ALPHONSE CHAPANIS
**Human Factors in Systems Engineering**

YACOV Y. HAIMES
**Risk Modeling, Assessment, and Management**

DENNIS M. BUEDE
**The Engineering Design of Systems: Models and Methods**

ANDREW P. SAGE and JAMES E. ARMSTRONG, Jr.
**Introduction to Systems Engineering**

WILLIAM B. ROUSE
**Essential Challenges of Strategic Management**

YEFIM FASSER and DONALD BRETTNER
**Management for Quality in High-Technology Enterprises**

THOMAS B. SHERIDAN
**Humans and Automation: System Design and Research Issues**

ALEXANDER KOSSIAKOFF and WILLIAM N. SWEET
**Systems Engineering Principles and Practice**

HAROLD R. BOOHER
**Handbook of Human Systems Integration**

# HANDBOOK OF HUMAN SYSTEMS INTEGRATION

# HANDBOOK OF HUMAN SYSTEMS INTEGRATION

Edited by

Harold R. Booher

WILEY-INTERSCIENCE

A JOHN WILEY & SONS, INC., PUBLICATION

This book is printed on acid-free paper. ♾



Published simultaneously in Canada.



For ordering and customer service, call 1-800-CALL-WILEY.

***Library of Congress Cataloging-in-Publication Data:***
Booher, Harold R.
Handbook of human systems integration / Harold R. Booher
p. cm. -- (Wiley series in systems engineering and management)
"A Wiley-Interscience publication."
Includes bibliographical references and index.
ISBN 0-471-02053-2
1. Human engineering. 2. Systems engineering. I. Title. II. Series.

T59.7.B66 2003
620.8′2--dc21 2002044604

Printed in the United States of America.

10 9 8 7 6 5 4 3 2 1

# CONTENTS

# FOREWORD

If the emergence of human factors out of the US–UK engineering psychology experiences of WWII was its first milestone in the US Defense community; and if the second was the deliberate broadening by Gen. Max Thurman (as Army Deputy Chief of Staff for Personnel) of the U.S. Army human factors program to Manpower-Personnel Integration (MANPRINT); then the publication of this collected work on HSI analysis principles and methods is surely the third major milestone.

From the WWII origin, the tools of the embryonic human factors profession were those of the first practitioners, experimental psychologists. The experimental method with human subjects, system design alternatives as levels of the independent variables, dependent performance measures crafted to illuminate the design differences, an Analysis of Variance framework, and results judged by standards of statistical significance ensured professional rigor.

The reality of the accelerating technological change is pushing the classic HF experiment toward obsolescence as a method of design influence and analysis. The time and other fiscal investments required for deliberate experimentation cannot keep pace with the rate of critical concept and design decisions early in the development of complex, military systems. And the MANPRINT expansion of human factors to include manpower and personnel and other domains raises concept design issues not amenable to people-in-the-loop experimentation. The commonality of the human factors expansion evident in the UK Ministry of Defence Human Factors Integration/MANPRINT program of the mid-1990s and the emerging US Navy interest in Human Systems Integration suggests a change of some permanence for defense human factors.

The organizational context, theoretical bases, and, especially, the concept evaluation tools described in this text give a new reality to this wider view of human factors. As the military services depend on the broadened HF scope to support materiel acquisition decisions, the HF practice will increasingly depend on professionals who are more "engineers"—applying science—than researchers. As such it readily fits into the industrial and systems engineering educational fields. This text is an ideal cornerstone for the education of the new HF professionals.

General Thurman's MANPRINT brought into materiel acquisition decision making all of the issues that might be "human resources" in commercial institutions, e.g., number of operators, maintainers, and support personnel, the relative costs of their abilities and skills, their training, as well as human engineering. Increasingly, we see perennial labor shortages in key sectors of the commercial economy. The next major HF milestone, the fourth, will be adoption of many of these HSI analysis methods for commercial practice where numbers of workers, the costs of necessary aptitude and skills, and training costs will then drive re-design of worker interfaces.

Milestones falling at prominent junctures become landmarks. This comprehensive compilation of HSI principles and methods will soon be viewed as a landmark in the evolution of the human factors discipline.

Robin L. Keesee

Human Research and Engineering Directorate
Army Research Laboratory

# PREFACE

Government and industry must change their systems design and development orientation from "technology" driven to "people–technology" driven. Global competition, demographic trends, and high-risk technology demand it. These three forces work together as economic levers to increase demands for products and services while helping to assure their quality and affordability. Systems designed and developed for both military and commercial applications are greatly affected by the interaction of such economic factors, but in the past both business and engineering cultures have tended to view advances in technology as not only the main way to improve systems capability, but for solutions to systems quality and affordability problems as well. For example, greater automation may be seen as the solution to high personnel costs for operating major military systems, but demographics may show personnel and training costs will rise due to the limited availability of skilled people in the work force to operate and maintain highly automated systems. Global competition simply raises the need for all organizations world wide that produce new systems to find the competitive edge which will make their products successful in the market place. Quality programs like Deming's Total Quality Management have raised the standard in commercial manufacturing practices, but have yet to have a major effect in military systems or in such fields as education and medicine. In industries employing high-risk technology as in aerospace, petrochemical, nuclear, and biological environments, the hazards of not fully comprehending the people–technology interfaces all too often result in tragic and costly unintended consequences.

Human Systems Integration is very attractive as a new integrating discipline that can help move business and engineering cultures toward a people–technology orientation. To be effective, however, a cultural change is needed which must start with organizational leadership. At the heart of the need for a cultural change in business and engineering is the fact that human factors engineering as a people–technology interface discipline has, by itself, been largely ineffective at changing ingrained attitudes in government and in most industries.

This point is made clear by Charles Perrow ("The Organizational Context of Human Factors Engineering," *Administrative Science Quarterly*, 1983; Normal Accidents, Basic Books, 1984, 1999) and from my experience with the Army MANPRINT program (discussed in Chapter 1). There is little question of the value of human factors engineering to producing safe and effective products and systems, but even though major human factors programs were introduced in each branch of the Department of Defense and in the Department of Transportation in the 1960s, the nuclear industry was almost completely unaware of the discipline until the Three Mile Island accident. Even when the benefits of human factors are fully appreciated by top leadership, the influence on systems acquisition will tend to erode with changes in leadership. The Army MANPRINT program provided

$3.29 billion cost avoidance on a major aircraft program from efforts initiated in 1985, but by 1994, MANPRINT nearly disappeared from the Army as a result of downsizing and changes in DoD acquisition policy.

I mentioned some of Perrow's findings in the preface to *MANPRINT: An Approach to Systems Integration*, one of which is worth repeating here. In searches to assign blame for accidents with systems employing high-risk technologies, Perrow urges us to seek deeper than the design engineer and to "take into account the pervasive social casual factors inherent in organizations which make and operate our machines." Considering that Perrow came at the problem from an organizational analyst point of view, I stressed "he reminds us that managers and professionals respond to 'rewards and sanctions and prevailing belief systems of top management.' There is nothing to prevent top management, if it wishes, from informing designers about human factors principles. Furthermore, it is top management who 'can require that these principles be utilized.' They alone 'can structure the reward system so that it encourages designers to take these principles into account'."

These organizational "social causal factors" affecting HSI advancement have not gone unnoticed at the congressional level of government. Thanks to Congressman Ike Skelton, who was concerned about the regression of MANPRINT in the Army (see the appendix to the Afterword) HSI began to see a new burst of interest throughout the military and in other sectors at the end of the millennium. It was during this period that the *Handbook of Human Systems Integration* was conceived.

The cover image (created by Heather DuMont) of the *Handbook of Human Systems Integration* symbolizes the theme of the book, which is to provide principles and methods that can help integrate people, technology, and organizations with a common objective toward designing, developing, and operating systems effectively and efficiently. If organizations are to change significantly to take full advantage of the benefits that HSI can offer, I believe this is most likely to be accomplished as an inherent part of systems engineering and management. The publisher, John Wiley and Sons, agrees with us by having the *Handbook* appear in its Systems Engineering Series rather than its Human Factors Series. Human factors and ergonomics are necessary fields for the successful implementation of HSI, as reflected in the large number of contributors to the *Handbook* from those fields. And if one were to try to obtain advanced education in HSI, they would most likely need to acquire it from institutions that teach human factors and ergonomics (see Chapter 5). Human factors and ergonomics are necessary but not sufficient, because they do not fully cover other important human domains that need representation and because of their inability generally to significantly influence organizational decision makers.

The organizations for systems engineering and management are already well institutionalized in government, industry and academia and have the common goal with HSI to produce high performing, safe, and affordable systems. The major component currently missing from systems engineering and management is a detailed description of the principles and methods of human systems. The intent of the *Handbook* is to provide that component.

There are three types of stakeholders in any organization that designs, develops, tests, evaluates, operates, or maintains systems sufficiently complex to employ systems engineering and management processes, who should benefit by using the *Handbook.* These are 1) the HSI practitioners who work with systems using the principles and methods described; 2) systems engineers and managers along with related disciplines such as safety engineering and integrated logistics support who provide the framework for HSI roles and interfaces; and 3) organization decision makers, including program managers who must weight the recommendations of the first two types in making systems acquisition decisions.

The *Handbook* begins where *MANPRINT: An Approach to Systems Integration* (1990) left off. *MANPRINT*... was a basic introduction to an Army concept of integrating various human systems disciplines and technologies in the systems acquisition process. In doing so, it presented the uniqueness of MANPRINT as a systems integration model which focuses directly on the human element both as a critical component of the system and as the primary reason for designing, developing and deploying the system. The original MANPRINT concept has gradually been incorporated into other government system acquisition organizations, both military and commercial, either under the name Human Systems Integration or Human Factors Integration. Those familiar with *MANPRINT*... will be pleased to see the advances made since its publication.

The *Handbook* scope is much broader than *MANPRINT*... covering both public and commercial processes; especially as they interface with systems engineering processes and it provides much greater depth, particularly in presenting the state of the art for tools, techniques, and methodologies utilized by each of the HSI domains. Ninety some contributors, technical advisors and reviewers make up the technical representation from government, industry and academia. Chapters provided by authors from the United Kingdom and Canada represent their government and industry. Three services of the Department of Defense are well represented along with the Federal Aviation Administration and the National Academy of Sciences. Many of the chapters cover both military and non-military applications. The *Handbook* is divided into four parts, which I summarize in the introductory chapter.

I am grateful to the numerous contributors, advisors, and reviewers listed on the pages that follow who are responsible for the bulk of the work that went into this *Handbook.* Without their selfless and timely efforts, a book of this complexity could not have been produced. There are a few individuals, some on those lists and others who are not, who made special contributions to the conception, planning, and execution of the book. I am especially grateful for the services of Robin Keesee, who not only did a technical review of the entire manuscript, but also encouraged a large number of his staff at the Human Research and Engineering Directorate of the Army Research Laboratory to write and review many of the chapters. I also greatly appreciate the additional efforts of Ed Smootz, Glen Hewitt, Bill Rouse, and Andy Sage who formed my inner circle of advisors in handling the numerous issues that arise from a project of this complexity. Others who were particularly important to this project were Nancy Dolan, Arch Barrett, Frank Petho, Bill Natter, Larry Lehowicz, and Jack Wade who helped in a variety of special ways like finding contributors, providing motivational support, and stimulating interests in the *Handbook.*

Milton Lee and Kim Booher provided me with critical intellectual property information without which, the book may not have been produced. Susanna Clay, Debra Clark, Rebecca Singer, and Jeff Landis provided all the editorial assistance that went into the three years of manuscript development. I can probably never repay them for their dedication and perseverance to assure the quality of this project. Finally I am most appreciative of the staff at John Wiley and Sons, in particular George Telecki for accepting the *Handbook* for publication and to Cassie Craig, Brendan Codey, and Christine Punzo for helping me through the submittal and production processes.

Harold R. Booher

Baltimore, Maryland

# CONTRIBUTORS

**Laurel Allender**, Ph.D., heads the Cognitive and Perceptual Modeling Team at the Army Research Laboratory Human Research and Engineering Directorate where her research is focused on human behavior representation in models and simulation. Dr. Allender is past chair of the Systems Development Technical Group of the Human Factors and Ergonomics Society, and of the Manned System Modeling Sub-group of the Department of Defense Human Factors Engineering Technical Advisory Group.

**Susan Archer** is Director of Operations at Micro Analysis and Design in Boulder, Colorado. She has led the development of numerous human performance modeling techniques for both military and commercial applications and is the program manager of a large scale government, industry, academic basic research effort to advance the state of the art in cognitive and computer sciences.

**Michael Barnes** is principal investigator for the battlespace visualization program of Human Research and Engineering Directorate, US Army Research Laboratory. Having a master's degree in experimental psychology, Mr. Barnes has worked as a researcher and human factors manager for the US Navy (Naval Weapons Center and Naval Air Development Center) and as General Electric (GE) Human Factor's unit manager for the Aegis Combat System.

**David Beevis**, MSc., P Eng., worked in the Canadian Defence Research and Development branch, Toronto for thirty years, where he was responsible for developing Canada's defense research program in human engineering and human systems integration. He is a member of the Ergonomics Society, the Human Factors and Ergonomics Society, and the Professional Engineers of Ontario.

**Kenneth R. Boff**, Ph.D., is Chief Scientist of the Human Effectiveness Directorate in the Air Force Research Laboratory and is founder and technical director of the Department of Defense Human System Information Analysis Center. Dr. Boff is the US National Coordinator and NATO chair for the human factors technology area and is a Fellow of the Human Factors & Ergonomics Society and the International Ergonomics Association.

**Catherine A. Booher**, M.D., is currently a Physician Advisor for Clinical Utilization and Case Management at Good Samaritan Hospital in Baltimore. She has worked previously as a Medical Director for Aetna where she participated on their Patient Safety Task Force, and as Faculty in the Department of Medicine at Union Memorial

Hospital where she served on the Performance Improvement Committee. She is board certified in internal medicine and is a member of the American College of Physicians.

**Harold R. Booher**, Ph.D., consults on Human Systems Integration and MANPRINT applications to systems design, development and assessment. Dr. Booher was the first Senior Executive (SES) Director of MANPRINT for the Department of the Army. Hal has more than 35 years experience in Human Factors and Engineering in government, industry and academic applications, is Editor of *MANPRINT: An Approach to Systems Engineering*, and is a Fellow of the Human Factors and Ergonomics Society.

**J. Jeffrey Crowson, Jr.**, Ph.D., is a human factors psychologist on the Operations Research faculty at the Naval Postgraduate School, Monterey, California, where he works on a variety of human systems integration and human factors projects. Previously, Jeff applied his background in experimental psychopathology and statistics to Navy personnel selection and classification issues in human systems integration.

**Lee Scott Ehrhart**, Ph.D., is a cognitive systems engineer with the Center for Innovative Computing and Informatics at MITRE Corporation. She has focused her more than 20 years research and experience on designing the interaction of human problem solvers, enabling technologies, and organizational processes to create *decision systems* for military command and control, counter-terrorism crisis management, and critical care medicine.

**Melanie J. Forster** works for the Centre for Human Sciences at QinetiQ where she is responsible for business development. Prior to her current appointment, Mel led the UK MoD Human Factors Integration programme working both in MoD Headquarters and the Defence Procurement Agency.

**Bruce E. Hamilton**, Ph.D., works for Northrop Grumman Ship Systems on the DD(X) HSI team. Previously, he provided human factors engineering support for NASA's Shuttle, International Space Station, and basic research program. His prior government contractor activities include crewstation design for the RAH-66 Comanche and the YAF-22 Falcon.

**Charles S. Harris**, Ph.D., is the associate dean and professor of sociology at Marymount University in Arlington, Virginia. A specialist in organizational change and in research methodology, his recent work includes studies of the career patterns of Army civilian employees and of policy-making processes at the Department of Defense.

**John Harrison** is Technical Director of Nickleby HFE Ltd in the United Kingdom. John has over 35 years in the defence industry, with a large portion of his experience directed toward applying human factors principles to systems problems. He is author of *Human Factors Integration: A Practical Guide for Integrated Project Teams.*

**Betty K. Hart** consults in culture change, diversity and financial services. Applying organizational change techniques, Ms. Hart has led successful change initiatives in the Departments of the Army and Navy civilian work forces. Betty has a Masters of Science degree from Stanford University.

**Donald B. Headley**, Ph.D., is a Research Psychologist in the Cognitive Sciences Branch, Human Research and Engineering Directorate, US Army Research Laboratory, where one of his principal activities has been conducting MANPRINT assessments of automated information systems and combat support systems. Don is also an Adjunct Professor for the Industrial Engineering Department at North Carolina A&T University, Greensboro.

**Lawrence J. Hettinger**, Ph.D., is Senior Human Factors Engineer for Northrop Grumman Information Technology. Larry has 22 years of experience on human factors design issues in training and support of complex tasks and in the design and use of virtual environment systems. He is a member of the Human Factors and Ergonomics Society and the American Psychological Society.

**Glen Hewitt** is a Scientific and Technical Advisor for Human Factors in Federal Aviation Administration Research and Acquisitions. Glen is a graduate of both the Army's Command and General Staff College and the Navy's Naval Command College. He holds a BS in Engineering from the United States Military Academy and an M.S. in Systems Management and Safety from the University of Southern California.

**Brian M. Kleiner**, Ph.D., is Director of the Macroergonomics and Group Decision Systems Laboratory and the Human Factors Engineering and Ergonomics Center at Virginia Tech. He is co-author of *Macroergonomics: An Introduction to Work System Design* and co-editor of *Macroergonomics: Theory, Methods and Applications.*

**John Klesch** is presently working for Computer Sciences Corporation, where he develops interactive multimedia instruction for training systems. Prior to this, John completed over 35 years of federal service with the US Air Force and US Army where he worked on job performance aids, improvement of technical manuals, new systems training development and distance learning.

**Robert M. Lindberg** is an active duty Major in the United States Air Force (USAF) supporting the 311th Human Systems Wing by leading Human Systems Integration for the USAF. Major Lindberg has a variety of experience in various military acquisition activities including acquisition logistics, aircraft flight test, aircraft engines, avionics, maintenance, manufacturing/production, and program management.

**John F. Lockett, III** leads a group of researchers engaged in application and development of Human Engineering Analysis Tools at the US Army Research Laboratory, Human Research and Engineering Directorate. John has over 17 years of human factors research and development experience; focused primarily toward applying workload analysis and human figure modeling technologies to the Army's MANPRINT Program.

**Anne Mavor** is the staff director for the Committee on Human Factors at the National Academy of Sciences/National Research Council. As a senior staff officer for the National Research Council since 1989, Anne's numerous projects have included air traffic control automation, modeling human and organizational behavior, and emerging needs in human factors.

**Nita Lewis Miller**, Ph.D., is the director of the Human Systems Integration Laboratory, conducts research, and teaches human factors engineering and human systems integration in the Operations Research and Systems Engineering Departments at the Naval Postgraduate School at Monterey, California. Nita holds a doctorate in Behavioral Science from the University of Texas and was a Postdoctoral Fellow at the USAF School of Aerospace Medicine.

**James Minninger** is a retired Army Aviation Warrant Officer and former RAH-66 Comanche Assistant Program Manager for MANPRINT. He is currently the U.S. Army Research Laboratory, Human Research Engineering Directorate Aviation Field Element Representative at Redstone Arsenal, Alabama.

**Jennifer McGovern Narkevicius**, Ph.D., is a cognitive psychologist serving as Senior Manager of Human Systems Engineering at ARINC Engineering Services, LLC. Jennifer has supported a variety of Navy, Air Force, and FAA aviation and avionics programs including Electronic Warfare/Electronic Attack, Air Traffic Management, Automated Systems, and Collaborative Knowledge, as well as web based applications and ground transportation initiatives.

**Stephen R. Olson** is a principal with Systems Management & Integration. Steve has over 30 years experience as a systems engineer with the US Navy, Texas Instruments and the Hughes Aircraft Company. He is a graduate of the Naval Academy and the Naval Postgraduate School and has completed all of the course work toward a Ph.D. in Information Technology at George Mason University.

**Glenn Osga**, Ph.D., is a Human Factors Psychologist with the Space & Naval Warfare Systems Center in San Diego, California. Glenn has contributed to numerous US Navy Command, Control and Combat Systems Human Engineering programs, including the development of decision-aids in use on USN Combatant Ships and acquiring patents on new user-interface interaction methods.

**Dino Piccione** is a human factors engineer in the Federal Aviation Administration (FAA) where he has been actively engaged in the integration of human factors into the development of air traffic systems and managing air traffic human factors research programs. Previously, he has conducted human factors test and evaluation of aircraft and aviation systems for the Army and worked as a human factors engineer for Boeing. He is a private pilot and Certified Professional Ergonomist.

**Linda Pierce**, Ph.D., is Acting Chief of the Human Factors and Integration Division of the Army Research Laboratory Human Research and Engineering Directorate, where her primary research areas are decision-making and teamwork in military command and control. She holds a doctoral degree in Industrial and Organizational Psychology from Texas Tech University.

**Jeffrey Powers** is the manager of Human Factors and Industrial Design at United Defense. Jeff is a Board Certified Ergonomist with special interests in workload, robotics, automation modeled after human behavior, crew station design, and quantification of man-in-the-loop system performance.

**Welford C. Roberts**, Ph.D., is an environmental and occupational health consultant and an Adjunct Associate Professor with Touro University, International. Dr. Roberts retired from the US Army as a Lieutenant Colonel where among his many assignments was Health Hazard Assessment Program Manager for the US Army Materiel Command and Assistant Professor at the Uniformed Services University of the Health Sciences.

**William B. Rouse**, Ph.D., is the H. Milton and Carolyn J. Stewart Chair of the School of Industrial and Systems Engineering at the Georgia Institute of Technology. Bill is a member of the National Academy of Engineering, a Fellow of the Institute of Electrical and Electronics Engineers, a Fellow of the Human Factors and Ergonomics Society, a

Fellow of the Institute for Operations Research and Management Science, and co-editor of the *Handbook of Systems Engineering and Management.*

**Andrew P. Sage**, Ph.D., is First American Bank Professor and *Founding Dean Emeritus* of the School of Information Technology and Engineering at George Mason University. Dr. Sage is editor of the John Wiley textbook series on Systems Engineering and Management and the INCOSE Wiley journal *Systems Engineering.* He is coeditor of *Information, Knowledge, and Systems Management*, and a Fellow of the Institute of Electrical and Electronics Engineers, the American Association for the Advancement of Science, and the International Council on Systems Engineering.

**Eduardo Salas**, Ph.D., is a Professor of Psychology and Director of the Applied Experimental & Human Factors Psychology Doctoral Program at the University of Central Florida. Dr. Salas also holds an appointment as Program Director for the Human-Systems Integration Department at the Institute for Simulation & Training at UCF. He is a Fellow of the American Psychological Association and the Human Factors and Ergonomics Society, and is Editor of the *Human Factors* journal.

**Joyce Shields**, Ph.D., serves as a senior leader of the Hay Group. Since joining Hay in 1985, Dr. Shields has directed and conducted consulting assignments in leadership development, executive coaching, human resource planning and development, change management, competency-based systems, selection and retention, and HR reengineering. Prior to joining the Hay Group, Joyce was the Director of the Manpower and Personnel Research Laboratory for the US Army Research Institute.

**Tonya L. Smith-Jackson**, Ph.D., is an Assistant Professor in the Grado Department of Industrial and Systems Engineering at Virginia Tech, where she is director of the Assessment and Cognitive Ergonomics Lab and co-director of the Environmental and Safety Lab. Her specialty areas are cultural ergonomics and safety, design of risk communications, and application of human information processing to system design and integration.

**Edwin R. Smootz**, Ph.D., has spent more than thirty years performing behavioral science and human factors research for the U.S. Army. Much of his career has been spent working to integrate human factors considerations into the test and evaluation process. His last position was as Chief of the Human Factors Integration Division within the Army Research Laboratory's Human Research and Engineering Directorate. He is now an independent consultant.

**William A. Stembler**, Ph.D., has been building training systems and solutions for over thirty years. For the past twenty years he has worked for Computer Sciences Corporation where he is founder of the corporation's Training Center of Excellence and the creator of the center's Courseware Factory.

**Donald W. Swallom** is a senior engineer for Science Applications International Corporation supporting the Army Aviation and Missile Command by overseeing system safety for the US Army's newest attack helicopter, the RAH-66 Comanche. In the United States Air Force, he served as a helicopter pilot and staff officer, where his last assignment was the chief of safety for the Arnold Engineering Development Center.

**Ronald A. Weiss**, Ph.D., is a human physiologist and bioengineer in the Survivability and Lethality Analysis Directorate of the Army Research Laboratory. His experience has focused on helping people to live and function in natural and hostile environments from under sea to outer space, from desert to polar regions, combat, radiological, biological, chemical, acoustic and motion environments.

**Christopher Wickens**, Ph.D., is a Professor of Psychology and Head of the Human Factors Division at the University of Illinois, Institute of Aviation. He has written two textbooks in Human Factors and Engineering Psychology, and served as chair of the National Research Council panel on the human factors of air traffic control automation.

**Richard Zigler** is the Ground Systems Mission Area Coordinator for the Survivability and Lethality Analysis Directorate of the U.S. Army Research Laboratory where he coordinates Army soldier survivability activities. Mr. Zigler is a principal contact for state of the art information on personnel survivability methods and techniques and a leading advocate for increasing awareness of personnel survivability issues in systems development and acquisition.

## TECHNICAL ADVISORS AND REVIEWERS

| | |
|---|---|
| Michael Allocco | Office of System Safety, Federal Aviation Administration |
| Archie Barrett | National Security Affairs, Naval Postgraduate School |
| Pam Bartlett | Office of Undersecretary of Defense (Personnel and Readiness) |
| Marilyn Sue Bogner | Institute for the Study of Human Error |
| Robert Bost | Human Systems Integration Directorate; Naval Sea Systems Command |
| Raymond G. Brandenburg | Manpower, Personnel, and Training; Mount Airy, Maryland |
| Janis A. Cannon-Bowers | Training Systems Division, Naval Air Systems Command Orlando, Florida |
| Paul H. Cunningham | Booz, Allen & Hamilton, Dayton, Ohio |
| Phil E. DePoy | Wayne E. Meyer Institute of Systems Engineering, Naval Postgraduate School |
| Nancy Dolan | Human Systems Integration, Chief of Naval Operations (Manpower & Personnel) |
| Michael Drillings | MANPRINT Directorate, Army Headquarters, Washington DC |
| Mary Dzindolet | Cameron University, Lawton, Oklahoma |
| Michael L. Fineberg | Booz, Allen & Hamilton, Falls Church, Virginia |
| Leonard M. Girling | Army Logistics Management College, Ft. Lee, Virginia |
| Robert Gross | Army Center for Health Promotion and Preventative Medicine |
| Patricia Hamburger | Combat Systems Department, Naval Sea Systems Command, Dahlgren, Virginia |
| W. Ian Hamilton | Human Engineering Limited, Bristol, England |
| David Harrah | Human Research and Engineering Directorate, Army Research Laboratory |
| Hal W. Hendrick | Hendrick & Associates, Englewood, Colorado |
| David Hoagland | Air Force Research Laboratory, Wright Patterson AFB, Dayton, Ohio |
| Mitchell A. Howell | Human Research and Engineering Directorate, Army Research Laboratory |
| Sherrie A. Jones | Training Systems Division, Naval Air Systems Command Orlando, Florida |
| L. Taylor Jones III | Battelle Memorial Institute, Huntsville, Alabama |

| | |
|---|---|
| Robin L. Keesee | Human Research and Engineering Directorate, Army Research Laboratory |
| Thomas H. Killion | Office of the Assistant Secretary of the Army for Acquisition, Logistics and Technology |
| Marcie K. Langelier | Crew Systems Department, Naval Air Systems Command, Patuxent, Maryland |
| Jan S. Lach | Jacobs Engineering, Oak Ridge, Tennessee |
| Ronald Lyons | Federal Data Corporation |
| James McGee | National Academy of Sciences, National Research Council |
| Richard McMahon | Human Research and Engineering Directorate, Army Research Laboratory |
| David Meister | Human Factors, San Diego, California |
| Thomas R. Metzler | Booz, Allen & Hamilton, Dayton, Ohio |
| James C. Miller | USAF Warfighter Fatigue Countermeasures, Brooks AFB, Texas |
| Christine M. Mitchell | School of Industrial and Systems Engineering, Georgia Tech |
| William H. Natter III | Committee on Armed Services, U.S. House of Representatives |
| Frank R. Paragallo | Human Research and Engineering Directorate, Army Research Laboratory |
| Richard W. Pew | BBN Technologies, Cambridge, Massachusetts |
| Jay A. Rachlin | Center for Devices and Radiological Health, Food and Drug Administration |
| Albert A. Sciarretta | CNS Technologies, Springfield, Virginia |
| Thomas B. Sheridan | Massachusetts Institute of Technology, Cambridge, Massachusetts |
| Ronald A. Spencer | Human Research and Engineering Directorate, Army Research Laboratory |
| Scott L. Smith | Office of Secretary of Air Force, Washington, DC |
| Steven D. Smith | Office of System Safety, Federal Aviation Administration |
| Michael H. Strub | Human Research and Engineering Directorate, Army Research Laboratory |
| Phil Sutton | Technology Development, Ministry of Defence, London, England |
| Scott VanBuren | Office of Architecture and Systems Engineering, Federal Aviation Administration |
| Harold P. Van Cott | Van Cott and Associates, Bethesda, Maryland |
| John P. Wilson | Army Logistics Management College, Ft. Lee, Virginia |
| Huw M. Walters | DERA Partnering Team, Ministry of Defence, London, England |
| Dennis White | Combat Systems Department, Naval Sea Systems Command, Dahlgren, Virginia |
| John P. Wilson | Army Logistics Management College, Ft. Lee, Virginia |
| Richard A. Young | Air Force HSI Office, Brooks AFB, Texas |
| Marjorie H. Zelco | MANPRINT Directorate, Army Headquarters, Washington, DC |

CHAPTER 1

# Introduction: Human Systems Integration

HAROLD R. BOOHER

## 1.1 BACKGROUND[1]

On March 28, 1979, the central control room operators were alerted to a loss of feed water to the pressurized water reactor (PWR) of the Three Mile Island Unit 2. The safety injection pumps came on automatically to pump in auxiliary water to the reactor vessel. The indicators for the pressurized vessel, however, showed a dangerous "water-solid" condition for the pressurized vessel. In an attempt to reduce the pressurizer water level, the operators turned off the safety injection pump. Consequently, the reactor core water cover became depleted, leading to a near meltdown. As a result, the nuclear industry radically changed its management and organizational approaches to nuclear safety but has never fully recovered from the effects of the accident or alleviated the public concerns with nuclear energy.[2]

In the Persian Gulf on July 3, 1988, the U.S. Navy crew of the *Vincennes* shot down an Iranian commercial jetliner killing all 290 civilians on board.[3] On February 16, 1996, Maryland Rail Commuter (MARC) train 286 collided with AMTRAC passenger train 29 near Silver Spring, Maryland. The 3 crewmembers and 8 of the passengers on the MARC were killed, and 26 people on the two trains were injured, in the derailment and subsequent fire.[4] October 31, 1999, Egypt Air flight 990 went into a rapid descent 30 minutes after takeoff from New York to Cairo, crashing into the Atlantic Ocean, killing all 217 people on board.[5] Although the cause is still undetermined, pilot suicide cannot be ruled out.

These are only a few of the more dramatic examples of the effects on society from failures in technology at the interface of people and machines. Internationally remote areas such as Bhopal, India, and Chernobyl, USSR, are now part of household vocabularies because of the tragedies there. The *Challenger* showed us even space is not immune from human-designed technological mistakes, and although safety is constantly improving in some technologies, it is still the leading cause of death in others. More people die in automobile accidents every year than the army casualties for the entire Vietnam war.[6]

*Handbook of Human Systems Integration*, Edited by Harold R. Booher.
ISBN 0-471-02053-2 

Although technology is constantly improving, the number of catastrophic incidents can be expected to rise as well, if for no other reason than the opportunities for both human and machine failures increase with complexity, and rapidly developing technologies involve greater and greater operational complexity (Perrow, 1999).

The loss of lives is not the only cost. The waste in productivity every year is astronomical. The Three Mile Island accident, where no lives were lost, cost General Public Utilities Nuclear over $1.3 billion in radioactive waste cleanup and borrowed electrical energy expense. The cost to the American industry from failure, rework, and waste resulting from substandard manufacturing has been estimated at over $600 billion a year—enough to pay off the national debt in less than 6 years![7]

Solutions to this modern societal problem of unnecessary losses in lives, productivity, and quality of life are extremely complex. People are both the cause and the solution. People are the benefactors and the victims. Through human error in design, operation, or repair of machines, people are hurt, killed, made unhappy, or, at the very least, inconvenienced. On the other hand, it is through human intelligence and unique human skills that equipment, organizations, and knowledge-enhancing products are designed and operated effectively, efficiently, and safely. The quality of any product produced by any industrial organization depends ultimately on the interplay of several primary factors, all under the control of people themselves. The manner in which people can assess how well they are controlling these factors depends on how well they can determine and manage cost, performance, schedule, requirements, and risk.

### 1.1.1 Focus on Human Element

It is a fundamental belief of this writer that through a focus on the human element it is possible to achieve both (a) dramatic reductions in waste and victims on the debit side of society's ledger and (b) dramatic increases in system performance and productivity on the credit side.

It is further believed that unless the human element is considered to be a critical component of the complex system, few of these benefits can be achieved by an organization that produces, buys, or operates complex systems. On the other hand, these benefits are most likely to be achievable if an organization's focus is first and foremost upon the people who, in some manner, will be directly exposed to the complex system.

The need to consider the human element as a critical component tends to be supported by the vast number of horror stories such as those sampled above. This recognition of the importance of the human element is generally accepted by systems engineering and systems management philosophies. People, technology, and organizations make up the three top-level components of any complex system (Sage and Rouse, 1999, p. 57). The idea that dramatic organizational benefits are *most likely* to be achieved through focus on people, however, is unique to the human systems integration (HSI) philosophy. While we may avoid some of the catastrophic outcomes from poorly human-engineered designs of systems or increase efficiency of poorly performing systems by recognizing that the human component is critical, it is not so obvious even to systems engineers and systems integrators that the *dramatic* reductions in costs, both human and financial, and the *dramatic* increases to performance and productivity are *most likely* to appear when the focus is upon the human element inherent in the system.

### 1.1.2 The Army HSI Program

The U.S. Army was the first large organization to fully implement and demonstrate the benefits of an HSI approach, by focusing upon the human element. In 1986, the army created a Manpower and Personnel Integration (MANPRINT) management and technical program designed to improve weapons systems and unit performance. The army leaders decided it was necessary to change the focus of equipment developers away from "equipment-only" toward a "total system" view—one that considered soldier performance and equipment reliability together as a system. The program was extremely broad, including all army management, technical processes, products, and related information covering the domains of manpower, personnel, training, human factors engineering, system safety, and health hazards. After the Gulf War, largely because of the fratricide incidents, the army added a new domain, soldier survivability, to MANPRINT (see Fig. 1.1).

The most unique aspect of the program was effective integration of human factors into the mainstream of system definition, development, and deployment. Organizationally, the MANPRINT domain functions were spread throughout the army with major roles being performed by Army Materiel Command, Training and Doctrine Command, Office of the Surgeon General, Army Safety Center, Army Research Institute, and the Human Engineering Laboratory. Responsibility for integrating these varied human factors functions into the materiel acquisition process was with the Deputy Chief of Staff for Personnel (DCSPER) on the Department of Army Staff. The policy that provided the formal definition and various roles and responsibilities was presented in Army Regulation

**Manpower**
The number of human resources, both men and women, military and civilian, required and available to operate and maintain military systems.

**Personnel**
The aptitudes, experiences, and other human characteristics necessary to achieve optimal system performance.

**Training**
The requisite knowledge, skills, and abilities needed by the available personnel to operate and maintain systems under operational conditions.

**Human Factors Engineering**
The comprehensive integration of human characteristics into system definition, design, development, and evaluation to optimize the performance of human-machine combinations.

**System Safety**
The inherent ability of the system to be used, operated, and maintained without accidental injury to personnel.

**Health Hazards**
The inherent conditions in the operation or use of a system (e.g., shock, recoil, vibration, toxic fumes, radiation, noise) that can cause death, injury, illness, disability, or reduce job performance of personnel.

**Soldier Survivability**
The characteristic of a system that can reduce detectability of the soldier; prevent attack if detected; prevent damage if attacked; minimize medical injury if wounded; and reduce physical and mental fatigue.

**Figure 1.1** MANPRINT domains.

602-2, Manpower and Personnel Integration (MANPRINT) in the materiel acquisition process (U.S. Army, 1990).

In the latest revisions to the MANPRINT policy (U.S. Army, 2001), several major roles have changed, most significantly with the Army Research Laboratory performing the technical work of all the domains except systems safety (Army Safety Center) and health hazards [Army Center for Health Promotion and Preventive Medicine (CHPPM)]. However, even though the basic philosophy laid out originally has not changed, implementation effectiveness has varied considerably over time (see Section 1.7).

### 1.1.3 Other HSI Programs

The desired objectives of the MANPRINT approach to systems integration and the human factors domains of the army program have both been adopted by the U.S. Department of Defense with its HSI program and in the UK Ministry of Defence with its human factors integration (HFI) program. The Federal Aviation Administration (FAA) has also implemented major portions of MANPRINT into its HFI program. The MANPRINT philosophy presented in the original MANPRINT book (Booher, 1990) is presented here as an HSI philosophy, with the understanding that essentially the same concepts and principles apply whether the term used is HSI, HFI, or MANPRINT. As the HSI philosophy evolves, newer HSI programs (U.S. Air Force, U.S. Navy, UK Ministry of Defence, The Netherlands Applied Scientific Research Organisation and FAA) are making significant contributions both in the development of advanced HSI technology and in procuring new systems with HSI in their systems engineering and management processes.

## 1.2 HSI CONCEPT

The HSI concept is described in several different ways:

1. The formal Department of Defense (DoD) definition
2. In terms of key benefits
3. As a top-level conceptual model

### 1.2.1 HSI Definition

Human systems integration is primarily a technical and managerial concept, with specific emphasis on methods and technologies that can be utilized to apply the HSI concept to systems integration. As a concept, the top-level societal objectives of HSI are to significantly and positively influence the complex relationships among:

1. People as designers, customers, users, and repairers of technology
2. Government and industrial organizations that regulate, acquire, design, manufacture, and/or operate technology
3. Methods and processes for design, production, and operation of systems and equipment

It is believed that most of the technical and managerial advances suggested by the HSI concept can be accomplished within an overall systems integration philosophy that places a special emphasis on how its roles and technology can be included within systems engineering and systems management processes. As such, the HSI concept is considered to be an important adjunct to the various levels of systems engineering and management described in the *Handbook of Systems Engineering and Management* (Sage and Rouse, 1999). The HSI concept is fully compatible with those systems engineering processes relevant to systems definition, development, and deployment and their life-cycle phases, as well as the systems engineering methods, tools, and technologies.

Organizations that have created programs that adopt certain aspects, wholly or in part, of the HSI concept provide their definitions in programmatic language. The army has defined its MANPRINT program as a comprehensive management and technical program "which focuses on the integration of human considerations into the system acquisition process to enhance human/system design, reduce life cycle ownership costs, and optimize total system performance. MANPRINT accomplishes this by ensuring that the human is fully and continuously considered as part of the total system in the development and/or acquisition of all systems" (U.S. Army, 2001, p. 1).

The DoD lists its requirements for HSI under its systems engineering requirements[8]:

> For all programs . . . the PM [program manager] shall initiate a comprehensive strategy for HSI early in the acquisition process to minimize ownership costs and ensure that the system is built to accommodate the human performance characteristics of the user population that will operate, maintain, and support the system. The PM shall work with the manpower, personnel, training, safety and occupational health . . . , habitability, survivability, and HFE [human factors engineering] communities to translate the HSI thresholds and objectives in the ORD [operational requirements document] into quantifiable and measurable system requirements. The PM shall include these requirements in specifications, the TEMP [test and evaluation master plan], and other program documentation, as appropriate, and use them to address HSI in the statement of work and contract. The PM shall identify any HSI related schedule or cost issues that could adversely impact program execution [U.S. Department of Defense, 2001, 5000.2-R, paragraph C5.2.3.5.9. HSI].[9]

### 1.2.2 HSI Benefits

Whether HSI professionals seek out their customers or the customer comes to them for assistance, both may be surprised at the wide range of HSI benefits possible depending on the customer's role in the organization. Table 1.1 lists three types of audiences who could benefit from HSI applications. One group includes those who have high levels of responsibility for systems decision. The second group includes those who are primarily concerned with the operational processes within the organization that form its day-to-day reason for existence. The third group comprises those who are closer to the technology and engineering disciplines with roles in assuring the organization actually produces something tangible and usable.

The cells show sample benefits that can be provided by the HSI professional for different audiences. Depending on the HSI theme, for example, "user focus," "integration," or "performance measurement," benefits can vary considerably. For example, if the leadership of an organization is open to knowing the value added from an HSI approach, the HSI

**TABLE 1.1 HSI Activities by Benefit Areas**

| Benefit Areas | User Focus | Integration | Performance Measures |
|---|---|---|---|
| Management and organization | Show leadership potential benefits of user focus and how to implement cultural change needed for their organization. | Show management how to integrate people, processes, and technology to achieve organization objectives. | Show decision makers how to assess risk from human error in their organization and how to measure costs and benefits of user-centered business decisions. |
| Operational processes | Develop new or revised operational processes that utilize user-centered techniques (such as functional and task analysis). These are compatible with systems engineering and operational analysis techniques. | Write policies and procedures for developing, changing, and conducting processes that integrate contributions from multiple skilled disciplines in human sciences and technology. | Measure effectiveness of processes with human–machine interfaces. Provide guidelines for changing processes to reflect user-centered performance and safety enhancements. |
| Product design | Develop and implement user-centered design procedures. | Utilize human factors skills and tools in the design and development of products. | Evaluate human performance, safety, and usability of products. |

professional can show top-level management ways to "integrate" people, processes, and technology in terms of the organizational objectives. It is always helpful to show the leadership ways in which HSI performance can be measured, as in assessing the risk from human error in their organization and how to measure costs and benefits from user-centered business decisions.

Benefits of interest for the operational process audience might include the development of revised procedures for the organization, which utilize such user-centered techniques as functional and task analyses. It would be helpful for the HSI specialist to point out how their techniques are compatible with systems engineering and operational analysis techniques.

When working directly with designers of products, the HSI specialist should be sufficiently familiar with the various human-related technologies and disciplines (sampled in the other chapters in this Handbook) to help the designer meet the organization's product goals and the user's needs without compromising cost, schedule, performance, or safety.

### 1.2.3 HSI Model

The model illustrated in Figure 1.2 describes HSI at its highest conceptual level. The HSI process inputs are systems that utilize all aspects of systems engineering and management, which can be summarized within the three fundamental stages of systems processing—systems definition, development, and deployment. The output of the HSI process is systems integration of people, technology, and organization. The reader might note that even without the HSI process being imposed, the principles and disciplines of engineering and operations would still define, develop, and deploy systems that would integrate people, technology, and organization. This process would be accomplished without full consideration, however, of the human requirements of the user or the expertise and technology offered by the HSI field, resulting in those very conditions of poor systems integration that organizations would like to avoid.

The HSI model, therefore, emphasizes two additional inputs that guide the HSI process as it influences the systems engineering and management processes. These inputs are (1) a user focus on all aspects of the systems definition, development, and deployment stages and (2) the combined application of human-related technologies and HSI disciplines. Any system, product, or equipment that passes through the HSI process modulated with these two inputs will have a high likelihood of having an adequate consideration of the people component.

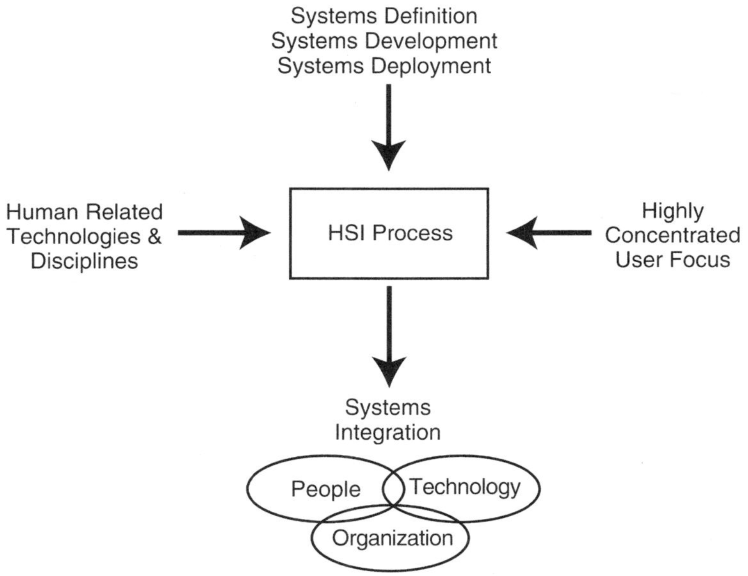


**Figure 1.2** Human systems integration model.

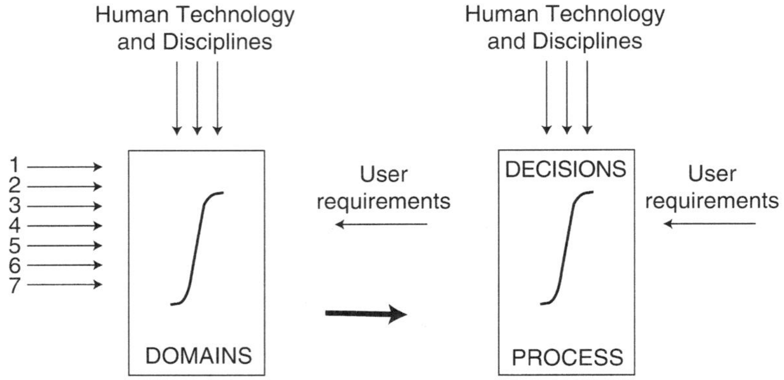


**Figure 1.3** HSI double-integration process.

### 1.2.4 Double-Integration Process

Figure 1.3 shows the HSI process behind the high-level concept. The HSI process is essentially a double-integration process where both integration steps are modulated by human-related technologies and disciplines and driven by user requirements. The first integration step creates a common focus for all seven HSI domains. Before MANPRINT in the army, each of the domains operated directly with the program manager (PM) for trade-off decisions regarding human concerns. Several of the domains, such as manpower, personnel, and training (MPT) were seldom considered at all in design trade-off decisions. The human factors engineer may have had the ear of the PM, but not early enough in the requirements definition process to make a design decision that affected MPT. Further, the HFE specialist had little expertise on these three domains. In other cases, perhaps several of the domains would raise a safety or health issue, but individually they were traded off, so that the issue was not given the visibility that MANPRINT provided. In the final analysis, the enormous increase in cost-effectiveness of a common HSI approach is critical to meet the system objectives. On one major aircraft program, it was demonstrated that the dramatic cost, performance, and safety objectives achieved with HSI were met with the same percentage of program funding as had been provided to past aircraft programs; yet in the earlier programs no such benefits were demonstrated. The difference was attributed to more effective use of funds when managed through HSI than when managed by domains independently.

At the second integration step, HSI contributions are fully integrated with the systems engineering and management processes. All the policies, procedures, guidelines, standards, specifications, and other documentation that relate to systems acquisition in the federal government and its contractors are part of the decisions process indicated in Figure 1.3. Such documentation must address all the HSI information needs and the roles and relationships of HSI with all the other systems engineering and management information needs, roles, and relationships.

For example, the types of user requirements (such as anthropometric measures or number and skills of personnel) might be reflected in a target audience description (TAD). The HSI technologies might include such items as man-in-loop simulation, human performance assessment, cognitive workload measurements, work space anthropometrics, or human systems trade-off methodology. Some of the disciplines closely associated with

HSI include human factors engineering, engineering psychology, manpower/personnel research, operations research analysis, systems safety engineering, and health hazards analysis.

## 1.3 SOCIOTECHNICAL SYSTEMS COMPLEXITY

The applicability of HSI varies with sociotechnical systems complexity. Table 1.2 shows sociotechnical systems ranging from very highly complex organizations with thousands of interactions between technology, people, and organizations to relatively small devices that make up functional systems (levels A to F). The HSI technology and disciplines are currently capable of aiding most directly in the design, development, and deployment of sociotechnical systems indicated as level D and below.

The greatest impact for life-cycle costs and performance improvements are centered on levels D and E. Examples include systems such as aircraft, tanks, cockpits, and other workstations for command and control operations. It is at these levels where the HSI state of the art is sufficiently developed to make major contributions. There are complex sociotechnical systems currently operating at level C and above, but HSI technology is not itself far enough along to make significant changes that will help make these levels perform more effectively. However, the needs of level C sociotechnical systems do suggest research and development (R&D) activities that should be taken to advance HSI technology. The military is already beginning to recognize that new HSI technology is needed to contribute to "technological system of systems" such as the navy's future aircraft carriers and the army's future family of battlefield vehicles. In a military "system of systems," the goal is to design complex sociotechnical systems made up of several diverse D level systems that will operate during war in a completely cohesive manner, analogous to individual systems.

Possibly the most complex operating environments outside the military that utilize current state-of-the-art HSI technology are human/machine control rooms for Air Traffic Control and National Aeronautics and Space Administration (NASA) Space Flight Centers. Nevertheless, because of the similarity of the items being controlled, and the design complexity of the technology, they are still level D sociotechnical systems. A hospital emergency center is conceptually a level C technological system because of the large number of activities (systems) going on simultaneously. An operating room may be considered a level D technology system because of the extreme diversity of procedures requiring people and technology interface. However, the "emergency and operating room system" is primarily made up of large numbers of level F systems of people and small devices. There are so few HSI sociotechnical applications for medical environments at the E level that it would be a great innovative leap to design a major technological system for an emergency or operating room.

The HSI applications to sociotechnical level A requires working with numerous other disciplines such as economics and political science in the formulations of policies, procedures, education, etc. The significant HSI changes to date have been directed at level B sociotechnical systems such as the DoD and FAA acquisition processes and their contractor organizations. It is believed that similar changes could benefit other mission areas such as health care, energy, and education with dramatic improvements in technological systems procured and/or regulated (see the Afterword). In two instances, level A organizations have provided the support and understanding necessary to introduce

**TABLE 1.2 Sociotechnical Systems—Levels of Complexity by Mission Areas**

| | Mission Areas | | | |
|---|---|---|---|---|
| Sociotechnical Systems | Military | Health Care | Energy | Transportation |
| A. Very highly complex organizations | | | | |
| Governmental agencies | Army department | | Dept. of Energy | Dept. of Trans. |
| Unpredictable environments | Warfighting units | | | |
| B. Highly complex organizations | | | | |
| Procurement/regulation agencies | DoD acquisition | Food and Drug Admin. | Nuclear Reg. Com. | Federal Aviation Admin., Federal Highway Agency |
| Product/service organizations | Large contractors | Hospitals | Nuclear power plant | |
| C. Complex organizations | | | | |
| System of systems | Aircraft carrier | Emergency room | | |
| D. Major technological system | Aircraft, tank, command & control | Operating room | Power generator control room | Train, car Air Traffic Control (ATC) room |
| E. Critical technological subsystem | Aircraft cockpit | | Controls/displays | ATC console |
| F. Small systems/devices (system parts) | Radio, radar (engine, wings) | MRI, monitors (tubes, cables) | Feed water pump (steam pipes) | Bicycle (tires) |

HSI into their B level organizations for systems acquisition. These organizations are the U.S. Department of the Army and the Ministry of Defence in the United Kingdom.

## 1.4 HSI UNIQUE ASPECTS

Table 1.3 lists unique aspects of the HSI philosophy characteristic of those organizations that adopt a people-oriented concept in their systems integration approaches to product design and manufacture. To begin, the benefits of a common focus of management is provided through the practice of top-down leadership and goal setting combined with bottom-up planning and execution. By giving high-level visibility to people-oriented concepts at all levels, the desired wide-sweeping changes have a realistic environment in which to grow. Understanding the concepts at the very top of an organization can bring focus on people throughout and sets up a reward system that instills competence and motivation in its employees.

The HSI concept recognizes the strengths and weaknesses of various disciplines. It may be wise to let the human factors engineer take the lead for integrating several other disciplines into any specific system design. But HSI would immediately remind the systems integrator of military systems, for example, that it is the enormous unnecessary MPT costs resulting from poor design that most influence the minds of government decision makers.

In HSI, decision makers and facilitators take advantage of technological developments in systems engineering and systems management. Inherent in several of these advances is the capability to quantify and measure human characteristics. These newer methods also allow better decisions to be made early in the design and development process where changes are relatively inexpensive to make.

The HSI process subscribes to the belief that investment in the front end on human factors will provide paybacks manyfold in the long term. But it goes beyond this to promise more immediate benefits as well. One of the problems of long-term payback is that the long-term rewards for the front-end decision makers often do not accrue for them

**TABLE 1.3 HSI Unique Aspects**

- High-level visibility of people-oriented concepts
- Focus throughout total organization on competence and motivation
- Top-down approach rather than bottom-up
- Multidisciplinary views of design
- Quantification of people variables
- Systematic early warning of human error considerations
- Provides trade-off techniques early in design
- Pushes technology and aids engineering advances
- Inherent part of system—not just supporting role
- Communicates in decision maker's language
- Encourages resources redirection rather than net increases
- Educates all people in the process
- Reduces demand for manpower, personnel, and training

personally. The HSI concept looks for more immediate productivity and cost avoidance measures first and then, as a bonus, points out the long-term advantages.

The HSI concept forces product technology to become more innovative. A company that recently adopted HSI into its helicopter engine design found out it had also produced an engine more competitive for commercial purposes. It did so at *no added cost* over its original design plan. Routinely, companies are finding that HSI provides the needed incentive to make the product not only user-friendly but also more reliable and cheaper to produce.

A fundamental concept of HSI is that people are considered part of any system being developed. At the same time, it is recognized that people issues and recommended actions must be described for decision makers in their own language. Frequently, these issues and actions are in terms of mission, resource, product, and/or process information.

Once introduced into an organization, a challenge for HSI is to remain viable during periods of budget constraint. It is easier to introduce new ideas when resources are increasing. But ideas that produce marginal returns during good times are vulnerable to extinction in bad times. HSI, therefore, encourages resource redirection rather than looking for a portion of a total net resource increase. Funds allocated to HSI and the disciplines it integrates frequently simply rise and fall proportionately with the fluctuation of the organization's budget. This comes from the belief that HSI is important, but no more so than any other activity contributing to the systems acquisition process. This belief is not above questioning as fiscally sound. In bad times, HSI may be one of the few areas where *increased* investment is warranted. In what other area can something be logically expected to produce more with less?

Education is absolutely essential to HSI. All people involved in the process from the top to the bottom must understand the concepts. Specialists are needed in certain areas, but to be successful HSI cannot, for example, rely solely on human factors specialists. New strategies are needed to provide HSI expertise to meet the increasing demand.

As the availability decreases (and consequently costs rise) for higher skilled operational and maintenance personnel, emphasis will be placed on simpler design. Systems and products that can be operated and repaired by fewer people, by lesser skilled people, and/or people with lesser training will be in greater demand. It is not an impractical expectation for the military, and probably in many commercial areas as well, to demand HSI designs that will allow reductions in all three areas—manpower, personnel, and training (MPT)—together. Too often in the past, cost reduction in one of these areas has merely shifted cost increases to one or both of the others. For example, when the army wanted to reduce total number of people, costs went up in recruiting higher skills. If it now tries to reduce cost of recruiting and retaining high-quality individuals, it will find increased training will be needed to maintain performance effectiveness. Improvement in equipment design is the most practical way to get major net reductions in MPT costs. It is a continual challenge for HSI researchers and engineers to develop methods and technologies that can help the HSI practitioner and systems engineer design systems that fully consider such MPT issues.

## 1.5 TEN HSI PRINCIPLES

During the past decade, 10 HSI principles have been identified that, to the degree they are applied, seem to assure that large organizations will capture the performance, cost, and safety objectives they desire for their systems. Conversely, to the degree any of these principles are left out entirely or a few are followed only marginally, large organizations risk their systems not meeting their desired system objectives. Moreover, specific systems programs that have followed these principles have been extremely successful, while those that have made compromises have made marginal progress. In some cases, programs have been canceled for failure to meet the system objectives; and in these cases, almost always the principles have been poorly followed as well. Several of these 10 principles have an obvious similarity to the unique aspects listed in Table 1.3. The main difference is that the HSI aspects are primarily stated in terms of special characteristics and benefits inherent in an organization that applies HSI—but no attempt is made to use them as criteria for building an HSI program or in assessing the program effectiveness—as is the purpose of the 10 principles.

The following 10 principles are crucial to effective HSI:

1. Top-level leadership
2. Focus on human-centered design (HCD)
3. Source selection policy
4. Organizational integration of all HSI domains
5. Documentation integration into procurement process
6. Quantification of human parameters
7. HSI technology
8. Test and evaluation/assessments
9. Highly qualified practitioners
10. Education and training program

### 1.5.1 Top-Level Leadership

Many programs in the past have attempted to bring greater focus to the human element in systems procurement using a bottom-up approach. Using this approach, organizations with human factors responsibilities and capabilities are assigned the responsibility to support other system design, development, and deployment activities. Historically, the bottom-up approach has had little positive effect on systems procurement. This is because the human factors disciplines operating in a support mode will find very few of the other HSI principles being applied throughout the systems acquisition organization. The HSI approach is completely reversed by having top-level leadership as its number one principle, meaning the leadership must infuse HSI throughout the organization, making HSI part of its culture, much in the Deming (1986) style that introduced total quality management in the commercial sector. Sponsorship of HSI by top-level leadership assures the organization adopts and continuously sustains the other nine HSI principles. This infusion should be done through implementing regulations with roles and responsibilities for HSI throughout the organization, actively encouraging HSI participation in top-level decision processes, providing wide-scale HSI education and training programs, and adopting HSI assessment, testing, and evaluation activities. Top-level sponsorship does not mean that large funding levels must be devoted to make the program successful, but it does mean that the advocacy is so visible that the nine other principles have a chance of being adopted. Without support from the highest management levels, the required organizational changes needed for effective HFI will not occur. If top-level sponsorship is withdrawn, support for the other principles will gradually wither.

### 1.5.2 Human-Centered Design Focus of Systems

This principle encourages the concept of defining a "system" more broadly than the hardware and software that industrial companies build. Procuring organizations should specify their requirements for a system in such terms as to include operators and maintainers as an inherent part of the "system." These requirements, which include the human element, should be translated quantitatively throughout the design, development, and testing processes in systems engineering measures of effectiveness and performance. Numerous examples of system failures in the past have been caused by failure to define a system in this broader sense. A very good illustration of how neglecting this principle of defining the system to include people is found in the U.S. Army's Stinger missile system. When the Stinger missile system was designed with a "probability of kill" at a certain level (without applying the HCD principle), the army found actual performance in the hands of the soldier was only one half of that expected. The designed performance was .6 PK (probability of kill), but actual performance (when the operator was included in the PK calculation) was only .3 PK. The designed performance had assumed human performance to be perfect and did not take into account the skill and training level of the operator. If the system probability of kill had been defined as "including the human operator," the procurement process would have been drastically different and training less costly.

### 1.5.3 Source Selection Policy

When the decision makers' attitude is "equip the man" rather than "man the equipment," the entire emphasis changes on what is actually being procured. For example, if in a military procurement the focus is on extending the capability of a fighter rather than fitting the fighter to a weapon, this creates a method of industry competition that is far more effective in cost and performance. To be successful, however, this concept has to reward those contractors who produce a better design at lower cost. This is done primarily by awarding contracts to those who demonstrate the best understanding and approach to designing and developing a system that will perform to the procuring activities performance requirements. When technologies and cost are close among competing proposals, HSI should be the discriminating factor in awarding the contract.

To help assure HSI can be a major factor in contract award, source selection policy for systems procurement should state that HSI evaluation factors will have the same *visibility* as technical and cost factors (as a major area) and will be evaluated in all *other relevant areas* as well. This source selection policy is a unique evaluation criterion requirement not specified similarly for any other factor. The HSI evaluation must not only show how well the contractor understands the HSI process (visibility) but also show that the contractor will use HSI technology and disciplines in the design of his equipment (other relevant areas).

### 1.5.4 Organizational Integration of All HSI Domains

A single focus for all HSI domains is necessary if any of the domains is to have substantial influence upon the procurement process and ultimately the quality of the product being procured. This single focus of the domains is the first integration step of the double-integration process (Fig. 1.3). The organizational integration principle can be a very difficult requirement for organizations to accept because the domains tend to be spread throughout large organizations within suborganizations that often have very little communication with one another. In 1986, the army chose the personnel organization (its DCSPER) as its integration center. The personnel organization became the representative of all six domains, even though manpower and personnel were the only domains specifically assigned to the DCSPER.

Although this decision worked well for the army, it is not necessary for HSI programs to have the domains integrated by personnel organizations. In fact, in some cases, the personnel organization may not be the best. The point and method of integration is highly dependent upon the implementation strategy for the organization as a whole. Regardless, expertise from each of the HSI domains should be provided to any system acquisition program with the idea of providing a common focus for HSI to the various systems engineering and systems integration working groups.

### 1.5.5 Documentation Integration into Procurement Process

The HSI model envisions a double-integration process, with the first integration taking place by bringing all the human factors domains together as described in principle 4 above. The second integration applies the information from the first integration directly into the procurement process. The HSI management tool for this principle for the DoD is the HSI Plan (HSIP) (U.S. Army, 2001). The HSIP is seen as the critical interface document feeding information into all other procurement documents and being fed by them. The quality of information in the HSIP depends on the quality of personnel assigned to the system joint working groups (SJWGs) and the tools and systems information available to the SJWGs.

### 1.5.6 Quantification of Human Parameters

The HSI process allows representation of all human factors domains in order to prescribe goals and constraints for the system being procured. Since the human is part of the system and the system is being designed to certain quantifiable specifications, the human aspects (to the degree possible) must be described quantifiably as well. Human parameters include data both from the perspective of the human as a measurable entity having such characteristics as body size and information processing capabilities and from a human performance perspective such as time and error performance on tasks. The quantification of human parameters database includes all characteristics, measures, and techniques that exist to describe and quantify a variety of human factors categories including anthropometrics, sensation and perception, mental abilities, social skills, and physiological attributes. Other representative quantitative measures include the levels of chemical, biological, and physical stressors that may adversely affect health and safety. In a systems approach to design one of the first set of human parameters to be quantified and provided to the contractor is the TAD. The U.S. military has compiled performance data for each occupational specialty (based on skill level and training) such that basic tasks can be analyzed for proposed weapons system designs. The HSI research community has a very strong role in providing human performance data that comprises cognitive as well as physical performance recorded in human reliability and human error terminology. This principle and the one following represent all the science and technology activities being performed by HSI professionals throughout government, academia, and industry.

### 1.5.7 HSI Technology

Conceptually, there are three different types of technologies, tools, and techniques important to HSI. First, each domain has tools that are either unique to it or shares commonality with one or more of the other domains. *Task analysis*, for example, is common to all the domains. Anthropometric design tools, however, are primarily HFE tools, although personnel, system safety, and health hazards would find use for them, as well. But domain tools are not generally envisioned to be trade-off technologies. The second and third types of HSI technology are classified under trade-off technologies.

The first of the trade-off technologies is represented in Figure 1.3 as the first integration. With this technology, the six (or more) domains of HSI can be traded off among each other to give the optimal system performance and cost. For example, if the total system affordability includes life-cycle costs [driven primarily by manpower (M), personnel, (P), and training (T)], then methodology is needed to assess such issues as the number of people who will operate and maintain the system throughout its life cycle, the occupational specialty of the people, and the type and amount of training for these people. A proposed system design might show increased automation to replace physical tasks but still demand a large number of high cognition tasks from the human operator. Trade-off techniques could show the advantages and disadvantages from a number of possibilities. For example, the number of operators and/or their skill levels might be reduced through increasing training. Or the number of operators and skill levels might be reduced through further automation. On the other hand, it might be desirable to increase the number of people to offset expensive specialty training or the projected casualty costs shown by the safety, health hazards, or soldier survivability domains.

The second trade-off methodology is used to trade off system capability against affordability based on HSI as an entity. For example, fewer systems requiring fewer personnel could be procured if the reliability of each system were increased because of better human performance. In such an exercise, human factors engineering plays a leading role by searching for design solutions that will assure the desired system performance while attempting to reduce the constraints imposed by the other domains.

### 1.5.8 Test and Evaluation/Assessments

Many of the principles addressed above, particularly those associated with staff requirements, statements of work, source selection, and proposal evaluation are concerned with creating a specification that meets the system buyer's needs and provides high assurance that the best supplier is chosen to meet the buyer's expectations. The HSI principles provide high assurance that the buyer and supplier fully understand the goals and constraints imposed by all the human factors associated with the system being procured. For each stage of design and development, it is just as important that a process is in place to evaluate how well the supplier is meeting the buyer's expectations. The HSI principle is to *crosswalk* all human performance requirements stated (or inferred) in the statement of work (with specifications) and evaluated in the proposal with the test and evaluation (T&E) process, as documented in the T&E plans and assessment procedures. Progress is reported in the official system T&E reports and through independent HSI assessment reports.

Representative users participate in operational T&E. If the system with these users fails the operational test, then the supplier's equipment has failed, not the user.

### 1.5.9 Highly Qualified Practitioners

Perhaps the most important principle of HSI is the requirement to use highly qualified practitioners. On the buyer's side such individuals will be found as domain representatives for the system working groups, as writers of requirements for statements of work, as proposal evaluators, and as assessors for the T&E process. It goes without saying that the supplier should employ equally qualified individuals. This requirement is often overlooked, for example, when the federal government introduces a large program and the needed skills are not immediately available or affordable. Consequently, the organization is tempted to try to implement wide-scale changes with insufficiently qualified practitioners. Such attempts will fail to be successful on anything other than a sporadic basis. The issues raised by HSI are nontrivial and not easily solved simply by imposing constraints on the system developer. HSI cannot influence design in any significant way by imposing requirements that cannot be defended by individuals conversant with the technology or operational complexity of the system. *Checklists cannot replace the technical judgment of personnel possessing the requisite formal education and on job experience that the domains require*. Most of the tools and techniques used by the domains and as HSI trade-off methodologies are applied best by experts in their field.

### 1.5.10 Education and Training Program

The HSI principle for education and training is to provide some aspect of HSI for everyone in the acquisition process, including government, industry, and academia. The implementation of this principle may appear so difficult and expensive that the organization will, as with some of the other principles, be tempted to ignore it, hoping benefits can accrue by a few policy changes and that industry will have incentives to provide more user-friendly products. But, wide-sweeping education and training is considered one of the most important principles for long-term institutionalization of HSI. Even if all the policy and procedures changes are implemented, systems will not be produced routinely that are significantly better if others throughout the procurement process do not "buy in" to the importance of human performance. There is a natural tendency to resist change, and if the future HSI program means doing still more with less, the resistance will be even greater than usual. Education and training is needed, therefore, not only for the practitioners, but also for the rest of those involved in the procurement and systems development processes. Fortunately, the cost to implement this principle is not prohibitive. Government agencies must make some investment to continue to assure a viable education and training program, but most education for the nonspecialist can be achieved by small modifications to other courses. The systems program manager in the DoD, for example, can be exposed to HSI during his already required defense systems management college course work. A one-semester graduate course in HSI could be added relatively easily to existing graduate programs in academic institutions for industrial engineering, systems engineering, or human factors.

## 1.6 HSI PRINCIPLES APPLIED TO SYSTEMS ACQUISITION

While it is easier to define, develop, and deploy systems with strong HSI influence in organizations with high HSI maturity ratings, it is possible for a few select systems to have acceptable quality HSI programs, even though the organization as a whole is weak on HSI. The phenomenon of a good HSI program within a weak HSI organization is because of the influence some PMs can have with procuring individual systems. This means it is extremely important for HSI professionals to be able to sell individual PMs on the cost and performance benefits HSI can offer.

A recent army study on HSI success factors for army systems concluded all of the HSI principles for organizations can be applied to specific systems procured by organizations (Booher, 1999). This study is described in more detail in Chapter 18, but some of the features of the HSI principles when applied to specific systems are briefly discussed here.

One of the study tasks was to identify critical factors resulting in MANPRINT cost and performance benefits for army systems. Ten representative army systems were selected and reviewed in this study. Most of the study systems had been recently reviewed by the U.S. Army Audit Agency (1997) to determine how well MANPRINT had been implemented on its newest systems coming into the army's inventory. Table 1.4 lists the systems reviewed and indicates their status in meeting the army's acquisition objectives at the time of the review. Six of the systems could be considered successful; two were marginal because of difficulties meeting soldier requirements, within cost, schedule, and performance objectives; one was fielded with reduced performance acceptance (degraded); and the army canceled one (failed).

The army systems were then rated on importance to systems success using three rankings: A, critical; B, important; and C, not important. When the same 10 principles for organizations are looked at from a PM's point of view, the principles can be described as 10 critical factors for systems acquisition. Each of the HSI system factors applies the corresponding principle as described below.

***Factor 1. Top-Level (TL) Leadership and Understanding*** This factor is the degree to which top-level management supports HSI concepts and practices for the specific system being developed. Top-level management includes the PM and the responsible decision makers he or she must report to in achieving program objectives. For large

**TABLE 1.4 Systems Reviewed for MANPRINT Involvement**

| System | Army Objectives |
|---|---|
| 1. Comanche Helicopter | Successful |
| 2. Longbow Apache Helicopter | Successful |
| 3. Javelin Antitank Guided Missile System | Successful |
| 4. Multiple Launch Rocket System—Extended Range | Successful |
| 5. Command and Control Vehicle (C2V) | Marginal |
| 6. Family of Medium Tactical Vehicles (FMTV) | Degraded |
| 7. Armored Gun System | Failed |
| 8. Crusader Artillery/Resupply | Successful |
| 9. Land Warrior | Marginal |
| 10. NBC Reconnaissance System (NBCRS–Fox) | Successful |

systems this factor is almost indistinguishable from principle 1 for the organization itself. For smaller systems, however, a few individuals can determine the strength of this factor. These individuals are those most responsible for the acquisition of small systems—the PMs themselves. Because of the rapid and controversial systems engineering trade-offs that often need to be made, it is important that the PM also understands HSI concepts and data as well as any other systems engineering concepts and data.

Factor 1 was found to be critical in determining system success or failure for 7 of the 10 systems evaluated. Top-level support and understanding was weak for the 2 marginal systems. Only in one case (Fox vehicle) was this factor considered unimportant to a successful program. (See Section 18.4.2 for the answer to this unusual finding.) Generally, where this factor is strong, the program is strong; and where it is weak, the program is weak.

***Factor 2. Human-Centered Design (HCD)*** Strong emphasis on HCD begins in the requirements stages. This factor generally follows the trend set by the PM on factor 1. Where it tends to differ is for nondevelopmental systems. The HCD concept is still important but can never be as central as with a full developmental program. Seven systems showed this factor to be either critical or important. The two systems that ultimately became degraded or failed considered HCD important but were unable to solve important MANPRINT issues along the way. This factor was not important to the success of two small systems.

***Factor 3. Source Selection (SS) Criteria*** Source selection criteria, which make HSI instrumental in determining who will win or lose a system procurement, is a necessary system success factor. Those army programs that were successful generally had strong MANPRINT source selection evaluation criteria. However, strong HSI source selection criteria without many of the other factors do not assure success. The marginal and degraded systems had adequate MANPRINT criteria. In other words, HSI source selection criteria are generally necessary but not sufficient to assuring a successful system.

***Factor 4. Domains Integration (DI)*** The integration of all the domains during the acquisition process was a factor that was critical or important on all 10 systems. In marginal, degraded, and failed systems the fact that this domain was active ensured the MANPRINT issues were properly identified. Once properly identified, the issues could be addressed in some cases, well enough to save the system, although in at least one system, it was too late.

***Factor 5. System Documentation Integration (SDI)*** The integration of MANPRINT into a system documentation process was a critical factor for 8 of the 10 systems. Adequate attention to this factor helped 6 programs to be successful. Weaknesses in system documentation integration made it a degradation factor for 3 of the systems. Once again the Fox vehicle was unique in that this factor was not important to its success.

***Factor 6. Quantitative Human Performance (QHP)*** For people to truly be included as part of the system, designers must be able to quantify their performance. A person can be treated as a system component and participate in trade-off methodology if the performance parameters are quantifiable. In 9 of the 10 systems this factor was either critical or important. The only exception was one system with little human involvement.

***Factor 7. HSI Technology (HT)*** HSI technology, which includes unique domain and trade-off methodology between domains, was critical or important for half of the systems. (See Fig. 1.3, the first stage of the double-integration process.) For those systems where several aspects are important to soldier performance but work against each other, trade-off methodology was essential. On the Javelin Antitank Guided Missile System, for example, a light-weight system is needed for single soldier portability, but many survivability characteristics needed by the soldier necessarily increase the weight. MANPRINT considers each of the domains, but frequently trade-offs among them will determine the best overall benefit to the soldier. Although the other half of the systems did not have trade-off issues among domains, this factor is still very important for aiding trade-offs between HSI and other systems engineering factors. (See Fig. 1.3, the second stage of the double-integration process.)

***Factor 8. Test and Evaluation (T&E)*** T&E was the only factor critical to all 10 systems. It is the one factor that must be present no matter what has gone before. It is the final and most reliable assurance factor for the army that the soldier will receive a safe and effective weapon before going into battle.

***Factor 9. Practitioners (PR)*** Skilled and available practitioners were critical to all systems except the one that did not have significant human involvement. Without this factor, several of the other factors could not be performed. [Notice, e.g., their importance to factors 4 (DI), 5 (SDI), 6 (QHP), 7 (HT), and 8 (T&E)]. Also factors 2 (HCD) and 3 (SS) need input from skilled practitioners.

***Factor 10. Education and Training (ET)*** The HSI education and training are essential to assure practitioners are qualified. For 9 of the 10 systems this factor was either critical or important to system success. The ET of nonpractitioners is almost equally important. In those systems that were marginal, degraded, or failed, the lack of appreciation of MANPRINT by nonpractitioners was one of the primary reasons for their deficiencies.

*Perhaps the most important finding from the study was that the 10 HSI principles provide 10 corresponding factors that are both necessary and sufficient for program management attention to assure system performance success.* Each of the 10 principles when applied to specific systems has unique characteristics, yet considerable connectivity with each of the others. In some cases the elimination of only one factor [such as factor 1 (TL)] will cause many of the others to degrade. On the other hand, one factor may be reduced, and the effect will be to create greater demand on one or more of the other factors, resulting in a domino effect among the factors. For example, if a PM decides to reduce the number of people representing the seven domains [i.e., reducing influence of factor 4 (DI)], a decrease in numbers of skilled practitioners [factor 9 (PR)] is created, which means the available practitioners must be skilled on more than one domain, which means greater pressure is applied on education and training [factor 10 (ET)] to produce higher skilled practitioners.

## 1.7 HSI ORGANIZATIONAL MATURITY

Attempts to make wide, sweeping changes in government organizations that either design and operate or influence the design and operation of technology are not new. As early as the mid-1960s, the air force, navy, and Department of Transportation initiated major human factors programs (Air Force, 1967; Fucigna, 1968; Little, 1966). In the 1980s, the Nuclear Regulatory Commission recognized the need for a comprehensive human factors program (Hopkins et al.; 1982; Moray and Huey, 1988). And in the 1990s, the air force and navy attempted to implement but soon shelved their IMPACTS and HARDMAN programs.

Although numerous specific examples of positive human factors influence can be cited, it is fair to conclude that past attempts to incorporate human factors as a primary consideration in government policy for the procurement or regulation of the nation's technology have been marginal at best. Human factors continued in the late 1990s to be viewed as a contributor to or supporter of design and operations that had not yet reached an equal footing with engineering or operations disciplines. The challenge for HSI in the twenty-first century is not only to reach an equal footing with these disciplines but also to actually surpass them in certain aspects; especially in organizational decision making about the purpose and approach to systems integration.

### 1.7.1 Assessment of Army MANPRINT Program

The 10 principles provide a means to assess the organizational maturity of HSI programs. In the 15 years since MANPRINT was first introduced to the U.S. Army there has been significant opportunity to refine the processes and techniques that are important to the HSI concept and show historically how well the army as an organization has applied the MANPRINT model to its military systems. Because MANPRINT has been continuously in existence since its inception, it provides the opportunity to evaluate an HSI program over time.

Three assessments of the U.S. Army MANPRINT program are shown in Figure 1.4 (1989, 1994, and 1999). The army HSI program was evaluated for level of maturity on each of the 10 HSI principles. A five-point scale defined on the figure was used for level of maturity. The HSI professionals familiar with the army MANPRINT program, from its inception until the present, performed the assessments. Major armywide studies on the effectiveness of MANPRINT implementation were primary sources for the assessments (Peters and Perkins, 1991; U.S. Army Audit Agency, 1997; General Officer Steering Committee, 1998).

The maturity assessments show it is very difficult to achieve and maintain an excellent rating on any of the HSI principles. On the other hand, excellence on all the principles is attainable. By 1989, the army MANPRINT program had achieved excellent ratings on 5 of the 10 principles and good ratings on two others. Curiously, two additional principles (7 and 8) that were among the lowest rated principles in 1989 are rated good by 1999. Only principle 2 (human-centered design focus of "systems") never reached better than average for any of the evaluation periods.

The assessments also show how easily support for an HSI organization can degrade. Within the 5 years between 1989 and 1994, all five of the excellent principles ratings fell at least three levels. Source selection (principle 3) fell four levels, all the way to the bottom. This drop was clearly not because no value was added from HSI, but rather because of the

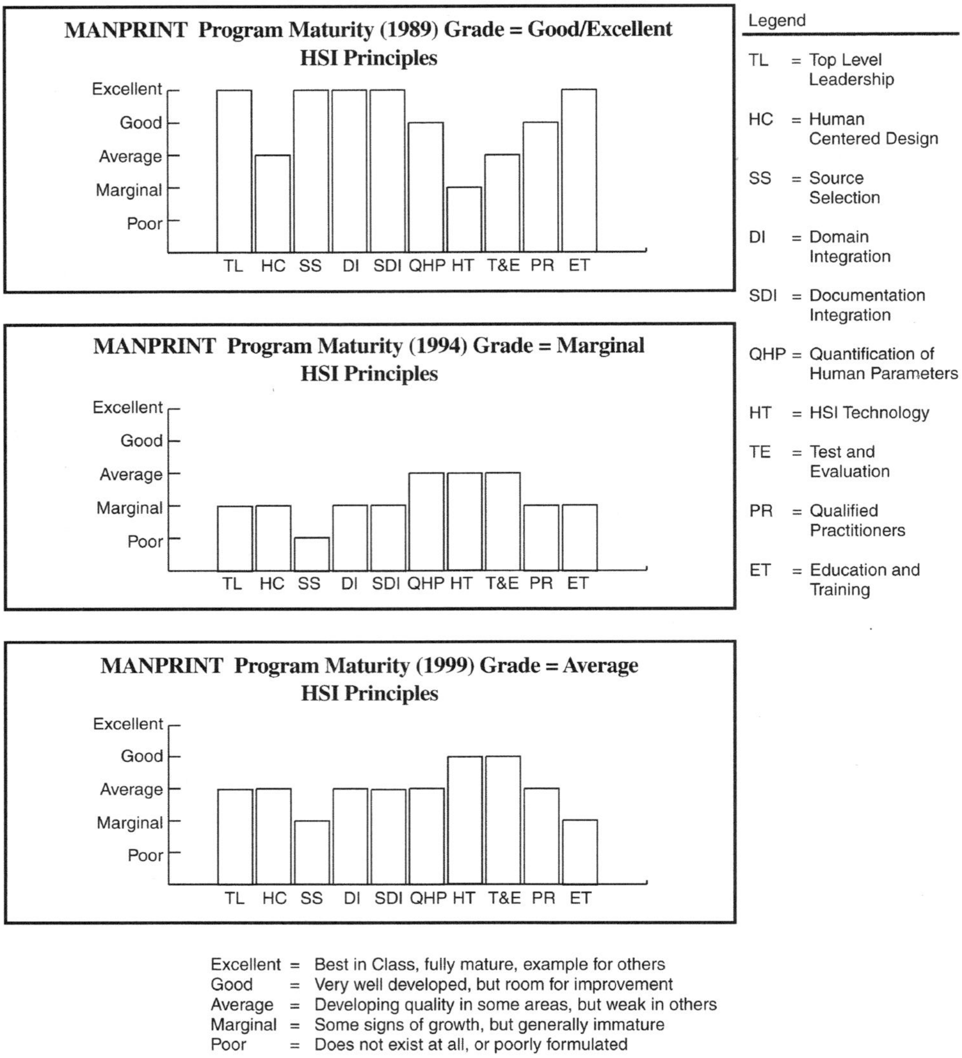


**Figure 1.4** U.S. Army MANPRINT program maturity.

turbulence within the army in changing to new acquisition strategies while undergoing major downsizing with its acquisition personnel. In 1998, once it became recognized that HSI was essential to accomplishing the new army objectives of fighting multiple missions with a reduced force, the army leadership initiated a revitalization effort for HSI, such that the army was able to bring back most of the principles' maturity to average by the time of the 1999 evaluation.

Also as mentioned above, the army was rated good on two principles (HSI trade-off methodology and T&E), which have continuously improved since they were first rated in 1989. This improvement is fortunate since these two principles are particularly critical in the new systems acquisition environment. The new acquisition environment relies upon performance-based requirements, advanced simulation and modeling, and T&E measures of performance to reduce risk and to determine systems acceptance criteria. This approach simply cannot be made to work without increased emphasis on the human component in defining requirements, simulating operational systems and environments, and measuring system performance in operational environments.

All 10 principles need to advance in maturity, however, if an organization wishes to make maximum use of the HSI approach to systems acquisition. If any systems acquisition organization would place an emphasis on HSI today as the army did on MANPRINT in the late 1980s, there is little doubt that the organization could achieve an excellent rating on all 10 HSI principles.

### 1.7.2 Assessment of Other HSI Programs

By the end of the millennium, a number of other government organizations had started to implement HSI programs. The most notable was the Ministry of Defence for the United Kingdom, which had implemented an HFI program for all three of its major services. In the United States, the navy and air force had provided a renewed thrust to HSI, and the FAA had also made significant strides in its HFI and systems safety programs. Based on assessments from HSI/HFI specialists within the various organizations depicted, Figure 1.5 provides a snapshot view of the maturity of these four programs on the 10 HSI principles in the year 2000. All assessors were familiar with the army MANPRINT program, which was used as the baseline for comparisons.

Two major *overall* comparisons among organizations are useful from the assessments summarized in Figures 1.4 and 1.5. These are the other organizations compared to the U.S. Army and to each other. Compared to the U.S. Army, the UK program was rated the same as the U.S. Army in 1999 (average), but with the U.S. Army still ahead on several of the principles. The FAA (marginal) was one step below the U.S. Army and the United Kingdom but farther along than either the U.S. Navy or U.S. Air Force in 2000. Since these assessments were made, the greatest HSI improvement activity has been shown by the U.S. Navy, which if assessed in 2003, would likely receive at least a marginal overall rating.

To make significant HSI maturity progress equal to the U.S. Army and UK programs, the most critical principle for the other organizations is top-level leadership. As that rises, other weak principles can receive more attention. For the navy, the most urgent attention needed (after leadership) in 2000 was for source selection, domain integration, qualified practitioners, and education and training. The same four weakest principles applied to the U.S. Air Force and two of the same (source selection and domain integration) applied to the FAA. The other weak principle for the FAA (HSI technology) does not receive the same attention as either the U.S. or UK military because of fewer manpower, personnel, and training trade-off opportunities.[10]

## 1.8 DISCUSSION AND SUMMARY

Human systems integration is a technical and managerial concept with specific emphasis on methods and technologies that can be utilized to apply the HSI concept to systems integration. As a concept the top-level societal objectives of HSI are to significantly and positively influence the complex relationships among: (1) people as designers, customers, users, and repairers of technology; (2) government and industrial organizations that

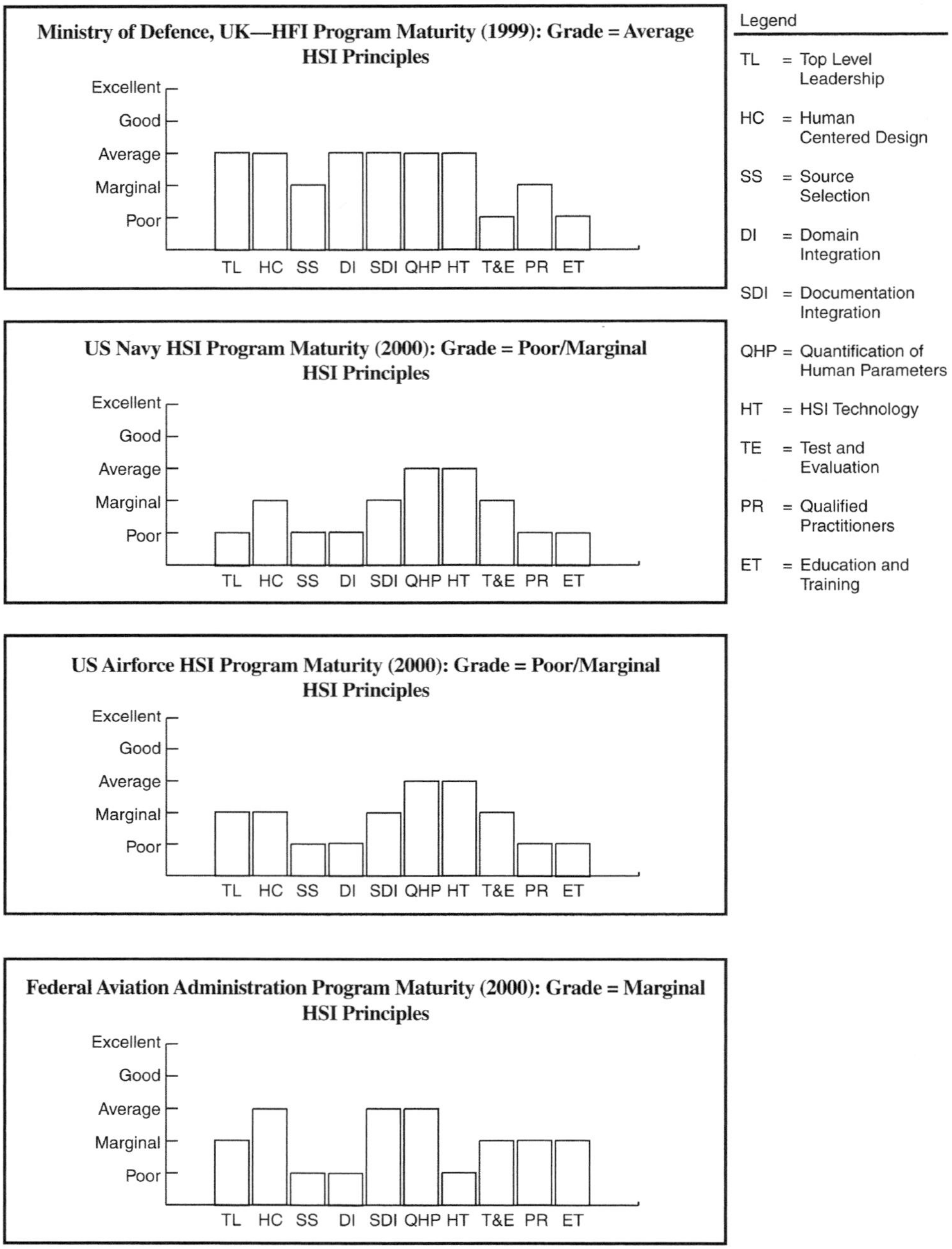


Excellent = Best in Class, fully mature, example for others
Good = Very well developed, but room for improvement
Average = Developing quality in some areas, but weak in others
Marginal = Some signs of growth, but generally immature
Poor = Does not exist at all, or poorly formulated

**Figure 1.5** Non U.S. Army HSI programs maturity.

regulate, acquire, design, manufacture, and/or operate technology; and (3) methods and processes for design, production, and operation of systems and equipment.

Most of the technical and managerial advances suggested by the HSI concept can be accomplished within an overall systems integration philosophy that places a special emphasis on how its roles and technology can be included within systems engineering and systems management processes. As a concept, HSI is fully compatible with those systems engineering processes relevant to systems definition, development, and deployment and their life-cycle phases, as well as the systems engineering methods, tools, and technologies. As a top-level model, HSI brings two novel features to the systems engineering model. These are (1) the highly concentrated user focus on all aspects of the systems definition, development, and deployment stages and (2) the application of the human-related technologies and the HSI disciplines throughout the systems engineering management and technical processes. No system, product, or equipment inputs can be considered as having had an adequate consideration of the people component if it does not pass through the HSI process modulated with these two inputs.

As a unique concept for integrating people, organizations, and technology, HSI can offer a wide range of benefits to an organization. Too often, non-HSI individuals do not appreciate these potential benefits because the benefits have not been communicated in a way that reflects most directly on their particular role in the organization. For example, people who have high levels of responsibility for systems acquisition decisions should be interested in HSI performance measures that help assess quantitatively the human error risk with operational systems. This can be contrasted with those primarily concerned with operational processes within the organization. The latter might be more stimulated by the ability of the HSI professional to help them develop people-oriented procedures that utilize user-centered techniques such as functional and task analyses.

The applicability of HSI varies with sociotechnical systems complexity. Sociotechnical systems can range from very highly complex organizations (such as the DoD) to critical technological human–machine subsystems (such as an aircraft cockpit). The HSI process needs support from the highest levels of an organization but is best applied as a concept to specific technological systems such as an aircraft or a control room. As HSI develops technologically, it will also become more relevant to systems design of more complex sociotechnical organizations that comprise a number of technological systems working in unison, such as an aircraft carrier or a hospital.

The HSI concept has unique aspects not fully demonstrated with any other human factors approach to systems integration. Such HSI aspects as "people-oriented processes," "focus throughout the organization on competence and motivation," and "educating all people in the process" are the same characteristics found in high-quality-oriented organizations. The HSI concept is closely aligned to top-level management and organizational objectives and as such aids technology decisions that benefit people while meeting these objectives. Most human factors (HF) advances have focused on the development of technologies and disciplines, which, if utilized, could produce high-performance systems with low human error rates. However, HF concepts have not traditionally played a significant role in the organizational decisions about what systems the organization should acquire. Neither has HF tended to influence the top-level decision makers about their policies toward people and education throughout the organization. The HF concept has begun to provide decision makers with sound economic arguments for human factors designs centered on the user; but these arguments are focused almost entirely on customers with specific product roles, not those in top-level management or operational processes roles. As a discipline, HF is making great strides toward "multidisciplinary views of design," making the human component an "inherent part of the system," and the "quantification of people variables"; but it does so largely in a support role to the engineering processes. As such, HF is not likely to emphasize "resources redirection rather than net increases" for a system acquisition, provide trade-off techniques among the HSI domains in design decisions, or be seen as the primary technology advancement for the system being developed.

During the past decade, 10 HSI principles have been identified that, to the degree they are applied, seem to assure that large organizations will capture the performance, cost, and safety objectives they desire for their systems. Conversely, to the degree any of these principles are left out entirely or a few are followed only marginally, large organizations risk their systems not meeting their desired system objectives. Moreover, specific systems programs that have followed these principles have been extremely successful, while those that have made compromises have made marginal progress. If the 10 HSI principles for organizations are looked at from a PM's point of view, they indicate 10 critical factors for system acquisition. In analyzing 10 army systems, Booher (1999) found the HSI factors that correspond to the 10 principles are both necessary and sufficient for program management attention to assure system cost, performance, and schedule success.

The 10 HSI principles are a blend between technical and managerial features. Some (such as top-level leadership, source selection, and domains integration) are purely management and organizational factors that can be raised or lowered in maturity through policy decisions. Others (such as quantitative human performance and HSI technology) are primarily technical factors. These tend to progress at the rate science and technology progresses for basic human performance knowledge and techniques. But still others are combinations of managerial applications and technical developments (such as HCD, skilled practitioners, and education and training). As technology advances, the organization can speed or impede progress depending on how well it understands and supports maturity development on these principles.

The 10 principles provide a means to assess the organizational maturity of HSI programs. In the 15 years since MANPRINT was first introduced to the army, there has been significant opportunity to refine the processes and techniques that are important to HSI concepts and show historically how well the army as an organization has applied the MANPRINT model to its military systems.

By describing the HSI model and its principles, this chapter provides the first step of a new movement within both public and private systems acquisition organizations to implement and improve their HSI capability.

In the chapters that follow, I asked the authors and myself to answer two questions about their contribution that relate to the HSI philosophy presented in this introduction. The first question was which of the 10 HSI principles relates most directly to their chapter. The second question was which of the principles relates to secondary or related information in their chapter. The matrix in Figure 1.6 shows our consensus. Each chapter indicated with a heavy dot corresponding to a principle will provide amplifying information bearing directly on that principle. Each chapter indicated with a white circle corresponding to a principle provides secondary or related information for the principle. Chapter 2, for example, provides information that addresses organization and management change concepts directly related to principle 1. Each chapter should therefore help the reader better understand some of the guidelines for top-level leadership that are highly beneficial to implementing HSI within an organization. Secondary or related information

| Chapters | Principles/Factors | 1. Top level Leadership | 2. Human Centered Design | 3. Source Selection | 4. Organizational Integration | 5. Documentation Integration | 6. Quantification–Human Parameters | 7. HSI Technology | 8. Test and Evaluation | 9. Highly Qualified Practitioners | 10. Education and Training |
|---|---|---|---|---|---|---|---|---|---|---|---|
| Part I | 2 | ● | ● | | ● | | | | ○ | ○ | ○ |
| | 3 | ● | ● | | ● | ● | | | ○ | ● | ○ |
| | 4 | ○ | ● | ● | ● | ● | ○ | ○ | ● | ○ | ○ |
| | 5 | ○ | ● | | | | ○ | ○ | | ● | ● |
| Part II | 6 | ○ | ● | ○ | ○ | ● | | | ● | | ○ |
| | 7 | ○ | ● | ● | | ● | ○ | ○ | ● | ○ | ○ |
| | 8 | | ● | ○ | ○ | ○ | ● | ● | ● | ○ | ○ |
| | 9 | ● | | ○ | ○ | ● | | | | | ○ |
| | 10 | ● | ● | | ● | ● | ● | ○ | ○ | ○ | ○ |
| Part III | 11 | ○ | ● | ○ | ○ | ○ | ● | ● | ○ | ● | ○ |
| | 12 | ○ | ● | ○ | ● | | ○ | ○ | ○ | ○ | ○ |
| | 13 | ○ | ● | ○ | | ○ | ● | ● | ○ | ● | ○ |
| | 14 | ○ | ● | ○ | | ○ | ● | ● | ● | ● | ○ |
| | 15 | ○ | ● | ○ | ○ | ○ | ● | ● | ● | ● | ○ |
| | 16 | ○ | ● | ○ | ○ | ○ | ● | ● | ● | ● | ○ |
| | 17 | ○ | ● | ○ | | | ● | ○ | | ○ | ○ |
| Part IV | 18 | ● | ● | ● | ● | ● | ● | ● | ● | ○ | ○ |
| | 19 | | ● | | | | ● | ○ | ○ | ○ | ○ |
| | 20 | | ● | | ○ | ○ | ● | ● | ● | ○ | ○ |
| | 21 | ○ | ● | | | | ○ | ○ | ○ | | ○ |
| | 22 | ○ | ● | ○ | ● | ○ | | ● | ○ | ○ | ○ |
| | 23 | ● | ● | | ○ | | ● | ○ | ● | | ○ |
| | 24 | | ● | | | | | | ● | | ○ |

**Figure 1.6** HSI principles by chapters.

will be found in Chapter 2 on principle 8 (measuring and evaluating organizational changes will indirectly aid systems acquisition processes) and principle 10 (if the guidelines for change are followed, the quality and amount of education and training devoted to HSI will increase).

## 1.9 BOOK OVERVIEW

The book organization is loosely modeled after that of the 1990 MANPRINT book. Chapters are categorized within four parts, but not meant to be restricted solely to the category where they are found. Any one of the chapters could provide valuable insight into each of the other parts. An attempt was made, however, to cluster chapter topics that provide information judged by the editor to fit best under the broad umbrella provided by each part as briefly described below. The book is constructed so that the reader can read from cover to cover, skip to parts, or read only pertinent chapters. The reader has an opportunity to decide on the relevance of chapter information before actually reading the full chapter. Each part briefly describes the chapters contained in the part, and each chapter begins with an introduction section.

Part I, *Organization, Management, and Culture*, discusses the engineering and management environments that affect HSI implementation, operation, and effectiveness. In particular it stresses those organizational, managerial, and cultural environments that procure, produce, and operate systems and equipment. To successfully apply HSI concepts, the organization's leadership, culture, and associated disciplines must at least tolerate, if not fully accept, the concepts. From the chapters included in Part I, the reader will find information that addresses the following types of questions:

- What is the role leadership must play for an organization that wishes to introduce HSI into its systems acquisition culture; particularly in motivating and managing change introduced by HSI concepts?
- What impact do cultural environments have upon implementation and operation of HSI programs?
- What are the roles and interfaces of HSI in systems engineering and management?
- What are key economic factors that drive decisions in the acquisition and systems engineering processes?
- What are the special needs and opportunities for HSI education and training for both the HSI professional and those other system acquisition stakeholders in an organization, newly introduced to the benefits of the HSI approach to systems integration.

Part II, *Systems Acquisition and Management Processes*, describes how HSI is involved throughout the major stages in acquiring a system, beginning with requirements determination (which have major personnel and training trade-offs), to system specifications (with extensive communication between the organization seeking a new system and the builder of the system), to system design and development, and, finally, to test, evaluation, and assessment of system performance. Part II includes:

- Descriptions of the systems acquisition model from both the government buyer's and the contractor seller's perspectives
- Guidance on how HSI requirements should be determined in the acquisition process
- Examples showing the importance of HSI system design trade-offs with personnel and training
- Descriptions of test and evaluation techniques that include HSI as part of system performance
- Special HSI issues associated with simulation architectures and procurement standards as applied in the modern acquisition process

Part III, *Methods, Tools, and Technologies*, describes the state-of-the-art for methods, tools, and technologies covered by the seven domains of HSI and those specially designed to integrate several domains. More specifically, Part III focuses on the description of tools, techniques, and technologies used by the HSI professional in the analysis and assessment of systems performance and integration issues. This part, more than any of the others, is designed to help the HSI professional acquire useful analysis and/or assessment information for system acquisition decisions. It addresses such questions as:

- Why conduct a particular type of analysis and why, or why not, employ a particular HSI method or tool?
- When should a tool, technique, or technology be used with respect to system development?
- What resources (time, money, computers, skills and qualifications of the analysts) are required for effective use of HSI methods and tools?
- What data are required to support a particular tool, technique, or technology?
- What are the critical interfaces among parameters and methods covered by the several domains?
- What methodologies are most applicable to cost–benefits analyses for HSI?

Part IV, *Applications*, gives us a wide range of examples that illustrate the methods and principles of HSI applied to systems from both the public and private acquisition processes. Many of the HSI systems applications presented in Part IV are drawn from military, aviation, and commercial environments that provide representative samples of the types of organizations, cultures, and technologies HSI professionals are likely to find themselves working in the future. Some of the more dramatic cost and performance benefits from HSI are demonstrated on major systems procurements such as those procured by large-scale public acquisition (DoD, NASA, FAA); but HSI can play an extremely important role in small systems developments such as appear with new commercial products, as well.

## NOTES

1. The background is an updated version of the background provided in Booher (1990, pp. 1–2). Much of the original material is repeated here with kind permission of Kluwer Academic Publishers.
2. The Three Mile Island nuclear accident description opened the MANPRINT book in 1990. The last sentence still applies 10 years later.
3. http://www.swishweb.com/Disasters/Aircraft/disaster03a.htm.
4. Railway Accident Report, NTSB Number RAR97/02. Collision and Derailment of Maryland Rail Commuter MARC Train 286 and National Railroad Passenger Corporation Amtrak Train 29. http://www.ntsb.gov/Publictn/1997/RAR9702.htm.
5. http://www.swishweb.com/Disasters/Aircraft/disaster04a.htm.
6. Total U.S. motor vehicle fatal crashes in 1999 were 37,043 (*Traffic Safety Facts 1999*); total motor vehicle fatalities in 1999 were 41,611 (*NHTSA 37-00*). The number of U.S. Army casualties (deaths) in Southeast Asia from 1957 to 1997 was 38,201.
7. Quality experts frequently state that 20 to 40 percent of payroll costs can be associated with waste, failure, and rework (Crosby, 1979; Deming, 1986; Juran, 1987). The $600 billion estimate is based on 20 percent of labor costs per annum, around $3 trillion (1989) and the national debt around $6.5 trillion (1999).
8. In addition to systems engineering, the DoD regulation also includes HSI requirements in the section for support strategy (C2.8.5. HSI).

9. DoD 5000 was canceled on August 29, 2002. Changes in the DoD acquisition process regulations are frequent; so references such as this in the *Handbook* cannot be depended on as the latest official regulatory policy. However, as of the publishing date, the June 2001 DoD 5000.2R document still provided the most relevant guidance information for HSI in military systems acquisition.
10. The distinction between human factors technology and HSI technology is discussed in other chapters (see especially Chapters 8, 11, 12, and 13).

## REFERENCES

Air Force Systems Command, Headquarters. (1967). *Handbook of Instructions for Aerospace Personnel Subsystems Design*, AFSCM 80-3. Washington, DC: Andrews Air Force Base.

Booher, H. R. (1999, May). *Acquisition Process Factors in MANPRINT*, Final Report for Army Research Laboratory, Human Engineering and Research Directorate. Reston, VA: Raytheon Systems Directorate, Instructional Systems Group.

Booher, H. R. (Ed.). (1990). *MANPRINT: An Approach to Systems Integration*. New York: Van Nostrand Reinhold.

Crosby, P. B. (1979). *Quality Is Free*. New York: McGraw-Hill.

Deming, W. E. (1986). *Out of the Crisis*. Cambridge, MA: Massachusetts Institute of Technology, Center for Advanced Engineering Study.

Fucigna, J. (1968). *Human Performance Development Program for the Naval Material Command*. Darien, CT: Dunlap and Associates.

General Officer Steering Committee (GOSC). (1998, July). *Report to the General Officer Steering Committee (GOSC) on the Army's Manpower and Personnel Integration (MANPRINT) Program*. Prepared for the Assistant Secretary of the Army (Manpower and Reserve Affairs). Washington, DC: Booz-Allen & Hamilton.

Hopkins, C. O., et al. (1982). *Critical Human Factors Issues in Nuclear Power Regulation and Recommended Comprehensive Long Range Plans*, Technical Report NUREG CR-2833. Washington, DC: Nuclear Regulatory Commission.

Juran, J. (1987). *On Quality Leadership*. Wilton, CT: Juran Institute.

Little, A. D. (1966). *The State of the Art of Traffic Safety: A Critical Review and Analysis of the Technical Information in Factors Affecting Traffic Safety*. Washington, DC: Automobile Manufacturers Association.

Moray, N., and Huey, B. (1988). *Human Factors Research and Nuclear Safety*. Washington, DC: National Academy Press.

Perrow, C. (1999). *Normal Accidents: Living with High-Risk Technologies*. New York: Basic Books.

Peters, J., and Perkins, M. (1991, May). *MANPRINT 2000: Program Assessment and Enhancement*. Science Applications International. McLean, VA.

Sage, A. P., and Rouse, W. B. (1999). *Handbook of Systems Engineering and Management*, New York: Wiley.

U.S. Army. (1990, May). *Manpower and Personnel Integration (MANPRINT) in the Materiel Acquisition Process*, Army Regulation 602-2. Washington, DC: Department of the Army.

U.S. Army. (2001, June). *Manpower and Personnel Integration (MANPRINT) in the Systems Acquisition Process*, Army Regulation 602-2. Washington, DC: Department of the Army.

U.S. Army Audit Agency. (1997, June 10). *Incorporating MANPRINT into Weapons Systems Development*, Audit Report: AA 97-205. Washington, DC: Army Audit Agency.

U.S. Department of Defense. (2001, June 16). *Mandatory Procedures for Major Defense Acquisition Programs (MDAPS) and Major Automated Information Systems (MAIS) Acquisition Programs*, DOD 5000.2-R. Washington, DC: U.S. Department of Defense.

PART I

# ORGANIZATION, MANAGEMENT, AND CULTURE

Organizations vary considerably to the degree their leaders (a) understand the human systems integration (HSI) concept, (b) are capable and willing to introduce HSI principles and methods into the organization's systems acquisition processes, and (c) create and sustain a culture where the HSI principles and methods thrive. Chapters 2 to 5 primarily focus on those aspects that can help the reader better understand the complex nature of creating and sustaining HSI in a systems acquisition culture.

Chapter 2, by Harris, Hart, and Shields, points out that while most organizational environments are not currently favorable to successful application of HSI, it is possible to transform them, provided senior leaders are willing to articulate the need for organizational change and build a structure to achieve an organizational culture favorable to HSI. To help leaders and HSI proponents better understand the transformation changes needed to achieve such a culture, Harris, Hart, and Shields first introduce the transformation changes process; then for the largest portion of the chapter they describe a four-stage implementation model (decide, guide, support, and sustain) accompanied with specific recommended tasks required at each stage; and finally they wrap up their effort by identifying barriers to change with strategies for overcoming them.

In Chapter 3 Hewitt and Piccione discuss the systems acquisition culture and HSI interactions from four perspectives. First is the broad perspective that addresses other surrounding cultures such as political and research environments that affect the systems acquisition culture that in turn affects HSI. Second is the historical view of those roles HSI has played in the past, and third is how systems acquisition organizations view themselves. Together these latter two views have the most influence currently in defining roles and responsibilities for HSI in systems acquisition. The fourth perspective is from the HSI culture itself, considering changes and trends within HSI, which can have ramifications for new expanded roles in the future. These four perspectives help provide a basis for understanding the distinction between the primary roles HSI players do have in government and commercial environments and the roles they perhaps should have. Hewitt and Piccione provide considerable detail on specific responsibilities, tasks, decisions, and interfaces for various HSI roles in systems acquisition, with particular emphasis on those

*Handbook of Human Systems Integration*, Edited by Harold R. Booher.
ISBN 0-471-02053-2 

roles considered critical system design roles that can positively influence systems performance and effectiveness. Such critical roles map the systems acquisition process starting with system concepts and requirements, running through design and development stages, and culminating in test and evaluations of those systems organizations wish to acquire.

Smootz in Chapter 4 describes the *defense acquisition management framework*, using it as a general model to show how and when human systems considerations should be integrated into the life-cycle process of any complex system. He emphasizes that it is only when key issues associated with variables such as manpower, personnel, training, and the objectives of the interface are addressed early, before fundamental decisions about requirements, investment strategy, and design approach are made, that HSI can have the impact that it needs to have to ensure that human performance contributes to, rather than detracts from, the cost and operational effectiveness of the total system. The acquisition framework will be important to remember and refer to when reading most of the other chapters in Parts II, III, and IV.

In Chapter 5, Kleiner and Booher present HSI for three types of reader. First are the HSI professionals who work on the seven domains of HSI. The education and training needs for the HSI practitioner are discussed in terms of what are the knowledge and skill requirements to perform quality HSI, what academic and other institutions provide relevant portions of these needs, and what is the vision for careers in HSI. Second are teachers and developers of programs and curricula for teaching advanced courses in HSI. Suggested topics and curricula for HSI education and training are presented, including a review of currently available academic programs. Third, is everyone associated with the systems acquisition process; from top-level decision makers, to program managers, to all those non-HSI individuals making input to the systems acquisition process. All three of these readers will play important roles in achieving and sustaining HSI in the various systems acquisition organizational structures and cultures described in Chapters 2, 3, and 4. This is the one chapter in the *Handbook* devoted primarily to principle 10 (education and training) and provides information for career paths in HSI adding to principle 9 (highly qualified practitioners) as well.

CHAPTER 2

# Leadership That Achieves Human Systems Integration

CHARLES S. HARRIS, BETTY K. HART and JOYCE SHIELDS

> The single most visible factor that distinguishes major cultural changes that succeed from those that fail is competent leadership at the top.
>
> —Kotter and Heskett, 1992

## 2.1 INTRODUCTION: BEYOND REDUCTIONISM

Stephen Jay Gould celebrated the February 2001 final release of data on the Human Genome Project as "a great day in the history of science and of human understanding in general." Compared to the number of genes in the humble fruit fly (13,000 to 14,000) or the round worm (just over 19,000), conventional scientific views had estimated well over 100,000 to 142,034 genes in *Homo sapiens* to account for the vastly greater complexity of humans. The Human Genome Project's findings were unexpected: We possess between 30,000 and 40,000 genes—fewer than half again as many as the tiny roundworm. Human complexity does not result from one-directional flow (one gene to protein) to generate our complexity; a single gene can create several messages because of the existence of coding (exons) and noncoding (introns) segments. As Gould exulted, this finding releases us from reductionism:

> From the late 16th century . . . science has strongly privileged the reductionist mode of thought that breaks overt complexity into constituent parts and then tries to explain the totality by the properties of these parts and simple interactions fully predictable from the parts. [This model] works triumphantly for simple systems—predicting eclipses or the motion of planets—but not [for] the histories of their complex surfaces. . . . The collapse of the doctrine of one gene for one protein . . . marks the failure of reductionism. . . . [T]he key to complexity is not more

*Handbook of Human Systems Integration*, Edited by Harold R. Booher.
ISBN 0-471-02053-2 

genes, but more combinations and interactions generated by fewer units ... and cannot be predicted.... So organisms must be explained in whole and not as a sum (of parts).

—Gould, 2001

While bureaucratic organizations and their transformation are not individuals, they have been the object of reductionist thinking. Most of the techniques for implementing change are directed at the individual within the organization's internal environment rather than that of the organization as a system (Hay Management Consultants, 1996). Too often system properties are disregarded and changes in individual variables (reward, advancement, and job satisfaction) are equated with changes in the organization itself.

The *structural* model of organizational change illustrates a second form of reductionism. Here the focus is on tangible, visible processes and technologies. Often, enterprises are viewed as a collection or conglomerate of parts and functions similar to a piece of complex machinery. Such an approach fails to consider explicitly the capacity of people to work and manage in new ways (Shields et al., 1999). The *behavioral* model of change seeks to avoid this form of reductionism, making explicit the changed behaviors that will be required of participants. Here it is members of the organization who will be tasked with making the processes, technology, and organizational changes happen.

A third form of reductionism can be found in *classical* and *neoclassical* organizational theory. While these models recognize system-level properties and the role of people in organizations, they define organizations as closed systems, acting independently of their environments (Baker, 1973). More contemporary writers avoid this pitfall by defining organizations as open systems, which exist in fast-changing turbulent environments. Mahortra (1993, p. 1) observes: "To conceptualize an organization as an open system is to emphasize the importance of its environment, upon which maintenance, survival and growth depend."

Whether they view them as open or closed systems, most students of organizations agree that organizations are not easily changed. Shields et al. (1990, p. 302) observe: "By their very nature organizations resist change. Typically, they are very conservative, highly segmented organizations with heavily regimented, strong hierarchical relationships and entrenched procedures." The elements for successful human systems integration (HSI) (Fig. 2.1) illustrate the complexity of changing a bureaucratic organization to achieve HSI. Rather than merely tinkering with an existing structure (a reductionist approach), change agents must succeed in having the organization adopt a new paradigm—one that changes its vision, alters its work culture, shifts awkward organizational components, and accomplishes related changes. As Figure 2.1 implies, these elements are necessary conditions; each must be in place if success is to be achieved.

In our view HSI is both a process and an end state. It is a process that places people at the center of change. Individuals, with their needs, abilities, and aspirations, shape the process of change and are both objects of and authors of that change. As an end state, an HSI-transformed organization is superior to its predecessor. It is more agile, integrated, and effective. It is an organization better able to respond to any challenges it encounters.

## 2.2 IMPORTANCE OF CULTURE

To understand change, we must first understand the status quo.

—Roger Martin, in Beer and Nohria, 2000

In addition to rejecting reductionism, the behavioral, structural, and open systems models share an interest in the place of culture in organizations. As used by social scientists, *culture* refers to a group's ways of thinking (beliefs, values, and other assumptions about the world) and doing (common patterns of behavior, including language and other forms of interaction). As Henslin (2001) observes, culture:

- Permeates deep into our thinking, becoming a taken-for-granted aspect of our lives.
- Serves as the lens through which we see our social world.
- Provides implicit instructions about what we ought to do in various situations, a fundamental basis for decision making.
- Creates "moral imperatives" or clear notions of what is right or wrong.
- Tends to persist over time, typically transmitted across generations.

In the world of work, culture represents everyday ways of "doing business" or patterns of behavior that new employees are expected to adopt. We term this form of culture *work culture*. The more visible characteristics of a culture are more malleable than deep-seated beliefs, which may be shaping the behaviors (Kotter and Heskett, 1992), but successful organizational change requires linking required behavioral changes to core values held by the individuals comprising the work culture.

Generally, most enterprises operate in a *functional work culture*. This classic, nineteenth-century industrial model derives from a time that emphasized control, conformity, and continuity. The industrial revolution, which was characterized by the uniformity driven by new industrial procedures and the revolutionary introduction of interchangeable parts, saw the creation of large hierarchical enterprises. The advantages of size generated requirements for reliability, increased work specialization, and continuity of processes

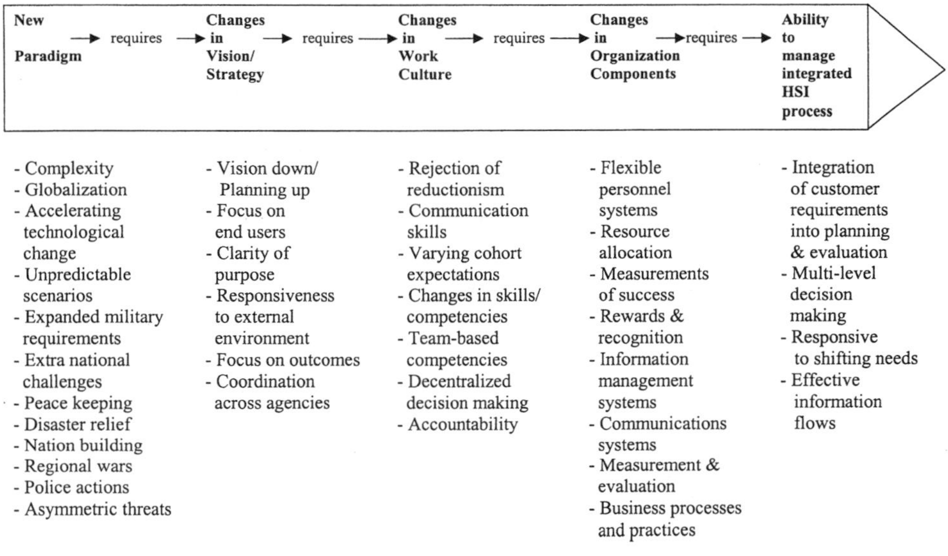


**Figure 2.1** Elements for successful HSI.

(in order to create viable products transported to ever more distant markets). A work culture that served emerging industrial enterprises well is ill suited to work that cannot be divided into repetitious, discrete jobs. Traditional management styles fail to capitalize on the total array of its peoples' competencies and creativity, which is necessary to provide the competitive edge for modern knowledge-based work cultures. Also traditional management practices tend to evoke compliance rather than commitment.

In our increasingly complex world, the functional work culture's emphasis on its internal processes and controls is not sufficient. Such a culture tends to encourage competition among functions with resultant overlap, redundancies, confusion, and fragmentation of effort. The competitiveness is reinforced by an American bias to emphasize individuality over collective efforts and to hail the aggressive, competitive individual as an ideal leader (Smith, 1995).

New work cultures focus outward on the end user—the soldier, sailor, airman, or customer—assessing the rapidly changing environment and quickly adapting to new circumstances and requirements. Interestingly, the Department of Defense (DoD) and the armed services, beginning in World War II, have traditionally used the so-called new work cultures for specific purposes, especially in wartime. Task forces, which by their very nature are illustrative of many new culture characteristics, have been an integral part of successful military actions. The shifting paradigm for organization success (Fig. 2.2) illustrates some of these changed requirements. Where the size of an organization was a measure of success in the past, speed of operation became the new measure. Flexibility, integration, and commitment replaced roles, specialization, and compliance.

For example, the reformation of the army's materiel acquisition process in the mid-1980s was a major cultural change, moving away from traditional behaviors. The army had introduced hundreds of new weapon and equipment systems into the force in the late 1970s and early 1980s. Force modernization was occurring just when the military went to an all-volunteer mode. The army encountered two special problems during this era. First, overall system performance did not always meet standards predicted during the engineering phase. Second, new complex systems failed to accommodate the "man–machine" interface when counting the requirements for specific types and numbers of operators, maintainers, and support personnel and impact on length and cost of the training cycle.

The development of the Bradley fighting vehicle during that period is a good illustration of the failure of traditional patterns of behavior. The mid-1970s prototype held one less soldier than the infantry squad was assigned. This was because designers had failed to accommodate the amount of equipment actually carried by each soldier. The solution was simply to eliminate one position in the squad. This was essentially deciding the number of

| **Old Success Factors** | **New Success Factors** |
|---|---|
| • Size | • Speed |
| • Role clarity | • Flexibility |
| • Specialization | • Integration |
| • Control | • Accountability |
| • "Need to Know" | • Open communications |
| • Process emphasized | • Outcome emphasized |
| • "Carrot and stick " rewards | • Internal motivation |
| • Compliance | • Commitment |
| • Internal focus | • External coordination |

**Figure 2.2** Shifting paradigm for organizational success.

infantryman after the fact—on a "space available" when fielding—rather than building a vehicle to fit the number of infantrymen determined early by army doctrine. In this case the army was paying for more infantrymen than could be carried to battle.

The newer, more encompassing systems development process focused attention on the ultimate customer: the user in the field. However, despite extensive documentation of problems and a committed senior leadership, the old culture rewarded program managers for moving a system forward on a tight time schedule and within costs. A number of leaders saw integrating human factors and manpower, personnel, and training (MPT) costs as increasing the "investment" costs and creating potential roadblocks to fielding systems on time. Many different parts of the bureaucracy reacted with concern, particularly where the various activities had been in competition so long they found it difficult to cooperate in an interdisciplinary approach. Nevertheless, the army did make major strides with its Manpower and Personnel Integration (MANPRINT) program, as indicated with the several specific examples in Chapter 18.

Successful cultural alignment for an integrated enterprise requires a shared mindset focusing on outcomes, mission, and requirements for change. The traditional functional work culture organized in a series of functional boxes and in vertical stovepipes must decentralize and also flatten the hierarchy in order to push decision making closer to action taking. Information must be shared laterally and upward as well as down through functional "stovepipes" in the organization. The more organic or behavioral approach showcases the importance of all organization members thinking and acting in a manner consistent with achieving a transformed enterprise. The behavioral approach also warns that such changes must transform the ways in which people in an organization think, act, and believe about the nature of their work.

## 2.3 LEADERSHIP MATTERS

> I tell people "what got us here won't keep us here."
>
> —Michael Ruettgers, in Hemp, 2001

Leadership is the critical element in creating a new work culture. In large part this is because it is leaders who communicate the behaviors, skills, and abilities necessary to the transformed organization. Leaders accomplish this feat through what they do *and* through what they say. Leaders are responsible for building the trust that will motivate people to follow them into the unknown. Competencies necessary to lead transformational change include vision, the ability to accept ambiguity while managing complexity, flexibility, and the ability to build and inspire a leadership team. A fundamental characteristic of successful change leaders is personal integrity, which is reflected by the ability to evoke trust in others by being consistent and open in words and actions. Such change leaders are open to and acknowledge the emotionalism and difficulties inherent in implementing organizational change. Machiavelli's *Prince* would not be able to lead in today's world unless he recognized the new paradigm for enterprise success.

For leadership development to work, the senior team must work together to make desired values and behaviors explicit, model these behaviors, and integrate them into the appraisal system. Successful change leaders also quickly remove or isolate high-ranking people who fail to "join the team." Major leaders such as Jack Welch of General Electric

(GE) have not hesitated to remove senior individuals who, although highly successful by objective standards, failed to "live the values" in their management of people. Kouzes and Posner (1995) assert that credibility requires *clarity* concerning guiding principles and capabilities necessary to success, *unity* on where the organization is headed and how values will be put into practice, and *intensity* or great consistency linking words and actions.

Our model reinforces Booher's HSI principle 1 (Chapter 1), which posits a top down approach to leadership. Successful leaders realize that, ultimately, only people transform organizations and that there must be "vision down, planning up." Upper level leaders institute change; middle and lower level leaders subsequently assure its success and sustainability. Leaders also assure that initiative is rewarded rather than thwarted at all levels.

Creating vision and overall strategy is largely inductive. Such directional guidance looks toward the future by assessing a broad array of information and looking for pattern. Management per se is fundamentally deductive, with the primary concern of producing orderly results. Planning, the manager domain, is necessary and complementary to vision but cannot substitute for it. Table 2.1 highlights these differences. For more information on comparisons between leader and manager roles see Kouzes and Posner (1987, Chapter 13) and Kotter (1990).

**TABLE 2.1 Leaders and Managers: A Comparison**

| | Leadership Roles | Management Roles |
|---|---|---|
| Approach | Inductive and analytic approach that looks toward future; assesses broad array of information, discerns patterns, risks, and opportunities. | Primarily deductive; focus on achieved results or predictable outcome; plan, organize, and control processes and systems. |
| Critical Tasks | Create vision, establish strategy, model values, motivate others; build constituent buy-in. | Plan, organize and control output/work/processes within a specific arena. Broader domains usually involve more planning whereas more limited spheres of influence tend to emphasize operational issues. |
| Assignment | Leadership roles assigned or assumed in more informal way, tend to be fluid. Usually given management assignments. | Formal assignment is part of management process. Some managers have leader roles either larger or smaller than management position. |
| Function through | Motivating others; modeling values; use of multiple communications channels; informal networks. | Planning, assigning routine responsibilities and authority; use of formal mechanisms outlining duties and means of resolving conflicts. |

## 2.4 TRANSFORMATIONAL CHANGE MODEL

> We are not interested in any change, but rather in change that produces results superior to the status quo.
>
> Roger Martin, in Beer and Nohria, 2000

Culture, leadership, and organizational components must all be incorporated into any change effort if enterprise-level transformation necessary to HSI is to succeed. Having outlined the key *elements*, we now turn our attention to the *process* of change. The main steps that earmark successful change are shown in Figure 2.3.

This four-phase approach to an integrated change process—decide, guide, support, and sustain—is prescriptive. If change is to have depth, scope, and sustainability, change agents must execute all phases in the process and incorporate all elements (culture, leaders, and components) for successful change. Omitting any single element will result in a change that is circumscribed or even aborted. Taken together, these four phases represent the way leaders can bring their vision of an HSI organization to reality.

In the following section we elaborate on the four-phase approach, describing the specific tasks that need to be accomplished to complete the process. Their scope and complexity underscore the challenges leaders face to implementing organizational change.

## 2.5 PHASE 1: DECIDE TO CHANGE

The decision to change initiates the change process (Fig. 2.4). It assumes that key decision makers have identified a compelling need to change, determined its magnitude, and estimated resources required. Leadership, while important to all four phases, is crucial here.

### 2.5.1 Task 1A: Link Vision to Core Values

With transformational change, the future is uncertain. Leaders are asking people to take a leap of faith into the unknown. An effective way to help people make this leap is to appeal to the organization's core values when communicating the vision. This approach allows

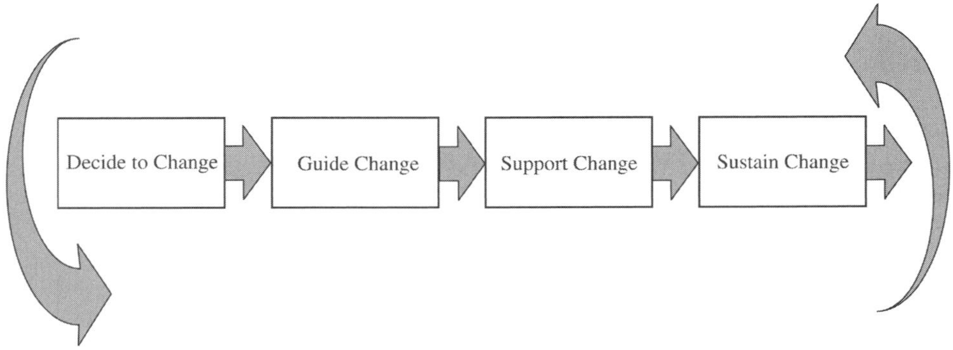


**Figure 2.3** HSI integrated change process.

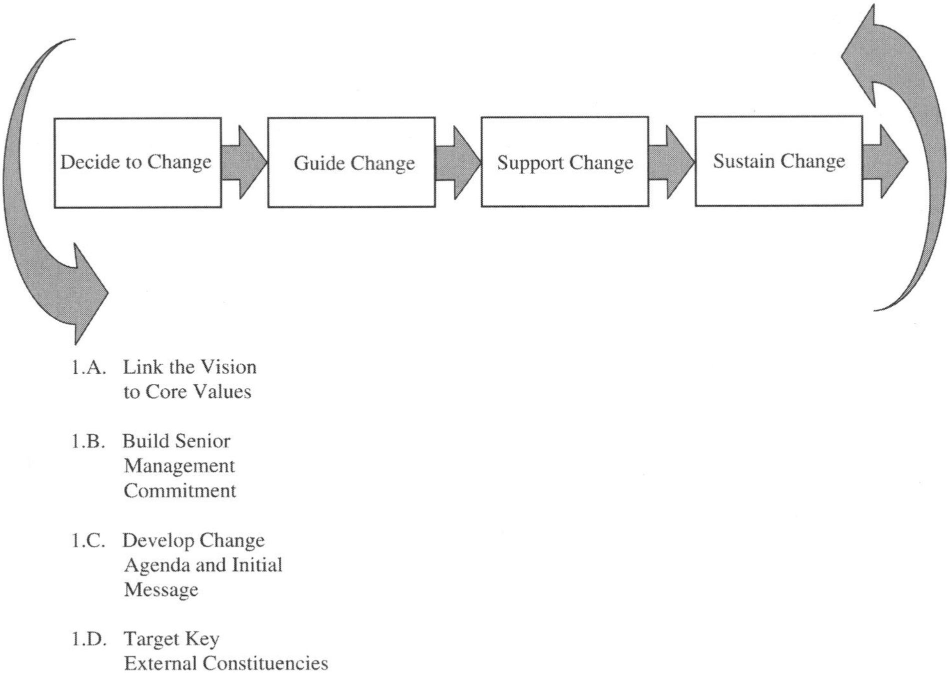


**Figure 2.4** Integrated change process—phase 1.

leaders to validate the past while focusing on the future. It provides a bridge for individuals to decide to change, helping them overcome their fear of the unknown.

The army's successful introduction of MANPRINT demonstrates major systems change that radically altered traditional stovepiped functional behaviors. MANPRINT marked the successful introduction of an integrated systems approach to the acquisition process for weapons and equipment in the armed services. The vision was to define a system *as an organization with people and equipment*, superseding the more parochial view that the system was just the equipment. The new comprehensive management system required continuous integration of six functional areas throughout the materiel and information system development and acquisition process. A major barrier facing the planners was the complexity of effort required combined with bureaucratic inertia. The effort was able to seize the moral high ground and simplify the concept by focusing on the fact that the army was going to design and build equipment with soldiers in mind. This was reflected in the slogan: "MANPRINT: Remember the Soldier."

At a different level, the key leadership at the Defense Printing Service (DPS) gave printers the opportunity to mourn the loss of their presses. Printers lost the tactile experience in producing finely crafted, crisp paper products when computer technology replaced century-old offset presses. The new vision tied the new technologies directly to the concept of continuing to take pride in providing the customer with highly effective and efficient service. This vision for DPS also built on the concept of each print shop opening the door to a "worldwide" team. This integrated approach enabled a leveraging of skills and capabilities to bring added technical resources to bear anywhere in the world. The DPS efforts captured the imagination and commitment of its employees, leading to successful change in the work culture.

Both DoD efforts—MANPRINT and DPS—succeeded in motivating their workforce by defining craft and technical success as that which enabled the ultimate user. The enterprise will only succeed when the ultimate user is effective. Therefore, the measure of success turned from process to outcome.

### 2.5.2 Task 1B: Build Senior Management[1] Commitment

Senior managers are the primary drivers of change. They often accomplish this by serving as role models for their followers (Burke and Litwin, 1992). The importance of followers' perception of leaders, their credibility and their competencies, cannot be overemphasized. Since leadership is a relational process involving two or more participants, leaders must inspire and motivate; only then will employees act in a manner both different from their conventional behavior and consistent with leadership expectations. Such leaders recognize the complexity of human behavior and are able to move employees to act "outside the box." These generic characteristics help inform our understanding of leadership in general. Successful change leadership may be defined as the process of mobilizing others to achieve a work culture that supports the desired vision and strategy.

When Donald Regan, then chief executive officer (CEO) of Merrill, Lynch, introduced the idea of a "cash management account", or CMA (the radical idea of linking money market and checking accounts to the traditional retail brokerage account), all the senior executives voiced serious and valid objections. Once Regan had gone around the table and listened calmly to each person, he sat back and simply said that, indeed, the difficulties, to include legal and regulatory, were genuine barriers. He then reframed the issue. Regan asked how they as an organization could resolve each of the issues. His leadership resulted in a major transformation across the traditional brokerage business (Pfeffer and Sutton, 2000, pp. 66–67).

One senior change leader in DoD did external benchmarking with a Fortune 50 corporation. When he inquired about managerial resistance to change, the executive stated: "We give management ten days to adjust their attitudes [to the change], or they're gone. Government change leaders do not have the same leeway as private corporations to dismiss their career civilians.[2] Instead, they devote significant personal time to individually working with career civilians in order to achieve support, retirement, or a transfer out of the organization" (Hay Management Consultants, 1996, p. 20).

Key elements for gaining senior management commitment include:

- Articulating a case for change
- Devoting personal time to senior managers, listening to their complaints and concerns in order to diffuse their opposition and provide an outlet for them to vent significant issues and negative reactions
- Assessing individual perspectives on the change required, then selecting only those managers to be members of the change leadership team who will be effective advocates[3]
- Initiating team building to build trust and open communication between the members of the senior team, particularly if the senior team includes people from very different

work groups; e.g. the three DoD work groups—political appointees, military officers, and career civilians

- Building support by including the senior management team in refining and further developing the vision

In the DPS, the new director had to grapple with a new high-level vision for an organization created by the consolidation of the air force, army, and navy printing establishments coupled with the loss of about one third of the personnel and grave fears that more severe reductions in force were to follow. The director built support for the new vision by involving his senior team in its further development. He pulled together the area directors in three different off-site facilitated conferences over a period of 6 months. This in itself was a new way of doing business because these people had never been in the same room together. Once the area directors began to coalesce into a team, they developed their vision, which included:

- "Going digital"—radically changing the nature of their business from printing to offering automating documents in computer-ready formats.
- Creating a "team concept" all the way down through the organization where employees were no longer "Army", "Navy", or "Air Force."
- Setting a new goal for communications, replacing the grapevine as the avenue for communicating change. When motivating for customer communication, the director stated "We wanted the bottom person in the organization as aware of and as enthusiastic about selling to customers as the top." (Hay Management Consultants, 1996, p. 27.) This also means when an organization is being radically down-sized, leaders should present the bad news directly throughout the organization, not try to soften with unrealistic optimistic forecasts.

### 2.5.3 Task 1C: Develop Change Agenda and Initial Message

Leaders often approach planning for organizational change in much the same way that they would approach project implementation planning. Reengineering efforts, in particular, have not realized the expected performance improvements because of a failure to address the challenges associated with change implementation, especially the challenges of managing the "people side" of change. Successful team members operate on the following principles when developing the change agenda or road map:

- Remain personally involved in the effort, particularly to make key decisions when people run into obstacles.
- Ensure that there is broad participation from all major constituencies in the organization in the planning, design, and implementation.
- Recognize that planning for transformational change needs to be iterative and exploratory. There is more variability in the process and less control; yet the senior leadership is still accountable for results. The agenda is a road map for change, rather than a detailed plan or blueprint.

Several common responses block a work culture's acceptance of the initial change message:

- Highly successful organizations resist the need to change by discounting the reality of the message.
- Change threatens the "comfort level" of individuals.
- Changes in customary "ways of doing business" disrupt power relationships.
- The notion that "everything may need to change" implies that the old roads to success may no longer be available (thereby discounting the senior person's success and increasing career uncertainty for midlevel personnel).
- The increasing amount of societal change has increased stress in people's lives, which contributes to resistance to further change.

To improve the chances of people accepting the change message, individuals must understand that their work lives depend on it. The messages must be concise, direct, and repeated. They must also link the urgency for change to core values. The initial message and subsequent communications about it need to provide honest assessment of how the proposed change will impact people. To maintain and build trust, the initial message should openly and frankly address job and position losses or other negative outcomes. The initial message does not seek to achieve "buy-in," it merely sets the stage for implementing change. Honest communications are central to issues of trust and acceptance.

### 2.5.4 Task 1D: Target Key External Constituencies

Senior management must also address its external environment. The external constituency includes all those individuals outside the purview of the senior management responsible for leading the change. For instance, the secretary of defense looks to elements outside the DoD—such as key congressional committees—whereas a defense agency may also look to elements in the office of the secretary of defense (OSD) (such as senior DoD officials including the secretary) and other players in the department. The corporate enterprise must look to shareholders, significant partners and competition, government regulators, Wall Street, and public opinion.

Senior managers have to build internal consensus in order to help keep down levels of discord. Discord and uncertainty often generate multiple letters to congressmen, unions, and other external political players. Providing clear, consistent communication helps avoid creating a backlash before the case for change is made to external constituencies who have the political influence to derail the process.

The approach for targeting key constituencies, particularly political influencers, includes the following actions:

- Identify key constituencies who will have concerns about the proposed change. The scope and level of effort spent here should mirror the extent of change planned and the size of organization undergoing the transformation.
- Incorporate plans to work affirmatively with members of the media. They provide a significant avenue for transmitting information and educating significant opinion makers in various constituencies. Such plans may be contingency-only.
- Develop a communications plan and use multiple means to influence targeted constituencies.

Table 2.2 provides a sample of those entities with interests in specific DoD components. A similar identification process can be used for any entity.

**TABLE 2.2 Identification of Key Constituencies within DoD**

| Entity Undergoing Change | Senior Team | Senior Official(s) | Agency Heads | Congress | Unions | Private Sector (defense contractors, trade-groups) | Media |
|---|---|---|---|---|---|---|---|
| Office of the Secretary of Defense<br>Joint Chiefs of Staff | Secretary of defense and key subordinates; chairman, JCS, service chiefs, and key subordinates | President and National Security Council | Senior appointees | Members, esp. committee chairs; other political leadership | National president; national conventions | CEOs, boards of directors | National media |
| Agencies and military services | Senior political appointees, SES, senior military | Secretary of defense, Office of Joint Chiefs of Staff | Senior appointees and career civilians | Congressional staff, committee staff | President and key subordinates | Key persons responsible for market and policy | Esp. trade publications |

While the DPS was undergoing significant change, the director identified the various constituencies and determined to target the union, Congress, and private contractors. At DPS, where unions are very significant, the union leaders were dismayed over the prospect of losing a high number of people. The director met with the national head of the employee union, and key DPS senior management worked with union members to ensure that a consistent message was delivered internally to employees and the union. The director and senior managers clearly articulated the changing environment and the need for moving forward with their change vision to internal employees and union officials. Congressional scrutiny of DPS's change effort was intense and hostile as a result of the downsizing and consolidation that was taking place. The director conferred with key committee staff with whom he had developed relationships in order to deliver, once again, a consistent message concerning the requirement for transformational change.

Printing contractors who had been printing massive amounts of DoD documents were concerned about losing business. Their concerns generated pressures from Congress and negative articles in trade publications. The director took the DPS case to the printing community using multiple channels of communication. He attended national conferences and talked with trade publications, emphasizing the fact that DPS needed to maintain certain core functions. Everything else would be put "on the table" for contracting to the private sector. As one senior DPS official noted, "The private sector printing community thought they could do things better. We said, 'You'll be a partner.'" (Hay Management Consultants, 1996, p. 42.)

Consider Boeing Industry, which attracted the world's attention, when it announced its relocation from Seattle to one of three potential locations—Denver, Chicago, or Dallas. This announcement served notice to a number of key constituencies both internal and external, from union to shareholders to cities eager to compete for the corporate presence. It broadcast a new vision for a Boeing corporation that had outgrown its prior image of a manufacturer of airplanes.

## 2.6 PHASE 2: GUIDE CHANGE

> [E]xemplary leadership and organizational change are impossible without the full inclusion, initiative and cooperation of followers.
>
> —Warren Bennis, in Beer and Nohria, 2000

The main objectives of phase 2 involve determining what new organization will emerge and the action necessary to achieve it (Fig. 2.5). This process is one of application and refinement, or trying out changes and readjusting, depending on the outcome. While the senior leadership is still the primary driver of change, participation begins to percolate throughout the organization, as more people are involved in starting to make the change vision a reality.

### 2.6.1 Task 2A: Develop Change Plan

> The First Principle: If you know by doing, there is no gap between what you know and what you do.
>
> —Pfeffer and Sutton, 2000

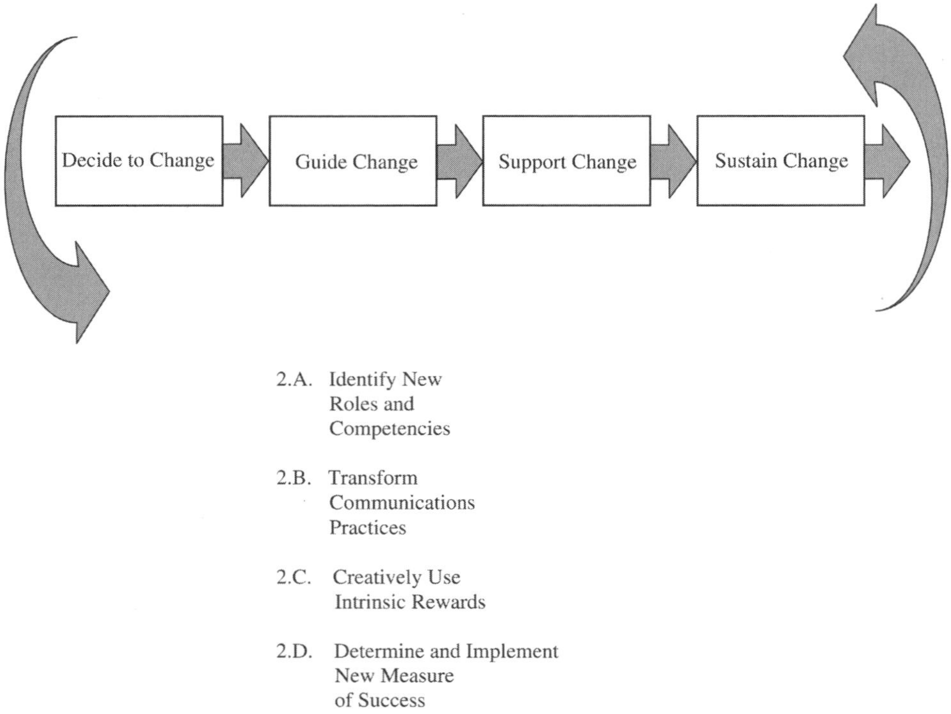


**Figure 2.5** Integrated change process—phase 2.

Public organizations commonly make only structural changes because they are relatively easy to implement compared to work culture behaviors. Change leaders need to ensure that the new structure supports changed business processes and the desired work culture. Pfeffer and Sutton (2000) cite the military as an example of learning and knowing by doing. They note the contradiction that "in many companies people are more likely to get ahead by talking smart than by doing smart and productive things." The challenge, then, is to both know and do. That is only possible with more doing than talking, planning, briefing, and critiquing.

Under the leadership of Jack Welch, GE became not only one of the most successful Fortune 50 companies across the past 20 years, it is also identified as the preeminent source of CEOs for other major corporations—demonstrated when the two "also rans" in the grooming for Welch's successor were named as CEOs of Home Depot and 3M. Welch has summarized his efforts as a "hardware" phase preceding the call to transformational change. His hardware phase pared a bloated bureaucracy, beginning with his immediate dismissal of a large central planning selling off or realigning 350 businesses into 13, and reducing the workforce by one quarter. Welch grasped that GE could continue to achieve growing productivity by using the initiative and knowledge of all its employees (Slater, 1994). His incentives have always demonstrated an awareness of the talents and capabilities residing in the entire workforce. In the mid-1980s Welch began a process called "work-out" aimed at bridging the knowing–doing gap. *Work-out* (1) focused on a business issue or key process within specific functional areas, as people recognized that issues and processes were cross functional; (2) brought together multifunctional/multilevel participation in small-group brainstorming; (3) presented all recommendations to business leaders at a town meeting; and (4) required immediate action by the manager (Pfeffer and Sutton, 2000). An action could be either accepted or rejected (with explanation), or if more information was needed, an action team and deadline for decision was established on the spot.

Work-out facilitated the transformation of GE's culture by overcoming functional specialization and hierarchical power differences that had inhibited information flow and action taking. The model helped all GE people learn how to present more specific, thought-out suggestions and develop initiative. It helped create a training ground for managers who displayed a bias for action, a willingness to listen, and who valued people "who dared to try new things." During the same era the development of "best practices" across functions led to "benchmarking" and to looking outward at other companies to see what practices or processes made them successful, independent of the products they made. While these initiatives may appear to be commonplace today, in many cases GE was breaking new ground with them.

When outcomes rather than processes became the imperative, organizations have to achieve greater flexibility in manpower utilization. Career development and internal selection also need to be recast to align new requirement with changed organizational needs. Through 20 years of leadership there have been only a handful of key initiatives that have driven GE excellence as a corporation and as a preeminent school of executives. During that period, there have been some variations in personnel systems, yet underlying each was the consistent theme of rewarding achievement, honoring risk taking, and ruthlessness in weeding out those who flout the organization's values.

In the federal sector the Defense Logistics Agency (DLA), a senior management group ("the gang of 10" assigned by the agency director to reengineer the headquarters), had 2 weeks to review contractor recommendations and create its reengineering blueprint. The DLA senior planning team redefined the headquarters role as policy, oversight, and resource management (Hay Management Consultants, 1996). Accordingly, the senior management group released certain headquarters executive service (ES) positions for transfer to the field. The design team put middle management back to work, assigning most headquarters (HQ) personnel to teams. This permanent team structure gave the reorganization more flexibility to meet changing business needs while giving managers the opportunity to better decision making by eliminating much of their previous focus on control-oriented tasks.

The senior team understood that individuals needed stability as well as opportunity when undergoing major change:

> We treated the teams like mountain climbing teams who establish camps all up the mountain, but with a "base camp" or a recognized "home room" or "base" doing training, time keeping, leave, and serving as the location where the position of record sits (Hay Management Consultants, 1996, p. 47).

Besides establishing stability at the top and bottom of the organization, DLA notified everyone that no changes in position descriptions or grades would occur for at least a year while the organization was going through the wholesale restructuring. This also provided a necessary element of stability. Senior management continued to concentrate time and energy on making the behavioral changes happen as the organization reengineered its work processes.

### 2.6.2 Task 2B: Design and Deploy Ongoing Total Communications

I wish we knew what we know at HP.

—Lou Platt, CEO Hewlett Packard, in Pfeffer and Sutton, 2000

Design and implementation of an effective communication strategy is a major challenge to teams directing change. Trust in the communicator is built through the integrity of the messages and the ways in which communications occur. Employees find it difficult to take actions on their own without access to accurate and timely information. Sharing information widely is a key hallmark of the new work cultures, particularly in times of transition. "Listening down and across" may be one of the most effective communication skills. Effective communication requires multiple channels—formal and informal, upward and downward. Policy announcements, reorganization plans, and procedural memos alone cannot do the job. Senior managers need to deliver change messages personally by, for example, calling an all-hands meeting.

### 2.6.3 Task 2.C: Create and Reward Early Successes

Change leaders must get people to "buy-in." Obtaining commitment to a new vision and new behaviors can be a daunting undertaking. Short-term successes provide concrete evidence that change is possible. Visible, early successes bring the change vision down to a tangible level while undermining the naysayers. Further, such visible successes establish the credibility of the project, the team, and the management that supports them.

In addition to creating early wins, building momentum requires a process of publicizing successes and rewarding the people involved. The work-out sessions in GE provided multiple opportunities for success, and, as Welch put it, allowed people to pick "the low-hanging fruit." This process reinforces behavior, sustains commitment, and communicates to the rest of the organization that there are personal benefits from the change. To summarize, leaders can capitalize on small wins and short-term successes by implementing early changes with the following characteristics:

- Visibility (changes in work processes that involve some form of organizational change are most visible).
- Strong likelihood for success defined in terms of desired outcomes.
- Targeted to organizational units where key managers are strong supporters of the change agenda.
- Success defined in terms of outcomes and behaviors consistent with the desired work culture.
- A number of people involved in the change.
- Recognition and rewards provided for those who achieve desired outcomes and behaviors.

Rewards and recognition can be designed to appeal to people's intrinsic motivation. *Intrinsic motivation* refers to psychological needs that drive people to perform, such as a desire to master skills that others recognize and appreciate, a desire to have control over one's work, and a desire to influence others (Steininger, 1994). In an environment where there is a requirement for massive change and significant ambiguity, individuals are naturally apprehensive and concerned. Issues about job security, power, control, and influence all become extremely important. An implementation plan that breaks the vision into doable pieces and provides an opportunity for people to experience short-term success helps to reestablish feelings of control and self-worth.

The army's key leaders incorporated planning for early success in their MANPRINT effort by integrating HSI into the army's weapons acquisition process. In order to expedite learning and minimize the impact of an early failure, a number of pilot projects were selected. These were chosen because they provided experience in each phase of the acquisition process and provided experience in procurement procedures.

The Light Helicopter Experimental (LHX) Program that produced the Comanche helicopter—now one of the weapons systems for the army—was selected as an army pilot program. Six functional areas (manpower, personnel, training, human factors, system safety, and health hazards) were identified and empowered to act as a team on this project. This process brought broad ownership, facilitated involvement, and *provided many opportunities for success*. Supervision of critical components was retained at a high level. The visibility of LHX demonstrated a high level of senior leadership commitment. The pilot project was the subject of an active information campaign ranging from keeping the senior leadership informed to routine briefings to newly assigned personnel. Such visibility not only sustained support but also allowed it to grow (Blackwood and Rivello, 1994).

## 2.7 PHASE 3: SUPPORT CHANGE

> If you are doing everything perfectly, you aren't trying hard enough![4]
>
> —Gordon Moore, 2001

During the third phase in the change process (Fig. 2.6), implementation is underway. The road map is in hand, implementation teams are at work, and the leadership team is tracking progress. The transformation effort now includes people at all levels. Supporting change requires efforts focusing on the redesign of various structural components that channel work flow and organizational structure so that the new work culture becomes the accepted norm. Structures help to anchor the change.

### 2.7.1 Task 3A: Identify New Roles and Competencies

Many reengineering efforts rely solely on training and education as the tool for turning the reengineered work processes into reality. The hard work of implementation, however, requires more than sending people to training sessions. As important as training is, change leaders must first identify the human resources they will need to perform each new or modified process—the new roles, responsibilities, and competencies people will need to do the work effectively. The process of identifying roles and competencies must deal with the characteristics of the *work required*, not the characteristics of *the people* who have been or will be doing the work. Organizations as disparate as GE and the World Bank are striving to create new opportunities for sharing this tacit knowledge.

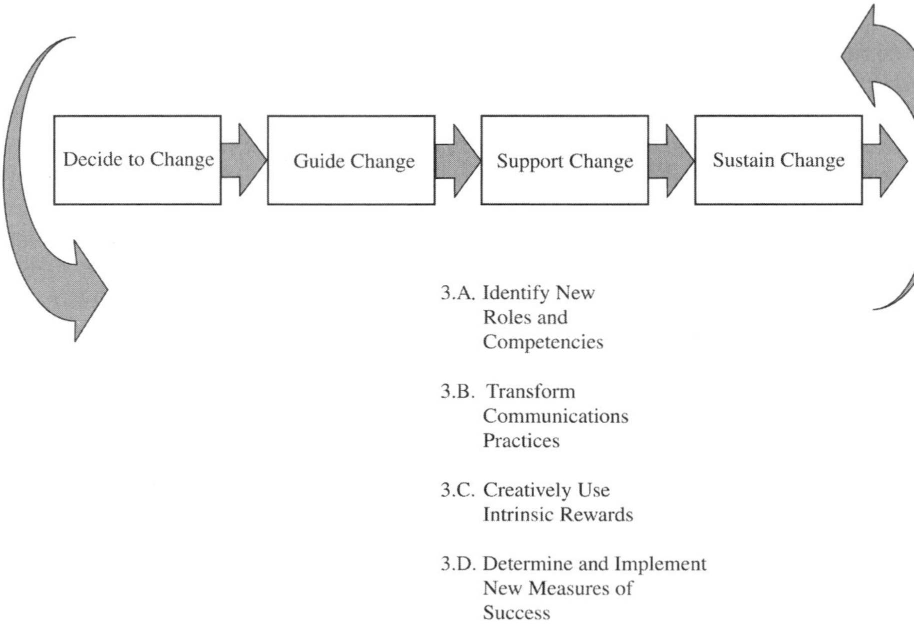


**Figure 2.6** Integrated change process—phase 3.

Change leaders must ensure that the following occurs in order to set the stage for effective change implementation:

- *Clearly articulate the new roles people will be required to play:* The new roles should support the new work flow, organizational structure, and work culture.
- *Identify what excellence looks like:* What are the specific responsibilities and competencies that will be needed for success? If the change involves moving to teams, identify the competencies (interpersonal skills, open communications skills) that would be required of all team members, leaders, and others.
- *Assess the existing talent pool:* This is needed to match people to the new roles and determine the extent to which the new competencies need to be developed. The assessment results become the basis for the organization's training and development plan.
- *Incorporate a bias for action:* Traditional work culture with its emphasis on routine, hierarchical deference, and tendency to hoard information seriously limits the ability to benefit from initiative, creativity, and on-the-spot knowledge. Preference for action taking and willingness to tolerate mistakes are critical to new work cultures.

Given the constraints of the public sector, some maintain that defining new roles and competency requirements would be difficult to achieve. However, many of the DoD change leaders interviewed by the authors have restructured their organizations, made greater use of teams, and created new team and individual roles for personnel.

A responsive, flexible workforce is a key component in all transformational change, regardless of the particular shape the work and organizational structure takes. In additional to technical competencies, this new workforce will need a broad spectrum of behavioral competencies. Communications, coaching, cooperation, and influence abilities are increasingly important for most of the workforce; more and more people are required to make decisions, work in teams, and negotiate across traditional functional boundaries. Unless organizations systematically develop these abilities in their people, old ways of "doing business" will persist.

GE and Allied Signal have taught thousands of managers and employees how to determine customer needs. This training helps provide organization members a common vocabulary, measures, approach to problem solving, and vision. Training programs are defined around competencies that align with business strategies.

Practice does not make perfect—*perfect practice makes perfect.* Acquiring behavioral skills requires hands-on training rather than education, which are traditionally more lecture, reading, and discussion driven. Those without the necessary skills or having very limited skills need to practice, correct, and practice again. The most effective way to build behavioral competencies is through practice, coaching and feedback, and the chance to apply new skills on the job. Often on-the-job, just-in-time opportunities for detail to other offices/directorates or opportunities for cross-functional training provide effective skills development techniques.

Historically, DoD training in behavioral competencies for civilians has been focused on career managers. For the most part, the system presumes behavioral competency in political appointees. In contrast, the military services have a longer history of shaping behavior through training (both for technical performance and leadership) throughout the ranks.

### 2.7.2 Task 3B: Transform Communications Practices

Communications is central to the issue of both sustaining and accelerating change in an organization. The Oracle Corporation sends change consultants to work with senior leaders at enterprises installing new Oracle systems so that the enterprise can both anticipate and shape the cultural ramifications (Hart, 2001). When an organization is reengineered around processes and has turned to a cross-functional (as opposed to a stovepiped) structure, the formal information flow will have been changed. Radical changes in the capture and flow of information will impact the culture. They will occur whether or not expected and planned for.

Prior to the development of computers and networking systems, gathering data and information was laborious and their distribution was difficult. A common characteristic of new work cultures is greater movement of information sideways and downward. When information is shared in this way, it short-circuits grapevines and rumors as a medium of exchange. Good information policies and practices:

- Are inclusive, rather than "need to know," reducing the potential for "ins and outs" generating internal tensions and competition.
- Provide the basis for decentralizing decision making and action taking by making the information necessary for good decisions at appropriate levels.
- Create a shared culture through shared vocabulary and common viewpoints.

Many classic avenues of communication have traditionally been exploited by work organizations. These strategies for transmitting knowledge and common understandings and acceptance of new processes can be improved upon and supplemented. This can be achieved by including the powerful informal transmission channels along with the formal ones. Good commanders as well as managers in the public and private sectors have historically used informal communication avenues to create an upward flow of information from subordinates. They have engaged in "walking around" and assessing morale. Senior managers should encourage their managers, supervisors, and team leaders to use this informal communications technique.

At the World Bank an "all-in-one" information kiosk allows teams to set up e-mail distribution lists so that progress reports, correspondence, and meeting notes can be easily shared and accessed by other members of the organization. SmithKline Beecham (Research and Development) uses groupware technology to set up databases by subject so that people from different geographic locations and other functions can contribute to information interchange.

GE, New York Life, Digital Equipment Corporation, Exxon, SmithKline Beecham, and the World Bank are organizations that have used town meetings as a means of removing vertical boundaries (Ashkenas et al., 1995).

In May 2001 IBM conducted a marathon brainstorming session titled "WorldJam" to which it invited all 320,000 employees. The *New York Times* reported:

> By the end of the three-day exchange . . . nearly 52,600 workers had logged in at one time or another. . . . The visitors generated more than 6,000 proposals and comments, and viewed five postings each on average. . . . IBM set out 10 broad problems to work on for a limited period in a setting that combined elements of moderated chat, electronic bulletin boards, and online polls.
>
> —Fedder, 2001, C1

It is too soon to say what the results will be in terms of knowledge transfer; however, one explicit goal IBM set was to study its potential for spreading ideas horizontally throughout the worldwide organization. Planning was already underway for a similar project spanning all of the sales forces later in 2001 (Fedder, 2001, C5).

### 2.7.3 Task 3C: Creatively Use Intrinsic Rewards

As Kouzes and Posner (1995) note, the role of the leader is to create the appropriate work culture wherein employees motivate themselves. Self-motivation or intrinsic motivation is variously defined as: "self-fulfillment" or sense of self-esteem and respect; or the necessity of having a clear role that gives meaning to life. Individuals are self-motivated when they are valued for their contributions. One theologian explains the vast difference between a mere job and meaningful work: a job is about economics, or extrinsic motivators, whereas "work comes from the inside out" (Fox, 1994, p. 120).

To use intrinsic rewards to modify behaviors, managers may have to first change themselves. In place of focusing on and controlling actions, they need to mobilize and motivate others. Command and control management involves close monitoring and controlling of subordinates' behavior and sends a message that the leader lacks trust in the ability and judgment of people within their purview. Command and control managers often fail to provide subordinates sufficient information for them to make effective decisions and initiate actions. As a result, control-oriented managers tend to discourage creativity and initiative. Effective leadership:

- Makes it safe to experiment or take risks.
- Encourages creative thinking.
- Avoids early negative feedback to new ideas.
- Rewards initiative and honors risk takers.

A tragic example of the consequences of failure in a traditional work culture is the United States' mistaken targeting of the Chinese embassy during the Kosovo conflict. The bombing coordinates were based on inaccurate and old data. At least one employee questioned the location but failed to aggressively call this to anyone's attention. The result was the death of two Chinese in their embassy. Ideally, a work culture that encourages subordinates to actively question discontinuities is also one where individuals are well informed about any given operation.

Effective senior managers also exhibit sensitivity to the cultural expectations of different work groups. For example, at DoD it has been customary to present awards for outstanding performance to service members when the individual is rotated from a given position or billet to the next assignment. This pattern differs from current practices surrounding a variety of civilian awards, which are given in connection with the annual performance appraisal or immediately upon the completion of a major project. As a consequence of the differences among the three work group cultures (career civilians, political appointees, and active-duty military), military supervisors sometimes fail to adequately manage the civilian awards system. Supervisors need to tailor rewards to different work cultures in all organizations, whether they are in the State Department or a Fortune 100 company.

### 2.7.4 Task 3D: Determine and Implement New Measures of Success

All organizations undergoing transformational change must establish a means to monitor and assess their progress toward achieving agreed-upon goals. This process can be encapsulated as "generating feedback mechanisms" (Nadler et al., 1995). These mechanisms are not separate activities; they are tailored to the specific change objectives and woven into the fabric of change management.

The team responsible for developing the new measures needs to follow a number of guidelines (Osborne and Gabler, 1992). These include:

- Recognize that there is a significant difference between measuring process and results. Process is relatively easy to capture; results are more difficult but more important to document.
- Be aware of the vast difference between documenting efficiency and effectiveness. Efficiency is a measure of the cost of production, while effectiveness measures the quality of output. Both phenomena need to be included in a comprehensive monitoring effort.
- Involve all participants in developing change measures. This guideline assures that employees will "buy into" the change measurement process. Initially it involves senior managers but will later broaden out to include all affected middle managers.

- Try to strike a balance between using too few and too many measures. If too few measures are used, not all change objectives will be measured. By creating too many, the power of all measures may be diluted and managers will be overwhelmed with information.
- Focus on maximizing the use of performance data. Performance measures assist managers in planning and conducting their work. Other measures, in contrast, are typically seen as burdensome and merely as reporting requirements.
- Plan to subject measures to annual review and modification. The fluid nature of organizational change means that measures, which appeared ideal when originally developed, may not work out as well in practice. While at this stage in the measurement design process such specifics are unknown, they still need to be anticipated.
- Realize that while "zero tolerance" may be an effective standard for scientific endeavors, it does not apply to human performance, especially when judgment is warranted.

The effectiveness of performance measures depends upon people's willingness to provide accurate data. People who are afraid of failure and its repercussions can distort or fail to provide accurate responses. Any new work culture must build on a foundation of trust and emphasize that the objective of measurement is to bring about continuing improvement.

One way to build this trust is to ensure that leaders deemphasize perfectionism in dealing with their people. The concept of "perfection," or absolute adherence to rules and regimen, are marks of success or failure in the functional work culture. Perfection, however, is the enemy of continuing change and improvement in the process, time-based, or network cultures. The pursuit of perfection destroys initiative and rewards the "we've always done it this way syndrome."

Some individuals simply may not be able to work with the new measurement system. One senior manager interviewed rewrote the performance standards for some of the key positions so that they were more outcome-oriented. People now had to relate budget expenditures to the overall new mission of the organization, whereas in the past, their performance standards related to how well they managed organizational spending. One of the affected employees refused to change his old approach, saying that he had not done business this way before and could not do it the new way; he left the organization.

## 2.8 PHASE 4: SUSTAIN CHANGE

When an organization has experienced significant change, the transformation will not be complete. Work flow may be different, behaviors may be cross-functional, and decision making may be more rapid, effective, and broad-based. These changes, however, may be short-lived. True transformation must be long-term, extending into the next generation. This phase addresses the steps HSI change leaders can take to sustain change (Fig. 2.7). They can achieve sustained change by assessing and readjusting direction, creating new and supportive policies and procedures. They can also work to dismantle existing barriers and prevent external constituencies from imposing new ones.

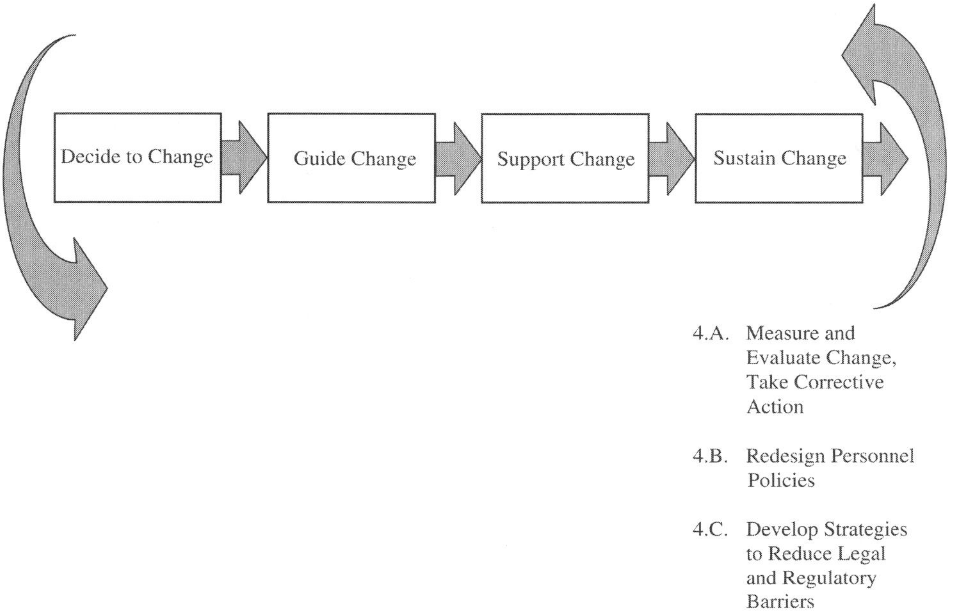


**Figure 2.7** Integrated change process—phase 4.

### 2.8.1 Task 4A: Measure and Evaluate Change, Take Corrective Action

In completing the previous task, the organization embraced measures documenting the full range of organizational change. To sustain change over time, however, leaders must use the information effectively and modify the measurement system as needed. Effective use of the rich lode of quantitative and qualitative information generated is another major challenge. Some of the issues that need to be resolved include:

- *Utility of Measures Adopted* Are the measures the organization now employ effectively capturing the kind of information needed or should others be substituted?
- *Frequency of Collection* Is there a good balance between timing of data gathered and the assessment of progress? Should some information be gathered more often or less often and why?
- *Resources Expended in Data Collection Effort* Is the magnitude, frequency, or timing of the data collection cost effective?
- *Extent of Use* Are the tools developed to assess change widely used, or are they restricted to a few work units or divisions?
- *Application to Decision Making* If information is to be useful, it should inform decision making and be used by management to direct actions that move the organization toward its goals. To what extent is decision making being driven by the information?

Any measurement effort that assesses change must also look at action taking, communication flow effectiveness, and participant behaviors. Additional questions to ask when assessing cultural change include:

**Action taking** Are managers and workers displaying a willingness to initiate actions as opposed to simply applying rules and procedures?

**Communications** Is information flowing more appropriately up and across the organization and is organizational information more accessible?

**Behaviors** Are cross-functional teams and behaviors becoming more standard than individual, functional, and technically stovepiped behaviors?

**Effectiveness** Do measures reflect the quality of output that supports the war fighter rather than focusing only on efficiency?

After a massive reinvention of itself that included consciously focusing on shaping the work culture, the Defense Mapping Agency (DMA) moved from a functional, work culture to one that is more process-based. DMA received the Vice Presidents' Hammer Award in January 1996. The process of change began in September 1994 when DMA established a nine-member Reinvention Task Force Team reporting to the agency's director. They found the following:

- The equivalent of 30 years' backlog at their current production rate
- Too much hierarchy
- A workforce isolated from its customers
- Functional stovepipes rather than an organization built around core processes

As a result of its "reinvention," the DMA reduced policy documents by 42 percent. By eliminating redundant or outdated forms, DMA achieved a 51 percent reduction in forms. The organization reduced its hierarchy from 11 to 3 levels, thereby achieving greater efficiency in action taking. More of the workforce was shifted to customer support teams. Most importantly, the agency now has a DMA report card from customers reporting how effective the organization is in producing "quick-fill" product requests. The outcome is more effective support to the war fighter. DMA now constantly evaluates performance, making change as necessary. "Reinvention doesn't stop. The Defense Mapping Agency is evolving with its customers in order to meet their evolving requirements, and that means continual improvement or change" (Hay Management Consultants, 1996, p. 70).

### 2.8.2 Task 4B: Redesign Personnel Policies

Although people by this stage are already performing tasks in new ways, change cannot be sustained into the next generation without addressing the formal personnel systems. For the new way of working to persist beyond the change leader's tenure, the changes that were implemented (in processes, organization structure, competencies, leadership, etc. discussed in phases 2 and 3) need to be institutionalized. This happens when senior management changes the systems and policies that drive the way people work together and are organized, evaluated, developed, promoted, and selected.

Addressing the federal personnel system poses a significant challenge. Implementing change in this setting involves identifying the specific personnel system changes that need to be made, the barriers to each, and whether the barriers are imposed internally or externally. For each required change, ranking their relative importance and ease of making the change can clarify which changes should be the top priority for action. For agency-imposed barriers (policies or regulations), effective use of negotiation and influence skills can bring success.

Making such changes can be a daunting task, and most of the change leaders interviewed discussed circumventing policy barriers rather than dismantling them. One DLA change leader interviewed described the strategy and actions he and his senior managers took to formalize personnel changes in this agency's senior ranks to assure the new work culture would be sustained after the leader left the organization. In DLA's change vision, headquarters was to focus on policy and the field on execution.

The leadership team took the following two actions to sustain the vision over time:

1. *Increased the span of control of Senior Executive Service (SES) positions*. This permitted an immediate transfer of some SES slots (and people occupying them) out of headquarters, emphasizing the decentralization of decision making for operations.
2. *Established a formal rotation process for a headquarters senior SES position*, whereby the agency's most senior civilian would rotate out of his/her position into another headquarters position every 3 years. The intended result was for the top SES position to rotate, with no single individual having the opportunity to become ensconced long-term in the top permanent civilian position. Another result was that the key change leader ensured that future "empire building" would be counter-productive. Further, no long-term civilian would be in a position to have sole influence over the selection or approval of the agency's senior cadre of permanent civilians—and would be less likely to be considered the *de facto* leader of the organization. Having a single permanent senior civilian as a deputy director can build a bias against change or at least toward a limited perspective on what the organization needs.

By deliberately departing from traditional HSI organizational structure, the agency change leaders were able to set up an operating concept that would persist in supporting a cross-functional perspective and openness to change that would not otherwise be possible.

### 2.8.3 Task 4C: Develop Strategies to Reduce Legal and Regulatory Barriers

In the earlier discussion, several objectives in the HSI change process (build senior manager commitment, target key political influencers, and creatively use intrinsic rewards) outlined strategies leaders can use to build internal and external support for change. As important as these actions are, they rely on the talent and energies of highly motivated leaders. If change is to be sustained, it needs to be further legitimized through internal policy redesign as well as by influencing the broader legislative and political environment. Externally imposed regulations that are irrelevant to DoD's primary mission reinforce risk-averse, compliance-oriented behaviors—typically viewed as "bureaucratic" behavior associated with the traditional functional work culture. In response to congressional or other agencies inquiries, it is easier for managers to explain they followed the rules than why their judgment led to a better deal for the government. Therefore, removing key regulatory and legal barriers can be of prime importance in ensuring that the new work culture persists over time.

Despite current procurement reform, the nature of our political system will likely continue to focus on equity over efficiency. The continued pressures to fund military programs over the DoD's and military departments' objections illustrate this point. Congress and the current administration are now perhaps more focused on government effectiveness than previously, but even this concern has many political elements. Therefore DoD's key political appointees have the responsibility for working with Congress to change legal barriers to change. The military and senior career civilian leadership in DoD has a responsibility for working affirmatively with other federal agencies to obtain regulatory change.

Good performance measures are an important tool for influencing Congress and other external constituencies. They help to enlist the support of members of Congress and to request consideration for pilot programs; setting new standards for performance measures that have been implemented effectively across the organization can be used proactively. Their presence can preempt the kind of General Accounting Office (GAO) and inspector general (IG) investigations that lead to the imposing of additional rules and procedures. Such impositions often thwart progress rather than adding value.

## 2.9 OVERCOMING CHALLENGES TO CHANGE

> Changing an organization is inherently and inescapably an emotional human process.
>
> —Duck, 2001

Collectively, the four phases of change and their associated tasks represent a model for designing, initiating, and implementing transformational change. The realization of such an enormous undertaking requires significant resources and high levels of commitment by the participants. Each *component of change* identified earlier in this chapter must be in place if the *process of change* is to succeed. As Figure 2.8 illustrates, when key elements of change are absent, planned changes are derailed. Adopting a static rather than a dynamic, responsive vision, for example, leads to diffuse, directionless actions and decision making by the organization and its leaders. Similarly, the failure to align the components of structure, process, technology, and measures results in a continuation of inefficiencies and an overall failure to sustain planned innovations.

Challenges and barriers to the process of change occur at all points. However, they are most significant at two watersheds—the decide to change and sustain change phases. The initial tasks of the former phase (build senior management commitment and link the vision to core values) provide the catalyst for initiating the macrolevel changes this report has outlined. It is the top leadership in any large organization that must initiate transformational change. Further, senior leaders are the source of the new vision. They must agree on what the vision comprises and then translate it into a message that resonates with the core values of the hundreds of thousands of men and women who will ultimately be responsible for making the vision real.

While the military culture dominates DoD, there are at least six distinct cultures that coexist within this complex organization—the four military service cultures, the civilian civil service culture, and the political appointee culture. If such diverse groups are to be

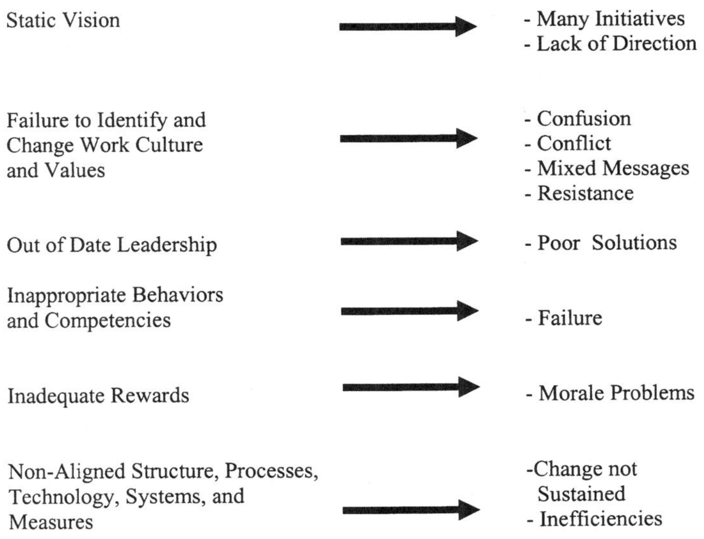


**Figure 2.8** Consequences of omitting change components.

persuaded that change is in their interest, leaders must tailor the message to each constituency. Leaders are faced with the challenge of respecting existing values while presenting an alternative way of conducting business. Furthermore, the scale of change required necessitates the early involvement of a senior leader cadre acting as a single voice, articulating the change message consistently. A considerable amount of energy and thought needs to be invested early in the change process, if the transformational change effort is to succeed. The change message must be broad enough to appeal to all culture groups while coalescing them around a new, shared vision.

At the latter phase (sustain change) a different set of challenges emerges. Now the organization has undergone major transformation and is better prepared to meet its changing mission. Recently minted, the changes instituted may be tenuous. If the organization is to be prevented from reverting to former, now inappropriate actions, renewed energy has to be devoted to maintaining the new structures. Leaders must assure that the organization continues to respond to changing circumstances—internal and external. A major component that emerges at this phase is the monitoring function (measure and evaluate change and take corrective action). The organization needs to quantify its progress in the areas of action taking, communication flows, and participant behaviors; multidimensional transformational change necessitates an ongoing monitoring of activities. Further, the information generated must be used to inform management decisions—a process that underscores the dynamic nature of HSI.

## 2.10 CONCLUSION

> In this and like communities, public sentiment is everything. With public sentiment, nothing can fail; without it, nothing can succeed.
>
> —Abraham Lincoln, First Lincoln–Douglas Debate

The Human Genome Project freed the scientific community from thinking that behavior could be reduced to genetic imperatives. In a similar vein, as students of organizations, we need to avoid viewing bureaucracies simplistically. In reality, work organizations are complex, multilayered entities that, by their very nature, resist change. Taking a simple formula for change and trying to apply it is merely tinkering with change. The approach outlined in this chapter builds on the premise that leaders can transform work environments through planned and systematic actions that touch all components of the organization. Leaders must make a case for change and paint a clear picture of what the transformed organization will look like. The process continues within the organization—involving more and more participants—and outside of it, as key constituencies are brought on board. Finally, the changes are sustained over time by implementing new structures and procedures.

While leaders are at the center of the process, it is the rank and file who determine whether or not an organization has been transformed. Do they now think and act the same way or differently? And, if the latter, do their actions support the new organizational culture? Only when these questions are answered can one determine whether or not successful organizational change has occurred. The chapters that follow will arm leaders with specific information to help decide, guide, support, and sustain the HSI culture.

## NOTES

1. In this and the narrative following, the term "senior management" is used to describe the most senior-level managers in the organization undergoing change, usually general officers and SES career civilians. "Middle management" is used to refer to the next level(s) of management to the supervisory level, generally major through colonel and GS-13 through GS-15.
2. Military officers and political appointees can more easily be removed from their positions than can career civilians at the GS-15 level and below.
3. Since DoD change leaders cannot make changes to the senior management team as easily as top leaders in the private sector can, some methods for constraining resistant senior managers include giving them assignments where their views will not be communicated downward, providing them with new performance standards, and helping them determine whether to stay or leave the organization.
4. ABC Nightly News, Interview with Gordon Moore (Chairman Emeritous of Intel Corp, known for "Moore's Law"), May 25, 2001.

## REFERENCES


Ashkenas, R., Ulrick, D., Jick, T., and Kerr, S. (1995). *The Boundaryless Organization: Breaking the Chains of Organizational Structure*. San Francisco: Jossey-Bass.

Beer, M., and Nohria, N. (Eds.). (2000). *Breaking the Code of Change*. Boston: Harvard Business School Press.

Blackwood, W. O., and Rivello, R. N. (1994, February). *Organizational Change: Lessons Learned from MANPRINT*. Washington, DC: Department of the Army.

Burke, W. W., and Litwin, G. H. (1992). A Causal Model of Organizational Performance and Change. *Journal of Management*, *18*(3), 523–545.

Duck, J. D. (2001). *The Change Monster: The Human Forces That Fuel or Foil Corporate Transformation and Change.* New York: Crown Business.

Fedder, B. J. (2001, May 28). IBM Meets with 52,600, Virtually. *New York Times*, pp. A1, C1.

Fox, M. (1994). *The Reinvention of Work: A New Vision of Livelihood for Our Time.* San Francisco: Harper.

Gould, S. J. (2001, February 19). Humbled by the Genome's Mysteries, Op-Ed. *New York Times*, section A.

Hart, B. K. (2001). Interview with Sandy Robinson, Organization Change Consultant for Oracle Corporation.

Hay Management Consultants. (1996). *Beyond Reengineering: A Behavioral Approach to Leading Change in the Department of Defense.* Arlington, VA: Hay Group.

Henslin, J. (2001). *Sociology: A Down to Earth Approach.* Boston: Allyn and Bacon.

Kotter, J. P. (1990). *A Force for Change: How Leadership Differs from Management.* New York: Free Press.

Kotter, J. P., and Heskett, J. L. (1992). *Corporate Culture and Performance.* New York: Free Press.

Kouzes, J. M., and Posner, B. Z. (1987). *The Leadership Challenge.* San Francisco: Jossey-Bass.

Kouzes, J. M., and Posner, B. Z. (1995). *Credibility: How Leaders Gain and Lose It, Why People Demand It.* San Francisco: Jossey-Bass.

Mahorta, Y. (1993). *Role of Information Technology in Managing Organizational Change and Organizational Interdependence.* Available: www.brint.com/papers/change (Retrieved January 10, 2001).

Osborne, D., and Gabler, T. (1992). *Reinventing Government: How the Entrepreneurial Spirit Is Transforming the Public Sector.* Reading, MA: Addison Wesley.

Pfeffer, J., and Sutton, R. I. (2000). *The Knowing-Doing Gap: How Smart Companies Turn Knowledge into Action.* Boston: Harvard Business School Press.

Shields, J., Harris, C. S., and Hart, B. K. (1999). Culture, Leadership and Organizational Change. In A. P. Sage and W. B. Rouse (Eds.), *Handbook of Systems Engineering and Management.* New York: Wiley.

Shields, J., Johnson, K. M., and Rivello, R. N. (1990). The Acquisition Decision Process. In H. R. Booher (Ed.), *MANPRINT: An Approach to Systems Integration.* New York: Van Nostrand Reinhold.

Slater, R. (1994). *Get Better or Get Beaten! 31 Leadership Secrets from GE's Jack Welch.* New York: Irwin Professional Publishing.

Smith, H. (1995). *Rethinking America.* New York: Random House.

Steininger, D. J. (1994). Why Quality Initiatives Are Failing: The Need to Address the Foundation of Human Motivation. *Human Resource Motivation. 33*(4), 601–617.

## RECOMMENDED READING

Baker, F. (1973). *Organizational Systems: General Systems Approaches to Complex Organizations.* Homewood, IL: Irwin.

Bennis, W., and Nanus, B. (1985). *Leaders: The Strategies for Taking Charge.* New York: Harper & Row.

Builder, C. (1989). *The Masks of War: American Military Styles in Strategy and Analysis.* Baltimore: Johns Hopkins University.

Farquar, K. W. (1995, Spring). Guest Editor's Note: Leadership Transitions—Current Issues, Future Directions. *Human Resource Management*, pp. 110–122.

Gilley, J. W., and Boughton, N. W. (1996). *Stop Managing, Start Coaching: How Performance Coaching Can Enhance Commitment and Improve Productivity.* New York: McGraw-Hill.

Hemp, P. (2001, January). Managing for the Next Big Thing: An Interview with EMC'S Michael Ruettgers. *Harvard Business Review.*

Nadler, D. A., Shaw, R. B., and Walton A. E. (1995). *Discontinuous Change: Leading Organizational Transformation*. San Francisco: Jossey-Bass.

Thomas, K. W. (2000). *Intrinsic Motivation at Work: Building Energy and Commitment*. San Francisco: Berrett-Koehler.

Wilson, J. Q. (1989). *Bureaucracies: What Government Agencies Do and Why They Do It*. New York: Basic Books.

CHAPTER 3

# Human Systems Integration Roles in a Systems Acquisition Culture

GLEN HEWITT and DINO PICCIONE

## 3.1 INTRODUCTION

What is a *system acquisition culture?* How does one recognize the system acquisition culture in which a program is operating? What effects do the acquisition cultural differences have upon a program, especially upon the human systems integration (HSI) element of the program? The degree to which answers to these questions are known and accommodated may determine the success of the acquisition program itself.

A system acquisition cannot be viewed as an isolated activity without interaction with its cultural environment or without the resultant ramifications of these interactions on the system being acquired. Nor can the role of the HSI specialist be assessed out of context with the complex environment of system acquisition. The impact of several dominant cultural influences must be considered as interacting with the systems acquisition process into which HSI must be immersed. The role of HSI is determined by these cultural influences. To approach the role of the HSI engineer without consideration of the cultural environment increases the difficulty in applying HSI best practices and in resolving the risks to the system's successful development and implementation. Consequently, this chapter focuses on the HSI practitioner's roles in the system acquisition process in light of the system acquisition culture within which they must work.

### 3.1.1 Acquisition Culture Defined

*Culture* has been defined sociologically as "the sum total of ways of living built up by a group of human beings, which is transmitted from one generation to another" (Barnhart and Stein, 1963 p. 327). Applied within the context of the acquisition environment, "culture" implies that there is a pattern in the "way of living" or some communicated and repeated way of conducting acquisition related business from one acquisition workforce generation

*Handbook of Human Systems Integration*, Edited by Harold R. Booher.
ISBN 0-471-02053-2 

to the next. *Culture pattern* similarly has been defined as "a group of interrelated cultural traits of some continuity" (Barnhart and Stein, 1963 p. 327). *Culture trait* may be referred to as "any fact in human activity acquired in social life and transmitted by communication." Recognizing an absence of precision in the definition, *acquisition culture* refers to the continuity and pattern of activities that reflect a habitual way of doing system acquisition business. The acquisition culture referred to here involves the buyers, sellers, and users of systems. The examples cited here are taken mainly from experience derived from the public sector where various agencies of the federal government are the buyers. The sellers are normally private companies (vendors) that engage in contracts with the government to provide the systems. However, the "sellers" during the development phase of acquisition may also be other government agencies, federally funded research and development centers, and other organizations that may act in the acquisition process in much the same manner as traditional system vendors. The buyers are normally government employees who act as agents for the user. Preferably users are "end users" of systems and equipment but may also be other participating individuals intermediate to the buyer and end user.

Each of these actors in the process has his or her own culture that often has cross-cultural ties. While it may be argued that there is a culture within a buyer's and seller's community that sets them apart from each other, the segments within each community have their own culture. As an example, let us assume that the U.S. Department of the Army is in the process of procuring a new aircraft. The culture within the U.S. Army's set of actors is certainly different from that of the culture within the aircraft vendor's community. However, the army program management staff may have more in common with the vendor's program management staff than with the army aviator in the field. Similarly, the vendor's program management staff may have a greater cultural tie to their army counterparts than to the technical design staff. The HSI practitioners on both sides of the army–vendor cultural divide are constantly called upon to jump the cultural gap and perform as advocates for the needs of the army aviators (the end user) in the field during the acquisition process. To deal with each cultural enclave successfully means that the culture must be identified and understood in terms of its patterns and traits.

### 3.1.2 Players in the Acquisition Culture

There are a number of players in the acquisition arena with which the HSI practitioner must deal. Each of these players has an influence on the role that HSI plays in the process of acquiring and fielding a system. One of the unique characteristics of the HSI practitioners is that they must straddle a number of different cultures and be comfortable speaking the language of the players in each. The primary player in this realm is the manager of the program that the practitioner supports. The HSI practitioner is normally a supporting member of the cast and often is admitted to the stage reluctantly.

The focus of program management is cost and schedule, with technical performance also a factor. Program management is a continual exercise in conflict resolution and resource allocation. Human systems integration is seen as a means to reduce the risk that cost, schedule, or performance goals are not met. Generally, program managers are "satisfiers," not optimizers. That is, their battle cry is "good enough is good enough." If the system minimally satisfies requirements and meets the cost and schedule goals, even though the system could experience substantial enhancement with a small additional expenditure, it is deemed successful. This success is narrowly defined through interpretations of the requirements documents and operational objectives that are specified in various documents and acquisition milestone directives. There is continual pressure from the operational community to expand the scope of the acquisition and enhance system capabilities.

The role of the HSI practitioner is often to understand the user's functional needs and the operational environment as a method to communicate the human performance component of system performance. The practitioner helps the program management staff realize what program risks might be encountered if the user community takes exception to various characteristics of the system. This role must be carefully orchestrated and performed in conjunction with the formal user representatives that are one of the primary stakeholders on the program manager's team. The practitioner must anticipate human performance issues for the system, define an HSI program to identify and control risk, and verify that the level of risk is acceptable to program management.

The operator and maintainer have their own culture that often clashes with that of program management. A prime directive for the HSI practitioner is to "Know Thy User," which includes the need for sensitivity to the user's culture. There is tremendous variability in these cultures that have an impact on the acceptability of any given system. Military air traffic controllers have a vastly different culture from their counterparts in the Federal Aviation Administration (FAA) who are represented by a union, even though the task demands and operational environment may be similar. This difference may result in the acquisition of a system that functionally meets a narrowly interpreted set of requirements that is acceptable to one user culture and not to the other.

Meister (1997) refers to the gap between HSI practitioners and researchers. Many HSI practitioners have been schooled in the fundamentals of experimental psychology and must deal with psychologists that perform research that the practitioner must alternatively direct, access, interpret, and apply to the system of interest at the moment. The practitioner must be able to influence research efforts so that the results can be transferred to the operational environment through its application in the design of the system. Researchers by their nature are academically oriented and focus on the process of defining a concept for investigation, proposing an investigative mechanism, controlling the experimental environment, and reporting the results. Much of the focus is on the experimental design and publishing a paper in the open literature. Success is often defined by the acceptance of the paper for publication in a refereed journal. The actual experimental result is of secondary importance. For the practitioner, the experimental results help to form what Meister (1997) refers to as "'workplace' knowledge." This is the grist for the HSI mill.

For the most part, researchers are reluctant to deal directly with engineers. Engineers demand quick, unequivocal answers to questions and want quantitative design input that can be directly applied to the design or its specification. Most engineering disciplines are used to dealing with objects that exhibit little variance and function in a predictable manner given a specified environment. If the environment changes, few measurements are needed to quantify the behavior of the object in the new environment. Engineers are generally intolerant of answers to design questions that contain the phrase "it depends." The HSI practitioner must quickly learn that the design of the system will proceed in most cases either with or without the influence of the benefits of the HSI program. The engineering staff is responsible for assuring the technical performance in the acquisition process, and the HSI practitioner must bridge the gaps between research results, user needs and demands, cost and schedule constraints, and his or her own sense of responsibility to all the parties that have a stake in the outcome. All these issues must be captured and communicated not only within the program manager's staff but also to the system vendor that ultimately must offer a product that meets the user's needs and helps to achieve the performance goals of the organization.

### 3.1.3 Influence of Culture

Just what are the ways that the acquisition culture affects HSI in an acquisition program? It is not within the scope of this chapter to explore the ways in which culture is developed, shaped, and transmitted from one system acquisition generation to the next. One merely needs to acknowledge that acquisition cultural influences do become established. We do know that experience, such as that from past acquisitions, shapes attitudes in ways that establish and develop patterns of doing business (see Fig. 3.1). These patterns create or influence tendencies or biases toward certain approaches which, when analyzed and properly understood, may predict trends for future acquisitions. It is important to recognize that knowledge about the acquisition culture can help not only to determine how a workforce may conduct an acquisition but also to modify the approach selected. Where there are HSI risks associated with certain patterns or trends in the acquisition culture, mitigating strategies can be employed to lessen the risks of these cultural traits.

## 3.2 COMMON CULTURAL INFLUENCES

What are the dominant cultural influences that interact with the human engineering process in an acquisition program? The key cultural influences that must be considered to determine their impact include the general business approach of the agency, the research and development culture, the influence of operational settings, and the strategic and tactical political environment. Each will be discussed.

### 3.2.1 Business Environment

For HSI, the cultural impact of the business environment is determined by the difference in responsibilities between the buyer and the seller. The purchaser has three responsibilities: (1) to define system/product requirements, (2) to conduct acquisition program support

**Experiences (past)**

↓

**Shape Attitudes**

↓

**Develop Culture**

↓

**Influence Tendencies/Biases**

↓

**Predict Trends (future)**

**Figure 3.1** Process of cultural development.

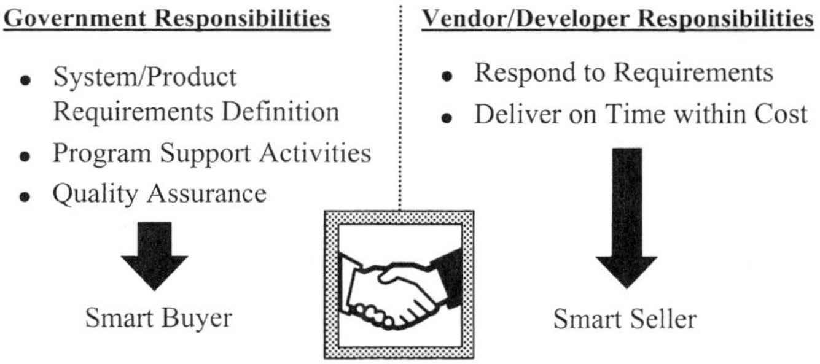


**Figure 3.2** HSI business cultural environment.

activities, and (3) to assure quality. From the vendor's viewpoint, there may be many competing influences upon the approach to be taken, but the basic responsibilities are to respond to the requirements and deliver on time and within cost. As for all components of the intended purchase, the HSI roles and responsibilities of the purchaser and developer/vendor reflect the relationships between being a "smart buyer" and a "smart seller." For the HSI component of the purchase, the central issue is how good the buyer and seller are at meeting HSI objectives in system acquisitions (see Fig. 3.2).

It is insufficient in the acquisition environment simply to have HSI expertise available to the acquisition community. The HSI practitioner must be prepared to address both the means needed to accomplish the HSI effort and the ends (resulting products and services) of the HSI endeavors. Many other "means" are essential to facilitate the required culture. That is, the management support, policies, processes, tools, and training must be in place to provide the supportive atmosphere for HSI to succeed. Figure 3.3 depicts some of the major elements of the HSI business measures that assist in the evaluation of the culture. Each of these measures contributes to the culture of the HSI environment. For example, if the vendor's acquisition workforce has HSI expertise that has positively influenced the design and development of past acquisitions, then the strategy of the buyer (and the roles of the HSI practitioner) differ significantly from that in which the vendor's ability or willingness to address human performance considerations is questionable. That is, in the

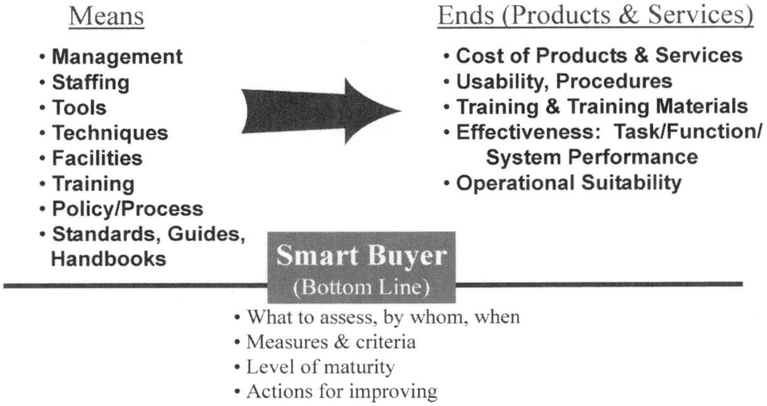


**Figure 3.3** HSI business measures.

first instance the purchasers' HSI participation in activities such as the design of user interface may be more limited (e.g., by emphasizing the role of quality assurance during test and evaluation). The HSI practitioner and the program management leadership must be responsive to the implications of the measures of business culture in general and the HSI culture specifically.

### 3.2.2 Research and Development Environment

A second key cultural influence that must be considered is the research and development (R&D) environment. In all effective system acquisitions, the purchase is made in an environment in which required HSI information is generated or otherwise acquired and then applied to the specific system (Wickens et al., 1997). The process by which this HSI information is acquired is usually referred to as "research" while the HSI activities are referred to as "application."

***Research Process Description*** Different organizations have various ways to describe the type of *research* (e.g., basic, applied). In all cases, there is both general (or *core*) research conducted to understand the fundamentals of the users' operational environment and targeted (or *specific*) research to understand how the fundamentals of operation are affected in certain systems, conditions, or applications. Regardless of the amount of or relationship between the core and specific research, the HSI practitioner's role will be affected by the infrastructure of the research program. For example, where research programs are well funded, expertly directed, and well coordinated with operational needs, the HSI participation is likely to be more influential on requirements definition and less dependent upon design, evaluation, and validation.

***Application Process Description*** Different organizations have various ways to describe the process, phases, or stages of *application* as well. In cases where the application process is mature, the information acquired is applied through a systematic process that includes requirements definition; proposed solutions; and evaluation, validation, and implementation. In the FAA, for example, the acquired information is applied through different processes for system acquisitions, air traffic services and operations, and regulation and certification functions (see Fig. 3.4). Whether these phases and processes are well defined or not, conducted intensely or not, specifically named or not, they are nevertheless sequentially (and often iteratively) implicated in all applications. The degree to which they are defined and institutionalized in the application imposes an influence upon the role of the HSI practitioner. For example, for applications during which a market analysis (conducted during the proposal of solutions) fully analyzes the human component, the HSI role will likely be better defined, funded, and integrated in the subsequent design and development.

***Relationship between Research and Application*** Within the R&D patterns that have been established, the relationship between the research component and the applications component of an acquisition also affects the HSI role and activities. There is a natural tendency for the cultures of research and application to differ. Research activities foster behaviors that are based upon "learning" mental models of the work environment. Assumptions related to this mental model tend to rely on outcomes that are dynamic, include many interdependencies, contain indirect influences, show continuing effects over

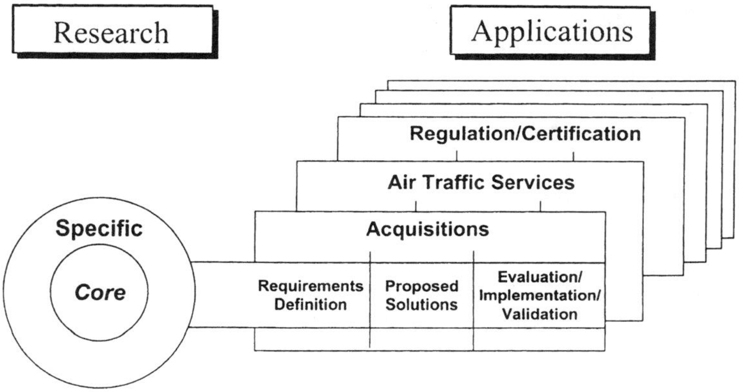


**Figure 3.4** HSI research and application cultural environment.

time, and view the human component less mechanically. Application activities motivate behaviors that are based more on "problem solving" mental models. Assumptions related to this mental model rely on outcomes that are decomposed into subproblems, focus on fixing a problem or achieving a goal, search for a root cause, and use mechanistic and physical metaphors for analysis. Table 3.1 provides a list of attributes related to each of these mental models (Carroll and Perin, 1995).

Acquisitions that bridge the gap between the differing mental models and find good connectivity between the research element of the agency and the application elements greatly reduce the risks associated with human performance in complex systems. Contrarily, cultures that fail to foster the direct linkage between research programs and application efforts likely will grow an agency research program that is disconnected with systems being acquired and fielded. Such a research effort will reinvent the required research (at a greater cost, with fewer options) during later phases of the application.

**TABLE 3.1 Attributes of Research and Applications Mental Models**

| Research: Learning Mental Model | Applications: Problem-Solving Model |
|---|---|
| Nonlinear | Linear |
| Integration | Decomposition |
| Dynamic | Cause–effect |
| Divergent | Convergent |
| Organic | Mechanistic |
| Human | Technological |
| Error expected | Error avoiding |
| Learning | Fixing |
| Relationships | Checklist audits |
| Understanding | Root cause |
| Collaboration | Specialties |

*Source:* Adapted from Carroll and Perin (1995).

***Linkages between Research and Application*** The HSI linkage between applications and research can take two forms. In one case, the operational community can specify needs that must be met or capabilities that are lacking. The articulation of these needs then provides the spark for the research effort that endeavors to provide the technology or nonmateriel solutions to the specified problem in response to operational demands. An alternative approach is for research to push operational capabilities by offering novel solutions to operational problems or providing operational capabilities that were previously not identified or articulated by the user community. Either approach can provide benefits and the HSI practitioner must be prepared to deal with providing a linkage between the research and acquisition communities in either mode. In the operational demand mode, the user community has already encountered or envisioned a shortcoming and often needs the solution in hand in the short term. The articulation of the need is often stated in the context of current procedures and technology that is deemed insufficient for a new environment, newly discovered threat, or new requirements for system throughput stated in terms of system capacity or safety. This leads to applied research that struggles to understand and define the criteria for success and the ultimate requirements for desired performance levels.

Many acquisition models specify that the acquisition cycle begins with a mission analysis that leads to subsequent activities, including the generation of a statement of operational requirements. These acquisition models often specify that the front-end documentation should specify the problem and not a solution. The models often state that both research and application activities should stem from mission needs. The challenge in this process is that the HSI practitioner is called upon to perform a portion of the mission analysis and must deal with an operational community that has learned to be pragmatic as a matter of training and daily survival. This pragmatism leads these constituents to be solution oriented, which can result in shortsighted requirements documents that are essentially shopping lists based on claims or demonstrations on the part of system vendors. Stating the HSI elements of mission needs and operational requirements is often difficult and requires an intense level of cooperation and communication to develop a visionary statement that directs research or development efforts. The HSI community is well served by the use of traditional mission and function analyses as well as human performance modeling. Such efforts often require the HSI practitioner to structure the analyses while the operational community provides the subject matter. The result can be a joint venture where each part of the community invests its own capital in the form of time and expertise to yield an analytical product. If the product is credible, the sense of ownership on the part of all the participants provides the motivation to form a team and sell the argument to organizational management for funding and development. By serving as a core element in the team, the HSI practitioner can be maximally effective.

In the cases where R&D efforts offer technology to push operational capabilities, the challenges for the HSI discipline are greater. Such offerings tend to be attractive to the management community since it potentially represents an opportunity to reduce risk and provide a system to the field with a shorter development cycle by sending the system to a field site to demonstrate the technology and get user buy-in. Once the user agrees that this technology is needed, the mission analysis and requirements documents reflect a need that can only be satisfied by this technology, and the focus shifts to delivery of technology rather than on the functional needs of the operational community. This leads to dissatisfaction during the test phase when a new group of users is exposed to the system and determines that the mission cannot be performed with the new system.

The role of HSI in a technology-push environment is to participate in the mission analysis and requirements definition process to assure that the critical needs of the user are articulated and the human interface risks associated with adopting the technology are defined. If additional investigations or other HSI activities are needed during the acquisition to assure that the mission needs are met, they must be identified at this point. Research programs that migrate to the acquisition cycle tend to focus on production versions of the research prototype with the assumption that if the system "worked" as a research product in a demonstration setting, it is good enough.

### 3.2.3 Operational Environment

Another cultural influence is the operational environment. That is, the culture of the operational environment will also bring a set of boundaries that will impact the role of the HSI effort. The considerations reflected in the HSI role by the operational environment include the operational mission approach, operational philosophy, and program operational emphasis. For example, consider the difference for the role of human safety in a business operation versus that of an aviation operation. Similarly, an operational philosophy such as "automation will be used extensively" has a different influence on the HSI role if one is considering the human on the deck of an aircraft carrier versus the human role in the control of the flight of missiles. Also, an acquisition that emphasizes the reduction of military staffing levels and manpower costs—such as a new naval warfare ship—will not present the same HSI role as one that emphasizes technological changes—such as the global positioning system application (Hughes and Dornheim, 1995).

As an illustration within the FAA, consider the operational environment of air traffic services (ATS) and airway facilities (AF) maintenance operations. Within ATS, there is a centralized emphasis on avoiding operational errors. Notwithstanding the importance of personnel safety within AF, decentralized operational effectiveness of the maintenance support function is the essential ingredient. The culture of these operational environments will become reflected in the strategies and approach to the development of procedures, training, system design, staffing guidelines, and other HSI roles within the acquisition programs. In another example, consider the appreciable differences in the HSI practitioner's role for the design and development of a small airport tower versus the design and development of a large, metropolitan tower where the complexities of the crew coordination, communications, interfacing procedures, program integration, and redundancy are paramount.

The operational environment can also levy an influence in terms of levels of technical sophistication in the workforce and the legacy systems with which the workforce must deal. While there is a significant movement in modern society to adopt new technology to resolve problems and advance productivity, not all workplaces can tolerate large changes in the technology base. Some workforces may be represented by powerful unions that will resist the rapid introduction of certain types of technology if it is perceived as a threat. The change may also be perceived by management as a threat if the proposed changes will entail risk to profits or revenue in the short term. This was a hallmark of the railroad industry in the 1970s. There were significant HSI and human factors challenges associated with enhancing safety and increasing the productivity of the rail system. Some of the enhancements were possible by combining advances in railcar braking systems and sensor systems associated with track-train dynamics to achieve enhanced levels of operator performance on long-haul freight trains. The legacy of the large amounts of rolling stock with 1950s-era braking systems coupled with resistance to change from both management and the labor force stymied the efforts of government agencies to apply the products of rail transportation research, including HSI and human factors efforts. The introduction of graphical displays to the locomotive cab at that time was met with skepticism. Thirty years later, the widespread use of computers in the workplace has eventually met with some degree of acceptance, as in the locomotive cab, although they still display the status of the same 1950s type of braking systems.

### 3.2.4 Political, Management, and Organizational

The last key cultural influence that affects the HSI role is the culture of the political, management, and organizational environment. Consider, for example, the implications upon the HSI acquisition functions that are set in an environment in which the workforce is unionized (such as for the FAA's air traffic control systems) versus one that is not [such as U.S. Department of Defense (DoD) weapon systems]. Consider one that is government (where political influence and oversight are direct) versus one that is nongovernment (such as the automobile industry, where market forces dominate). The context of these organizational forces materially affects the design and implementation of HSI endeavors (Wickens et al., 1998).

### 3.2.5 Impact of the Cultural Environment

Elements of the HSI role that are affected by the cultural environments include such parameters as the HSI management concept, organizational design, and practitioners' authority. For example, should the HSI effort be managed centrally by an HSI specialized office or should the effort be managed by a decentralized effort closest to the program or product? The answer to that question needs to be found within the cultural context of the business, R&D, operational, and political or organizational environment of the acquisition. One of the fundamental issues related to the management and execution of HSI support concerns the organizational structure of HSI professionals. Are they to be centralized and parceled out to engineering teams as the needs become evident? Or are they to be acquired by and for disparate system engineering applications without regard to centralization of the HSI engineering discipline? When making decisions on these topics, it is important to determine the degree to which domain expertise is needed on the programs. HSI practitioners tend to gain expertise in a particular area and often in specific subareas such as the "oceanic" domain of air traffic control, and the degree to which this expertise transfers to other domains varies. Working with subject matter experts and field personnel tends to be easier if their jargon and tasks are already known. Their confidence in HSI input, judgment, and assessments is heightened if the practitioner is already considered knowledgeable in the domain. Building this expertise and trust takes time that may not always be available. Closely tied to this issue is that of providing continuity among the HSI staff. Maintaining appropriate expertise may be especially difficult for projects that involve long-term studies, the application of domain-specific principles over a period of time, or an acquisition strategy that uses an iterative approach to building the system.

Other factors in defining and conducting the HSI role that are affected by the cultural environments include those in Table 3.2.

**TABLE 3.2 HSI Functions and Attributes Affected by Cultural Environment**

- *Management and execution concept:* centralized or decentralized HSI management and execution
- *Flexibility:* standardized support or tailored to organization being supported
- *Location:* separate or collocated with teams being supported
- *Organization:* matrixed support or structured under product leader
- *Responsibilities:* consultant/advisor or fully accountable for HSI portions of all program products
- *Authority:* provides suggestions and recommendations or has signature/approval for all HSI in plans, studies, analyses, documentation, reviews, reports, and tests
- *Coordination:* isolated program support component or representative to coordination and integration groups
- *Performance assessment:* independent in determination of performance or jointly appraised in conjunction with other program performance evaluations

## 3.3 HISTORICAL PERSPECTIVE OF CULTURE

A second perspective for cultural influences upon the HSI role is the traditional or historical role played by HSI practitioners and human factors engineers. It can be easily argued that the origins of human factors engineering have their antecedents in the fundamentals of twentieth-century management thought conceptualized by Henri Fayol in 1916 (Mittler et al., 1990). However, the practice of HSI has only recently seen widespread maturity in its application—and even now that maturity is sporadic. The history of HSI is certainly not long, and the description of the roles of the HSI practitioner continues to evolve. For example, in relation to software development activities for the computer–human interface, the software engineer and HSI practitioner increasingly communicate and work closely together in the process of rapidly developing prototypes. Notwithstanding the short history and dynamic nature of the HSI role, the influence of this historical and still emerging HSI role can be divided into three major categories: (a) the context of the HSI role in system acquisitions, (b) HSI roles in direct support of system acquisition activities, and (c) HSI managerial and oversight roles and responsibilities.

### 3.3.1 Context of HSI Roles in System Acquisition

The historical role of the HSI practitioner has been established and refined over the past two to four decades and is outlined here as an introduction. First it should be understood that the interfaces associated with the HSI role are many and varied. For example, historically the role of HSI and human factors engineering in the participation of equipment design activities is well documented in the literature of system engineering disciplines. This role is similar and complementary to the traditional role of the ergonomics engineer in which the issues related to knobs and dials, controls and displays, and fit and function are addressed. (A list of the common HSI issues is given in Table 3.3.)

The role of the HSI participant in the design and development of equipment (hardware and software) has also been extended to include other system development interfaces, that is, those interfaces related to safety and health, management and organization, cognition, or cooperation. For example, HSI participation in the system development process should entail consideration of the organizational interfaces. These interfaces go beyond the analysis of tasks and job functions by including the job design (i.e., how the functions

**TABLE 3.3 HSI Issues Common to Acquisition Programs**

- *Workload:* operator and maintainer task performance and workload
- *Cognitive decision making:* requirements for operator and maintainer tasks and decisions and related performance measures
- *Training:* minimized need for operator and maintainer training
- *Functional design:* equipment design for simplicity, consistency with desired human–system interface functions, and compatibility with expected operation and maintenance concepts
- *Computer–human interface:* standardization of computer–human interface to address common functions and employ similar user dialogues, interfaces, and procedures
- *Staffing:* accommodation of constraints and opportunities on staffing levels and organizational structures
- *Safety and health:* prevention of operator and maintainer exposure to safety and health hazards
- *Special skills and tools:* considerations to minimize the need for special or unique operator or maintainer skills, abilities, tools, or characteristics
- *Work space:* adequacy of workspace for personnel, their tools and equipment, and sufficient space for movements and actions they perform during operational and maintenance tasks under normal, adverse, and emergency conditions
- *Displays and controls:* design and arrangement of displays and controls that are consistent with operator's and maintainer's natural sequence of operational actions
- *Information requirements:* availability of information needed by operator and maintainer for a specific task when it is needed and in appropriate sequence
- *Display presentation:* ability of labels, symbols, colors, terms, acronyms, abbreviations, formats, and data fields to be consistent across display sets so that they enhance operator and maintainer performance
- *Visual/aural alerts:* design of visual and auditory alerts (including error messages) to invoke necessary operator and maintainer response
- *I/O devices:* capability of input and output devices and methods for performing task quickly and accurately, especially critical tasks
- *Communications:* system design considerations to enhance required user communications and teamwork
- *Procedures:* design of operation and maintenance procedures for simplicity and consistency with desired human–system interface functions
- *Anthropometrics and biomechanics:* system design accommodation of personnel (e.g., from 5th through 95th percentile levels of human physical characteristics) represented in user population
- *Documentation:* preparation of user documentation and technical manuals (including any electronic HELP functions) in a suitable format of information presentation, at appropriate reading level, and with required degree of technical sophistication and clarity
- *Environment:* accommodation of environmental factors (including extremes) to which user will be subjected and their effects on human–system performance

are performed), the management structure related to the job (i.e., how the job fits within the surrounding jobs and functions), and the organizational structure (i.e., how the job fits within the structure of other jobs, supervisors, and organizational relationships). Table 3.4 provides a list and brief description of the interfaces that have become elements of the HSI practitioners' role. Also provided are the performance dimensions and objectives that relate to the context of the enumerated human interface classes. These eight interfaces may be regarded as an integral part of the "total" system of equipment design and develop, but they are often and easily overlooked if not explicitly identified. It is this context of the HSI

**TABLE 3.4 HSI Interfaces in Sytems Acquisition**

| Human Interface Class | Performance Dimension | Performance Objective |
|---|---|---|
| 1. *Functional interfaces:* for operations and maintenance, role of human vs. automation; functions and tasks; manning levels; skills and training | Task performance | Ability to perform tasks within time and accuracy constraints |
| 2. *Information interfaces:* information media, electronic or hard copy, information characteristics, and information itself | Information handling/ processing performance | Ability to identify, obtain, integrate, understand, interpret, apply, and disseminate information |
| 3. *Environmental interfaces:* Physical, psychological, and tactical environments | Performance under environmental stress | Ability to perform under adverse environmental stress, including heat/cold, vibration, special clothing, illumination, reduced visibility, weather, constrained time, and psychological stress |
| 4. *Operational interfaces:* procedures, job aids, embedded or organic training, and on-line help | Sustained performance | Ability to maintain performance over time |
| 5. *Organizational interfaces:* job design, policies, lines of authority, management structure, organizational infrastructure | Job performance | Ability to perform jobs, tasks, and functions within management and organizational structure |
| 6. *Cooperational interfaces:* communications, interpersonal relations, team performance | Team performance | Ability to collectively achieve mission objectives |
| 7. *Cognitive interfaces:* cognitive aspects of human–computer interfaces (HCI), situational awareness, decision making, information integration, short-term memory | Cognitive performance | Ability to perform cognitive operations, e.g., problem solving, decision making, information integration, situational awareness |
| 8. *Physical interfaces:* physical aspects of system with which human interacts, e.g., HCI, controls and displays, workstations, and facilities | Operations and maintenance performance | Ability to perform operations and maintenance at workstations, work sites, and facilities using controls, displays, equipment, tools, manuals, etc. |

*Source:* Adapted from Federal Aviation Administration and Carlow International Incorporated.

role and the culture of the HSI and system acquisition effort that determines the extent to which these elements are adequately addressed.

### 3.3.2 HSI Roles in Direct Support of System Acquisition Activities

The roles of the HSI practitioner are as diverse as the applications they support. These roles and tasks begin early in the program with participation in mission analysis and requirements determination processes that may only serve to identify the major impacts and constraints of the human element upon the new capability being acquired. For example, early HSI activities may be initiated to limit the manpower and staffing requirements of predecessor systems, limit time-consuming and costly training requirements, simplify complex procedures that induce errors and increase task performance times, or all three. The HSI role continues through the verification of test and evaluation programs into the design of monitoring functions and data collection plans that serve to evaluate the degree to which human–system objectives are continuing to be met after system deployment and implementation (i.e., during the in-service management phase of a system). Note that the role of the HSI practitioner may change drastically if the vendor chosen for an acquisition has little or no HSI capability. In some cases, the vendor may reject the buyer's efforts to generate an HSI program that is integral to the design effort. The buyer's HSI practitioner must then determine a strategy that will reduce risk for the program and achieve the objectives without HSI support in the vendor's organization. While each organization may define their acquisition phases differently, Table 3.5 provides a list of the common major HSI roles in direct support of system acquisition activities.

**TABLE 3.5 Major HSI Roles in Systems Acquisition Activities**

HSI will perform, direct, or assist in conducting the following activities:

- Mission analysis and requirements determination (human impacts, constraints)
- Human–system interface considerations in market surveys/investigations/trade studies
- Generation and update of HSI plans
- HSI input to solicitation package preparation
- Identification and analysis of critical tasks performed by operators and maintainers
- Generation, refinement, and analysis of operational scenarios, human–system modeling, and human in loop simulations
- Development, demonstration, and evaluation of human–computer interface design requirements, prototypes, design, and development efforts
- Review/analysis of human engineering documentation
- Coordination of HSI working group activities
- Conduct of task performance analyses and coordination with training and logistics
- Conduct and coordination of safety and health hazard analyses
- HSI concepts, analyses, and assessments of engineering change proposals (ECPs) and design reviews
- HSI input to test and evaluation (T&E) plans, measures, criteria, and data collection efforts
- Design and evaluation of monitoring and data collection plans for postdeployment human–system performance

### 3.3.3 HSI Management and Oversight

The roles and responsibilities of the HSI practitioner are defined, in part, by the degree to which the HSI management and oversight responsibilities have been institutionalized (i.e., the "culture" of HSI management). Detailed descriptions of supervision and management of acquisition programs are covered in texts related to acquisition program management and general system engineering. However, four key areas of HSI management responsibility and authority determine the HSI acquisition culture: (a) policy, process, and procedures; (b) organization and infrastructure; (c) tools and training; and (d) integration activities.

***Policy, Process, and Procedures*** The culture surrounding the HSI effort will affect and be affected by the promulgation of policy, the definition of processes, and the practice of procedures related to HSI planning and implementation. That is, in agencies that publish and enforce strong HSI policies, the acquisition culture will be significantly more conducive to identifying and resolving HSI issues than those agencies without such policies. Those organizations that establish and exercise definitive HSI processes have proven to be more successful in the mitigation and resolution of human performance problems than those organizations without such processes. Similarly, in those agencies where the practice of HSI has become institutionalized, repeatable, and common, systems find better and cheaper solutions to HSI considerations. Policies, processes, and procedures that should be developed and institutionalized include those related to

- the importance and objectives of HSI;
- methods to coordinate HSI research with HSI engineering;
- the definition, scope, and role of HSI in the acquisition process;
- the process of conducting HSI activities and its relationship to other engineering disciplines and activities;
- documentation requirements related to the HSI activities; and
- mechanisms for the evaluation of HSI programs across the agency.

***Organization and Infrastructure*** Those charged with some responsibility for the HSI research and engineering functions within an organization should give significant consideration to the organization and staffing of the HSI organizational infrastructure. No element of the HSI program is more important than the number, qualifications, and organizational relationships of the HSI professionals supporting the acquisition efforts. No other indicator is more evident of the culture surrounding the HSI environment. If the status of the HSI personnel and organization reflects a weak investment, it is a probable indication of serious deficiencies in the HSI program at all levels of management and implementation. Establishing the HSI organizational infrastructure should be among the highest priorities for those intending to support HSI within the acquisition community.

***Tools and Training*** Many organizations attempt to develop some of their own HSI tools and training. Having the right set of tools and training readily accessible to the acquisition workforce will result in significant benefits to the acquisition program. However, precious HSI professional expertise can be wasted in the development of tools and training attempting to create capabilities that may already be available. It is necessary to evaluate where, how, and what kind of HSI capabilities should be acquired—whether purchased from outside the organization or grown from within.

***Integration Activities*** An adequate HSI culture is one that is able to promote the planning and execution of the appropriate engineering activities. Such a culture necessitates that the development of policy, process, and procedures; organization and infrastructure; and tools and training becomes integral to the acquisition process itself (Wickens et al., 1997). It may be obvious that development of this infrastructure and culture entails a considerable amount and continuity of effort. A list of HSI managerial and oversight roles and responsibilities is provided in Table 3.6.

***Sampling the HSI Culture*** The culture of the HSI environment will contribute to the definition of the management and oversight HSI responsibilities. The identification of some sample HSI responsibilities may serve to describe this HSI role more fully. Typical tasks assigned to an office with HSI responsibilities that should be considered within the cultural context of the acquisition environment include those in Table 3.7. This provides a list of typical tasks affiliated with the HSI management and oversight role.

### 3.3.4 Caution: Culture of Computer–Human Interface

More than a small number of HSI programs have suffered from the deleterious effects of failing to define the proper scope to the HSI effort. With the proliferation of management information systems and the predominance of software costs associated with the human–

**TABLE 3.6 HSI Managerial and Oversight Roles and Responsibilities**

Typical roles and responsibilities of a centralized HSI office include the following:

- Identifying an HSI point of contact for interaction with system product team/integrated product team (IPT)
- Providing guidelines for preparation of HSI plans and development and execution of HSI program
- Participating as member of human system integration working groups (HSIWGs)
- Coordinating and monitoring effectiveness of HSI processes and procedures
- Identifying and/or establishing standards and monitoring their application across IPT/product teams
- Advising IPT HSI coordinator of HSI risks and concerns associated with integration of systems across domains and recommending course(s) of action for their resolution
- Identifying and coordinating HSI and related research needed to address issues that cross products and/or cross IPTs that interact with domain
- Participating in technical interchange meetings with program/product HSI coordinators
- Reviewing acquisition and program-related documentation for proper inclusion of HSI considerations
- Providing HSI inputs to statements of work (SOWs), system specifications, and data item requirements, monitoring contractor activities, and reviewing contractor deliverables
- Providing information and participating in source selection activities, acquisition reviews, resource council briefings, and any other related program reviews
- Monitoring technological advances, marketplace trends, and relevant HSI research and analyses and sharing findings with HSIWG

**TABLE 3.7 HSI Management Tasks (Example)**

The HSI management and oversight tasks include the following:

- Identifying an HSI coordinator who serves as the focal point for coordinating all HSI activities among other organizational HSI elements and with other integrated product team (IPT) HSI representatives
- Preparing and executing IPT HSI plans that are compatible with organizational acquisition management system policy and guidance
- Identifying product HSI representatives responsible for integrating HSI considerations throughout product development
- Resolving human performance issues that occur during prototype development and testing
- Coordinating with HSI representatives to ensure HSI considerations within and across products are adequately addressed
- Establishing HSI coordinating groups that include HSI representatives from product teams and specialists from HSI areas of concern to serve as technical resources
- Establishing and monitoring effectiveness of HSI processes and procedures
- Establishing means for HSI representative to advise IPT of risks and concerns and recommending course(s) of action for their resolution
- Identifying HSI research needed to address issues common to product teams or that cross IPT boundaries and coordinating those research needs with other organizational elements
- Conducting periodic technical interchange meetings with HSI representatives to present and discuss HSI concerns and methods for mitigating them and to consider trade-offs among HSI technical areas for improving system performance or reducing cost
- Ensuring HSI considerations are addressed in all acquisition and program-related documentation [e.g., requirements documents, cost–benefit analyses, statements of work (SOWs), test plans]
- Ensuring HSI considerations are thoroughly addressed in transition plans for new systems and functions being integrated into environment
- Ensuring that HSI specialists actively monitor activities of prime contractors and subcontractors in all HSI technical areas as specified in SOWs, system specifications, and standards
- Providing HSI information for and participating in source selection activities, acquisition system and program reviews, resource council briefings, and any other related program reviews
- Monitoring technological advances and marketplace trends relevant to products and sharing pertinent findings with other HSI representatives

system interface, the importance of computer–human interface (CHI) or other similar terms such as human–computer interface (HCI) or man–machine interface (MMI) has been widely acknowledged. The interpretation of these various terms is often confused with the full scope of a proper HSI effort. That is, too often the HSI effort is confined to the important but limited considerations of screen design. The acquisition cultures that relegate HSI to this diminutive form are likely to flaw the program seriously.

The mundane and elementary factors associated with screen design can mask more serious concerns that center on the basic functions that can be performed by the system and the capabilities that the system operator can exercise in the performance of the mission or task. In many cases the functions and capabilities are centered on the technology and interests of the designers that coincidentally have an overlap with the needs of the user to perform the required task. A problem, for example, arises when the excess capability that is not needed requires maintenance and lies dormant in a system that is also deficient of needed capabilities. This creates a system mismatched to the user's needs. The unmet needs are often cited as "CHI" problems that then require extensive rework in the basic structure and mechanisms of the system. The "CHI" problems are really traceable to poorly articulated requirements or the inability of the designer to understand the details of the stated requirements and the operational needs. The HSI practitioner may in some cases unwittingly reinforce the concept that CHI relates to screen design by justifying the need for an HSI program that is primarily based on checklists, color guidelines, CHI principles, and CHI style guides. The design and management members of the team then see this as the main thrust of HSI and relegate the activities to the latter stages of development since these attributes can be easily reconfigured in most software intensive systems. Techniques, such as a heavy use of checklists, force a bottom-up approach to HSI where a more holistic approach may better serve the user and garner more support on the part of the acquisition team.

The scope of the HSI program and role of the HSI practitioner must be recognized for the macroergonomic as well as the microergonomic requirements. Table 3.8 provides a sampling of the differences between these two critical roles. For example, in addition to screen design, it is important to understand (and design) the cognitive requirements (e.g., memory requirements, calculations) associated with the screen's presentation of information. Also, while it is important to address the design of knobs and dials for the system, it is equally important to design the mapping of these controls to fit with the operators' and maintainers' tasks. In another example, the task performance of the individual user is paramount to good system design. No less important are the considerations of how these tasks fit into the design of crew performance considerations. In fact, well-designed crew resource management programs usually identify crew tasks or the sharing of tasks among individuals that may go unheeded in programs that solely or sequentially focus on the task of an individual.

## 3.4 CHANGING ACQUISITION CULTURE

Without regard to any particular effort to modify the general cultural environment within the acquisition community, there have been inescapable forces working to change it. Specific changes that have occurred over the past 10 years within the acquisition culture include (a) the increased intensity of HSI, (b) the elevated stature of usability, (c) the predominance of cognitive factors, (d) social and legal concerns for special accommodations, and (e) growing cross-cultural issues of interface design.

**TABLE 3.8 Microergonomic Versus Macroergonomic Dimensions of HSI: Examples of Range of HSI input**

| Microergonomic | Macroergonomic |
|---|---|
| Screen design | Cognitive requirements |
| Knobs and dials | Control/display-to-task mapping |
| Individual skills | Population attributes |
| Procedures | Job design and integration |
| Workload | Staffing and organizational design |
| Individual performance | Crew/team performance |
| Product usability | System usability |
| Training regimen | Skill acquisition and decay |

### 3.4.1 Intensity of HSI

No systems have been devised to operate without a human–system interface. Despite the consistency in the need for HSI support, the nature (intensity) of this interface has changed markedly, especially within the last two decades. Increasingly, more and more devices in the marketplace project an interface that personally and continuously interacts with the user. For example, an increasing number of household appliances contain information systems that provide feedback or instructions to the user and require input or response from the user. As miniaturization continues to create opportunities to embed tiny processors in the workplace equipment and personal possessions, an increasing number of applications contain a human–computer component. As the power of the embedded processors grows, so do the role and complexity of the HCI component. Because systems can be designed to be more responsive to their users (even individually tailored in their response), new systems find an increase in the authority and leverage of the user's interaction with the system. Naturally, this improvement in user authority reenforces one's expectations of systems to be responsive to unique and specialized use, which has resulted in an even greater proportion of the system's software being devoted to the human interface (Nielson, 1993). Thus, the density and intensity of the HSI component of systems necessitate a new (modern) cultural view of systems and their interface with the user. This new view involves one in which the HSI participant has a much larger share of the system's risk assessment, engineering, and budget.

### 3.4.2 Emphasis on Usability

*Usability*, in some respects, has become the modern buzzword and synonym for the proliferation of user–interface vernacular. Many authors have found "usability" to be an uncharged substitute for terminology that implies a cultural, organizational, or gender bias (e.g., ergonomics, MMI, CHI). Notwithstanding the natural utility of the term, its use has become more common simply because of increased emphasis on systems' usefulness. Contrary to a time not so many years ago when usability was primarily a concern for the user representatives during test and evaluation (as acceptance criteria), current acquisition programs are replete with concerns for system usability from the beginning of the program acquisition. Consideration of usability pervades the program's response to system stakeholders—from end users to senior management. While not every acquisition community has translated this growing emphasis on usability into a viable HSI organization or program, the increased emphasis does help to create a culture where HSI has a greater opportunity to flourish.

### 3.4.3 Predominance of Cognitive Factors in Design

Another new and growing impact upon the cultural environment of the acquisition community is the change from physical to cognitive attributes of the HSI effort in design and development. Still slow to receive the full acknowledgment of some members of the acquisition community, the design of systems involves a larger and larger contribution from the tasks related to the user's mental requirements. While the movement to cognitive tasks is clearly evident, the change continues with enough subtlety to cause a struggle in gaining full recognition from the acquisition community. Difficult as it may be to find adequate acknowledgment, system HSI design efforts increasingly reflect a predominance of things people must remember, recall, calculate, estimate, evaluate, analyze, interpret, recognize, or otherwise think about.

### 3.4.4 Special Needs and Accommodation

Human systems integration and its related disciplines (e.g., human factors, safety, CHI) have always been viewed as engineering for the user. Historically, the user has been described as a statistical proportion (e.g., 5th to 95th percentile of the population), a representative sample, or those who display typical operator or maintainer characteristics. Recently, customers of acquisition programs have grown to expect HSI to include previously disenfranchised portions of the population. New tools and user influences have emboldened HSI practitioners to design for all users including those with special needs. Federal law (e.g., Americans with Disabilities Act of 1990 and the Rehabilitation Act of 1973), protective standards (e.g., Uniform Federal Accessibility Standards), and other guidelines (e.g., Occupational Heath and Safety Act) promote expansion of public access requirements and design accommodations to assist those with special needs. Legal issues aside, public relations considerations (i.e., just plain community good will) cause public and private facilities, equipment, and services to meet wider population standards. Indeed, some commercial activities have found new markets in the accommodation of those with special requirements. In many cases, reaching beyond the traditional disabilities population (such as those in wheelchairs), designers have addressed unique access requirements and tailored workspaces to individual needs. As the public tolerance diminishes for designs that exclude even small portions of the population and as market share expands to meet tailored preferences, HSI has responded with tools and methods to incorporate these needs early in the requirements and development processes.

### 3.4.5 Cross-Cultural Issues of User–Interface Design

A major influence on the role of the HSI engineer and a determinant of the culture of the acquisition environment is the degree of sensitivity to cross-cultural considerations. There is no doubt that the expanded market to the international arena has imposed new design considerations upon the creators of equipment and systems. No longer can the manufacturing community ignore the economic opportunities inherent in appealing to distant populations. These populations contain cultural differences that must be considered at the earliest stages of development. Simplistically, for example, it is unlikely that American-made toys such as dolls or bicycles will meet with equal enthusiasm in Asia or Africa when made with only Americans in mind. Similarly, even the appearances and anthropomorphic elements of our new systems demand attention to cultural differences.

The well-recognized globalization of the marketplace brings with it a globalization of the user that goes beyond differences in size and shape. When the design and development community (including the HSI representatives) define the operator, past assumptions about the uniformity of the users' characteristics are no longer true. An aviation accident in which a commercial aircraft flew into a mountainside (an accident categorized as controlled flight into terrain, or CFIT) provides a salient example. In this accident, it has been implied that the Chinese pilot may not have understood the blaring aural alert directing the pilot to react. A recorded voice communication from one of the pilots to the other has been interpreted as "What does 'PULL UP' mean?" Some of these cultural differences are more subtle.

In another aviation example, the impact of culture has pervaded the philosophy of automation in the world's two largest commercial aircraft companies. It is well documented that Boeing and Airbus represent their cultural views very differently (Hughes and Dornheim, 1995). Where Boeing gives the pilot more degrees of flying freedom, Airbus designs tend to put automated boundaries around the pilot's safe-flying envelope. Pilots must be trained to react very differently for these two environments. Designers and especially HSI practitioners must consider the effects of variations in the technology, availability and impact of design tools, learning modalities, language, and management information methods among different cultures. Accommodating a global economy, the international markets, and intercultural communities of the user will continue to play a major role in HSI considerations.

## 3.5 TRENDS FOR THE FUTURE OF HSI

Changes and trends within the acquisition culture and HSI culture itself portend new HSI roles and relationships. The culture of HSI is responding to and must continue to respond to changes in (a) the relationship between hardware and software, (2) the use of off-the-shelf products, (3) the availability of HSI tools and technologies, (4) the dependence upon HSI compliance, and (5) approaches to program documentation. For a summary of these cultural trends, see Table 3.9.

### 3.5.1 Hardware/Software

Prior to the accelerated pace of the developments in the age of information (i.e., at least prior to the last decade or two), the acquisition community and HSI representatives focused upon the new hardware being procured. As operational platforms have become more expensive and opportunistic in capturing the advantages of increased information, software enhancements have become dominant. That is, the ratio of acquisition devoted to

**TABLE 3.9 Changes in HSI Trends**

| Historical | Future |
|---|---|
| 1. Hardware orientation on form and fit | 1. Software orientation with increased importance of procedures, cognitive tasks, and training |
| 2. Dependence upon communication to design engineer | 2. Greater dependence upon NDI/COTS solutions to human–systems performance requirements |
| 3. Limits of technology to integrate | 3. Increased rapid prototyping, modeling, simulation with human-in-the-loop |
| 4. Long lead time and iterative design approach | 4. Shorter acquisition time for consideration of human resources, human performance requirements |
| 5. Use of design guides and standards to assure level of quality | 5. Decreased use of compliance standards and specifications |
| 6. Larger documentation requirement | 6. Less dependence upon documentation |

the mechanical and hardware portion is decreasing relative to the software component. Not only are hardware systems embedding more software-intensive systems, but the number and size of the information management systems are increasing as well. Similarly, a greater percentage of the software development is devoted to the HCI component. In one study of several systems, Nielson found that 48% of the software was concerned with HCI, making it one of the most costly items of the product (Nielson, 1993).

As the software and HCI component become larger and more pervasive and intense in systems, the tendency and opportunity to make changes increase. The flexibility of system software promotes an acquisition environment in which systems may be (or at least appear to be) developed and revised more rapidly. This rapid development and software-intensive acquisition culture prescribes a role for the HSI representative that occurs earlier in the acquisition phases and is more intense.

Also, the larger share of software considerations (relative to hardware) implies a greater risk in the maintenance tasks as well as the operational functions. Software systems present more difficulty in the diagnostic capabilities and impose special considerations upon the HSI role for the design of maintenance operations, staffing, procedures, training, and the like. Responses from the HSI community will need to resolve demands for new technical skills in the HSI workforce, new methods and procedures to influence software HCI designs earlier, and new techniques and training to assist in this effort.

### 3.5.2 Development/Off the Shelf

The expense of long-term development programs and the need for faster acquisitions in order to meet the quick pace of changing technology have accentuated the seduction of buying off the shelf. This tendency to select acquisition strategies for nondevelopmental items (NDIs) or commercial-off-the-shelf (COTS) items changes the nature of the requirements by substituting the vendor's market demands for the purity and uniqueness of the users' needs and desires. While much can be said for the acquisition trend to follow the market, the lack of full control over the design and development modifies the risks of the program as well as the resultant role of the HSI practitioner. For example, instead of participating in each stage of the engineering design, the HSI representative assists in evaluating the vendors' alternatives for the human performance component. Changes in the acquisition "culture" may be realized relative to the HSI tools necessary to support NDI and COTS acquisitions, the HSI policies that acquisitions follow (e.g., the HSI role in source selection), and the processes and training of the acquisition workforce.

With respect to the user and maintainer interface with a system, the difference between COTS and NDI can be substantial. Program management is often tempted to dismiss the need for HSI support if the system is a COTS acquisition. The rationale is that the interface is "standard" and the marketplace has driven the vendors to produce a usable design. In the case of widely used products such as personal computer hardware, word processing software, or personal telecommunications equipment, this is sometimes a good rationale. However, most acquisitions for major systems that are to be used to perform essential missions for government agencies are actually NDI acquisitions using off-the-shelf modules and a customized human interface tailored to the functions contained in the system. These are unique systems that have not had the benefit of market scrutiny and feedback in the area of the human interface. These procurements run the risk of buying what the vendor wishes to sell without regard for user needs or the impact on human-in-the-loop system performance. Just as the engineering staff presses for such concepts as "open architecture" in a COTS/NDI procurement to avoid the pitfalls of proprietary components that create interface problems, the HSI staff needs to press for a "human–system architecture" that supports the user in the performance of various tasks and missions.

### 3.5.3 HSI Tool Technology

Harnessing the power of technology within the acquisition community has recently led to a new arsenal of tools available to the acquisition community, including the HSI elements. Despite the considerable efforts in past years to develop tools and techniques to support HSI, many of these endeavors provided only limited new capabilities. While the concepts for HSI tools have been sound, the limited computing power, high cost, or dependence upon huge databases often diminished their ability to influence timely acquisition decisions. As technology has enhanced processing power, increased the availability of data, and decreased cost, new opportunities have emerged for a proliferation of tools and technologies.

Powerful applications related to design alternatives (such as rapid prototyping techniques) are giving developers options and testers opportunities for evaluations well in advance of past acquisitions. New capabilities are providing the buyer methods to include visual specifications as government-furnished information, thereby decreasing the risk of poor interface designs. These capabilities are especially important to the HSI practitioner in the acquisition cases where there is little confidence in the ability of the vendor to provide a well-designed human interface for the emerging system.

New techniques (both high fidelity and low) are becoming easier to use and more widely distributed among HSI laboratories. Techniques that were labor intensive and lengthy are becoming more easily manipulated and able to be rerun repeatedly with revised parameters.

These new developments of HSI tools and techniques suggest that the future may lead to greater compatibility of tools across HSI domains (e.g., those devoted to workload and staffing, skill assessment and training, or error management) or even integration among various tools. These tools can potentially reduce program risk and have a positive impact on program budget and schedule by reducing the probability that the HSI aspects of the system will not achieve system objectives. The acquisition culture should be prepared for a future where HSI tools are more available, useful, readily taught to acquisition professionals, and integrated with acquisition processes.

### 3.5.4 Compliance

Decreased dependence upon government (or even commercial) standards and increased use of functional and performance specifications (replacing detailed specifications) are forcing acquisition programs and their assessments to be less compliance oriented. That is, acquisitions are moving away from reliance on design standards (e.g., military standards such as MIL-STD-1472) toward greater use of commercial standards, nonmandatory guidelines, or no standards at all. At the same time, design and program reviews, analysis, and test and evaluation are reverting from an approach that reviewed hundreds or thousands of specific items to one that assesses system performance. While not all acquisitions have met this new challenge, functional disciplines (especially the efforts related to HSI) are becoming less encumbered by "mind-numbing" tables of compliance requirements. Current trends show that the development of statements of work (SOWs) and associated system requirements are eliminating costly and sometimes misleading laundry lists of compliance requirements. The result of this trend for the HSI culture is to create greater dependence upon developing meaningful human performance specifications, monitoring programs more intensely to assure functional specifications are adequate, and collecting comprehensive performance data to provide exit criteria for operations and maintenance system interface. These exit criteria are to be reflected in the test and evaluation plans to determine if the system meets its HSI-related objectives. Changes in the nature of the HSI effort are evident in an earlier and stronger focus on performance data requirements and in the evolution of human–system performance evaluations.

### 3.5.5 Documentation

There are those who would argue that HSI professionals should not be overly bothered with preparing reports of engineering studies and analysis because no one reads them. No doubt, there is some truth in the statement because most acquisition programs rely on expeditious decisions once the research work is done. However, the value of documenting the HSI objectives, risks, strategies, plans, analyses, research and engineering studies, guidelines, and standards is well proven over time. Yet the trend in acquisitions is for less dependence upon the great volumes of information. These trends imply greater use of oral reports, shortened time frames for decision, and a shift from volumes of specifications to thin guidance and iterative rapid prototypes. Similarly, in many instances, acquisitions are switching from hard-copy to electronic formats. The HSI community must respond to this environment with an equal aptness to be less dependent upon voluminous, costly, and time-consuming documentation, while keeping in mind the undeniable need for written documentation throughout the acquisition program.

## 3.6 HSI CULTURAL MYTHS VERSUS REALITIES

The cultural biases about HSI that abound in the acquisition community have not served the HSI discipline or acquisition programs well. In order to dispel some of the injurious myths related to the conduct of HSI activities, 13 HSI program attributes and the related myths and realities are identified below and summarized in Table 3.10. In item 1, for example, unenlightened acquisition program participants contend that because we are human, identifying the HSI risks and constructing mitigation strategies and solutions can be done by anyone. This argument has led more than one program down the path of high risk until operational demonstrations or tests have illuminated serious user performance problems.

1. *HSI Expertise* HSI requires professional expertise. Acquisition experience has repeatedly demonstrated that HSI problems are only obvious in retrospect. Identifying and anticipating HSI issues and devising mitigation strategies and engineering solutions require the professional expertise of the HSI practitioner.

2. *HSI Research and Engineering Cost* Some argue that HSI is free or can be acquired at relatively little cost. There are those who suggest that because the process of applying and integrating human engineering is either negligible or very low in cost (compared to other budget items), budgets are not affected and budget planning is not

**TABLE 3.10 HSI Cultural Myths and Realities**

| HSI Attribute | Cultural Bias (Myth) | HSI Cultural Objectivity (Reality) |
|---|---|---|
| 1. HSI expertise | Since we are all human, anyone can do HSI. | Identifying HSI issues and devising mitigation strategies and engineering solutions require the professional expertise of the HSI practitioner. |
| 2. HSI research and engineering cost | HSI is free. There are seldom any significant costs to conducting HSI. | HSI is almost never free. However, to become a sustained and institutionalized activity at the individual project level, applying HSI must appear relatively inexpensive and be easy to obtain. |
| 3. HSI definition and scope | HSI success equals user acceptance. | User acceptance without rigorous performance criteria imposes risky criteria of user preferences. |
| 4. HSI research | HSI can be accomplished via quick and easy methods. | HSI research is a critical ingredient in most acquisition programs. But it is important to tailor the HSI effort to the time and resources available. Big benefits can often come from small rigorous studies. |
| 5. HSI requirements | HSI can be added once the requirements are defined. | HSI activities must be integrated at the earliest stages of the program to avoid human–system performance problems. |
| 6. Location of HSI practitioners | HSI professionals may be organized as an adjunct engineering discipline. | HSI people must be collocated with the product teams they serve and integrated as part of the team. |
| 7. Piecemeal participation | HSI should only be conducted as an activity where all the human elements are tied tightly together. | HSI problems are rarely simple enough to resolve all elements of the issue simultaneously. |
| 8. End of development testing | Because HSI tests are qualification and acceptance tests, they should be conducted at the end of the program. | HSI issues should be tested early and often and should contribute to the iteration of design. |
| 9. Use of controlled conditions | HSI facilities should be sterile laboratories where user performance is tightly controlled. | HSI engineering (similar to other engineering disciplines) benefits from the collaboration that occurs when it invites open technical interchange. |
| 10. Use of typical users | HSI studies should employ only the typical user populations to ensure valid results. | Users are important, but including management and other members of the acquisition team in the project may add increased understanding and credibility to the HSI role and function. |
| 11. Compelling HSI evidence | Only rigorous study in a controlled environment provides sufficient documentation for good HSI designs. | Anecdotal information of real problems may provide compelling evidence for the need of an HSI effort. |

(*continued*)

**TABLE 3.10** ***(Continued)***

| HSI Attribute | Cultural Bias (Myth) | HSI Cultural Objectivity (Reality) |
|---|---|---|
| 12. Documenting study results | Like rigorous research, engineering studies and solutions must be followed up with complete documentation. | Results of engineering studies should be kept as short as possible and be integrated directly into guidelines, tables, or standards. |
| 13. Dependence upon upper management support | HSI depends upon upper management support to succeed. | HSI success is at least equally dependent upon the project management leadership and engineering team members who will actually make it happen. |

required. In fact, HSI is almost never free but in many instances can be relatively inexpensive (especially compared to the benefits). Human systems integration is rarely without some costs, but the total program costs are greater and almost always decrease the program options when HSI is not done. A viable research and engineering budgeting process should account for the real costs and benefits of conducting HSI. Nevertheless, to become a sustained and institutionalized activity at the individual project level, applying HSI must appear relatively inexpensive and easy to obtain at the program level.

3. *HSI Definition and Scope* Some acquisition program management personnel suggest that HSI success should be equated to user acceptance of the product or system. This is a risky equation. Users provide an essential ingredient in the development and evaluation of HSI solutions. However, user acceptance without rigorous performance criteria relegates the success to the whimsical and risky criteria of user opinion and preferences.

4. *Ease of HSI Application* Some believe that HSI can almost always be accomplished with a quick and easy or "just-in-time" application. In truth, many acquisition programs require HSI studies and activities with tightly controlled and rigorous data collection and analysis. However, because the engineering community cannot tolerate an HSI solution that does not accommodate the realities of the acquisition schedule, prompt solutions are often sought. It is important to be responsive to the program and tailor the HSI effort to the time and resources available. Often, small studies beget big benefits and should not be neglected.

5. *HSI Requirements* Often, HSI has been applied as if it can catch up on the back end of a program or be added once the rest of the team has defined the requirements or solution adequately. In fact, for a successful program, HSI activities must be integrated at the earliest stages of the program. At a minimum, some form of "attention-getting" HSI requirements should be listed at the initiation of requirements. These early requirements add momentum, definition, and value as the program progresses to more mature states of development—usually at great benefit to the identification and mitigation of potential human–system performance problems.

6. *Location of HSI Practitioners* Sometimes HSI professionals are organized in an acquisition program as an adjunct engineering discipline and not fully integrated with the

other mainline engineers. To be effective, HSI people must be colocated with the product teams they serve. Informal questions and quick responses are likely to add value to the program and build HSI credibility. The HSI engineers need to be a part of the team and other team members need to learn to view them that way.

7. *Piecemeal Participation* Human systems integration involves an effort that considers the various complexities of operator and maintainer system interfaces. Because the HSI effort is best described as an integration discipline, some think that it should *only* be conducted as a whole activity in which all the human elements are tied intricately together. In fact, rarely are all the HSI problems simple enough to resolve all elements of the issues simultaneously. Furthermore, significant benefits will often accrue to the HSI program if other members of the engineering team view it as one in which problems are tackled as they emerge. Because projects generally follow an iterative development process, the benefits of working on HSI challenges "piecemeal" far outweigh the risks of solving only part of the problem at a time.

8. *End of Development Testing* Sometimes HSI tests are viewed as qualification and acceptance tests that should be conducted mostly at the end of the program development to avoid unnecessary costs in testing and evaluation. This is an erroneous assumption. No different from other engineering considerations, HSI issues should be tested early and often and should contribute to the iteration of design. Results of these efforts should be regularly reported, identifying the issues and their status.

9. *Use of Controlled Conditions* Because HSI addresses complex and extremely sensitive issues, it is sometimes assumed that HSI facilities should be sterile laboratories where HSI professionals can evaluate user performance in tightly controlled conditions. No doubt, every research program deserves an appropriate amount of control of the conditions. However, HSI engineering does not differ from other engineering disciplines that benefit from the casual collaboration that occurs when facilities are an open and inviting technical interchange among interested and competent participants.

10. *Use of Typical Users* Human systems integration studies must describe and understand all appropriate parameters of the user population. Consequently, it is often believed that these studies should employ *only* the typical user populations during their research, studies, analysis, and tests to ensure valid results. Of course, the user population (operators and maintainers) and their tasks must be properly identified for study. However, some studies do not necessarily require absolute fidelity in the participating population. Also, including management and other members of the acquisition team in the project provides first-hand knowledge of what HSI is and how HSI activities are conducted, thereby adding increased understanding and credibility to the HSI role and function.

11. *Compelling HSI Evidence* Because research of complex HSI issues requires a well-controlled environment, some suggest that *only* rigorous study and analysis will enable collection of the appropriate data for good HSI designs. It is true that carefully controlled experiments have importance in HSI research and in evaluating human performance. However, anecdotal information or short videos of real problems provide compelling evidence for the need of a well-defined and properly funded HSI program. These short vignettes can be powerful accomplices to the development of more comprehensive HSI efforts.

12. *Documenting Study Results* Engineering studies and solutions should be followed up with appropriate documentation to ensure similar programs benefit from past experiences. This documentation can be invaluable in developing guidelines and standards and in

avoiding repetition of earlier problems. Clearly, documenting progress and results is an essential ingredient in a scientific application of HSI. However, seldom are engineering reports read by a wide audience. Reports of engineering studies should be kept as short as possible. The lessons learned should be integrated directly into HSI guidelines, tables, or standards. Except for rigorous research and studies, one should avoid generating as much of the engineering paper trail as possible.

13. *Dependence upon Upper Management Support* Like many other fledgling initiatives, HSI depends upon some upper management support to succeed. The value of an HSI champion in upper management has been documented in several agencies. Despite the value of this support, in many practical applications HSI success is more dependent upon the project management leadership and engineering team members than on upper management. Teaching and demonstrating the value of HSI to the program participants at the middle management level are essential for success. These members of the team will help make it happen.

## 3.7 ROLES AND RESPONSIBILITIES

The roles of the HSI players for both government and commercial environments may differ significantly depending upon the culture of the organization and environment. However, the prescribed roles, responsibilities, tasks, decisions, and interfaces for HSI practitioners in an acquisition environment that has proven to be healthy for the HSI effort are often quite similar. While specific HSI timelines of decisions, roles, and downstream consequences vary depending upon the size, complexity, sponsorship, mission, cost, schedule, technological reach, and other factors in the system acquisition program, the sequence of the iterative activities is usually quite predictable. The extent to which HSI plays a part in the design decisions is dependent upon the timely accomplishment of the various HSI tasks. The Appendix provides a generic flow of how HSI roles support (and are supported by) the information, organization, and resources of the acquisition environment (U.S. Department of Transportation, 1998).

## 3.8 SUMMARY AND CONCLUSIONS

The elements that define the acquisition culture include the business approach and processes; research and development methods and structures; mission and operational considerations; and political, management, and organizational environments. The culture in which an acquisition team operates identifies the habitual patterns of how systems are acquired, affects the roles of the HSI practitioner, and influences the way HSI business is conducted. Knowing how to identify and mitigate the cultural aspects that may impose risks to the acquisition program enables the HSI practitioner to overcome some of the obstacles to achieving valued HSI and to bring the acquisition program increased success. Recent changes within the acquisition culture (especially within the HSI discipline) and trends for the future of HSI attest to the growing importance of addressing these acquisition cultural risks and dispelling the traditional HSI cultural myths.

## REFERENCES

Barnhart, C. L., and Stein, J. (Eds.). (1963). *The American College Dictionary*, 17th ed. New York: Random House.

Carroll, J. S., and Perin, C. (1995). Corporate Culture and Operational Safety. Paper presented at a symposium conducted by the MIT Sloan School of Management at the National Transportation Safety Board, Alexandria, VA.

Hughes, D., and Dornheim, M. A. (1995, January 30). Accidents Direct Focus on Cockpit automation. *Aviation Week & Space Technology*, *148*(4), 52–55.

Meister, D. (1997). In S. Casey (Ed.), *The Practice of Ergonomics: Reflections on a Profession.* Bellingham, WA: Board of Certification in Professional Ergonomics. Chap III Measurement and Research (p. 108).

Mittler, E., Hewitt, G., and Vehlow, C. (1990). Management Integration Methods. In H. R. Booher (Ed.), *MANPRINT: An Approach to Systems Integration.* New York: Van Nostrand Reinhold.

Nielson, J. (1993). *Usability Engineering.* Boston: Academic.

U.S. Department of Transportation (DOT), Federal Aviation Administration. (1998). *The Management of Human Factors in FAA Acquisition Programs.* Washington, DC: DOT.

Wickens, C. D., Mavor, A. S., and McGee, J. P. (Eds.). (1997). *Flight to the Future: Human Factors in Air Traffic Control.* Washington, DC: National Academy Press.

Wickens, C. D., Mavor, A. S., Parasuraman, R., and McGee, J. P. (Eds.). (1998). *The Future of Air Traffic Control: Human Operators and Automation.* Washington, DC: National Academy Press.

## APPENDIX: HSI ROLES AND RESPONSIBILITIES

| Action | Roles/Responsibilities | Program/ Project Sponsor | Program Project Team | HSI Expert | User Representatives |
|---|---|---|---|---|---|
| | *Activities Supporting All Phases of Acquisition Life Cycle* | | | | |
| Develop (HSI) policy | • HSI representative develops the HSI policy.<br>• Agency lines of business (i.e., program/product sponsoring organizations) and integrated product team/product teams (IPT/PTs) review and comment on policy. | X | X | X | |
| Develop HSI infrastructure | • Agency lines of business, IPT/PTs, and HSI representative coordinate to identify HSI infrastructure requirements (e.g., staffing, laboratory capabilities, equipment) and develop recommended solutions. | X | X | X | |
| Plan for budget for HSI activities and to support user involvement in HSI | • Agency lines of business and acquisition teams [e.g., IPT/PTs, mission analysis (MA) teams, concept exploration and investment analysis teams, postdeployment teams] identify funding requirements and confer with HSI representative in developing cost projections to support HSI activities and user involvement. | X | X | X | X |
| Coordinate and integrate research, engineering, procurement, and operations funded agency HSI activities/projects | • Agency lines of business and IPT/PTs develop integrated funding profiles and plans for programs/projects.<br>• HSI representative develops recommendations for coordinated HSI program and identifies potential duplication and/or disconnects in existing or proposed projects. | X | X | X | |

| | | | | | |
|---|---|---|---|---|---|
| Address cross-cutting HSI issues or HSI issues that exceed team empowerment boundaries | • Acquisition teams (e.g., IPT/PTs, MA teams, concept exploration and investment analysis teams, postdeployment teams) ensure that HSI issues are raised at the acquisition executive level, as appropriate, and for coordinating/consulting with the HSI representative.<br>• HSI representative is responsible for ensuring that mechanisms are in place to identify and resolve known and potential HSI risks. | | X | X | |
| Report on status of HSI | • Acquisition teams (e.g., IPT/PTs, MA teams, concept exploration and investment analysis teams) provide the HSI status on specific acquisition programs (e.g., at acquisition reviews, resource council reviews).<br>• HSI representative prepares and delivers status report on agency HSI to acquisition executives, program management team members, and HSI specialists.<br>• Acquisition executive, agency lines of business, IPT/PTs, and user representatives provide information and feedback. | X | X | X | X |
| | *Activities Supporting Mission Analysis (MA) Phase of Acquisition Life Cycle* | | | | |
| Establish MA teams | • The sponsoring agency line of business establishes and leads the MA team.<br>• MA team leader notifies the agency lines of business, user representatives, and HSI representative that MA team is being established.<br>• Agency lines of business and/or IPT/PTs provide domain-specific HSI expertise. | X | X | | |
| Designate HSI coordinator to support MA team | • Agency line of business coordinates with HSI representative to designate the HSI representative to support the MA team. | X | | X | |

*(continued)*

**TABLE** ***(Continued)***

| Action | Roles/Responsibilities | Program/ Project Sponsor | Program Project Team | HSI Expert | User Representatives |
|---|---|---|---|---|---|
| Identify and conduct HSI analyses and data collection to support MA | • Team HSI representative leads planning and provides technical direction or assistance.<br>• An HSI working group (HSIWG) may be established by the team HSI representative to coordinate HSI activities and conduct studies or analyses using a variety of HSI resources. | | | X | |
| Identify requirements for users to serve as subject matter experts in HSI analyses | • Team HSI representative identifies the requirements for a valid, representative sample in conjunction with MA team leader and user representatives. | | | | X |
| Develop and submit requests for user involvement in HSI during MA | • MA team prepares request with support from team HSI representative.<br>• The appropriate agency line of business processes requests. | X | | | |
| Develop HSI input for mission needs statement (MNS) | • Team HSI representative in conjunction with HSIWG and other members of the MA team draft HSI input.<br>• MA team responds to questions on MNS HSI issues with support from team HSI representative and other team members. | | X | X | X |
| Prepare final MNS and appropriate briefings and submit to decision authority | • MA team prepares and submits all necessary documents.<br>• MA team leader notifies the HSI specialists and user representatives when MNS decision meetings are scheduled. | | | | X |

| | | | | | |
|---|---|---|---|---|---|
| Establish requirements team to develop requirements document (RD) | • The agency line of business sponsoring the MNS establishes a requirements team.<br>• The leader of the requirements effort notifies user representatives, HSI representative, and IPT/PT that the requirements team has been established. | X | | X | X |
| *Activities Supporting Concept Exploration (CE) and Investment Analysis (IA) Phase of Acquisition Life Cycle* | | | | | |
| Establish investment analysis (IA) teams | • Office responsible for IA establishes IA team.<br>• IA team leader notifies appropriate agency line of business, IPT/PT, user representatives, and HSI representative that team is being established. | X | X | X | X |
| Designate HSI coordinator(s) to support IA team and integrated requirements team (IRT) | • CE and IA team leader designates an HSI representative to support the concept exploration and IA team.<br>• Requirements team leader designates an HSI representative to support the requirements team. | | | X | |
| Identify and conduct HSI analyses and data collection to support IA and requirements development | • Team HSI representative leads planning and provides technical direction or assistance.<br>• A HSIWG may be established by team HSI representative to coordinate HSI activities and conduct studies and analyses using a variety of HSI resources.<br>• Acquisition teams (e.g., IPT/PT) provide domain-specific HSI expertise. | | X | X | |
| Identify requirements for users to serve as subject matter experts in HSI analyses for IA and requirements development | • Team HSI representative identifies the requirements for a valid, representative sample in conjunction with team leaders and user representatives. | | | | X |

*(continued)*

**TABLE** ***(Continued)***

| Action | Roles/Responsibilities | Program/ Project Sponsor | Program Project Team | HSI Expert | User Representatives |
|---|---|---|---|---|---|
| Develop and submit official requests for user involvement in IA and requirements HSI | • IA team prepares and submits IA user request(s) with support from team HSI representative.<br>• Requirements team prepares and submits user request(s) with support from the team HSI representative.<br>• The appropriate agency line of business processes requests. | X | | | X |
| Develop HSI input to IA documentation | • Team HSI representative prepares HSI input for the investment analysis reports (IARs), the acquisition program baselines (APBs), RDs, and other acquisition documentation (e.g., market survey, alternative analyses, investment analysis plans) in conjunction with HSIWG and other members of the IA and requirements teams.<br>• IA and requirements teams respond to questions on HSI issues with support from team HSI representative and other team members.<br>• The IA and requirements teams prepare and submit all necessary final documents including appropriate briefings to decision authority.<br>• CE and IA team HSI representative provides input to APB on HSI resource requirements.<br>• The IA team leader notifies the HSI specialists and user representatives when investment decision meetings are scheduled. | | X | | X |

| | | | | | |
|---|---|---|---|---|---|
| *Activities Supporting Engineering and Development (E&D) Phase of Acquisition Life Cycle* | | | | | |
| Assign acquisition program to IPT after investment decision | • Decision authority makes IPT assignment. | | X | | |
| Notify user representatives and HSI representative of new acquisition program | • IPT/PT leader provides the notifications to these representatives. The tentative schedule for preparation of the acquisition strategy paper (ASP) and integrated program plan (IPP) will be provided to the user representatives. | | X | X | X |
| Assign HSI representative to support IPT/PT | • IPT/PT leader designates an HSI representative. | | X | | |
| Identify opportunities for HSI analyses and appropriate HSI data to be considered during the E&D phase | • HSI representative is responsible to lead this action. | | X | | |
| Identify and conduct HSI analyses and data collection to support the acquisition program, including operability assessments leading to formal test and evaluation activities | • HSI representative leads this action and provides technical direction or assistance.<br>• Analyses may be conducted by internal agency resources or through contractor support to the IPT/PT.<br>• Team HSI representative coordinates the processing of funding documents outlining HSI support needed. | | X | X | |
| Identify requirements for users to serve as subject matter experts in HSI analyses throughout E&D phase, including operability assessments | • HSI representative works in conjunction with other IPT/PT members to identify the requirements for a valid, representative sample. | | X | | |
| Develop and submit requests for user involvement in HSI analyses supporting acquisition program | • The IPT/PT prepares and submits request(s) with support from the HSI representative.<br>• The appropriate agency line of business processes requests for user participation. | X | X | | X |

(*continued*)

**TABLE (*Continued*)**

| Action | Roles/Responsibilities | Program/ Project Sponsor | Program Project Team | HSI Expert | User Representatives |
|---|---|---|---|---|---|
| Develop HSI input for E&D documentation | • HSI representative (in conjunction with other members of the team) prepares the HSI input for E&D documentation, including the ASPs, IPPs, requests for proposal (RFP), information requests, statements of work (SOWs), contract documents, etc.<br>• IPT/PT responds to questions on HSI issues with assistance from the HSI representative.<br>• IPT/PT prepares and submits all necessary final E&D documents to the decision authority. | | X | | |
| Develop and execute HSI/human performance plans | • HSI representative (with the appropriate IPT/PT members) develops and executes HSI/human performance plans and data collection for studies, analyses, assessments, test and evaluation, and other activities. | | X | | |
| Report on status of HSI | • IPT/PT reports on the HSI status on specific acquisition programs. | | X | | |
| | *Activities Supporting Post Deployment Phase of Acquisition Life Cycle* | | | | |
| Assign HSI representative to support service or product during postdeployment | • The postdeployment team leader designates an HSI representative. | | X | X | |
| Identify and conduct HSI analyses and data collection to support the program, including postdeployment operability assessments | • Postdeployment team HSI representative leads this action and provides technical direction or assistance.<br>• Actual analyses may be conducted by internal agency resources or through contractor support to the postdeployment team. | | | X | |

| | | | | | |
|---|---|---|---|---|---|
| Identify data and user requirements for analyses throughout the postdeployment phase | • Team HSI representative works in conjunction with other postdeployment team members to identify the data and user requirements.<br>• The appropriate agency line of business organization prepares and processes user study requests. | X | | | |
| Develop and execute HSI/human performance studies, analyses, and data collection plans and activities | • Team HSI representative works with the appropriate postdeployment team members to develop and execute the HSI plans and activities. | X | | X | X |
| Develop HSI input for lessons learned and other postdeployment documentation | • Team HSI representative prepares the HSI input in conjunction with other members of the team.<br>• Postdeployment team coordinates with appropriate user representatives for feedback. | X | X | X | X |

CHAPTER 4

# Human Systems Integration and Systems Acquisition Interfaces

EDWIN R. SMOOTZ

## 4.1 INTRODUCTION

In 1974 the U.S. Army Training and Doctrine Command, which is responsible for defining the U.S. Army's need for new systems, signed a letter of agreement with the U.S. Army Materiel Command to develop a remotely piloted vehicle system technology demonstration (U.S. Army, 1988). The intent, following a 1971 recommendation by the Defense Science Board, was to demonstrate how a remotely piloted vehicle could aid a ground commander to perform reconnaissance, acquire targets, designate targets with a laser, and adjust artillery fire. The system, later named Aquila, was to capitalize on emerging technology and give the U.S. Army a state-of-the-art multipurpose remotely piloted airborne system second to none. In 1988, after the expenditure of approximately $1 billion (Ladendorf, 1988), the army cancelled the program. The reasons for the cancellation are many, but surely failure to adequately address HSI-related requirements early in system development and integrate them fully into early testing and evaluation were among the problems that led not only to an inefficient expenditure of large amounts of money but also to a delay in fielding a needed system (Stewart et al., 1989). As the army entered the twenty-first century, it still did not have a widely fielded unmanned air reconnaissance system.

The acquisition of complex systems such as Aquila requires the successful execution of three functions: definition of requirements, design and development of the system, and deployment. This chapter will focus on processes required for the acquisition of complex systems and how to integrate human performance considerations into those processes in order to avoid more costly examples such as Aquila. It will begin with an overview of acquisition life-cycle models and then discuss various ways in which human systems integration (HSI) interfaces with the life-cycle process. The discussion will center around the U.S. Department of Defense (DoD) defense acquisition management framework. Since

*Handbook of Human Systems Integration*, Edited by Harold R. Booher.
ISBN 0-471-02053-2 

the purpose of the chapter is to show how human factors can be integrated into systems acquisition, attention will be paid to identifying acquisition documents that are normally part of the process and should be attended to and modified with human factors information if HSI is going to be successful.

An assumption underlying the discussion is that human factors personnel who are involved in this process must be well trained and educated in the field of human factors if they are going to have an impact on the systems acquisition process.[1] Just being interested in human factors is not enough. One must be cognizant of the human factors and human performance literature that comprises the field; must be facile with human performance measurement methods, modeling tools, and analysis techniques; and must be familiar with the functional area (e.g., vehicles, communications systems, individual soldier weapons, etc.) in which he or she is working. Without a strong grounding in these fundamentals, the human factors representative will have little to offer during the system development process and will be ignored.

Finally, throughout the chapter it will be noted that the author uses the term *operators and maintainers* when referring to the users of a system. In the past HSI has too often focused on issues associated solely with operating a system and neglected issues concerned with maintaining the system. But modern complex systems require a great deal of maintenance in order to keep them functioning. For example, the Aquila system mentioned above had such high maintenance requirements that it was impossible to keep the system at full operational status for more than several days with the maintenance manpower that had been allocated to it. Designing for the maintenance of complex systems can be just as important as designing for operation. Use of the term operators and maintainers is done to reinforce that point.

## 4.2 SYSTEMS ACQUISITION PROCESSES

A key element in the systems engineering of complex systems is a life-cycle model to lend structure and organization to the development process. In its simplest form, the process occurs in three phases: system definition, system development, and system deployment (see Fig. 4.1). During system definition, the requirements for the system are specified in accordance with the stated needs of the users of the system. In system development, the system is conceptualized, designed, built, and evaluated. Finally, in system deployment, the completed system is delivered to the end user.

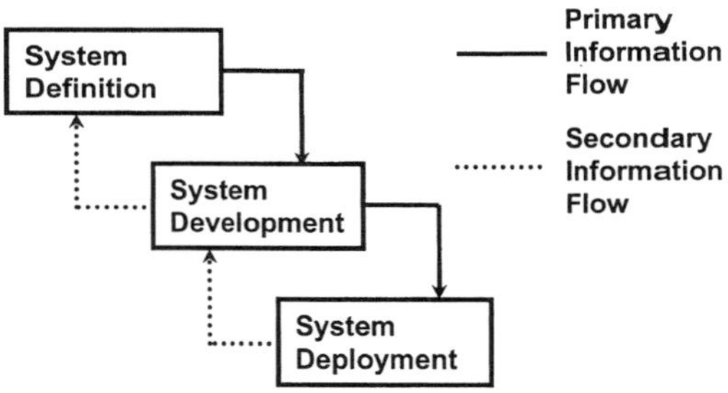


**Figure 4.1** Fundamental system engineering life-cycle phases [*Handbook of Systems Engineering and Mangement*, Sage, A. P. and Rouse, W. B. (Eds.), (1999). Reprinted by permission of John Wiley & Sons, Inc.]

### 4.2.1 DoD System Life-Cycle Model

A variety of life-cycle models have been produced by systems engineering professionals since 1970 [for overviews see Blanchard and Fabrycky (1998) and Sage (1992)]. One of the most well known is that produced by the DoD (2000b). The DoD has spent a great deal of effort modifying and adjusting its life-cycle systems model over the years, partly because the DoD is unique in being both the user and developer of its systems. This is required because many of the systems that it develops are for combat with distribution limited strictly to the armed forces of the United States and its allies. In fact, DoD system acquisition is strictly governed by a number of authorization documents, to include the law (e.g., Title 10 of the *United States Code*), directives of the executive branch of the government and the Federal Acquisition Regulation. Finally, because of the need to constantly improve the capability of the armed forces, the DoD is one of, if not the largest, acquirer of complex, customized systems in the world and thus needs an effective life-cycle systems model to help manage such a large endeavor. Because the DoD model is so ubiquitous, it will serve as the basis for most of the discussion in this chapter. Nevertheless, the principles and guidance espoused have general applicability to the acquisition of any complex system, regardless of what life-cycle model, document names, process identifiers, or other nomenclature are used.

The version of the DoD model presented in acquisition documents in effect until August 2002, as well as its immediate predecesssor, is shown in Figure 4.2.[2] It is useful to review the older version in order to understand the newer version. The older version has four phases with milestones (0 to III) leading into each phase. Following concept exploration is essentially each of the classic life-cycle phases shown in Figure 4.1. In past years system acquisition typically began with concept exploration and proceeded lock

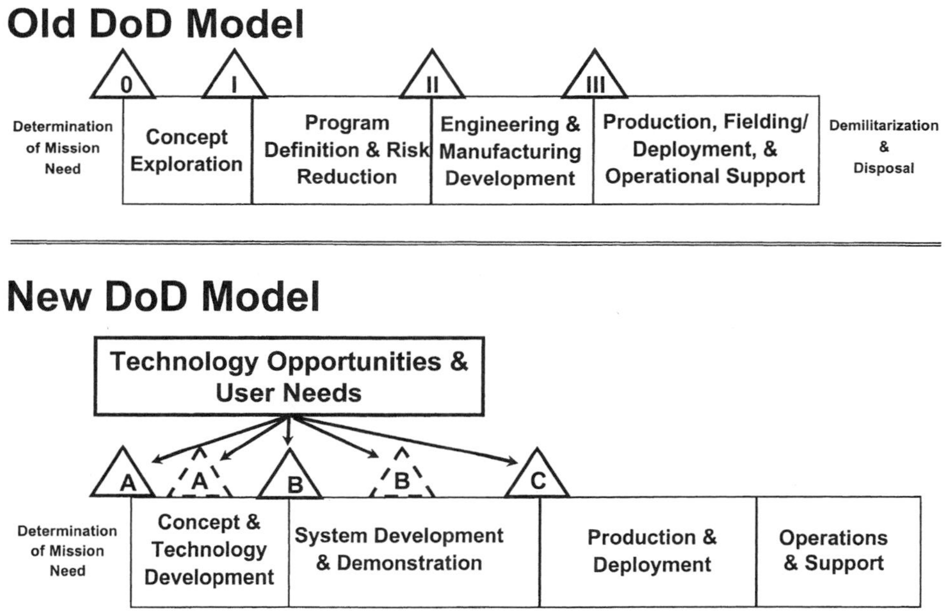


**Figure 4.2** Old and new DoD system life-cycle models.

step through the remaining phases. However, this resulted in rather long acquisition times, on the order of 10 to 15 years. It came to be recognized that the life-cycle model needed to be modified in order for the DoD to be able to capitalize on rapidly evolving technology characteristic of today's business environment. Consequently, the DoD revised its acquisition guidance regulations (DoD, 2002a, b, 2001) around a set of principles designed to make systems acquisition more rapid, affordable, and effective. In the process, the life-cycle model was modified to accommodate this new environment. As can be seen in Figure 4.2, it is still a four-phase process, but with the provision for entering the process at any point during the first two phases of *concept and technology development* and *system development and demonstration*, depending on the state of the primary technology or technologies that are forming the core of the new system.

The new life cycle, now called the defense acquisition management framework, is shown in more detail in Figure 4.3. It consists of three overarching *activities* (presystems acquisition, systems acquisition, and sustainment), which in turn are divided into a series of *phases* (concept and technology development, system development and demonstration, production and deployment, and operations and support). The four phases are each broken down into two *work efforts*. The nature of the work occurring in each work effort is obvious from the titles. It is important to note that the system acquisition process can begin at any point during the first two phases as long as an established materiel need can be satisfied by a technology at a state of development appropriate for that point in the cycle. To enter at a milestone B point would require a technology that is already mature and has been demonstrated in some application. To enter at milestone C would only be possible if there were already a system that had been developed for other markets and was fully capable of meeting a particular defense materiel need. This is not to imply that the DoD has never acquired previously developed technologies or systems; it has. But such

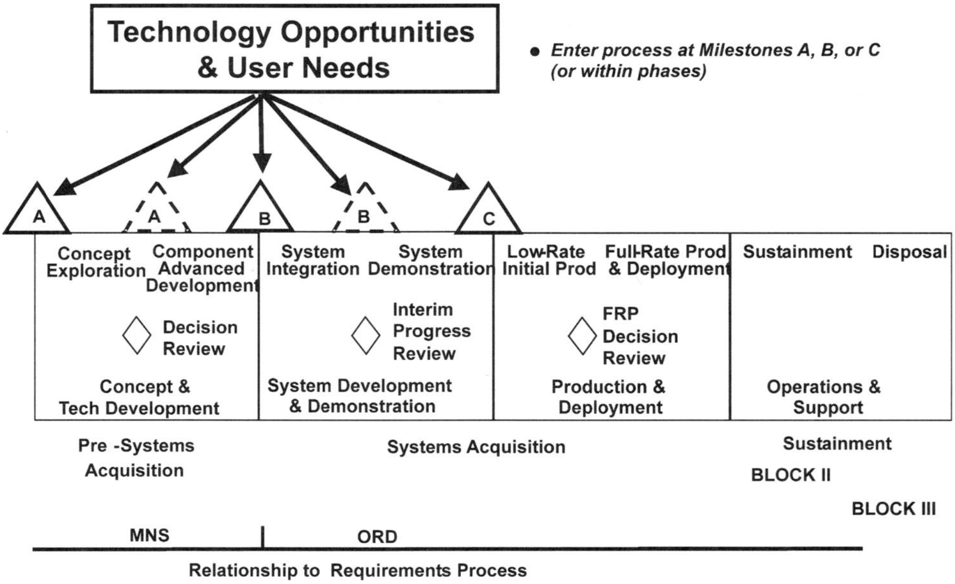


**Figure 4.3** Defense acquisition management framework.

acquisitions have been treated as exceptions to the acquisition process. Under the current defense acquisition management framework, they are part of the standard acquisition process.

Another advantage of the new life-cycle model is the accommodation of time-phased requirements in support of evolutionary acquisition. Under the evolutionary acquisition concept, an acquisition strategy is adopted for fielding a core capability with a modular open structure and the provision for future capability upgrades.[3] The new life-cycle model allows time-phased requirements to be introduced at various points in the life cycle, thus facilitating fielding, over time, systems of increasing capability. While this approach is not appropriate for all acquisition programs, it is essential for the timely acquisition of automated information systems in today's environment of rapidly evolving automation technology.

Another important emphasis in the DoD (2000a, b, 2001) 5000 series is a total systems approach to systems acquisition. The DoD Directive 5000.1 (which is the lead regulation in the 5000 series and provides overarching guidance) states: "Acquisition programs shall be managed to optimize total system performance and minimize total ownership costs by addressing both the equipment and the human part of the total system equation, through application of systems engineering" (DoD, 2000a, p. 5). This statement has profound implications for HSI, since it states up front, for the first time, in the highest level DoD regulation that governs system acquisition, that human performance considerations must be looked at in conjunction with equipment considerations when optimizing system performance and minimizing system cost. The directive goes on to state: "Program managers shall give full consideration to all aspects of system support including logistics planning; *manpower, personnel, and training; human, environmental, safety, occupational health, accessibility, survivability*, and security factors; and spectrum management and the operational electromagnetic environment."[4] The directive clearly indicates the importance of and need for HSI in the systems acquisition process.

### 4.2.2 HSI in the Life-Cycle Process

The question that arises at this point is how to accomplish the integration of human performance considerations into the life-cycle process to optimize system performance and minimize cost. How does one ensure that the human engineering, i.e., the design of the interfaces in the system, does not cause operators to make errors at critical times or does not cognitively overload the operator such that he or she makes repeated errors? How does one ensure that enough maintainers with the right skills are available to support the system when it is fielded without extraordinary amounts of training? An approach to doing this is to iterate an HSI cycle through the various phases and work efforts of the systems acquisition life cycle. Such an approach is shown in Figure 4.4. As can be seen, a total system concept must first be formulated. This is based on the requirements that have been specified for the system, its intended method of employment, notions of what technology will be used as the core of the hardware, and what parameters will characterize the operators and maintainers of the system. From this total system concept is derived requirements for human performance, such as numbers of operators required for operation; special cognitive, physical, or sensory skills required; estimates of the training that will be required; estimates of the workload under various scenarios; and so forth. From this estimate of human performance requirements, a judgment must be made as to whether the

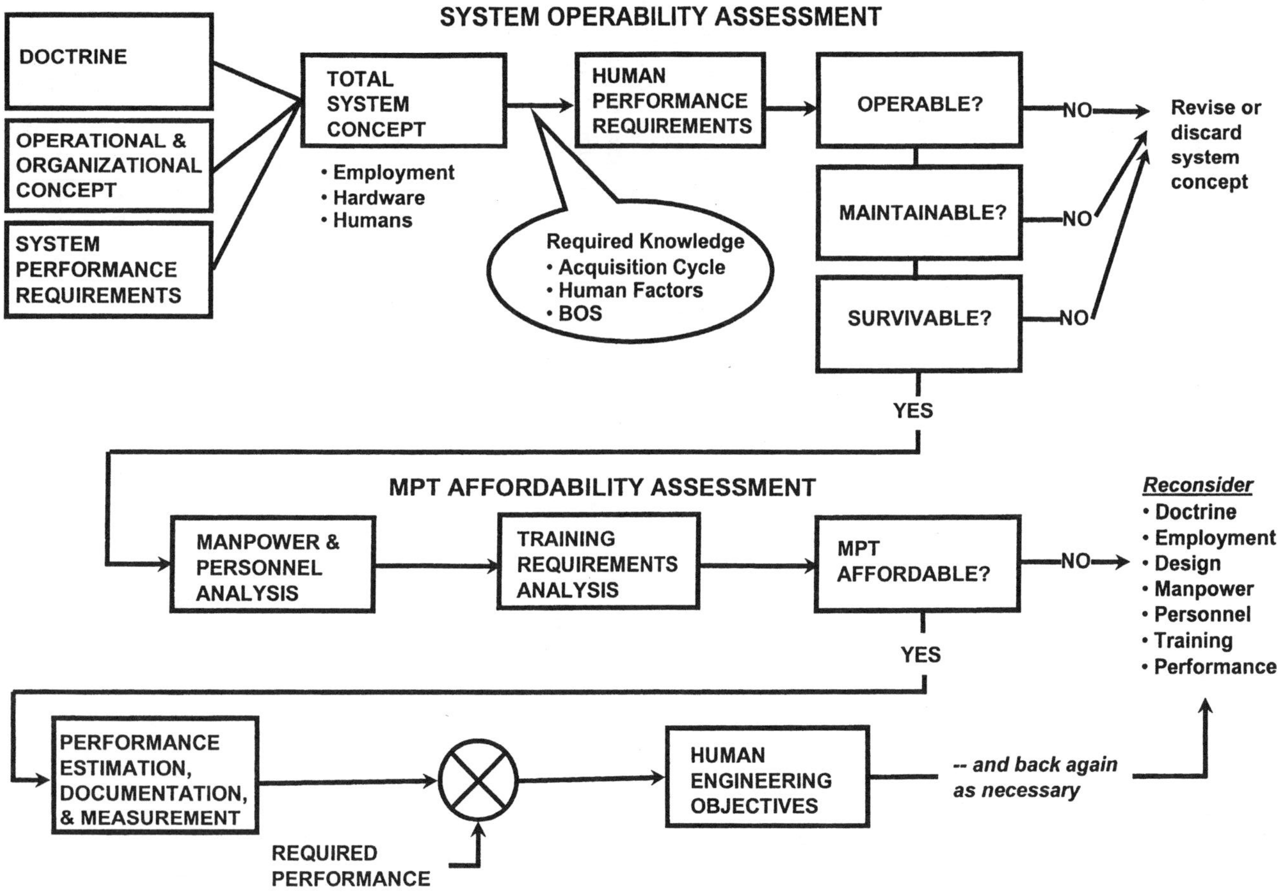


**Figure 4.4** HSI in the life-cycle process.

system is operable, maintainable, and survivable by the typical soldier, sailor, airman, or marine that is going to use it. This decision must be made jointly by the system program manager and the human factors professionals who have the knowledge of human performance in similar systems, the familiarity with the relevant human performance literature, and the skills for using the modeling and simulation tools that can help make such an evaluation.[5] If the decision is negative, then the system concept must be revised.

If the decision is positive (e.g., it is decided that the system as envisioned can be effectively operated and maintained by the typical humans who will be using it and the design will not put such a large workload on them that they will become fatigued and overwhelmed by all of the tasks that they are required to do, thus reducing their chances of surviving when coming under attack, etc.), then one proceeds to make a determination as to whether or not the human aspects of the system are affordable. This focuses on manpower, personnel, and training. The annual costs of recruiting, training, and paying personnel are among the largest costs in the DoD, typically exceeding 50 percent of the entire budget (Price, 1990). Costs related to manning and training on a system typically account for nearly 60 percent of total life-cycle costs (Graine, 1988). Therefore, this part of the analysis is critical to determining if the system if affordable. Reliable estimates must be made of the number of operators and maintainers that will be needed, whether any specialized aptitudes will be required (e.g., persons with extremely high cognitive abilities are expensive to recruit and retain), and how much training will be required. The training requirements analysis is particularly critical since training costs can escalate rapidly. Items that contribute to such costs are specialized facilities and training equipment, highly trained instructors, and the sheer length of time that a human sits in a training environment rather than being a productive member of the force. As above, human factors professionals must use their knowledge of manpower, personnel, and training needs of similar systems; their knowledge of the technical literature; and their skills with modeling and simulation tools to help the system program manager make a determination as to whether the manpower, personnel, and training costs will fit within budget constraints. If not, then trade-off analyses between manpower, personnel quality, training requirements, and design must be made in order to meet budget constraints.

It should be noted that the above decisions are not always made in as linear a fashion as Figure 4.4 implies. In the early phases of the system life cycle, decisions about operability and maintainability are made at a more general level so that affordability can then be addressed. The decisions are then refined as more detailed information becomes available during the system design process.

If the affordability decision is yes, then the next step is to make and document reasonable estimates of total system performance and costs and the contribution that human-related variables make to them. From these estimates one can make decisions as to how total system performance can actually be improved without increasing costs and set human engineering-related design objectives for accomplishing this.

This HSI cycle should be applied in the early phases of system acquisition and iterated during each phase. It will be most effective when it is applied early, i.e., during *concept exploration* and *component advanced development*. It is early in system development that human-related system development problems can be most easily and cost effectively addressed by changing the design of the system. By the later phases many aspects of the design have become fixed and are enormously expensive to change. Most life-cycle costs are determined by decisions made early in system development, even though the majority of life-cycle costs are not expended until production of the system is complete (Fig. 4.5).

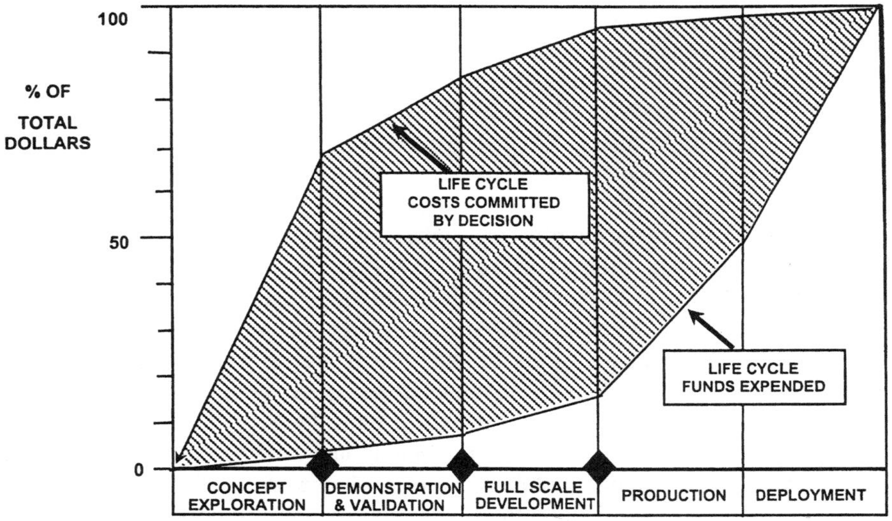


**FIGURE 4.5** Schedule of commitment decisions and life-cycle costs [Graine, G. N., "The Engineering Syndrome vs. the Manpower Personnel and Training Dilemma," *Naval Engineers Journal*, March 1988, p. 56. Reprinted by permission of the publisher].

One of the critical activities that is done early in the acquisition cycle is specifying requirements for the system. The DoD has emphasized in recent years that requirements be performance based; e.g., an armored vehicle must be able to traverse off-road terrain at a given speed, destroy an enemy armored vehicle at a given distance with a given probability, travel a given distance before requiring major maintenance, etc. It is critical that requirements related to the human factors of the system be specified at the same time. It is also critical that these requirements be adequately incorporated into the test and evaluation program that determines whether the system that is developed meets the requirements laid out for it. This has not always been done well in the past and has received renewed emphasis recently by the DoD. The 2001 version of Directive 5000.2-R (DoD, 2001, p. 83) states: "The program manager (PM) shall work with the manpower, personnel, training, safety and occupational health, habitability, survivability, and human factors engineering (HFE) communities to translate the HSI thresholds and objectives in the operational requirements document (ORD) into quantifiable and measurable system requirements. The PM shall include these requirements in specifications, the test and evaluation management plan (TEMP), and other program documentation, as appropriate, and use them to address HSI in the statement of work and contract." These two topics, HSI in requirements specification and testing and evaluation, receive particular attention in this chapter.

## 4.3 PRESYSTEMS ACQUISITION

The system requirements generation process typically begins before the system development process but overlaps its early phases. Adequate and comprehensive requirements generation is critical to the successful development of a system. If it is not clear what specific need a system is filling, it is unlikely that what is developed will fill any need satisfactorily. The DoD's requirements generation process is detailed in an instruction issued by the Chairman of the Joint Chiefs of Staff (DoD, 1999c). It provides for the creation of two basic documents; the mission needs statement (MNS) and the ORD. While these documents are military oriented, the principles underlying their development have generic application to requirements generation in general.

### 4.3.1 Needs Determination

To determine if there is a need to develop a new system, a series of analyses must be performed to match existing capability with what is needed to cope with the situation at hand. In the case of a private corporation that builds vehicles, for example, this might involve looking at how well its current vehicles satisfy the demands made by the marketplace as well as stack up against the competition. For the military, the process is a bit more complex. It requires an analysis conducted in the context of existing system capabilities, the current and projected future capabilities of the perceived threat, and how the four armed services plan to operate together in a joint fashion. From this analysis are derived needed future operational capabilities, which in turn give direction to future research and technology development and a set of assessments based on experimentation, modeling, and testing and evaluation to help determine how best to meet any deficiencies that have been uncovered [see TRADOC Pamphlet 71-9 (U.S. Army, 1999) for how the U.S. Army approaches this]. It should be noted that materiel solutions are not the first choice for addressing a deficiency. Rather, the following are considered first, in the following order: changes in doctrine, training, leader development, organizational structure, and soldier capabilities. Only after these have been considered and dismissed as not being able to address the deficiency is a materiel solution considered.

If a materiel solution is determined to be the best course of action, then a document is produced that describes the capability required and the rationale for it. Within the DoD this is called an MNS, and it drives the activities in the *concept and technology development phase* of the systems life-cycle process. An MNS does four primary things. First, it defines a need in terms of mission, objectives, and *general* capabilities (it purposely does not describe a need in terms of *specific* performance characteristics). The threat and threat environment to be countered are described, as are the shortfalls of current capability. Second, the reason why nonmateriel solutions are not acceptable is discussed. Third, potential material alternatives that already exist in other armed forces or in the commercial market are explored. Finally, constraints related to infrastructure support are discussed. These include manpower, personnel, and training constraints as well as available facilities, logistics support, and transportation, among other things.

The human factors professional has an important role to play in this process. The understanding that he or she has of the manpower, personnel, and training demands of similar systems in existence, based on experience with such systems, knowledge from reports and databases on those systems, along with access to modeling tools that can be used to conduct trade-off analyses on these variables, can make him or her an important contributor to the establishment of manpower, personnel, and training constraints. In addition, his or her expertise in the area of human performance can be of tremendous benefit in conducting the analyses and assessments needed to support the mission needs analysis and in identifying user–system interface issues associated with new technology.

### 4.3.2 Operational Requirements Development

Once a need has been documented and accepted by the appropriate decision authority at milestone A, it becomes necessary to state specific operational requirements for the new system. These are performance based and, within the DoD, are derived from efforts in the concept and technology development phase of the *defense acquisition management framework*. The concept exploration effort begins right after an MNS has been approved and validated and involves a series of studies that compare various system approaches to satisfying the needs stated in the MNS. The focus is on analyzing and evaluating the feasibility of alternative concepts and assessing their relative merits. This is followed by defining the most promising concepts in terms of broad objectives for cost, performance, interoperability, infrastructure, etc., after which the most promising concept for which there are available technologies is pursued. With this concept in hand, the component advanced development work effort begins wherein advanced technologies are further developed and demonstrated in relevant environments. The demonstrations are formal exercises that can become quite elaborate and involved and are of two types: advanced technology demonstrations (ATDs) and advanced concept technology demonstrations (ACTDs). The ATDs are designed to demonstrate that a technology is mature and has the potential for enhancing military operational capability in a cost-effective manner. The ACTDs determine the military utility of technology that is already proven and help develop the concept of operations that will optimize operational effectiveness. The information coming out of these technology demonstrations is then used to establish system architecture based on those technologies.

The results of these efforts also serve as the basis for specifying the requirements for a system in realistic operational terms. Within the DoD, these requirements are compiled into a formal document called the ORD (DoD, 1999c). Requirements are stated as operational performance parameters specific to a type of system (e.g., weapon, ground vehicle, communications system, etc.) and include system-level performance capabilities (e.g., probability of kill, maximum maintenance time per unit time of operation, range of transmission, etc.). They are written in output-oriented and measurable terms with both threshold (i.e., minimum) and objective (i.e., desired) criteria. Those parameters that are considered to be absolutely essential to successful satisfaction of the capabilities required by the MNS are called key performance parameters (KPPs). They are limited in number (usually eight or fewer), and failure to meet the threshold of even one of them can serve as the basis for terminating a program. They clearly serve as the drivers of the system development process.

### 4.3.3 Concept Exploration

In concept exploration efforts it is important to estimate operator and maintainer performance capabilities and limitations with various technologies. Models of total system performance can then be built that incorporate those estimates and be used to make reasonable comparisons of the effectiveness of the various technologies. Furthermore, such models can be used to support trade-off analyses by comparing the effectiveness and cost of various technologies under different levels of manpower, personnel capabilities, and training.

It is during this time that the human factors team should begin putting together an HSI plan for managing the human factors effort. A solid HSI management plan identifies and tracks the HSI issues and concerns that are identified for a new system and gives direction to the HSI effort by mapping a strategy for resolving those issues. To be successful, it must be fully coordinated with the PM or other individual who has oversight of the program at this stage in the life-cycle process.

### 4.3.4 Component Advanced Development

The role of HSI during component advanced development efforts is somewhat different than that in concept exploration. Since the emphasis is on demonstrations and experimentation, the need is for input into the design of the demonstrations and experiments to ensure that the right human performance data are collected so that the contribution that the human makes to total system performance can be accurately measured. Without reliable and valid data indicating what types of errors operators and maintainers are making or what critical tasks they are failing to complete within required time constraints, it is not possible to accurately evaluate the effectiveness of the operator or maintainer as part of the total system. The right human performance data must be collected and analyzed in the context of their contribution to total system performance in order for the data to be useful.

### 4.3.5 HSI Management Plan

It is critical that HSI considerations be fully incorporated into the ORD. The Joint Chiefs of Staff (DoD, 1999c) address the role of HSI in operational requirements generation, although it is in that part of the ORD called Program Support (pp. E-A-3 to E-A-6) rather than under the Capabilities Required section that includes the KPPs. Table 4.1 shows those

**TABLE 4.1 HSI Functions To Be Accomplished and Addressed in ORD**

1. Establish broad manpower constraints for operators, maintainers, and support personnel.
2. Identify requirements for manpower factors that impact system design (e.g., utilization rates, pilot-to-seat ratios, and maintenance ratios).
3. Establish broad cognitive, physical, and sensory requirements for the operators, maintainers, or support personnel that contribute to or constrain total system performance.
4. Establish requirements for human performance that will achieve effective human–system interfaces.
5. Identify requirements for combining, modifying, or establishing new military occupational specialties.
6. Describe the training concept to include requirements for the training support package (e.g., simulators, training devices, embedded training) and training logistics.
7. Include safety or health considerations that reduce job performance or system effectiveness, given the operational environment.
8. Identify critical errors that reduce job performance or system effectiveness, given the operational environment.
9. Determine objectives and thresholds for the above requirements, as appropriate.

*Source:* Adapted from U.S. Department of Defense, 1999c.

HSI considerations that must be addressed under Program Support. As one can see, the requirement is to cover the basic domains of HSI and to include manpower, personnel, training, and human factors engineering as well as safety and health hazards. However, it is important that the HSI professional closely examine the HSI issues and determine if any of them warrant being included in the capabilities section as a KPP. If the evaluations and analyses during concept exploration and component advanced development have revealed any human performance parameters that are likely to lead to system failure if they are not adequately considered during system development, then they need to be put forth as KPP candidates. This is where the existence of an HSI management plan that documents HSI issues and concerns can be helpful. It can serve as a ready source of issues for inclusion in the ORD.

The HSI management plan can also serve as the vehicle for introducing HSI considerations into plans for testing and evaluating the system. The DoD regulations require that the PM establish an overarching document, called the TEMP, early in the life cycle that lays out the plan for testing and evaluating a system during the course of its development. It is essential that human performance issues be integrated into that document if the human is to be adequately tested and evaluated as a contributor to total system performance. Directive 5000.2-R (DoD, 2001, p. 83) addresses these points. It indicates that "the program manager shall initiate a comprehensive strategy for HSI early in the acquisition process to minimize ownership costs and ensure that the system is built to accommodate the human performance characteristics of the user population that will operate, maintain, and support the system." It goes on to state: "The PM shall work with the manpower, personnel, training, safety and occupational health, habitability, survivability, and HFE communities to translate the HSI thresholds and objectives in the ORD into quantifiable and measurable system requirements." Finally, it states: "The PM shall include these requirements in specifications, the TEMP, and other program documentation, as appropriate, and use them to address HSI in the statement of work and contract." Clearly, an HSI management plan is needed to coordinate and satisfy these requirements.

A completed ORD with detailed performance specifications is required for entering the next phase of the life-cycle process. Once the appropriate decision authority has given a favorable decision at milestone B, then a new system acquisition is formally started and the process enters the next phase.

## 4.4 SYSTEMS ACQUISITION

At this point in the life cycle a system architecture has been specified based on technologies that are relatively mature and have a reasonable chance of contributing to successful system development, and the operational requirements have been specified in the ORD. The ORD, rather than the MNS, now drives the life-cycle process. The first phase of system development in the DoD life-cycle model is called system development and demonstration. As the name indicates, its purpose is to develop the system fully while reducing program risk, ensuring operational supportability and affordability, designing for producibility, and demonstrating system integration, interoperability with relevant systems, and utility. It consists of two work efforts; namely, system integration and system demonstration. If these are successful, the production and deployment work efforts follow.

### 4.4.1 System Development and Demonstration

During system integration the various technologies or subsystems are integrated into the overall systems architecture to produce a complete system. An initial prototype is built and demonstrated in an environment that is relevant to the system in which it is expected to operate. The focus of the whole effort is to ensure that the subsystem technologies can be made to work in an integrated fashion and reduce system-level risk. Following the successful demonstration of the initial prototype, the system demonstration work effort is begun wherein the system is fully developed and engineering development models are successfully demonstrated in the intended environment. It is during this work effort that the system is subjected to a variety of testing and evaluation to ensure that it is on track in meeting the specifications and requirements laid out in the ORD.

### 4.4.2 Request for Proposal

During the system development and demonstration phase it is important to refine and update the issues in the HSI plan by iterating through the HSI framework discussed previously and shown in Figure 4.4. In addition, there are a number of activities that are unique to this phase and that should have HSI input. It is at the very beginning of this phase that the PM issues a request for proposal (RFP), which describes to industry the requirements for a new system. This document contains numerous sections, but there are several in which HSI should be represented.[6] They include the statement of work (SOW), the system specifications, and the basis of award.

The SOW contains a description of the work to be performed by the contractor to include any efforts needed to produce required management and technical data. The HSI tasks that are appropriate for this section include the scope of the contractor's HSI effort, including the various HSI domains that need to be addressed, and specific HSI data collection efforts and analyses to be performed to ensure that human performance is effectively contributing to total system performance.

The system specifications section describes the performance requirements for the system to be developed and are based on the operational requirements found in the ORD. Key HSI issues in the ORD must be reflected in this section. For example, if there is a limit on the manpower available to operate and maintain the system, that should be stated clearly in the specifications.

The basis of award section describes how the technical proposals from the various vendors will be evaluated and how a winning vendor is selected. Proper attention to the inclusion of HSI in this section is critical if HSI is to be given meaningful attention by the vendor who is awarded the contract for system development. Typically, management, technical, and cost considerations have been given the highest weight in system proposal evaluations. However, the U.S. Army HSI community has pushed for and been successful in getting HSI to have equal weight with those factors. In the case of one system, the Comanche, HSI and training were given 17.5 percent of the total weighting; reliability, availability, maintainability, and integrated logistics support, which have clear HSI implications, were given 17.5 percent; and technical, which had numerous areas with HSI implications, was given 35 percent (Booher, 1997). The result was that HSI had as much or more weight in the final evaluation leading to source selection than any of the other evaluation factors. The Comanche development has been quite successful from an HSI point of view.

### 4.4.3 Design Support

Human systems integration has an additional role to play during system development and demonstration. It is during this phase of the acquisition process that the design becomes relatively fixed. The HSI information is needed to make rational design decisions, and much of this information comes from the knowledge of the HSI professional and the information he or she can derive from small focused human performance experiments and from modeling and simulation.[7] The DoD has in the past few years put increasing emphasis on the use of modeling and simulation as a means to reduce development costs and acquisition time. Human figure modeling tools[8] can give critical information about the adequacy of the physical layout of operator and crew stations, to include reach distances, fields of view, and display and control layouts. Task performance oriented modeling tools can give valuable information about the likelihood of operators and crews being able to effectively interface with the design.[9]

### 4.4.4 Testing and Evaluation

Another important function occurring during this phase is testing and evaluation (T&E) of the prototype system. While planning and some preliminary test and evaluation activities occur in the concept and technology development phase of the life cycle, most formal testing and evaluation are accomplished during the system development and demonstration phase and the production and deployment phase. There are two basic types of testing that are typically done to support the evaluation of a developing system: developmental testing (DT) and operational testing (OT). Developmental testing is engineering-oriented testing directed toward ensuring that adequate progress is being made technically. It is under the control of the PM and seeks to minimize design risks and, at the appropriate time, certifies that the system is ready for initial operational testing. On the other hand, OT is operations oriented and is done to ensure that the system can be operated and maintained by typical individuals and crews in mission environments. It is controlled by agencies that are independent of the PM. Developmental testing examines not only the system but also subsystems and components, whereas OT typically looks at the whole system. Within both DT and OT are a variety of tests that are executed, depending on the state of the system.

Human systems integration has a role to play in both DT and OT [see Meister (1986) and O'Brien and Charlton (1996) for detailed expositions on the role of HSI in T&E]. Both are opportunities to collect human factors data that can address the issues that have been laid out in requirements documents and the HSI management plan. The DoD military handbook 46855A (DoD, 1999b) lists the following HSI activities that should be accomplished in T&E:

- Verify that personnel can safely perform tasks to time and accuracy standards without excessive workload.
- Verify that system characteristics support effective and efficient operation.
- Verify that the human engineering design of the hardware supports human use.

- Determine the adequacy of human performance as a component of system performance.
- Determine the effects of operational variables on human performance.
- Determine the appropriateness of modifications.

In the past, HSI professionals have not attended to DT as much as to OT. This is partly due, perhaps, to the fact that the DT environment is not as operationally realistic as the OT environment. Nevertheless, DT has an advantage in that there is usually much more control over the total test environment, thus increasing the likelihood of the HSI professional being able to collect substantial amounts of reliable and valid data. In the future it would behoove the HSI community to take more advantage of the opportunities that DT has to offer.

Regardless of whether one is working in a DT or OT environment, there are a number of approaches that are used to collect human factors data relevant to HSI. These include taking physical measurements of equipment, the system environment, and the human; measuring task performance when operators and maintainers are using the system or components thereof; and administering instruments to obtain subjective assessments from operators and maintainers, such as questionnaires and interviews.

Information from physical measurements is the most standardized. Methods and instrumentation for obtaining such information [e.g., see Eastman Kodak Company (1983) and Meister (1986)] and associated standards (e.g., see DoD, 1999a) are well established and can be found in a variety of resources. It is obvious that humans operate best when various aspects of their work environment are within limited bounds, including air temperature, humidity, and quality (i.e., absence of noxious and toxic gases); illumination; sound (to include limits on the intensity of background noise and the relative intensity of communications and auditory signals); vibration; and physical characteristics of equipment (e.g., the weight of an item that has to be raised to shoulder height when setting up a system, the force required to activate a switch, etc.). Measurements on these variables are important not only because they contribute to operator and maintainer efficiency but also because extreme levels can be hazardous to health. Along these lines, measures of operator physiological state are sometimes taken to obtain indications of the extent to which he or she is experiencing stress (O'Brien and Charlton, 1996).

Measurement of operator and maintainer task performance consists of determining the critical tasks that have to be completed for successful accomplishment of a system's mission and then determining the extent to which these critical tasks are actually accomplished during a mission. The performance measures that can be taken include type of error made, time to accomplish a task (which can be a type of error in and of itself), and frequency of errors [see Gawron (2000) for a more detailed discussion of human error measurement and Reason (1990) for a theoretical treatment of human error]. Collection of error information can be done in a variety of ways. Trained observers can watch operators and maintainers perform their tasks and record and describe errors as they occur, either during the test itself or from videotape that was made during the test (the latter is usually more practical since there is often not enough room in crew compartments or operator stations for an extra person). Or information can be tapped from system databases to indicate the status of switches and controls relative to activities that are occurring during the test, thus revealing whether or not the operator took the correct actions at the appropriate time.

Error data are arguably the most important HSI information that can be collected, for without an objective, quantitative measure of human performance it is very difficult, if not impossible, to determine the effect the human is having on total system performance. When a system fails to meet a mission goal, the only meaningful way to determine if the operator or maintainer contributed to that failure is by linking the performance of the system at critical times to pertinent actions of the operator or maintainer. If such linkage shows that task errors partially or fully led to the degradation of system performance and mission failure, then the reasons for the task failure can be investigated and HSI-related changes made, such as redesigning the user–system interface, reallocating tasks among crew members, increasing the training of operators on that task, increasing the size of the crew, and so forth.

The most widely used approach to obtaining HSI information during T&E involves the use of subjective instruments. These include questionnaires, rating scales, and interviews. Such instruments, when properly designed, can provide a wealth of information to the HSI professional. However, they are most useful in providing information for explaining why errors have occurred or are likely to occur and as such are best used in conjunction with actual error measurement methods. The most common approach is the use of the questionnaire, most likely because it can be put together quite easily, rapidly tailored to the characteristics of the system under consideration, and easily administered [although see Babbit and Nystrom (1989) and Charlton (1996) for discussions of good questionnaire construction techniques]. But good questionnaires take a fair amount of thought and effort to ensure that the questions asked are well constructed, clearly understandable by the target audience, and focused on relevant issues. They are best supplemented by one-on-one interviews with respondents to clarify the answers given.

Other subjective techniques include various types of rating scales. Many of these have advantages over tailored questionnaires because they have been validated in a variety of situations, whereas a questionnaire is usually tailored to a given system, with the questions being written so that they have face validity; no independent validation is attempted. Several rating scales for obtaining subjective assessments of workload and situation awareness have become quite popular and have been widely used in recent years (e.g., see Gawron, 2000).

Regardless of the subjective technique that is used, it is important to recognize the point made several paragraphs ago—namely, that these are best used to provide supplemental information to direct performance measurements in order to explain why critical tasks are not being performed correctly or to provide information indicating that certain critical tasks are not likely to be performed correctly under certain situations, as in high workload, for example.

### 4.4.5 Production and Deployment

Once the relevant DoD decision authority has deemed a system mature enough for production (i.e., it has passed milestone C), it enters the production and deployment phase. In this phase it has to undergo a major initial operational test and evaluation (IOTE), which is a statutory requirement that must be performed on all major military systems by an agency independent of the developer. It is a large field test conducted under relatively realistic operational conditions with production representative systems. It is the final hurdle of system development before proceeding into full-scale production. A low-rate initial production of systems off of the assembly line is authorized for the test. Following a favorable evaluation, full rate production is authorized and the system is fielded.

The IOTE is usually the final opportunity to collect HSI-relevant data on the system during the formal acquisition process (although sometimes a system is required to have additional testing—called follow-on T&E—if some troublesome but not serious problems are uncovered in IOTE). A tremendous amount of time and effort is spent on IOTE, and HSI concerns and issues are typically prominent in the overall data collection plan, especially if they have been prominently identified in the ORD and are included in a KPP. The IOTE data collection effort is structured around critical operational issues and criteria (COICs) found in the TEMP. The COICs in turn are derived from critical issues and requirements in the ORD, most notable of which are the KPPs. But COICs are not the only issues addressed in IOTE; many additional issues are also addressed. What is important about IOTE with respect to HSI is that multiple copies of the system are being exercised as part of a unit with typical operators and maintainers in a relatively realistic operational scenario occurring over several days. This context gives the data that are collected more meaning than previously collected data, and inferences drawn about numbers, skills, and training of operators and maintainers needed for effectively operating and maintaining the system are likely to be more valid, as are conclusions about the user–system interfaces. However, the downside of this situation is that the design of the system is essentially complete, so any recommendations that the HSI professional might make for improving the design are not likely to be accepted unless they result in the correction of a design characteristic that is contributing to the substantial degradation of system performance. Thus, HSI must place substantial effort on the early stages of system design; efforts at that phase are likely to have the most impact.

## 4.5 SUSTAINMENT

The final activity in the life cycle is that of sustainment. During this period the fully fielded system is supported with maintenance activities and minor modifications that are needed to keep the system fully operational during its useful life. The HSI activities have not focused on this period in the past. However, much useful HSI data could be obtained here without a tremendous amount of effort. Perusal of maintenance records can reveal much about maintenance demands of the system over time, revealing high driver maintenance tasks that should be avoided in future upgrades to the system or in the development of similar systems. Questionnaires and interviews can be used to solicit information from operators and crews about operational problems they have experienced with the system, serving much the same role. It is impossible to thoroughly test a new system in all of the situations in which it is likely to be employed, and some flaws in a system may only be revealed when the system is in one of those types of situations. Such information that might be of use to the HSI community is lost unless specific focused efforts are made to retrieve it.

## 4.6 CONCLUSION

The life cycle of a system proceeds through a number of activities, to include defining what the system should be; designing, building, and evaluating the system; and sustaining it after fielding. Human systems integration has an important role to play in each of these periods, but its contributions with the most potential are in the early stages of the life cycle, when the system requirements are being established and the initial design is being derived. It is here that fundamental decisions are made about what functions the system is to perform and how it is going to be designed to satisfy them. The HSI professionals must step forth with their knowledge of human factors, their measurement methods, and their analysis and modeling tools and help program managers and system developers design systems that are operationally efficient and cost effective. While HSI activities in the latter phases of system acquisition are useful, their contribution by then is mainly in helping to make corrections to minor system problems and in providing information that may be of use in later upgrades of the system or in the development of future systems. It is only in the early stages of acquisition that substantial impact can be made. It is incumbent upon the HSI community to strive to make an impact during this early period to ensure that the systems that are developed can be used effectively, efficiently, and safely. The men and women of the armed forces who will use many of these systems in life-threatening situations deserve no less than our full and dedicated efforts in this endeavor.

## NOTES

1. See Chapter 5 for a discussion of education issues in HSI.
2. On 29 August 2002 DoD announced that it was revising its acquisition policy. Pending formal publication of the revised policy, interim guidance was issued in SecDef Memorandum, "The Defense Acquisition System," August 29, 2002, and in SecDef Memorandum, "Operation of the Defense Acquisition System," August 29, 2002. The interim guidance retained the newer DoD model shown in Figure 4.2.
3. See Chapter 10 for a discussion of evolutionary and incremental acquisition.
4. Italics has been added for emphasis.
5. See the chapters in Part III for a description of essential HSI tools and techniques.
6. See Chapter 7 for a contractor's perspective on this process.
7. See Chapter 9 for a description of simulation-based acquisition.
8. See Chapter 13 for a description of human figure models.
9. See Chapter 11 for a description of human performance models.

## REFERENCES

Babbit, B. A., and Nystrom, C. O. (1989, June). *Questionnaire Construction Manual*, Research Product 89-20. Fort Hood, TX: U.S. Army Research Institute for the Behavioral and Social Sciences.

Blanchard, B. S., and Fabrycky, W. J. (1998). *Systems Engineering and Analysis*, 3rd ed. Upper Saddle River, NJ: Prentice-Hall.

Booher, H. R. (1997, July). *Human Factors Integration: Cost and Performance Benefits on Army Systems*, AR-CR-341. Washington, DC: Army Research Laboratory.

Charlton, S. G. (1996). Questionnaire Techniques for Test and Evaluation. In T. G. O'Brien and S. G. Charlton (Eds.), *Handbook of Human Factors Testing and Evaluation.* Nahwah, NJ: Lawrence Erlbaum Associates.

Eastman Kodak Company, The Human Factors Section of the Health, Safety and Human Factors Laboratory. (1983). *Ergonomic Design for People at Work*, Vol. I. Belmont, CA: Lifetime Learning Publications.

Gawron, V. J. (2000). *Human Performance Measures Handbook.* Mahwah, NJ: Lawrence Erlbaum Associates.

Graine, G. N. (1988). The Engineering Syndrome vs. the Manpower, Personnel and Training Dilemma. *Naval Engineers Journal*, pp. 54–59.

Ladendorf, K. (1988, February 19). Lockheed Austin Loses Aquila Project: Lean 1989 Budget Axes Military Drone. *Austin American-Statesman*, pp. G1, G3.

Meister, D. (1986). *Human Factors Testing and Evaluation.* Amsterdam: Elsevier.

O'Brien, T. G., and Charlton, S. G. (Ed.). (1996). *Handbook of Human Factors Testing and Evaluation.* Mahwah, NJ: Lawrence Erlbaum Associates.

Price, H. E. (1990). Conceptual System Design and the Human Role. In H. R. Booher (Ed.), *MANPRINT: An Approach to Systems Integration* (pp. 161–203). New York: Van Nostrand Reinhold.

Reason, J. (1990). *Human Error.* Cambridge: Cambridge University Press.

Sage, A. P. (1992). *Systems Engineering.* New York: Wiley.

Sage, A. P., and Rouse, W. B. (Ed.). (1999). *Handbook of Systems Engineering and Management.* New York: Wiley.

Stewart, J. E. III, Smootz, E. R., and Nicholson, N. R. (1989, June). *MANPRINT Support of Aquila, the Army's Remotely Piloted Vehicle: Lessons Learned*, Research Report 1525. Alexandra, VA: U.S. Army Research Institute for the Behavioral and Social Sciences.

U.S. Army. (1988). *Remotely Piloted Vehicle: Force Development Test and Experimentation*, TRADOC Test and Experimentation Command Test Report FD 0186. Fort Hood, TX: U.S. Department of the Army.

U.S. Army. (1999). *Force Development Requirements Determination*, TRADOC Pamphlet 71-9. Fort Monroe, VA: U.S. Department of the Army.

U.S. Department of Defense (DoD). (1999a). *Design Criteria Standard, Human Engineering* (MIL-STD-1472F). Washington, DC: USDoD.

U.S. Department of Defense (DoD). (1999b). *Human Engineering Program Process and Procedures*, MIL-HDBK-46855A. Washington, DC: DoD.

U.S. Department of Defense (DoD). (1999c). *Requirements Generation System*, Chairman of the Joint Chiefs of Staff Instruction 3170.01A, August 10, 1999. Washington, DC: DoD.

U.S. Department of Defense (DoD). (2000a). *The Defense Acquisition System*, Directive 5000.1. Washington, DC: DoD.

U.S. Department of Defense (DoD). (2000b). *Operation of the Defense Acquisition System*, Instruction 5000.2. Washington, DC: DoD.

U.S. Department of Defense (DoD). (2001). *Mandatory Procedures for Major Defense Acquisition Programs (MDAPS) and Major Automated Information System (MAIS) Acquisition Programs*, Directive 5000.2-R. Washington, DC: USDoD.

CHAPTER 5

# Human Systems Integration Education and Training

BRIAN M. KLEINER and HAROLD R. BOOHER

## 5.1 INTRODUCTION

Human systems integration (HSI) requires highly qualified personnel applying their expertise to systems engineering and management processes if potential dramatic improvements in system performance, safety, and affordability are to be realized. Unfortunately, very few qualified people are currently available in the national pool for organizations to draw upon the needed expertise.[1] Three critical ingredients are required to increase the size and talent of the HSI personnel pool. First is greater demand for their talents, which is reflected in an increasing number of projects that mention HSI and jobs that require HSI qualifications. But demand for HSI jobs is not likely to increase until the need for the skills and methods outlined in this book are fully appreciated by organizations that procure, produce, and/or use systems. Second, as the demand increases for HSI expertise, the need for education and training programs to provide the necessary qualifications also increases (Van Cott and Huey, 1992). Currently, few education and training avenues exist to provide even minimal qualifications in HSI (Hollis, 1995; U.S. Army, 1995). The HSI workforce cannot increase substantially, therefore, without adequate education and training sources for developing HSI talent. Third, even if demand rises and education and training programs are available, something more is needed to motivate people to seek out such programs. To be sufficiently motivated to become a highly qualified HSI practitioner, individuals need to see the career potential of the HSI field. The demand for HSI practitioners and the development of qualification programs must progress together within an overall national vision of long-term careers for HSI specialists.

An army manpower, personnel, integration (MANPRINT) study on this topic recommended that a national workforce be established for HSI[2] (U.S. Army, 1993), which, if done, would bring together the second and third ingredients needed to solve the HSI

*Handbook of Human Systems Integration*, Edited by Harold R. Booher.
ISBN 0-471-02053-2 

personnel supply problem.[3] An essential first step would need to be made by the U.S. federal government (U.S. Army, 1993, p. ES-1):

> The need to integrate human issues or concerns into the engineering design process ... has been recognized by the Department of Defense (DoD) and other agencies of the Federal Government.... A series of the Congressional studies, General Accounting Office reports, and experiences of the [military] services ... identified a number of problems which illustrated the overriding fact [that] government agencies are designing, building, and buying systems ... [they] cannot easily use. A root cause of these problems is the lack of trained and qualified people in the workforce who understand how to integrate human issues into the process of research, design, development, and system implementation.

The army study recognized, however, that the federal government could not accomplish such a demanding challenge on its own: "Although, the Federal Government discharges a key and essential role in establishing and building a national [HSI] workforce, it is essential to involve academia and industry as well [as] the Federal Government" (U.S. Army, 1993, p. ES-1). The study also found that there was sufficient demand for the government to establish a formal job series and career progression within it. If this were done, it would be a major step toward attracting and retaining highly qualified personnel for both the managerial and technical work needed in HSI. Unfortunately, the third ingredient is so intertwined with the second that the federal government is reluctant to formalize its processes until education and training programs needed for a formal career in HSI are available.

The focus of this chapter therefore, is on the second ingredient, national education and training for HSI. Muckler and Seven (1990) first addressed the central issues of HSI education and training in the context of developing a national MANPRINT workforce: They considered "the kind of skills and knowledge required to conduct the MANPRINT effort"; examined "some of the ... institutional systems that educate and train many of the specialties of MANPRINT"; and outlined the various challenges and prospects facing the establishment of a national HSI workforce at the time (p. 519).

This chapter will build and expand upon the Muckler and Seven foundation while describing the current HSI requirements and institutional systems available to educate and train HSI specialties. A key purpose of the chapter is to amplify HSI principles 9 (qualified practitioners) and 10 (HSI education and training) described in Chapter 1.

More specifically, the objectives of the chapter are to:

- outline the HSI competencies needed for qualified HSI specialists,
- define what is available in academic settings to meet HSI qualifications,
- describe what is available in HSI practitioner training courses,
- summarize what HSI information is available in technical handbooks and textbooks,
- outline the education and training gaps between HSI needs and what is available,
- discuss special issues with establishing an HSI career path.

## 5.2 HSI COMPETENCIES NEEDED

Many of the competencies needed to perform HSI tasks in occupations have been identified. Past DoD studies (U.S. Army, 1993, 1995) provide a consensus of HSI practitioners and education and training specialists on job competency requirements for HSI personnel. These efforts investigated the types of work being conducted in the army MANPRINT jobs and identified related jobs in other federal agencies. The different contexts in which nearly all of the MANPRINT- or HSI-related work was being performed could be classified into four major categories (or functions):

- research,
- engineering,
- acquisition, and
- regulatory and policy.

Research jobs involve working in laboratory or field settings with responsibilities to determine such things as determining basic, quantitative effects of environmental stimuli on human behavior or performance. Research jobs also entail such work as the development of new HSI technology that is usually considered applied research. As such, HSI research is primarily concerned with producing results that advance the state of the art for HSI principles 6 (quantitative human parameters) and 7 (HSI technology).

Engineering jobs involve design, development, and testing of hardware and software. Acquisition jobs involve documenting and supporting a system or product program through the various acquisition stages discussed at length in several chapters in this book. For the HSI specialist, the engineering acquisition jobs categories are usually implemented through the systems engineering and management processes but may also be part of logistics support, training, safety, and test and evaluation processes, depending on the organization's engineering management structure. The regulatory and policy jobs frequently involve setting (or applying) standards and practices for technical HSI domains [usually human factors engineering (HFE), system safety (SS), or health hazards (HH) domains]. In HSI mature organizations, the entire HSI program and the other domains will also have HSI policy jobs.

Based on the typical task assignments and levels of competency required for these four job functions and on surveys of active MANPRINT practitioners, the U.S. Army (1993) developed lists of HSI task activities and personnel competencies. The following describes and updates these lists in discussing

- levels of competency and functional assignments, and
- core HSI competency requirements.

### 5.2.1 Levels of Competency and Functional Assignments

An HSI professional workforce can be envisioned as operating at three levels of competency. A career path for a HSI practitioner will tend to progress through three levels, with varying job requirements by level:

- Level I: HSI domain expert.
- Level II: System integrator; knowledgeable in all HSI domains.
- Level III: Overall HSI Manager and/or policymaker.

Specific job assignments can be defined at each level for each of the four job functions. Table 5.1 provides a matrix of job titles appropriate to each level and function.

The personnel competencies that are needed for the various jobs vary by both function and level. Table 5.2 shows some of the kinds of competencies that are frequently required for the four major functions.

Table 5.3 lists the types of tasks HSI practitioners might be asked to perform for a new system acquisition. Categorized by the seven HSI domains, most of these tasks were obtained from a MANPRINT practitioners survey conducted by the U.S. Army (1993). Although reflecting actual task domain requirements, the tasks itemized in Table 5.3 are only representative and not meant to be exhaustive of the tasks that can be required for the various domains. This listing of tasks does, however, indicate a number of characteristics of HSI jobs:

1. HSI requirements tend to mix management and technical tasks.
2. There is some overlap between domain requirements (e.g., both training and HFE do task analyses).
3. Some domains (e.g., HFE) show greater analytical and design requirements than others (e.g., manpower). This is more a function of domain maturity, than the need for the domain activity. For example, as HSI research and development provides more manpower analysis and design tools, the manpower domain plays a greater technical role.

In addition to specialized domain requirements, all HSI practitioners need to play a role in integration, both among domains and among the other acquisition process fields, such as systems engineering, safety engineering, integrated logistics support, etc. Consequently, an overall integration category needs to be added to the individual domains to cover the various analytical and managerial competencies listed in Table 5.2 below the domains.

### 5.2.2 Core HSI Competency Requirements

The army (Hollis, 1995; U.S. Army, 1993) developed an initial set of core requirements for the HSI professional to cover the broad functional competencies (Table 5.2) needed to do the sample HSI tasks (Table 5.3). We have modified the original list based on lessons learned from contributors writing this handbook, allowing us to compile a new list of core competencies, reflected in the key words of Table 5.4. Full descriptions of the topics are covered in the chapters that discuss them. (See, e.g., Chapters 4 and 10 for systems engineering and integration models, Chapter 6 for requirements determination, Chapter 8 for human performance measures, and Chapter 11 for HSI technology.)

Table 5.4 illustrates many of the competencies likely to be needed for those HSI practitioners working in one or more domains in the acquisition process. The major categories are (a) HSI domains and (b) systems engineering and integration. As might be expected, HSI competencies start with expertise in the HSI domains. Domain expertise at the entry level is required only for the domain in which the individual works, along with a familiarity of the particular systems engineering and integration functions associated with the entry job position. However, to progress in an HSI career path, expertise will be needed in more than one domain, and a working knowledge of most of the systems engineering and integration items listed in Table 5.4 will be required.

**TABLE 5.1 Job Titles by Level and Function**

| Research | Applied Engineering | Acquisition | Policy/Regulatory |
|---|---|---|---|
| | Level I | | |
| Research and development (R&D) laboratory staff | Junior engineer | Combat development (CD) staff | Regulatory agency staff |
| Research psychologist | Cost analyst | Training development (TD) staff | Acquisition policy staff |
| Anthropometrician | Safety engineer | Program management (PM) staff | Cost accounting standards staff |
| Operations research analyst | Industrial hygienist | Testing and evaluation (T&E) staff | Agency inspector |
| Operational medicine | Engineering psychologist | Integrated logistics support (ILS) staff | |
| | Industrial engineer | | |
| | Level II | | |
| R&D team leader | Design engineer | CD/TD branch chief | Regulatory agency branch chief |
| Senior scientist | Human factors engineer | Training system manager | Acquisition policy senior staff |
| | Test engineer | PM branch chief/senior staff position | |
| | | Senior ILS position | |
| | | T&E branch chief | |
| | Level III | | |
| R&D branch chief | System engineer | Program manager | Regulatory division director |
| R&D laboratory director | Program engineer | T&E director | Acquisition policy director |
| Chief scientist | HSI chief engineer | HSI program manager | |
| Technical director | | | |

**TABLE 5.2 Job Competencies by Function**

| Competencies | Research | Application Engineering | Regulatory | Acquisition |
|---|---|---|---|---|
| Domains | • | • | • | • |
| HFE and system safety | • | • | • | • |
| Manpower and personnel | • | • | • | • |
| Training | • | • | • | • |
| Health Hazards | • | • | • | • |
| Analytical techniques | • | • | • | • |
| Information management | • | • | • | • |
| Acquisition management | | • | | • |
| Regulatory management | | • | • | |
| Integrated logistic support | • | • | | • |
| System engineering | • | • | • | • |
| Research and development | • | • | • | • |
| Testing and evaluation | • | • | • | • |
| Procurement | | | | • |
| Cost analysis | • | • | • | • |
| Program funding | | | | • |
| Program management | | | | • |
| Organizational integration | • | | | • |
| Operations research | • | • | | • |

The core competencies are those needed collectively by all HSI jobs, levels I, II, and III and for all relevant functions—research, engineering, acquisition, and regulatory. These then become the basic list of knowledge, skills, and abilities (KSAs) for curriculum developers, education and training delivery systems, and students to focus upon in meeting HSI competency needs. A full-blown HSI career package would, of course, include all the advanced and specialized KSAs to cover all functional assignments. The academic coursework and special training courses to develop HSI expertise covered in the next sections will use the core competencies shown in Table 5.4 as the basic requirements for the HSI practitioner. It is understood, of course, that HSI is continually maturing and that this list is only something to start a dialog among the various stakeholders interested in enhancing the systems acquisition process and the HSI profession.

## 5.3 ACADEMIC EDUCATION

This section discusses HSI academic offerings. It begins with a review of currently available curricula that include coursework related to HSI requirements. It then covers two areas of special interest to HSI–human systems interface technology and new content trends—to help provide a broad perspective to the considerations involved in making up an HSI academic major. This is followed by a brief discussion on the content gaps between what is currently available and what is needed. Finally, perhaps the most important part of the chapter is a discussion of specific HSI course content. This is complemented by two hypothetical but plausible graduate-degree tracks focused on HSI.

**TABLE 5.3 HSI System Acquisition Tasks by Domain**[a]

| Domain | Task |
|---|---|
| Manpower | • Document changes to organizational structure caused by the introduction of a new system<br>• Determine numbers of required and authorized personnel for the units and types of personnel that will use, maintain, and support a new system<br>• Calculate whether a new system will require more personnel than is authorized or required currently |
| Personnel | • Specify human user, operator, or maintainer requirements (aptitudes and experience)<br>• Document changes to agency personnel, personnel management, and personnel policy caused by the introduction of a new system<br>• Develop, update, and maintain a description of the equipment operator, user, and maintainer |
| Training | • Prepare instructional or procedural documents<br>• Define instructional requirements<br>• Specify training objectives<br>• Assess the effectiveness of training (systems, courses, aids, simulators)<br>• Conduct training<br>• Design training aids<br>• Develop training content and instructional methods<br>• Design simulation systems<br>• Document the changes to agency training strategy, plans, policy, and procedures caused by introduction of a new system |
| Human factors engineering | • Assess mental workload<br>• Assess physical workload<br>• Analyze effects of environmental stressors<br>• Perform human reliability analyses<br>• Apply human factors criteria and principles<br>• Verify design conformance to human factors specification<br>• Design human–equipment interfaces<br>• Design workspace layouts<br>• Design software–user interfaces<br>• Prepare/review drawings for conformance to human factors specifications<br>• Develop, update, and maintain human factors management plans |
| System safety | • Develop analytical models and methods<br>• Collect data on errors, failures, or accidents<br>• Perform safety analyses<br>• Conduct root-cause analyses<br>• Perform failure-mode and effects analyses<br>• Develop and analyze fault trees<br>• Develop, update, and maintain system safety plans |
| Health hazards | • Assess performance risks from health hazards categories (noise, contaminants, etc.)<br>• Support product liability litigation<br>• Prepare product warnings<br>• Develop, update, and maintain health hazards prevention plans |
| Personnel survivability | • Conduct personnel survivability assessments<br>• Support casualty analyses<br>• Develop personnel survivability enhancement procedures |

[a]This table is a sampling of types of tasks assignable to the various domains for a new system acquisition.

**TABLE 5.4 HSI Core Competencies**

| HSI domains | Systems Engineering and Integration |
|---|---|
| • Statistics | • Acquisition process models |
| a. Experimental design | a. Traditional |
| b. Regression methods | b. Streamlined |
| c. Nonparametrics | c. Nondevelopmental items |
| • Sensory and perceptual processes | d. Materiel improvement |
| • Cognition and decision making | • Requirements determination |
| • Physical abilities and limits | a. Systems requirements analysis |
| • Anthropometry and work physiology | b. HSI issues and criteria |
| • Simulation methodology | c. MPT trade-offs |
| • Human system modeling | • Systems design and management |
| • Human performance measurement | a. Human-centered design |
| • Design of displays, controls, and workstations | b. Requests for proposal, proposal development, and evaluations |
| • Skill acquisition | c. HSI assessments |
| • Personnel selection | d. Program management |
| • Team performance | • Testing and evaluation |
| • Environmental health hazards | a. Measures of effectiveness and performance |
| • System safety | b. HSI in test design plans |
| • Human survivability in hostile environments | c. HSI in test reports |
| • Organization design | • HSI technology research and development |
| • Analytical techniques | • Operations research |
| a. Early comparability analysis | • Integrated logistics support processes |
| b. Manpower staffing analysis | • Safety engineering and management |
| c. Information requirements analysis | • Training approaches and methodologies |
| d. Task, function, and workload analysis | • Economic and cost analyses |
| e. Training effectiveness analysis | |
| f. HSI domain trade-off analysis | |
| g. Accident analysis | |
| h. Human error and reliability analyses | |

### 5.3.1 Currently Available Curricula

Due to its similarity with HSI content, a content analysis was performed on program descriptions of human factors/ergonomics (HF/E) academic programs in the United States and Canada [Human Factors and Ergonomics Society (HFES), 2000a]. As illustrated in Table 5.5, there are 83 programs listed by the HFES (2000a). These 83 programs reside on 72 campuses. Therefore, 11 campuses offer human factors programs in multiple departments. Several additional campuses have integrated their multiple offerings into a single program. Seventy-eight percent of the programs offer doctoral degrees. Thirty-six percent of the programs are in psychology departments while 36% are in industrial engineering departments. This is an interesting statistic because there are twice as many psychologists as engineers in the HFES, yet these disciplines have the same number of programs. The remaining 28% of programs are from either integrated departments or departments other than psychology or industrial engineering.

**TABLE 5.5 Graduate Programs in North America**

| Program | Number of Faculty | Department | Degrees Offered |
|---|---|---|---|
| Arizona St. | 8 | School of Design | MS |
| Auburn | 3 | Industrial and Systems Engineering | MS, MISE, PhD |
| Cal. St., Long Beach | 4 | Psychology | MAR |
| Cal. St., Northridge | 8 | Psychology | MA |
| Catholic | 5 | Psychology | MA, PhD |
| Central Michigan | 8 | Psychology | MS, PhD |
| Clemson | 12 | Psychology | MS |
| Cornell | 5 | Design and Env. Analysis | MS |
| Embry-Riddle | 7 | Human Factors and Systems | MS |
| FIT | 9 | Space Coast Center for HF Research | MS |
| Florida International | 2 | Industrial and Systems Engineering | MS |
| George Mason | 5 | Psychology | MA, PhD |
| Georgia Tech. | 6 | Psychology | MS, PhD |
| Iowa St. | 2 | Industrial and Manufacturing Systems | MS, PhD |
| Kansas St. | 2 | Industrial and Manufacturing Systems | MSIE, PhD |
| Kansas St. | 6 | Psychology | MS, PhD |
| Louisiana St. | 10 | Industrial Engineering | MSIE, PhD |
| Marquette | 5 | Mechanical and Industrial | MS, PhD |
| Marshall | 5 | Safety Technology | MS, MA |
| Mississippi St. | 3 | Industrial Engineering | MS, PhD |
| Mississippi St. | 9 | Psychology | PhD |
| MIT | 8 | Aeronautics and Astronautics | MS, PhD |
| Miami of Ohio | 10 | Psychology | MS, PhD |
| NJIT | 3 | Industrial and Manufacturing Engineering | MS |
| New Mexico St. | 7 | Psychology | MA, PhD |
| NYU | 5 | Occupational Therapy/ Environmental Medicine | MA, PhD |
| NC A&T | 1 | Industrial Engineering | MS |
| NC St. | 5 | Industrial Engineering | MS, MIE, PhD |
| NC St. | 10 | Psychology | MS, PhD |
| Northeastern | 4 | Mechanical, Industrial and Manufacturing | MS, PhD |
| Ohio St. | 10 | Industrial and Systems | MS, PhD |
| Ohio St. | 5 | Psychology | MA, PhD |
| ODU | 10 | Psychology | MS, PhD |
| Oregon St. | 3 | Industrial and Manufacturing Engineering | MS, PhD |
| Penn St. | 3 | Industrial and Manufacturing Engineering | MS, PhD |

*(continued)*

**TABLE 5.5** ***(Continued)***

| Program | Number of Faculty | Department | Degrees Offered |
|---|---|---|---|
| Purdue | 6 | Industrial Engineering | MS, PhD |
| RPI | 6 | Psychology, Philosophy and Cognitive Sciences | MS |
| Rice | 3 | Psychology | MA, PhD |
| San Jose | 8 | College of Graduate Studies | MS |
| SUNY Buffalo | 3 | Industrial Engineering | MS, PhD |
| Texas A&M | 5 | Nuclear Engineering | MS, PhD |
| Texas Tech | 3 | Industrial Engineering | MS, PhD |
| Texas Tech | 7 | Psychology | MA, PhD |
| Tufts | 2 | Mechanical Engineering | MS, PhD |
| Tufts | 3 | Psychology | MS, PhD |
| U. of Alabama-Huntsville | 1 | Psychology | MA |
| U. of Alabama-Tuscaloosa | 2 | Industrial Engineering | MS |
| U. of Calgary | 9 | Psychology | MS, PhD |
| U. California-Berkeley | 7 | School of Health/Bioengineering | MS, PhD |
| U. California-Berkeley | 7 | Vision Science | MS, PhD |
| U. Central Florida | 9 | Psychology | PhD |
| U. Cincinnati | 5 | Mechanical, Industrial and Nuclear Engineering | MS, PhD |
| U. Cincinnati | 24 | Psychology | MA, PhD |
| U. Conn. | 6 | Psychology | MA, PhD |
| U. Dayton | 7 | Psychology | MA |
| U. Houston | 5 | Industrial Engineering | MS, PhD |
| U. Idaho | 3 | Psychology | MS |
| U. Illinois-Urbana-Champaign | 8 | Psychology and Mechanical and Industrial Engineering | MA/MS, PhD |
| U. Iowa | 10 | Industrial Engineering | MS, PhD |
| U. Kentucky | 4 | Psychology | PhD |
| U. Louisville | 5 | Industrial Engineering | ME, PhD |
| U. Mass-Amherst | 13 | Mechanical and Industrial Engineering, Exercise Science and Psychology | MS, PhD |
| U. Mass-Lowell | 3 | Work Environment | MS, ScD |
| U. Miami | 13 | Industrial Engineering | MS, PhD |
| U. Michigan | 7 | Industrial and Operations Engineering | MS, PhD |
| U. Minnesota | 15 | Graduate Studies | MS, PhD |
| U. Nebraska-Lincoln | 4 | Industrial and Management Systems Engineering | MS, PhD |
| U. Oklahoma | 2 | Industrial Engineering | MS, PhD |
| U. South Dakota | 7 | Psychology | MA, PhD |
| U. Texas-Arlington | 8 | Industrial Engineering | MS, PhD |
| U. Toronto | 4 | Mechanical and Industrial Engineering | MEng, MS |
| U. Utah | 6 | Mechanical Engineering | MS, PhD |
| U. Virginia | 3 | Systems Engineering | ME, MS, PhD |

**TABLE 5.5 (*Continued*)**

| Program | Number of Faculty | Department | Degrees Offered |
|---|---|---|---|
| U. Waterloo | 6 | Kinesiology | MS, PhD |
| U. Waterloo | 4 | Systems Design Engineering | MS, PhD |
| U. Wisconsin-Madison | 9 | Industrial Engineering | MS, PhD |
| VPI & SU | 8 | Industrial and Systems Engineering | MS, PhD |
| WV U. | 7 | Industrial and Management Systems Engineering | MS, PhD |
| Wichita State | 4 | Industrial and Manufacturing Engineering | MS, PhD |
| Wichita State | 8 | Psychology | MA, PhD |
| Wright State | 6 | Biomedical, Industrial and Human Factors Engineering | MS, PhD |
| Wright State | 11 | Psychology | MS, PhD |
| York U. | 3 | Psychology | MA, PhD |

*Source:* Reprinted with permission from *Directory of Human Factors/Ergonomics Graduate Programs in the United States and Canada.* 

Programs that are offered in such departments as environmental design (e.g., Cornell) tend to concentrate on the human–environment interface. Many psychology programs tend not to focus on the larger system or holistic issues primarily because they have a sister industrial/organizational (I/O) psychology program that focuses on organizational issues. An example would be Clemson. There are still several HF/E programs that are offered in both psychology and engineering departments at the same institution. Examples include Clemson and North Carolina State. Many of these campuses do coordinate and collaborate with respect to their offerings. Programs that offer human factors or ergonomics as a concentration within industrial engineering tend to promote a systems approach. The analogous situation for psychology departments is to promote linkage with I/O programs or departments.

Many departments have integrated appropriate faculty from departments other than the primary degree-granting department. In some cases, as in the case of Iowa, these faculty members are referred to as "affiliated faculty." In other cases, the descriptions do not differentiate between affiliated and regular faculty, but the listing includes both. Still others operate by integrating several departments. An example would be the 15 faculty listed for the University of Minnesota who span eight different departments. Care therefore should be taken when evaluating the number of faculty per program. Indeed, it is difficult to evaluate the extent to which an institution has made a fixed-cost commitment to ergonomics.

It is also somewhat difficult to evaluate the human factors offerings per university by evaluating individual programs. In several cases (e.g., Wichita State), programs exist in multiple departments. Most typically, this involves a program in both psychology and engineering. At Wichita State, for example, psychology lists eight faculty positions and industrial and manufacturing lists four faculty members. Together, 12 faculty participants are quite respectable for an HF/E program at a university.

### 5.3.2 Human–System Interface Technology

Scientific disciplines can most readily and distinctly be defined by the nature of their unique technology (Hendrick and Kleiner, 2001). Based on its survey of HF/E internationally, the Strategic Planning Committee of the HFES identified the unique technology of HF/E as human–system interface technology. Included are the interfaces between the people portion of systems and the other sociotechnical system components. This includes jobs, hardware, software, internal and external environments, and work system structures and processes. As a science, HF/E involves the study of human performance capabilities, limitations, and other characteristics. These data then are used to develop human–system interface technology. The technology takes the form of interface design principles, guidelines, specifications, methods, and tools. As a practice, HF/E professionals apply the technology to the design, analysis, test and evaluation, standardization, and control of systems. The general objective of the discipline is to improve the human condition, including health, safety, comfort, productivity, and quality of life (HFES, 2000b).

Human–system interface technology has at least five clearly identifiable subparts, each with a related design focus (Hendrick and Kleiner, 2001, 2002) as follows:

- *human–machine* interface technology or hardware ergonomics,
- *human–environment* interface technology or environmental ergonomics,
- *human–software* interface technology or cognitive ergonomics,
- *human–job* interface technology or work design ergonomics, and
- *human–organization* interface technology or macroergonomics.

The first four of these technologies are focused on the individual or, at best, the subsystem level. Thus, they constitute the technologies of what often is referred to in the literature as *microergonomics*. The fifth is focused on the overall work system level and, accordingly, is the primary technology of *macroergonomics*.

### 5.3.3 Content Trends

Imada and Kleiner (2000) identified emerging educational topics for educating human factors engineering and ergonomics professionals (Table 5.6). These might have implications for the new HSI professional. Traditionally, a broad educational program will provide coverage of such basic areas as human–machine systems design, human sensation and perception, cognitive processing and decision making, psychomotor response, display and control layout/design, applied anthropometry and biomechanics, materials handling, environmental effects, control of dynamic systems, research methodology and experimental design, and workplace design. The demands imposed by modern systems, expansion of information technology, increases in government and other regulations concerning ergonomics and safety, increased environmental concerns, increased attention to special user populations, and constant pressure of litigation constitute some examples of why additional knowledge and skill might be needed (as illustrated in Table 5.6).

**TABLE 5.6 Content Trends for Educational Programs**

| |
|---|
| Information and communications systems design |
| Virtual reality |
| Usability testing and evaluation |
| Green design |
| Ergonomics standards and regulations |
| Forensics |
| Professional certification |
| Macroergonomics |
| Security |

With increases in digitization comes the need to accelerate human–computer interaction work. Whatever can be automated is being automated without particular regard to people. The impact of digitization will affect the HSI practitioner as well. One particular area of this is virtual reality (VR). Virtual reality and augmented reality (AR) systems are being used for training, product test and evaluation, even medical applications such as surgery. Again, human interface issues warrant attention to avoid technology-driven systems. One area that helps address these concerns is usability testing and evaluation. Training in the latest methods of usability testing is therefore helpful to the HSI specialist.

Environmentally conscious design is increasingly a concern for ergonomists. Thus, the HSI professional should be aware of how system design impacts the larger ecological balance. Related to the environment, but spanning well beyond its borders, are ergonomics standards and regulations. While these are changing in a dynamic political environment, the HSI professional needs to understand the relevant standards and regulations. Related to these issues is the frequency with which ergonomists find themselves involved with legal matters. Students need basic coursework information in tort law, professional ethics, forensic responsibilities of the ergonomics expert, and forensic research and data collection techniques. Professional certification is also an important goal for the HSI specialist to consider. Certification as offered in North America by the Board of Certified Professional Ergonomists (BCPE) can provide credibility to the HSI expert. Ergonomics curricula should include discussions of the need for professional certification and how to achieve it and encourage graduates to seek this goal as practicing ergonomists.

Ergonomists have for some time realized it was possible to do a good job of ergonomically designing a system's components, modules, and subsystems yet not attain relevant systems effectiveness goals because of inattention to the macroergonomic design of the overall work system (Hendrick 1984, 1986a, b). Macroergonomics has been shown to increase productivity, improve health and safety, and improve the competitiveness of organizations (Hendrick and Kleiner, 2001). In the laboratory, the science of macroergonomics continues to develop, with new knowledge being created about personnel, technological, and environmental work system factors. Several formal academic programs and courses exist, and given the compatibility with HSI content and philosophy, the HSI student is encouraged to gain as much exposure to macroergonomics as possible.

Finally, September 11, 2001, created a new trend—security. It is predicted that increasingly HSI will be seen as an invaluable approach to integrating the various system components necessary to create and maintain a level of safety and security in many types of critical systems.

### 5.3.4 What Are Gaps between Needs and What Is Available?

Most programs in human factors and ergonomics do not require and many do not offer the array of courses recommended for HSI specialization. Many programs, either intentionally or informally, specialize in a given subarea of ergonomics. Institutions are only beginning to establish tracks for HSI. Generally, the HSI specialist will need to piece together courses from existing programs to create the appropriate degree of specialization.

As for HSI content areas, the core areas listed in Table 5.4 are well represented across programs, but only a few are both broad and detailed enough to provide extensive coverage. Some programs actually specialize in taking a systems approach or provide allied programs such as macroergonomics, but these seem to be the exception, not the rule. It appears, then, that students interested in HSI ought to consider both the breadth and depth of the academic programs before making a selection.

Based on the core competencies needed for HSI, seven broad content areas (Table 5.7) are recommended for HSI education. Using these content areas as a guide, the student has a checklist for programmatic content that can be drawn from available course. Ideally, the curriculum would also be sensitive to the emerging trends outlined in Section 5.3.3.

### 5.3.5 Recommended Courses for HSI Major

An obvious and important question is what technical content is needed to be a practitioner or user of the HSI approach? This section describes in more detail the seven content areas listed in Table 5.7.

***Human Factors Engineering*** Human factors engineering content refers primarily to cognitive or behavioral issues related to design for human capability and limitation in the interest of better design of tasks, jobs, or related tools and equipment. We limit our operational definition here to single user–machine interactions and primarily cognitive demands.

Designing for human interfaces essentially postures the designer as a human advocate. The design criterion has to do with finding ways to support the system user to improve performance and well-being. The theory is by supporting operators and maintainers, performance and well-being will be maximized and errors will be reduced. The two principal ways to support a person are to change the system or change the person. Changing the system essentially involves system redesign. Changing the person essentially involves selection and/or training the person. When errors or accidents occur within this human–centered philosophy, the system has broken down in some way. Therefore, it is fruitless and misguided to blame the human operator or categorize mishaps as "human error." In reality, the system has failed and a systems approach is required to rectify the situation. System intervention can focus on system components and/or the interfaces between the human users or operators and the system.

When automating, special attention is paid to allocating function between human and machine. While automation is desirable for a number of tasks (e.g., repetitive, hazardous, etc.), as long as the human is needed within the system, it is important to proactively decide his or her role. When the human's role becomes too passive, he or she might not be able to appropriately operate or intervene should the conditions warrant an increased role at some point in the system's life cycle. Therefore, attention should be paid to building and

**TABLE 5.7 Recommended Content Areas for HSI Education**

| |
|---|
| Human Factors Engineering |
| Psychological capability and limitation |
| Sensation and perception |
| Workplace analysis and design |
| Cognitive ergonomics |
| Control/display design |
| Ergonomics |
| Physiological capability and limitation |
| Lifting and handling |
| Biomechanics and anthropometry |
| Training |
| Learning |
| Training transfer |
| Instructional design |
| Training technology |
| Training evaluation |
| Training models |
| Personnel |
| Skills, knowledge, and abilities analysis (SKA) |
| Selection |
| SKA/training trade-offs |
| Performance measurement and evaluation |
| Motivation |
| Reward systems |
| Health and Safety |
| System reliability analysis |
| Human error analysis |
| System safety planning |
| Safety training |
| Environmental stressors evaluation |
| Psychological stressors evaluation |
| Designing for health and safety |
| Protective equipment and gear |
| Controlling workplace hazards |
| Product reliability and liability |
| Forensics |
| Macroergonomics and Systems |
| Sociotechnical system design |
| Work system analysis and design |
| Participatory ergonomics |
| Function allocation |
| Organizational design |
| Collaborative/group work systems |
| Group task analysis |
| Methodology |
| Experimental design |
| Survey research |
| Simulation |
| Mission, function, and task analysis |
| Data analysis |

maintaining an appropriate mental model during system operation. (See Chapter 13 for greater discussion of HFE content.)

***Ergonomics*** Relative to the term *human factors*, we use the term *ergonomics* to cover physiological demands. Ergonomics content refers to physiological issues related to design for human capability and limitation in the interest of better design of tasks, jobs, or related tools and equipment. Physical work is typically the beneficiary of ergonomics intervention or prevention. For example, a worker injured through lifting heavy loads could lead to an ergonomic analysis of the job to identify risk factors. This analysis in turn could lead to a redesign effort. As in human factors engineering, intervention is likely to focus on changing the operator or changing the system. Again, selection and training are the major methods for changing the person.

The lower back is one targeted area for risk and injury reduction, especially in work that requires lifting and/or twisting. Repetitive-motion injuries such as carpal tunnel syndrome characterize maladies of the hands and wrists. Slips and falls relate to dynamic movement issues having to do with gait, balance, floor surface interaction, etc. In addition to physical injury and accidents, physical fatigue and workload are areas of concern as well.

***Training*** Training as used here refers primarily, but not exclusively, to skills training. It is recognized that knowledge acquisition is a necessary part of most skills training, but we are not referring to higher education per se, which is broadly covered throughout this chapter. It is recognized that most skills training involves some knowledge acquisition as well. There are various types of training approaches, training schedules, and other logistical issues to consider. One special consideration is team training. Here, one must be cognizant of both individual and group developmental needs. An assumption cannot be made that a properly trained individual will necessarily make a good team member. In fact, this highlights the distinction we can make between technical skills and social skills training. Both are needed to operate within an organization, especially a team-based organization.

If anything, the emphasis on technological systems for training has increased over time. Augmented reality and VR systems, for example, represent extremely sophisticated technologies currently being used in training. For example, distance learning technologies and computer-based training are now the norm for training rather than the exception. Web-based courses also exist, but their results are unknown at this juncture. What is known is that purely Web-based courses are quite difficult to develop and complex to deliver. (See Chapters 11, 12 and 22 for more information on training issues and techniques.)

***Personnel*** In many educational institutions, "personnel" and "manpower" are combined, thus converging much of the content for both domains. This content includes skill, knowledge, and ability attributes of people. Motivational issues are also included. "Manpower," or "person-power," as it is often referred to today, includes person-power demands and costs of acquiring, operating, and maintaining a system or systems.

Determining and maintaining manpower requirements is a staple skill in project and human resource management. What is less straightforward is how to develop and maintain a proper level of motivation in an organization. Traditional motivational approaches include extrinsic reward systems such as pay for performance approaches. However, these behavioral approaches have been scrutinized. The theory is that behavioral reward systems only lead to temporary compliance at best and feelings of inadequacy and punishment at worst. An alternative philosophy is to compensate people well, get their focus off pay and other extrinsic rewards, and then let the work itself intrinsically motivate people. Anything less than intrinsic motivation will be transient and unworkable in the long term. (See Chapter 11 for greater detail on the HSI tools and techniques used for manpower and personnel systems analysis.)

***Health and Safety*** Health and safety tend to be combined in academic courses. System safety content relates to system reliability predictions and the separation of human error from true systemic issues. Again, within a human-centered philosophy, while the actual error might have come from a person, we look to the system for causes and solutions. Human-related issues include human reliability, human error, problem personnel, motivation, etc. This content includes the inevitability of system failure and thus includes content related to emergency management, along with risk and risk perception. A systems approach necessarily also takes into account the management aspects of safety and safety programs as well. Military and civilian regulations and standards are also important to the management of health and safety. Liability and legislation are deemed important topics as well. Safety processes and procedures are typically the focus of the technical subsystem whereas the roles of safety personnel are the focus of the personnel subsystem.

From the health perspective, emphasis is on environmental health hazard issues where categories such as human effects from acoustics, chemical substances, and temperature extremes are covered. Such items as protective garments and gear are included. This category is broad, covering other relevant topics such as product liability and forensics. Hazard control and management might include such topics as determining hazards, providing safeguards, fail-safe designs, failure minimization, monitoring, warnings, failure minimization, etc. (See Chapters 14 and 15 for detailed coverage of health and safety.)

***Macroergonomics*** Conceptually, macroergonomics may be defined as a top-down sociotechnical systems approach to the design of work systems and the carry-through of the overall work system design to the design of the human–job, human–machine, and human–software interfaces in creating a fully harmonized work system (Hendrick, 1997; Hendrick and Kleiner, 2001).

Macroergonomics covers multiuser and organizational scenarios. The central purpose of macroergonomics is to optimize work system design (Hendrick and Kleiner, 2001). This is what distinguishes it from both microergonomics and organizational psychology, making it a particularly valuable contribution to HSI. Macroergonomics is accomplished through a systematic consideration of the relevant sociotechnical system variables in the ergonomics analysis, design, implementation, evaluation, and control process. These variables relate to three basic, empirically identified sociotechnical system components: the technical (technological) subsystem, social (personnel) subsystem, and the external environment. A fourth sociotechnical system element, job (and task) design, falls within the purview of microergonomics but is influenced by and interacts with design of the work system. In reality, macroergonomic design of the work system helps determine many of the characteristics that should be designed into individual jobs for joint optimization of the total system. Work system design, thus, includes both the structure and related processes of the entire work system.

As well as a top-down process, macroergonomics includes bottom-up and middle-out analysis, design, and evaluation processes (Hendrick and Kleiner, 2001). Top-down, an overall work system structure, may be prescribed to match the organization's sociotechnical characteristics. Middle-out, an analysis of subsystems and work process, can be assessed both up and down the hierarchy from intermediate levels and changes made to ensure a harmonized work system design. Bottom-up most often involves identification of opportunities by employees and lower level supervisors that result from higher level work system structural or process design. A true macroergonomics intervention effort thus requires employee participation at all levels of the organization.

Technically, macroergonomics is a human–organization interface technology. The empirical science underlying this discipline is concerned with factors in the technological subsystem, personnel subsystem, external environment, and their interactions as they impact work system design. As a perspective, certain guiding principles assist the ergonomist. These include participation, flexibility, joint optimization, joint design, system harmonization, and continuous improvement of processes.

***Methodology*** The last category in Table 5.7 is concerned with basic methods and techniques. Each content area previously described also has specific methods and techniques, but a special emphasis is needed for integration tools and simulation techniques for the HSI practitioner. Methodology is unique for HSI in at least two ways. The first is integration methodology that cuts across the various domains, especially manpower, personnel, and training (MPT); the second is integration methodology, which helps make HSI issues quantitatively visible and capable of being introduced into the acquisition process at very early stages. Chapter 11 describes the state of the art for MPT integration methodology, and Chapter 13 describes the manner in which human factors engineering methods and tools help quantify HSI issues for early acquisition process decisions.

The primary special methodological content needs for the HSI practitioner are in simulation and modeling methods and tools that include the human component. Use of modeling techniques that have predictive quantifiable results for human performance variables is most helpful in identifying problems early in the design process, thus allowing a range of possible solutions for consideration that would be unlikely options if introduced later in the acquisition cycle.

It is important that courses on simulation and modeling emphasize methods of quantifying MPT parameters in simulation and modeling processes. For example, methods and tools should demonstrate ways that MPT trade-offs in resources and system performance can be applied to system requirements, design, development, and test and evaluation.

### 5.3.6 Two Hypothetical HSI Graduate-Degree Tracks

To illustrate how an existing academic program could take the aforementioned recommended course content and develop a graduate program focused on HSI, we developed hypothetical HSI graduate-degree tracks from two institutions that currently teach large portions of the needed HSI content. These are Virginia Tech and the Naval Postgraduate School (NPS). Many of the institutions listed in Table 5.5 could undoubtedly produce acceptable HSI tracks, but these two are most familiar to the authors and have had extensive review for feasibility by faculty at these institutions. At the time of this writing, one institution has already approved a two-year master's HSI track, and the other expects to have an approved track in the near future. Should other educational institutions decide to create graduate degrees for HSI, the information provided here should be useful models for programs to meet military and commercial educational needs.

***Virginia Tech*** Most of the degree tracks from Virginia Tech appear in its published graduate manual (http//www.ise.vt.edu). At Virginia Tech, the M.S. and Ph.D. degrees in the HF/E option emphasize both methodology and content areas. Using the master's program as an example, students are required to complete five core courses, a minimum of four elective courses, and six hours of thesis work. The core courses are Human Information Processing, Human Factors System Design I, Human Physical Capabilities, Human Factors Research Design I, and Integrated Systems Design. Students then can select preapproved "tracks" in order to specialize beyond the core. Currently, these tracks include Cognitive Ergonomics, Human–Computer Interaction, Macroergonomics, Methods, Occupational Biomechanics, Sensory and Perception, Safety, Telecommunications, Transportation, and Work-Related Musculoskeletal Disorders.

In consulting Table 5.7, the core master's-level courses cover the following domains: human factors engineering, ergonomics, and methodology. The four electives would then be selected to provide coverage of the remaining content areas as well as provide a single course that covers the breadth of HSI. The other electives then could be in the areas of training, safety, and macroergonomics. Table 5.8 illustrates a currently hypothetical but potential program of study for an HSI master's-degree specialization. As an HSI program, some modifications from the Human Factors Engineering and Ergonomics (HFEE) option would be recommended. As examples, some of the most prominent modifications would include the following:

- One of the core courses (Integrated Systems Design) might be replaced with Usability Engineering.

**TABLE 5.8 Hypothetical HSI Masters Program—Virginia Tech**

| Course Type | Course Title | Credit Hours |
|---|---|---|
| Core | Human Information Processing | 3 |
| Core | Human Factors System Design I | 3 |
| Core | Human Physical Capabilities | 3 |
| Core | Human Factors Research Design I | 4 |
| Core | Usability Engineering[a] | 3 |
| Research | Research Thesis | 6 |
| Elective | Human Systems Integration[b] or Macroergonomics | 3 |
| Elective | Simulation Modeling[c] or Modeling Processes in Operations Research[c] | 3 |
| Elective | Training Systems or Human Computer Systems | 3 |
| Elective | System Safety, Industrial Health and Safety,[d] Occupational Safety, or HSI in Security Operations[e] | 3 |
| Total hours | | 34 |

[a]Replace current core course.
[b]Create new course.
[c]Add HSI Modeling.
[d]Add Personnel Survivability.
[e]Create new course.

- A new course on HSI principles and methods should be developed and offered as an elective.
- The personnel survivability domain is not represented currently by any Virginia Tech courses. Some content on personnel survivability could be added to one of the safety or health courses.
- The operations research courses in simulation modeling and modeling processes in operations research should be modified to include HSI trade-off models.
- A new course on HSI and national security could be developed and offered as an elective.

Table 5.9 provides descriptions of several representative courses drawn from the 2001 graduate manual of Virginia Tech. Suggested additions to existing courses for HSI modeling and personnel survivability are provided in italics but are not current offerings. The strength of these courses would be the application to industrial settings that could apply to both military and nonmilitary systems and products.

***Naval Postgraduate School*** Most of the degree tracks at the NPS are shown in its general catalog. Degrees include master of arts (MA), master of science (MS), engineer, and doctor's degrees. The MS options are most relevant for an HSI track. Depending on the course of study, MS degrees range from one year (four quarters) to two years (eight quarters). The minimum for any MS thesis degree is 48 quarter hours for a one-year program. Numerous courses are currently taught at NPS that, properly organized, could provide the HSI academic requirements. These are primarily in the Operations Research Department and the Systems Management Department. Two academic groups—Systems Engineering/Integration and Modeling, Virtual Environments and Simulation (MOVES)—also have courses of study that would be relevant to an HSI degree track. Table 5.10 provides a hypothetical one-year HSI master's program at NPS[4] that would draw from the Operations Service and Operations Analysis courses in the Operations Research Department, courses in the System Management Department, and courses from the two academic groups. At least six of the courses would be core requirements, with six others drawn from the electives. A new course on HSI principles and methods should be developed as part of the track.

Table 5.11 provides descriptions of several representative courses drawn from the 2001 NPS catalog. With the exception of a new survey course in HSI, the NPS has sufficient current courses that could be applied to an HSI track without major modifications. Its strength for DoD students would be the direct military applications in all of the coursework.

## 5.4 TEXTBOOKS

Another perspective on the information available for training is in terms of source books and texts. Muckler and Seven (1990) reviewed seven human factors textbooks for their coverage of the six MANPRINT domains. As stated in Muckler and Seven (1990), the comparisons reviewed here are presented to offer an alternative view of national education and training resources and are in no way intended to serve as a critical comparison of these texts.

**TABLE 5.9 HSI Course Descriptions—Virginia Tech**

| Course Title | Course Description |
|---|---|
| Human Information Processing | An examination of present human engineering design criteria, principles, and practices to achieve mission success through integration of the human into the system, subsystem, equipment, and facility design in order to achieve effectiveness, simplicity, reliability, and safety of system operation, training, and maintenance. |
| Human Factors System Design I | Human factors input into manned-system design, development, testing, and evaluation. Emphasis on the systems approach to human–machine interfacing, with discussion and application of specific methodologies and analytical techniques. Display and control design and selection fundamentals with engineering modeling of manual control systems. |
| Human Physical Capabilities | An examination of human physical attributes in human–machine systems with an emphasis on models of anthropometry and biomechanics, intero- and exteroceptors and on the work environment; force fields (transitory and sustained), sound, light, climate. |
| Human Factors Research Design I | Procedures for conducting human factors experiments, including research methodology, multifactor design alternatives, field research, designs for reducing data collection, empirical model building, and sequential research procedures. |
| Usability Engineering | An overview of the development process of interactive software interfaces, including iterative life-cycle management, systems analysis, design representation techniques, prototyping, and formative user-based evaluation. Integrative and cross-disciplinary approach with main emphasis on usability methods and user interaction development process. |
| Macroergonomics | The optimization of work system design through consideration of relevant personnel, technological, and environmental variables and their interactions. Emphasis is on the theoretical background, research methods, analysis, design, development and application of work system, and relationship between macro- and microergonomics. |
| Simulation Modeling | Introduction to discrete-event digital simulation, including development of simulation models, random-number and random-variable generation, model validation and testing, analysis of model output, and an overview of simulation languages. *Emphasizes the use of human systems simulation modeling in decision making through a series of projects involving manpower, personnel, and training trade-off problems.* |
| Modeling Processes in Operations Research | Introduction to and demonstrations of the phases and activities involved in the development, validation, and use of models in the solution of management decision problems. *Emphasizes the methods of quantifying manpower, personnel, and training parameters in modeling processes.* |
| Training Systems Design | A systems approach to the design and development of training with an emphasis on techniques to conduct training needs analysis, a survey of training methodology with an emphasis on computer-assisted techniques and training simulators, and procedures to evaluate training effectiveness. |

**TABLE 5.9 (*Continued*)**

| Course Title | Course Description |
|---|---|
| Human–Computer Systems | A survey of human factors procedures used in the design of computer-based systems. Consideration is given to the iterative interface design process, hardware interface design, software interface design, and workplace design. |
| Industrial Health and Safety | Addresses the identification, analysis, and control of biological, chemical, radiation, and fire hazards in industrial settings. *Instruction on personnel survivability would be included for HSI students.* |
| System Safety Analysis | Review of the systems safety analytical techniques and documentation requirements to protect against product liability and to provide proper design of equipment and systems. |

The texts reviewed by Muckler and Seven (1990) were Huchingson (1981), Oborne (1982), Kantowitz and Sorkin (1983), Wickens (1984), Salvendy (1987), Sanders and McCormick (1987), and Adams (1989). Salvendy's (1987) *Handbook of Human Factors* was a major source book for human factors at the time of the Muckler and Seven review, and with 68 technical chapters and nearly 2000 pages it is still widely regarded as the most

**TABLE 5.10 Hypothetical HSI Master's Curricula—Naval Postgraduate School**

| Course Type | Course Title | Credit Hours |
|---|---|---|
| Operations service (E) | Human Factors Engineering | 4QHR |
| | Manpower Requirements Determination | 4QHR |
| | Man–Machine Interaction | 4QHR |
| | Human Factors in Information Warfare | 4QHR |
| Operations service (R) | Manpower and Personnel Models | 4QHR |
| Operations analysis (R) | Human Factors in System Design I | 4QHR |
| | Human Factors in System Design II | 4QHR |
| Operations analysis (E) | Human Performance Evaluation | 4QHR |
| | Skilled Operator Performance | 4QHR |
| | Special Topics in Human Performance | 4QHR |
| | Test and Evaluation | 4QHR |
| | Cost Estimation | 4QHR |
| Operations analysis (R) | Human Systems Integration | 4QHR |
| Systems management (E) | Principles of Acquisition and Program Management | 4QHR |
| | Training Foundations | 4QHR |
| | Personnel Testing and Selection | 4QHR |
| Systems management (R) | Job Analysis and Personnel Selection | 4QHR |
| Systems engineering integration (R) | Systems Engineering I | 4QHR |
| MOVES (E) | Human Factors of Virtual Environments | 4QHR |
| | Training in Virtual Environments | 4QHR |
| Thesis (R) | — | 8QHR |

*Abbreviations:* R = required; E = elective; QHR = quarter hours; MOVES = modeling, virtual environments and simulation academic group; MS degree = 48 QHR plus thesis.

**TABLE 5.11 HSI Course Descriptions—Naval Postgraduate School**

| Course Title | Course Description |
|---|---|
| Human Factors Engineering | Introduction to human factors engineering. Designed to give an appreciation of human capacities and limitations and how these can affect optimum design of human–machine systems. Emphasis on integration of human factors into the system development cycle. |
| Manpower Requirements Determination | Utilize the tools of industrial engineering to determine quantity and quality of manpower required in military systems. Techniques include motion-and-time study, work sampling, time standards, work design and layout, materials handling, procedure review, and process design. |
| Manpower and Personnel Models | Covers the major types of manpower and personnel models for estimating the effects of policy changes on the personnel system. Topics include longitudinal and cross-sectional models, optimization models, and data requirements and validation. |
| Human Factors in Systems Design I | Provide foundation to understand, explain, and predict human behavior. Students develop techniques in understanding how to ask a question and methodological procedures for collecting data and interpreting results. Laboratory exercises and field surveys to determine and demonstrate human strengths and limitations in the workplace. |
| Human Factors in Systems Design II | Includes selected topics on human engineering and psychophysics in measuring human performance as part of overall systems performance. Investigate sensory perceptual deficits associated with simulators, virtual environments, and other human–machine devices. |
| Test and Evaluation | Relates the theory and techniques of operations research to the problems associated with system test and evaluation. Examples of exercise design, reconstruction, and analysis are examined. |
| Cost Estimation | Advanced study in the methods and practice of systems analysis with emphasis on cost analysis; cost models and methods for total program structures and single projects; relationship of effectiveness models and measures to cost analysis. |
| Principles of Acquisition and Program Management | Introduce fundamental principles of DoD systems acquisition and program management. Covers planning, programming, and budgeting processes; acquisition strategies; contractual decisions; and program management. Key functional areas explored are research and development, testing and evaluation, contracting, funding and budgeting, logistics support, systems engineering, and legal issues. |
| Personnel Testing and Selection | Study of methods for evaluating and predicting training and work performance in organizations. Special emphasis on testing concepts and models for Armed Services Vocational Aptitude Battery, equal employment opportunity, and selection decisions based on cost–benefit analysis. |
| Job Analysis and Personnel Selection | Study of job analysis and its use in determining training requirements. Consideration of instructional systems development and training pipeline management. Attention to cost–benefit issues involving training in regard to selection, equipment design, changing job requirements, and career development. |

**TABLE 5.11** ***(Continued)***

| Course Title | Course Description |
|---|---|
| Systems Engineering/ Integration | Study the systems engineering process and studies to include knowledge of systems design, development, and deployment; technical and economic trade-offs; human-in-the-loop issues; project management; systems acquisition; and the planning, programming, and budgeting system (PPBS). |
| Human Factors of Virtual Environments | Study issues in virtual environments on human performance, perception, cognition, multimodal interfaces, locomotion, wayfinding, object selection and manipulation, visualization, simulator sickness, and individual differences. |

comprehensive summary of the field (see revised edition: Salvandy, 1997). Muckler and Seven briefly summarized the other six books reviewed for their relative coverage of the six MANPRINT domains. They concluded that Huchingson (1981) provides a "highly applied human factors engineering point of view" showing strong coverage of human factors engineering and health hazards, but purposely excludes "selection and training of personnel in favor of design-oriented material" (Muckler and Seven, 1990, p. 536). Wickens (1984), on the other hand, provides a somewhat basic engineering psychology point of view with only a few of the domain content areas covered to any depth. Each of the other texts reviewed by Muckler and Seven fell somewhere between these two views with a fairly good presentation of HF/E content and scattered coverage of the other HSI domains. The principal findings of Muckler and Seven were as follows:

1. None of the texts showed "any apparent interest in the manpower area."
2. The texts covered "personnel variables only insofar" as they interacted "with human engineering design problems."
3. The texts showed concern with training devices, but so far as training variables in general, they were covered only if they interacted with system design variables.
4. Collectively, the texts concentrated on "three major areas: human factors engineering, system safety, and health hazards."

Our review for this chapter restructured the domains and content areas examined by Muckler and Seven, reexamined their texts, and added several newer texts. When we look at all the texts today, the first finding is still true—the human factors texts do not generally show interest in the manpower area. We have eliminated manpower, therefore, in our textbook examination. There does, however, seem to be some coverage of the personnel and training domains in several of the texts as well as the major areas of human factors engineering, system safety, and health hazards.

Table 5.12 presents our combined review of those texts examined by Muckler and Seven and 10 more recent textbooks. These newer texts include Kroemer and Grandjean (1997), Pulat (1997), Tayyari and Smith (1997), Wickens et al. (1998), Hollands and Wickens (2000), Niebel and Freivalds (1999), Konz and Johnson (2000), Hendrick and Kleiner (2001), Kroemer et al. (2001), and Hendrick and Kleiner (2002).

The text by Kroemer and Grandjean (1997) focused on the physical side of work but does contain some interesting related topics such as control and display design. Pulat

TABLE 5.12 **Textbooks for HSI Domains**

| Domains | H 1981 | O 1982 | K/S 1983 | W 1984 | S 1987 | S/M 1987 | A 1989 | K/G 1997 | P 1997 | T/S 1997 | W/G/L 1998 | N/F 1999 | H/W 2000 | K/J 2000 | H/K 2001 | K/K/K 2001 | H/K 2002 |
|---|---|---|---|---|---|---|---|---|---|---|---|---|---|---|---|---|---|
| Human factors engineering | 0 | 0 | 0 | 0 | 0 | 0 | 0 | 0 | 0 | 0 | 0 | 0 | 0 | 0 | 0 | 0 | |
| Psychological capability and limitation | × | × | × | × | × | × | × | | × | | × | | × | | | × | |
| Sensation and perception | | | | | × | | | | × | × | × | | × | | | × | |
| Workplace analysis and design | × | × | × | | × | × | × | × | × | × | × | × | | × | × | × | × |
| Cognitive ergonomics | | | | | × | | | × | × | | × | | × | | | × | |
| Control/display design | × | × | × | × | × | × | × | × | × | × | × | | × | × | | × | |
| Ergonomics | 0 | 0 | 0 | 0 | 0 | 0 | 0 | 0 | 0 | 0 | 0 | | | 0 | | 0 | |
| Physiological capability and limitation | × | × | × | × | × | × | × | × | × | × | × | | | × | | × | |
| Lifting and handling | | | | | | | | × | × | × | × | | | × | | × | |
| Biomechanics and anthropometry | | | | | | | | × | × | × | × | | | × | | × | |
| Training | 0 | | | 0 | 0 | | 0 | | 0 | | 0 | 0 | 0 | | | | |
| Learning | | | | | × | | × | | | | × | | × | | | | × |
| Training transfer | | | | × | × | | × | | | | × | | × | | | | × |
| Instructional design | | | | × | × | | × | | | | × | | | | | | × |
| Training technology | × | | | | × | | × | | | | × | | | | | | |
| Training evaluation | | | | | × | | × | | × | | × | × | | | | | × |
| Training models | | | | | × | | × | | | | × | | | | | | |
| Personnel | 0 | 0 | 0 | 0 | 0 | 0 | 0 | | 0 | | 0 | 0 | | | 0 | | |
| Skills, knowledge, and abilities (SKA) analysis | | | × | × | × | | × | | | | | × | | | | | |
| Selection | | | | | × | | | | × | | × | | | | | | |
| SKA analysis/training trade-offs | × | | | | | | | | | | | | | | | | |
| Performance measurement and evaluation | × | × | × | × | × | × | × | | × | | | × | | | × | | × |
| Motivation | | | | | × | | | | | | | | | | × | | × |
| Reward systems | | | | | × | | | | | | | | | | × | | × |
| Health and safety | 0 | 0 | 0 | 0 | 0 | 0 | 0 | | 0 | 0 | 0 | | 0 | 0 | 0 | | |
| System reliability analysis | | | | | × | | × | | × | | × | | × | | | | × |
| Human error analysis | × | × | × | × | × | × | × | | × | | × | | × | × | | | |
| System safety planning | | | | | × | × | | | | | | | | | | | × |

*(continued)*

**TABLE 5.12 (*Continued*)**

| Domains | H 1981 | O 1982 | K/S 1983 | W 1984 | S 1987 | S/M 1987 | A 1989 | K/G 1997 | P 1997 | T/S 1997 | W/G/L 1998 | N/F 1999 | H/W 2000 | K/J 2000 | H/K 2001 | K/K/K 2001 | H/K 2002 |
|---|---|---|---|---|---|---|---|---|---|---|---|---|---|---|---|---|---|
| Safety training | × | | | | × | | | | | | | | | | | | × |
| Environmental stressors evaluation | × | × | × | | × | × | | | | | × | | | | | | × |
| Psychological stressors evaluation | × | × | × | × | × | | | | | | | | × | | | | |
| Designing for health and safety | × | × | | | × | | | | | | × | | | × | × | | × |
| Protective equipment and gear | × | | | | × | | | | | | × | | | | | | |
| Controlling workplace hazards | × | | × | | × | | × | | | × | | | | | | | × |
| Product reliability and liability | × | | × | | × | | × | | | | × | | | | | | |
| Forensics | × | | × | | × | × | | | | | | | | | | | |
| Macroergonomics | 0 | 0 | 0 | 0 | 0 | 0 | 0 | | | | | | 0 | | 0 | 0 | 0 |
| Sociotechnical system design | | | | | × | × | × | | | | | | | | × | | × |
| Work system analysis and design | | | | | × | | | | | | | | | | × | | × |
| Participatory ergonomics | × | × | × | × | × | × | × | | | | | | | | × | | × |
| Function allocation | × | | | | × | × | | | | | | | × | | × | | × |
| Organizational design | | | | | × | | | | | | | | | | × | | × |
| Collaborative/group work systems | | | | | | | | | | | | | | | × | × | × |
| Group task analysis | | | | | | | | | | | | | | | × | | × |
| Methodology | 0 | | | | 0 | | | | 0 | | 0 | 0 | | | 0 | | 0 |
| Experimental design | × | | | | × | | | | | | | | | | × | | × |
| Survey research | | | | | | | | | | | | | | | × | | × |
| Simulation | × | | | | × | | | | | | | | | | | | × |
| Mission, function, and task analysis | × | | | | × | | | | | | × | | | | × | | × |
| Data analysis | | | | | | | | | × | | | × | | | × | | |

*Abbreviations:* 0 = some treatment given to major area; × = in-depth treatment given to subtopic identified. Column headings: A = Adams; H = Huchingson; H/K = Hendrick–Kleiner; H/W = Hollands–Wickens; K/G = Kroemer–Grandjean; K/J = Konz–Johnson; K/K/K = Kroemer–Kroemer–Kroemer; K/S = Kantowitz–Sorkin; N/F = Niebel–Freivalds; O = Oborne; P = Pulat; S = Salvendy; S/M = Sanders–McCormick; T/S = Tayyari–Smith; W = Wickens; W/G/L = Wickens–Gordon–Liu.

(1997) offers a wide variety of topics. Like the other texts reviewed, it has excellent resource information in the physical ergonomics area. However, this book also has very good cognitive information and other related topics. As is typically the case, little information is provided relative to macroergonomic or HSI issues.

The Tayyari and Smith (1997) text is heavily oriented towards physical ergonomics. However, the text also provides excellent treatment of some cognitive ergonomics issues such as design of controls and displays. It also uniquely has a section on the management of ergonomics programs, a topic usually associated with macroergonomics. Wickens et al. (1998) present a very broad, excellent introductory human factors perspective that covers several of the key informational areas important to establish the HSI knowledge base. The Hollands and Wickens (2000) book bears the same title as that of Wickens (1984) (*Engineering, Psychology and Human Performance*.) It is an excellent text for the domain of human cognitive performance and error and related subjects. While it covers manual control quite well, it does not seek to cover the more traditional topics in physical ergonomics or cover the more recent domains such as macroergonomics. Niebel and Freivalds (1999), as their title (*Methods, Standards and Work Design*) implies, focuses upon work design from a standards and methods perspective. While it does not directly address the more traditional topics associated with the HF/E aspects of HSI, the text does provide good information with respect to methods, data collection, and analysis. Konz and Johnson (2000) present another work design text that offers some interesting additional topics such as "cumulative trauma" and "managing change." Hendrick and Kleiner (2001) focus almost exclusively on macroergonomics in the first treatment of the subject in a textbook. Kroemer et al. (2001) offer a broad treatment of industrial ergonomics. Both Kroemer et al. and Konz and Johnson provide at least a brief treatment of macroergonomics. Finally, Hendrick and Kleiner (2002) provide an in-depth coverage of several topics within the macroergonomics domain but again do not directly address many of the HSI specialized topics.

## 5.5 HSI TRAINING COURSES

Academic institutions cannot provide all HSI specialist course needs, especially for short, non-degree-focused curricula. Much of the training needed by the HSI professional is in areas not HSI specific. In particular, these include non-degree courses to improve KSAs in acquisition management, systems engineering, logistics engineering and management, operations research, and safety engineering. The government, especially the DoD, provides a number of training opportunities for the HSI professional to acquire KSAs in these well-established fields. But in addition, specialized training is needed for HSI. This section summarizes the questions and findings of a recent training needs analysis conducted for HSI training (Booher, 2001).

***Who Needs to Be Taught HSI?*** Within the military, the target audience for HSI is the same general audience as has been in existence since the inception of the army's MANPRINT program. This audience is frequently divided into four major categories:

1. Organizational leaders (e.g., senior executive service (SES), generals/admirals, political appointees),

2. MANPRINT and HSI practitioners (or action officers),
3. MANPRINT and HSI managers, and
4. Non-MANPRINT and HSI acquisition process participants.

The major change needed for new HSI courses is to expand the target audience from being primarily army personnel to include the other services and government agencies that procure, operate, and/or regulate systems and equipment. Human systems integration is now part of the DoD requirements in the DoD 5000 series under Systems Engineering. All DoD government and contractor personnel who are affected by this regulation need some education and training in HSI.

***What Needs to Be Taught?*** The specific content for each course that should be developed to meet the target audience needs is beyond the scope of this chapter. However, models for courses both in the government and at academic institutions do exist. The best military example of coursework is that taught by the Army Logistics Management College (ALMC) for MANPRINT. This coursework provides short modules to address target audiences 2, 3, and 4 above. Numerous short courses are available for some of the domains of HSI (as human factors, safety), but no training courses outside the army are currently available. As discussed in the previous section, numerous educational academic institutions have coursework that provides various domain expertise, but institutions are only beginning to consider coursework in HSI specifically. It is assumed that the other services will need as a minimum the type of coursework currently available to the army.

***What Coursework on HSI Currently Exists?*** Since so little training coursework exists beyond that provided by army MANPRINT, we can for all practical purposes be confident that what exists for MANPRINT is all that is currently available for HSI training. The primary existing formal army coursework is provided by the ALMC. This comprises

1. an eight-day MANPRINT action officer's course;
2. three- to five-day MANPRINT applications course—tailored and derived from the action officer's course;
3. numerous two-hour modules for about six other army courses; and
4. CD ROM on MANPRINT for individual instruction.

In addition to ALMC coursework, materials for a short course on MPT is available from the Army Research Laboratory's Human and Research Directorate (ARL-HRED); an e-learning course on HSI is being developed by the U.S. Air Force; the U.S. Navy offers a one-day introduction to HSI[5]; and the Federal Aviation Administration (FAA) has developed a short course on human factors for acquisition. Also, this handbook has been designed to help fill gaps in HSI principles and methods training coursework.

***What Training Coursework Is Needed?*** The two principal needs for new training coursework identified by Booher (2001) are

1. expanding the current ALMC coursework to cover other agencies and
2. developing new specialized short courses for HSI practitioners—focused on known methods and principles. Table 5.13 shows a five-day HSI course curriculum that could be developed and taught by an institution such as ALMC to meet this need.

## 5.6 HSI CAREERS

As mentioned in the introduction to this chapter, individuals need to see the career potential of the HSI field if they are to become sufficiently motivated to develop the skills required. There are a number of actions the federal government would need to take in order to establish a career field for an HSI workforce. Those addressed in the HSI national workforce study (U.S. Army, 1993) include the following:

- HSI professional workforce vision,
- HSI personnel job series and skill code, and
- HSI certification/licensing.

### 5.6.1 HSI Professional Workforce Vision

As discussed above, the demand for HSI practitioners and the development of qualification programs should progress together. In order for this to happen, there needs to be an overall

**TABLE 5.13 HSI Applications Course**

| Curriculum | Course Hours |
|---|---|
| Day 1 | |
| HSI overview: HSI concept and model, 10 principles of HSI, case examples | 2 |
| Systems engineering overview: systems engineering framework; systems trade-off analyses | 2 |
| Systems acquisition process and HSI interfaces: requirements; testing and evaluation; performance measures; reporting documents | 2 |
| Day 2 | |
| Manpower, personnel, and training (MPT) domains: definitions and concepts; methods and tools; MPT trade-off analyses | 3 |
| MPT group exercises/case studies/demos | 3 |
| Day 3 | |
| Human factors engineering (HFE) domain: definitions and concepts; HFE methods and tools; program planning and execution | 3 |
| HFE group exercises/case studies/demos | 3 |
| Day 4 | |
| Health hazards domain: types of hazards; health hazards analysis | 2 |
| System safety: accident analysis; risk assessment | 2 |
| Personnel survivability: parameter assessment list (PAL); survivability case examples | 2 |
| Day 5 | |
| System integration domain tradeoffs exercise | 2 |
| Course Review | 2 |
| Total: | 28 |

national vision and implementation of a career path for HSI professionals. We believe that the vision and initial implementation must come from the federal government. It is encouraging that a practical approach to a career path has already been provided in the studies mentioned above. An overview of the concept for an HSI government career path is illustrated in Figure 5.1, which shows a hierarchy of HSI government jobs progressing from junior to midmanagement positions to senior manager, HSI integrator positions.

Personnel could come into HSI jobs from any of the job series indicated on the left side of the figure. These job series currently relate to various individual HSI domains (MPT, human factors engineering, health and safety) and related engineering and analysis fields. Entry could be at any level in the hierarchy. Minimum entry requirements for each level are shown in Table 5.14. Individuals would need specialized experience and education in at least one of the HSI domains. Entry could be from any federal agency or from outside the federal government, so long as minimum qualifications were met.

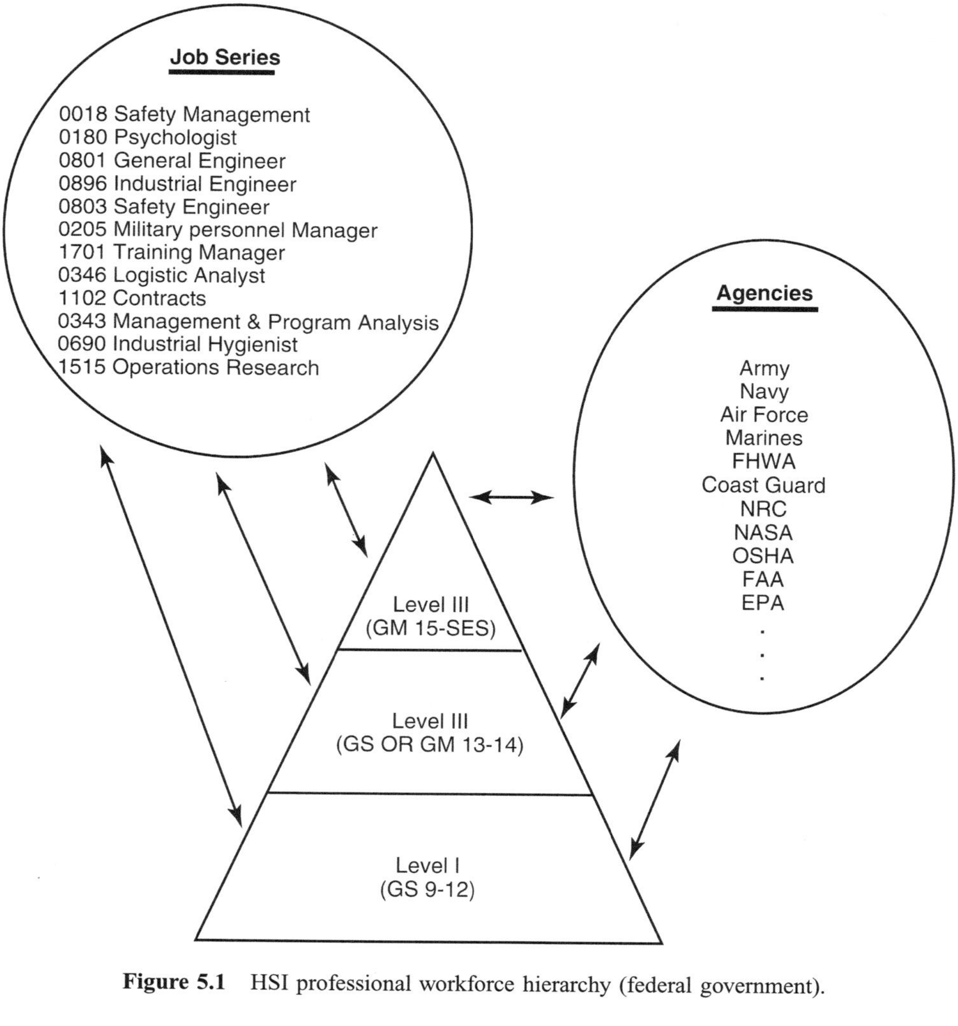


**Figure 5.1** HSI professional workforce hierarchy (federal government).

**TABLE 5.14 HSI Job Entry Requirements**

| Level | Grade | Experience | Education | Training |
|---|---|---|---|---|
| I | Junior to mid-technical, GS 9-12 | M—one domain, experience 1 yr<br>D—acquisition, experience 1 yr | D—baccalaureate degree | M—HSI basic course<br>M—certification<br>D—interdisciplinary courses<br>D—acquisition management courses |
| II | Mid to senior technical or mid-management, GS or GM 13/14 | M—HSI, experience 2 yr<br>D—acquisition or applied engineering, experience 2+ yr | D—master's degree | M—HSI advanced course<br>M—analytical techniques course<br>M—certification<br>D—interdisciplinary training<br>D—acquisition training<br>D—program manager course |
| III | Senior management, GM 15 and above | M—HSI, experience 4 yr, 2 of which in mid-management<br>M—acquisition, experience 2 yr | D—master's degree or doctoral | M—certification<br>D—executive training seminars |

*Note:* M = mandatory; D = desired.

Note that entry at each level requires a combination of experience, education, and training requirements. Some of the requirements are mandatory, whereas others are desired. Some trade-offs among the requirements would be expected. For example, an individual seeking to meet entry Level I requirements might have a bachelor's degree in one of the domains but no experience. It is likely the education (which is desired) could be substituted for the mandatory one year of experience for entry at the lowest level (GS-9).

Note also that HSI certification is considered highly important in the government vision. This is because there are currently no educational institutions that provide formal HSI degrees; thus, at least some of the KSAs needed to enter the field must be made up with government training. Successfully completing HSI training can be the primary criterion for certification.

The federal government study on establishing a national workforce proposed a comprehensive education option as part of the career path. We have slightly modified this option, as illustrated in Figure 5.2, to include training. The major features of a comprehensive education and training program for HSI would include the following major features:

- ***Continuing Education*** Primarily from nongovernment educational institutions in courses that relate to HSI domains and/or related engineering, analyses, and management fields, this would include courses for individuals completing or already having bachelor's degrees. Continuing education is shown as open to the largest portion of the workforce primarily for improving KSAs for level I jobs.
- ***HSI Specialized Education and Training*** The next largest portion of the workforce could participate in HSI specific courses, provided by either the government or civilian educational institutions. If HSI certification is required, it could be acquired

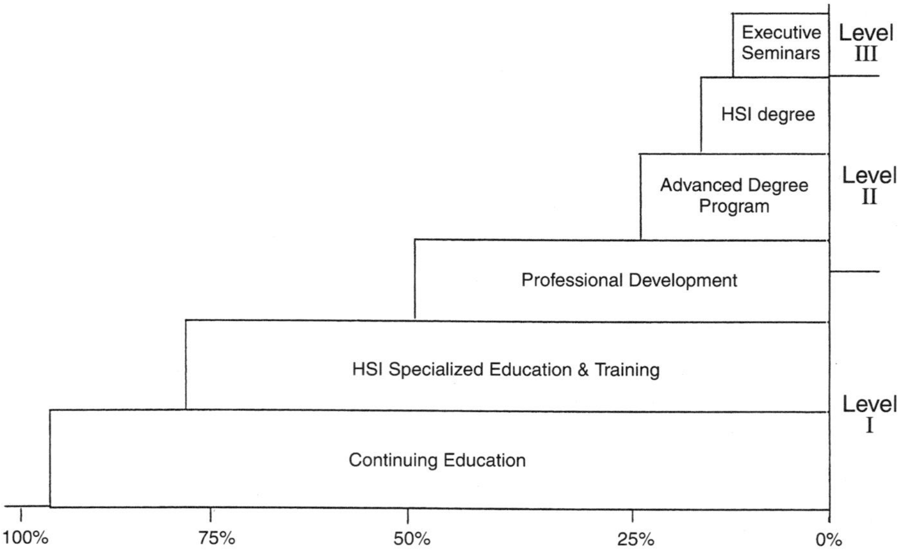


**Figure 5.2** HSI education and training by job level and percentage of workforce.

by successful completion of specialized HSI coursework. Nongovernment personnel would also be eligible to acquire certification through this coursework.

- ***Professional Development*** This represents a way to improve the KSAs for HSI practitioners, especially for new and mid-management supervisors. Many of the individuals who end up in supervisory jobs related to HSI may have specialized expertise in one or more domains but need professional development in the other domains and related engineering and management fields.
- ***Advanced-Degree Program*** Individuals who have increasingly greater responsibilities for HSI research, applications, and/or management could be selected as candidates for advanced degrees in related, existing fields, such as human factors engineering, systems engineering, or operations research and analysis. For every year spent in such a program at government expense, three years of government service in HSI would generally be required.
- ***HSI Degree*** A master's or doctoral degree in HSI from an accredited institution would be considered the highest educational achievement for the HSI professional. Ideally, when such degrees become attainable, they will serve as a criterion of excellence displayed by the highest level HSI professionals in the federal government.
- ***Executive Seminars*** These are currently available to government senior executives. However, such seminars do not usually have the discipline of HSI integrated into the coursework. For organizations wishing to provide a comprehensive, executive educational program, both specialized HSI seminars and seminars with HSI integrated into their curricula would be part of their educational program.

### 5.6.2 HSI Personnel Job Series and Skill Code

The study on establishment of a HSI career workforce (U.S. Army, 1993, p. 2-1) concluded that a viable career professional development program in the federal government must consist of the following personnel system components:

1. authorized full and part-time positions with job descriptions and grade requirements within the workforce structure;
2. a group of full-time HSI professionals that can be considered the HSI corps; and
3. a management system that facilitates recruiting new talent, providing them with required, sequential HSI-specific training and additional academic education, and assigning them to authorized positions.

As was shown in Figure 5.1, the HSI professional workforce career ladder hierarchy can have three levels. Typical job assignments for the levels in the four functional contexts are illustrated in Table 5.1. Currently, the military services have some full- and part-time positions that attempt to carry out HSI responsibilities. This is not done with any formal career structure, and none of the services has a personnel management system that facilitates the identification and assignment of individuals and their skills. There is no ability within the federal government to track personnel with HSI skills, so consequently it cannot evaluate any return on education and training investment to develop such skills.

In order to satisfactorily fulfill any of the three desired personnel system components, two fundamental decisions need to be made by federal agencies. The first is the determination of acceptable job series options for HSI personnel; the second is the development of approved job descriptions for HSI functions. The Office of Personnel Management (OPM) is the authority for establishing a new job series, but each agency has considerable latitude in utilizing existing job series and creating job descriptions.

The primary choices for HSI job series are (1) creating and assigning HSI personnel positions within existing job series as determined by the specific agency need; (2) creating and assigning HSI personnel positions within existing job series augmented by an HSI appendix;[6] (3) creating and assigning HSI personnel positions within existing job series, augmented by job skill codes;[7] or (4) creating a specialized HSI job series for HSI positions.[8] The national workforce study examined the advantages and disadvantages of each option. Essentially, the advantages of having a federal workforce that can be identified, tracked, and managed increase as one moves from option 1 to option 4. For example, as shown in Table 5.15, none of the existing job series is fully satisfactory to meet the job demands of an HSI position, making option 4 the preferred option. However, options 2 and 3 would be better than option 1 by providing greater controls over existing job series when used for HSI positions.

The disadvantages of implementing the options, however, increase in just the opposite direction (due to increased cost and coordination requirements). Currently, option 1 is the only choice being exercised.

Agencies can make some improvements in the career workforce simply by exercising some of the choices at each agency's disposal. For example, Table 5.15 shows that three of the existing job series do have strong technical requirements. These are engineering psychologist (180), general engineer (801), and operations research (1515). Options 2 and 3 could be exercised by augmenting these three job series either with an HSI appendix or a job skill code inserted into the position description. Table 5.16 illustrates a draft of what the national workforce study considered an acceptable job description for the HSI job series (option 4), but it could also be the basis for augmenting existing job series for any of the other options.

### 5.6.3 Certification and Licensing

The issues of certification and licensing of HSI practitioners have not changed since first examined by Muckler and Seven (1990). Their discussion on this topic was so enlightening that it is worth repeating in its entirety (pp. 540–542 with kind permission of Kluwer Academic Publishers):

> Once a labor pool of highly skilled professionals is established, a way of indicating to the public that they are (1) skilled and (2) skilled in something specific may be inevitable. The issue is certification and/or licensing of human factors/MANPRINT/[HSI] professionals where "certification" usually is the weaker form of regulation and restricts the use of a professional title. "Licensing," on the other hand, usually implies some specific kind of skill level and requires some form of examination.
>
> The basic purpose of certification and licensing is said to be the protection of the public. The practice goes back at least 3,000 years to Hammurabi and the Babylonians (Oates, 1979, pp. 180–183, on the practice of medicine). It should be understood that this is an extraordinarily controversial subject, and that there are those who argue that certification and licensing protects no one except guild members and may operate to the detriment of the public as well as practitioners (cf., Danish and Smyer, 1981; Gross, 1978; Koocher, 1979; Wiens and Menne, 1981; Herbsleb, Sales, and Overcast, 1985). It has been an issue of major concern to human

**TABLE 5.15 Pros and Cons of Job Series for HSI Careers**

| Job Series | Title | Pros | Cons |
|---|---|---|---|
| 0018 | Safety and Occupational Health Management | Strong integrated look at personnel health and safety | Lacks MPT knowledge |
| 0180 | Psychologist[a] | Most human factors engineers are now in this series | Restricted to those with psychology degrees and has no integration focus |
| 0205 | Military Personnel Management | Strong view of manpower and personnel issues | No technical requirements or engineering background |
| 0343 | Management and Program Analysis | Emphasis on integration and management | No technical requirements or background |
| 0346 | Logistics Analysis | Requires integration skills; relates to logistic part of acquisition | Too narrow a focus and weak in technical requirements |
| 0801 | General Engineering[a] | Engineering perspective and background helps interactions with design engineers | Lacks MPT background |
| 0803 | Safety Engineer | Engineering perspective and background helps interactions with design engineers | Lacks MPT background |
| 0845 | Industrial Engineer | Strong human factors orientation | Weakest in personnel areas |
| 1102 | Contracting | Relates well to acquisition process | Not technical enough |
| 1515 | Operations Research[a] | Good analytical skills that apply across all domains | Weak in human factors, health hazards, safety |
| 1701 | Training Manager | Strong human performance orientation | No technical or engineering background |
| XXXX | Human Systems Integration (new) | Tailored for HSI mission | Requires new job series |

[a]Series with technical grounding.

factors practitioners (cf. Siegel, 1980; Blanchard, 1985). Perhaps the best that can be done here is to describe some of the difficult issues raised by certification and licensing.

- It is possible to apply the process either to individuals or organizations. Normally, it is the individual who will be certified or licensed. An alternative would be blanket approval

**TABLE 5.16 Draft Definition of HSI Job Series**

Personnel in the HSI job series are primarily concerned with integrating the human component into the design, development, and acquisition of new systems or modifications of existing systems. System is defined in its broadest organizational context and includes equipment, people, and the procedures they use.

Individuals in this job series will perform in a number of environments, to include research and development, applied engineering, acquisition, and regulatory.

HSI professionals would perform the following principal functions:

- Develop HSI goals and constraints
- Develop and validate designs to meet HSI goals and constraints and performance requirements
- Trade off alternate design options to optimize (validate) compliance with requirements
- Validate design compliance
- Develop demographic and anthropometric description of workforce
- Identify critical usability issues for test and evaluation
- Prepare input to government solicitation documents
- Participate in source selection
- Identify issues for research and development
- Monitor contractor and/or government performance

Individuals in the HSI job series must be capable of conducting a variety of analyses in support of the procurement of hardware or software or the regulation of the end products. The analyses include but are not limited to the following:

| | | |
|---|---|---|
| Human–machine interface | Workload | Training |
| Organization workflow | Organization productivity | Anthropometric analysis |
| Manpower requirements | Health and safety analysis | Life-cycle analysis |

The results of these analyses are included in government regulation requirement and procurement documents (request for proposal, request for information, and request for quote) to which industry is required to respond.

HSI personnel are further responsible for evaluating the technical quality and consistency of industry response to government solicitations and requirements and supervising and evaluating contractor performance in the following functional areas:

| | | |
|---|---|---|
| Manpower | Human factors engineering | Occupational health and safety |
| Personnel | Health hazards | Personnel survivability |
| Training | Systems safety | Habitability |

for all those who graduate from an accredited educational program. What agency would do the accreditation has been the subject of disagreement.

- For individuals making their initial entry into a field, there is the question of what kinds of requirements should be established for either certification or licensing. For example, some fields require a formal, written, and often very rigorous, examination. Another alternative would be an oral examination of some form. Further, it may probably be assumed that some minimum level of experience and/or education would be required.
- There must be some written set of standards or ethics for what constitutes good and bad practice. There should be some form of statement as to what constitutes "harmful" activity. The American Psychological Association (APA) has a detailed code of ethics and defines at length acceptable and unacceptable behavior. The Human Factors Society

has also recently ratified a statement of ethics. [The Human Factors Society has changed its name to the Human Factors and Ergonomics Society. As of 2003, it still retains a statement of ethics for its members.]

- There have been many questions as to who should perform the certification and licensing. Paramount is the legal desire of the state. If the state deems the profession potentially harmful to the public, the state will probably provide formal mechanisms for certification and licensing. Even then, the state rarely does so without the approval and assistance of professional societies and organizations. In some cases, there will be both professional and public representation. For many professions (e.g., medicine) the state provides certification and licensing which may create some problems in jurisdiction when practicing in different states (cf., Henderson and Hildreth, 1965) or in standardization of requirements.
- One useful question is: What do the people in the field want? Rarely are current practitioners asked, but in one case they were. Siegel (1980) queried human factors professionals. His findings are interesting. First, current professionals preferred certification over licensing, principally on the practical grounds of the investment required to do licensing. Second, a master's degree was felt to be a sufficient level of education for certification (even then, 10% of the respondents felt that degrees were either not required or "immaterial"). Third, there was little agreement on the most appropriate background. Fourth, practitioners felt that experience should be substituted for formal education where appropriate. Perhaps the last two items reflect, in part, the fact that several major contributors to the field of human factors have not come from traditional educational backgrounds.
- A final issue is how frequently an individual should be assessed for his or her competency. For some, initial assessment is sufficient. But there has been a growing social feeling that more frequent assessments might be useful to see if competence is maintained. There appears to be no particular agreement on what form of assessment might be best. One could require formal, written, and/or oral examinations. Another method could be the submittal of performance and achievements to an assessment board. A third way could be a requirement to return to an educational institution at certain prescribed times for certain required number of courses.

## 5.7 HSI PROFESSIONAL PERSONNEL SUPPLY

Finally, an important aspect of establishing an HSI career workforce is having a reliable supply of professional personnel. One way of predicting the supply and composition of HSI professionals for the future is by examining the historical trends of groups having interest in applying human factors to the design of systems and equipment. Therefore, we wrap up our discussion on education and training of an HSI career workforce by examining the HFES membership by academic specialty and highest degree held (HFES, 2000b).

As can be seen in Figure 5.3, psychologists dominate the membership of the HFES (45 percent). This group includes general, experimental, industrial, engineering, cognitive, and other psychologists such as physiological, social, and clinical psychologists. The next largest group is comprised of engineers (23 percent). This includes general, industrial, mechanical, electrical, aeronautical/astronautical, and other engineers, including system engineering and biotechnology. The "other" category (discussed shortly) is almost as large as engineering (21 percent). Members with degrees in human factors or ergonomics constitute 11 percent of the HFES. This group includes general, psychology, and

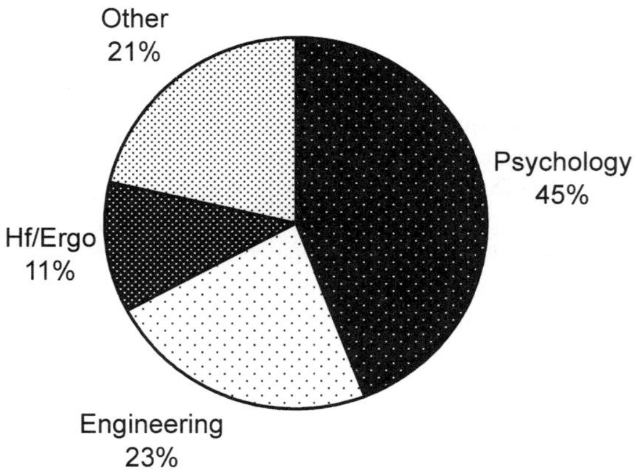


**Figure 5.3** Major academic specialties (adapted with permission from HFES, 2000b).

engineering specialists. Detailing the "other" category further, it becomes obvious that the HFES is diverse indeed. Figure 5.4 illustrates the composition of the 21 percent "other" in Figure 5.3.

As can be seen in Figure 5.4, there are many constituencies in HFES. The "other" category includes a variety of specialties such as physics, anthropology, sociology, architecture, industrial management, and operations research (HFES, 2000b). Academic specialties are not listed separately by HFES unless they constitute 0.8 percent of the total membership.

Regarding degrees of professionalism, the highest degree held analysis is interesting as well. As illustrated in Figure 5.5, there are almost as many members with master's degrees (37.7 percent) as those with Ph.D. degrees (38.4 percent). Members with a bachelor's degree account for 15.5 percent. Nondegreed members account for 0.75 percent of the total membership. Other undergraduates constitute 2.16 percent, and 5.54 percent did not declare.

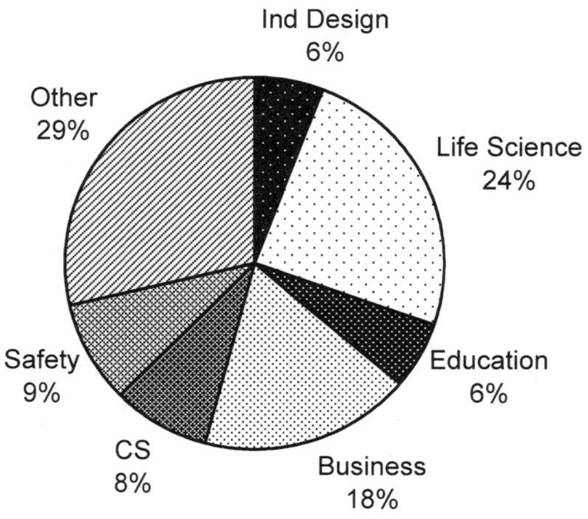


**Figure 5.4** Breakdown of "other" category (adapted with permission from HFES, 2000b).

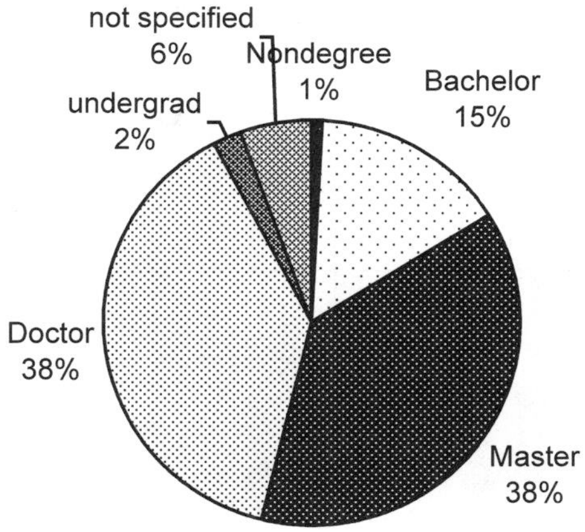


**Figure 5.5** Highest degree held (adapted with permission from HFES, 2000b).

As illustrated in Figure 5.6, the largest single membership group is comprised of general psychologists with Ph.D. degrees (12.11 percent). There are approximately half as many Ph.D.s in Industrial Engineering, the next largest group (6.53 percent). Those with a Master's in Experimental Psychology make up the next group (6.25 percent), followed by industrial engineers with MS degrees (5.70 percent) and those with a Bachelor's in General Psychology (4.90 percent). Together, these five groups comprise 35.49 percent of the membership. The majority of members is comprised of very small groups outside these five, reflecting the tremendous diversity of professional populations making up the HFES.

Figure 5.7 illustrates the growth of the HFES from 1960 until 2000. The HFES has grown from approximately 500 members in the 1960s to well over 5000 in 2000. Student membership has also increased but not so sharply.

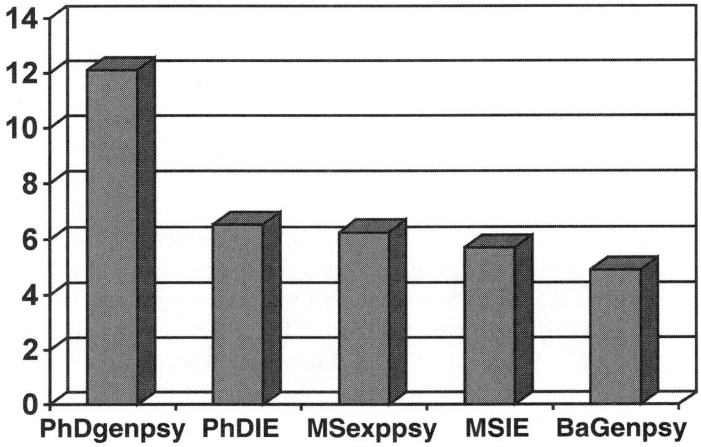


**Figure 5.6** Largest member groups in HFES (adapted with permission from HFES, 2000b).

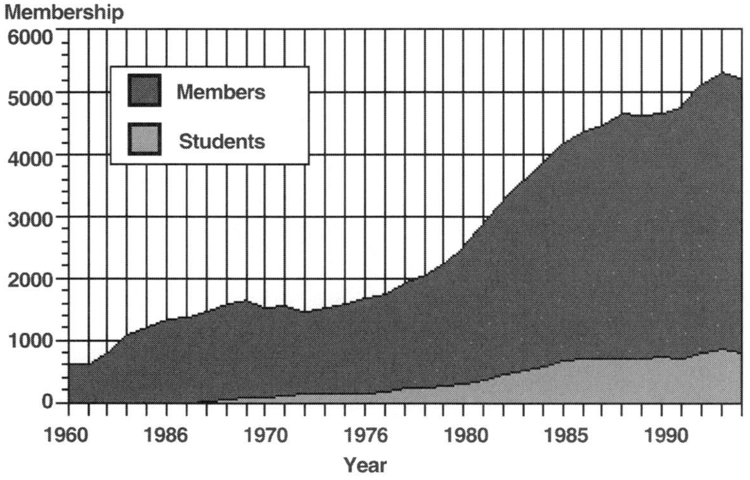


**Figure 5.7** Membership trends in 1960–2000 (HFES, 2001).

It is clear that the HF/E professional domain offers a diverse environment and home for the HSI professional. Since certification appears to be a trend that will continue, HSI professionals should consider BCPE certification or a comparable designation. Unless and until dedicated HSI degree programs are developed, multiple degrees, even from multiple disciplines, are appropriate. Potential degree programs include macroergonomics, systems engineering, industrial engineering, human resources, and I/O psychology.

## 5.8 SUMMARY AND CONCLUSIONS

In this chapter, we first outlined the core competencies needed for qualified HSI specialists. This included a discussion of levels of competency and functional assignments. We then explored the academic environment in which HSI study can be pursued. Current curricula in several institutions were reviewed to illustrate the state of the educational environment for HSI pursuit. We then identified the recommended content areas for HSI education and illustrated through two hypothetical examples how a student might pursue an advanced degree in HSI. Textbook coverage of HSI content areas and the status of practitioner-training coursework were also discussed. In addition to academic education, specialized training for the HSI professional is also important, so the issues and status of HSI training were addressed as well. Since education and training development of HSI professionals is closely linked to careers in HSI, several issues regarding the development of an HSI career path were discussed. These covered a professional workforce vision, personnel job series and job descriptions, and certification and licensing. Finally, we examined the composition of the most likely supply of HSI professionals through membership data from the HFES. We have attempted to address questions that would interest three readers: (1) HSI professionals, (2) teachers and developers of HSI courses and materials, and (3) those involved with the system acquisition process. Common to all readers is material on the issues and options involved in acquiring KSAs in HSI principles and methods. For the HSI professional, whether it is through emerging HSI courses or obtaining advanced degrees or on-the-job experience, the authors are hopeful these readers will pursue life-long HSI careers.

## NOTES

1. The U.S. Army (1993, 1995) national workforce studies for MANPRINT identified the situation then, and military downsizing since has further diminished the national pool of HSI professionals: "Few qualified people . . . having needed expertise" should not be construed that there are not a large number of individuals with many of the skills important to HSI. Neither does it mean that to be qualified requires an individual to have expertise in all the HSI domains. However, HSI as a profession is still immature and is discovering that it needs certain expertise (such as that which meets the unique management and technical integration requirements discussed in Chapter 1; see Table 1.1) not widely available. Currently, few people could have acquired full HSI qualifications through either experience or academic institutions. This is because the number of work opportunities has been low and the definition of unique educational requirements has not fully evolved.
2. The study was for a MANPRINT national workforce, but outside the army, the program is better known as HSI (see Chapter 1).
3. It is understood that there would be no need for the second and third ingredients if sufficient demand were not present.
4. The NPS does have an approved two-year master's HSI track.
5. The one-day navy HSI course, Human Factors and Safety Engineering, is sponsored by the American Society of Naval Engineers (ASNE) and sanctioned by Virginia Tech.
6. An HSI appendix would be information on HSI skills and functions that could be appended to each related job series descriptions in the government personnel manual (X-118). It would be available to managers to help in defining position descriptions.
7. Job skill codes are a way of formalizing experience or training that reflect qualifications needed for the job that are not reflected in the job series. For example, individuals could obtain a skill code in HSI by meeting specified criteria such as working in an HSI position, attending HSI training, or becoming certified in HSI. Skill codes are controlled by each agency and are not standard across agencies.
8. A separate job series for HSI could be designated from existing job series or created as a totally new series, but its functional contexts would include acquisition, regulation, research, and applied engineering. Although agency coordination across all federal agencies would be required, the creation of a new job series in HSI would have a very positive effect on individual career opportunities, motivating universities and industry to support the workforce, improving the availability and quality of HSI professionals, and allowing the management of the workforce.

## REFERENCES

Adams, J. A. (1989). *Human Factors Engineering*. New York: Macmillan.

Booher, H. R. (2001). *MANPRINT and HSI Technical Training: Task 2: Training Needs Analysis*. Aberdeen Proving Ground, MD: Army Research Laboratory, Human Research and Engineering Directorate.

Hendrick, H. W. (1984). Wagging the Tail with the Dog: Organizational Design Considerations in Ergonomics. In *Proceedings of the Human Factors Society 28th Annual Meeting* (pp. 899–903). Santa Monica, CA: Human Factors Society.

Hendrick, H. W. (1986a). Macroergonomics: A Conceptual Model for Integrating Human Factors with Organizational Design. In O. Brown Jr. and H. W. Hendrick (Eds.), *Human Factors in Organizational Design and Management II* (pp. 467–478). Amsterdam: North-Holland.

Hendrick, H. W. (1986b). Macroergonomics: A Concept Whose Time Has Come. *Human Factors Society Bulletin, 30*(2), 1–3.

Hendrick, H. W. (1997). Organizational Design and Macroergonomics. In G. Salvendy (Ed.), *Handbook of Human Factors and Ergonomics* (pp. 594–636). New York: Wiley.

Hendrick, H. W., and Kleiner, B. M. (2001). *Macroergonomics: An Introduction to Work System Design*. Santa Monica, CA: Human Factors and Ergonomics Society.

Hendrick, H. W., and Kleiner, B. M. (2002). *Macroergonomics: Theory, Methods and Applications*. Mahwah, NJ: Lawrence Erlbaum Associates.

Hollands, J. G., and Wickens, C. D. (2000). *Engineering, Psychology and Human Performance*. Upper Saddle River, NJ: Prentice-Hall.

Hollis, W. (1995, March 30–31). MANPRINT in Higher Education Workshop. Handout materials, DUSA(OR)–Deputy Undersecretary of the Army (Operations Research), Pentagon, Washington, DC.

Huchingson, R. D. (1981). *New Horizons for Human Factors in Design*. New York: McGraw-Hill.

Human Factors and Ergonomics Society. (2000a). *Directory of Human Factors/Ergonomics Graduate Programs in the United States and Canada*. Santa Monica, CA: Human Factors and Ergonomics Society.

Human Factors and Ergonomics Society. (2000b). *2000–2001 Directory and Yearbook*. Santa Monica, CA: Human Factors and Ergonomics Society.

Human Factors and Ergonomics Society. (2001). Available: http://hfes.org/About/Members.html.

Imada, A., and Kleiner, B. M. (2000). Macroergonomic Analysis of Work Systems in the 21$^{st}$ Century. Panel delivered at *IEA 2000/Annual Meeting of the Human Factors and Ergonomics Society.* Santa Monica, CA: The Human Factors and Ergonomics Society.

Kantowitz, B. H., and Sorkin, R. D. (1983). *Human Factors: Understanding People-System Relationships*. New York: Wiley.

Konz, S., and Johnson, S. (2000). *Work Design: Industrial Ergonomics*. Scottsdale, AZ: Holcomb Hathaway Publishers.

Kroemer, K., Kroemer, H., and Kroemer-Elbert, K. (2001). *Ergonomics: How to Design for Ease and Efficiency.* Upper Saddle River, NJ: Prentice-Hall.

Kroemer, K. H. E., and Grandjean, E. (1997). *Fitting the Task to the Human*. London: Taylor and Francis.

Mital, A., Ayoub, M. M., Casali, J. G., Kleiner, B. M., Fernandez, J. E., and Resnick, M. L. (1998). Ergonomics Issues beyond the Year 2000. In *Proceedings of the Industrial Engineering Research Conference*. Norcross, GA: Institute of Industrial Engineers.

Muckler, F. A., and Seven, S. A. (1990). National Education and Training. In H. R. Booher (Ed.), *MANPRINT: An Approach to Systems Integration*. New York: Van Nostrand Reinhold. pp. 519–549.

Niebel, B., and Freivalds, A. (1999). *Methods, Standards and Work Design*. Boston: McGraw-Hill.

Oborne, D. J. (1982). *Ergonomics at Work*. New York: Wiley.

Pulat, B. M. (1997). *Fundamentals of Industrial Ergonomics*. Prospect Heights, IL: Waveland.

Salvendy, G. (Ed.). (1987). *Handbook of Human Factors*. New York: Wiley.

Salvendy, G. (Ed.). (1997). *Handbook of Human Factors and Ergonomics*. New York: Wiley.

Sanders, M. S., and McCormick, E. J. (1987). *Human Factors in Engineering and Design*. New York: McGraw-Hill.

Tayyari, F., and Smith, J. L. (1997). *Occupational Ergonomics Principles and Applications*. London: Chapman and Hall.

U.S. Army, Office of the Deputy Chief of Staff for Personnel. (1993). Establishment of a MANPRINT National Workforce. Report by PRC and MTL Services, Contract No. MDA903-91-D-0031, DAPE-MR, Pentagon, Washington, DC.

U.S. Army, Office of the Deputy Chief of Staff for Personnel. (1995). Development of MANPRINT Planning, Programming, and Guidance Information; Task III: Implementation of MANPRINT in Higher Education. Final report by Hay Management Consultants and Remtech Services, Contract No. DASW01-94-C-0087, DAPE-MR, Pentagon, Washington, DC.

Van Cott, H. P., and Huey, B. M. (Eds.). (1992). *Human Factors Specialists' Education and Utilization: Results of a Survey*, Washington, DC: National Academy Press.

Wickens, C. D. (1984). *Engineering, Psychology and Human Performance*. Columbus: Charles E. Merrill.

Wickens, C. D., Gordon, S. E., and Liu, Y. (1998). *An Introduction to Human Factors Engineering*. New York: Addison Wesley Longman.

PART II

# SYSTEMS ACQUISITION AND MANAGEMENT PROCESSES

Human systems integration (HSI) deals with processes and methods to better understand and accommodate the role of the human being within systems. It is a thorough and comprehensive systems engineering and management strategy that is begun early in the process of system acquisition to ensure that all pertinent human concerns are addressed throughout the life-cycle process. Five chapters are provided that describe how HSI is involved throughout the major stages in acquiring a system, beginning with requirements determination, to system specifications, to system design and development, and finally to test, evaluation, and assessment of system performance.

The focus of Chapter 6 by Harrison and Forster is on the very earliest stages of the acquisition process. They rightly point out that decisions made in the concept stages will determine whether a project will proceed, define the key risks and issues to be addressed, and determine the allocation of resources for subsequent phases; yet human factors disciplines have traditionally been perceived as limited in their ability to contribute at this phase. Part of the perceived inability to contribute in the early stages comes from the fact that many tools and techniques available have been more suited to assessment of designs rather than addressing predesign concepts and analyses. Another part of the perceived limitations for human factors early in the acquisition process comes from cultural attitudes such as those described in Chapters 2 and 3. Together the limitations of tools and cultural attitudes have severely constrained the ability of HSI specialists to exert influence at perhaps the most critical stage in the system life cycle. Harrison and Forster present information that should be helpful in illustrating ways for the HSI professional to begin to reverse these past limitations. Their contribution takes two forms. First they discuss HSI activities that *should* be undertaken at early concept stages, including a general description of the requirements determination process, the types of HSI requirements and constraints that *should* be integrated into major project documents, and the roles of user information and target audience descriptions. Second, they present a promising new approach, developed within the United Kingdom (UK) Ministry of Defence's (MoD) Corporate Research Program. Known as the early human factors analysis (EHFA), the UK approach provides a mechanism to identify human-related risks and requirements early in

*Handbook of Human Systems Integration*, Edited by Harold R. Booher.
ISBN 0-471-02053-2 

an acquisition program and make an early cost and benefits assessment of the analysis results. Although this effort draws particularly on applications to UK defense acquisition, the principles described are generally applicable to other military and nonmilitary systems acquisition.

In Chapter 7 Hamilton describes the relationship between required HSI tasks and the acquisition process from the contractor's point of view. This is accomplished in three ways:

1. The critical contractor products and HSI tasks are described for each major stage of contract procurement activity—from contract award through the system life cycle up to testing and certification.
2. The principal documentation events of the contractor solicitation and selection process are discussed, which include the buyer solicitation announcement, the buyer request for proposal (RFP), the seller proposal, and the buyer source selection.
3. Guidelines are provided for contractor HSI practitioners on how to prepare proposals and plan and manage an integrated HSI program.

Chapter 8 by Barnes and Beevis, attempts to increase the reader's appreciation of the difficulty in designing modern complex systems by focusing on the system interactions among human, environmental, operational, and engineering components. It points out that design optimization through trade-offs entails measuring the performance not only of the various components but also of their interactions. Barnes and Beeves propose a systematic approach to measuring human performances trade-offs in terms reflecting system goals and subgoals. These models address human performance measurement not only in terms of the intended user but also as an integral part of the cost–benefit equation used to drive the design process. The chapter also discusses the advantages and limitations of various measurement techniques emphasizing the unique measurement problems that the human introduces into the process. The authors conclude that measuring complex systems requires some combination of modeling, hypothesis testing, and realistic simulation methods.

In Chapter 9 Olson and Sage discuss how HSI is affected by the increasing use of computer-based models and simulations within the system engineering and product acquisition domains. Although most of the chapters in this handbook focus on human involvement in the use of the system being engineered, this chapter focuses on another realm of human involvement—that of the human in the acquisition process itself. In summarizing the changing acquisition process and environment, the authors describe simulated-based acquisition as a potential model for the future and speculate on the role that humans will play in this new environment involving simulation-based acquisition.

In Chapter 10 Ehrhart and Sage present an HSI framework for user-centered systems engineering. The framework emphasizes methods for creating, structuring, and applying models and processes needed to identify and address HSI issues across all phases of the acquisition life cycle. The chapter discussion centers on the opportunities and challenges of the system acquisition life cycle, showing how HSI issues can be incorporated into systems engineering processes. More specifically, the framework provides a guide for the systems engineering manager to better appreciate and employ cognitive systems engineering methods to define problems, identify and represent cognitive task requirements, develop design goals, and implement and evaluate system designs for human–machine decision making.

CHAPTER 6

# Human Systems Integration Requirements in Systems Acquisition

JOHN A. HARRISON and MELANIE J. FORSTER

## 6.1 INTRODUCTION

Military effectiveness relies on having the capability to achieve military objectives in the face of equally determined opposition. The issue for "requirements" for systems engineering management and the human systems integration (HSI) specialist is how to express a capability need in a way that will ensure it is achieved or at least minimize the risk of its not being achieved.

Over the last century, the "edge" needed to achieve military capability has become reliant on ever more complex technology that in many cases has failed to deliver the desired result when in the hands of its intended users. This stimulated the birth of human factors (HF) and ergonomics as new scientific disciplines in the mid-twentieth century, but these activities were primarily applied as constraints on designs of equipment rather than addressing the full role of people integrated into system capability: The procurement process was still equipment centered rather than people centered.

Initiatives of the last decade have institutionalized consideration of human users within the procurement process. More recently, thinking about procurement as a whole has broadened to make explicit the acquisition of capability, of which procuring equipment is only a part. All of these moves affect requirements for the "system" being acquired.

A system is "an integrated composite of people, products and processes that provide a capability to satisfy a stated need or objective" [U.S. Department of Defense (DoD), 1992]. This system definition is central to HSI and must be central to system requirements thinking. Hitchins (1998, p. 195) describes a system as "an open set of complementary, interacting parts, with Properties, Capabilities and Behaviours (PCBs) emerging both from the parts and their interactions." This definition focuses on the interaction between parts. Hitchins also emphasizes that system engineering must concern itself with not just the "product system" (that will exist within the user organization) but the "process system"

*Handbook of Human Systems Integration*, Edited by Harold R. Booher.
ISBN 0-471-02053-2 

that creates it (and is often in a different organization). Both require integration, involving multiple interacting parts and people. Those with responsibility for HSI must therefore think not only about how end users interact with their equipment but also about how the results of HSI analysis will be integrated with the work of system engineers and others whose concerns, training, and mind set may be very different from their own. Traditionally, HSI has been more concerned with the product system, whereas the 10 principles of HSI listed in Chapter 1 strongly concern the process system.

The requirements and other key documents produced early in a project significantly determine what will eventually be fielded. For HSI to be truly effective, human-related issues and their consequences must be *woven into the fabric of the project* rather than appearing as stand-alone items. Developing systems is difficult work, with many trade-offs at all levels. It is important to remember that no "HSI product" is ever fielded or sees action. The equipment that is fielded, the people who operate, support, and maintain it, and the procedures that bind them all together are produced by other stake holders. Human systems integration must inform and add value to the output of these other players, whether it is hardware, software, trained people, or military handbooks. The ultimate test of that added value is how well human and non-human components work together, maximizing their own and each other's performance.

Requirements for a system are motivated by several different factors: what it must be capable of achieving and the constraints imposed on it by the environment in which it will operate, the systems with which it must interact, or the components it must include. At the *total system* level, humans are a vital system component, and they bring many constraints with them. At the slightly lower level of an equipment system (strictly a subsystem) to be procured, the humans represent another subsystem with which it must interact closely.

Requirements (in general) can be expressed in three different ways, often called the *three P's*:

- Product—attributes the product must have (e.g., height, load space).
- Performance—how well the product must do something (e.g., speed, accuracy, failure rate).
- Process—aspects of how the product must be developed or produced (e.g., quality procedures, design disclosure, road testing).

Human-related requirements are of all three types and are discussed further in Section 6.2.2.

Traditionally HSI requirements have been *requirements to conduct human factors tasks* (e.g., "the contractor shall do a task analysis") and requirements for desirable human-related attributes (e.g., "displays shall be easy to read"), but this undervalues the full role of HSI as a contributor to system requirements. Human systems integration can be applied to the full range and depth of system requirements. All aspects of a requirement affect the outcome and, hence, potentially the effectiveness of human to nonhuman integration. All parts of the requirement are legitimate candidates for HSI intervention.

**Example 6.1 Security and Operability** The feasibility study for a command system was facing many challenging issues, each being explored by respective specialists. The dominant HSI issue was the size of the command team as part of an overall drive for lean manning. One of the pressing technology issues was the need to handle a small amount of information at high security level and the consequent need not only to prove that the software was reliable but to do so in a way that would not greatly increase software development cost. After wrestling with this problem for some time, the system engineers presented a system architecture they believed would solve the problem. Their solution was to minimize the high-security software by dividing the system in two. However, the HSI manning analysis revealed the engineers' solution would add two people to the ship's complement, because the new architecture would split a critical operator role down the middle. This stark revelation, plus the relationship built up with the HSI leader helping the engineers to resolve smaller issues during the project, was enough to cause a rethink in the interpretation of the requirement and eventually come up with a solution to the software problem without increasing manpower.

## 6.2 HUMAN SYSTEMS INTEGRATION IN REQUIREMENTS

The requirement captures the reason the project exists. It is at the heart of the project. Downs et al. (1992, p. 93) state, "Projects start because people who are in some way significant in an organization feel that things are wrong, or at least that there is a reasonable chance that they could be made better."

This earliest phase of an acquisition program is most critical. The whole direction of the future project (or indeed whether the project goes ahead) will depend on decisions made during the preconcept and concept phase. The decisions will frame the overall requirement, define what risks are to drive the plan, and allocate resources. Human factors have traditionally been perceived as contributing little to this phase, yet failure to capture human-related issues as part of the initial requirement makes it harder to give the human dimension due weight in later phases.

### 6.2.1 HSI and the Requirements Process

Requirements engineering is a distinct discipline within the systems engineering community. It is the subject of active research with many unsolved problems, but there are established frameworks to which successful HSI must relate.

Requirements engineering has achieved prominence partly because of the widespread experience that it is difficult to do well, and partly because system problems are attributed to failures in requirements more than any other cause. Most phases of systems engineering transform one reasonably well defined entity into another. For example, designing turns a specification of what is required into a description of something that can be built; testing turns a hypothesis about performance into evidence. Each stage of the process expects to receive something well defined from the previous phase. This would mean a precise, unambiguous specification in the design stage and a well-defined set of test schedules and conditions and performance criteria in the testing stage.

The requirements process (as a whole) is more difficult than the design and development processes, since the inputs to requirements engineering are less well defined than those for the engineering processes that follow. The requirements process starts with loosely defined needs and progressively elaborates those needs into a much larger number of precise statements for something that can be contracted, built, and tested.

As a discipline, HSI has much to contribute to requirements. Much of the imprecision and uncertainty in the requirement sources stems from the need to involve people, such as operators and maintainers as components of the total system. But people are complex, hard-to-specify components. Moreover, people "own" the problems that the system is trying to solve. For these reasons, there is scope for productive cooperation between HSI and requirements engineering practitioners and considerable overlap between their research interests.

Figure 6.1 shows a simplified view of how a requirement for system capability is formulated and then split into requirements for the equipment and human parts of the system. Through a series of acquisition steps, the requirements are progressively transformed into a system composed of both people and equipment.

Much of the elaboration of requirements for the technical aspects of an equipment draws on a large pool of knowledge about the equipment domain, whereas many HSI requirements must draw on a very different and smaller pool—one that describes people

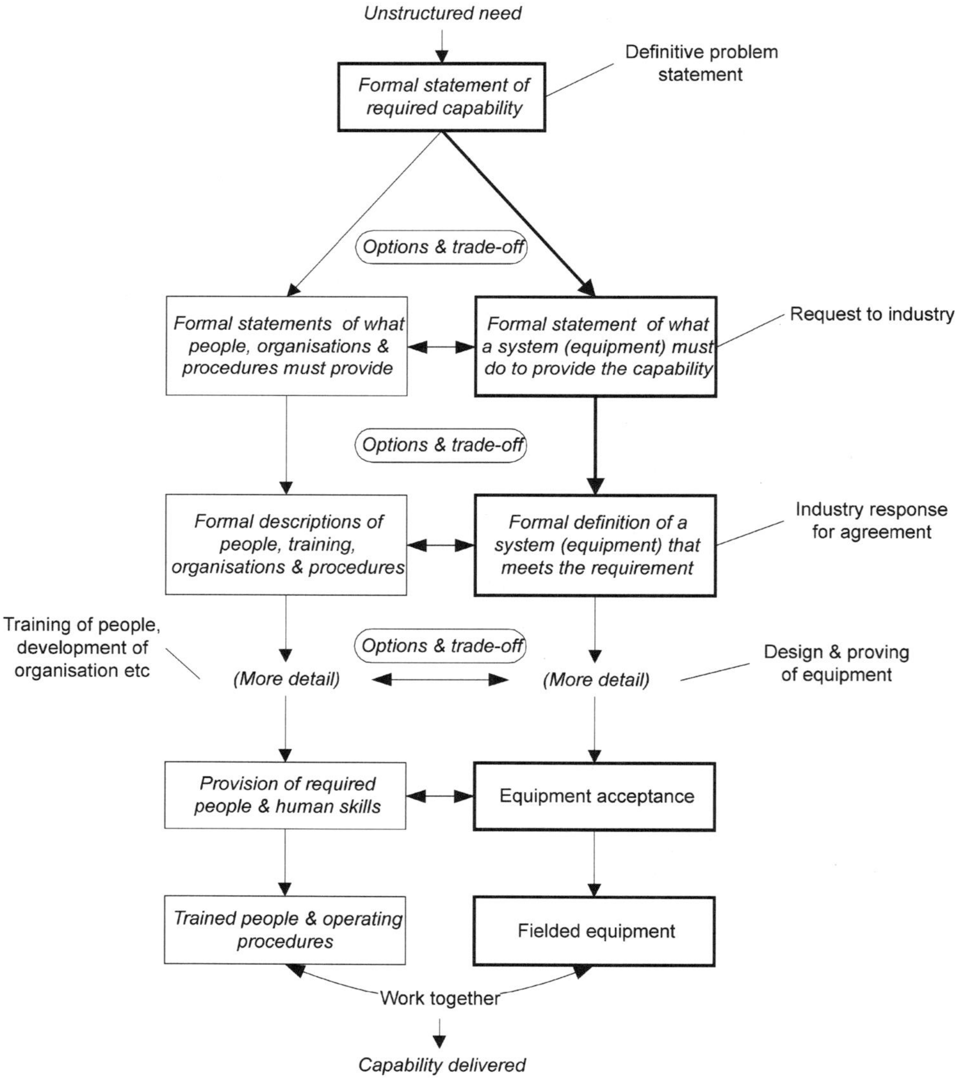


**Figure 6.1** Balanced view of systems acquisition process.

and how they behave. Specifiers and procurers can draw on the heavy investment in the underlying engineering science by equipment contractors, whereas for the corresponding human science the most recent HSI advances will generally come from smaller investments by government, academic, and HF consulting firms.

Another function of HSI is to act as the bridge between the human and equipment components of the system. Figure 6.1 shows a balance between the equipment and people paths of the system acquisition process. Without HSI involvement, the process tends to run along the equipment path, as illustrated by the thick-edged boxes along the right of Figure 6.1. From a procurement perspective, this is the "mainstream," but as a means of evolving from the statement of required capability to the fielded capability, it must interact with the people stream on the left, as illustrated in the balanced view. As the examples provided in this chapter illustrate, the HSI practitioner must often take the initiative to bridge human and equipment requirements in such a way that the system requirements can be met in a cost-effective manner.

In this more balanced view of things, the results of human-related disciplines in the left-hand stream complement the traditional activities in the right. This is not just "supplementary information" but ranks on a par with the evolving description of what the equipment must do. The unique role of HSI is to restore this balance and ensure that the two halves are properly coordinated at all stages.

**Example 6.2 Radar Display Clutter** The engineers designing a naval command system knew that clutter on radar displays made it hard for operators to see targets. They had the technology to suppress clutter. Land clutter was more difficult, especially at the edges, but with advanced digital technology they could suppress all land clutter. Everyone including the user representatives agreed this was a good idea, since it would make target identification easier. The system went to sea, and operators found they could not see the coastline on their radar screens, so they turned off the sophisticated suppression, thus making the advanced technology useless. Had a proper task description (routine with HSI practitioners) been developed in parallel with the ideas on equipment function, it would have become clear that operators need to see the land in some situations (e.g., where the coastline is) as well as enemy aircraft. The technology could have met all their needs at negligible cost and enhanced their performance, but instead a technology was bought that was not used, and performance capability was reduced.

Placing the emphasis of the initial formal statement on operational capability, rather than just on equipment performance (i.e., placing the first box in the center, rather than at the right of the diagram) is having a fundamental impact on procurement thinking, certainly in the United Kingdom (UK). An initial thinking stage has always existed, but traditionally it has been prior to the creation of the first formal requirement (which was for equipment). Tracing requirements to statements of capability, rather than to equipment needs, represents a major opportunity for HSI, since it makes the case for introducing human performance into the system performance equation.

### 6.2.2 Human-Related Contribution to Requirements

A *capability requirement* is the first to be formalized. Its production is likely to be *user led* rather than *system engineer led*. It will be a well-structured document that defines the required capability in fairly high level terms. Subsequent, more detailed requirements will be traced to it. Although such a high-level statement of military need might seem far from the detailed concerns of much HSI work, it is essential to ensure that human-related concerns are properly reflected in it, to provide the "hooks" that can subsequently be elaborated into greater detail.

The key point to remember is that it is people working with the equipment procured as a result of the requirement that will deliver the capability. The statement must cover, at an appropriate level, all aspects that will affect the ability to work effectively with the equipment. Performance targets must cover human as well as equipment performance. Overarching requirements stemming from legal and moral obligations to people at large as well as people within the system must also be covered.

The *system requirement* as supplied to contractors is mainly focused on what the proposed equipment will do. It is detailed and can be quite large. It will normally be managed with some sort of system engineering tool. Standard requirements practice recognizes the functional (F), nonfunctional (NF), and constraints (C) types of requirement statement. The NF requirements cover performance, quality of service, etc. This nonintuitive usage is well established in the system engineering community but can cause misunderstanding (e.g., the requirement for the speed of a car is NF). In an extreme case, everything that is not *functional* might be bundled together as *nonfunctional.*

The HSI requirements should fit within this framework, and indeed they do, but Table 6.1 shows two additional types—human performance (HP) and process (P)—that are necessary when specifying manned systems and especially the interface between the human and the equipment. Although technically HP and P are subdivisions of the NF type, it is helpful to differentiate them for HSI purposes.

The HP requirements are critical, since without them there would be no contractual check on whether the equipment really was operable and able to integrate the human component properly into the whole system. Making HP requirements explicit allows them to be tested reliably and allows contractors to focus on how to meet them, knowing that it will affect *that part of* acceptance.

The P requirements help to fill an important gap where the functional or performance needs cannot be clearly specified. The buyer specifies something for the contractor to do (e.g., demonstrate design options, explore the effect of certain trade-offs, or conduct pilot trials) without placing unwarranted constraints on the design solution. Process requirements normally appear in the statement of work (SOW) accompanying a contract, but they are mainly at high level and of broad scope, applicable to the system development as a whole (e.g., requirements to maintain records and traceability). Process requirement statements within a system requirement will normally be more specific, relating to a

**TABLE 6.1 Types of Requirements**

| | Requirement Type | Content |
|---|---|---|
| F | Functional | What the equipment must do |
| NF | Nonfunctional | How, or how well, it must do it (quality and performance) |
| C | Constraints | Limits on the solution |
| HP | Human performance | How well the user must perform tasks using the equipment |
| P | Process | Things the contractor must do |

**TABLE 6.2 HSI Contributions by Requirement Type**

| Requirement Type | Suggested HFI Contribution |
|---|---|
| Functional | Functions needed to support human operator or maintainer, enhance or compensate for human performance limitations, and provide for human safety and well-being |
| Nonfunctional | Requirements to support specified human tasks and required equipment responsiveness to human component |
| Human performance | Required human performance while using equipment (error and/or time) and requirements for legibility, comprehensibility, ability to manipulate controls, etc. |
| Constraints | Features to accommodate human characteristics or mode of working and ensure human health and safety and limitations of human mental and physical capability |
| Process | Required involvement of users to ensure design elaboration takes account of detailed user needs; requirements for operability prototypes, demonstrations, etc., for evaluation; and requirements for coordination (e.g., to ensure common operating principles between different equipments) and process visibility in any areas of uncertainty |

particular issue or need that cannot be properly covered at that stage as a functional or performance specification. Such requirements will subsequently be replaced by other requirement types as the detailed requirement becomes clearer and can be unambiguously specified.

Human systems integration should contribute requirements to address the implications for the human component of the system at a comparable level of detail to those from other areas. Several examples of HSI contributions applicable to the various requirement types are provided in Table 6.2.

**Example 6.3 Equipment Versus Task-Focused Requirements** The requirements for a command system were derived by reference to the predecessor system on an incremental basis. New functions needed were added, functions that did not work well were specified more closely, and redundant functions were dropped. Human issues were of concern during the system definition study, in particular the workload that would be imposed on members of the command team. A contractor studying user tasks noticed that a major fraction of the workload of one member of the team came from encoding and decoding messages using a slow pencil-and-paper method. This significant chore could have been removed by a very simple piece of software as part of the facilities provided by the new system. It was not, because there were no requirements to provide that function. Unfortunately, the requirement was based on replacing the predecessor system, not on enhancing human and equipment performance collectively.

### 6.2.3 User Interface Requirements

Requirements for the user interface can be the largest set of HSI-inspired system requirements. The user interface mediates much of the equipment's impact on its users and their behavior. The results of much HSI analysis (target audience description, task analysis, workload analysis, error analysis, etc.) eventually make their impact on the equipment design via requirements for how it will interact with users.

**TABLE 6.3 User Interface Requirement Structure**

| Section | Typical sub sections |
|---|---|
| Context of use | Scenarios, users, environment, assumptions, tasks to be supported |
| Generic requirements—that apply systemwide | Overall concepts, consistency, views, interaction, alerts, etc. |
| Function area specific requirements—that relate to individual functions (e.g., for command systems they might be tactical picture, weapon management, navigation, etc.) | Task supported, views used, information required, actions supported, feedback, constraints, alerts |
| Nonfunctional requirements—quality and performance | Responsiveness, human performance, health, safety, compliance with HFI policy and standards |
| Acceptance | Methods and criteria |

For information-rich systems, a good case can be made for producing the user interface requirement even earlier than the main functional requirement. Typically, the user interface requirement can be derived from the results of an interactive requirements prototyping exercise with users. A user interface requirement is often structured much as a system requirement, with contextual information, and F and NF requirements. Because of the extreme importance of overall coherence in a user interface, it is sensible to separate generic requirements from those for specific functional areas. Table 6.3 shows a typical user interface requirement structure.

### 6.2.4 Acceptance of HSI Requirements

*Acceptance* is the formal process to certify that contractual commitments have been met and that the deliverables meet their requirements. Equipment acceptance is based on evidence that the equipment has the required attributes and performs as specified in the system requirements document (SRD). This evidence is normally based on a combination of inspection (of the equipment and supporting documentation) and tests or trials (of the equipment). Evidence should be gathered incrementally during development and manufacture.

Where HSI requirements are expressed in terms of conventionally testable equipment attributes (size, weight, brightness, etc.) based on well-proven standards or the result of prior trials, their acceptance is no different from that of any other equipment attribute. Simple (yes–no) requirements are checked off, while quantitative ones are measured (with ruler, stop watch, photometer, etc.). (See the first two types of tests in Table 6.4.) Other HSI requirements cannot be expressed and tested in this simple way but require direct evidence about how the equipment and users interact with each other, as shown by the third and fourth types of tests in Table 6.4. These requirements (for operability, maintainability, trainability, and supportability) must be tested using people as "test instruments." Doing this reliably needs special procedures and techniques.

The first four tests shown in Table 6.4 are roughly in order of increasing cost. It is therefore good practice to ensure that items that can be simply tested are specified so as to

**TABLE 6.4 Types of Systems Tests**

| Type | Description | Comment |
|---|---|---|
| Design inspection (DI) | Formal inspection of design documentation, against principles and requirements in the specification | Offers a cheap and convenient supplement to other approaches for larger amounts of information, normally backed up by selective use of other tests |
| Functional demonstration (FD) | Formal controlled presentation of equipment and its working to assess the presence or absence of functionality | Suitable for HSI requirements that (through evaluation, experiment, or reference to standards) can reliably be expressed in terms of simple, testable equipment characteristics, and functions |
| Task walk-through (TW) | Formal controlled presentation of task support facilities; test criteria based upon ability to complete tasks and qualitative measures of user acceptability | Brings human interaction into the loop; semisubjective but controlled; more cost effective than full operability trials, especially with large amounts of detail where complexity and cognitive performance are more relevant than physical and skill-based performance |
| Operability trial (OT) | Formal controlled and structured data collection of human performance or subjective user reaction against agreed, predefined criteria | Ultimate test of human integration; comes closest to a real-life test and can also be applied to higher levels of requirements hierarchy but costly to implement |
| Process review (PR) | Review of evidence of development process (records, minutes, plans, etc.). Criteria for acceptance are based upon the existence of required evidence | Does not test system characteristics directly but provides confidence in the way they were derived, is auditable, and enables further scrutiny of supporting evidence if necessary |

permit this. The available trials budget and time should not be used up performing operability trials to check well-proven details. Operability trials (OTs) and task walk-throughs (TWs) should be used for areas of uncertain task interaction, task complexity, and overall integrated task performance. The OT and TW should also be used to test things such as performance degradation over extended periods and skill retention during periods of nonuse, as appropriate.

The matrix in Table 6.5 shows which types of test are most suitable for which types of requirement. The primary test for functional requirements is functional demonstration (FD), but in some cases design inspection (DI) or TW might be appropriate. For example, TWs can provide a useful check that the functions have been correctly interpreted from a task perspective. The NF requirements vary considerably, with corresponding diversity in the appropriate way to test them. The primary test for HP requirements is OT, with TW a

**TABLE 6.5 Tests Suitable for Type Requirements**

| | F | NF | HP | P | C |
|---|---|---|---|---|---|
| OT | — | ? | xx | — | ? |
| TW | x | x | x | — | ? |
| FD | xx | x | — | — | x |
| DI | x | x | — | x | x |
| PR | — | x | — | xx | ? |

*Note:* xx = primary, x = alternative, ? = Possible, — = unsuitable.

cost-effective alternative in many cases. Process requirements are primarily tested by process review (PR) i.e., reviewing the process evidence, but in some cases it will be more appropriate to inspect the design resulting from the process. Constraints normally relate to functions and design detail.

Combinations indicated with a question mark represent unlikely situations and in most cases could be handled in a different way. For example, meeting a constraint to accommodate users who are physically small could be tested by OTs to see whether small users could perform tasks or by PR to see that small users took part in the trials. Usually, it would be related to dimensions of the design that could be tested (possibly with anthropometric modeling tools). Likewise, a NF requirement for machine response time could be tested in an OT by seeing whether it undermined user performance, but it would be simpler and cheaper to measure it.

*Acceptance* is a contractual process based on evidence that the system meets its requirements. Traditionally "acceptance" has been equated with comprehensive functional testing of the completed system. Current acceptance practice recognizes that this is not a cost-effective approach and that evidence should be gathered incrementally over the period of development and manufacture within a quality-controlled process. Deficiencies can be corrected earlier, and everything is not tested twice—by both manufacturer and customer.

Evidence of operability should be gathered as early as possible using the most cost-effective type of test. The customer may retain the right (selectively) to repeat some of the tests, but in many cases, it is possible to use evidence from early trials to substitute less costly tests for confirmation later, for example, in production.

**Example 6.4 Display Legibility Requirement** Consider a requirement for legibility of information on a display. Standards such as MIL-STD-1472 (DoD, 1998) specify minimum angular subtense of characters, luminance contrast, and so on, but legibility can be degraded by vibration, content, and other task-related factors. Use of larger characters and symbols can offset the degradation, but at a cost; it reduces the screen capacity and might increase task complexity if information has to be spread over more screens. At the requirement stage, it might not be clear where the best trade-off lies, so specifying the character size is not appropriate. On the other hand, a performance requirement can be specified, with criteria for permissible error rates based on the nature of the task. Testing the requirement for acceptance is feasible but is costly in time and resources. The performance requirement does not immediately translate into designer action, leaving a risk that equipment might be presented for acceptance that did not meet it. When this happens, there is a high risk that marginal failures (or worse) will be traded away to avoid undermining the whole program, since changes late in development cause delay and can be extremely costly. Conducting appropriate trials as soon as representative displays are available and the information to be displayed is well enough understood can reduce this risk. The tests would build confidence that the performance requirement will be met. The trials could also provide a basis for additional requirements for character size, contrast ratio, etc., that would be less costly to test. The original performance requirement would still stand but should not need to be tested unless there was evidence to suggest that the substituted criteria had been invalidated, for example, by some change in the mode of use.

### 6.2.5 Human Systems Integration Process Requirements

Requirements for the conduct of HSI have an important role to play in defense contracting. It could be argued that competent contractors do not need to be told how to proceed but that overlooks two benefits that such requirements bring.

First, mandating processes to generate evidence about human aspects of the system can reduce risk by enabling better coupling of contractor results with the procuring authority's own internal HSI processes and program, especially during extended development periods. Such requirements should focus on the evidence to be produced, rather than the detailed technical processes to produce it. The requirements should make clear what is needed, in what form, and when. Such evidence can include, for example, documentary results of analysis, modeling, evaluation, and tests as well as demonstrations and user interaction sessions with prototypes.

Second, since the cost of HSI-related work can be a significant part of a contractor's project budget, mandating key evidence-producing activities puts all competitors on an even footing. Thus the procuring authority is less likely to be faced with an *apples-and-oranges* comparison between a more expensive development bid supported by comprehensive HSI and a cheaper one with no guarantee of such support.

As well as requirements to undertake HSI activities, it is also sensible to require contractors to be able to demonstrate how the design solutions offered have been influenced by the HSI results. For example, the Dunchurch report (MoD-Industry Human Factors Integration Group, 1995, p. 6) concluded. "where designs are produced, task based justification should be expected and should form part of the judgement of contractor competence."

An alternative approach to ensuring that appropriate processes will be deployed is to assess the *capability maturity* of the organization. The concept of a capability maturity model (CMM), developed for U.S. government procurement of large software systems, has spread to other disciplines. One of the earliest relevant to HSI was the usability management maturity grid (Flanagan, 1996). The UK Ministry of Defence (MoD) has recently co-sponsored work under the International Organization for Standardization (ISO) on quality-in-use processes and their integration (ISO, 2001). This is a CMM rooted in the concepts of ISO efforts on human-centered design processes for interactive systems (ISO, 1999).

## 6.3 HUMAN SYSTEMS INTEGRATION REQUIREMENTS ISSUES

The process of taking full account of the human dimension of systems within engineering and procurement has changed markedly during the last decade. Although the basic framework now seems clear, the process is still evolving, and issues are still being worked out.

Three issues are discussed here:

- the need to integrate with a requirements engineering process that is itself maturing and becoming more tool dependent;
- evolving ideas about how to formulate HSI requirements, especially the role of user interface requirements; and
- how the disciplines of HSI can adapt to an acquisition regime increasingly dependent on "off-the-shelf" equipment.

### 6.3.1 Working with System Requirements

In order to ensure full integration of HSI concerns into system requirements, it is necessary to understand the broad structures of a system requirement and some of the factors that influence it. System requirements invariably become large and are usually structured hierarchically, though other forms of structure can be imposed by the various software tools used to manage them, for example, to show traceability or dependencies. The use of such tools by system engineers imposes a constraint on those responsible for HSI who (to be fully integrated in the team) need to express requirements using the same tool and thus work within its constraints.

***Hierarchical Structures*** Traditional system requirements were either hierarchically structured textual documents or "flat" databases. The former are difficult to track and the latter are difficult to understand. More recently, requirements management tools have improved the situation with hierarchically structured databases that can automatically generate the structured documents. Even so, it is not uncommon for these to be used in such a way that only the headings are hierarchical, with all the "actual" requirements at the bottom level.

Good requirements engineering practice (Hunt, 1997) encourages the generation of a hierarchy of requirements statements, with a small number of "parent" requirements linked to "children" that describe contributory requirements, specify interfaces, or allocate performance budgets. For example, a high-level requirement might specify the rate at which aircraft will be able to take off. Lower level ones might specify the rate at which fuel can be delivered, the interaction between aircraft movement and ship safety, or the time to be allowed for moving aircraft between decks.

The hierarchical structure permits a much better understanding of how the whole requirement "hangs together" than a large collection of low-level statements. It also provides a better view of how requirements, especially *emergent properties*, of the system as a whole can be tested.

As illustrated in Figure 6.2, a "good" requirement will have a diamond-shaped profile, with a small number of requirement statements at the highest level increasing to a maximum at midlevel and reducing again at very low levels of detail. This differs from a *design* which typically has a more triangular shape with a lot of detail at the base. Traditional requirements databases often "captured" the content of well-thought-out documents only to produce very many (thousands) of low-level statements of detail.

Ideally, HSI requirements should be incorporated in the hierarchical structure of system requirements, not all isolated in a separate section. In many cases, HSI requirements will appear at most levels of the hierarchy, not just at the bottom but higher up as well. Indeed,

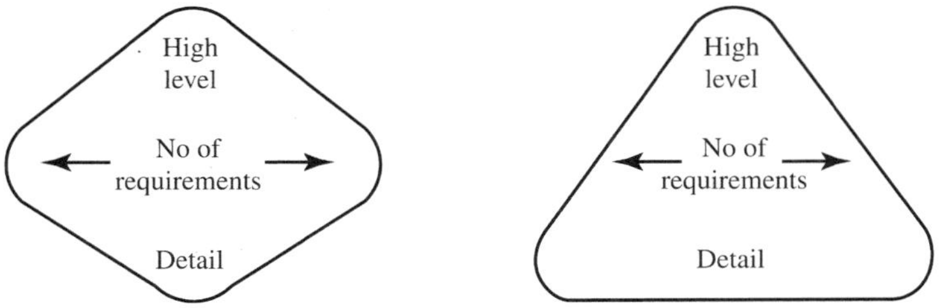


Good Requirement Profile

Bad Requirements Profile
(Good Design Profile)

**Figure 6.2** Requirements profiles.

many of the more challenging and important human performance requirements will relate to the performance of whole tasks or jobs, i.e., significant areas of functional support. Section 6.3.3 describes how a hierarchical structure could be used to provide coherence to a user interface requirement.

***Subject Matter*** The detailed structure and content depend on the type of system and mainly contains specific, itemized statements about what is needed. The HSI requirements should be subject to the same quality criteria as other requirements. There are many criteria to differentiate between *good* and *poor* requirements; some relate to individual statements and overall structure, whereas others are not always easy to apply. Criteria of particular relevance to HSI are as follows:

- *Justification* The wording should clearly show how HSI-inspired requirements affect system cost and/or effectiveness, so they do not appear either abstract or "merely common sense."
- *Verification* Requirements that cannot be verified are not taken seriously. Requirements such as ease of use are of less value than specified percentages of representative users being able to achieve designated tasks to a performance criterion. The statement of required capability is not the place for great detail, but it should provide the "hooks" from which more detail can be elaborated in the system requirement produced for industry.
- *Solution Avoidance* Avoiding solutions without being too vague to be effective can be difficult for user interface requirements. Specifying the need to support identified tasks, with appropriate performance requirements, is most effective at the high level. Specifying information and controls to be provided, with format relevant to the task, is appropriate at the low level.
- *Clearly Understood Need* The problem might be understood, but if what is needed to solve it is not, then it would be better managed as a risk, with corresponding activities in the plan to quantify it. Task or broad performance-based requirements can often act as "place holders" for such issues in the requirement, possibly with "to be determined" (TBD).
- *Comments* Some requirement structures allow the addition of comments. Although not part of the definitive requirement statement, an explanatory comment can often

help users of the requirement document to understand it more effectively. Understanding plays a key role in the way requirements are interpreted as well as the "legal" meaning of the words. Many of the people who use the requirement document will not be approaching it from a human-centered perspective.

- *Links to Other Requirements* Human systems integration should tie in with everything else. Note that there should be linkages to as well as from HSI requirements.
- *Overall Balance* The requirements should adequately reflect the importance of the human-related issues faced by the project, but it should not be dominated by too much detail. The level of expression is important.

Perhaps the most difficult aspect for HSI is testability. Section 6.2.4 described different ways to test human-related requirements, but in practice such things can be quite hard to specify. Well-meaning phrases such as "Displays must be clear" might be adequate as reminders or checklist items but will not pass the rigor of requirements engineering, and even if they do slip through, they will be of little help in acceptance. For marginal cases, the contractor will be unsure whether it has met the requirement, and if not, proving it will be difficult.

Focusing on testability has an unfortunate side effect. It can distort what is specified. Easy things to measure get included, while important but harder to measure things can be quietly forgotten.

**Example 6.5 Trivial Detail Can Dominate Requirements** An information system for a military headquarters involved many complex display screens. The details of all the screen layouts had been derived by analysis and signed off by the customer. Acceptance of the screens was determined by whether they conformed to the agreed screen definitions. To add precision, and because it was easy to test, the analysts had included dimensions on the drawings, i.e., the number of millimeters between the edge of the screen and data fields. During implementation, some of these dimensions were a few millimeters out. This was detected during inspection and the system had to be reworked, often to extremely tight deadlines, to correct the "deficiencies." However, this effort was of little use in determining whether the screens would be useful to the headquarters personnel who would be using them. There were no requirements related to whether the screens were legible or could be used to perform real tasks.

It is often difficult to specify and test what matters most, as in the example above. Traditionally, attempts to specify important higher order properties have resulted in well meaning but vague requirements such as ease of use that in the hard contractual world carry little weight. In most cases, human performance underlies the required quality and should be identified clearly in the early requirements. The expression should be as explicit as possible, specifying what it is that the target audience are required to be able to do and what criteria (time, error, accuracy, etc.) should be used to judge success. The actual thresholds for the criteria might not be known at this stage, but the contractors have a clear view of how they will be judged, and the procurer has a clear agenda of TBDs to be resolved by trial or other means as part of the HSI program.

Subsequent work must refine or elaborate these requirements. Some will lead to full acceptance criteria requiring operability trials with representative users and the associated procedures to ensure reliability of the results. This is costly and not practical for every detail. Recourse to standards, risk-based evaluation, and prototyping will allow some requirements to be replaced by equipment attributes that are easier to test (see Section 6.2.4 on acceptance).

***Different System Types*** Procurement organizations often classify systems in terms of the customer area or the technology, for example, aircraft system, electronics system, missile system, ordnance system, ship system, space system, and surface vehicle system. These are still broad classes and cut across many HSI issues. For example, a ship system could be a whole aircraft carrier, a propulsion system, or a navigation system. An electronic system could be a radar set or a command information system.

Some different types of systems will have a more significant impact on how HSI is handled:

- information-rich systems (also called software intensive systems),
- complex multiprocurement systems, (e.g., platforms), and
- manual intensive equipment.

*Information-Rich Systems (Software Intensive Systems)* Requirements may be structured in two separate parts, representing *infrastructure* and *applications* (Computing Services Association, 1992). Infrastructure covers physical aspects (equipment on which the software runs) and the underlying *software architecture* for communications and information storage. In software parlance, applications run on this "platform." Applications requirements are often the most numerous—mostly functional requirements split up by area (e.g., weapons, sensors, navigation).

The most direct HSI contribution to a requirement of this type is to specify requirements for the user interface. Many will relate to the applications and should be integrated with each functional area in the requirement structure. Other more generic user interface requirements should be identified separately, possibly integrated with the software infrastructure requirements. Physical aspects of the user interface (size of controls, force required, brightness of displays, etc.) fit logically as a subsection within the *physical characteristics* of the equipment.

Other HSI requirements will be located in the various nonfunctional sections, though there is a case for closer integration of some of these with the functional requirements that they qualify (see Section 6.2.2).

*Complex Multiprocurement Systems* Major procurements such as military platforms are managed as clusters of systems and subsystems under an overall umbrella procurement. There are separate but related requirement sets at each level, i.e., the whole platform, its major components (hull, propulsion, combat system, etc.), and various subsystems such as navigation.

The platform-level requirements should not duplicate lower level detail, but they must contain the definitive HSI requirements from which lower level requirements can be derived and to which they can be traced. Otherwise, there would not be adequate contractual incentives for them to be fully addressed at the subordinate system levels. Some HSI issues can seem abstract at the platform (or major system) level and may be difficult to address, but particular areas (i.e., whole system performance, work system boundaries, complementing requirements, and coherence) should be covered (see Table 6.6).

**TABLE 6.6 Platform-Level Requirements**

| | Description | Comment |
|---|---|---|
| Whole-system performance | Overall capability, subsequently translated into more concrete requirements that can be apportioned between the different component systems (in theory irrespective of whether the systems contain machines, people, or the usual mixture of both) | In practice, this can miss important aspects of performance that depend on people and in some extreme cases can lead to critical equipment that supports human functions such as communication, being almost "invisible" in formal measures of effectiveness |
| Work system boundaries | Contextual information to ensure that it flows down to each of the affected lower level requirements in a consistent way, especially where human roles, and hence the need for human integration, cross procurement boundaries | Work system boundaries, i.e., the job boundaries of individuals or teams, commonly differ from equipment boundaries |
| Complementing | Overall requirements for and constraints on the number and type of people who will operate, maintain, and support a system or a group of systems | Partly because of work system boundaries (as above), partly for platformwide issues such as damage control, accommodation, and watch keeping |
| Coherence | Requirements for consistency of operating practices, user interface conventions, labeling, etc. | Should be made explicit, whether standards, preestablished conventions, project specific "style guides," or merely aspirations |

**Example 6.6 Seating Is Part of the System** A warship procurement included a complex new command system. Human issues featured strongly in the command system procurement, with great efforts to take account of them. Console designers used anthropometric models and a full-scale mock-up to ensure operators would be able to see and reach all of the controls comfortably without risk to health and safety. At the platform and compartment level there was no overarching plan for HSI. The contract to supply the seating was let separately from the contract to develop the consoles, both independent of the ship builder, and with no exchange of information between the various contractors. As a result, the positioning of the seats relative to the consoles resulted in some operators having to twist their spines to use the consoles. This potentially costly risk to the health of the crew could have been avoided by HSI requirements at the platform level.

*Manual Intensive Equipment* This inelegant title covers the numerically large number of procurements of "hands on" equipment that are often less glamorous than the big complex projects. Two points are worth noting:

- Small items are often procured in large numbers. The number of people using basic items for personal use multiplies the impact of deficiencies in their human compatibility. Individual effects might not be "showstoppers," but the collective effect on force effectiveness can be significant.

- In some cases, incompatibility of simple items can become a showstopper because basic human interactions were not properly anticipated.

**Example 6.7 Integration on the Soldier** A helmet, body armor, and gun sight were separately procured and performed their specified functions. When integrated on the body of a soldier lying prone, the body armor pushed the helmet forward, making it impossible to see through the sight.

The requirements for such procurements are less complex than for larger systems, and the subject matter is likely to be differently structured. Nevertheless, the same principles apply, and the same type of HSI input is needed. The questions to be answered are the same:

- What must human and equipment be capable of doing (together and separately)?
- In what context must they do it?
- What tasks will the human need to perform?
- Under what conditions will the tasks be performed?
- How well must the human perform them?
- What equipment properties, capabilities, and behaviors will make this possible?

### 6.3.2 Risk-Based Approach to HSI Requirements

There is a conundrum underlying HSI. Human issues must be recognized very early to avoid major failures of human integration, and yet many of the overtly human-related interventions in equipment development appear to be of a detailed nature best suited for later stages. Traditionally this has led engineers and managers to delay consideration of human-related issues until detailed design, by which time it can be too late or too costly to remedy major shortfalls.

The UK MoD has developed early human factors analysis (EHFA), a simple, intuitive process to help project managers identify and assess human-related risks early in a project. Contractors working on systems with a human dimension can also apply EHFA, but the scope of the risks "owned" will depend on the contractual relationship. Ideally, client and contractor should jointly manage shared risks.

Taking a risk-based approach to HSI makes it easier to know what must be done early and what can be safely left until later. This approach should provide three links with the main system engineering effort:

- Human-related risks can be fed directly to the project risk register. They may need some aggregation to match the level at which other project risks are managed but should be better formulated and more comprehensive having emerged from a formal process.
- After assessment, most human-related risks will point directly to the need for actions to mitigate them or at least to quantify them (e.g., analysis or evaluation). Thus the risks will drive requirements for the HSI program (within the project plan), and the

formal assessment underpinning them should help to justify their priority in the queue for limited project resources.

- Some human-related risks, once identified, can be nominally removed by the creation of new system requirements.

**Example 6.8 Skill Practice Requirement** The risk that operational performance of an occasional task might be inadequate because of skill fade could be mitigated by a requirement for a built-in training facility to permit regular practice of the skill.

### 6.3.3 Traceability of User Interface Requirements

Table 6.3 outlines a basic requirements structure for user interface requirements, but having a separate user interface requirement (or section) does not solve all problems. To be valid, a user interface requirement must map onto the task needs of the users, and to be viable, it must map onto the functional requirements for the equipment.

The need to map between user interface requirements and other functional requirements can in principle be handled by using the same requirements tool for both and using it to link the two parts of the database. For example, a requirement for the user to view a particular parameter would be matched with requirements to measure or receive the parameter at a suitable rate and resolution and process it into a form suitable for viewing (e.g., by smoothing it).

In practice, the different levels to which different areas of requirement are broken down during early phases of acquisition can make this form of linking difficult. It also raises issues about tool compatibility between procurement authority, consultants, other agencies, and contractors. Such issues do not arise with simple documents or even flat databases, (see Section 6.3.1).

Mapping between user interface requirements and user task needs is a more complex problem. Traditionally the link has been via the analyst's understanding, but this is not readily amenable to automated checking or tracing to determine the impact of changed requirements. User task needs are traditionally documented in a task structure (often a hierarchy) separate from the user interface requirements. Recent work (Harrison, 1999) has suggested the possibility of linking the two and using the task structure as the organizing framework for the user interface requirement. This would only really be possible using a suitable requirements management tool. The concept is illustrated in Figure 6.3.

The goal level at the top of the task tree represents the small set of high-level responsibilities that define an operational role. Typically they represent the granularity at which tasks can be readily delegated.

Requirements attached higher up the tree would apply to the whole tree below (e.g., broad information needs, types of view required, and alerts relevant to responsibility). Those attached at the bottom would relate specifically to individual actions. Information needed for tasks, constraints, or feedback might appear at different levels depending on whether they cover a single action or a cluster of related actions.

Requirements for alerts would normally link directly to a goal, since it is the goal level responsibility that determines the "need to know" and justifies delivery of the alert to the individual role. Below the alert, both information and an action represent the needed response.

Of course, in some areas the task tree might be quite shallow, possibly with only a single level between goal and action.

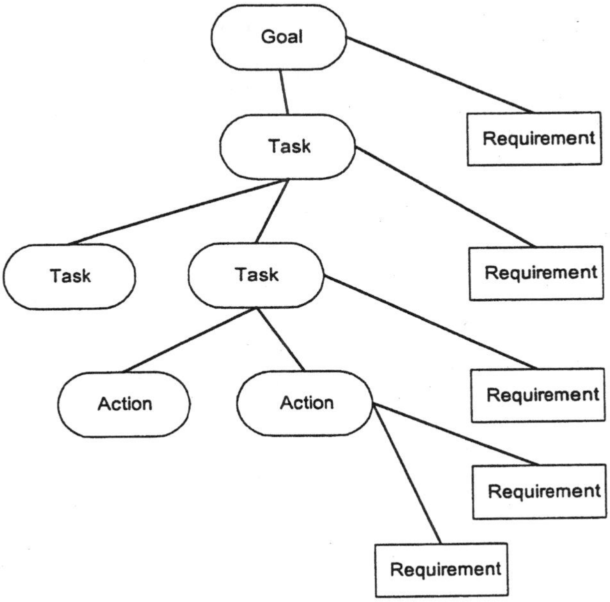


**Figure 6.3** Task structure as a framework for user interface requirements.

### 6.3.4 Implications of Commercial Off-the-Shelf (COTS) Equipment

The major influence of HSI is normally through design interventions, initially at the requirement stage and subsequently through development. Where for political, commercial, or other reasons it is intended to procure *off the shelf*, it is harder to see how the discipline of HSI can exert an influence, but it can, if properly focused. Presented below are a few points of clarification:

- *True COTS means that knowledge of performance can be used in selection.* Buying an item that exists offers the supreme advantage of "try before buy." Real performance information can be used to inform the selection, and to a considerable extent this should reduce the risk of not being able to make design interventions (to forestall unanticipated performance problems).
- *COTS is not nondevelopmental item (NDI).* The term COTS is sometimes wrongly used to describe a NDI. An NDI is not bought off the shelf but off someone else's drawing board. Despite the reduced cost of initial purchase, this offers the worst of both worlds—the risks of new development without the ability to make design interventions.
- *The whole system is not off the shelf.* Rarely is an equipment system bought entirely off the shelf. Normally one or more major components is, with others added or modified. In the extreme case where components are bought off "different shelves" and brought together, the system thus formed is new and requires design at the system level if it is to work. The fact that the system designer's hands are tied by the use of predesigned components is a major constraint that makes the design of the system difficult, but that is not the same as having no design to do.

- *The human component changes.* Even if the equipment system were bought entirely unaltered off the shelf, the system that will be expected to deliver the capability will still be new, because the human component will be different from that with which the equipment previously worked. People from a different force will be integrated with the equipment in a different tactical and possibly physical environment. They will be organized and trained differently, probably come from a different culture, and probably be required to deliver a slightly different operational capability.

From this it is clear that there is no such thing as an off-the-shelf system. The issue is how to design effective systems of people and equipment using major predefined components off the shelf. The principles of system engineering (and HSI) apply to systems in a COTS environment as they do to others, but the constraints are different, notably in the areas of trade-off and in how to manage the issues of human integration.

A particularly important difference between COTS (and other NDI) acquisitions from more conventional acquisitions is the restraint placed on conducting the most critical HSI design analysis. The following two points elaborate on this difference:

- In a COTS or other NDI acquisition, the most critical HSI design analysis is done in the somewhat iterative cycle of requirements statement (including, of course, the human performance requirements) and in the market survey.
- The requirement–market survey cycle occurs before any contract with the vendor is made.

At the level of capability requirement, there is no difference between the two approaches, but the dynamics of the subsequent process are different. A COTS (or part COTS) option will be considered because it offers potential advantages in one or more trade-off. Usually these include cost, time to deliver, and (equipment) development risk. A COTS option must be evaluated in a similar way to any other option to ensure that all the implied costs (initial and through life) are understood and that the risk of it failing to deliver the desired capability is acceptably low.

The difference between COTS-based options and those involving new development lies in the mechanisms available to intervene if some aspect proves unacceptable on initial evaluation. With new development, many limitations can be overcome by design intervention. There is normally a cost, but the cost can be modest if the intervention is made early. Detailed intervention at lower levels can often be left until later, provided such a contingency is planned. The scope for design intervention with a COTS component is severely limited, since changes to a developed item (even if possible) are usually very costly.

In practice, this means that any identified shortfall in total system performance or incompatibility between human and equipment components can only be mitigated by changes in the human component or by changes to any non-COTS elements that bind the major items together. By analogy, we might call the latter "glue." Such changes might indeed be able to make good the deficiencies, but neither should be assumed to be capable of easy change.

The human users come at the end of the line. If the COTS component is unchangeable and the glue used to join them together cannot be adapted to make good any incompatibilities, then the users will be left to try to make the whole system work. If they cannot (or if they can only do so with a penalty such as high risk of errors or risk of long-term health hazard), then the military customer might be faced with capability failure, high unplanned remedial costs, or both. The COTS-based systems represent a significant problem for HSI, because far more problems must be foreseen and forestalled (before commitment) rather than being detected and cured (during development).

The severe limitations on downstream intervention with a COTS-based system option make it essential to establish whether the system will be capable of the required performance and to identify any intervention needed to make it do so, either with non-COTS items or with the human component (training, selection, support, etc). Only when the true cost of the total system is known can a valid comparison with other options (COTS or not) be made. If the performance cannot be achieved, even with feasible interventions, then either the option must be rejected or the capability targets reduced, in which case other options might need consideration.

The HSI requirements play a key role in ensuring that COTS-based *systems* are selected on a sound basis, thus avoiding the risk of cheap COTS *equipment* leading to high downstream human-related costs and/or poor overall system performance.

Table 6.7 compares HSI activity relevant to COTS-based acquisition, with typical HSI activity for a new development, based on the system engineering activities of which they are a part.

Once a COTS selection has been made, the emphasis of HSI switches from foreseeing the human implications in order to influence the selection to system optimization within the remaining degrees of freedom (or damage limitation if the selection did not adequately account for human-related concerns). The scope for this remaining action is limited to

- any options within the COTS component (e.g., built-in customization facility or changes negotiated as a condition of the selection);
- influencing the design of system components other than the COTS items;
- identifying the need for task aids (e.g., cognitive props such as crib sheets, pocket calculators, overlays, or physical props such as lifting aids, supplementary tool boxes, vision aids);
- optimizing operating procedures;
- optimizing the training; and
- (exceptionally) making a more restrictive selection to increase the skill levels.

The last three options involve changes to the human element of the system. Some of these changes might prove untenable (e.g., tighter selection when manpower supply is already inadequate). All are likely to have a cost (which should have been foreseen and accounted for during the selection process). Changes to procedures and training might introduce less predictable side effects, such as increased errors caused by negative transfer from other equipment, the need to change selection criteria, or the need to reallocate work to different parts of an organization.

Given the difficulty of making major changes to the human component, the most effective way to obtain satisfactory human integration in a COTS-based system is to influence the initial option selection intelligently, by highlighting the true cost and performance of the different options, not just patching up the system afterward.

**TABLE 6.7 Comparison of Systems Engineering and HSI Activities for COTS-Based Systems**

| Systems Engineering Activity | HSI Activity | HSI Activity Relevant to COTS |
|---|---|---|
| Define required capability | Identify human issues implied by the capability. | As left (should be solution independent) |
| Identify and assess system options to provide it | 1. Identify human issues associated with predecessor systems.<br>2. Identify differences in context of use and predict impact on system options.<br>3. Assess human-related risks and requirements for each option. | 1. Identify human issues associated with COTS elements in current use, including user performance.<br>2. As left, informed by current use of COTS components.<br>2a. Seek evidence of compatibility of COTS equipment with intended target audience and operational tasks, drawing on existing service performance, comparability analysis, and evaluation of performance in relevant trials.<br>3. As left. |
| Define system options for comparison and selection | Ensure human parts of overall system (manpower, training, support, etc.) are adequately defined and costed. | 1. As left.<br>2. Identify and cost all additional equipment needed to make overall system work.<br>3. Identify and cost human interventions (selection, training, support, etc.) needed to make overall system work.<br>4. Identify and cost any performance shortfalls of overall system due to mismatch between equipment and people. |
| Select option | Take part in option trade-off across all system domains. | Inject above into option trade-off process. Focus on the total system, not just the COTS equipment. |
| Specify system requirements | 1. Identify human-related system requirements.<br>2. Identify human-related risks still to be addressed.<br>3. Plan activity to mitigate human-related risks. | 1. As left, but focusing on any freedom within COTS components, on glue components, and on performance requirements for overall system.<br>2. As left<br>3. As left. |

**Example 6.9 Aircrew Performance Requirements** When a replacement aircraft was ordered, the aircraft performance was specified, but inadequate attention was given to the performance of the crew. The aircraft was a derivative of a successful in-service product, but the operational tasks and the manning of the in-service item differed significantly from that intended for the new version. It rapidly became apparent to the HSI team that there would be major performance limitations in some operational conditions, but with contracts already let,

firm evidence would be needed to make any changes. That evidence could only be gained from flight trials. The contracted requirements for flight trials were based on demonstrating acceptable performance of the slightly varied aircraft, with no allowance for evaluating crew performance in the new role. Trials were on a tight schedule that could not be changed. As a result, the project continued heading toward expensive problems that no one could prevent, because requirements had not captured the human implications of operating the aircraft in a changed context of use, and there were no contractual requirements for trials to demonstrate crew performance as part of the total system.

## 6.4 UNITED KINGDOM HFI PROCESS

After running a UK version of manpower and personnel integration (MANPRINT) for land systems in 1990 and then a slightly modified HFI program for sea systems from 1991, the UK MoD adopted a triservice HFI policy that eventually became mandatory for all acquisition projects (UK MoD, 1998). Human factors integration inherited the six MANPRINT domains: manpower, personnel, training, human factors engineering, system safety, and health hazards.

### 6.4.1 HFI within Smart Procurement

Following the strategic defence review (SDR) in the late 1990s, the UK adopted an approach to capability acquisition commonly known as smart procurement (UK MoD, 2000). Smart procurement was motivated by far wider issues than HFI, but it is notable that many of the principles it embodies accord well with the changes that the HFI initiative seeks to achieve. Table 6.8 summarizes the key features of the Acquisition Management System (AMS) that implements smart procurement, and their relevance to HFI.

Focusing on capability directs attention toward effectiveness in use (output) rather than equipment performance (input). This provides a more secure basis for reasoning about the human contribution as a part of the solution, rather than something to be added separately when the equipment is in service. The option for a *nonequipment* solution also recognizes the potential to increase effectiveness by upgrading the human component (procedures, organization, training, etc.) in a more positive way than the *do-nothing* option under which such action would previously have been assessed.

Smart procurement picks up some of the HFI changes that were already occurring. For example, the role of HFI focus had already been mandated (UK MoD, 1998) but the greater autonomy of the integrated project team (IPT) and the more formalized inclusion of a wider range of stakeholders in the process should make it a more effective role. Perhaps the greatest challenge to HFI in smart procurement, as under any regime, is the demand it creates for broad-based, talented personnel to manage HFI effectively. The MoD has put in place guidance aimed at nonspecialists responsible for managing HFI (Defence Evaluation and Research Agency, 2001).

The manner in which HFI management should work within a smart procurement project is illustrated in Figure 6.4. The figure shows a high-level view of the HFI process during early development up to main-gate approval. The left-hand side of the diagram shows the HFI management roles of identifying and understanding the human-related issues, supported by analyses of various kinds, and the right-hand side shows how HFI information is used to enrich the key project outputs: requirements, plans, and solutions.

**TABLE 6.8 Smart Procurement Features and HFI**

| Feature | Comment | HFI relevance |
|---|---|---|
| Explicit focus on acquiring capability | MoD has always planned for capability, but previous equipment procurement regimes were focused on equipment rather earlier. | Capability delivered by a combination of people and equipment is the underpinning rationale for HFI. |
| Standing body within MoD is responsible for capability management | Capability working groups (CWGs) spawn new projects and act as the initial central customer. | The broader perspective should enable a better view of cross-project HFI issues. |
| Integrated project team (IPT) manages a project | Its members (core and associate) span a wide range of specialities. | The IPT manages requirements, ILS, (Integrated Logistic Support) and HFI. One of its members assumes the role of the HFI focus. |
| CWG works ahead of and then alongside new IPTs to create user requirement document (URD) | URD is the first formal statement of required capability. | The CWG should raise initial human-related issues. URD should include key human-related requirements (see Section 6.4.3). |
| IPT supported by CWG explores options for providing capability | Options lead to system requirements document (SRD), a formal requirement sent to industry to offer solutions. | SRD defines boundaries between human and equipment components with performance specified at the boundaries as well as for the overall system. |
| Streamlined acquisition life cycle with approvals at initial gate and main gate | Acceptable business case needed to pass gates. Otherwise IPT leader is free to act within agreed-upon limits of performance, cost, and time. | Opportunity for well-managed cost-effective HFI that can demonstrate added value. |
| IPT manages project through its whole life cycle | Concept, assessment, demonstration, manufacture, in-service (including incremental capability upgrades), and disposal. | IPT is responsible for development costs (which fund much HFI work) and downstream costs (where HFI can save cost of training, support, etc.). |
| Industrial collaboration encouraged from the start | Normally an initial period when contractors must collaborate with MoD while competing with each other. | Collaboration should ease HFI management, but the period of partial collaboration and partial competition represents a challenge. |

**TABLE 6.8** ***Continued***

| Feature | Comment | HFI relevance |
|---|---|---|
| After down selection chosen contractor becomes full member of IPT | | Coordination of HFI with industry should be easier, with better management of trade-off and less duplication, e.g., evaluations leading to acceptance. |
| Government-funded development not assumed | Off-the-shelf procurement, public–private partnerships, etc., should be considered. | The full implications of this shift have yet to be worked out but promise to take HFI into new areas. |

The whole HFI process depends on effective integration with other stakeholders, shown at the top of the diagram, while the bottom of the diagram shows explicit links to requirements and risk management.

Following main gate approval, the focus shifts more to implementation and evaluation, but the requirements process is revisited whenever there are changes, especially when upgrades are initiated.

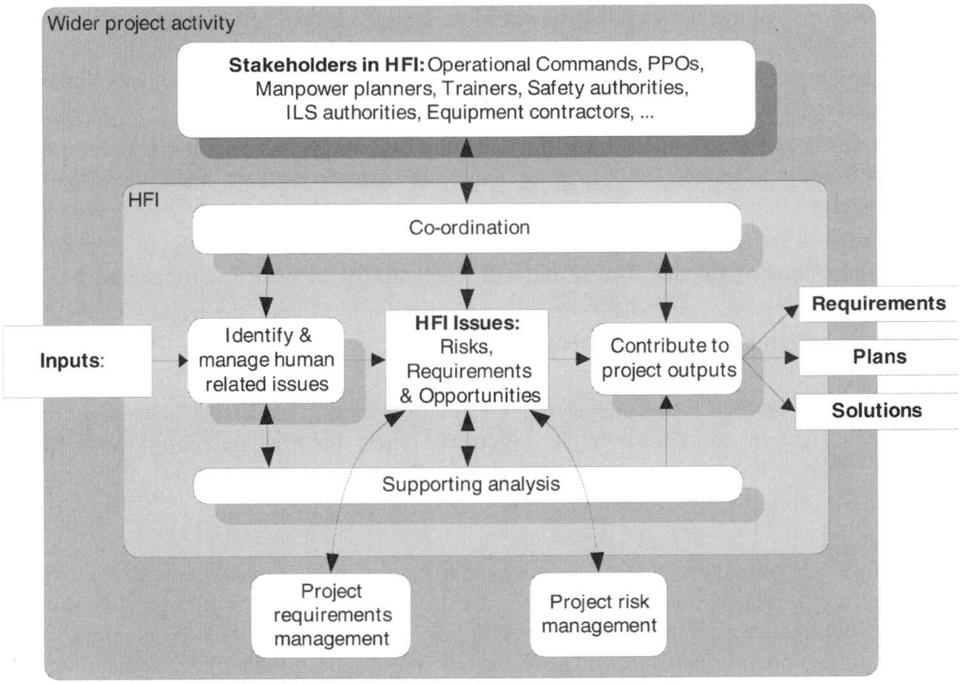


**Figure 6.4** Overview of the HFI management process.

### 6.4.2 Early Human Factors Analysis

Acquisition projects conduct an EHFA as soon as possible in the concept phase, with review at the start of subsequent phases and when initiating major capability upgrades during the in-service phase. A simple, intuitive process that need not require extensive resources, depending on the project scale, EHFA helps project managers to identify and assess human-related risks, for example: capability failure due to human performance limitations, difficulty using or maintaining equipment, the number and type of people required, health and safety problems, or training and maintenance costs. The outputs of EHFA can feed directly into project risk management, requirements engineering, and project planning.

Initially, EHFA was developed to "kick start" the process of HFI early in a project (Defence Evaluation and Research Agency, 1996). Being simple makes it easier to mandate, and by helping to focus on where most value can be added, it appeals to project managers. The need for EHFA grew out of the desire to help identify the key human-related issues early enough for them to be addressed effectively and to feed forward lessons learned from in-service equipment.

Early human factors analysis comprises seven steps:

1. *Document the project baseline* for the analysis (documents, concept options, requirements, constraints, etc).
2. *Document assumptions*, including those arising from baseline material as well as others that become apparent during the analysis.
3. *Identify concerns* that might represent risks to the project. Encourage stakeholders to make concerns explicit rather than taking them for granted or assuming they are not important.
4. *Review the concerns* that need further analysis and treat them as key issues or requirements.
5. *Analyze the key issues* to ensure that they are expressed unambiguously and are properly understood and that associated assumptions have been identified and checked.
6. *Estimate risks* associated with each key issue in terms of the likelihood and dimensions of impact. The result will feed into the project risk register.
7. *Identify strategies to reduce serious risks* in order to provide the basis for planning a work program to reduce human-related risks.

As shown in Figure 6.5 the underlying information model for EHFA is simple; EHFA groups substantive concerns into four general types, broadly reflecting the different motives of stakeholders who contribute them. The terminology reflects the terms in which stakeholder worry statements are often expressed. These are:

*It Must...* What the system must do is a natural way for stakeholders to express concerns, to retain good features of the predecessor they think might get lost, correct shortcomings in the predecessor they think might get overlooked, or meet more demanding operational needs. These may or may not be formally articulated in the requirements. When focusing on human issues, these statements are likely to be about things that look small overall but have a big impact on operability. Some of these might

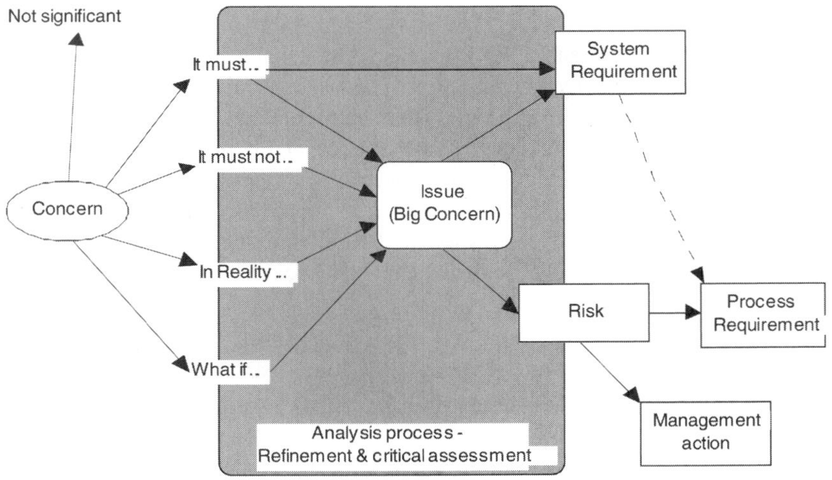


**Figure 6.5** EHFA underlying information model.

feed straight into requirements, but others will not, since the concern is not in the articulation of the need but reflects a belief that it might be subverted by other factors.

*It Must Not . . .* These statements are likely to relate to bitter experience of what has gone wrong in the past. They will not normally find their way into requirements (but they might subsequently with inverted wording, e.g., "the power shall be at least *x*" for "it must not be underpowered"). As such things have happened in the past, there is a natural concern that they will do so again.

*In Reality . . .* This covers a multitude of insights and information about the real-world context in which the system will be used. It is mainly motivated by a concern that the procurers and developers do not adequately understand operational realities. The insight could relate to the environment, the way things are used, the difficulty people have doing things, or the user's experience with the predecessor. These statements are likely to require more interpretation to generate either requirements or risk statements. Some might be better used to inform the use study, task analysis, or target audience description.

*What If . . .* This reflects the creative attempts of stakeholders to look beyond what they currently have and visualize future possibilities of equipment, scenarios, or situations, including emergencies and extreme conditions, or equipment being used in ways for which it is not designed.

Figure 6.5 shows two outputs from the central issues box. Except for the dotted line between system requirements and process requirements, there would be a clean divide between system requirements on the one hand and risks plus the processes and actions to manage them on the other. (The dotted line is included because requirements such as operability are often more effectively expressed, at least partly, in process terms; see Section 6.2.2.)

### 6.4.3 HFI in the URD

The user requirement document (URD) describes the capability needed. It is a relatively *high-level* document (typically several dozen pages) central to the business case for proceeding with the project. Subsequent system requirements will trace back to it, so it must include HFI "hooks" in appropriate places. HFI must inform requirements where people either are or might be involved, even if the human contribution to the capability itself is subsumed within more general statements. This will help avoid later difficulties when deriving the more specific SRD from the URD. The HFI content will depend on the human dimension of the capability requirement. One of the following will apply:

1. A human-free solution is not acceptable for some reason (ethical, moral, or legal).
2. Support to humans is the objective (transporting, housing, and providing them with information, etc.).
3. The most cost-effective (or only technically feasible) solution involves humans.
4. Constraints apply to humans used in the solution.

In cases 1 and 2, humans are inherent to the requirement and must be fully covered by requirements within the URD.

In case 3, human inclusion is a matter for the solution domain (i.e., the SRD). Therefore, the URD should not specify humans in the solution. It might however be sensible to word requirements in the URD in such a way that they can be readily interpreted in terms of human components if this is likely, in order to simplify later trace from SRD to URD.

In case 4, the human aspects are constraints on the solution, rather than requirements for what it must be able to do. These constraints (at a suitable level) must be covered by the URD unless (anticipating the SRD) humans can be categorically ruled out of the solution.

***General Description (Part 1)*** Table 6.9 suggests the HFI contribution, assuming a typical URD structure (UK MoD, 1999).

***Key User Requirements (KURs) (Part 2)*** The KURs form a small, high-level subset of requirements that epitomize the whole need. If they fall short, then the whole capability is undermined. Most military capabilities are critically dependent on human-related requirements. Whether they appear here as separate top-level items or are aggregated with other concerns will depend on the specific situation.

***User Requirements and Constraints (Part 3)*** Most HFI input will appear in the full set of atomized requirements and constraints. The structure will vary with the capability. The HFI contributions might include

- human functions—inherent human tasks that must form part of the capability;
- human support functions—specific equipment capabilities needed to enable effective performance and safety of the human component;
- user and organizational constraints, e.g., the expected availability, characteristics, and performance of the human component; and

- measures of effectiveness (MOEs)—ensure that the MOEs cover all the capabilities of the required system where capability is quantified, in particular those that depend on the performance of the human component.

***Context Documents (Part 4)*** These supplement and provide extra depth to help understand the user need, particularly for those less familiar with the background to the requirement (e.g., industry). They are vital for formulating validation and acceptance tests and trials. Context documents express the conditions under which effectiveness must be achieved. Key reports of HFI studies, particularly EHFA, should feature among the context documents. In some cases, there should be HFI contributions to other context documents, where the people issues feature strongly.

***Priorities*** The HFI requirements must fit within the overall priority structure, as shown in Table 6.10.

Performance requirements may be specified at different priority levels, e.g., maximum number of actions to perform a critical function might be (KUR) 4, (E) 3, (H) 2, and (D) 1.

### 6.4.4 HFI in the SRD

The SRD specifies a solution to the requirement in terms of what it will do. In most cases, the SRD also makes some high-level decisions about which components (equipment or human) will provide different functions. The boundary around those functions allotted to equipment will represent the contractual boundary for its procurement. The SRD must therefore be explicit about the interfaces between functions and the performance requirements (of human and equipment) at the interfaces.

**TABLE 6.9 Suggested HSI Content for Part 1 of Typical URD**

| Section | Suggested HFI Contribution |
|---|---|
| Background | State inherent human-related need (case 1 or 2) |
| Single statement of need | Human issues probably not explicit unless:<br>• Human performance a key driver (e.g., needs enhancing from what could be achieved with current equipment in future scenarios)<br>• Manpower cost a key driver (e.g., must be reduced while maintaining operational effectiveness) |
| Assumptions | Future manning, personnel, and training, projects that could share resolution of human-related issues, equipment with which human component must interoperate |
| Dependencies | Human component in related capabilities, outcome of trials |
| General constraints | Limits on manning, personnel deployment, and factors needed to avoid degrading performance and sustainability of human component |
| Users (of the capability) | Where capability users will interact with it directly, e.g., hands on or receiving information from it as part of their operational tasks, describe key characteristics (enough to help focus on concepts and options that match intended users) |

**TABLE 6.10 Requirement Priorities**

| Priority | Code | Definition | Example HFI Requirements |
|---|---|---|---|
| Mandatory | M | Must be met; requirements and constraints represent legal obligations | Health and safety at work, conditions of employment |
| Key user requirement | KUR | Mission critical to users of capability; functions not tradable; performance trade-off below a given level might require reendorsement | Factors affecting human performance of critical functions, factors affecting ability to man the solution |
| Essential | E | Subject to affordable technical capability; functions not traded without reference to user | Factors affecting manpower costs |
| Highly desirable | H | Tradable, but more important than desirable | Factors affecting working efficiency and noncritical error rates |
| Desirable | D | | Other human-related enhancements |

The SRD forms a key part of the business case for proceeding past main gate. Industry's response, and the ensuing contracts, will be traced back to the SRD, so it is essential to ensure that HFI requirements are included in all appropriate places. The SRD is more detailed and therefore larger than the URD but mirrors its structure, with the detailed internal structure depending on the nature of the system being specified. Tables 6.11 and 6.12 indicate the typical HFI content of an SRD.

Note that although a separate section on performance is usual, there is a case for including some performance requirements in the main functional sections, alongside the functions to which they relate. The contributions shown under human performance are in this category.

*Operability* is listed as another nonfunctional requirement in the SRD template, but it is fundamental to the effectiveness of systems involving people. Operability is about the equipment's effect on human performance. Human performance requirements predominate and should be testable by operator performance trials supplemented where appropriate by task walk-throughs. In areas where there are well-proven design rules to enhance operability, these should be specified and will generally be tested by inspection or demonstration of the corresponding equipment function.

All requirements should be accompanied by agreed-upon acceptance criteria. These might not be fully defined at initial issue but should be completed before main gate. Pass thresholds (values of criteria to be exceeded) should be agreed-upon before contract. See Section 6.2.4 on HFI in acceptance.

***Human Components in SRD*** The main documents for specifying the human components are the target audience description (TAD) and a high-level task description.

**TABLE 6.11 Typical HFI Content within an SRD (Equipment Component)**

| Equipment | Typical HFI-related content |
|---|---|
| Context | Human-related assumptions, reference to high-level task structure, reference to target audience description (TAD) (see below) |
| Functional | Functions needed to support human operator or maintainer, enhance or compensate for human performance limitations, and provide for human safety and well-being |
| Performance | Required equipment responsiveness to human users |
| Nonfunctional | |
| Reliability | Requirements to minimize risk and impact of human errors that could cause failure, e.g., inadvertent operational error (violation of safety rules, loss of information, etc.) or maintenance error (incorrect settings, things fitted wrongly, etc.) |
| Maintainability | Requirements to reduce demands on time and/or skill of maintainers (e.g., to reduce manpower or training costs) and reduce stress and/or hazard to maintainer (e.g., ease of access, visibility, ease of fitting) |
| Operability | Required human performance using equipment (error, time, accuracy); human interaction requirements (legibility, comprehensibility, ability to manipulate controls, performance over extended periods, etc.) |
| Safety | Requirements to mitigate adverse effects of system on people who come into contact with it (e.g., "standards", legislation, duty of care); requirements to avoid hazards caused by human action (e.g., erroneous use of confusing controls); requirements relating to operator safety, stress, fatigue, and boredom (might change dramatically with changed technology) |
| Security | Requirements to make human aspects of security effective (e.g. memorable passwords), requirements for support to human security enforcement (e.g., usable security audit tools), any emergency overrides and safeguards (if appropriate) |
| Engineering standards | Requirements to accommodate human size, strength, etc. (including future populations); requirements for consistency of use with other systems (e.g., style guides, conventions) |
| Environmental | Aspects that would affect performance or well-being of human component (e.g., vibration, air quality, heat efflux) |
| Support | Key human aspects of support requirements; constraints on size, weight, portability, etc.; requirements to simplify loading, assembly, setup, etc. |
| External interface | Interfaces to operators, maintainers, support personnel, or other people (detail will vary with the type of system) |

### 6.4.5 HFI Content of Requests for Industry Participation

The MoD will invite industrial participation in projects at different phases and in different ways, depending on the nature of the project. Such invitations specify the scope of what the contractor(s) will be required to do as well as system requirements for any equipment to be delivered as a result of the ensuing contract. This should include appropriate HFI requirements. The main types of invitation are as follows:

- *Expression of Interest* A preliminary invitation, published in the *MoD Contracts Bulletin* and the *Official Journal of the European Communities*, this provides a brief description of the project need.
- *Prequalification Questionnaire (PQQ)* This is issued to obtain evidence of contractor capability in order to select a short list. Where management of human-related issues and requirements will be needed, the PQQ should include questions covering relevant experience, resources, etc.
- *Invitation to Tender (ITT)* This formal statement of requirements prior to contract is normally supported by at least a (draft) SRD (or URD for a concept phase study) and a statement of work (SOW). The SOW should include requirements to provide HFI evidence in support of study results, designs, etc., and specify what HFI contribution will be available from the MoD and how the contractor is expected to collaborate with MoD HFI activity (joint working groups, demonstrations, provision of prototypes, etc.).
- *Invitation to Submit Outline Proposal (ISOP)* This is a formal statement of requirement whereby the MoD anticipates contractors will (partly) align commercial development work with an emerging MoD requirement. The HFI requirements are similar to above, but probably with less formal interaction.
- *Invitation to Negotiate (ITN)* Following from above, this occurs when the intention is to adopt a solution heavily based on a commercial development. The HFI requirements will focus mainly on providing evidence of operability, maintainability, safety, etc., to enable coordination with the MoD program.

**TABLE 6.12 Typical HFI Content within SRD (Human Component)**

| Human | Typical Content |
|---|---|
| Context | Human related assumptions, Task & role context |
| Human tasks | Tasks to be performed: inherent human functions, human functions that must be supported, human functions to support equipment, human tasks (outside system) that form part of role holder's job |
| Human performance | Required performance of key tasks |
| Target audience description (TAD) | Physical, sensory, and psychological, social, and cultural characteristics; organization, training, and career structure |
| Manning | Availability |
| Constraints | Policy, legal, and other constraints covering health, safety, well-being, etc. |

## 6.5 SUMMARY AND CONCLUSIONS

The increasing dependence of the armed forces on technology means that military capability must be delivered by a carefully integrated combination of people and equipment. The extreme demands often placed on both can undermine that integration unless specific measures are taken to ensure it is explicitly managed. Procurement based on equipment requirements alone cannot guarantee delivery of intended capability.

Requirements for equipment must therefore be seen as part of the requirement for a wider system, including the people, and this means that equipment requirements must be strongly influenced by human-related requirements. During the last decade, the knowledge needed to ensure effective integration of people and equipment has been institutionalized into the processes of acquisition and procurement, initially in the United States followed rapidly by the United Kingdom and others.

Different initiatives have emphasized different aspects of HSI. The initial approach of mandating HSI processes, with many formal HSI deliverables, achieved some notable successes by avoiding major project costs, but in more routine projects it often proved costly for the benefit gained or else the cost deterred its effective application on the ground. The UK introduced EHFA as an explicitly risk based process to address the cost-effectiveness issue, but cost-effective follow-through still relies on the skills and influence of individuals involved.

Recent thinking has recognized that system requirements are a powerful means to influence the eventual system through the mainstream system engineering activity. Injecting well-supported, human-related information into system requirements provides high leverage. In particular, making human performance requirements explicit provides a firm basis for later acceptance of operability and overall system effectiveness.

Contributing material into system requirements and integration with system engineering are not substitutes for the important analysis and trade-off in manpower, training, task analysis, etc. These need to be done, and using their output to help shape the system requirements directly can be a more effective means of system intervention than merely delivering HSI reports and hoping that someone else will take appropriate actions.

Human systems integration is a relatively young discipline, as is requirements engineering. Both are still evolving and there are many issues to work through. The HSI practitioners need to become more comfortable with system engineering methods and the routine use of system engineering tools, especially requirements engineering tools. One worthwhile goal will be to achieve integrated specifications of the equipment and people (task) aspects of a system in a single tool support environment, without making the result too complex for user stakeholders to understand.

The practice of HSI has been pushed forward on some types of systems more than on others. It is unreasonable to expect the HSI contribution to the requirements for all sorts of systems to look alike, and the HSI approach must adapt to suit different types of system requirements.

The risk-based approach to HSI in general and HSI requirements in particular is appealing to project managers but can appear to HF professionals as an excuse for cutting corners. The approach needs to mature before its full impact can be assessed.

One of the biggest challenges to HSI is the increasing pressure to base systems on COTS, with the dominant component often being predeveloped for a different context of use. In such cases, the role of HSI must engage early and influence the high-level choices of option rather than focusing on the development phase (which will not exist for the COTS components themselves). Equally, where political, economic, or other pressures force a compromise with human-related penalties, considerable creativity will be needed by HSI professionals to manipulate the few remaining variables to achieve a working system—before the "get well program."

Traditionally HSI requirements have been either too vague to be enforced or at too low a level of detail. Often what was easy to specify or measure took precedence over what really mattered to ensure that people could perform effectively and hence to achieving the desired overall capability. A proactive approach to the early identification of human-related risks and the systematic inclusion of human-inspired detail within the whole fabric of engineering requirements can help overcome these problems. On that foundation can be built systems of equipment that integrates properly with the human component to deliver the sought-for military capability.

## REFERENCES


Computing Services Association (CSA). (1992). *Views on the Procurement of Software-Intensive Defence Systems*, Annex to ASSC/330/3/117. London: CSA.

Defence Evaluation and Research Agency (DERA), Centre for Human Sciences. (1996.) *Early Human Factors Analysis*. Farnborough, UK: DERA.

Defence Evaluation and Research Agency (DERA), Centre for Human Sciences. (2001). *Human Factors Integration (HFI)—Practical Guidance for Integrated Project Teams*. Farnborough, UK: DERA.

Downs, E., Clare, P., and Coe, I. (1992). *Structured Systems Analysis and Design Method—Application and Context*, 2nd ed. Hemel Hempstead, UK: Prentice-Hall.

Flanagan, G. A. (1996). Usability Management Maturity, Pt 1—How Do You Stack Up? *SIGCHI Bulletin*, *28*(4).

Harrison, J. A. (1999). Validation of FC BISA—HCI Specification, B/C140/FD.7/01/01. Report to Defence Evaluation and Research Agency, Land Systems, Fort Halstead, UK.

Hitchins, D. K. (1998). Systems Engineering—In Search of the Elusive Optimum. *IEE Engineering Management Journal*. (Vol 7, No 3, pp. 114–116.)

Hunt, L. B. (1997). GMARC—Getting the Requirements Right—A professional Approach. Paper presented at the IEEE Conference on Software Technology and Engineering Practice (STEP 97).

MoD-Industry Human Factors Integration Group. (1995). Task Analysis and MANPRINT. Report of MoD-Industry workshop held at GEC Management College, Dunchurch, UK.

International Organisation for Standardisation. (1999). *ISO 13407 Human-Centred Design Processes for Interactive Systems*, TC 159/SC4. London: British Standards Institute.

International Organisation for Standardisation. (2001). *Ergonomics—Human-System Interface. Human-System Life Cycle Processes*, ISO/TC 159C 4 N HSL. London: British Standards Institute.

UK Ministry of Defence (MoD). (1998). Chief of Defence Procurement Instruction. *Managing Human Factors Integration*, CDPI Tech 330. Bristol, UK: MoD.

UK Ministry of Defence (MoD). (1999). *Smart Requirements Model. Acquisition Management System*. Bristol, UK: MoD.

UK Ministry of Defence (MoD). (2000). *The Acquisition Handbook—A Guide to Smart Procurement "Faster, Better, Cheaper,"* 3rd ed. London, UK: MoD.

U.S. Department of Defense (DoD). (1992). *Systems Engineering*, MIL-STD 499B. Washington, DC: DoD.

U.S. Department of Defense (DoD). (1998). *Department of Defense Design Criteria Standard: Human Engineering*, MIL-STD 1472E. Washington, DC: DoD.

CHAPTER 7

# Human Systems Integration and Acquisition: Contractor's Perspective

BRUCE E. HAMILTON

## 7.1 INTRODUCTION

Most of the information useful to the human systems integration (HSI) practitioner on the acquisition process has been developed by government organizations from the point of view of government activities and tasks. This chapter provides a different focus—that from the contractor's point of view. Because the contractor is constrained by the contractual language of specifications and standards, much of the emphasis will be on the contractual process that goes on between the buyer and seller. For example, typical HSI tasks required throughout a typical contract are identified not only generally but more specifically in terms of their relationship to contract milestones and required products.

The federal government acquisition process defines, requests, funds, and provides authority for effective (HSI) during product development. This process has changed considerably over the past 20 years in regard to the visibility and effectiveness of human factors in the design process. The change has generally progressed from a few domains of HSI being considered as something to be added to the basic design program to a completely integrated set of HSI domains being considered an inherent part of systems engineering and management throughout the acquisition process.

Several factors have contributed to this change. First, government buyers of systems began to realize that human capabilities were limiting the performance and effectiveness of major high-technology systems. Second, the personnel costs of maintaining and supporting military and space program systems were found to be prohibitive (accounting for more than 50 percent of the total life-cycle costs). Third, the complexity of new systems required multidisciplinary approaches to system design starting with the buyer's requirements process (the manner in which the buyer documents requirements and needs to its vendors), continuing through the contractor's design and development phases, and culminating in the system performance demonstrations prior to buyer acceptance. The HSI approach to

*Handbook of Human Systems Integration*, Edited by Harold R. Booher.
ISBN 0-471-02053-2 

systems integration not only provides skills and technology that provide positive effects to each of these factors but also its encouragement to focus on the human throughout the process has (for highly successful systems) become the "design driver."

The process by which products begin their life and ultimately are produced is called the system life cycle. A number of very good overviews of the current processes and phases of system life cycle have been described (Kirk, 1973; Clark et al., 1986; Blanchard and Fabrycky, 1990; Cushman and Rosenberg, 1991; and Kirwan and Ainsworth, 1992). Each overview has slightly different terminology, but their descriptions have more similarities than differences. The systems acquisition framework chosen for this chapter will closely follow the military systems life-cycle process shown in Blanchard and Fabrycky (1990) and as laid out in the military handbook *Human Engineering Program Process and Procedures* [U.S. Department of Defense (DoD) (1999)]. This is because military weapon systems procurements have driven the maturation of human factors from sideline commentator to design driver throughout the system life cycle. As more companies become certified to external, international quality assurance management system standards, such as ISO 9000, the process by which products are developed and manufactured will become more standardized such that the distinction between military and commercial processes will be reduced.

The following discussion will cover three major topics:

1. The stages of a procurement activity, from contract award through the system life cycle up to testing and certification. The critical contractor products and HSI tasks will be described for each contract major stage.
2. The principal documentation events of the contractor solicitation and selection process, which include the buyer solicitation announcement, the buyer request for proposal (RFP), the seller proposal, and the buyer source selection.
3. Guidelines for the contractor HSI practitioner attempting to plan and manage an integrated HSI program for the first time.

## 7.2 STAGES OF PROCUREMENT ACTIVITY

Up until October 2000, it was mandated that a new military system be acquired in four major phases, generally identified as phase I, concept exploration; phase II, program definition and risk reduction; phase III, engineering and manufacturing development; and phase IV, production and deployment. This framework is described in the DoD (1998) regulation 5000.2-R. This mandatory framework of procedures has recently been replaced by DoD Directive 5000.1, which provides guidance through a set of management policies and principles. The cancellation of the mandatory procedures does not diminish the utility of the acquisition that is still commonly used as a model for civil government, military and civilian acquisition. For example, the National Aeronautics and Space Administration (NASA) Johnson Space Center (JSC) Program Life Cycle and the System Engineering Process (JSC 49037; NASA, 1993) is the controlling document for the life cycle of a system, either acquired or built within the agency. This document mirrors that of DoD 5000.2-R and mandates the four-phase framework for the JSC facility. In time a new framework or paradigm for acquisition may emerge, but for the moment the old procedures remain the guiding framework. (See Chapter 4 for detailed discussion of DoD Directive 5000.1).

Consequently, the systems acquisition framework most useful for our discussion in this chapter derives from the four phases laid out in DoD 5000.2-R. The purpose of the concept exploration phase is to conduct the research necessary to support the programmatic decision that the technologies involved have firm scientific basis and have demonstrated application to the system being considered. The purpose of the program definition and risk reduction phase is to transition new critical technologies from the laboratory to practical demonstration. This leads to the programmatic decision that the necessary technologies and resources are mature enough for start of system development. The purpose of an engineering and manufacturing development phase is to design and implement the system to the point where technology risks between components and subsystems that could only be evaluated as part of a completed system have been tested and the maturity of the designs demonstrated. This leads to the programmatic decision that the system can begin phase IV, the production and deployment of a mature, tested system having the necessary characteristics as originally envisioned in phase I and as systematically modified in phases II and III.

Two approaches are important to helping the reader to better understand the role of HSI required throughout the system life cycle. One is to outline the differences in types of tasks and goals for each of the acquisition phases. For example, Table VI, human engineering (HE) analysis methods selection characteristics of (DoD, 1998), shows that during the concept exploration phase, HSI practitioners may be conducting mission analyses, while in the production and deployment phase, they may be conducting workload analyses. The other approach is to help the reader comprehend what it takes for successful accomplishment of the specific contract within each acquisition phase. The different HSI tasks and goals among the four phases are well addressed in the above reviews. Consequently, the emphasis in this chapter is upon the latter, stressing what is required for successful accomplishment of a contract within a system life-cycle phase.

Broadly, there are four key milestone reviews of contractor-developed products for a typical systems acquisition contract. The policies and principles presented in DoD Directive 5000.1 provide flexibility that may modify or even eliminate the framework phases, but it is anticipated that most specific contracts will have these primary milestones:

I. Program requirements review
II. Preliminary design review
III. Critical design review
IV. Testing and certification

Once a contract for work has been awarded, the contract has its own stages and milestones. Accomplishment of each of these contract milestones allows the program to proceed from its current life-cycle phase to the next life-cycle phase. These milestones tend to be constant across the broad gamut of products and possible acquisition strategies. While contracts can be modified, the basic pattern of contract stages tends to be that outlined in Figure 7.1.

The contract award starts a process in which the high-level requirements of the contract are decomposed into low-level specifics. The low-level specifics are then assigned to various disciplines and a preliminary design emerges. After approval of the integrated design, detailed production drawings are created and test articles produced. After certification and testing, production can begin.

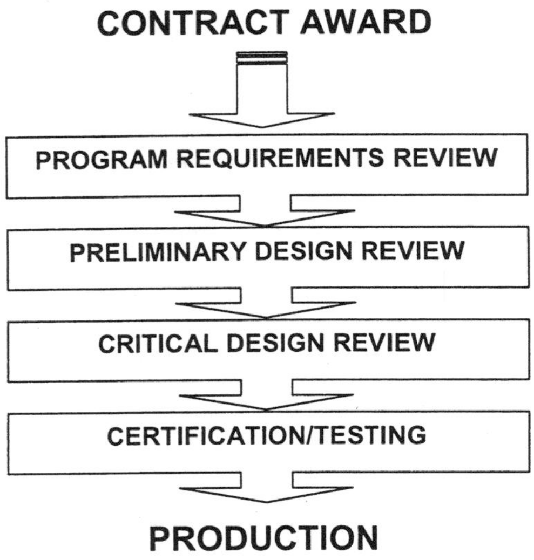


**Figure 7.1** Stages in development of a product: overview.

Control of this process is maintained by a hierarchy of documents and a specification tree. The specification tree starts with a single specification that documents what the product must do and what tests must be performed. These are the "parent" requirements to which successive levels of the tree must provide derived "children." The goal is to decompose general requirements into unique, testable design tasks that stand alone. In this manner, every high-level requirement can trace to a specific design task and each low-level can trace to a high-level parent. The sibling relationships can be understood and documented as interfaces.

The HSI tasks will occur in all stages of this process. The tasks will range widely dependent upon phase and task.

The first contract stage begins with contract award and ends with the program requirements review (PRR). At this review, the contractor demonstrates how he or she, as the seller, has accounted for the requirements of a contract and how the proposed product will ultimately demonstrate compliance. Thus at the PRR the contractor demonstrates that he or she understands the requirements of the contract for a product as it is to be designed and developed.

After successfully passing the PRR, the contractor begins to put together the design that will be used to satisfy the requirements of the contract. This second contract stage ends with a preliminary design review (PDR), which links design features to contract requirements. Subsequently, the contractor begins to make detailed designs and fabrication drawings.

The third contract stage begins after PDR and ends when approximately 90 to 95 percent of the drawings are done. This is when the critical design review (CDR) is held, at which the contractor is cleared to proceed to production of equipment and prototypes. The final contract stage is that of testing and certification. In this stage, the designs are tested to demonstrate that the specifications outlined during the PRR have been met. Additionally, some components may require certification when it is not adequate to merely demonstrate the component can accomplish its intended function. For instance, a component may be a critical safety component that is to last 100 hours in use. The contractor may be required to certify that this component will perform as designed for the intended life span. The following sections expand upon these activities and identify the tasks for the HSI program as a function of the milestone reviews of the contract.

### 7.2.1 Contract Award to Program Requirements Review

The initial contract stage starts upon contract award and ends with the PRR (see Fig. 7.2).

After the contract award, technology research and/or applied research may be needed in order to select material, processes, techniques, and/or technologies with the appropriate characteristics for use in satisfying the goals of the development. During this stage, basic decisions are made about how to approach the design and development of the product.

Type A specification, or system/system segment specification, supplemented by other referenced specifications, as necessary, are developed to specify (1) all essential functional characteristics, (2) necessary interface characteristics, (3) specific designation of the performance characteristics of key functional elements, and (4) all of the tests required to demonstrate achievement of each specified characteristic.

Typical HSI tasks include functional allocation, determination of control/display characteristics, preliminary human interface selection, establishment of personnel limitations, initial task and workload studies, development of target populations, lessons learned, and initial manpower estimates.

A PRR is conducted after functional analyses and preliminary requirements allocation studies are completed. These studies determine the initial direction and progress of the contractor's system engineering management effort for convergence upon an optimum and complete configuration (DoD, 1976). The total system engineering management activity

**Figure 7.2** Stages in development of a product: program requirements review.

**TABLE 7.1 Items for Review in a Program Requirements Review**

| | |
|---|---|
| Analyses | Plans |
| • Mission and requirements analysis | • Integrated test planning |
| • Functional flow analysis | • Data management plans |
| • System/cost effectiveness analysis | • Productibility analysis plans |
| • Logistics support analysis | • Preliminary manufacturing plans |
| • Program risk analysis | • Configuration management plans |
| • Human factors analysis | • Milestone schedules |
| • Life-cycle cost analysis | |
| • Manpower requirements/ personnel analysis | Specifications |
| | • Preliminary requirements allocation |
| | • Generation of specifications |
| Studies | |
| • Trade studies | Miscellaneous |
| • Specialty discipline studies | • Technical performance measurement |
| • System interface studies | • Engineering integration |
| • Value engineering studies | • System safety |

and its output are reviewed for responsiveness to the statement of work (SOW) and system requirements. The products to be reviewed may include any of the items in Table 7.1. Depending upon the major acquisition phase, the contract SOW (and deliverable items list) will require some subset of these products and describe the overall scope of the effort.

***Type A Specification*** The most significant product emerging from the initial-stage contract activities is the type A (or system/system segment) specification. The type A specification (DoD, 1995, p. 6) will

- state the technical and mission requirements for a system/segment as an entity,
- allocate requirements to functional areas,
- document design constraints, and
- define the interfaces between or among the functional areas.

When the requirements review is over, the system/system segment specification will outline the design to be produced, the interfaces, requirements, and test conditions. This document should both support the need for HSI and allow HSI requirements to apply to subsystems for implementation. If the type A specification does not recognize HSI requirements and does not require implementation in its subsystems, implementation of an HSI program will be difficult. Human systems integration must get its requirements integrated into the type A specification or face having its requirements viewed as optional.

***HSI Tasks for First Contract Stage*** There are three major HSI tasks in the initial contract stage. The first major task is compiling and classifying the HSI domain inputs to the system specifications. It is critical that HSI requirements be reflected in the type A specifications. If the requirements are in this document, the program is obligated to pass these requirements to the various subsystems, track that the requirement is implemented, and test for compliance. Chapanis (1996, p. 71) identifies seven classes of HSI input to the systems specification in the requirements stage (see Table 7.2).

The second major HSI task in the initial contract stage is to understand how to make the greatest HSI impact. Chapanis suggests this depends greatly upon recognizing the differences in two kinds of specifications known as "design to" and "build to." The design-to specification focuses on functional requirements whereas a build-to specification attempts to prescribe a proposed solution. Chapanis (1996, p. 71) states that the "distinction between the two kinds of specifications is important for the human-factors professional because the amount of human-factors detail to be supplied depends on the kind of specification for which that information is supplied." For example, a design-to specification might specify a 10-button keyboard be included for operator use while a build-to specification might specify the exact size, shape, and layout of keys on the board. The type A specification should be a design-to specification. The HSI practitioner preparing for a PRR should describe what is functionally required rather than a proposed solution.

The third major HSI task in the initial contract stage is to carefully read the contract and do the following:

1. Find every paragraph in the SOW that has HSI implications or explicit requirements and make a copy of them.
2. Find every deliverable item that HSI is either directly responsible for or provides inputs to and the dates the items are due.
3. Determine the HSI budget for accomplishing both of the above.

There is a significant amount of work for HSI during the period of time between contract award and PRR. To the degree both the buyer and seller have provided quality HSI inputs, it is in this stage that the greatest impact for the lowest cost can be made by HSI by virtue of early involvement. The buyer should identify specific HSI tasks in the SOW and deliverable items list. The seller should recognize an appropriate degree of HSI "front-loaded" activity and estimate man-hours, materials, and budgets accordingly.

***Example of HSI/System Activity for PRR*** At a PRR of a major U.S. Army helicopter program, a full day was devoted to demonstrating that the contractors under-

**TABLE 7.2 Suggested Human Factors Inputs to System Specification**

| | |
|---|---|
| Human-performance requirements<br>• Staffing, operating, maintaining, and support requirements<br>• Human–machine interfaces requirements<br>• Identification of areas in which human errors would be particularly serious | Dimensional and volume requirements<br>• Crew spaces<br>• Operator station layouts<br>• Ingresses<br>• Egresses<br>• Accesses for maintenance |
| Methods of operating the system | Maintainability requirements |
| Personnel requirements<br>Health and safety requirements | Training requirements |

stood how their proposed system was integrated into typical operations of the U.S. Army. The contractors started with a demonstration of how planning for the use of the system within a mission would be conducted and then walked the reviewers through the various operational activities, including

- mission planning,
- support processes to ensure aircraft and support material would be available,
- individual planning,
- mission conduct,
- after-mission briefing,
- aircraft maintenance, and
- the point where another mission cycle could begin.

This activity highlighted the effects of HSI, showing, for example, how crews would be able to meet timelines and performance criteria within the stated manpower goals and the likely available personnel skill capabilities.

### 7.2.2 Program Requirements Review to Preliminary Design Review

The next major milestone after the PRR is the PDR (see Fig. 7.3).

In the early stages of system design or development, functions are allocated to hardware, software, or people. Early decisions made with little regard to operator capabilities and limitations are likely to result in expensive training, staffing, or redesign of products.

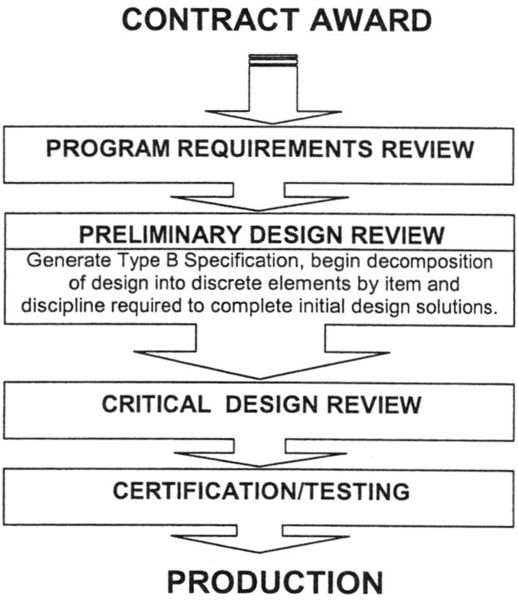


**Figure 7.3** Stages in development of a product: preliminary design review.

The type B specifications, or development specifications, state the requirements for the design or engineering development of a configuration item during the development period. Since the breakdown of a system into its elements involves identification, management, and control of configuration items of various degrees of complexity, it is desirable to classify by subtypes. Generally these are prime, critical, hardware, software, interface, noncomplex.

The HSI emphasis during this stage should be upon influencing the system engineering process. Considerations are how human performance affects system performance, identification of skill levels the target population will have, training impacts, performance measurement issues, human–machine allocation, and providing a common view of the human interface across elements and components.

The PDR ends the second contract stage. Each configuration item or aggregate of configuration items will have its own PDR. There are four primary purposes of the PDR. First, the PDR evaluates the progress, technical adequacy, and risk resolution (on a technical, cost, and schedule basis) of the selected design approach. Second, the PDR determines the compatibility of the proposed design with performance and any engineering specialty requirements of the hardware configuration item (HWCI) development specification. Third, the PDR evaluates the maturity of the design definition and assesses the technical risk associated with the selected manufacturing methods and processes. Finally, the PDR establishes the existence and compatibility of the physical and functional interfaces between the configuration item and other items of equipment, facilities, computer software, and personnel. The term *configuration item* is a contracting term used to denote hardware or software products whose design is being monitored and formally controlled by the contract. For computer software configuration items (CSCIs), the PDR focuses on the evaluation of the progress, consistency, and technical adequacy of the selected top-level design and test approach; the compatibility between software requirements and preliminary design; and the preliminary version of the operation and support documents.

The PDR is a formal technical review of the basic design approach for a configuration item or a functionally related group of configuration items. It should be held after hardware specifics are completed and preliminary drafts of supporting computer documentation are available. The list in Table 7.3 is an example of typical products reviewed in a PDR.

***Type B Specification*** The activities between the PRR and PDR mainly involve the generation of requirements documents and preliminary drawings. After the PRR and leading up to the PDR, the agreed-upon type A specifications (from the program requirements stage) are "decomposed" into lower level, more detailed specifications—the type B specifications. Type B (development) specifications state the requirements for the design or engineering development of a product during the development period. Each development specification should be in sufficient detail to describe effectively the performance characteristics that each configuration item is to achieve when an item has matured into a detail design. As shown in Table 7.4, there are five forms of type B specifications.

The systematic statement of a requirement and the generation of additional, detailed requirements necessary to implement the requirement are referred to as the *decomposition* process. Decomposition of higher, more general requirements into lower, more detailed requirements is at the heart of the specification effort. In a perfect program, each sentence of the type A specification, the high-level document, contains a single, unique, testable

**TABLE 7.3 Products for Review During Preliminary Design Reviews**

| | |
|---|---|
| Design<br>• Preliminary design synthesis of hardware development specification for item(s) being reviewed<br>• Equipment layout drawings and preliminary drawings<br>• Environment control and thermal design<br>• Power distribution and grounding design aspects<br>• Preliminary mechanical and packaging design of consoles, racks, drawers, etc.<br>• Interface requirements<br>• Mock ups, models, breadboards, or prototype hardware<br>• Transportability, packaging, and handling<br>• Standardization<br>• Human engineering and biomedical<br>• Safety engineering considerations<br>• Electromagnetic compatibility survivability/vulnerability | Data<br>• Pertinent reliability/maintainability/availability data<br>• Preliminary weight data<br>• Development test data<br>• Preliminary lists of materials, parts, and processes<br><br>Analyses<br>• Trade studies and design studies<br>• Functional flow, requirements allocation data, and schematic diagrams<br><br>Software<br>• Functional flows<br>• Storage allocation<br>• Control function description<br>• Software structure<br><br>Miscellaneous<br>• Development schedules<br>• Security<br>• Operation and support documents |

requirement for the performance of the item. Each requirement should spawn additional, more detailed requirements that transition from *what* needs to be done to *how* it is to be done. Ideally, each type A requirement has a direct trace to a type B requirement (the type B being the children of the high-level type A). Also no type B requirement should exist without a trace to a higher level parent. (A requirement without a higher level parent requirement is known as an "orphan" requirement.) Prior to the PDR, the HSI program provides requirements for the type A specification that, in the activities after the PDR and leading up to the CDR, are turned into design requirements. Thus the HSI program should be able to monitor the flow of requirements from high-level design-to specifications to the build-to requirements needed for production.

***HSI Tasks for Second Contract Stage*** During the time between the PRR and the PDR (the second contract stage), the HSI program should be active using rapid prototyping, simulators, mockups, and other techniques to understand mission, technology, and emerging designs.

In a recent DoD helicopter program, military flight crews flew simulated missions containing critical tasks in order to evaluate the proposed technological solutions during the development of the type B specifications. Round-table forums, with representatives from all technical disciplines, were then held in which HSI-generated requirements were evaluated for technological risk, availability of alternatives, and operational benefits. The time and money spent in this activity proved cost effective because developing the simulation matured the HSI requirements; flying the simulated missions put esoteric

**TABLE 7.4 Type B Specifications**

| Type | Title | Comments |
|---|---|---|
| B1 | Prime item specification | Any item that is so complex that it requires (1) formal acceptance by the contracting agency, (2) provisioning action required, (3) technical manuals or other instructional material required, and (4) quality conformance inspection of each item, as opposed to sampling |
| B2 | Critical item specification | Applicable when an item is deemed to be less complex but still has a critical nature |
| B3 | Noncomplex item specification | Applicable when an item is of relatively simple design that can be shown suitable for its intended use by inspection or demonstration, does not require acceptance testing but can use conformance to drawings instead, and is not software |
| B4 | Facility or ship specification | Applicable when the focus is upon a facility (building) or ship development that is an integral part of the system |
| B5 | Software specification | Applicable when software development specifications are required; can further be subdivided between software requirements specification and interface requirements |

technology into perspective for the users, and the technologists were able to identify potential risks that could be easily avoided.

The value of prototypes, simulators, and mockups cannot be overstated. It is very difficult for end users to assimilate design features on paper and apply their experience to the ultimate usability of the product. In an effort to design and build crew accommodations for the International Space Station (ISS), the typical design activities were augmented with early, full-size mockups, and crews were given the opportunity to use the mockups in a normal, earth gravity environment. The study was documented with pictures, video, and questionnaires, providing clear evidence that the basic requirements were incorrect and would require additional systems engineering analysis. As a result, it became clear that the contract was flawed. What was required would not meet the buyer's expectations. Ultimately, the contract was canceled and additional effort was made by the ISS program to define the requirements.

The outcome of such efforts is used to provide input to the type B specification development process. Shown in Table 7.5, Chapanis (1996, p. 74) provides a succinct list of the types of inputs made by HSI to type B specifications during the time between the PRR and PDR.

In summary, during the second contract stage, the HSI program should be engaged in monitoring the flow down of high-level requirements from design-to (type A) to build-to (type B) requirements; conducting simulations, evaluations, and testing to verify that allocations of function between machine, individual operators, and possible operator teams are correct and desirable; providing detailed interface definitions and requirements; providing inputs to documentation; and preparing for operational tests and validations.

**TABLE 7.5 HSI inputs to Type B Specification**

Detailed interface requirements

- Operating modes and functions performed at each station
- Displays and controls used at each station
- Exact formats and contents of each display, e.g., data locations, spaces, abbreviations, message lengths, special symbols
- Formats of all operator inputs
- Control and data entry devices, e.g., cranks, levers, pedals, keyboards, special function keys, cursor controls
- Status, error, and data printouts

Detailed requirements for tests and evaluations interface requirements

Verification of allocation of functions to operators to ensure that their capabilities are utilized and their limitations are not exceeded

Technical manuals and documentation coverage

### 7.2.3 Preliminary Design Review to Critical Design Review

The third contract stage covers the period from the PDR to the CDR. A CDR is conducted for each configuration item when detail design is essentially complete (see Fig. 7.4).

After the preliminary design is reviewed and approved, the process moves to detailed design of components and ultimately to CDR. In this phase, how tasks will be performed and what human interfaces will look like become determined.

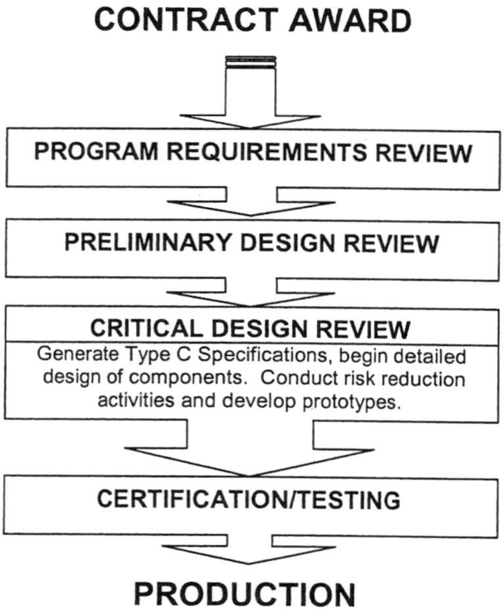


**Figure 7.4** Stages in development of a product: critical design review.

Type C specifications, or product specifications, establish the performance, design, test, manufacture, and acceptance requirements for a prime item. A type C may be a function specification when the contractor does not develop the item or a fabrication specification. Fabrication specifications state detailed part specification and assemblies, performance requirements and tests, and corresponding inspections.

The HSI emphasis during this stage continues to be on the system engineering process, but significant attention must be paid to the emerging design specifics. Task inventories and workload predictions become especially important during this stage. Iteration of design and communication between disciplines are key activities.

The purposes of the CDR are to (a) determine that the detailed design of the configuration item under review satisfies the performance and engineering specialty requirements of the HWCI development specifications; (b) establish the detailed design compatibility between the configuration item and other items of equipment, facilities, computer software, and personnel; (c) assess configuration item risk areas (on a technical, cost, and schedule basis); (d) assess the results of the productibility analyses conducted on system hardware; and (e) review the preliminary hardware product specifications.

For CSCIs, the CDR will focus on the determination of the acceptability of the detailed design, performance, and test characteristics of the design solution and on the adequacy of the operation and support documents (MIL-STD-1521). The items to be reviewed during the CDR per MIL-STD-1521 (DoD, 1976, pp. 54–56) include those listed in Table 7.6.

**TABLE 7.6 Items Typically Reviewed During CDR**

Hardware

- Adequacy of the detail design reflected in the draft hardware product specifications (type C specifications)
- Detail engineering drawings for the hardware, including schematic diagrams
- Adequacy of the detailed design in all areas, including:
  a. Manpower
  b. Personnel
  c. Training
  d. Human factors engineering
  e. System safety
  f. Health hazards
  g. Interface control drawings
  h. Mockups, breadboards, and/or prototype hardware
  i. Design analysis and test data
  j. System allocation document for hardware

Software

- Software detailed design, database design, and interface design documents
- Documentation describing results of analyses, testing, etc., by agreement
- Manuals and operation/support documents

Support equipment

- Requirements review
- Special equipment (problems, provisioning, reliability, logistics support)
- Calibration requirements

***Type C Specifications*** During the third contract stage, the PDR to CDR period of the contract, the emphasis is on the development of the engineering drawings and the type C production specifications. Type C production specifications are oriented toward either procurement of a product through specification of primarily functional (performance) requirements or primarily fabrication (detailed design) requirements (as per MIL-STD-490; DoD, 1995). Type C specifications take many forms due to the wide range of products or product design requirements. The variety of forms of type C specifications are shown in Table 7.7.

***HSI Tasks for Third Contract Stage*** The HSI program will continue monitoring the decomposition of requirements as the contract prepares to go to production. During this period of activity, the difficulties in transitioning technology, packaging and interfacing, meeting weight/space/power goals, and production costs become major drivers. The need for structure, support, bend radius, bolt sizes, and other manufacturing requirements intrudes upon the clean design described in the types A and B specifications. For example, even though size is supposed to be fixed by requirement, components may just not be capable of being packaged with as little structure as desired. Therefore, the size of individual objects may grow within a packaging footprint. Trade-offs and compromises abound during this stage as engineering struggles to make everything fit within weight, space, power, cooling, and other requirements. Often performance of the item is reduced to match what is capable of being produced given cost and schedule. This type of problem exemplifies why the HSI program in the earliest stages needed to spend so much time and effort documenting requirements. When these engineering struggles begin, the HSI requirements must be clean, and crisp and have a well-documented relationship to system function and, ultimately, to usability in order to battle against pressing weight, cost, and schedule demands.

Chapanis (1996, p. 75) states, "no other new human-factors inputs should be required at this time because they should all have been made prior to this point. At this stage, changes required to make equipment meet human-factors requirements would be extremely, perhaps prohibitively, costly." Chapanis does, however, indicate that the HSI program should review the detailed designs or drawings, schematics, mockups, or actual hardware and evaluate by checklists or other formal means the adequacy of designs with regard to items listed in Table 7.8. In addition, time/cost/effectiveness considerations and forced trade-offs of HSI design features should be thoroughly reviewed.

During this stage, engineering deals with trade-offs that affect the operator functions and capabilities, and the maintainer and supporter functions begin to be defined. The HSI program must pay careful attention to the details becoming available about how to maintain and support individual components of the product in order to assemble a clear picture of the manpower, skill sets, and timelines required. This information typically becomes available just before the prototypes of products are built and, owing to the maturity of the design, is very resistant to change. The opportunity for influencing these designs is very limited and will require significant HSI manpower to effect change, depending upon the size and complexity of the product.

Early in the contract, task inventories and workload predictions were "invented." That is, the best thoughts about how the design would eventually turn out and how the product would be used were captured in task inventories and workload predictions. As the program nears CDR, this information should be revisited given the data about the design that is now known. Tasks and workload obtained with simulations and prototypes should be compared

**TABLE 7.7 Type C Specifications**

| Type | Title | Comments |
|---|---|---|
| C | Product specifications | Applicable to any configuration item below the system level and may be oriented toward procurement of a product through specification of performance requirements or fabrication requirements |
| C1a | Prime item function | Applicable to prime items when a "form, fit, and function" description is acceptable |
| C1b | Prime item fabrication | Applicable to prime items when inclusion of a detailed design disclosure package is required |
| C2a | Critical item function | Applicable to a critical prime item when performance is of greater concern than interchangeability or control over design and a "form, fit, and function" description is adequate |
| C2b | Critical item fabrication | Applicable to a critical prime item when a detailed design is made available or where adequate performance can be achieved from the provided design |
| C3 | Noncomplex item fabrication | Applicable when procuring a noncomplex item to a provided detailed design |
| C4 | Inventory item | Applicable when procuring an item from an established inventory such as the DoD inventory |
| C5 | Software product | Applicable to a delivered computer software configuration item and is the "as built" software specification. It consists of the following: |
| | Software top-level design document | Describes how the top-level components implement requirements allocated from the software requirements specification |
| | Software detailed design document | Describes the detailed decomposition of upper level components into lower level components |
| | Database design document | Describes one or more databases(s) used by the configuration item |
| | Interface design document | Describes the detailed design of one or more configuration item interfaces |

**TABLE 7.8 Review Items for CDR**

| |
|---|
| Operator displays |
| Operator controls |
| Maintenance features |
| Anthropometry |
| Safety features and emergency equipment |
| Workspace layout |
| Internal environmental conditions |
| Training equipment |
| Personnel accommodations |

to early predictions, and deviations from predictions should be brought to the attention of management. Although late in the design process, there is still time to make corrections for critical issues.

While conducting reviews and comparisons of obtained tasks and workload to predictions, HSI can facilitate the process of communication and iteration of design details by maintaining a "big picture" focus. Specifically, HSI should focus on overall workload and mission/task workload rather than focusing on any one or two tasks. HSI can identify the "rough seams" between configuration items by identifying the confusion, increase in workload, or errors being made by users when they switch between components or disciplines within the design. For instance, the convention for representing something as ubiquitous as ON/OFF may differ in different functional areas. In some areas, status (*on* or *off*) may be indicated and in other areas the next state (*going to on* or *going to off*) may be indicated. This switch in convention could cause errors in performance. Smoothing these rough seams between design areas will require identification of the problem and an open mind on the part of all concerned to decide how best to resolve cost, schedule, and risk.

The ability of HSI to remain focused on the big picture can also help disparate disciplines resolve trade study issues by providing insight into the relative utility of design features. For example, in a recent military helicopter development program it became clear that the sensor manufacturer, controls and displays group, and software processing group were at odds as far as what they wanted to do. A series of meetings were held with all parties attending along with customer users. At these meetings, the various groups laid out their concerns and impacts with the HSI group, putting the trade-offs into the context of procedures and workload for the customer users. A consensus was reached as to which requirements had to be retained and those that could be modified or eliminated. This sort of meeting was also widely used in the development of the Boeing 777, albeit not as focused on specific technical capabilities as overall customer usability.

### 7.2.4 Critical Design Review to Testing and Certification

Once the design has been reviewed and all issues resolved, the fourth contract stage is initiated and configuration items start to be built. As configuration items are built, the product begins to be assembled. Whether the product being built is an aircraft, a cellular telephone, or a toaster, at some point in the process the product is tested to make sure that it performs as required or certified as meeting its specifications. The number of evaluations, tests, or certifications required depends upon the product and the contact requirements. In aviation, the requirements call for strict testing and certification due to obvious safety requirements. A hair dryer, on the other hand, may require casual functional testing but rigorous certification testing for achieving Underwriter Laboratories certification.

Five generic types of formal reviews occur after the CDR (DoD, 1976, p. 5). They are listed, along with a brief description in Table 7.9. Typically, one or more of these reviews are used to formally present the results of individual tests called for by the SOW or conducted as part of the development and verification process.

During product certification/testing, the HSI program assumes a major role of design monitor in providing design critique of the product (see Fig. 7.5).

After the CDR determines how tasks will be performed and what the human interfaces will look like, most efforts turn to fabrication and testing. As parts are finished, they are tested or certified for use. As subsystems are assembled, they too are tested. Eventually, the system is available for testing.

The number of tests and certifications required are set by the contract and by the requirements of the type A specification. The format typically is determined by the

**TABLE 7.9 Table of Post-CDR Reviews**

| Review Name | Description |
|---|---|
| Test readiness review (TRR) | Review conducted for each computer software configuration item to determine whether the software test procedures are complete and to assure that the contractor is prepared for formal testing; more generally, a meeting to assure that the contractor is prepared for any formal testing |
| Functional configuration audit (FCA) | A formal audit to validate that the development of a configuration item has been completed satisfactorily and that the configuration item has achieved the performance and functional characteristics specified in the functional or allocated configuration identification |
| Physical configuration audit (PCA) | A technical examination of a designated configuration item to verify that the configuration item "as built" conforms to the technical documentation that defines the configuration item |
| Formal qualification review (FQR) | The test, inspection, or analytical process by which a group of configuration items comprising the system is verified to have met specific contracting agency contractual performance requirements (specifications or equivalent) |
| Production readiness review (PRR) | Review intended to determine the status of completion of the specific actions that must be satisfactorily accomplished prior to executing a production go-ahead decision |

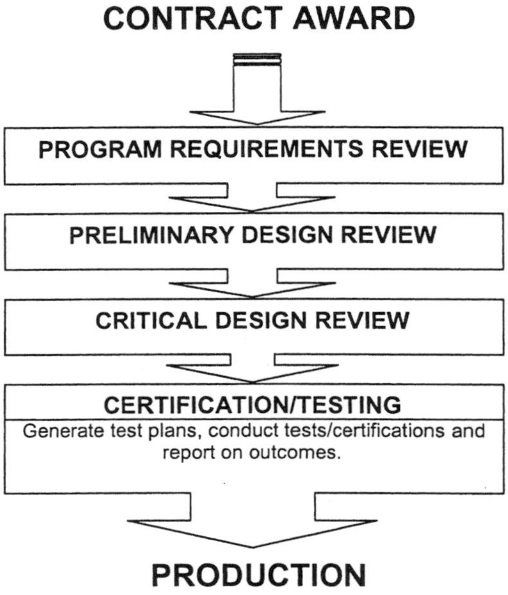


**Figure 7.5** Stages in development of a product: certification/testing.

contractor. A verification matrix is used to track specific points for testing and assign testing to disciplines and specific tests. Test plans and test reports are generally created and signed by the contractor and the monitoring technical community of the contract agency.

The HSI emphasis during this stage is to bring closure to the previously identified issues, prepare to test performance requirements compliance, and verify that the design meets identified criteria and standards. This is done through design analysis, generation of test plans, and participation in the certification planning process.

If an HSI program, as espoused in this chapter, has been followed, the HSI program should be using the results of its studies and analyses to aid in speeding product certification and testing rather than trying to force changes into the design. The workload, performance, form/fit/function studies conducted should now all support and verify previous studies, thus indicating that the design is suitable and ready for production.

By virtue of its continuing product studies during development, the HSI community is in a position to significantly enhance the certification and testing procedures. Final certification and testing will naturally focus on the product user. The HSI community can bring its knowledge and experience to the test communities envisioned scenarios along with predicted workload and timelines. This information can shortcut both the planning and execution time required for certification and testing. The HSI community's knowledge of predecessor systems and the target population can help focus testing on critical usability issues and provide a confident benchmark for performance with the new product. The HSI community can assist in generating training materials, user procedures, and lists of general design advantages. The HSI's database, gained over the course of the product development and focused on critical issues, can significantly help reduce the costs and timelines of testing.

The certification and testing of a design for the Warning, Caution, and Advisory System in a new military helicopter program provides a good illustration of the differences between the old ways of testing for system and component compliance versus the HSI method. Past aircraft would typically have simple sensors with dedicated lines to illuminate dedicated alert lamps in the cockpit for warnings, cautions, and advisory signs. For example, a magnetic oil drain plug sensor would have a dedicated line to a warning lamp on engine oil in the cockpit. The certification test was simply: "Use a screwdriver to mimic accumulation of metal shavings on the magnetic plug to see if the light would illuminate and therefore demonstrate the system is working." In the new aircraft, computers that processed sensor information data and controlled the display of information to the crew were dependent upon the content of all the sensors and their status. Everything in the new system (the computers, sensors, displays, and processing algorithms) had to be demonstrated as failure proof. Additionally, the specific sensor and crew display were also evaluated. Ultimately, the certification was awarded based in large part upon prior HSI studies of the Warning, Caution and Advisory System conducted in a full mission simulator as part of the developmental testing. The human factors certification became the validation method, showing that the flight software operated in a manner consistent with the simulator operations.

Installation of a computerized sensor and the Warning, Caution and Advisory System saved the aircraft considerable weight as well as reducing the maintenance burden that would have been imposed had a dedicated alert panel and individual wires been required. Transition to the new technology was feasible because early HSI studies and design involved the crew and the various technologists of the subsystems. High-fidelity simulations verified early testing and workload predictions. The simulations were used to focus certification efforts on the critical elements of the design and establish baselines to which certification testing could be compared. Without HSI, a technologically sound and obvious approach might not have been adopted due to concerns over verification. It is clear that the HSI approach to testing and evaluation helped the design team avoid a requirement for a heavier, redundant system that was conventional in all previous operational "glass" cockpits.[1]

## 7.3 PRINCIPAL DOCUMENTATION EVENTS OF ACQUISITION

The preceding section followed the execution of a generic contract from contract award through completion of the contract. The tasks of the HSI program were identified and placed within the various stages of the contract. The following section outlines the HSI program and associated documentation that should be developed and implemented in the acquisition process.

### 7.3.1 Definitions

*Acquisition* means the acquiring by contract with appropriated funds of supplies or services (including construction) by and for the use of the federal government, or comparable business entity, through purchase or lease, whether the supplies or services are already in existence or must be created, developed, demonstrated, and evaluated. Acquisition begins at the point when agency needs are established and includes the description of requirements to satisfy agency needs, solicitation and selection of sources, award of contracts, contract financing, contract performance, contract administration, and those technical and management functions directly related to the process of fulfilling agency needs by contract.

*Contract* means a mutually binding legal relationship obligating the seller to furnish the supplies or services (including construction) and the buyer to pay for them. These definitions are uniformly accepted and used by businesses worldwide [Federal Acquisition Regulation (FAR), 1997a].

### 7.3.2 Solicitation Process Summary

The first step in acquisition is to establish the "needs" of the buyer and solicit sources to fulfill the need; these are the sellers (contractors or vendors). Regardless of the life-cycle phase, the acquisition process of most buyers, including the government, follows a fairly common and consistent process for solicitation and award of contracts. The procedures, in short, are to announce that that the buyer is considering buying a product or services and publish a draft of the specifics of what is needed (work that is being requested and/or the performance requirements of items). This often starts in the form of a request for information (RFI), although the method could be a presentation at business conferences, public hearings, or presolicitation conferences. An RFI is used when there is no immediate intent to award a contract but there is a desire to obtain price, delivery, other market information, or capabilities for planning purposes.

After reviewing seller responses to the RFI, the buyer may make modifications to its requirements and release a RFP. This document communicates requirements and solicits formal proposals or offers to sell. Prospective sellers document how they intend to fulfill the conditions of the RFP, their design, and its features and submit this information, along with cost information, as their proposal. The buyer's technical staff conducts a formal technical evaluation of the competing proposals. As part of the evaluation, face-to-face negotiations may be held about the seller's intent and details of their design. After submission of revisions and/or cost updates, the buyer formally considers each of the proposals received, selects acceptable sources (contactors) and awards the contract.

### 7.3.3 Request for Proposal

The RFP is used to communicate requirements to prospective sellers and to solicit proposals. For competitive acquisitions, RFPs typically describe the requirements and anticipated terms and conditions that will apply, information required in the seller's proposal and factors, and significant subfactors that will be used to evaluate the proposal and their relative importance. The RFP is, in fact, where the HSI program formally begins for the both the seller and buyer HSI teams.[2] The buyer's HSI staff should be involved in the generation of the RFP. The buyer's HSI staff should be providing program and design requirements to shape the nature and extent of the HSI program that is believed necessary for a successful system development. Proper preparation of the RFP is by far the most significant single factor in ensuring that an adequate HSI program will be applied. When preparing the RFP, inputs must be clear, accurate, and concise, with rationale provided to justify all inputs. There is rarely a way to recover from a poorly constructed RFP; therefore, it is critical to invest the time to ensure a quality RFP. Ambiguity in an RFP forces the HSI practitioner or one's successor to live with that uncertainty. Since the acquisition process typically extends over many years, one may not be around to interpret issues that are ambiguous; therefore, it is better to take the time to be clear from the beginning (McCommons, 1987). HSI inputs to the RFP vary considerably depending on the size and nature of the procurement (DoD, 1999, Section 6.5).

***Human Factors Section*** The RFP typically has a section entitled Human Factors. This section is where the buyer's HSI community can state their requirements under the contract for the seller's HSI programs. The human factors section should make clear what the minimum program requirements are and what is expected of a fully implemented program. The buyer's HSI community needs to consider carefully what it requires from the winning contractor and create contract data requirements list (CDRL) items and data item definitions (DIDs) for products to be delivered. It should be kept in mind that formal data submittals by the seller require considerable labor to prepare and therefore drive the contract cost up. Deliverable items that are "nice to have" can incur significant cost penalties. However, critical items should be on the list regardless of cost. The reason for this list is that the buyer's contract officer monitors the items on the CDRL for compliance because reimbursement of charges may be withheld if the CDRL items are not provided or what is provided is inadequate. That is, the buyer's acquisition office monitors the contract to enforce delivery of the CDRL items. When funding becomes tight (not if, when), items on the CDRL will undergo significant pressure to justify value or face being deleted from the contract. Items not specifically required by the contract are subject to being interpreted as information or suggestions to the seller and can be deleted for cost avoidance. In other words, in order to save documentation money, the requirement might be interpreted as a requirement for doing a workload analysis but not as a requirement for delivering a formal document describing the analysis. Without an explicit requirement for the document, obtaining an adequate budget to document the result may be impossible. In general, complying with a discrete, testable requirement that has a monitored deliverable will provide a certain level of immunity to cost cutting.

In a recent major DoD helicopter competition, the released RFP had little in the way of HSI requirements within the human factors section and only had a total of seven direct crew interface requirements. Despite this paucity of requirements, HSI was a major consideration in determining contract award. The reason that HSI was considered so important was because the RFP required that a contractually binding human factors program plan be included as part of the proposed SOW. By not attempting to define a build-to human factors solution into the requirements, the DoD HSI team was able to help choose the process by which the solution would be engineered. In this manner, the DoD was able to select the contractor whose proposed solution held the best potential for fulfilling the HSI requirements.

***RFP Summary*** In summary, the RFP is one of the most important documents for the HSI community because competitive contractors will budget only for tasks that are explicitly required. In a review of the DoD HSI programs, Wright and Hall (1994, p. B-4) noted that underfunding of the front-end analyses conducted by HSI was a problem even in the relatively strong U.S. Army manpower, personnel, and integration (MANPRINT) program. They attributed this to "a lack of appreciation, among people responsible for funding FEA (front-end analyses), for the critical role FEA plays in the HSI Program during the very earliest phases of the new acquisition." Wright and Hall concluded, "Unfortunately, if the window of opportunity to make meaningful input to major system documentation that drives the design process is missed, there are no inexpensive ways to catch up later in the acquisition."

### 7.3.4 Seller's Proposal

The government's codification of procedures into a set of regulations has not gone unnoticed within the private sector. Many texts have been written on how the contractors (sellers) can best do business with the government. Alston et al. (1984) has produced a seminal work on obtaining government business. While it is primarily directed at how to conduct business with the U.S. government, the points made are completely appropriate for private-sector contracts.

A proposal is an offer by a seller to supply a product, perform a service, or provide a combination of the two. In some cases, the products or services are simple tasks and have been done before. In other cases, they are unique research-and-development efforts with a number of substantial state-of-the-art problems to be solved. From the seller's perspective, the function of the proposal is to sell the managerial and technical capabilities of the company to carry out the required work at a reasonable cost. The proposal is the point of sale; it should be considered the seller's most important selling tool. The proposal document must convince the buyer that the seller is offering an acceptable solution to the problem for a reasonable cost. It also communicates that the seller has an adequate organization and sufficient personnel and facilities to perform the required effort within the time specified. The seller has considerable latitude in conveying these messages.

The proposal must adequately cover three broad areas. The first area relates to the technical solution to the problem as defined in the solicitation. The second area describes how the seller will manage contract performance, and the third area addresses how much the proposed work will cost. The HSI community of both the seller and buyer should be interested in all three sections. The *technical section* of the proposal should demonstrate the seller's understanding of the problem and the proposed method of solving it. It is in this section that the technical elements of the HSI program belong. This section outlines the specifics of what tools, techniques, and procedures of HSI will be utilized to implement the HSI program. This section should also identify which technical elements will be responsible for delivery of the CDRL items and support other disciplines in fulfilling their CDRL requirements.

The purpose of the *management section* is to explain how the seller intends to manage the effort required under the proposed contract. The seller's HSI staff should provide details of where its management resides within the overall structure of the proposed program. If HSI is truly integrated into the program, then the management of HSI should be visible within the program management structure and in such a position that HSI is able to effect changes deemed necessary by their analyses. In other words, HSI management should be at a high level within the seller's organizational hierarchy.

The third section, *cost*, should also identify the proposed budgets of the seller's HSI effort. A budget within the program is necessary to assign labor and other resources to complete tasks. The scope of the effort described in the technical section should match the proposed costs. Unless the technical and cost are balanced, the management structure proposed will be without the means to accomplish the technical work and deliver the requested products.

Those attempting to respond to a proposal for the first time should consult MIL-HDBK-46855A (DoD, 1999). This handbook is an excellent document on how to decide what tasks and efforts are appropriate under a variety of circumstances. Another excellent source of guidance is DOT/FAA/RD-95/3 (Hewitt, 1995). This document is oriented more toward users and their problems than the more programmatic MIL-HDBK-46855.

The seller should also read and carefully consider both the explicit requirements in the human factors section of an RFP and those requirements that may be buried in other technical sections. For instance, human engineering requirements may be implied by specific information required in a particular test by the test and evaluation sections. Even if HSI is not responsible for a segment of the design or test, HSI involvement may be required that, if not properly documented, will require diverting core HSI program resources into support of other disciplines at unexpected times. For example, a subsystem test may require that sellers test their subsystem in "critical user tasks." While not explicitly in the human factors section, HSI inputs will ultimately be required to identify the critical tasks and evaluate user workload and acceptance.

As a general rule, the vendor should structure its offering such that the evaluators can easily find all the critical elements requested. If vendors have made up unnecessary and confused breakouts, evaluators may conclude that the proposal does not meet the buyer's intention. In extreme cases, sellers have been known to follow the RFP sentence by sentence, addressing each sentence in the RFP as a separate topic. In a government RFP, there are specific sections that spell out what the content of the proposal should include and the criteria by which the proposal will be scored.

The seller should look carefully at CDRL requirements outlined in the RFP. Formal deliverables tend to be labor intensive, subject to multiple reviews by management, and take longer to create than anticipated. There should be a logical relationship between work being accomplished in order to get to the desired design and what is deliverable. If a seller finds that new HSI work is being added to the proposal specifically to address CDRL requirements, the seller should revisit its proposed HSI plan, as the plan is probably not complying with the buyer's intended program. If the seller believes that their proposed HSI plan is fully compliant with the RFP intent and further believes that the CDRL is not required, an argument should be made for modification or deletion of the CDRL as a cost avoidance. Changing the CDRL can be accomplished in a variety of ways, from informal discussion with the buyer's HSI community to pricing a reduced-cost alternative without the CDRL.

In a recent response to a government, space-related RFP, a seller's inputs to a team proposal failed to address the technical questions posed. The staff responding to the assigned technical section decided to use the allotted proposal pages to market the company's background and capabilities. Their strategy was that the technical section submitted would convey the message that they had been successful in the past so they would be successful in the future. However, since technical information about the proposed design did not make it into the proposal, that company did not get the award.

A similar way to ignore the RFP and lose a contract is to adopt the approach that the buyers releasing the RFP do not know what they want or what is good for them. With this approach, the vendor responding to the RFP proposes something related but different than specified. What is proposed is usually a reworked design from either a slightly different but successful project or a previously failed proposal in which the company has already invested heavily. This typically results in yet another failed proposal.

Not all acquisition failures happen to the sellers. In a new space program, crew requirements for habitation were so delayed that a major negative impact was projected for the final product. Due to program delays and reorganizations, the need date for crew habitation accommodations had moved several years behind the rest of the program. This inadvertently caused basic requirements for the crews to be removed from the program since they were outside the contract scope and the seller selected could not legally work on "outside scope" issues. Eventually it became clear that crew requirements would have to be addressed and so the program, to buy time to decide what the "permanent" solution would be, adopted a "temporary" solution. This temporary solution was initiated and a contract awarded, but the winning company soon discovered that fundamental support requirements necessary for contract completion had not been made for the infrastructure (electrical power, computer connections, ventilation, etc.). Ultimately, the contract had to be canceled.

### 7.3.5 Source Selection

The U.S. Government FAR (1997b) defines proposal evaluation as "an assessment of the proposal and the offeror's ability to perform the prospective contract successfully." The FAR also provides the following guidance: "An agency shall evaluate competitive proposals and then assess their relative qualities solely on the factors and subfactors specified in the solicitation. Evaluations may be conducted using any rating method or combination of methods, including color or adjectival ratings, numerical weights, and ordinal rankings. The relative strengths, deficiencies, significant weaknesses, and risks supporting proposal evaluation shall be documented in the contract file" (p. 15-10).

Once an RFP has been released, sellers may respond with proposals. Typically, representatives of the requesting buyer's HSI community will be asked to serve on an evaluation board with the responsibility to read all proposals and ask the sellers for technical clarifications, omissions, errors, and deletions to their proposals. This may be done in writing or in face-to-face meetings with written commitments for any changes made as a result of meetings. Eventually, the technical work proposed will be scored. According to the instructions of the board, the buyer's HSI representative will be asked to provide ratings for each contractor's technical response to the RFP technical sections. Scoring should consider how well the proposed HSI program will satisfy technical requirements and be integrated into the product development. The buyer's HSI representative should also consider the proposed organizational structure. The proposed organization and management of the program give strong evidence of how well the HSI requirements spread throughout the RFP will be addressed and should reflect the technical expertise of the staff proposed to address the varied technical areas. This is important information to determine the quality of the HSI effort that will result after contract award. It can also help answer such questions as whether the seller's HSI practitioner will have organizational authority to push for HSI-desired features.

Another area of concern to the buyer's HSI community is cost. The HSI programs require significant manpower depending upon the phase of acquisition. An evaluation must be made whether the costs associated with the scope of work match the likely amount of work to be performed. For example, if the seller proposes to do task analysis and workload predictions for a major system in 12 months with two people, the costs might be suspect as being too low to credibly perform. On the other hand, if the contractor already possesses a significant portion of the data, this level of effort might be reasonable. Some sellers may demonstrate company information and prior work with direct application to the contract. This work may significantly reduce costs as well as showing appropriate skills are available to do the proposed work. This work may be the result of prior contracts—either on the current development or on predecessor systems—or may be the result of company-funded independent research and development. If the work is cited as having been independently developed, the buyer's HSI representative should consult with the acquisition authority to determine the status of rights to the data. Independently developed work may or may not become the property of the contract.

In evaluating the program plan and the scope of the planned work, the buyer's HSI representatives will come into contact with the seller's proposed design. The buyer's HSI representatives were presumably selected for their knowledge of predecessor systems and the task/mission being supported. This knowledge can be both a help and a hindrance in evaluating designs. Proposals for design similar to predecessor systems may evoke responses ranging from "its just like" to "its not like" the predecessor system. If the proposed design is a novel approach, the same range of responses may occur. The buyer's HSI representatives need to utilize their knowledge of the systems, tasks, and missions without trying to drive the solution toward one they are comfortable with or away from one they dislike. The evaluation should be on the quality of the proposed plan and the extent to which the design will meet the HSI criteria embedded within the technical sections and requested in the RFP.

## 7.4 HSI PROGRAM MANAGEMENT GUIDELINES

Acquisition has its own language, timing, and rules. Those who know the language, timing, and rules can promote, limit, or effectively block an HSI program. Section 2 was directed toward helping practitioners involved with the acquisition system understand its basic language, timing, and rules. Section 3 addressed the major events and products necessary to contract for an HSI program. This section provides basic information to help the HSI manager plan and implement an HSI program.

### 7.4.1 Specifications: From A to E

Throughout the life of an acquisition, the HSI program will remain "information starved." That is, HSI practitioners will constantly be trying to learn how various systems, subsystems, components, and items work and interact, at low and high levels, between components and the users. Since training manuals, descriptions, or overviews will not have been written while hardware/software items are in a design stage, the only place the practitioner can get specific information is from the engineer in charge of the system or subsystem. However, since it takes a significant amount of time to document how to operate a developing system or subsystem, it is likely that the engineer will point to his or her system's or subsystem's primary specification or controlling interface control document (ICD) as the only available information. The engineer would most likely ask the practitioner to review these documents before asking additional questions. This is a helpful suggestion, since much of the needed information should be in the specifications. Reading a specification, however, can be a daunting task, and several need to be read before the system design becomes clear. A few practical tips may help the practitioner get started.

The HSI practitioners should know that specifications run the gamut from top-level system specifications (type A discussed above) to D (process) and E (material) specifications. Type A is usually too general and type C too detailed except where very technical answers to very specific questions are concerned. Therefore, most of the practitioner's time is likely to be spent with type B specifications.

Each specification, regardless of type, follows a standard format. The practitioner generally needs to look in detail only at the section on requirements, specifically at the subsections called:

1. definition,
2. characteristics,
3. design and construction,
4. logistics,
5. personnel and training, and
6. characteristics of subordinate elements.

These sections are required in this standard format specifically to assist other disciplines in understanding how a system works. These documents are reviewed and accepted by the program either at PRR, PDR, or CDR. An HSI practitioner should have previously reviewed the contents of these sections and agreed that the level of detail and information was acceptable for later HSI use. If adequate information about the system is not included during the development phase, funding to revise the document and provide the information later is not likely.

In the past, these sections of the specifications were written by engineers for other engineers of the same discipline and reviewed by those engineers working on the system. The information contained was cursory and often assumed the reader would have significant knowledge of the technology and design. The HSI practitioners did not routinely see these documents until they had become information starved and then steered by engineering personnel to the specification. The cursory information typically found in the specifications made it impossible for the HSI practitioner or researcher to determine how the system or subsystem was intended to work. Tasks and workload predictions would lack definition, and the opportunity to influence design would be lost due to lack of information.

When a specification is in review, the adequacy of sections can be formally challenged and resolved if the information provided is too limited for later use. After review and release, it is difficult and expensive to change documentation, especially just for updating information rather than for a design change. Information will typically be provided only to the level demanded by other disciplines. If the HSI program does not or cannot force adequate documentation in the development of specifications, the HSI program will remain information starved. Later, HSI practitioners, attempting to revise tasks and workload, conduct simulations, or prepare for certifications, will require significant unplanned expenditures of time and money to become familiar with the details of the design. The alternative is to wait until the design is finished, but changes then cannot be economically accommodated.

### 7.4.2 Planning, Programming, and Budgeting

Planning, programming, and budgeting of the seller's HSI effort are program management requirements that, for HSI practitioners, tend to be learned by mentoring with an experienced manager. A single HSI practitioner can handle some programs such that little management is required. The task required, for instance, may be merely an upgrade to an existing product rather than development of a new function. Many programs, however, require multiple HSI practitioners with a variety of skills, all working across time in a coordinated fashion. Good program management will help make sure that required skills, financial resources, and time are available as required. Planning is used here to refer to the process of explicitly writing down the tasks, activities, and products that will fulfill the contractually required SOW throughout the period of performance. The seller's planning is a written list of all the major and minor tasks that have to be done in order to fulfill contractual requirements and commitments. Programming is used to describe the process of time phasing the planned work and identifying the interrelationships between the various work components. Programming describes what work must be done by what time in order to allow the next, dependent work to be done. It is the cascade or waterfall of work that leads to contract completion. The term *budgeting* is used to place the work to be done (planning) on a time-phased schedule (programming) in order to forecast how much the work to be done will cost over time. The result is a budget request to the program management.

To accomplish the work, staff (either in-house or contracted) must be available to do the work. To get staff to do work (implement the planning), funds (a budget) must be available at the right time. In other words, there must be programming of planned work with a budget to support staff specialists across time. If the management planning, programming, and budget are correct, the HSI program can progress by focusing on technical issues without the diversion of concerns about whether preliminary work was done or if follow-on work will happen. If work is not programmed correctly, the HSI effort will likely miss deliveries and reviews and not get tasks accomplished when needed. This will create a rolling wave of unfinished work moving to the right on the schedule. At some point, some tasks pass their effective window and never get done. At other points, the correct answer comes when it is too costly to implement. Below are some basic rules for planning, programming, and budgeting:

1. *Determine Deliverables and Milestones* Start planning from the end of the program by identifying what is to be delivered when the contract is completed. Work backward from the deliverables toward the start of the program to identify those tasks that must be done in order to complete the deliverables. Account for all deliverables (those items explicitly named in the contract as well as bodies of work that are to be conducted). Identify the date that each deliverable item is to be completed. Develop a timeline that incorporates each of the deliverables and dates under significant, discrete events called "milestones."
2. *Determine Work Tasks* Create a "bullet-sized" statement of work for each of the milestones. Identify all the significant tasks with a brief one- or two-line description of the task. Group the bullets into packages in which there are obvious start and finish activities. This is a task analysis of the HSI work. The HSI practitioners have missions, phases, tasks, and subtasks to complete the HSI program, just like the users of the vendor's products have missions, phases, tasks, etc. The HSI tools apply to HSI activities as well as others in the systems engineering and management process.
3. *Put Work into Monthly Packages* Since the budget comes in yearly packages, break the bullets of work into year groups and grossly identify start/stop points by month.

4. *Determine Labor Required and Costs* Identify how many people with what kind of knowledge and skills will be needed for a bullet and how many days it will take to get the bullet done. Generate a spreadsheet that has *month-by-skill-type* intersections. Fill the cells with the number of days (at 8 hours a day) times the number of people with that skill working (to get personnel hours per day) times the number of days in that month that those people will work on that task (personnel hours per month by skill). The contractor's finance people can help by using a labor rate (cost) for the skill indicated, adding overhead, fee, etc. (burdens), to assemble a budget for that work for that month.
5. *Determine Material Required* Identify any material items that will be needed and when needed. This is the budget for specialized computer software, hardware, mockup material, shop time, etc. For example, if a research assessment is planned using a mockup, the cost of building a mockup should be identified. It might be possible to use someone else's mockup or modify existing mockups to fit the need, but the HSI manager needs to start by identifying to higher management that material items will be required.
6. *Determine Travel Required* Identify any travel projected as needed to conduct the tasks, providing such information as when the travel will likely happen, to where, for how long, and with how many people.
7. *Determine Data Requirements* Identify what data will be needed from what other groups or companies. Summarize what is needed and when, then coordinate, in writing, with the source of the information to get agreement that it will be supplied.

Personal experience, company experience with prior programs, information from vendors, etc., all provide help with this process. Management tools are available to help as well, but there is no substitute for independent thinking about the problem and laying out the planning, programming, and budgeting in language understandable to both HSI and systems management. If the contract is with the government, it is likely that the contractor will be required to manage the work using earned-value accounting techniques. *Earned-value* techniques force the vendor to plan ahead and reduce the number of unpleasant surprises for both the buyer and seller.

### 7.4.3 Industry's Dilemma

The winner of an acquisition competition is typically the seller that promises significant technical advancement without introducing excessive technical risk, all for the best price. In other words, the winner proposes a design that appears to meet the requirements without reliance on technology that might not be available and at a price the buyer can afford. To win a competition that will result in a system that actually meets these competing criteria presents a dilemma for industry. On the one hand, if the seller promises too much technically—a very strong tendency when moving technology from the laboratory to production—any number of unforeseen problems in critical technologies may cause the following:

a. delays—which means using unplanned time and money (overruns and schedule slips) to resolve problems associated with bringing technology to production;

b. reworks—taking a different approach (new design and development costs for less risky technology) or additional design and development costs to address problems discovered;
c. unforeseen problems with support and logistics—e.g., new technology turns out to require significant maintenance that was unplanned and costly; and/or
d. failure to meet exit criteria for the contract—which always increases time and costs to complete.

On the other hand, promising too little advanced technology runs the risk that the buyer will perceive industry is not offering a solution that will provide significant improvement or benefit. It is generally a rare event when a seller can make this call correctly. Virtually all major acquisitions end with the seller having been either too conservative or too optimistic. An adequate discussion of lessons learned on this topic is outside the scope of this chapter. However, some common pitfalls for HSI can be identified as cautions as the seller struggles with this dilemma. Two fallacies typically applied by the seller are reducing work of "limited value" and assuming that the more work proposed, the better the product.

***Reducing Work of Limited Value*** One frequently used method of addressing the industry's dilemma is to reduce costs by cutting work that is of limited value compared with the "value" of advancing the state of the art. This equates to stripping money from the "support" tasks in order to offset technology risk. The technology is just as risky, but the budget can include costs for rework or risk reduction activities as a management risk reduction program. This reduces the perceived risk to the program of maturing the technology because budget is available to work the problem rather than having to overrun unexpectedly. For example, a contractor might choose to reduce up-front work (such as mission analysis) for design support organizations such as HSI or eliminate documentation in order to free up money for the risk management program.

Within this dynamic interplay, the HSI program can be overlooked and badly damaged. This is where the HSI community must be visible as a member of the design team and not viewed as just an after-the-fact reviewer or "grader" of the work of others. In a successful HSI program, work feeds upon prior work and information. Elimination of parts of the HSI program or reduction in the detail of program components can dramatically lower the ultimate value of specific tasks. Elimination of tasks may make prior lead-in work of no value in that there is no one to use the information (all the work has been done but no one is using it) or make it impossible to conduct follow-on work due to lead-in work not being done. Also, the addition of technology to enhance the chances of winning may place new, unknown burdens on the users that will not be noticed until much later in the program. Management of both the buyer and seller need to understand that *reducing work of limited value* poses a risk just as technology maturation is a risk.

***The More Work Proposed, the Better the Product?*** The simple view is that the more work a seller proposes, the better the proposed product will be, but the cost obviously goes up. If the cost is too high, the contract will be lost. If the amount of work being proposed is too little, the risk that cost overruns and schedule slips will occur while trying to bring the product to production may be viewed as too high and the contract can also be lost. Contracts are continuous, unrelenting trade-offs between cost, schedule, risk, and product quality. The HSI practitioner must understand this and be prepared to modify the

HSI program to maximize the effective contribution to the program. It has been said that many programs have failed because they could not afford to do it right. Balancing technology risks, costs, and schedule challenges seldom allows HSI all it needs, even with programs that attempt to fully apply the HSI principles outlined in Chapter 1. The HSI practitioners should be prepared to do the best they can within risk, schedule, and cost constraints.

### 7.4.4 Winning through Candor and Cooperation

Typically, a proposal will have a page count imposed that limits the contractor's input, making every word in the response important. Decisions on what to expend words on and how many words on a topic need carefully planning. The following are a basic set of rules to follow when generating a proposal:

- Read the SOW, specifications, and any instructions to the offeror that are provided for HSI content. Reviewers always appreciate the vendor being responsive to the questions and issues they ask about.
- State clearly what the HSI program will be and in what sequence it intends to resolve risk. Refrain from platitudes and marketing.
- Recognize that if all the answers were known, it is likely that the work would not be required. In other words, it is okay to acknowledge that the answers to the problems posed are unknown, but it is critical to show how the HSI program will identify and resolve problems throughout the contract.
- Identify the known items (and unknown factors) that pose risk or that will be difficult to accomplish. Articulate how the HSI program intends to approach and attempt to solve the problem(s) identified. In other words, define how the offeror intends to control risk.
- Coordination, cooperation, and data exchange between the buyer's and seller's technical staffs regarding the proposed HSI program are expected and encouraged by the buyer. Recognize this in the proposal and state how this interchange will be handled.
- Discussion of tools and techniques is helpful provided the seller identifies how the tools and techniques discussed will fit into the overall program. Do not just list tools that might be used. Identify which tools will be used for specific risk reduction, define what knowledge is to be gained, and discuss how that knowledge will solve problems or otherwise contribute to the HSI program.
- Read the proposal in draft form and ask whether it answers questions and/or leaves questions hanging. In reviewing the proposal, typical questions to be asked include "What is going to be done?" "How is it going to be done?" and "How does this contribute to the HSI program?" If the answers are not obvious, rework the text.
- It is especially beneficial for the seller to indicate how the program will assess its own progress and adjust to changing requirements. How will the HSI management team monitor itself to make sure that it is making progress?

## 7.5 SUMMARY

A new emphasis on HSI has been added to NASA and DoD federal government acquisition processes. Conscientious attention to HSI by contractors who provide services and systems to the government has been demonstrated to save significant money and enhance system performance (Booher, 1990, 1997). To aid the HSI practitioner from the contractor's perspective this chapter discussed three types of information:

1. the specifications and critical HSI tasks for each major stage of a contract,
2. the principal documentation events for the buyer and seller in the acquisition process, and
3. guidelines on planning, programming, and budgeting for an HSI program.

There are four major milestone reviews of contractor-developed products in a typical contract: I—the program requirements review (PRR); II—the preliminary design review (PDR); III—the critical design review (CDR); and IV—testing and certification. Each of these reviews ends a critical stage of work for the contractor that should have incorporated specific planned, programmed, and budgeted HSI tasks.

Successful integration of HSI into product development starts with the RFP. The buyer's RFP must have a clear requirement for the seller's HSI, including an explicit contribution to contract award; otherwise, the message is sent to the vendors that the user's performance with the proposed system is not particularly important.

After contract award, the HSI program begins with early (front-end) analyses in support of generation of the system/system segment specifications. When completed, the requirements and the design outline are reviewed at the PRR.

Once the high-level requirements and design have been agreed upon between the contracting agency and the contractor, work commences on the decomposition of the system/system segment specification requirements into the development specifications that state the requirements for the design or engineering development of a product during the development period. The development specifications indicate how the design-to specifications are to be built-to and ensure that all higher level specifications are addressed. This effort culminates in a PDR that validates the design requirements decomposition process and checks general design progress, technical adequacy, and risk resolution on a technical, cost, and schedule basis.

After the PDR is complete, work focuses on completing the product specifications. All these efforts are reviewed at the CDR during which the decision to proceed to production is made.

The final efforts are for certification and testing of the product. Specifications of a variety of types are the products of these reviews and control development of the product. The HSI practitioner should be involved in the generation, review, and implementation of the specifications.

A winning HSI program will have been well thought out before contract award. This will be reflected in the proposal with a clear description of what will be done and how it will occur. A well-thought-out program will also be reflected in a clear, defensible statement of costs to conduct the program, identify risks, and provide a means to assess risk, progress, and overall quality. The HSI program should be active from contract award through production and should be integrated into the mainstream development effort.

## NOTES

1. The term *glass cockpit* refers to an aircraft cockpit in which computer display(s) have been incorporated. These displays are typically cathode ray tube (CRT) or liquid crystal display (LCD)

and are made of glass—hence the use of the term *glass*. Generically, the term is used to describe a cockpit whose crew interface includes computer-synthesized displays and computer-mediated inputs.

2. There are, of course, months to years of analytical activity leading up to the issuance of an RFP.

## REFERENCES


Alston, F. M., Johnson, F. R., Worthington, M. M., Goldsman, L. P., and DeVito, F. J. (1984). *Contracting with the Federal Government*. New York: Wiley.

Blanchard, B. S., and Fabrycky, W. J. (1990). *Systems Engineering and Analysis*, 2nd ed. Englewood Cliffs, NJ: Prentice-Hall.

Booher, H. R. (1997). *Human Factors Integration: Cost and Performance Benefits on Army Systems*, DTIC No. AD-A330 776. Aberdeen Proving Ground, MD: Army Research Laboratory.

Booher, H. R. (Ed.). (1990). *Manprint: An Approach to Systems Integration*. New York: Van Nostrand Reinhold.

Chapanis, A. (1996). *Human Factors in Systems Engineering*. New York: Wiley.

Clark, D. W., Cramer, M. L., and Hoffman, M. S. (1986). *Human Factors and Product Development: Solutions for Success*. Amsterdam: Elsevier.

Cushman, W. H., and Rosenberg, D. J. (1991). *Human Factors in Product Design*. Amsterdam: Elsevier.

Hewitt, G. (1995). Human Factors in Systems Acquisition. In K. M. Cardosi and E. D. Murphy (Eds.), *Human Factors in the Design and Evaluation of Air Traffic Control Systems*. DOT-VNTSC-FAA-95-3. Washington, DC: FAA.

Kirk, F. G. (1973). *Total System Development for Information Systems*. New York: Wiley.

Kirwan, B., and Ainsworth, L. K. (1992). *A Guide to Task Analysis*. London: Taylor and Francis.

McCommons, R. B. (1987, February). McCommons' Laws. *MANPRINT Bulletin*, *1*(8), p. I.

National Aeronautics and Space Administration (NASA). (1993). *Program Life Cycle and the System Engineering Process*, JSC Document no. 49037. Houston, TX: NASA.

U.S. Department of Defense (DoD). (1976, June 1). *Human Engineering Program Process and Procedures*, MIL-STD-1521B. Washington, DC: DoD.

U.S. Department of Defense (DoD). (1995, August 31). *Human Engineering Program Process and Procedures*, MIL-STD-490A. Washington, DC: DoD.

U.S. Department of Defense (DoD). (1998, February 27). *Mandatory Procedures for Major Defense Acquisition Programs and Major Automated Information Systems Programs*, DoD 5000.2-R. Washington, DC: DoD.

U.S. Department of Defense (DoD). (1999, May 17). *Human Engineering Program Process and Procedures*, MIL-HDBK-46855A. Washington, DC: DoD.

U.S. Federal Acquisition Regulation. (1997a). Part 2—Definitions of Words and Terms, Section 2.101 Definitions.

U.S. Federal Acquisition Regulation. (1997b). Circular 97-02, Section 15.305, Proposal Evaluation.

Wright, W., and Hall, R. (1994). *U.S. Coast Guard Human Systems Integration (HSI) Process Model*, NTIS No. CG-D-26-94. Fairfax, VA: OGNEDN/ERC Government Systems.

CHAPTER 8

# Human System Measurements and Trade-offs in System Design

MICHAEL BARNES and DAVID BEEVIS

## 8.1 INTRODUCTION

The aim of human systems integration (HSI) for military systems is to ensure that "human considerations . . . shall be effectively integrated into the design effort for defense systems to improve total system performance and reduce costs of ownership" [U.S. Department of Defense (DoD), 1991a]. The purpose of this chapter is to discuss human performance measurement issues related to system design, particularly those issues that require trade-off decisions. The basic approach is to measure performance in terms of system goals and to choose measurement processes that reflect the context of the environment the system is being designed to operate within. This is similar to *use-centered* and *ecological interface design* philosophies that are currently being investigated in the cognitive engineering domain (Flach et al., 1998). That is to say, measurement is not concerned with human limitations per se or even environmental or technological problems but rather how these problems in concert affect the accomplishment of overall system goals (cf. interface design issues; Rasmussen and Vicente, 1989). The measurement problem is difficult because of the complexity of the design space, which includes not only volatile operational environments but also changing technological and human requirements as well. The solution is to develop a flexible measurement strategy that addresses system goals and top-level requirements while it adjusts to the various contingencies of the evolving design process. In effect, a successful measurement paradigm allows the design team to answer the "show me the payoff" question implicit in all design decisions.

## 8.2 HUMAN SYSTEM MEASUREMENT

Human system measurement processes consist of assigning numbers to events that have important design implications. The ability of the process to predict the effects of design

*Handbook of Human Systems Integration*, Edited by Harold R. Booher.
ISBN 0-471-02053-2 

outcomes crucial to operational objectives determines the value of a particular measurement procedure. By their very nature, HSI issues that pertain to complex systems have multiple operational requirements as well as future operating environments that are not well understood much less well defined. Because of this complexity, there is no unique measurement solution (Barnes et al., 1996). Various approaches, from traditional human engineering analyses to laboratory experiments to large-scale computer simulations, each having its own advantages and limitations, are required to measure those aspects of system performance impacting HSI during various phases of the design process. The criteria for choosing a specific approach depend on the operational context in which the system will be used. This includes its ecological validity, ease of use, cost, and insight it brings to particular design or crew issues in relation to functional requirements (Flach et al., 1998; Meister, 1986). In evaluating military systems, an important distinction that underlies all assessment is that between measures of performance (MOPs) and measures of effectiveness (MOEs). Starting with this distinction and adapting measurement paradigms developed by Meister (1985) and Erickson (1984), a general measurement strategy will be described below that allows the analyst to consider the ecological context and inherent complexity of future military systems while maintaining an appropriately eclectic approach for each design phase. Three topics of human system measurement help provide context for the rest of the chapter:

- human measures of performance,
- HSI measures of effectiveness, and
- human systems measurement problems.

### 8.2.1 Human Measures of Performance

The basic human-related MOPs are time to perform a task and the accuracy with which the task is completed, or its complement, the level of human error. The latter variable may be expressed in other terms, such as the probability of detecting a target. Most often, human-related MOPs are collected in laboratory experiments. The drawback to using such measures is that test results may have little impact because their effect on overall system performance is unknown. For example, Erickson (1984) argues that if the measurement process focuses on performance issues rather than overall systems effectiveness, the HSI design process is in danger of becoming marginalized. This is because the focus on local performance issues often fails to show a link between a proposed design change and overall system requirements. Indicating an operator's ability to resolve resolution lines on a display or measuring his or her comfort level may have little impact on the design process. The crucial question is not whether the operator is comfortable or the display is optimal but rather what effect these conditions have on system goals. In particular, what is the system cost of a poor interface or an uncomfortable seat in terms of overall effectiveness?

Human performance measures must be interpreted in the context of the operational situation, the state of the individual performing the task, and the implications of the measured performance on systems effectiveness (Charlton, 1996). Thus, human performance measures must be translated into what used to be called *systems-relevant criteria* but now more commonly are called *measures of effectiveness*. The MOEs permit the comparison of alternative systems in terms of functional objectives and mission needs.

Examples of MOEs include loss exchange results, systems saved, and tons delivered per day (DoD, 1991b). Chapanis (1960) provides early examples of the differences between human factors MOPs and MOEs. For example, he refers to a case where data on the human detection of a radar target (MOP) were translated into the detection range of the radar (MOE). Another example is translating measurements of human performance in an air traffic control task (MOP) into criteria such as the amount of fuel consumed by aircraft, the length of time the aircraft would be under control, and the separations that would be maintained between aircraft (MOE).

### 8.2.2 HSI Measures of Effectiveness

The corollary of the kinds of translations described by Chapanis (1960) is that the measures of human performance must be compatible with the system of measurement used to express MOEs. Erickson (1984) argues that system component and operator performance requirements are not explicit in the upper levels of any system analysis that is conducted early in concept development. There may be no direct relationship between operator task performance and system performance criteria unless the connection is made explicit by analysis. This becomes increasingly difficult as systems become more automated. Erickson describes an approach to developing a "capability hierarchy" starting with a functional analysis and decomposing the performance requirements from that level. He notes that it is necessary to go down at least two levels in the hierarchy before operator performance criteria become apparent (see Fig. 8.1).

Figure 8.1 depicts a useful paradigm for measuring the impact of design factors on system effectiveness. Erickson's paradigm captures the spirit of much of the above discussion. The process starts with mission goals (block 1) and uses a step-by-step process that encompasses a variety of measurement methods to build a utility model of the developing system within its intended military environment. Implicit in Erickson's discussion is the importance of the derivation of various combat costs and benefits that will be used to create MOEs within this environment. After determining mission goals, the analyst defines the system and develops the MOEs in relation to how a system interacts within the total operational environment. Descriptive modeling and task-analytic approaches are used to define the system and begin to understand the role of the crew within the operational milieu. Naikar and Sanderson (2000) have extended Erickson's approach by using work domain analysis (WDA), which is derived from an analysis of system goals and provides a framework for (1) evaluating the implications of detailed technical proposals for system overall functionality and (2) aggregating the implications of the lower level MOPs.

It is important that blocks 2, 3, and 4 be done as a team effort. The MOEs must be specified, and the critical system factors must be identified. The process of generating operational and systems data requires a team approach because no one engineering discipline or operational expert can understand all the ramifications of a complex system. When these steps are completed, we can then move to blocks 5, 6, and 7 to complete the evaluation of system effectiveness. Of particular note is the use of human performance data as inputs into constructing the framework for the modeling environment and human performance experimentation as an integral part of exercising the model. Erickson does not argue that human performance data are unimportant, only that data must be understood within its full operational context (i.e., or more generally its ecological context; Flach et al., 1998). Realistic simulation methods are the most likely confirmation of an effectiveness

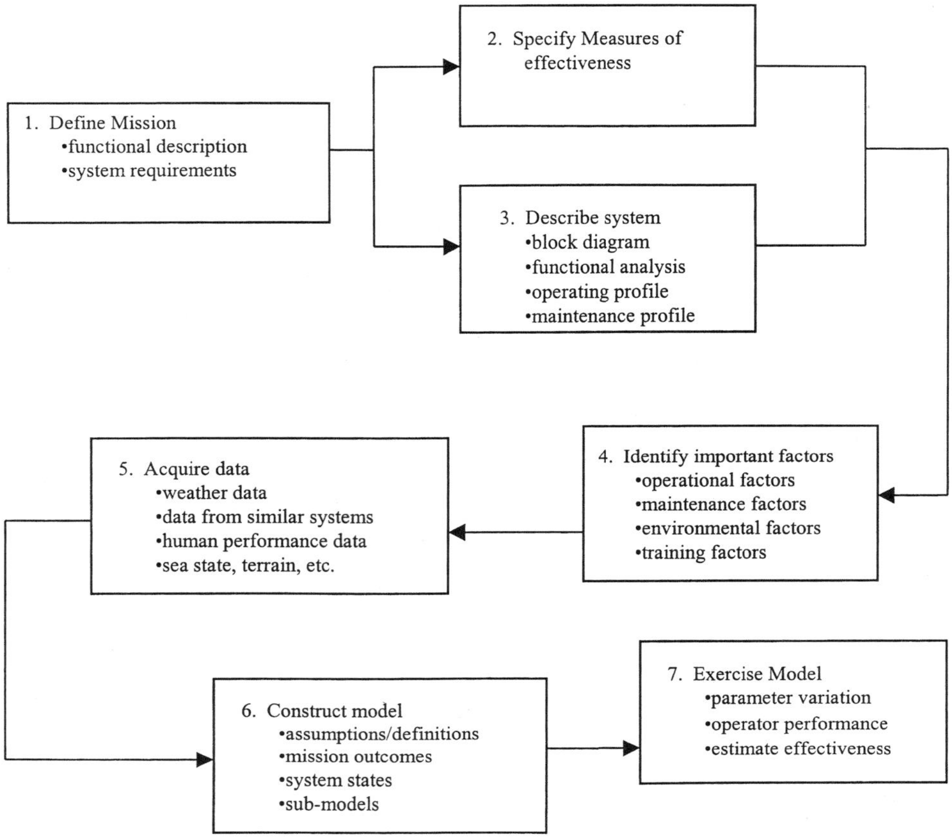


**Figure 8.1** Principal activities required to evaluate system effectiveness (after Erickson, 1984).

model given the current state of the art. The multistep, carefully integrated measurement process was considered necessary by Erickson because military environments are too complex to be represented by simple performance or analytical methods isolated from their effect on the total system.

### 8.2.3 Human Systems Measurement Problems

The principal drawback to systematic approaches such as Erickson's (1984) paradigm is the implication that we must know a great deal about the system before we can start the HSI process. The opposite is true; understanding the human role should proceed in parallel with the engineering design. In most military systems, the human role is paramount either in execution or decision making, and this role must be defined at least loosely before a working concept can be developed. Moreover, the more accurate the conceptualization of the human role is within the design process, the more likely the system is to meet milestones and cost goals. In attempting to understand the human role during the concept stage, the same basic approach can be used—certain overarching goals must be developed and human performance issues must be identified in terms of these goals. Methods must be used to try to measure or perhaps more realistically approximate the impact of the proposed human component on the nascent system even before detailed engineering specifications are available. Understanding the human role before knowing the exact parameters of the proposed system has always been a difficult problem; however, the advent of more cost-effective modeling techniques and simulation paradigms should make this problem more tractable.

In most cases, predecessor systems exist, but rapid changes in technology make direct comparisons of human performance issues for a new generation of systems difficult. For example, researchers at Fort Huachuca investigated generic crew roles for a family of unmanned aerial vehicles (UAVs) that were in various stages of conceptual development (Barnes et al., 2000). The training and doctrine system manager (TSM) was concerned with the skill level necessary to operate these new UAVs because this variable would influence all future UAV design decisions. Two flight operators were evaluated: the external pilot (EP) who flies the UAV for take-off and landing using a radio-controlled hand device and the air vehicle operator (AVO) in the ground shelter who flies the UAV at all other times. The crucial question was whether the flight crew needed to be flight certified. When a variety of test instruments and interviews with over 70 subject matter experts were used to answer the question, it was found that most of the flight safety problems took place during take-off and landing. As an example, Figure 8.2 shows that most of the skill-loading problems related to safety issues for the EP positions were the result of differences between experienced and inexperienced operators. In effect, experienced EPs used conceptualization skills about the qualities and characteristics of the air vehicle to anticipate problems for take-off and landing whereas the inexperienced operator used perceptual and psychomotor skills associated with direct control of the system to react to problems. This suggested that the problem was most likely a training issue. Accident reports and training performance data supported this hypothesis. The result of the analyses suggested a different training strategy rather than flight certification as the answer to improving EP performance. An indirect result was a greater emphasis on developing safe automatic take-off and landing technologies to circumvent this problem altogether.

The serial nature of the data collection effort in Figure 8.2 reflects the linear, compartmentalized design philosophy of the 1980s, which was both time consuming and cumbersome. Improved simulation methods and a determined effort by the military to infuse experimentation and concept exploration earlier in the design process have resulted in more of a spiraling concurrent engineering paradigm. Early approximations of the system are modeled and tested and even field tested before any mature design concept exists. This approach is highly iterative; the paradigm is closed loop with iterations often starting at the "define mission" stage because the design process is also being used to define mission elements. This process is still new, but the establishment of battlefield laboratories by all U.S. services, extensive use of warfighter experiments to validate new concepts, and an emphasis on using modeling and simulation during the design process all point to an evolving design philosophy. This approach, with emphasis on modeling and simulation tools, requires a more adaptable measurement paradigm. The same Erickson processes still occur but in parallel. Since a single model or simulation that is in any sense complete is usually not available early in the design process, the HSI practitioner must rely on a combination of modeling, human experimentation, and simulation for each stage of development. Erickson's approach still provides a valuable measurement framework. The same general philosophy is used in current programs, but the measurement process takes place in a more dynamic and iterative environment.

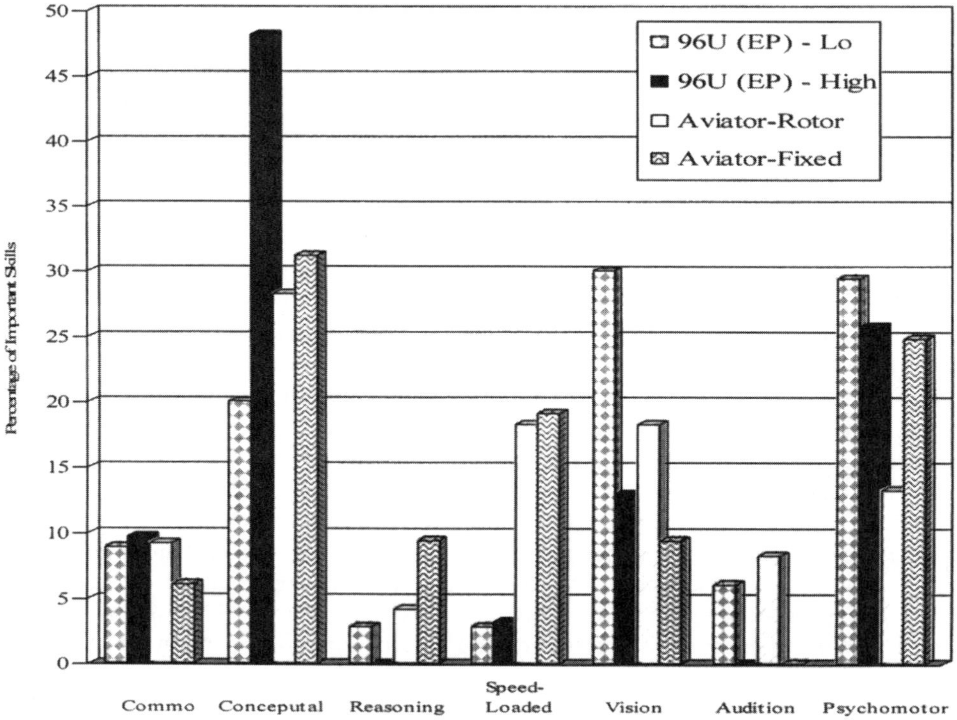


**Figure 8.2** Percentage of important skills used during emergencies, shown by external pilots (EPs) with high and low experience levels. Aviator information is shown for comparison.

## 8.3 GENERAL MEASUREMENT MODEL FOR HSI

The measurement process should be motivated by a cost–benefit paradigm. Measurement procedures and scales need to be chosen carefully to mirror important systems requirements and costs (Meister, 1985). All successful strategies have the same general components: some way of defining the system and its general impact on the stated requirements and methods for evaluating its impact to ensure the end products meet system goals (Gould, 1988; Meister, 1985; Whiteside et al., 1988). Data collection strategies start with system objectives (see Fig. 8.3), which lead to the definition of top-level system requirements and top-level definition of MOEs. From the bottom up there are three basic measurement protocols:

- analytical methods and models,
- human performance experiment, and
- realistic validation methods.

Each of these measurement methods is useful in helping to understand the impact of design decisions at various stages of the process. The MOPs developed from this combination of protocols feed into the MOEs that help determine critical design decisions.

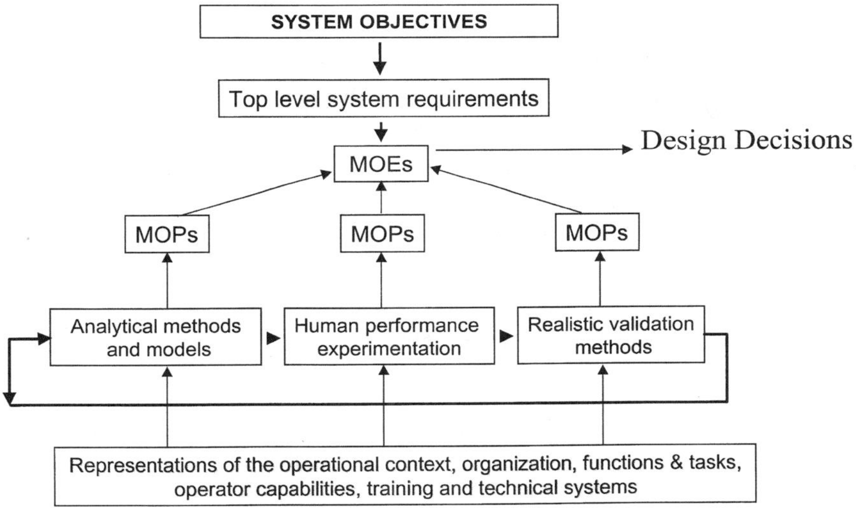


**Figure 8.3** General measurement model for HSI.

The importance of using these three classes of measurement is that they perform separate functions related to assessing system performance during the design process. The descriptive models and analytical techniques give the designer an understanding of system dynamics and MOEs and permit analysis of concepts before any actual design decisions are made. Human performance experimentation is the mechanism for evaluating specific design decisions or trade-offs important enough to be evaluated early to midway through the design process. Also, well-controlled experiments can be used to verify or update early modeling efforts and to understand the cognitive and performance issues evinced by the operational environment. The realistic simulations and exercises are usually the first time the design prototypes are exposed to complex operational environments. These exercises can be part of the formal testing cycle, but the exercises are usually done early enough in design to be sensitive to important operational and engineering issues. The purpose is not only to validate the design but also to allow the engineers to evaluate design options and trade-off decisions in their intended environment. Validation methods differ from human performance experimentation because of the greater emphasis on realism and the extensive use of closed-loop exercises to measure performance. The price of attaining realism is usually a decrease in experimental control and an increase in cost.

The various techniques provide different types of measurement scales, from nominal scales to ratio scales. (Nominal scales identify distinct entities, e.g., A not B; ordinal scales express differences as greater or less without reference to units; interval scales use equal intervals of measurement, such as degrees Fahrenheit, but without defining the end points; and ratio scales can be treated by all mathematical manipulations because they have defined zero points, such as a measurement of length or speed.) In the initial requirements definition and concept development stages, some MOEs may be clear and expressed on a ratio scale such as system range or payload, whereas others and human MOPs may be expressed on an ordinal scale of measurement. Since the various HSI techniques are applicable at different points in the acquisition design and development cycle, the scale of measurement that can be used changes throughout the system design and development process. For example, mission analyses, which are used to define the requirements for new systems, describe events and sequences of events that are nominal or ordinal data, yet mission analyses also include distances, times, and probabilities, which are ratio-scale data. In contrast, functional analyses, which are derived from mission analyses, provide descriptions of the functions to be performed but may not include performance requirements (Beevis, 1999).

To guide the HSI specialist in developing MOPs, the more generic HSI techniques, along with their measurement parameters and their relationships to system performance, are given in Table 8.1. The techniques are classified by domains (manpower, personnel, etc.) as described in Chapter 1. Table 8.1 shows some domain-specific considerations when measuring the human impact on complex systems.

The key to measurement is to match a specific technique to a particular problem during various phases of system development. The use of several of the above methods in concert is critical in the design process because there is no single research method that is flexible enough, cost effective, well controlled, and sufficiently realistic to address the multitude of problems associated with complex systems (Wickens and Hollands, 2000). In selecting a method, users should remember that many techniques are complementary and should be selected for their contribution to subsequent stages of measurement. For example, field observations, function and task analysis, and experimentation can be used successfully to define the requirements for complex human-in-the-loop simulation of advanced defense systems (Greenley et al., 1998). The literature suggests general guidelines for choosing methods based on past research and engineering experience (Barnes et al., 2000; Gould, 1988; Knapp, 1996; Meister, 1985, 1987):

- Evaluation early in the process is necessary because fixed designs are nearly impossible to change. Inexpensive analytic techniques (especially in the early phases) can uncover many HSI problems before they impact the design process.
- Evaluation must be conducted in a system context.
- Evaluation is iterative; results should verify previous analyses as well as suggest future exercises.
- Well-controlled simple procedures should always precede complex evaluation methodologies.

Early evaluation is important. For example, identifying the personnel and training requirements is crucial to early design decisions because the personnel skill mix necessary for a system is not only an important life-cycle cost but also an issue that affects the military organization as a whole. The supply of the necessary operators and maintainers has to be initiated years ahead of the delivery date for a new system. Therefore, it is important to identify and evaluate design features that affect such issues early in the project. Inexpensive analytic techniques often have a very high payoff if used early. To ensure that these methods do impact the design, they need to be tailored to specific design issues early enough in the process to be useful to the designer. Also, the HSI effort must be tightly coupled with the other engineering efforts. For example, operator workload analysis can be very useful as a design tool provided the results are applied to design problems such as crew task allocation or automation decisions (Barnes et al., 2000; Beevis and Essens, 1997) and the HSI team works closely with other design engineers to understand the

**TABLE 8.1 HSI Measurement Techniques and Their Relationship to System Performance**

| Technique | Measurement Parameters | Relationship to System Performance |
|---|---|---|
| | *Manpower Domain* | |
| Analysis and modeling—parametric estimates based on displacement, weight, power, etc. | Numbers of personnel | Assumes acceptable system performance and thus availability; may not reflect performance of future systems |
| Analysis by analogy (e.g., early comparability analysis) | Staffing estimates based on similar systems | Assumes acceptable system performance; may not reflect performance of future systems |
| Computer simulation of operator tasks | Numbers of personnel required at different points through a mission | Estimates can be related to system availability in normal and degraded modes of operation or maintenance. |
| | *Personnel Domain* | |
| Analysis by analogy (e.g., early comparability analysis) | Inventories of aptitudes, skills, and experience levels required and special physical requirements based on similar subsystems | Assumes acceptable performance based on skills required with existing systems |
| Analysis of skills | Inventories of aptitudes, skills, and experience levels and special physical requirements | Assumes acceptable performance based on analysis of categories of skills required |
| Computer simulation (e.g., FOOTPRINT) | Data on manpower, personnel, and training characteristics of each military occupation speciality. | Assumes acceptable performance based on analysis of skills required and available |
| Analysis and modeling—personnel cost estimates | Costs of operational, maintenance, and training personnel | Can be used as input to cost trade-offs for given level of system performance |
| | *Training Domain* | |
| Analysis by analogy with existing systems | Estimates of training system requirements based on existing systems | Defines training performance goals to be met for system to be effective |
| Analysis—training needs | Estimates of training system requirements | Defines performance goals for operator tasks required for system to be effective |

*(continued)*

**TABLE 8.1** ***(Continued)***

| Technique | Measurement Parameters | Relationship to System Performance |
|---|---|---|
| Computer simulation (IMPRINT) | Estimates of system performance as a function of skill levels | System performance is simulated as a function of workload and operator skills and task performance. |
| | *Health Hazards Domain* | |
| Analysis—health hazards assessment | Exposure times for personnel for health hazards | System effectiveness may be constrained by exposure times, e.g., limits to load carriage and number of rounds fired by a weapon. |
| | *System Safety Domain* | |
| Analysis—risk assessment | Probabilities of injury or system failure | System effectiveness can be predicted as a function of personnel and subsystem availability as predicted by failure analysis. |
| | *Human Factors Engineering (HFE) Domain* | |
| Analysis—missions and scenarios | Sequences of events, distances, ranges, times and environmental conditions | Scenarios for HFE should be developed from the scenarios in the mission needs statement and the operational requirements document. |
| Functional analysis of system[a] | Sequences and flows of functions required to perform system missions | Functional analyses define system and subsystem goals. System effectiveness is implicit in meeting those goals. Can be used to identify MOEs as suggested by Erickson (1984). |
| Task analysis[a] | Sequences and times of operator tasks, task tolerances and performance requirements and operator interface design requirements | Analyses reflect system performance requirements and establish criteria for experiments, rapid prototype evaluations, and field trial performance. |

**TABLE 8.1 (*Continued*)**

| Technique | Measurement Parameters | Relationship to System Performance |
|---|---|---|
| Cognitive task analysis | Diagrams showing the relationship of system goals to cognitive processes (e.g., information processing) and to lower level cognitive and physical tasks | Analyses reflect goals and knowledge-oriented behavior rather than data-oriented and interface manipulation behavior (Rasmussen, 1986). Emphasis is on top-down analysis. |
| Task network models and simulations | Times and probabilities of completing sequences of tasks | Simulation outputs must be translated to relate them to system effectiveness. Can produce estimates of workload and some MOEs and confirm assumptions about manning and personnel issues. |
| Laboratory experiments | Times, errors, measures of comprehension, retention of information, etc. | Results must be translated to be relevant to system effectiveness. |
| Rapid prototyping of operator: machine interface | Times, errors, measures of comprehension, and ease of use | Results must be translated to be relevant to system effectiveness. |
| Complex human-in-the-loop simulation | Performance of specific tasks, times, errors, measures of comprehension, retention of information, and ease of use | System MOEs can be measured directly if the simulation fidelity permits. Human MOPs must be translated into system MOEs. |
| Complex field trials and warfighter exercises | Times, errors, measures of comprehension, retention of information, and ease of use | MOEs can be collected directly, as well as MOPs. |

[a]A more detailed review of the links between HFE analyses and system performance is provided in Beevis, 1999.

various design and crew trade-offs necessary to reduce crew workload before a mature design is in place.

System context is vital—the measurement scale must reflect important system dynamics. For example, Knapp (1996) points out that for command and control (C2) systems the measure of interest is the information flow. Specifically, she felt it was important to measure the impact of operator workload and the resulting message flow in terms of its effects on command decision making and execution. Knapp was able to model the information flow for a number of C2 systems showing the costs and benefits of various design options for proposed future systems. Her measures were closely related to important system parameters. If she had simply reported her results in terms of system latency or overall workload, it is doubtful her results would have been as well received.

Knapp also delineated these effects in terms of the cognitive skills necessary to perform a particular task giving the design team additional insight into the quality of personnel needed for the various positions as different automation and design options were considered (Muckler et al., 1991).

Evaluation must be iterative because the overall design process is one of successive approximations. Once the design options are set in concrete, HSI analysis tends to be ignored. The good news is that the variety and usefulness of HSI tool sets have increased remarkably in the last 10 years, especially in their ability to evaluate iterative design options. Human systems integration modeling packages such as improved performance research integration (IMPRINT) tools allow multiple analyses using the same software environment. The chief advantage of these new modeling environments is that as more data are collected and new design options are proposed, it is relatively easy to revise the original model and rerun the analyses.

Well-controlled simple procedures can reduce the amount of effort required to investigate HSI issues. A simple analysis or experiment can uncover obvious flaws and permit the analyst to direct future efforts toward investigating the more subtle and complex trade-off issues. Simply requiring extensive HSI analysis or modeling does not guarantee that these tools or their results will be used in the design process. But well-controlled procedures can negate requirements for further complex techniques that add little to the design process.

## 8.4 ANALYTICAL AND MODELING TECHNIQUES EARLY IN DESIGN PROCESS

Most activities in the early stages of system design and development involve the analysis and synthesis of a design solution. Defining system components, operations, and crew issues is the initial step of any analytical effort for a new system. There are a number of traditional methods and data sources that will make the analyst's job easier. Various documents that are part of the design process such as mission needs statements, operational requirements, etc., are good starting points for the analyst to begin to understand the purpose and intended uses of the system early in the development cycle.

In many of the military programs, there are early exercises and "rock drills" (walk-through simulations of various doctrinal concepts) that define the doctrine being developed to counter future-threat profiles. Being part of these exercises is extremely useful because the exercises emphasize and elaborate the operational issues that the system is being designed to address. The importance of understanding the military purpose and intended environment before any analysis is attempted cannot be overemphasized.

Analytical techniques that are part of the traditional system engineering approach include functional-flow diagrams (Meister, 1985) and requirements modeling (Hatley and Pirbhai, 1987). These approaches are used during the design process to model the system and its various components and their interrelationships. These methods are quite helpful in developing a "blueprint" of the overall system and as such provide a good reference point for HSI analyses. However, if the models are developed solely for engineering guidance there are a number of drawbacks:

- These models often do not exist in the early conceptual stages and yet important HSI issues need to be addressed before a well-defined design is developed.

- System component and operator performance requirements are not explicit in the upper levels of any system analyses that are conducted early in concept development (Erickson, 1984).
- Often the human role in the system is not well defined in these representations (Beevis, 1987).
- Functional-flow diagrams that focus on engineering issues can unintentionally foster a myopic view of the system, especially in terms of how the system interacts as part of the total combat system.

After understanding whatever engineering documentation exists at that point (including HSI documents), the next step is to choose a method to represent the human component as part of the system. The purpose of these analytical and modeling methods is to understand and predict the impact of various important crew functions on early design concepts when there is little or no performance data for the new system. Choosing the correct method and measurement scale depends on the design issue.

### 8.4.1 HSI Analysis Techniques

Human factors engineering technology includes a suite of analysis techniques that are compatible with a number of systems engineering analysis methods. The generic forms of analysis correspond to those recommended in US-MIL-HDBK 46855 and include mission and scenario analysis, function analysis, function allocation, task analysis, performance prediction, and interface and workspace design (Beevis, 1999). These analyses can make a major contribution to the implementation of human factors in a project, particularly if coupled with other techniques such as experimentation, rapid prototyping, and human-in-the-loop simulation. For example, in the development of the F-18 aircraft, mission analyses were used to identify the likely use of aircraft systems, their operational modes, pilot tasks, and control and display requirements to establish an overall concept for the operator–machine interface. The results of the analytical effort were then refined and validated in a very extensive manned simulation of the aircraft (Merriman and Moore, 1984).

As described earlier, the various analytical techniques provide a range of predictive performance measures, some of which can be related to MOEs and some of which require additional analysis or quantification through experimentation, simulation, or trials. Function-flow diagrams provide the basis for developing performance specifications related to each system function as recommended by Erickson (1984) and for developing requirements specifications for the operator–machine interface. Because they describe the functions that must be performed by a system without reference to hardware, software, or humans, the early stages of human factors engineering (HFE) analysis can be reused and updated as technology improves.

Early analysis can reduce the potential range of design solutions to a point where options can be evaluated through user trials or experimentation. Early analysis can also identify areas where performance may be a significant factor that requires confirmation by experimentation, modeling, or simulation. For example, in the development of a targeting device for a shoulder-launched, ground-to-air missile, 18 possible combinations of display devices and formats for displaying gross and fine azimuth to the operator (a design option decision tree) were analyzed to identify the advantages, disadvantages, and performance implications of each one (see Fig. 8.4). From the analysis the two most

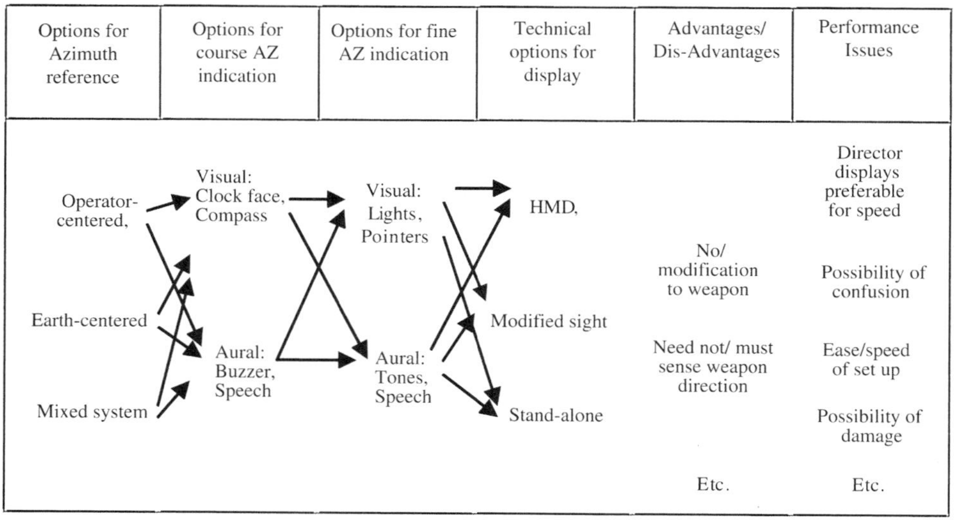


**Figure 8.4**[1] Design options decision tree used to identify most promising candidates for a display.

promising concepts were selected for a simple simulation experiment that confirmed the performance characteristics predicted by the analysis.

## 8.5 HUMAN PERFORMANCE EXPERIMENTATION

Measures of human performance are components of all the measurement techniques under discussion. The use of traditional experimental methods (Kirk, 1982) is an important source of obtaining these data, and these methods allow for a more precise understanding of the processes that affect system design. The problem with many of the more subjective methods and realistic simulations and field exercises is the lack of experimental control (McBurney, 1998). The causal relationships among variables may be impossible to untangle during a large field exercise, and many of the analytic techniques lack rigorous statistical verification. Experimental methods can be used both for parameter estimation and hypothesis testing. The problem with these methods is that the control conditions inevitably introduce a degree of artificiality into the measurement process. This is especially true because cost and pragmatic considerations limit the number of factors and ambient conditions that can be considered. For example, a 10-factor between-subject design to ensure no sequence effects would require a minimum of 2048 subjects if each factor had two levels to use conventional analysis-of-variance techniques.

More flexible experimental approaches are available (Williges, 1995) and are discussed below; however, in general, the tighter the experimental control, the more likely constraints are to be placed on realism. This is not an argument against experimental control; rather, it is a reminder that the true power of a good experimental approach is in hypothesis testing and not in measuring real-world processes.

The measurement model presented here is predicated on the use of initial modeling and realistic validation procedures being used in concert with true experiments. The role of the controlled experiment should be dictated by the initial analysis, which should point to the critical design decisions that need to be considered: the important crew-related factors that need to be investigated and the MOEs that need to be addressed. A good experiment is similar to a laser; it covers a small area of the response surface (the surface defined by the multiple regression equation that describes the relationship among all experimental variables), but it is very effective if it is directed toward the most critical factors. As any good experimentalist knows, it is the relevance of the questions asked and the experimental procedures used that determine the value of an experiment.

### 8.5.1 Person-in-the-Loop Simulation Experiments

Small-scale elegant experiments are rare during the design process because few of the problems being investigated for complex military systems lend themselves to such a paradigm. An example might be investigating a number of control-display options in the laboratory before narrowing the field to a few that can be evaluated during a more realistic simulation exercise. It is cost effective to eliminate deficiencies early in the design cycle using simple experimental methods wherever possible.

The initial experimentation involving either part-task simulations or more limited scenarios should be designed to optimize experimental control. These exercises are a bridge between the laboratory and more realistic simulations and field testing that follow. For example, a series of simulation experiments were performed at Redstone Arsenal to investigate fatigue and equipment factors for a variety of UAV options requiring day and night crews to rotate duty cycles every 12 of 24 hours over a 72-hour operational tempo (Barnes and Matz, 1998). The investigators were able to control most experimental factors and ensure that all participants received the same target sets, rest conditions, etc., to the point of controlling their rest periods when they were not on 12-hour shifts. The ground control station, flight parameters of the simulator, mission taskings, and target sets were all realistic. Important design requirements related to manning and console design were addressed by using measures of target acquisition and safety. The results were useful in that they indicated serious problems with single crew configurations and suggested crew fatigue problems related to circadian rhythms. Although the exercise was well controlled and realistic, it could not capture the actual stress, unpredictability, and interrupted sleep patterns of combat. Again, important compromises between experimental control and realism had to be made by the data collectors.

### 8.5.2 Experimental Designs for Complex Spaces

A number of experimental approaches such as response surface methodologies, confounded designs, and quasi-experimental methods have been developed specifically to measure complex environments (Box et al., 1978; Cook and Campbell, 1979). Response surface and confounded designs allow the investigator to measure a restricted portion of the response surface to estimate behaviors over the entire response surface.

A simple example is an experiment designed to model the effect of display size on 10 operational and sensor factors for a navy attack aircraft. The initial investigation used a screening technique (supersaturated fractional factorial design; Barnes, 1978) that identified four experimental factors and their second-order interactions as accounting for 50 percent of the variance in a target acquisition task. These four factors were used to create an orthogonal regression equation using conventional experimental designs. The fractional design permitted the investigation of a large response surface with few subjects and experimental conditions, and the results were used to investigate the most crucial factors in an unconfounded data space.

These designs are not panaceas; their utility depends on the particular measurement problem addressed. The measurement of complexity often results in the loss of experimental control, but this loss is worth the price if the response surface is better understood and the results lead to more predictive models or better experimental precision using so-called pure experimental designs.

The following sections describe a variety of simulation and virtual methods that are capable of reproducing much of the realism of field exercises and actual operational conditions. Complete control is impossible if for no other reason than the closed-loop nature of the real world and the bewildering array of variables involved. Under these conditions, the use of quasi-experimental methods and statistical controls is as important as the use of control techniques in classical experimental paradigms. The more sources of variation that exist, the more crucial it is to control as many of them as possible.

## 8.6 MODELING AND SIMULATION

A model is a representation of critical aspects of objects or situations. In the context of HSI, modeling refers to mathematical models of human performance and crew workload. Simulation is a method for implementing a model over time. Simulations can be "constructive" based on mathematical or parametric models: they can be "virtual" simulations in which operators use representations or rapid prototypes of systems and interfaces or they can be live simulations with users conducting trials with actual equipment. One advantage of the use of modeling and simulation is that work done for project planning or concept development can be carried over and reused and exploited in project development, project definition, and implementation. This permits MOPs and MOEs to be refined throughout the system's development process.

Operations analysis (OA) makes extensive use of models and simulation. However , it has proven very difficult to establish a link between the OA modeling activities and HSI modeling, despite the Military Operations Research Society having held several conferences on the subject. One reason for this is that combat models tend to focus on the outcome of engagements, whereas human factors models focus on performance of specific tasks. McMillan and Martin (1994) suggested that the human factors models can be used off-line to generate statistical distributions and performance shaping functions for use in OA models. It is not clear how the output of OA models can be used to focus HSI efforts without going through the kind of decomposition of performance requirements that is recommended by Erickson (1984).

Compared with other engineering disciplines, a review of the human factors and human engineering literature reveals a limited amount of human performance data to support modeling and constructive simulation. While it is true that much more effort is required in this area, a variety of models of human performance are available (McMillan et al., 1991). Some of these models are parametric, but several have been developed from first principles. Many of these models can be expressed as task network models for use in systems design and development. Task network modeling (TNM) represents the complex pattern of operator tasks as an interlinked network of simple human performance task models. One such task model reported by Card et al. (1986) is an information-processing representation of the human operator. The model, comprising perceptual, cognitive, and motor systems, includes times for a variety of types of information processing. Despite their seeming simplicity, such models can represent quite complex applications. For example, using an information-processing model and general principles of human performance, including assumptions of single-channel processing and task completion times estimated from the literature, a network model was developed for the tasks associated with air traffic control. Run as a constructive simulation, the model produced results for most performance parameters that were close to experimental observations of human performance obtained in a manned simulation of the same tasks (Burbank, 1994).

Probably the most adaptable approach to modeling human factors aspects of systems, TNM produces descriptive models of the human tasks and interrelationships in systems, and such models can be used as the basis for simulations. Such simulations can address a wide range of problems and are probably the most common application of TNM. They can be applied to very simple multitask models or to complex multioperator systems. The validity of such models is sometimes questioned. Because they contain less than complete detail, all models have limited validity, but many models can be constructed that are useful in the design and development process. The key to successful TNM simulations is the level of detail of the analysis on which the model is based. Function-flow diagrams must be decomposed to at least the fourth level because cases of feedback and coupling between operator-performed functions are seldom identified at higher levels of analysis. This contrasts with some of the models used to predict manpower requirements, which use information generated by the second or third level of functional decomposition.

### 8.6.1 Deterministic, Stochastic, and Hybrid Models and Simulations

Many task network models are deterministic because they are based on a scripted input (mission analyses and scenarios) and a predetermined sequence of tasks. If human performance terms such as times and probabilities of completing each task are added, task network models can be used to run constructive simulations of operator tasks. Such simulations are hybrids; some aspects of the simulation are stochastic, but the inputs and sequences of tasks are predetermined.

Implemented as a Monte Carlo simulation, where a random-number technique draws values for task completion times and probabilities of task completion, the simulation becomes stochastic. Further elaboration of such models can make the inputs subject to variation, for example, by using probabilistic mission events. The model of the operator's response to such inputs or factors such as task load can also be made probabilistic. Stochastic simulations avoid the criticism that operator performance is not deterministic and cannot be properly represented by models; however, they require many replications, typically at least 100, to collect the necessary distribution of outcomes to properly represent the operation of a proposed system. In fact, when simulating complex enterprises such as command and control systems, it is important to examine the distribution of possible outcomes as the system is simulated (RSG.19, 1999) so many replications are required. This is not a significant problem for computer-based constructive simulations; however, the need for many replications is a limitation for complex human-in-the-loop simulations. Establishing clear start and end points for complex sequences of probabilistic events is another difficulty and can preclude drawing generalizations across a wide range of simulation runs.

Any human performance that affects the time or probability of completing a task can be incorporated in TNM simulations. This makes the technique useful for HSI investigations because personnel factors such as skill level can be expressed in the time-and-error distributions used in such simulations. Thus, TNM simulations can be used to investigate HSI trade-offs involving manning levels, training or skill levels, and human factors engineering (Knapp, 1996). One extensive simulation used task network models to reflect the impact on crew tasks of four technical upgrades to a maritime patrol aircraft: data fusion technology, multifunction workstations, voice interactive technology, and electronic library applications. Simulation results were interpreted in terms of changes to the operators' workload and effectiveness and changes to the system effectiveness [Canadian Marconi Co. (CMC), (1995)]. Simulating human performance using task network models is seen as an important tool in achieving significant HSI cost reductions in future systems. For example, such simulations have been used to explore decreased manning levels in future naval systems (Campbell and Laughery, 1999), and such simulations can link job and task skill demands with system design, manning levels, and system performance (Middlebrook et al., 1999).

The outputs from task network simulations predict some characteristic of human performance, such as time to perform a sequence of tasks. By applying suitable algorithms to the task simulations, operator workload can be predicted as the sequence of tasks unfolds (Hendy et al., 1990). The simulation outputs can also be used to generate measures of a system's effectiveness. For example, the number of occurrences of specific tasks, such as verbal communication with a particular unit, can be used in conjunction with workload predictions and figures of merit derived from subject matter experts to generate MOEs for mission segments. Other transformations of operator MOPs to system MOEs can be calculated from task performance data. For example, the rates of processing messages and message-processing delays or backlogs can be calculated to give a MOE a C2 system (Middlebrook et al., 1999).

Early attempts at modeling human factors issues used commercially available software such as the General Purpose Simulation System (GPSS) originally developed by IBM (Overmayer, 1975) or custom-written software (McCann and Sweeney, 1976). Software tools are now available to support TNM. The chief tools are SAINT, developed by the U.S. Air Force (Wortman et al., 1978) and Micro Saint developed by Micro Analysis and Design and available commercially (Laughery and Drews, 1985).

The IMPRINT tool was developed from Micro Saint to incorporate nine separate tools for HSI analysis that had been developed previously by the U.S. Army Research Laboratory. IMPRINT can be used to model both crew and individual soldier performance for operator and maintainer tasks. Detailed operator–machine interface designs can be evaluated through the effects on task performance. For some analyses, workload profiles are generated so that crew–workload distribution, operator–system task allocation, and workload coping strategies can be examined. Maintainer workload can be assessed along with the resulting system availability. Using embedded algorithms, IMPRINT also models the effects of personnel characteristics, training frequency, and environmental stressors on the overall system performance. Manpower requirements estimates produced by IMPRINT can be used as the basis for estimating manpower life-cycle costs. The predecessor to IMPRINT, the hardware versus manpower (HARDMAN) analysis and simulation tool was subjected to verification, validation, and accreditation (VV&A) (Allender et al., 1995), and the key analysis capabilities of IMPRINT have also been subjected to VV&A. (See Chapter 11 for more information on IMPRINT.)

### 8.6.2 Complex Simulations and Warfighter Exercises

Human interaction with advanced systems is complex and emergent. It cannot be fully predicted because it emerges from the interaction between what the operator does with the machine and what the machine imposes as tasks or constraints on operator behavior (Taylor, 1959; De Greene, 1991). Because of this, in the early years of human factors research there was a strong emphasis on making observations in real-world conditions (Green et al., 1995; Moroney, 1995). The disadvantage of this approach is that the real world is not controlled. Thus, it may not be possible to make the observations required to evaluate system or operator performance without an unrealistic amount of observation (arranging situations to provide the required observations is the basis of most experimentation).

Compared to highly controlled laboratory experiments that have a strong theoretical basis but no close relationship to the real world, complex simulations and exercises are loosely controlled and are near the opposite extreme of techniques for measuring performance (Chapanis and Van Cott, 1972). This is primarily because of differences in the numbers of independent and dependent variables involved. Laboratory experiments involve few independent and dependent variables—sometimes only one dependent variable. In contrast, exercises and field trials involve a large number of both types of variables. Complex human-in-the-loop simulations lie between these extremes as they take place in quasi-operational conditions, are close to real-world observations, but are arranged to permit useful observations to be made as required.

While complex simulations, field trials, and exercises cannot replicate the stress of actual combat environmental stressors, the stress of sleep loss can be included as a factor when measuring performance. Other stressors can and may need to be simulated. For example, Muir et al. (1989) reported the effectiveness of using financial incentives in studies of emergency evacuation times from aircraft. They also reported that without such incentives behavior was not realistic and did not provide realistic MOPs.

At the outset of a development program, complex simulations, field trials, and warfighter exercises can provide empirical data for use in models or can validate models used in OA (Bryson, 1989). However, the primary use of complex simulations and field trials is to validate predictions about performance made earlier in the program. These human-in-the-loop simulation efforts can extend over several years during the development of a major weapon system and require firm commitments of personnel from operational units over that period. They typically require operational units to commit both operational and support personnel, weapons, and logistics required for the trial and require from months to years of preparation. In addition, analyses of the results from such trials typically require four to six times the amount of time required to make the observations, so results are not available immediately after the completion of the trial. Because of the time and effort required, large-scale field trials are easier to manage when conducted separately from specific system development projects to explore new system concepts or concepts of operations that can lead to development projects. When the trials are used to complement other measures of performance and to validate systems concepts, they must be anticipated, planned, and budgeted well in advance.

The requirements for time and effort preclude repeating field trials and exercises. Therefore, a key factor in organizing field trials is to arrange them so that the required observations can be made. To achieve this, such trials may need to be scripted to evolve in a particular way or to have operators perform specific mission segments. The selection of relevant performance measures that can be observed reliably in a field trial is a highly skilled activity. All measures will be affected somehow by the way that a field trial evolves, from almost negligible effects to measures that are strongly associated with the final states of the trials teams and systems. This is particularly true of the evaluation of C2 systems where commanders' decisions can have a significant effect on outcome measures (RSG.19, 1999). Table 8.2 provides some examples of the range of field trials and the corresponding script requirements.

In an effort to reduce the organizational and logistic requirements, field trials are sometimes arranged to "piggyback" on planned military exercises or other trials. This approach affords much less control over the trial. One consistent problem is that insufficient time may be scheduled by the operational units conducting the exercises for the trial troops to train and establish repeatable standard operating procedures (SOPs) for the new systems being evaluated (Poisson and Beevis, 2000). Usually, performance with a crew-served system increases with training until a plateau is attained (Towill, 1989). When

**TABLE 8.2 Measures and Scripting Required for Various Field Trials**

| Aim of Trial | Measurements | Script Requirements |
|---|---|---|
| Suitability of personal equipment and individual weapons | Operator evaluations of comfort or suitability when issued to operational personnel for use during regular duties | No script required (Webb et al., 1998) |
| Physical workload associated with new equipment or procedures | Physiological measures of thermal stress and physical performance in field conditions | Some scripting of physical tasks required (Tack, 1996) |
| Direct control of a destroyer during specific ship-handling maneuvers | Measurement of accuracy of ship's track-keeping; comparison with results of a mathematical control model | Affected by weather conditions and requires definition of the maneuvers (Lewis et al., 1966) |
| Low-level navigation in tactical aircraft | Measures of aircraft track, accuracy of navigation in normal and unusual circumstances | Affected by weather, day/night, and scenario; requires definition of scenario and reactions to becoming lost (Lewis et al., 1968) |
| Evaluation of new systems for an existing role, such as the use of a hovercraft for search and rescue | Evaluation against a given set of goals for the system; measures of completion of specific tasks | Affected by weather and scenario; requires definition of scenario and tasks (Lewis et al., 1967). |
| Digital battlefield equipment and procedures | Multiple measures of information flow, situation awareness, decision making | Affected by evolution of scenario; easy to lose "thread" between independent variables and outcome measures |

the operators are familiar with a crew-served weapon system and have established SOPs, performance times can be expected to conform to a learning curve. The advantage of this is that the final plateau of performance can be predicted from three observations early in the trials. If SOPs are changed, performance does not improve in a consistent manner and the plateau of performance is impossible to predict. In such a case, it is possible that the performance measurements could result in the selection from among several candidates of a system that is not the most effective.

## 8.7 INTERACTIONS AMONG HSI DOMAINS

The general measurement model described above is intended to support HSI efforts during the development process by focusing on system-relevant aspects of human performance. When planning human performance measures for a system under development, it is important to remember that the various HSI domains are interrelated. Changes in design to improve one domain nearly always affect other domains [Office of the Deputy Chief of Staff for Personnel (ODCSP), 1997]. Such changes must be considered when conducting design trade-offs. For example, in many weapon systems the operator's performance is adversely affected by having to wear protective clothing. Typical problems are hindrance or inability to perform tasks when wearing cold-weather gloves, inability to use weapon sights or other displays when wearing respirators, and reduced reach envelopes due to the bulk of clothing (Poisson and Beevis, 2000). Training can overcome some of these problems, but the most restrictive combinations of protective clothing and equipment may not be routinely included in training. Thus, from the viewpoint of system acquisition, it is important that the operator–machine interface and the training system be designed to accommodate all necessary combinations of protective clothing and equipment. The human factors engineering, health hazards, and training domains interact, and operator performance must be measured under conditions that represent those interactions.

The human systems integration plan for the defense acquisition process must address design trade-offs (DoD, 1991a). Manuals for HSI do not provide much guidance on HSI trade-offs; Booher (1990), for example, includes only two references to "trade-off". Domain interactions and HSI trade-offs that have been suggested include the following (ODCSP, 1997; Walters, 1992):

- increasing system costs through automation to reduce costs for manpower, or training, or reduced requirements for experience;
- increasing system costs through built-in test equipment to reduce the requirements for skilled personnel to avoid drawing from a classification that is projected to be under strength;
- increasing system costs through simplification of the user interface to reduce training time costs; and
- reducing costs for operator manpower by increasing support manpower and personnel requirements.

These trade-offs are binomial. However, it is clear that a change in one domain could interact with several other domains. Reducing manpower costs through automation or simplification of the user interface will increase system costs but might also increase the maintenance training requirement or lead to skill degradation. Also, automation entails its own set of performance issues and costs (Parasuraman et al., 2000).

Although the need is obviously great, there is no well-established body of knowledge on HSI domain trade-offs. Kennedy and Jones (1992) noted that weapon systems designers do not have the expertise nor the tools that are required to make MPTS trade-off decisions. For example, iso-performance curves have been recommended to support quantified trade-offs among personnel abilities and factors such as training time and training system effectiveness (Kennedy and Jones, 1992). The curves (see Fig. 8.5) show the relationship between personnel abilities measured on an aptitude scale and the time to train a given percentage of operators or maintainers up to an acceptable standard. An improvement in the ease of use of an item of equipment should result in a change in the iso-performance curve, because the new equipment requires less training than the predecessor to achieve the same level of proficiency. Unfortunately, Kennedy and Jones report that no data were archived by any of the armed services that would support the generation of iso-performance curves and that developing such curves must be done opportunistically.

### 8.7.1 The F-18 Example

To better understand interactions between the HSI domains and their effects on operator and system performance, Davidson et al. (1991) analyzed data from a review of human factors affecting flight safety and operational effectiveness of the F/A-18 Hornet aircraft operated by the Canadian Armed Forces. The data, derived from interviews with F-18 pilots, were categorized into the HSI domains and reviewed for interactions (Beevis, 1996). Figure 8.6 shows the interactions between 34 factors related to manpower, personnel, training, system safety, health hazards, and human factors engineering.

One thread of HSI domain interactions will be followed as an example. Through a landmark effort in human factors engineering (Merriman and Moore, 1984), the F-18 was designed and developed as a multirole aircraft that can be flown by one person. The F-18 replaced two-place interceptors on some squadrons, thereby reducing the manpower levels;

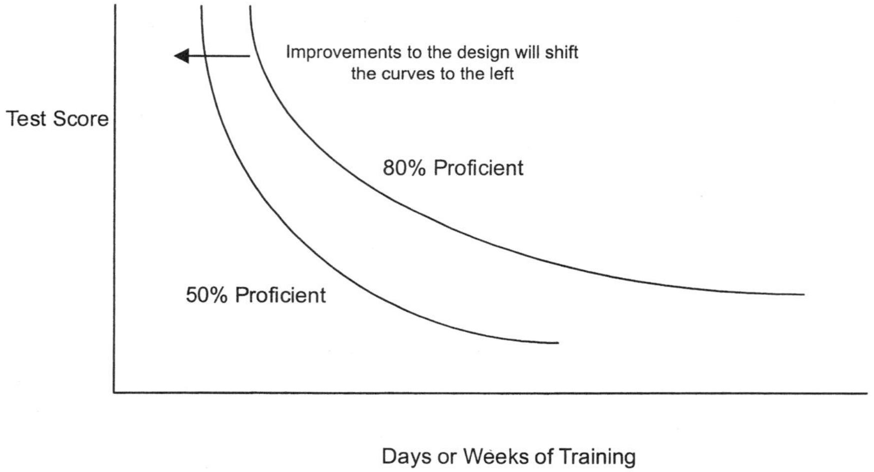


**Figure 8.5** Iso-performance curves for ability and training time (after Kennedy and Jones, 1992).

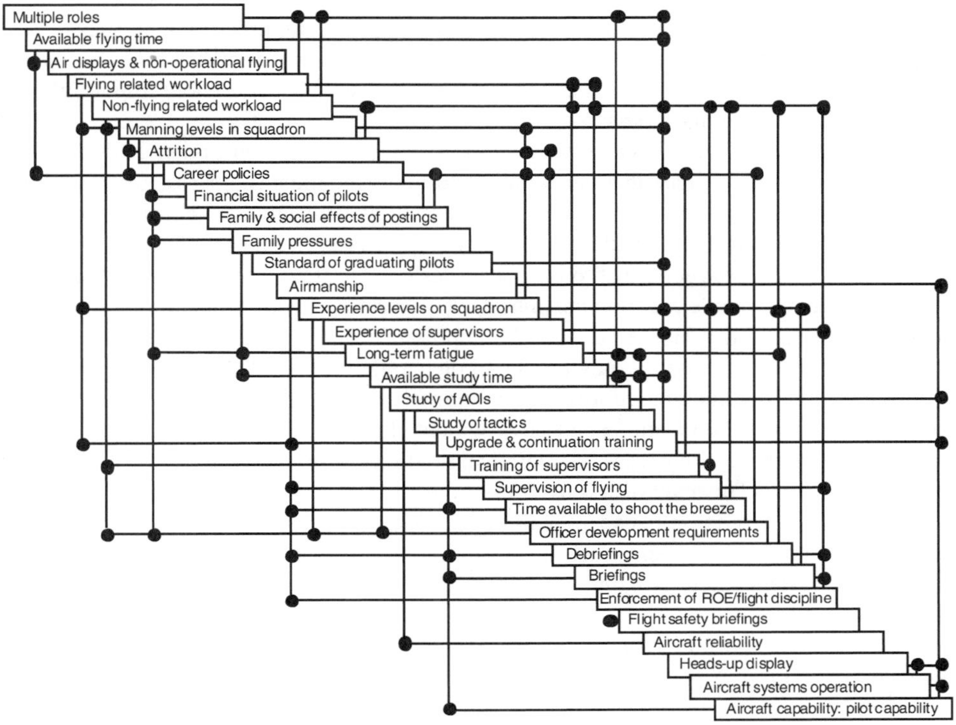


**Figure 8.6**[1] Factors affecting flight safety and operational effectiveness in F-18 aircraft (horizontal lines are outputs; vertical lines are inputs).

thus, the HFE domain interacts with manpower. The reduction in squadron manning levels increased the overall workload of nonflying activities on squadron personnel, including study time and classroom training; thus, the manning domain interacts with training. The capability of the F-18 made the study of tactics and aircraft systems more demanding, but the reduction in available study time made it more difficult for aircrew to do this; thus, training has internal interactions. In addition, lack of familiarity with aircraft operating instructions for emergency procedures affected flight safety. At the same time, on-squadron proficiency training and practice (upgrade and continuation training in Fig. 8.6) affected the standard of airmanship in squadron pilots. Airmanship affects the level of flight safety in the squadron as well as the quality of training once pilots are on-squadron; thus, training affects system safety in several ways. The U.S. Air Force has made similar observations. For example, an investigation of a fatal accident involving two combat aircraft concluded that the quality of crews had been hurt by too many deployments, which had precluded developmental training (Newsbreaks, 1994).

The interactions among the factors in the F-18 HSI study were analyzed using matrix algebra (the MICMAC method of Godet, 1991) to identify important direct and indirect interactions. Figure 8.7 shows the direct and indirect interactions among the HSI domains, derived from the relationships in Figure 8.6. It was concluded that even though the HSI domains of manpower, personnel, training, system safety, and human factors engineering do not seem to interact directly, they do have strong indirect interactions. Human factors engineering appeared to interact directly with only the training domain, whereas it interacted indirectly with manpower, personnel, and system safety factors. The results suggested that the interactions between these domains were probably more important than the interactions within the domains.

The personnel domain appeared to have the most interaction with other HSI domains. This seems reasonable given that personnel factors include the basic performance abilities of the humans in a system. Some important interactions between the 34 factors examined were a function of operational and organizational issues (e.g., career policy or operational commitments). This suggests that, once a well-designed system is introduced into service, operational requirements and policies may have a greater effect on operational effectiveness than the individual HSI domains. The importance of these factors is reflected in the conclusion of a NATO study group that the implementation of HSI would be facilitated by user descriptions that include information on individual units and organizational matters as well as the intended applications and environment (RSG.21, 1994).

Surprisingly, the health hazards domain had no direct or indirect interaction with any of the other domains. This may be a reflection of the particular application of the F-18, since there should be interaction between health hazards and system safety at a minimum. The seventh domain of HSI, personnel survivability, would also interact with both health hazards and system safety at a minimum, with desired interaction with HFE and probably training as well.

Overall, the analysis of HSI factors in F-18 squadrons showed that the pattern of the interactions is complex and does not lead to simple statements about trade-offs among the HSI domains. Rather than operating in isolation, operational practice, developmental training, manning levels, and the experience levels of personnel interact with the design resulting from the human factors engineering effort to affect the overall level of performance, effectiveness, and safety of an operational unit. For example, the thread of indirect domain interactions outlined above suggests that good human factors engineering can lead to deterioration in operational safety standards unless organizational measures are taken to avoid it.

### 8.7.2 Quantification of HSI Trade-offs

Many of the interactions examined in the F-18 case study provide qualitative information about trade-offs among the HSI domains (e.g., the need for protective gloves requires

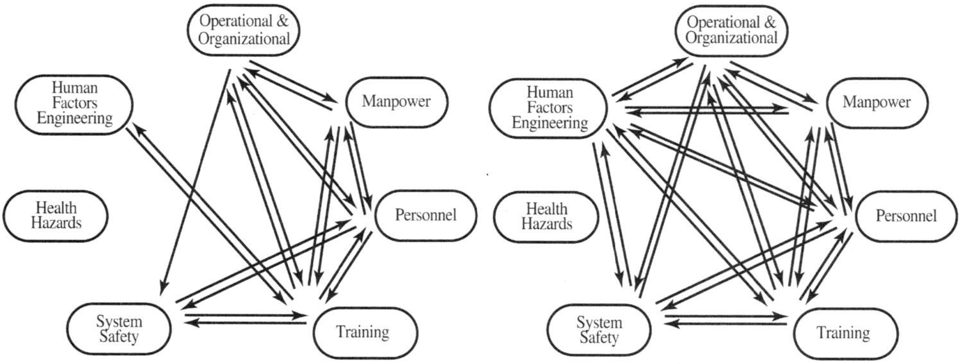


**Figure 8.7**[1] Direct (left) plus indirect (right) interactions among HSI domains for CF-18 aircraft.

additional human factors engineering effort to avoid decrements in performance). However, the design process requires quantitative information for trade-offs. Ideally, trade-off analyses should describe and compare either equal-cost or equal-capability options (DoD, 1991b). Thus, in a formal cost–benefit analysis all measurements must be transformed into an equivalent cost or performance delta.

Some costs associated with HSI issues are obvious, particularly those associated with manpower and personnel (DoD, 1991c). Several personnel cost models are available; for example, the Army Military–Civilian Cost System (AMCOS) is a database of active, reserve, and civilian manpower data that generates the manpower costs for the life cycle of a proposed system from "manpower-by-grade" information (Horne, 1987). System life-cycle cost models often include manpower and personnel costs. One model used for military system procurement is shown in Figure 8.8 (Kerzner and Bayne, 1991) with the most obvious personnel costs broken out. Reviewing its applicability to assist HSI trade-off analyses, the model developers concluded that it could be used to cost different system concepts, including ones with different manning levels or training costs. Procurement agencies have successfully used such cost models to compare HSI life-cycle costs across competing systems by requiring bidders to provide the data necessary to run the cost model.

In most cases, the life-cycle cost model approach does not help system developers and designers make the trade-offs required during the design process. First, during the design process, life-cycle costs are difficult to identify because the criteria are multivariate and elusive, including, for example, mission performance costs, safety costs, and logistic costs such as the supply, maintenance, and replacement of protective clothing and equipment. Both knowledge and tools are needed to identify such costs and to make appropriate trade-

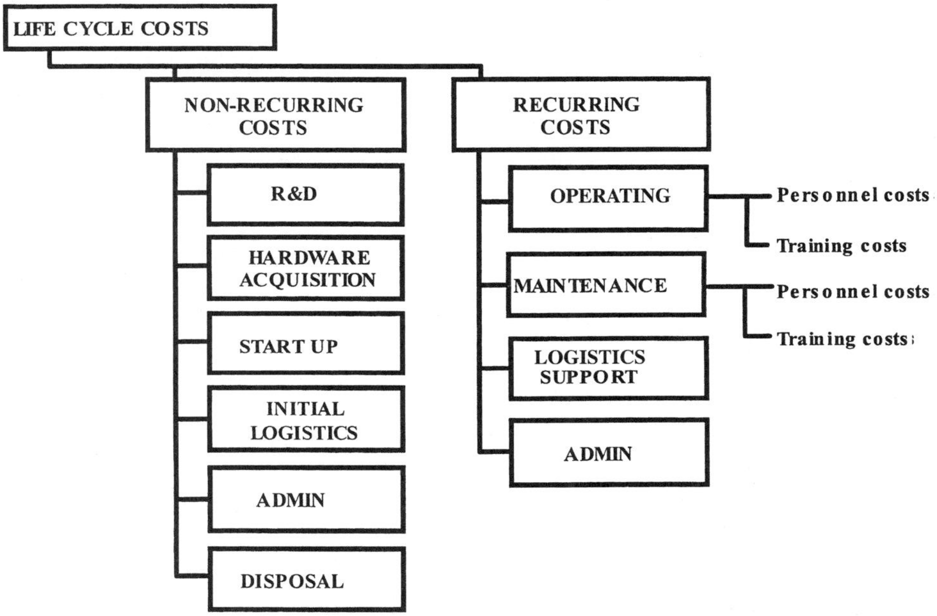


**Figure 8.8** Life-cycle cost model.

offs. However, detailed analyses of the costs in HSI domains do not appear to have received much systematic study. Second, cost models that require a designer to produce a complete system concept to which costs can be assigned are unlikely to be used by designers at the desk level. The HSI manuals do not provide cost trade-off tools and typically approach cost control by identifying high-driver tasks (ODCSP, 1997). Due to these problems, HFE specialists rarely perform a formal cost–benefit analysis. Much too frequently, human factors (HF) specialists on design teams leave cost considerations to specialists and focus on the measurements related to the performance of the human–machine system. Unfortunately, this limits the supply of information that would support studies of the cost effectiveness of HSI. To address that problem, at least one large-scale program has been started.[2]

## 8.8 FUTURE TRENDS

When Erickson (1984) produced his recommendations for linking human performance measures to system effectiveness measures, the predominant human factors technique was laboratory experimentation. This is changing rapidly with the advent of large warfighter exercises and the complex simulations discussed previously.

The DoD is developing several large simulation systems that allow constructive instantiations of developing systems to be evaluated collaboratively at diverse locations literally all over the globe. Other engineering disciplines are responding to the need to maximize system effectiveness, minimize life-cycle costs, and reduce development costs and times needs with a *revolution in business affairs* (RBA). This revolution places much greater emphasis on iterative design, integrated program management teams, and the use of technologies such as synthetic environments, which expands the use of modeling and simulation and computer-aided design through an integrated approach known as *simulation-based acquisition* (SBA).

Unless HSI processes and techniques are able to link into and exploit these processes, it will become increasingly difficult for HSI specialists to influence the eventual design solution. Thus, it can be expected that HSI activities will become more closely associated with constructive, virtual, and live simulations. Given the limitations of knowledge about human behavior needed for constructive simulations and the costs and lead times associated with live simulations, much more use is likely to be made of virtual simulations and experimentation. The use of virtual caves and other virtual representations is still more of a laboratory phenomenon than an engineering tool, but the pendulum is definitely swinging toward the use of virtual and other realistic simulation environments. That will present its own measurement challenges, because evidence to date is that human behavior in a virtual environment has some significant differences from that in the real world (Wickens and Baker, 1995).

Perhaps even more disturbing is the dramatic change in the military environments that will challenge the new systems under development. Technology may prove counterproductive in many of the nonlinear and asynchronous environments that will constitute modern battlefields and peacekeeping missions (Barnes and Fichtl, 1999). This puts an added burden on the evaluation process to ensure that military flexibility and the ability to operate in diverse environments are considered as part of the HSI design process.

Finally, systems themselves are becoming more complex, and the concept of a system of systems is becoming an accepted part of military doctrine. Evaluating systems in isolation will become increasingly more difficult to justify. The cost of ignoring complexity outweighs the considerable cost of investigating developing systems using the full panoply of measurement techniques that we have discussed above.

## 8.9 SUMMARY AND CONCLUSION

In summary, a wide range of variables must be considered and measured if human factors are to be integrated effectively into the design of complex systems to improve total system performance and reduce costs of ownership. Thus, a prime goal of any HSI program should be the measurement of human performance that can generate MOEs related to top-level design requirements. Measurements must be made in the context of the growing use of a spiral concurrent engineering effort where early approximations of the system are modeled and field evaluated before any mature design concept exists. Since available measurement techniques are applicable at different points in the acquisition design and development cycle, the scale of measurement that can be used changes throughout the system design and development process.

Three general approaches to measurement have been found applicable to the development and evaluation of defense systems: (1) analysis and computer simulations, (2) laboratory experiments, and (3) complex human-in-the-loop simulations combined with large-scale field trials. Many activities in the early stages of system design and development involve the analysis and synthesis of a design solution. Task network simulations can predict a range of characteristics of human performance. The simulation outputs can also be used to generate MOEs. Experimental methods are most useful for addressing specific design issues and for investigating specific cognitive and human performance questions related to these issues. Validation methods that include complex simulations and field exercises are essential in allowing the designer to evaluate design concepts in a realistic military environment. The human factors associated with the different HSI domains may have important interactions, but these are hard to predict and there is little quantitative information available to support trade-offs between domains. It is important that more quantitative data to aid trade-offs be developed. Without such data, it is difficult for life-cycle cost models to help system developers and designers make the necessary trade-offs.

The general conclusion is that a careful blend of measurement tools can and should be used during the design and development process to uncover the performance benefits of various design options that are impacted by human systems considerations. If the derived benefits can be related directly to design requirements and overall system goals, the payoff for performance-centered HSI trade-off studies is significant.

## NOTES

1. Figures 8.4, 8.6, and 8.7 are © HER MAJESTY THE QUEEN IN RIGHT OF CANADA (2003) as represented by the Minister of National Defence. Reproduced with permission.
2. See Defence Research and Development Canada. (2001). *Human Systems Integration Capability: Concept Description*. Ottawa, Canada: Defence Research and Development Canada, Director of Science and Technology for Human Performance.

## REFERENCES

Allender, L., Kelley, T. D., Salvi, L., Lockett, J., Headley, D. B., Promisel, D., Mitchell, D., Richer, C., and Feng, T. (1995). Verification, Validation, and Accreditation of a Soldier-System Modeling

Tool. In *Proceedings of the Human Factors and Ergonomics Society 39th Annual Meeting*. Santa Monica, CA: Human Factors and Ergonomics Society.

Barnes, M. J. (1978). *Display Size and Target Acquisition Performance*, Tech. Report No. 6006. China Lake, CA: U.S. Naval Weapons Center.

Barnes, M. J., and Fichtl, T. (1999). Cognitive Issues for the Intelligence Analyst of the Future. In *Proceedings of the 3rd Annual Federated Laboratory Symposium* (pp. 15–19). Adelphi, MD: U.S. Army Research Laboratory.

Barnes, M. J., Knapp, B. G., Tillman, B. W., Walters, B. A., and Velicki, D. (2000). *Crew Systems Analysis of Unmanned Aerial Vehicles Future Job and Tasking Environments*, ARL-TR-2081. Aberdeen Proving Ground, MD: U.S. Army Research Laboratory.

Barnes, M. J., and Matz, M. (1998). Crew Simulations for Unmanned Aerial Vehicle Applications: Shift Factors, Interface Issues, and Crew Size. In *Proceedings of the Human Factors and Ergonomics Society 42nd Annual Meeting* (pp. 143–148). Santa Monica, CA: Human Factors and Ergonomics Society.

Barnes, M. J., McClean, M., and La Nasa, F. (1996). Empirical and Analytical Methods for User Centered Design: A Synergistic Approach for the Downsized Ground Control Station. In *Proceedings of the 64th Annual Military Operational Research Society Symposium* (p. 149). Ft. Leavenworth, KS.

Beevis, D. (1987). Experience in the Integration of Human Engineering Effort with Avionics Systems Development. In *Proceedings of AGARD Symposium on the Design, Development, and Testing of Complex Avionics Systems, CP-417*, Vol 27, pp. 1–9. Neuilly sur Seine, France: Advisory Group for Aerospace Research and Development.

Beevis, D. (1996). Human Functions and System Functions. In D. Beevis, P. Essens, and H. Schuffel (Eds.), *Improving Function Allocation for Integrated Systems Design* (pp. 63–79). Wright-Patterson Air Force Base, OH: Crew Systems Ergonomics Information Analysis Center.

Beevis, D. (Ed.). (1999). *Analysis Techniques for Human-Machine System Design: A Report Produced under the Auspices of NATO Defence Research Group 8*. Wright-Patterson Air Force Base, OH: Crew System, Ergonomics Information Analysis Center.

Beevis, D., and Essens, P. (1997). The NATO Defence Research Group Workshop on Function Allocation. In E. F. Fallon, L. Bannon, and J. McCarthy (Eds.), *ALLFN'97 Revisiting the Allocation of Functions Issue: New Perspectives*. Louisville, KY: University of Louisville, IEA Press.

Booher, H. R. (1990). *MANPRINT: An Approach to Systems Integration*. New York: Van Nostrand-Reinhold.

Box, G. E., Hunter, W. G., and Hunter, J. S. (1978). *Statistics for Experimenters: An Introduction to Design, Data Analysis and Model Building*. New York: Wiley.

Burbank, N. S. (1994). *The Development of a Task Network Model of Operator Performance in Asimulated Air Traffic Control Task*, DCIEM-94-05. Toronto, ON: Defence and Civil Institute of Environmental Medicine.

Bryson, M. R. (1989). Experiments. In W. P. Hughes (Ed.), *Military Modeling*, 2nd ed. Washington, DC: Military Operations Research Society.

Card, S. K., Moran, T. P., and Newell, A. (1986). The Model Human Processor—An Engineering Model of Human Performance. In K. R. Boff, L. Kaufman, and J. P. Thomas (Eds.), *Handbook of Perception and Human Performance*, Vol. II (pp. 45-1 to 45-35). New York: Wiley.

Cook, T. D., and Campbell, D. T. (1979). *Quasi-Experimentation: Design and Analysis for Field Settings*. Chicago: Rand McNally.

Campbell, G., and Laughery, R. (1999). Modeling Human Performance on the Road to Manpower Optimization. *CSERIAC Gateway, X*(2), 12–13.

Canadian Marconi Company (CMC). (1995). *Aurora Life Extension Project, CP-140 Aurora Crew Efficiency Study: Report on the Preliminary Assessment of the Potential to Increase Crew*

*Efficiency*, Vol. 1, Department of Defence DSS Contract W8477-4-AC96/01-QC. Kanata, ON: CMC.

Chapanis, A. (1960). On Some Relations between Human Engineering, Operations Research, and Systems Engineering. In D. P. Eckman (Ed.), *Systems: Research and Design*. New York: Wiley.

Chapanis, A., and Van Cott, H. P. (1972). Human Engineering Tests and Evaluations. In H. P. Van Cott and R. G. Kincade (Eds.), *Human Engineering Guide to Equipment Design*. New York: Wiley.

Charlton, S. G. (1996). SITE: An Integrated Approach to Human Factors Testing. In T. G. O'Brien and S. G. Charlton (Eds.), *Handbook of Human Factors Testing and Evaluation*. Mahwah, NJ: Erlbaum.

Davidson, R. A., Beevis, D., Buick, F., Donati, A. L. M., Kantor, L., Bannister, S. R. H., Brook, E. A., Rochefort, J. A. P., and Turner, J. R. (1991). *Human Factors in the CF-18 Pilot Environment*, DCIEM Report No. 91-11. Toronto: Defence and Civil Institute for Environmental Medicine.

De Greene, K. B. (1991). Emergent Complexity and Person-Machine Systems. *International Journal of Man-Machine Studies, 3*, 219–234.

Erickson, R. A. (1984). *The Human Operator and System Effectiveness*, Tech. Publication 6541. China Lake, CA: Naval Weapons Center.

Flach, J. M., Vicente, K. J., Tanabe, F., Kazuo, K., and Rasmussen, J. (1998). An Ecological Approach to Interface Design. In *Proceedings of the Human Factors and Ergonomics Society 42nd Annual Meeting* (pp. 295–299). Santa Monica, CA: Human Factors and Ergonomics Society.

Godet, M. (1991). *Problemes at methodes de prospective—Boite a outils.* Paris: Edition Futuribles.

Gould, J. D. (1988). How to Design Usable Systems. In M. Helander (Ed.), *Handbook of Human–Computer Interaction* (pp. 757–785). North-Holland: Elsevier.

Green, R. J., Herschel, C. S., and Ellifritt, T. S. (1995). *50 Years of Human Engineering: History and Cumulative Bibliography of the Fitts Human Engineering Division*. Wright-Patterson Air Force Base, OH: Crew Systems Directorate, Armstrong Laboratory.

Greenley, M., Downs, G., and Espenant, M. (1998). Human Factors in an Armored Vehicle Simulation Project. In *Proceedings Interservice, Industry Training Simulation 48th Education Conference (IITSEC)*. Orlando, FL: IITSEC.

Hatley, D. J., and Pirbhai, I. A. (1987). *Strategies for Real-Time System Specification*. Chichester, UK: Wiley.

Hendy, K. C., Campbell, E. L., and Schuck, M. M. (1990). The Use of Human Information Processing Models for Workload Predictions Using Network Simulations. In E. J. Lovesey (Ed.), *Proceedings of the Ergonomics Society's 1990 Annual Conference, "Ergonomics-Setting Standards for the 90's"* (pp. 129–134). London: Taylor and Francis.

Horne, D. K. (1987). Army Manpower Cost System. *MANPRINT Bulletin, II*(4). (See also http://dticam.dtic.mil/hsi/index/hsi4.html.)

Kennedy, R. S., and Jones, M. B. (1992). *Simulating the Impact of MPTS Trade-off Decisions by Application of Isoperformance Methodology*. Orlando, FL: Naval Training Systems Center.

Kerzner, L. F., and Bayne, R. H. (1991). *A User's Guide to DND LCC 2.0*, ORAE Project Report No. PR 556. Ottawa: Department of National Defence, Operational Research, and Analysis Establishment.

Kirk, R. E. (1982). *Experimental Design: Procedures for the Behavioral Sciences*. Pacific Grove, CA: Brooks Cole.

Knapp, B. G. (1996). Task and Workload Analysis for Army Command, Control, Communications and Intelligence Systems. In D. Beevis, P. Essens, and H. Schuffel (Eds.), *Improving Function Allocation of Integrated Design Systems*. Dayton, OH: Crew Systems Ergonomic and Information Analysis Center.

Laughery, K. R., and Drews, C. (1985). Micro SAINT: A Computer Simulation System Designed for Human Factors Engineers. In *Proceedings of the Human Factors Society 29th Annual Meeting* (pp. 1061–1064). Santa Monica, CA: Human Factors and Ergonomics Society

Lewis, R. E. F., de la Riviere, W. D., and Logan, O. (1966). Sea Trials of Direct Control of Engines and Helm. In *Proceedings of the Ship Control Systems Symposium*, Vol. 1 (pp. VIII-A-1 to VIII-A-4). Annapolis, MD: U.S. Dept. of the Navy.

Lewis, R. E. F., de la Riviere, W. D., and Sweeney, D. M. (1968). Dual versus Solo Pilot Navigation in Helicopters and Low Level. *Ergonomics, 11*, 145–155.

Lewis, R. E. F., Storr, J. W., and Brewer, F. G. (1967). *Trials of an SK-5 Hovercraft for the Canadian Coast Guard*, DMRL Report 667. Toronto, ON: Defence and Civil Institute of Environmental Medicine.

McBurney, D. H. (1998). *Research Methods*. Pacific Grove, CA: Brooks Cole.

McCann, C., and Sweeney, D. M. C. (1976). *A Simulation to Study the Loading of Manpower and Facilities in the Data Interpretation and Analysis Centre for the LRPA, Part 3: A Guide to the Operation of the Mk.III Model*, DCIEM76-X-67. Toronto, ON: Defence and Civil Institute of Environmental Medicine.

McMillan, G. R., Beevis, D., Stein, W., Strub, M. H., Salas, E., Sutton, G., and Reynolds, K. C. (1991). *A Directory of Human Performance Models for System Design*, AC/243 TR/1. Brussels: NATO Defence Research Group Panel-8.

McMillan, G. R., and Martin, E. A. (1994). An Overview of Human Performance Models and Potential Applications to Combat Simulation. In C. A. Murtaugh (Ed.), *Human Behavior and Performance as Essential Ingredients in Realistic Modeling of Combat—MORIMOC III. Proceedings of the Military Operations Research Society Workshop, 27–29 May 1990*. Alexandria, VA: Military Operations Research Society.

Meister, D. (1985). *Behavioral Analysis and Measurement Methods*. New York: Wiley.

Meister, D. (1986). A Survey of Test and Evaluation Procedures. In *Proceedings of the Human Factors Society 30th Annual Meeting*, Vol. 9 (pp. 1239–1243). Santa Monica, CA: Human Factors Society.

Meister, D. (1987). Systems Effectiveness Testing. In G. Salvendy (Ed.), *Handbook of Human Factors*. New York: Wiley.

Merriman, S. C., and Moore, J. P. (1984). The F-18—A New Era for Human Factors. In *Proceedings of AGARD Symposium on Human Factors Considerations in High Performance Aircraft*, AGARD-CP-371, Vol. 19, pp. 1–5. Neuilly-sur-Seine, France: Advisory Group for Aerospace Research and Development.

Middlebrook, S. E., Knapp, B. G., Barnette, B. D., Bird, C. A., Johnson, J. M., Kilduff, P. W., Schipani, S. P., Swoboda, J. C., Wojciechowski, J. Q., Tillman, B. W., Ensing, A. R., Archer, S. G., Archer, R. D., and Plott, B. M. (1999). *Computer Simulation Models to Investigate Human Performance Task and Workload Conditions in a US Army Heavy Maneuver Battalion Tactical Operations Center*, ARL-TR-1994. Aberdeen Proving Ground, MD: Army Research Laboratory.

Moroney, W. F. (1995). The Evolution of Human Engineering: A Selected Review. In J. Weimer (Ed.), *Research Techniques in Human Engineering*. Englewood Cliffs, NJ: Prentice-Hall.

Muckler, F. A., Seven, S., and Akman, A. (1991). *Construction of a Military Occupation Specialty Taxonomy*, ARI Research Note 91–10. Alexandria, VA: U.S. Army Research Institute.

Muir, H., Marrison, C., and Evans, A. (1989). *Aircraft Evacuations: The Effect of Passenger Motivation and Cabin Configuration Adjacent to the Exit*, CAA Paper 89019. Cranfield, Beds, UK: Cranfield Institute of Technology Applied Psychology Unit.

Naikar, N., and Sanderson, P. M. (2000). Evaluating Design Proposals with Work Domain Analysis. In *Proceedings of the IEA 2000/HFES 2000 Congress* (pp. 202–205). Santa Monica, CA: Human Factors and Ergonomics Society.

Newsbreaks. (1994, October 17). *Aviation Week and Space Technology, 141*(16), 15.

Office of the Deputy Chief of Staff for Personnel (ODCSP). (1997). *A Handbook for MANPRINT in Acquisition*. Washington, DC: ODCSP.

Overmayer, R. W. (1975). *A Computer Model for Command and Control Analysis*, Report under Contract 00014-74-C-0324. Arlington, VA: Office of Naval Research, Naval Analysis Programs.

Parasuraman, R., Sherdian, T. B., and Wickens, C. D. (2000). A Model for Types and Levels of Interaction with Automation. *IEEE Transactions on Systems, Man, and Cybernetics*.

Poisson, R. M., and Beevis, D. (2000). *Human Factors Issues in Land Forces Weapons Systems Evaluations*, DCIEM Report TR 2000-032. Toronto, ON: Defence and Civil Institute of Environmental Medicine.

Rasmussen, J. (1986). *Information Processing and Human Machine Interaction*. New York: North-Holland.

Rasmussen, J., and Vicente, K. J. (1989). Coping with Human Error through Interface Design: Implications for Human Interface Designs. *International Journal of Man-Machine Studies, 31*, 517–534.

RSG.19. (1999). *Code of Best Practice (COBP) on the Assessment of C2*, RTO-TR-9. Neuilly-sur-Seine, France: NATO Research and Technology Organization.

RSG.21. (1994). *Liveware Integration—Final report*, NATO DRG Report No. AC/243 (Panel 8) TR/18. Brussels: NATO Defence Research Group Panel-8.

Tack, D. (1996). *Hot Weather CB Protective Clothing: Trial Concept and Outline Plan*, DCIEM-97-CR-36. Toronto, ON: Defence and Civil Institute of Environmental Medicine.

Taylor, F. V. (1959). *Human Engineering and Psychology*. Washington, DC: U.S. Naval Research Laboratory (unpublished preprint).

Towill, D. R. (1989). Selecting Learning Curve Models for Human Operator Performance. In G. R. McMillan, D. Beevis, E. Salas, M. H. Strub, R. Sutton, and L. van Breda (Eds.), *Applications of Human Performance Models to System Design*. New York: Plenum.

U.S. Department of Defense (DoD). (1991a). *Defense Acquisition. Part 7, Section B: Human Systems Integration*, Directive 5000.1. Washington, DC: DoD.

U.S. Department of Defense (DoD). (1991b). *Defense Acquisition. Management Documentation and Reports, Part 8: Cost and Operational Effectiveness Analysis*, Directive 5000.2. Washington, DC: DoD.

U.S. Department of Defense (DoD). (1991c). *Defense Acquisition. Management Documentation and Reports, Part 6: Manpower Estimate Report*, Directive 5000.2. Washington, DC: DoD.

U.S. Department of Defense (DoD). (1999, May). MIL-HDBK-46855A, Human Engineering Programs, Process, and Procedures. Washington, DC: DoD.

Walters, H. (1992). *Ministry of Defence: The MANPRINT Handbook*, 2nd ed. London: Controller, Her Majesty's Stationery Office.

Webb, R., Tack, D., and Gaughan, P. (1998). *DND Prototype Combat Helmet User Acceptance Trials*, HSI Report for DCIEM. Toronto, ON: Defence and Civil Institute of Environmental Medicine.

Whiteside, J., Bennett, J., and Holtzblatt, K. (1988). Usability Engineering: Our Progress and Evolution. In M. Helander (Ed.), *Handbook of Human Computer Interaction* (pp. 791–819). Amsterdam: North-Holland.

Wickens, C. D., and Baker, P. (1995). Cognitive Issues in Virtual Reality. In W. Barfield and T. A. Furness (Eds.), *Virtual Environments and Advanced Interface Design* (pp. 515–541). New York: Oxford University Press.

Wickens, C. D., and Hollands, J. G. (2000). *Engineering Psychology and Human Performance*, 3rd ed. Upper Saddle River, NJ: Prentice-Hall.

Williges, R. C. (1995). Review of Experimental Design. In J. Weimer (Ed.), *Research Techniques in Human Engineering*. Englewood Cliffs, NJ: Prentice-Hall.

Wortman, D. B., Duket, S. D., Seifert, D. J., Hann, R. L., and Chubb, G. P. (1978). *Simulation Using SAINT: A User-Oriented Instructional Manual*, AMRL-TR-77-61. Wright-Patterson Air Force Base, OH: Aerospace Medical Research Laboratory.

CHAPTER 9

# Simulation-Based Acquisition

STEPHEN R. OLSON and ANDREW P. SAGE

## 9.1 INTRODUCTION

The decade of the 1990s was a tumultuous time for the U.S. Department of Defense (DoD). It was a time that not only gave birth to a new defense posture for the country but also demanded change in the methods used to develop and produce large and complex systems for national defense. Brief comments on these two issues are of importance to simulation-based acquisition (SBA).

### 9.1.1 Background: U.S. Military Posture (1980 vs. 2001)

The 40-year Cold War (1950–1990) was a period of significant world tension and confrontation between the communist countries led by the Soviet Union and democracies generally led by the United States. In response to the posture of the USSR, the United States invested heavily in the development of increasingly complex aircraft, ships, tanks, and global surveillance systems. The Cold War arms race culminated in what is sometimes termed the "Reagan Build-up" of the early 1980s, the subsequent collapse of communism in Europe, and the disintegration of the Soviet Union by 1990. In practically a few months, the Cold War abruptly ended. There was much to be thankful for and the public sentiment in the United States quickly turned in favor of decreased spending on national defense. While spending commitments and long-term procurement plans prohibited an immediate collapse of expenditures, by 1993 spending on the development and procurement of military equipment had decreased by more than 50 percent from the peak in 1985 of about \$175 billion per year expenditures to about \$80 billion. It has remained essentially constant from 1993 through 2000. At the same time, the disintegration of Soviet influence in Europe and the Middle East gave rise to new sources of global instability and threats to regional peace. While different, these new threats confounded the complexity associated with the bilateral confrontation pattern of the Cold War. The U.S. defense establishment was confronted by the need to reconsider almost every aspect of how it planned and

*Handbook of Human Systems Integration*, Edited by Harold R. Booher.
ISBN 0-471-02053-2 

equipped for national defense. These changes in the world environment and the repercussions on national defense have come to be called the *revolution in military affairs* (RMA).

While rethinking the national defense strategy occupied many defense planners during this time, there was another equally complex problem confronting the DoD. During the Cold War the United States was a leader in developing and applying electronic technology to address defense needs. Initiatives funded by the DoD covered the spectrum from the manufacture of reliable chips and components to the design of "supercomputers" and the architecture of complex software and communications systems. As costs associated with these maturing technologies fell dramatically, their commercial potential blossomed, and the 1980s witnessed an extraordinary commercial growth in digital and communications technologies, a pattern that has continued into the twenty-first century. Commercial investment in these new-age information technologies rapidly outpaced that of the DoD. At first glance, this might seem a piece of good fortune, in that it became possible in principle for major cost savings to result from leveraging these commercial investments through use of commercial off-the-shelf (COTS) products. However, modern commercial business practices in contemporary high-technology industries generally bear little resemblance to the DoD business practices of the Cold War. Today, the great preponderance of commercial electronics and software and the associated information technology products have a life of only a few years. Furthermore, the information technology industry has developed a business model that makes it often insensitive to penalties associated with product defects and errors through continual release of upgrades and reengineered products, a situation not acceptable for military weapons systems. As the commercial market for digital components has expanded, manufacturers may have little interest in the relatively low production quantities required for unique military systems. Today, even a relatively new military system has components that are obsolete, with a very limited supply of compatible spare parts available in the market. Many defense systems, such as ships, aircraft, and tanks, experience operational lives of 20 to 50 years, and the effort and expense in maintaining the technological currency of these systems have proven to be a challenge. This problem has been compounded by the significant reduction in defense spending noted earlier. By 1993, the United States had practically ceased the acquisition of major new defense equipment, investing the bulk of DoD funds in the modernization and maintenance of existing equipment. To some degree, this strategy was acceptable, particularly given the large quantities of new equipment still in the production and delivery pipeline from the Reagan years. The result, however, by the year 2001 is that we had a rapidly aging fleet of ships, planes, and weapons of all types.

All of these issues have served to remind the DoD that the total cost of system ownership is dominated by operating, maintenance, and support costs that occur years after product acquisition (Buede, 2000). However, there is great difficulty in accurately projecting these future costs and in minimizing their impact during the product development phase. Furthermore, the need to change a product in an evolutionary manner after it is fielded is increasing, because maintenance of technological currency is essential to maintaining combat effectiveness, while the rate of technology "innovation" continues to escalate. These problems emphasize the need to anticipate the retrofit of new technology into fielded systems so that product upgrading can be planned and accomplished in a cost-effective and timely manner. The obvious question with all of these issues and with little or no prospect of funding ever returning to the levels of 1985 yet with demands for military presence and peacekeeping across the globe is, "How can the United States sustain its defense capability in a trustworthy manner?"

All of these factors have caused defense planners to realize that the DoD can no longer continue to develop and acquire systems as it has in the past. Just as the change in the world balance of power and U.S. military posture has generated the need for a RMA, this change in the business landscape caused many to declare that the United States also needed a revolution in business affairs (RBA), often referred to more modestly as *acquisition reform*.

### 9.1.2 Background: Motivation for Revised Acquisition Practices

A large number of problems have been encountered with "grand design" or waterfall life-cycle efforts traditionally used to engineer a system. Thus, there have been a number of efforts to extend developmental approaches beyond the classic waterfall approach (Sage, 1992, 1995; Sage and Rouse, 1999). Today, the classic waterfall approach is suggested only in those rare cases where user and system-level requirements are crystal clear and unlikely to change and where necessary funding for all life-cycle phases associated with the grand design is essentially guaranteed. This is rarely the case for major systems, especially those that are software intensive. Changing user needs and technology virtually guarantees that major systems cannot be developed using the grand design approach.

Two leading alternative approaches to the engineering of systems are termed incremental and evolutionary. Incremental development has as a plan to deliver the system in preplanned phases or increments, in which each delivered module is functionally useful. In such an approach, the overall system capability improves with the addition of successive modules. In such an approach, the desired system capability is planned to change from the beginning as the result of "build $N$" being augmented and enhanced through the phased increment of "build $N + 1$." This approach enables a well-functioning implementation to be delivered and fielded within a relatively short time and augmented through additional builds. This approach also allows time for system users to thoroughly implement and evaluate an initial system with limited functionality compared to the ultimately desired system. Generally, the notion of preplanning of future builds is strong in incremental development. As experience with the system at "build $N$" is gained, requirements changes for module $N + 1$ may be more easily incorporated into this and subsequent builds.

Evolutionary life-cycle development is similar in approach to its incremental complement; however, future changes are not necessarily preplanned. In this approach, we recognize that we are unable to initially predict and set forth engineering plans for the exact nature of these changes. The system is engineered at "build $N + 1$" through reengineering the system that existed at "build $N$". In this approach, a new functional system is delivered at each build, rather than obtaining "build $N + 1$" from "build $N$" by adding a new module. The enhancements to be made to obtain a future system are not determined in advance, as in the case of incremental builds. Evolutionary development approaches can be very effective in cases where user requirements are expected to shift dramatically over time and where emerging and innovative technologies allow for major future improvements. It is especially useful for the engineering of unprecedented systems that involve substantial risk and allows potentially enhanced risk management. Evolutionary development may help program managers adjust to changing requirements and funding priority shifts over time since new functionality introductions can be advanced or delayed in time in order to accommodate user requirements and funding changes. Open, flexible, and adaptable system architecture is central to the notion of evolutionary development. As a follow-on to this, it appears that evolutionary development of a system architecture has the potential to greatly decrease the risk and costs of excessive rework of a system of systems or a federation of systems after it has been initially engineered. Figure 9.1 indicates the general nature of the evolutionary life cycle. This can be represented as a continuing waterfall, with feedback across the life cycle or as a spiral. Much of what has come to be known as evolutionary acquisition is based upon an equivalent spiral life cycle (Boehm and Hansen, 2001).

The DoD has not been unmindful of these needs and the need for evolutionary life cycles. Incremental life cycles were recognized a decade ago and made a part of the DoD 498 standard, which is no longer operational due to the decision to use commercial standards whenever feasible. Acquisition reform is a major effort now and has been for much of the past decade. In the effort to reduce acquisition response time, the rewrite of the DoD 5000 series regulations (DoD, 2000a) calls for evolutionary acquisition to be the preferred method for future defense acquisition programs. It also calls for SBA to support this. Unfortunately, there is often considerable confusion over the meaning of these terms and life-cycle development methods that should be used in the pursuit of various evolutionary acquisition and simulation-based acquisition approaches. Some of this mystification is evident in the use of expressions such as evolutionary development, spiral development, spiral acquisition, evolutionary spiral development, and a host of other expressions where the meanings are not well understood and accepted across those using the terms.

There are a number of follow-on evolutionary acquisition efforts. Evolutionary acquisition strategies define, develop, and deploy an initial, militarily useful capability and a plan for subsequent definition, development, test, and production/deployment of increments beyond the initial capability over time. The scope, performance capabilities, and timing of subsequent increments shall be based on continuous communications among the requirements, acquisition, intelligence, logistics, and budget communities.

An excellent overview of evolutionary acquisition may be found in a Defense Systems Management College (DSMC, 1998) report. There it is indicated that evolutionary

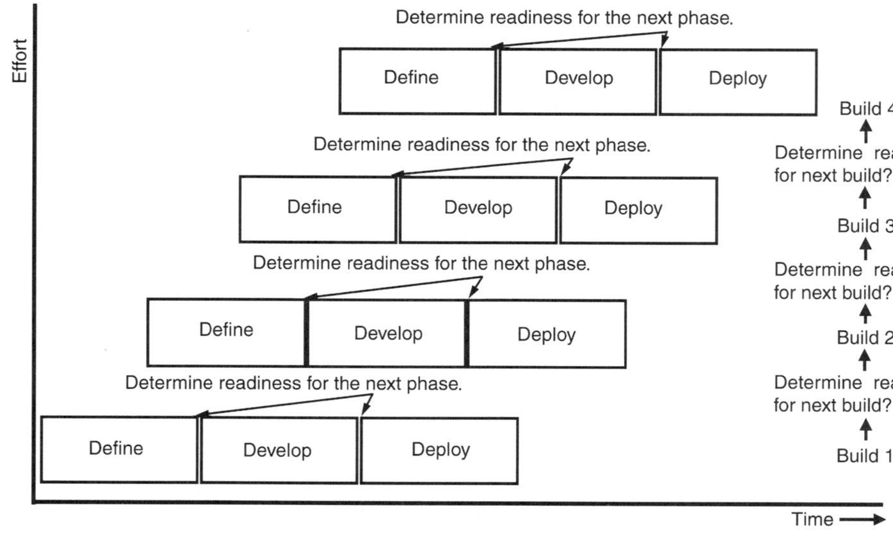


**Figure 9.1** Iterative life cycles in evolutionary acquisition.

acquisition is a strategy for use when it is anticipated that achieving the desired overall capability will require the system to evolve during development, manufacturing, or deployment. This appropriate definition provides a suitable linkage between the concepts of evolutionary acquisition and complex adaptive systems through use of the term *emergence*.

It is important also to recognize the complex adaptive nature of today's technologies and organizations. Interestingly, most studies of complex systems often run completely counter to the trend toward increasing fragmentation and specialization in most disciplines. It is not at all a large number of parts in a system that makes the system complex; it is the way that the parts interact. A product may consist of abundant parts, but if these parts interact only in a known, designed, and structured fashion; the system is not complex, although it may be big. Complexity exists when the interconnected parts of a system interact in unanticipated ways. One of the defining characteristics of complex systems is the property known as emergence. Here, the behavior of the overall system is different from the aggregate behavior of the parts, and knowledge of the behavior of the parts will not allow us to predict the behavior of the whole system. The emergence property is a form of control. It allows distributed agents to organize together to determine consequential higher order system behavior. In systems that are "complex," structure and control emanate or grow from the bottom up. Thus, the reductionist scientific approach generally does not work with complex systems. Virtually all organizational behavior in such systems is comprised of agents adapting to their environments and, in the process of so doing, affecting the environments of all other agents. In some situations, when systems are driven sufficiently far from equilibrium, *bifurcations* occur and chaotic behavior may result.

Clearly, there are many considerations involved in efforts such as these. The prototypical steps in building an experimental and exploratory model of a complex adaptive system might be described as follows:

1. Simplify the problem as much as possible, being sure to retain the essential features of the situation.
2. Identify a potentially appropriate model of the situation that represents agents that follow simple rules with specified interactions and randomizing elements.
3. Construct a simulation based on this model.
4. Run the simulation many times with appropriately different random variables, collect the data, and compute statistics from the different runs.
5. Identify how simple behavioral rules result in observed behavior.
6. Study the responses obtained by sensitivity studies and appropriate parameter changes to determine critical parameters, sources of behavior, and effects of different parameters on system responses.

There is a major role for modeling and simulation in the several activities suggested in this list. This creates a strong linkage between evolutionary acquisition and SBA as a way to study evolutionary acquisition and other acquisition phenomena.

## 9.2 OBJECTIVES FOR SBA

Simulation-based acquisition involves much increased use of computer-based models and simulations within system engineering and product acquisition life cycles. It is an acquisition process enabled by modeling and simulation technology integrated across acquisition phases and programs. It has the objectives of reducing the time, resources, and risk associated with the acquisition process while improving the trustworthiness and supportability of deployed systems.

Before proceeding further, it is useful for us to introduce some definitions appropriate to the "simulation world." There are a significant number of terms and associated definitions used in SBA. A relatively thorough list may be found in *Acquisition Modeling and Simulation Comprehensive Core Body of Knowledge* (Acquisition Functional Working Group, 1999), which also contains extensive references to the literature in this area.

For the most part, these terms agree with terminology often used in systems engineering. In particular, it is important to note that a model is a physical, mathematical, or logical representation of a system, entity, phenomenon, or process. A simulation is an implementation of a model such that the behavior of the model, generally over time, can be observed. Also, simulation is a technique for testing, analysis, or training in which real-world systems are used or where a model of these systems reproduces real-world and conceptual systems. There are three different classes of simulations: constructive, virtual, and live. Constructive simulations are solely resident in software. Engineers attempting to conceptualize, design, and implement various facets of a product or process most often use constructive models. They have the benefit of being repeatable and generally fast and can be run stochastically and repetitively, thereby providing a means to quantitatively assess the inherent uncertainty in some tasks and processes. Virtual simulations have a constructive component but also explicitly include a human-in-the-loop component, although in an artificial setting. These simulations may also include "real" operational software intended to run in the fielded product or physical hardware end items. Although virtual simulation repeatability and consistency are generally suspect, it is very useful for human factors engineering and individual training purposes. Because there is a human in the loop, virtual simulations generally run in real time, which adds to their expense and limits the amount of stochastic data that can be generated within complex scenarios. Live simulations have human "players" complemented with a broad mix of constructive models and operational hardware and software and include a "realistic" simulation environment. Live simulations are generally very costly to conduct but are considered essential to validate operational concepts and tactics and for unit and combined arms training.

Simulation-based acquisition recognizes the increasing role that these computer-based simulations and synthetic environments have in designing for and validating the changed acquisition process and environment and the role that humans will play in this new environment. Acquisition reform covers a broad spectrum but is largely focused on using information technology (IT) to bring efficiencies and commercial practices to DoD acquisition. The acquisition reform website (DoD, 2000b) presents a number of useful and current documents concerning this subject.

Numerous studies have illustrated that the early stages of a program involving system definition are when most of a program's life-cycle costs are really determined, and the ensuing rush to build something quickly is inevitably followed by a lengthy period of design changes and modifications in order both to get the "system right" and, more importantly, to get the "right system." Even then, the initial build may be so far from the right system that no amount of modification can redress the initial flawed approach. It is for this reason that evolutionary acquisition approaches are a common contemporary suggestion.

We must understand that as we approach what appears to be almost limitless computing capacity and speed, the time consumed in running simulations is shrinking dramatically. This is particularly the case for those simulations absent any "human in the loop." The issue is less the speed of model or simulation execution and more whether appropriate models can be built for evolutionary acquisition. Reduced product development life-cycle times must also assure the objective of superior system effectiveness with well-understood and manageable costs. Models and simulations are being used today within almost every functional domain of system acquisition. Unfortunately, these tools reflect a broad spectrum of adequacy: Some are derived from historical data of no modern relevance, some are employed outside their intended realm, and many are using specific product data inconsistent with data used for the same product within another functional domain. Few of these tools, or their underlying data that support their use, are integrated or interoperable with each other. Thus, it is very difficult to capture total system effectiveness and cost or to facilitate integration of systems to result in a system of systems. Thus, we have a potentially huge computational capacity to support modern and rapid product development but without a systems design approach that harnesses this power.

It is necessary to ask whether industry will spontaneously arrive at an SBA environment compatible with the government's interest in the presence of government inactivity in this regard. All evidence suggests this is unlikely. It would require an investment of discretionary funds at a time when the market capitalization of major defense firms has fallen sharply. Also, an individual company's strategy will surely be to create competitive advantage for the company. Significant investments have been made and are continuing to be made in engineering tools for a variety of functions. The tools that a company purchases are mostly of commercial origin and are usually tailored to the specific need of the organization. This often results in creation of a unique product environment where, once a vendor and that product are established, it becomes very difficult to substitute a new supplier for one initially chosen.

The phenomena of *path dependence* and *lock-in* are particularly present in products and services based on IT innovations (Shapiro and Varian, 1999). These invert the usual return to scale notions found in conventional products. The defense industry, as well as automotive companies and other manufacturers, has begun to take note of its costs and dependencies associated with its information technology suppliers. Defense companies may choose to ignore this issue because the costs are passed along to the DoD customer and the situation has the added benefit of creating barriers against future competition for developed products. The situation becomes even more pronounced when tools that have no commercial counterpart are involved. This is clearly the case for most models of combat capability or military vulnerability. These tools, while often unclassified, are crucially linked to classified data that characterize specific threats or friendly system behavior. Because of the high degree of complexity of modern military systems, many of these tools and simulations are themselves highly detailed and sophisticated. Even if we ignore the technology and data classification issues, serious users of these tools are only found within the defense industry, and they may constitute an insufficient market for speculative investment by companies. Often, these very tools are essential during concept development and architecting. These are the tools that a prime contractor must employ in order to conduct comprehensive trade-offs of virtual designs and conceptual architectures.

Consider two choices that the government may have with regard to these tools, especially the situation where we become more reliant on virtual product demonstrations and evaluations potentially brought about by SBA. The first choice is to let each major prime contractor develop its own set of tools and simulations. This approach will likely result in each contractor investing significant funds on tools to capture the operational environment but where each set of tools reflects the contractor's proprietary view of the military mission under consideration. There will then be a need to get government agreement that the individual representations are valid and that they also faithfully represent performance, supportability, maintainability, and cost characteristics of the specific products that an individual vendor is proposing. The government is then faced with validating each contractor's tools and accrediting them for use in a specific source selection process. The government will also be placed in the position of having to compare and trade off each competing representation against all others in order to make a source selection decision. This is a formidable task, particularly in the absence of a predetermined strategy for how such source selections will be conducted and with only "virtual" product results available at the time of selection. This suggests a very program-centric approach toward model development. Each procurement will produce procurement-specific and perhaps service-specific models, model environments, and associated simulations. There will be little incentive to generate shared approaches to modeling complex environments, particularly with individual service-dominated views of the battle-space and associated operational requirements. This outcome will likely be costly and ultimately result in budget- and time-constrained tools reflecting mediocrity in their comprehensive understanding of evolving requirements.

Alternatively, consider an environment in which the government and industry are encouraged to jointly agree on the development of common models for individual mission areas. This suggests that the number of models and their purpose be managed to produce collaborative model development environments and simulations. Contractors would participate in model development and be afforded the opportunity to contribute model improvements, even as proprietary model objects where competitive issues are at stake. The government would specify how the models are to be used during source selection and a contractor would be required to "protest" in advance if it believed the evaluation model incorrectly captured the salient features of an anticipated product development proposal. This model environment would afford some level of interoperability with a contractor's indigenous IT-based tool set, achieved through interoperability standards and procedures. The overall tool environment would reflect a comprehensive strategy of how the government intended to interface to a contractor's environment, addressing both data and the interoperability of models as well as data that may reach beyond the specific procurement under consideration. This latter requirement would facilitate the evaluation of a "system of systems," "family of systems," or "federation of systems" (Sage and Cuppan, 2001) and the integration of data to conduct higher levels of aggregated analysis. Finally, the data requirements and formats would be made known to all contractors, and all data required for subsequent competition would be available.

The difficult part is to create a SBA environment, such as the one just described, that is based as much as possible on commercial tools and environments. The government should not want to stifle competition where a viable commercial tool environment exists; it needs to develop "world-class" approaches for the subset of tools for which there is no commercial market, ensuring that these tools are available to its suppliers and compatible with its internal environments.

Simulation-based acquisition calls for the virtual development of a system through iterative improvement of its model representations of the system, beginning with the identification of system concepts, continuing with the selection of "best" concepts and the evaluation of those concepts against user life-cycle requirements, progressing through manufacture and deployment, and ending with system retirement. As these myriad representations mature, test artifacts may be used to validate model descriptions and to reveal instances in which models do and do not properly represent *real world* conditions.

To "build" a comprehensive digital representation of a system whose authenticity is accepted by all interested parties is a daunting task. It requires cooperation among all stakeholders, and it also requires an environment that supports and encourages this level of cooperation on a large scale. Ideally, SBA will go a long way in helping to realize this cooperation by capitalizing on the synergy between a vastly improved culture, process, and systems engineering environment to enable people and organizations to accomplish work in an integrated fashion.

## 9.3 SIMULATION-BASED ACQUISITION: STRUCTURE, FUNCTION, AND PURPOSE

The Office of the Secretary of Defense (OSD) has expressed strong support for the concept of SBA. The DoD's vision for SBA is "to have an acquisition process that is enabled by robust, collaborative use of simulation technology that is integrated across acquisition phases and programs. The purposeful objectives of SBA are to: reduce the time, resources, and risk associated with the acquisition process; increase the quality, military utility, and supportability of systems developed and fielded, and; enable integrated product and process development (IPPD) from requirements definition and initial concept development through testing, manufacturing, and fielding" (Sanders, 1997, p. 75).

Because SBA is an evolving concept, there are differing interpretations on its scope and method of implementation. In their book on SBA, Johnson et al. (1998) expanded the definition with a detailed explanation of a dominantly functional interpretation of SBA: "Simulation Based Acquisition is an iterative, integrated product and process approach to acquisition, using modeling and simulation, that enables the warfighting, resource allocation, and acquisition communities to fulfill the warfighter's materiel needs, while maintaining Cost As an Independent Variable (CAIV) over the system's entire life cycle and within the DoD's system of systems." The highlights of their definition are that "simulation based acquisition is . . . "

- "*. . . an iterative, integrated product and process approach to acquisition*"—Thus, SBA enables IPPD teams, in which the DoD and contractor organizations work internally and with each other in an integrated team effort, to converge on trustworthy solutions through use of an iterative design process that is based on a well-adjusted set of system requirements.
- "*. . . through modeling and simulation*"—Modeling and simulation activities make SBA possible through creating a synthetic environment that enables exercising the power of simulation to explore many more iterations of virtual designs than would be possible with physical prototypes. The associated level of increased user involvement leads to better learning and problem solving than obtained from the more traditional approach obtained from physical prototypes. The resulting increased communication and enhanced learning make team members more effective.

- "*. . . to enable the warfighting, resource allocation, and acquisition communities*" —A major objective of SBA is to integrate three principal acquisition support systems: the Requirements Generation System, the Planning Programming and Budgeting System (PPBS), and the Acquisition Management System (AMS) in support of the acquisition community and the related government and industry agents.
- "*. . . to fulfill the warfighter's materiel needs while maintaining Cost As an Independent Variable (CAIV)*"—The desire here is to maximize need satisfaction to the maximum amount possible within resource constraints. The cost as an independent variable concept (Brady, 2001) will, in principle, allow more trustworthy predictions of the costs of different alternatives and thereby enable better informed analysis of trade-offs.
- "*. . . over the system's entire life cycle*"—This suggests examination of all relevant facets associated with systems acquisition early in the acquisition life cycle and throughout acquisition of the system. These facets include the "ilities" associated with quality management of the acquisition process: affordability, availability, flexibility, interoperability, lethality, maintainability, manufacturability, mobility, reliability, supportability, survivability, and sustainability.
- "*. . . and within the DoD's system of systems*"—This suggests investigation of all significant interactions within and across the various systems that, collectively, result in the overall system of systems. This should enable total systems integration and, ultimately, expansion of the system-of-systems concept to include federation-of-systems concepts (Krygiel, 1999; Carlock and Fenton, 2001; Sage and Cuppan, 2001) that are needed in combined operations brought about by collaborative allied systems and programs.

The above reference to a system of systems enables the capture of important realities brought about by the fact that modern defense systems are not monolithic. Rather, they have 5 characteristics (Maier, 1998) that makes the system of systems (Krygiel, 1999; Carlock and Fenton, 2001; Sage and Cuppan, 2001) designation most appropriate:

1. *Operational Independence of the Individual Systems* A system of systems is composed of systems that are independent and useful in their own right. If a system of systems is disassembled into the component systems, these component systems are capable of independently performing useful operations independently of one another.
2. *Managerial Independence of the Systems* The component systems not only can operate independently but also generally do operate independently to achieve an intended purpose. The component systems are generally individually acquired and integrated, and they maintain a continuing operational existence that is independent of the system of systems.
3. *Geographic Distribution* Geographic dispersion of component systems is often large. Often, these systems can readily exchange only information and knowledge with one another and not substantial quantities of physical mass or energy.
4. *Emergent Behavior* The system of systems performs functions and carries out purposes that do not reside in any component system. These behaviors are emergent properties of the entire system of systems and not the behavior of any component system. The principal purposes supporting the engineering of these systems is fulfilled by these emergent behaviors.

5. *Evolutionary Development* A system of systems is never fully formed or complete. Development of these systems is evolutionary over time and with structure, function, and purpose added, removed, and modified as experience with the system grows.

We see that the operational concepts needed for a trustworthy SBA process are not at all simple. There are many needed elements. There is a distributed data repository that contains all of the data about the product under development. This is centralized in a virtual sense, and all of the different stakeholders have a shared responsibility to keep the repository up to date such that all have rapid access, throughout the life cycle, to information required to understand and define, develop, and deploy a trustworthy system.

It is very important to have confidence and trust in the models and simulations that comprise an SBA approach. When questions regarding confidence in modeling and simulation activities are raised, most often this relates to the notions of verification, validation, and accreditation (VV&A) of models and simulations. Appropriate definitions of these terms may be found in a variety of sources (Banks, 1998; NDIA, 1999; U.S. Navy, 2001):

*Verification* is the process of determining that a model or simulation implementation is transformed from one phase of development to another in a way that is consistent with the documented requirements and specifications. It is concerned with *building the model or simulation right.*

*Validation* is the process of determining the degree to which a model or simulation is an accurate representation of the real world from the perspective of the intended uses of the model. It is concerned with determining that the model or simulation behaves with sufficient accuracy relative to intended purposes or with *building the right model or simulation.*

*Accreditation* is the process of certification that a model or simulation is acceptable for use for a specific purpose. It represents official recognition that a *model or simulation produces credible results and is otherwise usable.*

In general, models and simulations are examined throughout the VV&A process from the users' application needs perspectives.

Many suggest that the SBA vision cannot be realized without an investment to develop the processes and architectures for the SBA way of doing business. They believe that specific strategies are needed to assure the appropriate level of data standardization and tool interoperability and that these strategies will not evolve spontaneously. Because of the continuing pressure on the defense budget, any suggestion of a new investment, no matter how small, comes under intense scrutiny within the Pentagon, and the funding to support SBA strategy development and execution generally requires a "business case" that will warrant the investment.

In developing this business case for SBA, it is necessary to understand the current state of product development and production. Models and simulations are today pervasive across all phases associated with engineering a system—definition, development, and deployment. One challenge is that tools used for modeling and simulation are generally not integrated and operate only on unique data that may be inconsistent across different views of the same product. Users of these tools can easily forget their limitations and may place unwarranted confidence in their results. A fully developed SBA environment will have integrated these models and addressed these concerns so that the procuring authority will, with confidence, understand the technical, design, cost, and operation performance risks of a product before any physical prototype of that product exists. This situation will be realized because the product will have been designed, tested, and operated in an integrated virtual environment that will, itself, be designed to illuminate the uncertainties of the integrated product knowledge as embodied in its interoperable models and simulations. Furthermore, modeling and simulation technologies will be applied to understand and project the trainability, maintainability, and supportability factors and costs for the equipment before it is produced or placed in the field.

Only when an SBA procuring authority has reached a satisfactory level of understanding of this virtual domain will it proceed into planning the next phases of prototype development, testing, and initial production. This planning will reflect a prototype and testing program that focuses on those issues in the virtual domain that revealed the weakest level of modeling and simulation "confidence," thereby demanding greater scrutiny before a final production decision. The resulting product prototyping and testing strategy should have, as a major objective, not just the validation of a point design, but the collection of sufficient data to improve the models and simulations, and should provide greater confidence in future virtual developments. Implicit in this process is that vendors will compete their designs in these virtual domains and the procuring authority will as a result have sufficient insight into both its own SBA environment as well as those of others in the supply chain in order to become and remain an informed procurer. This is particularly of significance when dealing with federated modeling and simulation issues (Nance, 1999; NDIA, 1999; U.S. Navy, 2001).

Even though the SBA concept is potentially appealing, there are a number of obstacles that need to be overcome. Much initial work must be accomplished before the first physical item is available using the SBA approach, especially in a distributed or federated environment. However, numerous studies have illustrated that the early stages of a program are when most of a program's life-cycle costs become determined. Premature cessation of the definition phase and proceeding to development with potentially volatile requirements and specifications in order to be able to build something quickly are inevitably followed by a lengthy period of design changes and modifications in order to obtain the right system. The initially configured requirements and specifications may be so flawed from appropriate ones that no amount of modification and associated expenditure can redress the initial flawed approach. Can we attach a cost savings to the solution of this problem through use of SBA? There is no shortage of anecdotal and factual data on the savings realized through modeling and simulation in defense acquisition. The Joint Simulation Based Acquisition Task Force (1998) has a 28-page discussion on SBA's *return on investment* (ROI) replete with examples of cost savings. The savings are impressive, but there may be difficulties in scaling the data up to the level of application envisioned for contemporary SBA processes and environments. The ROI calculations just attempt to do that, and one can indeed generate some numbers on the prospect of SBA that are so large they become very suspicious and apparently not believed.

While there is much discussion about SBA, there is very little in the way of formal guidance or a DoD-wide implementation plan for it at this time. Some have indicated that the SBA efforts are now the purview of the individual services and not the DoD and that it is not sufficiently emphasized in the new series 5000 regulations (Johnson, 2000). During the 1997 to 1999 time frame, industry attempted to describe the long-range vision for SBA. This was done by the SBA Industry Steering Group (SBA ISG) to the acquisition council, subordinate to the DoD Executive Council on Modeling and Simulation. Its ideas were developed in the ISG's SBA functional description document (FDD) (NDIA, 1999) and are summarized in the next few paragraphs.

The ISG believes that, when properly implemented, SBA can potentially make possible high-quality, enterprise-wide, collaborative decision making throughout the acquisition life cycle. Simulation-based acquisition is intended to be a process, culture, and environment whose use will result in more reliable and dependable assessments of the consequences of making acquisition decisions prior to funding commitments, thereby diminishing acquisition risk. This is to be accomplished by maximizing the use of relevant acquisition information while simplifying the process of capturing, managing, and assessing that information. In 1998, the ISG and OSD jointly declared the SBA vision as an Acquisition process in which the DoD and Industry are enabled by robust, collaborative use of simulation technology that is integrated across acquisition phases and programs. It was further stated that the goals of SBA are to substantially reduce time, resources, and risk associated with the entire acquisition process; increase the quality, military worth, and supportability of fielded systems while reducing total ownership costs throughout the total life cycle; and enable Integrated Product and Process Development (IPPD) across the entire acquisition life cycle. The apparent understanding here is that, as a new systems acquisition paradigm, SBA embraces the total system life cycle from initial realization of an unmet need, carrying all the way forward through system design production, operation, and retirement.

This paradigm is supported by three principal characteristics of the SBA process, as well stated in this FDD:

1. *SBA is an evolved culture* in which enterprise-wide and DoD-wide cooperation is the rule and individual technical contributions and innovations are encouraged and efficiently and effectively managed. This culture encourages needed changes, such as to lead to enhanced concurrent development and provision of incentives for organizations to provide tools and procedures for use by other organizations and without institutional or service-imposed barriers.
2. *SBA is a refined system acquisition process* that capitalizes on changes in the acquisition culture in order to facilitate collaboration by many integrated product teams (IPTs) across the entire system acquisition life cycle.
3. *SBA is associated with an advanced systems engineering environment* in which the application of various automated tools and methods supports all system life-cycle activities and encourages software reuse and interoperability maximization. This SBA environment provides a means to execute an extensible, tailorable, and repeatable acquisition process through creation of reusable product description repositories that can ultimately be used to cost effectively reengineer products for enhanced effectiveness and efficiency. This environment supports the seamless flow of data between acquisition, engineering, support, and training communities. This integrated SBA environment supports an evolutionary system acquisition process.

## 9.4 AN SBA APPROACH TO HUMAN SYSTEMS INTEGRATION

In the remainder of this chapter we will discuss approaches to be taken to achieve some of the benefits of SBA when addressing human systems integration (HSI) in contemporary acquisition environments. We will do so from the point of view of a "new" program.

Those concerned with integration of a legacy system as inherent in the new system being developed will need to tailor the suggestions made here in accordance with those systems integration needs that affect their program (Sage and Lynch, 1998). The approach presented here is based on experience in the DoD product development process and the precepts of SBA. It is written as a guideline for the HSI professional participating in the context of a large systems engineering development program.

In considering an SBA development strategy and related HSI concerns, it appears best to consider three different perspectives on process for engineering a system:

- program development objectives and related processes for engineering or acquiring systems;
- models, analysis, and data collection or methods and tools; and
- systems and program management.

### 9.4.1 Development Objectives and Processes

Texts on the topic of systems engineering and management (Sage, 1992, 1995; Sage and Rouse, 1999), devote a great deal of attention to the process of requirements definition, and it is the phase of the systems engineering life cycle for the ultimate product where virtual environments and simulations can have the biggest pay-off in the SBA context.

The requirements definition process is the first phase of any new program (including upgrade programs to existing equipment). For very large programs, such as acquisition category 1 (ACAT 1) programs, there will exist a mission needs statement (MNS), a capstone requirements document (CRD) generated by a commander in chief (CINC) of a unified command, and a supporting operational requirements document (ORD) generated by one of the military services. The derivation of these documents is an evolutionary process typically spanning a number of years. These requirements are very broad and are meant to be descriptive rather than prescriptive in nature. Nevertheless, they can have a dramatic impact on the human–system interface aspects of a design. For example, a top-level requirement of the navy's DD-21 program is to have a total crew complement of no more than 95 people, in contrast to the approximately 300 people normally found on a twentieth-century destroyer. Even if the HSI portion of the design was not exposed in any detail during the early requirements definition phase, it is important for systems engineers and systems managers to understand the source of the major system design requirements and how they came to be.

One must have a clear understanding of the operational purpose of the item being developed and the potential roles that HSI will play in various systems engineering design and development approaches that could satisfy operational needs. This understanding of operational requirements is fundamental to good system engineering practices. The next stage is to understand how different design concepts are to be assessed. It may be that there is either an explicit or implicit model of the system's value. This model could be expressed in terms of acquisition or operating cost or it could include a wide array of individual performance or cost metrics. In all likelihood, this model will address a large number of issues that involve, as well as some that are beyond, the purview of the HSI domain. It is important that we recognize this model, even if it is not explicitly expressed as a "model." Any set of requirements that is expressed in measurable design parameters is, in fact, equivalent to an assessment model for that system. If such a model is not explicitly stated, we must attempt to derive the model from implicit requirements. If the model exists or can be reasonably constructed from implied requirements, then our next task is to analyze the model from an HSI perspective and determine if it is appropriate. In some cases, the system requirements may be presented as unalterable, but that does not relieve us from establishing the model assessment priorities. As an HSI practitioner, the most important early step is to determine those aspects of the model that capture issues that are impacted by the HSI architecture of the solution.

We may find that there are no parameters in the model that appear to relate to the HSI "view" of the system. If this is the case, we have identified a potentially significant issue that needs to be addressed. In the "grand scheme" of things, if there are no HSI systems engineering parameters that impact the model that will evaluate the system, then a very strong argument exists that any investment in HSI is unnecessary. This is, of course, a foolish and untenable situation. If such a situation is detected, it will then be very desirable and necessary to establish the linkage between the evaluation model and the HSI parameters of the system that relate to the model in order to diagnose the issue and suggest potentially corrective measures. In most cases this should not be difficult. For example, if the probability of success is based on reaction to some external stimulus, then the issue is how the system to be engineered can decrease the reaction time, and this is often an HSI issue. If the linkage between the evaluation model and the human system design is not apparent, then it is up to the HSI designer to argue for the inclusion of the appropriate parameters. Failure to establish this connection must necessarily relegate the HSI part of the design to an insignificant part of the overall system design and related investment.

In addition to the assessment of the value of the HSI components of the system design, we must also consider the cost impact of HSI-related decisions. The question needs to be asked: Are these costs correctly portrayed in the cost estimating tools associated with the design? While we are collecting the needed information on the sources of the HSI requirements, we should also be developing an understanding of all the major system requirements and the technology and subsystem domains to which they have been allocated. In particular, we need to identify any domains that may impact or reflect HSI design decisions. We must identify the analysis and design environments in which major system requirements are to be assessed and whether their relationship to HSI parameters is properly handled. For example, in a situation where the performance of a system is highly dependent on a series of tasks in a platform with many workstations, the following questions are pertinent. Do other aspects of the design reflect assumptions about the performance of workstation tasks? How do they address uncertainty or human variability? Is there an associated risk that other parts of the design are assuming a best case approach and will be insensitive to subsequent changes in the HSI design as it evolves? One of the larger issues that SBA attempts to address is that of "harmonizing" and integrating different views of the system and to enable "real-time" incorporation of design changes and their implications across the entire design space. Until an IT environment is developed that can reliably perform that function, it is up to individual engineering teams to maintain this broad general awareness. They must constantly evaluate how their decisions relate to the performance requirements and objectives of the total system and the impact of these design decisions on domains outside the HSI field of regard.

### 9.4.2 Methods and Tools

While an effort is being made to understand the requirements world, we must also focus on the HSI engineering environment. In the SBA context this entails an evaluation of all of the

HSI data that are relevant to engineering the system and the tools, methods, and models that will be used to evaluate design trade-offs and analysis.

In this context, one might consider models in the manner described well by Blanchard and Fabrycky (1998, p. 91): A model is "a simplified representation of the real world which abstracts features of the situation relative to the problem being analyzed. It is a tool employed by an analyst to assess the likely consequences of various alternative courses of action being examined. The model must be adapted to the problem at hand and the output must be oriented to the selected evaluation criteria. The model, in itself, is not the decision maker but is a tool that provides the necessary data in a timely manner in support of the decision-making process. The extensiveness of the model will depend on the nature of the problem, the number of variables, input parameter relationships, number of alternatives being evaluated, and the complexity of operation. The ultimate objective in the selection and development of a model is simplicity and usefulness." The authors suggest that models should represent the dynamics of the system in a way simple enough to understand and use and close enough to reality to yield successful results; highlight those factors that are most relevant to the situation at hand and repress unimportant ones; be comprehensive through inclusion of all relevant factors; be reliable in terms of repeatability of results; be simple enough to allow timely implementation and use; and incorporate provisions for ease of modification or expansion to permit evaluation of additional factors that are not immediately apparent and that occur later. These are not necessarily trivial features to incorporate in a model. This suggests that models themselves should be adaptive and evolutionary.

In the SBA context, models of the system's behavior and the data describing the instantiation of the product being evaluated by those models represent the system description at that point in its evolution, wherever that point may be in the life cycle of engineering the system. Indeed, the adequacy of the models in confidently predicting the consequences and behavior of the system is a direct reflection of the overall risk of the conceptual system design itself. Any difficulty in modeling a system should be cause for a reevaluation of our own understanding of the factors that are relevant to its description. There is a general precept in the world of SBA that if we cannot model a system's behavior and interactions, then we have a poorly understood basis for engineering the system.

A strong word of caution should be injected at this point. When confronted with the challenge to produce an adequate model of a system, the immediate reaction is often to "go off and create one." This is often an inappropriate reaction. There are a lot of models that may potentially be used, so the first task should be a thorough survey and understanding of what is available. With that knowledge, we next should determine what is actually usable and appropriate in the situation at hand. Very often the best situation is one where a model has been developed and is available commercially. Such models may well be very responsive to requirements and are almost always cheaper than developing an in-house approach. Ultimately, we will have to make the choice about what models to use "off the shelf" and what to develop. However, developing a model from scratch carries a heavy burden of VV&A, as described earlier. Self-developed models should deservedly be met with customer skepticism until model accuracy and appropriateness have been fully demonstrated. Even if these conditions are met, engineering models are rarely static; they require a steady diet of funding since they need to evolve and maintain currency with the underlying technologies that they are to emulate.

Development of the SBA concept requires strong focus on the need to share data among different models and simulations. Ultimately, this may include real-time interaction among detailed engineering simulations and cost models operating "synchronously." This is an area that has the potential to yield great insight, but it will also doubtlessly result in a great deal of confusion as we try to make models work together through integration. Once we have the models identified, they can be used to analyze the system in a great variety of ways, some of which were not possible before the era of computer-implemented mathematical models that can be run at very great speed to enable experimentation with potentially complex adaptive behavior. Future models will be exercised over a very broad range of parameters that otherwise might remain unexplored.

Within any domain such as HSI or any subdomain, there may be many models and simulations that can be brought to bear, but we very often find that the use of these models is limited by the availability or reliability of data. Furthermore, when an engineering effort within a domain completes an analysis and creates new data relevant to the design, it is very often not clear how the data are distributed to the affected parties outside that engineering domain. There are important data interpretation and configuration management issues that must be addressed for SBA to be implemented successfully. Generally, the data problems for the SBA environment are of two types: (1) understanding the data as information and knowledge and (2) distributing and managing the data. Understanding the data refers to the need to share data with others, who may not know very much about the source of the data or their correct interpretation. This is a nontrivial problem when large systems are being analyzed, compared, and engineered. This often has to do with the underlying assumptions or conditions that existed at the time the data were created. Today, we often find ourselves with apparently useful data to address a question but with little confidence about some of the underlying attributes of the data. This is often because different engineering communities, technology domains, or cultures assume different things and terms have different meanings. In a world where the communication across technology domains and engineering teams is strictly controlled, the interfaces across boundaries can be managed to minimize this source of communication problem. However, SBA envisions an engineering environment where there are few boundaries and information can flow effortlessly and instantaneously across the engineering enterprise. Thus, there is a need for a more organized way to retain all of the important information about a data file so that others can properly interpret it. This is typically accomplished through the process of "data modeling," another growing field in the area of IT.

Data modeling attempts to create an unambiguous description of data and the relationships among data elements. Often this data model is "object flavored," and relationships between all data entities are mapped (e.g., parent–child relationships) and the attributes of each data entity are explicitly defined. Communities of interest have begun to develop their own data models to be published as international standards. For example, manufacturers have been developing standards for the exchange of parts data under the auspices of a global organization whose sole purpose is to establish standards, such as the International Organization for Standardization (ISO). There are many domains where there are no definitive terms and conventions for creating and managing data. Among those fields with no definitive data model are cost estimating and analysis and human–system interface design. Some aspects of human–system interfaces may already be accommodated by related engineering standards efforts, but data dealing with human system performance may have no common format or well-understood interpretation beyond the realm of a small subcommunity of practitioners. Setting data standards is not a panacea, and in some engineering domains and areas of research it may be premature or inappropriate. Furthermore, establishing and maintaining international standards can be a slow process.

Also, a difficulty with standards is that there are often so many from which to choose. Despite these shortcomings, the HSI community desirous of actively participating in an SBA program environment must come to grips with how data about the HSI aspects of a design are collected, stored, and shared. At a minimum, the HSI effort should include development of a data dictionary. This is an explicit definition of all the terms the team is going to use and apply to HSI data. Ideally it would also specify all of the attributes about a data file, model, or simulation that are considered essential and must be retained with that item. As discussed earlier, we must also identify all of those domains that will be providing data and those that will be recipients of HSI data. It is then necessary to understand all the issues about format, assumptions, and constraints on the data that apply. The significant benefit of a data model is that it can make all of these issues explicit and immediately visible to someone who is searching for or viewing the data. It may also be possible that a program could create its own data model, integrating individual pieces from the ISO or U.S. standards-setting bodies [e.g., American National Standards Institute (ANSI), Electronic Industries Association (EIA), and Institute of Electrical and Electronics Engineers (IEEE)]. This program-specific data model may be the best strategy to ensure existence of a data-modeling approach that can be shared across all programs.

Some may ask: "Why is it necessary to create a data model for a program? If we have to share data among two domains or applications, we will simply write a translator." This is indeed how many companies have been handling their data exchange needs. But that approach is becoming very costly, because it creates an $n$-squared problem. That is, if we have a relatively small program that uses 100 different applications that share different pieces of data (e.g., cost analysis spreadsheets, schedules, technical data, analysis results, simulation results, stored in spreadsheets, data files, text files, etc.), then we would need to write a data translator for each pair that needed to exchange information. Even if each application only exchanges data with 10 other programs, we would need 1000 translators or 500 bidirectional exchange translators. Furthermore, any time that an application changed, 10 translators would have to be inspected and potentially modified. The translation problem explodes exponentially for very large complex programs that may be sharing data between, e.g., contractors or clients/customers. If a virtual repository exists under one data model, then this translation/inspection only happens once for each application.

Another major issue about data in the SBA paradigm is that of configuration management (CM). Sage (1992) defines CM as the systems management process that identifies needed functional characteristics of a system early in the life cycle, controls changes to those characteristics in a planned manner, and documents system changes and implementation status. Determination and documentation of who made what changes, why the changes were made, and when the changes were made are the functional products of CM. Under SBA, the CM function must be maintained on all data, models, and simulations that impact a system or are used in the acquisition life cycle.

We will not expand this further here because CM is a well-recognized concern for programs and will only emphasize that it is even more important in the SBA context. The CM of the data about a product and the configuration of the models and simulations themselves are all very important SBA ingredients. This can become a complex problem when the effort is carrying forward multiple alternatives, each with its own data and each having unique performance characteristics predicted by the modeling and simulation environments. The CM system not only must keep track of data associated with each design but also should be able to track the configuration of a tool or simulation that produced the data or supported a specific design analysis. This is important because the broad sharing of data will only be successful if users of the data have confidence in its source and its "pedigree." Users must be able to look into a model or data file and learn all they need to about that data in order to determine that it is appropriate for use in some other application. Early in this chapter we introduced the notion of VV&A. The CM of data and models must include the VV&A attributes (e.g., date and source) of those items. Finally, we should remark that CM issues in an evolutionary acquisition context have yet to receive definitive study and consideration.

Once we have established how data can be understood and interpreted, you will have taken a major step in enabling the sharing of data across a program. The actual distribution of the data is more of a classical IT problem with many different approaches. Simulation-based acquisition simply recognizes that when models and data are going to be shared on such a broad scale, the underlying data repository and distribution system must appreciate the sophistication of the SBA paradigm. The most pressing requirement is that of data, and information, control and access. The fundamental goal of SBA is to achieve faster and more effective product development and support through the rapid exchange, understanding, and exploitation of all the data and information that exist about a product. But there may be valid reasons for restricting this information flow based on national security issues, company proprietary concerns, competition sensitivity, and other relevant factors. The IT communication and data backbone must satisfy these concerns.

### 9.4.3 Systems and Program Management

It is not difficult to take the position that implementing SBA suggests that we are "systems engineering" the acquisition process and the program management process from the perspective of creating a modeling and simulation environment that optimizes managerial effectiveness and problem solving. The models and simulations that we are referring to cover not just the domains of the engineering of systems but also the management functions of technical direction, planning, scheduling, and virtually all the tasks that are carried out under the systems engineering and systems management umbrella.

Topics of major interest for system engineering and management are the scope and type of development model implied by the evolutionary acquisition concept and the potential use of modeling and simulation in achieving this. The scope of SBA covers all the phases of a system's life cycle. Itemizing the list of subdomains based on the perspective of the user, owner, or builder would produce a lengthy list. Figure 9.1 presents an evolutionary life cycle that was comprised of three phases: definition, development, and deployment, and this life cycle can easily be expanded to yield a more realistic number of phases, such as shown in Figure 9.2. This figure represents a single build in Figure 9.1 as expanded into three life cycles: research, development, test, and evaluation (RDT&E); systems acquisition; and planning and marketing. A realistic systems engineering acquisition life cycle is necessarily associated with a life cycle for planning and marketing and a life cycle for RDT&E. In Figure 9.2, the life cycle for acquisition is expanded from the basic three phases of definition, development, and deployment to a more realistic seven-phase life cycle. Discussions of these expanded systems engineering life cycles and such related concerns as risk management may be found elsewhere (Sage, 1992, 1995; Sage and Rouse, 1999).

When we consider the engineering of a system, we also often find ourselves considering architectural views or perspectives. Many discussions of systems architectures

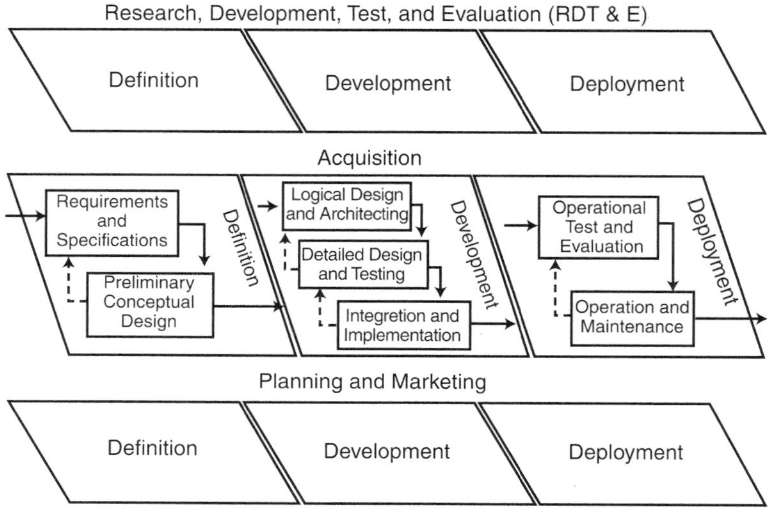


**Figure 9.2** Research, development, test, and evaluation.

focus on three primary architectural views. Here we will use *functional*, *physical*, and *implementation* to describe these (Sage and Lynch, 1998). The approximate corresponding DoD terminology in the Joint Technical Architecture (Sage and Lynch, 1998) is *operational*, *systems*, and *technical*[1]. Development of the implementation or technical architecture is the process during which the entire physical system design is integrated. This process also provides the raw materials for definition of the system's external and internal interfaces. Each of these activities in the design process is first completed at a high level of abstraction and correspondingly low level of detail. This results in an initial implementation or technical architecture for the system at a high level of abstraction. Then the entire process is repeated at a lower level of abstraction associated with greater detail for the next level of components. This repetition at lower and lower levels of abstraction and greater and greater detail is continued until ultimately the detailed implementation architecture is realized. The associated decisions and designs are reviewed, and changes are implemented at the higher levels of abstraction to the extent needed and then iterated downward. The implementation architecture integrates system requirements with the functional and physical architectures. This process also provides the raw materials for definition of the system's external and internal interfaces. These three architectures are first conceptualized at a high level of abstraction and correspondingly low level of detail. This first results in an implementation architecture for the system at a high level of abstraction. Then the entire process is repeated at a lower level of abstraction associated with greater detail for the next level of components. This repetition at lower and lower levels of abstraction and greater and greater detail is continued until ultimately an implementation architecture is realized. The associated decisions and designs are reviewed and changes are implemented at the higher levels of abstraction to the extent needed. Sage and Lynch (1998) describe a multistroke decomposition process for architecting a system and its interfaces that is roughly equivalent to this.

We have emphasized that systems engineering is a multiphase process. Each of these phases can be viewed at a number of levels: family of systems, system, subsystem, component, and part. These are generally defined in a functional block diagram structure for the system being engineered. At each of these levels, the various phases of the systems engineering process need to be enabled through identification of appropriate various work efforts. A work breakdown structure (WBS) or system breakdown structure (SBS) is an appropriate way to display this information. We may identify a two-dimensional matrix framework representation of the phases and levels in the form of hierarchical levels in the SBS, as shown in Figure 9.3. When we recall that this framework needs to extend across each of the three major systems engineering life cycles and the family of systems may be comprised of a large number of "systems," the complexity of the effort to engineer a system becomes apparent. This provides major encouragement for modeling and simulation as a part of the effort to successfully engineer a system.

A challenge for systems management is to determine how the investment in the infrastructure of data, models, and simulations is developed and evolved. There are no easy answers here, and it will be up to each program manager to determine the best strategy for a particular program. It is perhaps most important to begin with an open mind and the approach that many take toward quality. Neither quality nor a productive SBA environment is "free," but both bring the potential for far greater success in the long run if properly addressed and managed.

Relative to systems management, it is also important to transform the engineering and acquisition cultures in order to be able to accept the broad sharing of tools and data that are implicit in SBA. This has been suggested as the most important factor in implementing SBA. The need for cultural change focuses on the need to share data across different domains of the acquisition process, so that agents focusing on different modalities of the same design are using the same or consistent data. A fundamental objective is the appropriate and early involvement of stakeholders that today exist at the periphery of the acquisition process. This includes agents involved in the training and maintenance of a system as well as other systems with which the primary system must interoperate. The method of participation of these agents within SBA would be through models and simulations that portray the diverse key interests and unique cost and performance sensitivities appropriate to the system architecture and design.

| Level \ Phase | Definition | Development | Deployment |
|---|---|---|---|
| Family | $\text{Artifact}_{11}$ | $\text{Artifact}_{12}$ | $\text{Artifact}_{13}$ |
| System | $\text{Artifact}_{21}$ | $\text{Artifact}_{22}$ | $\text{Artifact}_{23}$ |
| Subsystem | $\text{Artifact}_{31}$ | $\text{Artifact}_{32}$ | $\text{Artifact}_{33}$ |
| Component | $\text{Artifact}_{41}$ | $\text{Artifact}_{42}$ | $\text{Artifact}_{43}$ |
| Part | $\text{Artifact}_{51}$ | $\text{Artifact}_{52}$ | $\text{Artifact}_{53}$ |

**Figure 9.3** Framework for activities by level and phase.

There are many impediments to the timely exchange of data and models to conduct these kinds of early trades, including issues of job security and the fundamental fear that someone else may use "my" data inappropriately. An example of this threat to current business practices is the impact on product testing during development and operational evaluation. Simulation-based acquisition suggests that as much testing as possible should be conducted in the virtual domain and a physical item test should only be scheduled after thorough evaluation of the shortcomings of models and simulations to address the risks being addressed by the physical test. Also, a major objective of any physical test should be to improve the models and simulations of the test parameters such as to reduce the number of future tests. Test and evaluation professionals need to understand, accept, and evolve this concept.

Cultural barriers include issues associated with sharing data between customers and suppliers and between teams competing relative to new opportunities. Whatever the environment for sharing, it must provide appropriate safeguards for the protection of proprietary information; however, it should not unnecessarily restrict the flow of data. The SBA initiative to date has largely been supported by those who have had a historically strong role in the evolution of modeling and simulation (M&S), specifically those supporting the development of simulators for training and wargaming and simulations used in performance trade-offs at the conceptual phase of product engineering. A major cultural challenge is to educate the engineering and support organizations that SBA is not just a classical M&S "fad," but a true initiative that requires that the broader engineering and management constituents of the acquisition process become major contributors and leaders of SBA practices. Thus, SBA can in no way be regarded as a replacement for systems engineering and management; it is an enhancer of good systems engineering and management efforts. It allows for wide-scope display of information and knowledge and thus supports the development of learning organizations. It does not represent a loss of responsibility and accountability or of security and competitive advantage. Rather, it is intended to enhance these for the betterment of all.

There have been a number of efforts to implement strong modeling and simulation capabilities in support of system acquisition. Particularly noteworthy among these is the U.S. Army (2001) Program for Simulation and Modeling for Acquisition Requirements, and Training (SMART). The Army Model and Simulation Office (AMSO, 2001) provides institutional support for SMART as the U.S. Army initiative that promotes the robust use of M&S efforts integrated across acquisition programs in an effort to reduce total ownership costs (TOCs), provide quicker delivery of products to the field, and simultaneously increase utility and worth of engineered systems. SMART is intended to more closely integrate the efforts of the requirements, acquisition, and training communities through the use of a variety of modeling and simulation approaches, including SBA. SMART is intended to foster collaboration across these three communities by integrating M&S beginning at the earliest phases in the acquisition process, thereby allowing better understanding of the process and enhancing its productivity and trustworthiness. SMART involves rapid prototyping to facilitate systems engineering so that the ultimately deployed systems meet users' needs in an affordable and timely manner with minimal and controlled risks. The intent is to enable collaborative environments across organizational and functional barriers among users, developers, testers, sustainers, and trainers. Analysis of alternatives (AOA) and CAIV are two of the analyses that support the decision process early in the life cycle.

SMART initiatives require that a comprehensive management and technical strategy for HSI be initiated early in the acquisition process in order to ensure that human performance factors are considered throughout the evolution of the system design. SMART requires that human factors engineering requirements be established in order to develop effective human–machine interfaces and to avoid system features that require extensive cognitive, physical, or sensory skills. Also, it requires that systems be designed for human interaction to minimize human errors in using deployed systems. Various M&S tools are suggested to support decisions and trade-offs through analysis of design suitability and prediction of the effects of alternative designs and architectures on human–system effectiveness. Two authoritative publications describe the current state of this comprehensive effort (U.S. Army, 2000, 2001), and the latter of these contains a comprehensive listing of resources, including websites, relating to M&S for system acquisition.

## 9.5 SBA QUALITY ASSURANCE QUESTIONS

Today, many smaller DoD programs are significant users of M&S technologies. Because of this, many assert that they are already executing SBA. In some cases they are motivated to take this position because they believe that SBA is the "current buzzword" and by asserting an SBA capability they will win both status and funding. We should be careful about how we view this position.

Programs that are effectively employing M&S are likely to be the biggest proponents of SBA. They are the "believers" and provide the demonstrations that the use of M&S is saving money and reducing risk for their programs. For these reasons, SBA advocates do not want to alienate these aggressive M&S adopters. But the aggressive use of M&S simply represents the evolutionary path to SBA, not the aggressive goals of the SBA vision, and we need to make this distinction. Program-specific M&S adopters will agree that they do not have the ability to readily exchange data across M&S environments. They also feel no commitment, because they have no associated funding, to support other programs or develop models or simulations with broader application than their own immediate needs. Finally, none of these programs set out with SBA as the kernel of their acquisition strategy and therefore have not invested in the infrastructure required to optimize the benefits of SBA.

But the revolution to SBA cannot be just for brand-new programs, and it cannot wait until the full infrastructure is in place. What should a program be doing while it is waiting for the revolution, and how will it know how well it is doing? With these questions in mind, we have prepared a *very preliminary draft* set of quality assurance questions that might ultimately be used for an SBA capability maturity model (CMM) assessment checklist based upon SBA desiderata established in a report of the Joint Simulation Based Acquisition Task Force (1998) and other readings. This SBA quality assurance checklist may help a program manager better understand where a program is and whether it is creating the opportunity to reduce program cost and risk through incremental implementation of the SBA vision. This checklist does not conform to the formal staged structure of the Software Engineering Institute's systems or software CMM or any of the related efforts, such as capability maturity model integration (CMMI) initiatives, with their progressive levels of maturity that may happen later. For the moment it is simply a set of questions that might be asked by a government program manager:

1. *Collaborative Environment* (*CE*) Is the program participating in a multiproduct CE in which the exchange of data is facilitated through standards and configuration management? What is the purpose of this collaborative environment? Are tools being shared?

2. *Distributed Product Description* (*DPD*) Do you have a description of your product that is maintained as the "virtual baseline" for your product? Is it used as the primary reference for design, development, and analysis? Is it under configuration management? Is it responsive to the needs of the collaborative environment? Has this distributed information been integrated such that it appears to users as a single integrated information repository and is this DPD consistent and coherent and with sufficient access control to protect classified and proprietary information?

3. *M&S Planning* Does the program have an M&S plan? This plan should identify the full spectrum of program M&S constituents. It should have a plan for prioritization of those models and simulations that will bring greatest benefit to the program, both individually and if their data are shared and interoperable. Does the plan address the collaborative use of M&S and how the CE will be supported? Have M&S shortfalls been prioritized? Does the M&S plan reflect an investment strategy for models and simulation purchases or upgrades that will bring the greatest cost benefit to the program? Does the plan provide explicit guidelines for VV&A for models and simulations? Does the plan address configuration management of M&S used within the program? Has the M&S plan been integrated into the system engineering management plan (SEMP)?

4. *Program Management* Is the potential of M&S fully identified and addressed in the program management plan (PMP)? Areas to look into include cost-estimating tools and decision analysis tools. The PMP should also have a statement regarding the certification and credibility of models to be used. It should reflect an understanding of the importance of VV&A for models that may have a key impact on program decisions. Does the PMP address the significance of configuration management of the tools and data employed by the program? Does the source selection plan address the use of M&S? Will all tools used in source selection be provided to potential offerors?

5. *System Engineering and Management* Is the potential of modeling and simulation fully identified and addressed in the SEMP. Were M&S tools used in the system requirements analysis? Are those tools in use today? Does the SEMP require a specific evaluation of the use of models or simulations to assess performance, reduce cost, and minimize program risk by all functional engineering domains? Are appropriate tools integrated and/or do they generate and use data interoperably with other key M&S elements in the program? Is the DPD used as the source/repository for all data on the current state of the baseline design and design alternatives? Does the program have an integrated approach to data management to support the system engineering process? Is this approach further integrated into the program information management system?

6. *Test and Evaluation* Is the potential of M&S fully identified and addressed in the test and evaluation management plan (TEMP)? Will M&S be used to identify the highest priority physical tests, i.e., those necessary to assess the most critical parameters for which M&S tools are inadequate? For all physical tests that will be required, has a specific analysis on why models or simulations cannot be substituted been conducted? Have anticipated physical test parameters and results been projected through models and simulations? If not, what result is expected from the test, and what is the basis for these expectations? Will test events generate information to upgrade/improve existing models or simulations? Will physical tests provide statistically relevant sample sizes? Will M&S be used to interpolate or extrapolate physical test results? Are the methods valid?

7. *Data Management* Does the program have an approach for participating in the appropriate integrated data environment (IDE) or for establishing a new IDE if necessary? Does the IDE reflect an integrated data-sharing approach for all engineering functional and life-cycle domains? Does the IDE identify who are the users and suppliers of data and how these data are to be shared? Will the IDE enable interoperation of tools and methods? Does the IDE reflect a plan for continuous data object/element ownership?

8. *Knowledge Management* Does the program have appropriate plans to enhance knowledge generation, transfer, and sharing? Does this lead to efficient and effective knowledge sharing across all elements of the DPD and among all concerned parties such that they have a mutually consistent and accurate understanding of the system, system of systems, and family of systems to be acquired.

9. *Technical Performance* Has the program identified and prioritized the technical performance measures (TPMs) for the program? (TPMs are those quantitative factors that represent the most important product features in the eye of the customer.) What analysis methods were applied in the requirements analysis to derive these TPMs? If models or simulations were used, are these tools still appropriate and are they being used to validate the current state of the design? Have these tools evolved as the product has evolved? Are other models or simulations being exploited to understand and mitigate performance risk and program cost associated with these TPMs?

10. *Cost Estimating and Analysis* How does the program assess cost across three domains: (1) cost of current program execution, (2) projected recurring cost of the prime items to be delivered, and (3) life-cycle cost of the complete system? What cost models are employed and how effective are these models at projecting cost? Are they integrated? The program manager must appreciate the importance of understanding life-cycle cost and its impact on system design. There are few programs today that do not have a major objective of minimizing TOC. Because we cannot collect historical cost on future systems, we have no choice but to use a model or simulation to estimate future costs. Are the tools the right tools? Can they assess CAIV? The program manager must appreciate that many cost tools use historical data and are therefore limited in their ability to project cost benefits of technology improvements and state-of-the-art changes in processes and methods. How do the program's cost models handle this? Do these models maintain currency with the existing design? (That is, are they synchronized with the product development process?) Does the customer understand and have confidence in the suggested cost-estimating methods?

11. *Source Selection* Will models or simulations be used to evaluate competitive offers? What tools will be used? Do the competitors have these models? If not, why not? Do the potential offerors trust these models? What is the plan for how contractors can propose model improvements and modifications to better reflect their potential offer? Is proprietary information safeguarded?

12. *Contractor Use of M&S* Does the program office staff have sufficient knowledge of those models employed by the contractor that can have a direct impact on the government's assessment of overall contractor performance? If not, how will the government measure and assess the contractor's performance under the contract?

13. *Program Schedule* Do the program schedule and major milestone events reflect key demonstrations of M&S maturity and capability?

14. *Customer Satisfaction* Does the customer understand the value of M&S in reducing cost and understanding and mitigating risk? Is the customer comfortable with the use of M&S? Do you understand where he or she is uncomfortable and do you have a plan to address the issues?

15. *Evolutionary Development* Have the systems engineering life-cycle process and the systems management efforts been configured to support evolutionary acquisition? Is the CM and configuration control board structured so as to cope with the various facets of evolutionary acquisition? Have appropriate modeling and support tools and processes for evolutionary acquisition been obtained and does the culture of the organization(s) support use of these?

16. *Human Systems Integration* Have appropriate planning and action been taken to ensure that relevant HSI concerns are addressed throughout the acquisition effort?

## 9.6 CONCLUSION

Simulation-based acquisition in some form is inevitable. It is simply the logical progression of the process with which we select and develop products to satisfy future needs. It is not at all merely using M&S to support systems acquisition. Highly sophisticated applications of M&S techniques already exist in the aerospace, automotive, and heavy-equipment industries. These provide demonstrations that product simulation and experimentation are powerful concepts. The SBA approach may be the only way that large, complex, and costly systems can be developed and tested before committing to production. Fundamentally, SBA is a risk mitigation strategy. It recognizes that all that we know can be modeled and what we do not know may be the best justification for intense model development as a method to focus our identification, pursuit, and resolution of the unknown risks.

Simulation-based acquisition has fundamental and profound implications for system engineering. System engineering in the SBA environment must be the focal point for architecting and creating the simulation paradigm of a system development and life-cycle support activity. But no one doubts that there is plenty of opportunity to spend money in the wrong places. Since SBA is a concept with much to be filled in, it offers the opportunity for invention and innovation in how we get there. It is a chance to "system engineer" the acquisition and product development process itself.

In closing this chapter, the authors recognize that we present an optimistic view of the potential of SBA to reform and improve the system engineering and acquisition environment. However, we recognize that there are at least three areas of concern that we have not dwelt on and that deserve far greater treatment that is beyond the scope of this chapter. These three areas are (1) commercial analogues to SBA, (2) simulation of combat or stressful environments, and (3) the shortcomings or "gaming" of models that misrepresent a system or preserve a point of view.

In this chapter we have emphasized SBA as an initiative within OSD, because that is where the abbreviation comes from and that is where it is being applied. The notion of SBA arising from the DoD is not inconsistent with the observation that, historically, many management and technology innovations that have broad commercial potential and utility have their origins in "state-of-the art" defense programs. That notwithstanding, there is a broad appreciation of the benefits of M&S in commercial industries. This is particularly true among vehicle manufacturers. For example, auto manufacturers are placing increasing reliance on M&S to assess driver comfort and dashboard design as well as precursors to crash tests and the assessment and design for occupant safety. The details of their approaches are worthy of more detailed investigation. We should point out that a big difference between DoD and commercial ventures in the implementation of SBA lies in the fact that most large-scale U.S. manufacturers have pervasive control over their processes. If they see the economic benefit to some aspect of SBA, they can generally change quickly. The DoD and the defense industry operate under a more pervasive set of constraints, limiting the ability to both initiate and fund process changes.

Perhaps the single greatest shortfall in the world of DoD simulations is the ability to model human performance under stress. This includes speed and accuracy at the console as well as the effectiveness of troops under fire and experiencing casualties. The authors know of no easy solutions here. Continued investigation is obviously required, but there may be a point where the best we can hope for is to bound the problem within some range of anticipated performance based on training, leadership, motivation, and other factors, which, admittedly, are rather difficult things to quantify.

The final issue is the inadequacy of some models or even the blatant misuse of models and simulations to protect a system or position. This is a real-world problem, often brought about because the best representation of a system's performance is often that provided by the program management office responsible for fielding the system. Indeed, because of the way money flows in the acquisition process, the program office may be the only source of funding for the construction of the system model/representations. This is a potentially very dangerous situation. The danger is greatest for sophisticated or complex systems that require significant understanding and model detail to capture their behavior. One of the early reviewers of this chapter pointed out the failure to adequately predict weapon susceptibility to enemy countermeasures. The authors believe that the failure is due less to diligent understanding of the system's shortcomings and more to the failure to have an independent and qualified team objectively assess and model the system. In any case, the fact that we can model a system behavior with great fidelity does not provide the guarantee that we will. This is one of those process and cultural issues that must be focused on in the evolution of SBA constructs. These are continuing concerns in the effort to fully develop simulation-based acquisition in the 21st Century (National Research Council, 2002).

## NOTE

1. Some authors, such as Buede (2000) use the term *operational architecture* to describe essentially what we call the *implementation architecture*. Either is an acceptable term. We use the term implementation primarily because it is the term used in the National Institute of Standards and Technology (NIST) systems integration and management architecture (SIMA) and to avoid possible confusion with the DoD joint technical architecture, where the term *operational* is used with a somewhat different meaning.

## REFERENCES

Acquisition Functional Working Group. (1999, December). *Acquisition Modeling and Simulation Comprehensive Core Body of Knowledge*. Department of Defense. Available: www.dmso.mil.

Army Modeling and Simulation Office (AMSO). (2001, April). *Simulation and Modeling for Acquisition, Requirements and Training (SMART) Reference Guide*. Available: www.amso.army.mil/smart/documents/ref-guide.

Banks, J. (Ed.). (1998). *Handbook of Simulation: Principles, Methodology, Advances, Applications, and Practice.* New York: Wiley.

Blanchard, B. S., and Fabrycky, W. J. (1998). *Systems Engineering and Analysis*, 3rd ed. Upper Saddle River NJ: Prentice-Hall.

Boehm, B., and Hansen, W. J. (2001, May). The Spiral Model as a Tool for Evolutionary Acquisition. *Crosstalk*, *14*(5), 4–11.

Brady, J. H. (2001). Systems Engineering and Cost as an Independent Variable. *Systems Engineering*, *4*(2), 233–241.

Buede, D. M. (2000). *The Engineering Design of Systems*. New York: Wiley.

Carlock, P. G., and Fenton, R. E. (2001). Systems of Systems (SoS) Enterprise Systems Engineering for Information-Intensive Organizations. *Systems Engineering*, *4*(4), 242–261.

Defense Systems Management College. (1998). *Joint Logistics Commanders Guidance for Use of Evolutionary Acquisition Strategy to Acquire Weapons Systems*. Fort Belvoir, VA: Defence Systems Management College Press. Available: www.dsmc.dsm.mil/pubs/gdbks/jlc_evol_acqstrat.htm.

Johnson, M. V. R. Sr., McKeon, M. F., and Szanto, T. R. (1998, December). *Simulation Based Acquisition: A New Approach.* Fort Belvoir, VA: Defense Systems Management College Press. Available: www.dsmc.dsm.mil/pubs/mfrpts/mrfr%5F1998.htm.

Johnson, M. W. (2000, June). Big Industry View of SBA. Paper presented at the National Defense Planning and Analysis Society (DPAAS) Conference on Simulation Based Acquisition. Available: www.dpaas.com/SBA%20M%20JOHNSON.PDF.

Joint Simulation Based Acquisition Task Force. (1998, December). *A Road Map for Simulation Based Acquisition*. Available: www.acq-ref.navy.mil and www.msosa.dmso.mil/sba/documents.asp.

Krygiel, A. J. (1999). *Behind the Wizard's Curtain: An Integration Environment for a System of Systems*. Washington DC, Command and Control Research Press.

Maier, M. W. (1998). Architecting Principles for Systems-of-Systems. *Systems Engineering*, *1*(4), 267–284.

Nance, R. E. (1999). Distributed Simulation with Federated Models: Expectations, Realizations, and Limitations. In *Proceedings of the 1999 Winter Simulation Conference*, Society for Computer Simulation International, San Diego, CA (pp. 1026–1031).

National Defense Industry Association (NDIA) SBA Industry Steering Group. (1999, February). *Simulation Based Acquisition Functional Description Document*. Available: www.msiac.dmso.mil/sba/documents.asp.

National Research Council, *Modeling and Simulation in Manufacturing and Defense Systems Acquisition*, National Academies Press, Washington, DC, 2002.

Sage, A. P. (1992). *Systems Engineering*. New York: Wiley.

Sage, A. P. (1995). *Systems Management for Information Technology and Software Engineering*. New York: Wiley.

Sage, A. P., and Cuppan, C. D. (2001). On the Systems Engineering and Management of Systems of Systems and Federations of Systems. *Information, Knowledge, and Systems Management*, *2*(4), 325–345.

Sage, A. P., and Lynch, C. L. (1998, November). Systems Integration and Architecting: An Overview of Principles, Practices, and Perspectives. *Systems Engineering*, *1*(3), 176–227.

Sage, A. P., and Rouse, W. B. (Eds.). (1999). *Handbook of Systems Engineering and Management*. New York: Wiley.

Sanders, P. (1997, September/October). Simulation Based Acquisition: An Effective, Affordable Mechanism for Fielding Complex Technologies. *Program Manager*, pp. 72–77. Available: www.dsmc.dsm.mil/pubs/pdf/pmpdf97/sanders.pdf.

Shapiro, C., and Varian, H. R. (1999). *Information Rules: A Strategic Guide to the Networked Economy*, Boston, MA: Harvard Business School Press.

U.S. Army. (2000, September). *Planning Guidelines for Simulation and Modeling for Acquisition, Requirements, and Training*. Available: www.amso.army.mil/smart/documents/guidelines/index.htm.

U.S. Department of Defense. (2000a, October 23). *The Defense Acquisition System* (Directive 5000.1) and *Operation of the Defense Acquisition System* (Instruction 5000.2). Available: ric.crane. navy.mil.

U.S. Department of Defense. (2000b). Acquisition Reform website: www.acq.osd.mil/ar.

U.S. Navy. (2001, February). *Modeling and Simulation Verification, Validation, and Accreditation Handbook*. Available: navmso.hq.navy.mil.

## CHAPTER 10

# User-Centered Systems Engineering Framework

LEE SCOTT EHRHART and ANDREW P. SAGE

## 10.1 INTRODUCTION

Human systems integration (HSI) is a systems engineering effort employed across the life-cycle process for the engineering of systems to ensure the incorporation of such critical human factors as usability, reliability, manning, training, and safety within the deployed system. Even when regulations require HSI plans[1] as part of the system acquisition process, process reviews within the U.S. Department of Defense (DoD) indicate a failure in making and implementing these plans (cf. DoD, 1994). Many traditional systems engineering efforts result in the generation of rich and precise hardware/software specifications and implementations with only meager representation and accommodation of users and their tasks (Ehrhart, 1994). Crucial information about users and the tasks they must perform is often lost somewhere between initial problem description and final detailed design specification. As a result, technologies introduced to streamline organizational processes and to facilitate other human activities often create new bottlenecks instead. Efforts in reviewing the literature in new product development and associated decisions notice few instances of concern for human issues in product development (Krishnan and Ulrich, 2001).

The recent emphasis on quality management in systems engineering (Sage, 1992, 1995; Sage and Rouse, 1999) reflects growing concern over the high cost of systems that either:

- Fail to adequately address the functional needs of the operational environment or
- Fail to support the users' successful access to that required functionality (Ehrhart, 1994).

The first issue noted is essentially one of system *utility*; the second issue is that of system *usability*. Both have critical implications for task performance and mission success. A

*Handbook of Human Systems Integration*, Edited by Harold R. Booher.
ISBN 0-471-02053-2 

goal of HSI activities is to orchestrate the application and introduction of new technologies to effectively support individual and team performance to meet organizational needs.

Successful implementation of systems for human users requires an understanding of what it means to provide technology in support of purposeful action in organizations. This includes the formulation, analysis, and interpretation of decision-aiding requirements and the engineering of human–computer cooperative systems to address the identified utility and usability requirements. Systems engineering process models and the associated architectures that lead to system development must address these requirements effectively to enable the design and engineering of useful and usable systems for human interaction (Sage, 1987). The needed efforts encompass various aspects of the problem domain and require evolving technological solutions, each with major human interaction and integration facets. Requirements documents are text-based models of the operational need; software and hardware designs are text and graphic models of the solution path proposed. Prototypes are also models, representing the current design of the system being developed. In between are many more models created as part of artifacts such as data structures, drawings, and charts. Structuring, evaluating, and refining these models highlights gaps in the requirements or conceptual design and alerts the systems engineering team responsible for requirements and conceptual design to critical human–machine factors that effect performance.

This chapter presents frameworks for user-centered systems engineering for HSI. We discuss frameworks for definition and development of systems that emphasize methods for creating, structuring, and applying models and processes needed to identify and address HSI issues across all phases of the development life cycle. Our hope is that this chapter will enable those responsible for HSI to address the wide scope issues that affect systems integration issues affecting humans, technologies, and organizations. Thus, we provide approaches that will enable determination of the value and impact of effective and ineffective user interfaces on systems integration. We address the diversity of users and tasks and their impact on the design of interfaces for HSI. We discuss different system development life cycles, including those particularly applicable to HSI, and show how HSI issues can be incorporated into systems engineering process life cycles. The references provide detailed information concerning sources available on these subjects.

### 10.1.1 HSI Players and Interactions

There are many stakeholders involved in HSI issues. Figure 10.1 illustrates five of these stakeholder groups and their roles. HSI methods and processes need to engage all the development stakeholders: operational end users of the system, as well as the organizations and enterprises for whom the system is to be engineered. These stakeholders also include those in systems engineering and management, who are responsible for technical direction and communications relative to the process of engineering the system, and the detailed implementation specialists responsible for detailed design production. Our concern is primarily with the first three groups. Their support needs are identified in Table 10.1.

The management, cognitive, and behavioral sciences include many advocates for holistic approaches to understanding the multiple facets of human–machine collaboration in organizations. These crosscut disciplinary interests and organizational functions may be called "transdisciplinary" endeavors (Somerville and Rapport, 2000). For example, enterprise management interests drive process modeling and improvement efforts for software process improvement (Humphrey, 1989), process reengineering (Hammer and Champy, 1993; Hammer and Stanton, 1995; Yu et al.; 1996, Sage 1995, 1999); and

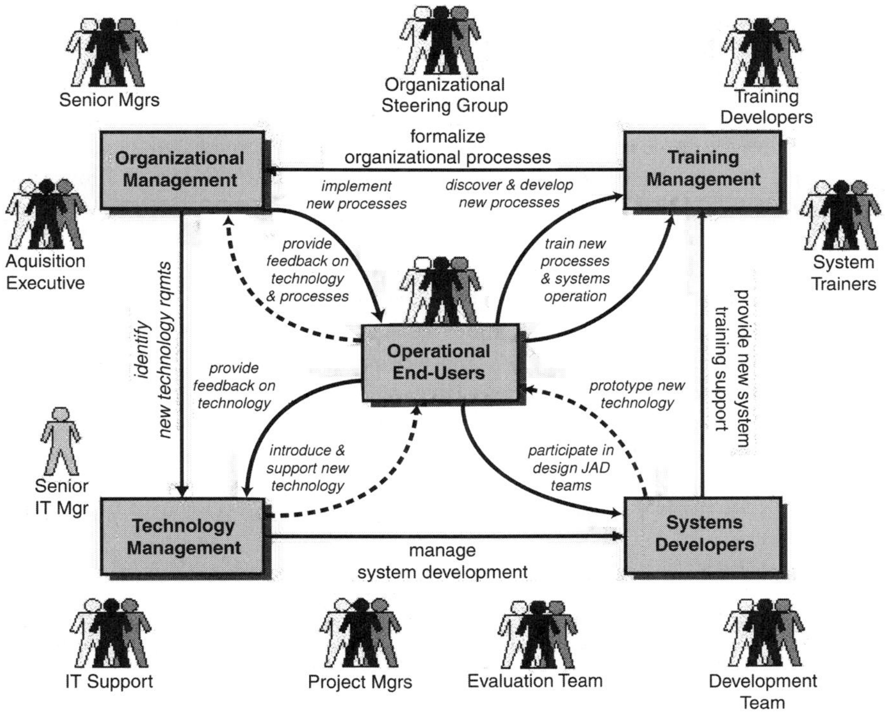


**Figure 10.1** Stakeholder roles in human systems integration.

**TABLE 10.1 HSI Application to Stakeholder Issues**

| Group | HSI Support Needs |
|---|---|
| Operational end users of deployed system | • Focusing on critical factors in decision tasks to make better decisions faster<br>• Understanding and managing information flows and cognitive workload<br>• Facilitating distributed collaboration and cooperative problem solving across multiple users and systems |
| Organizations and enterprises who acquire the system | • Identifying and expressing organizational requirements<br>• Co-evolving organizational processes with information technology introduction<br>• Synchronizing information operations across functional boundaries |
| Systems engineering and management for development effort | • Identifying and representing the human information processing and decision support requirements<br>• Identifying and addressing potential sources of error in the decision system<br>• Relating operational needs to system concepts for effective human–computer cooperative problem solving and decision making<br>• Identifying and managing development risks in evolutionary development of complex decision systems |

information technology enabled change (Manzoni and Angehrn, 1998). Training and education in organizations is supported by research on situated learning (Suchman and Trigg, 1991), action research (Checkland and Holwell, 1998), and learning organizations and knowledge management (Choo, 1998; Senge, 1990; Senge et al., 1994, 1999). In addition, the cognitive and behavior sciences have contributed user-centered design (Norman and Draper, 1986), decision-centered design (Andriole and Adelman, 1995; Ehrhart and Aiken, 1991; Woods and Roth, 1988), collaboration support (Olson and Olson, 1991), participatory design (Greenbaum and Kyng, 1991), and a broad range of approaches to enhance human information processing in systems and organizations (Sage, 1990). These approaches model humans and technology support as "organic" to information processing, knowledge-creating, and decision-making processes within organizations. Handbooks of human factors and ergonomics (Salvendy, 2001) and systems engineering and management (Sage and Rouse, 1999) generally address these issues. Although systems engineering and systems management necessarily cross boundaries to connect these disciplines, the state-of-practice is often still multidisciplinary, rather than interdisciplinary and transdisciplinary.

### 10.1.2 Cognitive Systems Engineering

An extremely important interface for HSI and systems engineering for decision systems is enabled through cognitive systems engineering (CSE). Cognitive systems engineering is often spoken of as the practice of the engineering user-centered, decision-focused, information-technology-based systems. The CSE concept provides approaches to the engineering of systems that have major human–machine cooperative problem solving and organizational decision-making requirements. There are three major imperatives:

- Model organizations as *decision systems* to better understand their aiding and training requirements
- Focus on the operational end users—their processes, organization, environment, technology support requirements, and training
- Drive the organizational decision system design to permit the co-evolution of organizational structure, advanced information technology, user and team tasks/processes, and the training design to ensure the successful integration of technology

The CSE approach synthesizes tools and methods across multiple disciplines—including artificial intelligence, cognitive science/psychology, sociology and organization science, systems engineering, and operations research—to provide both the scientific base and applied technologies necessary to support research and development. For both the designer and manager, incorporating CSE activities into the development process assures a better match to operational needs by capturing a more robust set of functional and nonfunctional requirements. This understanding supports informed decision making when design tradeoffs must be made during development life cycle.

### 10.1.3 Systems Engineering Life Cycle

The traditional systems engineering process is comprised of an iterative, multiphase process providing essential guidance in engineering effective systems. The essential phases

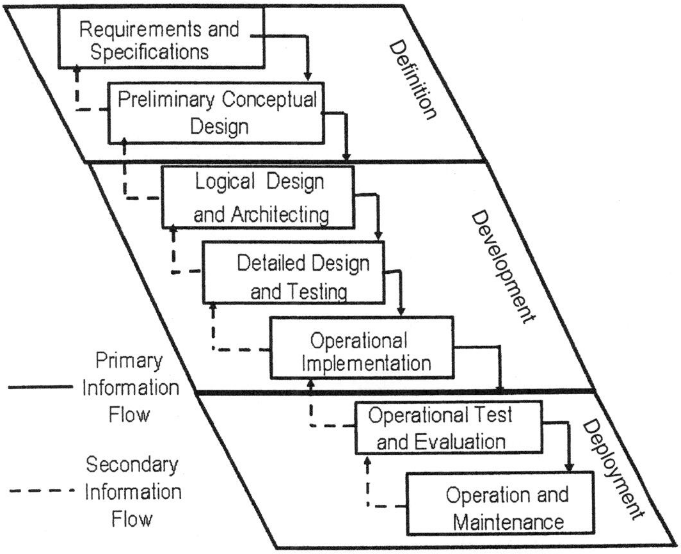


**Figure 10.2** Typical structure of a systems engineering life cycle.

in a systems engineering life cycle involve definition, development, and deployment as suggested in Figure 10.2 and may be described in terms of seven constituent phases as follows.

### *System Definition*

*1. Requirements and Specifications* The first part of a systems engineering effort results in the identification of user requirements and the translation of these into technological specifications for a product, process, or system. The goal of this phase is the identification of client and stakeholder needs, activities, and objectives for the functionally operational system. This means that information is a necessary ingredient and results in the mandate to obtain, from the client for a systems engineering effort, a set of needs and requirements for the product, process, or system that is to result from the effort. This information requirement serves as the input to the rest of the systems engineering process. This phase results in the identification and description of preliminary conceptual design considerations for the next phase. It is necessary to translate operational deployment needs into requirements specifications so that these needs may be addressed by the system design and development efforts. Thus, information requirements specifications are affected by, and affect each of the other design and development phases of, the systems engineering life cycle.

*2. Preliminary Conceptual Design and High-Level System Architecting* The primary goal of this phase is to develop several concepts that might work and are responsive to the specifications identified in the previous phase of the life cycle. The preliminary conceptual design selected must be one that is responsive to user requirements for the system and associated technical specifications. Rapid prototyping of the conceptual design is clearly desirable for many applications as one way of achieving an appropriate conceptual design. Several potential options are identified and then subjected to at least a preliminary evaluation in order to eliminate clearly unacceptable alternatives. The surviving alternatives are next subjected to more detailed design efforts, and more complete functional and physical architectures or specifications are obtained. It is at this phase that the enterprise, functional, and physical architectures are initially identified. Functional analysis approaches are particularly useful in this phase of effort.

### System Development

*3. Logical Design and Physical Architectural Specifications* This phase results in an effort to specify the content of the system product in question and to provide more detail to the associated high-level functional and physical architectures that were identified in the previous phase. Specifications are translated into detailed representations in logical form such that system development may proceed. This logical design product (sometimes called a functional architecture) and the product architectural specifications are realized in terms of the physical architecture (sometimes called engineering architecture) of the system that will ultimately be implemented.

*4. Detailed Design, Production, and Testing* The goal of this phase is a set of detailed design specifications that should result in a useful system product. There should exist a high degree of user confidence so that a useful product will result from detailed design, or the entire design effort should be redone or abandoned. Another product of this phase is a refined set of specifications for the operational deployment and evaluation phases of the life cycle. Again, design alternatives are evaluated and a final choice is made, which can be developed with detailed design testing and preliminary operational implementation. This results in the implementation architecture for the system. Utilization of this implementation, or detailed design architecture, results in the actual system. Preparations for actual production and manufacturing are made in this phase.

*5. Operational Implementation* An implementation contractor produces the system here, often in an outsourced manner. A product, process, or system is implemented or fielded for operational evaluation. Preliminary evaluation criteria for final acceptance of the system are obtained and then modified during the following two phases.

### System Deployment

*6. Operational Test and Evaluation (and Associated Modification)* Once implementation has occurred, operational test and evaluation of the system can occur. The system design may be modified as a result of this evaluation, leading, hopefully, to an improved system and, ultimately, operational deployment. Generally, the critical issues for evaluation are adaptations of the elements present in the requirements specifications phase of the systems engineering life-cycle process. A set of specific evaluation test requirements and tests are evolved from the objectives, and needs are determined in the requirements specifications. These should be such that each objective and critical evaluation component can be measured by at least one evaluation test instrument. If it is determined, perhaps through an operational test and evaluation, that the resulting systems product cannot meet user needs, the life-cycle process reverts iteratively to an earlier phase, and the effort continues. An important by-product of system evaluation is determination of ultimate performance limitations for an operationally realizable system. Often, operational evaluation is the only realistic way to establish meaningful information concerning functional effectiveness of the result of a systems engineering effort. Successful evaluation is dependent upon having predetermined explicit evaluation standards.

*7. Operational Functioning and Maintenance* The last phase includes final acceptance and operational deployment. Maintenance and retrofit can be defined either as part of this phase or as additional phases in the life cycle; either is an acceptable way to define the system life cycle for system acquisition or production. Maintenance can include reengineering of the product or system or retirement and phase-out.

With only cursory examination, this process would seem to confine evaluation activities primarily to the last stages of development. This is correct in the sense that it is at this phase that the formal operational test of the deployed system is conducted. However, configuration management efforts include evaluation and verification efforts at all phases of the life cycle. Adelman (1992) is among many that recognize that judgments and decisions pervade every phase of the systems engineering process. The results of analysis and evaluation, represented as iteration or feedback loops in most models, provide input to support development objectives at each phase and determine whether those goals have been achieved. This continuous evaluation is a critical component in requirements-driven design.

The early phases of system development are characterized by the greatest degree of uncertainty. As a result, as much as 80 percent of the mismatch between what the user wanted and what the developers delivered has been traced to shortfall in the definition of requirements (Boar, 1984). Barry Boehm's (1981, 2000) research indicates that the cost to fix these discrepancies may range as high as 100 times the cost had correct requirements been identified during the requirements analysis phase. Furthermore, empirical evidence from a number of studies reveals dramatic increases in error correction costs the later in the development cycle the error is found (Daly, 1977; Boehm, 1976; Fagan, 1974). The requisite rework leads to cost overruns and schedule slippage. Conversely, approaches to development that eliminate rework and postdevelopment modification promise productivity improvements from 30 to 50 percent (Boehm, 1987). For this reason, the search for cost-effective system performance improvement methods has focused on improving the quality of requirements identification and representation methods.

## 10.2 MODELS FOR HSI

As we have noted, there are three essential life-cycle phases in engineering a system: definition, development, and deployment. Models are especially useful in implementing these phases, especially in the definition phase and the early portion of the development phase. There are three generic types of models that are most useful here: conceptual models, requirement models, and prototyping models. Each of the phases of systems engineering is an iterative refining process in which formulation, analysis, and interpretation—and associated modeling and evaluation—interact continually. In the early phases, especially during system definition, models may be largely informal, conceptual expressions of the system engineer's architectural view of the system and its context. Evaluation of existing system operations supports the early stages of concept definition that, in turn, may form the first system model. Implicit in this model is some representation of the system's purpose as it relates to organizational goals and the identification of criteria by which the achievement of those goals is recognized.

As definition progresses, the current system model is analyzed in terms of the perceived deficiencies, or shortfalls, between what the system provides and what the organization needs. This process leads to the definition of yet another model—a set of requirements for the next-generation system and the criteria by which alternative architectures and designs, or system models, will be evaluated with respect to those requirements. Evaluation and modeling continue to play a key role in supporting decisions throughout the iterative process of architecting and systems design engineering. Even in the early life-cycle phases, evaluation is still being performed upon models in the form of system prototypes. Finally, evaluation of operational systems is accomplished in the early phases of system deployment based on the assumption that the evaluation criteria, established in the form of measures of performance (MOPs) and measures of effectiveness (MOEs), accurately represent (or model) the relationships between component, subsystem, and system performance and the larger purpose for which the system is intended.

The HSI approach uses models to conceptualize the user, tasks, and system supports. To effectively incorporate human systems engineering into systems engineering processes requires a framework for integrating and extending the multiple models that support understanding, representing, and translating the user's role in the human–machine system in terms of tasks performed, knowledge required, context of use, and organizational objectives. Ultimately, the level of detail chosen must be determined by the information required. Effective models may be characterized in terms of several key desirable characteristics:

- The level of detail is adequate to support evaluation of principal factors of interest at the current phase in the life-cycle process.
- The issue representation scheme and mode are appropriate to the question at hand.
- The assumptions regarding the nature and relationship of pertinent system variables can be supported by valid sources (historical data, acknowledged experts, output from other validated models).
- The resulting model is understandable to the responsible analysts and the critical reviewers.

A number of approaches for modeling based on human factors and systems engineering concerns are discussed in Salvendy (1997, 2001) and Sage and Rouse (1999). We will now turn our attention to a number of modeling issues in systems engineering as they specifically relate to HSI. These issues are covered for the following major sections: systems definition, system requirements, system conceptual and architectural design, prototyping and implementation, and system evaluation.

## 10.3 SYSTEM DEFINITION

### 10.3.1 System Definition Goals

The problem definition phase of systems engineering serves two purposes. First, the definition phase determines the scope of the proposed system in terms of what is needed and technically feasible. Second, this initial phase establishes the goals and objectives for the system development effort to follow. System definition is accomplished by examining three general types of information:

- *System Context* Who will use the system, what they are trying to do with it, under what conditions it will be used, etc.
- *Constraints* "Built in" requirements for inputs, outputs, interconnection, environmental tolerances, etc.
- *Technological Opportunities* Leverage points where technology may be applied with greatest benefit

During the initial portion of the system definition phase, the systems engineering team gathers information needed to understand the functional objectives of the user enterprise for the system to be engineered. Information drawn from various organizational documents and discussions with the user enterprise help to sketch the system boundaries and develop a profile of the system context as defined by:

- *Users* Experience, training, organizational roles
- *Tasks* High-level functions, performance goals, decision task characteristics (timing, criticality)
- *Organizational Context* Organizational goals, missions, control structures, communication modes
- *Environmental/Situational Context* When, where, how, and under what conditions will the system be used

This information comprises the operational needs that the new system must meet. The various dimensions of the system context each generate constraints on the system that must be explored during the requirements phase and addressed in the design. Moreover, constraints involving human performance, hardware, and software interact. For this reason, it is essential that the human factors unit of the systems engineering team coordinate with the other members of the team during these early definition efforts in order to consider these interdependencies. Initial decisions regarding the system concept trade off these technological opportunities (i.e., what *might* be done) against the system context and constraints (i.e., what *must* be done). The impacts of human user model both affect and are affected by the other hardware and software issues.

### 10.3.2 Models for System Definition

The early portions of the system definition phase provide the initial suggestions that guide the more detailed requirements identification and analysis and the subsequent technological specifications that follow. For this purpose, the most useful outputs from this early part of the definition phase are preliminary models, such as concept maps and functional decomposition diagrams, which define the central constructs of the system and indicate relationships between them. One of the most difficult aspects of the initial part of the system definition phase is the internal (and sometimes external) pressure to "define" in terms of solutions. Jumping to solution thinking during this phase may focus the subsequent requirements identification activities on a subset of the problem while neglecting other equally relevant aspects. This "tunnel vision" early in the development can lead to one of the most common sources of error—defining the wrong problem and then proceeding to solve it.

System definition activities focus on understanding the current ("as-is") organizational activity and developing goals and descriptions of the desired ("to-be") processes, functions, and technology support. System definition and the associated requirements identification activities vary widely in the granularity of representations that are required. The same system may use different modeling methods for different development efforts. Some models are suitable for extension and elaboration as the system concept evolves, while others are more narrowly focused with limited application. Several methods specifically address the semantic aspects of domain knowledge and are useful to the human systems engineer. For example, concept mapping is an informal technique for modeling relationships and interdependencies. The method was developed in the field of educational psychology and has been applied successfully to the acquisition and modeling of knowledge requirements for decision support systems (Seamster et al., 1997; Vennix et al., 1994; Klein, 1993b). Kieras (1988) developed a similar set of goal-task models to structure cognitive learning tasks. This method was used to identify and structure the cognitive requirements for embedded training in tactical information systems (Williams et al., 1989). Cognitive mapping (Montazemi and Conrath, 1986) is a more formal technique that evolved in the field of artificial intelligence. It focuses on modeling cause-and-effect relationships for process or behavior understanding and has been adapted to create computable cognitive architectures in neural networks (Senge and Sterman, 1994; Zhang et al., 1992). Soft systems methodology (SSM), developed by Peter Checkland (1981, 1998) in the 1970s, has been applied to action research and information systems development in medical, industrial, military, and other governmental organizations. The method uses informal models as a means to explore purposeful action within an organization and the necessary information support required.

Figure 10.3 presents a model, drawn from Ehrhart (1994), of a simple decision task in relating incoming information and the human information interpretation process. This example models aspects of the tanker duty officer's (TDO's) tasks in the Air Operations Center. The TDO is responsible for providing air-refueling support to all scheduled missions that require refueling. Replanning is required when new missions are created, existing missions rerouted, or air-refueling resources change. The TDO performs replanning tasks as indicated by his own assessment of the evolving situation and as tasked by other duty officers. For the HSI team, this model helps to identify the elements, or key variables, that need to be presented to the user such as current, planned, and required resources and operational situations. It also indicates that the user is basing part of the interpretation of this information on the potential change in information values across time. The model is annotated with HSI issues, such as potential error sources, experience, and aiding requirements.

Byrd et al. (1992) surveyed 18 requirements analysis and knowledge acquisition techniques that facilitate problem domain understanding in terms of information requirements, process understanding, behavior understanding, and problem frame understanding. They emphasize that none of the methods is suitable for eliciting and modeling all the dimensions of domain knowledge. The key to effective problem definition is finding a means for creating and relating *multiple* models, or views, of the problem. When the problem is complex and multidimensional, the design team needs methods specifically designed to facilitate interdisciplinary thinking. For example, multiperspective context

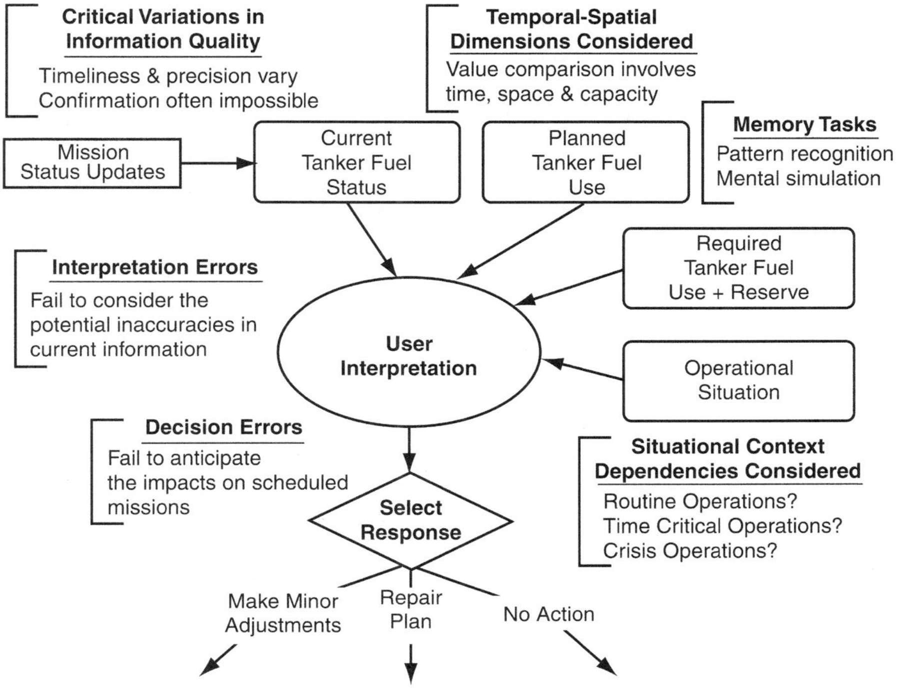


**Figure 10.3** Simple decision task with HSI annotations.

models, such as those described for problem analysis in Davis (1993), assist in creating informal models for review and iteration with the sponsors and operational users. Similarly, Zahniser (1993) describes the creation of $N$-dimensional views of the system developed by cross-functional development teams. The process is designed to encourage innovative thinking and bring multidisciplinary experience to bear on system development problems.

System definition models help to organize the system goals and objectives to guide the developers in the requirements identification phase. For the human engineering team, the most relevant issues are those aspects of the problem definition that address the functional roles and activities that are modeled for the human users. Using the initial high-level function allocation, the design team must begin to identify and analyze the human task requirements and the associated implications for HSI.

## 10.4 SYSTEM REQUIREMENTS

Although the deployment efforts to bring about a technical solution to problems are often cited and blamed for performance failures, the results of several studies of software-intensive systems traced the majority of errors in delivered systems to the *pre*deployment phases (Davis, 1993). Thus, the greatest leverage on improving the product integrity in human–machine systems is to be gained by adopting a systematic method for improving the predeployment or preimplementation processes and products. This may best be accomplished by obtaining a more comprehensive *understanding* of the users, tasks, and operational context; a more accurate *representation* of the technical requirements; and a more effective *translation* of requirements into development specifications, architectures, and system designs.

### 10.4.1 Requirements Identification, Analysis, and Representation Goals

During the portion of the definition phase that concerns requirements identification and analysis, the systems engineering team focuses on deepening and extending the knowledge represented in the system definition models with respect to the human users and their task support needs. System design and development requirements provide a focal point for integrating the information gathered on the users, problem-solving tasks, and the operational environment in order to guide design and development decisions. These requirements include not only the human–machine interaction requirements that define the operation of interfaces, but also the cognitive task requirements (CTRs) that define the supports for the user's decision task performance. Particularly in cases where the decision tasks are complex and must be performed in a dynamic, time-stressed environment, the operation of the interface must not distract the user from the primary tasks involved in accomplishing the organizational goals. The systems engineering architecture and design team uses the cognitive and interaction task information to determine the most beneficial human–machine task allocation, information representation modes, display formatting, and information interaction protocols.

During the requirements analysis phase of development, the CTRs can be identified and defined as part of the normal requirements identification activities. The goal during this phase is to gain an understanding of the functional tasks that the human user(s) must perform and how the user, organization, and situation define and impact those tasks. Using the high-level conceptual models from the early system definition activities and the evolving hardware and software requirements, the team develops models of information flows, task allocations, and organizational procedures for decision making. At this point, it is useful to observe the way the organization currently addresses the problem and interview representative users to expand and correct various preliminary functional, procedural, and dependency models.

***User-Centered Requirements Framework*** The CSE framework includes a user-centered requirements framework (Ehrhart, 1997) that expands upon information obtained during the system requirements analysis. This should be modeled to include a representation of the user's cognitive tasks, as implied by the information flows or prescribed by operational procedures, and the interpretation of analysis of that model with respect to the user's information requirements and the possible sources of cognitive errors. The CTRs are constructed through the process of evolving and relating models that profile the user and organization. They describe the environmental and situational contexts and define the various cognitive tasks involved in accomplishing the functional tasks assigned to the human–computer decision component, as shown in Figure 10.4.

### 10.4.2 Models for Requirements Capture and Analysis

A CTR represents either the nature of the *input* required for a human decision-making task or the content of the *output* required from that task. Thus, initial objectives in the

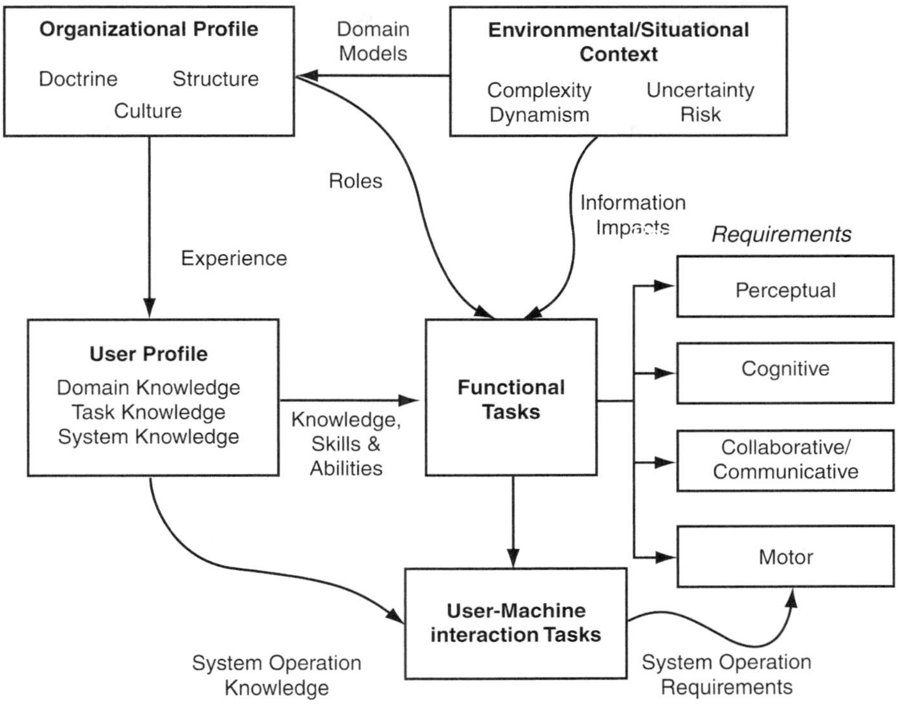


**Figure 10.4** Relationship between elements of user-centered requirements model.

requirements phase are to identify the kinds of cognitive tasks that the users may be required to perform and to examine the factors that may affect performance. If a task affects decision performance, it is necessary to find out what characteristics of the task do so. Meister (1981) identifies five task dimensions that may affect performance:

1. Functional requirements (cognition, perception, etc.)
2. Complexity
3. Mental workload
4. Temporal factors (pace, duration, sequence, etc.)
5. Criticality

Cognitive task taxonomies, such as those found in Fleishman and Quaintance (1984) and Rasmussen et al. (1990) can be used as a filter to identify and categorize basic cognitive tasks with respect to these dimensions. In addition, Andriole and Adelman (1995) present a taxonomic discussion of human information processing and inferencing tasks with respect to the potential cognitive errors associated with each.

As the team reviews the context diagrams, functional decomposition diagrams, and straw-person storyboards, descriptions of activities can be examined for verbal constructs that indicate human user actions. For example, in systems where the human user must *monitor* a situation and *interpret* evolving events, the software designers may view the inputs to the user as updates to a database. From the user's perspective, however, this requirement has implications not only for interface operation design but also for the information presentation design. In order to interpret those updates, the changes must not only be visible to the user but also presented within a meaningful context. Using the concepts of analogical representation and causal reasoning, this context might include some mapping of relationships between key factors, tracing of changes in relevant factors over time, and/or models of a goal state to which certain parameters should conform. At this point, the information presentation and interaction requirements continue to be identified from the user's perspective without specifying the design solution.

The simple model shown in Figure 10.4 raises numerous questions for the support system (aiding) design, as well as the training and staffing design, such as:

- How often must the status information be updated?
- How does the user need the information presented to comprehend the meaning of the change?
- Does the user ever need to know or review resource trends going back several updates? If so, is the current direction of the design implying that the user will retain this in his or her memory or keep notes off-line?
- Does the user make these interpretations routinely? Occasionally? Rarely?
- How does the change in current resources relate to the operational situation?
- Will the user have experienced a wide or narrow range of interpretation situations?
- What situational contingencies might negatively affect the user's accurate interpretation of these factors?
- How does the task/decision impact the mission? How critical is it? How rapidly must the decision be made? Where and how will it be disseminated?

These questions and others may need to be addressed in the design and coordinated with the other development teams involved in the effort to engineer the system. To answer them requires understanding not only the structure of information flows but also the way in which that information is used. Thus, it is important to address relevant issues identified in cognitive research regarding users, tasks, organizations, and situational context.

### 10.4.3 Profiling the Operational Context

More often than not, performance issues that affect organizational operations strongly relate to human performance issues (Stolovitch and Keeps, 1992). Models of the situational context, or decision environment, should capture and represent the conditions that impact decision making. They should also capture the effects of agents and events that are internal and external to the organization itself. The models in this section provide several perspectives for modeling situational context and interpreting potential impacts on decision making. Table 10.2 highlights important factors, associated characteristics, and human performance issues. Due to their considerable interaction with the decision tasks and users, similar issues are addressed with respect to the characteristics of the users, organizations, and tasks.

***Context Categories and Situational Response*** Meister (1991) presents a categorization of situational contexts in terms of four possible levels of determinacy that

**TABLE 10.2 Environmental/Situational Context Profiles**

| Factor | Characteristics | Human Performance Issues |
|---|---|---|
| Determinacy | • Ranges from *determinate* (highly predictable with only one probable outcome) to *indeterminate* (no outcome can be identified as significantly more probable).<br>• Describes degree to which all variables affecting given problem or situation are known and understood.<br>• Interacts with complexity and dynamics of environment. | • *Task allocation and aiding impacts:* dynamic demand on user's perceptual and cognitive resources<br>• *Information design:* level of detail; representation in discrete or symbolic formats; importance of structure in information presentation<br>• *Personnel assignment and training requirements:* more stochastic environments require expertise acquired through rich experience; experiential learning difficult in highly mutable environments |
| Structure and boundedness | • *Structure:* Describes extent to which crucial information for task performance is known, available, and quantifiable.<br>• *Boundedness:* Describes extent to which problem is constrained, may be represented in reliable fashion, and is tractable for human information processing. | • *Task allocation and aiding impacts:* tractability impacts on user cognitive workload; structural impacts on aiding concept<br>• *Information design:* unstructured environments may require symbolic representation to support understanding of the qualitative aspects of the task<br>• *Personnel assignment and training requirements:* reduced structure demands greater breadth of understanding; procedure-oriented training may not develop adequate task knowledge |
| Complexity | • Defined by number of interconnected components or aspects and degree of interdependence between them.<br>• Interdependence critically impacts "fault tolerance" of procedural designs. | • *Task allocation and aiding impacts:* task control may be distributed and require "what-if" tools to predict effects of proposed actions on related elements<br>• *Information design:* users need representations of dependencies to understand effects of possible events on chain of dependent factors<br>• *Personnel assignment and training requirements:* users need training to understand structure of operational context and requirement to examine "ripple effects" of their actions |

roughly equate to the degree to which the domain is well-bounded and predictable. The situational context may be considered *determinate* when the given situation or initial condition has only one significantly probable outcome. This highly predictable context for decisions includes common mechanical systems, some highly institutionalized social systems, and certain control systems. *Moderately stochastic* situations have only a limited number of qualitatively similar outcomes with a significant probability of occurrence. In this context, prediction of outcomes remains tractable as in the case of genetic processes or system variability due to variable dimensions in the component parts. *Severely stochastic* situations have a large number of qualitatively similar outcomes with a significant probability of occurrence. While event outcomes in these situations remain predictable, they are computationally intensive and beyond the range of unaided human computation. Severely stochastic situations involving human agents also have qualitative aspects that increase the difficulty of response and outcome prediction. *Indeterminate* situations provide so little information about possible outcomes that no outcome can be identified as significantly more probable. Meister cites psychotic human behavior and some political alliances as examples of indeterminate contexts.

These "environments" rarely exist in discrete form in practice; system users generally perform tasks simultaneously across a range of environments. For example, flying an aircraft requires interacting with multiple environments. The aircraft systems perform within determinate to moderately stochastic ranges. Air speed and altitude are absolute values with narrowly defined meanings for certain tasks. Other parameters (i.e., fuel consumption) represent calculated values for which there are ranges of accuracy. Outside the cockpit, the aircraft pilot must interact with severely stochastic weather conditions that may affect the aircraft in unpredictable ways. When the aircraft involved is a military aircraft, the pilot must also respond to the indeterminate environment of the battlefield.

In more determinate contexts, the operational goals focus on applying well-understood procedures to respond effectively to a highly constrained set of triggering events. The users seek to maintain operational consistency and control to meet routine performance demands. Errors occur when responses are too rigid to react to major changes and/or novel events or when users apply inappropriate procedures in a changed environment. Since control is maintained by manipulating key factors to create predictable outcomes, users need detailed information about situation inputs, user actions, and outcomes. As environmental/situational uncertainty increases, users must make efficient use of resources in a succession of varying short-term situations to rapidly and effectively exploit opportunities. The emphasis is on flexibility and adaptive, creative responses in the face of novel events. Errors occur when latency between recognition of situation and internal readjustment results in a lack of effective control. In addition, the high degree of uncertainty and ambiguity in novel events makes the application of experiential learning more difficult. Detailed data is often less valuable than symbolic representations of functional relationships to convey structure and assist users in recognizing and interpreting common aspects of novel events. Overviews and aggregated displays can support pattern matching and provide externalized mental models. Since extremely stochastic and indeterminate environments are often complex and dynamic, it is important to support users with multiple levels of abstraction to meet adaptive cognitive control requirements.

In addition to the determinacy of the situational context, it is useful to understand and model the degree of structure as well as the boundedness and complexity inherent in the situational context and typical decision tasks. Several researchers discuss the interaction of these factors (Fleishman and Quaintance, 1984; Meister, 1991; Rasmussen et al., 1994) and their implications for aiding the user. The structural characteristics of the decision context and tasks should be considered in the selection of the analytical methods that form the basis of the decision aid design as well as the interaction routines that facilitate the human–computer cooperation.

The degree of structure in a decision domain characterizes the typical situations and decision tasks in terms of the extent to which information on the key variables is available and quantifiable. For example, *highly structured* contexts are those where all critical information is readily available and quantifiable for accurate manipulation. In *semistructured* contexts, the key variables may be quantified without losing critical information or making difficult assumptions; however, often some of the critical information is unavailable. In this case, the uncertainty surrounding the decision involves "known unknowns" that may have to be inferred if further information cannot be obtained. Finally, *unstructured* contexts involve qualitative variables that may not be legitimately quantified. In addition, there may be "unknown unknowns," that is, critical information that is either not available or not represented in the user's model of the situation or task.

Closely related to determinacy and structure, "boundedness" incorporates the degree to which the key variables constrain the problem to make it tractable. The representativeness and reliability of the variables also contribute to the boundedness of the problem domain. A *closed* domain may be constrained and described accurately with variables that require minimal cognitive demands to manipulate. When the domain is *semibounded*, the variables may only be generally representative and reliable. The associated uncertainty is manageable only by highly trained and motivated experts. The *open*, or unbounded, context involves variables that may not be well-understood and/or reliable. The resulting uncertainty exceeds human ability to absorb and manipulate.

The degree of complexity characteristic in the domain is interwoven in the concepts of both structure and boundedness. Woods (1988) defines complexity in a domain or a system in terms of the number of interconnecting parts or subsystems and the degree of interdependence between them. Using a structural model of situational context, complexity may be further delineated with respect to the number of hierarchical levels (vertical complexity) and number of parts or subsystems per level (horizontal complexity). In *simple* domains, both the vertical and horizontal complexity is low and the critical variables in the situation do not interact. In a system context, this absence of interdependence results in component functioning unaffected by performance of other system parts. In *moderately complex* domains, the degree of vertical and horizontal complexity increases and there is greater interdependence between the variables involved. In moderately complex domains, performance of functions may be enhanced or degraded by the performance or nonperformance of other subsystems. *Complex* domains and systems involve many hierarchical levels extended by many interdependent parts and subsystems. The functions of a complex system cannot be performed if other subsystems perform poorly or not at all. The inherent complexity of the situational context plays a significant role in the user's ability to mentally simulate the consequences of a proposed response. From a design perspective, simplifying domain complexity may eliminate critical information with unpredictable results.

***Effects of Situational Context on Task Performance*** Situational context is an important variable in several models of human information processing and decision making. For example, Rasmussen's (1986) skills–rules–knowledge (SRK) model has three levels of cognitive control based upon situational contingencies and user knowledge. *Skill-based* control comprises the highly integrated, automatic sensory–motor responses that occur with little conscious effort. Efficient control in this mode is dependent upon experience and a predictable environment. In *rule-based* responses, the user is consciously aware of taking a sequence of steps to attain a goal that may not be explicitly formulated. As a result, the user can accurately describe the procedure or rule triggered by the situation, but often cannot explain the situational cues that triggered the rule. In novel situations or unfamiliar environments, the user does not have readily understandable cues to trigger procedural responses and must use additional cognitive resources to analyze the situation. Situation assessment in *knowledge-based*, or model-based, control is used to formulate an explicit goal and identify procedures to attain the goal. When reasoning identifies an appropriate rule or procedure, control drops back to the rule-based level. The decision-making effectiveness in this mode depends upon the quality of the user's "mental model" of the situational context.

Understanding the situational context can provide insight into the potential cognitive demand placed on the human user. For example, in simple, primarily determinate contexts, the decision-making efforts focus on optimizing the outcome by manipulating the initial conditions. This generally involves skill-based actions and some rule-based control. Semistructured, moderately stochastic contexts tend to induce attempts to manipulate initial conditions using primarily rule-based control. Since the possible outcomes are bounded, efforts often focus on optimizing the expected value of outcomes. In severely stochastic contexts, precisely manipulating initial conditions cannot control outcomes. Furthermore, detailed planning and reliance on preplanned procedures (rules) are less useful due to the unpredictability of complex evolving situations. In this case, combinations of knowledge- and rule-based response control efforts focus on preparation for unfavorable outcomes and maintaining an ability to recognize and rapidly exploit opportunities. Decision responses in a complex, indeterminate situational context rely primarily on knowledge-based control. Effective performance depends upon knowing enough about the situation and the domain to classify it. The highly unpredictable nature of these contexts requires an intuitive approach based upon well-developed mental models of the domain and environment to protect against disastrous response errors.

Situation assessment and mental models also drive Klein's (1993a) recognition primed decision (RPD) making model of expert decision making in dynamic situations. The RPD model describes decision-making behaviors comparable to the rule-based and knowledge-based behaviors described by the SRK model. When forced to respond quickly in an unfamiliar situation, the expert user attempts to identify aspects of the situation similar to previously experienced situations. In simple recognition situations, matching the current situation to a previously experienced analog automatically indicates the appropriate course of action in terms of the procedure to follow. In more complex recognition situations where there is no readily available analog addressing the key features of the situation, users must also reason about possible courses of action. This reasoning involves mental simulation of the possible outcome(s) of a particular course of action based upon the user's mental model of the situational context and ability to manipulate the network of interdependencies. The resulting cognitive demands lead to a satisfying, rather than optimizing, strategy in which the user selects the first course of action that appears to satisfactorily attain the goal.

Crises form a special case in situational contexts that impact the users, organization, and decision tasks. Hermann (1972) defines a *crisis* as a situation that:

- Presents a *threat* to one or more important goals of the organization.
- Permits only a very *short decision time* before situation changes significantly.
- Involves novel or unanticipated events that *surprise* the system users.

Threat or risk to the organization plays a central role in such domains as international politics, corporate management, and military operations. In each case, the situational context is dynamic and complex. The normal states of these environments range from moderately stochastic to indeterminate. The systems designed to cope with normal operations also must support rapid response to unanticipated events.

### 10.4.4 Profiling Organizations

The situational context surrounding a judgment or choice situation forms the external environment for decision making. The structure, function, and purpose goals of the organization provide the internal environment. The contingency task structure of a decision situation, which represents the external and internal environment surrounding a task needing attention and the experiential familiarity of the person or group undertaking judgment and choice with the task and the external and internal environment, will influence the mode of judgment and choice that is used. Another relevant variable, organizational culture, is dominantly influenced by the shared values and group behavior norms that shape human and organizational progress (Harrison and Huntington, 2000). Systems designed to support decision making within organizations must take into account not only the technical facets of the hardware, software, and communications architectures with which they cooperate but also the structure of the human organization in which they function. This involves understanding the organizational culture and how it directly, or indirectly, impacts and is impacted by the individual users, their tasks, and the contingency task structure surrounding decisions that must be made. Table 10.3 presents likely characteristics and human performance issues associated with leadership and authority, communication, and decision-making aspects of an organizational profile. Organizational policy, whether implicitly or explicitly communicated to a person or group attempting to exercise judgment, provides not only procedural guidelines for structured tasks but also conceptual perspectives and strategy objectives that must be considered. Organizational response to issues is generally evolutionary, emergent, and adaptive, and the resulting organizational systems share these characteristics. The engineer of organizational systems must be sensitive to the associated redefinition effects of new systems on the organization and its culture and doctrine. This subsection presents methods for profiling organizations and modeling the relationship of the organization to the other dimensions of systems engineering for human interaction.

***Methods for Profiling Organizations*** In a seminal work, Kotter and Heskett (1992) identify shared values and group behavior norms as the two major ingredients of organizational culture. Values are virtually invisible and are difficult to change. Norms that result from values are easier to identify and to change. However, any attempt to change norms, without an accommodating change in values, is likely to produce very unsatisfactory results. A multistage development approach illustrates how organizational cultures often emerge.

1. The top management in a new organization attempts to implement a strategic vision to support organizational strategy.
2. The deployment is successful and organizational personnel are guided by the new vision and strategy.

**TABLE 10.3 Organizational Profile**

| Factor | Characteristics | Human Performance Issues |
|---|---|---|
| Leadership and authority | • Describes span of control, degree of centralization, style of interaction, and flexibility of organizational authority structures. | • Less flexible organizational structures depend upon specific role assignments and narrowly focused training to ensure reliable task performance; more flexible organization structures require cross-training to ensure adaptable responses. |
| Communication | • Describes chain of interaction required to affect control and obtain feedback on actions taken. | • Organization's structural complexity impacts communication speed and may impact performance across all phases from planning through execution.<br>• Feedback delays due to complex communication chains may result in overcorrection when information on the results of actions is delayed. |
| Decision making | • Organizational "decision systems" require creation and sharing of common understanding of task domain, goals, and methods for achieving goals. | • Information presentation and interaction requirements must support creation and communication of shared mental models.<br>• Training must include an understanding of task domain, characteristics patterns, roles, and responsibilities within and across functional boundaries. |

3. The successful deployment leads to organizational success that continues over the years.
4. A culture results that reflects the vision, strategy, and experiences of organizational leadership.

Many have noted the fact that cultures develop when people interact over a sustained period of time and when they are successful in producing desired results. The longer the initial solution works, the deeper the particular culture becomes imbedded in the organization. Any number of external threats and opportunities may challenge the then prevalent organizational culture. The extent to which it can adapt to future needs determines the extent to which the organization will survive as an excellent organization. The notion of adaptation is key here. Figure 10.5 represents an extension of the ideas presented to illustrate adaptive and maladaptive organizational behavior.

In a landmark work, Edgar Schein (1992) identifies 10 phenomena that exist in a culture. On the basis of these and additional stability and integration requirements, he defines a group or organizational culture as a pattern of shared basic assumptions the group learned as it solved its problems of external adaptation and internal integration. A successful organizational culture is one that has worked well enough to be considered valid and taught to new members as the correct way to perceive, think, and feel in relation

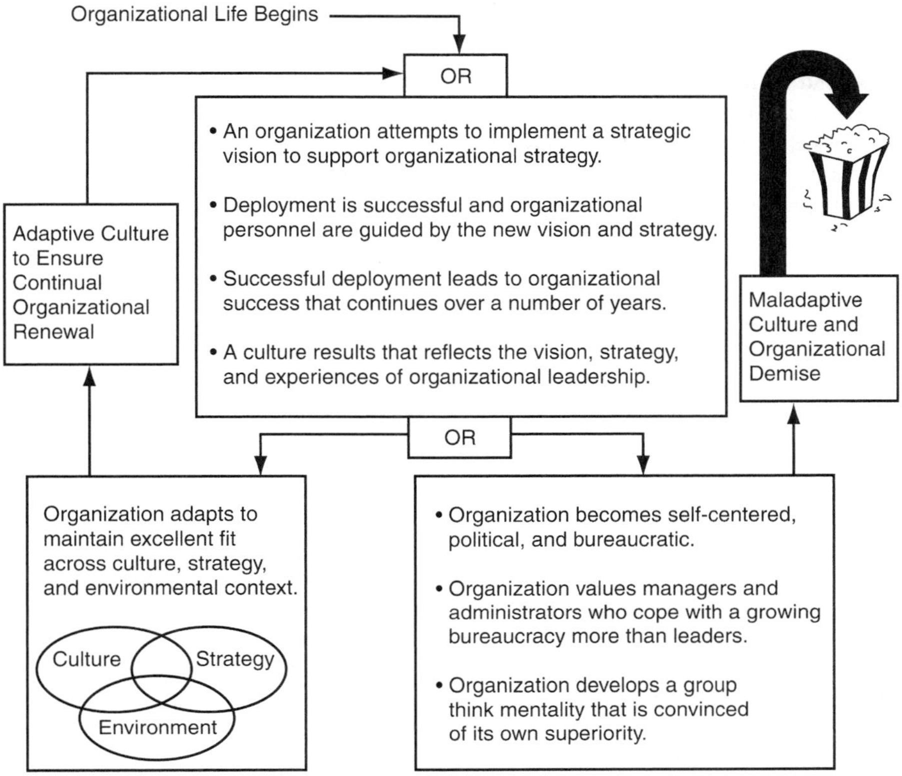


**Figure 10.5** Adaptive culture creation and emergence in organization.

to organizational problems. There are three major elements in this definition: *socialization issues*, including the process of how one learns; *behavior issues*; and *issues of subcultures*, and the extent to which they will develop. From this perspective there are causal dynamics involved in culture and leadership. Leaders initially create cultures when creating groups and organizations. Once a culture exists, it will determine the criteria for leadership. A leader in a dysfunctional culture must either change it, such that the group survives, or the culture will ultimately govern the leader.

Schein suggests three levels at which culture may be studied. At the unconscious level of basic underlying assumptions there are often unarticulated beliefs, thoughts, and feelings that represent the ultimate top-level source for the resulting values and organizational structures and processes. At the level of espoused values, formal statements of organizational objectives, purposes, and philosophies may be found. At the level of artifacts, which are comprised of organizational structures and processes, or functions, the organization attempts to implement its espoused values. We can most easily observe the cultural product that is embedded in an organization's structure. With potentially a little more difficulty, we can observe organizational artifacts as represented by processes or organizational functions. It is more difficult to examine the espoused value for the organization and even more difficult to determine the actual values from the espoused values. It is also difficult to measure cultural facets at the level of espoused purpose and inherent values. However, it is important that this be done. Measurements at the levels of process and structure possibly allow for good inferences about espoused values, but will not easily lead to information about actual values. To obtain these, we need observations of espoused values and the processes and structure implied by these, and the products of the actual value system in terms of processes, structure, and organizational products.

Six observable factors, or observables, are identified as primary means to embed culture in an organization. These factors are:

1. Critical factors that leaders measure and control
2. Approaches taken in response to crises and critical incidents
3. Observed criteria for resource allocation
4. Activities of leaders as role models, teachers, and coaches
5. Observed criteria, used in practice, to allocate organizational rewards and status
6. Observed criteria, used in practice, for recruitment of new organizational members and outplacement of existing organizational members

There are a number of supportive mechanisms that relate to organizational structure and processes, symbols and rituals, and formal statements of organizational purpose and philosophy. The formation of subcultures is an important aspect of organizational leadership. Subgroup and subculture formation is an inherent likelihood that results from the differentiation process that invariably occurs as an organization expands and grows. Differentiation may be functional, geographical, divisional, hierarchical, or may result across products or markets. Subcultures are, in no sense, always harmful. They may be supportive or harmful to an organization's mission depending upon how they are grown and how they mature. Thus, it is important to be able to characterize the factors that will lead to organizational growth, maturity, decline, and rebirth and the role of technology, especially the role of information technology, in supporting these changes. This creates a major need for organizational profiling.

In relevant work concerning profiling, Burton and Obel (1995) formalize the interactions between the organizational features, external environment, and technology use to generate prescriptive advice for organizational design. The underlying guidelines also may be used to project technology needs based upon the interaction of such organizational factors as structure, coordination and control, size, and strategy and such environmental factors as ambiguity, uncertainty, and complexity.

French and Bell (1973) present a hierarchical framework for developing an understanding of organizational functioning based upon information regarding organizational culture, climate, processes, and goals. The framework permits study of the organization as a whole and provides methods for examining and relating the subsystems, teams, and individual functional roles. At each level in the hierarchy, the analyst may select from a range of knowledge elicitation techniques to characterize activities and model the relationship of that level to the rest of the organization. At the top level of the hierarchy, investigation focuses on the organization as an entity with a common mission and power structure. It may also include the relevant external organizations, groups, or forces and lateral associations that control or interact with the organization. Investigation methods include questionnaires, interviews, focus groups, and examination of organizational documents that concern such relevant aspects as policies and standards.

There are a number of related studies. Salama (1992) reviews organizational "biographies," or histories of the development and activities of an organization, and shows that they provide insight into organizational culture. Questions related to culture, climate, and attitudes also are relevant at the team or group level functioning within an organization. The analysis reported here seeks to discover answers to such questions as:

- What are the major problems of a group or team?
- How can team effectiveness be improved?
- How well do the member/leader relationships work?
- How does the team relate to organizational goals? Do members understand this relationship?
- How well are team resources employed?

Individual interviews, using techniques such as concept mapping, followed by group review and discussion aid in identifying and refining models of team/group functioning. The models developed may be used in conjunction with more detailed cognitive task analysis to link team structure and function to the specifics of the task and environment.

Most models developed to describe organizational structure and functioning assume additional meaning when an understanding of the organizational culture augments them. Robbins (1990) identifies 10 dimensions that define organizational culture. These include *structural features* such as control, integration, interaction patterns, and rewards; *management characteristics* such as direction and support; *organization responses* such as conflict tolerance and risk tolerance; and *individual characteristics* such as initiative and identification. A strong organizational culture communicates the organization's model of appropriate behaviors to the individual members and increases their identification with the organization. An organization is said to have a strong culture when the core values of the organization are clearly understood, intensely held, and widely shared. The resulting unit cohesion prevents breakdowns in procedures in high-stress, crisis situations and is critical for effective performance. For this reason, technologies introduced into an organization must facilitate and not interrupt the flow of communication and interaction that supports team cohesion. A strong organizational culture also can have negative effects on decision making, such as the social pressure for uniformity and failure to question weak arguments common in "groupthink" situations (Janis, 1982, 1989).

The concepts of *collective cognition* and the *collective mind* propose to describe the purposeful interaction characterizing team performance in situations requiring a high level of continuous reliability (Weick, 1995; Weick and Roberts, 1993). The collective mind is evidenced by the manner in which the team members structure and coordinate their actions with respect to a shared mental model of the system. The research of Weick and Roberts (1993) examined the effects of variations in the individual models and coordination of actions in aircraft carrier flight deck operations. As team members increased the conscious interrelating of their actions within the system, they improved their comprehension of unfolding events and reduced the incidence of error. The researchers present a model of collective cognition that relates actions (contributions), the shared mental model (representations), and the coordination of actions within the system (subordinations). In related research, Schneider and Angelmar (1993) investigated collective cognition in organizations and proposed a cognitive framework based on structure, process, and style that is applicable to the individual, group, and organizational levels of analysis.

Examining the formal and informal lines of communication in an organization provides additional information on the means by which control is exercised in an organization. Harrison (1985) discovered that patterns of interaction defined through communication between the hierarchical levels of an organization establish a shared understanding about levels of influence in decision-making processes and how such influence may be exercised. Moreover, the definition of participation through interaction dominated the perceptions of subordinates, regardless of the management style reported by their superiors. The results indicate the importance of actively supporting interaction between levels of the organization where decision-making effectiveness depends upon intraunit participation. Thus, it is not surprising to see a variety of literature in this area relating to the culture of work organizations and organizational design and functioning (Trice and Beyer, 1993) and works on organizational design in the management science literature appearing over the years (Nystrom and Starbuck, 1981; Galbraith, 1977, 1995; Nadler and Tushman, 1997). The subject also plays a prominent role in works that concern such subjects as systems engineering (Sage, 1992) and systems management (Sage, 1995).

***Organizational Responses to Situational Contexts and the Role of Contingency Task Structures*** Organizations and systems must be designed for effective response in both routine operating conditions and problem situations (Meister, 1991). Organizations develop routine (or standard) operating procedures to guide responses in relatively stable, predictable environments. Although specific tasks may involve some risks there is usually low threat and adequate response time. In this context, users respond to problems arising in their sphere of responsibility according to specific guidance from superior authority. These procedures permit a high degree of control and consistency across all organizational levels to ensure organizational objectives are met. The longer decision horizons permit subordinate users to defer responses when situations exceed the scope of their responsibility. The reduced threat allows users to reduce their workload through the use of various cognitive shortcuts or heuristics. Janis (1989) suggests that the cognitive shortcuts used in routine decision making provide more efficient responses than the conscious pursuit of more precise decisions through formal reasoning efforts. Efforts such as this were at the heart of the cognitive mode model of Janis and Mann (1977), which indicated that individuals search for information to:

1. Enable recognition of a potentially challenging opportunity.
2. Enable determination of potential losses if the present course of action is continued.
3. Enable determination of potential losses if a change is made to a new but familiar course of action.
4. Enable determination of whether it is reasonable to find a better course of action than the familiar ones already considered and initially dismissed as improper.
5. Ascertain if familiar courses of action not previously considered are acceptable.
6. Ascertain whether the remaining time until the decision must be appropriate to formal rational deliberation.
7. Support a formal formulation, analysis, and interpretation of the issues and the resulting vigilant search, processing, and deliberation.

Crisis conditions trigger shifts in organizational communication and control patterns (Hermann, 1972; Meister, 1991). Organizations designed to operate effectively in dynamic, high-threat environments must adapt rapidly to crisis conditions and novel situations. Communication delays may impair information gathering and decision implementation. For this reason, users must respond to novel problems arising in their sphere of responsibility during a crisis with only general guidance from superior authority. There is some evidence that more loosely coupled organizational structures with built in redundancy and informal interaction are necessary to respond effectively in complex, dynamic, high-threat environments (Pew, 1988). With training and experience in crisis operations, users gain experiences to develop a wide range of creative responses; however, their focus on the immediate problem may result in a satisfying response that does not meet organizational objectives.

Hermann (1972) describes the effects of crisis situations on three organizational dimensions: leadership and control, communication, and decision making. The leaders' attitudes toward rank and authority are critical determinants of subordinates' willingness to raise issues that appear to challenge the prevailing hypothesis. Conversely, weak or inexperienced leaders may be influenced in crisis situations by subordinates to make incorrect decisions (Janis, 1989). In crisis operations, there is typically a marked increase in communication with internal and external agencies. The increased intrateam communication may lead to a general air of confusion (and potentially panic) and increase the impulse to action.

When routine operations constitute the majority of organizational experience, users have little opportunity to develop a wide range of responses and may be ill prepared for sudden shifts in the environment. This can have disastrous effects for response coordination. For example, Helmreich (1988) cites National Aeronautics and Space Administration (NASA) and National Transportation Safety Board (NTSB) studies implicating crew coordination in more than 70 percent of aircraft crashes. Often such cases involved minor malfunctions, simple errors, or erroneous assumption that were compounded through inattention or incorrect judgments by a team into nonrecoverable crises. Human–human and human–machine miscommunications, poor use of available support resources, and inadequate situation assessment are the major contributing factors to the resulting failures (Helmreich, 1988).

Designers are rarely able to observe the functioning of organizations during crisis or intense periods of activity. Research indicates that organizational performance during crisis operations may be enhanced through aiding designs that support improved situation assessment and facilitate communication based upon shared mental models (Orasanu and Salas, 1993). The organizational models developed to guide design should explore the human user requirements associated with both crisis and routine operations. The knowledge acquired through these models is used to determine appropriate human–machine task allocation, design information presentation, and develop interaction routines. The organizational models should also provide structures to link user and task profiles. When utilized for the engineering of systems, these principles should lead to appropriate designs for cognitive task performance (Orasanu and Shafto, 1999).

### 10.4.5 Profiling Users

The functional roles of system users within an organization often are developed in conjunction with the profile of the organization. The HSI systems engineering team also needs to develop a profile of typical users' knowledge and experience. In certain organizations, such as military units, this information may be assumed in part by the functional definition of the position. For example, an aircraft commander may be assumed to have a minimum number of flying hours, to have completed specific training, and passed certain qualifying examinations. System designers need information that may not be assumed automatically from job descriptions. To design the information presentation and interaction routines that coordinate the performance of human–computer cooperative decision making, the design team must develop a profile of the user's experiential familiarity and knowledge of the domain, tasks, and systems involved or the contingency task structure.

Dreyfus and Dreyfus (1986) identify six levels of knowledge that a user may progress through in developing expertise. These levels (novice, advanced beginner, competent, proficient, expert, and master) provide a more detailed picture of the role of expertise in cognitive tasks. Intended in large part initially to support the design of training aids, a subject of much contemporary interest (Salas and Cannon-Bowers, 2001), the Dreyfus model describes the differences that various knowledge/skill/expertise levels make in influencing the mental functions employed in decision-making tasks and the associated mental attributes of the person exercising judgment. The mental functions involved in decision-making tasks include: differences in ability to recognize similarity in environmental and task features, differences in the way task components are conceptualized and recognized, and differences in the decision strategies employed. Figure 10.6 is an interpretation of how the various transitions occur at various levels of proficiency in this model.

The ability to make similarity judgments is essential for rapid recognition of prototypical situations and analogical reasoning for unfamiliar situations (Beach, 1992; Klein, 1993a). Tasks and situations are perceived as decomposed attributes at lower levels of

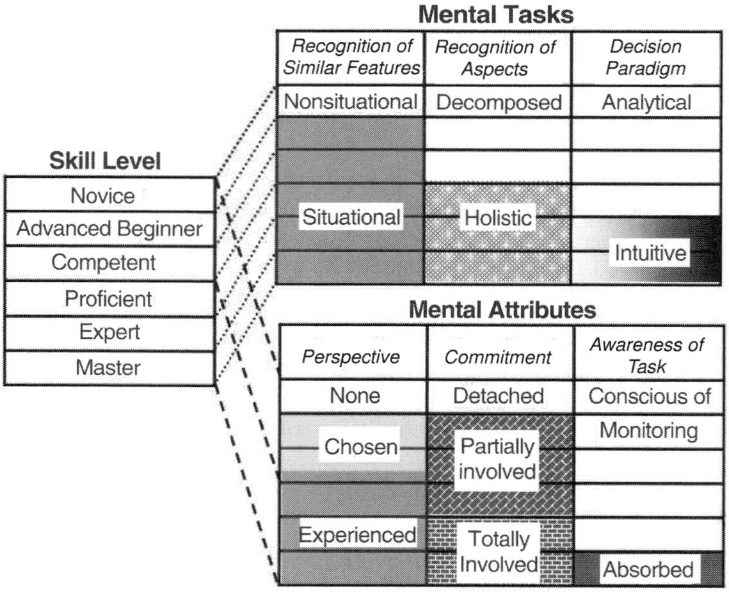


**Figure 10.6** Interpretation of the Dreyfus model of decision style.

proficiency or as a whole at the highest levels. Expertise also factors in the decision strategy employed. Lower levels of expertise usually require analytical strategies to manage the problem perceived as parameters or attributes. The holistic models that characterize higher levels of expertise facilitate intuitive strategies. Hammond (2000) has long been concerned with a cognitive continuum model where cognition varies from analytical to intuitive as a function of experiential familiarity with the contingency task structure. Thus, decision strategy selection is based upon the attributes of the task and task situation and expertise of the decider. Clearly, these models are related and address the combination of factors that determine decision strategy.

Several resources are available to guide the system designer in modeling human users (Meister, 1991; Senders and Moray, 1991), and psychology is often viewed as a requisite science of system design (Meister, 1991; Carroll, 1997). Table 10.4 characterizes user experiential familiarity and knowledge in each dimension as low, medium, and high and provides a simple representation of the continuum of knowledge and experience that usually exists as a mixture of expertise—deep in some areas and broad in others. The same system user may have different expertise and knowledge levels across domain, task understanding, and systems control ability.

The user's knowledge of the specific functional tasks to be performed generally interacts with domain knowledge. For example, a user may have considerable knowledge and experience with the situational contexts that characterize the domain but may have never performed the specific tasks now assigned. In such cases, the user may understand intuitively *what* must be done to accomplish a goal but not know *how* to do it. If a system user is not able to distinguish between relevant and irrelevant information needed to perform a task, the associated cognitive workload will increase. A moderate level of task knowledge supports task performance, and the amount of sustenance is based on the users' experience and familiarity with the task at hand and the associated facility with procedures for system use in accomplishing this task. This knowledge permits the user to trade-off performance quality in order to maintain a reasonable workload and still attain the desired goal. At the highest levels of task knowledge, the user demonstrates flexible, intuitive task performance. Depending upon the level of their domain knowledge, the users can rapidly recognize prototypical situations and adapt their task performance in an appropriate resourceful response to task requirements.

There can be several manifestations of low system knowledge and expertise. For example, when new technology is introduced, the user may have knowledge and experience with the domain and functional tasks but may have had little or no experience using the new system itself. Depending upon their role in the organization, a user may only have used a few system functions while remaining largely unaware of its other capabilities. Both of these knowledge levels result in a limited, often fragmented, knowledge of system operation. As a result, the user usually has an insufficient mental model of the system and may be confused by any errors in system operation that result from its use. The resulting increase in cognitive workload may greatly impair performance in tasks at which the user is otherwise proficient.

The competent user has a moderate knowledge of system functions and the interaction routines required to exercise those functions. The user understands the operation of commonly used system features and can operate various interfaces in order to accomplish the required tasks. The competent user's mental model of the system provides an adequate foundation to allow them to learn from operational errors. The master, or "power" user, has a strong, accurate mental model of the relationships between himself, the machine, and

**TABLE 10.4 User Knowledge and Experience Profiles**

| User Knowledge | Low | Moderate | High |
|---|---|---|---|
| Contingency task structure (domain) | • Limited, fragmented models of domain<br>• Very limited ability to recognize prototypical situations or interpret novel situations | • Recognizes some prototypical situations and can use reasoning to respond to unfamiliar situations. | • Rapidly recognizes prototypical situations and can intuitively interpret novel situations based on similarities to other prototypical situations. |
| Functional task understanding | • Adequate for most routine operations; may be too brittle to handle novel and crisis situations | • Adequate for all routine operations and some novel situations. | • Intuitively able to interpret task outcomes in novel situations. |
| Systems control ability | • Novice or casual user; limited, fragmented knowledge of system operation | • Competent user; understands operation of all commonly used system features; successfully learns from operational errors. | • Master ("power user"); strong, accurate mental model of system relationships between self, machine, and tasks to perform. |

the tasks each performs. This system model permits the user to coordinate fluid operation of the interface such that the system operation tasks are "transparent." The user is, thus, freed from the additional cognitive load associated with system operation and is able to focus directly upon the functional tasks at hand. This level of facility is critical in situations where tasks must be performed rapidly and under pressure.

User knowledge is a function of training, experience, and level of interaction with the system. Even well-trained, experienced users who rarely interact with the system cannot maintain their fluency with the system due to this infrequent interaction. When technology plays a crucial role in the organization's mission, the information presentation and interaction routines selected for the human–computer interface (HCI) must support the anticipated variation in user knowledge across all three dimensions. Where performance reliability is critical, the HCI design must make up the deficit in the user's system knowledge. Depending upon technological feasibility and the goals set for the system, it may also attempt to address deficit knowledge of the domain and tasks. Finally, the system's HCI design should provide the means for the users to extend their knowledge and improve their performance.

### 10.4.6 Profiling User Tasks

The task analysis is the usual focal point of HSI requirements models; however, there is no general method for capturing and analyzing tasks that fully addresses the range of task factors and questions. The process that leads to profiling user tasks involves application of task analysis findings to the engineering—including design, development, and evaluation—of the target system. The specific features of the task analysis choices should drive the selection of a suitable method for these analyses. Stammers et al. (1990) identify a range of task analysis methods defined by such representation techniques as hierarchical, network, and flowchart methods or by such content entities as cognitive and knowledge description, taxonomies, and formal grammars. Table 10.5 adapts the definitions from Stammers et al. (1990) and Meister (1985) to depict the advantages, disadvantages, and examples for several methods that can be used to compare task analyses.

It is important to note that every cognitive task performed by the human–computer cooperative decision system and supported by system design is impacted by the user, the organizational structure and goals that define the role of the system user, and the situational environment that provides the context for user judgments and decisions. During the identification and analysis phase, the HSI team must gain sufficient knowledge about the multiple dimensions of the requirements in order to be able to model their interactions and implications for system design. The activities involved in capturing and modeling the situational context, the organizational user, and task profiles are not necessarily discrete or sequential. These analyses occur largely in parallel and often represent shifts in focus as the system evolves over time, rather than separate discrete efforts.

***Modeling Tasks to Determine Requirements*** From the perspective of system users, the functional tasks encompass the activities that the human user performs to fulfill their roles in supporting the organization's mission. Functional tasks include not only the human–machine cooperative tasks and decision-making activities but also human–human communication activities. These tasks are separate from the system operation (user–machine interaction) tasks that constitute the focus of most traditional human factors

**TABLE 10.5 Task Analysis Methods Comparison**

| Method | Examples | Advantages | Disadvantages |
|---|---|---|---|
| Hierarchical methods | Hierarchical task analysis (HTA); GOMS (goals, operators, methods, and selection) (Card et al., 1983); GQM (goal, question, metric) (Basili and Weiss, 1984) | • Provides systematic and complete structure for analysis<br>• Broadly applicable<br>• Well-documented; easy to learn and apply | • Difficulties with representation of parallel activities<br>• Limited representation of cognitive factors |
| Network methods | Time event charting, link analysis (Chapanis, 1959); petri nets (Perdu and Levis, 1993) | • Models temporal order; can represent parallel activities<br>• Analysis may be performed quickly given right data<br>• Relatively easy to learn | • Limited applicability and narrow focus<br>• Does not consider underlying cognitive or behavioral relationships |
| Knowledge description and cognitive methods | Cognitive task analysis (Seamster et al., 1997; Rasmussen et al., 1990); task analysis for knowledge description (TAKD) (Johnson et al., 1984) | • Provides methods for characterizing cognitive tasks not found in other TA methods<br>• Consistent structure for representing task information | • Difficulty assuring completeness in highly cognitive tasks<br>• Requires considerable analyst skill in knowledge elicitation<br>• Expert sources may not be able to adequately verbalize knowledge |
| Taxonomic methods | Fleishman and Quaintance (1984) provide a descriptive survey of many human behavior and performance classification methods. | • Explicit categorization of task information; variety of uses<br>• Well-documented; relatively easy to learn and apply | • Difficult to assure completeness and definition of mutually<br>• Potential for inconsistent allocation of task elements |
| Formal grammar methods | Task action grammar (TAG) (Payne and Green, 1986) | • Rigorous, formal specification of procedural task information<br>• Provides a mapping of tasks to actions for HCI dialog | • Narrowly focused, inflexible structure<br>• Does not model relationships between task elements |
| Flowchart methods | Job process chart (Tainsh, 1985); decision models | • Models parallel user/system tasks and information flows<br>• Well-documented; relatively easy to learn and apply | • Difficult to assure completeness and definition of mutually exclusive categories<br>• Potential for inconsistent allocation of task elements |

engineering. For example, an air traffic controller has functional tasks that include using computer-based support to track aircraft in flight and on the ground, making decisions about control options and communicating directly with the aircraft personnel. Each of these broad categories of functional tasks must be considered in the development of the computer system that supports the controller.

The early problem definition models provide an initial framework for task definition. During the requirements modeling phase, tasks are iteratively defined using a combination of top-down and bottom-up analysis methods. Andriole (1986) and Ehrhart (1993) describe a variety of task analysis methods useful for investigating both the functional and interface operation tasks in decision-aiding systems. There are a number of other resources that describe techniques for capturing and modeling the cognitive aspects of decision making (Sage, 1992; Andriole and Adelman, 1995; Klein, 1998; Senge and Sterman, 1994; Zachary, 1988). Task profiling and requirements identification activities focus on four areas:

1. Identification and modeling of the sequencing and dynamics of the tasks.
2. Identification and characterization of decision-critical information regarding the situation elements external to the system (support systems, physical environment, threats, etc.).
3. Identification of the ways that users interact with all of this information to explore situations, develop hypotheses, generate options, make choices, and implement their decisions.
4. Identification of the information presentation and interaction requirements of the alternative analytical methods proposed to support tasks and decision processes.

### 10.4.7 Functional Tasks in HSI

The general characteristics of the functional tasks involved in HSI are very important as they lead to identification of the cognitive characteristics of decision-making tasks. One of the principal goals of the task analysis models is to identify and characterize the key variables in the task inputs, outputs, and feedback that define the tasks and affect task performance. The characteristics of these variables and their interrelationship have implications for task allocation, flow of control, information presentation and interaction design, as well as hardware, software, and communications requirements. As the tasks and their associated variables are identified, the individual variables must be characterized vis-á-vis these various dimensions and related in order to model the dependencies, information flows, etc. The relationships defined then provide building blocks for design of information presentation and interaction design.

***Functional Task Characteristics*** The HSI design efforts often begin with information from organizational job descriptions or the functional role models developed in the organizational profile. In addition to the individual task parameters discussed previously, the designer must develop a profile of the overall shape and flow of various tasks. This profile considers such principal human functions as discrimination, communication, and interpretation that are required in order to complete the tasks. In addition, the combination, or cumulative effect, of tasks is examined in terms of such factors as complexity, loading, pacing, and criticality. The overall complexity of the human users' tasks is a function of the number of interdependent factors or subtasks involved in the overall task. The level of complexity is highly correlated with task difficulty (Meister, 1991). Also related to complexity, the overall task load describes the demands placed on users by such factors as the number of concurrent tasks, interactions, and their sequencing. Meister distinguished task "load" from task "stress" by the absence of an element of fear or anxiety. Finally, both the pacing and criticality of task performance must be understood to assess their impacts on timing, accuracy and precision, prioritization, and attention requirements.

In addition to the overall profile of the functional tasks, the HSI design team must also discover and model the relationships between the elemental aspects of tasks, such as variables, constants, actions, and processes. A simple system model interrelating task elements allows the designer to categorize each element in terms of whether it is:

1. An input to the task
2. A response activity
3. A process output from the task
4. Feedback on action(s) that have been taken

This broad categorization helps to identify the characteristics of elements that are relevant to the task flow. To be especially useful, these elements must be defined further.

***Input Characteristics*** The task input characteristics incorporate the concepts of triggering events (stimuli) and task information that the user must sense, perceive, attend to, and interpret in order to generate a response. For instance, the task stimuli or input information may vary over time in a predictable or random fashion. This variation can affect not only stimulus detection but also the user's ability to recognize and identify the stimulus. Stimuli with numerous patterns of variation task users' long-term memory and create additional cognitive workload as they attempt to match features against remembered patterns. The duration of the stimulus relative to the task time and other tasks occurring simultaneously has ramifications for the user's attention and short-term memory resources. When the stimulus occurs only briefly or changes while occurring, it may be necessary to store and redisplay stimuli for examination. When users can neither control nor predict the occurrence of stimuli, they may fail to detect occurrence or recognize significance. Moreover, where task-relevant stimuli are mixed with such irrelevant stimuli as a "noisy" environment, the user may fail to detect the relevant stimuli or mistake irrelevant stimuli as relevant and thereby experience a "false sensation." In addition to the added workload, an abundance of irrelevant stimuli can create confusion and seriously degrade performance (Meister, 1991). This provides strong motivation for proper provision of cognitive support for decision making (Lerch and Harter, 2001).

***Response Characteristics*** The requirements and characteristics of the user's response are closely related to the output characteristics. For example, task allocation strategies and feedback design depend upon how often the user must respond and what the response frequency and precision must be. The difficulty of attaining these response goals becomes a function of the number of component elements incorporated in the task output unit and the output workload. Very low levels for goal attainment difficulty may affect the user's attention and interest (Meister, 1991). In contrast, very high levels of difficulty may indicate tasks out of the range of human performance. These factors also have emotional consequences in terms of motivation, frustration, and stress. The cognitive demands associated with the content of the decision-making task are discussed in the section on decision task characteristics.

Once the broad tasks are identified, the designer must look at the subtasks or procedures that comprise those tasks. For example, the number and interdependency of the procedural steps required in the response to produce one task output unit also impacts task complexity. The precision required in responding has implications for both the information presentation precision and the means by which the user formulates the response. Tasks and subtasks that must be performed more or less simultaneously create extra demands on attention and cognitive resources (Wickens and Hollands, 1999). This issue must be addressed in task allocation strategies if these are to be appropriate. Finally, in addition to task precision parameters, the designer needs to take into consideration how closely the user must adhere to prescribed procedures. Tasks requiring absolute adherence to a strict procedure may be candidates for automation. At the very least, the sequencing of valid actions will have to be controlled in the HCI design through the use of constraints and affordances (Norman, 1986).

***Output Characteristics*** From the system's perspective, the functional tasks are the outputs of a human–system cooperative response process. For this reason, the identification and analysis often begins with desired outputs. One of the first issues to be resolved is what constitutes an output unit. An output unit may be a single task, such as the assignment of a single entity to a service unit, or a composite task composed of a number of elements or component tasks, such as would result from planning a series of activities for multiple actors. Task volume, or throughput, is measured in terms of the number of output units produced during a period of time. In some cases, the duration of the output unit is also an issue. For example, an operator may have to maintain some signal or machine state for a set period of time or until an appropriate feedback signal is received. In this case, the workload associated with task output is a function of the number of task output units produced during a set time period and the duration the output is maintained.

The HSI designer must be concerned with several issues brought about by the task output characteristics. For example, the output volume required has implications for human attention, workload, and short-term memory capacity that must be considered in human–computer task allocation decisions (Ehrhart, 1990; Gardiner and Christie, 1989; Huey and Wickens, 1993). The number and format of elements composing the output task unit have implications for the level of detail that must be addressed, manipulated, and then sent as output. The duration over which the task output unit must be maintained also impacts attention, memory, and workload by limiting resources available to respond to incoming tasks; therefore, it must be considered in task allocation schemes. Finally, the level of workload associated with output requirements affects not only the task allocation design but also impacts the cognitive resources required to maintain the level of vigilant performance required.

***Feedback Characteristics*** Feedback during task performance informs the user on the appropriateness and efficacy of the response. In continuous tasks, feedback becomes part of the input for the next response cycle. Feedback on task performance may be characterized in terms of pacing factors such as feedback lag and the ratio of reaction time to feedback lag. When there is no feedback or feedback is greatly delayed, task performance may be impaired (Rasmussen, 1986). In addition, the absence of usable feedback impedes experiential learning (Gardiner and Christie, 1989). Delayed feedback is often misinterpreted or incorrectly associated with the wrong response causing the user to construct invalid causal models of the task and domain (Brehmer, 1987; Reason, 1990). When the user's reaction time must be faster than the feedback returned, the delay in feedback may lead to overcorrection in the mistaken belief that the response had no effect. Feedback is also important with respect to the number of subtasks involved in making choices based on feedback on the outcome of the previous response. When feedback is variable in quality or delayed, the effects propagate through a network of dependent choices making the reliability of task performance unpredictable.

Table 10.6 summarizes the impacts of the functional task characteristics. The HSI team can use information developed to raise issues regarding task allocation between user and system, team interaction designs, staffing cycles, training designs, and knowledge and experience required for effective team and individual performance.

**TABLE 10.6 Impacts of Functional Task Characteristics**

| Characteristics | Elemental Task Features | Potential HSI Issues |
|---|---|---|
| Input | • Input variability<br>• Input duration<br>• Occurrence regularity<br>• User's control of input | • Stimulus detection and identification; long-term memory<br>• Impacts on attention and short-term memory requirements<br>• Stimulus detection; attention<br>• Stimulus detection and identification; work-load and frustration |
| Response | • Goal attainment difficulty<br>• Response precision<br>• Response frequency<br>• Number simultaneous subtasks<br>• Number and interdependency of procedural steps<br>• Degree of procedural<br>• Degree of procedural adherence required | • Task complexity factor, impacts on user frustration and motivation levels<br>• Impacts information display precision and response input mechanisms<br>• Demand on attentional resources; response input mechanisms; task allocation strategies<br>• Demand on attentional focus; response input mechanisms; response feedback design<br>• Task complexity; impacts on short-term memory<br>• Impacts on level of autonomous control extended to user; attention requirements |
| Output | • Number output units<br>• Number elements/output unit<br>• Duration output unit maintained<br>• Output workload | • Task allocation; short-term memory; attention<br>• Identification of appropriate level of infor-mation detail<br>• Impacts on attentional or short-term memory resources; task allocation design<br>• Impacts of extended vigilance on attentional or short-term memory resources |
| Feedback | • User's control of response lag<br>• Feedback lag<br>• Reaction time/feedback lag ratio<br>• Number of choice subtasks | • Task allocation strategies<br>• Attention; impacts on short- and long-term memory<br>• Impact of feedback on performance quality<br>• Impact of feedback on performance quality; short-term memory |

### 10.4.8 Cognitive Decision Task Characteristics and Error Sources

Many authors have identified cognitive tasks in decision making. Although specific terminology varies across authors, these cognitive tasks are commonly described in terms of the following four generic activities:

1. *Information processing*—to collect and organize decision information
2. *Inferencing*—to interpret information for situation assessment
3. *Judgment*—to identify a suitable response
4. *Mental simulation*—to plan the execution of the chosen response

Each activity has further cognitive implications in terms of attention and memory demand or workload and such potential errors as biased interpretation or inappropriate use of heuristics. For example, overloading human attentional and memory resources impacts situational awareness, triggers accuracy/effort trade-offs, and influences judgment and choice strategies (Andriole and Adelman, 1995; Janis, 1989; Payne et al., 1993; Reason, 1990; Svenson and Maule, 1993). Decision task profiling helps to identify aspects of task and task sequence that must be supported in the design of information presentation and interaction routines.

As noted, decision tasks may be characterized and modeled using a variety of methods (Andriole and Adelman, 1995; Sage, 1992; Ehrhart, 1993; Fleishman and Quaintance, 1984). One of the most commonly used general models for decision making in complex, dynamic situations is Wohl's (1981) stimulus–hypothesis–option–response (SHOR) model that was initially proposed for tactical air combat decision making. The SHOR model's four generic elements, representing the aspects of the decision cycle, are subdivided into the cognitive functions or activities involved in each:

1. *Stimulus*—the detection/recall, manipulation, display, and storage of the decision data (i.e., situational context and variable inputs)
2. *Hypothesis*—the creation, evaluation, and selection of alternative perceptions or interpretations of the stimulus
3. *Option*—the creation, evaluation, and selection of feasible response alternatives to the hypotheses
4. *Response*—the planning, organization, and execution of the selected response option

The SHOR model provides a useful framework for identifying the characteristics, potential sources of error, and support requirements associated with decision tasks. Using the task characteristics identified in Table 10.7, the designer identifies potential decision task-related design issues, such as task allocation, information presentation, decision/task aiding, training, and staffing. We discuss each of these four characteristics further since they are central to the implementation of the SHOR model.

***Decision Stimulus*** The decision stimuli constitute the primary inputs into hypothesis generation and evaluation for situation assessment efforts. The stimulus phase of decision making is concerned with initial data gathering and processing. Performance during this phase is determined by the quality of monitoring, focus of attention, and such processing

**TABLE 10.7 Impacts of Decision Task Characteristics**

| Characteristics | Elemental Task Features | Potential HSI Issues |
|---|---|---|
| Stimulus | • Vigilance level required | • Impacts on attention; short-term memory; task allocation and workload; staffing cycles |
| | • Stimulus detection difficulty | • Stimulus alerts and display |
| | • Stimuli and decision cue detail | • Information display and interpretation; error characteristics |
| | • Qualitative vs. quantitative characteristics | • Information display and interpretation |
| | • Rate and volume of incoming information | • Impacts on short-term memory; dynamic task allocation |
| | • Reliability and representative-ness of stimulus variables | • Problem perception; judgment and reasoning errors; training and experience required |
| Hypothesis | • Situation novelty | • Problem perception; long-term memory, judgment and reasoning errors; training and experience required |
| | • Number of possible hypotheses | • Problem perception; cognitive work-load; judgment and reasoning errors; training and experience required |
| | • Decision horizon (time allowed) | • Attention, memory, workload, judgment and reasoning errors; aiding require-ments |
| | • Inferencing required | • Cognitive workload, reasoning errors; training and aiding required |
| Option | • Number of possible options | • Attentional focus; memory; information processing; judgment and reasoning errors |
| | • Option evaluation tractability | • Memory and information processing workloads; judgment and reasoning errors; aiding design; training and experience required |
| | • Potential for goal shifts or conflict | • Memory load; feedback required; judgment and reasoning errors; templating for rapid recognition of new goal requirements; training for adaptive response |
| | • Option value assessment difficulty | • Aiding for option understanding and comparisons; training and experience required |
| | • Outcome uncertainty | • Feedback information requirements; inferencing requirements; aiding required |
| Response | • Planning required | • Memory; reasoning; decision horizon; training and aiding requirements |
| | • Coordination required | • Memory; organizational structure; decision horizon; staffing and training requirements; communication design |
| | • Execution control required | • Memory; organizational structure; decision horizon; staffing and training requirements; monitoring aids |

activities that bring meaning to data gathered as filtering, aggregation, and correlation. In addition, performance depends upon memory of the evolving context, previous experiences, and training to identify relevance and code stimuli.

In addition to the pacing and volume characteristics of the inputs discussed in the previous section, the data inputs to the decision task must be examined in terms of their impacts on attention, memory, cognitive workload, and information processing. Situational awareness requires varying levels of vigilance depending upon the dynamics of the environment. Therefore, the attentional requirements associated with a decision task may require little active monitoring, monitoring at intervals, or continuous monitoring of the situation. The low monitoring requirements of typically stable or very slowly changing situations may result in poor situational awareness when the stimulus event occurs. When continuous monitoring is required, human fatigue can result in loss of attentional focus. Monitoring at set or random intervals incurs additional cognitive workload as the user may be required to maintain a working memory of the sequence of signals or events monitored in order to create an accurate mental model of the evolving situation. Monitoring at intervals is often involved in divided attention tasks and may require a rapid mental reorientation each time attention is refocused (Wickens, 1987). Attention is also related to the degree of difficulty in detecting the stimuli. Stimuli that are very difficult to detect, either due to inherent characteristics or the presence of other stimuli, may not attract attention during monitoring. In these cases, stimuli may require machine monitoring for detection or enhancement to facilitate perception or focus attention.

Over and above the cognitive resources demanded by the attention requirements, the pacing and volume of incoming decision data place demands upon the user's short-term memory. The designer must evaluate these impacts in terms of whether the typical memory demands exceed the capability of proposed users. At the lowest levels, the pace and volume of incoming information are manageable by the average trained user. As the demands increase, only highly motivated experts can manage the flow of information. The expert uses domain and task knowledge to cluster information in meaningful "chunks" rather than as discrete elements (Badre, 1982). At the highest levels, the volume of information overloads human ability to absorb and manipulate. At this point, machine monitoring and preprocessing is required to aggregate information into more manageable forms.

One of the key issues the design team must examine is the appropriate level of abstraction, or the proper level of detail, that is required in information presentation in order to permit the user to effectively interpret the decision data. Rasmussen and his colleagues (1986, 1994) categorize three levels of abstraction for decision inputs: signals, signs, and symbols. Signals are sensed information directly representing time–space data about the environment. Signs are indirect representations of the state of the environment derived from the pattern of physical signals. Signs serve to trigger learned behaviors or rules for response. Symbols are conceptual, rather than physical, structures that represent functional properties and relationships. Signs, or indicators, carry with them a context that triggers not only interpretation but also expectation. When the situational context differs from the learned context, as in novel situations, it may not be possible to correctly interpret the available information as signs. Symbols represent the more abstract conceptualization of domain relationships necessary in causal reasoning to interpret unfamiliar situations. Forcing users to work with information at the wrong level of abstraction can either overburden them with unmanageable detail or provide them with insufficient information to adequately assess the situation.

Conceptual foundations have been developed recently for human–machine interface design, based primarily on supporting these three cognitive levels. In general, humans use skill-based knowledge, rule-based knowledge, and formal-reasoning-based knowledge in an attempt to keep processing effort at the lowest cognitive level that trustworthy performance of the task requires. The ecological interface design construct attempts to minimize the difficulty of controlling a complex system while, at the same time, supporting the entire range of activities that specific users may require. Vicente and Rasmussen (1992) suggest that the usual approach to interface design, which is generally based on a direct manipulation interface (DMI), fails to consider that:

1. Practical problem solving can take place at various levels of abstraction in a hierarchical problem domain representation.
2. The same interface can be interpreted in different ways.
3. The way in which information is interpreted triggers qualitatively different modes of information processing, each requiring a different type of computer support.

Vicente and Rasmussen (1992) indicate that human errors may be related to: problems of learning and adaptation, interference among competing control structures, lack of resources to avoid error, and entrenched human variability. These four categories of human error are impacted somewhat differently as a function of whether skill-based, rule-based, or formal-reasoning-based approaches to performance are used. The ecological interface construct suggests that an interface design must take these factors into account if it is to be a viable aid that supports human interaction. Ecological interfaces are related to direct manipulation interfaces, direct perception interfaces, object displays, and graphics-based displays. To ensure that these requirements are met satisfactorily, interface designers must be concerned with the best way of describing or representing the complexity of the domain and the best way of communicating this information to the system operator. These concerns translate directly into more specific questions—one relating to the problem side of design and one relating to user resources for the designer of systems. They may be stated as follows: Problem side—What is the best way of presenting the complexity inherent in the problem or issue at hand? User side—What is the most effective resource that the operator has for coping with complexity and how can this be best utilized?

The reliability and representativeness of the input information affects the extent to which the variables may be understood and correctly interpreted. Moreover, when information is incomplete or ambiguous, users may focus on irrelevant information and inappropriate causal explanations (Reason, 1990). Users may be unaware that critical information is missing and need reminders or models that call attention to missing, imprecise, or ambiguous values in relevant stimuli. Strategies for analytical support and information presentation require an understanding of which data elements may vary in information reliability and how potential variation may affect interpretation.

***Hypothesis Formation*** During hypothesis formation, the user seeks to bring an order to the information collected by creating, evaluating, and selecting a causal explanation or assessment of the possible situation that would account for the collected data. Several factors characterize the decision tasks during the hypothesis phase. First, the degree to which decisions are made in familiar or unfamiliar conditions affects the reasoning that must be supported and extent to which functions may be automated. For example, routine situations may be handled with procedural reasoning or automated to reduce workload. In contrast, decision making in highly uncertain environments requires support for interpreting unfamiliar situations. In complex, dynamic environments, human decision-making errors often stem from failure to consider processes across time, such that evolving and emerging trends are neglected and form a tendency toward thinking in causal tree representations rather than causal network representations.

The decision tasks should also be characterized in terms of the number of feasible hypotheses that commonly may be generated to explain the available information. In well-bounded domains with few possible hypothesis alternatives, situation assessment is usually performed with rule-based, procedural reasoning. Errors in hypothesis evaluation in such instances result from selecting an inappropriate evaluation rule or a flawed evaluation rule (Reason, 1990). In situations where the number of feasible explanations for stimuli may be large, users may use cognitive shortcuts to rapidly reduce complex hypothetical relationships into loosely integrated general hypotheses. In such cases, the hypotheses may never be adequately integrated for evaluation purposes, and the evaluation will be consequently flawed.

Another dimensional characteristic of the hypothesis phase tasks that must be identified is the time allowed for hypothesis generation, evaluation, and selection. Planning and forecasting tasks have longer decision horizons and do not require rapid hypothesis evaluation; however, the delays in feedback can affect the quality of the causal models used to interpret decision inputs. The shortened decision horizon in time-critical tasks increases the effects of user experience, attention, and workload. The more robust mental models developed with experience increase the user's ability to focus attention on relevant information, reducing workload to evaluate complex stimuli in shorter periods of time (Shanteau, 1992; Rouse et al., 1992, 1993). Real-time decision making may require almost instantaneous situation assessment. In addition to experience level and attention focus, decision performance may depend upon vigilance levels maintained and the speed of feedback (Edland and Svenson, 1993; Janis and Mann, 1997; Janis, 1989).

The stress associated with shorter decision horizons results in general narrowing of perceptual focus ("tunnel vision") or issue fixation, rendering decision makers less capable of dealing with multiple stimuli/issues (Helmreich, 1988; Janis, 1989; Orasanu and Salas, 1993). This tends to result in a decrease in the number of information sources used in situation assessment and the number of alternative courses of action considered. In addition, there is often a failure to critique the microdecisions that aggregate to a larger, central decision. The frequency of action or decisions increases as users feel "impelled" to action.

The nature and amount of inference that is required to interpret situational data impacts the quality of hypothesis evaluation. Situational and presentation contexts affect not only the detection of stimuli but also their cognitive interpretation. In cognitive tasks, the context in which stimuli occur appears to have greater significance than its physical attributes. For example, Lockhead (1992) found context and sequence were the primary factors affecting similarity judgments in recognition and categorization tasks. In other research, Edgell et al. (1992) discovered a context effect in the perception of cue salience for probability judgments. The sequence, or presentation order, of decision stimuli has also been found to affect their interpretation in expert situation assessment tasks (Adelman et al., 1993). In a series of experimental studies, researchers found that experts constructed different causal explanations for event sequences depending upon presentation order. The explanations provided indicated that the significance experts attached to a particular decision cue differed based upon its sequential context.

The human ability to perceive and interpret information based upon context is an essential strength in situation assessment. When decisions must be made in high-threat, dynamic environments, contextual interpretation permits the user to make accurate assessments intuitively and respond rapidly. Context, however, has also been a factor in misinterpretation and disastrous decisions. For example, the erroneous shooting of the Iranian Airbus in 1988 by the USS *Vincennes* was, in part, due to the context under which the available information was interpreted (Helmreich, 1988; Klein, 1998). Similarly, Pentagon investigations revealed that the April 1994 shooting of two U.S. Army UH-60 Black Hawk helicopters by U.S. Air Force F-15C fighters occurred when the fighter pilots misidentified the helicopters as Russian-made Hind helicopters flown by the Iraqis. Expectation may have been a contributing factor in the misidentification. The fighter pilots had not been briefed that allied helicopters would be in the area (Harris, 1994). Other cues, such as negative identification friend or foe (IFF) response and Airborne Warning and Control System (AWACS) communication, increased the expectation that the helicopters were either unknown or hostile and may have influenced visual identification.

***Option Generation, Evaluation, and Selection*** The objective of the option generation, evaluation, and selection phase is to seek a feasible response to the hypothesized situation. Several characteristics of tasks during option generation, evaluation, and selection bear examination during task modeling. Many of the same factors affecting hypothesis generation and evaluation, such as situational context, boundedness, and tractability, also influence the performance of the option phase tasks. The number of potential responses to a situation affects the boundedness of option evaluation. Furthermore, when there are many feasible options to a situation, users may shift from option to option without sufficient evaluation or attempt to oversimplify (Janis and Mann, 1977; Dörner, 1987; Janis, 1989). Information volume and problem boundedness also affect tractability and may cause the workload in the option evaluation task to exceed human manipulation abilities. The goal variability inherent in the environment impacts option evaluation based on the rapidity and predictability of the variation and resulting option conflicts. In multistage, evolving decisions, a change in goals may supersede previous subchoices. Such shifts require rapid reprioritization and reevaluation of current options against higher level goals (Klein, 1993b). Feedback timeliness also becomes more critical as goals shift rapidly.

The difficulty of option evaluation tasks is judged by the extent to which outcome values are well understood and easy to determine. In bounded and semibounded domains with well-understood outcome values, users may employ rule-based evaluation. Higher levels of evaluation difficulty become less tractable for unaided evaluation. At this point, the decision making may be unacceptably delayed as users wrestle with the possible consequences of possible courses of action. Inference is required where outcome values are uncertain. In complex environments, the network of uncertainties rapidly becomes intractable for human evaluation, leading users to simplify with insupportable inference leaps (Dörner, 1987; Hogarth, 1987). Users may also avoid committing to any option, often waiting to see if changing events force or suggest a choice (Janis and Mann, 1977; Janis, 1989).

***Response Planning and Execution*** The response planning and execution phase involves planning, coordination, and execution control as required to carry out the course of action option that has been selected. Plans are essentially hypotheses based on a network of causal assumptions about the sequence of steps that will bring about the desired goal. Simple responses based on experiential familiarity involve little or no planning. Skill-based control evokes reactive responses based on experiences with similar situations; rule-based control triggers procedural plans; and formal reasoning based control will usually require explicit use of analytic procedures. Moderate levels of planning feature manageable levels of effort using ad hoc or prepackaged plans. Complex responses usually require extensive planning or replanning involving the reevaluation of goals and adjustment of control structures.

Reason (1990) categorizes cognition-based plan failures that result from properly implemented action plans not accomplishing the intent that led to their implementation as mistakes,[2] that is, errors of intention. Reason suggests three basic sources of these planning failures, including: errors in the working database, such as stimulus phase errors; errors in mental operations, such as hypothesis and option phase errors; errors in the properties of the schema or a misguided act of poor planning. Reason traces these errors to characteristics of the human planner based upon limits of attention and memory and a powerful urge to accept seemingly rational explanations that bring order to complex, chaotic situations. The response coordination requirements are determined as a function of the size, complexity, and dispersion of the network of the agents that must be coordinated. These elements depend upon the organizational factors discussed previously and the time available for a response. Coordination tasks are communication intensive. The effectiveness of coordination is dependent upon experience, training, shared task and situational models, and flow of communication.

Execution control is defined in terms of the number and interdependency of the actions required in the planned response. As such, control is closely related to coordination. Multiphased, interdependent responses increase the coordination effort required to track the status of the evolving response. Moreover, the network of dependencies increases the difficulty of tracking all the possible consequences or "ripple effects" of actions taken. If feedback is delayed, it may be associated with the wrong phase and result in confusion and overcorrection (Meister, 1991). Finally additional cognitive resources, requiring attention and memory, are demanded to handle the wider range of control and potential goal shifting in multiphase responses.

### 10.4.9 Relating Cognitive Task Characteristics to Task Models

Investigating the situational, organizational, user, and task dimensions helps to identify the specific aspects of the decision tasks that should be considered in the design of the decision information presentation and interaction routines. The user's cognitive tasks emerge as part of describing the sequence of steps involved in performing a task or procedure. As discussed in the previous section, the tasks in decision making involve such generic cognitive functions as information processing, inference formation, judgment, and mental simulation. The situational context, organizational structure and culture, the user's experience and training, and the inherent features of the task influence each of these functions.

As the task models are developed, the designer can begin to explore the cognitive requirements involved in successful task performance. As the design team refines models of the problem domain and tasks, the issues raised may be compiled for later distillation and structuring. Figure 10.7 is based on the example TDO decision tasks model presented earlier and summarized in Figure 10.4. It indicates some cognitive support issues that might surface during requirements identification and modeling. For example, if identification of the status updates to current resources reveals a variation in the timeliness and reliability of the data, this fact must be considered in presentation of that information to the user. The reliability will also be a consideration in determining the analytical method used to track and compare the change in resources. Additionally, since current and projected status of refueling assets, missions, and available fuel are also multidimensional constructs involving time, location, and capacity/range, the combination of those dimensions must be presented in a form that is meaningful to the user and representative of the underlying relationship. The situational context involving operations of a routine, exercise, or crisis nature is also a factor in the decision to reassign, reroute, or cancel refueling missions. Where constraint-based planning tools are employed, the user will generally need support for modifying the constraints to meet operational circumstances. This is the case, since projecting changes in the complex network of tanker and fuel-receiving missions exceeds human short-term memory capabilities and requires aiding to trace the ripple effects of change across the schedule. The task models also provide means for projecting possible errors related to human performance (Fields et al., 1995; Reason, 1990).

An understanding of the TDO's tasks also suggests requirements for training and staffing. For example, the proposed operational process the new system must support incorporated assumptions about the TDO's knowledge and experience both in the air-

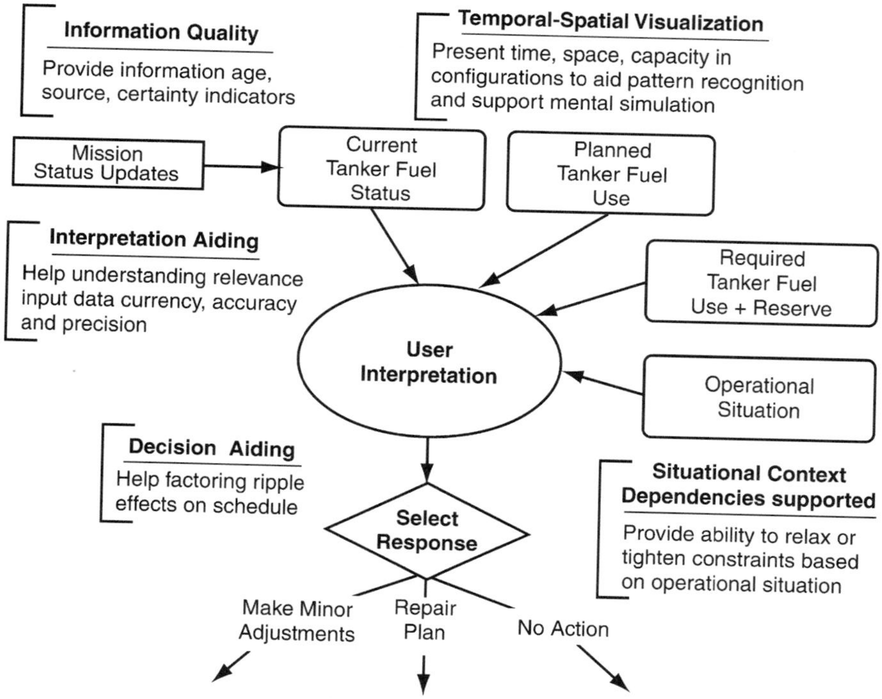


**Figure 10.7** Potential impacts of user/task/context on task aiding.

refueling domain and in performing the execution control tasks. These assumptions must be feasible with respect to both personnel projections and training regime. The user will need training both on the new task processes and roles and the capabilities and limitations of the aiding technology. Figure 10.8 indicates possible training and staffing impacts suggested by the task model.

The models represented in Figures 10.7 and 10.8 provide the means to characterize various aspects of the decision domain and tasks in terms of readily observable, broad criteria. Location of the domain and tasks within certain parameters suggests possible sources of cognitive demand and user error. These potential problems are evaluated in terms of system support and expressed as cognitive task requirements. Figure 10.9 presents an example of how the issues raised while analyzing the decision task requirements suggest cognitive task requirements (CTRs).

Contemporary system design teams have a range of software tools that facilitate the creation of rich representations of task requirements. For example, modeling software that permits hypermedia links provides the means for "annotating" the basic task models with such additional models as one based on situational context, text-based descriptions, audio clips from interviews, and even field video of task performance under realistic conditions (Ehrhart and Aiken, 1990). The process of building and reviewing these models helps to identify the cognitive characteristics of each task. These cognitive characteristics, in turn, raise performance and HSI issues that should be included in the requirements documentation to assure their inclusion in the design and implementation of the system.

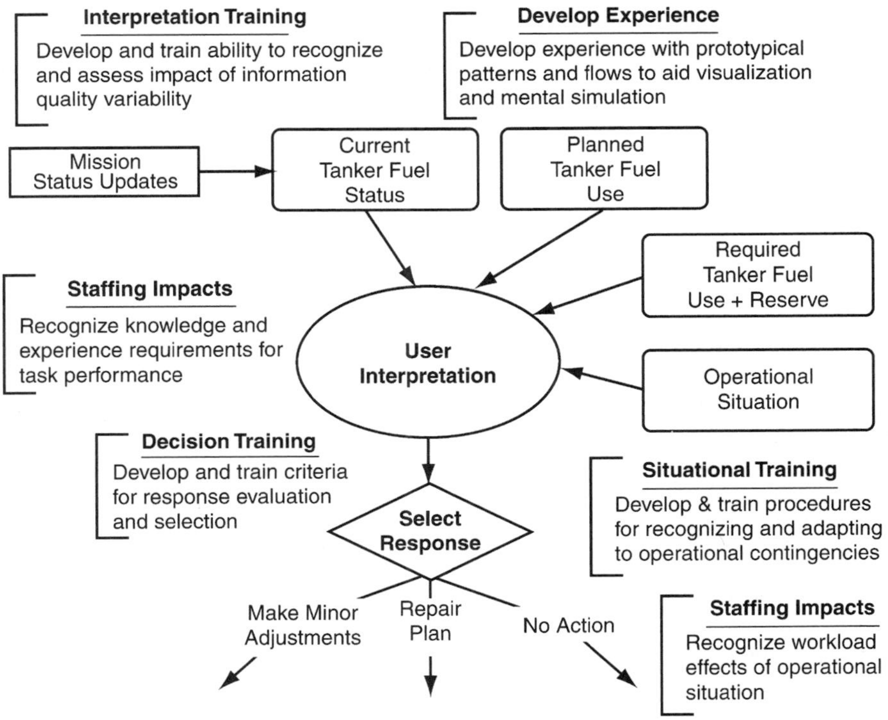


**Figure 10.8** Potential impacts of user/task/context on training and staffing.

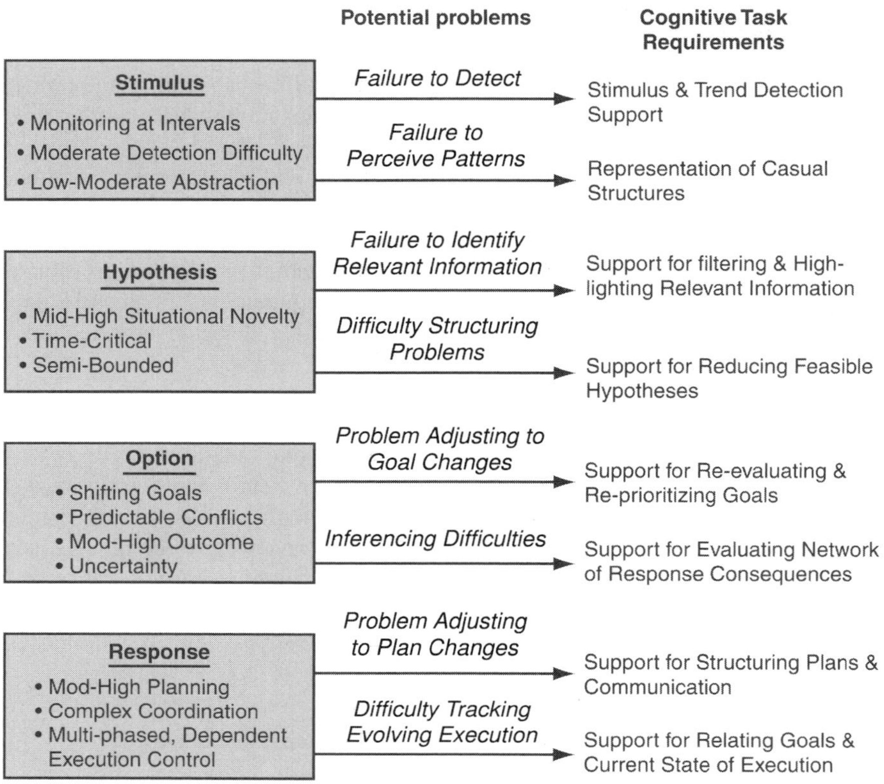


**Figure 10.9** Mapping decision task characteristics to cognitive task requirements.

## 10.5 SYSTEM CONCEPTUAL AND ARCHITECTURAL DESIGN

During the conceptual and architectural design phase, the systems engineering team begins to interpret the cognitive task requirements in terms of cognitive systems engineering design principles, such as those presented in Gardiner and Christie (1987) and Rasmussen et al. (1994). This allows interpretation of the cognitive demands that characterize certain tasks and task situations in terms of the impacts on information presentation and human–computer interaction. When coupled with basic human factors guidelines for system design, the cognitive systems engineering design principles help to identify technological solutions that support the cognitive task requirements and which also conform to the identified hardware and software requirements and specifications. For example, selectively focusing attention is a coping strategy invoked when the user is overwhelmed by large amounts of information. This information processing strategy may be associated with such biases as fixation on one problem element or overemphasis of cues that support the current hypothesis. The design principles that address "selective attention" provide reminders of the "larger world" to avoid tunnel vision and means for directing the user's focus to the most relevant information. The design goals for implementation of these principles include:

- Providing an overview or "establishing shot" to expand the decision-makers perspective
- Exploiting common representational analogies, such as maps and models to highlight the relationships between domain factors

As new requirements and related design "goals" are identified and understood, they can be integrated into the developing system concept. Rather than occurring in a rigid sequence, this process continues iteratively as requirements surface and prototype concepts are proposed. In this fashion, the prototype design evolves as the incarnation of the designers' hypotheses regarding the decision-making activities and interaction requirements.

Figure 10.10 illustrates formulation of design goals based on informal requirements knowledge as embodied in situational, organizational, user, and task models; formal requirements specification; and standards and guidance literature concerning cognitive systems engineering principals and human factors. The resulting conceptual design and architecture concept is a configuration of features including the information presentation methods, interaction routines, and the hardware and software technologies that support them. Each feature must be traceable to the requirements and specifications documentation. The specific incarnation of the feature and its configuration in the design should be traceable to the higher level design goals, principle(s), and guideline(s) that defined or suggested it. This dual traceability ensures that the proposed design adequately meets requirements and helps the systems engineering design team make better use of the technology options available to them.

For purposes of generalization, the discussion of design goals presented here, as well as the principles and guidelines underlying them, is restricted to the higher level design goals

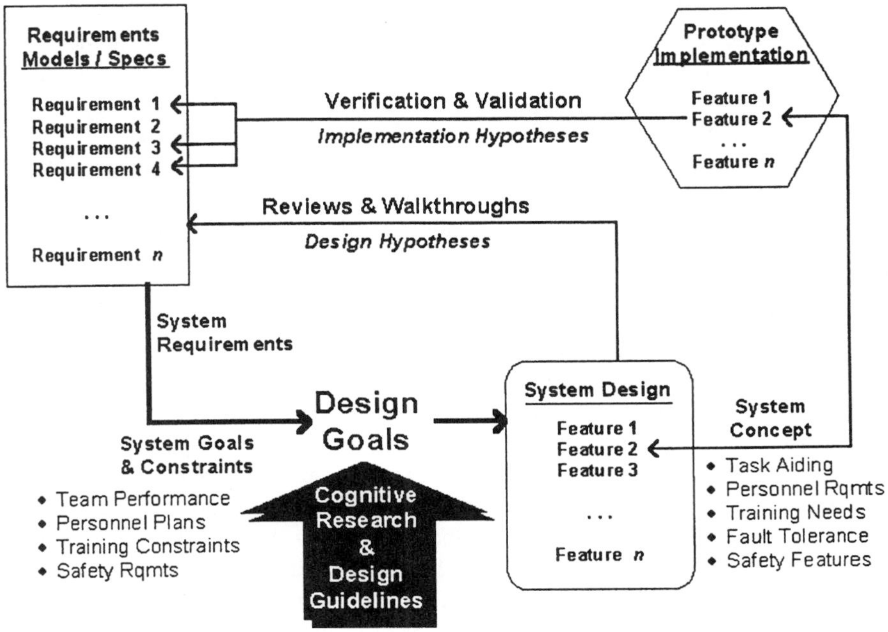


**Figure 10.10** Tracing user/task requirements.

as contrasted with detailed design considerations needed to implement a physical realization of the system. The design practitioner is directed to the human factors and decision support (Sage, 1991) literature for more detailed presentations. Among more recent efforts, the following references are particularly noteworthy:

- *Principles and Guidelines* Salvendy (1997), Wickens and Hollands (1999), Sheridan (2002), Hackos and Redish (1998), Gardiner and Christie (1987), Preece et al. (1994), Rasmussen et al. (1994), Shneiderman (1997), and Smith and Mosier (1986).
- *Empirical and Experimental Evaluations* Andriole and Adelman (1995), Brannick et al. (1997), Guzzo and Salas (1995), and Svenson and Maule (1993).
- *Theoretical Foundations* Card et al. (1983), Dreyfus and Dreyfus (1986), Janis (1989), Klein et al. (1993), Meister (1991), Norman and Draper (1986), Rasmussen and Vicente (1989), Sage (1991, 1992), and Senders and Moray (1991).

The remainder of this section surveys the cognitive systems engineering principles and design guidance that relates to the situational, organizational, user knowledge, and task characteristics developed in the profiles. Each category is discussed in terms of information requirements, support for potential performance errors, and possible systems engineering design goals.

### 10.5.1 Design Goals Associated with Situational/Environmental Context

Vicente and Rasmussen's (1992) ecological interface design (EID) model presents two environment-related design goals based on Rasmussen's (1986) model of cognitive control. First, the interface design should not force the user to use a higher level of cognitive control than required by the task. Empirical evidence suggests that the skill-based and rule-based levels of cognitive control produce the most efficient response, provided the user has correctly interpreted the situation through possession of sufficient experiential familiarity. In addition, there is evidence that users attempt to reduce task demand by relying on the cognitive shortcuts provided by the lower levels of control (Hammond, 2000; Rasmussen, 1993; Rastegary and Landy, 1993). Second, the interface should support all three levels of control: skill-based, rule-based, and formal-knowledge-based. This goal reflects the user's requirement to operate in the multiple environments that make up complex domains.

In determinate environments, the principal design goal is providing support for users to help them rapidly select an effective response to a relatively unchanging and predictable environment (Meister, 1991; Rasmussen, 1986). In such an environment, the limited highly structured set of cause-and-effect relationships permits response automation in situations when very rapid response is required. Users generally need detailed displays that present specific values for relevant parameters, such as altitude and air speed in aircraft. When these values must be considered together, the display should either integrate them or present them in sufficiently close proximity that the user can compare the readings almost simultaneously (Vicente and Rasmussen, 1992). Interactions should be designed to allow the user to act directly on the display to manipulate the time–space signals in an appropriate and timely manner.

In moderately stochastic environments, the user needs to understand the effects of variability in some parameters and the interaction of the parameters. In some cases, the display of some individual parameter values may be integrated into a single display for interpretation as signs rather than as signals. There is empirical evidence that indicates the use of "configural displays" improves performance by allowing users to extract critical data relationships from both the low-level parameter values and the high-level constraints (Bennett et al., 1993). Woods and Roth (1988) indicate that the strength in configural displays lies not only in the economy of representation but also in the emergence of certain domain features. It is important, however, that displays representing complex domains not reduce the complexity below the level of the fundamental parameters and their interdependencies. This is in keeping with Ashby's law of requisite variety (Ashby, 1956).

In severely stochastic and indeterminate environments, the design goals focus on providing the means to make most efficient use of resources in a succession of varying, short-term situations. Users must be able to rapidly develop creative, adaptive responses to effectively exploit opportunities and avoid disasters. This requirement suggests the need for displays that represent the causal relationships and make use of goal-relevant domain models. The representation of causal networks provides externalized mental models that relieve the user of the cognitively demanding tasks involved in comprehending the causal factors underlying a situation and the network of consequences associated with options (Rasmussen and Vicente, 1989). As such, these displays help to support the mental simulation required for the intuitive response patterns suggested in Klein's recognition primed decision-making model (Klein, 1993a). Table 10.8 summarizes the design goals related to the situational and environmental contexts that the human–computer cooperative decision system must operate.

### 10.5.2 HSI Design Goals Associated with Organizational Contexts

Response selection and coordination within an organizational context involves synchronizing multiple perspectives, synthesizing intraorganizational information, and recognizing relevant patterns in evolving situations in order to formulate an appropriate response. The design goals associated with the organizational context focus on: responding to interdependencies of organizational structure, facilitating communication, incorporating accepted doctrine, and supporting the shared mental models required for effective organizational response.

In organizations characterized by complex interdependent structures, the performance of one unit or subsystem affects the performance of the others. The extent of this effect

**TABLE 10.8 Design Goals Summary—Situational/Environmental Context**

- Support all three levels of cognitive control: skill-based, rule-based, and knowledge-based.
- Support skill-based control with displays and interaction methods that allow users to directly manipulate the signal-level parameters of the problem.
- Support rule-based control with displays that map structure and constraints of environment. Model structural relationships and make domain variables salient through design and highlighting.
- Support knowledge-based (or model-based) control with domain models that help to relate problem parameters to goals. Model causal relationships and make goal-relevant information salient through design and highlighting.

may range from enhancing or degrading functional performances to a tightly coupled relationship where one function cannot be performed if the other fails. In either case, the users responsible for the performance of a function within an interdependent structure must maintain some awareness of the organizational functions that support their functional responsibility, as well as the organizational functions that are affected by their decisions. Users must consider these causal factors in contexts where knowledge-based control is used to adapt to complex, dynamic environments. Depending upon the tasks supported and degree of interdependence within the organization, design goals for organizational structure may include models that relate the dependent network of supporting functions for diagnostic reasoning and situational awareness. In addition, causal models can provide reminders of the potential consequences of decisions for other organizational functions. Finally, models may present the flow of coordination and control involved in implementing decisions within the organization.

The shifts in organizational response that occur during crisis situations present challenges that may also require attention in conceptual system design. For example, if decision making is performed in a distributed environment, the user may have to cope with failure of communication links that provide updates to critical information. The design of information presentations must provide indications of the data elements affected. The interaction design may include methods for reorganizing the display of information that may be needed when there are changes in data reliability. The system design may also have to accommodate shifts in decision-making autonomy under crisis conditions. In these cases, the standard operating procedures and channels of authorization may be replaced by a set of high-level goals and constraints, such as military rules of engagement, to permit faster, semiautonomous responses. Based upon the information gathered in requirements, the information presentation design and interaction control should be adaptable to these conditions.

Wellens (1993) presents an information-processing model for multiperson and human–machine decision making in a distributed decision-making environment that addresses some of the problems of communication design. This model incorporates the concept of communication bandwidth, representing degree of richness in communication, associated with the modes of interaction and communication that are supported by the system. For example, video-conferencing should provide all the cognitive content of face-to-face discussion but should "filter" out some behavioral facets that may be counterproductive if retained. This "filtering" is not at all due to any function of the electronic medium; rather it is due to the participants' tendency to focus on rational presentation of factual information without additional emotional behaviors. Despite the intuitive appeal of increasing communication bandwidth, Wellen's experimental research with dynamic situational awareness in team decision making indicated that increases in information richness were not always associated with improved situational awareness. This result seems to be due largely to the time pressures and additional filtering required in an information-rich media.

The design goals for supporting communications in organizational contexts should lead to an understanding of who must share information, what information and knowledge must be shared, and how it must be communicated. Knowledge management and knowledge sharing are very important contemporary research issues (Von Krogh et al., 2000; Sage and Small, 2000), and this provides yet another important dimension to the complexity of HSI issues. Within this high-level construct, information interaction design concepts should strive to maintain an appropriate distance and directness in the communication between members of the team or organization. As such, the design should facilitate the integration of users who must cooperate and not interfere with their cooperative tasks.

Much of the strength in shared mental models appears to be task, training, and communication dependent. Rouse et al. (1992) state that the current empirical evidence is insufficient to form a coherent theory of team-based design. In fact, there seems to be some evidence that technology interferes with shared mental models. For example, Duffy (1993) cites the loss of "backchannel communication" as a potential negative effect of introducing technology in team processes. The communication that occurs in the background of the primary communication provides team members the opportunity to question, clarify, and confirm their understanding of the situation. This secondary communication is a critical part of avoiding errors due to miscommunication. As another illustration of this, the investigation of the Black Hawk helicopter shooting indicated that some of the members of the AWACS team knew before the shooting that the helicopters were U.S. Army Black Hawks, but the information did not get communicated to the pilots of the F-15Cs (Harris, 1994).

Group or team situational awareness involves sharing of common perspectives between two or more individuals regarding current environmental events, including their meaning and anticipated future (Wellens, 1993). The HSI designs to support shared mental models should incorporate not only the advantages of multiple perspectives but also the power of shared knowledge and training. This shared knowledge includes doctrinal concepts and common representations of both abstract and concrete organizational information (Kahan et al., 1989). Table 10.9 summarizes the design goals associated with the organizational context.

### 10.5.3 Design Goals Associated with User Profiles

The user profile characterizes predicted levels of knowledge, experience, and training that the users are likely to have with respect to three knowledge areas: the domain, the functional tasks, and the operation of the system. The effects of this knowledge generally conform to models of beginners with low amounts of experiential familiarity and expertise, competent practitioners with moderate levels of these, and experts with high-level knowledge and expertise. Individual system users will typically demonstrate a range of competency across the three knowledge areas. The three knowledge levels have a number of common features, regardless of the area of knowledge involved. As with cognitive control, the predicted knowledge levels of the prototypical user must be supported for each area. Each knowledge level is discussed below with design goals for each area of

**TABLE 10.9 Design Goals Summary—Organizational Context**

- Provide models of interdependencies in organization to aid user in assessing causes of situations and effects of choices.
- Provide means for users to adapt to shift in organizational response during crisis situations. Encourage consideration of organizational doctrine through use of goal- and constraint-based displays.
- Facilitate all necessary and useful communication between decision participants with information display and interaction concepts that support team interaction.
- Support sharing of team or unit mental models to foster effective task coordination.

knowledge. The HSI design goals identified here were synthesized from Dreyfus and Dreyfus (1986), Rasmussen (1986), Rasmussen et al. (1994), and Senders and Moray (1991).

At the lowest level of expertise, the user may not recognize critical cues regarding the situation, task, or system state. In addition, the user usually has only limited ability to reason about the cues provided. In novel situations this limitation may induce confusion and error. The beginner often lacks confidence and may be slower to respond and reluctant to commit to action. Finally, lower expertise is associated with a limited goal framework that increases the probability of errors of intent.

Where domain knowledge is low, users benefit from displays that are formatted as accepted domain models such as to present situational information in context and to map causal relationships. Constraints, supports, and reminders help to guide domain understanding and increase confidence in situation assessment in these low-knowledge situations. In addition, templates of prototypical domain constructs with relevant cues highlighted can assist the user in making comparisons and developing responses in novel situations.

Low task knowledge often results in an inability to handle shorter decision horizons and heavy information loads. Additional time may be lost reviewing irrelevant information or inappropriate options. As a result, the beginner has difficulty maintaining performance quality under increased task workload. Lower levels of task knowledge are characterized by limited response option generation and evaluation capabilities. Finally, the beginner has difficulty prioritizing tasks. Display and interaction supports for functional tasks are similar to those discussed for low domain knowledge. To support the beginner in developing task knowledge, the HSI design should allow the user to query the constraints and affordances that have been built into the task models. Automation strategies should be explored to relieve the beginning user from excessive cognitive workload. When feasible, adaptive "intelligent" decision aids may be appropriate to filter displays and propose options. Where this type of aiding is infeasible, organizational structures may provide the same kinds of error trapping, error flagging, and redundancy afforded in machine design.

Low system knowledge is addressed in most fundamental guidance for systems design (cf. Preece et al., 1994; Shneiderman, 1997; Wickens and Hollands, 1999). Several general guidelines apply to help reduce errors and foster system learning. First, the information presentation design should provide overview screens to help users develop a mental model of the system resources available and understand where they are in a process. Moreover, the human–machine communication should make the current state of the system implicit and the available options visible. The interaction design should include built-in constraints to prevent an unrecoverable error and alert the user to the nature of their error and current response options. Finally, Norman (1986) encourages designers to make use of natural or domain knowledge in the interaction symbology in order to allow the user to interact with the task in the most familiar terms.

Moderate levels of expertise lead to performance errors based on misinterpretation of cues due to limits of the user's domain, task, or system models. Alternatively, errors can occur when the user fixates on the most available models. Moderately experienced users have limited ability to resolve conflicts between multiple models. Finally, moderate expertise is characterized by a reliance on learned procedures and a limited ability to reason at higher levels of abstraction in unfamiliar situations.

Moderate domain knowledge may be supplemented with displays formatted as accepted domain models to present situational information in context and map causal relationships.

The associated interface and interaction design efforts should support construction of more robust mental models by providing the option to view deeper levels of explanation. Since users may fail to recognize the degree and impacts of uncertainty in situational cues, displays and interaction routines are required that make the sources and extent of domain uncertainty explicit.

Moderate knowledge of the functional task requires some of the same support described for lower knowledge levels. For example, the user's task knowledge may not be sufficiently robust to understand the effects of subtask uncertainty. Displays and interaction design should help the user to understand the source of uncertainty and explore the potential effects on task performance. Moderate levels of task knowledge also benefit from designs that make task constraints and affordances visible. In high information volume situations, the user may not have adequate schema to distinguish relevant information. The system design should provide goal or decision-oriented displays in order to focus attention on relevant information and to provide strong encouragement for error control.

Moderate system knowledge is characterized by response mode errors that are based on incorrect assumptions about the current system state. For this reason, system state, available options, and similar information should be visible or available on demand. It is also beneficial to minimize the use of similar interaction sequences that vary in effect given different operational modes. Moderate levels of system operation expertise may not provide sufficient procedural information to respond to unexpected system behavior. In addition, the competent user may become lost in complex, linked sequences of displays. Overview displays and interaction routines that help the user to trace recent steps help the user maintain orientation (Woods, 1984). Interaction and interface designs for moderate system operational knowledge should still facilitate error recovery with "undo" commands and similar recovery devices. Finally, these designs should feature multiple levels of help in order to allow the user to select an appropriate presentation depth for the information desired.

The highest levels of expertise are also associated with errors in the selection and interpretation of information and judgments regarding appropriate responses. Although users have expert levels of domain knowledge, they may exhibit inconsistencies in combining situational cues. In addition, the multiple models in their repertoire may compete, with selection triggered by availability rather than reasoned choice. Experts benefit from the option to use domain model displays or customize displays and interaction routines to match their mental models. As with the moderately experienced user, experts require designs that support the continued development of mental models and that provide the option to view deeper levels of explanation. Expert users may display overconfidence in their situational interpretation or response choice. Displays that make explicit the sources and extent of domain uncertainty continue to be useful at this level.

High task knowledge may also be associated with, and in fact plagued by, overconfidence. This stems in part from insensitivity to the potential for aggregated errors in subtasks and microdecisions performed in a multistage decision fashion and a failure to revise decisions as needed in light of new information. Due to these realities, experts will continue to benefit from being presented with constraint representations that allow for error control. They will also benefit from having optional supports and reminders available during situation assessment. These may be provided in goal-oriented displays or displays and interaction routines user-customized to match their mental models. These displays also help to promote understanding the causal network of contributing causes and consequences of action. Finally, the difficulties that expert users may have in adequately considering domain uncertainties may be reduced through use of displays that make the sources and extent of uncertainty in key variables explicit.

The users with high levels of expertise in system operation can still be confounded by illogical interaction and interface designs. In general, the rules for consistent design of information presentation and interaction routines discussed for the lower levels of expertise apply to the expert. Several additional considerations apply primarily to the higher levels of system operation knowledge. For example, expert system operators are usually very intolerant of being forced to use lengthy procedures to accomplish a simple task. Thus, appropriate designs should allow the user to tailor the interface to optimize for best performance. However, when users reach expert levels in system operation, their ability to bypass some operational sequences may result in unintended actions. For this reason, experts also benefit from the error tolerant design guidelines suggested for lower levels. Table 10.10 summarizes the design goals associated with the user's knowledge and experience in the domain, tasks, and system operation.

### 10.5.4 Design Goals Associated with the Task Requirements

It is very difficult to generalize about tasks outside of such very broad categories as planning and situation assessment. While broad categories provide a general framework for discussing common error sources and failure modes, the details of system design remain tied to the specifics of the actual task to be supported. Woods and Roth (1988) describe cognitive systems engineering as "problem-driven and tool-constrained." In this systems-oriented view, the requirements analysis process describes the cognitive tasks and performance context and then attempts to trace the causal factors associated with both satisfactory and unsatisfactory performance. The goal of this process is to raise issues for addressing these causal factors in both the system and the associated interaction and interface design. While there remains no adequate theoretical basis for a prescriptive approach to design, the empirical literature provides some insights into broad categories of causal factors relating to such issues as attention span, memory, and workload. Unfortunately, the response of specific designs to these factors are highly task and context dependent and, thus, often do not generalize to other tasks or contexts. A full explication of the possible system design responses is beyond the scope of this work. Instead, this section focuses on the identification of high-level design goals in the support of cognitive task performance and the information presentation and interaction solutions suggested by the empirical and experimental literatures in human factors, decision science, and cognitive psychology.

**TABLE 10.10 Design Goals Summary—User Knowledge and Experience**

- Provide support for predicted levels of user knowledge and experience with domain, functional tasks, and system operation.
- Provide less experienced users with reminders to support performance, constraints, and recovery routines to prevent serious errors and embedded models to promote learning.
- Provide moderately experienced users with goal- or decision-oriented displays to aid in reasoning with multiple models.
- Provide highly experienced users with ability to take shortcuts and adapt system to meet response goals.

Norman (1986) defines the means by which designers and users understand and interact with computer-based systems in terms of the construction and use of multiple models. The designer develops a conceptual model of the target system that accurately, consistently, and completely represents that system. Although Norman does not discuss user or system requirements, the designer's conceptual model is built upon a mental model of the domain and task requirements that is often imprecise, inconsistent, and incomplete. The interface design presents a system image intended to convey information about system operation. The user constructs a mental model of the system based upon interaction with the system and system image as represented in the interface. The user's mental model of the system may not match the designer's conceptual model. Note also that a user's mental model of the task domain is determined by training and experience and, thus, varies in accuracy, consistency, and completeness from one system user to the other.

This work highlights the points at which the translation of requirements to design breaks down. The completeness of a designer's conceptual model of the system is a function of his understanding of the system architecture, not the requirements of the task domain. Thus, the users' mental models of the system, constructed through interaction with the system and the interface (system image), may or may not compliment their model of the task domain. The "transparency" of the interaction, that is, the degree to which the users perceive themselves to be interacting directly with their tasks, is determined by the convergence of these various models. Mismatches in interface and interaction design reduce transparency such that the user is more occupied with the operation of the system than the performance of the task.

One of the fundamental strengths of human users is their ability to conceptualize or construct mental models of causal relationships. This ability lies at the heart of intuitive and analogical reasoning. The concept of a "mental model" appears with various definitions, taxonomic structures, and applications in the cognitive process literature. Johnson-Laird (1983, 1999) discusses mental models as analogical representations for deductive inference tasks. One of the principle contributions of this work was to emphasize the semantic aspects of thought. Gentner and Gentner (1983) propose a "structure-mapping" theory to explain the cognitive processing of analogies. Their research employed protocol analysis and experimental manipulations to demonstrate the difference in domain understanding resulting from differing causal explanations of physical phenomena. Carroll and Olson (1988) review the mental model literature and offer a practical definition of mental models. In their definition, a mental model: (1) incorporates a full and detailed structure; (2) involves an understanding of what the system contains in terms of structure, how it works in terms of function, why it works that way in terms of purpose; and (3) provides a way to simulate actions mentally before choosing the appropriate action option to perform. The concept of "running" a mental model is roughly analogous to the mental simulation activity described in Klein's (1993a) RPD making model.

When used as an analog, a mental model serves as an "advance organizer" for the interpretation of novel concepts (Mayer, 1979; Mayer and Bromage, 1980). Anderson (1983) suggests that analogy and the creation of mental associations may be the only way that people learn. Bott (1979) found that users will generate their own analogies in order to explain system behavior if none is provided for this purpose. Research indicates that an inaccurate mapping between the user's model and the actual functioning of the system can increase task complexity and result in performance errors (Carroll et al., 1988). Lehner and Zirk's (1987) experimental studies involving expert system users found that an accurate mental model of system processes was key to cooperative problem-solving performance.

Moreover, the high performance attained with an accurate mental model continued even when the user's problem-solving method was different than the expert system behavior, as will often be the case.

Decision models such as Klein's (1993a) RPD making model and Rasmussen's (1986) SRK model propose conceptualization and analogical reasoning as the means by which users respond to novel situations. Similarly, conceptualization is a common factor in all four phases of the SHOR decision paradigm. The analog selected serves to reduce cognitive demand by identifying and structuring the relevant information and filtering out the irrelevant information. The mental models associated with the proposed analog then provide the means for mentally simulating the potential outcomes of the available options. Although this model appears to explain much of what makes for expert decision performance, there are several potential pitfalls. For example, the selection of an analogy may be affected by its availability in memory due to its vividness or recent experience. The selection of and adherence to an incorrect analogy may blind the user to relevant information, generally information that contradicts the working hypothesis. Subsequent mental simulations built upon these incorrect assumptions could mislead users with respect to the potential effects of their actions. Finally, the ability to "run" complex mental models is constrained by the limitations in human working memory and information-processing capability. In highly complex domains with extensive interactions among the various factors, the mental simulation required may be intractable.

The cognitive science literature presents numerous descriptive theories and empirical studies that attest to the existence of mental models and their use in judgment and decision making (Staggers and Norcio, 1993; Mellers et al., 1998; Hastie, 2001). However, there remains no systematic method for satisfactorily harnessing the power of mental models to guide the design of interactions and interfaces for decision support. Several difficulties in the practical application or manipulation of mental models negatively impact their prescriptive value in design. In practice, mental models are fragmentary and lack discrete boundaries or formalized definitions. The incomplete and disconnected aspects of a mental model permit the incorporation of contradictory, nonrational, and invalid concepts. Furthermore, mental models of rarely used systems or procedures can deteriorate over time due to forgetting.

In complex, dynamic environments, the interaction models required for human–computer cooperative decision making must assist the user in maintaining situational awareness and understanding the short- and long-term consequences of decisions. This implies a framework of models in the mind of the user that must be represented in the interaction and interface design. These include:

- *Task Interaction Models* Representation of the current state of the target domain (situational awareness) means for acting on the domain (task variables) and means for predicting the consequences of actions on the domain (outcome simulation).
- *System Interaction Models* Representation of the current state of the system and the means to understand the actions required to perform tasks using the system.

Carroll (1987, 1997) proposes a structured methodology for designing effective interface metaphors that provides a useful starting point for developing interaction models. Extending this method to the design of interaction and interfaces for decision aiding suggests the following basic activities:

- Identify potential task domain models, such as network models for route planning.
- Describe the match between models and the domain in terms of user task scenarios, such as constraints and affordances implied by the analogy.
- Identify the potential mismatches and their implications, such as identification of the gaps or breakdowns in the analogy.
- Determine the appropriate design strategies to help users manage unavoidable mismatches.

This task profile characterizes functional tasks along four dimensions (input, output, response, and feedback) and decision tasks in terms of four decision phases in the SHOR paradigm (stimulus, hypothesis, option, and response). Although Table 10.11 also suggests potential impacts of these dimensions, coherent design for information presentation and interaction cannot be derived from the assemblage of individual "fixes." Woods and Roth (1988) refer to this as the "prosthesis approach" to design. In contrast, they suggest that the design goals of cognitive systems engineering focus on extending the human user's conceptual abilities. This concept is consistent with Zachary's (1988) approach to the design of the knowledge representation, data management, and analytical methods for decision support systems. Toward this end, the tables serve to identify the cognitive areas that the individual characteristic may impact, such as attention and situational awareness, and supports making suggestions regarding the appropriate design features that address these areas.

The human user's ability to meet the cognitive demands of decision tasks is determined by both the quality of their conceptual skills and the cognitive resources, such as attention and memory, that they bring to the task. Design goals for supporting decision tasks fall into two general categories: those related to enhancing human users' understanding, and those related to reducing the negative effects of human cognitive limits. These categories are actually two sides of the same coin, rather than distinctly different constructs. The first category involves what Woods and Roth (1988) term "conceptual tools." The conceptual tools are those features of the design that enhance the user's ability to structure the problem, formulate the goal, select a solution path, and implement the selected response. The second category addresses the limits of human attentional and memory resources that interfere with effective use of the human cognitive strengths that aid conceptualization. The attempts of a system user to cope with their own cognitive limits often result in erroneous problem formulation and option selection. On the other side, enhancing conceptualization reduces certain aspects of cognitive demand and, thus, reduces the

**TABLE 10.11 Design Goals Summary—Task Requirements**

- Structure problem representation to enhance the user's perception and understanding of values and relationships between key task variables.
- Provide multiperspective conceptualization aids that make abstract or nonvisible concepts and relationships visible.
- Provide decision- or goal-oriented perspectives for organizing and prioritizing tasks.
- Support mental simulation with representations of network of problem dependencies for situational assessment, option evaluation, and response coordination.
- Allocate tasks between user and computer to reduce cognitive workload and support human user's adaptive, intuitive cognitive abilities.

cognitive resources required. The HSI design goals proposed here also play such dual roles.

One of the most effective means of enhancing a user's conceptualization is to structure the problem representation to highlight the values and relationships between the relevant task variables. Woods and Roth (1988) propose that the extent to which designers can successfully structure representation is a function of three factors:

1. The designer's ability to anticipate the decision tasks and situational variables
2. The characteristics of the representation that influence decision performance
3. The degree of domain variation in the relationship between key criteria and decisions

Task analysis helps to identify the decision variables and map their relationship in the decision process. Representation impacts the user's ability to monitor, perceive, combine, and relate data to assess the situation and formulate an appropriate response. In this way, structure of the representation makes the semantics of the domain visible, such as obtained in the ecological interface designs proposed by Vicente and Rasmussen (1992). The third factor addresses the issue of representational economy and variety. In complex, dynamic environments the users' requirements for problem views may change given a variation in the situational context, such as from routine operations to crisis. In addition, representational structures may have to provide multiple perspectives on the problem.

The use of representations to aid conceptualization lies at the heart of several approaches to structuring decision variables. Treu (1992) presents examples of several structural primitives and composite structures. Each primitive is considered with respect to its effects on cognition and memory and its representation in computer-based systems. For example, node and arc structures may imply paths, scripts, spatial location, or distance. When combined with the vertical hierarchy primitive that suggests concepts of rank, ordering, and levels of abstraction, the composite structure conveys a tree or object hierarchy.

Configural or integrative displays combine the low-level syntactic data to form a high-level semantic representation. The goal of integrative displays is to facilitate the user's holistic perception of domain or situational features that are not apparent when the data elements are separated. Initially, the concept of configural displays focused on the benefits of data proximity in object displays (Carswell and Wickens, 1987). It appears to be the "emergent features" of the configural display, rather than the mapping to a recognizable object, which determines the benefits in performance (Sanderson et al., 1989). The value of integrative displays has been questioned where users must also attend to individual data elements (Bennett and Flach, 1992). Bennett et al. (1993) empirically demonstrated the benefits of configural displays to promote extraction of both high-level and low-level data.

In their EID method, Vicente and Rasmussen (1992) also incorporate integrative displays derived from an abstraction hierarchy of the work domain. Based on improvements in decision performance in an experimental study, they suggest that the representation based on the abstraction hierarchy provides a better match to the user's mental model of the work domain. These findings support the benefits of multiple problem perspectives for decision making. Support for the multiple perspective design also applies to the research in integrative displays. Coury and Boulette (1992) investigated the effects of configural displays on diagnostic tasks in conditions involving time pressure and uncertainty. Their findings suggest that accurate and timely situation assessment under all conditions of time stress and uncertainty requires both integrated and separated displays.

In cases where representation requires multiple screens, Woods (1984) proposes that the integrative construct is the level of "visual momentum" the information presentation and interaction supports. When visual momentum is low, information processing occurs in a series of unintegrated data views, and this requires the system user to reorient and search for relevant information in each view. This may add considerably to the user's cognitive workload and, as a result, may degrade performance. Woods suggests several structural features to increase visual momentum, including the "long shot" or overview display, perceptual landmarks, display overlap, spatial organization, and spatial cognition. Overviews, landmarks, and overlap provide information about the location of one display with respect to another and support the multiple perspective aspects of the Rasmussen (1986) abstraction hierarchy. Spatial organization uses such spatial orientation entities as hierarchies, paths, and maps to serve as preorganizers and aids for exploring the domain or situation. Spatial cognition refers to the use of analogical representations to provide a map to all the features of the underlying process. Woods suggests that increasing visual momentum reduces mental workload, improves data sampling behavior and identification of relevant information, and improves the cooperation between the human user and the computer-based support.

Representation structure also affects the cognitive demand associated with decision tasks. Norman et al. (1986) present strategies for cognitive layouts in windowing designs that address selective attention, multicue integration, and variable levels of cognitive control. The layout of windows to reduce the demands of monitoring multiple activities is based on a model of attention that suggests that attention works as a dynamic filter. When multiple signals are present, attention is focused on one signal while the remaining signals are attenuated. The shift of attention may be voluntary, such as when the user actively searches for necessary information, or involuntary as in the case of flashing alarms. Spatial layouts for multicue integration may achieve all or part of the integration, possibly through use of integrative displays, or much of the necessary integration may be left to the user. Although integrative displays are quite powerful, effectiveness depends upon the extent of domain-criteria variability. Leaving integration to the user is the most flexible approach for the designer but unfortunately places the burden of integration entirely on the user. Finally, the use of spatial layouts may be used to provide multiple perspectives of the problem or domain from the syntactic to the semantic, in terms of signals, signs, and symbols.

Another approach to representation is a decision-oriented or goal-oriented display. In decision-based representation, displays present problem information structured such as to aid interpretation. Similar to the concepts in integrative displays, these display paradigms shift much of the cognitive demand in data integration and interpretation from the human user to the computer. In contrast to data-oriented displays that present all available information, decision-oriented displays provide only the information that is relevant for the decision task. In an experimental study involving multiphase decisions in a complex, dynamic environment, MacMillan and Entin (1991) found that decision-oriented displays resulted in faster decisions with fewer errors. Goal-oriented displays represent the domain or situational structures that relate to desired goal or system state. These can take the form of goal and subgoal hierarchies or diagrammatic views of the system or process. Kieras (1992) employed diagrammatic displays for diagnostic tasks in control system management. Experimental investigations indicated that the causal structures and one-to-one mapping of component state to the diagram produced better diagnostic performance than the more traditional representation in which the diagram and component state values were separated.

Goal-oriented representation may also be used to support the mental simulation required to identify causes for situation assessment, evaluate consequences of options, and plan for response coordination and implementation. Woods and Hollnagel (1987) present a methodology for constructing goal–means networks that incorporate the task goals, functions (the means to achieve goals), and requirements that instantiate new goals based on what the function needs to accomplish the higher goal. Woods and Roth (1988) propose goal-oriented displays for evidence processing, situation assessment, and planning. Bainbridge (1988) discusses problem representation as structure–function and goal–means networks. These graphic representations use hierarchies and cause–effect links to support pattern recognition, planning and prediction, and semantic organizers. Bennett et al. (1997) describe the representation aiding approach to display design in some detail and provide a useful tutorial as well as a description of four alternative approaches: aesthetic, psychophysical, attention based, and problem solving and decision making.

Mental simulation is important not only for time-pressured situations but also where feedback is delayed due to the inherent response latency of the system. Hoc (1989) describes the problems that are unique to long response latencies. In such environments, the diagnosis and response to changes in a system cannot be affected in direct cause-and-effect relationships. The user cannot directly manipulate the goal variable but must manipulate it indirectly through causally related variables. Planning is complicated by uncontrolled and unanticipated interventions in the causal network and long delays in response effect and feedback. The mental simulation required for planning in this context rapidly becomes intractable without appropriate aiding.

Ball and Ord (1983) present a graphic planning aid to support the mental simulation required to predict the consequences of options in an air traffic control task. Their aid presented two problem views: the current situation with radar and a predictive display of the planned response. Bell and Ord emphasize the problem of dealing with the multiple realities of the present and predicted situational displays. Their planning aid handled these representations as discrete displays and featured both manual and computer-generated options. Experimental studies with air traffic control teams revealed decision-making problems associated with requiring the controllers to relate information from both the planning and situation displays. This design forces the user to choose between maintaining situational awareness and evaluating the consequences of his or her response.

In a cooperative decision system, the design of the cooperation in the allocation of tasks between the human user and the computer-based support is a key factor in reducing the user's cognitive task workload. This is most often accomplished by automating the attention-intensive monitoring tasks; rapid, memory, or computation-intensive tasks; or time-constrained response tasks. Task allocation also attempts to assign to the human user those tasks, such as inference and judgment that involve adaptive, intuitive, cognitive abilities. For example, in Ball and Ord's (1983) air traffic control aid, the human controller and computer shared responsibilities for monitoring and planning. The activities within those responsibilities were allocated based on the different strengths of the human controller and the computer. The human user's tasks involved pattern recognition and maintaining situational awareness; the computer was assigned responsibility for continuous updating of the situational data and detailed trend analysis.

Most cooperative decision-making task allocation strategies involve some form of static allocation where some or all of the tasks are directly assigned to either the human user or the computer support. For example, Ball and Ord's air traffic control system featured static allocation. Other research in cooperative decision making also features dynamic task allocation (Andriole and Adelman, 1995; Andriole and Ehrhart, 1990; Vanderhaegen et al., 1994). Vanderhaegen et al. (1994) also present a design for human–computer cooperative decision making that involves a dynamic activity regulation strategy based on a model of "horizontal cooperation." The concept of horizontal cooperation attempts to avoid the negative performance effects sometimes encountered with the passive user in vertical, or master–slave type, cooperation tasks (Roth et al., 1988). Horizontal cooperation places both the human and computer on the same hierarchical level and allows explicit and implicit dynamic task allocation in much the same fashion as the human–human cooperation in team decision making. One particularly intriguing feature of this design is the dynamic task demand estimation capability modeled on workload and performance assessment. Rather than attempting the more subjective task of estimating mental workload, the task demand estimator employs a weighted additive model of the functional task decomposition. In the work cited, expert controllers determined weights empirically. This task demand-modeling concept would appear to also have utility in the determination of task loading during the design phase.

This section has presented several models for interpreting the task-related aspects of HSI-related design. The high-level decision task goals proposed provide:

- A starting point for integrating the situational, organizational, and user goals
- Signposts to the determination of the more detailed design goals associated with the specific tasks

Table 10.11 summarizes the design goals that support the decision task requirements. The next section discusses implementation of the current HCI design concept in prototype form for evaluation and iterative modification.

The high-level design goals need to consider the effects of situational context, organizational context, and user knowledge and experience on the cognitive requirements of the tasks that the human–machine cooperative system must perform. Each high-level goal maps to a deeper layer of task and situation-dependent design goals. In essence, these goals provide a checklist for design, implementation, and evaluation. The design concept is a configuration of information presentation and interaction strategies that represent the designer's resolution of these high-level and specific design goals.

### 10.5.5 Evaluating Designs for Usability

One of the most often used terms in human–machine-related areas, especially with respect to design evaluation, is "usability." Narrow definitions of the term limit usability to the mechanics of operating the interface. Nielsen's (1993) usability heuristics exemplify this narrow definition. In a somewhat broader definition, usability may be seen as the measure of the system design's ability to support the user in accomplishing their tasks (Mayhew, 1999). This model of usability incorporates the interface operation tasks as a subset of an overall measure of the effectiveness and ease of use of the system.

Several researchers have proposed the use of so-called discount usability evaluation methods to identify areas for improvement early in design (Nielsen, 1993; Wright and Monk, 1989, 1991). Nielsen's (1993) *heuristic evaluation* approach is based upon such accepted design guidance as "use simple and natural dialog and also provide adequate feedback." It converts such heuristics into checklists of nine usability properties. Heuristic evaluation may be performed by 3 to 5 evaluators and does not involve interaction with users. Empirical evaluations using as many as 77 evaluators indicated that aggregating the responses of as few as 5 evaluators resulted in the capture of 55 to 90 percent of usability problems (Nielsen and Molich, 1990). This research also pointed out the relatively poor performance of individual evaluators. The fundamental limitation of Nielsen's heuristics is their focus on design aspects of interface operation. As designed, the checklists do not provide the means to examine the extent to which the design addresses the cognitive task requirements. A recent work by Henneman (1999) discusses skills (human factors, multimedia interaction design systems engineers) and tools (design laboratories, usability standards, and guidelines) that support a user-centered process for engineering usable systems. The author's conclusion that the key to improving the usability of new systems and products lies in the development staff and the organizational environment in which they work, appears undeniable.

### 10.5.6 Evaluating System Designs for Reliability

To be effective, systems must be reliable. From the user's perspective, this means the system is available upon demand with current, accurate information. The best case is 100 percent reliability; the worst case is multiple failures in critical systems. The most likely case is that there will be some disruption of services and delays in information updates. Systems fail in a variety of ways. Van Gigch (1991) lists five types of system failures:

1. Failures of structure and control, which often results from reliance on faulty controls built into the structure of the system or expecting other parts of the system to catch mistakes or take care of problems
2. Failures of technology, due to technology that does not perform as expected and that provides incorrect, incomplete, and/or imprecise information
3. Failures of decision processes through flawed assumptions and information-processing biases that effect judgment and choice
4. Failures of behavior, which generally occur through doing the wrong thing
5. Failures of evolution, due to rigid, nonadaptive human behavior

Design for reliability is, as a consequence of these failures, very important. Pecht (1995, 1999) identifies eight tasks, each requiring full systems engineering and systems management commitment:

1. Define realistic system requirements.
2. Define the system usage requirements, including the environment in which the system must operate.
3. Identify potential system failure modes and mechanisms.
4. Thoroughly characterize the system component materials and the manufacturing and assembly, or integration, process.
5. Design reliable systems within constraints posed by these.
6. Certify these processes.

7. Monitor and control these processes.
8. Manage the life-cycle process for the system to be engineered such as to improve reliability, quality, and cost effectiveness of the system to be engineered through use of this process.

The classic engineering response to reliability issues is often to build in "graceful" degradation so that failure of one subsystem does not propagate and lead to multiple failures. Information about the effects of outages in such systems is often provided in cryptic form for system administrators, but the system users are left to fend for themselves. Users need clear, understandable information about the extent to which their current information may be impaired by system outages or delays. As an illustration, it is often very helpful to provide appropriate feedback such as to reduce negative impacts on decision-maker confidence of:

- Information currency indicators
- Summary of update times and content
- Overview diagrams of systems that are affected by delays and failures

Operators may need assistance in identifying what information must be restored to bring the system up-to-date. Finally, decision makers need to be alerted when systems or networks are unavailable.

Ideally, this information would also be represented in the certainty factors for information in dependent systems. For example, if the intelligence systems supporting the enemy situation displays were impaired, the predicted or last known location could be displayed with a change in the icon that indicated its position was not based upon direct sensing or recently updated information. Without those uncertainty indications, the user may misinterpret the data provided. There are a large number of variables that affect and impact system reliability. The interaction between reliability and the closely related subject of maintainability is of major concern in implementing trustworthy systems of all types.

## 10.6 PROTOTYPING AND IMPLEMENTATION

A prototype is a physical manifestation of the configuration of information presentation and interaction methods, and functional capabilities and technologies, which have been proposed in the system conceptual design and architecture as potentially satisfying the user requirements for the system to be engineered. Developing prototypes during the early phases of system development provides a low-risk means for evaluating both the conceptual design and architecture and the system implementation hypotheses. At each stage in the system development effort, the represented design can be reviewed against the current version of requirements. In this way, sponsors and operational users can respond to the prototyped design to refine the requirements base and assess the utility and usability of the proposed interface for the decision tasks. Prototypes vary widely in scope and definition, from preliminary paper storyboards to functional interfaces to data. The choice of the form of prototype depends upon the questions that must be answered at the current phase of system development. For example, early in the development a prototype may be no more than a set of sample screens sketched on paper or a cardboard mockup of a control panel. More commonly, the term "prototype" is applied to early functioning versions of software and hardware.

### 10.6.1 Prototyping Design Concepts

Assessing the appropriateness and effectiveness of the proposed system to support the complex interactions among humans, equipment, environment, and information within the organization often requires some form of interactive prototype. Using an interactive prototype also provides useful insight for the overall development effort. The HCI design embodies most of the system concept that is "available" to the user to guide his or her mental model of the system. For example, the HCI design incorporates such critical system design factors as:

- The representation of information regarding the situational elements external to the system, such as environment and external threats and opportunities
- The representation of system states and feedback to the operator based on results of actions taken
- The allocation of tasks between the human user and the physical system as determined by the dynamics of the situation and the requirements of the analytical methods selected to support decision processes
- The modes in which users may interact with all of this information to explore situations, develop hypotheses, generate options, select among alternatives, and implement their decisions

In a requirements-driven design process, the judgments and decisions made during each phase determine the objectives of the analyses and evaluations required to support those decisions. Table 10.12 presents the relationship of prototyping objectives and the associated scope and boundaries of the prototyping effort. During each phase, the system design is considered in the context of the organizational and environmental factors that impact performance; however, these factors are represented at varying levels of detail depending upon the phase requirements. For example, during the problem definition phase and early in the requirements identification, the system design in question is modeled at a relatively high level of abstraction. The desired performance is expressed primarily in qualitative terms; the nature of the interaction with other support systems and the external environment is modeled in very low detail. As development proceeds to later phases, the specification of requirements increases in detail with respect to the system itself and its interaction with other systems in the organization and external environment. This specification, in turn, dictates the inclusion of more precise quantitative and qualitative analysis to assure the system design meets both engineering specifications and organizational requirements.

### 10.6.2 Prototyping Strategies

The software engineering and information systems development literatures suggest a wide variety of approaches to prototyping (cf. Andriole, 1990; Arthur, 1992; Connell and Shafter, 1989; Sage and Palmer, 1990). The selection of prototype form should be based on the goals of the current development phase and the information that must be derived

**TABLE 10.12 Prototyping Goals for System Development Life-Cycle Phases**

| Design Phase | Prototyping Objectives | Prototype Characteristics |
|---|---|---|
| Problem definition and requirements identification | • Determining desirable system and HCI characteristics<br>• Determining existing system capabilities and deficiencies<br>• Selecting "best" of alternative system definition | • System represented at high level of abstraction<br>• Qualitative analysis<br>• Organizational, environmental interactions represented in minimal detail |
| Requirements specification and design | • Developing requirements specifications and design alternatives<br>• Determining "best" design | • System represented in moderate to high detail<br>• Qualitative and quantitative analyses<br>• Organizational and environmental interactions modeled in moderate detail |
| Implementation | • Determining whether developmental prototype meets specifications<br>• Providing feedback on detailed design | • System modeled in moderate to high detail<br>• Qualitative and quantitative analyses<br>• Organizational and environmental interactions modeled in moderate to high detail |
| Testing and evaluation | • Determining whether proposed design as prototyped meets system and organizational requirements | • Qualitative and quantitative analyses<br>• High detail in system and context modeling |

from the prototype. Nielsen (1993) identifies the trade-offs in prototype implementation in terms of depth of functionality (vertical prototyping) versus breadth of features (horizontal prototyping). Vertical prototyping is used in "functional" prototypes that permit the user to interact with real information; however, only a narrow range of system features is represented. In contrast, horizontal prototyping permits the presentation of the full range of system features but without the functional capability to interact with real data.

Another common prototype classification involves the extensibility of the prototype. "Throw-away" prototypes, such as paper storyboards and mockups, are used in early definition phases often before the target hardware and software have been identified. The name conveys a pejorative image of sunk costs; however, the throw-away prototype facilitates communication between development teams, system designers, sponsors, and end users. The information gathered not only contributes to design but can also be used to develop instruments for the evaluation phases. "Evolutionary" prototypes involve incremental development that attempts to represent the breadth of the system with functional depth evolving incrementally. The term, rapid prototyping, is generally used to refer to an evolutionary prototype. Interactive storyboards are commonly used as throw-away prototypes. In situations where commercial off-the-shelf (COTS) programs and computer-aided systems (software) engineering (CASE) tools may be used for development, interactive storyboards become the early forms of rapid, evolutionary prototypes. These four general approaches to prototyping are discussed in further detail below and summarized in Table 10.13.

**TABLE 10.13 Prototyping Procedures**

| Method | Advantages | Disadvantages |
|---|---|---|
| Paper storyboards | • Low-cost, low-risk method for exploring requirements.<br>• Scenarios can be reused for later evaluations of design.<br>• Storyboards and scenarios can later be incorporated into interactive storyboards. | • Verbal descriptions in scenarios are not as vivid as visual representations.<br>• Paper storyboards support very limited exploration of interaction.<br>• May have less utility in identifying potential human errors. |
| Mockups | • Low-cost method for verifying physical layout of custom interaction hardware.<br>• May be useful in simulating environment for exercises where full interaction is not required. | • Limited to representing surface features.<br>• Full capture of ergonomic aspects of performance requires more expensive representation (pushable buttons, turnable knobs, etc.). |
| Interactive storyboards | • Useful for refining requirements and identifying potential human errors.<br>• Provides low- to medium-fidelity environment for performing usability trials.<br>• May be developed with low to moderate cost using COTS software. | • Will not identify throughput or information overload problems associated with data volume.<br>• Designers must be careful to present only *feasible* design options within given hardware/software constraints. |
| Integrated rapid prototyping | • Useful (within limits) for evaluating performance with actual or simulated inputs.<br>• May help prevent premature "freezing" of design. | • Moderate to high cost (some costs reduced when CASE tools provide easily modified prototypes).<br>• Increasing fidelity is costly. |

***Paper Storyboards*** Paper storyboards provide a relatively low-cost, low-risk method for getting a preliminary feel for how the system would be used in terms of typical tasks and situations. Storyboards may be annotated, reordered, or even redesigned during requirements definition interviews. Paper storyboards are limited to representation of a set scenario with little possibility of exploring the range of interaction possible with the given design. The technique presents the sequence of screens but does not capture potential interaction errors or the cognitive workload associated with a particular design. These aspects are better addressed with interactive storyboards.

***Mockups*** Mockups encompass a variety of nonfunctioning physical representations ranging from cardboard models of single control panels to full-scale control centers with turnable knobs and flippable switches. They are primarily used for studying the ergonomic impacts of equipment layout of physical task performance. In many cases, physical mockups are unnecessary for studying the implications of system designs since most of the visible features of interest are incorporated in interactive storyboards or prototype systems.

Where custom interaction hardware is required for user input or users must perform other physical tasks while operating the system, mockups assist in doing early evaluations of the potential workload associated with the set of system design alternatives.

***Interactive Storyboards*** Interactive storyboards serve as a powerful means for exploring design alternatives without incurring the expense of developing a working prototype. This is particularly advantageous when the investigation is focused on evaluating several advanced interaction technologies rather than supporting the design of a specific system. Interactive storyboards are also useful for working with experts or end users to refine requirements. Subjects interact with a computer-based storyboard simulating the actual operation of the system. Interaction may take the form of informal exploration or subjects may be presented with tasks to perform using the simulated system. In the latter case, the storyboard provides a low to medium fidelity environment for assessing usability and identifying potential human errors. Verbal protocol methods may be used to elicit the cognitive processes involved in the interaction.

Where storyboards are used in requirements definition and refinement, care must be taken not to present something in storyboard form that is infeasible within the technological and resource constraints likely to be present in the operational working system. Although this method can be used to identify problems with cognitive workload due to the allocation of tasks between the human operator and the physical system, it may not task the overall system sufficiently to enable delineation of user or computer performance problems related to throughput or information overload. These issues may be addressed with operational prototypes that accept real-time data.

***Integrated Rapid Prototypes*** Although developing prototype versions of a system is not a new concept, until recently software prototyping tended to be restricted to semioperational *beta* versions of systems under construction. As such, they represented a considerable investment in time and effort, and major changes to the design were highly discouraged. Furthermore, it was not uncommon for a cost-conscious sponsor to stop development with the prototype. If the prototype offered most of the functionality of the completed system, the sponsor would take delivery on the prototype and cancel further development. Similarly, if the prototype indicated major problems with the design or development effort, the sponsor might consider it good management to cut his or her losses at that point. For obvious reasons, developers grew reluctant to show prototypes to their clients.

The introduction of fourth-generation languages and CASE tools dramatically changed the role of prototyping in system design and development. Using the toolboxes provided in COTS system prototypes with complete interactive displays using windows and pull-down menus can now be developed very rapidly for UNIX, DOS, Macintosh, and other environments. This rapid development capability and the corresponding ease with which the software may be modified or even substantially redesigned or reengineered, makes it possible for designers to develop and use prototypes during the earliest phases of design. These early prototypes provide many of the features of interactive storyboards while reducing the possibility of presenting the user with an infeasible system concept. Nevertheless, until the system is tasked with the full volume of data expected in the target setting, actual system performance and its impacts on the users would not be fully apparent. This has important implications for reliability and validity of system and design evaluations.

## 10.7 SYSTEM EVALUATION

With the growth of interactive computing and its application in support of complex decision support systems of humans and machines, conceptual design prototyping has become an important tool in capturing and analyzing user requirements. Figure 10.11 illustrates potential prototype evaluation benefits. In iterative design and development processes, prototype evaluation aids in verifying and validating the working design against the requirements. Each prototyping phase culminates with some form of evaluation, and the evaluation goals vary depending upon the current development phase. Early evaluation provides a means for extending requirements and task analyses to the evaluation of the procedures embedded in the current design solution. In this manner, evaluation provides a means for acquiring information about the current version of the system design with respect to the performance characteristics and capabilities of the human–machine cooperative decision system. Finally, this process of iterative design, prototype implementation, and evaluation supports the project management planning and control processes that ensure the overall development effort stays on track both with respect to the delivery of a quality product and within the cost and schedule parameters. The cost effectiveness of incorporating an evaluation method depends not only upon the size and complexity of the project but also at which point during development the prototype evaluation is conducted (Mantei and Teorey, 1988).

Information feedback during prototyping enables iterative and evolutionary system development course correction. Early evaluation allows design modification during the initial life-cycle phases when the cost to modify a system is much lower than it will be at later phases. For the systems engineering development team, evaluation is also a discovery process. Findings from the evaluation provide input for requirements and design modification and help to set MOPs and MOEs (Sproles, 2000, 2001), benchmark targets for later system-level evaluations. Evaluation feedback informs not only the design of the

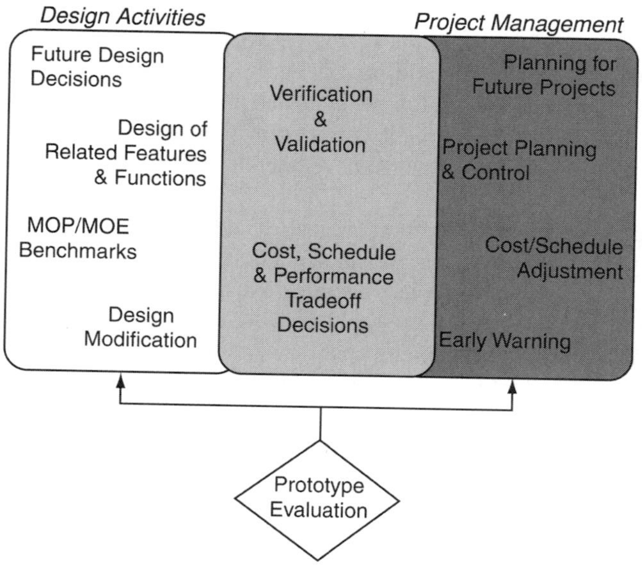


**Figure 10.11** Benefits of prototype evaluation feedback to development.

particular functions and features considered, but also provides input for the design of related components. For the project manager, evaluation feedback is a critical part of project planning and control. Early evaluation flags potential problems that may require cost, schedule, or, in some cases, contract modification.

The introduction of new technology into a complex organizational system will modify its processes and the related structures and subtasks. This organizational evolution must also be mapped into the evolving system and development process. Defining cognitive requirements and evaluating their implementation in support systems is a critical part of ensuring the effectiveness of new systems. Often, of course, new technologies are introduced into organizations to cope with recognized needs for reengineered organizational processes. These statements illustrate the interactive relationships across humans, organizations, and technologies as well as relationships between these and the internal and external environment. User-centered design is an approach that enables and enhances this integration by embracing three basic principles:

- Design of aiding technology embodies the relationship of human users and computer-based aids in achieving organizational goals.
- Decomposition of functions, processes, and tasks provides measurable indicators of the extent to which specific designs fulfill system objectives.
- The utility of evaluation to the system design process depends upon the application and interpretation of performance measures in the context of a valid framework of objectives, functions, processes, and tasks as appropriate MOPs and MOEs.

Thus, the qualitative aspects of support to decision making must be included in the earliest evaluations. Designs for the complex systems supporting decision making derive conceptual requirements from models of organizational processes. The doctrine incorporated in these models, and the missions defined by the organization provide the context for identifying the functional and task requirements that structure the relationships of humans and machines. These requirements, in turn, help to determine the appropriate MOPs and MOEs that form the selection criteria for aiding designs. These can be applied through a combination of checklists, expert reviews, end-user walkthroughs, and heuristic evaluation. As early as possible, developers need the input of "real users" using the system under the most representative conditions. Pilot exercises and experiments generate a wealth of information on the complex interactions of users, processes, and system supports that can be used to assess the development paths of future systems.

### 10.7.1 Setting Evaluation Goals

The evaluation goals of systems engineering practitioners are often quite distinctly different from those of cognitive science researchers. System interfaces and interactions, and associated human–machine cooperative decision-making tasks, involve highly complex constructs. As discussed, the evaluation of a conceptual design and architectural prototype should track to the associated requirements. Two principal evaluations should be conducted at each level of prototyping:

- Verification of implementation of user requirements by the system
- Validation of design implementation's effectiveness in terms of interface usability and utility

Depending upon the systems engineering phase under consideration, the evaluation scope, and the level of detail in the prototype, evaluation may range from designer-reviewed checklists and rating scales to empirical evaluations with representative users.

Computer-based interactive prototypes provide opportunities for direct observation of the human–computer decision performance. Several methods are available for examining interaction processes through automated capture and analysis of interaction protocols to facilitate the rapid data analysis required for design iteration (Smith et al., 1993). The empirical study approach builds information in a data-intensive, bottom-up fashion. While empirical evaluations can be used to determine performance benchmarks, they do not permit direct insight into the performance *requirements*.

These requirements evolve from a top-down analysis based upon the organizational and system objectives, functions, and tasks identified with those functions. An analytical framework for empirical evaluation is desirable as, without an analytical framework, the measures collected in empirical studies lack context and can misdirect decision makers. In this context, Rogers (1992) questions the desirability of the microanalysis and theoretical rigor that characterize research in cognitive psychology. Rogers suggests that applied research and, by extension, design evaluation benefits much from a macrolevel analysis that allows a parallel, symbiotic relationship with the theoretical aims of the research. This provides much support for case-study-based evaluation (Yin, 1994; Stake, 1995).

Rasmussen and Pejtersen (1993) conceptualize the well-balanced evaluation of a cognitive systems engineering design product as a combination of top-down analytical evaluation and bottom-up empirical assessments. System design often evolves through the top-down analysis of the intended purpose and identified functions. Functions are then decomposed into the procedures and tasks allocated to the machine and user, culminating in the design that maps the system's form. Bottom-up empirical evaluations first address the lower level human factors issues associated with fundamental usability and continue by evaluating the support of the cognitive requirements involved in the tasks. These human requirements interact with the system's allocation of functional requirements and the capabilities afforded by the design.

Despite some variations in terminology, this prescription for a combination of top-down analytical and bottom-up empirical evaluation is consistent with similar discussions in Meister (1985, 1991) and Adelman (1992). Meister (1985) presents a series of human performance questions grouped by development stage and indicates the various analysis and evaluation methods that supply answers. The balance between analytical and empirical evaluation approaches shifts depending upon the stage of the life-cycle process that leads to the system itself. For example, in the early stages of planning and design, there is a strong reliance on top-down analysis methods supported by the available objective data and subjective judgments. The later phases of detail design and prototype testing employ more rigorous empirical evaluation methods and well-structured subjective measures to assess performance in terms of the functional requirements outlined in earlier phases of development. Figure 10.12 illustrates the relationship of efforts at the various phases of engineering a system and evaluation objectives.

### 10.7.2 Selecting Evaluation Methods

Nielsen's (1993) text on usability engineering discusses usability heuristics and heuristic evaluation but does not present example checklists or sufficient information to guide the conduct of heuristic evaluation. Ravden and Johnson (1989) present a comprehensive

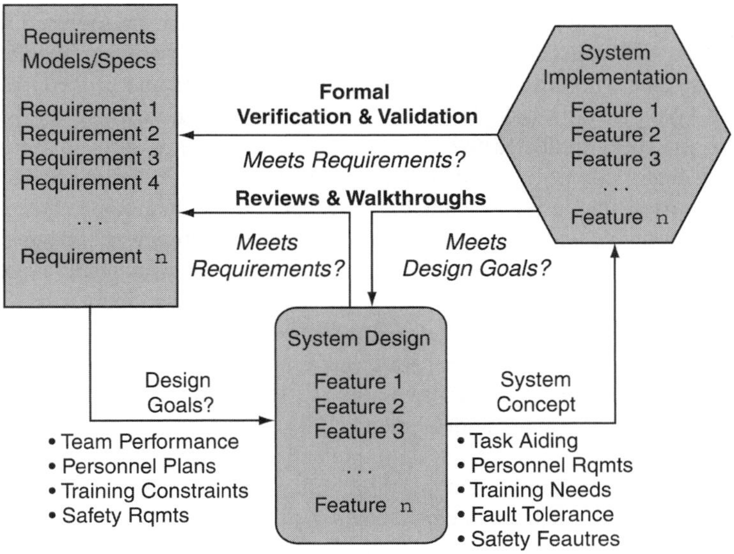


**Figure 10.12** Relationship of development process inputs to evaluation goals at each development phase.

evaluation that employs nine usability criteria. Each criterion is addressed in 10 to 12 questions that may be amended to address the specific evaluation goals. Evaluators include the designer(s), representative end users, and such other technical professionals as human factors experts. The members of the evaluation team complete the checklists individually as they perform a predetermined set of exemplary tasks. The principal advantage of Ravden and Johnson's method is the potential for rapid analysis and the ready conversion of the subjective data into quantitative measures for comparison. The most significant source of overhead is in the selection and development of interaction tasks. Depending upon the goals of the evaluation, the development of simple tasks or task scenarios may entail extensive preparation.

Wright and Monk (1991) avoid some of the shortfalls in heuristic evaluation while retaining its low cost and effort features. Although they acknowledge the value of careful quantitative evaluation, they suggest that qualitative evaluation provides more cost-effective guidance for the early phases of design. Their approach, intended for the design practitioner, involves designers and users in a *cooperative evaluation* using think-aloud protocols and verbal probes. Analysis in this early phase is highly focused to capture the relevant information within cost and schedule requirements. Wright and Monk (1989) indicate that evidence in the form of either critical incidents or breakdowns is sufficient to identify system design problems. A *critical incident* is some user behavior that fails to use the functionality of the system efficiently. A *breakdown* designates any point in the interaction where the user's focus on the task is broken due to the demands imposed by the system, such as when the interface ceases to be "transparent." This approach is similar to, but in many ways different from, the retrospective analysis technique used by decision researchers (Klein, 1989).

The ability to use the systems engineering designer as the evaluator allows for the speedy, inexpensive evaluation necessary for iteration in the early stages when the conceptual design is evolving rapidly. Experimental investigations performed with design trainees indicate that satisfactory rates for detecting design problems may be achieved quickly by designers with little or no human factors background and limited training in the method itself (Wright and Monk, 1991). Rather than merely endorsing their own designs, the results of the study indicated that these designers were better at evaluating their own systems using this method than similarly experienced evaluators not associated with the design. Furthermore, the designer-evaluators uncovered more unanticipated problems than the evaluators not involved in the design. The principal limitations in the cooperative evaluation method include problems with the task altering aspects of think-aloud protocols and the potential for bias in the single designer-evaluator model. Similar to the usability evaluation method of Ravden and Johnson (1989), cooperative evaluation also requires the preparation of meaningful tasks that provide the context for the evaluation sessions.

Departing from the design-phase orientation of the classic system development life-cycle (SDLC) model, Gardiner and Christie (1987) examine the role of prototypes in addressing questions on four human system interaction design levels: conceptual, relating to the system concept; semantic, relating to the interaction concept; syntactic, relating to the interaction form; and lexical, relating to interaction details. In related work, Ehrhart (1993) presents a survey of evaluation methods useful for assessing system designs to support human–machine cooperative decision making. Gardiner and Christie's model provides some useful guidelines for trading off the time and expense required for developing a prototype against the functionality and performance achieved. In addition, it indicates the extent of evaluation support possible with a relatively small investment in so doing. Table 10.14 combines the suggestions of Gardiner and Christie (1987), Nielsen

**TABLE 10.14 Prototyping and Evaluation to Match Human–System Design Requirements**

| Design Level | Design Evaluation Focus | Prototyping Support | Evaluation Tools and Techniques |
|---|---|---|---|
| Conceptual | • System and conceptual design concept (architecture)<br>• Appropriate for user requirements | • Written descriptions and scenarios<br>• Storyboards<br>• Interactive storyboards | • Focus groups<br>• Walk-through<br>• Predictive models<br>• Heuristic methods |
| Semantic | • Interaction concept<br>• Broad definition of interaction, error feedback, and user support | • Interactive storyboards<br>• Hardware mockups<br>• Partial prototypes | • Informal user tests and observation<br>• Walk-through<br>• Checklists and rating scales |
| Syntactic | • Interaction form<br>• Dialog parameters and interaction sequences | • Interactive storyboards<br>• Partial (developmental) prototypes | • Formal and informal user tests<br>• Walk-through<br>• Controlled laboratory tests<br>• Field tests and observation |
| Lexical | • Interaction detail<br>• Specification of human–system interaction and interfaces | • Partial (developmental) prototypes<br>• Functional prototypes | • Formal user tests<br>• Gaming and simulation<br>• Field tests and observation |

(1993), and Ehrhart (1993) for linking the proposed evaluation focus, prototyping support, and evaluation techniques appropriate at each design level.

## 10.8 SUMMARY AND CONCLUSIONS

This chapter presents a framework for employing cognitive systems engineering methods to define problems, identify and represent cognitive task requirements, develop design goals, and implement and evaluate system designs for information presentation and interaction in human–machine cooperative systems. To improve HSI planning and implementation in program management requires effort among all the stakeholders. The acquisition process must be revised to require the definition and tasking of HSI responsibilities in all program management directives and acquisition program plans (DoD, 1994). System program managers must have adequate training to understand and direct HSI efforts. The HSI effectiveness should be an element in system program managers' performance ratings. Program funding must provide sufficient resources for implementation of HSI practices during all phases of the engineering of a system. These life-cycle phases must include reviews with organizational stakeholders, including representative end users, training designers, and support personnel.

## NOTES

1. To improve and support HSI activities, the Defense Information Systems Agency (DISA) maintains an annotated directory of HSI design support tools and techniques developed by U.S. government agencies, NATO countries, academia, and private industry (Dean, 1998). The DoD also provides a searchable set of mandatory and discretionary HSI guidelines as part of the *Acquisition Deskbook* (DoD, 2001).
2. The other types of cognitive errors are mental memory lapses and physical action slips, which are errors that cause improper implementation of action plans.

## REFERENCES

Adelman, L. (1992). *Evaluating Decision Support and Expert Systems.* New York: Wiley.

Adelman, L., Tolcott, M., and Bresnick, T. (1993). Examining the Effect of Information Order on Expert Judgment. *Organizational Behavior and Human Processes*, *56*, 348–369.

Anderson, J. R. (1983). *The Architecture of Cognition.* Cambridge, MA: Harvard University Press.

Andriole, S., and Adelman, L. (1995). *Cognitive Systems Engineering for User-Computer Interface Design, Prototyping, and Evaluation.* Hillsdale, NJ: Erlbaum.

Arthur, L. J. (1992). *Rapid Evolutionary Development: Requirements, Prototyping & Software Creation.* New York: John Wiley.

Ashby, W. R. (1956). *Introduction to Cybernetics.* New York: Wiley.

Badre, A. N. (1982, April). Selecting and Representing Information Structures for Visual Presentation. *IEEE Transactions on Systems, Man, and Cybernetics*, *12*(4), 495–504.

Bainbridge, L. (1987). The Ironies of Automation. In J. Rasmussen, K. Duncan & J. Leplat (Eds.), *New Technology and Human Error.* London: John Wiley.

Ball, R. G., and Ord, G. (1983). Interactive Conflict Resolution in Air Traffic Control. In M. J. Coombs, and J. L. Alty, (Eds.), *Designing for Human-Computer Communication*, London: Academic.

Basili, V., and Weiss, D. M. (1984, November). A Methodology for Collecting Valid Software Engineering Data. *IEEE Transactions on Software Engineering, 10*(6), 728–738.

Beach, L. R. (1992). Epistemic Strategies: Causal Thinking in Expert and Nonexpert Judgment. In G. Wright and F. Bolger (Eds.), *Expertise and Decision Support*, New York: Plenum.

Bennett, K. B., Nagy, A. L., and Flach, J. M. (1997). Visual Displays. In G. Salvendy (Ed.), *Handbook of Human Factors and Ergonomics* (pp. 659–696). New York: Wiley.

Bennett, K. B., Toms, M. L., and Woods, D. D. (1993). Emergent Features and Graphical Elements: Designing More Effective Configural Displays. *Human Factors, 35*(1), 71–97.

Boar, B. (1984). *Application Prototyping: A Requirements Definition Strategy for the 80s*. New York: Wiley.

Boehm, B. W. (1976, December). Software Engineering. *IEEE Transactions on Computers, 25*(12), 1226–1241.

Boehm, B. W. (1981). *Software Engineering Economics*. Englewood Cliffs, NJ: Prentice-Hall.

Boehm, B. W. (1987, September). Improving Software Productivity. *IEEE Computer*, pp. 43–57.

Boehm, B. W. (1988, May). A Spiral Model of Software Development and Enhancement. *IEEE Computer, 21*(5), 61–72.

Boehm, B. W. (1999, March). Making RAD Work for Your Project. *IEEE Computer, 32*(3), 113–117.

Boehm, B. et al. (2000). *Software Cost Estimation with COCOMO II*. Upper Saddle River, NJ: Prentice-Hall.

Bott, R. (1979). *A Study of Complex Learning: Theory and Methodology*, CHIP Report 82, La Jolla, CA: Center for Human Information Processing, University of California at San Diego.

Brannick, M. T., Salas, E., and Prince, C. (Eds.). (1997). *Team Performance Assessment and Measurement: Theory, Methods, and Applications*. Mahwah, NJ: Erlbaum.

Brehmer, B. (1987). Models of Diagnostic Judgment. In J. Rasmussen, K. Duncan, and J. Leplat (Eds.), *New Technology and Human Error*. New York: Wiley, 1987.

Burton, R. M., and Obel, B. (1995). *Strategic Organizational Diagnosis: Developing Theory for Application*. Boston: Kluwer Academic.

Byrd, T. A., Cossick, K. L., and Zmud, R. W. (1992, March). A Synthesis of Research on Requirements Analysis and Knowledge Acquisition Techniques. *MIS Quarterly*, pp. 117–138.

Card, S. K., Moran, T. P., and Newell, A. (1983). *The Psychology of Human-Computer Interaction*. Hillsdale, NJ: Lawrence Erlbaum.

Carroll, J. M. (1997). Human Computer Interaction: Psychology as a Science of Design. *Annual Review of Psychology, 48*, 61–83.

Carroll, J. M., Mack, R. L., and Kellogg, W. A. (1988). Interface Metaphors and User Interface Design. In M. Helander (Ed.), *Handbook of Human-Computer Interaction*. New York: Elsevier.

Carroll, J. M., and Olson, J. R. (1988). Mental Models in Human-Computer Interaction. In M. Helander (Ed.), *Handbook of Human-Computer Interaction*. Amsterdam: Elsevier.

Carswell, C. M., and Wickens, C. D. (1987). Information Integration and the Object Display. *Ergonomics, 30*, 511–527.

Chapanis, A. (1959). *Research Techniques in Human Engineering*. Baltimore, MD: Johns Hopkins Press.

Checkland, P. (1981). *Systems Thinking, Systems Practice*. New York: Wiley.

Checkland, P., and Holwell, S. (1998). *Information, Systems and Information Systems—Making Sense of the Field*. New York: Wiley.

Choo, C. W. (1998). *The Knowing Organization: How Organizations Use Information to Construct Meaning, Create Knowledge, and Make Decisions*. London: Oxford University Press.

Connell, J. L., and L. Shafter. (1989). *Structured Rapid Prototyping: An Evolutionary Approach to Software Development*. Englewood Cliffs, NJ: Prentice Hall.

Coury, B. G., and Boulette, M. D. (1992). Time Stress and the Processing of Visual Displays. *Human Factors*, *34*(6), 707–725.

Daly, E. (1977). Management of Software Development. *IEEE Transactions on Software Engineering*, *3*(3), 229–242.

Davis, A. M. (1993). *Software Requirements: Objects, Functions, and States*, rev. ed. Englewood Cliffs, NJ: Prentice-Hall.

Dean, T. F. (Ed.). (1998, October). *Directory of Design Support Methods*, Technical Report ADA355192. San Diego, CA: Defense Technical Information Center. Available: http://dticam.dtic.mil.

Dörner, D. (1987). On the Difficulties People Have in Dealing with Complexity. In J. Rasmussen, K. Duncan, and J. Leplat (Eds.), *New Technology and Human Error*. New York: Wiley.

Dreyfus, J. L., and Dreyfus, S. E. (1986). *Mind Over Machine: The Power of Human Intuition and Expertise in the Era of the Computer*. New York: Free Press.

Duffy, L. (1993). Team Decision-Making and Technology. In N. J. Castellan, Jr. (Ed.), *Individual and Group Decision Making*. Hillsdale, NJ: Erlbaum.

Edgell, S. E., Bright, R. D., Ng, P. C., Noonan, T. K., and Ford, L. A. (1992). The Effects of Representation on the Processing of Probabilistic Information. In B. Burns (Ed.), *Percepts, Concepts and Categories: The Representation and Processing of Information*. Amsterdam: Elsevier Science.

Edland, A., and Svenson, O. (1993). Judgment and Decision Making under Time Pressure: Studies and Findings. In O. Svenson and A. J. Maule (Eds.), *Time Pressure and Stress in Human Judgment and Decision Making*. New York: Plenum.

Ehrhart, L. S. (1993, March). *Tasks, Models and Methods: System-Level Evaluation of Human-Computer Interaction (HCI) for Command and Control Decision Support*, Technical Report, Contract No. F30602-92-C-0119. Fairfax, VA: $C^3$I Center, George Mason University.

Ehrhart, L. S. (1994, August). Cognitive Systems Engineering: Human Computer Interaction Design for Decision Support. Ph.D. Dissertation, School of Information Technology & Engineering, George Mason University, Fairfax, VA.

Ehrhart, L. S. (1997, July). Integrating Multimedia Information for Joint Theater Command and Control. In *Proceedings of the 49th IEEE National Aerospace and Electronics Conference*, Dayton, OH. New York: IEEE Press.

Ehrhart, L. S., and Aiken, P. H. (1991). Cognitive Engineering for Intelligent Control System Design: Preserving User Models in Requirements Analysis. In *Proceedings of the 1991 American Control Conference*, Boston, MA, June 26–28, 1991. Evanston, IL: American Automatic Control Council.

Fagan, M. (1974, December). *Design and Code Inspections and Process Control in the Development of Programs*, Report IBM-SDD-TR-21-572. New York: IBM.

Fields, R. E., Wright, P. C., and Harrison, M. D. (1995). A Task Centered Approach to Analyzing Human Error Tolerance Requirements. In P. Zave and M. D. Harrison (Eds.), *Proceedings of the 2nd International Symposium on Requirements Engineering (ICRE'95)* (pp. 18–26). Los Alamitos, CA: IEEE Computer Society.

Fleishman, E. A., and Quaintance, M. K. (1984). *Taxonomies of Human Performance: The Description of Human Tasks*. New York: Academic.

French, W. L., and Bell, Jr. C. H. (1973). *Organization Development: Behavioral Science Interventions for Organization Improvement*. Englewood Cliffs, NJ: Prentice-Hall.

Galbraith, J. R. (1997). *Organizational Design*. Reading, MA: Addison Wesley.

Galbraith, J. R. (1995). *Designing Organizations*. San Francisco, CA: Jossey-Bass.

Gardiner, M. M., and Christie, B. (1987). *Applying Cognitive Psychology to User-Interface Design*. New York: Wiley.

Gentner, D., and Gentner, D. R. (1983) Flowing Waters or Teeming Crowds: Mental Models of Electricity. In D. Gentner and A. L. Stevens (Eds.), *Mental Models*. Hillsdale, NJ: Erlbaum.

Greenbaum, J., and Kyng, M. (Eds.). (1991). *Design at Work: Cooperative Design of Computer Systems*. Hillsdale, NJ: Erlbaum.

Guzzo, R. A., and Salas, E. (Eds.) (1995). *Team Effectiveness and Decision Making in Organizations*. New York: Jossey Bass.

Hackos, J. T., and Redish, J. C. (1998). *User and Task Analysis for Interface Design*. New York: Wiley.

Hammer, M., and Champy, J. (1993). *Reengineering the Corporation: A Manifesto for Business Revolution*. New York: Harper Business.

Hammer, M., and Stanton, S. A. (1995). *The Reengineering Revolution: A Handbook*. New York: Harper Business.

Hammond, K. R. (2000). *Judgments Under Stress*. New York: Oxford University Press.

Harris, J. F. (1994, July 14). Downing of Copters Blamed on Blunders. *Washington Post, 117*(221), A1, A10.

Harrison, L. E., and Huntington, S. P. (Eds.). (2000). *Culture Matters: How Values Shape Human Performance*. New York: Basic Books.

Harrison, T. M. (1985, Spring). Communication and Participative Decision Making: An Exploratory Study. *Personnel Psychology, 38*(10), 93–116.

Hastie, R. (2001). Problems for Judgment and Decision Making. *Annual Review of Psychology, 52*, 653–683.

Helmreich, R. L. (1988, October). *Testimony before the U.S. House of Representatives, Committee on Armed Services on the Subject of the USS VINCENNES Downing of Iranian Air Flight 655*.

Henneman, R. L. (1999, Summer). Design for Usability: Processes, Skills, and Tools. *Information, Knowledge, and Systems Management, 1*(2), 133–144.

Hermann, C. F. (1972). Threat, Time, and Surprise: A Simulation of International Crisis. In C. F. Hermann (Ed.), *International Crises: Insights from Behavioral Research*. New York: Free Press.

Hoc, J.-M. (1989). Strategies in Controlling a Continuous Process with Long Response Latencies: Need for Computer Support to Diagnosis. *International Journal of Man-Machine Studies*, 30, pp. 47-67.

Hogarth, R. (1987). Judgement and Choice: the Psychology of Decision, 2nd Edition. New York: John Wiley.

Huey, B. M., and Wickens, C. D. (Eds.). (1993). *Workload Transition: Implications for Individual and Team Performance; Panel on Workload Transition, Committee on Human Factors, Commission on Behavioral and Social Sciences and Education, National Research Council*. Washington, DC: National Academy of Sciences.

Humphrey, W. S. (1989). *Managing the Software Process*. Reading, MA: Addison-Wesley.

Janis, I. L. (1982). *Victims of Groupthink*. New York: Free Press.

Janis, I. L. (1989). *Crucial Decisions: Leadership in Policy Making and Crisis Management*. New York: Free Press.

Janis, I. L., and Mann, L. (1977). *Decision Making: A Psychological Analysis of Conflict, Choice, and Commitment*. New York: Free Press.

Johnson, P., Diaper, D., and Long, J. (1984). Tasks, Skills and Knowledge: Task Analysis for Knowledge Based Descriptions. In *Interact'84*, Vol. 1. London: IFIP.

Johnson-Laird, P. N. (1983). *Mental Models*. Cambridge: Cambridge University Press.

Johnson-Laird, P. N. (1999). Deductive Reasoning. *Annual Review of Psychology, 50*, 109–135.

Kahan, J. P., Worley, D. R., and Stasz, C. (1989) *Understanding Commander's Information Needs*. [Technical Report R-3761-A]. Santa Monica, CA: The RAND Corporation.

Kieras, D. E. (1988). Towards a Practical GOMS Model Methodology for User Interface Design. In M. Helander (Ed.), *Handbook of Human-Computer Interaction*. Amsterdam: Elsevier.

Klein, G. A. (1993a). A Recognition-Primed Decision (RPD) Model of Rapid Decision Making. In G. A. Klein, J. Orasanu, R. Calderwood, and C. E. Zsambok (Eds.), *Decision Making in Action: Models and Methods*. Norwood, NJ: Ablex.

Klein, G. A. (1993b, April). *Naturalistic Decision Making: Implications for Design*, Technical Report: CSERIAC SOAR 93-1. Wright-Patterson AFB, OH: Crew System Ergonomics, Information Analysis Center.

Klein, G. A. (1998). *Sources of Power: How People Make Decisions*. Cambridge, MA: MIT Press.

Kotter, J. P., and Heskett, J. L. (1992). *Corporate Culture and Performance*. New York: Free Press.

Krishnan, V., and Ulrich, K. T. (2001, January). Product Design Decisions: A Review of the Literature. *Management Science*, *47*(1), 1–21.

Lehner, P. E., and Zirk, D. A. (1987). Cognitive Factors in User/Expert-System Interaction. *Human Factors*, *29*(1), 97–109.

Lerch, F. J., and Harter, D. E. (2001, March). Cognitive Support for Real Time Dynamic Decision Making. *Information Systems Research*, *12*(1), 63–82.

Lockhead, G. R. (1992). On Identifying Things: A Case for Context. In B. Burns (Ed.), *Percepts, Concepts and Categories: The Representation and Processing of Information*. Amsterdam: Elsevier Science.

MacMillan, J., and Entin, E. B. (1991). Decision-Oriented Display Design and Evaluation. In *Proceedings of the 1991 Symposium on Command and Control Research*, June 24–25, 1991, National Defense University, Ft. Lesley J. McNair, Washington, DC. McLean, VA: SAIC, December.

Mantei, M. M., and Teorey, T. J. (1988, April). "Cost/Benefit Analysis for Incorporating Human Factors in the Software Lifecycle." *Communications of the ACM*, 31(4), pp. 428–439.

Manzoni, J. F., and Angehrn, A. A. (1998, Winter). Understanding Organizational Dynamics of IT-Enabled Change: A Multimedia Simulation Approach. *Journal of Management Information Systems*, *14*(3), 109–140.

Mayer, R. E. (1979). Can Advance Organizers Influence Meaningful Learning? *Review of Educational Research*, *49*, 371–383.

Mayer, R. E., and Bromage, B. K. (1980). Different Recall Protocols for Technical Texts Due to Advance Organizers. *Journal of Educational Psychology*, *72*, 209–225.

Mayhew, D. J. (1999). *The Usability Engineering Lifecycle: A Practitioner's Handbook for User Interface Design*. San Francisco, CA: Morgan Kaufmann.

Meister, D. (1981). *Behavioral Research and Government Policy: Civilian and Military R and D*. New York: Pergamon.

Meister, D. (1985). *Behavioral Analysis and Measurement Methods*. New York: Wiley.

Meister, D. (1991). *Psychology of System Design*. Amsterdam: Elsevier.

Mellers, B. A., Schwartz, A., and Cooke, A. D. J. (1998). Judgment and Decision Making. *Annual Review of Psychology*, *49*, 447–477.

Montazemi, A. R., and Conrath, D. W. (1986, March). The Use of Cognitive Mapping for Information Requirements Analysis, *MIS Quarterly*, *10*(1), 45–56.

Nadler, D. A., and Tushman, M. L. (1997). *Competing by Design: The Power of Organizational Architecture*. New York: Oxford University Press.

Nielsen, J., and Molich, R. (1990). Heuristic Evaluation of User Interfaces. In J. Carrasco Chew and J. Whiteside (Eds.) *Proceedings of ACM CHI '90 Conference. Seattle, WA, 1–5 April 1990*. New York: ACM Press, pp. 249–256

Nielsen, J. (1993). *Usability Engineering*. San Diego, CA: Academic.

Norman, D. A., and Draper, S. W. (1986). *User Centered System Design*. Hillsdale, NJ: Erlbaum.

Nystrom, P. C., and Starbuck, W. H. (Eds.). (1981). *Handbook of Organizational Design*. New York: Oxford University Press.

Olson, G. M., and Olson, J. S. (1991). User-Centered Design of Collaboration Technology. *Journal of Organizational Computing*, *1*, 61–83.

Orasanu, J., and Salas, E. (1993). Team Decisionmaking in Complex Environments. In G. A. Klein, J. Orasanu, R. Calderwood, and C. E. Zsambok (Eds.), *Decision Making in Action: Models and Methods*, Norwood, NJ: Ablex.

Orasanu, J., and Shafto, M. G. (1999). Designing for Cognitive Task Performance. In A. P. Sage and W. B. Rouse (Eds.), *Handbook of Systems Engineering and Management* (pp. 629–658). New York: Wiley.

Payne, J. W., Bettman, J. R., and Johnson, E. J. (1993). The Use of Multiple Strategies in Judgment and Choice. In N. J. Castellan, Jr. (Ed.), *Individual and Group Decision Making*. Hillsdale, NJ: Lawrence Erlbaum.

Pecht, M. G. (1995). *Product Reliability, Maintainability, and Supportability Handbook*. Boca Raton, FL: CRC Press.

Pecht, M. G. (1999). Reliability, Availability, and Maintainability. In A. P. Sage and W. B. Rouse (Eds.), *Handbook of Systems Engineering and Management* (pp. 303–326). New York: Wiley.

Perdu, D., and Levis, A. H. (1993, September). Requirements Determination Using the Cube Tool Methodology and Petri Nets. *IEEE Transactions on Systems, Man, and Cybernetics*, *23*(5), 1255–1264.

Pew, R. W. (1988, October). Testimony before the U.S. House of Representatives, Committee on Armed Services on the Subject of the USS VINCENNES Downing of Iranian Air Flight 655.

Preece, J., Rogers, Y., Sharp, H., and Benyon, D. (1994). *Human-Computer Interaction*. Reading, MA: Addison-Wesley.

Rasmussen, J. (1986). *Information Processing and Human-Machine Interaction: An Approach to Cognitive Engineering*. New York: North-Holland.

Rasmussen, J., Pejtersen, A. M., and Schmidt, K. (1990, September). *Taxonomy for Cognitive Work Analysis*, Technical Report Risø-M-2871. Roskilde, Denmark: Risø National Laboratory.

Rasmussen, J., Pejtersen, A. M., and Goodstein, L. P. (1994). *Cognitive Systems Engineering*. New York: Wiley.

Rasmussen, J., and Vicente, K. J. (1989). Coping with Human Errors through System Design: Implications for Ecological Interface Design. *International Journal of Man-Machine Studies*, *31*, 517–534.

Rastegary, H., and Landy, F. J. (1993). The Interactions among Time Urgency, Uncertainty, and Time Pressure. In O. Svenson and A. J. Maule (Eds.), *Time Pressure and Stress in Human Judgment and Decision Making*. New York: Plenum Press.

Ravden, S. J., and Johnson, G. I. (1989). *Evaluating Usability of Human-Computer Interfaces: A Practical Method*. Chichester: Ellis Horwood.

Reason, J. (1990). *Human Error*. Cambridge, UK: Cambridge University Press.

Robbins, S. P. (1990). *Organization Theory: Structure, Design, and Applications*. Englewood Cliffs, NJ: Prentice-Hall.

Rogers, Y. (1992). Mental Models and Complex Tasks. In Y. Rogers, A. Rutherford and P. A. Bibby (Eds.), *Models in the Mind: Theory, Perspective and Application*. New York: Academic Press.

Roth, E. M., Bennett, K. B., and Woods. D. D. (1988). Human Interaction with an "Intelligent" Machine. In E. Hollnagel, G. Mancini, and D. D. Woods, (Eds.), *Cognitive Engineering in Complex Dynamic Worlds*. London: Academic.

Rouse, W. B., Cannon-Bowers, J. A., and Salas, E. (1992, November/December). The Role of Mental Models in Team Performance in Complex Systems. *IEEE Transactions on Systems, Man, and Cybernetics*, *22*(6), 96–138.

Rouse, W. B., Edwards, S. L., and Hammer, J. M. (1993, November/December). Modeling the Dynamics of Mental Workload and Human Performance in Complex Systems. *IEEE Transactions on Systems, Man, and Cybernetics*, *23*(6), 1662–1671.

Sage, A. P. (1991). *Decision Support Systems Engineering*. New York: Wiley.

Sage, A. P. (1992). *Systems Engineering*. New York: Wiley.

Sage, A. P. (1995). *Systems Management for Information Technology and Software Engineering*. New York: Wiley.

Sage, A. P. (1999). Systems Reengineering. In A. P. Sage and W. B. Rouse (Eds.), *Handbook of Systems Engineering and Management* (pp. 825–932). New York: Wiley.

Sage, A. P. (Ed.). (1987). *System Design for Human Interaction*. New York: IEEE.

Sage, A. P. (Ed.). (1990). *Concise Encyclopedia of Information Processing in Systems and Organizations*. Oxford: Pergamon.

Sage, A. P., and Palmer, J. D. (1990). *Software Systems Engineering*. New York: Wiley.

Sage, A. P., and Rouse, W. B. (Eds.). (1999). *Handbook of Systems Engineering and Management*. New York: Wiley.

Sage, A. P., and Small, C. T. (2000, October). A Simulation Perspective on Knowledge Management and Sharing, and Conflict and Complexity in Social Systems Management. In *Proceedings of the IEEE Conference on Systems, Man, and Cybernetics*, Nashville TN (pp. 536–541).

Salama, A. (1992, Autumn). The Use of an Organization's Biography as a Research Method for Investigating Organizational Development. *Management Education and Development*, *23*(3), 225–233.

Salas, E., and Cannon-Bowers, J. A. (2001). The Science of Training: A Decade of Progress. *Annual Review of Psychology*, *52*, 471–499.

Salvendy, G. (Ed.). (1997). *Handbook of Human Factors and Ergonomics*. New York: Wiley.

Salvendy, G. (Ed.). (2001). *Handbook of Industrial Engineering: Technology and Operations Management*. New York: Wiley.

Sanderson, P. M., Flach, J. M., Buttigieg, M. A., and Casey, E. J. (1989). Object Displays Do Not Always Support Better Integrated Task Performance. *Human Factors*, *31*(2), 183–198.

Schein, E. H. (1992). *Organizational Culture and Leadership*, 2nd ed. San Francisco, CA: Jossey-Bass.

Schneider, S. C., and Angelmar, R. (1993). Cognition in Organizational Analysis: Who's Minding the Store? *Organizational Studies*, *14*(3), 347–374.

Seamster, T. J., Redding, R. E., and Kaempf, G. L. (1997). *Applied Cognitive Task Analysis in Aviation*. Brookfield, VT: Avebury.

Senders, J. W., and Moray, N. P. (1991). *Human Error: Cause, Prediction, and Reduction*. Hillsdale, NJ: Lawrence Erlbaum.

Senge, P. M. (1990). *The Fifth Discipline: The Art and Practice of the Learning Organization*. New York: Doubleday Currency.

Senge, P. M., Roberts, C., Ross, R. B., Smith, B. J., and Kleiner, A. (1994). *The Fifth Discipline Fieldbook: Strategies and Tools for Building a Learning Organization*. New York: Doubleday.

Senge, P. M., and Sterman, J. D. (1994). Systems Thinking and Organizational Learning: Acting Locally and Thinking Globally in the Organization of the Future. In J. D. W. Morecroft and J. D. Sterman (Eds.), *Modeling for Learning Organizations*. Portland, OR: Productivity Press.

Senge, P. M., Kleiner, A., Roberts, C., Ross, R. B., Roth, G., and Smith, B. J. (1999). *The Dance of Change: The Challenges to Sustaining Momentum in a Learning Organization*. New York: Doubleday.

Shanteau, J. (1992). The Psychology of Experts: An Alternative View. In G. Wright and F. Bolger (Eds.), *Expertise and Decision Support*. New York: Plenum Press.

Sheridan, T. B. (2002). *Humans and Automation: System Design and Research Issues*. New York: Wiley.

Shneiderman, B. (1997). *Designing the User Interface: Strategies for Effective Human-Computer Interaction*, (3rd ed.). Reading, MA: Addison-Wesley.

Smith, S. L., and Mosier, J. N. (1986, August). *Guidelines for Designing User Interface Software*, Report ESD-TR-86-278. Bedford, MA: MITRE Corporation.

Somerville, M. A., and Rapport, D. J. (Eds.). (2000). *Transdisciplinarity: Recreating Integrated Knowledge*. Oxford: EOLSS Publishers.

Sproles, N. (2000, February). Coming to Grips with Measures of Effectiveness. *Systems Engineering*, *3*(1), 50–58.

Sproles, N. (2001, May). The Difficult Problem of Establishing Measures of Effectiveness for Command and Control: A Systems Engineering Perspective. *Systems Engineering*, *4*(2), 145–155.

Staggers, N., and Norcio, A. F. (1993). Mental Models: Concepts for Human-Computer Interaction Research. *International Journal of Man-Machine Studies*, *38*, 587–605.

Stake, R. E. (1995). *The Art of Case Study Research*. Thousand Oaks, CA: Sage.

Stammers, R. B., Carey, M. S., and Astley, J. A. (1990). Task Analysis. In J. R. Wilson and E. N. Corlett (Eds.), *Evaluation of Human Work: A Practical Ergonomics Methodology*. New York: Taylor and Francis.

Stolovitch, H. D., and Keeps, E. J. (Eds.). (1992). *Handbook of Human Performance Technology: A Comprehensive Guide for Analyzing and Solving Performance Problems in Organizations*. San Francisco, CA: Jossey Bass.

Suchman, L. A., and Trigg, R. H. (1991). Understanding Practice: Video as a Medium for Reflection and Design. In J. Greenbaum and M. Kyng (Eds.), *Design at Work: Cooperative Design of Computer Systems*. Hillsdale, NJ: Erlbaum.

Svenson, O., and A. J. Maule (Eds.) (1993). *Time Pressure and Stress in Human Judgment and Decision Making*. New York: Plenum Press.

Tainsh, M. A. (1985). Job Process Charts and Man-Computer Interaction within Naval Command Systems. *Ergonomics*, *28*, 555–565.

Treu, S. (1992). Interface Structures: Conceptual, Logical, and Physical Patterns Applicable to Human-Computer Interaction. *International Journal of Man-Machine Studies*, *37*, 565–593.

Trice, H. M., and Beyer, J. M. (1993). *The Culture of Work Organizations*. Saddle River, NJ: Prentice-Hall.

U.S. Department of Defense (DOD). (1994, June 8). *Human Systems Integration Requirements for Air Force Acquisition Programs*. OAIG-AUD, Report No. 94-124. Arlington, VA: Office of the Inspector General.

U.S. Department of Defense (DOD). (2000, October 23). *The Defense Acquisition System (Directive 5000.1) and Operation of the Defense Acquisition System (Instruction 5000.2)*. Available: ric.crane.navy.mil.

U.S. Department of Defense (DOD). (2001). *Acquisition Deskbook*. Washington, DC: Office of the Secretary of Defense. Available: www.deskbook.osd.mil.

van Gigch, J. P. (1991). *System Design Modeling and Metamodeling*. New York: Plenum.

Vanderhaegen, F., Crevits, I., Debernard, S., and Millot, P. (1994). Human-Machine Cooperation: Toward an Activity Regulation Assistance for Different Air Traffic Control Levels. *International Journal of Human-Computer Interaction*, *6*(1), 65–104.

Vennix, J. A. M., Andersen, D. F., Richardson, G. P., and Rohrbaugh, J. (1994). Modeling Building for Group Decision Support: Issues and Alternatives in Knowledge Elicitation. In J. D. W. Morecroft and J. D. Sterman (Eds.), *Modeling for Learning Organizations*. Portland, OR: Productivity Press.

Vicente, K. J., and Rasmussen, J. (1992, July/August). Ecological Interface Design: Theoretical Foundations. *IEEE Transactions on Systems, Man and Cybernetics*, *22*(4), 589–606.

Von Krogh, G., Ichijo, K., and Nonaka, I. (2000). *Enabling Knowledge Creation: How to Unlock the Mystery of Tacit Knowledge and Release the Power of Innovation*. New York: Oxford University Press.

Weick, K. E. (1995). *Sensemaking in Organizations*. Thousand Oaks, CA: Sage.

Weick, K. E., and Roberts, K. H. (1993). Collective Mind in Organizations: Heedful Interrelating on Flight Decks. *Administrative Quarterly*, *38*(3), 357–381.

Wellens, A. R. (1993). Group Situation Awareness and Distributed Decision Making: From Military to Civilian Applications. In N. J. Castellan, Jr. (Ed.), *Individual and Group Decision Making*. Hillsdale, NJ: Lawrence Erlbaum.

Wickens, C. D. (1987). Attention. (1987). In P. A. Hancock (Ed.), *Human Factors Psychology*. Amsterdam: North-Holland.

Wickens, C. D., and Hollands, J. G. (1999). *Engineering Psychology and Human Performance*, 3rd ed. Englewood Cliffs, NJ: Prentice-Hall.

Williams, K. E., Reynolds, R. E., Carolan, T. F., Anglin, P. D., and Shrestha, L. B. (1989, December). *An Evaluation of a Methodology for Cognitively Structuring and Adaptively Sequencing Exercise Content for Embedded Training*. Technical Report TR89-035. Orlando, FL: University of Central Florida, Institute for Simulation and Training.

Wohl, J. C. (1981). Force Management Decision Requirements for Air Force Tactical Command and Control. *IEEE Transactions on Systems, Man, and Cybernetics*, SMC-11(9), pp. 618–639.

Woods, D. D. (1984). Visual Momentum: A Concept to Improve the Cognitive Coupling of Person and Computer. *International Journal of Man-Machine Studies*, *21*, 229–244.

Woods, D. D., and Hollnagel, E. (1987). Mapping Cognitive Demands in Complex Problem-Solving Worlds. *International Journal of Man-Machine Studies*, 26, pp. 257–275.

Woods, D. D. (1988). Coping with Complexity: The Psychology of Human Behavior in Complex Systems. In L. P. Goodstein, H. B. Andersen, and S. E. Olsen (Eds.), *Tasks, Errors, and Mental Models*. London: Taylor and Francis.

Woods, D. D., and Roth, E. M. (1988). Cognitive Systems Engineering. In M. Helander (Ed.), *Handbook of Human-Computer Interaction*. Amsterdam: Elsevier.

Wright, P. C., and Monk, A. F. (1989). Evaluation for Design. In A. Sutcliffe and L. Macaulay (Eds.), *People and Computers V.* Cambridge: Cambridge University Press.

Wright, P. C., and Monk, A. F. (1991). A Cost-Effective Evaluation Method for Use by Designers. *International Journal of Man-Machine Studies*, *35*, 891–912.

Yin, R. K. (1994). *Case Study Research: Design and Methods*, 2nd ed. Thousand Oaks, CA: Sage.

Yu, E. S. K., Mylopoulos, J., and Lesperance, Y. (1996, August). Modelling the Organization: New Concepts and Tools for Re-Engineering. *IEEE Expert*, pp. 16–23.

Zachary, W. W. (1988). Decision Support Systems: Designing to Extend the Cognitive Limits. In M. Helander (Ed.), *Handbook of Human-Computer Interaction*. Amsterdam: Elsevier.

Zahniser, R. A. (1993, October). Design by Walking Around. *Comm. of the ACM*, *36*(10), 115–123.

PART III

# METHODS, TOOLS, AND TECHNOLOGIES

From the point of view of the human systems integration (HSI) practitioner, this part is the heart of the *Handbook*. Collectively the following seven chapters provide the state of the art for those methods, tools, and technologies needed by practitioners to effectively participate in the systems acquisition culture (Part I) and processes (Part II). All of the HSI domains have at least one chapter devoted to the methods, tools, and technologies specific to their domain, but the contributors have made special efforts to indicate areas where there is overlap among their disciplines and point out those methods, tools, and technologies that are designed to integrate several domains. Most of the chapters in this part provide amplifying information that is directly related to principle 7 (HSI technology).

Chapter 11, by Archer, Headley, and Allender presents the state of the art for manpower, personnel, and training (MPT) integration methods and tools. It begins with a description of MPT factors important in analysis and assessment issues, covers examples of tools developed by the U.S. military services, the United Kingdom, and Canada to support MPT trade-off processes, illustrates examples of nonmilitary MPT tools usage, and ends with the key challenges facing future developers of MPT integration technology. This chapter is the most comprehensive chapter in the *Handbook* for illustrating the technical results of military and commercial investment in HSI technology for the past decade.

In Chapter 12, Hettinger describes training as one of the most critically important disciplines involved in the safe and effective operation of complex human–machine systems. He recognizes, however, that the training domain is perhaps the least developed HSI domain conceptually, methodologically, and culturally. Although training, when utilized to prepare individuals to function within the context of existing systems, is well established, it has been largely unsuccessful at achieving the benefits that should be expected from an HSI approach to systems acquisition. The objective of training from an HSI perspective is not simply to instill knowledge, skills, and abilities (KSAs) into personnel adequate to satisfactorily operate and maintain developed systems. Training should also (a) influence systems requirements, design, development, test, and evaluation throughout all stages of the system acquisition process and (b) incorporate knowledge from other technical domains when conceiving training requirements and strategies. Because of the relative immaturity of the training domain at influencing the systems

*Handbook of Human Systems Integration*, Edited by Harold R. Booher.
ISBN 0-471-02053-2 

acquisition process, the primary objective of Hettinger is to provide a state-of-the-art review on training from an HSI perspective. He does this by providing a new training model as a framework for integrating training within the systems acquisition process; discussing the key training issues and challenges currently facing the training domain; and summarizing some of the more pressing research and applications questions needing answers to help steer training into a more effective HSI domain.

With Chapter 13, Lockett and Powers present a much needed overview of the broad area of human factors engineering (HFE) methods and tools. Although there is extensive information on methods and tools throughout the human factors literature, it is difficult for the beginning HFE practitioner working as part of an HSI team to employ HFE methods and tools effectively. Lockett and Powers organize their chapter so that it begins with a brief discussion of the purpose and scope of human factors engineering, followed with a description of HFE basic methods accompanied with useful information on how to find, select, and employ HFE technologies and tools. They describe the major classes of tools and methods available to address HFE issues as a part of HSI. Examples of tools and methods covered range from guidelines, standards, user juries, and mockups to task network modeling, human figure modeling, and work domain analysis. Near the end, they provide a section on common pitfalls to avoid in HFE that has proved especially useful for HFE program managers. Selected references are provided to guide the reader through the vast field of HFE.

Chapter 14, by Swallom, Lindberg, and Smith-Jackson, provides an introduction to the system safety (SS) domain for HSI. System safety includes system safety engineering and management and involves both the organizational practice of a systems approach and the design of the work environment to be consistent with a systems approach. System safety engineering deals with the tools of the trade, the principles and methodology of analyzing the hazards of system components, subsystems, and interfaces. System safety management deals with how system safety decisions are made based on the system safety engineering analysis. Swallom, Lindberg, and Smith-Jackson designed their chapter to provide information useful not only to SS practitioners but also to be helpful to those disciplines having interfaces with SS in conducting system development efforts. Their chapter includes key safety definitions and references, basic risk assessment models, a comprehensive list with selected examples of system safety methods and techniques, and an outline of the system safety process. The specific methods and techniques described include various hazard analyses, event and fault tree analyses, failure mode and effects analysis, and cause–consequence analysis. The chapter provides sufficient detail of the SS process steps (identify, assess, mitigate, verify, accept, and track SS requirements) so that the chapter can act as a general guideline for system safety applications in the systems acquisition process.

In Chapter 15, Roberts presents the state of the art on health hazards (HH) categories applicable to military and commercial environments. He describes each HH category in considerable detail, including the hazard definition, its health consequences, typical hazard sources, typical subject matter experts, and current tools and techniques used to assess and control the hazard. The HH categories are: acoustic energy, biological substances, chemical substances, oxygen deficiency (ventilation and high altitude), radiation (non-ionizing and ionizing) energy, shock (nonelectrical), temperature extremes (hot and cold), trauma, and vibration. Roberts also provides an extensive reference list for detailed sources on each HH category.

Perhaps the most unique domain for HSI is personnel survivability. Unlike the other six domains, it is not considered a formal discipline having a career field such as human factors, safety, health, personnel, and training. There are no institutions that grant degrees in personnel survivability. It is, however, an extremely important HSI domain. Personnel survivability is the HSI domain that addresses those system acquisition concerns of how to help personnel survive during combat and other hostile events. With Chapter 16, Zigler and Weiss introduce the reader to the personnel survivability domain by describing the systems survivability concept along with personnel survivability methodology available for use by survivability analysts. Zigler and Weiss discuss the systems acquisition issues involved in trying to increase the likelihood that equipment and people operating as a system can survive hostilities and illustrate these issues with numerous examples drawn from actual events. Their chapter defines the six components that make up personnel survivability and discusses each of the components in the context of a survivability analyses process. Survivability analysis tools such as the MANPRINT soldier survivability (SSv) "parameter assessment list" and the interactive survivability considerations chart are described to illustrate how the HSI analyst proceeds through the survivability analysis process.

Rouse and Boff acknowledge in Chapter 17 that assessing and trading off economic and noneconomic factors in HSI can be very difficult. The combination of such uncertainties as both tangible and intangible benefits, multiple stakeholders, and inherent unpredictability make cost–benefit analyses for HSI a considerable challenge. Rouse and Boff address this challenge by first reviewing alternative frameworks for cost–benefit analysis and then proposing an overall methodology for situations with these characteristics. Use of the methodology is illustrated by its application in three investment problems that involve technologies for aiding, training, and ensuring the health and safety of personnel in military systems. This chapter is one that perhaps could have fit into one or more of the other parts as well as Part III. It was placed here, however, because of its methodological approach and to expose HSI methods, tools, and technology developers to the subtleties and complexities of making a convincing cost–benefits argument for program managers and acquisition decision makers.

CHAPTER 11

# Manpower, Personnel, and Training Integration Methods and Tools

SUSAN ARCHER, DONALD HEADLEY, and LAUREL ALLENDER

## 11.1 INTRODUCTION: WORKFORCE CHALLENGES

A key concept in personnel staffing and systems integration is determining, acquiring, training, and retaining the proper number of people with the right skills for the jobs required to operate and maintain systems. Traditionally, most organizations attempt to match the number and skills of people necessary to meet acceptable performance at minimum cost. More recently, organizations have begun to recognize that the introduction of new technology—ranging from information technology, process monitoring and control, and robotic manufacturing to weapons technology—can significantly increase the difficulty of maintaining a proper mix of numbers and skills of people in the workplace. Some technology may help to reduce numbers and skills required as well as reduce the workload (both physical and mental) on employees. In other cases technology may, through its sophistication, cause an increase in the need for, and therefore the cost of, skilled individuals to operate the systems. Also, technology may not reduce workload but simply shift it from physical workload to mental workload. The technology–people trade-off in the workplace is a job design issue that can be addressed via the human systems integration (HSI) approach. The primary objective of this chapter is to describe the state of the art for HSI methods and tools particularly useful for analysis and assessment of manpower, personnel, and training (MPT) issues on system design and development programs.

The complexity and universality of the workforce problem facing organizations that procure, manufacture, and use systems and products can be appreciated from the following overview of workforce challenges facing the military, other government activities, and commercial industry.

*Handbook of Human Systems Integration*, Edited by Harold R. Booher.
ISBN 0-471-02053-2 

### 11.1.1 Military Manpower and Personnel

The military is facing serious manpower and personnel challenges as we enter the twenty-first century. Problems in meeting recruitment quotas due to competition from industry and college and an increasing sophistication of equipment equate to a need for soldiers with skills and aptitudes in advanced technologies. Also, jobs must be performed within the environment of a battlefield "office," which is further complicated by stressors such as long-term operations, fatigue, night operations, temperature extremes, protection against the nuclear–chemical–biological threat, noise, precipitation, crowding, rough terrain, and fear of the enemy strength. Now with an ever-increasing use of computing power on the battlefield in weapon, vehicle, and communication systems, warfighters must also cope with operating and maintaining complex systems and dealing with an information-rich tactical environment. Decision making will need to take place under conditions of high cognitive workload, perhaps to the point of "information overload," coupled with information uncertainty and time pressure. Special consideration is required to match the skills required to successfully perform these challenging jobs with the skills and abilities that the warfighters possess.

Simply throwing technology at the problem is not the solution. This notion is aptly stated in *SAILOR 21: A Research Vision to Attract, Retain, and Utilize the 21st Century Sailor*: "Many in Congress, the Department of Defense, and the Navy believe that if we have newer, bigger, more high-tech weapons systems, we don't need to worry about people. These new technologies may require fewer people, but those same people must be more capable, able to learn more, faster, and perform a much broader range of tasks" (Keeney and Rowe, 1998, p. 5).

### 11.1.2 Government Workforce

In spite of recent moves to reduce the federal workforce, the federal government is still the largest employer in the United States. The vast majority of the workforce is employed under the executive branch (about 98 percent) and is mostly distributed across 14 cabinet departments and 90 independent agencies. The U.S. Office of Personnel Management (OPM) reported in March 1999 that the executive branch employed nearly 2.8 million civilians. Of that number, 918,000 were in the Department of Defense. The remainder, slightly less than 2 million, are distributed among the other government cabinet departments and agencies.

There are significant differences between the federal working force and the workforce as a whole. In particular, almost half of all federal workers perform professional or managerial jobs. This rate is nearly twice as high as the remainder of the U.S. workforce, in which the largest group of workers is engineers, including chemical, civil, aeronautical, industrial, electrical, mechanical, and nuclear engineers.

The outlook for federal employment is bleak; it is projected to decline by 9 percent through 2008. Due to the competitive benefits and perceived stability of federal employment, this will translate to particularly aggressive competition for the remaining jobs. While this will lessen somewhat the pressure to recruit new employees, concerns about retaining high-quality workers will increase in order to maintain a desirable skill and experience balance across the workforce. In sharp contrast, 707,000 new government jobs are expected to arise in state and local government, reflecting growth in the population and its demand for public services.

These statistics describe the pressures in achieving a manpower and personnel balance in the government workforce that supports dynamic adjustment in the mix of quantities and skill levels needed for each job.

### 11.1.3 Commercial Workforce

There were about 7 million business establishments in the United States in 1997, providing close to 128 million jobs. This workforce is projected to reach nearly 150 million by 2008, not including approximately 12 million self-employed workers. Commercial organizations, including manufacturing, service, and financial organizations, are diverse and have a wide variety of manning and personnel challenges and must align with the vast range in workforce demographics. Broad disparities in age, experience, education level, and even motivation require a correspondingly broad range of manning and personnel policies. There are also some similarities to the federal sector and military: Recruiting is typically needed to replace workers in existing jobs, rather than to fuel employment growth.

While the operational environment of the military is much riskier than the environment typically experienced in commercial industries, nonmilitary organizations can also have unique challenges. A good example of this is a university hospital in California that was required to define and justify the skill levels of its medical staff before it would be considered for grant funding (Hager et al., 1998). Addressing manpower and personnel issues and selecting employees that are qualified and prepared for the jobs in a fast moving technological society is a nontrivial effort, and dealing with the gaps between the skills employees have and those that they need is quite difficult.

### 11.1.4 Meeting the Workforce Challenges

The solution to these identified skill-requirements gaps in industry is similar to the solutions pursued by the military and government in that if the selection process does not succeed in supplying the needed skills and aptitudes, the gap must be filled through implementing a training solution, a materiel solution, or a process reengineering solution. Training solutions are fairly easy to understand in an industrial setting because the structure of skills and abilities by worker specialty is not usually as rigid as in the military (although some labor unions do have similar structures). In industry, however, procuring training often requires extensive effort since competitive forces typically do not support sharing training courses across employers.

Materiel (i.e., equipment) solutions include adjusting the system and equipment to be simpler to use or including job aids that provide advice or assistance on the most challenging tasks. This solution is probably the most commonly used in response to addressing manpower and personnel gaps. One example from the commercial world is the advanced automated call centers that involve a wide range of speech recognition and automated message generation capabilities. Another example is the advanced production management systems installed in many production facilities that include robotic and electronic machine control tools. These tools have not only decreased the need for manual, low-skill tasks but also have allowed artificial intelligence techniques to replace some of the decision-making effort that used to require highly experienced worker involvement.

Process reengineering solutions can encompass the other types just discussed but can also extend to rethinking an existing process in order to improve it. Many examples of successful business process reengineering (BPR) efforts can be found in the literature

(Malhotra, 1998; Caron et al., 1994). One specific case is of a major consumer bank. In order to increase its competitive position, this bank instituted an effort to examine the flow of work inside its organization. In this study, bank managers recognized that no part of the process or the organization was "sacred" and that a well-done BPR effort had to have license to examine and question every aspect of the work flow (Grover et al., 1995). Every step of the process flow was examined to ensure that it added value to the product or service, and that it was performed as efficiently as possible. In this case, jobs were redesigned in order to streamline existing processes, impacting staffing, training, and selection policies.

The above examples are typical illustrations of how MPT issues affect government and commercial organizational decisions for systems and product acquisition. In the following sections, discussion of MPT domain factors, with the emphasis on the military, along with examples of the analysis and assessment methods currently available, are presented to illustrate how HSI technology can help address MPT challenges relevant to a wide variety of organizations. At the end of the chapter, a number of challenges for HSI technology are identified.

## 11.2 MANPOWER, PERSONNEL, AND TRAINING DOMAINS

A primary function of the various HSI tools and techniques developed to date is the integration of HSI domains, which generally refers in particular to the interaction of manpower, personnel capabilities, and training, and also human factors engineering. Figure 11.1 along with the following description illustrates the basic system integration model for HSI:

> A particular system design concept determines the human tasks that are required to operate, maintain, and provide logistical support to the system. The tasks in turn drive the requirements for quantitative manning, required characteristics and innate abilities of personnel, and needed training. Human performance is the product of the interactions of tasks with manpower, personnel, and training. The combination of human performance with the system design, in terms, for example, of lethality, mobility, vulnerability, reliability, maintainability, and availability, drives system performance. —Hay Systems, 1991, pp. 1, 3

In this section we describe the MPT domains individually and address the distinguishing elements of analysis for each element. A further guide to MPT issues and risks is via an electronic booklet developed by the U.S. Army Total Personnel Command and available at https://www.perscomonline.army.mil/DCSOPS/DCSOPS_MANPOWER.htm.

### 11.2.1 Manpower

The manpower domain focuses on establishing the number of people needed to operate, maintain, and support the system. These numbers include military and civilian resources.

***Typical Issues and Questions in the Manpower Domain*** While considered the most costly, manpower is perhaps the easiest domain to define and understand. It solely concerns the number of people and does not attempt to describe the people. In HSI language this is often referred to as the "spaces" problem, whereas considerations relating

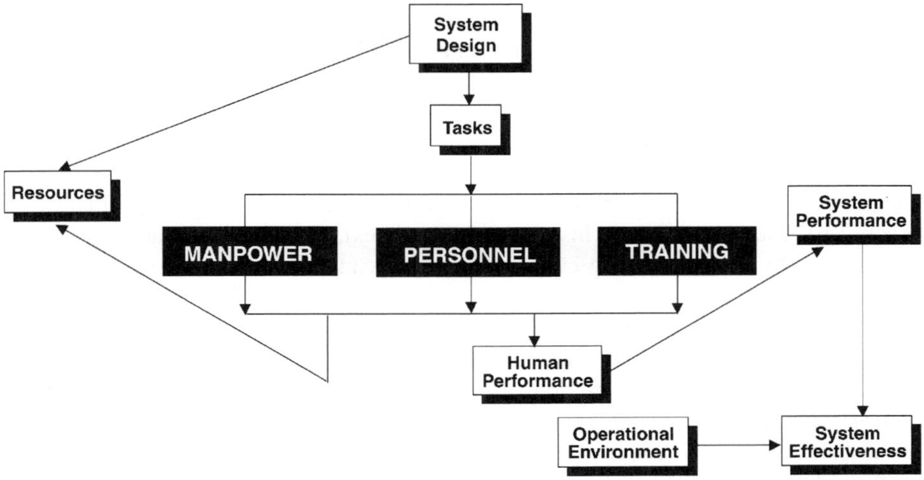


**Figure 11.1** MPT domains within a system structure (Hay Systems, 1991).

to the characteristics of the people are often referred to as the "faces" problem. This latter challenge belongs to the personnel domain.

A sample of the questions that can be answered by a manpower analysis include:

- How many maintainers do I need at each maintenance level in order to achieve the required system availability?
- Do I need two operators in my new system, or will advanced levels of automation allow me to reduce the crew size to one?
- Will the new concepts for bolt-on armor enable my crew to be able to transition an air-dropped vehicle to battle-ready status in less than one hour?
- Does the system require round-the-clock or sustained operations?
- Has special test equipment been identified that will require maintenance?
- Do sufficient human resources exist in the units that will receive the new system to conduct operations at the identified tempo?
- How many semiautonomous robotic systems can a single operator control? Or, conversely, how many operators are required to control a single robot?

In order to address each of these questions, we must use some method of estimating how much work needs to be done and dividing it by how much work a single person can do. The solution to this equation is the number of people you need. While this sounds extremely simple, both the numerator and the divisor can be difficult to estimate with precision. The remainder of this section provides examples of techniques that have been used to successfully assess these values.

***Manpower High Drivers*** A high driver is anything about the system that is resource intensive and could lead directly or indirectly to system performance problems. The system characteristics that drive manpower requirements are all linked to the amount of attention that a system needs in order to remain operational. The common element across all types of manpower analysis is the system's operational tempo. The operational tempo specifies the intensity and number of missions that must be performed by the system. This value drives operational, maintenance, supply, and support requirements of the system.

Manning requirements, as shown earlier in Figure 11.1, are driven by system design, but those requirements must be evaluated within the full operational context. For operational manning requirements especially, that context is described as MET-T (mission, environment, terrain, and tactics.) Those factors drive the frequency, difficulty, and criticality of task performance. In a battlefield context, the speed and accuracy of task performance are crucial. Making even a small error can have serious consequences.

Maintenance manpower requirements are driven by operational tempo as well as component reliability (i.e., how often the system must be repaired), maintenance concept (i.e., how many spares are available), and combat damage levels. Additionally, the shift lengths of the maintenance crew and the availability of special skills across each shift matched with the operational mission schedule can play a large role in determining maintenance manpower requirements. The notion of the operator-maintainer, while appealing, bears special consideration since both sets of drivers must be included in the process of manpower requirements determination.

Support manpower requirements are driven by the geographic relationship between the supply unit and the operational unit, as well as the physical requirements of the tasks. For example, if each round of ammunition must be manually loaded and is a two-man lift rather than loaded via an automated process, this will have a dramatic effect on support manpower requirements. Likewise, if resupply is required frequently and the distance to the ammunition storage point is great, then support manpower requirements are again increased.

Regardless of the specific manpower element of interest, it is clear that there is a strong and direct relationship between system design and manpower requirements. This provides the basis for the HSI community's need to apply manpower analysis tools very early in the weapon system acquisition process. It is through this process that the HSI community can influence system designers to implement design elements that reduce manpower requirements, leading to a more affordable system.

***Key Manpower Analysis Elements*** In the military environment, reaching an understanding between "what do we need" and "what can we have" for numbers of personnel (manpower) is an important part of the early planning process. An equally key determination is identification of the *bill payer* for manpower slots (i.e., the spaces). In a zero-sum environment, if a new or revised system is to add people, then the cost involves paying for these new slots by taking away from elsewhere. This consideration requires that a comprehensive manpower analysis system necessarily include force-level assessments. This is equally important for operators, maintainers, and support manpower.

The analysis elements for operational manning requirements typically begin with a classic task analysis. First, the functions that the system must perform in order to accomplish all of the missions in the operational requirement are identified. Once this is done, the functions can be decomposed into subfunctions and then allocated to people or machines. Once they are allocated to people, these subfunctions become operational tasks and form the basis for a detailed, job-oriented task analysis. This type of task analysis delineates the step-by-step process that must be performed to achieve a function, the estimated performance parameters (e.g., time, accuracy, workload) for the steps, and the consequences of incorrect performance. Once this information is gathered, the individual tasks and groups of tasks can be analyzed in order to determine where workload, or effort, peaks during the mission and function processes. These peaks can drive the work allocation among crew members and, thus, can determine the operational crew size required to successfully achieve the missions. Specific methods for performing this analysis are described later in this chapter.

Assessing the force-level implications of operator crews is usually straightforward. Once crew size for a single system is established (using comprehensive task analysis and workload modeling techniques discussed in Section 11.2.2 and Chapter 13), crew ratios (i.e., the number of crews per single system) are used as multipliers to extend the operational crew requirement to the various unit levels.

The analysis of maintenance manning requirements requires linking performance of maintenance tasks to system reliability data. An example of a comprehensive maintenance analysis tool is the Improved Performance Research Integration Tool (IMPRINT) developed by the Army Research Laboratory Human Research Engineering Directorate (Archer and Adkins, 1999; Archer and Allender, 2001).

IMPRINT predicts manpower requirements by simulating the maintenance requirements of a unit as the systems are sent out on missions, then to maintenance (as required), and then placed back into a pool of available systems. This process must continue for a significant period so that components with high reliabilities have an opportunity to fail, providing a realistic sample of maintenance requirements. There are, of course, a number of complexities involved in this process. For instance, when a system comes into maintenance, it is prioritized and scheduled for repair based on the pools of maintainers with specific specialties available in the particular maintenance level needed. Therefore, if the manpower pools are very tightly constrained, the maintenance will take longer, since fewer repairs can be performed in parallel. This will have an adverse effect on the system availability. If the system is not available to be sent out on other missions, it will actually accrue less usage because it is in maintenance, rather than performing missions on the battlefield. Oddly enough, this will result in less maintenance over time since the components are not accruing as much wear and tear. Other issues that affect the manpower required to maintain a system include spare availability, combat damage, maintenance shifting, and the criticality of individual component failures.

The process of developing a maintenance analysis, running it, and analyzing results is analogous to designing, executing, and analyzing an experiment. In this vein, the first step is to determine the questions that analysis must answer. A comprehensive maintenance manpower analysis capability can answer questions such as:

- How many people of each specialty are needed in order to meet the system availability requirement?
- Which pieces of equipment (i.e., subsystems) are the high drivers for maintenance?
- How should each organizational level be staffed?
- How sensitive is the maintenance manpower requirement to the failure rates of individual components?

After defining the questions that need to be answered, the dependent and independent variables must be selected. After these items are determined, the analyst is equipped to conduct a study that is designed to address the relevant questions.

To conduct an IMPRINT analysis of the maintenance man-hour requirements to support a particular system, four basic activities must be performed:

1. Prepare a description of the maintenance requirements for the system by specifying the following information for each component system:

   a. How often the component needs to be maintained (i.e., rounds fired, time operated).
   b. The type of maintenance task that needs to be performed (remove and replace, repair, inspect, troubleshoot, etc.).
   c. The type and number of maintainers that are needed to perform the maintenance task, including selection of specific soldier specialties.
   d. How long it will take to perform each maintenance task.
   e. Whether the maintenance is scheduled or unscheduled.
   f. The maintenance organizational type at which the task needs to be performed; up to three levels of organizations can be specified including factors of geographic dispersion.
   g. Whether a contact team could perform the maintenance.

2. Build a simulation scenario that defines the conditions that will be used for the system being modeled and the amount of usage the components in each system will incur. Usage can be described in time or distance units and also as the amount of ammunition fired, which permits greater simulation fidelity. Missions that are relevant to the scenario will determine system usage and probabilities for combat damage. Each scenario can contain multiple missions.

3. Define the unit configuration and support parameters for each scenario. These parameters include:

   a. Operational crew (per system)—This is an optional parameter and the information defaults to an empty set of operational crew members.
   b. Maintenance shift manning (size, type)—This parameter defaults to the minimum possible shift manning, as well as one shift per day that is 8 hours long. IMPRINT calculates the minimum shift manning by examining each maintenance task to find the minimum number of people in each specialty that will enable any given task to be performed.
   c. Spare parts (availability, wait times)—This is also an optional parameter and is specified at the subsystem level. This parameter defaults to 100 percent availability and a zero wait time.

4. Assess the results after executing the simulation for a sufficient number of scenario days so that both low- and high-reliability parts have a chance to fail. IMPRINT provides reports that identify the direct maintenance man-hours needed to achieve a specific reliability and availability or readiness level for the input parameters described in the previous three elements. These man-hours are converted into "bodies" by considering the required crew sizes to perform individual tasks and the annual maintenance man-hour availability at each organizational level. IMPRINT also outputs reports to help assess the subsystems and specialties that are "high drivers" in terms of maintenance requirements, so that users can evaluate trade-offs between reliability, maintenance concepts, manning, and operational capability.

If IMPRINT is used to evaluate the maintenance requirements for a proposed new system in the acquisition cycle, the component maintenance parameters can be entered from scratch, a system design, or contractor-provided logistics and reliability data [e.g., from a Logistics Support Analysis Report (LSAR)]. However, it is more likely that an analyst would begin by copying maintenance parameters from a similar library system and then modifying existing components and/or adding new components to reflect the system being evaluated. It is also possible to use IMPRINT for a different purpose, such as to address unit or organizational design questions. In this case, the analyst will probably copy a library system and use it as is. In this manner, the analyst can use the same components and maintenance actions and modify only the types of maintainers, maintenance levels, or other parameters for the existing components.

As mentioned above, the maintenance analysis also requires a mission schedule or operational profile. This information is used to determine the intensity of the scenario. The intensity, or operating tempo, drives the distance the system travels, the number of rounds each weapon system fires, and the number of operating hours that are accrued during each day of the scenario. This information, in turn, controls when the individual components will fail. Often, data that help define the operational profile are available in the operational requirements document (ORD), test and evaluation reports of a similar predecessor, system, or from subject matter experts (SMEs). If data elements exist for which there are no sources, multiple analyses can be performed using the most likely and worst-case values for these elements. The analysis can then be performed iteratively to determine how sensitive the maintenance manpower results are to the variability of these data items. If the results are not very sensitive to the values in question, then it is probably not necessary to invest more resources in finding better data. If the results are very sensitive, then it is important to improve the quality of the data, or to provide a decision maker with information on the most likely and worst-case results.

Once optimal maintenance manning decisions are made through an analysis process such as the one described, they are added to the operational and supply manpower requirements at the force level, and the total is used as a basis for calculating any necessary additional directed manpower. Directed manpower is usually ratio-based. That is, it is a proportional allotment of manpower slots needed to account for other support personnel (e.g., chaplains, administration). A system's total manpower burden can then be expressed in terms of numbers of people by specialty and organizational level.

### 11.2.2 Personnel

The personnel domain focuses on assessing the types of people needed to operate, maintain, and support the system. The experience, aptitudes, and physical characteristics can all be used to describe the personnel requirements.

***Typical Issues and Questions in the Personnel Domain*** In conducting a personnel staffing capabilities assessment for decision makers, the HSI analyst typically looks for "disconnects" between what the system is supposed to do and the capabilities of the men and women who will operate, maintain, and support that system. Example issues and questions facing military organizations and decision makers in the procurement process when personnel capabilities and new technology do not connect include:

- Do the skills of the target audience match up with the system requirements?
- Will the system be usable for the target audience?
- Does the new system require jobs that:

  a. Are difficult to recruit because of high entrance requirements?
  b. Are difficult to retain because of ample public-sector opportunities?
  c. Require a Top Secret security clearance?

The remainder of this section describes the U.S. Army's method of assessing personnel staffing capabilities that would be required for a system being planned or revised.

***Personnel High Drivers*** One very useful result of a personnel staffing capabilities assessment is to project as early as possible what likely *high drivers* will occur with the system when it is implemented in the organization. Typical items include requirement of specialty personnel, time constraints for system development, high physical or cognitive workload requirements, an undefined or hard to define target audience, and a wide-scoped target audience.

Elaborated by Headley (in press) and illustrated in Figure 11.2, the personnel capabilities assessment model follows a four-step process in determining high drivers. Headley derived the process from guidance provided by a number of sources (Guerrier et al., 1991; U.S. Total Army Personnel Command, 1991; Herlihy et al., 1990; Archer and Adkins, 1999).

Steps 1 and 4 of Figure 11.2, although important and often time-consuming, are obvious and need little discussion here. To aid in step 1, beneficial contacts, documents, and web sources for military systems and personnel background information are provided by Headley (2002) and Herlihy et al. (1990). For step 4, the end result of the assessment methodology, the primary feature to appreciate is that it is a formal report that states the MPT issues with associated *concern, major*, or *critical* ratings and that when these ratings

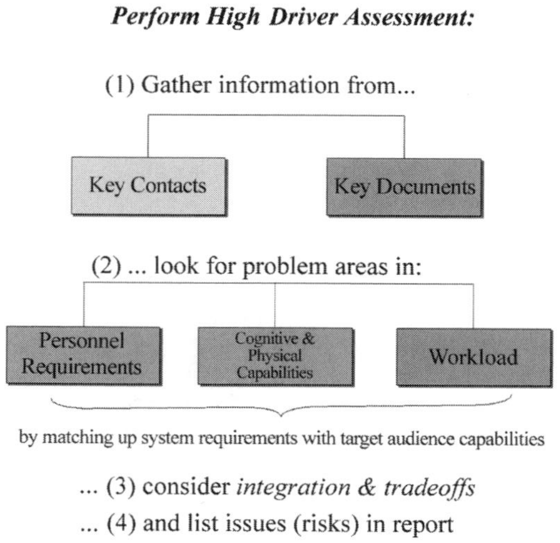


**Figure 11.2** Key steps in performing personnel staffing assessments.

are determined by an organization that has an acceptable level of HSI maturity (see Chapter 1), a system will not be allowed to continue in the acquisition cycle until major and critical issues are resolved (Headley, in press). Step 2 is described in the section immediately following, and step 3 is elaborated for manpower, personnel, and training in Section 11.2.4.

***Key Personnel Analysis Elements*** Any organization that wishes to improve its capabilities through the introduction of new technology must consider the personnel staffing capabilities needed to operate and maintain the technology and determine whether the personnel costs of this capability is affordable. Criteria to make this judgment include the skills of people needed to perform the jobs and tasks both generated and eliminated by the new technology. Assessing whether the required capabilities are available and affordable is a difficult task and cannot be done effectively with large, complex operations such as military systems acquisition without decision aids.

Critical information in determining personnel requirements is found in describing the *target audience*, which includes the types of systems used, key statistics on the personnel pool, and descriptions of relevant occupational specialties. Given the typical constraints of no increases in personnel, no newly created specialties, and no increase in training burden, designing for usability by a defined audience is important. Sometimes the target audience is narrow and easily defined, such as an infantryman. But for some systems, the user community is vast and the ORD may include the term *general-purpose user* (GPU). An example of a true GPU system would be the Army Distance Learning Program, which is building digital training facilities (CD-ROM courses, two-way televised instruction, e-mail, etc.). Any soldier or Department of the Army employee is potentially a student in such a classroom. As a result, the digital training facilities can be designed with this very broad target audience in mind. Note, however, that even with a very focused target audience, when the mission broadens, as with the U.S. Army's vision for the Objective Force Warrior, and the skills and abilities required also broaden, the description of the target audience is still essential and, in fact, may need to be filled in with greater detail.

Key statistics describing the active enlisted personnel that comprise typical enlisted operators and maintainers can be found in a U.S. Army study (Department of the Army 1997):

- 50 percent are in the age range 21 to 29; 94 percent in the range 17 to 39; median age is 26 years.
- 96 percent of the enlisted force have high school diplomas.
- Average years of service is eight.
- 15 percent are females.
- The most frequent rank is that of specialist (E4), comprising 25 percent of the enlisted force.

Many systems are meant for operation by both genders. This is especially important to know from a design point of view if anthropometrics is a key factor, necessitating designing for the "5th percentile female to the 95th percentile male" soldier on stated parameters. However, users on some systems will be all males, as will be the case for those specialties closed to women. This exclusion is due to the Direct Combat Probability Coding Policy, which proscribes that jobs routinely exposed to direct combat are off limits to women. Currently, 40 U.S. Army jobs are closed to women for this reason; example jobs are infantryman, armor crewman, cavalry scout, M1 Abrams tank turret mechanic, and combat engineer (Department of the Army, 1999).

The armed services place great emphasis on recruiting and retaining the proper skill base to attain operational readiness. Given that the state of the skill base is always in some degree of flux, there is a need to examine how the available base will tie in with the quality of training and usability of the system interface. Some jobs are typically hard to fill, especially those requiring the smarter soldiers (e.g., intelligence analyst) as defined by high entrance exam scores. This consideration gains importance in filling the need for more "digital warriors," that is, soldiers with basic skills in operating and maintaining computerized systems and also higher cognitive abilities for complex multitasking and attention management. The U.S. Army's Office of the Deputy Chief of Staff for Personnel routinely maintains a list of the top 25 recruiting priorities for entry-level soldiers. An HSI analyst can use this list to identify high-demand and low-supply skills.

*Cognitive and Physical Requirements* Entry into most workplaces in our society requires meeting certain standards. For jobs, whether in the academic, business, or military communities, an applicant is expected to possess or have the potential to acquire selected skills and abilities. The military requires meeting strict standards in these areas:

- Educational
- Medical
- Height, weight, and size characteristics
- Moral
- Physical requirements
- General muscular strength and endurance and cardio-respiratory fitness
- Strength requirements for performance of job-specific physically demanding tasks
- Aptitudes related to job requirements

Of particular note to the HSI analyst is that an applicant for a given job must meet specific aptitude criteria, as established in Army Pamphlet 611-21, "Military Occupational Classification and Structure." (Department of the Army, 1999). Examples of requirements for two army jobs from this document are:

| 11B Infantryman | 96B Intelligence Analyst |
|---|---|
| • A minimum score of 90 in aptitude area "combat"<br>• Color discrimination of red/green<br>• Occasionally raises and carries 160-pound person on back. Frequently performs all other tasks while carrying a minimum of 65 pounds, evenly distributed over entire body | • A minimum score of 105 in aptitude area "skilled technical"<br>• Eligibility to meet requirements for top secret security clearance and sensitive compartmented information access<br>• Occasionally lifts 37 pounds and carries 50 feet as part of a multiperson lift |

Cognitive aptitude comes into play for recruiting, training, and interface design of systems. The cognitive aptitude component is stated in the form of a minimally acceptable score on 1 of 10 *aptitude areas*. In the examples above, the cutoff score for an *infantryman* is 90 on the aptitude area *combat*, and the cutoff for an *intelligence analyst* is 105 on the area *skilled technical*. Aptitude area scores are derived from scores earned on subtests of the recruiting test battery called the Armed Services Vocational Aptitude Battery (ASVAB). The better the performance on this test, the more jobs for which one is eligible. In addition to the aptitude area scores that are specific for different job categories, ASVAB subtests are used to derive the more general Armed Forces Qualification Test (AFQT) score. Four ASVAB subtests (arithmetic reasoning, mathematics knowledge, word knowledge, and paragraph comprehension) contribute to the AFQT. As stated by the Office of the Assistant Secretary of Defense (1999), the AFQT score is representative of:

- General measure of trainability
- Predictor of on-the-job performance
- Primary index of recruit aptitude

Test scores on the AFQT have been shown to be predictors of military performance. In one study, performance on tasks pertaining to communications networks highly correlated with the average AFQT score of three-soldier teams (Winkler, 1999). Another study found consistent predictability between AFQT scores and performance on a number of written and hands-on tests by soldiers in missile-system-related job series (Horne, 1986). Modest correlations were found between ASVAB scores and vehicle identification performance (Heuckeroth and Smith, 1990).

Prior to induction, recruits are rated according to a physical profile made up of six factors: physical capacity, upper extremities, lower extremities, hearing, and ears, eyes, and psychiatric (PULHES). These categories are shown in Table 11.1.

Each U.S. Army job lists a profile for these physical abilities on a scale that ranges from 1 (high level of medical fitness) to 4 ("medical conditions or physical defects of such severity that performance of military duty must be drastically limited"; AR 40–501; Department of Army, 1998). For induction the practical limit is a 2 on any given ability for most U.S. Army jobs. Most jobs have a *physical demands* rating that indicates the level of body strength needed to perform the tasks. For example, the infantryman rating is "very heavy," which is defined as "lift on an occasional basis over 100 pounds with frequent or constant lifting in excess of 50 pounds" (Department of the Army, 1999). A legal specialist's job is rated as "light" ("lift on an occasional basis a maximum of 20 pounds with frequent or constant lifting of 10 pounds"). The analyst should assess whether physical tasks marked for a given job are within the already set rating.

**TABLE 11.1 Some Examples of Required Fitness Levels**

| Ability | Infantryman | Recruiter | Cavalry Scout |
|---|---|---|---|
| Physical capacity or stamina | 1 | 1 | 1 |
| Upper extremities | 1 | 3 | 1 |
| Lower extremities | 1 | 2 | 1 |
| Hearing and ears | 2 | 2 | 1 |
| Eyes | 2 | 2 | 2 |
| Psychiatric | 1 | 1 | 1 |

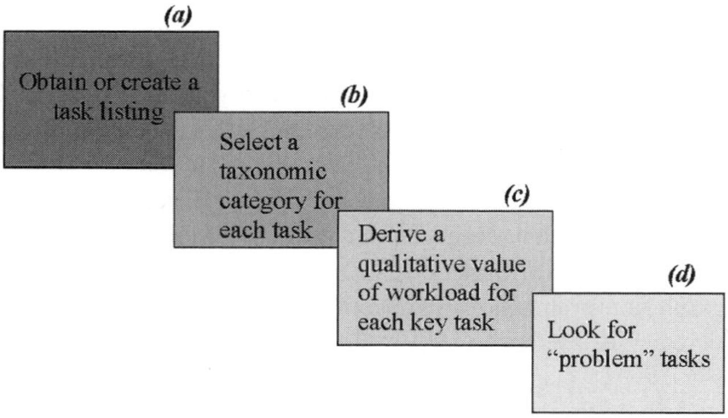


**Figure 11.3** Strategy for task workload assessment.

*Task Workload* Workload is a measure of the amount and quality of effort required to perform a set of tasks and is used here to refer to workload involving perceptual, cognitive, and fine motor processes rather than gross motor abilities. Many theories exist that define the unit of workload measurement and the threshold of workload that is acceptable. However, most theoreticians agree that one or more tasks of a job could cause high cognitive or physical workload such that overall system performance is affected. Solutions to high workload could be job redesign, interface redesign, or increased crew size (the latter, of course, is typically to be avoided).

Figure 11.3 illustrates typical procedures for identifying high workload tasks. These are conventional task analyses exercises and can be found discussed more fully in Headley (in press) and Chapters 10, 13, and 19.

As illustrated in step (b) in Figure 11.3, it is helpful to classify a given task by the channel(s) (visual, auditory, etc.) required. This categorization aids later diagnosis of how to alleviate a high workload condition. For example, if visual workload is high at the same time that auditory workload is low, the HSI analyst should consider whether some visual information can be converted to auditory cues, possibly through a speech synthesis system. SMEs are a good source for these data. Three convenient taxonomies are listed in Table 11.2.

**TABLE 11.2 Workload Taxonomies**[a]

| VACP Model[b] | NASA TLX Model[c] | SWAT[d] |
|---|---|---|
| Visual | Mental demand | Time load |
| Auditory | Physical demand | Mental effort load |
| Cognitive | Temporal demand | Psychological stress load |
| Psychomotor | Performance | |
| | Effort | |

[a]Abbreviations: VACP, visual, auditory, cognitive, psychomotor; NASA TLX, National Aeronautics and Space Administration Task Load Index; SWAT, Subjective Workload Assessment Technique.
[b]McCracken and Aldrich (1984).
[c]Hart and Staveland (1988).
[d]Reid et al. (1989).

A final note about the key personnel analysis element of task workload and cognitive requirements has to do with the capability to examine this in detail. For task workload, the effort required to perform a task is rated. Similarly, the cognitive skill required to perform a task can be rated. Such ratings are important for both personnel and training analysis and one well-known approach, the Job Assessment Software System (JASS) (e.g., Knapp and Tillman, 1998) is briefly described in Section 11.2.3. Another approach is available within IMPRINT, where an abbreviated taxonomy of nine task types is used to describe individual tasks. For each task, when an operator is assigned to perform the task, a specific U.S. Army military occupational specialty (MOS) is also selected. Along with that MOS selection comes the associated ASVAB-based aptitude area and AFQT scores. If a proposed task or set of tasks is suspected to require a different set of skills and abilities than those generally available with that MOS, an analysis can be conducted. For example, in order to evaluate whether an increase in personnel abilities is required, and if so, how much of an increase, a higher score cutoff is selected and the resulting effect can be determined at the level of the individual task performance time and accuracy or aggregated at the overall system mission level. However, it may also be the case that a particular task type does not benefit from an increase in ability level, in which case, no appreciable effect is seen and some other approach to bridging the personnel-performance gap such as training must be examined.

### 11.2.3 Training

The training domain is concerned with assessing the likelihood that the stated system instructional plan will provide personnel with the job skills and knowledge required to properly operate, maintain, and support a system. Notice that this domain addresses the training *re-quirements* for a system and does not typically include the development of training methods and techniques per se. (Note, however, that this distinction is blurred when the system being developed is itself a large-scale training system such as the U.S. Army's Combined Arms Tactical Trainer.) Often training requirements are presented in terms of gaps or differences between the current training program and the new system. Once this is accomplished, this information can be used to determine the training needs for the new or modified system.

***Typical Issues and Questions in the Training Domain*** In conducting a training requirements assessment for decision makers, the HSI analyst typically looks for "disconnects" or "leap aheads" between how the selected operators, maintainers, and support crew are currently prepared to perform their jobs and what they must do for the new system. Example issues and questions facing military organizations and decision makers in the procurement process when current training plans and new technology do not connect include:

- How do the current knowledge and skills of the target audience as provided by the current training [i.e., program of instruction (POI)] match up with the tasks the crew members must perform on the new system? What is the basic ability to attain the skills that will be trained?
- Does the new system require jobs that:

  a. Are hard to train because of high aptitude requirements?
  b. Are expensive to train because they are unique?
  c. Are difficult to train because experienced performers do not exist?

The remainder of this section describes the U.S. Army's method of assessing training requirements that would be required for a new or modified system.

***Training High Drivers*** Historically, the training burden implications of a new or modified system have been overlooked or underestimated. One reason for this is that the bill payer for training is not always the system developer, which indirectly influences the developer to use training as a panacea for usability problems. One very useful result of the training domain of HSI is that it requires a detailed assessment of training requirements early in the design process, to include the tasks and equipment elements that are contributing to the requirement.

High drivers that emerge as a result of a detailed training requirements assessment will likely fall into three general categories—the frequency, difficulty, and criticality of task performance. If a task is performed with a high degree of frequency, even if it is a relatively easy, not cognitively demanding task, it is a training high driver. Tasks also may be deemed high drivers if they are inherently difficult; for example, if they require high cognitive skill and will be performed under time pressure. Or it may be that some aspects of task performance are hard to remember, say a task with a large number of substeps that must be performed in sequence without a memory aid. [See Rose et al. (1985) for a detailed discussion.] Other possible high driver tasks are those that will be performed under high physical workload or stress or performed as a part of a team. These sorts of tasks are critical, yet difficult or costly to train. Finally, tasks are training high drivers if they are critical to the overall mission.

***Key Training Analysis Elements*** Unfortunately, analysis tool development to support the training domain has received less attention than either the manpower or personnel domain. This is probably due in part to a dearth of empirical data to support quantitative, performance-based analysis. Training-related data are more likely to pertain to the effectiveness of a training method or technique rather than to training requirements determination.

Analyses conducted to support this domain provide important data necessary to ensure that a system can be successfully fielded within the schedule and the life-cycle cost constraints. The results of this element of the HSI analysis process provide the inputs for the training development process. In most system designs, this particular element of the analysis will focus on tasks and functions in which new and unique skills and knowledge are required of the users—those "leap ahead" requirements.

The training analyst accepts as input the knowledge, skills, and abilities (KSA) elements attached to each task and the equipment required to perform each task. These KSAs encompass the cognitive and physical workload aspects identified to support the personnel domain, and extend to broader considerations of task performance. One tool that has been successfully used to identify KSAs needed to perform new tasks is the U.S. Army's JASS (Knapp and Tillman, 1998; Barnes et al., 2000a). The individual skills and abilities in the JASS taxonomy are shown in Figure 11.4 (see Fleishman and Quaintance, 1984, for foundational work on JASS categories).

A typical training analysis process is shown in Figure 11.5, in which each task is scored using benchmarks aligned to basic performance skills (e.g., communication, analysis, information processing, gross motor, fine motor, etc.). The tasks are then assigned to a targeted specialty. The current POI for that specialty and the personnel characteristics drive the KSA profiles for the soldiers. The two profiles can be directly compared across the eight major categories of skills yielding skill gaps (i.e., skill categories for which the

**Cognitive Skill and Experience Clusters**

| Communication | Conceptual | Reasoning | Speed-Loaded |
|---|---|---|---|
| 1. Oral Comprehension | 5. Memorization | 13. Inductive Reasoning | 19. Time Sharing |
| 2. Written Comprehension | 6. Problem Sensitivity | 14. Category Flexibility | 20. Speed of Closure |
| 3. Oral Expression | 7. Originality | 15. Deductive Reasoning | 21. Perceptual Speed and Accuracy |
| 4. Written Expression | 8. Fluency of Ideas | 16. Information Ordering | 22. Reaction Time |
| | 9. Flexibility of Closure | 17. Mathematical Reasoning | 23. Choice Reaction Time |
| | 10. Selective Attention | 18. Number Facility | |
| | 11. Spatial Orientation | | |
| | 12. Visualization | | |

**Perceptual-Motor Ability Clusters**

| Vision | Audition | Psychomotor | Gross Motor |
|---|---|---|---|
| 24. Near Vision | 31. General Hearing | 34. Control Precision | 41. Extent Flexibility |
| 25. Far Vision | 32. Auditory Attention | 35. Rate Control | 42. Dynamic Flexibility |
| 26. Night Vision | 33. Sound Localization | 36. Wrist-Finger Speed | 43. Speed of Limb Movement |
| 27. Visual Color Discrimination | | 37. Finger Dexterity | 44. Gross Body Equilibrium |
| 28. Peripheral Vision | | 38. Manual Dexterity | 45. Gross Body Coordination |
| 29. Depth Perception | | 39. Arm-hand Steadiness | 46. Static Strength |
| 30. Glare Sensitivity | | 40. Multi-Limb Coordination | 47. Explosive Strength |
| | | | 48. Dynamic Strength |
| | | | 49. Trunk Strength |
| | | | 50. Stamina |

**Figure 11.4** JASS categories.

allocated tasks exceed the skill levels currently supplied by the training and selection criteria for that MOS) and high driver tasks (i.e., specific tasks that cause unique or exceptionally high skill levels). The skill gaps become training or possibly personnel selection challenges. The high driver tasks become design and integration challenges.

This approach enables the HSI analyst to assess training impacts very early in the system design process so that training development can address unique needs of the new or modified system and can progress in parallel with the system integration process. It provides a clear link between the tasks that are driven by technology selections and integration decisions and the skills needed to perform those tasks. Finally, it compares those skill requirements to skill availability in selected MOSs and identifies skill gaps, many of which become training requirements and can be passed on to the training development team.

### 11.2.4 Trade-offs Among Domains

The HSI initiative helps an analyst identify issues within separate domains that affect system design and development. However, much of the value of HSI is in how it identifies issues that interact across domains. For military applications, this notion is directly stated

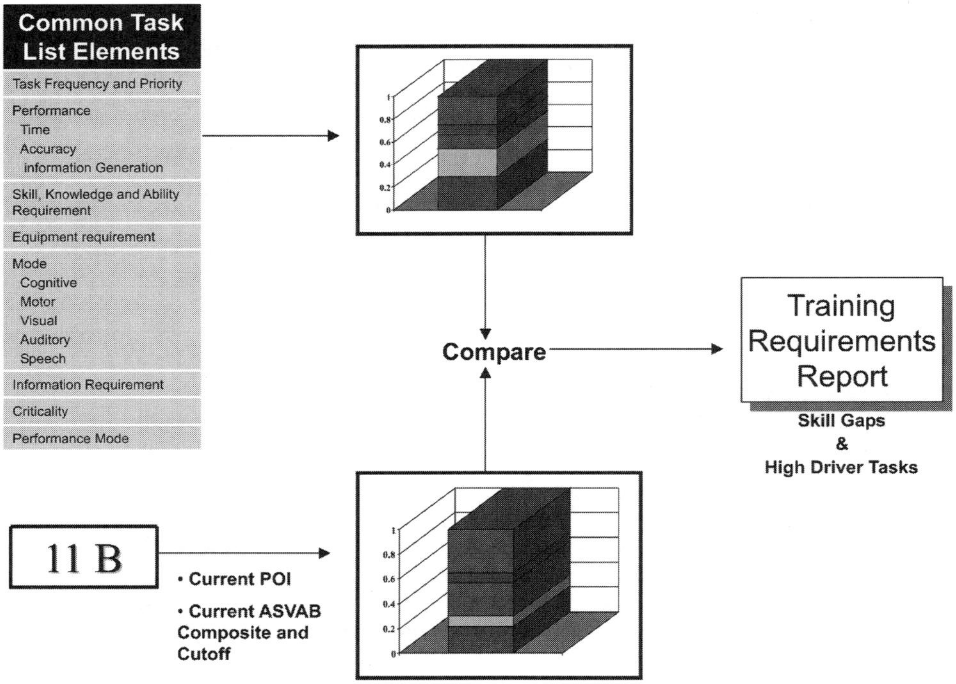


**Figure 11.5** Task data drive skill requirements and enables direct comparison to POI and personnel aptitude data to reveal specific training or selection needs.

in the Military Handbook *Human Engineering Program Process and Procedures* [MIL-HDBK-46855A; (U.S. Department of Defense, 1999)]: "With few exceptions, the MPT areas are predominant sources of life cycle cost of major systems. Since initial MPT constraints limit design, while equipment, software, and task design place demands on MPT resources, it is imperative that the HE and MPT communities harmonize their work from requirements-setting to total system design and evaluation" (p. 23).

Frequently, domains interact with one another such that successful resolution of a high driver in one domain might cause a significant negative impact in another domain. Acceptable manpower limits, the aptitudes and characteristics of user personnel, the training burden, the design, and acceptable performance criteria must all be considered together. For example, take the interaction between manpower and personnel. If manpower requirements go up (more people are needed), personnel quality will go down. Similarly, when the human factors engineering of a system is poorly done, the training requirement will increase. The multiway trade-offs among training, personnel, and equipment design are often a major HSI consideration in minimizing the effects of high drivers. For example, if a relatively poor design were locked in place, and the target audience were relatively unsophisticated in terms of the system at hand, then the only variable to manipulate is training in order to assure acceptable performance. In the remainder of this section, we describe some of these trade-offs in more detail in order to set the stage for a subsequent discussion on MPT integration tools.

Table 11.3 highlights the principal areas of trade-off applicable to the manpower and personnel capabilities domains (see also Dynamics Research Corporation, 2000). The three examples below illustrate more specifically the variety and importance of making proper trade-offs in design decisions:

***Trade-offs between Aptitude-Training-Performance*** Often the most desirable system design desired is one that allows acceptable system performance with low-aptitude operators and minimal training. However, this outcome is seldom possible even with financial resources available to assist the design effort. If the system design results in a more complicated system than expected, but not enough high-aptitude operators are readily available, then enhancing the training package might be required. As shown in Figure 11.6, different combinations of aptitude and training can result in different design concepts.

**TABLE 11.3 Trade-off Areas**

| Areas | Comment |
|---|---|
| • Operators versus automation | Functional allocation of task to people or equipment |
| • Aptitude versus training | Lower aptitude people have greater training requirements |
| • Aptitude versus decision aids | Relationship between aptitude and best use of decision aids |
| • Design versus target audience (5th–95th percentile) | Consider cost and increase in personnel availability between a 10–90 versus a 5–95 percentile design |
| • Manpower versus built-in test/built-in test equipment | Embedded diagnostics can reduce maintenance times and personnel required (but, consider cost and reliability of equipment) |
| • Design versus manpower | System's design influences workload that in turn influences numbers of operators and maintainers |
| • Design versus aptitude | More complex tasks that are not automated will require higher aptitudes in order for them to be performed at required level |
| • Maintenance manpower versus support manpower | Watch for trade-offs that reduce burden in one area but end up increasing burden in other areas |
| • Manpower versus training | Using maintenance as an example, if specialist positions are required, training time goes down but overall manpower is increased; if generalist positions are planned (e.g., a suite of systems is to be maintained by one specialty), then the number of positions is decreased, but training time will likely need to be longer |
| • Manpower versus aptitude | Fewer personnel with high aptitudes may be required to perform a given set of tasks to time and accuracy standards; but, such people may be in limited availability |
| • Personnel characteristics versus safety | Example: A visual warning in red is required; a constraint now exists to allow only operators who are not color blind |
| • Health hazards versus personnel characteristics | Army constraints of operator's capabilities (e.g., strict visual requirements) reduce risk also reduce the availability of the personnel pool |

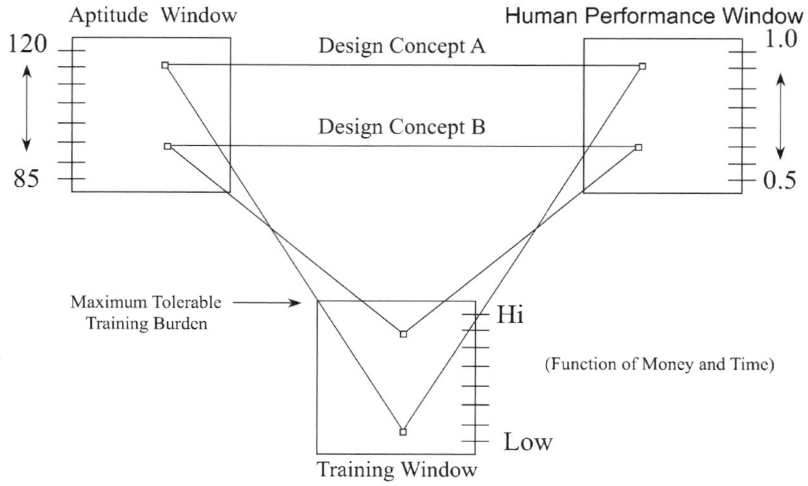


**Figure 11.6** Aptitude training, and performance trade-offs (adapted from Guerrier et al., 1991).

***Computer Systems*** Achieving usability of automated information systems is typically accomplished through balancing an interaction of the quality of screen designs (i.e., navigation, consistency, layout, labeling, etc.), the type of training required, and the background the target audience brings to the system. Unlike the standard practice of the mid-1980s when a box of software included a paper user's guide, today much commercial-off-the-shelf (COTS) software is shipped without the manual; rather, a reduced set of useful information is embedded in the software. This is often, but not always, successful. When successful, this achievement is due in part to designing to the low-skill end of a wide target audience (i.e., the least common denominator); therefore, the more usable the product the greater the potential for increased sales.

Military software is often customized and built for a more specific user group. It may also be more complicated than a typical COTS product. Therefore, it is important to consider how the interface, operator aptitudes and experience, and required training interact. Three minimally acceptable scenarios for combinations of the three factors are shown in Figure 11.7. Poor interface design will likely require both complex training and a smart target audience. Conversely, an excellent interface design may allow simple training and a "not so smart" target audience. If the interface design is medium quality, it is possible to have a more medium training and medium target audience combination for acceptable usability.

***Aptitude Areas*** The importance of aptitudes to the HSI analyst is that some skills are hard to obtain and therefore are constantly in high demand throughout the military. Thus, there is a relationship between aptitude area and recruiting, training, and retention. This relationship is made all the more dramatic if an aptitude area cutoff score is changed to either bring smarter personnel into a given job to handle added complexities of a new system (i.e., cutoff is *raised*) or to bring more people to the job (i.e., cutoff is *lowered*). The following is likely to happen for each of these two scenarios, respectively (concepts taken from Warner and Knapp, 1999):

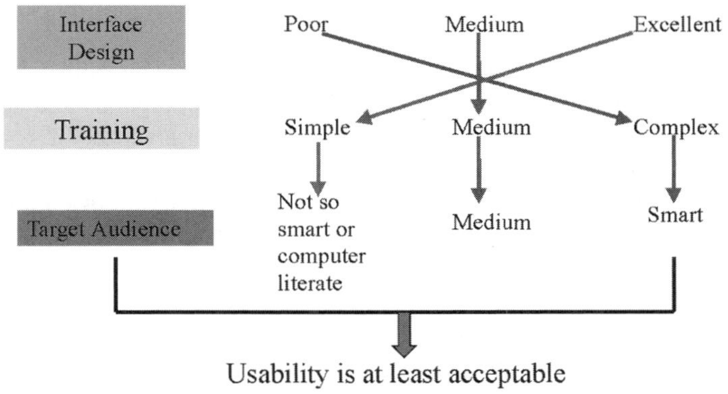


**Figure 11.7** Three scenarios for domain trade-offs in automated information systems.

- A higher aptitude area cutoff score will result in higher quality graduates of an advanced training course. Also, because of the higher caliber student, the number of academic drop-outs will be fewer. However, the total number of graduates for the aptitude area will not be as many because the number of qualified recruits for entry into the school will drop.
- Lowering an aptitude area's cutoff score means more personnel will be qualified to enter the class, and as a result the graduating class will be larger, but the training expense incurred will be greater due to the cost of many academic drop-outs and of additional material that might have to be presented.

With a better understanding of the high-level personnel capabilities considerations provided by HSI assessment methodology and seeing the types of personnel data needed to exercise the tools useful to HSI analysts, we can now review the status of MPT systems integration tools that have been developed by the U.S. Federal Government (primarily military services) and other countries.

## 11.3 MPT SYSTEMS INTEGRATION TOOLS

Most systems integration tools have been developed for use in military systems acquisition. Each military service has a slightly different perspective on MPT integration, borne of their different missions and the different environments in which they operate, and sometimes capitalizing on different research findings. This has led to the separate development of analytical tools and techniques that support the differences in service organization focus. In this section we have selected a tool from each of four different organizations [U.S. Army, U.S. Air Force, U.S. Navy, and UK Ministry of Defence (MoD)] in order to highlight the various similarities and differences with the major HSI integration tools. This is not intended to be a complete discussion of the tools, techniques, and technologies currently available or being developed for HSI.

### 11.3.1 U.S. Army

The U.S. Army took the early lead in HSI tool development and recognized that the power of these tools lay in their ability to support quantitative, unambiguous trade-offs. The U.S. ARL-HRED has been active in performing HSI analysis on a variety of systems and has sponsored recent work in this area. ARL-HRED developed a modeling and analysis tool named IMPRINT. The IMPRINT tool grew out of common U.S. Air Force, Navy, and Army MPT concerns identified in the mid-1970s. These concerns centered about two key questions: How to estimate MPT constraints and requirements early in system acquisition and how to enter those considerations into the design and decision-making process.

To address these questions, the U.S. Navy first developed the HARDMAN (hardware vs. manpower) comparability methodology (HCM). The U.S. Army then tailored the manual HCM, which became known as HARDMAN I, for application to a broad range of weapon systems and later developed an automated version, HARDMAN II. In HARDMAN I and II, however, there was no direct link between MPT and performance. To directly remedy this shortcoming, the U.S. Army began the development of a set of software analysis modules in the mid-1980s (Kaplan et al., 1989). This set of modules was called HARDMAN III, and although the name was the same, it represented a significant advance in the field through using a fundamentally different approach for addressing MPT concerns than previous methods. It provided an explicit link between MPT variables and soldier–system performance. IMPRINT is essentially an integrated and refined version of HARDMAN III in the Windows environment.

The mechanism for the MPT performance link is task network modeling provided by the commercially available Micro Saint task network simulation modeling engine, PC software designed for describing and analyzing task networks. The modeling capability offered in IMPRINT can be further characterized based on three distinctions (Law and Kelton, 2000): (1) static versus dynamic, (2) deterministic versus stochastic, and (3) continuous versus discrete. A static model does not address system effects over time, whereas a dynamic model represents a system as it changes with time. A deterministic model does not represent any probabilistic or random elements, whereas a stochastic model does encompass random elements and produces output that contains random error. A discrete model refers to instances where the variables characterizing the system change instantaneously at separated points in time. A continuous model is the converse, with variables that change continuously with time. In some instances, systems can be treated as either discrete or continuous, depending on the objectives of the analysis.

Using these definitions, IMPRINT can be described as a dynamic, stochastic, discrete-event modeling tool. When certain assumptions hold, namely, (1) that the system of interest can be adequately described by task activities and networked sequencing, (2) that dynamic processes and random variability are of interest, and (3) that any continuous tasks can be fairly transformed into discrete tasks, then IMPRINT is an appropriate tool to use to represent and analyze soldier–system performance.

The basic modeling capability in IMPRINT requires the decomposition of a system mission into functions, which, in turn, are decomposed into tasks. The functions are linked together into a network describing the flow of events. The network can include various types of branching logic such as parallel branches, probabilistic branches, and repeating branches. Within each function, the tasks are sequenced using the same types of branching logic options. At the task level, estimates of task performance time and accuracy means and standard deviations are input along with the consequences of the failure to perform a

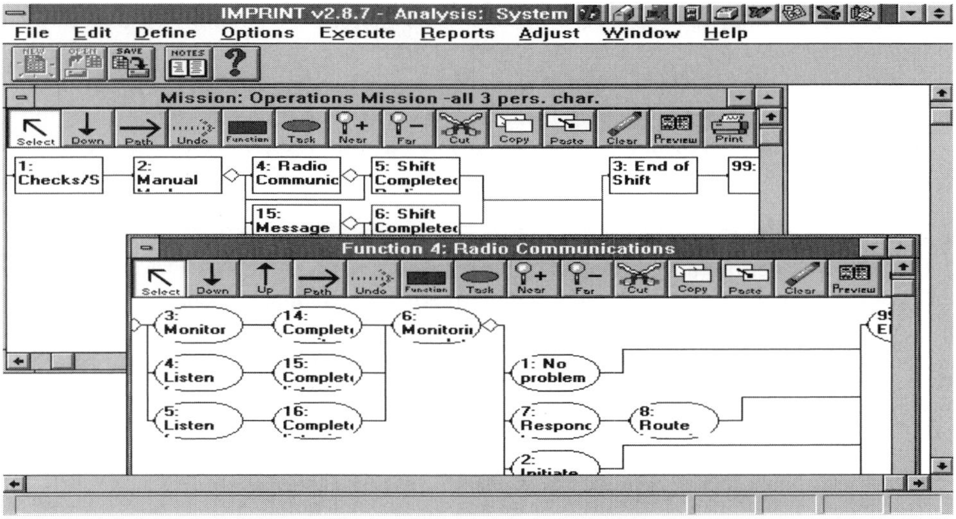


**Figure 11.8** Sample IMPRINT screen.

task accurately enough. The available failure consequence options are: (1) no effect, (2) total mission abort, (3) repetition of that or some other task, or (4) subsequent degradation of some other task. The data entered are assumed representative of performance under "typical" or baseline conditions. In addition, standards of performance can be entered to provide benchmarks for performance adequacy at the mission, function, and task levels. A sample IMPRINT screen depicting both the function and task level networks is shown in Figure 11.8.

IMPRINT executes a mission model task by task by first drawing a task time from the distribution as defined by the mean and standard deviation input for each task. Then it calculates the probability of success for the task based on the accuracy inputs. Next it determines, for this instance, whether there is an accuracy failure. After checking for a given task, IMPRINT proceeds through the task and function networks in accord with the established branching logic and analyzes the output according to the standards. When the model execution is completed (which can consist of several repetitions), reports of estimated performance at each of the three levels are generated along with the comparisons to the standards. Although any given model and its associated assumptions must be scrutinized, this approach is particularly useful for comparisons across systems or system conditions.

Several aspects of IMPRINT are unique. First, IMPRINT provides a method through which users can assess the effectiveness of the performance (i.e., "how successful was our system") as well as the more traditional efficiency assessment (i.e., "how busy were the soldiers" and "how many soldiers do we need"). The question of "how busy" can be answered in greater detail by using the embedded mental workload scales, either a straightforward assessment of visual, auditory, cognitive, and psychomotor workload channels or an assessment using a more advanced scale that includes a calculation of single task demands and intertask conflicts.

Second, IMPRINT includes specific algorithms to assess performance under diverse environmental conditions. Recall that the task performance data entered in the baseline model are assumed to represent performance under "typical" conditions. The embedded environmental stressors automatically adjust performance to account for the changes expected under different levels of the stressors. Currently, IMPRINT includes five environmental stressors: (1) protective clothing (i.e., mission-oriented protective posture, or MOPP); (2) heat; (3) cold; (4) noise; and (5) hours since last sleep (see Fig. 11.9). The application of a stressor will result in either less accurate task performance, longer times to complete the task, or both. Stressors may be applied to an individual task or to all the tasks assigned to a particular job or MOS for the mission. When the model is rerun, the new, or "stressed," task performance time and/or accuracy are used as the task estimates that are "rolled up" in the task, function, and mission reports are compared against the standards. Importantly, the results can also be compared with the baseline model predictions. [See Archer and Adkins (1999) for more complete documentation.]

The third unique aspect of IMPRINT is embedded data to enable users to adjust performance based on personnel characteristics (i.e., ASVAB scores) of the performing soldiers. This capability is a key element of the integration analysis, for it ties the variables from the personnel HSI domain into the system performance prediction.

IMPRINT is truly an integration analysis tool. Manpower (the number of crew members), personnel (the aptitudes of those soldiers), training (frequency and recency of practice for tasks), and human factors engineering (the design of the crew station) are all well represented in the total system performance estimate. A number of applications of IMPRINT to system acquisition and design issues provide ample evidence of integrated HSI analysis (Allender, 2000).

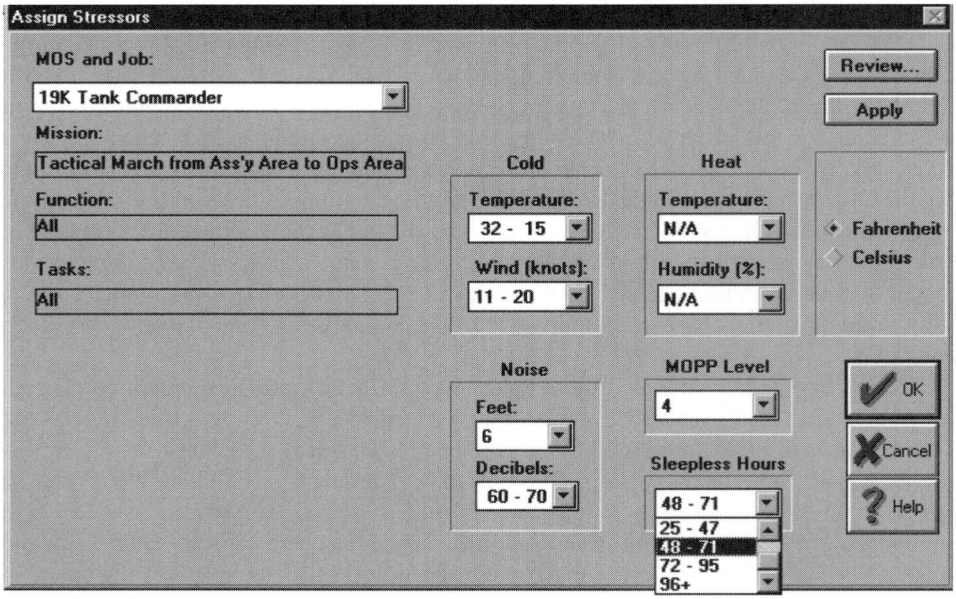


**Figure 11.9** Environmental stressors.

### 11.3.2 U.S. Navy

The U.S. Navy has a unique HSI problem to solve. Since most of its platforms are shipboard systems, the operators, maintainers, supply, and support personnel are typically the same sailors, most of who live on the system. This difference, coupled with recent emphasis on dramatic reductions in crew size and increased concern about each sailor's quality of life, has driven the U.S. Navy to focus on properly allocating tasks between crew (either aboard the system while underway or in port) and automation devices. This emphasis has driven development of tools that are attentive to the skills required by new equipment and faithful representation of sailor training and selection considerations in order to determine whether those skills are likely to be available.

The Systems Engineering Analysis Integration Tool (SEAIT) is a small business innovative research (SBIR) project managed by NAVSEA Dahlgren, Dahlgren, Virginia. The purpose of SEAIT is to evaluate the effects of reduced shipboard manning on ship design, system performance, and cost. A primary strength of this tool is that it supports a flexible analysis approach through which a system designer can apply varying levels of fidelity to the analysis of manning and automation alternatives. The scope of the functional analyses includes shipboard operations, unplanned corrective maintenance, and support functions. The tool allows for evaluation and trade-off analyses of ship manning during both normal and emergency or special conditions for a variety of operational activities (e.g., underway in open water, entering port, heavy weather, man overboard, fire, combat operations). A unique and innovative capability of this discrete event-based simulation tool is a module that solves for the best crew manning strategy based on the user's goal.

The SEAIT tool assists designers in assessing the impact of reduced manning levels on performance in various dimensions of the systems. These include the levels of automation required, the allocation of tasks to human operators of the system, the workload of the reduced crew, and subsequent risk associated with degraded performance due to excessive workload and other performance stressors. Users of SEAIT evaluate and trade-off these factors to determine the ultimate affordability of the new system. Costs associated with a new system are limited to the dollar cost of developing the system, including new automation, and the manpower costs of the required crew.

Several aspects of SEAIT make it unique among HSI tools. In SEAIT, users are prompted to provide limitations regarding crew size and the flexibility of voyage functions and also for the analytical goal (i.e., minimize number of jobs, minimize hardware cost, minimize skill gaps between existing and new jobs). SEAIT will run multiple iterations of the task network model in order to find the solution that best accomplishes the goal, within the constraints.

The SEAIT tool incorporates a method through which users can specify the types and levels of skills necessary to perform tasks. These skill scales are taken from work by Fleishman (Fleishman and Quaintance, 1984) and were originally included in a tool developed for the U.S. Army and discussed earlier in this chapter, JASS. The SEAIT method contains the JASS skill taxonomy as a library table that provides information on the skills and their levels that are available within the navy personnel inventory. These embedded data allow SEAIT to perform three tasks. First, the tool can help users allocate tasks appropriately through providing guidance to the user when tasks are assigned to jobs, by listing all the existing jobs that contain the requisite levels of the necessary skills. Second, SEAIT can allocate tasks to jobs "on the fly" whenever competing (parallel) tasks exceed available crew members for specific jobs. Finally, SEAIT can determine whether a task has a unique skill (i.e., one that is not available in the crew). This task would then be a candidate for redesign, automation, or possibly a training solution.

The interface available to users behind the Skills/Automation tab is shown in Figure 11.10. If users indicate that this function will not be automated, they can use the Define Required Skills buttons to change the list of skills attached to each function. If users click on the Define Required Skills button, the interface shown in Figure 11.11 is presented.

When users add required skills to a function or task, they can select the skills they want from the complete list or from skill categories (e.g., Communication). The list is presented in a spreadsheet interface and includes a description of the skill. Once they select a skill and click on the Score Skills button, the interface shown in Figure 11.12 is displayed.

The Assign Skill Levels interface helps users determine the level of a skill needed for the function or task. As shown in Figure 11.12, the skill scale name and description are displayed on the interface. Below these, the benchmarks for this skill scale are shown, with scores ranging from 0 to 70. For example, "Understand a lecture on navigation in space" is assigned a score of 63 from a maximum possible 70 under the category Oral Comprehension. Users can move the slider on the scoring scale to set the score, or alternatively, can type in a score in the text box underneath the scale. The control at the bottom left of the screen lets users move through the skill scales and score them without returning to the previous screen and selecting them individually from the list of skills.

Skills and required levels are combined and assigned to tasks by the SEAIT discrete event simulation engine. This engine generates a composite of the skill requirements over time by each crew member of the system. Figure 11.13 provides an example of this type of

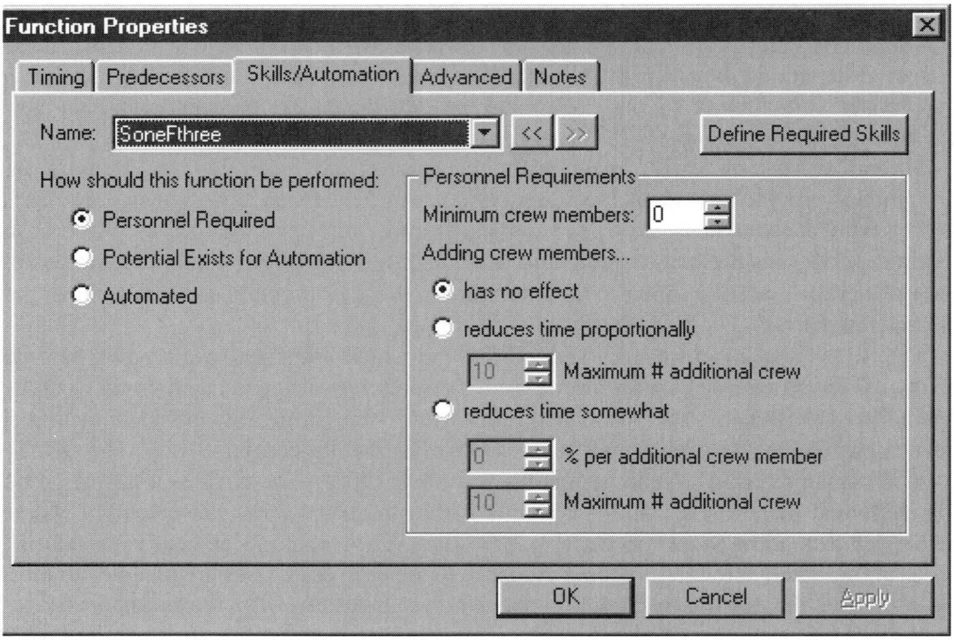


**Figure 11.10** SEAIT Skills/Automatic tab.

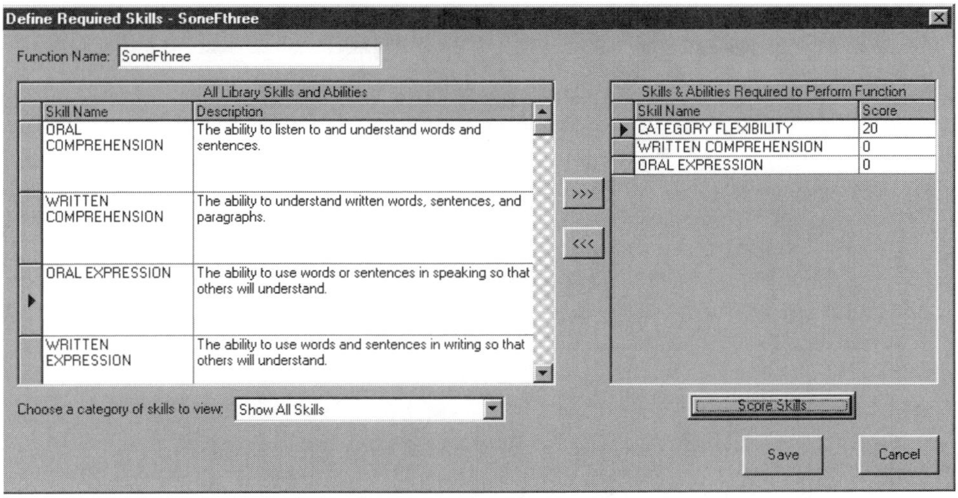


**Figure 11.11** SEAIT Define Required Skills.

output. This result can be used to determine whether a specific set of tasks or operational requirements combine to drive the skill requirements of the system and to help the user identify fruitful task allocation or automation strategies.

SEAIT became available in late 2000 and includes some system data from fielded ships. Follow-on plans include significant augmentation and integration with other U.S. Navy analysis environments.

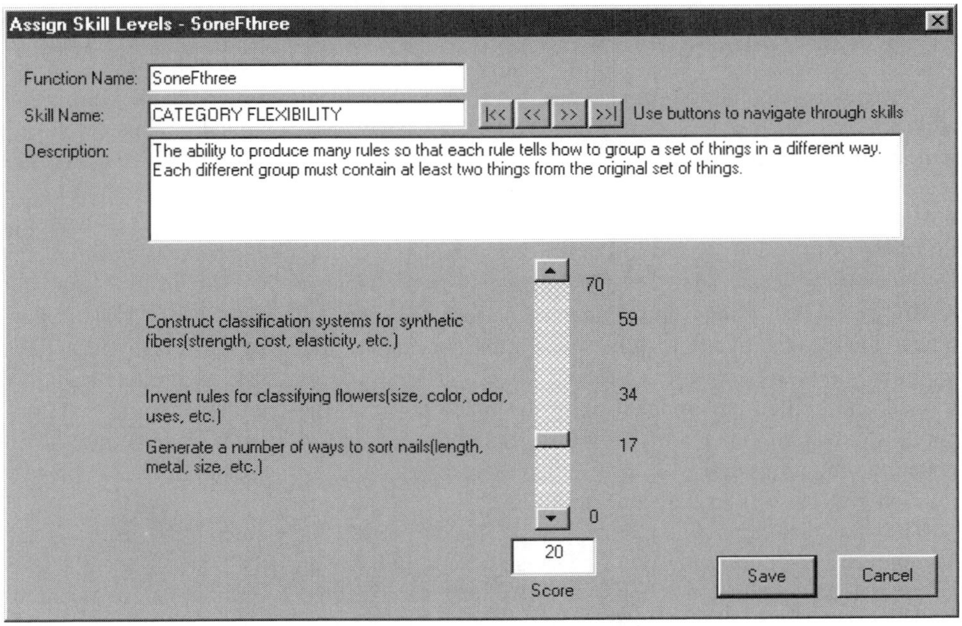


**Figure 11.12** SEAIT Assign Skill Levels.

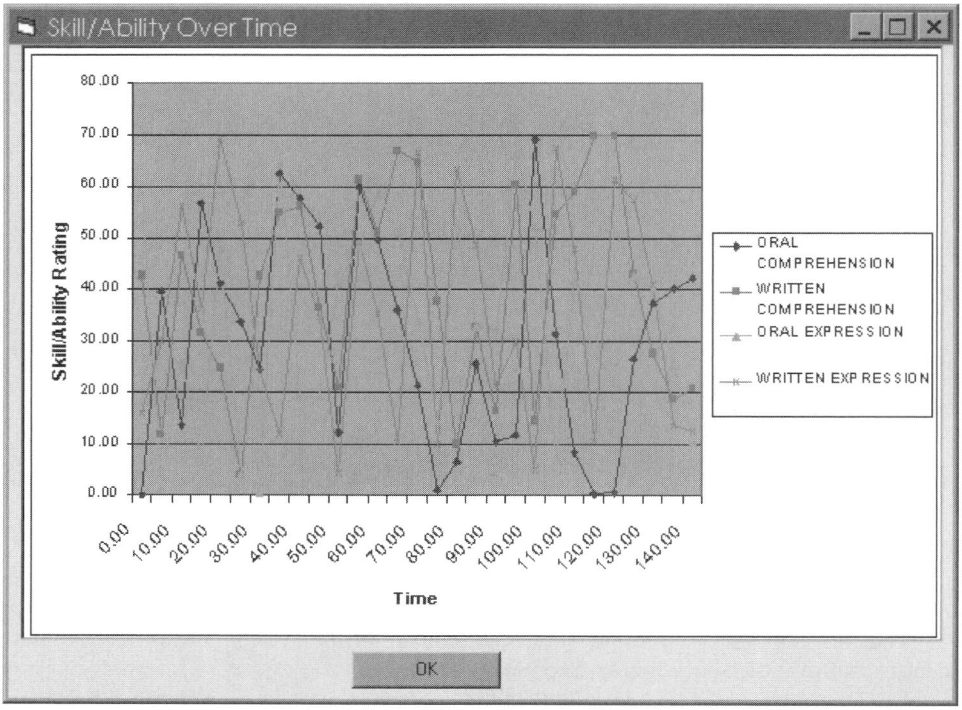


**Figure 11.13** SEAIT output example.

### 11.3.3 U.S. Air Force

U.S. Air Force tool development has traveled a slightly different path than the other services. While the U.S. Army and the U.S. Navy have typically considered the maintenance and operational HSI analysis of a system together, the U.S. Air Force has tended to separate the analysis of the maintenance requirements of the system from the operational requirements. This is partly attributable to the traditional organizational divisions of the analysts and to the different emphasis of the two analytical requirements for aviation systems. The need for weapon system maintenance is driven by equipment reliability under various operational scenarios. The resulting maintenance task performance then leads directly to sortie generation rate, one of the most important and highest visibility readiness measures in the air force. Air force operators are exceptionally highly trained and skilled personnel. Unlike the other services, the operator crew (i.e., pilots, copilots, navigators) performs very little maintenance, so there is little overlap in terms of resource limitations and workload requirements. This reduces the requirement to consider the two sets of tasks in the same analysis.

The tool that best characterizes the U.S. Air Force's MPT integration effort is the Manpower, Personnel, and Training Decision Support System (MPT DSS). The MPT DSS helps analysts conduct the complex MPT analyses required to support Department of Defense (DoD) HSI requirements. MPT DSS provides a non-simulation-based analytical environment that has several unique capabilities.

MPT DSS is composed of a number of tools that communicate through a central database, enabling users to maintain data integrity and a clear audit trail of data elements and their resulting outputs as the design for the analyzed system matures. This design also enables the software to direct users toward specific tools that fit their analytical goal and level of expertise.

One of the most unique elements of MPT DSS is its ability to allow users to import MPT data from existing databases. Through a tool called the database integration (DBINT) tool, users can import data from the logistics composite model (LCOM), the comprehensive aircraft support effectiveness evaluation (CASEE), F forms 611/612 training cost data, U.S. Navy (USN) catalog of navy training courses (CANTRAC), and the programmed technical training manpower standards (PTTMSs). This helps the user begin to populate the analysis with reliable data. While this ability is powerful and useful, it does create a maintenance requirement for MPT DSS in order to stay current with updated data formats of the external data sources.

The individual analysis tools of MPT DSS are designed to address reasonably separable aspects of the HSI analytical picture. As the user progresses through the tools, the inputs and the outputs are integrated into the master database in order to support the trade-off capability that is central to an integration tool.

The tool elements of MPT DSS can be placed into three categories. The first category contains tools that help the user develop a representation of the system design.

*Database Integration* (DBINT) helps users import the data needed to conduct MPT analyses.

*System Definition Systems* (SDSs) help users refine the equipment and tasks to be included in the MPT analysis. The LCOM and CASEE methods provide existing aircraft work unit codes and maintenance task reliability and maintainability data.

The second category includes six tools used to perform selected, somewhat domain-specific analysis.

*Manpower Estimation* (ME) helps users determine the direct maintainer manpower required to support a squadron.

*Force Structures* (FS) help users aggregate squadron manpower and calculate indirect manpower within an organization structure. The FS categorizes and outputs manpower as specified by the Manpower Estimate Report and includes various cost categories.

*Training Task Analysis* (TTA) helps users identify the tasks that require training, determine the instructional setting in which the tasks should be trained, and determine the time required to train each task adjusting for skill and knowledge similarity.

*Training Resources and Requirements* (TRR) help users define operator and maintainer courses including length, training devices, and comparable cost.

*Life-cycle Cost* (LCC) applies standard cost factors, including inflation, to the MPT resources generated by the other tools to produce the MPT portions of LCC.

The third category includes three tools that are used to support comparison and trade-off analyses between versions of an analysis and across domains. Three tools have been designed to combine the results of the individual analysis tools into an overall assessment of the system.

*The high driver tool* identifies the parts of the weapon system that are the major contributors to a key measure of effectiveness. For example, the high driver tool can rank order equipment items in terms of the maintenance man-hours they require.

*The comparison tool* allows users to compare MPT analysis results between different versions of the same system. The comparison tool presents differences in both tabular and graphical reports.

*The trade-off tool* helps users conduct trade-off and sensitivity analyses of key MPT parameters. The trade-off tool can vary two parameters systematically, rerun the analyses needed to assess their impact on a particular measure of effectiveness, and graphically depict how variations in these parameters impact the measure of effectiveness.

The MPT DSS central database maintains a representation of the input and output associated with each process. During the trade-off analysis process, this representation allows users to change MPT parameters and automatically rerun the analyses needed to assess impacts on key measures of effectiveness.

Originally developed for U.S. Air Force (USAF) systems, the MPT DSS has been expanded to include the analysis of U.S. Navy (USN) and U.S. Marine Corps (USMC) systems to support the Joint Strike Fighter Program Office (JSFPO). The USAF has also played a leading role in meeting the challenge of integrating human performance models (HPMs) with other constructive simulations. The Air Force Research Laboratory/Human Effectiveness Directorate (AFRL/HED) supports research development for the air force in the areas of human/system design. AFRL/HED is leading the development of a methodology and a suite of computer simulation tools to evaluate crew system designs from the perspective of crew performance as they affect weapon system mission effectiveness. The capability will allow users (e.g., system program offices, industry, etc.) to impact design decisions much earlier in the acquisition process in ways that historically have only happened later in the design phase. Being able to affect the design of crew interface and associated aircraft subsystems (avionics, weapons, etc.) much earlier permits greater inclusion of human factors and crew systems engineering trades.

The software tool currently being developed by the USAF is called Combat Automation Requirements Testbed (CART). CART helps analysts, operation researchers, and engineers develop HPMs that are realistic in their behaviors and can also interact with external mission and engineering-level simulations. Rather than begin anew, AFRL/HE built upon the U.S. Army's proven IMPRINT human performance modeling environment. There were two major modifications made to IMPRINT that provided this new capability. The first was goal orientation, and the second was integration with external simulations.

Based on Jens Rasmussen's abstraction hierarchy concepts, the hierarchy of goal-to-task is key to CART's translation of real-world actions and events into usable operator models (see Fig. 11.14). CART permits the user to decompose a mission (i.e., destroy a time–critical target) into high-level goals (threat evasion, attack a target, etc.). Once these goals are established, the user can create a series of high-fidelity operator tasks that support each higher level goal. Creation of the lower level tasks in the model can be based on real-world experience, engineering analysis, interviews with SMEs, or simply assumptions about how the operator will interact with the crew interface (should a physical form not yet exist). As the task network model is being built, users can input key performance features such as time, accuracy, variability, etc.

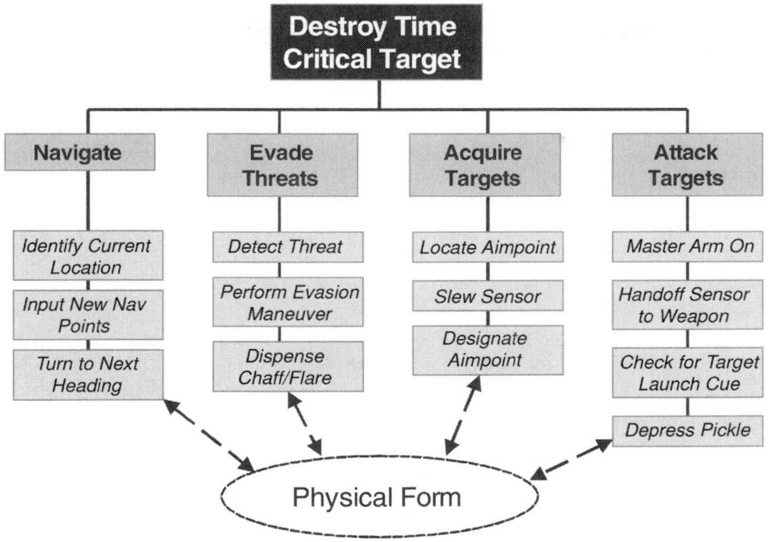


**Figure 11.14** CART goal orientation.

The second major extension to IMPRINT under the CART program was the addition of a network communication interfaces (NCI) layer or "sockets" that permit data to be exchanged with an external simulation running in parallel but outside of the CART simulation. In this mode, the external simulation passes target location, threat identification, and other key environmental factors to the task network model built by the CART user using the NCI. The task network model in CART "perceives" the real world through a set of crew interfaces (displays, audio) and visual cues that have also been modeled. Based on the user's construction of the task network model, the interaction between the HPM and the external simulation will vary—much like real pilots under different environmental conditions. The variability within and between the external simulation and the CART model can be measured whereby validation of the operator model becomes straightforward. In practice, AFRL has demonstrated the efficacy of the CART software to faithfully represent operator tasks in highly dynamic time-critical targeting missions.

By using these features (goal orientation and integration with external simulations), CART users are able to create realistic models of human behavior while avoiding the costs associated with fabricating and running cumbersome human-in-the-loop simulations. In effect, this capability provides the acquisition community with a tool that relates the effect that operator performance has on system lethality and survivability.

### 11.3.4 Other Developers (Non-U.S.)

The Integrated Performance Modeling Environment (IPME) is an integrated environment of models intended to help the human factors practitioner analyze human system performance. IPME builds from a discrete event modeling environment similar to that embedded in IMPRINT but adds modules to expand the descriptions of selected aspects of human behavior. The development of IPME has been a collaborative development effort among the United Kingdom's Defense Evaluation Research Agency's Centre for Human

Sciences (DERA CHS), Canada's Defense and Civil Institute for Environmental Medicine (DCIEM), and Micro Analysis and Design Inc. (MAAD).

The IPME uses a process-oriented modeling approach and builds upon an SME's accounting of how operator activities are organized or may be organized to meet operational objectives. Operator responsibilities and goals can be recorded at a high level of abstraction (such as "prepare for mission") that can be decomposed into a hierarchy of functional blocks (such as "prepare met brief") until the analyst has reached a level of granularity (such as "read current weather map") appropriate to study a given problem.

The key IPME features are:

- *Environment Model* The analyst can model environmental factors or what behavioral scientists refer to as performance-shaping factors. These include environmental variables such as temperature, humidity, time of day, etc.
- *Operator Characteristics* Operator traits and states are simulated. Traits are variables such as mental ability, susceptibility to motion sickness, time since trained, etc. States are variables such as fatigue, hunger, etc. The operator state is dynamically updated during a simulation. Therefore, each operator in the simulation can have unique characteristics.
- *Performance-Shaping Functions (PSFs)* These user-defined functions dynamically modify individual operator task "time to perform" and "probability of failure" values. These PSFs define how performance-shaping factors (environment variables or operator characteristics) affect operator performance. The PSFs are linked to individual tasks through a task taxonomy allowing one PSF to be dynamically applied to any similar task in a model. Since PSFs can use operator states as expression variables, simulations can be built that have two operators performing the same task type with different, and therefore more realistic, "time to perform" and "probability of failure."
- *Prediction of Operator Performance (POP) Scheduler and Workload Measurement* A new algorithm for estimating operator workload developed by the British Centre for Human Sciences has been built into IPME and can be used to evaluate when operator task demands exceed capacity (Farmer et al., 1995).
- *Information Processing (IP) Scheduler* The IP scheduler is a new scheduling approach that establishes an "operator load" rather than a "task load." The operator load is defined as execution of the tasks that *can* be simultaneously completed within a human's resource capacity. Thus, the scheduler emulates expected human performance under loaded conditions. The IP mode of IPME establishes time criterions, structural and resource contention, and human memory limits for each task. As the simulation executes, a time pressure is calculated for each operator within the simulation based upon the slack time established for task execution. For simultaneously executing tasks, the scheduler determines if there are conflicts between structural resources such as hands, vision (fovea and peripheral views), and cognitive conflicts. In addition, it emulates the concept of prospective (or short-term) memory limits and the resulting effects from task overloading and attention distractions. These features produce realistic human behaviors, both for simulation-based acquisition and training applications.

- *Measurement Suite* Through a measurement suite, the user is able to set up experimental runs using independent variables that can be set to different initial values for each experimental run. Multiple experimental runs can be defined *and* multiple simulation runs (or iterations) can be specified for each experimental condition. Blocked experimental designs are supported.

The IPME is based on the proven Micro Saint simulation engine with the human operator simulator (HOS) extensions. It is a discrete event Monte Carlo simulation engine with a graphical user interface (GUI). The GUI provides a drawing space where network diagrams defining man and machine tasks are constructed using visual components. Network element sequence is defined by connecting model components with mouse point, click, and drag operations. Micro Saint supports several types of human decision models and queues to allow the representation of complex operations.

The HOS extensions provide a mechanism to define a workspace associated with a task network. This workspace can contain work zones or work surfaces, operators, and positional markers. Work surfaces can contain work controls with which the operator would interact. These controls can include things such as keyboards, mice, dials, knobs, etc.

IPME contains a simple socket protocol to allow passing variable information from external applications. External applications can be processing anywhere from which a connected socket can communicate. This means the IPME simulator can interact with other simulators on the same machine, or other machines connected via an intra- or internet.

IPME continues to have new features added under collaborative funding. The Canadian DCIEM continues to transition its simulated operator loading evaluation (SOLE) methodologies into the IPME environment. The final target is to have a complete human factors analysis capability that implements methodologies consistent with MIL-HDBK-46855A (Hendy and Farrell, 1997). The Centre for Human Sciences continues to advance the capabilities of the IPME simulation tool.

### 11.3.5 Summary of Tool Characteristics

The tools considered in this chapter are summarized in Table 11.4. They are IMPRINT, SEAIT, MPT DSS, CART, and IPME. The following five basic characteristics of each tool are included in the table: (1) principal purpose, (2) features supporting MPT requirements analysis, (3) additional analytic and data capabilities, (4) platform, and (5) distribution. It should be noted that many of these tools are still in active development, and the features are changing in order for the tools to continue to meet user needs. One particularly fertile area is in the development of links between these tools and other analysis tools and databases. The development of standards such as high-level architecture (HLA) have fueled this work and will eventually provide a cost-effective method for the tools to share algorithms, methods, and data.

### 11.3.6 Technical Gaps and Emerging Technologies

Although admittedly not quite perfected, and, unfortunately, not even in every-day use, the MPT tools available today are far more than simple manpower calculators or training days

TABLE 11.4 **Summary of Tool Features and Capabilities**

| | IMPRINT VACP | IMPRINT Advanced | IMPRINT Maintainer | SEAIT | MPT DSS | CART | IPME |
|---|---|---|---|---|---|---|---|
| *Proponent* | | | | | | | |
| | Army | Army | Army | Navy | Air Force | Air Force | United Kingdom |
| *Principal Purpose* | | | | | | | |
| Level of analysis | Individual & small team (systems of systems in development) | Individual & small team | Single system type/force-wide roll-up | Navy manpower analysis ship-wide | Air Force & Navy aircraft maintainer manpower | Individual & small team | General modeling & individual performance |
| Acquisition phase | All | Postconcept | All | All | Postconcept | All | All |
| HSI domains | MPT & HFE (performance) | M & HFE | MPT | MPT | MPT | MPT & HFE (performance) | MPT & HFE (performance) |
| *Manpower Requirements Analysis Features* | | | | | | | |
| Analysis of staffing or team composition | Yes | Yes | Yes | Yes | Yes | Yes | Yes |
| Embedded workload algorithm | McCracken-Aldrich | Wickens' MRT | Utilization | Busy/not busy | Utilization | McCracken-Aldrich | POP and IP/PCT, VACP, W/Index |
| Dynamic operator assignments | No | Triggered by workload overload | Triggered by utilization | Triggered by workload overload | Triggered by utilization | Triggered by task environment | Available through user-defined simulation expressions |

| | | | | | | | |
|---|---|---|---|---|---|---|---|
| Embedded stressors | Sophisticated noise, heat, cold, MOPP, sleep deprivation | User calibrated soldier quality, fatigue | Sophisticated noise, heat, cold, MOPP, sleep deprivation | No | No | Sophisticated noise, heat, cold, MOPP, sleep deprivation | User defined |
| Embedded shift work and circadian rhythm | No (Circadian algorithm in development) | No | Shift definition | Limited – work week/ overtime tracking | Shift definition | No | No |
| *Personnel Requirements Analysis Features* | | | | | | | |
| Performance variation by personnel quality | Yes, by ASVAB score | User defined | Yes, by ASVAB score | No | Yes | Yes, by ASVAB score | User defined |
| Skill requirements estimation | Yes – detailed | Yes – limited | Yes – limited | Yes – very detailed | Yes | Yes – detailed | Yes – limited |
| *Training Requirements Analysis Features* | | | | | | | |
| Performance variation by frequency of practice | Yes (sophisticated algorithm in development) | User defined | Yes | No | No | Yes | User defined |
| Training resource requirements estimation | No | No | No | No | Yes | No | No |
| *Additional Analytical & Data Capabilities* | | | | | | | |
| Task analysis data used as input | Yes | Yes | Yes & logistics data | Yes | Yes | Yes | Yes |
| Sample models | Army-specific models | Army-specific models | Army-specific models | Navy data elements | May be available on request | Air Force models also available | Limited set of military & commercial models |

(*continued*)

**TABLE 11.4 (*Continued*)**

| | IMPRINT VACP | IMPRINT Advanced | IMPRINT Maintainer | SEAIT | MPT DSS | CART | IPME |
|---|---|---|---|---|---|---|---|
| Number of task decomposition levels | Unlimited | Two | N/A | Two | N/A | Unlimited | Eight |
| Performance time distributions | Normal, gamma, lognormal | Normal, gamma, lognormal | Normal | Normal, gamma, lognormal | Normal | Normal, gamma, lognormal | Normal, gamma, exponential, rectangular, lognormal |
| Automated task accuracy & failure consequences | Yes | Yes | Yes | Yes | Yes | Yes | Yes |
| Embedded human performance micro models | Yes | Yes | No | No | No | Yes | Yes |
| Link to cognitive modeling architecture | Yes | No | No | No | No | Yes | No |
| Representation of work center (x, y, z coordinates affect model outcome) | No | Can be implemented manually | No | Can be implemented manually | No | Can be implemented manually | Automated |
| Automated optimization techniques | No | No | No | Limited | Trade-offs, high drivers, comparison | No | No |

| | | | | | | | |
|---|---|---|---|---|---|---|---|
| Automated data collection | Yes | Yes | Yes | Yes | Yes | Yes | Yes |
| Measurement suite to support experimental designs | In development | In development | In development | No | Version comparison capability | No | Yes |
| Graphical point & click model building interface | Yes | Yes | N/A | Yes | N/A | Yes | Yes |
| Animation | Task network | Task network | N/A | Task network | N/A | Task network | Task network |
| Debugging/ tracing capability | Yes | Yes | No | Yes | Yes | Yes | Yes |
| Simulation to simulation communication | Yes, HLA compatible | No | No | No | No | Yes, HLA compatible | Yes |
| Life-cycle cost | No (link to tool provided) | No | No | No | Yes | No | No |
| | | | *Platform* | | | | |
| Windows 2000 | Yes | Yes | Yes | Yes | N/A | Yes | HFAT only |
| Unix–Sgi | No | No | No | No | N/A | No | Yes |
| Linux | No | No | No | No | N/A | No | Yes |
| | | | *Distribution* | | | | |
| Availability | Immediately | Immediately | Immediately | 12/00 | Not currently available | Immediately | Immediately |
| Limitations | Free to U.S. govt. and contractors | Free to U.S. govt. and contractors | Free to U.S. govt. and contractors | Free to U.S. govt. and contractors | N/A | Free to U.S. govt. and contractors | Free to UK govt. and DCIEM |

estimators. *The "so-what" question has been answered.* The effect of MPT factors on task and system performance has been firmly established. IMPRINT, for example, has been used to apply stochastic, task network modeling, and mental workload assessments to a number of systems, resulting in significant system design impacts (see Allender, 2000, for a summary). At the same time, the need for MPT integration and assessment continues to grow in both breadth and depth.

Fortunately, research results and new technologies are now becoming available to address these expanded MPT integration and assessment needs. The HSI community is building on advances in computing technology and human performance modeling techniques (Laughery and Corker, 1997). In 1998, Pew and Mavor published their comprehensive review of the state of human behavior modeling in the military, which has proved to be a touchstone for both the modeling and the HSI communities. In the Human Systems Integration Technologies, Tools, and Techniques Symposium (2000), sponsored by ARL-HRED, the discussions of emerging expectations were all discussions of what modeling can offer: "Cognitive Modeling: Does HSI Need a Brain?," "Integration of Models: Is It the Holy Grail?," "Team Modeling: Can Human Teams be Engineered?," "High-level Architecture (HLA) and HSI: Do They Need Each Other?," and "Joint HSI: Are the Models Color Blind?" In addition to these advances came the formal recognition from the defense community that an authoritative representation of human behavior is essential for military simulations (U.S. Department of Defense, 1995). In this section, a few of the key advances that support expanded MPT integration and assessment needs are highlighted.

***Computers and Computing Techniques*** Advances in computing have boosted and will continue to boost the performance of MPT integration tools. Computers are faster, and at the same time, faster computers are more accessible. On the "low end," the latest PC technology, which is actually quite powerful, sits on virtually every HSI practitioner's desk. On the high end, for example, the ARL's Major Shared Resource Center lists dozens of high-performance computers and programming languages that are available to government and government-affiliated researchers, either on-site or via remote access.

Somewhat distinct from the variety and number of available computers are what might be termed computing techniques coming from the fields of computer science, artificial intelligence, and mathematics. For example, in their review, Pew and Mavor (1998) singled out neural networks as having particular promise for the representation of human behavior. Both neural networks and genetic algorithms, in addition to their surface "biological" or "physiological" appeal, have potential for human behavior representation in that they "learn" or "evolve" in ways that may reasonably represent learning, although both also require large amounts of data to feed or train the software. Bayesian networks are another way to represent learning and growth and can be used in two ways of interest here for: embedded decision-aiding or embedded within another modeling environment to represent an aspect of cognitive processing (e.g., Anderson and Lebiere, 1998). Another development is agent-based programming, and while the uniquely defining characteristics are still being debated within the computer science community (Bradshaw, 1997), for our purposes here, suffice it to say that agents are another way to structure software that may be value added for simulating human–human or human–system collaboration. [See Zhang et al. (2001) for an example of tactical operations staff collaboration.] While only a few techniques are mentioned briefly here, detailed descriptions abound in the scientific literature and even in the popular press.

***Cognitive Modeling*** The requirement to understand and predict soldier–system performance has expanded beyond the classic sort of task analysis where a task is generally an observable unit of behavior that takes seconds to minutes, even hours, to perform. Now, there is a need to know not just *what* the tasks are but *how* they are performed. System developers and designers need to know what cognitive mechanisms are invoked as a part of task performance. For example, displaying information on a helmet-mounted display does not simply increase the amount of information available to a soldier. The change in technology changes the perception, memory, and processing of the information as well as the kinds of errors associated with it. Therefore, truly understanding the MPT implications of a system requires an understanding not only of the obvious and observable interface features but also the cognitive interface and resulting cognitive demands.

Cognitive modeling capabilities, originating with academic research, are maturing to meet this need (Pew and Mavor, 1998). The 2001 International Conference on Cognitive Modeling even included a special panel on government interests and opportunities in the area (Gluck et al., 2001). The U.S. Air Force has sponsored the Agent-based Modeling and Behavior Representation (AMBR) project that examined several cognitive modeling approaches all modeling the same task of air traffic control (Gluck and Pew, 2001). The National Air and Space Administration (NASA) has initiated a human error modeling effort utilizing a number of cognitive models to describe and predict pilot errors with current equipment and practices as well as planned future equipment enhancements. The U.S. ARL has recently conducted work in-house using the Atomic Components of Thought-rational (ACT-R) cognitive modeling architecture (Anderson and Lebiere, 1998) to model aspects of soldier behavior (e.g., Kelley et al., 2001).

Among the most well-known cognitive modeling architectures are the ACT-R (Anderson and Lebiere, 1998), Executive-Process/Interactive Control (EPIC) (Meyer and Kieras, 1997), Soar (Laird et al., 1987), and COGnition as a NEtwork of Tasks (COGNET) (Zachary et al., 1992); but also include the various versions of the HOS (e.g., Hood and Allender, 1993) or goals, operators, methods, and selection rules (GOMS) concepts (e.g., Card et al., 1983). Whereas task network modeling, such as is resident in IMPRINT, is essentially atheoretical with respect to cognition, cognitive modeling approaches typically derive from a specific theoretical basis. ACT-R grew out of research in basic learning and memory and the findings in that research are "built-in" as constraints on the modeling. The development of EPIC drew on work in perception and psychomotor skills; Soar can be considered a product primarily of the artificial intelligence community with its reliance on rule-based productions; and COGNET combines several aspects of psychology and the more applied approach of task analysis. A fundamental concept of HOS was to take "laws" of the basics of human performance and make those available to aggregate into larger descriptions of performance. Fitts' law pertaining to reaction time is a classic example.

As development has continued on individual cognitive modeling approaches, some aspects of that development have drawn from other approaches. For example, ACT-R now includes a perceptual-motor component in ACT-R-PM. Another aspect of development has been to address the software interface, the "ease-of-use" question. Training classes are offered routinely for COGNET, ACT-R, and Soar and changes to their interfaces are being considered to enhance usability. In sum, there are cognitive modeling tools available to the HSI community that can provide significant predictive and explanatory power even though they are not yet turn-key operations.

***Decision Modeling*** Human decision making has been considered within the HSI context most simply by using flow diagrams or operational sequence diagrams to show possible decisions, and, more robustly, by using task network modeling. Within task network modeling, both decision variability and the time to make a decision can be addressed. While both diagrams and network modeling are useful, there are limitations, at least in the current formulations.

One way in which limitations in decision modeling have been addressed is by bringing the outside world in. Links to other simulations have increased the richness of the available decision environment. Specifically, this means that the scenario against which a model runs does not have to be preset within the model. The CART program described earlier was a significant advancement in this regard. The goal orientation capability implemented in CART permits modeling of decision-making strategies as goal switching with the "triggers" for switching goals coming from events in another simulation. A test of the CART methodology in a Joint Strike Fighter testbed reported by Hoagland et al. (2001) showed very good correspondence between the model-in-the-loop decision making with respect to use of controls and displays and that of actual pilots-in-the-loop in the same simulation.

Many of the outcome measures of human performance modeling are predictive of efficiency, that is, the time to perform. One effort (Wojciechowski et al., 2001) recently expanded the effectiveness measures to include quality, where decision quality is determined as a function of information quality, operator factors such as fatigue, experience, training, stressors, and team performance. This framework has been implemented in a field artillery sensor-to-shooter model and has shown great promise as a way to provide a measure of decision quality and the resulting influence on overall system performance over and above the more typical efficiency measures.

Recognition-primed decision making (RPD) (Klein, 1989) has been suggested as a more appropriate way to represent the way humans actually make real-world decisions than criterion weighting, pure memory strength, or adherence to predefined strategies. Recently progress has been made to represent the RPD approach via modeling, such as the effort reported by Warwick et al. (2002) where task network and cognitive modeling capabilities are combined with the promise of increased descriptive and diagnostic power.

***Model Integration and Model Federations*** The recent growth in model integration and the advances in model federations have been motivated by all of the developments just mentioned in this section. Computing advances have made it possible for different models to "talk to each other" and share information more easily. This includes military model communication standardization efforts such as distributed interactive simulation (DIS) and high-level architecture (HLA). The increasing desire, or need, to include accurate and appropriate representations of human cognitive and decision-making behavior in military models and simulations—and the reality of having to do it cost effectively—require assembling models using the best of what is available and reusing models across multiple environments.

One example of model integration is found in Lebiere et al. (2002). In that effort, IMPRINT and ACT-R were integrated in order to capitalize on the strengths of IMPRINT for modeling task sequences with flexible branching logic and the richness of ACT-R for representing the influence of memory limits and competing information on decision making. Craig et al. (2002) report on a CART (or IMPRINT)-ACT-R integration, again building on their individual strengths, although via a slightly different communication protocol. This sort of "hybrid" integration holds great promise.

The appearance of human behavior representation in large-scale military combat models and simulations is not new. The Conference on Computer-Generated Forces and Behavior Representation has been meeting since 1992. However, what is new is the escalating emphasis on a *truly* authoritative representation of psychologically plausible behavior as opposed to, say, simple movement rates. Within the CART program, Hoagland et al. (2001) built a model of a pilot that "flew" and reacted to targets in a jet fighter simulation, a simulation that could also be run as soldier-in-the-loop. This federated, model-in-the-loop configuration was used to evaluate design and tactics options. The U.S. Army's Joint Virtual Battlespace (JVB) federation has funded the integration of Micro Saint–based models of a robot controller and of a field artillery staff for the purpose of helping to evaluate Future Combat Systems (FCS) concepts. In this way, an accessible, standalone human performance modeling environment can take inputs from other models and simulations, which stimulate and affect the human performance model—including cognition, decision making, and the actual goal-oriented behavior of the human model—and in turn, the human performance model can exert an influence on the course of the overall, federated simulation, that is, on the combat outcome.

## 11.4 COMMERCIAL APPLICATIONS

There is tremendous pressure in commercial industry to reduce the cost of delivering products and services. As with the military, a key factor in product cost is manpower cost. Therefore, competitive advantage can be gained through reducing the numbers of people needed in the process, or through reducing the skills required by the workforce, while maintaining target production levels. The techniques used in industry to attain this reduction often include some version of the four-step process described in Section 11.2.2. Unlike the military, however, the documentation and requirements process does not have a common structure across companies, even within the same industry. While this is not surprising, the sharing of techniques is also thwarted by the need for companies to protect proprietary information for competitive purposes. However, one common element across firms is that the manpower and personnel questions are addressed by a combination of the human resources, facilities planning, and technical staffs.

While the level of analytical power brought to the questions varies widely between companies; in general, a well-structured approach is typically used when a firm is considering a change that could affect staffing. In this process, the first step is to reach a detailed understanding of the system in which the work is being performed. The second step is to identify viable alternatives. The third step is to test those alternatives in a realistic environment. And finally, the selected improvement must be implemented. In this section, we will discuss two examples describing how this process was successfully conducted.

### 11.4.1 Example 1: Automation and Manpower Trade-offs

Gates Rubber is a large manufacturing corporation that produces rubber products for the automotive industry. Many years ago, they recognized that their production throughput could probably be improved and they began a process of updating their traditional manufacturing facility from a straight-line production process to a cell-based manufacturing process (Harshell and Dahl, 1988). They hoped that this would allow them to maximize the utilization of large pieces of capital equipment, such as cranes and ovens. However, there were many unanswered questions that they wanted to examine before making this dramatic change. First, they needed to understand how much the new process would increase throughput. This would enable them to assess the financial payoffs of the new equipment purchases. Second, as a unionized organization, they needed to understand the manpower and personnel implications in order to fulfill their obligations to their workforce. Finally, they needed to perform detailed analysis that would allow them to balance their production line through the removal of bottlenecks.

To address these questions, Gates Rubber embarked on an analytical process. Because many of their questions needed to be answered in relative terms, as to how much the new process was different from the existing process, the first part of the process was to develop a reliable and accurate baseline of the existing process. As with many manufacturing organizations, Gates Rubber possessed very detailed records of how much time each step in their existing process would take. These records of time were collected using a combination of empirical data collection and motion–time–method (MTM) techniques. Collectively, these data became their labor standards upon which negotiations with union personnel regarding staffing levels were based. Because of these existing data, it was a relatively simple process to benchmark the existing manufacturing production data, providing a reliable basis to which the manpower and personnel implications of the changes could be compared.

This information was used to develop a discrete event simulation model of the baseline process. This type of simulation is commonly used in industry, and most industrial engineers have some knowledge of these simulation techniques. In this case, the simulation model was constructed using Micro Saint, which is based on a task network modeling approach.

To develop the baseline models, a flow diagram was constructed that described the flow of the product through the manufacturing line. This diagram was complicated because the current product mix consisted of 40 different product types, resulting in different flows through the line, depending on the orders that were being processed in a given batch. Figure 11.15 shows a portion of the baseline task network.

Each node in the flow diagram represented a step in the manufacturing process. Associated with each step was performance information, such as the time it took to perform the step, the requirement for resources or manpower, and the potential for rework.

Once the model was fully populated, it was executed against a production schedule. In this way, the same process could be tested against different product mixes, lot sizes, and manning and personnel solutions. As the model ran, an animated depiction of the manufacturing floor could be viewed, as shown in Figure 11.16. This animation allowed the design team to quickly evaluate bottlenecks in the process, manpower problems, and flow rates. Additional output from the baseline model consisted of data files that provided a record of product throughput and manpower utilizations that were used to assess the allocation of workload among the manufacturing staff resulting from the baseline layout.

The next step in the process was to identify viable alternatives that could potentially increase or maintain production throughput while decreasing cost. Ideas were collected using a wide range of techniques, from management-driven initiatives to suggestions from floor production staff. Many alternatives were considered, which ranged from changes in automation levels, to reengineering of the process itself (through a reorganization of the

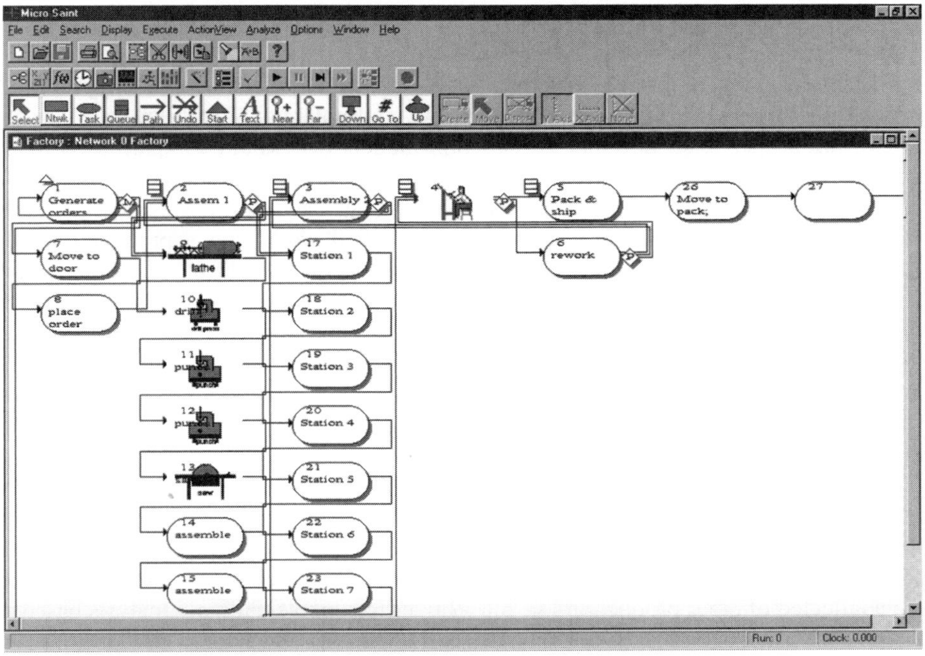


Figure 11.15 Gates Rubber baseline task network.

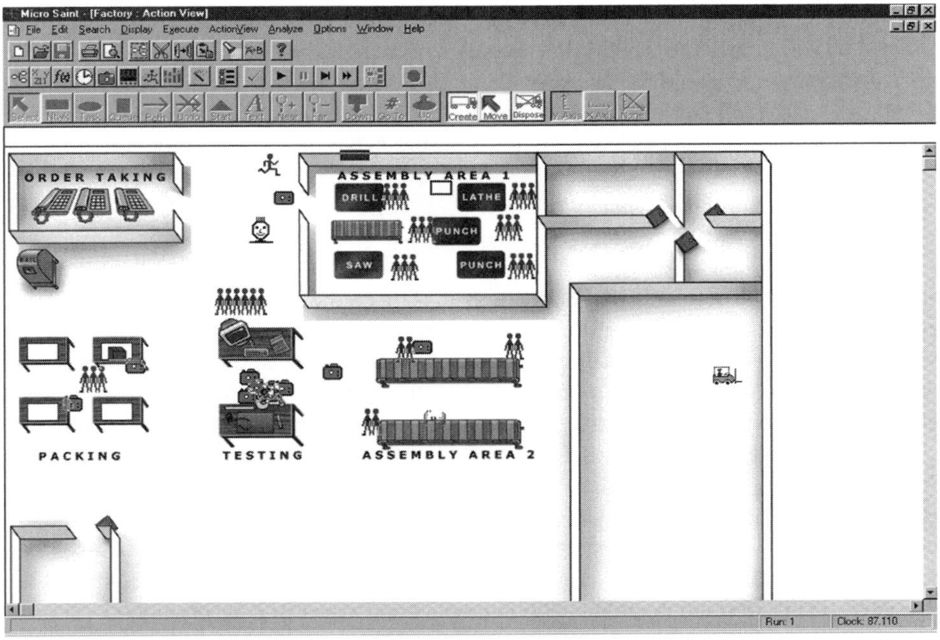


Figure 11.16 Dynamic view of the Gates Rubber manufacturing model.

production stations and adjustment of the work-in-progress levels), to implementing advanced automation that could reduce motor and cognitive workload levels. Each of these alternatives was described in terms of how they would affect the existing process in terms of time, worker involvement in the process, and potential rework levels.

Next, the project team developed simulation models that would compare each of the alternatives to the existing process. The output of the discrete event simulation model provided measures of worker utilization (i.e., how "busy" each worker was throughout the shift), the number of tasks that required highly skilled work, and a prediction of the throughput of the process. Pay rates, fully loaded with overhead and fringe benefit amounts by experience level for each worker, were combined with the cost of each alternative process to generate a prediction of the cost per unit of the product for each alternative. This output provided quantitative, unambiguous comparisons of the effects of trading off manpower and personnel adjustments against materiel solutions, and clarified for the project team the facts associated with making these decisions in a system that is controlled and operated by humans.

One less obvious benefit of the work was that the entire team gained a detailed understanding of the system, illuminating many other areas that could potentially be improved. This project did not attempt to discover the "optimal design" for the cell. Rather, it only identified one of an entire family of solutions that would work. Later studies were conducted of other production line alternatives to support trade-off analysis based on operator utilization balanced against operator costs, training, and skill levels. These studies were also designed so that machines and capacity could be traded against the cost of the equipment, the operator training and installation requirement, and the processing time. Optimization was measured using cost–benefit analyses that balanced labor and facility costs against production rates.

The final step in this process was to implement the chosen change to the production line and to attempt to validate the data predicted by the simulation model. This step was critical in that it would determine whether the use of simulation to evaluate process alternatives was trustworthy and could be used for additional reengineering studies. The validation process was extremely successful, and showed that the predicted measures were within 96 percent of the production levels experienced on the newly redesigned production line.

### 11.4.2 Example 2: Health Care

To stay competitive in the service industry, it is becoming more and more critical to accurately predict customer need while remaining cost efficient. At the same time, the market demands instant, high-quality service and support. Nowhere is this challenge more apparent than in the health care industry. Health care professions must find ways to become more efficient and effective in order to keep up with varying patient needs.

One of the determining factors for health care facilities is the need to properly plan staffing of new or renovated facilities. Prior to the 1980s the methodology used to assess manning needs was based on a ratio formula involving patient volume and length of stay. During the 1980s, a computerized model using probability theory and the Poisson mathematical formula became a popular method for determining obstetric bed need and associated manpower and personnel (i.e., numbers and types of care providers). Unfortunately, neither of these methods could incorporate the impact of scheduled procedures (inductions and caesarian births) or seasonal variability.

In the early 1990s, discrete event simulation became a popular method for attacking this problem. Since simulation can take into consideration the dynamic variability of patient arrival rates, length of stay, and service times, as well as the individual characteristics of the level of care needed for particular procedures, it would provide needed predictive accuracy and power.

Figure 11.17 shows a diagram of the top level of an obstetric model that has been used by Smith Hager Bajo, a health care consulting firm, to predict resource needs (Hager et al., 1998). Development of this model began with eliciting descriptions of the various scenarios that could impact staffing decisions from knowledgeable experts. These scenarios were used to develop a task network model of the medical process. The network is hierarchically organized, with the rectangles representing networks of tasks. The patient arrival rates are modeled using existing patient scheduling data, and the flow of each patient through the process is determined by stochastically generated patient profiles, representing historical procedure records.

This effort resulted in a simulation model that can be used as an analytical tool through which staffing requirements and the impact of training and skill levels on bed need can be assessed. Figure 11.18 provides sample output taken from a snapshot in time during the execution of a scenario in this model.

The obstetric model has been used by over 50 facilities since it was first developed. The projects have ranged from facility planning for bed need, to staffing analysis, to decision making regarding practice changes. Several users have documented significant savings in remodeling costs dues to the detailed analysis supported by this tool.

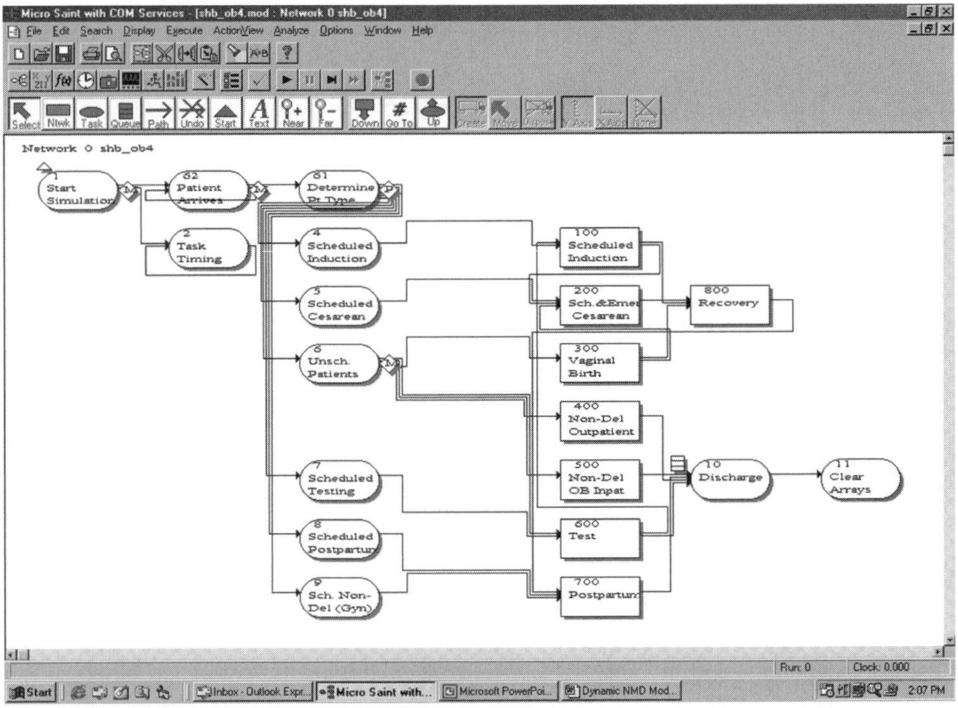


**Figure 11.17** Obstetric model.

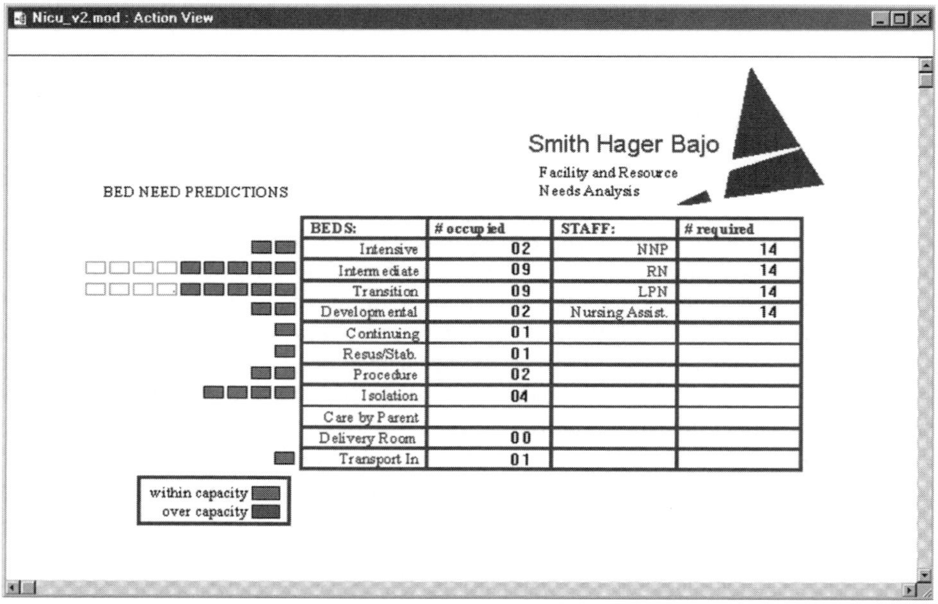


**Figure 11.18** Sample outputs from the obstetrics analysis simulation tool.

### 11.4.3 Summary of Commercial Systems

Emerging systems are relying on advanced automation at an unprecedented level. Decision aids and adaptive automation devices are becoming common solutions to the challenge of decreasing personnel costs. However, the insertion of these technologies can dramatically change the character of the tasks assigned to the people in the system. It is necessary that designers influence system design with manpower and personnel considerations in order to ensure the people can operate, maintain, and support the system.

Commercial industry must attack issues very similar to those experienced by the military in which the role of humans within a system can be adjusted, with concomitant analysis of the resulting implications to system efficiency and effectiveness. Commercial organizations appear to have a very detailed grasp of the financial bottom line and use that grasp as the final objective measure of whether a solution should be implemented. Similar projects to the ones discussed in this section have been conducted across a variety of industries including banking, financial management, fleet vehicle maintenance, automotive production, airport and transportation center design, and food service. In each of these examples, similar questions had to be answered that addressed the role of the people in the system and how the roles impacted system performance and return on investment.

## 11.5 CONCLUSION: CHALLENGES FOR MPT INTEGRATION TECHNOLOGIES

Changes in military and commercial operational environments have contributed to the need to better understand how to take advantage of the increased information available in today's systems. The challenge is to attain this advantage in spite of the limitations in MPT resources and how they affect our capabilities to work with these systems. As designers respond by moving toward distributed, network-centric systems, the research and development (R&D) community must support that move by understanding the human's role in this new environment. Additionally, the R&D community must communicate in ways that designers and system engineers value so that the potential of these technological advances is not wasted.

As a conclusion, we provide a brief survey of some of the most pressing challenges for MPT integration technology development.

***Challenge 1: To Better Understand the Role of the Human as a Component of a Robotic System*** Limitations in manpower and personnel as well as a continuing desire to reduce the exposure of humans to risky environments increase the interest in robotic systems. Because these systems are typically thought of as "unmanned," the implications of human interaction are often overlooked or undervalued. We must develop tools to help robotic system designers understand the payoff of instituting human-centered design processes into the control and maintenance of robotic systems.

Advanced robotic systems are typically endowed with task managers that provide the system with some level of autonomy (Endsley and Kaber, 1999). It is often difficult for the human controller to understand when to intervene in the robot's task and when to let automation take over. The level to which the human trusts the automation and understands the robot's limitations is a critical issue. Yeh and Wickens (2000) are just one team of many researchers who have done work to increase our understanding in this area. Alerting humans to the state of the system, levels of potential error and uncertainty, and possible remediation for automation missteps is a complex issue and must be accommodated by properly designed user interfaces and operator training programs.

It is important that we select robotic technologies, and implement decision-aiding systems (e.g., task managers, intelligent agents) in a thoughtful way, with an eye toward improving total system performance. This is an extensive challenge and engulfs many interesting research questions. Research in the areas of visual perception, multimodal displays, user-adaptive interfaces, and intelligent agents must all be conducted in order to ensure that we are making well-reasoned choices in the design of command and control interfaces for robotic systems.

***Challenge 2: Engage in the Design and Development of Effective Collaborative Tools*** Tools that enhance the ability for distributed teams to collaborate and share a "common operating picture" have great promise. However, anecdotal evidence indicates that these tools have not yet lived up to their promise to reduce workload in times of stress and uncertainty.

Effective collaborative tools rely on vast streams of data and limited bandwidth places practical limits on the application of such tools (Darken et al., 2001). It is necessary that we develop an understanding of the trade-offs between hardware capability and the ability of humans to use information successfully so make wise choices in developing collaborative systems.

***Challenge 3: Continue to Explore How People Make Decisions in Realistic, Complex, Stressful Environments*** Many researchers have made significant progress in developing an understanding of the decision-making process (Klein, 1998; Orasanu and Connolly, 1993; Zsambok, 1997). This understanding has been applied to a variety of operational contexts (Warwick et al., 2002; Peterson et al., 2001) to ensure that the theory makes sense in military environments. The promising results will allow us to design systems that provide a "decision-friendly" environment. This influence spans user interface development, job and task design, and training program development. This work must continue, and further, computational models of human decision making must be developed so that we can predict the decisions a human will make in particular situations. This will enable us to design systems that are error–tolerant.

***Challenge 4: Intelligently Control Information Flow*** Digitization technologies and the increased data flow in everyday life, as well as on the battlefield, have increased the importance of applying visualization technologies in order to help people turn data into knowledge that helps them perform their tasks. Innovative work in this area is being performed (Barnes et al., 2000b) and should be continued.

In a related area, while technologies can be developed to help humans "see" more, problems with limited bandwidth, workload overload, etc. do not always enable them to "notice" more. It is theoretically possible to develop technologies that will augment cognition (Schmorrow et al., 2001). Instead of simply displaying video to a decision maker and allowing the decision maker to identify items on the video that require attention, rule sets and intelligent agents could be developed to "prescreen" the images so that the decision maker's attention is directed toward potential items of interest. Not only will this enhance performance, but it also simplifies many of the tasks associated with the perceptual challenges of a data-rich environment.

***Challenge 5: Develop a Quantitative Understanding of the Links Between Training and Performance*** Some efforts have been made to understand the quantitative links between training and performance (Archer et al., 2002). This area of work is critical if MPT analysts are to make defensible trade-offs between numbers, quantities, and training levels of the humans in the system. As crew sizes decline and tasks become more cognitive in nature, gaps between the skills available to perform tasks that are increasingly complex will create training needs that must be well understood.

***Challenge 6: To Continue to Work Toward "Speaking the Same Language" as Other Engineering Disciplines*** Significant leaps in human performance modeling techniques (Sargent and LaVine, 2000) have provided a way for HSI analysts to participate in large-scale distributed simulation efforts through integrating models of the soldier with models of other system components. The combined simulation places the human model in a realistic context, interacting with changes in the simulated environment, the operating tempo of other computerized components, including enemy forces, and to changes in the state of the system (e.g., sensor outputs, weapon status). The outputs of the composite integrated simulation system are measures of performance and effectiveness that inherently include the variability of the human. These improvements have enabled considerations of human capability to impact system design and have gained credibility for an MPT analyst's ability to provide design information early in the acquisition process.

While these recent advances have been quite successful, much work remains. We must continue to work toward improving the research base so that we can accurately predict the elements (or "first principles") of performance that are instantiated in high-fidelity human performance models. Particularly interesting work is being conducted in developing models of perception and cognition (Lebiere, 2001), and work in this area will improve the credibility of HSI analysts and the applicability of and acceptance for our products.

## REFERENCES


Allender, L. (2000). Modeling Human Performance: Impacting System Design, Performance, and Cost. In M. Chinni (Ed.), *Proceedings of the Military, Government and Aerospace Simulation Symposium, 2000 Advanced Simulation Technologies Conference* (pp. 139–144). San Diego, CA: The Society for Computer Simulation International.

Anderson, J. R., and Lebiere, C. (1998). *The Atomic Components of Thought*. Mahwah, NJ: Lawrence Erlbaum Associates.

Archer, S., and Adkins, R. (1999, April). *Improved Performance Research Integration Tool (IMPRINT) User's Guide, Version 5*. Prepared under Contract DAAL01-95-C-0122 for the U.S. Army Research Laboratory, Human Research and Engineering Directorate, Aberdeen Proving Ground, MD. Boulder, CO: Micro Analysis and Design.

Archer, S., and Allender, L. (2001). New Capabilities in the Army's Human Performance Modeling Tool. In M. Chinni (Ed.), *Proceedings of the Military, Government, and Aerospace Simulation (MGA 2001) Conference* (pp. 22–27). Seattle, WA: The Society for Computer Simulation International.

Archer, R., Oster, A. and Walters, B. (2002). Incorporating Training and Stressors into Computer Generated Force Behaviors In M. Chinni (Ed.), *Proceedings of the Military, Government, and Aerospace Simulation (MGA 2002) Conference* (pp. 53–58). San Diego, CA: The Society for Computer Simulation International.

Barnes, M. J., Knapp, B., Tillman, B., Walters, B., and Velicki, D. (2000a). *Crew Systems Analysis of Unmanned Aerial Vehicle (UAV) Future Job and Tasking Environments*, ARL-TR-2081. Aberdeen Proving Ground, MD: Army Research Laboratory.

Barnes, M. J., Wickens, C. D., and Smith, M. (2000b). Visualizing Uncertainty in an Automated National Missile Defense Simulation Environment. In M. E. Benedict (Ed.), *Proceedings of the 4th Annual FedLab Symposium: Advanced Displays and Interactive Displays* (pp. 117–122). Adelphi, MD: U.S. Army Research Laboratory.

Bradshaw, J. J. (1997). *Software Agents*. Cambridge, MA: AAAI Press/MIT Press.

Card, S. K., Moran, T. P., and Newell, A. (1983). *The Psychology of Human-Computer Interaction*. Hillsdale, NJ: Lawrence Erlbaum Associates.

Caron, M., Jarvenpaa, S. L., and Stoddard, D. B. (1994, September). Business Reengineering at CIGNA Corporation: Experiences and Lessons Learned from the First Five Years. *MIS Quarterly*, pp. 233–250.

Craig, K., Doyal, J., Brett, B., Lebiere, C., Biefeld, E., and Martin E. A. (2002). Development of a Hybrid Model of Tactical Fighter Pilot Behavior Using IMPRINT Task Network Modeling and the Adaptive Control of Thought—Rational (ACT-R). Paper presented at the Eleventh Conference on Computer Generated Forces and Behavior Representation, Orlando, FL, pp. 233–241.

Darken, R. P., Kempster, K., and Peterson, B. (2001). Effects of Streaming Video Quality of Service on Spatial Comprehension in a Reconnaissance Task. Paper presented at the Interservice/Industry Training and Education Conference, Orlando FL. Arlington, VA: NDIA.

Department of the Army. (1997, September). *Army Demographics*. Office of the Deputy Chief of Staff for Personnel, Human Resources Directorate, Demographics Unit. Available: http://www.odcsper.army.mil/directorates/hr/demographics/front.asp.

Department of the Army. (1998, February 28). *Standards of Medical Fitness*, Army Regulation 40-501. Available: http://www.usapa.army.mil/pdffiles/r40_501.pdf.

Department of the Army. (1999, March 31). *Military Occupational Classification and Structure*, Department of the Army Pamphlet 611-21. Available: http://www.usapa.army.mil/pdffiles/p611_21.pdf.

Dynamics Research Corporation (DRC). (2000, January 18). *National Missile Defense (NMD) Human Systems Integration (HSI) Technical Metrics Report*, Delivery Order 0001BF for Contract No. DAAL01-95-C-0115, Human System Integration Office, Brooks Air Force Base, TX. Andover, MA: DRC.

Endsley, M. R., and Kaber, D. B. (1999). Level of Automation Effects on Performance, Situation Awareness and Workload in a Dynamic Control Task. *Ergonomics*, *42*, 462–492.

Farmer, E. W., Belyavin, A. J., Jordan, C. S., Bunting, A. J., Tattersall, A. J., and Jones, D. M. (1995, March). *Predictive Workload Assessment: Final Report*, DRA/AS/MMI/CR95100/1.

Fleishman, E. A., and Quaintance, M. K. (1984). *Taxonomies of Human Performance: The Descriptions of Human Tasks*. New York: Academic.

Gluck, K., Allard, T., Allender, L., Chipman, S., Glinert, E., Tangney, J., and Warren, R. (2001). Panel on Government Interests and Opportunities in Cognitive Modeling. In E. M. Altmann, A. Cleesemans, C. D. Schunn, and W. D. Gray (Eds.), *Proceedings of the 2001 Fourth International Conference on Cognitive Modeling* (p. 33), Fairfax, VA: Lawrence Erlbaum Associates.

Gluck, K., and Pew, R. W. (2001). Overview of the Agent-Based Modeling and Behavior Representation (AMBR) Model Comparison Project. In *Proceedings of the 10th Conference on Computer Generated Forces and Behavior Representation* (pp. 3–6), Norfolk, VA.

Grover, V., Jeong, S. R., Kettinger, W. J. and Teng, J. T. C. (1995). The Implementation of Business Process Reengineering. *Journal of Management Information Systems*, *12*(1), 109–144.

Guerrier, J. H., Lowry, J. C., Jones, R. E., Guthrie, J. L., Barber, J. L, and Miles, J. L. (1991, April). *Handbook for Conducting Analysis of the Manpower, Personnel, and Training Elements for a MANPRINT Assessment*, U.S. Army Research Institute for the Behavioral and Social Sciences Research Note 91-43, DTIC No. AD-A23-5430. Alexandria, VA: U.S. Army Research Institute for the Behavioral and Social Sciences.

Hager, J., Bajo, K., Smith, J., Archer, R., and LaVine, N. (1998). The Next Generation of Simulation: Obstetric Bed Need/Staffing Projections. In *Proceedings of the Healthcare Information Management Systems Society Conference*, Orlando, FL.

Harshell, J., and Dahl, S. G. (1988, December). Simulation Model Developed to Convert Production to Cellular Manufacturing Layout. *Industrial Engineering*, *20*(12), pp. 40–45.

Hart, S. G., and Staveland, L. E. (1988). Development of NASA-TLX (Task Load Index): Results of Empirical and Theoretical Research. In P. A. Hancock and N. Meshkati (Eds.), *Human Mental Workload*. Amsterdam: North-Holland.

Hay Systems. (1991, December). *Generic MANPRINT Analysis Adjunct Lessons Learned Technical Reports on MPT in Army MANPRINT Analyses: Executive Summary*, Delivery Order No. 0031, prepared for U.S. Army Training and Doctrine Command, VA, Contract No. DABT60-87-D-3873, DTIC No. AD-A258 639. Washington, DC: Hay Systems.

Headley, D. B. (in press). *Methods for Assessing Human Systems Integration Issues in the Manpower and Personnel Domains*. U.S. Army Research Laboratory, Human Research and Engineering Directorate Technical Report, Aberdeen Proving Ground: MD.

Hendy, K., and Farrell, P. S. E. (1997, December). *Implementing a Model of Human Information Processing in a Task Network Simulation Environment*, DCIEM No. 97-R-71. Defence and Civil Institute of Environmental Medicine, Toronto, Ontario, Canada.

Herlihy, D., Bondaruk, J., Nicholas, G., Guptill, R., and Park, J. (1990, May). *Hardware Versus Manpower Compatibility Methodology*, Vol. 1: *Overview and Manager's Guide* (DTIC AD-A225 122); Vol. 2: *Systems Analysis* (AD-A225 745); Vol. 3: *Manpower Requirements Analysis* (AD-A225 746); Vol. 4: *Personnel Pipeline Analysis* (AD-A225 747); Vol. 5: *Training Resource*

*Requirements Analysis* (AD-A225-748); Vol. 6: *Impact Analysis* (AD-A225 733); Vol. 7: *Tradeoff Analysis* (AD-A225 732). Andover, MA: Dynamics Research Corporation.

Heuckeroth, O. H., and Smith, N. D. (1990, March). *Relationship between Vehicle Identification Performance and the Armed Services Vocational Aptitude Battery (ASVAB)*, U.S. Army Research Institute for the Behavioral and Social Sciences Technical Report 882, DTIC AD-A221 558. Alexandria, VA: U.S. Army Research Institute for the Behavioral and Social Sciences.

Hoagland, D., Martin, E., Anesgart, M., Brett, B., Lavine, N., and Archer, S. (Fall, 2001). Representing Goal-Oriented Human Performance in Constructive Simulations: Validation of a Model Performing Complex Time-Critical-Target Missions. Paper presented at the Simulation Interoperability Workshop, Orlando, FL. Arlington, VA: NDIA.

Hood, L., and Allender, L. (1993, August 8–13). Micro Saint/HOS—A Computer Modeling Tool for Evaluating the Human-Computer Interface. In M. J. Smith and G. Salvendy (Eds.), *Proceedings of the 5th International Conference in Human-Computer Interaction* (jointly with the 9th Symposium on Human Interface), (p. 252). School of Industrial Engineering, Purdue University, West Lafayette, IN.

Horne, D. K. (1986, April). *The Impact of Soldier Quality on Performance in the Army*, U.S. Army Research Institute for the Behavioral and Social Sciences Technical Report 708, DTIC AD-A173 946, Alexandria, VA: U.S. Army Research Institute for the Behavioral and Social Sciences.

Human Systems Integration Technologies, Tools, and Techniques (HSIT3) Seminar Presentation CD. (2000, September 25–26). Building 196, 2261 Monahan Way, Wright Patterson Air Force Base, OH: AFRL/HEC/HSIAC.

Kaplan, J. D., Dahl, S. G., Laughery, K. R., Archer, R. D., O'Brien, L., Connelly, E., Conroy, J., Cherry, W. P., Thompson, D., Proegler, L., Roerty, D., and Roberts, D. (1989, June). *MANPRINT Methods, Monograph: Aiding the Development of Manned System Performance Criteria*, Technical Report 852. Alexandria, VA: U.S. Army Research Institute for the Behavioral and Social Sciences.

Keeney, W. M., and Rowe, M. (1998, December). *SAILOR 21: A Research Vision to Attract, Retain, and Utilize the 21st Century Sailor*. Available: http://www.nprst.navy.mil.

Kelley, T. D., Patton, D. J., and Allender, L. (2001). Predicting Situation Awareness Errors Using Cognitive Modeling. In M. J. Smith, G. Salvendy, D. Harris, and R. J. Koubek (Eds.), *Proceedings of Human-Computer Interaction International 2001 Conference*. Vol. 1: *Usability Evaluation and Interface Design: Cognitive Engineering, Intelligent Agents and Virtual Reality* (pp. 1455–1459). Mahwah, NJ: Lawrence Erlbaum Associates.

Klein, G. A. (1989). Recognition-Primed Decisions. In W. B. Rouse (Ed.), *Advances in Man-Machine Systems Research* (pp. 47–92). Greenwich, CT: JAI Press.

Klein, G. A. (1998). *Sources of Power: How People Make Decisions*. Cambridge, MA: MIT Press.

Knapp, B., and Tillman, B. (1998). Job Assessment Software System (JASS). In *Proceedings of the Human Factors and Ergonomics Society 42nd Annual Meeting* (pp. 1319–1322). Chicago, IL. Human Factors and Ergonomics Society, Santa Monica, CA.

Laird, J. E., Newell, A., and Rosenbloom, P. S. (1987). Soar: An Architecture for General Intelligence. *Artificial Intelligence*, *33*, 1–64.

Laughery, K. R., Jr., and Corker, K. (1997). Computer Modeling and Simulation. In G. Salvendy (Ed.), *Handbook of Human Factors and Ergonomics* (pp. 1375–1408). New York: Wiley.

Law, A. M., and Kelton, W. D. (2000). *Simulation Modeling and Analysis*, 3rd ed. McGraw-Hill Series in Industrial Engineering and Management Science. New York: McGraw-Hill.

Lebiere, C. (2001). A Theory-Based Model of Cognitive Workload and Its Applications. In *Proceedings of the 2001 Interservice/Industry Training, Simulation and Education Conference*. Held at Orlando, FL. Arlington, VA: NDIA.

Lebiere, C., Biefeld, E., Archer, R., Archer, S., Allender, L., and Kelley, T. D. (2002). IMPRINT/ACT-R: Integration of a Task Network Modeling Architecture with a Cognitive

Architecture and Its Application to Human Error Modeling. In M. Chinni (Ed.), *Proceedings of the Military, Government and Aerospace Simulation Symposium, 2002 Advanced Simulation Technologies Conference*. San Diego, CA: The Society for Computer Simulation International.

Malhotra, Y. (1998, Fall). Business Process Redesign: An Overview. *IEEE Engineering Management Review, 26*(3), 27–31.

McCracken, J. H., and Aldrich, T. B. (1984). *Analyses of Selected LHX Mission Functions: Implications for Operator Workload and System Automation Goals*, TNA ASI479-024-84, (DTIC No. AD-A232330). Fort Rucker, AL: Anacapa Sciences.

Meyer, D. E., and Kieras, D. E. (1997). A Computational Theory of Executive Cognitive Processes and Multiple-Task Performance. Part 1. Basic Mechanisms. *Psychological Review, 104*(1), 2–65.

Office of the Assistant Secretary of Defense (Force Management Policy). (1999, November). *Population Representation in the Military Services: Fiscal Year 1998*. Available: http://dticaw.dtic.mil/prhome/poprep98.

Orasanu, J., and Connolly, T. (1993). *The Reinvention of Decision Making*. In G. A. Klein, J. Orasanu, R. Calderwood, and C. E. Zsambok (Eds.), *Decision Making in Action: Models and Methods* (pp. 3–20). Norwood, NJ: Ablex.

Peterson, B., Boswell J., and Darken, R. (2001). Collaborative Navigation in Real and Virtual Environments. Paper presented at the Interservice/Industry Training and Education Conference, Orlando, FL. Arlington, VA: NDIA.

Pew, R. W., and Mavor, A. S. (Eds.). (1998). *Modeling Human and Organizational Behavior Application to Military Simulations*. Washington, DC: National Academy Press.

Reid, G. B., Potter, S. S., and Bressler, J. R. (1989). *Subjective Workload Assessment Technique (SWAT): A User's Guide*, Technical Report No. AAMRL-TR-89-023. Wright Patterson Air Force Base, OH: Harry G. Armstrong Aerospace Medical Research Laboratory.

Rose, A. M., Radtke, P. H., Shettel, H. H., and Hagman, J. D. (1985). *User's Manual for Predicting Military Task Retention (Revised June 1985)*, ARI Research Product 85-26. Alexandria, VA: U.S. Army Research Institute for the Behavioral and Social Sciences.

Sargent, R. A., and LaVine, N. D. (2000, April). Combat Automation Requirements Testbed (CART)—Using Simulation Interoperability for Improved Simulation Based Acquisition. In M. Chinni (Ed.), *Proceedings of the Military, Government, and Aerospace Simulation Symposium, 2000 Advanced Simulation Technologies Conference* (pp. 114–119). San Diego, CA: The Society for Computer Simulation International.

Schmorrow, D., Worcester, L., and Patrey, J. (2001). Augmented Cognition: New Design Principles for Human-Computer Symbiosis. In *Proceedings of the International Applied Military Psychology Symposium*. Prague, Czech Republic.

U.S. Department of Defense. (1995). *Modeling and Simulation Master Plan*. Available: https://www.dmso.mil/public/sitemap.

U.S. Department of Defense. (1999, May 17). *Human Engineering Program Process and Procedures*, MIL-HDBK-46855A. Redstone Arsenal, AL: U.S. Army Aviation and Missile Command. Available: http://www.hf.faa.gov/docs/46855a.pdf.

U.S. Total Army Personnel Command. (1991, June). *Early Comparability Analysis (ECA) Procedural Guide*. Alexandria, VA: U.S. Total Army Personnel Command.

Warner, J., and Knapp, B. (1999, July). *What Is the Relationship of ST [Skilled Technical] to MOS Success in Training?* Briefing slides. Fort Huachuca, AZ: U.S. Army Research Laboratory—Human Research and Engineering Directorate Field Element.

Warwick, W., McIlwaine, S., and Hutton, R. J. B. (2002). Developing Computational Models of Recognition-Primed Decisions: Progress and Lessons Learned. *In Proceedings of the 11th Conference on Computer Generated Forces and Behavior Representation* (pp. 479–485). Orlando, FL.

Winkler, J. D. (1999). Are Smart Communicators Better? Soldier Aptitude and Team Performance. *Military Psychology, 11*(4), 405–422.

Wojciechowki, J. Q., Wojcik, T., Archer, S., and Dittman, S. (2001). Information-Driven Decision Making Human Performance Modeling. In M. Chinni (Ed.), *Proceedings of the Military, Government, and Aerospace Simulation (MGA 2001) Conference*, (pp. 3–8). Seattle, WA. San Diego, CA: The Society for Computer Simulation International.

Yeh, M., and Wickens, C. D. (2000). Effects of Cue Reliability, Realism, and Interactivity on Biases of Attention and Trust in Augmented Reality. Paper presented at the IEA 20000/HFES 2000 Congress, San Diego, CA.

Zachary, W., Ryder, J., Ross, L., and Weiland, M. Z. (1992). Intelligent Computer-Human Interaction in Real-Time, Multi-Tasking Process Control and Monitoring Systems. In M. Helander and M. Nagamachi (Eds.), *Human Factors in Design for Manufacturability*. New York: Taylor and Francis.

Zhang, Y., He, L., Biggers, K., Yen, J., and Ioerger, T. R. (2001). Simulating Teamwork and Information Flow in Tactical Operations Centers Using Multi-Agent Systems. In *Proceedings of the Tenth Conference on Computer Generated Forces and Behavior Representation* (pp. 529–539). Norfolk, VA.

Zsambok, C. E. (1997). Naturalistic Decision Making: Where Are We Now? In C. E. Zsambok and G. Klein (Eds.), *Naturalistic Decision Making* (pp. 3–16). Mahwah, NJ: Lawrence Erlbaum.

## CHAPTER 12

# Integrating Training into the Design and Operation of Complex Systems

LAWRENCE J. HETTINGER

## 12.1 INTRODUCTION

A challenging, new state of affairs is confronting public- and private-sector organizations engaged in the design and deployment of complex sociotechnical systems.[1] Chapter 1 aptly summarizes this by noting that systems and products that can be operated and repaired by fewer people, lesser skilled people, and/or people with less training will be in greater demand. It is not an impractical expectation for the military, and probably for many commercial areas as well, to demand human systems integration (HSI) designs that will allow reductions in all three areas—manpower, personnel, and training (MPT)—together.

If current trends hold, these organizations will be increasingly challenged to accomplish more with significantly diminished financial and personnel resources. Specifically, they will be called upon to develop and field systems that are "better" (more operationally effective) than their predecessors, even though their design and operation will be supported with fewer traditional resources (e.g., funding, manpower, etc.). Advanced military systems, such as the U.S. Navy's planned CVNX aircraft carrier, are already being approached with these expectations and constraints firmly in place. The CVNX is intended to be far more operationally effective than earlier Nimitz-class carriers but with an approximate 27 percent reduction in crew size (Smith and Driscoll, 2000). Similarly, the planned DD-X class of new naval destroyers is projected to operate with an even greater percentage reduction in crew size—and, again, with greatly improved performance capabilities. Indeed, it seems apparent that the U.S. military's strategic paradigm is shifting away from a reliance on overwhelming force (in terms of sheer numbers of personnel and weapons platforms) toward increased reliance on superior technology, stealth, flexibility, and speed. And while critical resources underlying system development and deployment may be diminishing, performance expectations certainly are not. Clearly, the development of large-scale military systems has entered an era in which intelligent, coordinated,

*Handbook of Human Systems Integration*, Edited by Harold R. Booher.
ISBN 0-471-02053-2 

multidisciplinary cooperation among design disciplines will be required to produce high-performance technologies.

An analogous shift is occurring in many nongovernment sectors of the economy, those in which some combination of personnel cutbacks, budgetary pressures, and increased performance demands are present.[2] The fact that these constraints are also typically accompanied by significant timeline pressures (e.g., bringing products to market before one's competitors in the private sector; adhering to aggressive timelines for system development and deployment in the public sector) has led to significant interest in the development and application of an efficient and effective means of designing and deploying new systems.

The HSI approach provides a set of effective guidelines for accomplishing these complex and frequently conflicting objectives. The principles and methods of the HSI approach reflect (1) an unwavering concentration on the user as the focal point of any sociotechnical system in order to achieve high levels of safe and effective system performance and (2) the application of a coordinated, multidisciplinary approach among the core elements of the systems engineering process during all phases of acquisition from concept formulation to deployment. As described elsewhere in this book (e.g., Chapter 18), the HSI approach has led to the development of effective and safe systems in a manner whose efficiency is reflected in the life-cycle cost savings as well as in enhanced usability and effectiveness of the systems themselves.

Despite several successful applications to date, the HSI model is still a new approach with a number of important areas remaining for growth and elaboration. Nowhere is this more evident than in enhancing training to reach its full potential as a critical HSI domain.

### 12.1.1 Role of Training in System Design and Deployment

The vital importance of training in supporting the effective performance of systems has long been recognized. Obviously, effective training has been, is, and will continue to be a critical element of successful systems for the foreseeable future and beyond. However, an HSI approach to systems acquisition affords an important, new role for training within the total context of an *integrated, multidisciplinary* approach to systems design and deployment. It also addresses the apparent workforce reality that resources for training will continue to experience significant pressure and the training community (like other systems engineering components) will continue to be called on to accomplish more with less.

The HSI design philosophy emphasizes three key aspects of training:

1. the role training plays in supporting the effective performance of deployed systems;
2. the role that training considerations and expertise play throughout all phases of the *design* and *test* of systems, well before they are deployed; and
3. the positive impact that the HSI approach can have on the training community itself, primarily by facilitating interactions with other domains whose concerns are principally with optimizing aspects of human performance.

The first point corresponds to training's traditional role—one in which a specific, fully developed system is taken as a starting point and within which training applies its trade toward equipping people with the knowledge, skills, and abilities (KSAs) and devices necessary to interact with the system. This is the area in which training specialists have most commonly devoted their time and attention. Adding the second and third points is the challenge posed by HSI.

As one of the oldest and most prestigious domains of applied psychology, training is an area of impressive scholarship and real-world accomplishment, and there are several excellent, contemporary reviews of this work (e.g., Patrick, 1992; Salas and Cannon-Bowers, 2001; Swezey and Andrews, 2001). While I will discuss some of the main elements of this literature and have relied heavily on these reviews throughout this chapter, my main interest will be to review several key areas of training research and development in order to analyze how the various issues and challenges relate to an HSI training approach. Specifically, how can we best apply the considerable expertise contained within the training community to better support the *entire* system design and deployment process? How can we intelligently design systems so that training requirements will not be as onerous once they are built and deployed, thereby enabling more efficient and effective use of limited training resources? And how can training itself be improved through more thorough integration with other HSI domains and procedures?

To approach this problem, we will need to examine the nature of the potential interactions between training and the other elements of the HSI approach and how the products of these interactions can be used to enhance the total system acquisition process. For instance, training and personnel selection are two areas with considerable functional overlap. Clearly, it is vital for training and personnel selection specialists to efficiently interact with one another—for instance, when training informs personnel of projected personnel attributes and characteristics that will be needed to support specific systems or when personnel informs training of a deficit of such people in the personnel assignment pool. However, it is also likely that training expertise and knowledge can inform and benefit the activities of human factors engineers, safety and health specialists, and others involved in the up-front development and testing of a new system. Similarly the expertise of these groups, in turn, can likely inform and benefit the activities of training experts.

One of the purposes of this chapter is to show examples where considerable benefits to the systems acquisition process can be derived from application of training expertise earlier in the process and through greater interaction with the other functional domains. These benefits can be expected not only in the form of cost and time savings through early anticipation of training problems but also through more innovative designs. It is not unreasonable to expect many creative ideas to appear when training specialists interact with other HSI professionals and systems engineering disciplines on a consistent, daily basis. The outcome of these synergistic processes will take the form of design and test insights that will, in all likelihood, significantly enhance the overall performance of new systems.

### 12.1.2 Overview of Chapter

The objective of this chapter is to discuss issues and challenges associated with effectively applying the training domain throughout the system life cycle using the HSI principles and methods described throughout this book.

The chapter covers the following topics:

1. *Traditional Training Model* This section covers the objectives and fundamental activities of training and provides an overview of the systems development process depicting the traditional training role.

2. *New HSI Training Model* A new training definition is provided, synergies between training and the other HSI domains are discussed, and considerations for integrating training within the systems development process are identified.
3. *Issues and Challenges* This section reviews some of the key issues that have been examined by training specialists over the years, identifies key challenges currently facing the training domain, and considers the relevance of these topics for an HSI approach to systems design.
4. *Conclusions and Recommendations* In order for training to be most effectively integrated within an HSI approach, a number of important research questions must be addressed. In addition, there are important issues more directly related to the culture and organization of complex systems design and deployment. The former topics are more empirical and scientific in nature, while the latter are more related to programmatic and administrative concerns. This section will describe issues and make some recommendations relevant to both types.

## 12.2 TRADITIONAL TRAINING MODEL

What is the purpose of training? At the most general level, this question is usually addressed by noting that training exists to promote the acquisition, retention, and transfer of specific sets of skills and abilities (e.g., Patrick, 1992, pp. 13–14). Training, it has often been noted, is not the same as education. The two domains have traditionally been differentiated by emphasizing training's concentration on very specifically defined sets of skills as opposed to education's more global purpose of "broadening the mind" and developing the intellect. However, as a result of the profound changes impacting the nature of contemporary work environments, the traditional differences between training and education are perhaps becoming less distinct and less meaningful than in the past. As we attempt to prepare individuals to become adept at coping with rapid and significant change in work environment characteristics, much may be gained by broadening the scope of training to include skills associated with "learning to learn."[3]

Clearly, one of the major purposes of training is to identify and devise practical methods and technologies for imparting the skills that people need to successfully function within complex systems. However, within an HSI perspective the objective of training is expanded to include a concern with the development of KSAs to handle *all* aspects of the sociotechnical system. Simply put, the new objective of training is to promote the design and operation of highly functional systems by

1. promoting awareness of training needs and requirements at all phases of the system acquisition process and
2. incorporating knowledge from other technical domains within its conception of system training requirements and approaches.

### 12.2.1 Fundamental Training Activities

Figure 12.1 provides a conceptual illustration of the role of training in the systems design process implied by the traditional definition. Four fundamental activities are seen as

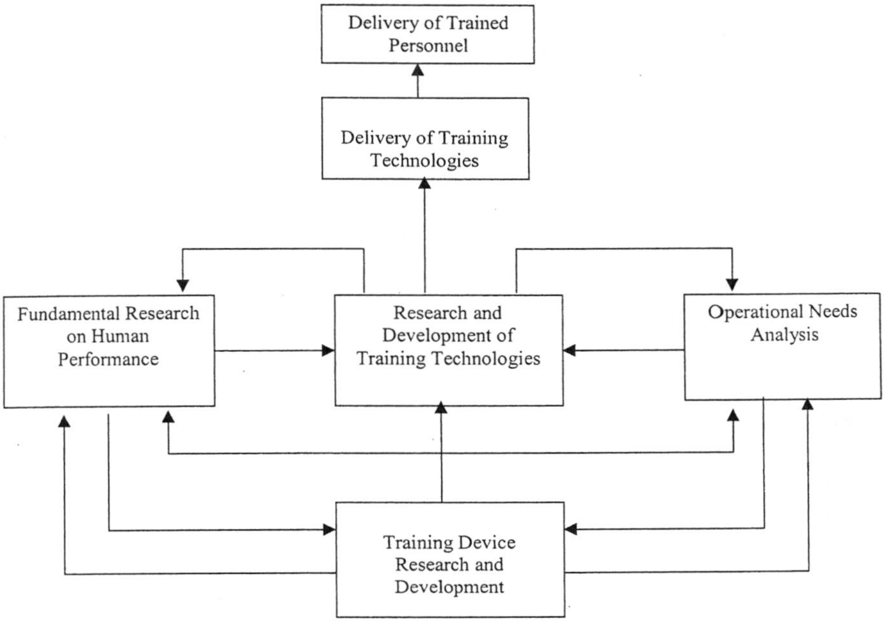


**Figure 12.1** Schematic diagram representing traditional flow of training expertise within systems design and deployment process.

forming the basis of the role of training in complex systems design and deployment, particularly with respect to large-scale military procurements:

1. *Fundamental Research on Human Performance* Research within the broad area of human performance, including domains such as cognition, learning, problem solving, decision making, psychomotor performance, etc., plays an important role in the identification of (1) the types of cognitive, perceptual, and motor abilities that underlie skilled performance; (2) the means by which humans acquire these skills; and (3) the means by which the presence or absence of these abilities can be reliably assessed. In the realm of training research and practice, these are perhaps the most scientifically based areas of investigation.

2. *Operational Needs Analysis* This analysis is an empirical determination of system performance requirements, particularly with respect to the nature of the role played by human operators. Techniques such as cognitive work analysis, hierarchical task analysis, knowledge mapping, etc., are widely used to extract information about global characteristics of operators' tasks as well as more detailed information such as the timing of information delivery and corresponding human performance requirements in operational environments.

3. *Training Device Research and Development* This activity reflects the engineering and human factors research conducted on the design and application of new technical approaches to training (e.g., advanced simulation devices, virtual environment systems, etc.). Typically, this area of training research and practice is devoted to the development of new training hardware and software for eventual deployment into the operational training

environment and includes work in the area of simulation and virtual environment systems as well as technologies for delivering training during system deployment.

4. *Training Technology Research and Development* The output of the above three activities is manifest in the form of training technologies—specifically, training concepts, curricula, and devices that best represent current knowledge about training needs, training devices, and human skills acquisition. When combined with specific knowledge about the operational needs and constraints of the targeted training environment, the output of this area of training research and practice is the delivery of training technologies to the operational community.

As depicted in Figure 12.1, none of these areas operate independently of one another. Nor is there (or should there be) a strictly linear relationship between these activities. One does not feed directly into another without some reciprocal influence. Certainly it is in the best interest of large-scale training programs to have all these areas effectively interacting with one another. Particularly in an era of reduced research and development (R&D) budgets it would be wise, for example, for activities within the *fundamental research on human performance* area to be closely linked with current and future requirements identified by the *operational needs analysis* area. Otherwise, the former risks becoming estranged from the real-world requirements identified by the latter.

Figure 12.1 illustrates a model that makes sense and has worked well [particularly within large-scale Department of Defense (DoD) training programs], but it has a serious flaw with respect to the interaction of training programs with other disciplines involved in system development. Specifically, within the traditional model, training takes the form of a very "stovepiped" activity. In other words, while certainly not operating in a complete vacuum, training tends to be isolated from other mainstream system disciplines.[4]

### 12.2.2 Training in the System Development Process

Figure 12.2 provides a simplified schematic of the systems design and deployment process in which the traditional training role is depicted. Four major system acquisition cycle activities are illustrated:

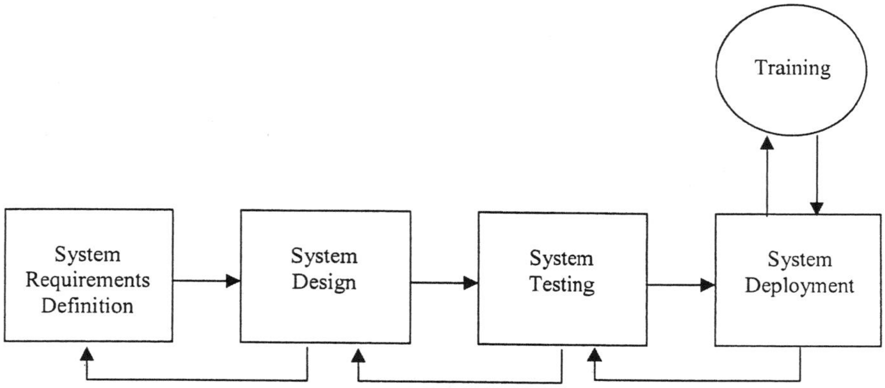


**Figure 12.2** Schematic representation of system development process featuring the traditional role played by training.

1. *System Requirements Definition* The origin of any new system results from the establishment of some need in the user community. In the case of military systems, this need may arise from the level of the individual warfighter (e.g., a stated need for improved sighting mechanisms on small arms) or from a more strategic level (e.g., a stated need for a high-performance destroyer to operate in *littoral* environments with greatly reduced crew sizes—the DD-X). Quite often, of course, the identification of needs has relevance for both levels of analysis. It is in this stage that the overall objectives of the system are identified and performance specifications are determined.

2. *System Design* Once a need has been established and the necessary approval granted to proceed with development of candidate systems,[5] the design process can begin in earnest. With particularly transformational technologies (such as the U.S. Navy's DD-X and CVNX programs, the U.S. Air Force's F-22 program, and the multiservice Joint Strike Fighter program) this stage of the process quickly becomes extremely complex. Large-scale, *transformative* systems acquisition projects can lead to the initiation of extensive basic and applied R&D programs involving large teams of scientists, engineers, and management and support personnel. While these teams are often widely dispersed in a geographical and organizational sense (and are often in competition with one another for resources and influence), their overall mission is to produce a working prototype of the system under consideration.

3. *System Testing* While testing of system components is almost always conducted as part of the system design process, ultimately, a form of acceptance testing must be conducted to determine whether or not the entire finished product meets stated specifications. Acceptance testing of systems in the field typically includes assessment of overall system performance, generally assessed in terms of a battery of operationally relevant performance metrics, as well as a variety of tests of subsystem components.

4. *System Deployment* Following successful completion of the initial three phases of design, the system is ready to be deployed. At this point and throughout the system's effective life cycle, changes may occasionally be made based on changing strategic needs, advances in technology, etc. However, for all intents and purposes the design cycle is effectively ended once the system is deployed.

As the model in Figure 12.2 implies, input from the training community has traditionally been solicited only toward the end of the design cycle. In the worst-case scenario, training specialists are called upon to begin their work once a system has more or less reached its final design and is ready for deployment, at which point they obviously will not have had any input on requirements, design, and testing issues that may have facilitated the eventual design of training programs and enhanced the overall design of the system itself. In a somewhat less dire scenario, training inputs may have been solicited at these earlier stages in system development, but in a way that effectively isolated training specialists from other technical domains (i.e., stovepiping), thereby limiting the effectiveness of the system development process in ways described above.

## 12.3 HSI TRAINING MODEL

The most compelling feature of the traditional model is that training R&D efforts are devoted exclusively to very specific training objectives. In other words, training researchers and practitioners work on well-defined training issues to achieve very specific training objectives. On the face of it, this may not seem like such an unreasonable proposition; indeed, it is almost a tautology. One might also argue that concentrating on very specific training objectives should produce more sharply focused and effective solutions to training problems.

However, there are at least three major problems with this approach: (1) it perpetuates organizational stovepiping with its many associated problems, (2) it prevents other HSI domains from benefiting from training expertise, and (3) it cuts off training experts from benefits that could be gained from more regular interaction with other HSI domains.

The commonly understood objectives of training (i.e., supporting the effective and efficient acquisition and retention of functional KSAs) characterizes training as a process that is unnecessarily restricted to a narrow range of activities in the life cycle of a given sociotechnical system. The HSI approach encourages the incorporation of training expertise and requirements at all stages of the system development process.

### 12.3.1 New Training Definition

A new definition of training is proposed to take into account a broader, more integrated perspective: *Training* is concerned with promoting the safe and effective performance of sociotechnical systems by facilitating the acquisition, retention, and transfer of user KSAs through the design of effective curricula and training technologies and through influencing system design, development, test, and deployment in such a way as to effectively integrate knowledge about requisite user skills, abilities, and performance requirements throughout all phases of the system life cycle.

Figure 12.3 provides a conceptual illustration of the role of training in the systems development process as implied by the new definition of training. The figure illustrates the

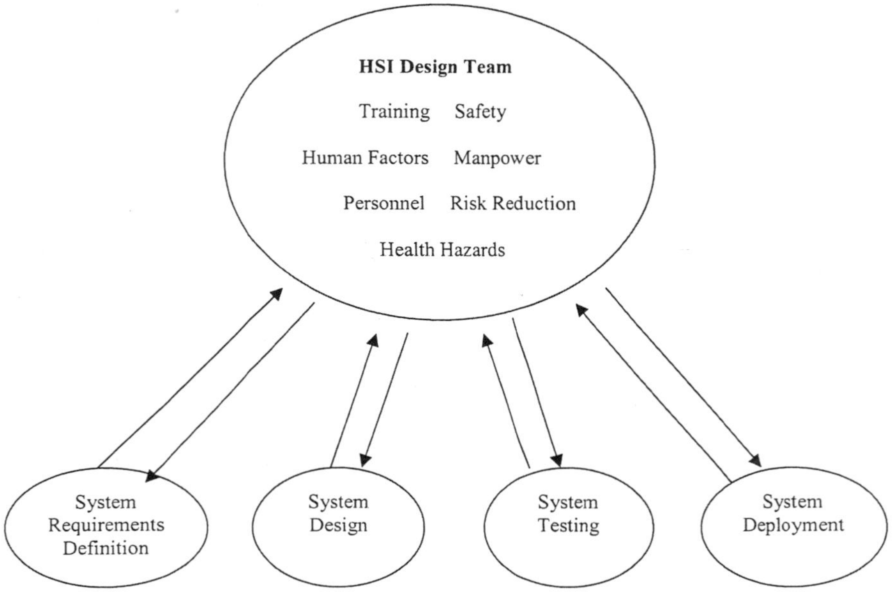


**Figure 12.3** Schematic diagram representing flow of training expertise in HSI approach.

concept of training, along with the other HSI domains, working in close cooperation with one another throughout all major phases of system development.

The satisfactory application of HSI principles to complex system design requires not only that appropriate and up-to-date knowledge about designing and implementing training programs be employed but also that training considerations be factored into the discussions of system design at all stages and in conjunction with all of the other HSI disciplines. For example, although training research and practice have played a very important role in the successful application of human–machine technologies, its role has been unnecessarily restricted to already developed or nearly developed systems.

Also, many aspects of training, as traditionally conceived, can have a negative impact on overall system performance. These include poor training (e.g., training delivered by ineffective and/or poorly qualified individuals, training delivered using defective or outdated materials, etc.), inappropriate or nonfocused training (e.g., training delivered at an inappropriate level or with insufficient attention given to the specificity of the material to be covered or the desired outcome), training delivered at the wrong time, and many others.

These problems are all clearly *risk factors* in designing training for any given sociotechnical system. However, they are more than just training risks and are best understood as *system risks*, because to the extent that such risks are present, the functionality of the entire system will be adversely affected. Training, it might be said, is too important to be left to training specialists alone—just as human factors, personnel selection, safety and health, etc., are too important to be left to their respective specialists alone. The implications of risks not being adequately addressed in any of these areas have profound consequences for an entire system, and therefore, a systems approach such as the HSI methodology is needed.

### 12.3.2 Synergies between Training and Other HSI Domains

One of the main areas of emphasis of the HSI approach is the need for all functional entities within the system development process to work together in a consistent and coordinated fashion. This is critical not only from the perspective of eliminating costly redundancies or discrepancies in effort and other negative effects of misunderstanding and miscommunication but also to enhance the introduction of creative, user-centered approaches to system design. In order to better appreciate the benefits of such synergies, several examples of interactions between training specialists and those from the personnel selection, human factors, and safety and health domains are provided below.

***Training and Personnel Selection*** As noted earlier in the chapter, the synergistic relation between the domains of training and personnel selection is one of the most clearly evident. To a great extent, the primary mission of each of these areas is largely identical—providing individuals who are well suited to operate systems safely and effectively. The difference, of course, is that one group has primary responsibility for selecting individuals from the pool of possible candidates while the other group has primary responsibility for providing these individuals with the necessary KSAs to perform their assigned tasks.

One can easily imagine the inefficiencies that can result when these two groups operate in relative isolation from one another. From a more positive perspective, however, the benefits of these groups working in close coordination are also clear. Those responsible for personnel selection concern themselves on a daily basis with the nature of the pool of candidates from which they must select those individuals who will eventually, following training, be called upon to operate and/or function within the various systems that comprise a given organization. Consistent interaction between these specialists and those with training responsibility can help the latter prepare training technologies and curricula that are optimally suited to the types of trainees that are most likely to be made available to them. Conversely, personnel selection specialists will come away from this regular interaction with a much more accurate concept of the types of KSAs that are going to be required for recruits, job candidates, etc., to succeed in the work environment.

***Training and Human Factors Engineering*** Training and human factors specialists approach system design problems in ways that often seem largely complementary to one another. Human factors engineering (HFE) attempts to design human–machine systems that are sufficiently intuitive and functional so that the need for extensive training is diminished. Training specialists design training methods to compensate for the inability of HFE to perfectly achieve its goal. While interactions between these groups are certainly far from uncommon, to the extent that they are kept separate within the context of stovepiped system development efforts, many potential design and training benefits can still be lost.

For instance, since a large part of training's traditional role involves preparing individuals to successfully operate human–machine systems, it would be advantageous for training personnel to have the opportunity to provide inputs on the design of these systems. For all of the various reasons described above—expertise in human performance, skill acquisition (e.g., what has and has not "worked" in the past), and analysis and assessment of tasks—training personnel bring a unique perspective to human–machine interface design problems. Similarly, human factors specialists can reciprocate with their insights on how humans interact with novel technologies and the types of human performance failures (and successes) that they have observed as part of their human–machine system design process. Additionally, if training specialists are assigned a meaningful role along with human factors engineers (and hardware and software specialists, etc.) in the early design of human–machine system concepts, then the possibility of designing in effective "on-duty" training technologies is greatly enhanced. By permitting operators to exercise opportunities for training using the very human–machine systems they use to perform their tasks, greater efficiencies in the use of training resources (time and money) can be achieved. Greater training effectiveness and personnel readiness can also be achieved as needs for training are identified by means of on-line performance assessment followed by carefully targeted training to address performance deficits as they are identified.

***Training and Personnel Safety and Health*** Personnel safety and health specialists concern themselves with all aspects of system design and operation that impact on operators' physical and psychological well-being. Within large-scale industrial settings, such as the oil and chemical industries, the day-to-day work of training specialists is nearly always very directly concerned with safety and health issues, and the same is also true of military settings. However, personnel and health specialists often find themselves in the same situation as their training counterparts—they are called upon to deal with the constraints of a system well after it has reached the final stages of development. To the extent that these two groups work with one another at all stages of the development process, their complementary insights on the acquisition of safe and effective work procedures are likely to shed significant light on how such systems should be designed originally.

Each of the discussions above is analogous to what might be referred to as a *two-way interaction* statistical analysis. Two-way interactions are useful in illustrating the HSI concept, but it should be understood that three-, four-, five-, and six-way (and more) interactions can occur, providing even finer synergies when all HSI disciplines interact consistently with one another.

### 12.3.3 Integrating Training within the HSI Approach

The key challenges involved in overcoming the inherent limitations in the model depicted in Figure 12.3 include establishing organizational structures that

1. overcome the effects of stovepiping,
2. facilitate consistent and close coordination between all HSI disciplines, and
3. assure a lead HSI administrator/engineer serves at a very high level within the design team's organization.

Although HSI administrators need not be training specialists, they must report directly to the overall program manager and have people responsible for all HSI areas reporting directly to them. As Chapter 1 notes, without ongoing commitment to and participation of HSI at the very highest organization levels, the support and performance of specific HSI tasks will suffer. In other words, it is absolutely critical that the structure of the system development organization be such that an overall HSI leader is positioned at a very high level reporting directly to the program manager.

Another challenge involves changing the traditional organizational mindset or "culture" to reflect a greater commitment to the sort of coordinated "systems" approach reflected in the HSI model. A systems approach to training is not a new idea among training experts themselves. For example, Patrick (1992) states, "training . . . can be viewed as a system which interacts with other systems, such as personnel selection, ergonomics, etc." (p. 14). However, as things currently stand outside the training community, training considerations, as well as those of other HSI domains particularly with regard to their integration, are all too rarely taken into account during the early phases of complex system development. Nevertheless, it is at this stage that the implications of alternative designs can be examined with respect to trade-offs between personnel requirements, human–machine interface design characteristics, and training requirements while there is still time to influence key, early acquisition cycle decisions.

Effective and thorough integration is key to the successful application of HSI methodology and, ultimately, to successful system design, deployment, and maintenance. As with many other organizational entities whose strength lies in the synthesis of partially independent, partially overlapping areas, the "whole" of HSI is much greater than the sum of its parts. As Chapter 1 points out, HSI is a technical *and* managerial concept (emphasis mine). One of the greatest challenges to successful incorporation of an HSI approach therefore involves overcoming the many organizational and managerial sorts of obstacles that stand in its way.

Organizations will occasionally attempt to conduct an *HSI approach* to the design of a new system, particularly when explicitly instructed to do so by their customers (e.g., the government), but fail to grasp that unless there is strong coordination and near daily interaction between *all* of the various elements that make up the HSI approach, then much of its potential benefit will be lost. Therefore, it is not sufficient to have one group working on "training issues" while another independent group works on "human factors" or "personnel selection" issues. There must be integration of these efforts, and they must be coordinated from the highest organizational levels. What is lost if integration is not successfully accomplished are the types of synergistic insights that can arise when people of quasi-independent areas of expertise come together with a common focus—in this case, efficient and effective human-centered systems design. A related problem, particularly common in complex system development projects performed by large and (often) geographically far-flung organizations comprised of numerous system specialties, involves the significant probability of miscommunication and wasted or duplicated effort.

How can we tailor training to better support overall system goals of effective and timely performance? More importantly, how can we effectively *integrate* training with other aspects of an HSI approach to design? The model depicted in Figure 12.4 is proposed as an approach to overcoming the types of problems and limitations in current system development models that have been described throughout this chapter. It is also proposed as a means to facilitate the types of economic and synergistic benefits, also described above, of adopting a more comprehensive user-centered HSI approach to system development.

Figure 12.4 actually presents an overview of a fully integrated HSI approach to system development from the point of view of the training community. Specifically, training is involved with all phases of the system development process and interacts on a continual, day-to-day basis with each of the other user-centered HSI domains. Issues and advantages of this approach with respect to each of the four key areas of system development are described below.

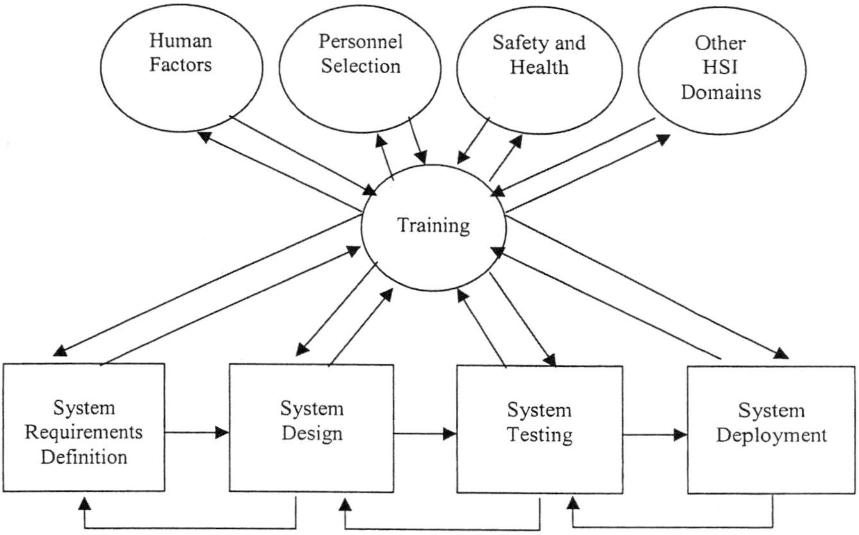


**Figure 12.4** Schematic representation of system development process featuring expanded role for training.

***System Requirements Definition*** Training expertise is required in the earliest phase of systems design—system requirements definition—in order to accomplish several specific objectives. First, it is important for training personnel to begin to anticipate any unusual training demands that might be forecast as a result of emerging characteristics of the new system. For instance, early on in the DD-X system definition process it became clear that greatly reduced crew size was going to be one of the most radical new features of this system. After decades of training crews of 300 or more to perform destroyer-based tasks, suddenly the training community was faced with being required to do the same with a crew size closer to 100. Given the enormous challenges that this represents, it is vital that training personnel are involved at this phase in order to (1) begin to prepare themselves to meet the challenge sooner rather than later, (2) avoid delays in the system development process, and (3) advise other elements of the system design process of the feasibility of being able to meet new challenges of this sort.

The advantage of having not only training personnel present in this phase but also training personnel interacting directly and consistently with all other user-centered and technical domains is that problems in the early system definition problem can be approached from a multidisciplinary perspective, providing a more creative approach to defining a system that has a chance of success. For instance, given the greatly reduced crew size requirement of the DD-X and CVNX, what looks like an insurmountable training problem at first might become much less so when training and human factors personnel interact. Such interaction could perhaps design onboard human–machine systems that can carry much of the training load that might otherwise need to be assumed by traditional on-shore training centers.

***System Design*** The system design phase refers to the phase of design in which work begins to develop concrete designs, prototypes, concept demonstrations, etc., to provide concrete but innovative approaches to meeting the specifications and requirements of the system as specified in the system definition phase. Traditionally, system design has been heavily dominated by engineering considerations, with HSI concerns either treated as something of an afterthought or relegated to a lower level within the system development hierarchy. Chapter 6 and Chapter 20 address problems associated with this traditional design paradigm and discuss methods for overcoming it. Within the HSI approach, it is important to specify the role that each user-centered discipline plays at this stage of system development and define how they can optimally integrate with one another. This is largely an organizational issue and requires that an administrative structure be put in place that (1) recognizes HSI as a discipline on an equal setting with all other design disciplines and (2) mandates the consistent, daily interaction of all HSI domains—working in close cooperation with engineering design elements to produce design concepts.

What, specifically, does it mean to incorporate training requirements and expertise into the design process? First, it means defining the constellation of KSAs that operators will need to acquire in order to successfully use the system. This knowledge emerges as characteristics of the system's human–machine components begin to emerge and allows training specialists to begin to structure their related technologies and curricula accordingly. However, it also enables training specialists to provide designers of human–machine systems with information about these KSAs for the purpose of helping to (1) design some of these features "into the system" in order to decrease training requirements on the human,[6] and (2) examine methods of designing training technologies and curricula that are embedded in the human–machine systems, enabling on-duty training to take place.

At this stage of system development training specialists must also assess the ability of the *training system* (e.g., the training centers and schools, the simulation technologies, etc., that support training) to support the needed skill acquisition. Can the system handle it? If not, what is needed to get the training system up to speed so that deployment of the operational system is not delayed?

***System Testing*** Complex systems undergo testing at many different levels and at many different periods of time during their life cycle. When initially designed, systems are tested to ensure that they meet the functional specifications set for them earlier in the system development process. Later, as new subsystems are developed and used to replace older legacy subsystems, testing is often repeated to ensure that these new elements conform to expectations and specifications.

For any system or subsystem that requires human intervention at some point in its life cycle (either in the form of regular operational intervention—as in the case of a control room display panel—or in the form of maintenance or repair), training is a concern. In preparation for initial system testing, training specialists will need to focus on several key tasks. First, initial training procedures will need to be developed in order to prepare a crew to exercise either the entire system or subcomponents of the system. Clearly, this is a daunting task that is most efficiently approached using a multidisciplinary approach such as that proposed by the HSI model. The importance of a smooth development of the test procedure (in terms of maximizing the effectiveness of resources expended) is just as important in the test phase as in the other phases of system development. Therefore, it is very important that training preparations reflect the full input of all other relevant operational domains.

System tests also afford the opportunity to collect human performance data that are of interest. For example, such data can shed light on the effectiveness of the design of human–machine interfaces and can also provide information about the effectiveness of training procedures (including any innovative on-line/on-duty training procedures). The collection of human performance data is vital at this stage to evaluate the effectiveness of many aspects of the system design, including the training technologies and curricula.

***System Deployment*** Training has always had its broadest and most important application at the system deployment stage, and in spite of the value it can add at earlier phases, this will probably be the case for many years to come. Once a system is deployed, it is the responsibility of the training community, in conjunction with their colleagues in the personnel selection domain, to supply a steady stream of qualified individuals to operate and/or function within it. While there is nothing new in this requirement, there are new ways that the training community can accomplish this important objective provided they have been able to participate in the early system design stages. This may, for example, include the development of novel training technologies that can be "embedded" within the design of human–machine systems. Continued progress in the area of human performance assessment (an area in which collaboration between training and human factors specialists would be beneficial) could hasten the development of "adaptive" training technologies in which individualized training regimens can be designed and applied on an as-needed basis as a function of deficits in operators' task performance.

In the current operational environment, the biggest challenge facing training is operating at this final stage of the system life cycle with reduced resources and enlarged expectations. This challenge, along with the several others identified in this chapter, can best be met by a more thorough integration of training expertise at all levels of the system development process, especially by helping to design system components and procedures that require less training for operators to master while at the same time incorporating required training elements into their design.

## 12.4 ISSUES AND CHALLENGES

Because of the immaturity of the training domain at influencing system design, a major objective of this chapter is to provide a state-of-the-art review of training from an HSI perspective. Such a review was found necessary in order to provide a basis for better understanding what needs to be involved culturally, organizationally, and technically to bring the training domain into an effective, efficient HSI discipline. This section, therefore, presents the findings of that review. The categories considered are

- dominant themes in the history of training research and practice,
- emerging issues in training research and practice,
- technical challenges, and
- cultural changes.

### 12.4.1 Dominant Themes in the History of Training Research and Practice

The scholarly and technical literature on training is vast and daunting. A thorough review of this literature is beyond the scope of this chapter, but it is important to examine several of the major themes in this body of work. The reasons for doing so are (1) to examine the relevance of this knowledge base to the broader "design team" philosophy underlying the HSI approach to system development and (2) to examine the extent to which the proposed broader definition of training and its role in the system development process may require the field to expand upon the principles and knowledge contained within these areas. Indeed, in many cases, very important new sets of questions and research issues for the training community may appear.

The themes from the training literature briefly reviewed are (1) individual versus team training, (2) measurement of training effectiveness, (3) the analysis of tasks and related aspects of skill, and (4) the design of training programs. These topics by no means comprise an exhaustive list of themes from the training literature of potential relevance to the HSI design philosophy; they merely represent an exemplary sampling of topics whose discussion is intended to highlight the relevance of the area as a whole to all aspects of complex system design and deployment.

***Individual versus Team Training*** In recent years the attention of training specialists and researchers has increasingly turned from considerations of training *individuals* to a more concentrated focus on training *groups* of individuals or *team training* (e.g., Cannon-Bowers and Salas, 1997, 1998; Entin and Serfaty, 1999; Salas and Cannon-Bowers, 2000; Swezey and Salas, 1992). In the human factors and training literature, a team is defined as a set of two or more individuals who must interact and adapt to achieve specified, shared, and valued objectives (Salas et al., 1992). Team training is clearly a key issue in the development and successful application of sociotechnical systems whose successful operation relies on the skilled performance of coordinated groups of operators. It is becoming increasingly rare for operators to perform functions within complex systems that have little or no direct relevance for the activities of other operators. Therefore, the training of efficient group or team skills is becoming increasingly important.

For many years researchers seemed to avoid studying team training, perhaps because the problems involved in training and reliably assessing the performance of *individuals* were daunting enough, let alone *teams* of individuals. Nevertheless, this area has begun to receive extensive research attention in recent years, and the work that has been accomplished to date has significant implications for all aspects of system development. The concerns that face training specialists in designing approaches to team training and the knowledge they have obtained from studying the problem are of great potential value to other design team members. Conversely, the training community's understanding of team training is likely to be greatly aided by the inputs of others who deal with teams in settings other than training.

For example, a topic of great current interest in the team training field involves the concept of the *shared mental model* (e.g., Kraut et al., 1999; Stout et al., 1996). Simply put, this notion reflects the hypothesis that a team's performance is likely to be significantly enhanced to the extent that individuals within the team share common concepts (mental models) relating to the factors that underlie their work performance. In other words, in order to function cohesively and effectively, it is generally thought that team members should share a very similar set of functional concepts about (1) the team's long- and short-term objectives within any given situation and (2) the means for accomplishing them. Additionally, of course, individual team members must possess the requisite sets of skills needed to successfully achieve these objectives.[7] To perform at a high level, individuals within teams should be able to accurately and easily perceive and comprehend the current operational state of affairs with respect to the performance of the team and their role within it. Furthermore, they must also be able to accurately anticipate the state of affairs in the operationally relevant near future. In short, the team should share a common, high-level *situation awareness* (e.g., Endsley, 1999) in order to function effectively.

Clearly, training must concentrate on developing the sets of skills and abilities that will produce effective team performance, and much work remains to be done to refine the methods needed to accomplish this important goal. The identification of team-based KSAs as well as the design of methods with which to impart them through training continues to be a topic of intensive research in the training community. However, by factoring these training considerations into the design phase of system development, it should be possible to design human–machine systems that will facilitate team training and team performance objectives while simultaneously reducing the overall demand on training resources.[8] By doing so, overall system performance objectives can be enhanced within a more efficient and cost-effective design setting.

A few relevant skeptical questions at this point might be: In operational terms, in terms of the success or failure of the total system, how vitally important is this concept of a shared mental model? Does it make sense to devote a significant portion of our limited resources to fostering it? Will it really make a difference in terms of the overall performance of the system? The answers to these questions are not always clear and are likely to be heavily situation dependent,[9] although evidence (much of it from the commercial aviation literature) strongly suggests that many accidents occur as a result of breakdown in team coordination (e.g., Helmreich and Foushee, 1993). In general, however, if the answer to the first question is that it is very important to promote effective team performance and to contribute to this goal by doing what we can to create a consistent, shared cognitive mission perspective, then it is critical that the development of a shared mental model among team members become a broad rather than a narrow concern.

In general, one might say that this is true of all such human performance issues. If the system cost associated with a breakdown in human performance is high, and if factors can be identified that might lead to such a breakdown, then it is vitally important for the broad design team as a whole to be concerned with them and not just one narrow subdiscipline. For instance, while training experts have become increasingly familiar with the shared mental model concept, others involved in different aspects of system design have not. Therefore, in a stovepiped approach to system design, this potentially critical issue becomes little more than a *training* objective when in fact it must be a total *system* objective.

What can be done to address this organizational deficiency? Involving training personnel in the system design phase, as a start, can help place emphasis on the design of human–machine systems that facilitate the acquisition and maintenance of shared mental models. This can be accomplished through the design of displays that provide individual team members with intuitive information concerning the activity of the team and the team's role in the overall structure of ongoing events. Consistent interaction (early in the system design phase) between training, human factors, and human–machine systems design personnel is an obvious first step toward obtaining the objective of "designing" in shared mental models. A failure to take a coordinated approach of this type means that training personnel must take whatever they are given in the way of a completed system design and work with it to promote effective team training. Without their inputs in the design phase, this will almost certainly be a more arduous and time-consuming (and, in the end, less effective) process.

To revisit an important theme, in the HSI approach to system design, the focus of the design team is first and foremost on the *user* of the system. The goal of system design is shifted away from a primary emphasis of achieving purely *engineering* objectives to achieving *system performance* objectives on the assumption that the human user is the most critical element of the system. This can only be successfully accomplished by incorporating those with knowledge of human performance requirements at all phases of the system design process. As illustrated above, such an approach can result in more functional systems that achieve their objectives in a coordinated and efficient manner.

***Measurement of Training Effectiveness*** The ability to reliably assess the effectiveness of training has long been an area of central concern to the training community. When substantial time and resources are devoted to the development of training techniques and technologies, it is important to be able to determine if they are achieving their objectives. Training researchers have devoted substantial attention to identifying valid and reliable metrics of training effectiveness (how *well* task-relevant KSAs are acquired and retained) and efficiency (how *quickly* they are acquired and with what level of resource commitment) (e.g. Damos, 1988; Lintern, 1991). Consumers of training technologies are of course very interested in being able to gauge their return on investment for training resources spent.

Training involves ongoing changes in underlying cognitive skills but also manifests itself as behavioral change (e.g., Salas et al., 1999). In other words, effective training enables trainees to *think* more effectively about their jobs, and it also enables them to *perform* those jobs more effectively. Therefore, a key challenge faced by training researchers has been to devise measures of training effectiveness that tap into both of these key outcomes.

The ability to determine training effectiveness relies heavily on the more fundamental ability to reliably assess the dimensions of human performance that underlie the KSAs of interest. In other words, metrics (usually quantitative) of human performance are used to assess the effectiveness of training and allow researchers, trainers, and consumers to make informed decisions on the adequacy of various training approaches and technologies.

Given the progress that has been made in assessing the human performance of complex tasks (e.g., Boff et al., 1986; Lane, 1987), a logical extension of the training domain within the HSI approach might involve the implementation of on-line human performance assessment to indicate when an operator or user of a system may need further training in one or more specific areas. Once a particular performance deficit has been identified, appropriate individualized training could then be prescribed for the operator. In essence, then, a continual monitoring or sampling of operator performance (particularly in tasks where the costs associated with human error are high, such as those involved in managing complex weapons systems or nuclear power plant control) can be conducted to assess the degree to which a particular operator requires additional or refresher training.

While the measurement of training effectiveness with respect to individuals remains a critical research area, an even more complex challenge involves the determination of factors influencing the effectiveness of team training. The determination of appropriate team performance metrics could also lead to the same sort of possibilities for real-time/on-line assessment of team performance and subsequent identification of just-in-time or refresher training requirements. In either case, the key point is that system designers can build on this important training research to develop human–machine systems that possess the capability to effectively monitor the performance of individuals and teams, identify performance deficits, and prescribe and deliver training to address those deficits.

Additionally, knowledge about desired human/team performance outcomes is (or ought to be) critical in the early phases of system design. Human performance metrics should be among the most carefully examined factors when designing and evaluating human–machine components of complex systems. Clearly, the role of training experts in this phase of the system development process is vital as they represent a significant portion of a design team's expertise on the topic of targeted human performance "goals" that the system must meet in order to achieve success.

***Analysis of Tasks and Related Aspects of Skill*** Many methods have been developed over the years in an attempt to describe the tasks that people execute when performing a given job and to identify the underlying KSAs needed to successfully perform those tasks. The most common and best known are generally referred to as job or task analysis techniques (e.g., Schraagen et al., 2000) and are based on the observation of work and the interviewing of subject matter experts (e.g., workers, operators, supervisors, etc.), resulting in the production of detailed, schematic representations of the spatiotemporal characteristics of the work environment as well as the perceptual, cognitive, and motor requirements associated with task performance. *Cognitive work analysis* (e.g., Roth and Woods, 1989; Vicente, 1999) and *cognitive task analysis* (e.g., Schraagen et al., 2000) represent more recent variants on traditional task analysis and are more focused on identifying and understanding perceptual and cognitive tasks as they relate to the costs and constraints of the task environment. Other related techniques include *knowledge mapping* or *concept mapping* (e.g., McNeese et al., 1995), techniques that seek to explicitly delineate the hypothesized connections between events, costs, and constraints of the task environment and the operator's related cognitive and perceptual tasks.

These techniques have been used over the years to help identify requisite KSAs underlying the performance of operational tasks in support of the development and application of training technologies and programs. Clearly, in order to train effectively, one must have a focused awareness of the aspects of skilled performance that are relevant in any given situation, and task analysis techniques are specialized for uncovering this type of information. However, the utility of a detailed specification of cognitive, perceptual, and motor task requirements understood with relation to the operational environment of interest extends far beyond the training domain. For example, as exemplified by the role that cognitive task analysis has played in *ecological interface design*, there is a significant role for this type of information in the design of human–machine interfaces (Vicente and Rasmussen, 1992). The utility of this type of information in other domains such as manpower and personnel selection is also clear.

Three areas of training research and practice—team training, the assessment of training effectiveness, and the analysis of tasks and related aspects of skill—exemplify the types of expertise characteristic of training that have much broader application within the total system development process. These areas provide a baseline of existing training information that should be continually examined for extension to broader system questions beyond the traditional applications.

### 12.4.2 Emerging Issues in Training Research and Practice

Having reviewed some of the major issues from the literature on training for potential relevance to HSI in system development, we can now examine several emerging issues for training. The purpose of this section is to illustrate the nature of some of the key challenges that the training domain can expect to face in the near future and how an HSI approach to system development can facilitate their solution.

Environments within which people perform their jobs have changed dramatically over the past several decades, and the pace of change appears to be continually accelerating (e.g., Gleick, 1999). Clearly, the technology has changed immensely, primarily in the direction of providing individuals and organizations with greatly enhanced "power" in terms of the ability to extract, process, display, and act upon information. Powerful and portable computers, the Internet, widespread wireless telecommunications, advanced display and control technologies (e.g., virtual environments), personal digital assistants, and many other information-based technologies are commonplace tools that in many cases were still in the conceptual phase no more than 10 years ago.

However, many important social, economic, and political trends have accompanied the so-called information revolution, most notably (for purposes of this chapter) many that relate to the structure of the work environment.[10] Downsizing, increased workload, and increased requirements for multitasking and concomitant multiple skill sets are several of the emerging realities of workplaces (military or otherwise) that impact strongly on training and the relationship of training to other elements of system development.

The ubiquitous nature of significant technical and societal change has profound significance for all aspects of complex system development, including training. How, for instance, can training technologies and curricula be designed and implemented to retain relevance for more than a short period of time? How can we train individuals to accommodate to and thrive within a rapidly changing sociotechnical environment? And how can we incorporate knowledge from research on these questions into more efficient design of complex systems? These and many other questions must be successfully resolved to ensure that training fulfills the important role implied in the expanded definition of the field used in this chapter. Before proposing approaches that might answer some of these questions, it will be helpful to consider how training has accommodated similar circumstances in the past.

The military has been interested in systematic, large-scale training for a long time, perhaps longer than any other area of human endeavor.[11] For instance, the development of fundamental fighting skills and abilities (handling a spear, swordsmanship, archery, etc.) has been a subject of military interest since ancient times. However, coincident with the onset of World War I, the emphasis on military training began to dramatically shift in at least two important ways. First, training (and related disciplines such as personnel selection) began to transform its focus from the development of as many competent foot soldiers as possible toward the more challenging need to prepare individuals to deal with human–machine systems (e.g., airplanes, tanks, increasingly complex artillery systems, etc.) whose complexity and sophistication were much greater than anything previously encountered. Second, the nature of training began to be a topic of serious scientific inquiry, particularly among applied psychologists of the day. Training moved from being a purely "seat-of-the-pants" enterprise practiced more or less along the lines of the journeyman-apprentice model to a much more scientifically based approach. This change of focus was partly necessitated by the requirement to train large numbers of people in a hurry and concomitant increased government funding of training research. In essence, training was forced to change in order to accommodate to the vast strategic, technical, and scientific changes characteristic of the time, which it must again do now.

Obviously, a great deal has changed since World War I, although many demands and constraints on the training community remain very much the same. Training still largely focuses on preparing large numbers of individuals to become skilled users of complex human–machine systems, and the level of involvement of psychological scientists in matters of training is still high. However, the human–machine systems themselves as well as the situations under which they are used have changed dramatically.

The two emerging areas of greatest concern for the future of training are (1) effects of new technologies on sociotechnical system performance and (2) cultural changes, including the effects of personnel downsizing on system performance.

Among the many emerging challenges facing training and systems development in general, the remarkably rapid change in the technological capabilities of systems is certainly one of the most daunting. As the number of years that it takes to transition from one "generation" of technology to the next continues to decrease, workers and military personnel are continually at risk of finding that their hard-earned KSAs that were appropriate in one technical "era" have become obsolete. Those responsible for the smooth life-cycle operation of large-scale systems (i.e., some military weapons platforms are designed with the goal of being operable for 30 years or more) must concern themselves not only with the periodic upgrading of technology but also with the ability of training programs to be able to smoothly transition personnel from one generation of technology to the next. In addition, many new technologies present challenges (and opportunities) for the training community in that they may require new forms of skilled human performance that have received little attention in the past. Some of these technologies are described in Section 12.4.3.

The second major emerging challenge relates to general concerns with changes in the "culture" of work and in particular the effects of downsizing. Downsizing in and of itself does not necessarily guarantee a negative impact on overall system performance. Whether it does or not, and the degree to which it does or not, depends strongly on (1) organizational expectations concerning the impact of personnel downsizing on system performance (i.e., is organizational "output" expected to decrease in a manner commensurate with the loss of personnel, remain at a constant, pre-downsizing level, or even increase in spite of the loss of personnel resources?) and (2) whether or not technical tools and other resources are introduced to offset the loss of personnel resources. Indeed, the general area of *resource downsizing* is a related, major area of concern as not only personnel but also budgets and key resources are cut back.

Those who must successfully operate a particular system with fewer colleagues frequently experience the collateral effects of personnel downsizing. Increased demands in the form of multitasking and the need to perform under conditions of potentially increased stress and fatigue are among the challenges they face. What role does training have in preparing individuals to cope with these changes characteristic of the workplace?

These areas of technical and socioeconomic change are hardly independent. The nature of their interactions is important for the training and larger system development community, as will be discussed in Section 12.4.4. Before that, however, we will examine a number of technical challenges facing the training community.

### 12.4.3 Technical Challenges

Currently there are four technical challenges that are particularly critical to the progress of training in an HSI environment:

1. increased reliance on automation,
2. novel human–machine interface technologies,
3. enhanced functionality of legacy systems, and
4. complexity.

***Increased Reliance on Automation*** In order to enhance the overall performance of sociotechnical systems in the face of reduced manpower, it is apparent that machines will be called upon to accomplish more tasks (that were previously performed by humans) and to do so with increased speed and reliability. The increased use of automation is one system design strategy that will be heavily relied upon to accomplish this goal. As has been frequently noted (e.g., Wickens, 1999), as automation becomes a more ubiquitous characteristic of sociotechnical systems, the role of the human increasingly shifts from one of active operator to one more accurately described as active monitor. As has also frequently been noted, automation is a double-edged sword (e.g., Funk et al., 1996). Assuming that automation is introduced into a system for reasons other that, economic cost savings (i.e., there is strong reason to expect increased safety and/or effectiveness of system performance), there are still many potential design and training problems that interact with one another. For instance, if automation is not designed properly, then it may lack the functionality or performance desired by operators in all relevant situations, necessitating "workarounds" when it fails to perform as anticipated. Automation may not control processes or systems the way an operator might, which suggests that the gap between the operator's mental model of the system and the actual system model itself must be successfully bridged to avoid potential difficulties associated with the operator being "out of the loop" when a problem arises.

Obviously, to the extent that the design of automation is not "human centered," the demands on training will be greatly increased. However, even under ideal design conditions training has to be able to address the unique human performance demands of this new role of the human in the functioning of sociotechnical systems.

***Introduction of Novel Human–Machine Interface Technologies*** Another method of enhancing the overall performance effectiveness of current and future systems is to introduce effective, new human–machine interface technologies. Emerging concepts such as virtual environment technology (e.g., Durlach and Mavor, 1994) and adaptive interface technology (e.g., Bennett et al., 2001) hold a great deal of promise for developing more effective systems by producing human–machine interfaces that are far more intuitive than current approaches. However, the introduction of such novel technologies raises a number of issues for system developers that require the application of training expertise.

First, the design and functional integration of new technologies such as virtual and adaptive interfaces are best approached from a user-centered design framework, one in which the technical aspects of the design are in close alignment with the needs, capabilities, and limitations of the user. As discussed earlier, the training domain has developed significant expertise over the years in domains such as individual and team performance assessment, task analysis, and assessment of the impact of novel technologies on human performance and skill acquisition. Along with human factors specialists, training personnel have a great deal to offer in helping to design technologies to match the needs of the user.

Second, novel technologies present novel challenges for training specialists within their more traditional role of providing trainees with the KSAs needed to safely and effectively interact with them. Early, direct, and consistent involvement of training personnel with technology development and testing will result in greater training effectiveness and efficiency by allowing training specialists to (1) anticipate training requirements of new technologies early enough to permit training to take place as soon as possible upon system deployment and (2) participate in the development of the technology itself, specifically with an eye toward helping to build in on-line just-in-time/refresher training capabilities into the technologies while they are still in the design phase.

Finally, it should be pointed out that these technologies offer significant new opportunities for training (Durlach and Mavor, 1994). This topic will be discussed more fully in Section 12.5.

***Enhanced Functionality of Legacy Systems*** While in many cases operators may perform their tasks at workstations that are not appreciably different from legacy systems in terms of the amount, type, and appearance of the equipment involved, the equipment itself may be enhanced to control a significantly increased number of processes and subsystems. The greater amount of activity to be performed by an operator in the same unit period of time is a significant factor to be accounted for in the design of these systems and in training operators to use them.

***Complexity*** One of the most profound technical issues impacting the future effectiveness of systems and one that each element of the system development process will be called upon to address is complexity, a phenomenon that cuts across all aspects of the technical challenges discussed above. As stated by Pool (1997), the safety and effectiveness of complex systems depend not just on their physical characteristics but also on the people and organizations operating them. Complexity creates uncertainty, and uncertainty demands human judgment (a problematic situation for applications of automation). As noted by Perrow (1984) and others, complex systems can be very sensitive to even small changes or perturbations, so that a minor mistake or malfunction can quickly deteriorate into a major accident. As Pool (1997, p. 250) notes: "Such uncertainty and sensitivity make it impossible to write out procedures for every possible situation—and attempts to do so can backfire. Operators who are trained to always go by the book may freeze up when an unexpected situation arises. Or worse, they may misinterpret the situation as something they're familiar with and take exactly the wrong action."

Training individuals to cope with complexity, particularly in the operation of high-risk systems such as weapons platforms, oil and chemical refineries, and nuclear technologies, presents challenges that, in the end, almost certainly cannot be addressed by training alone. Indeed, the technical issues described above present significant challenges not only for training but also for all aspects of complex sociotechnical system development. However, adding the overall level of complication is another emerging set of trends whose individual elements can perhaps best be classified under the umbrella of *sociocultural* change.

### 12.4.4 Cultural Changes

A number of sociocultural trends were described above whose presence represents an emerging challenge for training as traditionally conceived and training as recast within the HSI approach to system development. Chief among these in terms of the breadth of its influence is downsizing, the reduction of manpower/personnel that has characterized both the public and private sectors in recent years and is anticipated to continue at least into the foreseeable future. Downsizing results in a number of issues for training and system design, challenges that require extensive coordination between HSI domains in order to be satisfactorily met. Among these are the increased need for cross training and the need to train individuals to accommodate to and thrive within an environment of change.

***Increased Need for Cross Training and Multitasking*** A key challenge in downsized and reduced manpower environments is the need for cross training and, analogously, the requirement for individuals to be able to perform well in a multitasking environment. As with the emerging technical issues described above, an efficient and satisfactory approach to this problem will require a multidisciplinary approach involving training, human factors, and personnel selection specialists at a minimum. While each field, if it were to attempt to address the problem in isolation, would undoubtedly develop useful approaches and concepts, it is very likely that a combined approach would result in a more satisfactory, cost-effective solution.[12]

***Training to Adapt to Change*** There is little doubt that the environment and overall culture within which people live and work have changed dramatically within the course of the last several generations. One of the most central features of our current culture is the ubiquity and rapidity of *change*. This sociocultural characteristic has important implications for the design of training approaches and technologies as well as their integration within an overall system development approach.

Levy (1998) has noted that society's relationship to information and knowledge has changed dramatically since World War II and to an even greater extent since the seventies: "Until the second half of the twentieth century, a person could utilize skills learned during his youth throughout his career. More important, he was able to transmit this knowledge, nearly unchanged, to his children or apprentices. Today this pattern has become obsolete" (Levy, 1998, p. 70). Many people, perhaps the majority, are now required to change their skills several times throughout their life. As observed by Levy, even within a given "trade," knowledge has an increasingly shorter life span (e.g., three years or less in computer technology). It has therefore become difficult to design and train "basic" skills in a given field. The relevance and importance of KSAs are now far less durable than at any time in the past, being subject to change as a function of the emergence of new technologies (e.g., the personal computer) or new socioeconomic conditions (e.g., downsized work environments). As a result, society appears to have made a transition from the dominance of stable skill sets to what Levy refers to as "a condition of permanent apprenticeship."

## 12.5 CONCLUSIONS AND RECOMMENDATIONS

Despite its importance and the many decades of work devoted to understanding it, there is still a great deal of misunderstanding about training. Salas et al. (1999) have recently summarized many of the current "myths, misconceptions, and mistaken assumptions" about training, and their discussion is well worth studying. Perhaps the greatest myth about training is that its applicability is limited to the design of training technologies and procedures. Those who accept the HSI approach to systems integration have little doubt that training expertise can significantly enhance other key aspects of user-centered system design and that such interaction is vital to achieve enhanced system performance in a reduced resource environment.

A key point emphasized in this chapter is that training considerations need to be thoroughly interwoven with all other HSI considerations throughout the life cycle of major systems. The means of determining how to do this is perhaps much less an empirical or scientific question than a socioeconomic-political question. As noted in Chapter 1 and elsewhere in this book, it is incumbent upon HSI professionals to effectively "sell" the HSI approach to those in charge of major programs. In addition, there is a need to develop a systematic approach to incorporating training as part of a total life-cycle approach to system procurement. This process must involve systematic and periodic reexamination of mission requirements, task requirements, personnel requirements, etc.

However, there are also many technical areas in which the area of training must continue to develop to meet the challenges of system design in the twenty-first century. For instance, as noted by Patrick (1992), there is still considerable work to be done to identify principles that will permit the development of training programs that are not problem specific but that generalize across a range of training problems. This is a critical area in light of the reduced resource constraint. While there will always be a need to tailor training approaches to specific system requirements, there is little doubt that funding agencies and organizations are going to continue to exert pressure on the training community to develop generalized training paradigms and methodologies that can transfer relatively easily from one domain to another.

As these pressures mount and both system developers and training communities recognize the need to make better use of HSI principles throughout systems acquisition, strategies to address the following considerations will become more important.

1. *Recognize the Change in Focus for Training Specialists* Clearly, the types of suggestions raised in this chapter, were they to be implemented, would involve many profound changes in the traditional role of the training specialist. Chief among these is the much broader role to be played throughout the entire system design process. The realization of this broader, more interactive role will require the pursuit of new avenues of research and new developments in theoretical approaches to training.

The primary challenge will be to envision a much broader role for today's *training specialist* as tomorrow's *system design specialist* with particular expertise in user knowledge, skill, and ability acquisition and retention. This change in focus amounts to a fundamental redefinition of the role of the training expert within the system design process, one that acknowledges the relevance of training expertise at each stage of system acquisition and that also acknowledges the reciprocal influence of all other HSI domains on training.

2. *Explore the Question "Can We Train People to Adapt?"* In past centuries, people could apprentice themselves to a master of some trade or craft and very effectively learn a skill, safe in the knowledge that its basic components would not change appreciably over the course of their lifetime or career. Obviously this is no longer the case. The ubiquity of rapid technical change in nearly all career fields—indeed, the rapid appearance and disappearance of entire job specialties—requires that successful individuals be able to adapt quickly and effectively to changing sociotechnical work environments.

The author has observed control room operators in oil refineries who were either unwilling or unable to fully adapt to the presence of computer-based control and/or maintenance request systems that had recently been introduced to replace more traditional and familiar systems. There appeared to be many factors underlying operators' apprehensive approach to these systems, but for the most part they related to a compelling sense of discomfort with change on the job in general as well as the introduction of requirements for new skills that some operators felt they could not easily master. In response to such situations, a variety of coping behaviors may be adopted, e.g., delaying actions until there is "enough time" to sit down and devote the necessary attention to the task or asking someone more adept with the new technology to perform the task. In either case, these coping mechanisms can obviously produce performance and safety problems.

It is insufficient to merely train operators in the specifics of new technologies and new systems every time one comes along. Rarely is it the case that an entire work environment is replaced or upgraded; rather, individual components or subsystems of the work environment tend to get replaced. The rapidity and unpredictability of such change in itself create training challenges for operators. In addition, the possibility that new subsystems may not function predictably or well with existing components creates training and performance challenges of a different sort. Training specialists must concentrate on developing approaches to enable operators to cope with change. Furthermore, training specialists must be continually involved in the evolutionary development and upgrading of existing systems in order to help assure that new components and subsystems will function well within existing operator skill sets.

3. *Consider the Effects of Downsizing* As noted above, downsizing is one of the most compelling of the "new realities" facing designers of new systems as well as those responsible for the smooth and safe operation of existing systems. There is little doubt that significant reductions in the number of individuals available to operate aircraft carriers, oil and chemical refineries, and other complex and risky technologies will pose tremendous challenges for those responsible for their design and operation. Responsible approaches to

downsizing must be based on examining the trade-offs involved in reducing personnel while maintaining overall system performance at effective and safe levels. Tools for performing these sorts of analyses do not currently exist, and the training community must be called upon to provide reliable inputs on changes in training that will be needed to empower a smaller workforce to perform at an acceptable level.

4. *Explore the Effects of New Human–Machine Technologies* Many radically new approaches to human–machine technology are being developed and can soon be expected to appear more frequently in major systems. For instance, the development of virtual environments, adaptive interface, and alternative control technologies (to name just a few) will radically change the relationship between the operator and mechanical/computer system being controlled (e.g., Durlach and Mavor, 1994). In each case, these technologies present new challenges in terms of identifying new skills that need to be trained, new training regimens, etc. However, they also present exciting new opportunities to promote more effective training. For instance, the increased computational power that makes such technologies possible, combined with their underlying requirement to monitor human performance in real time, can provide training specialists with highly detailed, up-to-the-minute profiles of operator performance and skill levels. Virtual environments, in particular, can provide very versatile and realistic training environments, but exactly how these new technologies can best be utilized in training personnel remains to be demonstrated.

5. *Measure Team Performance* Cannon-Bowers and Salas (1997), in noting the importance that team training has assumed within recent years, state that the issue of team performance *measurement* has also grown in importance: "Without accurate, reliable measures of team performance, it is difficult to select or train team members or to manage team performance. Unfortunately, little research exists that provides theoretically based guidance to those interested in assessing team performance" (p. 45).

The need to consider team training in the development of modern sociotechnical systems is clear. This need is not limited to the training of teams but extends to the design of the very systems that teams use in the performance of their tasks. However, in order to realize the many potential benefits of training and designing for well-coordinated teams, their performance must be measured in a valid and reliable fashion. Without this ability, the impact of our interventions will remain unclear.

## NOTES

1. This chapter refers to a variety of systems, each of which involves humans interacting with technology and/or with one another. For the sake of convenience these are called *sociotechnical* systems. The use of this term is intended to emphasize that in each case the system under consideration is characterized by complex interactions between humans and machines as well as humans and other humans.
2. The medical field is a prime example. In spite of persistent budgetary problems and personnel shortages in key areas such as nursing, there is tremendous pressure to develop and implement sociotechnical systems that improve overall system performance, particularly with respect to the reduction of medical error and the promotion of greater cost and performance efficiency.
3. The goal of an undergraduate and graduate education is often described in terms of developing the student's ability to learn, rather than simply "loading the student up with facts and figures." As the

author's graduate school advisor once opined: "Having a Ph.D. doesn't prove you are smart, it proves you are educable." To the extent that training begins to focus on broad "skills" associated with learning, then perhaps a similar criterion can be applied to identify successful trainees.

4. The problem of stovepiping is one that is widely acknowledged as contributing to cost and timeline inefficiencies in the development of complex sociotechnical systems. Inefficiencies attributable to stovepiping include conflict between ostensibly coordinated but often in fact competitive organizations over resources and scheduling priorities, a lack of cohesive awareness of large-scale organizational goals and plans, and severe constraints on communication resulting in further inefficiencies. A large portion of the HSI design philosophy is geared toward the elimination of stovepiping and its negative effects on complex system development.
5. This "approval" process is in itself, of course, a highly complex and treacherous process characterized by generally Byzantine political and economic machinations. It is also one that persists throughout the entire system acquisition cycle as programs can be terminated or radically redirected at any point. While I do not want to minimize the fundamental importance of these processes, they are beyond the author's expertise and scope of this chapter.
6. For example, designing in automation features that can perform specific role functions faster and more accurately than humans can, offloading performance responsibility from the operator and diminishing overall training load.
7. Often the required skills of one team member are complementary to those of others on the team, which is just one of the factors that make holistic assessment of team performance and training so difficult.
8. Conversely, the early design of human–machine systems intended to support operational teams (e.g., the command center of a destroyer) can be assessed and, if necessary, modified on the basis of the assessment of team performance during design-phase simulation tests of prototypes.
9. One rule-of-thumb answer might be as follows: To the extent that the cost associated with a breakdown in effective team performance is high (i.e., results in the creation of dangerous or catastrophic conditions), the importance attached to creating and maintaining a consistent, shared mental model should be high.
10. The degree to which these societal changes have in fact been caused by the proliferation of information technology is a matter of debate outside the scope of this chapter. For our purposes, it is sufficient to note that these vastly influential trends have occurred at the same time.
11. While other areas, such as agriculture, medicine, the arts, and preindustrial trades (metallurgy, weaving, etc.), also have long training histories, the military domain has been historically unique in its need to train large groups at the same time and to train them to work as coordinated, unified entities.
12. *Cost effective* in this sense refers to cost savings associated with (1) a more efficient problem solution process and (2) a more efficient and durable solution.

## REFERENCES

Bennett, K. B., Cress, J. D., Hettinger, L. J., Stautberg, D., and Haas, M. W. (2001). A Theoretical Analysis and Preliminary Investigation of Dynamically Adaptive Interfaces. *International Journal of Aviation Psychology, 11*, 169–196.

Boff, K. R., Kaufman, L., and Thomas, J. P. (Eds.). (1986). *Handbook of Perception and Human Performance.* New York: Wiley.

Cannon-Bowers, J. A., and Salas, E. (1997). A Framework for Developing Team Performance Measures in Training. In M. T. Brannick, E. Salas, and C. Prince (Eds.), *Team Performance Assessment and Measurement: Theory, Methods, and Application.* Mahwah, NJ: Erlbaum.

Cannon-Bowers, J. A., and Salas, E. (1998). Individual and Team Training Under Stress. Theoretical Underpinnings. In J. Cannon-Bowers, and E. Salas, (Eds.), *Making Decisions Under Stress: Implications for Individual and Team Training* (pp. 17–38). Washington, DC: American Psychological Association.

Damos, D. (1988). Determining Transfer of Training Using Curve Fitting. In *Proceedings of the Human Factors Society Annual Meeting*, Santa Monica, CA, (pp. 1276–1279).

Durlach, N. I., and Mavor, A. S. (1994). *Virtual Reality: Scientific and Technical Challenges.* Washington, DC: National Academy Press.

Endsley, M. (1999). Situation Awareness in Aviation Systems. In D. G. Garland, J. A. Wise, and V. D. Hopkin (Eds.), *Handbook of Aviation Human Factors.* Mahwah, NJ: Lawrence Erlbaum Associates.

Entin, E. E., and Serfaty, D. (1999). Adaptive Team Coordination. *Human Factors*, *41*, 312–325.

Funk, K., Lyall, B., and Riley, V. (1996). *Perceived Human Factors Problems of Flightdeck Automation.* Grant 93—G-039, Phase I Final Report. Washington, DC: Federal Aviation Administration.

Gleick, J. (1999). *Faster: The Acceleration of Just About Everything.* New York: Random House.

Helmreich, R. L., and Foushee, H. C. (1993). Why Crew Resource Management? Empirical and Theoretical Bases of Human Factors Training in Aviation. In E. Wiener, B. Kanki, and R. Helmreich (Eds.), *Cockpit Resource Management* (pp. 3–45). San Diego, CA: Academic.

Kraut, R. E., Fussell, S. R., and Lerch, F. J. (1999, April 30–May 2). Shared Mental Models in Work Groups: A Controlled Field Study. Paper presented at the Annual Meeting of the Society for Industrial and Organizational Psychology (SIOP), Atlanta, GA.

Lane, N. E. (1987). *Skill Acquisition Rates and Patterns: Issues and Training Implications.* New York: Springer-Verlag.

Levy, P. (1998). *Becoming Virtual: Reality in the Digital Age*, New York: Plenum.

Lintern, G. (1991). An Informational Perspective on Skill Transfer in Human-Machine Systems. *Human Factors*, *33*, 251–266.

McNeese, M. D., Zaff, B. S., Citera, M., Brown, C. E., and Whitaker, R. D. (1995). AKADAM: Eliciting User Knowledge to Support Participatory Ergonomics. *International Journal of Industrial Ergonomics*, *15*, 345–364.

Patrick, J. (1992). *Training: Research and Practice.* London: Academic.

Perrow, C. (1984). *Normal Accidents: Living with High-Risk Technologies.* New York: Basic Books.

Pool, R. (1997). *Beyond Engineering: How Society Shapes Technology.* New York: Oxford University Press.

Roth, E. M., and Woods, D. D. (1989). Cognitive Task Analysis: An Approach to Knowledge Acquisition for Intelligent System Design. In G. Guida and C. Tasso (Eds.), *Topics in Expert System Design.* New York: Elsevier.

Salas, E., and Cannon-Bowers, J. A. (2000). The Anatomy of Team Training. In S. Tobias and J. D. Fletcher (Eds.), *Training and Retraining: A Handbook for Business, Industry, Government, and the Military* (pp. 312–335). New York: Macmillan.

Salas, E., and Cannon-Bowers, J. A. (2001). The Science of Training: A Decade of Progress. In S. Fiske, D. Schacter, and C. Zahn-Waxler (Eds.), *Annual Review of Psychology*, Vol. 52. Palo Alto, CA: Annual Reviews.

Salas, E., Cannon-Bowers, J. A., Rhodenizer, L., and Bowers, C. A. (1999). Training in Organizations: Myths, Misconceptions, and Mistaken Assumptions. *Research in Personnel and Human Resources Management*, *17*, 123–161.

Salas, E., Dickinson, T. L., Converse, S. A., and Tannenbaum, S. I. (1992). Toward an Understanding of Team Performance and Training. In R. W. Swezey and E. Salas (Eds.), *Teams: Their Training and Performance* (pp. 3–29), Norwood, NJ: Ablex.

Schraagen, J. M., Chipman, S. F., and Shalin, V. J. (2000). *Cognitive Task Analysis.* Mahwah, NJ: Lawrence Erlbaum.

Smith, J., and Driscoll, K. (2000). *CVNX Analysis of Alternatives: Part 3, Aircraft Carrier Study 5.* Available: www.dt.navy.mil/pao/excerpts%20pages/2000/cvnxjan.html.

Stout, R. J., Cannon-Bowers, J. A., and Salas, E. (1996). The Role of Shared Mental Models in Developing Team Situational Awareness: Implications for Training. *Training Research Journal, 2,* 85–116.

Swezey, R. W., and Andrews, D. H. (2001). *Readings in Training and Simulation: A 30-Year Perspective.* Santa Monica, CA: Human Factors and Ergonomics Society.

Swezey, R. W., and Salas, E. (Eds.). (1992). *Teams: Their Training and Performance.* Norwood, NJ: Ablex.

Vicente, K. J. (1999). *Cognitive Work Analysis: Toward Safe, Productive and Healthy Computer-Based Work.* Mahwah, NJ: Lawrence Erlbaum.

Vicente, K. J., and Rasmussen, J. (1992). Ecological Interface Design: Theoretical Foundations: *IEEE Transactions on Systems, Man, and Cybernetics, 22,* 589–606.

Wickens, C. D. (1999). Automation in Air Traffic Control: The Human Performance Issues. In M. W. Scerbo and M. Mouloua (Eds.), *Automation Technology and Human Performance: Current Trends and Research.* Mahwah, NJ: Lawrence Erlbaum Associates.

CHAPTER 13

# Human Factors Engineering Methods and Tools

JOHN F. LOCKETT, III and JEFFREY POWERS

## 13.1 INTRODUCTION

In popular culture, human factors engineering (HFE) has become synonymous with the terms *ergonomic* and *user friendly*. Even popular radio show hosts have a sense of these terms (Magliozzi and Magliozzi, 2000). But what do they really mean? How does one make something ergonomic and user friendly?

Ergonomics is the study of the principles of work. Taken literally, this definition is not too helpful, but we get a sense that how people use technology to accomplish work is important. The definition for HFE adopted by the U.S. Army manpower, personnel, and integration (MANPRINT) program provides a bit more insight. (Available on line at http://www.manprint,army.mil/manprint/index.htm) The definition is "the integration of human characteristics into system definition, design, development, and evaluation to optimize human–machine performance under operational conditions." From this definition we get the sense that we need to consider the physical and mental limits, biases, behaviors, health, and safety of the people who will be using technology when we decide how that technology (which can range from simple hand tools to a complex multimodal interface in a manufacturing plant control room) should be designed and what roles humans and machines should play. Some of these limits may seem obvious, such as body size or ability to lift, but others are more esoteric, such as those relating to human information processing. Some other definitions of HFE have included phrases such as "designing for human use" and an "approach that fits systems (or machines, jobs, processes) to people and not vice versa" (Wilson and Corlett, 1995).

The key to realizing this user-centered approach begins with seeing the design of new technology from the point of view of the full range of people who will ultimately use it—the target audience. Norman (1988) has written a very accessible and popular book on design that helps the reader recognize poor application of technology resulting from a lack

*Handbook of Human Systems Integration*, Edited by Harold R. Booher.
ISBN 0-471-02053-2 

of user-centered design. He proposes a set of design principles to use in evaluating a design and illustrates each with examples to which everyone can relate. Once we have developed the ability to view design from a user perspective, we realize that HFE applies to almost everything that humans create (machines, jobs, or processes), including consumer products, computer interfaces and websites, maintenance equipment and tasks, workstations, manufacturing assembly lines, buildings, vehicles, weapons systems, and medical devices. Given this wide range of application, what methods do we use to conduct an HFE program?

## 13.2 HUMAN FACTORS ENGINEERING METHODS

Over the past 80 or so years, many useful methods and techniques have been developed to accomplish user-centered design. These methods encompass research, design, and evaluation. It is impossible to describe all of them within the confines of this chapter [e.g., Wilson and Corlett (1995) lists more than 50 subcategories], but some of the basics will be highlighted. For more comprehensive information readers should refer to texts devoted solely to HFE topics and methods such as Karworski (2001), U.S. Department of Defense (DoD, 1999a), Wickens et al. (1997), Wilson and Corlett (1995), Weimer (1995), Sanders and McCormick (1993), and Salvendy (1987). On-line descriptions are also available. For example, see Nomos Management AB for usability methods at http://www.nomos.se/about/methods.shtml.

Based on methodical groupings from comprehensive texts on ergonomics methods, we propose a basic list of HFE methods. These methods encompass research, design, and evaluation:

- Time-and-motion analysis;
- Link analysis and operational sequence diagrams (OSDs);
- Task analysis, function allocation, and workload analysis;
- Accident and incident analyses;
- Anthropometric and biomechanical analyses; and
- Field study, survey, and usability analysis.

All of the methods involve developing an understanding of how the system you are designing will be used, by whom (the target audience), under what conditions, and what actions they will have to take. Each of the core methods provides different data and perspectives about the system, and the results of one method may serve as input to another. For example, time-and-motion analysis data are often used in both task and link analyses. Because of this, multiple methods are often used as part of an effective human engineering program.

### 13.2.1 Time-and-Motion Analysis

Time-and-motion analysis is one of the oldest methods in human and industrial engineering. Time-and-motion analysis involves observing a person using a system (or its predecessor) and recording the duration of each action performed. In an era of assembly line production and job specialization, engineers focused on improving the efficiency of job performance believed that any job could be performed more efficiently if unnecessary time and motions were eliminated. The first step in attacking this waste is to identify unnecessary motions (e.g., hand movements, back tracking, etc.) and the time required to perform them. Slack time and high-risk points for repetitive-motion injuries can also be identified. Choke points can be highlighted and the line rebalanced to eliminate them. Time-and-motion analyses may be carried out with the participant in the actual work setting (preferred) or a simplified representation of it. The recorder may be physically present with the participant or observations may be made from a time-stamped video recording. Care must be taken to clearly define start and stop points for each motion.

Time-and-motion analysis data are often used as input to other human engineering analysis methods. Data from multiple repetitions of the task are usually collected and often analyzed, simply by computing means and examining those means to identify key performance factors. If enough data are collected, the data can be described in terms of various sampling distributions (e.g., normal distribution with a mean $X$ and a standard deviation SD). Figure 13.1 shows time-and-motion analysis data from a low-resolution study about loading bags of material in a chemical plant. We can see that event 6 has the greatest waiting time and event 5 takes the most time to accomplish. Further examination of these events should improve the loading time.

A weakness of many time-and-motion data collection efforts is the lack of control or assessment of the motivation of the participant performing the task. Human behavior,

| Event | Description | Waiting Time | Time for Operation | Reach | Move | Grasp | Release | Preposition | Search | Inspect |
|---|---|---|---|---|---|---|---|---|---|---|
| 1 | Bags of material are filled by a dispensing machine | 35 | 120 | | | X | X | X | | |
| 2 | Bags are carried to a box | 20 | 110 | X | X | X | X | | | |
| 3 | Boxes are sealed and stacked into columns | 10 | 150 | X | X | X | X | X | | X |
| 4 | Boxes are moved onto a pallet | 20 | 30 | X | X | X | X | | | |
| 5 | Pallet is lifted and carried to the loading dock | 20 | 240 | X | X | | X | | | |
| 6 | Pallets are lifted and placed onto trucks | 320 | 45 | X | X | X | X | | | X |
| 7 | Pallets a........ | | | | | | | | | |
| 8 | | | | | | | | | | |

**Figure 13.1** Time-and-motion analysis data-loading bags of material in a chemical plant.

| Task Number | Task Action | Action Type | Modality Type | Elapsed Time | Beginning Criteria | Ending Criteria | Compare Standard to Recorded | Comments | Suggested Solutions |
|---|---|---|---|---|---|---|---|---|---|
| 1 | Load Printer Paper | P R L | V PF M | 30 | Paper is available, operator is trained, written material in native language, 9th grade reading level proficiency | Reloaded paper cassette, ready to print. | Standard is set at 20 seconds, task recorded at 30 seconds | Standard was exceeded due to difficulty removing cassette tray | Quick release lock to enhance removal task |
| 2 | Load Printer Ink Cartridge | P R L | V PF M A | 42 | Cartridges are available, operator is trained in cartridge replace, written instructions in native language, 9th grade reading level | Installed ink cartridges. Tone sounds upon completion. | Standard set at 30 seconds, task recorded at 42 seconds | Errors were experienced in placing the cartridge properly. Task repetition was the result. | Increase training required for operators |
| 3 | etc... | | | | | | | | |

Action Key
P= Prepare
R= Read
L= Load
S= Search

Modality Key
V= Visual
P= Psychomotor
F= Fine Motor
M= Manual
A= Audible
S= Speech

**Figure 13.2** Time studies and product usability.

including the speed, efficiency, and activity to perform tasks, is variable and dependent on motivation. These variables must be carefully assessed and controlled during a time-and-motion study.

Figure 13.2 demonstrates the use of time studies on product usability. In this case, the manufacturer of an office computer printer can identify the wasted efforts printer users are finding with the reloading of paper and ink cartridges. Extra time and steps can create a level of dissatisfaction and negative brand recognition. The study also provides data for factors such as the reading levels demanded by support material, labels, and product legends as well as the product training requirements. Niebel (1993) is a good source for detailed information on time-and-motion analysis.

### 13.2.2 Link Analysis and OSDs

Link analysis and OSDs are also focused on efficiency and are used to help identify the optimal placement of workspace infrastructure. Both use links and data from time-and-motion studies. A link may represent any relationship between a person and machine, between one person and another, or between one machine and another. Links can be characterized as

- Communication (visual, auditory, touch),
- Frequency (how often a person looks at something, movement from one place to another),

- Sequence of use (order of use or movement),
- Control (person to equipment), and
- Movement (eyes, hands, feet, whole-body location).

Once links have been identified and data collected, analysis is often facilitated through graphic representations. Graphical link data representation types are

- Link tables—summarize "importance of relationship" and "reason for proximity" per component pair;
- Adjacency layout diagrams—used to represent the frequency or importance of links (e.g., movements, functional connection, etc.);
- Spatial OSDs—describe the actual sequence of use; and
- Combination—provides elements of adjacency and spatial sequential use.

The diagrams are usually drawn in one plane (two dimensional) (see Fig. 13.3), but three-dimensional representations are possible. The diagrams are analyzed to try to arrange components according to sequence of use (place components in order of temporal use) or frequency of use (place most frequently used components in most convenient location). If links represent sequence or frequency, then this analysis primarily involves minimizing the distance between the strongest links.

Other factors should also be considered in the location of workstation components. These include placing important items in prominent positions, grouping components that

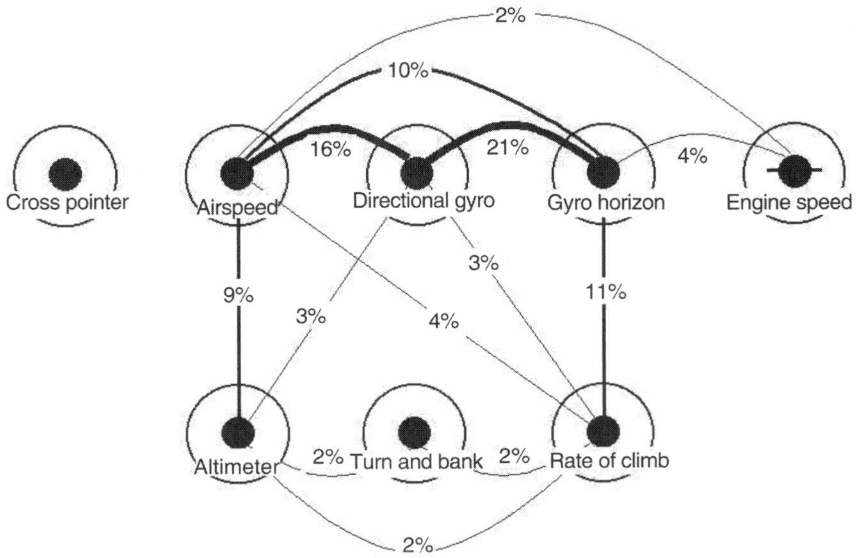


**Figure 13.3** Two-dimensional link analysis: eye fixations of aircraft pilots (adapted from Jones et al., 1949).

are functionally related in the operation of the system, consistency, clutter avoidance, and control-display compatibility or collocation (see, e.g., Wickens et al., 1997). As mentioned earlier, many inefficiencies and obstacles to optimal performance can be identified and resolved by studying the sequence of the activities involved. In link analysis and OSDs, activities are normally tracked with the focus on the operator; however, operations of the larger system can also be examined. Operators assigned to tasks but not optimized as a team may self-organize based on parameters not essential to optimum performance, such as the personality of other team members, seniority, traditions, etc. Operational sequence diagrams are one part of ensuring optimum team performance. This type of analysis allows the analyst to define the functions, tasks, assignment of tasks to operators, and task sequences in an optimized way.

The first step is to determine the appropriate sequence and order of operations. This will be based on requirements such as those that drive the human and system performance. Performance criteria in the form of error rates and time to perform should also be considered.

Once the activity is formatted in the proper order with the sequential relationships established, it can be readily analyzed for human effectiveness. The key to designing a system, product, or process to be optimally operated by a human is to design the task around the operator. Therefore, OSDs should be structured from the human's perspective. This may be different from other analytic methods that are focused on constraints that are external to the human and the immediate task. Kirwan and Ainsworth (1992) is a good source for more information about OSDs.

When designing a computer interface or a computer-controlled system, OSDs can be very useful for considering the sequence of information required by the automated systems. The sequence should include what user input is required as well as actions taken by the automated system. Computer systems may have very strict information, time, and sequence requirements, but human capabilities and expectations do not always match these requirements. Thus it is important to consider the human user when designing computers and computer-controlled systems. For example, human perception of sight and sound is limited and variable. Stimuli (e.g., the flash of a pixel on a screen or a noise) must be of sufficient intensity and duration to register and be perceived. This assumes that the user's attention is not directed elsewhere and that the stimuli are understood after they are perceived. So the computer interface and sequence must present stiumuli that the user can perceive and that will attract his attention.

The interface must also consider the cognitive profile of the target audience such as the user's experience, expectancy, and ability to recognize and apply metaphors. Many of these items are very dependent on cultural and educational backgrounds. Context sensitivity is important in the usability of computer systems. How users arrived at a point in a user interface as well as their previous experience in using that and other user interfaces can affect how they expect the computer system to respond. The user will draw conclusions from information presented and make decisions at nonreversible decision points depending on the goal of the task.

Perhaps the most important factor for handling this problem is elimination of sensitivity to sequence on the part of the hardware and software system. In situations where elimination of sequence sensibility cannot solve all of the issues or when it introduces unacceptable complications, the presentation of information in dialog boxes along the way can help assist operators in choosing appropriate sequences. Operational sequence diagrams can be helpful with both of these methods.

### 13.2.3 Task Analysis, Function Allocation, and Workload Analysis

*Task analysis* refers to a listing and examination of the basic actions a person must perform to accomplish a job. In terms of detail, HFE tasks usually are defined as an action (e.g., turn, lift, push, toggle) performed on an object (e.g., a wheel, lever, button, crank). Task analysis begins with a task list that may be generated using any number of techniques such as document review (user's manuals), interviews (users or designers), observation of personnel using the system, or cognitive walk-throughs and may represent physical as well as mental tasks. The method used and the amount of detail depend on the analysis questions that will be asked (i.e., the reason for the task analysis).

One of the most popular methods is simply a top-down hierarchical method. In this method, a general mission is broken down into increasingly detailed actions required to perform the mission. Until the mission breakdown reaches the level of an action on an object (task) assigned to either a person or a machine, the elements are referred to as functions. Examples of functions are drive, communicate, and engage target. Examples of tasks are turn steering wheel, apply brake, and push "talk" button. It is usually more efficient to gather time data (time-and-motion study), sequence information (OSDs and link analysis), task demands, skills required, etc., at the same time that the task list is generated.

Another popular method is *cognitive task analysis*. In contrast to traditional task analysis, this method is aimed at understanding the underlying thought processes required to perform observable tasks. In some cases the aim is to understand the knowledge, user experience, or biases that went into an observable action. Cognitive task analyses are conducted for many purposes such as design of computer systems and training. The idea is to support performance of tasks such as decision making and control of complex systems by understanding the mental aspects of how the tasks are performed. Those interested in cognitive task analysis may wish to visit the Cognitive Task Analysis Resource website at http://www2.ctaresource.com.

Once the task list is generated, it must be analyzed to be useful. Task data form the basis of many HFE methods, but the purpose of a task analysis is often for function allocation or workload analysis. When observed in the context of a task analysis, goals (what you are trying to do), human behaviors (how you are trying to do it), and environment (conditions under which you are performing) associated with each task are identified and reviewed for compatibility with each other. If there is a significant incompatibility among the physical or mental abilities of the target audience, established goals, or environment and design, the discrepancy becomes more obvious since the details of task goals, performance, and conditions have been specified. For example, if the design requires a person to reach for three widely spaced switches at a height of 8 feet all at the same time, this is not likely to be a reasonable requirement. Most people are not able to reach to 8 feet, and no one has three arms to activate three switches at the same time. The most effective forms of task analysis include improvement recommendations that are traceable to specific issues identified via the task analysis. In later phases when test and validation are conducted, the findings of the task analysis can help guide test issues to either confirm or refine the issue. Kirwan and Ainsworth (1992) provide numerous methods and examples of task analyses.

*Function allocation* is aimed at deciding which jobs (groups of tasks) should be assigned to humans (and which human if there are several available) and which to machines. The idea is that people are better at certain task types (e.g., dynamic decision making) and machines at others (e.g., tedious, repetitive tasks). But the state of the art in automation is advancing rapidly, causing significant changes to the manner in which tasks can be performed as well as affecting the types of tasks that can be automated. Current thinking (Parasuraman et al., 1996) is that adaptive aiding in which the function allocation is dynamic depending on the needs of the human may be required for performance in high-paced, information-rich, highly automated environments.

*Workload analysis* is used to ensure that people do not have too many (overload) or too few (underload) tasks and information to handle at any given time. Workload analysis considers the fact that a person's mental processing capacity is limited and not all tasks are purely physical. Accomodating these mental processing limitations is an important design constraint. This analysis may be as simple as counting the number of tasks that a person must perform at any given moment. However, more complex analyses consider the nature of the task (e.g., visual, auditory, cognitive processing, psychomotor control, or physical) and the difficulty in performing different types of tasks simultaneously. Other analyses consider novel versus highly learned tasks (i.e., higher workload until well learned) and dynamic management of workload (i.e. strategies used by people to attempt to keep workload at a comfortable level). See Damos (1991) for more information on workload analysis.

### 13.2.4 Accident and Incident Analyses

The study of incidents and accidents is a key HFE method. Performed properly, the analysis will be a valuable indicator of system behavior. The focus is on unplanned-for and unanticipated, undesirable system performance. Incidents typically involve near misses, accidents, and full-blown system failure. Many analytical constructs are devised and used to forecast the possible outcomes of a functioning system. However, no technique or approach can be all encompassing and anticipate all uses and combinations of variables. For those situations, accident and incident analyses will be the most useful methods, because the focus is on how the system is actually used and how that use led to a near miss or system failure.

This type of study is focused on finding important information from man–machine systems that are not doing what they were intended to do. It can be a system that is (1) under performing compared to the design performance specifications, (2) producing defective results, or (3) yielding undesired side effects. Frequently, these analyses are performed on a specific system with recurring problems or on a design that is in multiple locations and is experiencing one or more common problems.

When possible, conclusions should be drawn from multiple occurrences of similar incidences. The selection of sampling techniques and data collection process across the incidences is very important. Data structures and resolution should be tailored to assure that there is sensitivity in areas that are being studied. Additional attention should be paid to isolating those variables that are fixable within the context of the available solution set.

Emphasis must also be placed on the individuals performing the investigations and data collection. Collection of data by individuals that can be personally affected by the outcome of the investigation should be avoided. In other words, if the facts surrounding the incident will yield findings attributable to the behavior or performance of an individual or his or her direct associates, then this person should not be involved in the data collection and reduction. This will prevent possible reporting distortions and inappropriate leading of the investigation.

In addition, there should be a level of anonymity to many of the sources of the data collected. Information will be much more objective and useful if those providing it know that there will be no punitive outcomes regardless of personal involvement. This is a factor where the information providers might try to protect individuals from an organization or power structure based on perceptions of fairness. To avoid these effects, anonymity will assure there is no linkage between data collected and the personal well-being of the information sources or their direct associates.

*Incident analysis* is one of the most telling of the analytical methods. It allows the analyst to construct a specific scenario where the outcome was undesirable (i.e., an accident, near miss, etc.). By decomposing the factors contributing to the undesirable outcome, the analyst can identify specific causal factors and initiate solutions for prevention. While seemingly clear, there is some level of complexity to all incidences where human behavior is present. Therefore, measured trial-and-error fixes may be required before the optimal solutions are actually achieved. Flanagan (1954) was the first to use incident analysis in his study of near misses in aircraft. (See Chapter 14 for more information on causal analysis.)

A number of risk and human reliability assessment techniques have been developed. These techniques can be useful in analyzing accidents and incidents in more detail. See Wilson and Corlett (1995) and Kirwan (1988) for specifics on particular techniques and evaluation factors for selecting among them.

### 13.2.5 Anthropometric and Biomechanical Analyses

Anthropometric and biomechanical analyses are used here to refer to a group of methods and principles all dealing with the application of information about the size, shape, and physical abilities of people to the design of workstations, products, and jobs. Information about size and shape refers to reliable measurements of a person's body such as overall stature, limb lengths, functional reaches, and girth of body parts. These measurements are (ideally) collected according to very specific procedures and landmarks. Data are often summarized according to populations surveyed and reported as means, standard deviations, and percentiles. (Note that there is some controversy over the value of using percentiles for anthropometric analyses. See Section 13.6.)

Recent trends are to use three-dimensional laser scanning to collect detailed body surface measurements. Three-dimensional scanning works particularly well for obtaining girth and contour measurements for an individual in a static posture. A notable example of an anthropometric survey using three-dimensional scanning technology is the Civilian American and European Surface Anthropometry Resource (CAESAR) coordinated by the Society of Automotive Engineers. Physical ability data refer to characteristics such as strength (e.g., lift, push, pull) and range of motion for a particular body part.

The idea behind anthropometric and biomechanical analyses is to use these data to set design limits that will fit the target population and will not exceed their capabilities. The importance of knowing the anthropometric and biomechanical characteristics of the target population cannot be emphasized enough, especially when the characteristics are very different from those of the designer. For example, if the designer of an airplane seat is a Dutch male who is thin and strong and has long legs, he might not consider seat breadth to be as important as leg clearance. If the target population includes older North American females (greater hip breadth and shorter buttock–knee length than Dutch males), seat breadth is likely to be as critical as leg clearance. The designer might also underestimate the criticality of being able to reach an overhead bin without having to assume a weak posture (lifting arms over one's head). But if the plane were being designed for a population that includes Japanese women (shorter on average, and women on average have less upper body strength than men), this would be a very important design issue.

Insensitivity to user characteristics is often a problem when the target population includes children, the elderly, or disabled. If this designer did realize that these dimensions were important, how would he or she know what their limits should be? Designers can access anthropometric and biomechanical data sets to establish statistical limits so that a specified portion of a population (e.g., 90, 95, and 99 percent) will likely be accommodated by the design. Texts such as Sanders and McCormick (1993) and Wilson and Corlett (1995) provide more detail on traditional anthropometric and biomechanical analysis accommodation methods. Many traditional anthropometric data sets exist. One large survey is the 1988 anthropometry survey of U.S. Army personnel (Gordon et al., 1989). The *Directory of Databases Part I—Whole Body Anthropometry Surveys* (1996) lists whole body anthropometric surveys and provides current sources for the survey raw data and summary statistics.

### 13.2.6 Field Study, Survey, and Usability Analysis

*Field studies* are quasi-experiments (quantitative, yet without true randomization to control for various threats to validity) and are conducted in a setting as close as possible to the actual conditions under which the final product will be used. They are valuable because they allow the system (people and equipment) to be tested quantitatively against system performance requirements in a realistic environment. The price for this realism, however, is a lack of control over sources of bias and error. Field studies yield powerful insight into the situations and the environment of the human activity, process, or product that is being considered. Keep in mind that subject matter experts of the product, process, and environment will most likely be ill equipped to analyze their situation in a way that can be applied analytically to the overall target audience. Those conducting field studies are often faced with changes in conditions, participants, equipment, etc., on very short notice and must be knowledgeable of basic experimental design methods to minimize threats to validity when conducting the field study and responding to these changes. Only a combination of trained sensitivity to human issues, experimental design, and direct observation will produce valid and useful data.

*Surveys* are designed to ask people directly about their attitudes, opinions, or behaviors regarding some activity, product, or system. Surveys usually take the form of a questionnaire. It is easier to design a bad survey than a good one, which typically happens by introducing bias or too much length. Questions should be unambiguous and only those that are needed should be included. It is critical that a trained analyst plans and creates the questionnaire. It is also important that the administration of the survey be performed using a controlled, unbiased process. Charlton (1996) provides an excellent summary of questionnaire techniques. When survey data are analyzed, one should keep in mind that the data are subjective. Improper execution of the survey can easily bias results and lead to a reinforcement of preconceived notions if conducted under uncontrolled conditions.

*Usability analysis* typically involves the mock operation of a product by a target audience member or a subject matter expert. This analysis is a critical activity for comparing the suitability of a product or process with the target audience and environment. Used extensively in the development of consumer products and in the computer industry, it is a process of mocking up or simulating the entire environment that a user or customer will experience. This allows the user to experience a more complete representation of the product (in terms of factors that could affect performance) than is typically present in classic human experiments.

Classic experiments seek to control sources of error by keeping some aspects of the experiment (variables such as time of day or test environment) equal or constant, but important factors may be left out through this control. Usability analysis is favored by marketing organizations as a way of obtaining information about corporate and brand images of products. In this analysis, trade-offs of performance versus image are sometimes made in favor of advertising opportunities such as those commonly found on web pages. Subjective and objective results will be evaluated later and decisions will be made based on trade-off criteria of the usability project. Frequently, performance data in terms of errors and intervals to achieve an objective are not the primary goal of the activity. Instead, information such as desired features, use cases, and unanticipated behaviors may be of more interest. A user jury, if properly conducted, can be considered a form of usability analysis.

Usability laboratories are typically used to conduct these studies. A usability laboratory usually consists of two compartments. The first compartment contains the subjects, and the second compartment contains the test evaluators. The test subjects are positioned in a manner typical of how they would use the product being tested. The evaluators have visual and audible access to the test and also have test equipment, data collection, and other test apparatus at their disposal. A controlled training session will precede the activity and varies vastly among practitioners in terms of rigor and documentation. Sufficient documentation must be recorded to assure repeatability as well as to provide the scientific basis for assessing the training demands of the product. After being trained and prepared, the subject(s) operates the product under the surveillance of evaluators. In addition, video data recording equipment may be used to collect moving images and allows for time-sensitive recording of associated data. The evaluators and the associated equipment are invisible to the test subjects.

Some usability practitioners ask the subjects to talk through the problems as they occur. This is not recommended for systems or products that have cognitive workload constraints. Verbalizing actions will alter the instantaneous workload of the subjects and have a confounding effect (i.e., artificially increase their workload). Postsessions (i.e., review with subject following testing) can be helpful in identifying specific problem areas without impacting the operational session. Follow-up questionnaires may yield additional anecdotal and open formatted information. Postsessions should be conducted away from subject waiting and testing areas to avoid prejudicial effects on pending trials. The resulting data will be of several types (subjective and objective), and media formats should be reduced, analyzed, and compiled by an experienced practitioner in order to assure clarity, accuracy, and validity of the results.

Hix and Hartson (1993) are one source of additional information about ensuring usability of products, particularly software user interfaces. See also, Part III of Wilson and Corlett (1995).

## 13.3 HFE TOOLS AND TECHNOLOGIES

As HFE methods evolved, a variety of tools and technologies were developed to make application of the methods easier. The tools range from paper-and-pencil checklists to graphically sophisticated computer-based modeling tool suites. These tools and technologies generally are matched to one or more of the basic methods discussed in the previous section.

Before we discuss classes of tools and technologies, the point must be made that none of these tools do the thinking for you. A trained human factors practitioner is required to use them properly. So what good are these tools and techniques? What they do is structure, organize, describe, and provide the capability to visualize your system, but they do not interpret the results and typically do not tell you how the system must be changed to improve it. Even the best hammer and chisel will not produce a great carving in the hands of an unskilled craftsman. Just as it would be unwise to expect a psychologist to use a finite-element analysis tool to analyze a structure, it is unwise to expect a computer scientist or engineer to use an HFE tool with no training. The tools, especially those that are computer based, contain assumptions and qualifications that impact validity of results if misinterpreted or ignored. Many contain technical terms that may be familiar only to those in human factors or subfields of psychology. These tools should not be used unless the analyst understands the basics of the method and techniques underlying them (i.e., could perform the analysis by hand if given sufficient time).

Table 13.1 provides classes of tools in a list that helps provide some structure for our discussion and is not intended to be either comprehensive or orthogonal.

### 13.3.1 Guidelines and Standards

There are many guidelines and standards that apply to human engineering. The difference between guidelines and standards is that standards are usually mandatory and compliance is required while guidelines are generally only recommended practice. Many human engineering guidelines and standards have been developed and maintained by industry standardization groups [e.g., American National Standards Institute (ANSI), the Society of Automotive Engineers (SAE), and the International Organization for Standardization (ISO)]. Several standards groups also exist within the U.S. government [e.g., National Institute of Safety and Health (NIOSH), DoD, and National Aeronautics and Space Administration (NASA)]. Occupational health and safety standards are also covered in Chapters 14 and 15.

**TABLE 13.1 Classes of HFE Tools**

| |
|---|
| Guidelines and standards |
| Checklists |
| Subjective assessment tools |
| Simulation—unmanned |
| • Task network tools |
| • Perceptual models |
| • Congnitive process models and architectures |
| • Graphical human models |
| • Integrated tools |
| • Human behavioral representations (HBRs) in simulation federations |
| • HFE tools embedded in computer-aided design/computer-aided engineering (CAD/CAE) suites |
| Simulation—human in the loop |
| Miscellaneous analytical tools |

Some important human engineering standards are

- MIL-STD-1472F (DoD, 1996), DoD design criteria standard: *Human Engineering*;
- NASA-STD-3000 (NASA, 1995), *Man-Systems Integration Standards (MSIS)*; and
- MIL-STD-1474D (DoD, 1997), DoD design criteria standard, *Noise Limits*.

Examples of human factors–related guidelines include

- MIL-HDBK-759C (DoD, 1998), *Human Engineering Design Guidelines*;
- Numerous guidelines related to office ergonomics such as ergonomic requirements for office work with visual display terminals and ANSI B11, (ANSI, 1994) *Ergonomic Guidelines for the Design, Installation and Use of Machine Tools*[1]; and
- *The Human Factors Design Guide for Acquisition of Commercial Off-the-Shelf Subsystems, Non-Developmental Items, and Developmental Systems* (Wagner et al., 1996).

One problem with guidelines and standards is that not all cases and combinations of factors can be anticipated and their appropriate resolution specified. Also, the source and assumptions for some of the recommendations can be buried or lost so the HSI practitioner will not know how applicable the recommendation is for his purpose. For example, he or she may find a standard for the size of lettering you need so that a sign is readable at a distance of 30 feet but the standard might apply only to 20/20 corrected vision under ideal (clear) atmospheric conditions for a person (reader) standing still. It is unlikely that this standard will be appropriate for a person reading the sign from a moving vehicle on a foggy day.

### 13.3.2 Checklists

These tools consist of paper or computer-based lists of issues or design parameters that should be evaluated in the course of a human engineering program. These lists are based on prior experience and are often an attempt to capture human engineering subject matter expertise. Checklists may also take the form of "lessons learned" documents or branched question-and-answer tools and may even be labeled as guidelines. Examples of human engineering checklists are *Human Factors Evaluation Checklist for Tanks* (Clingan and Akens, 1986) and some aspects of the Cornell University ergonomic guidelines for arranging a computer workstation (http://ergo.human.cornell.edu/ergoguide.html).

Similar to guidelines and standards, checklists are limited in their ability to anticipate and cover all combinations of variables and conditions that may apply to a given design problem. They cannot capture the variability and dynamics of human performance. They are usually shorter than guidelines and standards and are generally geared to quicker evaluations. As such, they may not be detailed enough to capture very specific design problems. Their utility lies more in guiding inexperienced practitioners through a quick basic evaluation. A few checklists are quite elaborate and include references to more detailed analyses such as one developed by Kearney (1998). Therefore, they cross into the realm of process guidelines.

### 13.3.3 Subjective Assessment Tools

These tools are typically dependent (performance effect data) measures used during the conduct of a study. Those cited here are used most often during simulator or field studies. The best of these tools include guidance on how to administer and score them and then interpret results. Subjective assessment tools commonly involve feedback or ratings from participants. Examples of subjective assessment tools are questionnaires, workload measures such as the subjective workload assessment technique (SWAT) (Reid and Nygren, 1988); NASA task load index (TLX) (Hart and Staveland, 1988); the modified Cooper–Harper workload scale (Wierwille and Casali, 1983); and situation awareness (SA) measures such as the cognitive compatibility situation awareness rating technique (CC-SART) (Taylor et al., 1997) and the situation awareness global assessment technique (SAGAT) (Endsley and Garland, 1999). Objective metrics are usually measures of performance such as reaction time and error rate or the physiological state of participants that have been correlated with changes in performance of various task types. Examples are heart rate, eye blink, blood pressure, hand steadiness, and electromyogram. Performance assessment batteries that combine various measures from the subjective and objective categories have been developed to provide multidimensional insight into performance. Examples of performance assessment batteries are the complex cognitive assessment battery (CCAB) (described in Kane and Kay, 1992), COGSCREEN™ (described in Kane and Kay, 1992), and the delta battery (Turnage and Kennedy, 1992).

### 13.3.4 Simulations

Simulation offers the ability to create virtual elements of a future situation before they are readily available. This provides important answers about the situation, process, or product in a time frame when the design is still being formed. For example, in traditional product development programs, many months of designing would precede the availability of a prototype. Using simulations before the physical prototype is fabricated can lead to many important and timely discoveries.

A key to effective simulation is to find the right degree of simulation (include critical elements) and fidelity (model those elements to the correct level of accuracy) to make it representative but not more so than is necessary. The important items that require the maximum fidelity should be identified in the planning stages. Those findings should be forwarded to the simulation specifications. If the simulation has too much in it, it will be excessively costly and take too much time and resources to accomplish the objective. This will undermine a key benefit of using the technique (i.e., cost savings).

Human engineering simulations fall into two main categories: manned and unmanned. Manned simulations will include a real human as part of the execution of the model. Unmanned will have a part of the software that represents human activity.

Unmanned simulations are usually computer programs in which models are built. The models represent the environment within which the operator performs, contains a task or task network, and also has some representation of the human. The representation will depend on the purpose of the modeling activity. Physical and cognitive human behavior will be represented.

Exercising the model will yield results that forecast the output of the man–machine system. If the output is not satisfactory, then factors related to the environment, tasks, or human attributes can be modified. Such modifications are made to determine sensitivity of elements and will result in solutions to the system design problems. Usually the design is modified based on early runs. Subsequently, the model is updated to the new design and rerun in the simulation to validate and quantify the improvement. This process is most effective when schedules and availability allow for multiple iterations and collaboration between the members of the design team.

***Unmanned Simulation Models*** There are several types of models that belong in this category. Below, we briefly describe several of these types. More information can be found in other chapters of this book, as well as in the publications that are referenced.

*Task Network Modeling Tools* Task network modeling is a technique that allows predictive modeling of activities that can be subdivided into discrete elements (or tasks). Once defined, estimates of performance ranges (e.g., time, accuracy, workload) are attached to the lowest level of the decomposed hierarchy. The tasks are then simulated using a discrete-event simulation process and are typically subjected to a range of scenario events in order to trigger unique combinations of tasks. The end result of the simulation is a system-level performance estimate that is applicable to a broadened range of scenarios. Tools in this class are used for task analysis, function allocation, and workload analysis. Several tools exist that provide task network modeling environments, including Improved Performance Research Integration Tool (IMPRINT), WinCrew, and the Integrated Performance Modeling Environment (IPME). These tools are described in Chapter 11.

*Perceptual Models* These are models of how people register input stimuli from the environment. Perceptual modeling tools can help designers compare these stimuli to what their target audience can perceive. These sorts of models are often referred to as "first-principle models" and usually do not have an embedded sense of time. For this reason, they are not typically considered simulation models but are more often mathematical algorithms designed to help designers predict what a person can see or hear in a specific environment.

*Cognitive Process Models and Architectures* This family of models focuses on describing and predicting cognitive behavior and often includes a representation of memory. Currently, these models are best suited to modeling very detailed and short (several seconds) tasks, simply because they require a significant amount of effort and are not intended to predict psychomotor performance. Very few models in this category have been commercialized, and they require a great deal of expertise in cognitive psychology and, in most cases, computer science to use effectively. Several examples of these models and architectures are provided by Pew and Mavor (1998) (e.g., see atomic components of thought—rational (ACT-R), executive-process interactive control (EPIC), cognition as a network of tasks (COGNET), and Soar).

*Graphical Human Models* This class of models provides unique capabilities in visualization and is extremely helpful in evaluating and communicating the "fit" of the human into an existing or notional crew or workstation (anthropometric analysis). The tools typically have some limited CAD capability for creating an environment to represent the crew or workstation. Usually, the item being evaluated is imported from a more sophisticated CAD package, and if not well planned, this process may consume significant project resources. The most well known were developed to run on UNIX platforms such as

SGI (Silicon Graphic Inc) machines, but many have begun to migrate to more powerful personal computer platforms, and this is clearly the trend for the future. Common features of the tools include creation of different-sized figures from various anthropometric databases (including male and female), ability to see through the eyes of a figure, positioning figures into a limited set of predefined postures, limited predefined animated behaviors using scripting (e.g., walking a level path identified by start and end points), and specification of range-of-motion of joints. Most of the models have some level of embedded biomechanical representation. Some employ techniques such as inverse kinematics so that the body parts may be positioned with less user input and several can use data from various motion-tracking systems to replicate human movement. Biomechanical definition may include various degrees of simulation in the number of joints, degrees of freedom about each joint, and range of motion. Some of the models include modules that use algorithms for analyzing strength and lifting. Examples of graphical human models are Jack, Safework® Pro™, Ramsis, and Mannequin Pro (refer to the Directory of Design Support Methods and company websites for more detail[2]). See also Chaffin (2001) for examples of application of graphical human models.

*Biomechanical Models* Some graphical human models are primarily biomechanical models. They are typically used to predict occupant motion in crash test or ejection seat simulations. Primary examples of whole-body, biodynamic models include mathematical dynamical model (MADYMO) (Happee et al., 1998, and http://www.madymo.com) and articulated total body (ATB) (see Cheng et al., 1998, and http://www.atbmodel.com). Other biomechanical models represent specific parts of the body (e.g., spine, bones, joints, shoulder) in more detail. Similar to graphical human models, there are biomechanical models (some graphical and some just parameterized algorithms) that are intended to help predict the acceptability of a lift. Examples are the 3D Static Strength Prediction Program™ (3D-SSPP) (http://www.engin.umich.edu/dept/ioe/3DSSPP) and the NIOSH lifting equation (Waters et al., 1994). (See http://www.industrialhygiene.com/calc/lift.html for an on-line version of the equation.)

*Integrated Models* At a workshop held in 1985 (Kroemer et al., 1988), the National Research Council made several recommendations toward establishing an integrated ergonomic model. Their recommendation provided clear indication that models of the task environment or work process (i.e., task network models) would benefit from combination with theoretically correct models of cognition and perception. Additionally, graphical human models could be used to view a dynamic representation of the human interacting with the simulated environment. Since this report was published, a comprehensive human model has not been developed, but some progress toward that end has been made. One of the earliest and most ambitious integrated model efforts is Man–Machine Integration Design and Analysis System (MIDAS). MIDAS was designed primarily to answer questions related to the design of aviation cockpits and includes representations of pilot perception, cognition, and anthropometry (Hart et al., 2001). More recently, the U.S. Army Research Laboratory Human Research and Engineering Directorate (ARL HRED) has supported the development of a crewstation design tool intended to put HFE tools and models on the desktop of systems designers. A unique aspect of this effort is the inclusion of a library of controls and displays, indexed by human resources (e.g., visual, auditory) and associated with HFE standards and guidelines. In addition, NASA has made great progress toward integrating task network and cognitive models under their human error modeling program (e.g., IMPRINT and ACT-R), and this area of product development shows much promise.

*Human Behavior Representation (HBR) in Simulation Federations* In a class of models closely related to integrated tools, discussed above, another approach is to develop separate models that excel in one particular aspect of a problem and then share modeling results from one model to another to improve the degree and fidelity of all the simulations. This type of model has been used extensively in high-fidelity, force-level simulations. A particular and interesting challenge of this environment is the challenge of clock synchronization. The combination of task network models (which are typically discrete events in which the clock "jumps" from event to event at irregular intervals), system model, (which are typically continuous in which the clock "ticks" along at regular intervals), and first-principle models (which usually do not have any internal concept of time) is a complex and difficult effort. Nonetheless, many organizations are showing significant progress toward this end aided by higher level architecture (HLA) compliance. Examples in the human factors area are IPME and the combined Combat Automation Requirements Testbed (CART) IMPRINT effort (Martin et al., 1999).

*HFE Tools Embedded in CAD/CAE Suites* A more recent trend in human factors simulation tools is the inclusion of tools such as graphical human figure models as modules in larger computer-aided engineering (CAE) tools suites. This has positive and negative aspects. One of the largest risks in this approach is in exposing the quality of the human figure model database, which has direct consequences on the quality of the computer-aided design (CAD) assessment of human "fit." If the human figure model is not valid (e.g., torso is too long or too thin, arms are not proportional), then the end result could be an unfortunate combination of good intentions and misinformation. In fact, it could appear as though a thorough human factors analysis was performed when, in fact, the human factors assessment was completely lacking.

***Simulation—Human in the Loop*** Simulations are a key tool in the design of future products, processes, and almost any development activity. Simulations consist of hardware and software that are configured to reproduce a set of circumstances or an environment under which a task or activity is performed. The environment is constructed specifically to replicate a realistic and often complex set of conditions. Manned simulations usually consist of mock-ups, hot mock-ups, desktop simulators, or full simulators. In manned simulation, it is critical to have an appropriate experimental design, rigorous experimentation controls, and subjects that represent the target audience. It is difficult but extremely valuable to test simulated manned systems and get valid results. This activity is best left to trained human factors engineers or other professionals with the experience and awareness of experimental design to control confounding variables, especially those introduced by last minute changes.

### 13.3.5 Miscellaneous Tools

Some tools do not fit into the classification structure well, because they address either a specific method or a technique not covered in this chapter. Examples are protocol analysis tools, Locate II, and the Work Domain Analysis Workbench (WDAW). Protocol analysis tools such as MacSHAPA (Sanderson et al., 1994) are designed to support encoding and analysis of multiple sources of task performance data such as verbal communication, equipment logs (e.g., keystrokes), task sequence, and observable physical action. Typically data are encoded with a time stamp of who performed the action or who received the information. Then the tools can be used for data filtering and visualization, which are particularly useful in time-and-motion studies and task analysis. Locate II was developed specifically to facilitate using link analysis type data to arrange workstations within a two-dimensional (one plane at a time) workspace such that good movement patterns and visual, audible, and tactile communication are facilitated (Hendy, 1989, and http://www.interlog.com/~jle). The efficiency of layouts is compared using cost function values. The WDAW (developed with the support of the U.S. Air Force Research Laboratory and then the Australian Defense Sciences Technology Organization) was developed specifically to conduct work domain analyses (WDAs). (For information on WDA, see Rasmussen et al., 1994.) The tool uses a graphical interface to help an analyst with a working knowledge of cognitive engineering and cognitive work analysis to perform a WDA to examine issues such as potential conflicts (e.g., in information needs) across system or subsystem goals and functions. For information on the WDAW, see Sanderson et al., (1999) or http://www.it.swin.edu.au/schil/WDAW/wdaw.html. Note also that risk assessment tools and techniques are covered in more detail in Chapter 14.

Any written reference will be out of date before it is published due to accelerating developments, particularly with computer-based tools. Fortunately, several on-line databases of tools exist as resources to update information about HFE tools. One of the better (comprehensive and frequently updated) databases is the Directory of Design Support Methods and Liveware Survey maintained by the Manpower and Training Research Information System (MATRIS) Office of the Defense Technical Information Center (DTIC). Another is the Manning Affordability Website: www.manningaffordability.com

## 13.4 SELECTING TOOLS AND TECHNOLOGIES

How are HSI practitioners supposed to find the right tools for the job and determine whether or not a marketed or proposed tool or piece of software is really appropriate for the project?

The first step in tool selection is to identify the right class of tool from the methods and techniques discussed previously. Then, a search for currently available tools is performed (possibly through one of the on-line tools databases referenced earlier) to develop a list of candidates. Then, the following questions should be used to sort through the list of candidates:

1. *Original Purpose* Is the tool or model sensitive to the parameters that you are interested in varying? Can it be used or adapted to address your primary areas of interest?
2. *Degree of Accuracy* How accurate does my answer have to be? What is the tolerance for error? The answer will depend on the type of system being analyzed (e.g., the manufacturing tolerance in the cockpit of an airframe is much higher than in a workstation on a ship) and how early it is in the product life cycle (i.e., the more mature the design, the less tolerance for error).
3. *Level of Resolution* What is the scale of resolution? When tools are developed, a level of detail in problem investigation is often assumed. For example, a tool designed to help optimize the location and type of switches on a particular control panel in a cockpit might be ill suited to investigate the layout of workstations on an aircraft carrier.

4. *Validity* Are the data and algorithms valid for my application? Models are limited by the data from which they were developed. Care must be taken to match those data to the target audience. For example, it is inappropriate to use a database on the size of Japanese females for an analysis on stature accommodation of Dutch males. This may sound obvious, but often the underlying assumptions and databases of a model are not well documented or easy to trace.

5. *Realistic Resource Requirements* If the tool is software based, what does it take to run and use it (compared to resources available for your project)? Resources to consider include personnel with skills to learn the model and related packages (e.g., UNIX, CAD packages, basic programming), time, and computer platforms.

6. *Information Availability and Data Format* What data are required as input? Do you have access to these data or sufficient resources to develop them? If you have large amounts of data, consider whether the tool can read the data in as an electronic file. If so, what file formats are supported?

7. *System Compatibility* If the tool is software based, what platform(s) does it run on? This is becoming less of a problem as tool developers work to make their products multiplatform compatible, but not all applications run equally well on all platforms. For example, very large files on a Windows NT platform might overwhelm a graphical human model originally developed to run on a high-end UNIX platform.

8. *Cost* The cost of tools ranges from no cost for paper-and-pencil government-developed methods to over $70,000 for sophisticated human figure modeling software packages. Other cost issues to consider are as follows: Is other software necessary to use the tool? If you do not have that software, you will have to buy it as well. Are you buying one seat (one copy limited to use on one machine at a time) versus a site license (permission to use multiple copies throughout your organization)?

9. *Output Format* What output is needed? In what format? Consider what type of information you will need for output. Does the tool produce this output in a preformatted report or visualization or will you have to generate it yourself? Does the tool produce the data necessary to feed the report? If the output is needed in electronic form, does the tool produce files of that type? Incompatibilities in file format or having to generate them yourself can use up valuable resources in a project.

10. *Software Compatibility* If the tool is software based, does the model need to run in real time and is dynamic interaction with other simulations necessary? A stand-alone tool that runs faster than real time (e.g., Monte Carlo simulation) may be sufficient, but several standards for communication between models have been established so that each one will not have to be comprehensive in representing the system(s) and the environment in which it operates.

11. *Verification and Validation* Is the tool developer committed to on-going verification and validation? Verification, in particular, is only good for a given version number and must be repeated with each software release. Are model assumptions well documented so that you can produce a model or analysis that you can defend?

Care should be taken to match the version number of the tool with the information being used in your evaluation and the version you intend to purchase. In other words, do not base your decision on a description of a previous version because the tool's capabilities and underlying assumptions may have changed. Current users of the tool should be polled to determine whether or not marketing claims are accurate. There are expert systems such as HOMER (although not fully implemented), WCField, and OWLKNEST (AGARD, 1998; *Directory of Design Support Methods and Liveware Survey*, 2002) that may be useful in helping the analyst select tools and technologies. (They ask many of the same questions that we have listed above.) However, unless these systems are updated, it is possible for their recommendations to be inaccurate.

## 13.5 PLANNING FOR ANALYSIS

Perhaps the most important step in preparing to conduct a human engineering analysis is understanding how to get the most out of the available analysis resources. This requires the analyst to (1) carefully consider what type of output is required to identify issues, (2) provide the right level of data to evaluate solutions, and (3) provide the impetus necessary to get design changes implemented. Once required output has been determined, methods and tools that support the output and necessary input data can be identified. Method, tool selection, and availability of input data drive resource requirements.

Determining required output is not easy. One way to start is to think about what output is required to address the issues important to the project. This begins by asking what alternatives are being considered as part of the system design. Sometimes initial questions are articulated by the customer but the questions are incomplete, are not user centered, or do not get to the root of system performance. It is up to the human factors engineer to use the clues in those questions to determine human factors issues.

For example, a project might have an initial question about whether a head-mounted display provides better vision than a cathode ray tube (CRT) monitor. The analyst might ask what types of data in which format will best explain what the problems are and what the solutions should be? Examples of possible formats are time to perform tasks with the displays, type and number of errors in performing tasks, two-dimensional graphs showing visibility of a reference object at various distances, three-dimensional CAD files showing field of view, and relative changes in workload or a specific situation awareness (SA) measurement.

Deciding what type of output is needed will affect the modeling and analysis. For example, if it is known that a project is considering the value of a speech detection system, the analyst's model should have tasks modeled to a level in which activation, feedback, etc., of the speech detection system are represented. Otherwise, one may only need to represent communication in a broader sense such as "communicate internal to vehicle" and "communicate external to vehicle." Similarly, with CAD-based analyses, equipment, workstations, clothing constraints, and even humans can be modeled in varying degrees of resolution. Evaluation of preliminary concepts may be fairly low resolution due to the immaturity of exact equipment measurements. Items that are not of high tolerance or criticality may be modeled with less resolution. More mature designs or workstations in which high tolerances are critical (e.g., a helicopter cockpit redesign) require higher resolution input and output.

Once the required output and the method to produce it are determined, one will have a good idea of the input data required. At this point, the availability of the data should be checked. If the data will result from another part of the design process (e.g., CAD), it should be asked to be in a format as close to that which will be needed as possible. This can be as simple as specifying a file type or format (e.g., vrml, .xls comma delimited, .stl) or as complex as giving specific instructions on documentation of assumptions and sources, grouping of CAD parts, resolution, etc. There are two advantages to asking for the data as early as possible. First, the analyst has a better chance of getting what he or she wants without having to waste resources generating or reformatting it. Second, the analyst will know if the data will be available at all. Asking early helps scope the resources required to obtain the required input for analysis. Often, getting the right input data is the most expensive part of the HFE process.

A good mechanism for asking for the input data is to include it as a contract requirement. Requests for proposal can include specific data elements useful in evaluating the human engineering merits of the proposal. Likewise, contracts should include requirements for data useful to evaluate design options both during system development and for reuse on product improvements.

There is a wide assortment of methods and tools available to perform a wide variety of human engineering analyses. Each method and tool can be effective and useful as a stand-alone activity. However, the most powerful results can be achieved by using them in combination with each other. Once methods and tools have been selected to aid in human engineering for a program, a flow or management scheme for their application should be developed to maximize synergies among them. Basic texts on HFE (e.g., Wilson and Corlett, 1995; Sanders and McCormick, 1993; Wickens et al., 1997) should be consulted for standard approaches. Depicted in Figure 13.4, the following multiphase approach is recommended:

- Alpha phase—planning and task analysis;
- First phase—workload analysis;
- Second phase—anthropometric analysis;
- Third phase—human-in-the-loop simulation; and
- Final phase—design recommendation and documentation.

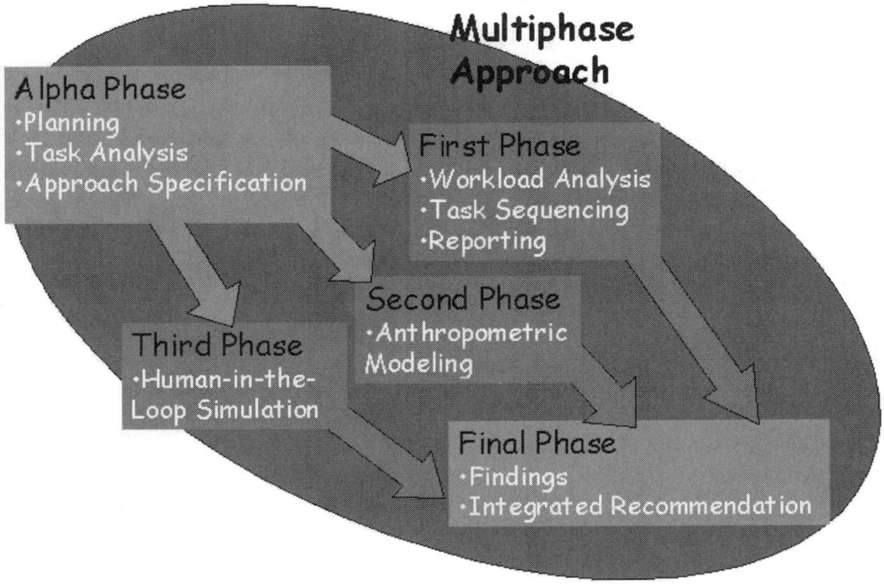


**Figure 13.4** HFE analysis phases.

### 13.5.1 Alpha Phase: Planning and Task Analysis

The alpha phase is the key to a multiphase approach. It is characterized by precision planning and task analysis conducted as a concurrent first step. It results in a clear direction with a documented specification that indicates the inputs and outputs from each of the planned activities, establishes links between the activities, and creates a network of objectives that are optimized for efficiency and effectiveness. It involves extensive collection and analysis of several information sources. Part of the specification is an allocation of tools and methods mapped against the specific challenges presented by the tasks implicit in the product's operation and use.

The first step is to collect and manage the knowledge related to the project. Examples are

Product mission, purpose of the product, process, or system;
Functions the operator(s) and jobs the product must perform;
Acceptable and desired expectations for product performance;
Schedule available for changes to the design;
Budget available for changes to the design;
Schedule available to provide human engineering inputs;
Budget available for human engineering;
Simulation infrastructure, experimental design capability, and availability;
Anthropometric model software and the skill level of the available analyst;
Workload software and skills of the analyst;
Existing models that relate to the product;
Scenario(s) the product will operate under;
Target audience description data; and
Lessons learned from related projects.

During the task analysis, the analyst will list questions that will need to be answered to assure that each studied task will be performed to an optimum or prescribed standard. Each question will be allocated to one of the dimensions for analysis, resolution, and validation. After the alpha phase, the activity splits into a triad, and thus the multiphase aspect is in effect. The economies achieved come from the targeted addressing of the issues with the most appropriate tool or process.

### 13.5.2 First Phase: Workload Analysis

The first dimension is the workload model building, execution, and analysis. The activities in this phase are tailored to addressing those workload situations that are fixable within the context of the project and within the sensitivity range of the modeling tool and model.

Based on the task analysis from the first phase, an experienced analyst will start by reviewing and updating the plan that was created in the alpha phase. This will assure that the analyst is sensitized to the task sequencing and workflow problems that were identified. Combining this knowledge with an understanding of the modeling tool (such as WinCrew or IMPRINT), a vision of possible solutions should be generated. The solution set should be a generalized notion or zone of solutions that fit within the constraints of the project scope. The specific solutions will come later as a result of the use of the tools.

The workload analyst then builds the model to a scope somewhat larger than the focused activity would dictate. This is to assure that other elements not identified in the planning phase will be captured. It can be anticipated that some items will emerge from product development, operational scheme changes, and a variety of other areas due to project immaturity at the time of planning.

As the workload modeler builds and executes the model, the tasks and the respective interface issues should be identified and recommendations documented. Recommendations for mitigation should be accompanied with quantification of the problem and estimates of the likely improvement potential. Subsequent runs with postulated notional design improvements are essential parts of the process and should be preplanned in the original project scope. Also, tasks that cannot be fully modeled or solutions that are dependent on accurate representation of the target audience should be identified and forwarded to the manned simulation activity phase. This prescreens the problem set that will be simulated. The desired result is for analysts to address the maximum number of the issues in this phase where costs and schedules are most favorable.

### 13.5.3 Second Phase: Anthropometric Modeling

This activity is focused on achieving the maximum results from the modeling tool (perhaps using one of the graphical human figure models mentioned earlier) while expending the minimum resources. During task analysis, those tasks and sequences of activities that are relevant to anthropometry were identified and prioritized. For example, driving tasks may have been allocated to use of foot pedals in a driver's station. This is in contrast to workstations that have no foot-operated controls or have passenger-type seating space. Since the documented results from the task analysis phase will specify requirements for the anthropometric phase to focus on areas most likely to give interesting results, it will steer the anthropometrist toward detailed evaluation of foot space when foot controls are part of the workstation. In contrast, it will also prevent wasted efforts for detailed foot space evaluations in passive and passenger applications where a static foot does not contribute to the task. This is a simple example, but it is important that use of each evaluation tool and method be based on the task performed.

The result will be a tailored modeling activity that has high fidelity and attention to detail for those areas that are most sensitive to dimensional human accommodation and that can have an impact on the design process within the project budget and schedule constraints. See Chaffin (2001) for anthropometric modeling case studies and Green (2000) for procedures to follow in conducting a human model–based analysis.

### 13.5.4 Third Phase: Human-in-the-Loop Simulation

Based on the results of the task analysis, an update is made to the simulation plan during the onset of this phase. Only those areas that are most conducive to the benefits of simulation will be included in the final simulation plan.

The tasks that were allocated to simulation in earlier phases should be reviewed. Each task that requires issue resolution or design recommendation should be reexamined for alternative processes and reallocated back if appropriate. Once the list is set, the tasks to be simulated will drive the level and scope of the simulator fidelity.

Preexisting simulation facilities may not be designed for human engineering activities. Even cases where there is an available simulator with an operator's station represented, it may in fact be inadequate for some levels of scientific human experimentation. An analyst experienced in experimentation and simulation should evaluate the simulator with an eye toward the ability to collect and record errors and response times as well as log activities performed. The facilities should be evaluated for a variety of baseline capabilities and also for capabilities that can be added for the project at hand.

In addition to dimensional and feature inclusion, aspects such as image generator latency and simulator reliability should be considered. Image generator latency can lead to confusing results. The tasks in the plan should be evaluated for response time sensitivity, if latency is anticipated. Also, unplanned downtime during an experiment is very costly and will affect the schedule. Long simulation sequences with combined tasks will be much more vulnerable to simulator lockups than short snippets. Preplanning based on simulator performance and reliability is essential to the success of this phase.

Since the maximum number of issues were allocated to the other dimensions in the earlier phases, these experiments can now be short and of limited scope and fidelity. This contributes to optimal solutions.

### 13.5.5 Final Phase: Design Recommendation and Documentation

The last developmental phase is a documentation and recommendation summary. Each phase should result in some recommendations, but it is here that the issues are brought back together after being allocated to the appropriate phase. In many cases, issues can transcend more than one dimension. It is at this point where the interaction and combining of results must take place. This summary provides the glue that holds together a comprehensive and integrated solution set and body of recommendations.

The final phase and the alpha phase are the only phases requiring a specific sequence. Final must be last, and alpha must be first. The other dimensions can be conducted concurrently or in a tailored order to accommodate the specific needs of the project. This approach further supports the concept of schedule compression and cost minimization.

In summary, these approaches have a planning foundation on which each module is built and modified. Using these elements together and creating synergy between them will yield the maximum benefit at the minimum expenditure of resources.

### 13.5.6 Case Study on Importance of Method Sequence

During the development of an operator station for a loader backhoe, a sequential finding resulted in saving more than $30,000 in retooling costs. In the initial anthropometric analysis, before sequential analysis was performed, a leg space dimension was derived based on the largest operator in the target audience (95th percentile U.S. male, 1988). This operator station requires that the operator swivel the seat 180° from facing forward for the loader to facing rearward for the backhoe. When this leg space dimension was swept front to rear to allow the transition, an arc was described on the right side of the operator's station. This resulted in the design of a single concavity into the tooled console to accommodate knee clearance.

The problem with this arc was that it was based on an implied sequential assumption that was inaccurate. The assumption made was that the large operator would swivel the seat from the rearmost position on the track in both directions. Sequential analysis indicated that operators set the seat first in order to operate and then transition. This discovery led to the requirement for two arcs required in the console. The resulting design solution was improved to allow two different swivel positions on the seat track for the 95th percentile operator, one from each operator facing position as a starting point.

Corrections to the design before the tooling was released saved tooling costs. It also illustrates the critical link between sequence task modeling and anthropometric analysis, which is usually considered a static construct. This saving represented a total return of the cost of the human engineering activity for this project.

## 13.6 COMMON ERRORS IN PERFORMING HFE

Although not intended to be a comprehensive list, the following are common pitfalls for some of the more popular technology developments.

1. *Human Factors Is Just Common Sense* If this were true, then no one would have problems figuring out how to program a VCR or have difficulty reaching an ATM from their car. Norman (1988) shows just how prevalent poor design is. In reality, without training in a user-centered approach to design, engineers would not know that they should factor in parameters from physiology, psychology, anthropology, and other studies of humans when designing new technology. A related common problem is that many engineers and designers assume that just because they are human and know how to design and are logical, they are qualified to do human engineering. The misconception here is that there is no special knowledge of data or methods required to perform human engineering. We have shown in this chapter that that is not true. There is tremendous variation in the behaviors, expectations, and physical and mental capabilities of people. People and products are often part of much larger, complex systems. Consideration of all of these factors requires knowledge of the data that characterizes people and specific methods for making use of it. Above all, it is important to avoid the trap of assuming that because the engineer, designer, or developer is part of a target audience they can anticipate the concerns of the entire target audience. It is best to focus development decisions on data, findings, and analysis rather than solely on the opinions of those who are experts in the systems. They may know much about how *they* perform in the system but may know little about how to accommodate the entire range of the target audience.

2. *Anthropometric Analysis and Workstation Design* The phrase "accommodate the 5th to 95th percentile soldier" often appears in military specifications. What is intended is to accommodate the central 90 percent of the target audience and not worry about the 10 percent of people who fall at the extremes of the population (e.g., smaller, taller, weaker, stronger). The problem is that not all body measurements are perfectly correlated so that using a figure with many dimensions sized to the 90th percentile actually represents an extreme much greater than 90 percent. Testing workspaces with figures sized by setting each body segment to a uniform "percentile" length is misleading and invalid (Bittner and Moroney, 1975; Meindl et al., 1993). This issue is, however, poorly understood in the engineering and design communities. Approaches that have been used to address this issue are principal component analysis, boundary mannequins, and Monte Carlo simulation (Robinette and McConville, 1981).

Another common error is to create and use an "average" figure sized to represent the 50th percentile user. Again, body dimensions are not perfectly correlated, so a person having sizes matching many 50th percentile dimensions is unlikely to exist. While it is possible that an "average-sized" figure may be useful for some applications such as animation or work flow analysis, 5th, 50th, and 95th percentile figures are never correct and the nomenclature is misleading. Percentiles are meaningless unless they refer to a specific dimension.

Still, a common error is workstation design around static postures. Especially when using noninteractive human figure models or templates, engineers may position the human in one static posture (e.g., seated at a workstation) and optimize the location of components around that posture. The problem is that the workstation user may have to change position to reach for components or get in and out of the workstation (possibly even during an emergency such as a fire) but the workstation design makes it impossible to do so.

Figure 13.5 shows a driver trying to exit a vehicle crew station through a hatch. The crew station was designed for use with a night vision device. The night vision device is the object hanging down in front of the driver's face. When the driver is in a static seated driving position, the layout of the crew station is adequate. However, when the driver tries to exit the vehicle through the hatch and dynamics and motion come into play, the night vision device location becomes a serious obstacle. Exiting the vehicle through the hatch must be performed quickly in an emergency.

3. *Task Analysis* Not all task analyses are appropriate for all uses. The main reason for this is that task analyses may be performed to answer different types of questions. The resulting task list may differ in terms of detail, area of focus, etc., depending on what that analysis question was. So, a task list developed for use in developing a training program for a system may be useless for feeding a workload analysis of the same system. Another

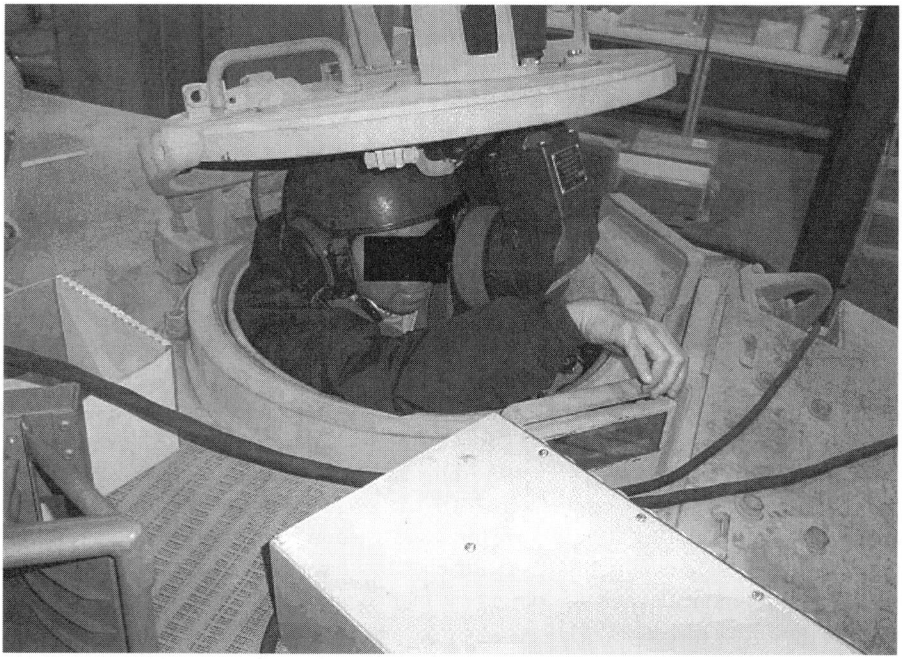

**Figure 13.5** Driver egress difficulty.

mistake is assuming that tasks must be listed serially. Many analysis methods that use task data as input can handle concurrent task performance. For some of them, such as workload analysis, this timing information is crucial. If task analysis data will be used to track and test compliance with system performance requirements, then the task must be clearly defined (clear definition of beginning and end points) and tied to relevant factors in design (e.g., what equipment is being used under what conditions) and measurement criteria (e.g., how fast, to what degree of accuracy, etc.)

4. *Use of Models* Care must be taken in applying models. By definition, a model is a simplification of the real world. The simplifications represent assumptions about which aspects of the real world are important. The assumptions on which a model is based should be identified and examined for compatibility with the target system and audience before the model is applied. For example, a human figure model developed for use by the U.S. military may use algorithms developed from a population of U.S. Army males. It is unlikely that this model will be equally valid for application to a product designed for Japanese females. This is one reason why it is meaningless to discuss the validity of a model outside the context of its intended application—just because a model has been "validated" does not mean that it is valid for use on any particular application.

Problems also arise when assumptions between modeling elements (within one modeling tool or when using multiple models) conflict. For example, the sizing data for a human model might be based on a military population but the feasibility of a lift analysis model might be derived from a civilian population. In some cases, this discrepancy may invalidate results. As stated in Section 3.5, models should not be applied without first determining the analysis questions that they will be used to answer. If these trade-off parameters and output measures were well thought out prior to modeling, issues related to the appropriateness of a model would be easier to resolve. For example, if we need to determine the feasibility of an arm lift for a military population but the model is based on a civilian population, we can check to see if the difference in lifting strength between the military and civilians is within an acceptable range for our analysis. This requires investigating the demographics of our target population (military) with the population used to derive the model (civilians) on the attribute for which we are interested (lifting strength of arm).

5. *Optimizing the Parts Instead of the Whole* When system design is broken down into subsystems and assigned to departments and disciplines, it is possible for total system requirements to be forgotten in favor of subsystem requirements. The result is a system with components that may work well when used separately but do not work at all when the system is used as intended. For example, during one phase of development of a new hatch and commander's weapon station for a main battle tank, the design of the vision blocks around the hatch was optimized for maximum viewing. This resulted in very tall blocks. Hatch operation was improved so that it allowed for easy adjustment into several positions, including one in which the hatch was partially open. A new machine gun mount just outside the hatch was also designed. Each individual component met design constraints but there was a system level requirement for the machine gun to be aimed when the commander was using the hatch in the partially open position. Because the view blocks were so tall and he could not get higher up due to the partially open hatch, the commander was unable to reach over the view blocks to the machine gun handles to aim at ground targets (see Figure 13.6).

6. *Misuse of the "User Jury"* A great deal of confusion seems to exist among human engineers regarding differences among user juries, experiments, and tests. The

**Figure 13.6** Commander difficulty reaching machine gun.

most important differences among them lie in their purpose and the method used to achieve that purpose. A properly run *user jury* or focus group can help elicit information about the appropriateness of design concepts or directions from potential users. From them, engineers can get a general sense of the factors important to users, the operating environment constraints, and the reaction to expect from specific design options. User juries are *not* appropriate for determining the best design from all possible choices because it is impossible to present all choices for consideration. Andre and Wickens (1995) found that subjective feedback from users might fail to predict design features that improve performance. *Experiments* are aimed at controlled proof of hypotheses using the scientific method. Proper experimentation can define relationships among design parameters and performance to help determine the best possible design choice. In contrast, the purpose of *tests* is to determine whether or not a given design meets the system requirements. The test is generally set up as pass or fail against a specific performance requirement threshold. Tests are best at evaluating whole-system performance against a realistic use scenario. An example of a test is whether or not I can fire an arrow without injuring my son into the apple on his head at 20 paces when the wind is blowing at 5 knots on a clear day using the new and improved crossbow that I just designed.

7. *Poor Experimental Methodology* When conducting usability analyses, user juries, and field studies, it is very easy to violate principles of good experimental methodology. This is a particular danger for engineers and physical scientists not trained in experimentation involving human participants. Human participants introduce numerous variables that are subtle and difficult to control in a field setting. Examples of common problems include too few subjects, not enough trials, poor control of motivation, order effects, and

experimenter-introduced bias. Typical sources of introduced bias are overtrained subjects, subjects with a personal stake in the outcome of the study (such as the designer), and subjects that have been coached. The worst candidates to conduct human system experiments are system designers because of the bias in favor of their own design. Martin (2000) is an excellent reference for readers unfamiliar with experimentation involving human participants. Weimer (1995) also covers experimental methods and discusses them in the context of human engineering application areas.

8. *Surveys* Questionnaires are very popular mechanisms for collecting subjective data, but a good questionnaire is difficult to find. The most common mistakes in developing questionnaires are

- Asking too many questions or "nice to know" questions (consider what you will do with the answers; how will they be analyzed?);
- Asking questions that are so vague that respondents are not sure what is being asked (e.g., how do you feel about the design?);
- Asking questions to which the answer can be misinterpreted (e.g., compound questions; did the respondent mean yes to both parts or just one part?); and
- Biased wording in questions (e.g., how much do you *like* the new design?).

9. *Focusing on System Operation and Ignoring Maintenance* When the system is analyzed, it is often the case that tasks required for maintenance actions are overlooked. Mental workload and time may not be an issue, but maintenance tasks may involve limited physical access to parts or openings, awkward postures, or heavy lifting. These tasks and analyses may have an effect on safety, error, and health hazards and should not be ignored. Biomechanical and anthropometric analyses are particularly useful investigating these issues. Problems with maintenance tasks may result in excessive costs (including manpower) to field a system, product liability, or unnecessary system downtime.

10. *Misunderstanding SA* Situation awareness refers to the user's level of awareness of his or her operational environment while performing a task or job. Designers sometimes think of SA as a static, one-dimensional aspect of the system they are designing, but SA levels are dynamic and can vary between individuals. As a concept, SA is context specific, so there is no one measure of SA that is appropriate all of the time and in all cases. Because of this, better measures of SA probe an operator's momentary awareness of a specific aspect of a specific parameter applicable to the context under which they are performing. For example, we might ask whether a pilot was aware of his altitude at a specific time or whether an automobile driver noticed a particular pedestrian crossing the street. To develop these questions that probe SA, it is often necessary to perform task and cognitive task analyses to understand the requirements of a job in context. Examples of context-specific SA metrics can be seen in the various versions of the situation awareness global assessment technique (SAGAT), e.g., air traffic control, air-to-air tactical, and commercial aircraft operation versions (Endsley and Garland, 1999).

## 13.7 BENEFITS OF MODELING FOR HFE

Modeling offers many benefits for the human engineer and for the success of system development projects. Proper implementation of an HFE modeling program should result in reduced resource expenditure, earlier identification and remediation of usability issues, and more readily accepted input from human factors engineers.

Most of the costs associated with an HFE modeling effort are incurred at the beginning of the effort, and the cost to produce derivatives of those models is often less. This is because once a model of a system, workstation, or product has been built, the data and labor needed to make modifications to that model to perform new analyses, answer new questions, or evaluate other design options are reduced greatly. This is especially true if future uses for the model are anticipated so that data reuse and increases in model fidelity are planned. For example, if a company uses modeling and simulation to develop a new system prototype on the speculation that it may be of interest to the DoD, the company will have those files to use as evidence of the soundness of its design in a proposal to the government. If the company wins a contract to develop the system, those same files can be modified and used to evaluate increasingly detailed designs as the system goes through the concept development, demonstration and validation, full-scale development, production, and product improvement phases. The cost of producing usable models is reduced for each phase because data and labor from previous designs have been leveraged and the corporate knowledge about those designs has been embedded in the models.

Human factors engineering modeling is also useful in evaluating the feasiblity of system performance requirements and is very effective in combating human engineering "requirements erosion." To better understand what is meant by requirements erosion, consider the following scenario. A design engineer asks for a lift constraint, (e.g., what is the heaviest I can make this part and still have a man lift it?), and an answer is given by the human factors engineer. The design engineer says she cannot meet the constraint, but how about going over by 5 pounds? A few months later, as weight constraints for the entire system get tighter, the engineer asks for a small 6 pounds on the lift constraint. The process is repeated and each time the engineer asks for a small compromise it may appear insignificant compared to the current limit, but when viewed against the original constraint, a small increase may be significant. The HFE specialists must remember to measure and present all requests for requirement, leniency against the original requirement, not against the previous number. Models can be used to document and illustrate the original constraint and each successive compromise.

Once the initial design phase begins, modeling benefits human factors engineers by allowing them to be proactive (rather than reactive) when identifying and resolving HFE design issues. If timed properly, use of predictive modeling results in identification of problems earlier in the design cycle before so many design constraints are set, thus limiting the options for problem resolution. In addition, human engineering modeling tools enable human factors engineers to participate more fully in concurrent engineering efforts. Instead of waiting for hard copies of design drawings or physical prototypes to evaluate for usability issues, human factors specialists can use modeling tools to evaluate computer-based designs as they are being developed by other engineering disciplines. Recommendations for modification can then be specified through changes to the same computer-based design files delivered to them rather than in written (and more easily ignored) text reports.

Models also benefit HFE by providing quantifiable results. Program managers use trade-off analyses that require numerical input and models help provide that numerical input. Even if the numbers output by the model are not accurate to the *n*th degree, these results help bound the problem and may be just as accurate as some cost or design parameter projections produced by other disciplines.

As stated earlier, in some cases, graphical output is more useful for acceptance of HFE recommendations than numerical or text reports. It is often difficult to convince other design and engineering disciplines that a problem is severe enough to warrant change. Sometimes it is difficult just to explain the problem. Graphical human engineering models illustrate these problems. Graphical human figure models are especially effective in showing inability to reach controls, fit in a workspace, or see displays and controls. Most people have no trouble putting themselves in the place of the human figure model and seeing the problem from the viewpoint of the model. Nonanthropomorphic graphical models such as task network models may not be as easy to relate to, but they are just as important to good design. Such models can help illustrate bottlenecks in task or information flow.

## 13.8 SUMMARY

Human factors engineering (HFE) and ergonomics are disciplines that focus on designing systems around users (i.e., user-centered design) and employing technology that acknowledges and complements human limitations and capabilities rather than forcing them to adapt to the technology. Well-designed, user-centered systems require relatively less training and aptitude to operate and maintain. They should also produce less errors when used. Human factors engineering is an engineering discipline that extends well beyond the application of common sense to design. Many HFE methods have been developed that center around understanding the thoughts and actions of the target audience of users. Several of the most commonly used methods are presented in this chapter. Tools (many computer based) have been developed to aid in application of HFE methods. Classes of HFE tools as well as factors to consider when deciding which are appropriate for a project are also described. The HFE tools are becoming increasingly integrated to more comprehensively represent the physical, mental, and behavioral aspects of human performance. Analysis templates and wizards are being incorporated into tools to aid in conducting HFE analysis and design. Despite the availability of adequate methods and an ever-improving set of tools, their application is not without problems. Ten common errors in application are identified to help program managers or those new to the discipline to avoid making these errors. Once the methods have been chosen and appropriate tools selected to apply the methods, it is useful to develop an HFE program plan to maximize use of HFE resources. A typical project flow is described to aid in development of an HFE plan. Finally, the chapter concludes with several points outlining the benefits of modeling to a successful HFE program.

## NOTES

1. Examples of these guidelines can be found at http://www.iso.org/iso/en/ISOOnline.frontpage http://risk.das.state.or.us/ergoguid.htm, and http://www.usyd.edu.au/su/ohs/ergonomics/welcome.html.
2. Human figure modeling software changes frequently. For Jack, see http://www.plmsolutions-eds.com/products/efactory/jack. For Safework®, see http://www.safework.com/safework_pro/sw_pro.html. For Ramsis, see http://www.hs.tecmath.de/english/ramsis_eng.shtml. For ManneQuin Pro, see http://www.nexgenergo.com/ergonomics/mqpro.html.

## REFERENCES

AGARD. (1998, December). *A Designer's Guide to Human Performance Modeling*, Advisory Report 356.) Neuilly-Sur-Seine, France: North Atlantic Treaty Organization.

American National Standards Institute. (1994). ANSI B 11 TRI Ergonomic guidelines for the design, installation and use of machine tools. American National Standards Institute, NY.

Andre, A. D., and Wickens, C. D. (1995, October). When Users Want What's Not Best for Them. *Ergonomics in Design*, pp. 10–14.

Bittner, A. C., Jr., and Moroney, W. F. (1975). The Accommodated Proportion of a Potential User Population: Compilation and Comparisons of Methods For Estimation. In *Proceedings of the18th Annual Meeting of the Human Factors Society. October 1974, Huntsville, Ala.* Santa Monica, CA: Human Factors Society.

Chaffin, D. B. (2001). *Digital Human Modeling for Vehicle and Workplace Design*. Warrendale, PA: SAE International.

Charlton, S. G. (1996). Questionnaire Techniques. In T. G. O'Brien and S. G. Charlton, (Eds.), *Handbook of Human Factors Testing and Evaluation*. Mahwah, NJ: Lawrence Erlbaum Associates.

Cheng, H., Rizer, A. L., and Obergefell, L. A. (1998). *Articulated Total Body Model Version V User's Manual*, Technical Report AFRL-HE-WP-TR-1998-0015. Wright-Patterson Air Force Base, OH: Air Force Research Laboratory. Available: http://www.atbmodel.com/pages/atbusrguide.pdf.

Clingan, J. N., and Akens, R. C. (1986). *Human Factors Evaluation Checklist for Tanks*, Technical Note 8-86. Aberdeen Proving Ground, MD: U.S. Army Human Engineering Laboratory.

*Cornell University Ergonomic Guidelines for Arranging a Computer Workstation*. (2001, November 26). Available: http://ergo.human.cornell.edu/ergoguide.html.

Damos, D. L. (Ed.). (1991). *Multiple-Task Performance*. London: Taylor & Francis.

*Directory of Databases part I—Whole Body Anthropometry Surveys*. (1996). SAE Aerospace Information Report SAEAIR5145. Warrendale, PA: Society of Automotive Engineers.

*Directory of Design Support Methods and Liveware Survey*. (2002, January 21). Available: http://dtica.dtic.mil/ddsm/hsi/index.html.

Endsley, M. R., and Garland, K. J. (1999). *Situation Awareness Analysis and Measurement*. Mahwah, NJ: Erlbaum.

Flanagan, C. (1954). The Critical Incident Technique. *Psychological Bulletin*, *51*, 327–386.

Gordon, C., Bradtmiller, B., Churchhill, T., Clauser, C., McConville, J., Tebbetts, I., and Walker, R. (1989). *1988 Anthropometry Survey of U.S. Army Personnel: Methods and Summary Statistics*, Technical Report Natick/TR-89/044. Natick, MA: U.S. Army Natick Research, Development and Engineering Center.

Green, R. F. (2000). A Generic Process for Human Model Analysis. SAE Document Number 2000-01-2167. In *Proceedings of the Digital Human Modeling Conference, June 2000, Dearborn, MI.* Warrendale, PA: SAE International.

Happee, R., Hoofman, M., van den Kroonenberg, A. J., Morsink, P., and Wismans, J. (1998). A Mathematical Human Body Model for Frontal and Rearward Seated Automotive Impact Loading. SAE Report Number 983150. In *Proceedings of the Forty-Second Stapp Car Crash Conference*. Warrendale, PA: SAE International.

Hart, S. G., Dahn, D., Atencio, A., and Dalal, K. M. (2001). Evaluation and Application of MIDAS v2.0. SAE Document Number 2001-01-2648. In *Proceedings of the Advances In Aviation Safety Conference & Exposition, September 2001, Seattle, WA.* Warrendale, PA: SAE International.

Hart, S. G., and Staveland, L. E. (1988). Development of NASA-TLX (Task Load Index): Results of Empirical and Theoretical Research. In P. A. Hancock and N. Meshkati (Eds.), *Human Mental Workload*. Amsterdam: North Holland.

Hendy, K. C. (1989). A Model for Human-Machine-Human Interaction in Workspace Layout Problems. *Human Factors*, *31*, 593–610.

Hix, D., and Hartson, R. H. (1993). *Developing User Interfaces: Ensuring Usability through Product & Process*. New York: Wiley.

Jones, R. E., Milton, J. L., and Fitts, P. M. (1949). *Eye Fixations of Aircraft Pilots: IV. Frequency, Duration and Sequence of Fixations during Routine Instrument Flight*, U.S. Technical Report 5975. Wright-Patterson Air Force Base, OH.

Kane, R. L., and Kay, G. G. (1992). Computerized Assessment in Neuropsychology: A Review of Tests and Test Batteries. *Neuropsychology Review*, *3*, 1–117.

Karworski, W. (Ed.). (2001). *International Encyclopedia of Ergonomics and Human Factors*, Vols. I–III. London: Taylor & Francis.

Kearney, D. S. (1998). *Ergonomics Made Easy: A Checklist Approach*. Rockville, MD: Government Institutes.

Kirwan, B. (1988). A Comparative Evaluation of Five Human Reliability Assessment Techniques. In B. A. Sayers (Ed.), *Human Factors and Decision Marking*, (pp. 87–104). Oxford: Elsevier.

Kirwan, B., and Ainsworth, L. K. (Eds.). (1992). *A Guide to Task Analysis*. London: Taylor & Francis.

Kroemer, K., Snook, S., Meadows, S., and Deutsch, S. (Eds.). (1988). *Ergonomic Models of Anthropometry, Human Biomechanics, and Operator-Equipment Interfaces: Proceedings of a Workshop*. Washington, DC: National Academy Press.

Jones, R. E., Milton, J. L., and Fitts, P. M. (1949). *Eye Fixations of Aircraft Pilots: IV. Frequency, Duration and Sequence of Fixations during Routine Instrument Flight*, U. S. Technical Report 5975. Wright-Patterson Air Force Base, OH.

Lockett, J., and Clingan, J. (1990). *Human Factors Assessment of Block II, M1 Improved Commander's Weapon Station (ICWS)*, TM 12-90. Aberdeen Proving Ground, MD: U.S. Army Human Engineering Laboratory.

Magliozzi, T., and Magliozzi, R. (2000). Blatant Ergonomic Blunders. In *Our humble Opinion: Car Talk's Click and Clack Rant and Rave* (pp. 42–47). New York: Perigee Books.

Martin, D. W. (2000). *Doing Psychology Experiments*, 5th ed. Pacific Grove, CA: Brooks/Cole.

Martin, E. A., Brett, B. E., and Hoagland, D. G. (1999). Tools for Including Realistic Representations of Operator Performance in DOD Constructive Simulations. In *Proceedings of the 1999 AIAA Modeling and Simulation Technologies Conference and Exhibit*.

Meindl, R. S., Zehner, G. F., and Hudson, J. A. (1993, March). *A Multivariate Anthropometric Method for Crew Station Design*, Technical Report AL-TR-93-0054. Wright-Patterson Air Force Base, OH: Crew Systems Directorate, Human Engineering Division, Armstrong Laboratory.

National Aeronautics and Space Administration. (1995). *Man–Systems Integration Standards (MSIS)*, NASA-STD-3000. Available: http://jsc-web-pub.jsc.nasa.gov/fpd/SHFB/Msis/msis_home.htm (retrieved October 28, 2001).

Niebel, B. W. (1993). *Motion and Time Study*. Burr Ridge, IL: Richard D. Irwin.

Norman, D. (1988). *The Psychology of Everyday Things*. New York: Basic Books.

Parasuraman, R., Mouloua, M., and Molloy, R. (1996). Effects of Adaptive Task Allocation on Monitoring of Automated Systems. *Human Factors, 38*, 665–679.

Pew, R. W., and Mavor, A. S. (Eds.). (1998). *Modeling Human and Organizational Behavior: Application to Military Simulations*. Washington, DC: National Academy Press.

Rasmussen, J., Pejtersen, A., and Goodstein, L. (1994). *Cognitive Engineering Concepts and Applications*. New York: Wiley.

Reid, G. B., and Nygren, T. E. (1988). The Subjective Workload Assessment Technique: A Scaling Procedure for Measuring Mental Workload. In P. A. Hancock and N. Meshkati (Eds.), *Human Mental Workload*. Amsterdam: North Holland.

Robinette, K., and McConville, J. (1981). *An Alternative to Percentile Models*, SAE Tech. Paper Series 810217. Warrendale, PA: Society of Automotive Engineers.

Salvendy, G. (Ed.). (1987). *Handbook of Human Factors*. New York: Wiley.

Sanders, M. S., and McCormick, E. J. (1993). *Human Factors in Engineering Design*. New York: McGraw-Hill.

Sanderson, P. M., Scott, J. J. P., Johnston, T., Mainzer, J., Watanabe, L. M., and James, J. M. (1994). MacSHAPA and the Enterprise of Exploratory Sequential Data Analysis (ESDA). *International Journal of Human-Computer Studies*, *41*(5), 633–681.

Sanderson, P., Eggleston, R., Skilton, W., and Cameron, S. (1999). Operationalising Cognitive Work Analysis with the Work Domain Analysis Workbench. In *Proceedings of the 43rd Annual Meeting of the Human Factors and Ergonomics Society. Houston, TX, 27 September–1 October*. Santa Monica, CA: Human Factors and Ergonomics Society.

Taylor, R. M., Vidulich, M. A., and Haas, M. W. (1997). CC-SART: Validation and Development. In *Proceedings of the Workshop on the Validation of Measurements, Models and Theories*, Report No. TTCP/HUM/97/006 (pp. 29–35). Washington, DC: Technical Cooperation Programme.

Turnage, J. J., and Kennedy, R. S. (1992). The Development and Use of a Computerized Human Performance Test Battery for Repeated-Measures Applications. *Human Performance, 5*(4), 265–301.

U.S. Department of Defense (1997, February). MIL-STD-1474D *Noise Limits* Washington, DC: U.S. Department of Defense.

U.S. Department of Defense. (1998, March). *Human Engineering Design Guidelines*, MIL-HDBK-759C. Washington, DC: U.S. Department of Defense.

U.S. Department of Defense. (1999a, May). *Human Engineering Program Process and Procedures*, MIL-HDBK-46855A. Washington, DC: U.S. Department of Defense.

U.S. Department of Defense. (1999b, August). *Human Engineering,* MIL-STD-1472F. Washington, DC: U.S. Department of Defense.

Wagner, D., Birt, J., Snyder, M., and Duncanson, J. (1996). *The Human Factors Design Guide for Acquisition of Commercial Off-the-Shelf Subsystems, Non-Developmental Items, and Developmental Systems*. Atlantic City, NJ: FAA Technical Center.

Waters, T. R., Putz-Anderson, V., and Garg, A. (1994). *Applications Manual for the Revised NIOSH Lifting Equation*. Available: http://aepo-xdv-www.epo.cdc.gov/wonder/prevguid/p0000427/p0000427.asp (retrieved January 25, 2002).

Weimer, J. (Ed.). (1995). *Research Techniques in Human Engineering*. Englewood Cliffs, NJ: Prentice-Hall.

Wickens, C. D., Gordon, S. E., and Liu, Y. (1997). *An Introduction to Human Factors Engineering*. New York: Addison-Wesley.

Wierwille, W., and Casali, J. (1983). A validated rating scale for global mental workload measurement application. In *Proceedings of the Human Factors Society 27th Annual Meeting* (pp. 129–133). Santa Monica, CA.

Wilson, J. R., and Corlett, N. E. (Eds.). (1995). *Evaluation of Human Work: A Practical Ergonomics Methodology*, 2nd ed. London: Taylor & Francis.

CHAPTER 14

# System Safety Principles and Methods

DONALD W. SWALLOM, ROBERT M. LINDBERG and TONYA L. SMITH-JACKSON

## 14.1 INTRODUCTION

System safety is perhaps the most familiar of all the human systems integration (HSI) domains to the general population as the discipline that helps in the design of equipment to avoid accidents. System safety is the first (and sometimes the only) HSI domain to be involved whenever major accidents occur in defense, medicine, transportation, manufacturing, and energy. No one wishes people to die in accidents, but it is especially upsetting when the accidents are avoidable. System safety specializes in preventing accidents through helping design systems that are safe. System safety is both a philosophy and a practice that focuses on designing in ways to prevent accidents.

The philosophy behind system safety is based on the medical model of primary prevention (referred to as *loss prevention* in the safety arena), which means that the main emphasis is on the complete removal of hazards from the environment. System safety uses secondary prevention, or *loss control*, as a last resort if primary prevention is not feasible. In loss control, it is understood that the hazard cannot be removed from the work environment, but the system (including the human component) can be protected in such a way that exposure to hazards is less likely or the consequences of exposure are less severe. Loss control is a prevention approach in which it is relatively difficult to determine how much protection to apply.

System safety faces a continual problem in demonstrating how to increase system safety without decreasing system performance to unacceptable limits or making the system unaffordable. For example, heavier aircraft may provide greater crashworthiness than lighter ones, but greater weight can also decrease system performance and increase system cost in development and operation. Automation may remove the hazard from the environment, but complex and sometimes expensive technologies must become part of the system. Although not always the case, sometimes the cost of safety outweighs the accident prevention advantages.

*Handbook of Human Systems Integration*, Edited by Harold R. Booher.
ISBN 0-471-02053-2 

The objective of the system safety discipline is to achieve a minimal level of risk within the constraints of operational effectiveness, time, and cost. System safety practitioners apply system safety principles and methods to accomplish such activities as hazard identification, hazard elimination, and risk control in the systems engineering process.

System safety and safety engineering extend as far back as 2100 BC, the estimated date of the first safety engineering manual, the Code of Hammurabi (Deitz et al., 2002; Kohn et al., 1996). This ancient Babylonian code focused on ship design, construction, loss control, and even specified the behavior of ship personnel, particularly when goods or lives were lost at sea. In the year 1743, the European-born doctor, Ulrich Ellenborg, identified lung diseases among builders that were caused by asbestos and identified other toxic substances that undermined the health of mine workers [Kohn et al., 1996; Occupational Safety and Health Administration (OSHA) online document]. The National Safety Congress convened in 1912 to organize efforts to protect the safety of the public [Kohn et al., 1996; National Safety Council (NSC), 2002]. This group later became the National Safety Council. The U.S. military beginning in World War II has also contributed to the development of the system safety discipline and, particularly, has developed specific methods and practices relevant to risk assessment.

There are a number of other historical contributors to system safety as well as safety engineering. Major attention was directed toward the protection of workers with the ratification of the Occupational Safety and Health Act in 1970 (OSHAct). OSHAct requires employers to adhere to standards of health and safety and provides regulatory authority to OSHA. More importantly, the passage of OSHAct and the establishment of OSHA forced employers to organize efforts within industry to protect the health and safety of workers, and, consequently, companies began to understand the importance of system approaches. For example, OSHA not only provides specific regulations as they apply to such system components as scaffolding, confined spaces, and materials handling, but it also addresses training practices, accident investigation, and process safety, all of which require careful integration with existing subsystems to be effective and compliant.

Today, system safety concepts are practiced within a wide range of industries, including: military, transportation, mining, manufacturing, nuclear, automotive, chemical processes, construction, and health care. Both federal and international standards have been developed that require system safety programs and methods to meet the objectives of comprehensive loss prevention and loss control.

The system safety engineer's primary job is to determine *how* the system can fail and cause death, injury, occupational illness, damage to or loss of equipment or property, loss of data, or damage to the environment. Knowing a system's potential for harm leads to the system design question: What can be done to eliminate or reduce that potential for harm?

The purpose of this chapter is to provide an overview of the analytical aspects of the system safety domain by discussing exemplary models, methods, and processes used by the system safety engineer to help identify and mitigate the potential harm from accidents. Before covering the details of these analytical approaches, we first define a number of terms familiar to system safety personnel. They include:

- Key safety definitions
- System safety engineering and management
- Safety groups and plans

### 14.1.1 Key Safety Definitions

There are several definitions that are useful to our discussion of system safety. Described in Table 14.1, they include:

- Safety
- Accidents
- Mishaps
- System
- System safety
- Hazards
- Risk
- Mishap risk
- Hazard severity
- Hazard probability
- Exposure

**TABLE 14.1 Safety Definitions**

*Safety* is condition in which there is low probability that harm will occur. Safety shares that definition with "security." However, security tends to mean freedom from harm from hostile person or group. Safety is more concerned with forms of harm from nonpersonal sources. The harm one might experience includes death, injury, occupational illness, damage to or loss of equipment or property, loss of data, or damage to the environment.

An *accident* is undesirable event or a series of undesirable events that result in harm.

A *mishap* is an accident. Mishap is terminology frequently used in DoD but is seldom used in commercial system safety practice.

A *system* is collection of things that work together. Military Standard 882, which delineates the Department of Defense practice of system safety, defines a system to be "an integrated composite of people, products, and processes that provide a capability to satisfy a stated need or objective" (DoD, 2000).

*System safety* is "the application of engineering and management principles, criteria, and techniques to achieve acceptable mishap risk, within the constraints of operational effectiveness, time, and cost, throughout all phases of the system life cycle" (DoD, 2000).

*Hazards* are the conditions or events in a system that can result in harm.

*Risk* is likelihood and severity of a loss. Conditions that require risk management include those that create significant risk of "death, injury, acute/chronic illness, disability, and/or reduced job performance of personnel who produce, test, operate, maintain, support, or dispose of the system" (DoD, 2001).

*Mishap risk* is expression of the impact and possibility of a mishap in terms of potential mishap severity and probability of occurrence (DoD, 2000). In life-cycle terms, mishap risk is the expected cost of mishaps stemming from a particular hazard over the life of the system.

*Hazard probability* is likelihood that adverse consequences from a specific hazard will occur.

*Hazard severity* is assessment of consequences of hazard. It is amount of harm that could potentially occur in one mishap due to specific hazard. It is degree of injury, occupational illness, property damage, equipment damage, or lost data in that mishap.

*Exposure* is time interval over which hazard occurs. Increasing exposure interval changes the probability of occurrence—as exposure interval increases, so does probability of occurrence.

### 14.1.2 System Safety Engineering and Management

*System safety engineering* deals with the tools of the trade, the principles and methodology of analyzing the hazards of system components, subsystems, and interfaces. One popular definition provided by Malasky (1982) states that system safety is:

> an optimum degree of safety, established within the constraints of operational effectiveness, time, and cost, and other application interfaces to safety, that is achievable throughout all phases of the system life cycle. (p. 17)

System safety should also be viewed as a systematic process to identify, eliminate, and control hazards. Figure 14.1 illustrates the overall process and specifically identifies opportunity windows that support system modification.

*System safety management* (or risk management) deals with how the decisions are made based on the analysis done by the system safety engineers to eliminate or reduce the associated *mishap risk*. Other aspects of system safety management include defining and allocating the resources required for the safety effort and providing system safety interfaces with other system development efforts. Generally, system safety engineering and management provide decision makers with information to ensure mishap risk is evaluated in a reasoned and balanced way.

Department of Defense (DoD) regulations require the program manager to comply with environment, safety, and occupational health (ESOH) regulatory requirements, which in general is to prevent or avoid ESOH hazards, where possible, and manage those hazards where they cannot be avoided[1] (DoD, 2001).

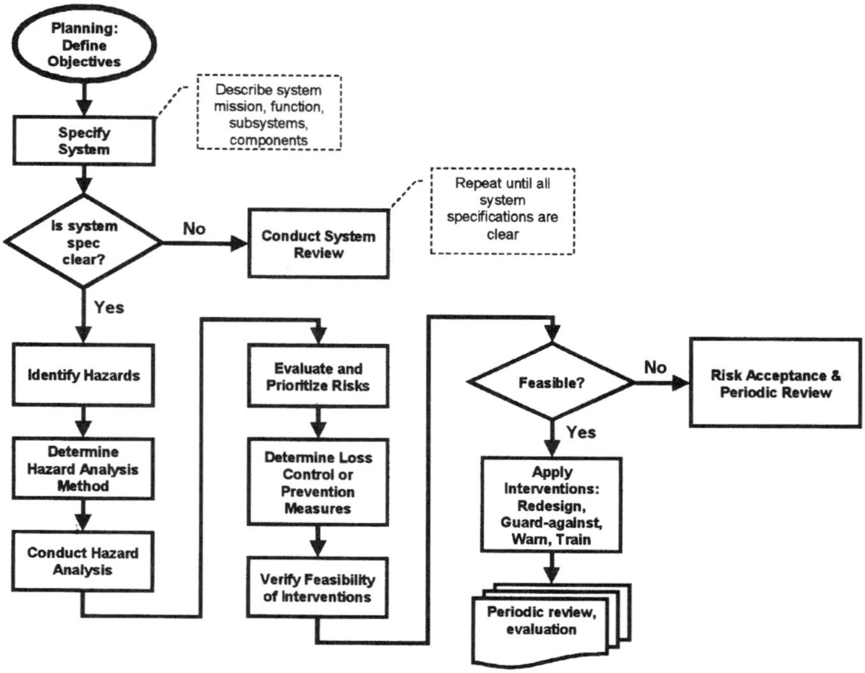


**Figure 14.1** Conceptual model of the system safety process.

OSHA addresses system safety management in a number of regulations. For example, 29 CFR 1910.333 specifically addresses robot system safety. The standard outlines work management practices for continuous attended operation, maintenance, and repair. OSHA's lock-out/tag-out standard (LOTO; 29 CFR 1910.147) defines an "authorized employee." Employers cannot make designations of "authorized" employees for LOTO that conflict with the definition given in the OSHA standard. Other government agencies such as the Federal Aviation Administration (FAA)[2] and the National Aeronautics and Space Administration (NASA) also require compliance with safety directives.[3]

### 14.1.3 Safety Groups and Plans

A *system safety group* (SSG), *system safety working group* (SSWG), or a *system safety integrated product team* (SSIPT) is a formally chartered group of persons representing organizations involved in a system's design, use, and management. Organizations that specifically practice total safety management (TSM), may integrate their safety groups into a quality circle (Goetsch, 2002). These teams are organized to assist the program manager in achieving acceptable mishap risk. The system safety plan spells out how each group functions, the processes that will be used to determine acceptance of mishap risks, and how to obtain additional resources to eliminate or reduce risk. The name of the group depends on the level of responsibility. For example, an SSG may function at one level of management while the SSWG may work at a lower level in the organization.

A *system safety program plan* (SSPP) is based on the principles of safety and any government or company systems safety policies. The SSPP lays out how the organization will reduce mishap risk to an acceptable level and still achieve the program objectives. The plan includes organizational resources, responsibilities and relationships, methods of accomplishment, milestones, depth of effort, and integration with other program activities and related systems. The plan also spells out how the system safety team functions. For DoD programs, a *system safety management plan* (SSMP) delineates how the government will manage system safety. For contractors, the SSPP outlines contractual responsibilities for system safety through the use of company methods and processes.

### 14.1.4 Chapter Outline

The chapter covers three major topics as reflected in the following sections:

- Risk assessment model
- System safety methods and techniques
- System safety process

## 14.2 RISK ASSESSMENT MODEL

Almost all accident event sequences can be traced back to a process failure between the system, environment, and human interfaces. In a majority of accident investigations, human error (operator, maintainer) is determined to be the root cause of the system mishap. However, even if there is an equipment failure, the question still remains whether someone failed to design, test, or produce the equipment correctly. This is critical for HSI because, carried to the extreme, almost all accident event sequences can be traced back to a human failure. Since people are not perfect, the system safety analysis process starts with the question, "What would a 'reasonable' worker know and do in the place of that person who failed?" Could the failure have been foreseen and a better decision made to prevent or minimize the mishap? But even reasonable people do not always foresee the results of their decisions. System safety, in that sense, is an effort to systematically provide knowledge to reasonable people so they can make the best decisions in the design of systems.

Although root causes of accidents are often attributed to human error somewhere within the system, it must be understood that the integration of system safety and HSI supports a slightly different view of "human error," as is found in other disciplines. In the case of the integrated system safety and HSI perspective, human error occurs because of design problems within the entire sociotechnical system, which includes such factors as training, management practices, machine design, human information processing, and even psychosocial stimuli such as stressors or culture. In addition, when investigations occur in environments that are applying a system safety/HSI approach, it is understood that accident causation can be explained by multiple factors that interact to produce hazardous situations.

### 14.2.1 Range of Outcomes

The first principle of the system safety model states that there is a range of possible outcomes from a hazard. One reason for the range of outcomes is that the hazard could manifest itself as a mishap at different times during the operation of the system. For example, if an aircraft or helicopter engine quits running while the vehicle is on the ground, the only damage is to the component and/or other engine parts. If the engine quits during cruise flight, engine debris may cause damage to the aircraft. If the engine fails during a critical phase of takeoff, the potential result is destruction of the aircraft and loss of life. A production line may introduce hazards anywhere from raw materials entry to export and waste disposal. Another reason for the range of outcomes is the interaction of more than one hazard in the mishap. If an electrical component fails and begins to arc, it may just burn up that component. If it arcs in conjunction with a fuel leak, it could result in a serious fire and loss of the whole system. Examples of the range of outcomes that could happen in a mishap are:

1. *Human Injury* Ranges from minor injury resulting in no days missed from work to death.
2. *Equipment Damage* Ranges from minor component damage requiring little repair to total system destruction.
3. *Environmental Damage* An example is a chemical spill that could range from a minor hazardous material spill requiring no reporting to a major environmental catastrophe.
4. *Health-Related Mishaps* These can range from short-term health impairment with 100 percent recovery to a lifetime health disability.
5. *Business-Related Mishap* These can range from loss of one computer file to the loss of an entire data storage site.

### 14.2.2 Risk Analysis

Risk analysis makes use of both quantitative and qualitative methods to assess risk. As with all forms of risk, one quantitative measure is dollars. Risk is the cost of mishaps stemming from a particular hazard over the life of the system. This is a very important concept and one that is often misunderstood. If the system is operated long enough with no changes, the risk will indeed be realized as an actual cost at some future point in time. Risk of a hazard has two components, *hazard severity* and *hazard probability.* Sometimes a third component, *exposure*, is identified (see Table 14.1 for definitions).

Since there is a range of possible consequences of a hazard, the likelihood of the worst consequence may be far less than those of lesser consequence. Figure 14.2 graphs this relationship in a \$20 million system. *As the severity of the hazard increases, the hazard probability decreases*. However, it should be noted that in real-life situations this paradigm is not always true.

Often the *risk curve* for a particular hazard follows the relationship where the severity ($S$) times the probability ($P$) is a constant ($S \times P = C$). From the Figure 14.2 one could ask: what is the probability of this hazard resulting in a loss of exactly \$10,000? That probability might be 0.00001 occurrences during a life cycle of the system. So the risk is \$10,000 per occurrence times 0.00001 occurrences per life cycle of the system equals \$1. This is done for every \$1 increase in mishap severity all the way up to a \$20 million mishap. When the risks are added for each level of severity mishap, the total risk from the hazard is identified. Mathematically, this is the area under the risk curve. If this system operates for the whole life cycle, the likelihood of having a mishap that is exactly \$10,000 is unlikely since the probability is 0.00001 or 1 in 10,000 life cycles. However, the total cost of the mishap should come close to the total calculated risk for the hazard. If 10 of the systems operate, the cost of the hazard per system will be closer to the calculated risk. If there are a thousand systems, it will be even closer. If there are an infinite number of systems, the cost per system will be the calculated risk.

There are several graphical representations used to illustrate risk levels and their relationship to probability and severity within different contexts. These same illustrations are used to aid in decision making regarding risk categorization. Figures 14.3 to 14.6 are examples. Figure 14.3 shows different levels of risk. Hazard 1 is high risk; hazards 2 and 3 are medium risk; and hazard 4 is low risk. Figure 14.4 depicts risk curves on logarithmic

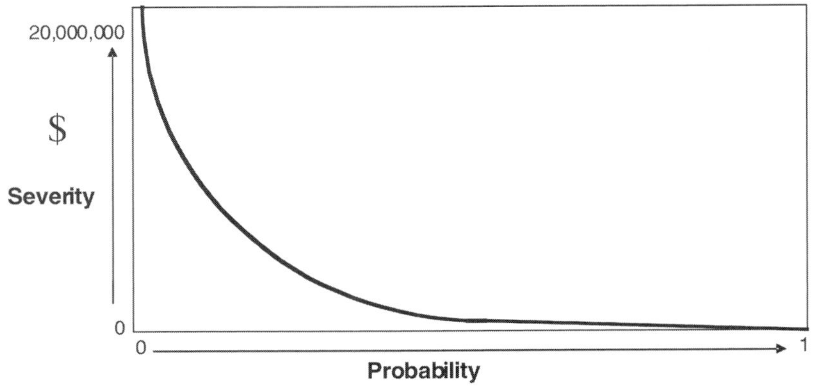


**Figure 14.2** Relationship of severity and probability.

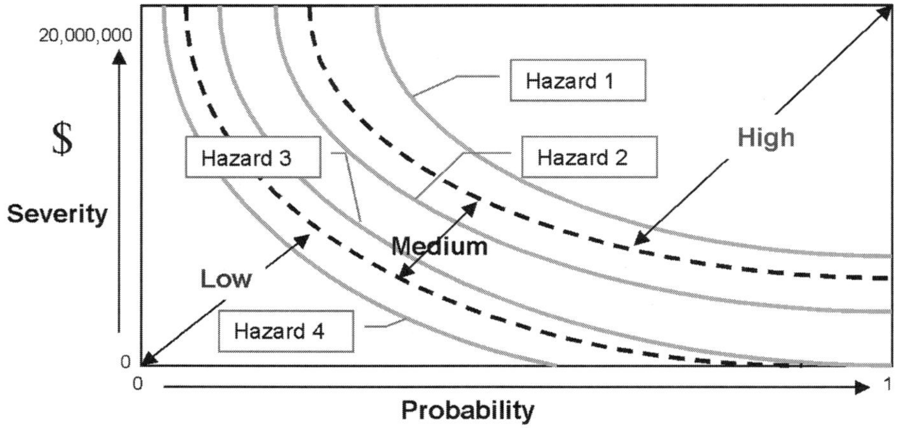


**Figure 14.3** Risk curves on linear scales. [Adapted with permission from Clemens and Simmons (1998, pp. II-3 to II-6).]

scales. Note the curves now are closer to being straight lines. Figure 14.5 shows the hazard assessment matrix developed by DoD in 1993 as part of the MIL-STD-882. The hazard assessment matrix is a hybrid that is both quantitative and qualitative. The approach is to use qualitative descriptors of the severity and likelihood of a hazard to assign a hazard risk index (HRI). The HRI is a quantitative descriptor of risk. The matrix can be simplified to something like that shown in Figure 14.6.

The above discussion illustrates the theory behind the risk matrix, but in practice a system safety engineer will take the worst credible consequence of a hazard to assign a risk assessment code from the matrix instead of dealing with the entire risk curve. The "worst credible consequence" is the most severe outcome of a hazard that can reasonably be expected to occur during the life cycle of a system. From Figure 14.6, the risk assessment code assigned could be 1A through 4F. The risk of the worst credible consequence then

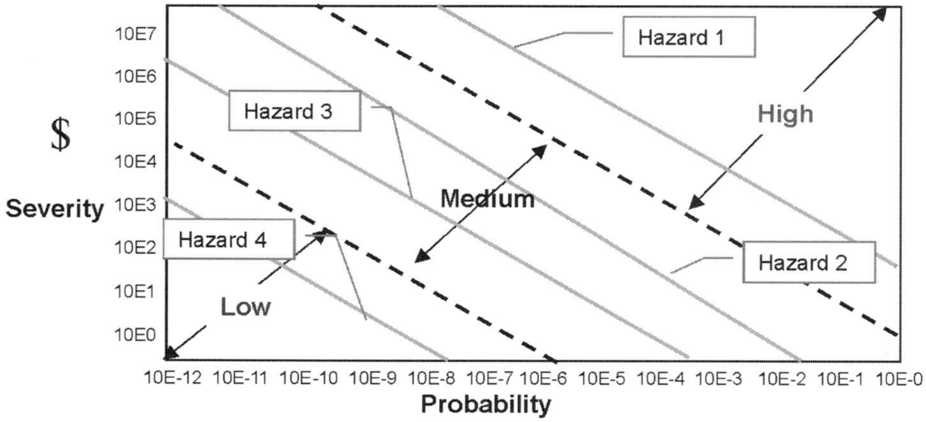


**Figure 14.4** Risk curves on log scales. [Adapted with permission from Clemens and Simmons (1998, pp. II-3 to II-6).]

Hazard Classification

Hazard Likelihood Categories

| FREQUENCY OF OCCURRENCE | | | Frequent | Likely | Occasional | Seldom | Unlikely |
|---|---|---|---|---|---|---|---|
| SEVERITY | Catastrophic | I | 1E | 1E | 2H | 2H | 3M |
| | Critical | II | 1E | 2H | 2H | 3M | 4L |
| | Marginal | III | 2H | 3M | 3M | 4L | 4L |
| | Negligible | IV | 3M | 3L | 4L | 4L | 4L |

Hazard Risk Index

**Abbreviations:** Extremely high (E), High (H), Moderate (M), and Low (L)
Hazard Risk Index Labels : 1 = Unacceptable, 2 = Undesirable with management waiver required, 3 = Acceptable with management review, 4 = Acceptable without review.

**Figure 14.5** Hazard assessment matrix with hazard risk indices (HRI) embedded. **Abbreviations:** Extremely high (E), high (H), moderate (M), and low (L). Hazard risk index labels: 1 = unacceptable, 2 = undesirable with management waiver required, 3 = acceptable with management review, 4 = acceptable without review. [Adapted from Kohn et al. (1996), p. 205; DoD (1993); Roland and Moriarty (1990), pp. 200, 204.]

represents the whole risk curve. In this manner, a system safety engineer can assign a risk code early in the program before there is a mature design or any substantial analysis. This early estimate of the risk helps allocate resources for further analysis and risk reduction.

Table 14.2 describes each probability level, ranging from "frequent" to "improbable," for a DoD aircraft program. These probability levels are also reflected in Figure 14.6.

Table 14.3 describes each severity level, ranging from 4, "negligible" to 1, "catastrophic" for a DoD aircraft program. These severity levels are also reflected in Figure 14.6.

Using these tables, a safety engineer can assign a risk assessment code that can be used to prioritize risk or hazard mitigation actions and determine if the risk is acceptable. The

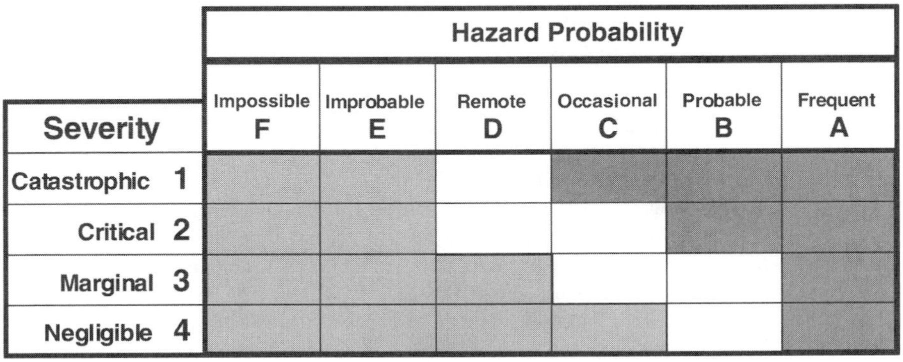


**Figure 14.6** Simplified risk matrix. [Adapted with permission from Clemens and Simmons (1998, pp. II-3 to II-6).]

**TABLE 14.2 Hazard Probability Levels**

| Level | Description | Specific Individual Aircraft | Fleet | Probability [Occurrences per Flight Hour ($p$)] |
|---|---|---|---|---|
| A | Frequent | Likely to occur often in life of aircraft | Continuously experienced | $p > 10^{-1}$ |
| B | Probable | Will occur several times in life of aircraft | Will occur frequently | $10^{-1} \geq p > 10^{-3}$ |
| C | Occasional | Likely to occur some time in life of aircraft | Will occur several times | $10^{-3} \geq p > 10^{-5}$ |
| D | Remote | Unlikely but possible to occur in life of aircraft | Unlikely but can reasonably be expected to occur | $10^{-5} \geq p > 10^{-7}$ |
| E | Improbable | So unlikely, it can be assumed occurrence may not be experienced | Unlikely to occur, but possible | $10^{-7} \geq p$ |
| F | Impossible | Cannot occur | Cannot occur | $10^{-9} \geq p$ |

*Source:* Adapted from *Air Force System Safety Handbook* (2000), p. 22.

assigned risk assessment code is likely to change as the program progresses. It may be that an analysis of the design shows that the initial risk assessment was too optimistic and a higher risk code should be assigned. If all goes well in the system safety effort and the redesign of the system, the hazard will be assigned a lower risk code that will be more acceptable to the risk acceptance authority.

*Total system risk* is the sum of the known and unknown risk of all system hazards. *Residual risk* is the risk that remains after all risk reduction efforts have been brought to

**TABLE 14.3 Hazard Severity Levels**

| Category | Description |
|---|---|
| 1 | Catastrophic: Death or permanent total disability; system loss or mishap damage greater then or equal to $1 million. |
| 2 | Critical: Severe injury or severe occupational illness (permanent partial disability); mishap damage greater than $200,000 but less then $1 million. |
| 3 | Marginal: Minor injury or minor occupational illness (no permanent effect); mishap damage greater than or equal to 20,000 but less than $200,000. |
| 4 | Negligible: Less than minor injury or occupational illness (no lost workdays); mishap damage less than $20,000. |

*Source:* Adapted from *Air Force System Safety Handbook* (2000, p. 22).

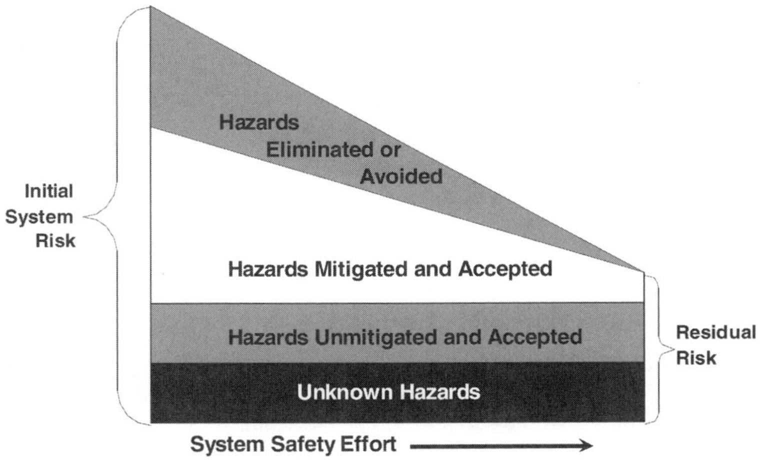


**Figure 14.7** Residual risk.

bear on a hazard. If a hazard is not eliminated, then some mishap risk still "resides" in the system. Figure 14.7 illustrates this concept. The initial system risk is the risk before the system safety effort began. The four components of risk are: hazards that are eliminated or avoided, hazards that are mitigated and then accepted, hazards unmitigated and accepted, and hazards that are never discovered.

## 14.3 SYSTEM SAFETY METHODS AND TECHNIQUES

There are a large number of methods and techniques available to help the system safety analyst. To aid analysis efforts, the System Safety Society (http://www.nm-esh.org/sss/ssshdbk.html) documents 101 safety analysis methods and techniques (Table 14.4) in the *System Safety Analysis Handbook* (Stephans and Talso, 1997). From this extensive tool kit, an experienced system safety engineer can select appropriate methods or techniques to identify and assess the risk of system hazards.

Although Table 14.4 identifies a large number of tools, they quickly reduce to a manageable number in actual systems application. Many of the methods and techniques are variations of one another or are methods specifically adapted to a particular type of system (nuclear, aircraft, facility, etc.) or type of hazard (human factors, explosive, electrical, fire, confined space, etc.). Some methods have been found to be less reliable than others and, consequently, are used less often. Further, many of these techniques are common to other HSI domains and are covered in the chapters for those domains (see, e.g., Chapter 13 for human factors analyses and Chapter 15 for health hazards assessment).

One caveat when conducting risk analyses is to first select the group that will conduct the analysis. No single individual should conduct a risk analysis because the quality of an analysis can be undermined by biases or oversights. Thus, a group analysis increases the chance that a comprehensive analysis is conducted.

**TABLE 14.4 Summary of System Safety Techniques and Methodologies**

| No. | Technique | Purpose | Application |
|---|---|---|---|
| 1 | Accident analysis | Evaluate accident scenarios | In conjunction with PHA or SSHA |
| 2 | Action error analysis | Analyze interactions between machines and humans | Human interface with automated or other processes |
| 3 | Barrier analysis | Analyze unwanted flow of hazardous energy | Systems analysis, occupational safety reviews, and accident analysis |
| 4 | Bent pin analysis (BPA) | Represent failures within cable connections | Electrical cable systems |
| 5 | Cable failure matrix (CFMA) | Represent failures within cable assemblies | Electrical cable assemblies; use with BPA |
| 6 | Cause–consequence analysis | Evaluate accident consequences | Similar to FTA or ETA |
| 7 | Change analysis | Examine potential effects of modification | All systems |
| 8 | Checklist analysis | Identify hazards using list of known deficiencies and accident situations | Evaluate compliance to standards |
| 9 | Chemical process Quantitative risk analysis (CPQRA) | Quantitative risk assessment within chemical process industry | Processes of all types |
| 10 | Common cause analysis | Identify common causes of accident sequences | All systems; extensively used in nuclear power industry |
| 11 | Comparison to criteria (CTC) | Structured format to guide compliance review | Any system designed to standards |
| 12 | Confined space safety | Systematic evaluation of spaces with limited egress | Implements OSHA requirements; supports PHA or SSHA |
| 13 | Contingency analysis | Prepare for emergencies by identifying potential accidents and measures to mitigate | All systems wherein advance preparation is needed |
| 14 | Control rating code (CRC) method | Produce safety effectiveness ratings | Systems, facilities, and equipment |
| 15 | Critical incident technique | Use historical information to identify and ameliorate hazards | Any system with human operators |
| 16 | Criticality analysis | Rank potential failure modes | Used with FMEA |
| 17 | Critical path analysis | Network modeling and analysis | Control and monitor complex safety management efforts |
| 18 | Cryogenic systems safety analysis | Specifically examine cryogenic systems | Use with PHA or SSHA |

**TABLE 14.4 (*Continued*)**

| No. | Technique | Purpose | Application |
|---|---|---|---|
| 19 | Damage mode and effects analysis | Provide early criteria for damage or vulnerability assessment | Uses results of FMEA |
| 20 | Deactivation safety analysis | Identify significant safety and health concerns integral to facility deactivation process | Facilities clean up and deactivation |
| 21 | Digraph utilization within system safety | Model failure effect scenarios | Model complex systems similar to FTA |
| 22 | Electromagnetic compatibility (EMC) analysis and testing | Prevent EM interference and protect form EMR | Any system requiring electrical circuit protection |
| 23 | Energy analysis | Evaluate safety through "energetics" | Any system that contains, stores, or uses energy in any form |
| 24 | Energy trace and barrier analysis (ETBA) for hazard discovery and analysis | Safety analysis through meticulous tracing of energy flow | Any system; often used with MORT and STEP |
| 25 | Energy trace checklist | Evaluate safety through "energetics" and lists of known energy hazards | Used with PHA or SSHA; defines system hazards |
| 26 | Environmental risk analysis | Assess risk of environmental noncompliance | Any system that produces potentially toxic or hazardous materials |
| 27 | Event and causal factor charting | Reconstruct accident event and determine root cause(s) | Any accident or mishap |
| 28 | Event tree analysis (ETA) | Organize, characterize, and quantify potential accidents | All systems wherein unwanted events can be anticipated |
| 29 | Explosive safety analysis | Evaluate potential effects of hazards involving handling, storing, and working explosives | Any situation involving gram to ton quantities of explosives |
| 30 | External events analysis | Focus attention on events outside the system under examination | In conjunction with PSAR or FSAR |
| 31 | Facilities system safety analysis | Apply system safety to a facility and its operation | Used to comply with OSHA 1910.119 |
| 32 | Failure modes and effects analysis (FMEA) | Determine and evaluate effects of subcomponent failures on system | Any system, subsystem, component, procedure, interface, etc. |

(*continued*)

**TABLE 14.4 (*Continued*)**

| No. | Technique | Purpose | Application |
|---|---|---|---|
| 33 | Failure modes, effects, and criticality analysis (FMECA) | Tabulate all system failure modes | Essentially a reliability tool |
| 34 | Fault hazard analysis | Systematically examine a system or facility using inductive analysis | Any system, subsystem, component, procedure, interface, etc. |
| 35 | Fault isolation methodology | Safety analysis of computer-controlled unmanned systems | Large electromechanical hardware/software systems |
| 36 | Fault tree analysis (FTA) | Postulate undesirable end event and examine contributing events | All systems wherein undesirable end events can be foreseen |
| 37 | Fire hazards analysis | Examine fire hazards using system safety techniques | Any system with fire safety concerns |
| 38 | Flow analysis | Evaluate effects of flow of fluids or energy | All systems that transport or control flow of fluids or energy. |
| 39 | Hazard analysis | Application of quantitative methods to solve safety problems | A generic technique that can be applied to chemical processes and similar systems. |
| 40 | Hazard and operability study (HAZOP) | Group review using structured brainstorming | Began with chemical industry. Any process or product using brainstorming |
| 41 | Hazard mode effects analysis | Introductory technique to determine if further safety analysis is necessary | Any project with safety concerns |
| 42 | Hardware/software safety analysis | Integrated hardware/software safety analysis | Used with PHA or SSHA |
| 43 | Health hazard assessment (HHA) | Detailed review of hazardous materials | Any system |
| 44 | Human error analysis | Evaluate any system where human error is of concern | Any system with human interfaces |
| 45 | Human factors analysis | Evaluate functions, tasks, resources among humans and machines | Any system with active human involvement |
| 46 | Human reliability analysis (HRA) | Assess factors of human reliability | Any system with active human involvement |
| 47 | Interface analysis | Identify potential hazards occurring due to interface incompatibilities | All systems with subsystems or components |
| 48 | Job safety analysis | Assess efficient and safe ways of task performance | Human operator functions |

**TABLE 14.4 (*Continued*)**

| No. | Technique | Purpose | Application |
|---|---|---|---|
| 49 | Laser safety analysis | Assess hazards associated with nonionizing radiation | All laser operations |
| 50 | Management oversight and risk tree (MORT) analysis | Analyze system to determine detailed information | All systems and processes |
| 51 | Materials compatibility analysis | Analyze physical degradation due to materials incompatibility | Aerospace, military, nuclear, marine, and chemical systems and processes |
| 52 | Maximum credible accident/worst case | Determine upper bounds of potential accident environment | All systems |
| 53 | Modeling | Create visual representation of complex safety program or process | Large and complex safety programs wherein a review tool is desirable |
| 54 | Naked man | Evaluate basic system to determine need for controls | Any system; particularly applicable to confined spaces |
| 55 | Network logic analysis | Examine a system in terms of Boolean mathematical representation | All systems that can be represented in bimodal elemental form |
| 56 | Nuclear criticality analysis | Ensure nuclear safety by eliminating possibility of a nuclear reaction | All facilities that handle fissile material |
| 57 | Nuclear explosives process hazard analysis | Identify high consequence (nuclear) activities to reduce possibility of nuclear explosive accident | Nuclear or similar high-risk activities |
| 58 | Nuclear safety analysis | Implement safety analysis requirements for nuclear facilities | All nuclear facilities and operations; DOE and NRC have rigid requirements |
| 59 | Nuclear safety cross-check analysis | Verifies software designs associated with nuclear systems | At present applies to military nuclear weapon systems |
| 60 | Operating and support hazard analysis | Identify and evaluate hazards associated with system operation | Operational phase of the systems acquisition cycle |
| 61 | Operational readiness review | Demonstrate the safety of startup or restart of a nuclear facility | DOE requirement; systematic approach to any complex facility |
| 62 | Petri net analysis | Model system components at an abstract level | Software control systems |
| 63 | Preliminary hazard analysis (PHA) | Initial analysis at early stages of system design | All systems |

*(continued)*

**TABLE 14.4 (*Continued*)**

| No. | Technique | Purpose | Application |
|---|---|---|---|
| 64 | Preliminary hazard list | List hazards at early stages of system design for management | Used with PHA or SSHA |
| 65 | Probabilistic hybrid analytical system evaluation | Describes the potential for failure and help in weighing cost–benefit analysis | Modeling where inputs lack precise definition of have dependence |
| 66 | Probabilistic risk assessment (PRA) | Quantified analysis of low probability, high severity events | Initially nuclear power industry, now any system with catastrophic accident potential |
| 67 | Procedure analysis | Step-by-step review of operational tasks | Systems involving human operators |
| 68 | Process hazard analysis | Management of highly hazardous chemicals | Requirement of 29 CFR 1910.119 for chemical process industry |
| 69 | Production system hazard analysis | Identify hazards associated with the manufacturing process | Transition from development engineering to production process |
| 70 | Prototype development | Modeling or simulation analysis of preproduction product | All manufacturing systems |
| 71 | Radiological hazard safety analysis | Structured approach to characterization and categorization of radiological hazards | Broadly applicable to all facilities engaged in managing radioactive materials |
| 72 | Relative ranking | Rank hazardous attributes (risk) of process | Any system wherein a ranking approach exists or can be constructed |
| 73 | Repetitive failure analysis | Model recurring events that prevent system from performing its function | Currently used in nuclear industry; potential for transfer to other fields |
| 74 | Risk-based decision analysis | Efficient approach for making rational and defensible decisions in complex situations | Applies to a wide spectrum of safety and economic analysis |
| 75 | Root cause analysis | Identify causal factors relating to a mishap or near- miss incident | Any system; widely used in aerospace and nuclear industries |
| 76 | Safety review | Generic assessment process | Any existing system |

**TABLE 14.4** (*Continued*)

| No. | Technique | Purpose | Application |
|---|---|---|---|
| 77 | Scenario analysis | Evaluation by postulating accident scenarios | All systems, particularly novel systems where there is little experience |
| 78 | Seismic analysis | Ensure structures and equipment resist failure in seismic event | Physical structures and equipment |
| 79 | Sequentially timed events plot (STEP) | Define and assess systems (accident analysis) | Any system that can be modeled |
| 80 | Single-point failure analysis | Identify those failures that would result in catastrophic events | Hardware and software systems and formalized human operator procedures |
| 81 | Sneak-circuit analysis | Identify unintended paths or sequences | Control and energy-delivery circuits of all kinds |
| 82 | Software failure modes and effects analysis (SFMEA) | Identify software-related design deficiencies | Any software process |
| 83 | Software fault tree analysis | Identify root causes(s) of undesired software events | Predominantly, software controlled hardware systems |
| 84 | Software hazard analysis | Eliminate software hazards during development process | All software development processes |
| 85 | Software sneak circuit analysis (SSCA) | Identify program logic that could cause undesired events | All software programs |
| 86 | Statistical process control | Understand and control variations in process | Any process where sufficient data can be obtained |
| 87 | Structural safety analysis | Validate mechanical structures | Any physical entity with a structural design |
| 88 | Subsystem hazard analysis | Identify hazards as a result of subsystem design | Any component (or group of components) at the less-than-system level |
| 89 | System hazard analysis (SHA) | Concatenate the results of SSHA | Any complex program |
| 90 | Systemic inspection | Review or audit a process or facility | Virtually without limit |
| 91 | Systematic occupational safety analysis | Evaluate facility from an OSHA standpoint | Any operation with personnel involved |
| 92 | Task analysis | Safety analysis of operation on task-by-task basis | Any operation with personnel involved |
| 93 | Technique for human error prediction (THERP) | Provide quantitative measure of human operator error | Any operation with personnel involved |

(*continued*)

**TABLE 14.4 (*Continued*)**

| No. | Technique | Purpose | Application |
|---|---|---|---|
| 94 | Test safety analysis (TSA) | Ensure safe environment during systems and prototype testing | Any test program |
| 95 | Threat hazard analysis | Evaluate potential threats (enemy) and self-induced (accident) throughout life cycle | Weapons systems; mandatory requirement of Mil-STD-2105B |
| 96 | Time/loss analysis (T/LA) for emergency response evaluation | Evaluate loss outcomes resulting from mishaps | Emergencies of all types |
| 97 | Uncertainty analysis | Identify the incertitude of result based on confidence levels | Any quantified safety analysis |
| 98 | Walkthrough task analysis | Determine and correct direct/root causes of unplanned occurrences | Any operation or process |
| 99 | What-if analysis | Identify hazards through a brainstorming approach | Any operation or system |
| 100 | What-if/checklist analysis | Logical identification of hazards combing two techniques | Any system |
| 101 | Wind/tornado analysis | Analysis of hazards resulting from all types of winds | All structures and buildings |

*Source:* Stephans, R. and Talso, W., (Eds) *System Safety Analysis Handbook*, 1997, pp. 3–4 to 3–7. Reproduced with permission System Safety Society.

A detailed discussion of all of these techniques is beyond the scope of this chapter. However, a few techniques that are unique to the system safety HSI domain are described below. Items discussed are:

1. Preliminary hazard analysis
2. Event tree analysis
3. Fault tree analysis
4. Failure mode and effects analysis
5. Fault hazard analysis
6. Subsystem hazard analysis
7. System hazard analysis
8. Cause–consequence analysis

### 14.3.1 Preliminary Hazard Analysis

The preliminary hazard analysis (PHA) activity is a safety engineering and software safety engineering function performed to identify the system hazards and their preliminary causal factors during system development. The hazards are formally documented to include information regarding the description of the hazard, causal factors, the effects of the hazard, and preliminary design considerations for hazard control by mitigating each cause. This analysis is preliminary and is used to provide early design considerations that may or may not become design requirements. The PHA activity can be used even before the system has been physically designed. For example, during the conceptual design phase of a system (when no prototypes or mockups exist), a PHA can be conducted using a team of safety personnel associated with the design of that system. Performing the analysis includes assessing hazardous components, safety-related interfaces between subsystems, environmental constraints, operation, test and support activities, emergency procedures, test and support facilities, and safety-related equipment and safeguards. The hazard analysis can start with a listing of hazards and a simple worksheet analysis, or can be conducted by using a series of "what if" scenarios. Figure 14.8 shows a sample Preliminary Hazard List and Preliminary Hazard Analysis Worksheet (U.S. Army, 1990). The actual hazard analysis process can become quite involved. Figure 14.9 outlines an example process flow for conducting PHAs (Clemens and Simmons, 1998, p. III-6).

The PHA becomes the springboard documentation to launch the subsystem hazard analysis (SSHA) and system hazard analysis (SHA) analyses as the design matures and progresses through the development life cycle. Preliminary hazards can be eliminated (or officially closed through the SSWG if they are deemed to be inappropriate for the design. For more comprehensive information readers should refer to texts devoted solely to system safety techniques and methods such as Li (1999), Clemens and Simmons (1998), Alberico et al. (1999), and Stephans and Talso (1997).

### 14.3.2 Event Tree Analysis

The event tree analysis (ETA) is an analytical tool that can be used to organize, characterize, and quantify potential accidents in a methodical manner. An event tree models the sequence of events that results from a single initiating event. The ETA concept uses forward logic; in other words, events are graphed from an initiating event (starting

*Preliminary Hazard List*

| Part | Hazard | Cause | Effects | Hazard Category | Comments |
|---|---|---|---|---|---|
| | | | | | |

*Preliminary Hazard Analysis Worksheet*

| Part | Hazard | Cause | Effects | Hazard Category | Corrective or Preventive Action |
|---|---|---|---|---|---|
| | | | | | |

**Figure 14.8** Sample preliminary hazard list and preliminary hazard analysis worksheet.

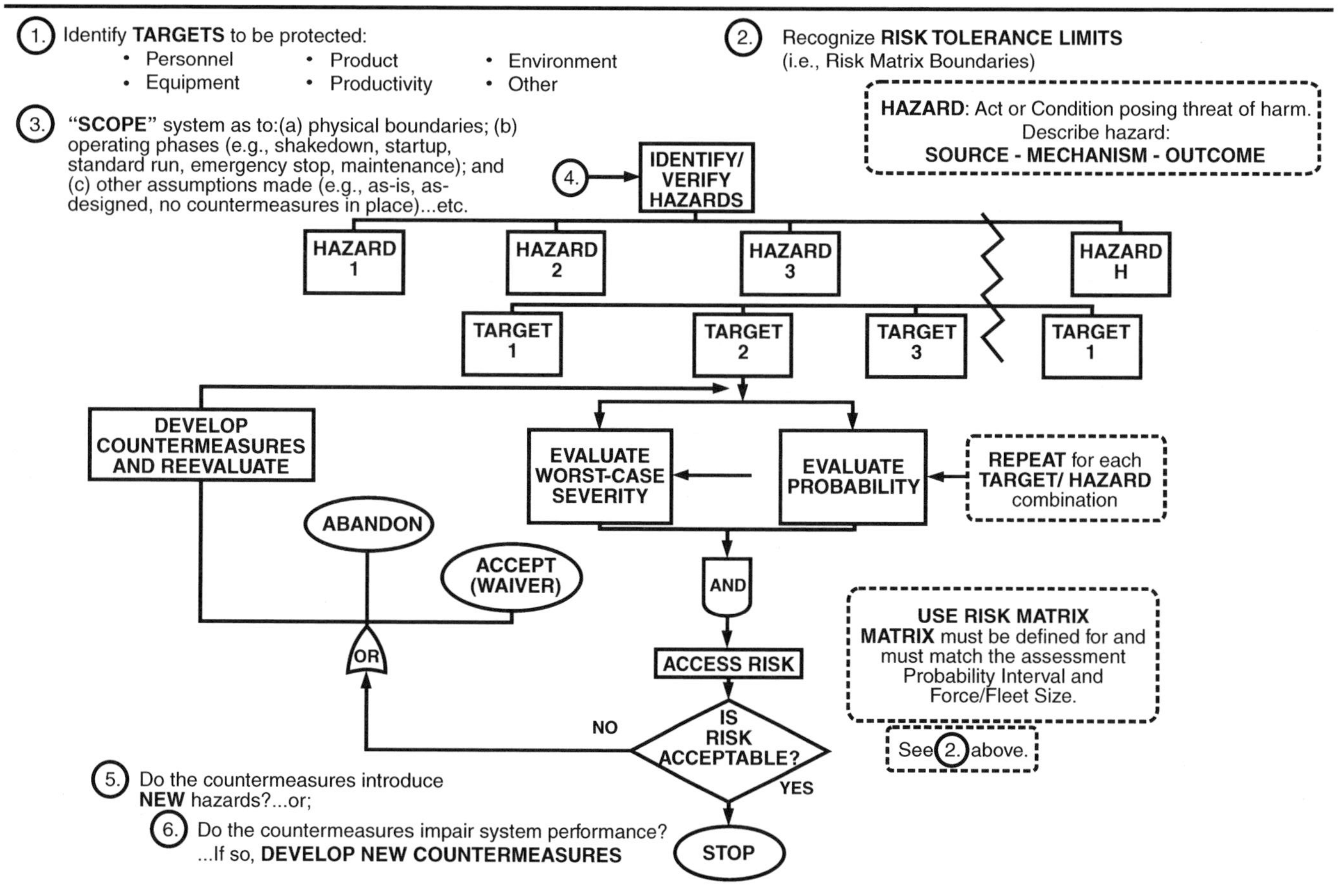


**Figure 14.9** Preliminary hazard analysis process flow.

point) to the consequent or resulting events. This logic approach is inductive, which means that the logic flows from the specific to the general. The process begins by selecting initiating events, both desired and undesired, and developing consequences through consideration of system/component failure-and-success alternatives. Identification of initiating events may be based on review of the system design and operation, the results of another analysis such as a failure modes and events analysis (FMEA), or personal operating experience acquired at a similar facility. The safety professional should then postulate the success and failure of the mitigating systems and continue through all alternate paths, considering each consequence as a new initiating event. Figure 14.10 is an example of an ETA using a building fire.

### 14.3.3 Fault Tree Analysis

The purpose of a fault tree analysis (FTA) is to assess a system by identifying a postulated undesirable end event and examining the range of potential events that could lead to that state or condition. The FTA can model the failure of a single event or multiple failures that lead to a single system failure. The FTA is a deductive approach meaning that the logic flows from general to specific, or moves from an event that is a result to the events that produced the result. The method identifies an undesirable event and the contributing elements (faults/conditions) that would precipitate it. The contributors are interconnected with the undesirable event, using network paths through Boolean logic gates. Figure 14.11 demonstrates a basic graphical depiction of the relationships between events and conditions that are associated with a car and a train on the section of a track simultaneously. See Stephans and Talso (1997) for specifics on particular FTA techniques.

The box with the words "Car and Train on Track at Same Time" in Figure 14.11 represents the top event that involves a fictional scenario in which an accident occurred

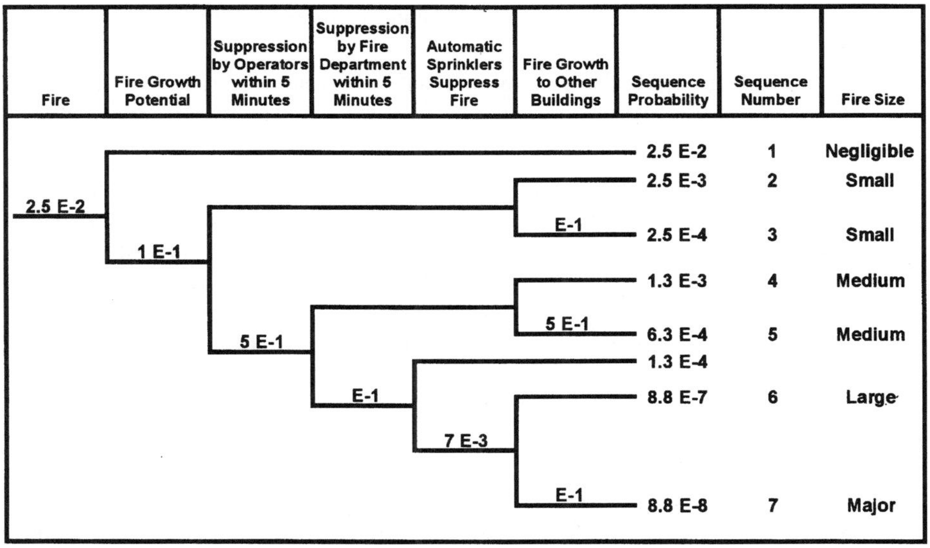


**Figure 14.10** Example of an event tree analysis for a building fire. (Reprinted with permission, System Safety Society, Stephens and Talso, 1997.)

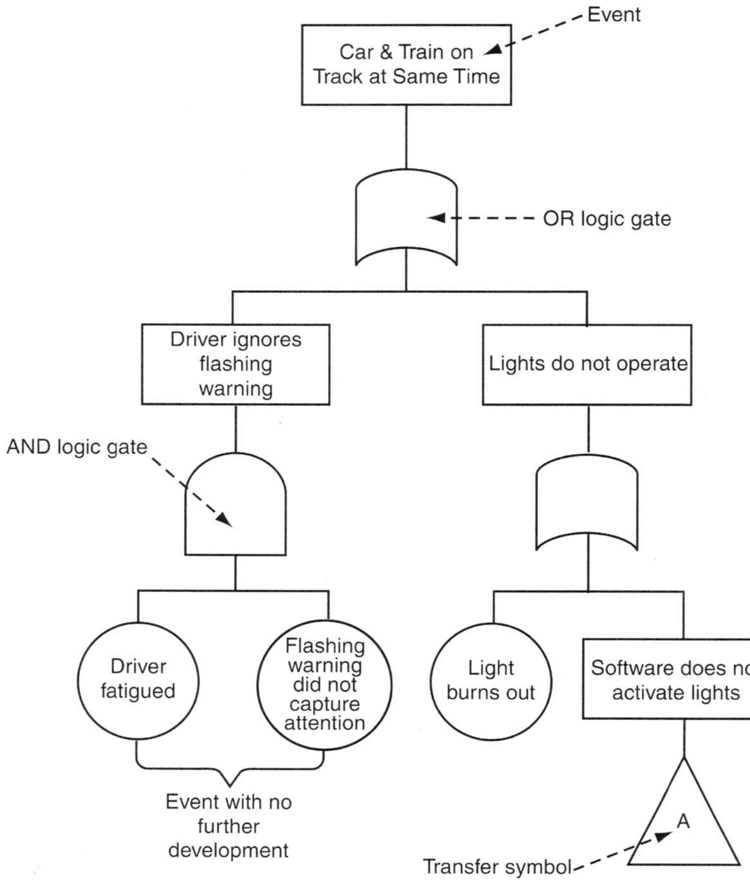


**Figure 14.11** Fault tree of a car and train accident scenario.

due to a train collision with a car. The circles represent those basic events that are not analyzed further. The object that looks like a quarter moon on its side is an OR gate. It indicates that any of the events or conditions that lead to it can cause the mishap. Another symbol is the AND gate, which looks like a half moon lying on its flat side and indicates that all the conditions below it must exist for the next event to take place. Transfer symbols are used to indicate continuation to or from another analysis. Fault trees are useful for helping to focus efforts on safety critical areas, visually displaying logic, and showing the relationships between conditions and events. Further, FTA helps the safety engineer to completely understand the subsystem or component being analyzed and help identify the root causes of the top event. If probability data is available for the basic events, then the probability of the top event can be mathematically determined.

However, FTA has limitations. One problem is that the top event of a fault tree must be clearly defined and limited in scope to be effective. If the top event is not clearly defined the analysis will become confusing and possibly misleading. Another limitation is that human factors failures are difficult to model with this type of analysis. It is also important to understand that FTA is subjective; thus, no two fault trees will be the same when done by two different assessors. Still, another limitation is that FTA can be expensive because of costs in obtaining data. Finally, a fairly mature systems design is required before FTA can be used effectively.

### 14.3.4 Failure Mode and Effects Analysis

Another common analysis method is the failure mode and effects analysis (FMEA). This analysis is a qualitative reasoning approach best suited for reviews of mechanical and electrical hardware systems. The FMEA technique (1) considers how the failure modes of each system component can result in system performance problems and (2) ensures that appropriate safeguards against such problems are in place. The system is divided up into different units in the form of a block diagram. Figure 14.12 illustrates the functional block diagram on four components (U.S. Coast Guard, 2001, p. 9-5). Failure modes are identified for the various units. Conceivable causes, consequences, and the significance of failure are assessed for each failure mode. An investigation is made into how the failure can be detected. Recommendations for suitable control measures are made. To document the findings, the FMEA record sheet addresses the following: identification, failure mode, failure cause, failure effect, failure detection, possible action, and probability and/or criticality level. A quantitative version of FMEA is known as failure modes, effects, and criticality analysis (FMECA). Table 14.5 provides a vessel-based FMEA record sheet example from the U.S. Coast Guard (Walker, 2000). The terminology used is similar to that mentioned above.

### 14.3.5 Fault Hazard Analysis

The fault hazard analysis (FHA) method is a basic inductive method of analysis that is used to perform an evaluation that starts with the most specific form of the system and integrates individual examinations into the total system evaluation. The purpose of the FHA is to systematically examine a facility or system and to identify hazards and their effects. The FHA methodology, like the FMEA, is to examine the system, element by

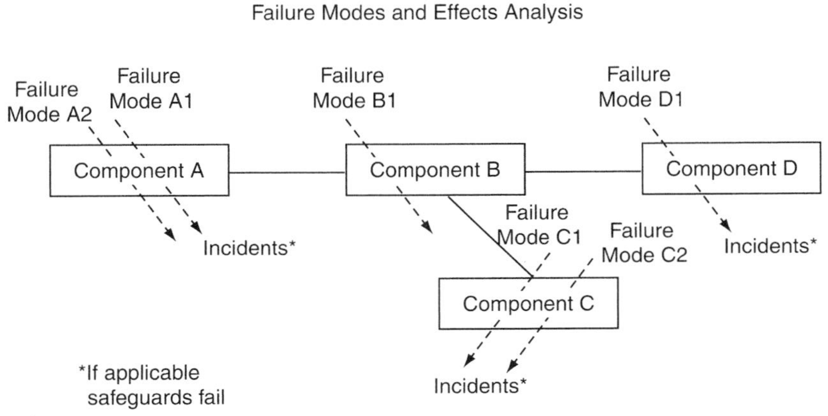


**Figure 14.12** Failure modes and effects analysis.

**TABLE 14.5 Example Failure Modes and Effects Analysis (FMEA)—Functional Failure-Based**

| Functional Failure | Loss Scenario (Effect) | Percent of All Reportable Marine Events | Dominant Causes | Applicable Inspection Activity | Inspection Effort | Criteria | Change in Risk |
|---|---|---|---|---|---|---|---|
| | | | Function: Providing Start Air for Engines | | | | |
| No or insufficient volume of start air provided to engines | No engine start, which can lead to loss of propulsion and a disabled vessel | ~25 | Condensation in bottles (62%) | Blowdown bottles during inspection | <10 min | Do not have to do on variable pitch propellers | Current practice (high negative impact if not performed) |
| | | | | Verify that regular blowdowns are scheduled and occurring (by record review) | <5 min | See above | Possibly high positive impact |
| | Could possibly lead to a grounding, collision, etc. | | | Communicate importance of blowdowns to crew | <5 min | See above | Current practice (high negative impact if not performed) |

| | | | | | |
|---|---|---|---|---|---|
| | Disabled compressors (multiple compressors) (5%) | Operation verification (measure discharge pressure) | <10 min | See above | Current practice (high negative impact if not performed) |
| | | Visual inspection for leaks, gauges functioning, obvious defects, etc. | <5 min | See above | Current practice (high negative impact if not performed) |
| | | Communication with pilots about known problems during transit | <5 min | See above | Current practice (high negative impact if not performed) |

element. Modes in which each element can fail are then identified. Finally, the effects to the system for each failure mode are determined, taken both singly and in combination with others (some variations classify effects according to their severity). The FHA method is very similar to a PHA and is a subset of the FMEA technique. Figure 14.13 provides an enhanced example of the FHA, which is very similar to the PHA. See Stephans and Talso (1997) for more information on FHAs.

### 14.3.6 System Hazard Analysis

The system hazard analysis (SHA) provides documentary evidence of safety analyses of the subsystem interfaces and system functional, physical, and zonal requirements. As the SSHA identifies the specific and unique hazards of the subsystem, the SHA identifies those hazards introduced to the system by the interfaces between subsystems, man–machine, and hardware–software. It assesses the entire system as a unit and the hazards and failure modes that could be introduced through system physical integration and system functional integration.

The SHA is accomplished in much the same way as the SSHA. That is, hazards and hazard causal factors are identified, hazard mitigation requirements communicated to the design engineers for implementation, and the implementation of the safety requirements are verified. However, several differences between the SSHA and SHA are evident. First, the SHA is accomplished during the acquisition life cycle where the hardware and software design architecture matures. Second, where the SSHA focused on subsystem-level hazards, the SHA refocuses on system-level hazards that were initially identified by the PHA. In most instances, the SHA activity will identify additional hazards and hazardous conditions because the analyst is assessing a more mature design than that which was assessed during the PHA activity. And third, the SHA activity will put primary emphasis on the physical and functional interfaces between subsystems, operational scenarios, and human interfaces. Figure 14.14 (Alberico et al., 1999) demonstrates the concept with a propulsion system. For further insight refer to Alberico et al. (1999) and Stephans and Talso (1997).

In the example illustrated in Figure 14.14, the fault tree approach is used to analyze a system-level hazard “Loss of Thrust Actuation.” The hazard is depicted as the top event of the fault tree. The SHA activity analyzes all causes to the hazard, including the software branch that is a branch of the OR gate to the top-level event. This particular hazard has hardware causes (actuator control arm failure), human error causes (pilot commands shutdown of control unit), and software-induced errors causes.

Further, “Thrust Actuation” is a function of the propulsion system and administratively controlled by the propulsion Integrated Product Team (IPT) of contractor A. The computer hardware and software controlling the thrust actuators are also within the engineering boundaries of the same IPT. However, the software safety analyst has determined, in this case, that a fault condition in the computer operating system (OS) is the primary causal factor of this failure mode. This OS fault did not allow actuator sensor data to be read into sensor limit tables and allowed an *overwrite* to occur in the table. The actuator control algorithm was utilizing this sensor data. In turn, the actuator control computer software component functional architecture could not compensate for loss of credible sensor data that transitioned the system to the hazardous condition. In this example, the actuator and controlling software are designed by contractor A; the sensor suite and throughput data bus are designed by contractor B; and the computer OS is developed by contractor C.

| SYSTEM: Launch Car | | | | SUBSYSTEM: Stabilization | | | | | | Revision 1 as of 18 April 1993 | |
|---|---|---|---|---|---|---|---|---|---|---|---|
| 1 Number | 2 Unit/ Item | 3 System Event Phase | 4 Brief HAZ Description | 5 Effect on System | 6 Severity | 7 Probability | 8 Index | 9 Category | 10 Recommended Control Requirement/Action | 11 Effects of Recommended Action | |
| SB10X01 | Stab Sys | Deploy | Mechanical energy of suppressed load | LSR lower while train is in motion resulting in rail car damage or derailment | I | D | 8 | X | Provide positive restraint of LSRs during transport | Reduces probability of occurrence I=12 (E - probability) | Design reviews and verification of incorporation into system design |
| SA10X02 | HPU | Deploy | Chemical energy of hydraulic fluid | Spilled hydraulic fluid could result in a fire | I | D | 8 | X | Control leakage of fluid thru fittings of boundary failure. Select fire tolerant fluid | Reduces probability of occurrence I=15 (E - probability) | Open pending subsequent design reviews and verification of incorporation into system design |
| | SCU | Maint | Electrical Energy - Electrical Potential | Electrocution/shock to personnel from contact with energized system components | I | D | 10 | X | Design per MIL-STD-454, Req 1 | Reduces probability of occurrence | Open pending subsequent design reviews and verification of incorporation into system design |
| | | | Electrical Energy | Erroneous signals from the SCU could result in improper leveling and MLC overturning during stabilization or erection operations | I | D | 8 | A | Provide two fault tolerance in the design | Reduces probability of occurrence | Open pending subsequent design reviews and verification of incorporation into system design |

**Figure 14.13** Fault hazard analysis example (reproduced with permission, Systems Safety Society, Stephans and Talso, 1997).

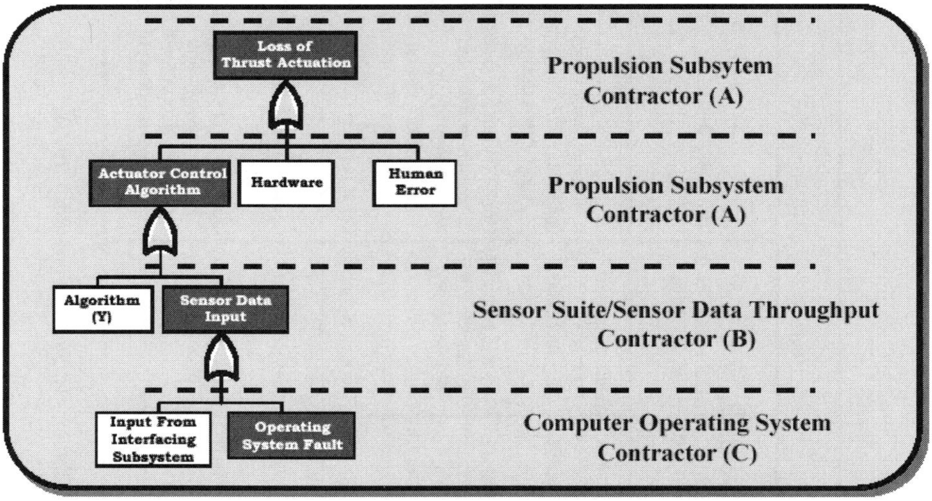


**Figure 14.14** Example of a system and subsystem hazard analysis.

The safety analysis performed by contractor C is demonstrated in this example. If contractor C is contractually obligated to perform a safety analysis (and specifically a software safety analysis) on the computer OS, the ability to bridge (bottom-up analysis) from an OS software fault to a hazardous event in the propulsion system is extremely difficult. The analysis may identify the potential fault condition but not identify its system-level effects. The analysis methodology must rely on the "clients," of the software OS, or contractor A, to perform the top-down analysis for the determination of causal factors at the lowest level of granularity.

### 14.3.7 Subsystem Hazard Analysis

The hazard analysis performed on individual subsystems of the (total) system is the subsystem hazard analysis (SSHA). This analysis is "launched" from the individual hazard records of the PHA that were identified as a logically distinct portion of a subsystem. Although the PHA is the starting point of the SSHA, it must be only that—a starting point. The SSHA is a more in-depth analysis of the functional relationships between components and equipment (this also includes the software) of the subsystem. Areas of consideration in the analysis include performance, performance degradation, functional failures, timing errors, design errors, or inadvertent functioning.

### 14.3.8 Cause–Consequence Analysis

The cause–consequence analysis (CCA) combines the inductive reasoning features of ETA with deductive reasoning features of FTA. The result is a technique using six steps that relates specific accident consequences to their many possible causes. The first step selects an event or type of accident situation to be evaluated. The various accident paths are then constructed based on the chronological successes and failures of the appropriate safety functions (systems, operator actions, etc.) that influence the course of the accident resulting from the event. The next step develops the accident paths resulting from the event through an ETA. Through the use of an FTA, the analyst develops the initiating event and the safety function failure event to determine their basic causes. The accident sequence is composed of a sequence of events, each of which is a top event for a fault tree that is part of the cause–consequence diagram. For an accident sequence to occur, all of the events in the sequence must occur. Evaluating the results of the CCA is a two-step process. First, the accident sequences are ranked based on their severity and importance to plant safety. Then, for each important accident sequence, the accident sequence minimal cut sets can be ranked to determine the most important basic causes. The final step in performing a CCA is to document the results of the study.

## 14.4 SYSTEM SAFETY PROCESS

The system safety process operates within the context of the systems acquisition process as illustrated in Figure 14.15. When the system is conceived, the designers take the operational environment, lessons learned from the past, technology, and the doctrine of how to achieve success, to determine the requirements for the system. As these requirements become the design, system safety engineers take these requirements, identify the hazards, and determine the safety requirements for the system. Testing that is conducted on the system also helps identify hazards. As the design of the system matures, safety engineers generate reports on the safety of the system to document the risk. The design and risk acceptance authorities use these reports to decide whether to accept the mishap risk and approve production of the design or allocate more resources for mishap risk reduction.

In order to determine the safety of a particular system, there are a number of important questions that need to be answered to define the system for which safety is a concern:

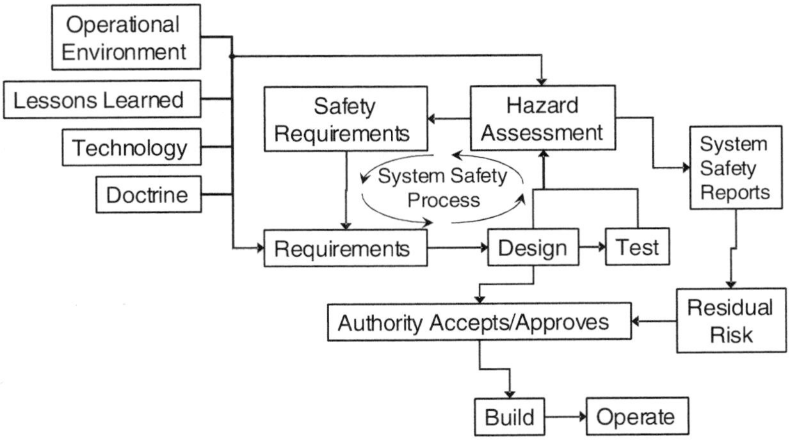


**Figure 14.15** System safety process within the systems acquisition process.

- What are its boundaries?
- What are the people, machines, and processes that make up the system?
- What are the needs and objectives that the system must fulfill?
- How do the components of the system interact?
- What are the interfaces with other systems?
- What parts of the system can we control and redesign to make them safer and more effective in the desired functions of that system?

Regardless of the methods used to conduct risk analysis within the system safety process, organizations must have a system safety program and plan. (Refer to Fig. 14.1 as the general model for the following steps.)

***Step 1. Develop System Safety Plan*** The first step of a system safety program is to develop a system safety program plan integrated with other program planning documents. OSHA addresses the components of a system safety program plan. Figure 14.16 shows how the elements of the system safety program are coordinated with other program efforts to ensure appropriate safety data is available at decision points in the program.

The preliminary hazard list (PHL) is developed at the beginning of the design and entered into the hazard tracking system (HTS). Following the preliminary design review, system safety starts a functional hazard analysis (FHA) to help determine what safety requirements need to be included in the requirements documentation. Systems safety further develops the FHA into an SSHA and SHA. The software hazard analysis and the safety assessment report are used at the critical design review to determine whether the design is ready for production. In conjunction with the critical design review, the residual risk will need to be accepted for each hazard that has not been eliminated. In addition, the safety data will be used to develop the test program. The data from the test program will be used to identify additional hazards and verify that mitigation measures are effective. Figure 14.16 also shows that meetings of the SSWG are scheduled to support program milestones and other requirements.

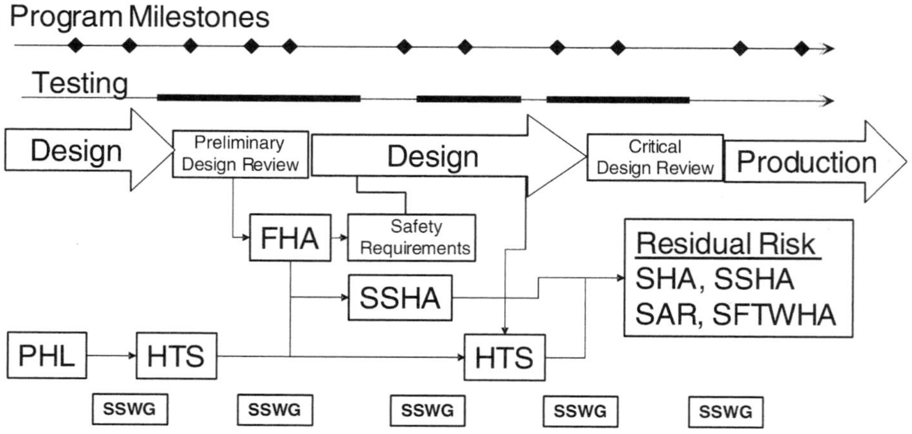


**Figure 14.16** System safety program planning.

This all points out the importance of system safety understanding the program needs and writing the system safety plan to ensure that resources are allocated and personnel are assigned to support the program schedule. Some things that might go into the plan are the system safety program purpose, safety policies, responsibilities, hazard assessment plan, and system safety specific information on the product.

Table 14.6 shows the typical content items of the system safety plan. For example, critical items for the system safety plan are the policy, objectives, and risk management methodologies.

*Policy* The safety analysis should spell out the management's policies regarding system safety and include specifics as needed for the program. Some ideas for this include:

1. Proactively identify all hazards that could cause personal injury or equipment damage.

**TABLE 14.6 Content of a System Safety Program Plan**

*References* List organizational documents that govern safety program in general and system safety specifically. List applicable government documents as well as company directives. List industry standards that will be used to give guidance to system safety group and other teams in system safety methodologies and techniques that will be used.

*Scope* Delineate what plan applies to and what general areas of effort plan covers.

*Policy* Spell out in simple terms management's system safety policies pertaining to effort. Restate organizations policies regarding system safety and include specifics as needed for program.

*Objectives* State in clearest terms final objectives of system safety program

*Task Objectives* Here is where specific tasks for various members of system safety group and supporting teams are delineated. Make sure responsibilities are clearly delineated and that open communication between members occurs.

*Risk Management Methodologies* Describe how safety issues are handled and what the process is for identifying and entering hazards into tracking system. Describe how hazards will be classified. Identify when and how hazards will be closed (see Figure 14.17). Describe the process for residual risk acceptance and how acceptance will be documented. Identify items that should be included in hazard tracking system. Define severity levels.

*Safety Integration with Other Disciplines* Describe lines of communication and information exchange with other teams, working groups, and other supporting organizations.

*Schedules* Describe how system safety program events interact with overall program schedule. Include timing of reports and other deliverable products. This could be detailed appendix to plan.

*System Safety Group Charter* This can be an appendix to system safety plan or stand-alone document. Membership should be updated as necessary when there are major program reorganizations, the program passes a milestone, and significant personnel changes occur.

*System Safety Document Examples* This could include a sample risk analysis formats, hazard tracking formats, and risk acceptance documents.

*Glossary of Abbreviations, Acronyms, and Terms* Even though trained and experience system safety engineers and managers speak language of system safety, it is good idea to include a glossary to ensure all those who will work with system safety team fully grasp meaning of terms. For those who have not had experience or training in system safety, there is often confusion as to meaning of "risk," "hazards," etc. A good glossary of terms may help prevent confusion and help avoid rabbit trails in discussions on risk of particular hazard or when hazard is ready to close. It also helps to standardize terminology when numerous vendors are subcontracting on particular program.

2. Evaluate the risks associated with system hazards.
3. Eliminate or mitigate hazards to the lowest possible level consistent with operational requirements and resource constraints.
4. Ensure identified hazards and management controls are examined with respect to all applicable design standards and accepted design practices to include operational scenarios and environments.
5. Report residual hazards and associated risk to the appropriate risk decision authority.
6. Document all hazards and risk management decisions throughout the program life cycle.

*Objectives* The safety analyst should state the final objectives of the system safety program, such as:

1. All potential hazards associated with the program are identified and formally tracked for the life of the system and that risks associated with those hazards are properly managed and resolved.
2. No known residual hazard is accepted without formal documentation of associated risks. The appropriate authority shall make risk acceptance decisions.
3. System safety maintains a two-way interface with the HSI program and all design integrated product teams.
4. Historical safety data is included in the system safety program. Significant safety data are documented as "lessons learned" and will be entered in appropriate data banks and submitted as proposed changes to applicable design handbooks and specifications.
5. Safety measures consistent with system requirements, technical feasibility and cost are included in the system safety planning, development, production, and fielding.
6. Retrofit actions required to improve safety are minimized through the timely inclusion of safety features early in the life cycle of the program.
7. Changes in design, configuration, or mission requirements are accomplished in a manner that maintains acceptable safety-related risk levels.
8. Maximum operational readiness and mission protection will be achieved through accident prevention.
9. Safety consideration is given to system design, production, fielding, and ease of disposal for all hazardous materials.

*Risk Management Methodologies* The safety analyst should describe such features as how safety issues are handled, how hazards will be classified, and how they will be closed. For example, hazard management tools such as shown in Figure 14.17 might be used in the hazard closure process. The flowchart describes such a process on a U.S Army system. It tracks the hazard process from the point a hazard is identified until the risk is accepted and the hazard closed by the program manager (PM) if the hazard is a low risk, by the program executive officer (PEO) if a medium risk, and the army acquisition executive if a high risk.

***Step 2. Identify Hazards*** The next step after developing the system safety plan is to identify the hazards. The system safety process follows the iterations of the system engineering process. As design requirements are identified in the conceptual phase and

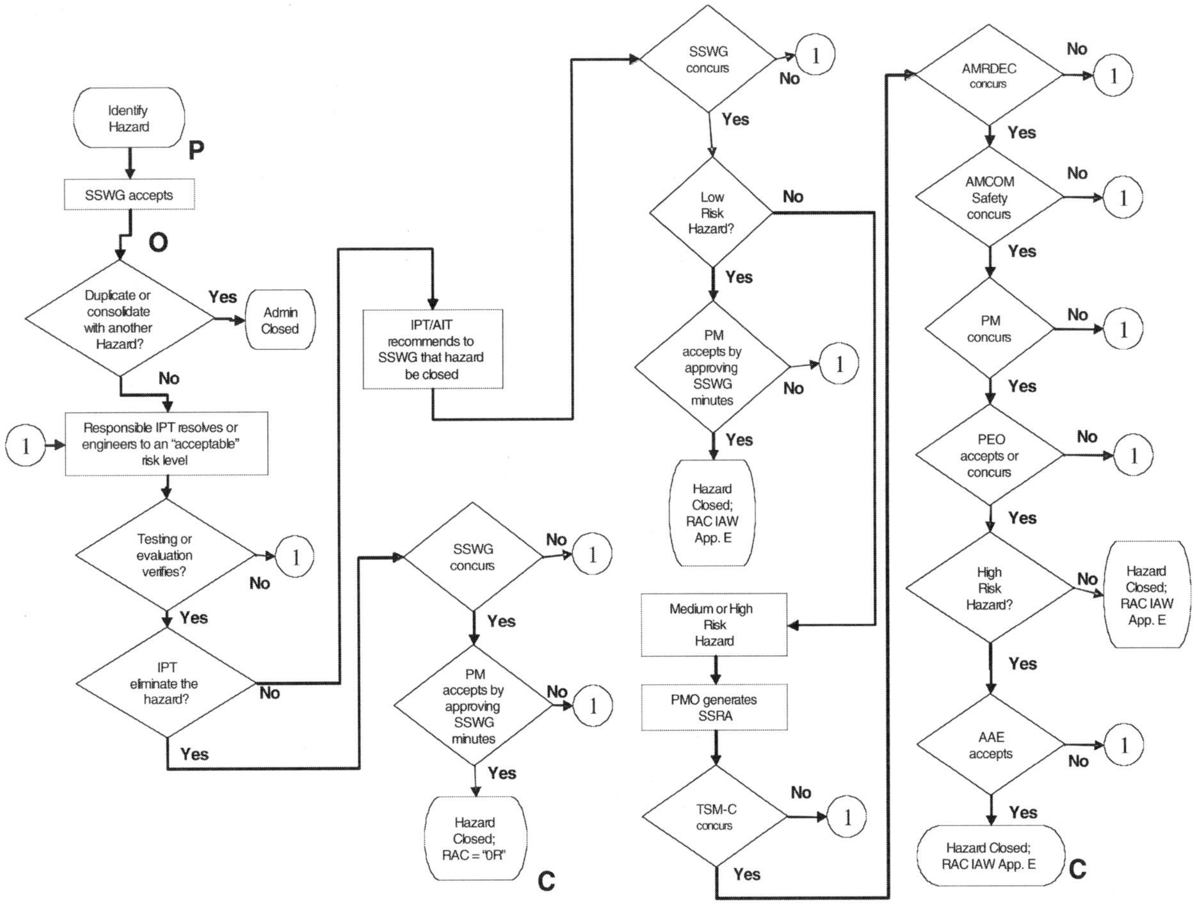


**Figure 14.17** Flowchart of a hazard closure process.

continuing throughout the entire system life cycle, so are hazards. One fundamental concept understood by safety engineers is that all mishap risk is never fully identified. There are always some undiscovered conditions that can cause subsystems and components to interact in undesirable ways that create hazards. So never be surprised to find new hazards as the system design matures, is tested, and is fielded.

One of the best places to start looking for hazards is the legacy system for which the new system is comparable or replacing. For example, the DoD reviewed hazard lists from current fighter aircraft prior to initiating the joint strike fighter effort. The basic subsystems and components of a fighter aircraft are similar, and many of the interactions between the airframe, engines, flight controls, avionics, life support system, etc. produce the same hazards. Of course, these aircraft also introduce new concepts, technologies, and materials along with the introduction of new hazards.

Another method is a hazard checklist. While the checklist can never be all-inclusive, the framework provides a starting point to catch hazards that may have otherwise been overlooked.

Still another method of unearthing hazards is to perform analyses on the functions of the system to see what functions the overall system, subsystems, and components must do to operate properly. The analyst should start with a list of system functions. These may be derived from the work breakdown structure, brainstorming ideas, or requirements documents for the system. As an example, Figure 14.18 shows a diagram of the top-level functions for a military helicopter.

As each function is studied ask the following, "What harm could come if this system, subsystem, or component fails to function correctly?" "What would happen if this failure is not detected?" "How would it be different if it is detected?" All the functions depicted must be present for the helicopter to work effectively.

The functional hazard analysis process produces a very comprehensive listing of hazards. This method is effective because it is a top-down process of identifying hazards and mitigation measures based on what the system must do and not just on the current design of the hardware. The analysis supports assessments on component criticality and hazard severity. With data supplied by the reliability engineers, a determination on the probability of the hazard resulting in a mishap can then be made.

It is also very useful in analyzing the system software in order to determine that the software functions correctly and what would happen if it failed. Just like functional hazard analysis on hardware, this information helps software designers to better understand the system requirements and design in order to make a better product. A major difference between hardware and software is that the former is visible while the latter is not. This makes software system safety analysis especially difficult.

The functional hazard analysis needs to be updated any time a function is added, deleted, or changed or when another type of analysis reveals additional failure modes. As soon as design requirements are generated, the functional safety analysis should begin.

As a system matures, hundreds, even thousands, of hazards could be identified depending on the complexity of the system. A PHL would be created only in the very early stages with hazards undergoing a PHA when appropriate. The PHA, usually the first analysis, can be very valuable because it identifies and characterizes possible hazards early in the design phase when needed redesign is least costly. It identifies known hazards, such as explosives, radiation sources, pressure vessels or lines, toxic materials, and high voltage, and specifies where each will occur in the system and the method to be used to eliminate the hazard or control the associated risk.

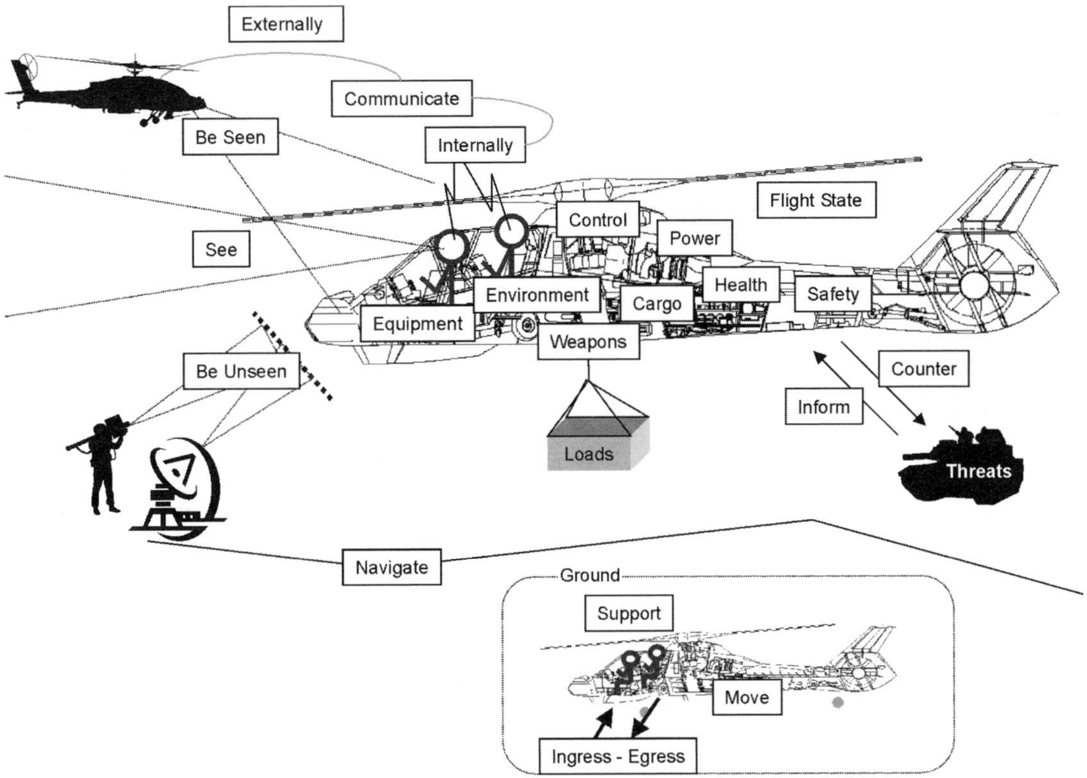


**Figure 14.18** Functions of a military helicopter.

Finally, the analyst needs to ensure the hazard is thoroughly described in the hazard description. The narrative of the potential hazard contains three elements that express the threat: a source, a mechanism, and an outcome. A source is an event or a condition that serves to initiate the chain of events in the mishap. A mechanism is the means by which the source can bring about the harm. An outcome is the harm that will be suffered in the mishap. If a hazard cannot be described using a source, a mechanism, and an outcome, it likely is not a hazard. A complex hazard may have multiple sources, mechanisms, and outcomes and may require a diagram in addition to a narrative to fully describe the hazard.

***Step 3. Assess Mishap Risk*** The third step in the system safety process is assessing the risk of the hazard in question. The most basic way to do this is to select a risk assessment code from the hazard matrix based on the hazard description and knowledge of the system. For example, what would be the risk of a military helicopter striking wires during flight? The severity would be "1," catastrophic, based on the description of catastrophic in Table 14.3. The worst credible outcome would be "Death or permanent total disability; system loss or mishap damage greater then or equal to $1,000,000."

But what is the probability? One way is to examine the mishap experience of a similar existing aircraft. The aircraft could be similar in mission and operating environment. Let's say the existing aircraft has a catastrophic mishap at a rate of 2.1 times $10^{-7}$ times per flight hour. That plots out to a "D, Remote" in Figure 14.6, since the probability falls in the range of $10^{-5}$ to $10^{-7}$ occurrences per flight hour. The resulting assessment based on design differences determines how much better or worse the new helicopter will perform. Perhaps the new helicopter will be able to see the wires better with sensors on board, or perhaps will have better or worse wire strike protection systems on board. Thus, the assessment provides decision makers with comparative information indicating whether the new system will be safe or not.

The final element in the accident analysis is cause–consequence analysis. This step evaluates the effect of the postulated accident on the workers, the public, and the environment. For some facilities, consequence analysis may also include health effects assessment, accident frequency estimates, or safety goal comparisons. Figure 14.19 is an example that highlights causes, preventive features, mitigation features, potential impact, and risk determination. For further information concerning qualitative consequence analysis, see U.S. Department of Energy (1997) for more information on workload analysis.

***Step 4. Identify Mitigation Measures*** The fourth step in the system safety process is to identify those measures that will eliminate or mitigate a hazard. To accomplish this most effectively, system safety engineers use the "system safety order of precedence." Elimination of all hazards would be ideal, however, not practicable from a programmatic point of view and is often impossible. Figure 14.20 illustrates the concept that will be discussed in greater detail below as adapted from the *Software System Safety Handbook: A Technical & Managerial Team Approach* (Alberico et al., 1999).

First in the system safety order of precedence is to design for minimum hazard. Although not always possible, designing to eliminate hazards is preferable to procedures or training to avoid them. In every design there are options, some of which avoid or eliminate the potential for the hazard. If elimination of a hazard cannot be accomplished, the next step is to at least reduce the risk to an acceptable level. The acceptable risk for a hazard can be based on the performance of legacy systems. The acceptability of risk will be refined in

| | |
|---|---|
| Causes: | Mixing of incompatible chemicals due to personnel error, container leakage, or improper maintenance of equipment |
| Preventive Features | |
| Design: | Ventilated storage cabinets and/or storage areas provided with sumps for spill containment |
| Administrative: | Segregation of non-compatible chemicals, regular inspection of containers and storage areas, instruction of personnel in proper handling techniques |
| Method of Detection: | Smoke and ionization detectors for fire conditions, personnel observation, appropriate alarms |
| Mitigation Features | |
| Design: | Fire suppression equipment (sprinklers, portable fire extinguishers), laboratory fume hoods, ventilation design |
| Administrative: | Employee training, safety procedures, automatic fire department response, emergency medical technicians available on site |
| Potential Impact: | Physical damage to affected area, potential water damage, potential injury to personnel from burns, explosions, or inhalation of toxic materials, partial shutdown of operations |
| Risk Determination: | |
| Probability Level: | Low |
| Consequence Level: | Medium |
| Risk Level: | Low |

**Figure 14.19** Example category 3 qualitative consequences analysis (uncontrolled chemical reaction).

an iterative review of the design to find an optimum balance of safety and other performance objectives.

The next activity in the order of precedence is to incorporate safety devices. If identified hazards cannot be eliminated or the associated risk adequately reduced through design selection, further risk reduction efforts are required by using fixed, automatic, or other protective safety design features or devices; for example, the addition of air bags and daytime running lights on vehicles.

If the hazard still presents a problem, warning (aural and visual) devices should be added to the system. Incorporate these devices to detect conditions related to the hazard to produce a warning signal for alerting personnel. Make sure the warning device(s) design minimizes incorrect reactions caused by nuisance warnings (false alarms). Work closely with human factors engineering to select visual, audible, or tactile warnings that are not ambiguous and cannot be confused with other warning mechanisms.

Finally, develop procedures and training. The reason procedures and training are listed last is that they rely on humans to provide the safety. People make errors in following procedures when distracted or bored. Training requires continuous monitoring on seldom-

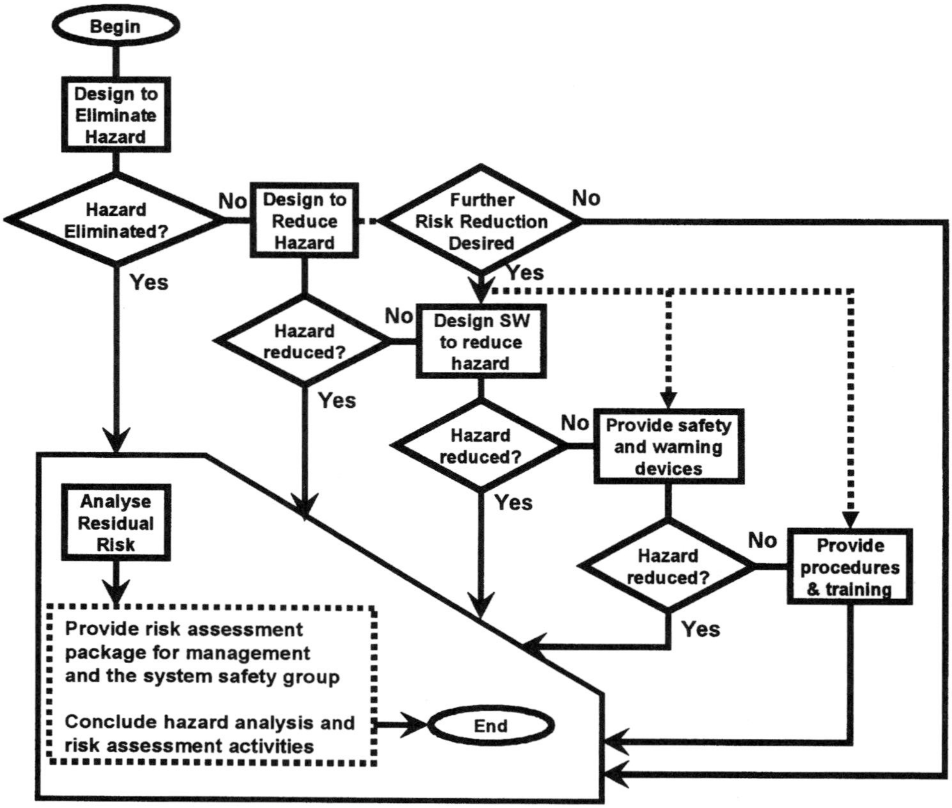


**Figure 14.20** System safety order of precedence.

used procedures, like emergency responses, since the system is unable or it is undesirable to provide actual situations to reinforcement training.

It is very important for the safety engineer to work closely with the other engineers in this step of the system safety process. Look for opportunities to apply lessons learned from other systems and for ways to apply new methods and technologies.

***Step 5. Verify Mitigation Effectiveness*** The next step in the process is to establish mechanisms to verify that mitigation measures are in place as designed and effectiveness is verified in actually eliminating or mitigating the hazard. This may involve reviewing drawings to see if proposed design changes were included, inspecting components, subsystems, and the system by observing fabrication and test activities, and by reviewing test reports. These verification mechanisms should be included as part of the hazard tracking system that will be discussed below. Remember that people do fail so verification mechanisms are important. Work closely with system engineering, configuration management, testing, and quality assurance to make sure effective safety-related design changes do make it into the final production configuration.

***Step 6. Accept Residual Risk*** As design decisions are finalized, program management must also begin to formally accept the residual risk. Often these decisions will be informally addressed prior to the formal acceptance of risk at design reviews and other decision meetings. Make sure the system safety plan clearly spells out how the process works and who must review the risk documentation before the final acceptance decision is made and signed off.

A decision maker may ask, “How do I know when to accept risk?” The best answer is depicted by the “bathtub curve.” As depicted in Figure 14.21, the total cost of safety is the sum of the cost of mishaps and the cost of safety mitigation measures. As the resources are expended on safety, the cost of mishaps decreases and the cost of mitigation increases. There comes a point where the cost of one more dollar of mitigation results in just one dollar of mishap cost reduction. The next dollar that is spent will only save 99 cents. This is the optimum level of risk and that is where spending money on risk reduction efforts yields no additional system benefit. This concept can be applied to mitigation measures for a specific hazard or can be applied to all the hazards of the system.

Another question that a decision authority may ask is, “If I have a limited amount of resources to spend on safety, how can I best spend those resources?” The answer is to prioritize the mitigation measures being considered for the entire system based on which ones produce the most reduction in mishap cost for the dollar expended. Usually, those mitigation efforts that bring the system closest to optimum level of risk are where limited resources should be spent.

Often, the critical issue is determining the cost to human life, those injuries and deaths due to mishaps in risk assessment and risk acceptance. The short answer is to determine the costs in terms of replacing trained and experienced people plus any additional cost for worker’s compensation and death benefits. For example, the DoD assigns a value to a military aircraft pilot of $1.1 million. There are obviously intangible costs to organizations for injuries and deaths. There are costs in terms of the suffering of families and co-workers. There is lost productivity and effectiveness due to poor morale. There are public relations impacts and legal requirements related to the provision of workers compensation benefits packages. There is time spent dealing with lawsuits and investigations. However, in practical terms, human life is worth what decision makers are willing to spend to protect it. Appropriate authorities must make judgments (whether or not based on quantified values) on how much to spend for risk mitigation to protect human lives. The role of the

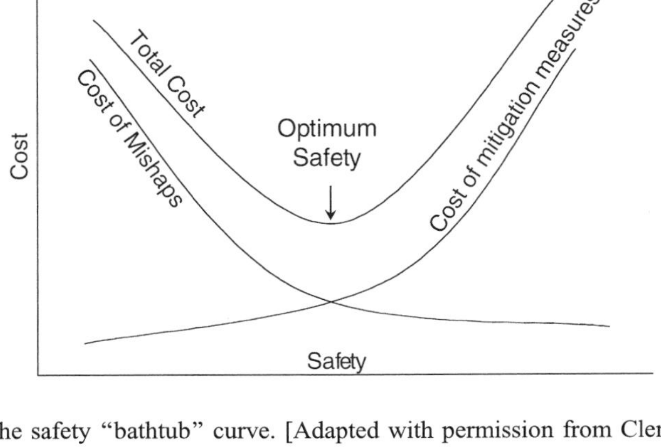


**Figure 14.21** The safety “bathtub” curve. [Adapted with permission from Clemens (2002, Sheet 00-6).]

safety engineer is to provide the best information as to the residual risk in terms of dollars, injuries, and deaths over the system life cycle as well as the cost of proposed countermeasures.

***Step 7. Track Hazards through Life Cycle*** The last step in the system safety process is to track the hazards throughout the life cycle of the system. In a complex system there will be thousands of hazards identified over time. The hazards will be published in documents such as the SHA, SSHA, operating and support hazard analysis (O&SHA), or the safety assessment report (SAR). As systems change due to changes in operating the system within a certain environment and planned product upgrades, previously identified hazards may resurface or new hazards may be introduced. Continuous risk reduction efforts will be required.

The best way to track hazards and risk reduction efforts is a computer database. By using such a database, reports can be easily generated to track the progress of risk reduction. Documents can be generated for risk acceptance. Mitigation measures can be tracked as well. Each hazard should have a record in the database. This record may be referred to as hazard tracking record (HTR), an SAR, or another name that fits the organization. The record fields should be described in the system safety plan and include those indicated in Table 14.7.

For example, the status of the hazard is important to record. A possible status could include:

- *Proposed* The hazard has not yet been accepted by the system safety group.
- *Open* The hazard was accepted by the system safety group with no corrective action plan in place.
- *Monitor* The hazard is accepted by the systems safety group with a corrective action plan in place.
- *Recommend Closure* The contractor determines that there will be no further elimination or mitigation of risk by design changes and proposes closure to the systems safety group.
- *Pending Closure* The systems safety group has concurred that the hazard should be closed and has forwarded the closure document to the risk acceptance authority.
- *Closed* The hazard has been eliminated or all corrective actions have been completed.
- *Verified* The appropriate decision authority has accepted any residual risk.
- *Administratively Closed* The hazard was determined by the system safety group as describing a hazard that is already identified in another hazard tracking record or it has been incorporated in the scope of another hazard tracking record.

If a hazard tracking system is thoughtfully designed, a variety of hazard reports and decision documents are available to the organization. The HTS can also be used to present information to the SSG and other teams directly from the database.

## 14.5 CONCLUSION

System safety focuses on providing the foundation for designing safe systems around human capabilities and limitations rather than reacting to unacceptable situations. By

**TABLE 14.7 Example Hazards Tracking Database**

*Unique Number* This number should, if at all possible, be based on system that helps users determine with what part of system hazard is associated. By convention, hazard is assigned to originating subsystem. However, this may be difficult to determine because often hazards are related to interface of two or three subsystems. To which subsystem will the hazard be assigned? This example highlights the need for system safety engineering to coordinate its activities with systems engineering and design teams.

*Hazard Topic* This is a short phrase used to quickly differentiate hazard from other hazards in hazard list.

*Hazard Description* A hazard description is brief narrative of potential mishap attributable to the hazard having three elements that express the threat: a source, a mechanism, and an outcome.

*Status* Tracking record should state where hazard is in process of analysis and risk acceptance. Possible status list of codes could include

- Proposed
- Open
- Monitor
- Recommend closure
- Closed and verified
- Administratively closed

Tailor the list of status codes to level of complexity dictated by system. This could range from two or three status codes to dozen or more.

*Risk Assessment Code* This is code assigned showing the worst credible consequence of hazard and its probability.

*Subsystem, Component, and Part Number* If hazard is associated with specific subsystem, component, or part, there should be a field to enter that data.

*Severity* Include reasoning used to assign severity of risk assessment code.

*Probability* Include rationale used to assign probability of risk assessment code.

*Special Considerations* Track whether hazard involves radioactive material, explosives, munitions, health hazard, or involves system requirement.

*Risk Reduction Alternatives* List here all reasonable alternatives for risk elimination or reduction. By listing all alternatives, creativity of design engineers to find even better alternative is enhanced.

*Recommendations* List here mitigation measures recommended by system safety engineer or system safety group. Individually track these recommendations as part of system safety management system.

*Consequences of Risk Acceptance* List here costs of risk acceptance if no further risk reduction is funded. This should include how many deaths and serious injuries may occur over life cycle of system. What are financial costs in terms of damage or other forms of loss such as data loss or environmental impact?

*Dates of Status Changes* Tracking dates provides understanding of hazard history at glance.

*System Description* Short description of subsystem or components involved with hazard helps users of hazard tracking system understand how hazard fits into the system.

*Sources and References* List design standards, safety standards, safety analyses, requirements documents, and other related documents and references related to hazard.

*Actions* Track actions taken and decisions made related to hazard individually in database within system safety management system. These should be dated to provide history of hazard.

understanding system safety concepts, principles, and elements, managers transform system requirements into operational systems through a comprehensive, iterative technical management process. It is an activity that must be done throughout the entire life cycle of the system, from "cradle to grave," and is a concurrent approach to both product and process development. If safety personnel are involved early in the design concept, they will be better able to identify hazards, avoid risk, and develop reliable countermeasures to those hazards that cannot be eliminated. The earlier system safety efforts are funded in a program, the more cost effective those dollars will be in reducing the mishap risk of the system. In order to aid decision authorities, a listing of 101 techniques and methods used by safety personnel throughout the entire life cycle was presented in the chapter. In particular, system safety engineering deals with the tools of the trade, the principles and methodology of analyzing the hazards of system components, subsystems, and interfaces. Whereas, system safety management deals with how the decisions are made based on the analysis done by the system safety engineers in order to eliminate or reduce the associated mishap risk. Through interaction between engineering and management, hopefully an acceptable level of risk can be achieved within the constraints of operational effectiveness, time, and cost. In order for system safety to be effective, the integrated product team must agree on a system safety plan that will identify hazards, assess mishap risk, identify elimination and mitigation measures, verify mitigation effectiveness by reducing risk, accept residual risk, and track hazards though the life cycle.

## NOTES

1. For acquisition work within the DoD, DoD (2001) 5000.2-R, paragraph C5.2.3.5.10.1 indicates, "All programs, regardless of acquisition category and throughout their life cycle, shall comply with this [the Environment, Safety, and Occupational Health (ESOH)] section. The PM [program manager] shall ensure a system design that can be tested, operated, maintained, repaired, and disposed of in accordance with ESOH statutes, regulations, policies, and, as applicable, environmental treaties and agreements (collectively termed regulatory requirements) and the requirements of this section." As such, paragraph C5.2.3.5.10.6.3 identifies that "Pub. L. 91-596 (1990) (reference (dddd)) [Public Law 91-h;596, "Occupational Safety and Health Act of 1970," as amended by Public Law 101-552, Section 3101, November 5, 1990] makes Federal Occupational Safety and Health Act standards and regulations applicable to all federal (military or civilian) and contractor employees working on DoD acquisition contracts or in DoD operations and workplaces. In the case of military-unique equipment, systems, operations, or workplaces, Federal safety and health standards, in whole or in part, shall apply to the extent practicable."
2. FAA Order 8040.4 (U.S. Department of Transportation, 1998) requires that "The FAA shall use a formal, disciplined, and documented decision making process to address safety risks in relation to high-consequence decisions impacting the complete product life cycle. The critical information resulting from a safety risk management process can thereby be effectively communicated in an objective and unbiased manner to decision makers, and from decision makers to the public. All decision making authorities within the FAA shall maintain safety risk management expertise appropriate to their operations, and shall perform and document the safety risk management process prior to issuing the high-consequence decision. The choice of methodologies to support risk management efforts remains the responsibility of each program office."
3. NASA (2000) NPG 8715.3 states, "This NASA Safety Manual is the central Agency document containing procedures and guidelines that define the NASA Safety Program. This document serves as a general framework to structure the more specific and detailed requirements for Headquarters, Program, and Center Directors."

## REFERENCES

*Air Force System Safety Handbook.* (2000, July). Kirtland Air Force Base, NM: Air Force Safety Agency. Available: http://safety.Kirtland.af.mil/AFSC/ROBMS/Training/ssm-hndbk.pdf.

Alberico, D., Bozarth, J., Brown, M., Gill, J., Mattern, S., and McKinlay VI, A. (1999, December). *Software System Safety Handbook: A Technical & Managerial Team Approach.* Dahlgren, VA: Joint Services Computer Resources Management Group. Available: http://www.egginc.com/dahlgren/files/ssshandbook.pdf

Clemens, P. L. (2002). *System Safety Scrapbook* 9th ed. Tulahoma, TN: Jacobs Sverdrup. Available: http://www.sverdrup.com/safety/scrapbook.pdf.

Clemens, P. L., and Simmons, R. J. (1998). *System Safety and Risk Management*, NIOSH Publication 96-37768. Cincinnati, OH: National Institute for Occupational Safety and Health. Available: http://www.sverdrup.com/safety/riskmgt/riskmgt.shtml (retrieved July 16, 2002).

Dietz, T. R., Frey, S., and Rosa, E. (2002). Risk, Technology and Society. In R. E. Dunlap and W. Michelson (Eds.), *Handbook of Environmental Sociology* (pp. 562–629). Westport, CT: Greenwood.

Federal Aviation Administration. (2000, December). *System Safety Handbook: Practices and Guidelines for Conducting System Safety Engineering and Management.* Washington DC: Department of Transportation. Available: http://www.asy.faa.gov/Risk/ (retrieved July 14, 2002).

Goetsch, D. L. (2002). *Occupational Safety and Health for Technologists, Engineers, and Managers*, 4th ed. Upper Saddle River, NJ: Prentice-Hall.

Kohn, J. P., Friend, M. A., and Winterberger, C. A. (1996). *Fundamentals of Occupational Safety and Health.* Rockville, MD: Government Institutes.

Li, T. S. (1999, December). *Principles of Risk Management.* Hong Kong: Hong Kong University of Science and Technology. Available: http://www.ab.ust.hk/sepo/esst/Riskmgt-notes.pdf (retrieved June 26, 2002).

Malasky, S. W. (1982). *System Safety: Technology and Application.* New York: Garland STMP.

National Aeronautics and Space Administration. (2000 January). *NASA Safety Manual*, NPG 8715.3. Washington, DC: National Aeronautics and Space Administration.

National Aeronautics and Space Administration. (2002, June 26). *NASA Guidebook for Safety Critical Software—Analysis and Development*, NASA-GB-1740.13-96. Available: http://swg.jpl.nasa.gov/resources.

National Safety Council. (2002, June). *About the Council.* Available: http://www.nsc.org/insidensc.htm.

Occupational Safety and Health Administration. *Overview of Occupational Safety and Health/Industrial Hygiene.* Available: http://www.osha-slc.gov/SLTC/smallbusiness/sec5.html.

Roland, H. E., and Moriarty, B. (1990). *System Safety Engineering and Management*, 2nd ed. New York: Wiley Interscience.

Stephans, R. A., and Talso, W. W. (Eds.). (1997). *System Safety Analysis Handbook*, 2nd ed. Albuquerque, MN: System Safety Society.

U.S. Army. (1990, June). *System Safety Management Guide*, Pamphlet 385-16. Washington, DC: Department of the Army.

U.S. Coast Guard. (2001). *Risk-Based Decision-Making Guidelines*, 2nd ed. Washington, DC: U.S. Coast Guard. Available: http://www.uscg.mil/hq/g-m/risk/e-guidelines/html/index.htm.

U.S. Department of Defense. (DOD) (1993). *System Safety Program Requirements*, MIL-STD 882C. Washington, DC: U.S. Department of Defense.

U.S. Department of Defense. (DOD) (2000, February). *Standard Practice for System Safety*, MIL-STD 882D. Washington, DC: U.S. Department of Defense.

U.S. Department of Defense. (DOD) (2001, June). *Mandatory Procedures For Major Defense Acquisition Programs (MDAPS) and Major Automated Information System (MAIS) Acquisition Programs*, DOD 5000.2-R. Washington, DC: U.S. Department of Defense.

U.S. Department of Energy (DOE). (1997, September). *Hazard Categorization and Accident Analysis Techniques for compliance with DOE Order 5480.23, Nuclear Safety Analysis Reports*, DOE-STD-1027-92, Change Notice 1. Washington, DC: U.S. Department of Energy.

U.S. Department of Transportation (DOT). (1998, June). *Safety Risk Management*, FAA Order 8040.4. Washington, DC: U.S. Department of Transportation.

Walker, D. A. (2000, May). *Field Demonstration Workshop on Performance-Based Inspection of Vessels Entering The St. Lawrence Seaway (Establishing Specific Inspection Plans)*, LR-101-11.3-1B-94. Washington DC: U.S. Coast Guard. Available: http://www.uscg.mil/hq/g-m/risk/e-guidelines/html/vol4/Volume4/Tool-spec_Rec/Failure%20Modes%20and%20Effects%20Analysis%20(FMEA)/FMEA.Htm.

## RELATED READINGS

Bahr, N. J. (1997). *System Safety Engineering and Risk Assessment: A Practical Approach*. Philadelphia, PA: Taylor and Francis.

Brauer, R. L. (1994). *Safety and Health for Engineers*. New York: Wiley.

Briscoe, G. J. (1997, June). *Risk Management Guide*, SSDC-11. Idaho Falls, ID: EG&G Idaho.

Clark, R., Benner, L., and White, L. M. (1987, March). *Risk Assessment Techniques Manual*. Oklahoma City, OK: Transportation Safety Institute.

Grose, V. L. (1987). *Managing Risk: Systematic Loss Prevention for Executives*, Arlington , VA: Omega Systems Group.

Kije, L. T. (1963). *Residual Risk*. Rusee.

Layton, D. M. (1989). *Systems Safety Including DOD Standards*, Weber Systems.

Leveson, N. G. (1995). *SAFEWARE: System Safety and Computers, a Guide to Preventing Accidents and Losses Caused by Technology*. Reading, MA: Addison-Wesley.

Rodgers, W. P. (1971). *Introduction to System Safety Engineering*, New York: Wiley.

Saaty, T. L. (1996). *The Analytic Hierarchy Process: Planning, Priority Setting, Resource Allocation*, 2nd ed. Pittsburgh, PA: RWS Publications.

Society of Automotive Engineers (SAE). (1996a). *Certification Considerations for Highly Integrated or Complex Aircraft Systems*, Aerospace Recommended Practice 4754. Warrendale, PA: SAE.

Society of Automotive Engineers (SAE). (1996b). *Guidelines and Methods for Conducting the Safety Assessment Process on Civil Airborne Systems and Equipment*, Aerospace Recommended Practice 4761. Warrendale, PA: SAE.

Stephenson, J. (1991). *System Safety 2000, a Practical Guide for Planning, Managing, and Conducting System Safety Programs*, New York Van Nostrand Reinhold.

Tarrants, W. E. (1980). *The Measurement of Safety Performance*, New York Garland STPM.

Vincoli, J. W. (1993). *Basic Guide to System Safety*, New York Van Nostrand Reinhold.

CHAPTER 15

# Environmental Health Hazard Analysis and Assessment

WELFORD C. ROBERTS

## 15.1 INTRODUCTION[1]

Health hazards associated with both military and civilian systems and equipment involve various aspects of the human–environment interface. The relationship between the environment and human health has been recognized for centuries and has become even more evident as society and technology advance. Often these relationships are considered in very compartmentalized ways such as in the workplace, home, or ambient (outside) environment. In the civilian sector, on the federal level, concerns about workplace health hazards are the responsibility of the Occupational Safety and Health Administration (OSHA). Responsibilities for environmental health hazards outside the workplace are assigned to a variety of other federal agencies such as the Environmental Protection Agency (EPA), agencies within Department of Health and Human Services, Food and Drug Administration (FDA), and others. These responsibilities are similarly divided at state and more local governmental levels. In the military services, programs that address these types of concerns have a variety of descriptors [e.g., environment, safety, and occupational health (ESOH); environmental and occupational health (EOH); and environment, health, and safety (EHS)]. There are specialized professionals and special detection and monitoring equipment and techniques to address issues associated with compartmentalized spaces. For example, an industrial hygienist may only deal with workplaces and use OSHA or National Institute of Occupational Safety and Health (NIOSH) sampling and analytical methods and guidelines to detect and characterize occupational health concerns. An Environmental Health Scientist may only deal with the ambient environment and use EPA sampling and analytical methods and guidelines to detect and characterize health concerns.

For simplicity, we will consider the health concerns that are addressed in this chapter to be within the discipline of *environmental health*. *Environment* is defined as all factors that

*Handbook of Human Systems Integration*, Edited by Harold R. Booher.
ISBN 0-471-02053-2 

are external to the human body. Thus, environmental health represents the relationship of the human–environment interface. When all environmental factors are considered collectively, they are considered to be the *macroenvironment*. Other, more specific human–environment interfaces, such as workplace exposures, indoor air quality issues, confined spaces, and others, can be considered to be *microenvironments*. When one assesses environmental health impacts, all exposure sources should be considered. Therefore, the risk assessment process (i.e., hazard determination, dose–response assessment, exposure assessment, risk characterization) should include the exposures from all microenvironments.

Many products that we use in our society can have potential dangers that may be inherent in their composition or may result from the way that they are used, stored, or disposed. For example, vehicles that use combustion engines and fossil fuels can generate a variety of potentially dangerous exhaust products such as carbon monoxide. Some vehicles (e.g., large trucks, aircraft, etc.) also can emit high noise levels that may damage hearing. Even the clothes that we wear may be treated with chemicals that can be harmful under certain conditions (e.g., pesticide-impregnated military garments) or they may be a factor associated with the development of temperature-related diseases. The potential dangers can be minimal or negligible if the hazardous component is very small, if people's contact with it is very limited, or if it is shielded to prevent human exposure. In the civilian community federal laws and federal and industry guidelines and practices provide standards and criteria for the safe design of many commodities.

When military equipment and systems are developed or acquired, concerns about many environmental health issues may converge. All of the military services—army, air force, navy, and marines—have formal programs to develop and acquire equipment and systems. And all services focus on health concerns for the people who must use, maintain, or otherwise handle military equipment. These people may be exposed to a variety of environmental health hazards that are chemical, biological, or physical in nature. As part of the human systems integration (HSI) program, the military services are required to have a formal process to evaluate new and modified military systems and equipment for potential health hazards and to acquire recommendations for eliminating or minimizing such hazards during design [U.S. Department of Defense (DoD), 2001].

The following discussion describes the health hazard analysis process applied by the U.S. Army as one example of how a DoD service identifies and mitigates environmental health hazards associated with military equipment and systems during the materiel acquisition process.

### 15.1.1 U.S. Army Health Hazard Assessment Program

The U.S. Army has consolidated the health aspects of HSI under a single centralized program, the Health Hazard Assessment (HHA) program. The HHA program is part of the army's Preventive Medicine Program; therefore, it applies a public health perspective to prevent soldiers from being exposed to harmful levels of chemical, physical, and biological agents.

The HHA program has a specific responsibility and support structure that promotes interaction between the combat development, materiel development, and medical communities. The U.S. Air Force, U.S. Navy, and U.S. Marine Corps also provide medical consultation, review, and support to their materiel developers but in a more decentralized manner. This discussion focuses upon the centralized army HHA program as a convenient method of organizing and presenting information about the various health categories. Even though the focus is on the army HHA process, it is the intent of this chapter to reflect state-of-the-art assessment practices for the various health categories. Therefore, from this approach one can extrapolate the technologies to health assessments being performed by other military services for similar or the same health hazards. It should be noted that occasionally there are joint developments where the army, air force, navy, and marines combine resources and efforts to develop materiel to be used by all services. During joint developments the Army may take the lead for the health issues and use the HHA program to identify and assess health concerns.

Combat scenarios and conditions are expected to be extreme, stressful, and harmful to the health and well-being of military personnel. In fact, we have come to expect that in combat, and, to a lesser extent, during military training, some soldiers will be killed or injured from causes other than enemy aggression. These effects are classified as Disease and Nonbattle Injuries (DNBIs) and are reported to account for a significant numbers of casualties (Lynch et al., 1999). Also, as exemplified in the next paragraph and as presented in more detail by Gaydos (1988, 1993), past history shows that military forces have been exposed to stresses from their own weapons and equipment causing adverse health effects and casualties. The adverse health consequences associated with the use of military weapons and equipment contributes to DNBI casualty statistics. Even if disease or death does not occur, such stresses can adversely affect soldiers' ability to perform their mission. The purpose of the HHA program is to eliminate (where possible) or (if not) to minimize such adverse conditions. In doing so, HHA may provide benefits ranging from enhancing soldier capability and mission effectiveness to preventing disease and loss of life.

Since the beginning of the U.S. Army its medical officers, and subsequently the Army Medical Department (AMEDD), have been providing commanders with informal health hazard information about military weapons and equipment. This informal process continued through the Civil War, World War I, and World War II until the late 1970s with issues involving exposure to toxic gases and vapors arising from guns and combustion engines, especially when used in confined spaces such as armored tanks (Gaydos, 1988, 1993). In 1976, blast overpressure hazards were identified during the development of the M198 towed howitzer, and the army surgeon general was asked for help to overcome this hazard (Gross and Broadwater, 1993; Gaydos, 1988). Also during the late 1970s the AMEDD became involved with the assessment of carbon monoxide health hazards from exposure in the Bradley fighting vehicle (Gross and Broadwater, 1993; Gaydos, 1988). In both of these situations the medical community was not involved with the development process until its later stages. The M198 howitzer and Bradley health issues should have been identified and addressed early in the conceptual stages of the acquisition process "to preclude costly and even unacceptable changes" (Gaydos, 1988). These events eventually led to the development of the formal HHA program during the early 1980s. In 1983 U.S. Army Regulation 40-10, *The Army Health Hazard Assessment Program in Support of the Materiel Acquisition Decision Process*, was published and marked the formal beginning of the army's HHA program. Since that time the HHA program has made great strides in providing formal health hazard support to the army's materiel acquisition decision process. For more detail about the history, background, or specific procedures associated with the HHA program, the reader should consult Gross and Broadwater (1993) and Gaydos (1988, 1993).

## 15.1.2 Military and Equipment Health Hazards

There are nine health hazard categories typically addressed by the HHA program: acoustic energy, biological substances, chemical substances, oxygen deficiency, radiation energy, shock, temperature extremes and humidity, trauma, and vibration. These hazards are listed in Figure 15.1 along with a brief description. A more detailed description of each category and subcategories and examples of the types of equipment and systems that may produce them are provided later in this chapter.

**Acoustic Energy**
The potential energy that transmits through the air and interacts with the body to cause hearing loss or damage to internal organs,

**Biological Substances**
The exposure to microorganisms, their toxins, and enzymes.

**Chemical Substances**
The hazards from excessive airborne concentrations of toxic materials. Exposure occurs through inhalation, ingestion, and skin or eye contact.

**Oxygen Deficiency**
The displacement of atmospheric oxygen from enclosed spaces or at high altitudes.

**Radiation Energy**
*Ionizing*: The radiation causing ionization when interfacing with living or inanimate matter
*Nonionizing*: The emissions from the electromagnetic spectrum with insufficient energy to produce ionization of molecules.

**Shock**
The mechanical impulse or impact on an individual from the acceleration or deceleration of a medium.

**Temperature Extremes & Humidity**
The human health effects associated with high or low temperatures, sometimes exacerbated by the use of a materiel system.

**Trauma**
*Physical*: The impact to the eyes or a body surface by a sharp or blunt object
*Musculoskeletal*: The effects to the system while lifting heavy objects.

**Vibration**
The contact of a mechanically oscillating surface with the body.

**Figure 15.1** Health hazards assessed through the army health hazard assessment program.

This chapter presents the health aspects of HSI by discussing the U.S. Army's perspective as reflected in its HHA program. Each of the nine heath hazard categories and pertinent subcategories are defined and further discussed in terms of the technology to detect, measure, analyze, and assess them. Examples of equipment and systems that produce each hazard are presented in addition to the many professionals who may be involved with the assessment process. There also are discussions about some of the common tools that support the overall HHA program and that are associated with all hazard categories. These include risk assessment and risk assessment codes, the exposure control hierarchy, the medical cost avoidance model, and the hazard tracking database.

## 15.2 HEALTH HAZARD CATEGORIES

There are nine health hazard categories routinely assessed by the AMEDD in support of the HHA program. This section defines and describes each hazard category, and when necessary subcategories, and provides examples of the types of equipment or systems that may have such hazards.

### 15.2.1 Acoustic Energy

The health hazard category of acoustic energy refers to the effects that result when people are exposed to *steady-state noise*, *impulse noise*, and *blast overpressure*. This discussion is limited to the health effects that can occur from exposure to such phenomena in air and not in any other media (e.g., water, other fluids, soil, and other solid material).

Acoustic energy is that energy caused by pressure waves that propagate through the air to interact with the body. The risk of injury from the various forms of acoustic energy is directly proportional to the amount of energy that forms the pressure wave. More specific detail concerning the physics of sound (e.g., pressure, intensity, power, frequency, and propagation) may be found in Bruce et al. (1998).

The terms *sound* and *noise* sometimes are used synonymously to refer to acoustic energy. Noise typically is defined as an unwanted or unnecessary sound [American Industrial Hygiene Association (AIHA); 1975]. There also may be psychological and social components to the reaction to noise (e.g., residents in neighborhoods close to airports, military installation, race car tracks, etc.); however, these are not discussed in this chapter.

Sound may interact physically with the body in a manner that can cause an adverse effect. There are two categories of noise that are of interest to the heath hazard assessor: steady-state noise and impulse noise. Blast overpressure is a form of impulse noise that also is addressed here.

***Steady-State Noise*** Steady-state noise is a variation in air pressure around the ambient atmospheric pressure. The duration of the variation exceeds 1 second (U.S. Army, 1996). Steady-state noise can be continuous, intermittent, or fluctuating. The primary health effect that is caused by steady-state noise is hearing loss. Noise-induced hearing loss is progressive, and its onset is generally imperceptible. Initially, individuals may be unaware of any hearing loss and may not have problems in quiet listening situations. Noise-induced hearing loss is characterized by reduced hearing sensitivity at frequencies above 2000 hertz (Hz). Other symptoms may include ringing in the ears (tinnitus), a temporary muffling of sound after noise exposure, and a sensation of fullness in the ears. Continued, unprotected exposure to hazardous noise levels causes progressive hearing loss into the lower frequencies and a marked loss in communication ability. Individuals with a high-frequency noise-induced hearing loss may complain that they can hear people talking, but they cannot understand their words. Repeated long-term exposure to high-intensity steady-state noise causes permanent hearing loss. Other health effects from exposure to steady-state noise (e.g., tissue heating) is physically possible but has not been encountered at the noise levels generated by equipment evaluated for military use.

There are numerous sources of steady-state noise. Examples found in both military and civilian settings include wheeled and tracked vehicles, self-propelled artillery, aircraft (rotary and fixed wing), communication headsets and speakers, alerting or warning signals, power generators, training simulators, maintenance tools and equipment, gas torches, and compressed air or gas (U.S. Army, 1994a; Gross and Broadwater, 1993; Leibrecht, 1990). People may be exposed to steady-state noise by operating this equipment or simply by being in close proximity of the hazardous noise sources.

***Impulse Noise*** Impulse noise is a variation in air pressure above the average atmospheric pressure lasting less than 1 second (U.S. Army, 1996). This phenomenon is referred to as impulse noise when it causes auditory effects. When there are non auditory effects, it is referred to as blast overpressure (discussed in the next section).

The primary health effect that is caused by impulse noise is hearing loss. At low noise levels, impulse noise does not cause adverse health effects. However, as the noise level increases, it produces a recoverable loss in hearing sensitivity referred to as a temporary threshold shift (TTS). If the TTS is small and recovers rapidly, long-term consequences are minimal. However, the short-term consequences could have adverse effects on performance when the detection of faint sounds is important. A large TTS could indicate an inner ear injury requiring more than 24 hours to recover. An injury of this type could be cumulative and could lead to a permanent hearing loss in addition to an adverse effect on performance that is dependent upon hearing ability. Impulse noise at higher levels can cause larger losses of hearing sensitivity that never completely recover to produce a permanent threshold shift (PTS). The PTS is indicative of a permanent organ injury. A PTS adversely affects hearing-dependent performance. At very high levels, impulse noise can cause tympanic membrane (eardrum) rupture and damage the ossicular chain and inner ear.

A few examples of items that produce impulse noise include pistols, machine guns, grenades, mortars, cannons, tank guns, howitzers, recoilless rifles, rockets, missiles, nuclear explosives, training simulators, and impact tools and equipment (U.S. Army, 1994a; Gross and Broadwater, 1993; Leibrecht, 1990). Some of these items are found in both the civilian and military environments.

***Blast Overpressure*** Blast overpressure is a special case of impulse noise produced generally by the rapid burning of material. A blast wave is characterized as variations in ambient pressure over time (pressure–time history). This increase in ambient pressure is called blast overpressure. The level of blast overpressure in a specific location depends on the energy of the explosion, the distance from the point of detonation, the elapsed time since explosion, and the measurement technique. Fuel air mixtures produce large overpressures with long durations while weapons produce modest peaks with shorter durations.

Blast overpressure can produce nonauditory injuries to gas-containing organs in the body. These include the middle ear, upper respiratory tract, lung, and gastrointestinal tract.

The health effects that occur from blast overpressure exposure can range from those that are transient and insignificant to death. They include trivial surface petechiae, rupture of the visceral pleura, pneumothoraces and hemothoraces, gross hemorrhage, and pulmonary edema.

Examples of typical sources that produce blast overpressure are mortars, cannons, tank guns, howitzers, recoilless rifles, rockets, missiles, explosives, and nuclear warheads (U.S. Army, 1994a; Gross and Broadwater, 1993; Leibrecht, 1990).

### 15.2.2 Biological Substances

Biological substance health hazards are those that are associated with living organisms or emanate from them. The organisms include poisonous plants and animals and a wide variety of human pathogenic microorganisms and/or their associated toxins. A biological hazard may produce disease, death, or a debilitating condition when the agent is introduced to a susceptible host in sufficient quantity to cause an adverse effect. It may be introduced by ingestion (e.g., contaminated food or water), inhalation (e.g., viruses, mold/fungi spores), skin contact (e.g., poison ivy or oak), or injection (e.g., a snake bite or bee sting).

Biological hazards may come from a broad spectrum of sources that may include microorganisms (e.g., bacteria, viruses, rickettsia, molds, etc), parasites, venomous insects, and other animals (e.g., reptiles, amphibians, marine animals, etc.), insect vectors, and poisonous and toxic plants. More detailed information concerning these types of hazards may be found in Russell (1996), Norton (1996), Kotsonis et al. (1996), U.S. Army (1994b), and Beneson (1990). Some specific examples of biological hazards are

- diseases transmitted to humans by various animal species, primarily insect and rodent vectors, such as malaria, encephalitides, hemorrhagic fevers, Lyme disease, leishmaniasis, plague, and rabies;
- various communicable diseases and allergic-type reactions to animal dander, which cause adverse respiratory or gastrointestinal symptoms resulting in noncombat lost time during training exercises and wartime activities;
- exposure to toxic plants such as poison ivy, poison oak, and numerous other common species;
- exposure to stinging and biting insects and arthropods such as bees, wasps, ticks, flies, mosquitoes, fire ants, spiders, scorpions, and centipedes;
- exposure to species of poisonous lizards and snakes such as Gila monster, coral snake, and pit vipers, including rattlesnakes, water moccasins, and copperheads;
- exposure to bloodborne pathogens such as hepatitis B and human immunodeficiency virus (HIV) as a result of work injury or improper medical waste disposal;
- diseases and debilitating ailments resulting from substandard levels of personal hygiene and sanitation such as dermatitis and other skin diseases, body, head, and crab lice, and scabies; and
- poor sanitation and personal hygiene practice (which includes potential hazards associated with operation of food service facilities and management of field rations), microbiological quality of water supply, solid and liquid waste disposal, management

of sewage disposal, infectious and medical wastes, pest management, graves registration, and field sanitation and personal hygiene practices.

The following are examples of military equipment-related situations encountered during HHAs that require analysis of biological hazards:

- new rations and field-feeding systems (e.g., methods of cooking, packaging, and delivering foods; equipment evaluation of food safety criteria; field cleaning and sanitizing processes; solid and liquid waste disposal; and pest management considerations);
- field water treatment and packaging systems (e.g., microbiological analysis, water treatment methodology, distribution system including plumbing, storage and packaging, and backflow and cross-connection control);
- field waste collection and disposal systems (e.g., field expedient disposal methods, wastewater disposal, recycling requirements, hazardous waste, effluent discharge and runoff, pest management considerations, and pollution prevention);
- field housing units and support facilities (e.g., tents and shelters, laundry and bathing facilities, personal hygiene standards, and pest management considerations); and
- new field uniforms (e.g., construction materials and design, field laundry operation, insecticide impregnation, snake and insect bite protection, and handling if exposed to bloodborne pathogens).

### 15.2.3 Chemical Substances

Chemicals are prevalent and ubiquitous in the environment. They can occur naturally in the environment; however, many are man made (synthetic) and introduced into the environment through anthropogenic ways. Typically, the HHA program assesses chemicals that are synthetic and/or processed (e.g., mined, refined, etc.) or their by-products. As defined for the HHA program, the chemical substances health hazard category focuses on toxic liquids and solids and excessive airborne concentrations of mists, gases, vapors, fumes, or particulate matter (Gross and Broadwater, 1993).

People or the environment may be exposed to potentially harmful chemicals during the development, use, storage, and ultimate disposal of military equipment. People can be exposed through several routes to include inhalation, ingestion, dermal (skin) absorption, or direct injection (parenteral).

The sources of chemical concerns typically addressed in the HHA process include both military-unique and nonmilitary-unique operations and equipment. Examples of such sources include wheeled and tracked vehicles, vessels, aircraft, weapon systems (e.g., guns, rifles, rockets, missiles, etc.), smokes and obscurants, chemical agents, and maintenance and logistics operations. Table 15.1 presents some examples of military systems that have been evaluated and their associated chemical hazards.

***Military-Unique Chemical Standards*** Some chemical substances are found more frequently in a military environment and/or the nature of the exposure pattern (e.g., frequency, duration, or concentration) is significantly different from that which occurs in non military settings. Therefore, the AMEDD, through its research efforts and in coordination with other military organizations (e.g., the materiel and combat development communities), developed military-unique exposure levels for such substances.

**TABLE 15.1 Examples of Military Systems and Associated Chemical Hazards**

| Military System | Chemical Hazard(s) |
|---|---|
| Avenger | Hydrogen chloride |
| Landing Craft Utility | Degreaser, welding and soldering fumes, sanding/grinding particulates |
| M43A1 Protective Mask | Bromobutyl natural rubber |
| Theater High Altitude Area Defense System (THAAD) | Diesel engine exhaust, rocket motor propellant and oxidizer, fire-extinguishing agents, nuclear–biological–chemical agents, off gassing |
| M-109A6 Paladin 155-mm Self-Propelled Howitzer Munitions | Lead |
| JAVELIN Advance Antitank Weapon System | Lead |

Carbon monoxide (CO) is one example of a chemical substance that has unique military concerns. It is a chemical substance most frequently identified as a potential hazard by the HHA program (U.S. Army, 1996). Its military-unique exposure standard is based upon the level of carboxyhemoglobin (COHb) that develops in the blood. Blood levels are determined by taking test data collected from weapon systems (i.e., measured CO concentrations) and applying it to an equation, which also has a computerized format, that was developed to estimate COHb levels (MIL-HDBK-759B, 1992; MIL-STD-1472D, 1989). The equation is also used to determine maximum allowable consecutive executions (MACEs) for firing from weapons systems (e.g., tanks and howitzers) and the number of minutes required between firings to allow predicted COHb levels to drop below allowable guidelines. Additional information concerning the military-unique CO standard can be found in several military references (MIL-HDBK-759C, 1995; MIL-STD-1472F, 1999; U.S. Army, 1996).

There are specific military guidelines for fog oil (U.S. Army, 1990b), some nerve agents (U.S. Army, 1990c), and mustard exposure (U.S. Army, 1991b). More detailed discussions concerning military-unique aspects of carbon monoxide, lead, various types of smokes and obscurants, diesel engine exhaust, and chemical warfare agents can be found in the *U.S. Army Health Hazard Assessors Guide* (U.S. Army, 1996).

The National Research Council's Committee on Toxicology (NRC-COT) develops military-unique exposure limits for a number of airborne contaminants for the DoD and National Aeronautical and Space Administration (NASA). The NRC-COT recommends exposure limits that allow army personnel to function in emergency situations with the unlikelihood of suffering from irreversible health effects. These limits are known as emergency and continuous exposure guidance levels (EEGLs) and short-term public emergency exposure limits (SPEGLs) (NRC, 1984a–c, 1985a,b, 1986, 1987, 1988). The specific chemicals are listed in Table 15.2.

### 15.2.4 Oxygen Deficiency (Ventilation)

The oxygen deficiency category addresses a variety of situations and conditions that affect the availability of oxygen in breathable air. This section discusses low oxygen concentrations as they relate to confined spaces, enclosed spaces, and ventilation needs.

**TABLE 15.2 Specific Chemicals Addressed in Emergency and Continuous Exposure Limits for Selected Airborne Contaminants (EEGLs) and Short-Term Public Emergency Exposure Limits (SPEGLs)**

| Volume | Chemicals |
|---|---|
| 1 | Acetone, acrolein, arsine, carbon disulfide, chloroform, fluorine, mercury vapor, methane, ozone, sulfuric acid |
| 2 | Chlorine; chlorine trifluoride; ethanolamine; fluorocarbon 11, 12, 21, 113, 114; isopropyl alcohol; phosgene; sodium hydroxide; sulfur dioxide; vinylidene chloride; xylene |
| 3 | Bromotrifluoromethane (Halon 1301) |
| 4 | Aluminum oxide, carbon monoxide, ethylene glycol, hydrogen sulfide, methanol, nitrogen dioxide, nitrogen tetroxide, nitrous oxide |
| 5 | Hydrazine, monomethylhydrazine, 1,1-dimethylhydrazine |
| 6 | Benzene, ethylene oxide |
| 7 | Ammonia, hydrogen chloride, lithium bromide, toluene |
| 8 | Lithium chromate, trichloroethylene |

An *oxygen-deficient atmosphere* is an atmosphere that contains less than 19.5 percent oxygen by volume (29 CFR 1910.146). *Ventilation* is one of the principal methods to control health hazards and may be defined as causing fresh air to circulate to replace contaminated air (Olishifski, 1985).

A *confined space* is an enclosed area that is large enough to accommodate a person's body, has limited means for entry and exit, and is not designed or intended for continuous human occupancy (Schroll and Harris, 1998). A confined space may create conditions that can affect the nature of its atmosphere. Oxygen deficiency may occur when gases or vapors exceed their upper explosion limit and when oxygen is consumed by chemical reactions (e.g., rusting) or biological reactions (e.g., biological breakdown of organic materials). Also, oxygen can be displaced by an inert gas (e.g., nitrogen) or by the operation of an internal combustion engine (Schroll and Harris, 1998; Todd, 1998). In addition to causing oxygen-deficient atmospheres, confined spaces also may concentrate air contaminants to hazardous levels, cause dangerous oxygen-enriched environments, and have an internal configuration that can cause its contents to trap, crush, and/or asphyxiate someone (NIOSH, 1979). Additional information concerning the characteristics of confined spaces may be found by referring to Dinardi (1998), NIOSH (1979), and OSHA standards (29 CFR 1910.146).

Similar to confined spaces, the military considers the health aspects of *enclosed spaces*. Enclosed spaces differ from confined ones in that they are designed for detail work or occupancy for extended periods of time and are designed to receive adequate ventilation (MIL-HDBK 759B, 1992).

Examples of confined spaces include but are not limited to storage/holding tanks, vessels, silos, pits, sewers, pipelines, tank cars, boilers, septic tanks, and utility vaults. Examples of enclosed spaces that are frequently encountered in the military include mobile vans, shelters, crew compartments, and vehicle cabs (MIL-HDBK 759C, 1995).

### 15.2.5 Ambient Pressure Changes

The human body can experience oxygen deficiency effects that result from decreased barometric pressure. Reduced atmospheric pressure decreases the rate at which oxygen diffuses into the blood. Thus, people who are exposed to low-pressure environments suffer the effects of a variety of hypobaric health hazards, which may include hypoxia, Benign Acute Mountain Sickness (Benign AMS), AMS, and Chronic Mountain Sickness (Monge's Disease) (Popendorf, 1998).

Hypobaric hypoxia can affect mental performance, judgment, sleep, and physical work capacity (Fulco and Cymerman, 1988; Houston, 1984). Specific information concerning, AMS can be found in a variety of references (Fulco and Cymerman, 1988; Hackett and Roach, 1987; Houston, 1984; Young and Young, 1988; Hackett et al., 1976, 1989; Hackett and Rennie, 1977; Montgomery et al., 1989; Reeves and Schoene, 1991; Tso, 1992). Based upon these references, the following paragraphs briefly describe AMS-associated effects and conditions. For more detail, the reader should consult the aforementioned references.

Benign AMS can occur when people move to higher elevations in a short period of time. Headache, anorexia, nausea, insomnia, labored breathing, and general weakness characterize its effects. These may occur as one ascends, but typically the effects occur 6 to 48 hours later, and they usually disappear in 3 to 5 days. Acute mountain sickness is rare below 2400 m (8000 ft) elevation but is very common above 3600 m (12,000 ft).

Acute mountain sickness (other than benign) is characterized by high-altitude pulmonary edema (HAPE) and/or high-altitude cerebral edema (HACE). The progressive symptoms of HAPE include fatigue, labored breathing on exertion, nonproductive cough, labored breathing at rest, a cough progressing from producing frothy white to slightly bloody sputum, and coma followed by death within 6 to 12 hours. High-altitude cerebral edema can occur in mountaineers who ascend rapidly to altitudes higher than 2400 m. Often occurring in conjunction with HAPE, HACE is a complication that results from rapid exposure to very high altitudes. It is characterized by mental dysfunction (hallucinations, bizarre behavior) and neurological abnormalities (ataxia, paralysis, cerebellar signs) and may progress to coma and death.

Monge's disease is similar symptomatically to benign AMS and may exhibit characteristics of HACE. However, it is rare and occurs after chronic (years of) exposure (Popendorf, 1998).

Decompression sickness (DCS) also can occur at high altitudes and is similar to that associated with diving (Popendorf, 1998). Even though it is not a cause of oxygen deficiency, DCS is considered when assessing barometric risks.

Environments that have decreased barometric pressure are found at high terrestrial elevations or are simulated in a hypobaric chamber. Typical occupations that encounter the elevation hazard include high-altitude construction, mining, and aviation (Popendorf, 1998). The military deploys its forces to terrestrial altitudes that are greater than 2500 m (8200 ft) where their health and performance can be affected adversely. Many strategic areas of the world, including the Middle East, Asia, and South America, contain land areas with elevations greater than 3000 m (10,000 ft) (U.S. Army, 1975). Military personnel deployed to these elevations are exposed to the hazards of hypobaric hypoxia and the AMSs presented in this section.

A hypobaric chamber facility can simulate high-terrestrial-altitude conditions by reducing the chamber pressure using vacuum pumps. These chambers are used to observe and determine the effects of hypobaric conditions on people in laboratory studies.

### 15.2.6 Radiation Energy

Radiation is electromagnetic energy that can be divided into two broad categories—that which causes matter to ionize (i.e., *ionizing radiation*) and that which does not cause matter to ionize (i.e., *nonionizing radiation*). Both nonionizing and ionizing radiation can be further subclassified by wavelength, frequency, and energy (Fig. 15.2). The types of nonionizing radiation and fields include ultraviolet, visible light, infrared, microwaves, radar, television, radio waves, extremely low frequency, electric fields, and magnetic fields. Ionizing radiation types are X-rays, gamma rays, and cosmic rays. This section presents and discusses health hazard issues associated with both nonionizing and ionizing radiation.

***Nonionizing Radiation*** The following discussion is organized by first presenting information that generally applies to all forms of nonionizing radiation. Subsequent subparts of this nonionizing section provide information specific to laser/optical radiation and radio-frequency radiation (RFR), respectively. Even though the nonionizing radiation spectrum includes several types of radiations, the focus of the HHA radiation energy health hazard category, and subsequently this subsection, is on laser optical radiation. In addition to information about RFR and laser, discussions about other forms of nonionizing radiation and other forms of optical radiation can be found in Hitchcock et al. (1998) and Moeller (1997). The American Conference of Governmental Industrial Hygienists (ACGIH) (2002) presents Threshold Limit Values (TLVs)® for RFR and laser sources, in addition to TLVs for magnetic fields, microwave radiation, light and near-infrared radiation, and ultraviolet radiation.

Injury from exposure to nonionizing radiation occurs when energy is absorbed by biological tissue. The mechanism and type of tissue injury varies and depends upon the part of the electromagnetic spectrum that is involved. Photochemical effects dominate when energy is absorbed from ultraviolet and short-wavelength visible radiation exposure. Thermal effects dominate when there is exposure to visible and infrared optical radiation. For most of the nonionizing radiation spectrum, the effect is limited to superficial organs such as the skin and eyes.

Nonionizing radiation does not create ions when interacting with matter. Because ions are charged particles, they are chemically more active than their electrically neutral forms. Chemical changes that occur in biological systems are cumulative and can be detrimental or even fatal. By contrast, the biological effects of nonionizing radiation are caused primarily by thermal stress (i.e., the accumulation of heat). When heat is dissipated, the effects do not persist (they are not cumulative). When the thermal stress is extreme, however, persisting injuries such as erythema, cataracts, and burns can occur.

*Laser Optical Radiation* Lasers emit light in a very narrow bandwidth (i.e., a single wavelength or color (monochromatic)) that propagates in a highly directional manner characterized by low divergence (beam spread) (Hitchcock et al., 1998).

Optical (laser) radiation occupies that portion of the electromagnetic spectrum that is the optical radiation region (Fig. 15.2). The eye is the most susceptible organ system to laser radiation in the visible and nonvisible infrared region of the spectrum, because the incident energy is focused to a small spot on the sensory retina.

Optical radiation is emitted by a large variety of army systems. These include communications systems, combat surveillance systems, fire control systems, and target acquisition systems. In addition to laser radiation, some systems also are high-intensity light sources. Optical radiation in this spectral region also may be emitted secondarily from rocket exhausts, the detonation of explosives, and warm bodies.

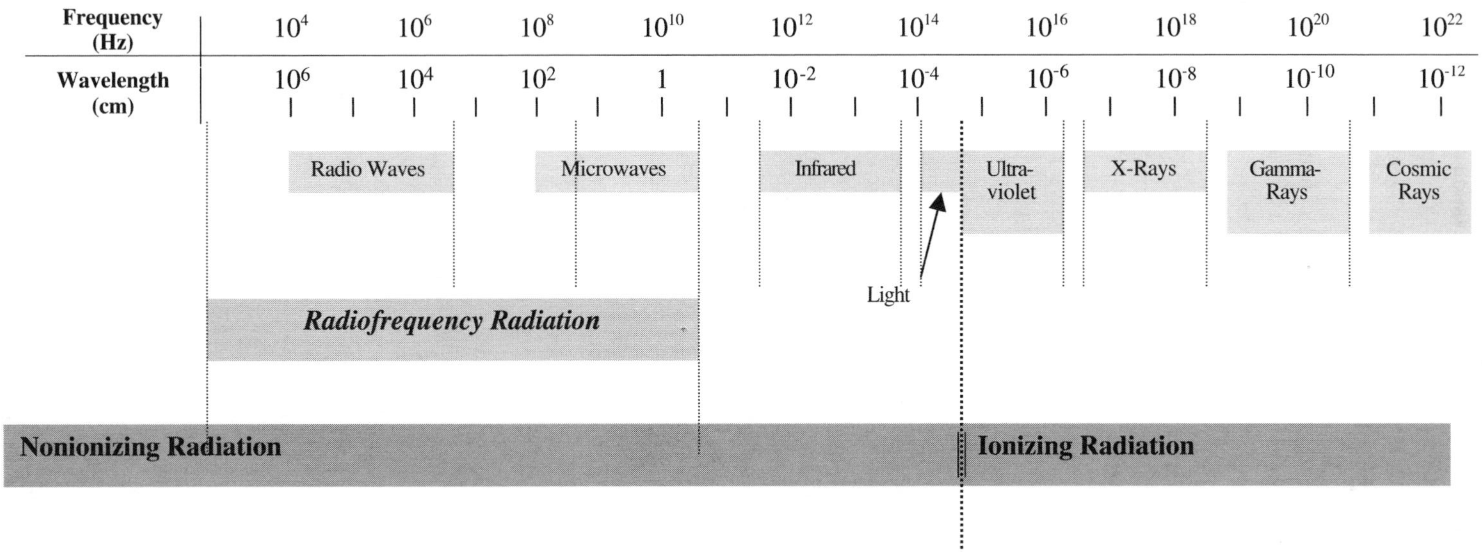


**Figure 15.2** Electromagnetic spectrum showing approximate frequencies in cycles per second ($H_2$) and wavelengths in centimeters (cm) of selected types of radiation.

*Radio-Frequency Radiation* The approximate location of RFR on the electromagnetic spectrum can be seen in Figure 15.2. It can heat tissue to cause a radio-frequency (RF) burn, which can be internal and life threatening. Another cause of biological effects is the induction of RF current in the body. This RF current can stimulate nervous tissue resulting in shock effects.

When RFR is absorbed by biological tissue and converted to heat, if the amount of energy absorbed exceeds the body's ability to dissipate heat, thermal stress or injury can occur. The site of energy absorption varies depending upon the orientation of the individual and the frequency of the energy. In the upper frequency bands (used by radars and satellite/communication sets), the effect is limited to external organs such as the eyes and skin. Internal organs may be affected at lower frequencies as the result of deep-body heating or induced currents. Other factors that influence individual sensitivity to RFR are the individual's unique physiology (especially height, weight, and gender) and the external environment (such as temperature and humidity). Furthermore, energy deposition is not uniform throughout the body but is a function of the dielectric characteristics of various body tissues.

During the operation of most RFR sources, users are exposed to low levels of RFR; the amount of RF energy absorbed does not stress the thermoregulatory system. Consequently, no effects are observed and individuals cannot perceive the RF energy being absorbed. There are no known long-term health effects from chronic exposure to low-level RFR. The perception of RFR does not imply that an injury has occurred, especially when most of the energy is deposited near the surface of the skin where temperature sensors abound. It is expected that many systems' operators will at some time in their career perceive a mild warming sensation near an RFR source. A few systems are capable of producing painful RFR intensities if the exposure is long enough. Although there may be no damage, the exposed individual will likely avoid repeating the encounter. In extreme cases, the incident could affect the operator's performance. For increasing intensities and exposure times, the threat of tissue damage increases. Acute effects such as cataracts and burns are possible. Ultimately, the injury may become life threatening or cause a permanent disability.

For RFR less than 100 megahertz (MHz), RF shock and burn may result from induced current. Two interactions may result relative to one or both of these effects: either a spark discharge, when one is close to an RF energized conductor, or a current or flow to ground while one is making contact with an RF energized conductor. The spark discharge phenomenon is comparable to what happens after walking on a carpet and touching a grounded object—a spark is drawn. A burn results if enough current enters the body through a small cross section. Spark discharges are expected near an antenna; however, these discharges are unacceptable coming from the transmitter chassis. The threshold for RF current perception is a function of frequency, surface area of the contact point, and individual sensitivity. The current is perceived as heat if it is above 100 kilohertz (kHz). It is perceived as a tingling sensation if the current is below 100 kHz. Although the limits are intended to prevent perception, there is no threat associated with it. The next effect is annoyance resulting from mild shocks. The individual will act to avoid repetitive shock. Individuals startled by an RF shock could injure themselves or someone else while jerking away from the source of the shock. Several serious effects may occur below 100 kHz and at sufficient current densities. Life-threatening situations can result when individuals are unable to release the conductor; they may be unable to breathe or their heart may fibrillate. Fortunately, there are very few tactical systems operating in this frequency range.

Interference with electronic life support systems (e.g., pacemakers) is an indirect effect of RFR exposure. The FDA tests all such devices to ensure that they are not susceptible at the RFR levels frequently encountered. Due to advances in pacemaker technology, the potential for pacemaker interference is virtually nonexistent.

Radio-Frequency electromagnetic waves are emitted by a large variety of army systems for communications, combat surveillance, fire control, and target acquisition. These waves are produced by converting electricity into RFR using various types of generators. It has nothing to do with radioactivity, and natural sources of RFR are inconsequential. Furthermore, unless the generator is operating, RFR will not be present. In fact, many systems are designed to minimize the amount of radiation emitted to prevent detection. These sources of RFR include the following types of systems: radars, radios, satellites communications set, industrial sealers, and electronic countermeasures sets [see U.S. Army (1996) for additional information concerning these systems].

***Ionizing Radiation*** Ionizing radiation is electromagnetic or particulate radiation capable of producing ions, directly or indirectly, in its passage through matter. Examples of ionizing radiation are alpha and beta particles, X-rays, gamma rays, neutrons, and heavily charged ions. The reader should consult the HHA *Assessor's Guide* (U.S. Army, 1996), McCarthy (1998), and Moeller (1997) for additional information concerning ionizing radiation measurement units and typical natural and iatrogenic exposures. The ACGIH (2002) presents TLVs for ionizing radiation.

The absorption of ionizing radiation in biological material may lead to excitation or ionization. In humans this may be demonstrated by genetic and somatic effects. The induction of cancer is the primary somatic effect. Other somatic effects include effects on growth and development, cataract of the eye lens, life shortening, fertility, and sterility. Exposure *in utero* may induce cancer during childhood. The genetic effects through mutagenesis are expressed, not in the irradiated individuals, but in their immediate or remote offspring.

There are electronic devices that are capable of emitting ionizing radiation (U.S. Army, 1996). Examples of these are X-ray machines, linear accelerators, electron microscopes, cyclotrons, RF generators that use klystrons or magnetrons, and other electron tubes that produce X-rays. There also are materials or combinations of materials that emit ionizing radiation. Such materials may be special nuclear material such as plutonium or enriched uranium; source material such as uranium or thorium; by-product material such as any radioactive material yielded in or made radioactive by exposure to radiation incident to the process of producing special nuclear material; naturally occurring or accelerator-produced radioactive material (NARM), such as radium, classified as source material; and materials containing induced or deposited radioactivity.

Ionizing radiation is used directly in army materiel systems as calibration and check sources for radiation, detection, indication, and computation (RADIAC) or other survey-type instruments, as a source of radio luminescence in meters and gauges, as an ionization source in various devices, and as radiographic sources. Indirectly, ionizing radiation may be emitted from an army materiel system as natural radioactivity or radioactivity incorporated into material or a component of the system.

### 15.2.7 Shock (Acceleration, Deceleration)

Shock is a health hazard category that has been evaluated infrequently in the HHA program. Consequently, its concepts, hazard evaluation, risk concerns, and implications toward soldiers have not been fully developed. In 1995, the health hazard community developed an assessor's guide to describe various health hazards associated with military systems and to document how the risks from being exposed to such hazards were estimated (U.S. Army, 1996). Shock was not included in this guide and was to be included at a later date. However, some concepts concerning this area of health hazard concern were assembled. These concepts are presented here but are not intended to be the definitive approach to evaluating shock in the HHA process. Rather, they are presented only to give the reader some idea about some of the factors that need to be considered when shock is assessed as a health hazard. The following information is compiled from a number of references to include the Society of Automotive Engineers standards (SAE, 1986, 1995).

*Shock, impact*, and *impulse* are terms used to describe the rapid and violent application of mechanical forces to the human body. These forces are characterized by short durations and high magnitudes. Impact injury may be described by the deformation of body tissues in excess of their failure limits, resulting in the destruction of their anatomical structures and, more importantly, the disablement of their physiological functions. The parameters of impact and the physical response of the human body are the focus of biomechanics research that attempts to understand the mechanisms of injury and to determine limits of human tolerance to impact.

Impact is considered to be *blunt* when the forces are distributed over some area of the body and do not cause penetrating injury. Examples of extremely short blunt impacts, 5 to 10 milliseconds (msec) in duration, are those produced by the rear surface of a body armor as it defeats a bullet or by the striking of the exposed head against rigid surfaces. When the direct blunt impact is cushioned by soft tissues of the body, such as blows to the abdomen, or when the striking surface deforms under impact, as in the case of energy-absorbing steering columns in modern automobiles, forces of 10 to 50 msec in durations are generated in the interaction.

Indirect impacts where the forces are generated as a result of sudden whole-body *accelerations* or *decelerations* are generally the longest type of impact, with durations ranging from 50 to 250 msec. For example, the crewmember in a ground vehicle is subjected to whole-body deceleration when the vehicle is brought to a sudden stop. Another example of whole-body deceleration is exposure of the seated pilot during a vertical helicopter crash. Examples of whole-body impact acceleration include a mine blast under a truck, the ballistic impact of a missile with a tank, and the seat ejection of a fixed-wing aircraft.

Penetrating injuries, which are produced by high-speed missiles or sharp objects, involve the concentration of forces over a small area of the body. Because the magnitudes and durations of such forces are difficult to measure, the severity of the impact is generally characterized by the energy of the striking object.

Impact to the human body may occur as a result of aircraft crashes, ground vehicle accidents, mine explosions, parachute opening shocks, landing falls, weapon recoil, and other interactions between the soldier and his or her environment during training or battle missions.

### 15.2.8 Heat Stress

Exposure to excessive heat levels can cause heat stress, which can lead to heat strain. This section briefly describes and characterizes these conditions and the factors associated with them. The information presented here is based on selected references (U.S. Army, 1980a, 1996; ACGIH, 2002; Burr, 1991; Gagge and Gonzales, 1996; Levine et al., 1995; Sawka et al., 1993, 1996; Stephenson and Kolka, 1993; Bishop, 1998; Ramsey and Beshir, 1998; NIOSH, 1986). The reader should consult these references for additional detail.

Heat stress is the product of an interaction of work activity (e.g., a military mission) and environmental factors with physiological factors (U.S. Army, 1996). Work activity can be characterized by factors such as clothing, load carried, terrain, and work rate. Environmental factors include temperature, humidity, solar load, and wind speed. Examples of physiological factors include fitness, hydration, acclimatization, rest, nutrition, medication, and health.

Heat stress can lead to heat strain. Heat strain is characterized by one or more of the following: hyperthermia, increased sweating rate (this decreases heat stroke), dehydration, compromised cardiovascular control, and an increased heart rate (U.S. Army, 1996). Heat strain can result in heat-related injuries such as heat cramps, heat exhaustion, and heat stroke. Mental and physical performance decrements can occur at dehydration levels that are higher than and/or hyperthermic levels lower than those that cause injury. Heat cramps and heat rash also can develop from excessive exposure to heat (Bishop, 1998).

Heat strain is caused by the interaction of work activity (e.g., a military mission) and environment with individual physiological factors (U.S. Army, 1996). The type, intensity, and duration of physical work required by an activity directly affects metabolic heat production. Clothing, especially chemical-protective (CP) clothing and equipment (e.g., heavy backpacks, heavy and/or awkward equipment, etc.), can have a profound affect on bodily heat production and storage. Chemical warfare (CW) treatment drugs such as atropine also increase heat storage by inhibiting sweating.

Environmental factors, such as ambient temperature, humidity, radiant-heat load, and wind speed, affect heat balance. If the ambient temperature (measured as dry-bulb temperature) is sufficiently hot (i.e., greater than or equal to body temperature), it will prevent direct heat transfer away from the body by convection/conduction and radiation. Evaporation will be the only route available for heat loss. Wind speed can aid evaporative heat loss, but clothing, vehicles, and shelters will impede evaporation. If the ambient humidity [measured as dewpoint temperature, wet-bulb temperature, vapor pressure or relative humidity (RH)] is high (i.e., greater than 50% RH), evaporative heat loss is compromised, and heat production by or heat transfer to the body cannot be dissipated.

Individual physiological factors that affect thermoregulation and heat balance include the following: acclimation status, aerobic fitness, hydration and nutrition, general health, use of pharmaceuticals, skin disorders, febrile illness, sleep status, age, gender, and anthropometric factors (i.e., body fat, size, surface area) (U.S. Army, 1996). For more detailed discussion on heat stress, thermoregulation, metabolic heat, and environmental factors associated with heat stress, the reader should consult Bishop (1998), Ramsey and Beshir (1998), and the HHA *Assessor's Guide* (U.S. Army, 1996).

Shelters, vehicles, and clothing are examples of military materiel systems that may cause heat stress. Shelters may cause heat stress if adequate air ventilation and air conditioning are not maintained. Consideration must be given to the added heat load of heat-generating equipment, such as computers, when assessing the heat stress or workspaces. Vehicles also may cause heat stress if air ventilation and/or air conditioning is inadequate. Again, consideration must be given to any additional heat-generating sources such as engines, which should be adequately insulated from the crew compartment.

Garment systems, especially CP garments, can interfere with thermoregulation by impeding evaporation of sweat, the most important physiological mechanism for dissipating heat in hot environments.

### 15.2.9 Cold Stress

Exposure to low temperatures can cause cold stress and injuries. The following paragraphs briefly describe and characterize these conditions and factors associated with them and are based upon selected references (Burr, 1993; U.S. Army, 1976; DoD, 1988; Freund et al., 1994; Young, 1988; Young et al., 1992a,b, Bishop, 1998; Ramsey and Beshir, 1998; ACGIH, 2002). The reader should consult these references for additional detail.

Cold stress is the product of an interaction between work activity and environmental factors (U.S. Army, 1996). Work activity can be characterized by factors such as type of physical work and its associated metabolic rate, work duration and intensity, clothing worn, equipment used, and exposure to wetness. Environmental factors include ambient air temperature, humidity, water temperature (if immersion occurs), humidity, radiant or solar load, terrain (including snow consistency and depth), exposure to wetness, and wind speed. Cold temperatures interact with wind to enhance cooling power, which commonly is referred to as *windchill.* Physiological factors (e.g., rest—sleep status), nutrition, dehydration, general health, medication, anthropometry, age, gender, and training also interact to affect the susceptibility to cold injury. Using CW treatment drugs also affects susceptibility to cold stress.

Cold temperatures can cause a general decrease in body temperature and/or affect specific body areas (U.S. Army, 1996). Generalized cold injury is called *hypothermia*, which is the reduction of body-core temperature. Hypothermia can affect the cardiovascular system and victims may lose consciousness (ACGIH, 2002). Other effects include loss of manual dexterity and fine motor skills, decreased visual acuity, and some psychological responses (Ramsey and Beshir, 1998).

Cold temperatures also can cause freezing injuries (frostnip and frostbite) and nonfreezing injuries (chilblains, trench foot, and immersion foot) to exposed skin and peripheral extremities (e.g., hands, fingers, feet, toes, nose, etc.). Cold stress causes shivering. It also affects peripheral and superficial (skin) blood vessels by causing them to constrict, especially in the extremities, nose, and ears. Other responses to cold stress include skin cooling and reduced blood flow to the hands and feet that can lead to blunted sensations of touch and pain and loss of dexterity and agility during cold exposures longer than an hour. Dehydration can impair performance and increase the risk of injury. It is a response to cold-induced diuresis and/or inadequate fluid intake or nutrition.

Shelters, vehicles, and clothing are examples of military materiel systems that may cause cold stress (U.S. Army, 1996). Shelters may cause cold stress if adequate heating is not maintained. Vehicles also may cause cold stress if heating is inadequate. Clothing systems can cause cold stress and injury if they are not properly insulated, layered, and ventilated. Nonfreezing injuries may occur if clothing restricts blood circulation and the hands or feet get wet.

### 15.2.10 Trauma (Physical and Musculoskeletal)

Trauma is a health hazard category that has been evaluated infrequently in the HHA program. Consequently, its concepts, hazard evaluation, risk concerns, and implications toward soldiers have not been fully developed. In 1995, the health hazard community developed an assessor's guide to describe various health hazards associated with military systems and to document how the risks from being exposed to such hazards were estimated (U.S. Army, 1996). Trauma was not included in this guide and was to be included at a later date. However, some concepts concerning this area of health hazard concern were assembled. These concepts are presented here but are not intended to be the definitive approach to evaluating trauma in the HHA process. Rather, they are presented only to give the reader some idea of some of the factors that need to be considered when trauma is assessed as a health hazard. The primary focus of this section is limited to health consequences addressed by the U.S. Army's ergonomics program. Consequently, much of this presentation is taken directly from the army's pamphlet that addresses the ergonomics program (U.S. Army, 2000). DiNardi (1998) provides specific details about work-related musculoskeletal disorders (WMSDs), work method evaluations, and epidemiological evidence. Other references that the reader may consult for additional information include those by Snook and Ciriello (1991, 1974), Snook and Irvine (1968), Ciriello and Snook (1978, 1983), Snook et al. (1970, Snook (1985), Garg and Ayoub (1980), Ayoub et al. (1980a,b), Asfour et al. (1986), and Ayoub (1991).

The ACGIH identifies ergonomics as "the term applied to the field that studies and designs of the human-machine interface to prevent injury and illness and to improve work performance" ACGIH (2000, p. 109). DiNardi (1998, p. 727) describes ergonomics as "the science of fitting workplace conditions and job demands to the capabilities of the work population." The American Industrial Hygiene Association recognizes that ergonomics is "a multidisciplinary science that applies principles based on the physical and psychological capabilities of people to the design or modification of job, equipment, products, and workplaces" DiNardi (1998, p. 727). Other terms for WMSDs include cumulative trauma disorders (CTDs), repetitive-motion illnesses (RMIs), and repetitive strain injuries (RSIs) (ACGIH, 2002).

The WMSDs are caused or aggravated by repeated biomechanical stress and microtrauma (U.S. Army, 2000). Over time, repeated microtrauma can evolve into a painful, debilitating state involving muscles, tendons, tendon sheaths, and nerves. Some examples of WMSDs include tendonitis, tenosynovitis, bursitis, chronic muscle strain, and nerve entrapment syndromes (e.g., carpal tunnel syndrome).

There are specific workplace conditions that can contribute to the development of WMSDs (U.S. Army, 2000). These are considered to be occupational risk factors and include the following: repetitive motions (especially during prolonged activities), sustained or awkward postures, excessive bending or twisting of the wrist, continued elbow or shoulder elevation (e.g., overhead work), forceful exertions (especially in an awkward posture), excessive use of small muscle groups (e.g., pinch grip), acceleration and velocity of dynamic motions, vibration, mechanical compression, restrictive workstations (e.g., inadequate clearances), improper seating or support, inappropriate hand tools, machine-pacing and production-based incentives, extreme temperatures, and extended exposure to hazardous or annoying noise. The combined effect of several risk factors in one job or workstation may lead to a higher probability of causing a WMSD.

There are many jobs that can cause WMSDs. In addition to the workplace conditions identified in the army ergonomics pamphlet (previous paragraph), DiNardi (1998) also lists some of the typical job activities associated with common CTDs of the upper extremities. These activities are diverse and include functions such as turning screws, grinding, buffing, polishing, hammering, surgery, typing, keying, wiring, etc. Given such diversity, it is not difficult to conclude that most military systems and equipment, either by their use or maintenance, are potential candidates for causing or aggravating WMSDs.

### 15.2.11 Vibration

When a vibration phenomenon involves the entire body, it is known as *whole-body vibration (WBV)*. When only specific parts of the body are involved, it is described as *segmental vibration*. Segmental vibration usually occurs to the hands, wrist, and arms and also may be described as *hand-transmitted vibration*. The primary focus on vibration hazards in the HHA program has been with WBV associated with vehicles. Consequently, the majority of this presentation is on WBV.

Vibration is an oscillatory motion characterized by alternate increase and decreases in displacement (U.S. Army, 1996). Oscillatory motions or vibrations in the human usually occur through physical contact with a vibrating source. Whole-body vibration occurs when oscillatory motions are transmitted to the entire body through contact with a vibrating source at the feet of a standing individual, at the buttocks of a seated individual, and along the entire side of the body of a supine individual.

*Segmental vibration* occurs when a specific body segment is in contact with a vibrating source, but the vibrations are not typically transmitted to other parts of the body. The major area of concern for segmental vibration is the hand–arm system; therefore, this also is referred to as *hand-transmitted vibration*. Griffin (1990) identifies disorders related to hand-transmitted vibration to include those associated with the vasculature, bones and joints, peripheral nerves, musculature, central nervous system, and the whole body. Additional information concerning hand-transmitted vibration can be found in Bruce et al. (1998) and ACGIH (2002).

*Resonance* is a factor that affects the hazard potential of vibration exposure. Resonance describes the interaction between the human body and a vibration source such that the body causes the vibration to amplify. This resonant vibration can cause large displacements in the body, which can result in damage (Bruce et al., 1998).

Signatures obtained from military ground vehicles operating over secondary and cross-country routes suggest that the traditional methods of defining oscillatory motion and evaluating the effects of vibration during operation of these vehicles may not be adequate (U.S. Army, 1996). These signatures are categorized as *repetitive impact* or *repeated impact* and are defined as broadband vibrations with embedded shocks. Vibration with high crest factors (ratio of peak acceleration to the root-mean-square acceleration) is considered to be repetitive impact. Military research suggests that evaluation criteria separate from that of International Organization for Standardization (ISO) 2631 (ISO, 1985) is required to assess repetitive impact for military equipment and scenarios (U.S. Army, 1996; Village et al., 1995a–c; Cameron et al., 1995).

The effects associated with exposure to vibration include physiological changes, discomfort, performance decrements, pain, and degenerative processes (U.S. Army, 1996). Examples of physiological effects include increases in heart rate, respiration rate, cardiac output, mean arterial blood pressure, pulmonary ventilation and oxygen uptake, and hyperventilation (U.S. Army, 1996). Discomfort and pain have been observed to affect the lower back, gastrointestinal and stomach areas, and thoracic area (U.S. Army, 1996; Henzel et al., 1966; Temple et al., 1965). Low-back discomfort can ultimately lead to clinical diagnosis of degenerative diseases, including herniated discs, osteochondrosis, spondylosis, and other disorders of the spinal column (U.S. Army, 1996; Bruce et al.,

1998). Among the field studies, back disorders were by far the most widely reported illness or injury associated with WBV (U.S. Army, 1996).

Back pain has been associated with the prolonged and repeated operation of both air and ground military vehicles, and vibration exposure is considered to be a factor in the generation of these symptoms (U.S. Army, 1996). Military helicopter pilots have reported symptoms of low-back pain and general discomfort after about 4 hours of flight (VanIngen-Dunn and Richards, 1992). Some air force and army helicopter pilots have reported that back pain disappears immediately upon removal of the vibratory stress associated with helicopter flight. Others have reported that it takes four to five days for the pain to disappear.

There are few studies on vibration conducted with women subjects. Those that exist have found that women are more sensitive to higher frequencies, but there appear to be no reported significant differences in comfort (U.S. Army, 1996). However, there are reports of increased gynecological and menstrual disorders in female drivers exposed to WBV (Böhm, 1964; Brovko, 1975).

The primary source of WBV is from transportation vehicles, including ground, air, and water vehicles. Vehicle vibrations can be generated by exposure to specific environmental conditions (e.g., ground terrain, wave conditions) and/or vibrations occurring by design (e.g., engines, rotor blades, etc.). These vibrations are transmitted to the operator or the passenger. Other sources of WBV include heavy machinery and buildings and vibrations that are transmitted directly through air or water. For example, military ground crews can be exposed to WBV resulting from exposure to aircraft propeller washes.

The primary source of segmental vibration occurs with the operation of hand tools. Tools that produce the most severe vibrations include chain saws, jackhammers, and tools used in a factory environment.

Whole-body vibration exposure and its consequences are unique in the military environment as compared to the civilian community (U.S. Army, 1996). The exposures are typically much longer due to the need for extended operations, and the vibrations are more severe due to the adverse conditions encountered in some military environments. This is a consequence of the greater maneuverability and speed required of these vehicles for combat readiness, effectiveness, and survivability. The wide range of operational capabilities required of military vehicles is the primary reason that military-unique standards will be developed and recommended for assessing WBV health hazards.

## 15.3 TOOLS AND TECHNIQUES

Regardless of the hazard category that is evaluated, health hazard assessors rely upon various tools and techniques to help estimate and characterize potential risks. Generally, these tools and techniques can be characterized as those that are used to detect and quantify potential hazards, methods to assess potential health impacts, and control measures to eliminate or reduce hazards to acceptable levels. This section presents information about the various tools and techniques that are applied to the various HHA categories.

### 15.3.1 Acoustic Energy

The tools and techniques that are applied to assess acoustic energy health hazards are associated with measuring noise sources and acquiring data and the interpretation of the measurements. Based upon the measured levels and the potential risk, various controls can be recommended to eliminate or reduce hazardous noise levels.

There are several types of devices that can be used to measure noise. These include sound-level meters, noise dosimeters, sound intensity meters, narrow-band analyzers, tape recorders, and graphic-level recorders. Some sound-level meters are able to record a "peak" response and measure impulse noise. Refer to discussions by Bruce et al. (1998) for detailed information concerning instruments typically used to measure and assess noise levels.

When noise levels are recorded, the unit of measurement typically is in decibels (dB), a logarithmic scale. Sound can be measured on several response scales depending upon its nature (i.e., steady state, impulse, etc.). The frequency scale may be a flat response or it may be weighted. For example, when steady-state noise is measured with a sound-level meter, it is set on a scale that approximates how the human ear responds to sound. This is an A-weighing network; therefore, the measurement units are expressed as dBA. Noise data also may and should include an octave-band analysis to assess the contribution of various frequencies.

The measurement requirements for military equipment are detailed in a military standard (MIL-STD-1474D, 1991). Typically, measurements are taken at each operator and crew position and at representative positions where individuals are likely to be located during typical system operation. The measurements are made with the system and all auxiliary equipment operating in a normal mode. The data acquired from measuring noise levels are applied to algorithms that allow the health hazard assessor to recommend exposure durations and levels (e.g., the number of rounds that can be fired from a weapon in a given period of time) that will prevent soldiers from being harmed. Details about the nature and application of such algorithms and possible recommendations can be found in the *U.S. Army Health Hazard Assessor's Guide* (U.S. Army, 1996).

Measuring and assessing blast overpressure involves recording noise data and estimating risk by applying the data to a computer model (INJURY) developed by the U.S. Army. This model predicts the probability of injury to the tympanic membrane, upper respiratory tract, and lung due to insults from blast waves (see U.S. Army, 1996).

Control strategies for eliminating or minimizing the potential hazards from noise are the same as those for other occupational hazards. The ideal control is to design and build systems so that they do not produce hazardous noise levels. When this is not feasible, other control measures may be used. Typical measures for noise control include use of personal protection equipment and limiting the exposure duration. Earmuffs and earplugs are the types of personal protection often used to reduce steady-state and impulse noise exposure. Blast overpressure usually is controlled administratively (e.g., limiting the number of blast events) or by controlling the distance between people and the blast site. Several other additional references should be consulted for details concerning personal protection equipment and other measures for controlling and minimizing noise hazards (Donahue and Ohlin, 1993; U.S. Army, 1996; Bruce et al., 1998).

**Example 15.1 Acoustical Hazard Assessment of a Mortar System** The M-120 series 120-mm *Battalion Mortar System (BMS-120)* is an example of a system that was assessed for acoustical hazard (U.S. Army, 1995b). The BMS-120 is a smoothbore, muzzle-loading, indirect fire mortar system. The system consists of the M-120 Towed Mortar, transported by a one-quarter-ton truck, and the M121 carrier configuration mounted in a modified M113 Armored Personnel Carrier. Its 7000-plus-meter (m) maximum range, high rate of fire, and excellent stability give the weapon a high trajectory and allow it to be fired from behind high cover. Even though a Blast-Attenuating Device (BAD) was developed to reduce exposure at crew locations, an assessment of the BMS-120 revealed blast overpressure as a potential health hazard. Impulse noise levels in excess of 140 decibels peak (dBP) is hazardous and will cause permanent hearing loss. The BMS-120 can produce noise levels in excess of the recommended limits. The HHA report required that 120-mm mortar crews wear earplugs (preferably E-A-R™ brand) during firing. When the mortar is fired, all personnel within 200 m wear approved hearing protection, medically trained personnel check crew members to assure proper fit of the E-A-R™ earplugs, and soldiers are informed about the significant risk of hearing loss from the BMS-120 and the proper use of hearing protection.

### 15.3.2 Biological Hazards

The process of measuring and evaluating biological health risks involves several steps (U.S. Army, 1996). These include evaluating the design and operation criteria from the mission needs statement, the operational requirements document, and the detailed test plan; comparing design and operation criteria against existing military and other reference standards; determining any untested criteria by comparing the detailed test plan to army standards and other reference standards; and developing test protocols to fill any information gaps.

Evaluating biological hazards is complex, requiring knowledge and skills in many scientific disciplines. This process consists of collecting (sampling), identifying, and measuring (quantifying) the potential hazardous organism or substance. Many sampling and analytical techniques to identify and measure biological hazards are available to scientists. Some examples of these are presented in Table 15.3.

Once sampling and analysis are completed, the risk to the target population is characterized. The health implications and outcome of biological substances may not always be clear. In some circumstances, a level of subjectivity and professional judgment is necessary. The assessment of such hazards may simply be to determine compliance with consensus standards or guidelines [e.g., drinking water standards (Safe Drinking Water Act), food sanitation standards (Model Food Code)]. If there is no such standard, then applying a risk assessment and management process may be necessary to estimate the impact to human health.

Protecting the soldier from exposure to biological substance(s), which can result in illness or loss of system effectiveness, requires installing and using multiple and overlapping controls. These may range from design and engineering solutions that prevent exposures to the use of personal protective equipment to block biological agents and prevent them from entering the body. Examples include immunizations, prophylactic drugs, personal hygiene, design and maintenance of soldiers' uniforms, screening and bed netting, ventilation, insecticidal sprays and repellents, and the application of sanitation standards in the areas of food and water.

**Example 15.2 Biological Hazards Evaluation of an Environmental Control System** The *Electrical Generator/Environmental Control System (EG/ECS)* was evaluated for biological hazards (U.S. Army, 1995b). The EG/ECS provides electrical power and heating and air conditioning for components of the Deployable Medical System (DEPMEDS). Each DEPMEDS is comprised of tentage arranged with Rigid Wall Tactical International Standards Organization Shelters. The EG/ECS consists of a 100-kilowatt generator system, a power distribution center, and an Environmental Control Unit (ECU). The ECU is set up outside the

**TABLE 15.3 Examples of Sampling, Analytical, and Assessment Methods for Biological Hazards**

| Biological Hazard | Examples of Methods |
|---|---|
| Animal or plant allergen | Sampling—collecting specimens<br>Analysis—sensitivity reaction assay (reference, Hayes), patch test<br>Assessment—risk assessment |
| Insect vector poisonous arthropod, poisonous snake | Sampling—trapping<br>Analysis—taxonomic identification by morphological characteristics<br>Assessment—risk assessment |
| Microorganism | Sampling—surface swipes, air-filtering techniques<br>Analysis—taxonomic identification by morphological and biochemical (metabolic) characteristics, immunoassay techniques, DNA assay<br>Assessment—compliance with consensus standards, risk assessment |
| Poisonous plant | Sampling—collecting specimens<br>Analysis—taxonomic identification by morphological characteristics, chemical analysis<br>Assessment—risk assessment |
| Sanitation | Sampling—methods associated with microorganisms, animals, and insects<br>Analysis—methods associated with microorganisms, animals, and insects<br>Assessment—compliance with consensus standards: time-temperature controls (e.g., food handling), physical construction and type of material (e.g., food preparation and storage equipment, drinking water treatment and storage devices) |
| Toxin | Sampling—grab and bulk sampling of neat substances or contaminated media (e.g., water, food, soil, etc), surface swipes, air-filtering techniques<br>Analysis—chemical analysis (direct-reading equipment, chromatography, spectometry, infrared and UV detection, etc.; reference, current chemical analytical text, NIOSH/OSHA), a variety of acute and chronic toxicity assays (reference, Hayes)<br>Assessment—compliance with consensus standards, risk assessment |

shelter and delivers conditioned air through a 9-ft supply duct to a nylon plenum running the length of the ceiling. Return air is recirculated to the unit through a second duct located at floor level. The ECU dehumidifies the return air by passing it across the evaporator coil where heat is extracted by the refrigerant. As the air is cooled, water vapor condenses and collects in the drip pan underneath the coil. Ideally, the condensate should drain from the pan and empty outside the ECU through the drain holes located on the right and left sides. An HHA of the system revealed, however, that without drain lines into the drip pan the condensate collects in the pan and causes the water to stagnate. Stagnant water in air-cooling or heating systems is an ideal growth medium for thermophylic *Actinomyces*, the causative agent of hypersensitivity pneumonitis and for *Legionella pneumophila*, the causative agent of legionellosis. Exposures to air contaminated with these microorganisms may induce severe health problems. The HHA report, therefore, recommended that suitable fittings and drain hoses be installed from the condensate drip pans of the ECU and routine inspections of the drip pan and drain hoses be performed to assure condensate is emptying from the ECU.

### 15.3.3 Chemical Hazards

When evaluating chemicals, there are several steps: identifying, sampling, analyzing, quantifying, assessing hazard and risk, and recommending controls. There are tools and techniques for each of these events. The interrelated tasks of chemical identification, sampling, and analysis are done using similar or related tools and techniques but also may require some that are unique to the task. For example, chemical identification may be done in the field using direct-reading instruments. However, some chemicals or field conditions may require that a sample of the chemical or environmental media (e.g., air, water, soil, etc.) be collected and taken to a laboratory for analysis.

Examples of direct-reading instruments are prominent in the assessment of air contaminants. Technologies include colorimetric indicators (e.g., detector tubes), airborne particulate analyzers (e.g., optical, electrical, piezoelectrical, and beta attenuation analyzers), and gas and vapor analyzers (e.g., electrical, radioactive, thermal, electromagnetic, chemi-electromagnetic, and chromatographic analyzers) (Lioy and Lioy, 1983; Todd, 1998).

Laboratory analytical techniques for chemical substances are as varied as, and sometimes similar to, the direct-reading technologies. Examples of chemical analytical methods applicable to the HHA process are reviewed by Draper et al. (1999) in their discussion of industrial hygiene chemistry. Some of the technologies that they discuss include spectrometry (e.g., infrared spectrometry, atomic absorption spectrometry, inductively coupled plasma spectrometry, ultraviolet/visible spectrophotometry, and mass spectrometry), and chromatography (e.g., gas chromatography, liquid chromatography, high-performance liquid chromatography, and ion chromatography). The review also lists several other analytical methods, including microscopy, X-ray spectroscopy, electroanalysis, immunoassay, and surface analysis. Other references that should be consulted concerning sampling and analytical techniques for chemicals in a variety of environmental media are published by the EPA [1983, 1988a,b, 1990–1995 (there are others; these are a few examples)], the Occupational Safety and Health Administration (OSHA, 1996), the NIOSH (NIOSH, 1994), and others [e.g., American Water Works Association (AWWA), 1995].

Once chemicals are identified and their presence in the environment is quantified, their hazards must be determined and the risk to people and the environment estimated. This involves the process of risk assessment where hazard determination, dose–response assessment, and exposure assessment are integrated to provide a risk characterization (NRC, 1983).

Chemical hazards often can be determined by searching references and scientific literature. There are a variety of commercial references (e.g., Clayton and Clayton, 1991; Klaassen, 1996; Hayes, 1994; Lewis, 1999), government reports and references (e.g., NIOSH criteria documents, NIOSH/OSHA guidelines for chemical hazards, EPA criteria and health effects assessment documents, the Agency for Toxic Substances and Disease Registry Toxicological Profiles, and others), and professional organization sources [e.g., The ACGIH, and the American Industrial Hygiene Association (ACGIH, 2000; AIHA,

2001)] that present summary or detailed information about the hazards of specific chemicals and the basis for recommended exposure levels. Also, there are a number of databases that can be accessed through the Internet. Examples of such databases can be found at the National Institutes of Medicine, National Library of Medicine Internet site and include the Hazardous Substances Data Bank (HSDB), Integrated Risk Information System (IRIS), Chemical Carcinogenesis Research Information System (CCRIS), GENE-TOX, Environmental Mutagen Information Center (EMIC), and Developmental and Reproductive Toxicology and Environmental Teratology Information Center (DART/ETIC).

When chemical hazard information is not available or is determined to be insufficient, toxicological and/or epidemiological studies may be required to produce the information. In the military medical research and support organizations [e.g., the U.S. Army Medical Research and Materiel Command, the U.S. Army Center for Health Promotion and Preventive Medicine (USACHPPM), the Naval Medical Research Institute, the Naval Health Research Center Toxicology Detachment, and the U.S. Air Force Institute for ESOH Risk Analysis] often are the organizations that would conduct such studies.

Epidemiology studies may be either *descriptive* or *analytical*. Descriptive studies assess the amount and distribution of health outcomes within a population or community by focusing on person, place, and time. Analytical studies may be either *experimental* or *observational*. In experimental studies an investigator controls exposure to an agent of interest. For example, a group of people with similar characteristics would be identified; then some would be exposed to a chemical and subsequently compared to others that were not exposed. Obviously, for ethical reasons, this is done rarely. Therefore, most epidemiology studies are analytical, which may be either *retrospective* or *prospective* studies. Retrospective studies assess effects from past exposures. Prospective studies determine how people are exposed currently and then monitor them through time to determine if future exposure effects occur. Biostatistical measures are applied to the study data to summarize them and to determine their significance. More detailed discussion about standard epidemiology research and study methods may be found in references by Tyler and Last (1998) and Mausner and Kramer (1985). Biostatistical methods are presented in references by Daniel (1983), Klienbaum and Kupper (1978), and Ott (1988).

When human epidemiology studies are not available or to further support existing studies, laboratory toxicological studies may be performed. Frequently these experimental studies are conducted on laboratory animals (e.g., rats, mice, dogs, monkeys, rabbits, and pigs), and the results are extrapolated to humans. Toxicology studies may be designed to evaluate effects that occur from an acute exposure by a single dose or multiple doses in a 24-hour period. Short-term studies (e.g., subacute and subchronic exposures) may range in durations from 14 days to 2 years. Long-term studies (e.g., chronic and lifetime exposures) may extend through the life of the test animals (Moeller, 1997). A variety of other types of studies that do not use laboratory animals also may provide supporting information to confirm toxicological end points or mechanisms. Some examples include the use of microorganisms (e.g., *Salmonella* sp.) and insects (*Drosophila*) to elucidate mutagenic potential, cell cultures (e.g., for neurotoxicity), and enzymes (e.g., for hepatic and renal toxicity). Hayes (1994) presents detailed discussions of various toxicological assay methods.

As with many of the other health hazards, the risk associated with chemical exposures may be based upon existing standards and criteria. If there are no existing standards or criteria, then risk assessment and management procedures are applied to establish recommendations for soldier protection.

Control measures to protect against the harmful effects from chemical exposure include the range of options presented toward the end of this chapter.

**Example 15.3 Chemical Hazard Assessment of the Paladin Self-Propelled Howitzer** The *M109A6 Paladin* is a self-propelled armored and full-tracked howitzer suitable for worldwide deployment with heavy division forces. It is an example of a chemical hazard assessment (U.S. Army, 1995b). Previous HHAs of the howitzer noted exposure to lead during the firing of munitions. The Howitzer Improvement Program (HIP) Program Manager, therefore, requested an additional assessment to address the health hazards associated with the substitution of inorganic tin foil for the lead foil decoppering agent in the propelling charge of the M109A6 munition. Trials of munitions using lead in the propelling charge and trials using tin as a replacement for lead in the propelling charge were conducted and compared. The trials with lead showed concentrations at various crew positions as high as 10 times the permissible exposure level (PEL) for lead. In contrast, in trials using tin as a replacement for lead, the exposure levels never reached one-tenth of the tin PEL. Because the testing data indicated that overexposure to tin when firing munitions is remote, the HHA report recommended substituting inorganic tin foil for the lead foil decoppering agent in the propelling charge of the M109A6 Paladin howitzer.

### 15.3.4 Oxygen Deficiency—Ventilation

Confined spaces should be tested for oxygen levels and other contaminants prior to entry and monitored continuously during occupancy (Schroll and Harris, 1998). Because of the need for instantaneous information, especially when confined spaces are occupied, direct-reading instruments are used to measure and monitor atmospheric oxygen levels. The instruments that are used typically are colorimetric indicators or electrochemical sensors (i.e., potentiometric and coulometric analyzers) and heat-of-combustion detectors. More detailed information about these technologies and specific products can be found in Todd (1998), Nader et al. (1983), and Saltzman (1983).

*Ventilation* is one of the principal methods to prevent or eliminate oxygen-deficient atmospheres in confined and enclosed spaces. It can be accomplished by supplying forced air to either dilute or displace air contaminants (Schroll and Harris, 1998). Exhaust ventilation, coupled with fresh makeup air, also can be used to remove oxygen-reducing point-source contaminants. Ventilation rates and air exchanges are determined by fan capacity, duct size, and type of contaminants to be removed. Instruments that can be used to determine air exchange rates, airflow, and air capacity include a variety of products that measure pressure, volumetric flow rate, and air velocity. Examples of such instruments include the U-tube manometer, pitot tube, vane anemometer, thermal anemometer, inclined anemometer, aneroid gauges, smoke tube, and tracer gas. Additional details about industrial ventilation system design, measurement techniques, and instrumentation are presented and detailed in the ACGIH *Industrial Ventilation* manual (ACGIH, 2001).

Enclosed spaces are not routinely tested and monitored for oxygen levels because ventilation normally is part of the design for the space. For example, there are specific ventilation requirements for personnel enclosures (e.g., shelters, armored vehicles, etc.) and vehicle cabs (MIL-STD 1472F). However, when the ventilation system is not functioning properly, oxygen deficiency should be considered as a possible outcome and assessed prior to returning the system to normal operation.

Sometimes it is necessary to remove oxygen from a confined space to prevent a fire or explosion. This can be done by *inerting*, which requires introducing a nonreactive (inert) gas (e.g., nitrogen, argon, or carbon dioxide) to displace the oxygen (Schroll and Harris, 1998). Individuals who enter this type of space must wear a supplied air respirator. A supplied air respirator also can be worn if ventilation cannot be used or if it will not correct an oxygen-deficient atmosphere.

**Example 15.4 Health Hazard Assessment of Ventilation on a Landing Craft** An HHA that involved ventilation concerns was performed on the *Landing Craft Mechanized (LCM-8)* (U.S. Army, 1995b). The LCM-8 is a U.S. Navy–designed, welded-steel, twin diesel-powered watercraft approximately 73 ft long and capable of carrying 60 tons. The vessel is designed to transport personnel and cargo in resupply or waterborne tactical operations. The army has a fleet of approximately 96 vessels, each with a crew of six enlisted soldiers. The Service Life Extension Program (SLEP) is a product improvement program intended to upgrade the engine and transmission performance of the existing fleet of LCM-8 Mod-1 vessels and extend their service life 20 years. A review of the design and description of the LCM-8 Mod-1 SLEP noted several health hazards, including confined spaces and the use of fire-extinguishing agents. Fatalities occur in confined spaces as a result of encountering one or more potential hazards. Liquid fuel fires on the LCM-8 are frequently extinguished by discharging carbon dioxide ($CO_2$) from portable fire extinguishers directly on the burning material. When $CO_2$ is discharged in a confined space or room, an oxygen-deficient atmosphere may result due to air displacement. The HHA report recommended that when a $CO_2$ discharge starts, personnel should consider the space oxygen deficient and employ appropriate confined-space entry procedures; enclosed spaces, crew spaces, and spaces containing diesel fuel tanks comply with current U.S. Coast Guard regulations for ventilation; and all personnel be trained in the hazards associated with confined-space entry and work procedures. (The reader should note that $CO_2$ has toxic properties that should be considered in addition to its ability to displace oxygen.)

### 15.3.5 Ambient Pressure Changes

The physical principles associated with the gas laws—Boyle's law, Dalton's law, and Henry's law—are key to the understanding of the nature of hypobaric conditions and the relationship with oxygen. These laws relate factors such as pressure, volume, temperature, mass, molecular weight, and others to gas properties and their effects. Gas principles and laws can be reviewed in Popendorf (1998) or any college-level general chemistry textbook.

Boyle's law can be applied to the expansion and contraction of gases in bodily organs (e.g., lung, ear, sinuses, and gastrointestinal tract) due to external pressure changes. These effects can cause pain and physical trauma in affected organs. Dalton's law (the law of partial pressures) addresses the significance of partial pressures exhibited by individual gases in a mixture and the summed effect of all the gases. Henry's law can be used to predict the body's absorption of gases from lung alveoli, their transport rate in blood, and the amount that may concentrate and be stored in various tissues in the body.

Ambient total pressure changes can be predicted by using an equation that is based upon various gas law principles. Popendorf (1998) discusses how this equation is derived and shows how the predicted pressures can be compared to values in various physiological tables to estimate health outcomes.

Techniques discussed by Popendorf (1998) to control, prevent, or minimize occupational health hazards in hypobaric environments reflect the range of options available in other typical workplaces. These include engineering controls (e.g., increasing air pressure in aircraft cockpits and cabins), using personal protective equipment (e.g., equipment similar to supplied air respirators that increase the availability of oxygen by increasing its molar fraction in the breathing air), and acclimatization. An acclimatization timetable is required to allow individuals to adapt to hypobaric conditions at high elevations. If acclimatization does not occur, work activities, especially complicated tasks, may be jeopardized by serious health problems. Military guidelines for high-altitude operations suggest a deployment timetable that permits some degree of acclimatization prior to mission execution (U.S. Army, 1975).

### 15.3.6 Nonionizing Radiation

There are dose–response relationships for a wide range of exposure conditions to include wavelength, pulse duration, pulse repetition frequency, source size, exposure, and dose. Response criteria include clinically visible response cutaneous erythema, minimally visible retinal lesion, microscopic cellular change, or permanent or temporary changes in visual function. From such dose–response relationships, comprehensive PELs were established. However, many exposure conditions inherent to new military developmental systems lack sufficient biological basis to assess the hazards. Generally, these situations are addressed by extrapolating from existing PELs.

There are several regulatory standards and criteria for the use, control, and exposure to radiation. These are enumerated in the army's HHA *Assessor's Guide*, which includes federal laws and regulations, national and international guidelines, DoD requirements, and army regulations and guidelines (U.S. Army, 1996).

***Laser Radiation*** Laser radiation control measures are based upon limiting access to the beam. This is achieved by a number of methods. The primary protective method is to block the beam by materials opaque to the laser wavelength(s). The greatest difficulty has been in providing laser-protective materials, which provide transparency for vision. Currently, there are no universally acceptable, general-purpose, eye-protective filter materials that do not block the visible wavelengths required for vision. Range restriction is another method to limit access to the beam when PELs are exceeded. Control of laser operation by engineering controls, such as enclosures and safety interlocks, can eliminate or control the hazard (see MIL-STD 1425A, 1991).

***Radio-Frequency Radiation*** The sequence of events and considerations for assessing and controlling potential RFR hazards from military systems are detailed in the HHA *Assessor's Guide* (U.S. Army, 1996). These include performing a nonionizing radiation protection study, taking field measurements, comparing measurements to established exposure limits, recommending control strategies, and assigning a risk assessment code (RAC).

Hitchcock et al. (1998) describes several types of instruments that can be used to measure RF fields, body currents, and contact currents. These include instruments that make densitometric measures, current monitors, personal monitors, and frequency counters.

There are engineering administrative control strategies to minimize exposures and health hazards associated with RFR (U.S. Army, 1996). Examples of engineering controls include using system sector blanking to prohibit radiation in certain areas, incorporating dummy loads, shielding high-voltage power supplies, and incorporating warning devices when the sources are activated. Administrative controls are publishing warning messages in technical manuals; providing safety training and briefings; periodically inspecting waveguides, interlocks, etc.; and installing barricades, fences, signs, and warning devices to prohibit access to unauthorized areas.

As dictated by army regulations (U.S. Army, 1980b, 1990d), RFR source safety evaluations typically are performed by USACHPPM scientists during various stages in the concept, development, and fielding of army materiel systems. Most of these systems are evaluated in the concept, research, and early development testing phases and usually are completed before the initiation of a HHA.

Exposure limits documented in U.S. Army Regulation 40-5 (U.S. Army, 1990d) and DoD standards are routinely applied for evaluating potential health hazards from new materiel. In the absence of exposure criteria, the USACHPPM consults with the U.S. Army Medical Research and Development Command's Walter Reed Army Institute of Research (WRAIR). The USACHPPM staff professionals will conduct an initial system evaluation and special and periodic installation RFR surveys. The information acquired from the special studies and routine surveys are entered into a database for evaluating potential health hazards from new material. The database consists of technical reports maintained by the USACHPPM. All reports and related publications, especially those that include recommendations for corrective action, are peer reviewed. The USACHPPM staff professionals make a number of recommendations for alleviating and reducing exposures to levels beyond PELs.

Risk assessment codes are assigned for non compliance with the recommendations of the study. The hazard severity is assigned based on the RFR exposure levels. The matrix that is used as a guideline for determining hazard severity is shown in the HHA *Assessor's Guide* (U.S. Army, 1996). The hazard probability is a subjective determination based on the experience and knowledge base of USACHPPM engineers. The assignment of an RAC is also subject to the review of an USACHPPM physician and is, therefore, a medical decision.

**Example 15.5 Health Hazard Assessment of RFR Source on an Air Defense System** The *HAWK Air Defense Guided Missile System (ADGMS)* exemplifies a RFR source for which an HHA was developed (U.S. Army, 1995b). The HAWK ADGMS defends against low- and medium-altitude attacking aircraft. The phase III product improvement program provided a highly mobile air defense system that can search for, detect, and designate hostile targets. The primary search, detection, and designation equipment consists of one continuous-wave acquisition radar and one high-power illuminator radar. The primary health concerns are exposure to RFR during tactical operations, field maintenance, and depot-level maintenance activities. The HHA identified several situations in which excessive exposures could occur and recommended the following actions to control those exposures: (1) publish warning messages in all appropriate technical and field manuals; (2) enroll maintenance personnel in a medical surveillance program; and (3) use warning lights, signs, barricades, and alarms to prevent soldiers from entering potential exposure areas in the field.

### 15.3.7 Ionizing Radiation

The HHA methodology for radiation hazards from military systems includes an identifying process and an evaluation process, which are detailed in the HHA *Assessor's Guide* (U.S. Army, 1996). The evaluation process consists of a hazard identification step and an exposure assessment step. The exposure assessment step involves the use of selected exposure standards (U.S. Army, 1995b; 10 CFR 20) and computer dosimetry models (e.g., the army's radiological bioassay and dosimetry program, the REMEDY program[2], or the medical internal radiation dose (MIRD) to estimate internal dose.

Ion particle–counting instruments and dose-measuring instruments are types of instruments used for detecting and measuring ionizing radiation. The principles of ion particle–counting instruments are used both in portable radiation–detection instruments and in laboratory instrumentation for measuring radioactivity in environmental samples and bioassay specimens. Examples of ion particle–counting instruments are the gas-filled particle counter (e.g., the ionization chamber, proportional counter, and Geiger counter) and the scintillation detector. The response of a dose-measuring instrument is proportional to the energy it absorbs. Examples of dose-measuring instruments are the free-air ionization chamber (laboratory instrument), the air-wall ionization chamber (condenser-type pocket dosimeter), the ion current chamber (cutie pie), and the thermoluminescent dosimeter. Since neutrons are not directly ionizing, they must interact with another medium to produce a primary ionizing particle, and therefore, require refinements in detectors for measurements and dosimetry. The army HHA *Assessor's Guide* (U.S. Army, 1996) and McCarthy (1998) have additional details concerning instruments for measuring ionizing radiation.

There are several regulatory standards and criteria for the use, control, and exposure to ionizing radiation. These are enumerated in the army's HHA *Assessor's Guide* and include federal laws and regulations, national and international guidelines, DoD requirements, and army regulations and guidelines (U.S. Army, 1996).

### 15.3.8 Shock

Known human tolerances to impact are summarized in a Society of Automotive Engineers report (SAE, 1986). The report focuses on automotive crash injuries; however, the tolerances may generally be applied to aviation and other combat vehicle impact loading conditions. The following information is derived primarily from literature associated with automotive crash injury assessments.

Injury ratings of automotive crashes traditionally are measured on the abbreviated injury scale (AIS), which ranges from zero to five. This scale may not be entirely appropriate for soldiers in a combat environment, because many injuries rated as serious but survivable (i.e., an AIS of 3) can cause temporary incapacitation to the soldier and may lead to further life-threatening exposures. Therefore, it is necessary to adjust some injury threshold levels (ITLs) commonly applied in the automotive environment to reflect the secondary threat resulting from crew incapacitation in a combat environment.

Several types of test manikins have been used to evaluate and certify new automobiles and are intended primarily to simulate a seated driver or passenger in automobile crashes. These manikins are not able to test spinal load and lateral chest impacts, which are relevant concerns for some military systems. However, these are the state-of-the-art manikins and must be used in the absence of better test surrogates in axial (vertical) and lateral (transverse) impacts.

In order to assess risk, an injury assessment value (IAV) is measured and compared to the lower bound of an ITL. The numerical relationship between the known ITL and the measured IAV defines an *injury assessment criterion*. Injury criteria have been established for the prediction of head (closed-brain) injuries; chest trauma; cervical, spinal, and lumbar spinal column fractures; and pelvis and lower extremity injuries. These criteria translate engineering measurements, such as forces, accelerations, and deflections, into probabilities of injury occurrence.

The engineering measurements needed for valid injury assessment must be obtained by following standard and accepted instrumentation and signal processing guidelines. Measurements are obtained from transducers that are mounted in the test surrogate (manikin) and attached to electronic data recorders. High-speed cinematography can be used to produce a visual record of the impact response of the test surrogate and to complement electronic data recordings.

Head injury predictions are based upon acceleration signals, measured at the head center of gravity, in the forward ($A_x$), leftward ($A_y$), and upward ($A_z$) directions. Neck injury predictions are based on measures of the neck axial ($F_z$), shear ($F_x$) forces, and the neck pitch ($M_y$) moment. Chest injury is predicted from chest acceleration signals that are measured in the forward ($A_x$), leftward ($A_y$), and upward ($A_z$) directions. Lower spine (lumbar) injury predictions are based on measures of forward ($F_x$) and leftward ($F_y$) shear, upward ($F_z$) axial forces, and fore/aft ($M_y$) and lateral ($M_x$) bending moments. Lower extremity injury prediction is based on strength data of femur and tibia bones, under compressive ($F_z$), shear, and bending loading. The reader should refer to the appropriate (1995) standard SAE for specific details concerning these and other measures.

**Example 15.6 Health Hazard Assessment for Shock from a Parachute** An example of a system that received an HHA for shock is The *tactical assault personnel parachute (TAPP)* (U.S. Army, 1995b). The TAPP was designed for use in low-altitude mass tactical assault airborne operations in order to lower the rate of descent and reduce the potential for landing injury. An initial HHA of the TAPP identified musculoskeletal trauma resulting from excessive opening forces and impact velocity as a potential adverse health effect to personnel during training and combat airborne operations. Although additional data were needed to fully assess the TAPP, the Initial HHA recommended including the current and improved paratrooper helmets in the TAPP test program to evaluate the effect of helmet mass on neck loads during opening shock and crushable foam on reducing deceleration in the head during parachute landing falls.

### 15.3.9 Heat Stress

There are several heat stress indices that are used to estimate the potential for people to develop heat illness. They combine measures of temperatures, humidity, and air velocity to produce a single numerical indicator of heat stress potential. Examples of these indices include the wet-bulb global temperature (WBGT), the wet-globe temperature (WGT), and the heat stress index. The formulas for these indices and details concerning their application are offered by Ramsey and Beshir (1998) and the ACGIH (2002).

There are a variety of instruments that measure temperature, radiant heat, humidity, and air velocity, and some electronic devices integrate the individual measures to produce a single heat index. Examples of each of these are presented by Ramsey and Beshir (1998). Examples of heat-measuring devices include liquid-in-glass thermometers, bimetallic thermometers, resistance thermometers, and thermocouples. Radiant heat can be measured with a radiometer or globe thermometer. Humidity can be measured with a psychrometer or hygrometer. Air velocity can be measured with a vane or thermal anemometer. Both a NIOSH publication (NIOSH, 1986) and the army's heat injury medical bulletin (U.S.

Army, 1980) show and describe how to assemble a device for determining the WBGT using a wet-bulb thermometer, a shielded thermometer, a dry-bulb thermometer, and a black-globe thermometer.

A psychrometric chart can be used to determine dry-bulb temperature, wet-bulb temperature, relative humidity, vapor pressure, or dew point (Ramsey and Beshir, 1998). Heat strain can be predicted by use of a computerized model (Gonzalez and Stroschein, 1991; Pandolf et al., 1986, 1971).

Engineering and administrative controls and personal protective equipment can be employed to prevent or minimize the occurrence of heat stress. Engineering controls may include air conditioning, ventilation, and isolation of heat sources. A microclimatic cooling vest is an example of personal protective equipment. Application of a work–rest schedule based upon a heat indicator level and planned consumption of adequate quantities of drinking water are examples of administrative controls. Detailed information concerning heat stress control and injury prevention strategies can be found in references by Bishop (1998), the ACGIH (2002), and the U.S. Army (1980a). Design considerations for military systems (e.g., vehicles and shelters) are found in MIL-HDBK-759A (1981) and MIL-STD-1472F (1999).

**Example 15.7 Heat Stress Assessment of the Pedestal-Mounted Stinger** A heat stress HHA example is the *pedestal mounted stinger (AVENGER)* weapon system (U.S. Army, 1995b). The AVENGER consists of a Stinger missile and.50-caliber machine gun pedestal that is turret mounted on a high-mobility multipurpose wheeled vehicle (HMMWV). Used against enemy fixed- and rotary-wing aircraft, the AVENGER is operated by a two-person crew (a driver and a gunner). The AVENGER is used for training and combat missions in a variety of environmental conditions, including hot weather. An HHA of the system identified heat stress as a potential adverse health effect to AVENGER personnel. When AVENGER fire missions of only 60 minutes or less were conducted with outside temperatures ranging from 82 to 85°F, test personnel reported that the gunner's station in the turret and the driver's station in the vehicle cab became uncomfortably hot. AVENGER crew members dressed in chemical protection suits (mission-oriented protective posture gear) during a chemical scenario to endure an even more significant heat load. Elevations in the driver's and gunner's core body temperatures may cause performance decrements or heat illness, such as cramps, exhaustion, and heat stroke. Since actual fire missions may last as long as 12 hours, the HHA report recommended installing a microclimatic cooling system in the AVENGER for use at all normally occupied crew positions. Administrative guidelines, such as work/rest/maximum work periods and water requirements information, were provided for interim use until the cooling system could be installed.

### 15.3.10 Cold Stress

Windchill is expressed as an *equivalent chill temperature*, which reflects the cooling power of wind on exposed skin (Ramsey and Beshir, 1998; ACGIH, 2002). Ramsey and Beshir (1996) show how the equivalent chill temperature (windchill index, WCI) can be calculated from the relative air velocity and the air temperature. The ACGIH (2002) presents a table that shows the resultant windchill index at given wind speeds and temperatures. The table also indicates whether there is "little danger," "increasing danger," or "great danger" of a given WCI causing exposed skin to freeze. The windchill chart also may be found in the army's cold injury medical bulletin (U.S. Army, 1976) and references by Jones et al. (1993) and Young et al. (1992a,b). Instruments for measuring air temperature and velocity were presented earlier in the heat stress section and can be found in the reference by Ramsey and Beshir (1998).

Clothing systems designed for cold temperatures must be layered, provide adequate insulation, and fit properly (U.S. Army, 1996). Loose clothing layers with air spaces between them, under a wind- and water-resistant outer garment, and insulated boots play key roles in preventing cold injury. Ramsey and Beshir (1998) discuss and present an equation to calculate an *index of required clothing insulation (IREQ)*. The IREQ is an international standard (ISO, 1993) that allows one to calculate the amount of insulation required for the body to maintain thermal equilibrium.

Engineering and administrative controls and personal protective equipment can be employed to prevent or minimize the occurrence of cold stress. Engineering controls may include provision of adequate heating sources in vehicles and shelters. Design considerations for military systems (e.g., vehicles and shelters) are found in MIL-HDBK-759C (1995) and MIL-STD-1472 F (1999). The ACGIH (2002) provides guidelines for a work–warming regimen for people who work in extremely cold environments ($-26°C/-15°F$ and below).

Detailed information concerning cold stress control and injury prevention strategies can be found in references by Bishop (1998), the ACGIH (2002), and the U.S. Army (1976).

### 15.3.11 Trauma

The tools and techniques of ergonomics are associated with hazard analysis, prevention, and control. These techniques and examples are listed in the army's ergonomics manual (U.S. Army, 2000) as follows:

> Complete a detailed analysis to further evaluate those jobs or worksites having WMSD risk factors as determined by systematic passive and active surveillance. The analysis should systematically consider the concept of multiple-causation and the degree of WMSD risk. Trends, including age, gender, work task, and time of injury should be considered. Work tasks or portions of the process that contain risk factors should be identified. Both problems and solutions should be identified.(p.4)

As a part of the analysis, data should be reviewed and analyzed. There are established data, analytical tools, and methods that may be helpful during a detailed analysis. Examples of these include incidence and severity rates (e.g., a log of federal occupational injuries and illnesses or equivalent); accident and injury reports and lost work time or absenteeism reports by job, unit, department, or facility; checklists, questionnaires, and interviews (see U.S. Army, 2002); and direct observation, videotape analysis, and job analyses (see U.S. Army, 2002). Assessment methodologies may include static and dynamic strength testing, timed activity analysis, biomechanical analysis, and cardiovascular measurements. The NIOSH provides guidance for assessing lifting (NIOSH, 1998).

There is a hierarchy of hazard prevention and control strategies that include process elimination, engineering controls, substitution, work practices, administrative controls, and personal protective equipment. Details and examples of these strategies can be found in the army ergonomics manual (U.S. Army, 2000). Both DiNardi (1998) and the ACGIH (2002) present discussions concerning control strategies and methods to prevent or reduce the risk of developing WMSDs. The ACGIH (2002) also presents a TLV for the hand, wrist, and forearm that is based on hand activity level and peak hand force.

**Example 15.8 Trauma Assessment for an Air Defense System** The HHA developed for the *Vulcan Air Defense System (VADS)* is an example of a trauma assessment (U.S. Army, 1995b). The VADS provides air defense against low-altitude threats and ground coverage against stationary or moving targets, such as personnel, trucks, and lightly armored vehicles. The system includes an M168 20-mm cannon, capable of delivering selected rates of fire of 1000 or 3000 rounds per minute. An HHA identified the potential for musculoskeletal trauma when crew members are required to lift ammunition boxes containing 100 rounds and weighing 97 pounds per box from an auxiliary ammunition carrier to the ammunition feed chute of the weapon. The height of the lift varies depending on the height of the system's carrier but does not exceed 3 ft. Since human engineering design criteria specifies 87 pounds as the maximum allowable weight for a one-time, one-person, two-handed lift to 3 ft, the HHA recommended (1) keeping the distance of the lift between the ammunition carrier and the ammunition feed chute to less than 3 ft or requiring a two-person lift and (2) coordinating with the U.S. Army Human Engineering Laboratory to investigate the need for alternate work practices or engineering design modifications to mitigate the lifting hazard.

### 15.3.12 Vibration

The current accepted method for assessing the effects of WBV is that as described in ISO (ISO, 1985). The U.S. version of ISO 2631 is American National Standards Institute (ANSI) S3.18 (ANSI, 1979). Transducers (e.g., accelerometers) and recorders are used to acquire vibration data. Bruce et al. (1998) describe the equipment and its placement for measuring vibration. The primary method for quantifying vibration is to define the motion in three translational axes (ISO, 1985). The translational motions include the fore-and-aft direction ($x$), lateral direction ($y$), and longitudinal or vertical direction ($z$).

There are several steps followed in order to apply the data derived from the ISO 2631 process to the HHA process (e.g., establishing hazard probability and severity and risk assessment codes):

- establishing an operational test matrix,
- collecting operational WBV data,
- developing the test condition probability classification for each mission, and
- defining the test condition severity classification for the vehicle based upon the WBV exposure criteria presented in ISO 2631.

The specific details for each of these steps are described in the army HHA *Assessor's Guide* (U.S. Army, 1996).

The methods for reducing WBV exposure can be through design or by administrative controls. Design methods may involve the vehicle suspension system and/or the seating system (U.S. Army, 1996). The vehicle suspension characteristics should allow for a wide range of vehicle loads but avoid a natural frequency in the region of human resonances. Seat cushioning and suspension mechanisms can attenuate the transmission of vibration to vehicle occupants. Administrative controls can include rest periods or intermittent participation in specific mission profiles.

The ACGIH has recommended TLVs for both WBV and hand-transmitted vibration (ACGIH, 2002). Bruce et al. (1998) provide additional details concerning exposure criteria, measuring techniques, and control measures for both WBV and hand-transmitted vibration.

**Example 15.9 Vibration Assessment for the Fast Attack Vehicle** An HHA addressing vibration was developed for the *fast attack vehicle (FAV)* (U.S. Army, 1995b). The FAV is a lightweight, all-terrain vehicle capable of high-speed, cross-country travel with high maneuverability and agility. The vehicle serves as a weapons or communications carrier/platform for antiarmor, reconnaissance, deep attack, and other missions that require speed, agility, or the negotiation of rough terrain. A review of the system's performance specifications and development testing identified excessive levels of whole-body vibration. Portable ride meters affixed to the driver's seat demonstrated that the FAV could not be driven for more than 1 minute over a rough surface course at speeds greater than 40 mph. More importantly, the predominant acceleration for the FAV occurred in the 3–5-Hz range—frequencies at which the body's internal organs may resonate. In addition, during development testing, 50 percent of the personnel reported kidney and back-related injuries. These injuries were attributed to the shock and vibration sustained by soldiers due to inadequate cushioning of the seats. The HHA report, therefore, recommended entering soldiers who operate the FAV into a medical surveillance program for whole-body vibration, giving special attention to the genitourinary and musculoskeletal systems and improving the shock absorbency of the FAV's seats and suspension system.

## 15.4 HEALTH HAZARD ASSESSMENT EXPERTISE

There are numerous scientists, engineers, and technicians who specialize in the health and engineering sciences that support the HHA program. Generally, these individuals are members of the AMEDD. The major army medical organizations that are involved with the HHA program include the USACHPPM (Aberdeen Proving Grounds, Maryland) and the U.S. Army Medical Research and Materiel Command (Fort Detrick, Maryland). These organizations and others involved with the program are listed in an HHA procedures guide (U.S. Army, 1994a). Table 15.4 lists various scientists, engineers, and physicians who may do HHAs or provide information to support them. The following paragraphs provide additional detail about the specific roles of some of these professionals.

***Acoustical Energy (Noise and Overpressure)*** A variety of professionals may be involved with acquiring and interpreting noise data and making recommendations to control hazard sources. These may include industrial hygiene and environmental health professionals, audiologists, and acoustic engineers. In the recent past, army physicians were involved in the assessment of blast overpressure to apply and validate a model to predict the potential for developing adverse health effects.

***Biological Substances*** Many scientific disciplines are involved in evaluating biological hazards. Several types of scientists and health professionals are versed in the nature and characteristics of microorganisms. Some examples include microbiologists, virologists, and mycologists. These specialists and others (e.g., physicians and veterinarians) also may be versed in aspects of communicable, infectious, and zoonotic diseases.

***Chemical Substances*** Multiple specialists may be involved with assessing chemical hazards. There are several factors that may influence who may be involved with the assessment. These include identifying, sampling, and analyzing the chemical(s); quantifying the potential exposure; determining hazard and risk; and recommending appropriate control measures. The skills and expertise of analytical chemists, industrial hygienists,

**TABLE 15.4 Examples of Technical Experts Who May Support or Provide Health Hazard Assessments**

| Hazard Category | Technical Experts |
|---|---|
| Acoustic energy | Industrial Hygienist, Environmental Health Scientist, Audiologists, Acoustic Engineers, Physicians (Blast Overpressure) |
| Biological substances | |
| Animal or plant allergen | Physician, Immunologist |
| Insect vector | Entomologist, Zoologist, Biologist |
| Microorganism | Microbiologist, Virologist, Mycologist, Physician, Veterinarian, Biologist |
| Poisonous arthropod | Entomologist, Zoologist, Biologist |
| Poisonous snake | Zoologist, Herpetologist, Biologist |
| Sanitation | Sanitarian, Environmental Health Scientist, Environmental Engineer, Biologist |
| Toxin | Toxicologist, Physician, Biologist |
| Chemical substances | |
| Chemical identification | Analytical Chemist |
| Chemical sampling | Analytical Chemist, Industrial Hygienist, Environmental Health Scientist |
| Chemical analysis and quantification | Analytical Chemist<br>Analytical Chemist, Industrial Hygienist, Environmental Health Scientist |
| Hazard and risk assessment | Industrial Hygienist, Environmental Health Scientist<br>Toxicologist, Risk Assessor, Industrial Hygienist, Environmental Health Scientist |
| Control(s) recommendation | Industrial Hygienist, Environmental Health Scientist |
| Shock | Physicist, Engineers (Automotive Safety Engineer, Biomechanical Engineer), Physicians |
| Temperature extremes and humidity | Physiologist, Physician, Environmental Health Scientist, Industrial Hygienist |
| Trauma | Ergonomist, Biomechanical Engineer, Human Factors Engineer, Industrial Hygienist, Physician, Environmental Health Scientist |
| Vibration | Biomechanical Engineer, Human Factors Engineer, Industrial Hygienist, Physician, Environmental Health Scientist |

toxicologists, other environmental health professionals, toxicologists, and risk assessors may be employed to assess chemical hazards.

***Oxygen Deficiency (Ventilation)*** Professionals who may measure and assess oxygen-deficient environments typically include industrial hygienists, gas-free engineers, and environmental health professionals. These are individuals who are trained to operate direct-reading instruments that measure atmospheric oxygen concentrations and are knowledgeable about the risks associated with oxygen-deficient atmospheres.

***Oxygen Deficiency (High Altitude)*** Physicians, especially aviation medicine and diving medicine specialists, and certain physiologists are knowledgeable about the dynamics and effects of oxygen deficiency due to reduced pressure atmospheres. Some industrial hygienists also are specialized in this area.

***Radiation Energy*** Health physicists are the primary professionals who measure and assess radiation hazards. Also, some industrial hygienists and other environmental health professionals may be specially trained in radiation assessment.

***Shock*** People who would be proficient to assess shock include professionals who are skilled in physics and engineering. This would include physicists and engineers, especially automotive safety engineers and biomechanical engineers. Physicians also may be involved with helping to define and establish when detrimental health effects occur to body organ systems.

***Temperature Extremes and Humidity (Heat Stress)*** There are several professionals who may evaluate and assess the potential for people to develop heat stress. Physiologists and physicians may be involved in predicting the types of health effects that may occur at various levels of heat exposure and subsequent to the development of heat-related medical conditions. Environmental health professionals and industrial hygienists typically monitor ambient and work environments and recommend control measures to prevent heat-related illness.

***Temperature Extremes and Humidity (Cold Stress)*** There are several professionals who may evaluate and assess the potential for people to develop cold stress and injury. Physiologists and physicians may be involved in predicting the types of health effects that may occur at various levels of cold exposure and subsequent to the development of cold-related medical conditions. Environmental health professionals and industrial hygienists typically monitor ambient and work environments and recommend control measures to prevent cold-related illness.

***Trauma*** People who would be proficient to assess ergonomic concerns would include ergonomists, biomechanical engineers, human factors engineers, industrial hygienist, some physicians, and some environmental health professionals. The Department of the Army considers "trained ergonomics personnel" to be "health care, industrial hygiene, environmental (health) science, safety, or engineering personnel with approved training in ergonomics" (U.S. Army, 2000, p.17).

***Vibration*** People who would be proficient to assess vibration hazard potential would include biomechanical engineers, human factors engineers, some industrial hygienists, some physicians, and some environmental health professionals.

## 15.5 HEALTH HAZARD ANALYSIS PROCESS

In this section, the overall HHA process is described in terms of HHA requirements and general guidelines.

### 15.5.1 Why and When to Do an HHA

The objective of HSI is to develop military equipment and systems that will "fit the soldier, sailor, airman" by improving the interface between the person and the system. This improvement includes protecting the health of the soldier/sailor/airman by eliminating or minimizing stresses that could occur from health hazards. Thus, the primary objective of the HHA program is to identify, assess, and eliminate or control health hazards associated with the life-cycle system management of weapon systems, munitions, equipment, clothing, training devices, materiel systems, and information systems (U.S. Army, 1995b). Other specific supporting objectives are to protect the serviceman, enhance mission effectiveness, and contain costs. These are to preserve and protect health, reduce performance decrement, enhance system effectiveness, reduce system design retrofits needed to control or eliminate health hazards, enhance readiness, and reduce personnel injury and illness compensation (U.S. Army, 1995b).

The HHA process should begin early in the life cycle of a system. This could occur as early as during the identification of requirements and needs, usually a combat developer's responsibility, and subsequent generation of documentation. Alternatively, the process could begin when the materiel developer is identified and formulates a development plan. However, the HHA process never should be delayed beyond concept exploration. When initiated beyond concept exploration, the process suffers because of competition for resources and a compressed development timeline. The HHA report (HHAR) (to include the initial and updated HHARs) should be provided to the materiel developer early in the acquisition process so that it can be considered during various decision stages (milestones). The HHAR also should serve as a source document to influence other aspects of the acquisition and development process (e.g., test plans, market investigations, safety releases, and system technical and training publications).

The HHA process also supports the acquisition of commercial off-the shelf (COTS) items and nondevelopmental items (NDIs). Even if a commercial product has been evaluated for health concerns, the HHA process can determine if the commercial or any other health assessment is relevant to the intended U.S. military use.

### 15.5.2 HHA Program Directives and Guidelines

The HHA program was developed by the U.S. Army in response to many years of addressing health hazards associated with the use of military weapons, equipment, and other systems. It became obvious that it was better to anticipate and address such issues when systems were being conceptualized and developed rather than after they were fielded and in use. Thus, in 1981 the army surgeon general formalized the HHA program by developing and coordinating the publication of Army Regulation (AR) 40-10, *Health Hazard Assessment in Support of the Army Materiel Acquisition Decision Process* (the current version is U.S. Army, 1991a). Since then the program has been integrated with DoD provisions that require a variety of human factor concerns to be addressed during system acquisition. Also, the requirement to integrate the HHA program into the materiel acquisition and development process has been incorporated into a number of army medical, personnel, safety, and other regulations. Examples of these are given in Table 15.5.

### 15.5.3 Military-Unique Hazards

One of the roles of the HHA program is to address situations that are unique to the military and do not have a direct civilian correlate (Gross and Broadwater, 1993). When military design, specifications, or requirements render compliance with civilian health standards infeasible or when no regulatory standard exists for such military application, DoD components can develop and publish special military standards, rules, or regulations prescribing occupational safety and health measures (DoD instruction 6055.1). Conse-

**TABLE 15.5 Selected Department of Defense Publications and Army Regulations that Address Integration of Health Hazard Assessment Program into Materiel Acquisition Decision Process**

| Publication Number | Publication Title |
|---|---|
| | *Department of Defense Publications* |
| Regulation 5000.2-R | *Mandatory Procedures for Major Defense Acquisition Programs (MDAPS) and Major Automated Information System (MAIS) Acquisition Programs* |
| Instruction 6055.1 | *DoD Occupational Safety and Health Program* |
| Instruction 5000.36 | *System Safety Engineering and Management* |
| MIL-HDBK 759A | *Human Factors Engineering Design for Army Materiel* |
| | *U.S. Army Publications* |
| Regulation 40-10 | *The Army Health Hazard Assessment Program in Support of the Materiel Acquisition Decision Process* |
| Regulation 40-5 | *Preventive Medicine* |
| Regulation 602-2 | *Manpower and Personnel Integration (MANPRINT) in the Materiel Acquisition Process* |
| Regulation 602-1 | *Human Factors Engineering Program* |
| Regulation 15-14 | *System Acquisition Review Council Procedures* |
| Regulation 70-1 | *Systems Acquisition Policy and Procedures* |
| Regulation 70-10 | *Test and Evaluation During Development and Acquisition of Materiel* |
| Regulation 70-15 | *Product Improvement of Materiel* |
| Regulation 70-142 | *Materiel Release, Fielding, and Transfer* |
| Regulation 200-1 | *Environmental Protection and Enhancement* |
| Regulation 385-16 | *Systems Safety Engineering and Management* |
| Regulation 70-75 | *Survivability of Army Personnel and Material* |
| PAM 700-142 | *Instructions for Materiel Release, Fielding, and Transfer* |
| Available from USACHPPM | *U.S. Army Health Hazard Assessment Program Strategy* |
| Available from USACHPPM | *Health Hazard Assessment Manual—Procedures Guide* |
| Available from USACHPPM | *Health Hazard Assessor's Guide* |
| Available from USACHPPM | *Materiel Developers Pocket Guide to Systems Health Hazards* |

quently, some of the health standards presented earlier in this chapter were developed to be applied uniquely to military situations.

### 15.5.4 U.S. Army Health Hazard Assessors Guide

The *U.S. Army Health Hazard Assessor's Guide* (U.S. Army, 1996) was a major reference for the contents of this chapter. This guide was developed and authored by scientists from the Army Medical Department, especially the U.S. Army Center for Health Promotion and Preventive Medicine and the U.S. Army Medical Research and Materiel Command (USAMRMC), and some from other organizations (see acknowledgment note at end of chapter). It was created to be a cumulative technical reference document and to serve as a guide to the medical criteria and standards used to assess health hazards associated with army systems. The guide's developers intend that it be updated routinely in order to incorporate new trends and technology and remove outdated information. The guide presents key information used by health hazard assessors in the various hazard categories to include definition of the hazard, hierarchy of criteria and standards, methods of developing army-specific criteria and standards, and methods for measuring hazards and interpreting health risks.

## 15.6 TOOLS THAT SUPPORT THE OVERALL HEALTH HAZARD ASSESSMENT PROCESS

There are several tools that are applied in the HHA process to all of the various hazard categories. They either complement the hazard assessment directly or the HHA process in general.

### 15.6.1 Risk Assessment and Risk Assessment Codes

One purpose of the HHA program is to convey to the materiel developer the risks associated with the operation and maintenance of military systems. The NRC described a process that involves risk assessment and risk management that is applicable to the HHA process (NRC, 1983). In the HHA process the Army Medical Department, through the USACHPPM, is the health *risk assessor*. The materiel developers (acquisition community, e.g., program and project managers) are the *risk managers*. Risk assessment generally is a four-stage process that evaluates the potential for people to develop disease or die from exposure to biological, chemical, or physical agents. The risk management process is separate from but includes the risk assessment. Managing risk includes consideration and integration of a variety of other factors (e.g., technology, economics and funding, politics, social concerns, military needs and requirements, and others) that influence the outcome of design and production decisions. Risk assessment and management paradigms are discussed in a variety of references [Roberts and Abernathy, 1995; Abernathy and Roberts, 1994; NRC, 1983; Presidential/Congressional Commission on Risk Assessment and Risk Managemet (PCCRARM), 1997a,b].

The health hazard assessor (or independent medical assessor as identified in AR 40-10 U.S. Army, 1991a) estimates the health risk potential and provides this information, qualitatively and quantitatively, to the developer. The four basic steps in the risk assessment process are hazard identification, dose–response assessment, exposure assessment, and risk characterization. Hazard identification is the process of determining the type of adverse health effects that a biological, chemical, or physical agent may cause. Dose–response assessment relates the severity of an adverse health effect in response to exposure to specific amounts or quantities of agents. Exposure assessment determines who will be exposed and how they will be exposed; the medium (e.g., air, water, soil, food, etc.), the routes (e.g., inhalation, ingestion, skin contact), the duration, the amount, etc. The risk characterization is the quantitative and/or qualitative expression of risk that combines the hazard identification, dose–response assessment, and exposure assessment.

The risk characterization is the tool that allows the independent medical assessor to convey the health risk and recommend to the materiel developer exposure levels to agents that should not cause adverse health effects to soldiers who use or maintain military systems. The assessment of such hazards may simply be to determine compliance with consensus standards or guidelines. If there is no such standard, then applying a risk assessment and management process may be necessary to estimate the impact to human health and develop a criteria or guideline to protect the health of soldiers.

The materiel developer then incorporates the health risk information (HHA) into the materiel acquisition development process. The developer is expected to make design and/or process changes that will incorporate the HHA recommendations to produce a system that will not cause any adverse health problems. However, it is not always possible or feasible to totally eliminate all conditions that can cause adverse health problems. Realistically, the developer must make design and production decisions based upon a variety of factors. Thus, the developer should integrate the health and medical recommendations into the overall decision process. When adverse health conditions cannot be totally eliminated, they should be reduced to some acceptable protective level. This level may be difficult to define and its pursuit should promote coordination and interaction between the developer and the HHA community.

In the military the estimated degree of risk that is associated with each health hazard is assigned a RAC. The RAC is an alphanumeric index that is based upon a hazard's probability (A, frequent; B, probable; C, occasional; D, remote; E, improbable) and its severity (I, catastrophic; II, critical; III, marginal; IV, negligible). Detailed discussion of RACs and their implications can be found in Gross and Broadwater (1993) and U.S. Army (1990a, 1991a) regulations. This alphanumeric designation is an example of how risk characterization is communicated to materiel developers.

### 15.6.2 Exposure Control Hierarchy

A key element of the HHA program is to identify and recommend strategies that will eliminate or decrease exposures to hazardous agents. Generally, there are several ways that any one potential hazard can be controlled. Throughout this chapter there are various controls discussed for each of the hazard categories. Various controls are presented here to show the reader that there is a hierarchy of possible control strategies. Such strategies may range from completely eliminating the hazard from the system or process to removing the person from the system or process. Generally removing the hazard and engineering controls is the best strategy because they do not rely upon an individual to comply with an activity or practice. Thus, there is a hierarchy of control strategies that can be recommended to control health hazards. The hierarchy of control strategies, in order of preference from a health perspective, includes engineering controls, work practices, personal protection, and administrative controls. These strategies and some examples

**TABLE 15.6 Hierarchy of Control Strategies and Examples to Eliminate or Control Chemical, Physical, and Biological Hazards**

| | |
|---|---|
| A. Engineering controls | C. Personal protection |
| • Elimination<br>• Substitution<br>• Isolation<br>• Enclosure.<br>• Ventilation<br>• Process change<br>• Product change | • Respiratory protection<br>• Gloves<br>• Apron<br>• Eye goggles<br>• Ear muffs and plugs |
| B. Work practices | D. Administrative controls |
| • Housekeeping.<br>• Dust suppression<br>• Maintenance.<br>• Sanitation<br>• Work practices<br>• Education<br>• Labeling and warning systems<br>• Waste disposal practices | • Work–rest cycles<br>• Exposure time limits<br>• Environmental monitoring<br>• Medical control<br>• Management program |

are provided in Table 15.6. The reader should consult Corn (1984) and the HHA procedures guide (U.S. Army, 1994a) for additional detail and examples of control strategies.

### 15.6.3 Medical Cost Avoidance Model

Historically, it has been difficult for public health practitioners to quantify the economic impact of preventive programs. Generally, it is difficult to predict how much disease, illness, or death public health programs will prevent and then relate it to economy. However, given that such programs require funding to operate, they must compete for economic resources. Funding resources often are allocated based upon priority of need, and economic impact (e.g., cost savings) is one factor that is considered in decisions to allocate funds. Therefore, public health and preventive medicine practitioners need ways to quantify their economic impact. Bratt et al. (1997) developed a *medical cost avoidance model (MCAM)* that estimates medical costs for health hazards based on risk assessment codes. This model estimates total medical costs for unabated health hazards based on the costs for clinic visits, hospitalization, lost time, disability, rehabilitation, and death. It quantifies health hazard costs, improves the understanding of a stated health risk, and assists materiel developers with making risk management and trade-off decisions concerning corrective actions.

### 15.6.4 Hazard Tracking Database

As the HHA program has matured, an electronic database maintained at the USACHPPM was developed and evolved to record and monitor the health hazards associated with military equipment and systems. The current database tracks results of HHAs and provides reporting designed to assist the HHA program manager in daily activities. Also, it is a resource for medical planners and advisors to use that can identify and estimate potential hazards that soldiers may encounter as they train and conduct missions. When new materiel systems are being considered for development, the database can be queried to provide information about health hazards associated with similar systems. Also, when existing systems are being considered for product improvement or modifications, the database can be queried to provide information about their existing health hazards. Additional information about the database, its history, and capabilities can be found in a discussion by Murnyak et al. (2002).

## 15.7 SUMMARY

By presenting the U.S. Army's HHA program, this chapter defines the typical hazards associated with military equipment and systems and demonstrates how the military services address the health component of the DoD HSI requirements. The reader should recognize that the sciences applied to and the process of conducting health assessments for military equipment are detailed and complex. Generally, these systems and equipment are evaluated by several health scientists in a multidisciplinary manner. Materiel developers should plan for and integrate this process into their acquisition and development plans and allow sufficient time for early and later HHAs to include the time for acquiring necessary data. Therefore, it should be evident that there is a need to integrate health concerns into the development and acquisition process during the early stages (e.g., during concept exploration). Early integration can help avoid costly retrofits to correct or eliminate hazardous conditions. The military services continue to apply state-of-the-art science, technology, and medical knowledge to assess and control military health hazards in order to protect and preserve the health of U.S. military forces and to enhance the military mission.

## NOTES

1. Much of this chapter was either based upon or inspired by the contents of the *U.S. Army Health Hazard Assessor's Guide* (U.S. Army, 1996), which was developed and authored by scientists from the Army Medical Department, especially the U.S. Army Center for Health Promotion and Preventive Medicine and the U.S. Army Medical Research and Materiel Command. While it is recognized that this guide is a military publication and a public domain document, the contributions of these scientists and engineers should be recognized for their contributions and dedication to the health hazard assessment program and military public health and preventive medicine. Therefore, I wish to acknowledge the following individuals for their contributions to the guide (the organizations shown are the ones where they worked during the development of the guide): Major John Albano, U.S. Army Aeromedical Research Laboratory; Major (retired) David Alberth, U.S. Army Center for Health Promotion and Preventive Medicine; Dr. Nabih Alem, U.S. Army Aeromedical Research Laboratory; Lieutenant Colonel Gregory Argyros, Walter Reed Army Medical Center; Lieutenant Colonel (retired) Gary Bratt, Office of the Surgeon General; Major Barkley Butler. U.S. Army Aeromedical Research Laboratory; Lieutenant Colonel James Carroll, U.S. Army Medical Research and Materiel Command; Mr. John DeFrank, U.S. Army Center for Health Promotion and Preventive Medicine; Mr. Jim Devine, U.S. Army Research Institute of

Environmental Medicine; Mr. Harris Edge (retired), U.S. Army Center for Health Promotion and Preventive Medicine; Lieutenant Colonel Tom Fitzpatrick, Walter Reed Army Institute of Research; Mr. Robert Gross, U.S. Army Center for Health Promotion and Preventive Medicine; Ms. Jennifer Houser, U.S. Army Center for Health Promotion and Preventive Medicine; Dr. Adolph Januskiewicz, Walter Reed Army Institute of Research; Lieutenant Colonel (retired) Roland Langford, U.S. Army Medical Research and Materiel Command; Ms. Leslie Levine, U.S. Army Research Institute of Environmental Medicine; Colonel Maria Mayorga, Walter Reed Army Institute of Research; Mr. Tom McNeil, U.S. Army Center for Health Promotion and Preventive Medicine; Mr. Ben Mozo (retired), U.S. Army Aeromedical Research Laboratory; LTC David Mukai, U.S. Army Center for Health Promotion and Preventive Medicine; Lieutenant Colonel (retired) George Mumyak, U.S. Army Center for Health Promotion and Preventive Medicine; Dr. James Patterson, U.S. Army Aeromedical Research Laboratory; Mr. Brad Roberts, U.S. Army Center for Health Promotion and Preventive Medicine; Lieutenant Colonel (retired) Welford C. Roberts, U.S. Army Materiel Command; Mr. Felix Sachs, U.S. Army Center for Health Promotion and Preventive Medicine; Dr. David Sliney, U.S. Army Center for Health Promotion and Preventive Medicine; Dr. Suzanne Smith, Wright-Patterson AFB; Mr. Bruce Stuck, Walter Reed Army Institute of Research; and Mr. Maurice Weeks (retired), U.S. Army Center for Health Promotion and Preventive Medicine.

2. REMEDY is a registered trademark of Scientific Application International Corporation (SAIC).

## REFERENCES

Abernathy, C. O., and Roberts W. C. (1994). Risk Assessment in the Environmental Protection Agency. *Journal of Hazardous Materials*, *39*, 135–142.

American Conference of Governmental Industrial Hygienists (ACGIH). (2000). *Documentation of the Threshold Limit Values and Biological Exposure Indices*, 7th ed. Cincinnati, OH: ACGIH.

American Conference of Governmental Industrial Hygienists (ACGIH). (2001). *Industrial Ventilation, a Manual of Recommended Practice*, 24th ed. Cincinnati, OH: ACGIH.

American Conference of Governmental Industrial Hygienists (ACGIH). (2002). *Threshold Limit Values for Chemical Substances and Physical Agents and Biological Exposure Indices.* Cincinnati, OH: ACGIH.

American Industrial Hygiene Association (AIHA). (1975). *Industrial Noise Manual*, 3rd ed. Fairfax, VA: AIHA.

American Industrial Hygiene Association (AIHA). (2001). *Emergency Response Planning Guidelines and Workplace Environmental Exposure Level Guides Handbook.* Fairfax, VA: AIHA.

American Industrial Hygiene Association (AIHA). (2001). *WEEL 2001 Update Set.* Fairfax, VA: AIHA.

American National Standards Institute (ANSI). (1979). *Guide for the Evaluation of Whole Body Vibration*, ANSI S3.18. Washington, DC: ANSI.

American Water Ways Association. (1995). *Standard Methods for the Examination of Water and Wastewater*, 19th ed. Washington, DC: American Public Health Association.

Asfour, S., Ayoub, M., Genaidy, A., and Khalil, T. (1986). A Database of Physiological Responses to Manual Lifting. *Trends in Ergonomics/Human Factors, III*, 801–809 (Abstract).

Ayoub, M., Mital, A., Bakken, G., Asfour, S., and Bethea, N. (1980a). Development of Strength and Capacity Norms for Manual Materials Handling Activities: The State of the Art. *Human Factors*, *22*(3), 271–283.

Ayoub, M. M. (1991). Determining Permissible Lifting Loads: An Approach. In *Proceedings of the Human Factors Society 35th Annual Meeting* (pp. 825–829). Santa Monica, CA: Human Factors Society.

Ayoub, M. M., Mital, A., Asfour, S. S., and Bethea, N. J. (1980b). Review, Evaluation, and Comparison of Models for Predicting Lifting Capacity. *Human Factors*, *22*(3), 257–269.

Beneson, A. S. (1990). *Control of Communicable Diseases in Man*. Washington, DC: American Public Health Association.

Bishop, P. A. (1998). Applied Physiology of Thermoregulation and Exposure Control. In S. R. DiNardi (Ed.), *The Occupational Environment – Its Evaluation and Control* (pp. 629–660). Fairfax, VA: American Industrial Hygiene Association.

Böhm, F. (1964). The Effect of Vibrations on the Genitalia of Female Driving Personnel. *Zeitschrift fur die gesamte Hygiene und ihre Grenzgebiete*, *10*(10), 720–736.

Bratt G. M., Doganiero, D. M., and Spencer C. O. (1997, Fall). Estimating the Health Hazard Costs of Army Materiel: A Method for Helping Program Managers Make Informed Health Risk Decisions. *Acquisition Review Quarterly*, pp. 443–471.

Brovko, E. I. (1975). On the Disturbance of Menstruation of Female Tram Drivers and Conductors. *Gigiena Truda i Professional 'nye Zabolevani*, *20*(11), 182–185.

Bruce, R. D., Bommer, A. S., and Moritz, C. T. (1998). Noise, Vibration, and Ultrasound. In S. R. DiNardi (Ed.), *The Occupational Environment—Its Evaluation and Control* (pp. 424–488). Fairfax, VA: American Industrial Hygiene Association.

Burr, R. E. (1991). *Heat Illness: A Handbook for Medical Officers*, USARIEM Technical Note 91-3. Natick, MA: US Army Research Institute of Environmental Medicine.

Burr, R. E. (1993). *Medical Aspects of Cold Weather Operations: A Handbook for Medical Officers*, USARIEM Technical Note 93-4. Natick, MA: US Army Research Institute of Environmental Medicine.

Cameron, B. et al. (1995). *Development of a Standard for the Health Hazard Assessment of Mechanical Shock and Repeat Impact in Army Vehicles. Phase 4 Experimental Phase*, USAARL CR 96-1. Fort Rucker, AL: U.S. Army Aeromedical Research Laboratory (as referenced in U.S. Army, 1996).

Ciriello, V., and Snook, S. (1978). The Effects of Size, Distance, Height, and Frequency on Manual Handling Performance. *Proceedings of the Human Factors Society*, pp. 318–322.

Ciriello, V., and Snook, S. (1983). A Study of Size, Distance, Height, and Frequency Effects on Manual Handling Tasks. *Human Factors*, *25*(5), 473–483.

Clayton, G. D., and Clayton, F. E. (Eds.). (1991). *Patty's Industrial Hygiene and Toxicology*, Vols. 1–3, 4th ed. New York: Wiley.

Code of Federal Regulations (CFR), Title 10, Chapter 1, *Nuclear Regulatory Commission*, Part 20, *Standards for Protection Against Radiation*. Washington, DC: Government Printing Office.

Code of Federal Regulations (CFR), Title 29, Part 1910.146, *Permit-Required Confined Space*. Washington, DC: Government Printing Office.

Corn, M. (1984). Role of Control Technologies in Preventing Occupational Disease. *Archives of Environmental Health*, *3*(39), 235.

Daniel, W. W. (1983). *Biostatistics: A Foundation for Analysis in the Health Sciences*, 3rd ed. New York: Wiley.

DiNardi, S. R. (1998). Ergonomics. In S. R. DiNardi (ed.). (1998), *The Occupational Environment—Its Evaluation and Control* (pp. 726–774). Fairfax, VA: American Industrial Hygiene Association.

Donahue, A. M., and Ohlin, D. W. (1993). Noise and the Impairment of Hearing. In D. P. Jenkins (Ed.), *Textbook of Military Medicine, Part III—Disease and the Environment*, Vol. 2: *Occupational Health, the Soldier and the Industrial Base* (pp. 207–252). Washington, DC: Office of the (Army) Surgeon General—Borden Institute.

Draper W. M., Ashley, K., Glowacki, C. R., and Michael, P. R. (1999). Industrial Hygiene Chemistry: Vol 71 No 12, Keeping Pace with Rapid Change in the Workplace. *Analytical Chemistry. 71*, 33R–60R.

Freund, B. J., O'Brien, C., and Young, A. J. (1994). Alcohol Ingestion and Temperature Regulation During Cold Exposure. *Journal of Wilderness Medicine*, *5*, 88–98.

Fulco, C. S., and Cymerman, A. (1988). Human Performance and Acute Hypoxia. In K. B. Pandolf, M. N., Sawka, and R. R. Gonzales (Eds.), *Human Performance Physiology and Environmental Medicine at Terrestrial Extremes*. Indianapolis, IN: Benchmark.

Gagge, A. P., and Gonzalez, R. R. (1996). Mechanisms of Heat Exchange: Biophysics and Physiology. In M. J. Fregly, and C. M. Blatteis (Eds.), *Handbook of Physiology*, Vol. 1 (pp. 45–84). New York: Oxford University Press.

Garg, A., and Ayoub, M. M. (1980). What Criteria Exist for Determining How Much Load Can Be Lifted Safely? *Human Factors*, *22*, 475–486.

Gaydos, J. C. (1988). A Historical View of Occupational Health for the Soldier. *Medical Bulletin of the US Army Medical Department*, *2*, 4–6.

Gaydos, J. C. (1993). Historical Review of the Need for Military Toxicology and the U.S. Army's Response. In H. J. Clewell III (Ed.), *Proceedings: Conference on Chemical Risk Assessment in the Department of Defense: Science, Policy, and Practice, Wright-Patterson Air Force Base, Ohio, 1992.* Cincinnati, OH: American Conference of Governmental Industrial Hygienists.

Gonzalez, R. R., and Stroschein, L. A. (1991). Computer Modeling: Predicting the Soldier's Heat Transfer and Work Capabilities in MOPP. In *Proceedings of the 33rd Annual Conference of the Military Testing Association*, San Antonio, Texas, October 28–31, 1991 (pp. 553–558).

Griffin, M. J. (1990). *Handbook of Human Vibration.* London, Academic.

Gross, R. A., and Broadwater, W. T. (1993). Health Hazard Assessments. In *Textbook of Military Medicine, Part III—Disease and the Environment*, Vol. 2. *Occupational Health, the Soldier and the Industrial Base*, (pp. 165–205). Washington, DC: Office of the (Army) Surgeon General—Borden Institute.

Hackett, P. H., and Rennie, D. (1977). Acute Mountain Sickness (Letter). *Lancet*, *7*, 1491–1492.

Hackett, P. H., Rennie, D., and Levine, H. D. (1976). The Incidence; Importance; and Prophylaxis of Acute Mountain Sickness. *Lancet*, *2*, 1149–1151.

Hackett, P. H., and Roach, R. C. (1987). Medical Therapy of Altitude Illness. *Annals of Emergency Medicine*, *76*, 980–986.

Hackett, P. H., Roach, R. C., and Sutton, J. R. (1989). High Altitude Medicine. In P. S. Auerbach, and E. C. Geehr (Eds), *Management of Wilderness and Environmental Emergencies*, (pp. 1–34). St. Louis, MO: C. V. Mosby.

Hayes, A. W. (Ed.). (1994). *Principles and Methods of Toxicology*, 3rd ed. New York: Raven.

Henzel, J. H. et al. (1966). *Effects of Anterior Intercostal Nerve Block on the Threshold of Thoracic Pain Associated with* $g_z$ and $g_x$ *Vibration*, AMRL-TR-65-68. Wright-Patterson Air Force Base, OH (as referenced in U.S. Army, 1996).

Hitchcock, R. T., Murray, W. E., Patterson, R. M., and Rockwell, R. J. (1998). Nonionizing Radiation. In S. R. DiNardi (Ed.), *The Occupational Environment—Its Evaluation and Control.* (pp. 490–579). Fairfax, VA: American Industrial Hygiene Association.

Houston, C. S. (1984). Altitude Illness. *Emergency Medicine Clinic of North America*, *2*(3), 503–512.

International Organization for Standardization (ISO). (1985). *Evaluation of Human Exposure to Whole Body Vibration—Part I. General Requirements*, ISO 2631/1. Geneva, Switzerland: ISO.

International Organization for Standardization (ISO). (1993). *Evaluation of Cold Environments: Determination of Required Clothing Insulation (IREQ)*, ISO/TR 11079. Geneva, Switzerland: ISO.

Jones, B. J., Rock, P. B., Sawka, M. N., Modrow, H. E., Lindsay, G. C., Petruccelli, B., Mays, M. Z., O'Mara, M. A., Young, A. J., Stroschein, L. A., and Krueger, G. P. (Eds.). (1993). *Sustaining Soldier Health and Performance in the Former Republic of Yugoslavia: Guidance for Small Unit Leaders*, USARIEM Technical Note 93-6. Natick, MA: U.S. Army Research Institute of Environmental Medicine.

Klaassen, C. D. (Ed.) (1996). *Casarett and Doull's Toxicology, the Basic Science of Poisons*, 5th ed. New York: McGraw-Hill.

Klienbaum, D. G., and Kupper, L. L. (1978). *Applied Regression Analysis and Other Multivariable Methods.* Boston, MA: Duxbury.

Kotsonis, F. N., Burdock, G. A., and Flamm, W. G. (1996). Food Toxicology. In C. D. Klaassen, (Ed.), *Casarett and Doull's Toxicology, the Basic Science of Poisons*, 5th ed. (pp. 909–949). New York: McGraw-Hill.

Leibrecht, B. (1990). *Health Hazard Assessment Primer.* U.S. Army USAARL 90-5. Fort Rucker, AL: Aeromedical Research Laboratory.

Levine, L., Sawka, M. N., and Gonzalez, R. R. (1995) *General Procedure for Clothing Evaluations Relative to Heat Stress*, USARIEM Technical Note 95-5. Natick, MA: U.S. Army Research Institute of Environmental Medicine.

Lewis, R. J., Sr. (1999). *Sax's Dangeous Properties of Industrial Materials*, 10th Ed. New York: Wiley.

Lioy, P. J., and Lioy, M. J. (Eds.). (1983). *Air Sampling Instruments for Evaluation of Atmospheric Contaminants*, 6th ed. (pp. T12–T14, T1–V118). Cincinnati, OH: American Conference of Governmental Industrial Hygienists.

Lynch, L. C., Elliott, C. W., and McMurry, P. (1999). Disease and Nonbattle Injury Forecasting. *Army Medical Department Journal*, pp. 3–8.

Mausner, J. S., and Kramer, S. (1985). *Epidemiology—An Introductory Text.* Philadelphia, PA: W. B. Saunders.

McCarthy, M. E. (1998). *Nonionizing Radiation.* In S. R. DiNardi, (Ed.), *The Occupational Environment—Its Evaluation and Control* (pp. 580–602). Fairfax, VA: American Industrial Hygiene Association.

Moeller, D. W. (1997). *Environmental Health.* Cambridge, MA: Harvard University Press.

Montgomery, A. B., and Mills, J., and Luce, J. M. (1989). Incidence of Acute Mountain Sickness at Intermediate Altitude. *Journal of the American Medical Association*, *267*, 732.

MIL-HDBK-759C. (1995). *Human Factors Engineering Design for Army Materiel (Metric).* Redstone Arsenal, AL: U.S. Army Missile Command.

MIL-STD-1472F. (1999). *Human Engineering Design Criteria for Military Systems, Equipment, and Facilities*. Washington, DC: Department of the Army.

MIL-STD 1474D. (1991). *Noise Limits for Military Materiel*. Washington, DC: Department of the Army. (Change Number 1 published 1993).

MIL-STD 1425A. (1991). *Safety Design Requirements for Military Lasers and Associated Support Equipment*. Washington, DC: Department of the Army.

Murnyak, G. R., Spencer, C. O., Chaney, A. E., and Roberts, W. C. (2002). The Evolution of a Health Hazard Assessment Database Management System for Military Weapons, Equipment, and Materiel. *Military Medicine*, *167*(4), 331–342.

Nader, J. S., Lauderdale, J. F., and McCammon, C. S. (1983). Direct Reading Instruments for Analyzing Airborne Gases and Vapors. In P. J. Lioy and M. J. Lioy (Eds.), *Air Sampling*

*Instruments for Evaluation of Atmospheric Contaminants*, 6th ed. (pp. V1–V4). Cincinnati, OH: American Conference of Governmental Industrial Hygienists.

National Institute of Occupational Safety and Health (NIOSH) (1979). *Criteria for a Recommended Standard: Working in Confined Spaces*, DHHS (NIOSH) Publication No. 80-106. Washington, DC: Department of Health and Human Services, National Institute of Occupational Safety and Health.

National Institute of Occupational Safety and Health (NIOSH) (1986). *Criteria for a Recommended Standard: Occupational Exposure to Hot Environments*, DHHS (NIOSH) Publication No. 86-113. Washington, DC: Department of Health and Human Services, National Institute of Occupational Safety and Health.

National Institute of Occupational Safety and Health (NIOSH) (1994). *NIOSH Manual of Analytical Methods (NMAM)*, NIOSH Publication No. 94-113. Washington, DC: Department of Health and Human Services, National Institute of Occupational Safety and Health.

National Institute of Occupational Safety and Health (NIOSH) (1998). *Applications Manual for the Revised NIOSH Lifting Equation*, NIOSH Publication No. 94-110. Washington, DC: Department of Health and Human Services, National Institute of Occupational Safety and Health.

Norton, S. (1996). Toxic Effects of Plants. In C. D. Klaassen (Ed.). (1996). *Casarett and Doull's Toxicology, the Basic Science of Poisons*, 5th ed. (pp. 841of 853). New York: McGraw-Hill.

National Research Council (NRC). (1983). *Risk Assessment in the Federal Government: Managing the Process*. Washington, DC: National Academy Press.

National Research Council (NRC). (1984a). *Emergency and Continuous Exposure Guidance Levels of Selected Airborne Contaminants*, Vol. 1. Washington, DC: National Academy Press.

National Research Council (NRC). (1984b). *Emergency and Continuous Exposure Guidance Levels of Selected Airborne Contaminants*, Vol. 2. Washington, DC: National Academy Press.

National Research Council (NRC). (1984c). *Emergency and Continuous Exposure Guidance Levels of Selected Airborne Contaminants*, Vol. 3. Washington, DC: National Academy Press.

National Research Council (NRC). (1985a). *Emergency and Continuous Exposure Guidance Levels of Selected Airborne Contaminants*, Vol. 4. Washington, DC: National Academy Press.

National Research Council (NRC). (1985b). *Emergency and Continuous Exposure Guidance Levels of Selected Airborne Contaminants*, Vol. 5. Washington, DC: National Academy Press.

National Research Council (NRC). (1986). *Emergency and Continuous Exposure Guidance Levels of Selected Airborne Contaminants*, Vol. 6. Washington, DC: National Academy Press.

National Research Council (NRC). (1987). *Emergency and Continuous Exposure Guidance Levels of Selected Airborne Contaminants*, Vol. 7. Washington, DC: National Academy Press.

National Research Council (NRC). (1988). *Emergency and Continuous Exposure Guidance Levels of Selected Airborne Contaminants*, Vol. 8. Washington, DC: National Academy Press.

Occupational Safety and Health Administration (OSHA). (1996). *Sampling and Analysis*, OSHA Technical Manual (OTM). Washington, DC: OSHA, U.S. Department of Labor.

Olishifski, J. B. (1985) Fundamentals of Industrial Hygiene, In Edition Chicago, IL: National Safety Council.

Ott, L. (1988). *An Introduction to Statistical Methods and Data Analysis*, 3rd ed. Boston, MA: PWS-Kent Publishing Company.

Pandolf, K. B., Stroschein, L. A., Drolet, L. L., Gonzalez, R. R., and Sawka, M. N. (1971, August). Prediction Modeling of Physiological Responses and Human Performance in the Heat with Application to Space Operations. In *Proceedings of the Seventh Annual Workshop on Space Operations, Applications, and Research (SOAR, 93), NASA Lyndon B. Johnson Space Center*, NASA CR-1855.

Pandolf, K. B., Stroschein, L. A., Drolet, L. L., Gonzalez, R. R., and Sawka, M. N. (1986). Prediction modeling of Physiological Responses and Human Performance in the Heat. *Computers in Biology and Medicine*, *6*, 319–329.

Popendorf, W. (1998). Barometric Hazards. In S. R. DiNardi (Ed.), *The Occupational Environment—Its Evaluation and Control* (pp. 605–626). Fairfax, VA: American Industrial Hygiene Association.

Presidential/Congressional Commission on Risk Assessment and Risk Management (PCCRARM). (1997a). *Framework for Environmental Health Risk Management; Final Report*, Vol. 1. Available: http://www.riskworld.com.

Presidential/Congressional Commission on Risk Assessment and Risk Management (PCCRARM). (1997b). *Risk Assessment and Risk Management in Regulatory Decision-Making; Final Report*, Vol 2. Available: http://www.riskworld.com.

Ramsey, J. D., and Beshir, M. Y. (1998). Thermal Standards and Measurement Techniques. In S. R. DiNardi (Ed.), *The Occupational Environment—Its Evaluation and Control* (pp. 661–693). Fairfax, VA: American Industrial Hygiene Association.

Reeves, J. T., and Schoene, R. B. (1991). When Lungs on Mountain Leak: Studying Pulmonary Edema at High Altitudes. *New England Journal of Medicine, 325*, 1306.

Roberts, W. C., and Abernathy, C. O. (1995). Risk Assessment: Principles and Methodologies. In A. M. Fan and L. W. Chang (Eds.), *Toxicology and Risk Assessment: Principles, Methods, and Applications*. New York: Marcel Dekker.

Russell, F. E. (1996). Toxic Effects of Animal Toxins. In C. D. Klaassen (Ed.), *Casarett and Doull's Toxicology, the Basic Science of Poisons*, 5th ed. (pp. 801–839). New York: McGraw-Hill.

Saltzman, E. (1983). Direct Reading Colorimetric Indicators. In P. J. Lioy and M. J. Lioy (Eds.), *Air Sampling Instruments for Evaluation of Atmospheric Contaminants*, 6th ed. (pp. T12–T14). Cincinnati, OH: American Conference of Governmental Industrial Hygienists.

Sawka, M. N., Wenger, C. B., and Pandolf, K. B. (1993). *Human Responses to Exercise—Heat Stress*, USARIEM Technical Report T94-3. Natick, MA: U.S. Army Research Institute of Environmental Medicine.

Sawka, M. N., Wenger, C. B., and Pandolf, K. B. (1996). Thermoregulatory Responses to Acute Exercise—Heat Stress and Heat Acclimation. In M. J. Fregly and C. M. Blatteis (Eds.), *Handbook of Physiology*, Vol. I (pp. 157–185). New York: Oxford University Press.

Schroll, C. R., and Harris, M. K. (1998). Confined spaces. In S. R. DiNardi (Ed.), *The Occupational Environment—Its Evaluation and Control* (pp. 1125—1150). Fairfax, VA: American Industrial Hygiene Association.

Snook, S., and Ciriello, V. (1974). Maximum Weights and Work Loads Acceptable to Female Workers. *Journal of Occupational Medicine, 16*(8), 527–534.

Snook, S. and Irvine, C. (1968). Maximum Frequency of Lift Acceptable to Male Industrial Workers. *American Industrial Hygiene Association Journal, 29*(6), 531–536.

Snook, S., Irvine, C., and Bass, S. (1970). Maximum Weights and Work Loads Acceptable to Male Industrial Workers. *American Industrial Hygiene Association Journal, 31*, 579–586.

Snook, S. H. (1985). Psychophysical Considerations in Permissible Loads. *Ergonomics, 28*, 327–330.

Snook, S. H., and Ciriello, V. M. (1991). The Design of Manual Handling Tasks: Revised Tables of Maximum Acceptable Weights and Forces. *Ergonomics, 34*(9), 1197–1213.

Snook, R. H., Campanelli, A. R., and Hart, W. J. (1978). A Study of Three Preventive Approaches to Low Back Injury. *Journal of Occupational Medicine, 20*, 478–481.

Society of Automotive Engineers (SAE). (1986). *Human Tolerance to Impact Conditions as Related to Motor Vehicle Design*. Warrendale, PA: SAE.

Society of Automotive Engineers (SAE). (1995). *Instrumentation for Impact Test—Part 1—Electronic Instrumentation*, SAE J211/1. Warrendale, PA: SAE.

Stephenson, L. A., and Kolka, M. A. (1993). Thermoregulation in Women. In *Exercise and Sport Science Review*, Vol. 21 (pp. 231–262). Baltimore, MD: Williams and Wilkins.

Temple, W. E. et al. (1965). *Man's Short-Time Tolerance to Vibration*, AMRL-TR-65-96. Wright-Patterson Air Force Base, OH (as referenced in U.S. Army, 1996).

Todd, L. A. (1998). Direct-Reading Instrumental Methods for Gases, Vapors, and Aerosols. In S. R. DiNardi (Ed.), *The Occupational Environment—Its Evaluation and Control* (pp. 177–210). Fairfax, VA: American Industrial Hygiene Association.

Tso, E. (1992). High Altitude Illness. *Emergency Medical Clinics of North America*, *10*(2), 231–247.

Tyler, C. W., Jr. and Last, J. M. (1998). Epidemiology. In R. B. Wallace (Ed.), *Public Health and Preventive Medicine* (pp. 5–33). Stamford, CT: Appleton and Lange.

U.S. Army. (1975). *Medical Problems of Man at High Terrestrial Elevations*, TB MED 288. Washington, DC: Department of the Army.

U.S. Army. (1976). *Cold Injury*, TB MED 81/NAVMED P-5052-29/AFP 161-11. Washington, DC: Department of the Army.

U.S. Army. (1980a). *Prevention, Treatment and Control of Heat Injury*, TB MED 507/NAVMED P-5052-5/AFP 160-1. Washington, DC: Department of the Army.

U.S. Army. (1980b). *Ionizing Radiation Protection (Licensing, Control, Transportation, Disposal, and Radiation Safety*, Army Regulation 385-11. Washington, DC: Department of the Army.

U.S. Army. (1990a). *Systems Safety Engineering and Management*, Army Regulation 385-16. Washington, DC: Department of the Army.

U.S. Army. (1990b). *Exposure Standard for Fog Oil*, Technical Report No. 9010. Fort Detrick, MD: U.S. Army Biomedical Research and Development Laboratory.

U.S. Army. (1990c). *Occupational Health Guidelines for the Evaluation and Control of Occupational Exposures to Nerve Agents GA, GB, GD, and VX*, DA Pamphlet 40-8. Washington, DC: Department of the Army.

U.S. Army. (1990d). *Preventive Medicine*, Army Regulation 40-5. Washington, DC: Department of the Army.

U.S. Army. (1991a). *Health Hazard Assessment Program in Support of the Army Materiel Acquisition Decision Process*, Army Regulation 40-10. Washington, DC: Department of the Army.

U.S. Army. (1991b). *Occupational Health Guidelines for the Evaluation and Control of Occupational Exposures to Mustard Agents H, HD, and HT*, DA Pamphlet 40-173. Washington, DC: Department of the Army.

U.S. Army. (1994a). *U.S. Army Health Hazard Assessment Manual, Procedures Guide.* Aberdeen, MD: U.S Army Center for Health Promotion and Preventive Medicine. Available from Defense Technical Information Center, Ft. Belvoir, VA.

U.S. Army. (1994b). *Guide to Poisonous and Toxic Plants*, U.S. Army Environmental Hygiene Agency Technical Guide TG 196. Aberdeen Proving Grounds, MD: U.S. Army Center for Health Promotion and Preventive Medicine.

U.S. Army. (1995b). *Occupational Ionizing Radiation Personnel Dosimetry*, Army Regulation 40-14. Washington, DC: Department of the Army.

U.S. Army. (1995b). *U.S. Army Health Hazard Assessment Program Strategy.* Fort Belvior, VA: Defense Technical Information Center.

U.S. Army. (1996). *U.S. Army Health Hazard Assessor's Guide.* Aberdeen, MD: U.S Army Center for Health Promotion and Preventive Medicine.

U.S. Army. (2000). *Ergonomics Program*, DA Pamphlet 40-21. Washington, DC: Department of the Army.

U.S. Army. (2002, June). *Ergonomics in Action*, Technical Guide No. 220. Aberdeen Proving Ground, MD: U.S. Army Center for Health Promotion and Preventive Medicine.

U.S. Department of Defense. (1988). Cold Injury. In T. E. Bowen and R. F. Bellamy (Eds.), *The Emergency War Surgery NATO Handbook*, 2d U.S. Revision (pp. 57–73). Washington, DC: U.S. Government Printing Office.

U.S. Department of Defense. (1998). *Department of Defense Instruction*, DODI 6055.1, DoD Safety and Occupational Health (SOH) Program. Washington, DC: U.S. Department of Defense.

U.S. Department of Defense. (2001). *Mandatory Procedures for Defense Acquisition Programs and Major Automated Information Systems Programs*, DOD 5000.2R. Washington, DC: U.S. Department of Defense.

U.S. Environmental Protection Agency (EPA). (1983). *EPA 100-400 Series—Methods for Chemical Analysis of Water and Wastes*, EPA-600/4-79-020. Washington, DC: EPA.

U.S. Environmental Protection Agency (EPA). (1988a). *EPA 500 Series—Methods for the Determination of Organic Compounds in Drinking Water*, EPA/600/4-88/039. Washington, DC: EPA.

U.S. Environmental Protection Agency (EPA). (1988b). *EPA 600 Series—40 CFR, Part 136, July 1, 1988.* Washington, DC: EPA.

U.S. Environmental Protection Agency (EPA). (1990). *EPA 500 Series—Methods for the Determination of Organic Compounds in Drinking Water.* Supplement I, EPA/600/4-90/020. Washington, DC: EPA.

U.S. Environmental Protection Agency (EPA). (1991). *EPA 200 Series—Methods for the Determination of Metals in Environmental Samples*, EPA-600/4-91-010. Washington, DC: EPA.

U.S. Environmental Protection Agency (EPA). (1992). *EPA 500 Series—Methods for the Determination of Organic Compounds in Drinking Water.* Supplement II, EPA/600/R-92/129. Washington, DC: EPA.

U.S. Environmental Protection Agency (EPA). (1993). *EPA 100-400 Series—Methods for the Determination of Inorganic Substances in Environmental Samples*, EPA-600/R-93/100. Washington, DC: EPA.

U.S. Environmental Protection Agency (EPA). (1994). *EPA 200 Series—Methods for the Determination of Metals in Environmental Samples*, Supplement 1, EPA-600/R-94/111. Washington, DC: EPA.

U.S. Environmental Protection Agency (EPA). (1995). *Test Methods for Evaluating Solid Waste: Physical/Chemical Methods*, EPA/530/SW-846. Washington, DC: EPA.

VanIngen-Dunn, C., and Richards, M. K. (1992). *Feasibility of Reducing the Incidence of Low Back Pain in Helicopter Pilots Using Improved Crewseat Cushions*, AL-SR-1991-0009, Final Report. Phoenix, AZ: Simula.

Village, J. et al. (1995a). *Development of a Standard for the Health Hazard Assessment of Mechanical Shock and Repeat Impact in Army Vehicles. Phase 1*, USAARL CR 95-1. Fort Rucker, AL: U.S. Army Aeromedical Research Laboratory (as referenced in U.S. Army, 1996).

Village, J. et al. (1995b). *Development of a Standard for the Health Hazard Assessment of Mechanical Shock and Repeat Impact in Army Vehicles. Phase 2*, USAARL CR 95-2. Fort Rucker, AL: U.S. Army Aeromedical Research Laboratory (as referenced in U.S. Army, 1996).

Village, J. et al. (1995c). *Development of a Standard for the Health Hazard Assessment of Mechanical Shock and Repeat Impact in Army Vehicles. Phase 3 Pilot Tests*, USAARL CR 95-3. Fort Rucker, AL: U.S. Army Aeromedical Research Laboratory (as referenced in U.S. Army, 1996).

Young, A. J. (1988). Human Adaptation to Cold. In K. B. Pandolf, M. N. Sawka, and R. R. Gonzales (Eds.), *Human Performance Physiology and Environmental Medicine at Terrestrial Extremes* (pp. 401–434). Indianapolis, IN: Benchmark.

Young, A. J., Roberts, D. E., Scott, D. P., Cook, J. E., Mays, M. Z., and Askew, E. W. (1992a). *Sustaining Health and Performance in the Cold: Environmental Medicine Guidance for Cold Weather Operations*, USARIEM Technical Note 92-2. Natick, MA: U.S. Army Research Institute of Environmental Medicine.

Young, A. J., Roberts, D. E., Scott, D. P., Cook, J. E. Mays, M. Z., and Askew, E. W. (1992b). *Sustaining Health and Performance in the Cold: A Pocket Guide to Environmental Medicine Aspects of Cold Weather Operations*, USARIEM Technical Note 93-2. Natick, MA: U.S. Army Research Institute of Environmental Medicine.

Young, A. J., and Young, P. J. (1988). *Human Acclimatization to High Terrestrial Altitude.* In K. B. Pandolf, M. N. Sawka, and R. R. Gonzales (Eds.), *Human Performance Physiology and Environmental Medicine at Terrestrial Extremes.* Indianapolis, IN: Benchmark.

CHAPTER 16

# Personnel Survivability Methodology

RICHARD N. ZIGLER and RONALD A. WEISS

## 16.1 INTRODUCTION

Survivability is the ability to exist and function through and after exposure to hostile situations or environments. This can apply to both personnel and equipment. With personnel survivability, the application is focused on the human. In the civilian sector, a realistic example would be living through an automobile collision even though the participants may have received major injuries. In the military sector, survivability can be illustrated in many different ways, from living through pitched battles on land, on and under the sea, and in the air to exploring hostile regions of the world. While survivability is an issue for the system designer in both civilian and military environments, this chapter concentrates on the military environment because of its critical need to include survivability in designing of equipment and training personnel and the existence of formal survivability analysis programs within the U.S. Department of Defense (DoD) that are not present in civilian environments.

### 16.1.1 Definitions

Within the DoD, *Mandatory Procedures for Major Defense Acquisition Programs (MDAP) and Major Automated Information System (MAIS) Acquisition Programs* (U.S. Department of Defense, 1999, p. 6-F-2) defines survivability as "the capability of a system and crew to avoid or withstand man-made hostile environments without suffering an abortive impairment of its ability to accomplish its designated mission." The army defines the characteristics for survivability more specifically in both system and personnel terms:

- *System Survivability of Army Personnel and Materiel* (U.S. Army, 1995) states it as "the characteristics of a system that can reduce fratricide, as well as reduce detectability, prevent attack if detected, prevent damage if attacked, minimize medical injury if wounded or otherwise injured, and reduce physical and mental fatigue." (p. 7)

*Handbook of Human Systems Integration*, Edited by Harold R. Booher.
ISBN 0-471-02053-2 

- *Personnel Survivability of Army Personnel and Materiel*, (U.S. Army, 1995) states it as "those characteristics of humans that enable them to withstand (or avoid) adverse military action or the effects of natural phenomena that would result in the loss of capability to continue effective performance of the prescribed mission." (p. 7)

### 16.1.2 MANPRINT Domain

The U.S. Army established 'soldier survivability' (SSv) as the seventh domain of its Manpower and Personnel Integration (MANPRINT) program in 1994 (see U.S. Army. 1994). As an institutionalized concept, the U.S. Army SSv program is referred to extensively throughout the chapter since the U.S. Navy, U.S. Air Force, and U.S. Marine Corps do not yet have a specific, cohesive area of human systems integration (HSI) for personnel survivability coverage. However, examples from each of the services and civilian life are used frequently to show the universal applicability of incorporating personnel survivability into equipment design and strategies for system operational utilization.

The military personnel survivability domain is built around the following six components:

- Reduce fratricide.
- Reduce detectability.
- Reduce probability of being attacked.
- Reduce damage.
- Minimize injury.
- Reduce mental and physical fatigue.

In the personnel survivability domain, it is assumed the warfighter is integral with his or her equipment during combat. Damage to that equipment or improperly functioning component due to an enemy or fratricide action may endanger the warfighters' well-being and place them immediately into a life-threatening situation. The effects on the equipment are then evaluated to determine the potential further effects on the personnel manning the specific system. Although personnel and equipment appear to be separate areas in the real world, they both fight together as a single intertwined unit, and reality dictates that they be evaluated together.

*Fratricide* is the unforeseen and unintentional death or injury of personnel (and of damaged and destroyed equipment systems) resulting from friendly forces employment of weapons and munitions. Personnel systems and weapons systems should contain improved antifratricide systems, such as identification of friend or foe (IFF) and situational awareness (SA) systems.

*Reducing detectability* considers a number of issues to minimize and possibly eliminate detectability of friendly personnel and equipment by confounding visual, acoustic, electromagnetic, infrared/thermal, and radar signatures and methods that may be utilized by enemy equipment and personnel. Methods of reducing detectability could include camouflage, low-observable technology, smoke, countermeasures, signature distortion, training, and/or doctrine.

*Reducing probability of attack* concentrates on a number of issues revolving around two primary concepts: (1) avoiding the appearance of similarity to a high-value target and (2) actively preventing or deterring attack. For a hardware system or personnel manning the system, these issues can range from determining whether there are warning sensors to assessing the ability to deflect attack by use of active countermeasures.

*Reducing damage if attacked* addresses many issues to answer the following concerns: (1) the system's ability to protect the operator(s) or crew member(s) from attacking weapons, (2) the effects of the methods and tactics of the system's field operation on the system's and unit's survivability, (3) the system's ability to protect the crew from on-board equipment (e.g., fuel, munitions, etc.) hazards in the event of an attack, and (4) the system's ability to minimize the risk to supporting personnel if the system is attacked. Subject matter experts (SMEs) in areas such as nuclear, biological, and chemical (NBC) warfare, ballistics, electronic warfare (EW), directed energy, medical treatment, human factors, and information assurance can add additional issues.

*Minimizing injury* explores (1) combat enemy weapon-caused injuries, (2) the system's potential sources and types of injury to both its crew and supported troops as it is used and maintained in the field, (3) the system's ability to prevent further injury to the fighter after being attacked, and (4) the system's ability to support treatment and evacuation of injured personnel. Combat-caused injuries or possible injuries are addressed in this portion of personnel survivability, along with the different perspectives on potential mechanisms for reducing damage.

*Reducing physical and mental fatigue* considers tasks that may have a direct bearing on the occurrence of combat-related mistakes. The primary thrust is reducing the complexity of combat-related or combat support tasks, where negative effects by operator error can be directly traced to fatigue. Associated human factors-related tasks need to be simple and not mentally fatiguing, knowing that stress and sleep deprivation make the mental processes for operating equipment increasingly difficult while mistakes in processing and judgment become more prolific.

## 16.2 PARAMETER ASSESSMENT LIST

When SSv was established in 1994, a parameter assessment list (PAL) (Tauson et al., 1995) was developed to provide a common tool for diverse individuals to perform SSv assessments. The PAL comprises approximately 170 different issues associated with soldier combat survival. Provision is made to maximize flexibility by allowing removal of those issues that do not apply to the particular program/project/product being assessed while providing for addition of other system-specific issues that may apply. The PAL was developed to aid a multidisciplinary approach using a number of subject matter authorities. A thorough understanding of SSv issues is necessary to do a competent job. The PAL can be used by the combat developer to construct a set of potential issues to be considered for inclusion in an operational requirements document.

When the SSv assessor is notified of a program start, the first step is to assign required system performance criteria for each issue. These criteria should result from consensus among the assessors, program manager (PM), proponent agency, and user community. Sources of required system performance levels may come from operational requirements documents, the system's concept of employment, and the evaluation of SMEs. Once required system performance levels are defined, they are compared to actual (testing, experimentation, technical analysis, etc.) system performance for each issue. Sources of information on system performance may include modeling output, performance from similar or predecessor systems, engineering plans, task analysis and crew workload data, and test data. In earlier milestones, this may constitute a best guess, with more substantive information becoming available later in the acquisition cycle.

The comparison of the required and the actual system performance leads to a rating for each issue. An issue may be assigned a deficiency rating of critical, major, minor, none, or does not apply. The rating is based on the magnitude of the difference between required and actual performance and the potential effect on injury to the soldier, mission completion, loss of the system, inability of the system to complete its mission, probability of occurrence, and unacceptable impact on other HSI domains.

The primary difference between personnel survivability and other HSI domains is that personnel survivability addresses issues involving enemy and friendly combat weapons-induced injuries and the inherent hazards to the human under threat/combat conditions. Under normal noncombat environment operating circumstances, some related issues would be considered in the human factors engineering, systems safety, and health hazards domains of HSI. When potential combat weapon-induced threat exposure is included, these issues are reevaluated differently as part of the survivability domain.

For the U.S. Army, the SSv assessment provides a service to the materiel and combat developers by providing an overall integrated technical review of the hardware/human system, the combat weapon-induced threat, the resulting program survivability issues, participation in resolving those issues, and technical support for milestone decisions. This SSv assessment assists in providing coverage and assurance that issues will be or are being addressed in such diverse areas as EW, NBC survivability, individual ballistic protection, directed-energy weapons, smoke/obscurants and atmospheric effects, physiological effects, and heat stress. Survivability work performed on a program can immediately be incorporated into addressing the SSv issues and efforts. The personnel survivability specialist can assist a program team in producing a survivability outline and strategy as well as creating developmental and operational test plans, live-fire test plans, and issues to be undertaken by threat working groups, while providing valuable information and insights for the combat developer.

## 16.3 SURVIVABILITY ANALYSIS PROCESS

Survivability analysis is a process that evaluates personnel susceptibility to attack and physical injury. It focuses on the effects of threats that might reduce the ability of a friendly system to complete its mission. Part of the survivability analysis determines if a potentially destructive matter (i.e., bullet, fragment, high-powered energy device, chemical/biological agent, environmental situation, etc.) can affect the system and to what extent. A survivability program and analysis should be established for every new equipment program that may be used in combat situations or indirectly affected by such effects.

A typical survivability analysis is diagrammed in Figure 16.1. The need for the system, its operational requirements and specifications, and types of missions are first reviewed. The combat or small-scale contingency threat(s) to the system are also reviewed. From this information, survivability system requirements are established. A complete understanding of the development through production hardware/human operator system is acquired, and the computerized target solid geometry of the system is prepared for simulation testing, if necessary. A susceptibility analysis is then performed to identify the inability of the system or its components to avoid the weapons (present, anticipated, potential, and creative usage) and other elements that make up the hostile environment (Ball, 1985). About the same

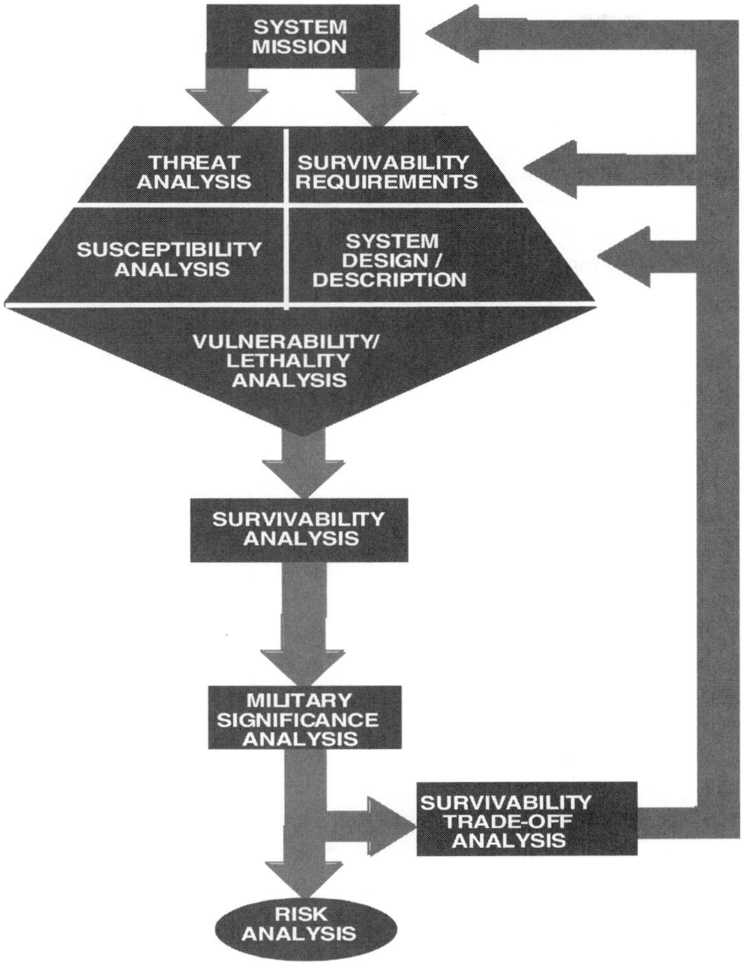


**Figure 16.1** Survivability analysis process.

time a vulnerability/lethality analysis is performed to determine the inability of the system and components to avoid or withstand the damage caused by the hostile environment. The vulnerability and susceptibility analyses together examine the criticality of each component and the effect of any degradation of a component to the overall system as well as determine the degraded condition or remaining capabilities of the system after it is attacked. The basic data to conduct these analyses can be generated from a variety of sources to include computer models, laboratory, hardware-in-the-loop investigations, or field experiments.

The survivability analyst uses information from these previous analyses to assess the battle damage of components and the entire hardware/human system. Also essential in conducting a survivability analysis are studies of battle damage assessment and repair (BDAR) capabilities, spare-parts availability, and logistics support provided for that system. Survivability deficiencies must be determined not only on the basis of individual system but also on a more global perspective. This includes the effect of system degradation on the surrounding equipment and overall missions, whether it is a long-range reconnaissance mission or a full-scale battlefront.

No system can be made invulnerable to all threats. When survivability deficiencies are identified, a military significance analysis is conducted to determine whether these deficiencies are mission critical. If the deficiencies are mission critical, a trade-off analysis may be recommended to determine if they can be rectified by changing doctrine, tactics, training, introducing survivability enhancements, redesigning the basic system, or developing an entirely new system. The effectiveness of the various alternatives is analyzed. In other cases, where the cost to correct the survivability deficiency is unacceptably high, with respect to the level of operational effectiveness gained, the army may decide to accept the deficiencies. Therefore, the ultimate value of the survivability/lethality/ vulnerability analysis is that it provides decision makers with the quantified technical information to understand and effectively manage systems.

Survivability efforts in the acquisition process are the key to maintaining a fighting force's size and effectiveness when going from one combat mission to the next. Highly trained personnel are key to all manned weapon systems performing effectively. Improvements in combat capabilities will occur by improving/enhancing personnel survivability in two primary ways: (1) by designing and producing increasingly combat-effective systems for land, sea, and air operations and (2) by ensuring all weapon systems incorporate systems design characteristics to enhance personnel survivability. The survival of a soldier, sailor, marine, or airman and his or her equipment depends on the type of mission conducted, the amount of friendly support available, and the effectiveness of the hostile combat environment encountered.

To perform a thorough survivability analysis of both the hardware system and the personnel using that system, the interrelated considerations diagrammed in Figure 16.2 must be evaluated iteratively. For example, the type of military mission must first be selected and then one of the military and natural fighting environments must be determined. Knowing the mission and fighting environment, an analysis is then conducted to determine the likely physiological and psychological states of the warfighters as well as how well they will be protected. After that analysis, the same mission is performed with a different military and natural environment, physiological and psychological state of the personnel, and availability of the protective equipment. This iterative process is continued until all combinations are considered and analyzed. Although this seems like a daunting task, it is really not that difficult, because many of the considerations may not apply for each system. However, before discarding a consideration, a rationale to discard it must be developed. By doing this type of analysis on each human/hardware system, a thorough understanding of that system's capability, limitations, and survivability will be established. This will make it easier for the materiel developer and the combat user to understand the system's functional boundaries and limitations.

## 16.4 PERSONNEL SURVIVABILITY COMPONENTS

Each of the six focal components for the personnel survivability analysis will now be considered in more detail, in light of the iterative review process. Note that all six components are interlinked in the real world of survivability. For clarity, they have been separated artificially in a continuing time sequence. All the components are based on

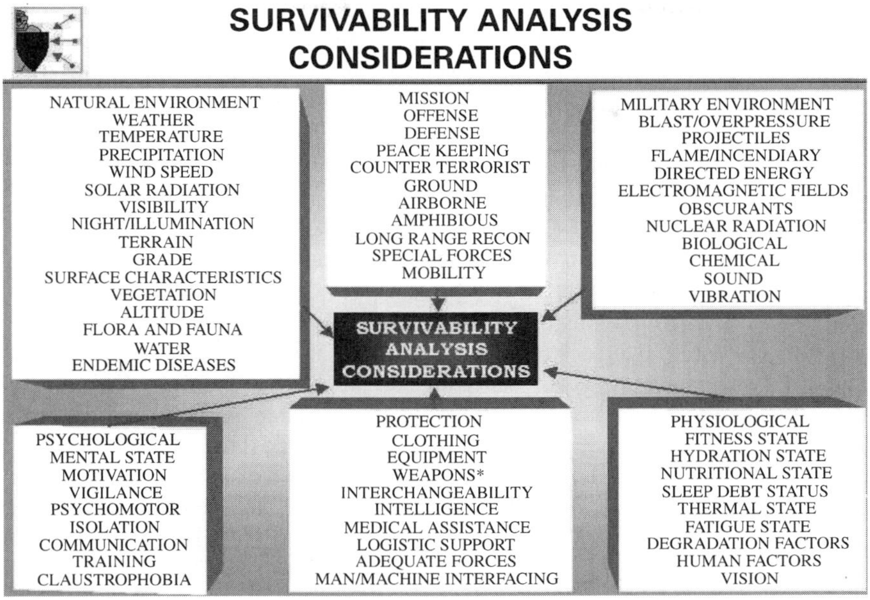


**Figure 16.2** Survivability considerations.

avoidance or minimization of an event and the correcting activities if the avoidance activity fails.

### 16.4.1 Reduce Fratricide

Unfortunate as it is, fratricide does occur under the normally less-than-perfect conditions of combat. Example 16.1 illustrates some historic battles where fratricide events played significant roles.

**Example 16.1 Fratricide in Historic Battles** During the Battle of Waterloo, the late afternoon approach of the Prussian allies to link up with the Duke of Wellington's left wing resulted in fratricidal situations because the Dutch–Belgian forces could not be recognized. Having fought with the French forces under Napoleon until his abdication the year before, the Dutch–Belgians still retained their blue and white uniforms, which the Prussians incorrectly recognized as French. However, those Prussians that arrived slightly later in the same vicinity as the Scottish troops had no recognition problems when they saw the kilts, feather bonnets, and scarlet coats. Well-known incidents occurred during the Civil War when Confederate Lieutenant General (LTG) Thomas "Stonewall" Jackson was returning from a quick night reconnaissance on the Chancellorsville battlefield after collapsing the Federal army's right flank. Hand and arm wounds inflicted by his own troops forced him from the battle and eventually led to his death. A year later, Confederate LTG James Longstreet was similarly hit in the arm by Confederate soldiers during daylight while returning from a reconnaissance after successfully attacking on and rolling up the Federal army's left flank during the Battle of the Wilderness.

Numerous attempts have been made in this century to reduce the risk of fratricidal incidents as evidenced by the large national insignias on aircraft of the world wars, multicolor roundels on the wings and fuselages of the British, French, and American aircraft, and crosses on German aircraft. Other nationalities used various markings, such as the Poles with their four white and red alternating squares. One of the more memorable attempts at reducing fratricidal risk were the big black and white "invasion stripes" on allied aircraft for D-Day and beyond.

***The 2 Percent Rate*** There have been several studies (Shrader, 1982; Dupuy, 1990) on casualties, with a commonly referred to 2 percent rate of fratricidal casualties compared to combat casualties. The following paragraph is intended to illustrate that fratricide can be a serious problem for the force strength overall. It is not intended to indicate what an "acceptable" percentage rate is for fratricidal casualties.

Although there was undoubtedly a great deal of research involved in arriving at the 2 percent figure, there is some question as to the accuracy of the data used. (See Example 16.2 for studies showing higher rates.) World War I is a good example since generally only the coarsest types of casualty information are reported and large portions of records were lost through time.

**Example 16.2 Studies Showing Greater Than 2 percent Fratricide Rate** An excellent article, entitled "Dealing Realistically with Fratricide," by Steinweg (1995) reveals some very detailed studies providing results at much different fratricidal rates. During World War II, battalion surgeon Captain James Hopkins of the 5307th Composite Unit (Merrill's Marauders) was interested in the effects of body armor and thus maintained very detailed records of wounds and interviews with patients, obtaining a figure of 14 percent of total casualties resulting from fratricide. For Vietnam, over 125 personnel were involved in a wounds data and munitions effectiveness team (WDMET). This team produced an evaluation of wounds data and munitions effectiveness in Vietnam from a 1967 to 1969 study of over 7800 casualties that showed fractricide accounting for 14 percent of total killed in action through rifle fire, 11 percent of total killed in action through fragments, and 11 percent of total wounded.

***Five Fratricide Issues*** How does one apply "reduce fratricide" to personnel survivability? The PAL (Tauson et al., 1995) provides a list of fratricide-associated issues that serve as a guide in evaluating the equipment and the soldiers in this regard. Examples of some of these issues are as follows:

1. Is the related IFF or target identification system effective to ranges at least as long as the weapons' ranges?
2. Is the system's signature (visible, electromagnetic, etc.) similar to potential threat vehicles? Is the system compatible with IFF receiver or identifier systems, such as active (question-and-answer) IFF receivers?
3. Is the IFF system a noncooperative target recognition system (i.e., if an enemy tries to target you to find your position, does the system refuse to cooperate so as not to give any information to the enemy)?
4. Does the self-location equipment provide sufficient resolution to reduce fratricide?
5. Is the system's ability to distinguish between friendly and enemy targets compatible with mission-oriented protective posture level IV (MOPP IV) (NBC individual protective equipment) conditions?

Elaboration of these five issues provides insight into how vulnerable the warfighter may be to fratricidal incidents and will provide the impetus to raise and correct noted deficiencies.

The first issue concerning weapon range being greater than identification (ID) capability is a constant struggle. Most sights are currently not capable of providing enough detail of a potential target at a great enough range for a gunner to positively identify friend or foe. Moving close to a potential target to provide positive ID is normally out of the question in combat situations (see Example 16.3).

**Example 16.3 The Chevron ("V") ID Sign** The chevron, or V, used in Operation Desert Storm was an attempt at positive ID. (It could be applied in several ways: horizontally laying on its side, vertical, and up-side down.) But, if the chevron were not large enough or not made of the recommended material, it would be nearly impossible for the platform sight's pixels to pick up the sign and provide it to the gunner to positively identify. In addition, if a platform's chevron were on the sides of the platform but not on the front, there would be an angle of approach to the targeting sight where it would not be able to pick out the ID sign.

As for the weapon ranges, although the tank gunner's primary sights can pick out targets at long ranges, they cannot (by themselves) pick up enough identifying features to positively identify friend or foe at extended ranges, which are increasing rapidly. During Operation Desert Storm, an ABRAMS tank's 120-mm cannon scored a kill at 3.4 km (Carhart, 1994), well beyond normal long-range gunnery expectations, while during Britain's equivalent, Operation Desert Sabre, a CHALLENGER tank scored a kill even further, at 5.1 km (Houck, 1994).

For the second issue, a friendly weapons platform having a signature(s) similar to or identical to a potential enemy's platform signature(s) can place the friendly platform and its crew in danger due to chaotic imperfect conditions, round-the-clock operations, and aggressive friendly forces (see Example 16.4).

**Example 16.4 Similar Tanks of Opposing Forces** During Operation Desert Storm, the Syrian and Kuwaiti contingents of the coalition forces employed T-series (Syria, T-62's and T-72's; Kuwait, M-84 Yugoslav-built T-72-design tanks) (DeBay, 1991) main battle tanks that were practically identical to those of the opposing Iraqi forces. During the coalition drive into southern Kuwait, the Syrians were to be assigned to align between U.S. 2nd Marine Division on their right and the Egyptians and Kuwaitis on their left (U.S. News and World Report, 1992). After a number of tense and frustrating meetings between the U.S. Marines and Syrians about how the marines would recognize the Syrian T-62's and T-72's to be friendly forces, the Syrians "refused to assume their assigned position and had then moved west to follow the Egyptians and Kuwaitis into Kuwait." The marines' left flank was open throughout the ground war, although the army's 1st Cavalry Division's Tiger Brigade was caused to be assigned to strengthen the 2nd Marine Division's left flank due to this potential fratricidal situation. Here, even positive visual ID of the type of tank (T-62 or T-72 series) would still leave one wondering as to whether it is a friend or foe. Unfortunately, a "friend" firing at you while you hesitate to fire to get confirmation of friend or foe is still a dangerous situation, no matter how friendly the opponent may be in other circumstances.

The Desert Storm coverage and publicity via television coverage of fratricidal incidents—giving rise to the third and fourth issues—have provided impetus to the U.S. Army to develop IFF systems, similar in concept to those available to the U.S. Air Force and U.S. Navy. The difficult part will be designing these systems to be noncooperative with potential enemy detection systems and to maintain security on them. This can be a very tough proposition; for example, fighter pilots and helicopter pilots on covert missions would be reluctant to turn on an active IFF system that broadcast a "Here I am!" signal to the world. In addition, making an IFF system small and light enough to be carried by dismounted soldiers will be one of the ultimate design challenges.

The fifth issue on MOPP IV addresses one of the potentially most confusing and frightening scenarios, where a soldier, marine, or air police may undergo an NBC attack with smokes and obscurants with very limited visibility (it is worse at night). In a close-combat situation, the individual will potentially be faced with situations where any other individuals in their protective clothing must be immediately recognized. Limited visibility, increased personal stress, and heat retention are situations that will have to be addressed for successful combat and survival. If available, increased SA capability can help the individual with orientation, personal location, and locating friendly forces.

### 16.4.2 Reduce Detectability

Detectability is an important issue of survival during wartime. Reducing the range at which an enemy can finally detect an individual or weapon system often improves the chances of surviving an engagement or performing a mission. If a combat sniper's position is detected at long range, he or she is immediately placed in danger; therefore, snipers are taught carefully conceived camouflage and concealment techniques that require an enemy force to approach to within a very short distance before the sniper can be detected. The same is true for other platforms. The moment that a combat aircraft or ship is detected by an opposing force, the specific platform and its crew members are both immediately placed into a state of vastly increased danger.

Personnel survivability is a vital concern and must be addressed to reduce detectability for both personnel and weapon systems in a wartime setting. A weapons system's signature is a function of its inherent characteristics and/or its interaction with the physical environment. All weapons systems and personnel possess signatures in the visible part of the electromagnetic (EM) spectrum and emit and/or reflect EM radiation (see Fig. 16.3). To the extent that these emissions or reflections can be associated with the existence of and ID of the person or system, they are referred to as a "signature."

What issues can be directed at reducing signatures while simultaneously reducing detectability and the dangers associated with both? The SSv PAL lists 21 probable issues for army systems, and SMEs are likely to raise more issues depending on the system's characteristics. Issues that would have to be studied and addressed for either a helicopter or a dismounted soldier system include the following:

- Is the system likely to be detected by threat forces because of its visible static signature?
- Is the system likely to be detected by threat forces because of its thermal (infrared) signature?
- Is the system likely to be detected by threat forces because of its radio-frequency signature?
- Have any electro-optical or optical components on the system been hardened to reduce optical cross-sectional measurements that are the cause of wide-angle and at-range detection?

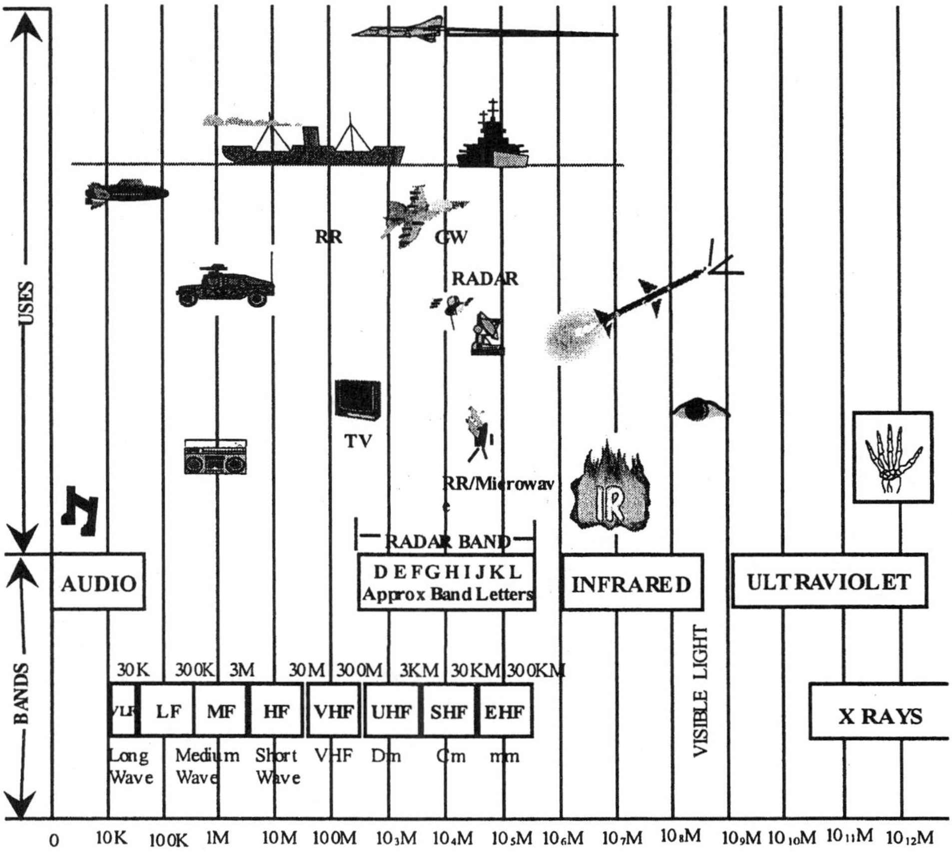


**Figure 16.3** Electromagnetic spectrum and representative sources for specific signatures.

- Will threat forces' use of obscurants prevent the system from detecting approaching enemy systems?

Failure to knowingly address these and other issues during the acquisition process will probably cause combat danger to the weapon system and put the military personnel in or near the system in potentially mortal danger. Environmental effects such as the diurnal temperature cycle can make the difference whether the combatant will become a casualty (see Example 16.5).

**Example 16.5 Diurnal Temperature Cycle** The diurnal temperature cycle (U.S. Army Research Laboratory, 1999) was a topic of concern during Operation Desert Shield, with many military personnel concerned about the potential signatures of enemy forces and their equipment. The diurnal cycle is a daily cycle where twice during the day (see Fig. 16.4) objects that change temperatures faster than the background sand will become the same temperature as the sand for a short period of time. For example, a cool tank's temperature will rise in the morning and become hotter than the desert, and the tank will then cool down faster during evening than the sand so that it is colder at night. The difference in temperature is the method used in thermal sights to find targets. The danger for systems (i.e., tanks, helicopters,

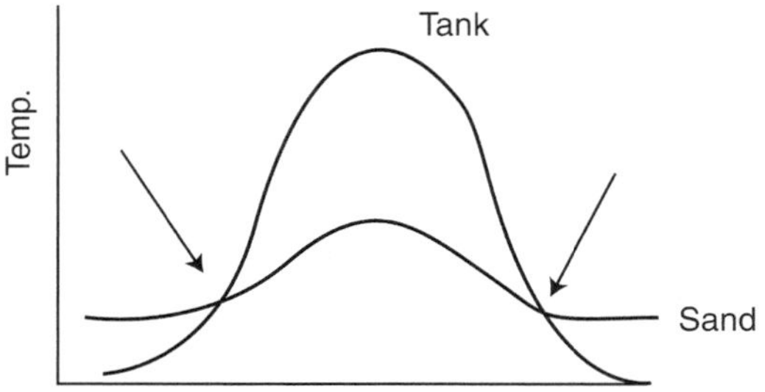


**Figure 16.4** Desert ambient temperature cycle (imposed on generic tank signature to show points where vehicle can "disappear" for short periods of time during 24-hour day).

aircraft, etc.) using thermal sights in their fighting is that once in the morning and once in the late evening thermal sights will not be able to see an opposing tank because its temperature will be indistinguishable from the background sand and therefore "disappearing." The Battle of 73 Easting in Operation Desert Storm (Carhart, 1994) was affected by this phenomenon when the lead U.S. tanks crested a ridge and stopped because they could not see expected enemy forces. What they could see appeared to them as "bowling balls" in the distance. As the lead troop studied the bowling balls, and as the other tanks came up alongside their position, one of the bowling balls moved upward, revealing arms and a torso. With that signature identified, the guidance was given to aim below the enemy tank commander's thermal signatures given off by their heads and thus fire at what would be armored vehicles below. The U.S. forces could have been in mortal danger when they could not see the enemy force, but fortunately they found and used the unintended enemy signatures to their advantage. The diurnal cycle itself can vary with wind speed, temperature, relative humidity, and turbulence, but it can be determined and provided to those combatant forces who need to know when it will affect their gun sights.

***Signatures Detection Technology*** Friendly weapons systems and military personnel throughout the combat theater will be subject to continuous enemy efforts to locate and identify them for deception, destruction, or avoidance. In this effort, the enemy will employ a variety of intelligence, surveillance, and reconnaissance (ISR) techniques, special operations forces teams, devices, and platforms oriented toward exploiting the signatures of individual weapons systems. (The dismounted soldier is now also being considered as a weapons system.) Acoustic, radar, sonar, infrared, and visual detection, tracking, and guidance devices are designed to sense EM radiation emitted or reflected from the system. Using counter-ISR means, such as camouflage, concealment, deception, and emission control, places an additional burden on soldiers and may reduce their operational effectiveness.

Detection avoidance includes technologies and methods used to suppress sights, sounds, and images associated with friendly weapons systems and military personnel. Suppressing these signatures, so that personnel and their weapon systems are indistinguishable from their background, provides the weapon systems with the ultimate advantages of battlefield surprise and protection. Making weapon systems harder to find increases their survivability. While some of the greatest gains in survivability may be achieved through detection avoidance technologies, they can sometimes be the most expensive to develop and integrate. Technology programs that apply to detection avoidance include acoustic, radar, seismic, thermal, and visual signature reduction and/or masking achieved by using advanced materials and coatings and materials that distort the apparent shape of the equipment.

***Obscurants*** Adverse atmospheric and obscurant conditions (U.S. Army Research Laboratory, 1998) (see Fig. 16.5) can be valuable battlefield countermeasures for defeating or degrading threat detection, acquisition, tracking, and terminal guidance functions. Early knowledge concerning atmospheric and obscurant parameters in developing a weapon system can provide decisive advantages in that system's use. Similarly, studying threat systems' vulnerabilities to meteorological conditions and atmospherics can help defeat those threat systems.

Obscurant materials, ranging from visual to the far infrared, are a valuable battlefield countermeasure for defeating or degrading threat detection equipment. An atmospheric and obscurant analysis would include subanalyses such as the diurnal temperature cycle, far-infrared attenuation, and weather effects analysis.

### 16.4.3 Reduce Probability of Being Attacked

This section addresses hit avoidance and countermeasures instituted to frustrate an enemy attempt to attack friendly forces. It is generally accomplished by reducing opponent target acquisition and guidance systems' ranges of operation and sometimes is accomplished by intimidation. This is a critical area, as failure will likely increase the numbers of casualties. The SSv PAL lists 17 probable issues for army systems, and SMEs are likely to raise more issues depending on the system's characteristics. A few of the issues that would have to be studied and answered, whether for a helicopter or for a dismounted soldier system, are as follows:

- Is the system able to deflect attack by the use of electronic jamming or spoofing of a munition's sensors?
- Is the system able to deflect attack by the use of active ballistic interdiction to deflect or destroy incoming munitions?
- Has any microprocessor code on the system been protected from the presence or insertion of malicious code ("back-doors," viruses, etc.)?

**Figure 16.5** Use of smoke/obscurants to hide battle field movement.

- Does the system present a unique or highly recognizable signature (visual, thermal, etc.)?

***Hit Avoidance*** Hit avoidance refers to technologies that reduce the probability of being hit by a weapon after being detected by the enemy. Hit avoidance includes both avoiding acquisition and tracking by enemy fire control and avoiding being struck by enemy weapons that have been fired. Hit avoidance technologies protect with sensors and countermeasures. The sensors detect incoming threats, and the countermeasures confuse, physically disrupt, deflect, or destroy those incoming threats. Examples of hit avoidance technologies and doctrinal countermeasures are early warning systems, such as missile and radar warning receivers, smoke and obscurants, jammers (optics, laser, and radar), decoys (laser, flares, and chaff), rapid mobility, and counterfire.

***Detection of Susceptibility*** System performance evaluators, decision makers, materiel and combat developers, and the eventual users of these systems benefit from knowing any limitations of these systems in degraded environments before they are fielded. Detecting susceptibilities early enables better management of a program and greatly reduces the risk of surprise findings in any developmental and operational testing or, more importantly, during actual combat. Proactively identifying potential counter-countermeasures that will negate or reduce these susceptibilities is also a benefit to these programs and to the combat forces that will use them.

The common infantryman's signature amid technology changes is an intriguing but easily understood example of reducing the probability of being attacked (see Example 16.6).

**Example 16.6 Military Clothing and Equipment** Concepts of military clothing and equipment have changed greatly over the last 100 years. During the Anglo-Boer War of 1899 to 1902, the British Army learned at great human expense the value of khaki-colored uniforms that allowed them to blend into the terrain. This was in stark contrast to the brilliant and colorful multicolor uniforms with shiny metal accoutrements that were the style used in most western armies of the time and up through the first months of World War I. During the Russo–Japanese War of 1904 to 1905, the Japanese Imperial Army (Burnett, 1913) learned the value of uniforms and equipment that would reduce acoustic noise down to almost nothing. This improvement in uniforms was especially effective for surprise when conducting mass night attacks against Russian positions and for the night "Banzai" attacks used later during World War II. The German Waffen SS and Wehrmacht units of World War II developed the flecks-of-spots (flecktern) camouflage patterns for battle-dress smocks and uniforms to help them blend in more with the surrounding terrain and growth during daylight. The starlight scopes of the U.S. forces in Vietnam started to take away the opportunity and advantage for infantry to close with an enemy without being detected during the dark of night. Image intensifiers and now thermal weapon sights are becoming prolific to the extent that the advantage of night movement, and even hiding under camouflage in daylight, is removed to some extent.

***Electronic Warfare*** Successful EW is a carefully integrated program of actions and countermeasures. Reconnaissance, firepower, communications, signal intelligence, jamming, direction finding (DF), and deception elements can be used by an enemy to attack priority networks, nodes, and links to nullify, limit, or delay command, control, communication, and computer (C4) systems while protecting their own operational capability. Airborne, sea-based, and ground-based platforms can allow an enemy to intercept, locate, and jam tactical communications systems from the high-frequency (HF) to superhigh frequency (SHF) portions of the radio-frequency (RF) spectrum. The EW threat will continue to grow as technologies are employed to affect low probability of intercept.

Knowing the susceptibility and vulnerability of combat and combat support systems to the full spectrum of EW effects is critical to the survival of soldiers and to their ability to fight effectively on the battlefield.

### 16.4.4 Reduce Damage

The personnel vulnerability component of "reduce damage" is one area where the life and death of military personnel are most directly affected. This component recognizes the moment when an attack is made, and the last line of defense of system and operators' personal equipment must stand up to this attack. Complicating this realization is the fact that weapon overmatch of system defenses often occurs, and efforts to minimize or preclude injury are very important. In PAL for SSv, there are 33 listed potential issues. This is the area where SMEs are most likely to raise a large number of additional system-specific and threat-related issues affecting survival of the system and the military personnel operating it. Some issues are as follows:

- Does the system adequately protect the crew from direct- and indirect-fire munitions through the specific damage mechanism of spall?
- Does the system provide crew protection from secondary explosions of on-board munitions if the system is attacked, by means of separation of ammunition storage in a compartment isolated from the crew?
- Does the system provide adequate crew protection from directed-energy weapons such as lasers?
- Does the system provide adequate warning and protection for the crew in a chemically or biologically contaminated environment?
- Will the system be able to operate in the presence of external electromagnetic environmental effects without affecting crew members and other military personnel?

***Ballistics Protection*** The ballistic impact of an overmatching munition of any type occasions significant effects behind the armor of a weapon system. If the projectile perforates, a residual of the penetrator or some portion of a shaped-charge jet (see Fig. 16.6) will exit behind the armor. Behind-armor debris (BAD) is then also generated. This may include residual penetrator pieces, armor plug, scab, small pieces of these objects if they break up, many armor fragments, and a fine spray of metallic particles referred to as spall. The result is the BAD cloud of fragments being projected forward at high speed, generation of blast overpressure, potential fire hazard, and combustion by-products, causing a great deal of damage and injury. In many cases, most of the lethality from a ballistic impact is caused by BAD. In addition to casualty-producing effects, debris often kills not only primary but redundant components and systems as well. A major success story has been the development of spall liners that are often attached to the interior walls of armored vehicles. These spall liners tremendously reduce the number and narrow the cone

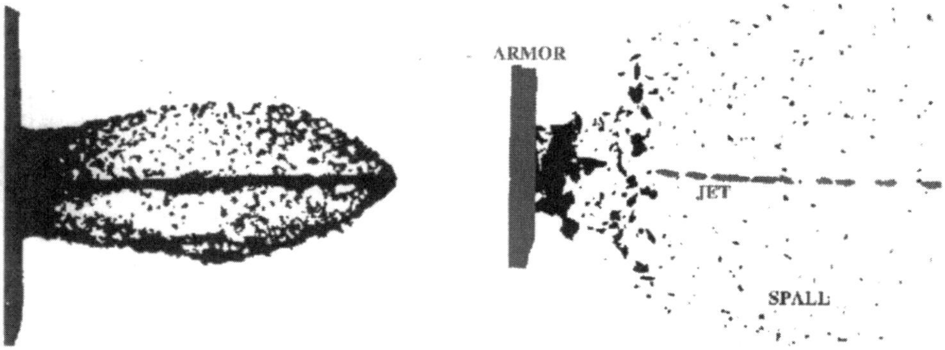


**Figure 16.6** X-ray photographs of shaped-charge weapon's molten metal jet penetrating armor and propelling spall armor fragments at very high velocity into a vehicle interior.

angle of spall fragments that can harm both onboard operating personnel and materiel systems within the vehicle.

Federal law (U.S. Code, Title 10, Section 2366, n.d.) directs that military survivability testing shall begin at the component, subsystem, and subassembly level, culminating with tests of the complete major system or program, or major product improvement, configured for combat. This means that live-fire test and evaluation (LFTE) is really a process, rather than a discrete event. A major system, major munitions, a missile program, or a product improvement to any of these programs may not proceed beyond low-rate initial production (LRIP) until realistic survivability or lethality testing is completed, and the report required by statute is submitted to the prescribed congressional committees. *Realistic survivability testing* means testing for vulnerability of the system under combatlike conditions by firing various weapons and munitions.

Fully dressed and equipped mannequins representing soldiers are often placed in the LFTE systems to determine injuries. They are placed to measure stress and strain effects and to note injuries that may be inflicted upon crew members. If a test munition is anticipated to overmatch a platform's armor to penetrate into the crew compartment, the mannequins will be constructed of wooden forms to hold the soldier-worn equipment in proper locations. If the personal equipment and the mannequin are penetrated, the mannequin will then act as a "witness" for examination and analysis in showing the shot-lines of any munition, spall, and associated fragments that hit it.

***Directed Energy*** Electronic equipment can be defeated or impaired by irradiation from directed-energy weapons (DEWs). Degradation can range from temporary upsets in electronics subsystems, permanent circuit deterioration, or permanent destruction due to burnout or electrical overload. Humans can be wounded, impaired, and, in extreme cases, killed by DEWs. These weapons produce casualties and upset or damage equipment by focusing energy on the target. The three principal divisions of DEWs include lasers (low and high energy), high power radio frequency (HPRF), and particle beams (charged and neutral).

- *Lasers* The presence of laser devices in the inventories of major armies is increasing, and any device, such as a target designator or a laser range finder, can be employed as a weapon if it is pointed at a target it can damage. In the near term, the most probable targets of laser weapons are electrical and electro-optical systems (specifically, fire control devices such as sights and the person behind the sights) and personnel. Most systems that contain optical components, including direct-view optics such as vision blocks, periscopes, laser protection goggles, and telescopes, along with cameras (television), laser range finders, image intensifiers, forward-looking infrared (FLIR) systems, trackers, or seekers are potentially susceptible to laser threats currently fielded or under development. At a minimum, protection of the soldier is required, and depending upon the particular system, hardening with regard to laser jamming susceptibility may be required for the system to pass a milestone III (production) decision.

- *Particle Beam* The particle beam weapon is the newest of the developing threats, using radiation in the form of accelerated subatomic particles focused on a target by magnetic fields. A particle beam weapon can inflict damage on a target by removal of material in the proximity of a hit, by detonation and ignition of explosives and fuel, or by radiation damage to vulnerable critical components.

***Weapons of Mass Destruction*** Weapons of mass destruction are an increasing threat against the U.S. military and civilian population and institutions. DoD Directive 5000.1 (U.S. DoD, 1996), U.S. Army Regulation AR 70-75 (U.S. Army, 1995), and many requirements documents require PMs to make their systems NBC survivable. These documents include requirements for nuclear, biological, and chemical contamination survivability (NBCCS) and nuclear weapons effect (NWE) survivability.

*NBC Contamination* The NBCCS requirements encompass three areas: hardness, compatibility, and decontaminability. Hardness is defined as the ability of a system to withstand both an NBC attack and the decontamination process without mission-essential equipment functioning unacceptably or failing. Compatibility is the ability of equipment to be operable by personnel wearing NBC-protective clothing and equipment. Decontaminability is the ease and degree of ability to restore equipment to safe levels of cleanliness so that personnel may remove burdensome protective equipment without fear of ill health from residual agent effects. Incomplete NBCCS processing of the system may cause casualties at a later date when crew members or maintenance personnel are maintaining the system.

*Nuclear Weapons Effects* The NWEs (see Fig. 16.7) include blast (overpressure), thermal, electromagnetic pulse (EMP), and initial nuclear radiation (INR). An EMP is a secondary effect of a nuclear weapons detonation.

Initial nuclear radiation consists of both neutrons directly emitted from a nuclear detonation and ionizing radiation (rays) caused directly or indirectly by the nuclear detonation. The INR ends within seconds of the detonation and affects both personnel and equipment. For tactical situations (e.g., the effects of a nearby detonation on ground equipment), semiconductor parts can be upset or burned out, and optical materials can have a significant temporary or permanent change in optical properties.

The thermal pulse consists of visible, infrared, and ultraviolet radiation emitted from the fireball. Thermal radiation can cause severe burns and flash blindness to soldiers and can cause many effects on materials, including melting of material, charring of surface

**Figure 16.7** Nuclear weapons effects can damage/destroy systems in many ways.

coatings, ignition of materials, debonding of laminates, optical obscuration, and degradation of material properties.

***Information Operations*** Information operations (IO) is defined as actions taken to achieve information superiority by affecting adversary information, information-based processes, information systems, and computer-based networks while defending friendly information, information-based processes, information systems, and computer-based networks (U.S. Army, 1999). The threats to the information infrastructure come from individuals and groups motivated by ego, curiosity, military, political, social, cultural, ethnic, religious, or personal or industrial gain. The information warfare (IW) threat focuses on intercepting, exploiting, corrupting, and/or destroying data in existing databases or data being exchanged between databases. Attacks can be designed with a delayed effect, such as corrupting a database or controlling program, as well as immediate actions to degrade or physically destroy. Examples include the following:

- unauthorized access to classified or sensitive military information;
- insertion of malicious software to cause a computer to operate in a manner other than that intended by its users (this category includes computer viruses, worms, logic bombs, and programs designed to bypass protective programs);
- corruption of data through use of malicious software or alteration of data;
- sowing disinformation;
- lengthening the command-decision cycle;
- misdirection of U.S. forces, weapons or sensors;
- delay or prevention of the development or deployment of military information systems;
- withholding battlefield or other situational information.

***Electromagnetic Environmental Effects*** Electromagnetic environmental effects (E3) is a broad term used to define the general effects of several diverse EM phenomena. The E3 address the performance, safety, and reliability of a system to be stored, transported, and operated in an EM radiation environment without suffering any detrimental effect. The E3 encompass that portion of the EM spectrum from very low frequencies (VLF band) to extremely high frequencies (millimeter-wave band).

The E3 can range from simple static interference or upset to permanent damage or burnout of electronic components and possibly catastrophic system failure. Examples of different effects in a military environment that can be caused by undesired EM energy are as follows:

- temporary or permanent injury of personnel;
- corruption of stored computer instructions or data;
- performance degradation of receiver signal processing circuits;
- ignition of electrically initiated devices, flammable materials, or explosives;
- operation of electromechanical equipment, electronic circuits, or components; and
- burnout or voltage breakdown of electronic components, antennas, or circuit cards.

Example 16.7 illustrates the need for emphasis on E3.

**Example 16.7 U.S.S. Forrestal Fire** An example from military experience illustrates the need for emphasis on E3. This incident occurred aboard the aircraft carrier *USS Forrestal* (U.S.S. Forrestal Museum, Inc., n.d.) in 1967 (see Fig. 16.8). During rearm and refuel activity for a strike operation over Vietnam, a 5-inch-diameter Zuni missile mounted under the wing of an F-4 Phantom that was preparing to fly its second strike of the day was accidentally launched up-deck and struck and knocked off the fuel tank and a 1000-lb bomb of an A-4D Skyhawk. The ensuing fire and explosions of ordnance resulted in 134 dead, 27 aircraft destroyed, and over $70 million damage to the carrier. The accident investigation concluded that the inadvertent coupling of EM energy from either a ship-mounted transmitter or a power generator located next to the aircraft wing most likely caused the accident. This E3 accident remains today as the greatest single incident of loss of life in the U.S. Navy since World War II.

### 16.4.5 Minimize Injury

Medical injury is one of the major components affecting the probability of survival. Figure 16.9 shows the percentage of wound cases and their anatomical distribution for some major wars fought in the twentieth century. Fragmentation and bullets accounted for 85 percent of the wound cases. These wounds were generally received in the upper body regions (upper extremities, head and neck, and chest and abdomen). The greatest percentage of wounds in a single region was inflicted in the lower extremities. The type of mission and the prevailing equipment used dictated the susceptibility of wound location. During the trench warfare activities of World War I, soldiers standing in those trenches exposed the head and neck, upper torso, and upper extremities to rifle fire and artillery fragmentation barrages. The only protection provided for an infantryman was a thin steel helmet that did not always stop a bullet or fragment. During the Vietnam War, most of the soldiers in the field were issued torso-covering body armor that provided good protection

**Figure 16.8** Fire aboard the carrier *U.S.S. Forrestal* resulting from an E3 incident.

COMBAT WOUNDS

Wound Causes | Wound Distribution

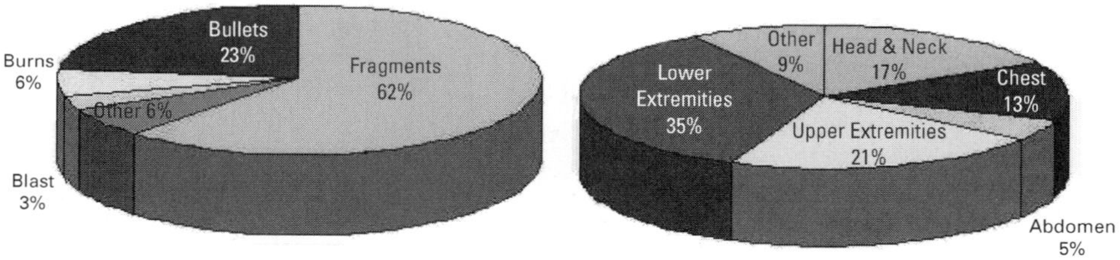


WWI, WWII, Korea, Vietnam, Middle East

SOURCE:
- Helmets & Body Armor in Modern Warfare, B. Dean, Yale Univ. Press, 1920
- Battle Casualities, G. Beebe Atal, C. Thomas Publ., 1952
- Wound Ballistics, Office of Surgeon Gen., 1962
- MLRR 257 (AD 29480), 1954; 144 (AD B954368), 1952; 221 (AD 28447), 1953
- WDMET, EASP 100-67, Vol 3, 1969
- MED Bulletin of the US Army, Europe, Vol 35, No. 8, Nov/Dec 1978
- USAIA 2201 1129 77, 1977
- WSEG 237, Vol 3, Part 3, 1975
- ISRAELI Journal of Med. Sciences, Vol 20, 1984

**Figure 16.9** Causes and anatomical distribution of wounds compiled from some major conflict studies.

against fragmentation weapons but not against bullets. Lack of protection of the extremities and lower torso still led to the many fragmentation wounds in those areas.

Today's scientific and highly technical battlefield environment encompasses much more than bullets and shrapnel. Air assault, both in the dark and in daylight, can result in broken bones due to falling from heights. Moving vehicles such as tanks, wheeled vehicles, and helicopters can produce skeletal injuries when detonating buried mines or accidents when the vehicles roll over or crash. Radiation injury may result from exposure to nuclear material. The use of chemical and biological warfare materials can produce nerve and soft tissue damage and possibly death. Directed-energy weapons can generate eye injuries, blindness, and burns. Electromagnetic effects, electric shock, and burns are potential medical injuries from short-circuited, capacitor-discharged, and ungrounded electronic equipment. In many cases, these injuries may be prevented or ameliorated when survivability is addressed in the initial hardware design and development stages.

The SSv PAL focuses on three general areas of minimizing medical injury to improve survivability:

- Identify the potential sources for personnel injury in the system design and when the soldier and equipment are functioning in the field.
- Assess the system's ability to prevent further injury to the soldier after being attacked or exposed to a hostile environment.
- Assess the ability of the system to support treatment and evacuation of the injured.

***Bail-Out Ejection*** Bail-out ejection from a high-performance aircraft illustrates how a system analysis using a PAL could identify both potential sources of medical injury from ejection and requirements to prevent further injury after leaving the aircraft (see Example 16.8).

**Example 16.8 Bail-Out Ejection Analysis** The purpose of the ejection seat is to get the aircrew member as far away as possible from a damaged/troubled aircraft to avoid injury to the crew member. Ejection is required because the speed and potential attitude of these modern aircraft precludes jettisoning the canopy and stepping on the wing to jump clear of the aircraft to execute a parachute jump. The design of the aircraft must include a method of rapid, complete removal of the canopy prior to ejection to avoid crew member injury as the rocket- or mortar-propelled seat or system moves up its guide rails and out of the aircraft. After leaving the aircraft, the seat must be thrust sufficiently high above the vertical stabilizer and horizontal tail surfaces to prevent them from striking the crew member or the seat to which the crew member is attached. In some aircraft, the instrument panel is immediately above the rudder pedals, feet, and lower legs. To prevent contact injury to the feet and lower legs by striking the underside of the instrument panel during ejection, the crew member must wear straps on his or her flying boots that will automatically retract the lower legs and feet against the seat and out from under the panel when the ejection process is activated. The seat pan must be sufficiently long to preclude the momentum "G" force of ejection and the weight of the crew member's legs from breaking both thigh bones during the rapid upward acceleration. The activation mechanism of the ejection system must be placed in a position so it can easily be reached and activated by one hand, if necessary, without placing the spinal column or the arms in a position where they can be injured. A timing mechanism must be in the seat design so that, once the ejection propellant is activated, enough time will elapse to clear the tail plane surfaces and separate the seat away from the crew member to preclude further seat contact injury when descending and landing.

During canopy separation and crew member/seat separation, depending on altitude and air speed, the crew member must have an eye shield and an oxygen mask that are functional. The eye shield is necessary to prevent facial injury from exposure to wind blast, rain, and particulates in the air traveling at speeds up to several hundred knots per hour. Depending on altitude, an easily activated oxygen bottle connected to the flight mask must be part of the survival kit to prevent hypoxia during the several-minute parachute descent. If the descent is into water or onto rocky terrain, a means of quickly disconnecting the parachute must be available to prevent bodily injury from being entangled in the parachute that is collapsing around the crew member or being dragged over the ground. If the crew member becomes entangled in the chute when in the water, he or she may not find their way out of the entanglement and drown. The crew member should also be wearing some sort of flotation device to keep the head above water until rescue is achieved.

While the potential sources of medical injury described in Example 16.8 will cover most high-performance aircraft, several other potential medical injury situations must be considered. If ejection is required at zero speed/zero altitude while the plane is sitting on the runway, hovering, or initiating take-off, this same ejection sequence must be used to clear the aircraft and fully open the parachute before the crew member comes in contact with the ground to minimize injury.

In some aircraft, emergency positional stability requires full-time monitoring and control. The slightest distraction may result in the aircraft rolling upside down. In this event, time is not available to sequentially jettison the canopy and eject. Ejection must go through the closed canopy. To prevent head and neck injury by having the crew member's head serve as the canopy penetrator, the back of the seat is higher than the tallest crew member using it. This seat back then absorbs the impact of canopy penetration rather than the head, neck, and body. In some cases, a pointed spike is mounted on the seat top to facilitate penetration. In the past, some preliminary tests of this mode of ejection using dummies indicated the potential for breaking and/or amputating shoulders and arms because the hole created in the canopy by the seat was not sufficiently large to allow the body to pass through cleanly. This led to canopies being scored in several places to facilitate canopy penetration by the seat and breakage of an adequate hole to allow crew member ejection through it without injury.

In the event of a crew member bailing out and landing with injuries involved, the ability to provide rescue and treatment must be considered. If the incident happens near a fixed installation, fire engines, ambulance, and emergency medical crews may be available to provide these services. If this same event happened over water or in a remote location, trained helicopter rescue teams and special equipment may be necessary to find, treat, and evacuate the downed crew member. Methods (procedures, equipment, and trained personnel) must be strategically located to minimize contact time.

***Fire Suppression*** Another example of minimizing injuries by designing soldier and system survivability into the hardware system can be illustrated with the fire suppression subsystem installed in a small enclosed compartment such as a barge engine room, tank turret, or vehicle crew compartment. In each situation, a fire detection system must be present to detect that a fire is actually present and give warning to unsuspecting occupants to leave the area. The detector must be functional for that particular environment. A combination of a thermocouple and sensor to observe the EM spectrum for specific flame signatures is often required to detect fires in these environments without false alarms. Once an alarm is given, the occupants of the compartment must quickly understand it. An acoustic signal must be heard over background noise, and a visible alarm light must be observed and recognized from any location in the compartment. More than one exit route must be available to flee the fire and potential explosion. In many systems, it is not possible to provide an easily accessible escape exit for each individual. [For example, the ABRAMS tank has two turret hatches on top, one each for loader and track commander (TC). The gunner is seated in front of the TC, and in an emergency, the gunner must wait for the TC to go through the hatch before being able to escape. If the tank has flipped over onto the top of its turret, the turret basket side screens and some equipment on the basket floor will have to be removed before the three turret crew members will be able to escape through the driver's hatch in the hull glacis plate.] The choice of fire-extinguishing material must be quick and effective (desired speed of extinguishments is measured in milliseconds for armored combat vehicles) on the burning material and non toxic to the crew or ample escape time must be provided to minimize toxic exposure. If a built-in fire suppression system is utilized, means to minimize compartment air exchange during the fire suppression treatment are needed for the treatment to work most effectively.

Under ordinary circumstances, halogen gas in high concentration would be used to suppress a fire by both reducing oxygen concentration and lowering the temperature of the ignited material. For short exposures during flight to safety, the gas is safe to breathe in high concentrations (Sass-Kortsak et al., 1985). Typical symptoms reported in confined spaces are light headedness, headache, and disorientation. These are all symptoms usually found with breathing low-oxygen concentrations and retaining carbon dioxide, not specific toxic materials. Because of the ban on the use of halogens due to ozone depletion, other agents must be used. Carbon dioxide in concentrations from 25 to 62 percent by volume (dependent upon the fuel source) is effective against fire in most situations. Unfortunately, these concentrations are extremely toxic to the crew within the compartment or confined space (see Example 16.20). Only a few breaths of these high concentrations can incapacitate the crew to the extent that they may not be able to make it out of the area on their own. If the fire must be fought with hand-held extinguishers, safety masks with carbon monoxide filtering materials must be used to protect those fighting the fire. Water as a fire suppressant cannot be used around electrical fires because the water will short out other electrical equipment and cause electrical shock hazards for the crew as they fight the fire.

Once the fire is out, how can the crew safely reenter the compartment to determine the extent of the damage and their ability to repair it? If either the suppressing substance did not sufficiently cool down the burning material or the cause of the fire was not isolated and shut off (e.g., leak of flammable hydraulic fluid, overloaded electric circuit, etc.), the action of a crew member opening the hatch to take a look could reignite the fire from the renewed supply of oxygen. The person opening this hatch could be severely burned by this flashback. Toxic combusted substances such as carbon monoxide and nitrous oxide may be present in the air within the confined space. These substances must be flushed out of the space before reentry can be safely attempted. Unless instruments are available to determine the presence of adequate oxygen to support life and toxic materials are absent or present in sufficiently low acceptable levels, reentry should not be attempted without a self-contained breathing apparatus.

***Evacuating the Wounded*** If the crew is injured in the confined space, how can they be extracted rapidly and safely? The use of loops attached to the shoulder area of their crew members' uniforms can assist in the lifting of injured, unconscious, or nonmobile crew members through narrow hatches. After a fire, pure oxygen breathing apparatus must be available to flush the crew members' lungs of soot particles and restore blood hemoglobin oxygen saturation levels. Where should the extracted personnel be laid out for examination, treatment, and recuperation until they can be evacuated to a more substantial medical treatment facility? In one situation, it was determined that the black plastic surface of a barge deck ordinarily was so hot from absorbing a solar heat load that sailors working on that surface periodically had to go inside to cool their feet. If injured personnel were laid directly on that hot surface, they could suffer additional burns and dermal injury beyond what they had been exposed to in the confined space during the fire.

Injuries resulting in loss of mobility or loss of utilization of the hands and arms can drastically limit the chances of survival if the soldier is alone, even though sometimes the tactical situation calls for the wounded to fend for themselves for a period of time before they can be retrieved, such as what occurred during the British campaign in the Falkland Islands. Besides the inability to travel safely, soldiers sometimes cannot get to water and food to sustain or protect themselves until rescue. Loss of blood and burns over large areas of the body would put individuals in shock and increase their dependence on water and nutrient electrolytes.

Overall mission performance degradation as the result of physical and mental wounds must also be considered. Potential combat-caused injury or the possibilities of injury are generally not addressed in the HSI health hazard assessment and system safety analysis. However, they are included in the reduction of the medical injury component of SSv.

***Wearing Protective Equipment*** There are a number of different needs for combatants or other personnel to wear specially designed protective equipment. Example 16.9 describes some of these special needs. Example 16.10 describes helmet design in particular.

> **Example 16.9 Special Needs for Protective Equipment** When soldiers are in combat, they will usually wear protective equipment to minimize personal injury when engaged with an opponent. This will usually include helmet and body armor. In some cases, a face shield may be worn. If a chemical attack is anticipated, the soldier may also wear a chemical respirator mask and protective clothing that will completely enclose and separate the body from the ambient environment. Also, the combatant as well as those not engaged in combat, working or residing in areas where biological weapons may be employed, may be inoculated with a vaccine against a specific potential threat. During assembly of bridges to cross wide bodies of water, engineers will wear life vests to protect themselves from drowning if they fall into the water.

An operational "cost" is associated with each of these protective systems. For example, wearing chemical protective clothing may reduce the potential exposure to chemical injury but increases the opportunity for heat stress injuries that may pose medical problems and death. Breathing through a gas mask will limit the ability to perform heavy work at a rate equal to doing it without a mask. Body armor capable of protecting against rifle and small arms fire is very heavy. To limit the amount of extra weight while still maintaining near-normal body mobility, it is generally only used to cover the upper torso (chest and back). Flexible body armor can protect most of the torso, but this protection is limited to shrapnel and fragmentation protection. The extremities are not protected from any of these ballistic projectiles.

**Example 16.10 Helmet Design** The current Army Personal Armor System for Ground Troops (PASGT) helmet was a design improvement over the World War II plastic helmet liner and steel outer shell by offering more protection to the ears, side of the head, and neck while making it lighter and more impervious to shrapnel. By covering the ears of some troops, however, a reduction of the ability to hear sounds (and their direction) occurred, jeopardizing their survival on patrols. Mounting night vision goggles and heads-up displays on the helmet requires that the helmet be anchored tightly by a chin strap to prevent "motion sickness" when looking through these devices. During World War II, it was found (Beyer et al., 1962) that when a helmet was held in place by a chin strap and artillery rounds or bombs exploded nearby, the resulting shock waves would get under the helmet, lifting it upward at some angle. Since the strap held the helmet fast to the head, the upward and angled motion could break the wearer's neck. Once this was learned, the soldiers during that war and the Korean War did not fasten helmet chin straps. Because artillery and bombs were seldom used in heavy concentration against our troops during the Vietnam and Desert Storm wars, and most of our current group of military designers have never really been exposed to combat, this lesson of certain combat medical injuries may have been forgotten.

### 16.4.6 Reduce Mental and Physical Fatigue

The component for the reduction of physical stress and mental fatigue was developed to address tasks that may have direct bearing on the occurrence of combat-related mistakes. Emphasis of this component is on the ID and reduction of complexity of combat-related or combat-support tasks, where negative effects by operator error can be directly traced to stress and fatigue. Associated human factors–related tasks need to be made simple and not mentally tasking, considering that stress and sleep deprivation make the mental processes for operating equipment increasingly difficult as time goes on while mistakes in processing and judgment become more prolific.

***Stress*** Stress can have both positive and negative effects on personnel. Both the anticipation as well as the experience of exposure to a hostile environment can cause a toll on functional ability. Personnel in a defensive position waiting for an attack to come will experience the "fight-or-flight" reactions described physiologically by Cannon (1932) and psychologically by Selye (1956). To some personnel, primarily those who have previously experienced the situation, their senses are sharpened and movements made more fluid, making them more prepared to protect themselves, defeat their attacker, and survive the event. Other personnel, generally new and inexperienced to the event about to unfold, will be frightened to the point of near panic and/or paralysis when protecting themselves and may wish to flee as the only method of survival.

For at least the last half century, military organizations have tried to place their recruits through training that simulated realistic fighting environments the individual was likely to experience in order to reduce personal fear and stress when the individual would be exposed to the same condition in a combat situation. The experience of being fired upon, going through unfamiliar/hostile territory at night, and then fighting the enemy according to a prearranged plan and timetable (with alternate courses of action in case of unforeseen circumstances), with a group of friends they have come to rely on to protect one another, is all training to condition the troops that the event is not that difficult if one does not let fear get the best of the mind. Stress should never be completely eliminated because stress is required to develop the mental "edge" needed to survive.

***Fatigue*** *Military combat is extremely fatiguing.* Those who have experienced this environment have learned to take their rest when and where they can. While fire-fights are short (usually less than an hour), the troops may be required to be on alert for long periods extending to days. It is not unusual for a patrol to be on the move all day long, and then while most of the group is trying to rest, some of the troop must remain awake as sentries. They frequently resume the patrol the next day before the sentries have had a chance to rest or sleep. Even though the next night sees different sentries remaining awake, eventually all of the patrol becomes fatigued from lack of sleep. The situation of being constantly on the alert is dangerous due to fatigue causing an individual to lose focus on his or her objectives.

***Stress and Fatigue Examples*** The SSv parameter assessment list initiates a review of a hardware system and the personnel who will operate the system in combat by focusing on the following five stress and fatigue areas:

- the physical constraints and workload placed on the soldier by this system,
- the cognitive constrains and workload placed on the soldier by this system,
- the system's ability to minimize the effect of the environmental stressors on the soldier,
- the system's ability to minimize the effect of mechanical (system-produced) stressors on the soldier, and
- the system's compatibility with crew life support and continuous operations.

Several examples of routine military and civilian activities that are very stressful and fatiguing in themselves are provided to illustrate how the additional stress of combat or lack of sleep can change the activities into life-threatening situations (see Examples 16.11 to 16.14).

**Example 16.11 Weapons Operations** The physical constraints placed on a weapon operator can be severe. Several decades ago, a weapon was developed for infantry reconnaissance platoons to protect themselves against helicopter attacks. Once an approaching enemy helicopter was identified by sound, the crew carrying the weapon was to stop, unpack the weapon, assemble its many component parts, acquire, aim, and fire it as the helicopter came within the crew's line of sight. To do all these tasks before the helicopter was able to find, identify, and destroy them and their weapon required about 1.75 min under the best of circumstances by crewmen who were in the upper 10 percent of their high school graduating class. The battery was only good for one firing. If the weapon misfired or the fired projectile missed, a second electric battery and missile had to be unstowed, attached, and checked and the weapon fired again. The speed of the workload on the crew and the knowledge and skill required to maintain and operate the weapon as well as the weight of the weapon created such a stress on the soldiers that an analysis of the fielded system provided an example that went to Congress and became part of the impetus in developing the MANPRINT program in the army.

**Example 16.12 Deep-Sea Diving** One civilian and military occupation that places very high cognitive and physical restraints on an individual is that of a deep-sea diver. The individual is confined to an ambient isolating suit that is only slightly larger than his or her body. The diver depends on a crew of knowledgeable people up to half a mile away to provide a blended mixture of breathing gas at a pressure equal to the diver's changing ambient pressure. This gas must be supplied in sufficient quantity to allow the diver to do heavy work without causing the diving suit to balloon and preclude the individual's movement. It must be passed through a 0.5-in.-diameter line that can be easily snagged and cut by features of the diver's surrounding work site. Beyond the 100-ft depth, the working environment is so dark that the diver must use a high-powered light to be able to see a distance as short as 3 ft away from the face. One method of communication with the top-side crew is a two-way wired telephone to enable the diver to clearly hear and understand the top-side crew. The diver, unfortunately, cannot be heard distinctly below approximately the 150-ft depth because the breathing gas mixture distorts the sound as it flows over the vocal cords so that the voice is changed in pitch and speed. The backup communication method is tugging on the lifeline a given number of pulls in a coded format. In many cases, the length of line damps out the communication being sent in either direction. The ambient environment contains predatory creatures, unseen currents to prevent the diver from staying at the workstation, and uneven working areas preventing standing or functioning in a comfortable position. The diver must usually work alone because of the danger involved. He or she must be skilled at several different occupations, including plumbing, welding, naval architecture, oceanology, and respiratory physiology. The diver knows that a cut or crimp in the breathing hose, a cut in the suit, loss of neutral buoyancy by the intentional or unintentional dropping of a weight, improper use or loss of tools can immediately cause death. Every body movement must be considered and the consequences weighed for the diver to survive.

When the diver attempts to return to safety upon the completion of work, the process must be very slow if he or she is to live. If the descent was to below 200 ft of depth for more than 12 min, the diver must ascend at the rate of only 1 ft of depth every 20 min in accordance with U.S. Navy diving tables. A dive to 900 ft to fix an oil well "Christmas Tree" can take more than 12 days to return to the ocean surface. To ease the return ascent, allow the diver to drink and eat, and protect the body from the constant extremely cold water (usually about 32°F), a diving chamber is sent near the bottom to provide a refuge during the return. Although isolated from other personnel for the full ascent time, there is access to food, water, warmth, and personal hygiene needs. Such an occupation can produce sensory overload and increased physiological and psychological fatigue. Unless the individual puts complete trust in his or her own ability, the ability of both the supporting top-side crew and the equipment, and the stability of the ambient environment above and below the ocean where the individual is working, the deep-sea diver may not survive the mental and physical stresses to which he or she is subjected.

**Example 16.13 High Climatic Temperatures** High climatic temperatures affect the ability to perform many different tasks, both military and civilian. Pepler (1958) measured the performances of British soldiers receiving Morse code, detecting simulated radar signals, tracking a moving pointer, and making decisions from rapidly changing visual displays. Performance started to deteriorate in temperatures above 81°F. The accuracy of West African soldiers conducting Morse code communication duties and tracking mission activities for 3-hour periods also deteriorated above 86°F according to Watkins (1956). Bursill (1958) conducted experiments to determine heat fatigue on simulated tracking. He had the men sit in both a 65 and a 96°F environment for 2 hours and then had them respond to small randomly timed lights appearing erratically in their peripheral field of view while they were tracking a moving target and point immediately in front of them. More peripheral lights were not observed in the higher temperature than in the lower temperature, suggesting that fatigue produced by heat, repetitive task, and possible boredom focused their awareness toward the central visual field. It is well known that a pilot looking directly ahead through the cockpit window without any cloud variation or ground to focus on for a period of time will gradually focus on a point of ocular convergence and miss any object in the surrounding visual field.

**Example 16.14 Heat and Sleep Deprivation** Pepler (1959) also conducted studies on serial tracking tasks associated with combinations of heat and lack of sleep. The effects of lack of sleep alone were different from the effects of heat alone. Loss of sleep reduced the subject's responsiveness to the required tasks but not the accuracy. The elevated temperature reduced the accuracy but not the responsiveness to the task.

## 16.5 SOME "LESS-THAN-OBVIOUS" EXAMPLES

For those working in the areas previously described, it is normal for many to make a studied observation of a given system based upon their expertise, list the issues of concern, and then work hard at addressing those particular issues without continually looking further. Periodically, issues arise after the equipment has been fielded, causing concern and potential disruption. A few examples are presented to suggest that an assessor continually needs to watch for the "flaw" that is often not obvious.

### 16.5.1 NBC-Protective Equipment

A person who has performed any physical task in a MOPP IV chemical protective ensemble and NBC-protective respirator immediately recognizes that this set of protective equipment attempts to address simultaneously survivability, human factors, and health hazards issues. Looking beyond the obvious concerns when individual protective equipment is used in the usual temperate environment, there are some potentially serious issues (see Examples 16.15 to 16.18).

**Example 16.15 Arctic Clothing in NBC Environment** Consider use with Arctic cold-weather clothing in below-freezing conditions. In the recent past, NBC protection was to be worn *under* the troops' Arctic clothing. This is not an obvious consideration until one realizes that after a chemical exposure the soldiers must remove their contaminated cold-weather clothing and MOPP clothing and equipment prior to entering a shelter. While going through the decontamination process outdoors, the soldiers could suffer severe hypothermic injury before entering the protective shelter. Once inside, the soldiers are not able to go back out into the contaminated environment unless the shelter has supplies of cold-weather clothing and MOPP protective equipment. Even if the environment is decontaminated, clean, cold-weather clothing must be provided for activity outside the shelter.

**Example 16.16 Shaving While Wearing NBC Mask** To properly utilize an NBC respirator, the male military person must take extra care when shaving and do so on a daily basis since NBC mask seal (as presently designed) leakage can be caused by a single whisker.

**Example 16.17 NBC Clothing and Stevedoring** Another NBC tangential issue can be that stevedoring (loading and unloading ships' cargoes) in the tropics may cause serious concerns. When under threat of missile attack with chemical weapons, the wearing of MOPP IV can be a big "monkey wrench" in normal personnel requirements. Working in MOPP IV in the tropics may limit work performed by personnel down in ships' holds to as little as 15-min/hr shifts, due to the existing intense humidity and temperature being severely intensified within ships' holds.

**Example 16.18 NBC Clothing and Reading Wristwatch** For a simpler example, consider the front-line leader wearing MOPP IV. Combat-related movements and actions are to be made at specified times. Has the leader realized that one cannot pull back the sleeve under contaminated conditions to look at the wristwatch for the time? Will the leader be thinking ahead and wear a favorite wristwatch over the sleeve and be willing to dispose of it if exposed to chemical/biological agents? Has logistics considered stocking replacements in sufficient quantity to meet the anticipated demand?

### 16.5.2 Single-Eye Viewer

Another example to consider is an opaque heads-up display for use by only a single eye (see Example 16.19).

**Example 16.19. Single-Eye Viewer** The display is periodically pulled down in front of that primary eye while the other eye is intended to maintain surrounding situation awareness (SA) in a ground or aviation environment. A display of this type, when put into position, is in the visual field of only the primary eye. Unrealized by many, this concentrated active utilization of one eye affects the vision of the other eye, causing this other eye to "shut down" its visual activity over a very short time period. This situation can tremendously reduce a soldier's visual awareness of his peripheral surroundings until the display is repositioned away from the single eye and both eyes are again to be used to view the surroundings. The soldier will then experience a short time lapse before the shut-down eye fully reactivates.

### 16.5.3 Fire Extinguisher Agents

Another less-than-obvious concern comes from a past consideration for fire-extinguishing agents. Due to environmental concerns for ozone effects, Halon manufacture was discontinued in the United States. One version of Halon was used in the automatic fire suppression systems of military vehicles due to its ability to quickly extinguish fires at a low concentration in crew compartments while simultaneously allowing the crew to breathe. Carbon dioxide was proposed as a substitute for Halon. (See Example 16.20 for the limitations of carbon dioxide as a fire-extinguishing agent.)

**Example 16.20 Carbon Dioxide as Fire-Extinguishing Agent in Closed Quarters** On the surface, carbon dioxide seems like a good candidate due to its common use in fire extinguishers and its ability to extinguish many sources of fire. Unfortunately, although common carbon dioxide fire extinguishers can be used to good effect in large rooms, outdoors, or in compartments without personnel, carbon dioxide does have serious effects on the human body when used in heavy concentration. Carbon dioxide needs to be in a concentration of 25 percent (minimum) to 62 percent (Cote, 1997) of total air volume dependent upon fuel source(s) to put out flame. Consider the little-known physiological fact that if humans are in a closed compartment (whether in a cockpit, ship, or armored vehicle), a lower than required (for extinguishing a blaze) 10 percent by volume atmospheric level (Bender et al., 1994) of carbon dioxide is subjectively intolerable to breathe, while slightly higher (12 to 15 percent) carbon dioxide levels produce unconsciousness and, eventually, death. For those personnel successfully evacuating the high-carbon-dioxide environment to fresh air, they experience severe headaches, nausea, vomiting, the smell of ammonia (which may lead a military person to believe that a chemical attack is underway), and potential loss of consciousness. At a carbon dioxide level of 25 percent or greater (minimum required to extinguish flame in a closed compartment), the crew members may immediately experience effects such as impaired mental ability, violent respiratory movements and convulsions, inability to take steps for self-preservation, loss of consciousness, and potentially stopping of the heart. These human body responses to high levels of carbon dioxide may only allow a three- to four-breath reaction time before collapse. Used in combat, a vehicle might be saved from a fire during combat, but members of the crew would likely perish from the extinguishing agent named in this example.

Personnel survivability investigations must constantly consider many diverse possibilities to find the lesser known and hidden issues until they are detected and pointed out. The effective use of the survivability considerations presented in Figure 16.2 by knowledgeable multidisciplined SMEs should minimize the less-than-obvious concerns if this information is incorporated in system design during the concept and development stage of a new or modified program.

## 16.6 CASUALTY ASSESSMENT TOOLS

Combat mathematical and computerized models are often used to project probabilities of war-fighting outcomes. At this level, combat models are used to feed assumptions into the analysis of alternatives (AOA). Although these models are used to project personnel casualties as well as weapon system losses during combat, they are unlikely to have utilized human performance data or taken into consideration SSv factors. For example, soldiers operating a tank in actual combat may become casualties even though the tank survives an attack.

### 16.6.1 The ORCA Model

In the past, casualty assessments relied on separate applications of several stand-alone models, each of which dealt with a specific battlefield insult (injury mechanism). These tools serve a crucial function by permitting the computation of human vulnerability measures for personnel serving as individual combatants or as "components" who contribute to the overall vulnerability of a weapon system. For example, the army's ComputerMan (Clare et al., 1980; Saucier, and Kash, 1994) model was often used to evaluate penetrating injuries, while BURNSIM (Knox, and Perry, 1993), an air force code, was frequently used to assess the likelihood of skin burns from thermal exposures. The Operational Requirements-based Casualty Assessment (ORCA) model (Neades et al., 1999) incorporates the best features of these and several other existing models and combines them in a way that allows consistent assessment of casualties across virtually all platform, task, and threat types (see Fig. 16.10). It provides a new methodology for assessing the antipersonnel effects associated with various munition-produced mechanisms. Development of this model was prompted by concern at the Office of the Secretary of Defense (OSD) level over the lack of standardized service methodology for computation of user casualties.

The ORCA computer code allows the analyst to calculate anatomical damage and the effect on individual performance as a result of exposure to kinetic energy threats. It will also tell what happens with other insults such as thermal, chemical, directed-energy (laser), blast, and accelerative loading threats, but, at the time of this writing, ORCA does not yet incorporate pressure–time histories for these latter threats. In each case, the effect of a computed injury is characterized by the predicted impairment of each of 24 human

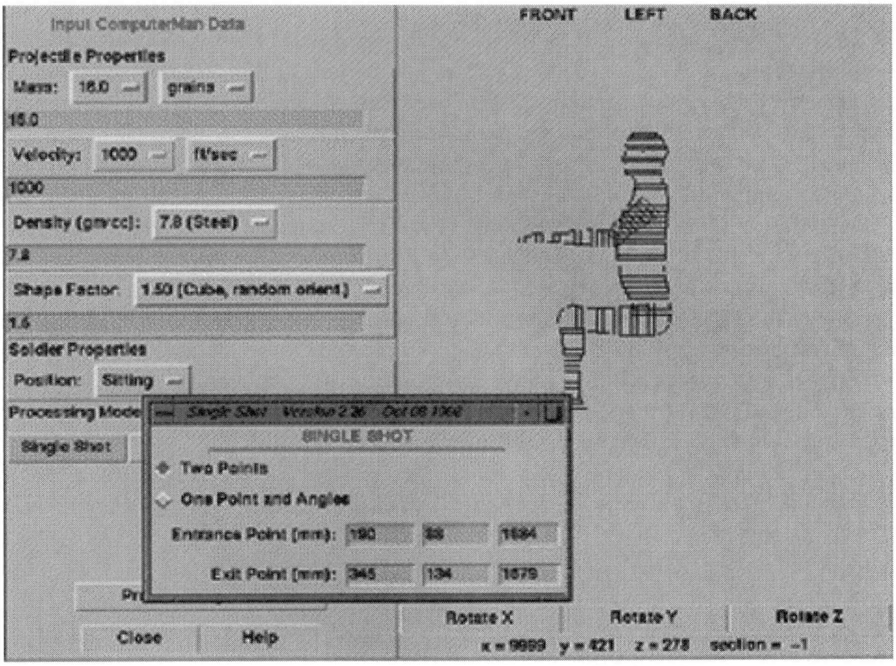


**Figure 16.10** ORCA analysis—computer screen showing areas of inputs and the human figure in one of the available aspects and configurations.

elemental capabilities (e.g., vision, cognition, and physical strength) as a function of time after injury. Postinjury capability is then compared to capability requirements associated with the individual's military job task or mission to determine if he or she is an operational casualty. ORCA has code outputs for the following: (1) discrete exposures (e.g., a single fragment impact) including a physical damage summary; (2) details of any deleterious processes (e.g., blood loss); (3) abbreviated injury score (AIS) (American Association for Automotive Medicine, 1998); (4) elemental capability status; and (5) remaining performance capability (comparable to incapacitation) as a function of time after wounding (six time periods ranging from immediate to 72 hours). In addition to addressing discrete simulations with single threats, ORCA can also be run in grid or batch mode to produce results that reflect a range of exposure conditions. ORCA users can specify the operational requirement for a military job, task, or mission by selecting from a database library of 20 army, navy, air force, and marine military occupational specialties, specific military tasks, or predefined mission scenarios. Although the determination of medical casualties is not within the charter of the DoD Joint Technical Coordinating Group for Munitions Effectiveness' Crew Casualty Working Group, it is essential, to the degree that medical and operational casualty factors are common, that ORCA be consistent with the needs of the combat casualty analytical medical community. To this end, significant care has been taken to define and record injuries in a way that serves future medical analysis needs. In particular, ORCA determines and tracks each injury's AIS severity score, an injury characterization system common throughout the civilian medical community.

### 16.6.2 Weapons Systems Effects Models

Beyond modeling of human effects, it is vitally important in SSv to understand effects of weapons on combat systems and equipment. What affects the platform or equipment provides the analyst with information on what to protect the human against. For example, a tank commander may have his head and shoulders out of his hatch to observe and give movement commands to avoid being hit by incoming SAGGER antitank guided missiles (ATGM). Based upon actual occurrence, the tank commander in this position will experience tremendous heat and overwhelming pressure from the detonation of the missile against the tank, at a minimum. The analyst must be knowledgeable of the potential weapon's and the platform's interaction effects to understand what the human being will potentially experience. An analyst must realize that soldiers by themselves would almost never be the targets of an ATGM, but, because of the tanks they are manning, they may be exposed to the ATGM's weapon effects upon a tank and its armor. Therefore, SSv must be viewed as interwoven with system survivability.

There are a fairly large number of computerized models that provide invaluable insights as to weapon system effects (U.S. Army Natick Research, 1992; National Defence Research Establishment, n.d.; U.S. Army Research Laboratory, 1996, 1997). Generally, they incorporate algorithms based upon actual test data and provide quite accurate outputs.

### 16.6.3 Current Limitations

A word of caution is due on the use of models and simulations. Models and simulations have limited value, depending upon their accuracy in the codes and the data used. For example, one particular chemical-related model does not take into consideration the heat stress on the body due to wearing MOPP IV clothing and its weakening effect before the chemical attack is even applied. This same model does not take into consideration the dehydrated state of the soldier in a hot desert environment or the fact that the individual may have taken pyridostigmine as a prophylactic before exposure. It completely escapes the analyst using the model that pyridostigmine will cause the soldier to have a hard time aiming a rifle to protect him- or herself unless the user is familiar with the pharmacology of the prophylactic. The model does not incorporate the protective layers of clothing and their stopping action. In many cases the models have been looking at the wrong route of entry for bioagents according to epidemiological studies. That said, models and simulations can add a great amount of value in an analysis and as guides useful in applying knowledgeable insight and experience.

An additional cautionary note is on time limitations of some models, as they may use 5-to 10-minute minimum time intervals. An illustrative example would be improving the breathing resistance of a gas mask. A soldier under fire is taught never to expose her- or himself to the enemy for more than 15 seconds when running from protective cover to protective cover. If the breathing resistance of the mask is improved 200 percent, the soldier will be able to get to protective cover faster (less than 15 seconds). This improvement is important for survivability but will not show any improvement in a simulation model where the shortest time segment is 5 minutes. This interval may often be too long to notice major technical improvements in individual equipment. Also, a wounded military person's variability of his or her physiological and psychological state sometimes causes extreme responses, different than those predicted. Models generally do not take into consideration the excited state of the body (being pumped up for trouble before the event), as they generally use model data of an individual physiologically at rest.

## 16.7 SUMMARY AND CONCLUSIONS

Personnel survivability most often deals with an inseparable combination of equipment and personnel. If military communications or other equipment becomes a target of a threat force, the personnel manning that equipment are also parts of the target. If a terrorist places a bomb in a plane, the personnel occupying the plane are also targets. There often is no simple means to disconnect the human from the machine for combat effects investigation and analysis. Therefore, personnel survivability must view man–machine interaction and effects as a whole with the ultimate focus on the survival of the human. For good investigations to occur, a number of different disciplines are required. A good personnel survivability assessment can bring together a medical combat care provider, an EW SME, an NBC assessor, a human factors professional, and a materiel developer. They may *all* be needed to make one integrated, threat- and combat-oriented HSI assessment.

A personnel survivability assessment may furnish a valuable service to the materiel developer and others by providing an overall integrated technical review of the hardware/human system. This includes appreciation of the combat weapon–induced threat and the resulting program survivability issues while encouraging HSI participation in resolving those issues. This type of assessment assists in providing coverage and assurance that issues will be, or are being, addressed in such diverse areas as EW, ballistics, DEWs, NBC warfare, physiological effects, and heat stress. A comprehensive personnel survivability list of issues can be tailored to assist a materiel development team in producing a survivability outline and strategy while simultaneously providing valuable information and insights for the combat developer.

The HSI practitioner should be able to utilize this method with some modifications to help the civilian who fights in hostile environments as well. The personnel survivability examples provided in the chapter on fire-extinguishing agents, deep-sea diving, high climatic temperatures, heat and sleep deprivation, single-eye viewers, and wearing protective clothing in hostile environments are as applicable for civilians as for military combatants. The kinds of civilian operations where personnel survivability becomes especially critical are situations such as bombs in airplanes and trains, chemicals placed by terrorists in water supplies, mask and special clothing utilization in NBC civilian environments, police firearms and armor, and fire fighter protection.

It is important to distinguish those human factors that might be routinely addressed in the systems safety and health hazards domains from those in the personnel survivability domain, which adds a new dimension to the general HSI arena. Whether a warfighter or a civilian, personnel survivability applies to those situations where the environment is so hostile that those to be protected are working on the outermost edge of their physical and cognitive abilities where at any moment they could become casualties. It is believed that the methods and examples described provide information that can play a positive role in reducing the likelihood of such casualties.

## REFERENCES

American Association for Automotive Medicine. (1998). *The Abbreviated Injury Scale 1998 Revision (AIS-98).* Des Plaines, IL: Committee on Injury Scaling.

Ball, R. E. (1985). *The Fundamentals of Aircraft Combat Survivability Analysis and Design.* New York: American Institute of Aeronautics and Astronautics.

Bender, N. K., Slonim, N. B., and Slonim, N. B. (Eds.). (1994). *Environmental Physiology.* Saint Louis, MO: C. V. Mosby.

Beyer, J. C., Enos, W. F., and Holmes, R. F. (1962). Personnel Protective Armor. In *Wound Ballistics* (Chapter XI). Washington, DC: Office of the Surgeon General, U.S. Army Medical Department.

Burnett, C. (1913). *Training in Night Movements, Based on Actual Experiences in War.* (Translated from the original Japanese.) Port Townsend, WA: Loompanics Unlimited.

Bursill, A. E. (1958). The Restriction of Peripheral Vision During Exposure to Hot and Humid Conditions. *Quarterly Journal of Experimental Psychology, 10*, 113.

Cannon, W. B. (1932). *Wisdom of the Body.* New York: Norton Publishing.

Carhart, T. (1994). *Iron Soldiers, How America's 1st Armored Division Crushed Iraq's Elite Republican Guard.* New York: Pocket Books.

Clare, V. R., Ashman, W., Broome, P., Jameson, J., Lewis, J., Merkler, J., Mickiewicz, A., Sacco, W., Sturdivan, L., Lamb, D., and Sylvanus, F. (1980). *The ARRADCOM Computer Man: An Automated Approach to Wound Ballistics*, ARCSL-TR-80021. Aberdeen Proving Ground, MD: U.S. Army Chemical Systems Laboratory.

Cote, A. E. (Ed.). (1997). *National Fire Protection Association Handbook*, 18th ed. Quincy, MA: National Fire Protection Association.

DeBay, Y. (1991). *Blitzkrieg in the Gulf, Armor of the 100-Hour War.* Tsuen Wan, New Territories, Hong Kong: Concord Publications Company.

Dupuy, T. N. (1990). *Attrition: Forecasting Battle Casualties and Equipment Losses in Modern War.* Fairfax, VA: Hero Books.

Houck, K. (1994). *Politics of War: Tank-Plinking in the Gulf.* Available: http://www.lbbs.org/zmag/articles/july94houck.htm.

Knox, F. S., III, and Perry, C. (1993). *User's Manual for BURNSIM: A Burn Hazard Assessment Model*, USAARL Report No. 93-13. Ft. Rucker, AL: U.S. Army Aeromedical Research Laboratory (also available from U.S. Air Force Laboratory, WPAFB, OH).

National Defence Research Establishment (NDRE), Department of NBC Defence. (n.d.). *Chemical Attack Simulation: Protection and Risk (CASPAR)*, S-901 82. Umea, Sweden: NDRE.

Neades, D. A., Klopcic, J. T., and Davis, E. G. (1999). New Methodology for the Assessment of Battlefield Insults and Injuries on the Performance of Army, Navy, and Air Force Military Tasks, RTO-MP-20 AC/323(HFM)TP/7. In *North Atlantic Treaty Organization Research and Technology Organization Proceedings 20, Models for Aircrew Safety Assessment: Uses, Limitations and Requirements*.

Pepler, R. D. (1958). Warmth and Performance: An Investigation in the Tropics. *Ergonomics*, *2*, 63.

Pepler, R. D. (1959). Warmth and Lack of Sleep: Accuracy or Activity Reduced. *Ergonomics*, *52*, 446.

Sass-Kortsak, A. M., Holness, D. L., and Stopps, G. J. (1985). An Accidental Discharge of a Halon 1301 Total Flooding Fire Extinguishing System. *American Industrial Hygiene Association Journal*, *46*(11), 670-3.

Saucier, R., and Kash, H. M., III. (1994). *ComputerMan Model Description*, U.S. Army Research Laboratory Technical Report No. ARL-TR-500. Aberdeen Proving Ground, MD: U.S. Army Research Laboratory.

Selye, H. (1956). *The Stress of Life.* New York: McGraw-Hill.

Shrader, C. R. (1982). *Amicicide: The Problem of Friendly Fire in Modern War.* Fort Leavenworth, KS: Combat Studies Institute, U.S. Army Command and General Staff College.

Steinweg, K. K. (1995). Dealing Realistically with Fratricide. *Parameters, U.S. Army War College Quarterly, 25*(1), 4.

Tauson, R. A., Doss, N. W., and Zigler, R. N. (1995). Methodology for Performing Soldier Survivability Assessments. *MANPRINT Quarterly, III*(2), 4–5.

U.S. Army Natick Research, Development, and Engineering Center. (1992). *Integrated Unit and Soldier System Survivability.* Available: http://www.natick.army.mil:80/soldier/m&a/IUSS.htm.

U.S. Army Research Laboratory, Survivability/Lethality Analysis Directorate. (1996). *BRL-CAD®.* Available: http://ftp.arl.mil/brlcad.

U.S. Army Research Laboratory, Survivability/Lethality Analysis Directorate. (1997). *Modular UNIX™-Based Vulnerability Estimation Suite (MUVES).* Available: http://www-slad.arl.army.mil/Services/SL-Modeling-MUVES.html.

U.S. Army Research Laboratory, Survivability/Lethality Analysis Directorate. (1998). *Atmospherics/Obscurants.* Available: http://www-slad.arl.army.mil/Services/Obscurants.html, http://www-slad.arl.army.mil/Services/Obscurants-Services.html.

U.S. Army Research Laboratory, Survivability/Lethality Analysis Directorate. (1999). *Atmospherics/Obscurants Analysis.* Available: http://web.arl.mil/Services/Obscurants-Analysis1.html.

U.S. Army. (1999). *Information Operations*, FM 100-6. Washington, DC: Department of the Army.

U.S. Army. (1994). *Manpower and Personnel Integration (MANPRINT) in the System Acquisition Process*, AR 602-2. Washington, DC: Department of the Army.

U.S. Army. (1995). *Survivability of Army Personnel and Materiel*, AR 70-75. Washington, DC: Department of the Army.

U.S. Code, Title 10, Section 2366. (n.d.) *Major Systems and Munitions Programs: Survivability Testing and Lethality Testing Before Full-Scale Production.* Washington, DC: Government Printing Office.

U.S. Department of Defense (DoD). (1996). *Defense Acquisition*, Directive 5000.1 (incorporating Change 1, 1999). Washington, DC: DoD.

U.S. Department of Defense (DoD). (1999). *Mandatory Procedures for Major Defense Acquisition Programs (MDAP) and Major Automated Information System (MAIS) Acquisition Programs*, Regulation 5000.2. Washington, DC: DoD.

U.S. News & World Report. (1992). *Triumph without Victory, the History of the Persian Gulf War.* New York: Times Books, a Division of Random House.

U.S.S. Forestal Museum. (n.d.). *The Tragic Fire July 29, 1967.* Available: http://forrestal.org/fidfacts/page13.htm.

Watkins, E. S. (1956). *The Effect of Heat on Psychomotor Efficiency with Particular Reference to Tropical Man.* Liverpool: University of Liverpool.

## RECOMMENDED READING

Bailey, H. (1944). *Surgery of Modern Warfare*, Vols. I and II. Baltimore, MD: Williams and Wilkins Company.

Burns, N. M., Chambers, R., and Hendler, E. (1962). *Unusual Environments and Human Behavior.* New York: Free Press of Glencoe.

Edholm, O. G., and Bacharach, A. L. (Eds.). (1965). *Physiology of Human Survival.* London: Academic Press.

English, J. A., and Gudmundsson, B. I. (1994). *On Infantry*, Rev. ed. Westport, CT: Praeger.

Glenn, J. F., Burr, R. E., Hubbard, R. W., Mays, M. Z., Moore, R. J., Jones, B. H., and Krueger, G. P. (1990). *Sustaining Health and Performance in the Desert: A Pocket Guide to Environmental Medicine for Operations in Southwest Asia.* Natick, MA: U.S. Army Institute of Environmental Medicine.

Isby, D. C. (1988). *Weapons and Tactics of the Soviet Army*, 2nd ed. London: Jane's Publishing.

Ivarsson, U., Nilsson, H., and Santesson, J. (Ed.). (1992). *A FOA Briefing on Chemical Weapons: Threat, Effects and Protection.* Sundybyberg, Sweden: Forsvarets forskningsanstalt (FOA).

Johnson, J. C., and Thaul, S. (Eds.). (1997). *An Evaluation of Radiation Exposure Guidance for Military Operations.* Washington, DC: National Academy Press.

Lawrence, T. E. (1935). *Seven Pillars of Wisdom.* New York: Doubleday, Doran Publishing.

Macksey, K. (Ed.). (1981). *Tank Facts and Feats*, 3rd ed. New York: Sterling Publishing.

Marshall, S. L. A., and Smith, P. (1947). *Men against fire, the Problem of Battle Command in Future War.* Norman, OK: University of Oklahoma Press.

McDonough, J. R. (1985). *Platoon Leader.* Novato, CA: Presidio.

Morris, C., and Morris, J. (1992). *The American Warrior.* Stamford, CT: Longmeadow.

Nesbitt, P. H., Pond, A. W., and Allen, W. H. (1959). *Survival Book.* New York: Van Nostrand Company.

Newburgh, L. H. (Ed.) (1949). *Physiology of Heat Regulation and the Science of Clothing.* Philadelphia, PA: Saunders.

Ogorkiewicz, R. M. (1991). *Technology of Tanks.* Surrey, United Kingdom: Jane's Information Group Limited.

Pandolf, K. B., Sawka, M. N., and Gonzalez, R. R. (Eds.). (1988). *Human Performance Physiology and Environmental Medicine at Terrestrial Extremes.* Indianapolis, IN: Benchmark.

Roberts, B. (1993). *Biological Weapons, Weapons of the Future.* Washington, DC: Center for Strategic and International Studies.

Salvendy, G. (Ed.). (1987). *Handbook of Human Factors.* New York: Wiley.

Sanders, M. S., and McCormick, E. J. (1987). *Human Factors in Engineering and Design*, 6th ed. St. Louis, MO: McGraw-Hill.

Slonin, N. B. (1974). *Environmental Physiology.* Saint Louis, MO: C. V. Mosby.

Survivability/Vulnerability Information Analysis Center (SURVIAC). (n.d.). *McPTD 2.1—Radar Cross Section (RCS) Computation Based on Physical Theory of Diffraction.* Booz, Allen & Hamilton. Wright-Patterson AFB, OH. Available: http://iac.dtic.mil/surviac/prod_serv/model_guide/mcptd.html.

Survivability/Vulnerability Information Analysis Center (SURVIAC). (n.d.). Booz, Allen & Hamilton Inc. Wright-Patterson AFB, OH. Available: http://iac.dtic.mil/surviac.

Swinton, E. D. (1986). *Defences of Duffer's Drift.* Wayne, NJ: Avery Publishing Group.

Thibodeau, G. A., and Patton, K. T. (1992). *The Human Body in Health and Disease.* St. Louis, MO: Mosby-Yearbook.

Tromp, S.W. (1963). *Medical Biometeorology.* Amsterdam: Elsevier.

U.S. Army. (n.d.). *Survival Manual*, FM 21-76. Washington, DC: Department of the Army.

U.S. Department of Defense (DoD). (1995). *Bosnia Country Handbook*, DOD-1540-16-96, Washington, DC: DoD.

Wilkinsom, N. B. (1983). *Explosives in History.* Wilmington, DE: Wilkinsom, Hagley Museum.

Winslow, C. E. A., and Herrington, L. P. (1949). *Temperature and Human Life.* Princeton, NJ: Princeton University Press.

## CHAPTER 17

# Cost–Benefit Analysis for Human Systems Integration

WILLIAM B. ROUSE and KENNETH R. BOFF

## 17.1 INTRODUCTION

The past decade has been a period of very serious scrutiny of the activities of most enterprises. Business processes have been reengineered and enterprises have been downsized or, more popularly, rightsized. Every aspect of an enterprise must now provide value to customers, earn revenues based on this value, and pay its share of costs. Aspects of an enterprise that do not satisfy these criteria are targeted for elimination.

This philosophy seems quite reasonable and straightforward. However, implementation of this philosophy becomes rather difficult when the "value" provided is indirect and abstract. When anticipated benefits are not readily measurable in monetary units and only indirectly affect things amenable to monetary measurement, it can be very difficult to assess the worth of investing in such benefits.

There are a wealth of examples of such situations. With any reasonable annual discount rate, the tangible discounted cash flow of benefits from investments in libraries and education, for example, would be so small as to make it very difficult to justify societal investments in these institutions and activities. Of course, we feel quite justified arguing for such investments. Thus there obviously must be more involved in such an analysis than just discounted cash flow.

This chapter addresses types of human systems integration (HSI) investments that have these intangible characteristics in addition to more tangible attributes. One type is research and development (R&D). This type of investment is often made for the purpose of creating long-term value. It will certainly require years and may take decades before returns are fully realized. It is easy to see how R&D can be difficult to justify in terms of impacts on, for instance, this year's sales and profits or current operational readiness (Rouse and Boff, 1998).

*Handbook of Human Systems Integration*, Edited by Harold R. Booher.
ISBN 0-471-02053-2 

Another type of investment with these intangible characteristics involves products and services that enhance human effectiveness and thereby HSI. This includes selection, training, system design, job design, organizational development, health and safety, and, in general, the wide range of things done to ensure and enhance the effectiveness of people in organizations ranging from businesses to military units. In particular, investments focused on increasing human potential, rather than direct job performance outputs, are much more difficult to justify than those with near-term financial returns (Rouse et al., 1997).

This chapter also addresses the complex interaction of R&D investments in human effectiveness as a central element of HSI. This is done by building on previous efforts by the authors and many others that address the two elements of this interaction. Investing in R&D to enhance human effectiveness presents a confluence of difficulties related to representing and quantifying benefits as well as attributing costs. Nevertheless, there is a widely shared sense that such investments are socially and economically important. It is difficult, however, to justify particular projects on the basis of such perceptions.

A primary difficulty involves the trade-off between the relatively short-term payoffs of direct improvements in job performance and the inherently long-term benefits of R&D efforts aimed at enhancing human effectiveness and HSI. Short-term investments usually involve less uncertainty and fewer risks. In contrast, revolutionary high-payoff innovations usually emerge from much earlier R&D investments. Thus small, certain, near-term returns compete with large, uncertain, long-term, and potentially very substantial returns. The methodology presented in this chapter enables addressing both types of investments.

In general, several issues underlie the difficulties in justifying the aforementioned types of long-term investments. As just noted, a fundamental issue concerns the associated uncertainties. Not only are the magnitudes and timing of returns uncertain; the very nature and characteristics of returns are uncertain. With R&D investments, for instance, the eventual payoffs from investments are almost always greater for unanticipated applications than for the originally envisioned applications (Burke, 1996). Further, organizations that make the original investments are often unable to take advantage of the eventual returns from R&D (Christensen, 1997).

Another central issue relates to the preponderance of intangible outcomes for these types of investments. For example, investments in training may enhance leadership skills of managers or commanders. Investments in organizational development can improve the cohesiveness of "mental models" of management teams or command teams and enhance the shared nature of these models. However, it is difficult to fully capture such impacts in terms of tangible, "bottom-line" metrics.

It is important to differentiate between intangible outcomes and those that are tangible but difficult to translate into monetary benefits or costs. For example, an investment might decrease pollution, which is very tangible, but it may be difficult to translate this projected reduction to estimated economic gain. This is a mainstream issue in economics and not unique to cost–benefit analyses.

A further issue concerns cost–benefit analyses across multiple stakeholders. Most companies' stakeholders include customers, shareholders, employees, suppliers, communities, etc. Government agencies often have quite diverse sociopolitical constituencies who benefit or stand to lose benefits in a myriad of ways depending on investment decisions. For example, government-sponsored market research may be part of a regional economic development plan or may be part of a broader political agenda focused on creating jobs. In general, diverse constituencies are quite likely to attempt to influence decisions in a variety of ways. These situations raise many basic questions relative to the importance of benefits and costs for the different stakeholders.

Yet another issue concerns the difference between assessing and predicting costs and benefits. It is certainly valuable to know whether past investments were justified. However, it would be substantially more valuable to be able to predict whether anticipated investments will later provide benefits that justify the initial investments. Of course, limits of our abilities to predict outcomes are not unique to cost–benefit analysis.

The types of investment problems addressed in this chapter are rife with many uncertainties, intangibles, and stakeholders and associated unpredictability. These issues are explored in this chapter in the context of alternative frameworks for performing cost–benefit analyses. This leads to clear conclusions about how best to methodologically handle these types of investments. Application of the resulting methodology is then illustrated in the context of three investment problems involving technologies for aiding, training, and ensuring the health and safety of personnel in military systems.

## 17.2 COST–BENEFIT FRAMEWORKS

There are a variety of frameworks for scrutinizing and justifying investments:

- *Cost–Benefit Analysis* Methods for estimating and evaluating time sequences of costs and benefits associated with alternative courses of action.
- *Cost–Effectiveness Analysis* Methods for estimating and evaluating time sequences of costs and multiattribute benefits to ensure that the greatest benefits accrue for given costs.
- *Life-Cycle Costing* Methods for estimating and evaluating costs of acquisition, operation, and retirement of alternative solutions over their total cycles of life.
- *Affordability Analysis* Methods for estimating and evaluating life-cycle costs compared to expected acquisition, operations, and maintenance budgets over the total life cycle of an alternative investment.
- *Return-on-Investment Analysis* Methods for projecting the ratio, expressed as a percentage, of anticipated free cash flow to planned resource investments.

This chapter focuses on cost–benefit analysis in a broad sense that includes many aspects of the other approaches. For more traditional treatments of cost–benefit analysis, as well as worked examples, see Layard and Glaister (1994) and Gramlich (1997).

Cost–benefit analyses are very straightforward when one considers fixed monetary investments made now to earn a known future stream of monetary returns over some time period. Things get much more complicated, however, when investments occur over time, some of which may be discretionary and when returns are uncertain.

Further complications arise when one must consider multiple stakeholders' preferences regarding risks and rewards. Additional complexity is added when returns are indirect and intangible rather than purely monetary. These complications and complexity are more common than are situations where the straightforward cost–benefit analyses are applicable. This section discusses alternative frameworks for addressing cost–benefit analyses and compares these alternatives relative to their abilities to address the issues considered in Section 17.1.

### 17.2.1 Traditional Economic Analysis

The time value of money is the central concept in this traditional approach. Resources invested now are worth more than the same amounts gained later. This is due to the costs of investment capital, which must be paid, or foregone, while waiting for subsequent returns on the investment. The time value of money is represented by discounting the cash flows produced by the investment to reflect the interest that would, in effect at least, have to be paid on the capital borrowed to finance the investment.

Equations (1–3) summarize the basic calculations of the discounted cash flow model. Given projections of costs $c_i$, $i = 0, 1, \ldots, N$, and returns $r_i$, $i = 0, 1, \ldots, N$, the calculations of *net present value (NPV), internal rate of return (IRR)*, or *cost–benefit ratio (CBR)* are quite straightforward elements of financial management (Brigham and Gapenski, 1988). The only subtlety is choosing a *discount rate* (DR) to reflect the current value of future returns decreasing as the time until those returns will be realized increases.

$$\mathrm{NPV} = \sum_{i=0}^{N} \frac{r_i - c_i}{(1 + \mathrm{DR})^i} \tag{1}$$

$$\mathrm{IRR} = \mathrm{DR} \quad \text{such that} \quad \sum_{i=0}^{N} \frac{r_i - c_i}{(1 + \mathrm{DR})^i} = 0 \tag{2}$$

$$\mathrm{CBR} = \frac{\sum_{i=0}^{N} c_i/(1 + \mathrm{DR})^i}{\sum_{i=0}^{N} r_i/(1 + \mathrm{DR})^i} \tag{3}$$

It is quite possible for *DR* to change with time, possibly reflecting expected increases in interest rates in the future. Equations (1)–(3) must be modified appropriately for time-varying discount rates.

The metrics in Equations (1)–(3) are interpreted as follows:

- The NPV reflects the amount one should be willing to pay now for benefits received in the future. These future benefits are discounted by the interest paid now to receive these later benefits.
- In contrast, IRR is the value of DR if NPV is zero. This metric enables comparing alternative investments by forcing the NPV of each investment to zero. Note that this assumes a fixed interest rate and reinvestment of intermediate returns at the internal rate of return.
- The CBR simply reflects the discounted cash outflows divided by the discounted cash inflows, or benefits.

### 17.2.2 Multiattribute Utility Models

Cost–benefit calculations become more complicated when benefits are not readily transformable to economic terms. Benefits such as safety, quality of life, and aesthetic value are very difficult to translate into strictly monetary values. Multiattribute utility models provide a means for dealing with situations involving mixtures of economic and noneconomic attributes.

Let cost attribute $i$ at time $j$ be denoted by $c_{ij}$, $i = 1, 2, \ldots, L$ and $j = 0, 1, \ldots, N$, and benefit attribute $i$ and time $j$ be denoted by $b_{ij}$, $i = 1, 2, \ldots, M$ and $j = 0, 1, \ldots, N$. The values of these costs and benefits are transformed to common utility scales using $u(c_{ij})$ and $u(b_{ij})$. These utility functions serve as inputs to the overall utility calculation at time $j$, as shown in Equation (4) (Keeney and Raiffa, 1976):

$$U(\underline{c}_j, \underline{b}_j) = U[u(c_{1j}), u(c_{2j}), \ldots, u(c_{Lj}), u(b_{1j}), u(b_{2j}), \ldots, u(b_{Mj})] \quad (4)$$

which provides the basis for an overall calculation across time using

$$U(\underline{C}, \underline{B}) = U[U(\underline{c}_1, \underline{b}_1), U(\underline{c}_2, \underline{b}_2), \ldots, U(\underline{c}_N, \underline{b}_N)] \quad (5)$$

Note that the time value of benefits depicted in Equations (1)–(3) is included in Equations (4) and (5) by dealing with the time value of costs and returns explicitly and separately from uncertainty.

An alternative approach involves assessing utility functions for discounted costs and benefits, possibly discounted as represented in Equations (1)–(3). With this approach, streams of costs and benefits are collapsed across time before the values are transformed to utility scales. The validity of this simpler approach depends on the extent to which people's preferences for discounted costs and benefits reflect their true preferences.

The mappings from $c_{ij}$ and $b_{ij}$ to $u(c_{ij})$ and $u(b_{ij})$, respectively, enable dealing with the subjectivity of preferences for noneconomic benefits. In other words, utility theory enables one to quantify and compare things that are often perceived as difficult to objectify. Unfortunately, models based on utility theory do not always reflect the ways in which human decision making actually works.

Subjective expected utility (SEU) theory reflects these human tendencies. Thus to the extent that one accepts that perceptions are reality, one needs to consider the SEU point of view when one makes expected-utility calculations. In fact, one should consider making these calculations using both objective and subjective probabilities to gain an understanding of the sensitivity of the results to perceptual differences.

Once one admits the subjective, one needs to address the issue of whose perceptions are considered. Most decisions involve multiple stakeholders, in other words, people who hold a stake in the outcome of a decision. It is therefore common for multiple stakeholders to influence a decision. Consequently, the cost–benefit calculation needs to take into account multiple sets of preferences. The result is a group utility model as shown in Equation (6) (Keeney and Raiffa, 1976; Kirkwood, 1979):

$$U = U[U_1(\underline{C}, \underline{B}), U_2(\underline{C}, \underline{B}), \ldots, U_K(\underline{C}, \underline{B})] \quad (6)$$

where $K$ is the number of stakeholders.

Formulation of such a model requires that two important issues be resolved. First, mappings from attributes to utilities must enable comparisons across stakeholders. In other words, one has to assume that $u = 0.8$, for example, implies the same value gained or lost for all stakeholders, although the mapping from attribute to utility may vary for each stakeholder. Thus all stakeholders may, for instance, have different needs or desires for safety and, hence, different utility functions. They also may have different time horizons within which they expect benefits. For example, stakeholders of different generations, some perhaps not yet born, have different time horizons within which they expect to receive benefits. However, once the mapping from attributes to utility is performed and utility metrics are determined, one has to assume that these metrics can be compared quantitatively.

The second important issue concerns the relative importance of stakeholders. Equation (6) implies that the overall utility attached to each stakeholder's utility can differ. For example, it is often the case that primary stakeholders' preferences receive more weight than the preferences of secondary stakeholders. The difficulty of this issue is obvious. Who decides? Is there a super-stakeholder, for instance? Do the groups of stakeholders, or their representatives, simply vote on who gets how much weight? Such a procedure has its own theoretical problems, which cannot be addressed here.

Beyond these two more theoretical issues, there are substantial practical issues associated with determining the functional forms of $u(c_{ij})$ and $u(b_{ij})$ and the parameters within these functional relationships. This also is true for the higher level forms represented by Equations (4)–(6). As the number of stakeholders ($K$), cost attributes ($L$), benefit attributes ($M$), and time periods ($N$) increase, these practical assessment problems can be quite daunting.

### 17.2.3 Option-Pricing Theory

Many investment decisions are not made all at once. Instead, initial investments are made to create the potential for possible future and usually larger investments involving much greater benefits than likely for the initial investments. For example, investments in R&D are often made to create the intellectual property and capabilities that will support or provide the opportunity to subsequently decide whether or not to invest in launching new products or services. These launch decisions are contingent on R&D reducing uncertainties and risks as well as further market information being gained in the interim between the R&D investment decision and possible launch decision. In this way, R&D investments amount to purchasing options to make future investments and earn subsequent returns. These options, of course, may or may not be exercised.

Amram and Kulatilaka (1999), Boer (1998, 1999), Lint and Pennings (1998), and Luehrman (1998) advocate using option-pricing theory to analyze investments involving such contingent downstream decisions. Option-pricing theory focuses on establishing the value of an option to make an investment decision, in an uncertain environment, at a later date. Equations (7)–(9) summarize the basic calculations as outlined by Luehrman.

One of two central elements in option pricing is the value of the returns from the contingent investment decision should one choose to make this decision, i.e., exercise the option. This value is represented by the NPV (net present value) quotient. As indicated in Equation (7), this quotient is formed as the ratio of the present asset value, that it, the traditional NPV of the free cash flow projected to result from exercising the option, and the present value (PV) of the investment required to acquire these assets, i.e., the option exercise price, $X$. As shown by Equation (8), the latter present value decreases as the risk-free rate of return, $r_f$, increases and/or the time until the option must be exercised or expire increases:

$$\text{NPV quotient} = \frac{\text{present asset value}}{\text{PV}(X)} \tag{7}$$

$$\text{PV}(X) = \frac{\text{present required investment value}}{(1 + r_f)^t} \tag{8}$$

The use of a risk-free rate is premised on the assumption that PV($X$) will be invested now and accrue interest at rate $r_f$ for $t$ time periods so that the exercise price, $X$, will be available when the option can be exercised. The risk-free rate is used because these funds are not at risk until investors decide to exercise the option. If they choose to let the option expire, they retain $X$ for other purposes.

The second central element in option pricing is the cumulative volatility expressed, as shown in Equation (9), as the product of the standard deviation of returns per period times the square root of the number of periods:

$$\text{Cumulative volatility} = \sigma\sqrt{t} \tag{9}$$

where $\sigma^2$ is the variance of returns per time period (expressed as a fraction) and $t$ equals the number of time periods. The inclusion of volatility in option-pricing models is central to realistically representing investments where it is seldom the case that future returns are certain.

It might be imagined that estimating $\sigma^2$ is quite difficult. However, it is common to use numbers in the 0.30 to 0.60 range (Luehrman, 1998) and sensitivity analysis to determine the extent to which particular option values depend on these estimates. It is important to keep in mind that the goal is making a well-informed investment decision, *not* making highly precise estimates of variables that only coarsely affect the decision at hand.

The values of the NPV quotient [Equation (7)] and cumulative volatility [Equation (9)] are used to ascertain Black–Scholes values. These values are computed from the partial differential equation that is the Black–Scholes option-pricing model (Black and Scholes, 1973). For certain classes of option-pricing problems (e.g., constant $\sigma^2$ and fixed $t$), Black–Scholes numbers have been precomputed and can be found tabulated in various publications (e.g., Amram and Kulatilaka,1999; Luehrman, 1998).

The Black–Scholes values, expressed as percentages, increase with increasing NPV quotient and increasing cumulative volatility. This percentage is multiplied times the present asset value in Equation (7) to determine the value of the option. Thus, in general, the value of an option to later decide on an investment increases with $r_f$, $\sigma^2$, and $t$. In particular, in the presence of high volatility and high risk-free returns, the longer one can wait to decide, the more valuable the option. However, the present asset value in Equation (7) decreases with time. Thus, depending on the specific cash flow and investment projections as well as the parameters chosen, the option value may increase, decrease, or possibly increase to a maximum and then decrease. Sensitivity analysis is a good way to gain an understanding of this range of possibilities.

The resulting option value is totally premised on the assumption that waiting does not preempt deciding later. In other words, the assumption is that the decision to exercise an option cannot be preempted by somebody else deciding earlier. In typical situations where other actors (e.g., competitors) can affect possible returns, it is common to represent their impact in terms of changes of projected cash flows (Amram and Kulatilaka, 1999). In many cases, competitors acting first will decrease potential cash flows that will decrease the option value. It is often possible to construct alternative competitive scenarios and determine an optimal exercise date.

A central attraction of this model is the explicit recognition that the purpose of an investment now (i.e., purchasing an option) is to ensure the option to make a subsequent investment later (i.e., exercise the option). For example, one invests in creating new technologies for the option of later incorporating these technologies in product and service lines. The significance of the contingent nature of this decision makes an option-pricing model a much better fit than a traditional discounted cash flow model.

However, not all long-term investment decisions have substantial contingent elements. For example, one may invest in training and development to later have the option of selecting among talented managers for elevation to executive positions. There are minimal investments associated with exercising such options; almost all of the investment occurs up front. Thus option-pricing models are not useful for such decisions.

### 17.2.4 Knowledge Capital Approach

Tangible assets and financial assets usually yield returns that are important elements of a company's overall earnings. It is often the case, however, that earnings far exceed what might be expected from these "hard" assets. For example, companies in the software, biotechnology, and pharmaceutical industries typically have much higher earnings than companies with similar hard assets in the aerospace, appliance, and automobile industries, to name just a few. It can be argued that these higher earnings are due, for example, to greater knowledge capital among software companies. However, since knowledge capital does not appear on financial statements, it is very difficult to identify and, better yet, to project knowledge earnings.

A recent article by Mintz (1998) summarizes a method developed by Baruch Lev for estimating knowledge capital and earnings. This article in *CFO* drew sufficient attention to be discussed in The *Economist* (1999) and reviewed by Strassman (1999). In general, both reviews applauded the progress represented by Mintz's article but also noted the shortcomings of his proposed metrics.

The key, Mintz and Lev argue, is to partition earnings into knowledge earnings and hard asset earnings. Equation (10) accomplishes this by first projecting normalized annual earnings from an average of three past years and estimates for three future years using readily available information. Earnings from tangible and financial assets are calculated from reported asset values using industry averages of 7 and 4.5 percent for tangible and financial assets, respectively. Knowledge capital is then estimated by dividing knowledge earnings by a knowledge capital discount rate, as shown in Equation (11). Based on an analysis of several knowledge-intensive industries, Mintz and Lev use 10.5 percent for this discount rate.

$$\begin{aligned}\text{Knowledge earnings} &= \text{normalized annual earnings} \\ &\quad - \text{earnings from tangible assets} \\ &\quad - \text{earnings from financial assets} \end{aligned} \tag{10}$$

$$\text{Knowledge capital} = \frac{\text{knowledge earnings}}{\text{knowledge capital discount rate}} \tag{11}$$

Using this approach to calculate knowledge capital, Mintz compares 20 pharmaceutical companies to 27 chemical companies. He determines, for example, knowledge capital–book value ratios of 2.45 for pharmaceutical companies and 1.42 for chemical companies. Similarly, the market value–book value ratios are 8.85 for pharmaceutical companies and 3.53 for chemical companies. Considering this correlation between knowledge capital and market value, Strassman (1999) points out that Mintz's estimates do not fully explain the full excess of market values over book values.

The key issue within this overall approach is being able to partition earnings. While earnings from financial assets should be readily identifiable, the distinction between tangible and knowledge assets is problematic. Further, using industry average return rates to attribute earnings to tangible assets does not allow for the significant possibility of tangible assets having little or no earnings potential. Finally, of course, simply attributing all earnings "left over" to knowledge assets amounts to giving knowledge assets credit for everything that cannot be explained by traditional financial methods.

Nevertheless, the knowledge capital construct appears to have potential application to investments involving, for example, R&D or training and development. The purpose of these two types of investments seems to obviously be that of increasing knowledge capital. Further, companies that make investments for this purpose do seem to create more knowledge capital. The key for cost–benefit analyses is being able to project investment returns in terms of knowledge capital and, in turn, project earnings and separate these earnings into knowledge earnings and hard earnings. Further, one needs to be able to do this for specific investment opportunities, not just the company as a whole.

### 17.2.5 Comparison of Frameworks

Table 17.1 provides a comparison of the four frameworks just reviewed. It is important to note that this assessment is not really an apples-to-apples comparison. Multiattribute utility theory provides much more of a general framework than the other three approaches, which emphasize financial metrics. Nevertheless, these four approaches represent the dominant alternatives.

Traditional economic analyses are clearly the most narrow. However, in situations where they apply, these analyses are powerful and useful. Most of the investment situations addressed in this chapter do not fit these narrow characteristics. For example, if R&D investments in the human effectiveness aspects of HSI are viewed within a traditional framework, with typical discount rates, no one would ever invest anything in such R&D. But people do make such investments and, thus, there must be more to it than just NPV, IRR, and CBR. [In fact, Cooper et al. (1998b) have found that companies relying solely on financial metrics for R&D investment decisions tend to be the poorest performers of R&D in terms of subsequent market success.]

One view is that R&D reduces uncertainty and buys time before committing very substantial resources to productization, process development, etc. Option-pricing theory seems to be a natural extension of traditional methods to enable handling these complications. As noted earlier, several authors have advocated this approach for analyses of R&D investments.

The knowledge capital approach provides another, less mathematical way of capturing the impacts of R&D investments in human effectiveness aspects of HSI. The difficulty of this approach, which is probably inherent to its origins in accounting and finance, is that it does not address the potential impacts of alternative investments. Instead, it serves to report the overall enterprise score after the game.

Multiattribute utility models can, in principle, address the full range of complications and complexity discussed thus far. Admittedly, the ability to create a rigorous multi-attribute utility model depends on the availability of substantial amounts of information regarding stakeholders' preference spaces, probability density functions, etc. However, in

**TABLE 17.1 Comparison of Cost–Benefit Frameworks**

| Issue/Framework | Traditional Economic Analysis | Multiattribute Utility Models | Option-Pricing Theory | Knowledge Capital Approach |
|---|---|---|---|---|
| Representation of uncertainties | Focuses on expected revenues and costs without consideration of variances. | Probabilistic uncertainties and stakeholders' preferences regarding uncertainties are central to models. | Volatility of returns is a central construct within this model. | Focuses on actual and expected earnings without consideration of variances. |
| Intangible vs. tangible outcomes | All outcomes must be converted to monetary units. | Preferences regarding intangible outcomes can be incorporated. | All outcomes must be converted to monetary units. | All outcomes must be converted to monetary units. |
| Multiple stakeholders in costs/benefits | One-dimensional nature of costs and benefits implies one stakeholder. | Formulations for multiple stakeholders are available and limitations are understood. | One-dimensional nature of costs and benefits implies one stakeholder. | One-dimensional nature of costs and benefits implies one stakeholder. |
| Assessing vs. projecting costs/benefits | Depends on abilities to project monetary costs and benefits. | Depends on abilities to project attributes of utility functions. | Depends on abilities to project monetary costs and benefits. | Difficult to project impact of particular investments. |

the absence of such information, a much more qualitative approach can be quite useful, as is discussed later in this chapter.

The value of the multiattribute utility approach also depends on being able to compare overall utilities of alternative investments, which in turn depends on being able to compare different stakeholders' utilities of the alternatives. This ability to transform a complex, multidimensional comparison into a scalar comparison is laden with assumptions. The saving grace of the approach, in this regard, is that it makes these assumptions quite explicit and, hence, open to testing. This does not, of course, guarantee that they will be tested.

Expected-utility calculations serve to show how one alternative is better than another, rather than providing absolute scores. Thus, differences of expected utilities among alternatives are usually more interesting than the absolute numbers. In fact, the dialog among stakeholders that is often associated with trying to understand the sources of expected-utility differences can provide crucial insights into the true nature of differences among alternatives.

Overall, one must conclude that multiattribute utility models provide the most generalizable approach. This is supported by the fact that multiattribute models can incorporate metrics such as NPV, option value, and knowledge capital as attributes within the overall model; indeed, the special case of one stakeholder, linear utility functions, and NPV as the sole attribute is equivalent to the traditional financial analysis. Different stakeholders' preferences for these metrics can then be assessed and appropriate weightings determined. Thus, use of multiattribute models does not preclude also taking advantage of the other approaches; the four approaches, therefore, can be viewed as complementary rather than competing. For these reasons, the multiattribute approach is carried forward in the remainder of this chapter.

## 17.3 COST–BENEFIT METHODOLOGY

Cost–benefit analysis should always be pursued in the context of particular decisions to be addressed. A valuable construct for facilitating an understanding of the context of an analysis is the value chain from investments to returns. More specifically, it is quite helpful to consider the value chain from investments (or costs), to products, to benefits, to stakeholders, to utility of benefits, to willingness to pay, and finally to returns on investments. This value chain can be depicted as follows:

| | | |
|---|---|---|
| Investments (costs) | to | Resulting products over time |
| Products over time | to | Benefits of products over time |
| Benefits over time | to | Range of stakeholders in benefits |
| Range of stakeholders | to | Utility of benefits to each stakeholder |
| Utility to stakeholders | to | Willingness to pay for utility gained |
| Willingness to pay | to | Returns to investors |

The process starts with investments, which result or will result in particular products over time. Products need not be end products; they might be knowledge, skills, or technologies. These products yield benefits, also over time. A variety of people—or stakeholders—have a stake in these benefits. These benefits provide some level of utility to each stakeholder. The utility perceived or anticipated by stakeholders affects their willingness to pay for these benefits. Their willingness to pay affects their "purchase" behaviors, which result in returns for investors.

The central methodological question concerns how one can predict the inputs and outputs of each element of this value chain. This question is addressed elsewhere in some detail for R&D management (Rouse et al., 1997; Rouse and Boff, 1997) and for human effectiveness (Rouse and Boff, 1997). Briefly, a variety of models have been developed for addressing this need for prediction. These models are very interesting and offer much potential. However, they suffer from a central shortcoming. With few exceptions, there is an almost overwhelming lack of data for estimating model parameters as well as a frequent lack of adequate input data. Use of data from baselines can help, but the validity of these baselines depends on new systems and products being very much like their predecessors. Overall, the paucity of data dictates development of a more qualitative methodology whose usefulness is not totally determined by availability of hard data. The remainder of this section outlines such a methodology.

As indicated in the earlier comparison of four frameworks for addressing cost–benefit analysis, the most broadly applicable of these alternatives are multiattribute utility models. The remainder of this section describes the following seven-step methodology:

- Step 1: Identify stakeholders in alternative investments.
- Step 2: Define benefits and costs of alternatives in terms of attributes.
- Step 3: Determine utility functions for attributes (benefits and costs).
- Step 4: Decide how utility functions should be combined across stakeholders.
- Step 5: Assess parameters within utility models.
- Step 6: Forecast levels of attributes (benefits and costs).
- Step 7: Calculate expected utility of alternative investments.

It is important to note that this methodology is by no means novel and builds upon works by many others related to multiattribute analysis (e.g., Keeney and Raiffa, 1976; Sage, 1977; Hammond et al., 1998; Matheson and Matheson, 1998; Sage and Armstrong, 2000).

***Step 1: Identify Stakeholders*** The first step involves identifying the stakeholders who are of concern relative to the investments being entertained. Usually this includes all the people in the value chain summarized earlier. This might include, for example, those who will provide the resources that will enable a solution, those who will create the solution, those who will implement the solution, and those who will benefit from the solution.

***Step 2: Define Benefit and Cost Attributes*** The next step involves defining the benefits and costs involved from the perspective of each stakeholder. These benefits and costs define the attributes of interest to the stakeholders. Usually, a hierarchy of benefits and costs emerges, with more abstract concepts at the top, e.g., viability, acceptability, and validity (Rouse, 1991), and concrete measurable attributes at the bottom.

***Step 3: Determine Stakeholders' Utility Functions*** The values that stakeholders attach to these attributes are defined by stakeholders' utility functions. The utility functions enable mapping disparate benefits and costs to a common scale. A variety of techniques are available for assessing utility functions (Keeney and Raiffa, 1976).

***Step 4: Determine Utility Functions Across Stakeholders*** Next, one determines how utility functions should be combined across stakeholders. At the very least, this involves assigning relative weights to different stakeholders' utilities. Other considerations such as desires for parity can make the ways in which utilities are combined more complicated. For example, Equation (6) may require interaction terms to assure all stakeholders some utility.

***Step 5: Assess Parameters of Utility Functions*** The next step focuses on assessing parameters within the utility models. For example, utility functions that include diminishing or accelerating increments of utility for each increment of benefit or cost involve rate parameters that must be estimated. As another instance, estimates of the weights for multistakeholder utility functions have to be estimated. Fortunately, there are a variety of standard methods for making such estimates.

***Step 6: Forecast Levels of Attributes*** With the cost–benefit model fully defined, one must next forecast levels of attributes or, in other words, benefits and costs. Thus, for each alternative investment, one must forecast the stream of benefits and costs that will result if this investment is made. Quite often, these forecasts involve probability density functions rather than point forecasts. Utility theory models can easily incorporate the impact of such uncertainties on stakeholders' risk aversions. On the other hand, information on probability density functions may not be available or may be prohibitively expensive. In these situations, beliefs of stakeholders and subject matter experts can be employed, perhaps coupled with sensitivity analysis (see step 7) to determine where additional data collection may be warranted.

***Step 7: Calculate Expected Utilities*** The final step involves calculating the expected utility of each alternative investment. These calculations are performed using specific forms of Equations (4)–(6). This step also involves using sensitivity analysis to assess, for example, the extent to which the rank ordering of alternatives, by overall utility, changes as parameters and attribute levels of the model are varied.

***Use of the Methodology*** Some elements of the cost–benefit methodology just outlined are more difficult than others. The overall calculations are quite straightforward. The validity of the resulting numbers depends, of course, on stakeholders and attributes having been identified appropriately. It further depends on the quality of the inputs to the calculations.

These inputs include estimates of model parameters and forecasts of attribute levels. As indicated earlier, the quality of these estimates is often compromised by lack of available data. Perhaps the most difficult data collection problems relate to situations where the impacts of investments are both uncertain and very much delayed. In such situations, it is not clear which data should be collected and when they should be collected.

A recurring question concerns the importance that should be assigned to differences in expected-utility results. If alternative $A$ yields $U(A) = 0.648$ and alternative $B$ yields

$U(B) = 0.553$, is $A$ really that much better than $B$? In fact, are either utilities sufficiently great to justify an investment?

These questions are best addressed by considering past investments. For successful past investments, what would their expected utilities have been at the time of the investment decisions? Similarly, for unsuccessful past investments, what were their expected utilities at the time? Such comparisons often yield substantial insights.

Of course, the issue is not always A versus $B$. Quite often the primary question concerns which alternatives belong in the portfolio of investments and which do not. Portfolio management is a fairly well-developed aspect of new product development (Cooper et al., 1998a; Gill et al., 1996). Well-known and recent books on R&D/technology strategy pay significant attention to portfolio selection and management (Roussel et al., 1991; Matheson and Matheson, 1998; Boer, 1999; Allen, 2000). In fact, the conceptual underpinnings of option-pricing theory are based on notions of market portfolios (Amram and Kulatilaka, 1999).

Most portfolio management methods rely on some scoring or ranking mechanism to decide which investments will be included in the portfolio. Expected utility is a quite reasonable approach to creating such scores or ranks. This is particularly useful if sensitivity analysis has been used to interactively explore the basis and validity of differences among alternatives.

A more sophisticated view of portfolio management considers interactions among alternatives in the sense that synergies between two alternatives may make both of them more attractive (Boer, 1999; Allen, 2000). Also correlated risks between two alternatives may make both of them less attractive. A good portfolio has an appropriate balance of synergies and risks.

In principle, at least, the notions of portfolio synergy and risk can be handled within multiattribute utility models. This can be addressed by adding attributes that are characteristics of multiple rather than individual alternatives. In fact, such additional attributes might be used to characterize the whole portfolio. An important limitation of this approach is the likely significant increase in the complexity of the overall problem formulation. Indeed, this is an issue in general when multiattribute utility models are elaborated to better represent problem complexities.

Beyond these technical issues, it is useful to consider how this cost–benefit methodology should affect decision making. To a very great extent, the purpose of this methodology is to get the right people to have the right types of discussions and debates on the right issues at the right time. If this happens, the value of people's insights from exploring the multiattribute model usually far outweighs the importance of any particular numbers.

The practical implications of this conclusion are quite simple. Very often, decision making happens within working groups who view computer-generated, large-screen displays of the investment problem formulation and results as they emerge. Such groups perform sensitivity analyses to determine the critical assumptions or attribute values that are causing some alternatives to be more highly rated or ranked than others. They use "What if...?" analyses to explore new alternatives, especially hybrid alternatives.

This approach to investment decision making helps to substantially decrease the impact of limited data being available. Groups quickly determine which elements of the myriad of unknowns really matter—where more data are needed and where more data, regardless of results, would not affect decisions. A robust problem formulation that can be manipulated, redesigned, and tested for sanity provides a good way for decision-making groups to reach defensible conclusions with some level of confidence and comfort.

## 17.4 THREE EXAMPLES

Human effectiveness concerns enhancing people's direct performance (aiding), improving their potential to perform (training), and ensuring their availability to perform (health and safety). These are central issues in HSI. Investments in human effectiveness also have the potential of increasing returns on other investments by, for example, enabling people to take full advantage of new technologies.

Three examples of aiding, training, and health and safety investments are discussed in this section—VCATS (aiding), DMT (training), and PTOX (health and safety). These examples focus on enhancing human effectiveness and HSI in military systems, particularly Air Force systems. The applicability of these technologies, and the relevance of the following analysis of the impacts of these technologies, to other military services and to non-military problems should also be readily apparent.

### 17.4.1 Visually Coupled Targeting and Acquisition System

The Visually Coupled Targeting and Acquisition System (VCATS) provides aiding to military aircraft pilots. The VCATS includes a helmet-mounted tracker and display (HMT/D), associated signal processing sensor/transducer hardware, interchangeable panoramic night vision goggle with head-up display (PNVG-HUD), and extensive upgrades to the aircraft's operational flight program software (Rastikis, 1998). The VCATS enables the pilot to cue and be cued by on-board and off-board systems, sensors, and weapons as well as be spatially and temporally coupled with the control processes implemented with the HMT/D and PNVG-HUD. The system is particularly effective in helping pilots to cue weapons and sensors to targets, maintain "ownship" formation situation awareness, and avoid threats via provision of a real-time, three-dimensional portrayal of the pilots' tactical and global battlefield status. In general, the VCATS enables pilots to acquire targets and threats faster. This results in improvements in terms of (1) how far, (2) how quickly, and (3) how long—for both initial contacts and countermeasures.

To a great extent, the case for advanced development has already been made for the VCATS and current support is substantial. However, the transition from advanced development to production involves ensuring that the options created by the VCATS and validated by combat pilots are exercised. The case has also been argued for ongoing investments in basic research and exploratory development to ensure that the VCATS has future technology options, particularly for migration to multirole fighter aircraft. The maturity of the program should help in making this case in terms of benefits already demonstrated. However, in the current budget climate, there is also substantial risk that VCATS research may be viewed as essentially "done." This raises the potential for negative decisions regarding further investments.

### 17.4.2 Distributed Mission Training

Distributed mission training (DMT) involves aircraft, virtual simulators, and constructive models that, collectively, provide opportunities for military pilots to gain experiences deemed important to their performance proficiency relative to anticipated mission requirements (Andrews, 2000). The desired training experiences are determined from competencies identified as needed to fulfill mission requirements. These competency requirements are translated to training requirements stated in terms of types and durations of experiences deemed sufficient to gain competency.

The case to be made for DMT involves investments to address research issues and technology upgrades of near-term capabilities. The primary options-oriented argument is that investments in R&D in DMT will create contingent possibilities for cost savings in training due to reduced use of actual aircraft. More specifically, DMT options, if exercised, will provide cash flows of savings that justify the investments needed to field this family of technologies.

A much more subtle options-oriented argument concerns the training experiences provided by DMT that could not otherwise be obtained. Clearly, the opportunity to have relevant training experiences must be better than not having these experiences. The option, therefore, relates to proficiency versus possible lack of proficiency.

As straightforward as this may seem, it quickly encounters the difficulty of projecting mission impacts—and the value of these impacts—of not having proficient personnel. One possible approach to quantifying these benefits is to project the costs of using real aircraft to gain the desired proficiences. While these costs are likely to be prohibitive, and thus never would be seriously considered, they nevertheless characterize the benefits of DMT.

### 17.4.3 Predictive Toxicology

Predictive toxicology (PTOX) is concerned with projecting the impacts on humans from exposure to operational chemicals (individual and mixtures). The impact can be characterized in terms of the possibility of performance decrement and consequent loss of force effectiveness, possible military and civilian casualties, and potential long-term health impacts. Also of concern are the impacts of countermeasures relative to sustaining immediate performance and minimizing long-term health impacts [Office of Science and Technology Policy (OSTP), 1998].

The case to be made involves investment in basic research and exploratory development programs, with longer term investment in an advanced development program to create deployable predictive toxicology capabilities. The requisite R&D involve developing and evaluating models for predicting performance and health impacts of operational chemicals. Advanced development will focus on field sensing and prediction, termed deployment toxicology. The nature of the necessary models is strongly affected by the real-time requirements imposed by deployment.

### 17.4.4 Applying the Methodology

The remainder of this section primarily addresses steps 1 to 4 of the cost–benefit methodology in the context of these three examples related to human effectiveness aspects of HSI. These steps constitute the "framing" steps of the methodology, rather than the "calculation" steps. Appropriate framing of cost–benefit analyses is critical to subsequent calculations being meaningful and useful.

***Step 1: Identify Stakeholders*** This step involves identifying people, usually types of people, and organizations that have a stake in costs and benefits. All three of the examples involve three classes of stakeholders: warfighters, developers, and the public. A key issue concerns the relative importance of these three types of stakeholders. Some would argue that warfighter preferences dominate decisions. Others recognize the strong role that developers and their constituencies play in procurement decisions. Yet another argument is that the dominating factor is value to the public, with the other stakeholders being secondary in importance.

Warfighters as stakeholders include military personnel in general, especially for PTOX. Warfighters of particular importance include aircraft pilots, personnel who support flight operations, and military commanders. Developers as stakeholders include companies and their constituencies, e.g., stockholders, employees, and communities. The public's interests are represented by several agents, including Congress, the executive functions within the military services, and the military procurement establishment. Pilots and other military personnel are users of the technologies of interest, developers are the providers, and the public's agents are the customers for these technologies. There are obvious trade-offs across the interests of users, providers, and customers.

***Step 2: Define Benefit and Cost Attributes*** Benefits and costs tend to fall in general classes. Examples of benefits for military organizations and contractors include the following:

| | |
|---|---|
| Enhanced impact | Increased lethality, survivability, and availability |
| Enhanced operability | Decreased response time and increased throughput |
| Enhanced design | New techniques and larger pool of experienced people |
| Increased opportunities | New tactics and countermeasures |

Example cost attributes applicable to military procurement include the following:

| | |
|---|---|
| Investment costs | Capital investments and R&D costs |
| Recurring costs | Operating and General and Administrative (G&A) costs |
| Time costs | Time from development to fielding to competent use |
| Opportunity costs | Other costs/benefits foregone |

These general classes of benefits and costs can be translated into specific benefit and cost attributes for the three classes of stakeholders in VCATS, DMT, and PTOX. Benefits for warfighters (users) include enhanced performance (e.g., response time), confidence in performance, and health and safety in varying combinations for the three examples. Costs for these stakeholders include learning time and changing their ways of doing things to ensure compatibility between new and legacy technologies.

Benefits for companies and their constituencies (providers) include R&D funds received, subsequent intellectual property created, and competitive advantages that result. Also important are jobs and economic impacts in the community. Direct costs include bid and proposal costs as well as opportunity costs. Less direct costs include, for instance, economic development resources and incentives provided to the companies by their communities.

The primary benefit sought by the public's agents (customer) is mission performance per dollar. It can easily be argued for all three examples that mission performance is increased. Unfortunately, it is difficult to attach a value to this increase. For example, what is the value of being able to generate 5 percent more sorties per time period? The answer depends on whether more sorties are needed.

Few would argue with the importance of successfully meeting mission requirements. However, if the types of innovations represented by these examples enable exceeding mission requirements, what are such increases worth? This is a politically sensitive question. If better performance is of substantive value, why was not this level of performance specified in the original requirements?

A good way to avoid this difficulty is to take mission requirements as a given and determine how much money could be saved in meeting these requirements by adopting the technologies in question. For example, could requirements be met with fewer aircraft, pilots, and support personnel? As shown in Table 17.2, the cost savings due to these decreases can be viewed as benefits of the technologies. It also might be possible for the VCATS, DMT, or PTOX to enable meeting mission requirements with less capable systems, rather than just fewer systems. This possibility provides substantial opportunities for increased benefits due to these technologies.

Note that this philosophy amounts to trying to provide a given level of defense for the least investment. Another approach might be to attempt to provide the most defense per investment dollar. However, this immediately begs the question of how much defense is enough. Unlike the business world where value is defined by the marketplace and, hence, can provide a basis for optimization—see Nevins and Winner (1999) for a good example—there is no widely agreed-upon approach to measuring military value and optimizing accordingly.

The rationale for the benefits indicated in Table 17.2 for each of the three examples includes the following:

- The VCATS enables pilots to compete with threats, increase the number of wins versus losses, and counter threats (e.g., missiles) in ways that they could not do otherwise. Consequently, it must be possible to meet *fixed* mission requirements with fewer aircraft and associated infrastructure. These benefits can be translated into financial returns in terms of cost avoidance.
- Distributed mission training provides opportunities to practice behaviors that would not otherwise be practiced, for the most part due to the costs of practice. This decreases the probability of not performing acceptably given inadequate training. Also, DMT provides training experiences that would not otherwise be possible. For example, in the DMT environment, pilot "kills" actually disappear. In contrast, field exercises often "reuse" kills because of the costs of getting adversaries into the exercise in the first place.
- Predictive toxicology enables larger proportions of deployed forces to be fully functional and less dependent on medical surveillance or medication and earlier intervention before the onset of problems. In principle, this should enable reducing the size of deployed force, which is critical for increasingly likely expeditionary military missions (Fuchs et al., 1997).
- In addition, PTOX provides cost avoidance due to downstream health impacts. The ability to predict the "body burden" of toxicity during deployment should enable removing personnel from risk once the burden is approaching predetermined limits. These capabilities are likely to also be very important for nonmilitary operations such as disaster clean-up.

It is *not* essential that the savings indicated in Table 17.2 actually occur. For example, it may be that the number of aircraft is not decreased, perhaps due to factors far beyond the

**TABLE 17.2 Public Benefits and Costs for Three Examples**

| | VCATS | DMT | PTOX |
|---|---|---|---|
| Benefits | Fewer aircraft and associated personnel to meet mission requirements due to better performance and fewer aircraft losses | Fewer aircraft and associated personnel to meet mission requirements due to better performance, fewer aircraft losses, and fewer aircraft for training | Fewer personnel to meet mission requirements and decreased medical costs due to fewer people affected by toxic materials, fewer people lost to toxic effects, fewer people to care for people affected, and decreased downstream medical costs |
| Costs | Initial investment (option price) for proposed R&D costs and, later, contingent investment (exercise price) for subsequent fielding of technology | Initial investment (option price) for proposed R&D costs and, later, contingent investment (exercise price) for subsequent fielding of technology | Initial investment (option price) for proposed R&D costs and, later, contingent investment (exercise price) for subsequent fielding of technology |

scope of these analyses. However, one can nevertheless attribute to these technologies the benefits of having provided opportunities to meet mission requirements in less costly manners. Technologies that provide such opportunities are valuable; the extent of this value is the extent of the opportunities for savings.

This argument puts all three examples on common ground. The benefits of all alternative technologies can be expressed as reduced costs to meet requirements. From an options-pricing perspective, these savings can be viewed as free cash flow returned on investments in these technologies. The "option price" is the R&D costs. The "exercise price" is the subsequent costs of fielding the technologies. Thus, assuming costs savings can be projected (albeit with substantial volatility), the options value of investing in these technologies can be calculated.

***Step 3: Determine Stakeholders' Utility Functions*** Different stakeholders' preferences over the benefit and cost attributes will vary substantially with specific situations. However, there is a small family of functional relationships that captures most, if not all, expressed preferences (Keeney and Raiffa, 1976). Thus while context-specific tailoring is needed, it can be performed within a prescribed (and preprogrammed) set of functions, both within and across stakeholders. Similarly, alternative parameter choices can be prescribed in terms of choices of weightings.

An important aspect of cost–benefit analyses, as advocated in this chapter, is the likely nonlinear nature of utility functions. In particular, diminishing returns and aspiration levels tend to be central to stakeholders' "preference spaces." In other words, while linear functions imply that incremental increases (or decreases) of attributes always yield the same incremental changes in utility, nonlinear functions lead to shifting preferences as attributes increase (or decrease). Figure 17.1 portrays a range of example utility functions; *a* and *d* illustrate linear relationships; *b* and *e* show accelerating relationships; and *c* and *f* show diminishing relationships.

To illustrate how these types of relationships can be employed to represent the preferences of users, providers, and customers, the general forms of each type of

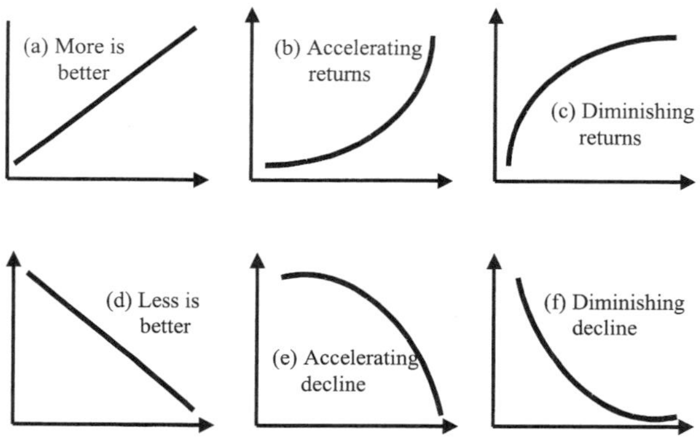


**Figure 17.1** Example utility functions: (*a*) More is better, (*b*) accelerating returns, (*c*) diminishing returns, (*d*) less is better, (*e*) accelerating decline, and (*f*) diminishing decline.

stakeholder's utility function are shown in Equations (12)–(14):

$$U_{\text{user}} = U[u(\text{performance}), u(\text{confidence}), u(\text{cost of change})] \tag{12}$$

$$U_{\text{provider}} = U[u(\text{resources}), u(\text{advantage}), u(\text{cost of pursuit})] \tag{13}$$

$$U_{\text{customer}} = u(\text{option value}) \tag{14}$$

where, as noted earlier, users are primarily concerned about impacts of investments on their performance, their confidence in their performance, and the costs of changing their ways of performing; providers are concerned with the investment resources supplied to develop the technologies in question, the competitive advantages created by the intellectual property created, and the costs of pursuing the investment opportunities; and, finally, customers are focused on the financial attractiveness of the investments as reflected in the option values of the alternatives, which are based on projected cash flows (i.e., costs savings), volatility of cash flows, magnitudes of investments required, and time periods until returns are realized.

Considering the elements of Equations (12)–(14), the appropriate functional forms from Figure 17.1 are likely to be as follows:

- *u*(performance) is an accelerating returns function (Fig. 17.1*b*):
  - The VCATS is least concave since relatively modest performance improvements are of substantial utility.
  - Distributed mission training is moderately concave since training on otherwise untrained tasks must produce substantial improvements to yield high utility.
  - Predictive toxicology is most concave since major decreases in performance risk are needed to ensure high utility increases of personnel availability.
- *u*(confidence) is a linear function (Fig. 17.1*a*) since greater confidence is always better but there are unlikely to be significant thresholds.
- *u*(cost of change) is an accelerating decline function (Fig. 17.1*f*) since low to moderate costs are easily sustained while larger costs present difficulties.
- *u*(resources) is an accelerating returns function (Fig. 17.1*b*) since moderate to large resources are needed to make opportunities attractive.
- *u*(advantage) is a linear function (Fig. 17.1*a*) since greater advantage is always better but there are unlikely to be significant thresholds.
- *u*(cost of pursuit) is an accelerating decline (Fig. 17.1*f*) since low to moderate costs are easily sustained while larger costs present difficulties.
- *u*(option value) is a linear function (Fig. 17.1*a*) since customers will inherently gain the expected value across a large number of investments.

It is important to note the importance of this last assumption. If customers'—that is, the public's—utility function were not linear, it would be necessary to entertain assessing the specific form of their function. Unlike users and providers, the public is not so easily identified and interviewed.

With the identification of the stakeholders (step 1) and framing of the cost–benefit attributes (step 2), the process of determining the form of stakeholders' utility functions (step 3) can draw upon considerable standard "machinery" of decision analysis. The specific versions of the functional forms discussed above are likely to vary with the VCATS, DMT, and PTOX. However, the overall formulation chosen is quite general.

***Step 4: Determine Utility Functions Across Stakeholders*** Another important aspect of the utility functions is their typical lack of alignment across stakeholders. Specifically, either different stakeholders care about different things or possibly they care about the same things in different ways. For example, customers may be very price sensitive while users, who seldom pay prices themselves, are usually much more concerned with impacts on their job performance

For the types of investment problems considered in this chapter, preferences typically differ across time horizons and across people with vested interests in different investment opportunities. Thus far in the formulation of the three examples, the stakeholders do not have attributes in common. However, they are nevertheless likely to have competing preferences since, for example, the alternative providing the greatest performance impact may not have the largest option value.

Differing preferences across stakeholders are often driving forces in pursuing cost–benefit analyses. These differing preferences can be aggregated, and traded off, by formulating a composite utility function such as

$$U = U[U_{\text{user}},\ U_{\text{provider}},\ U_{\text{customer}}] \tag{15}$$

Often Equation (15) will be linear in form with weights assigned to component utility functions to reflect the relative importance of stakeholders. Slightly more complicated are multilinear forms that include products of component functions, e.g., $U_{\text{user}} \times U_{\text{customer}}$. Multilinear formulations tend to ensure that all stakeholders gain nonzero utility because, otherwise, zero in either term in a product yields zero overall.

Considering trade-offs across stakeholders, it is important to note that the formulation of the analysis can often be usefully expanded to include a broader set of stakeholders. These additional stakeholders may include other entities that will benefit by advances of the technologies in question, although they may have little or no stake in the immediate application for the technology. It is also quite possible that stakeholders such as "the public" have multiple interests, e.g., military effectiveness and public safety from toxic risks.

Broadening the analysis in this way is likely to have differing impacts on the assessment for the three examples due to the natures of the technologies and issues being pursued. The three examples differ in this regard in the following ways:

- The VCATS addresses a rather esoteric set of issues from the public's perspective.
- Distributed mission training addresses an issue with broad general support from the public but narrower specific constituencies.
- Predictive toxicology addresses strong cross-cutting health and safety issues of substantial concern to the public.

These differences suggest that PTOX would gain a larger $\Delta U$ than DMT, and DMT would in turn gain a larger $\Delta U$ than the VCATS, by broadening the number of stakeholders and issues. Quite simply, the "spin-off" benefits of PTOX are likely to be perceived as much greater by a larger number of stakeholders.

However, if the formulation is further broadened to consider the likelihood that the desired technologies will emerge elsewhere if investments are not made in these efforts, the $\Delta U$ impacts will likely be the opposite. The PTOX research and development are being pursued by several agencies. Distributed mission training has broad applicability for both military and nonmilitary applications and consequently is being pursued by other parties. The VCATS, in contrast, is highly specialized and is unlikely to emerge from other sources.

These two possibilities for broadening the formulation, in terms of stakeholders and issues, clearly illustrate the substantial impact of the way in which cost–benefit assessments are framed. If the framing is too focused, important spin-off benefits may not be included. On the other hand, framing the analysis too broadly may raise issues that are difficult to quantify, even roughly, and includes stakeholders whose preferences are difficult to assess. Of course, many modeling efforts face such difficulties (Sage and Rouse, 1999).

***Steps 5–7: Calculation of Overall Cost–Benefit*** The remaining steps of the cost–benefit methodology involve assessing parameters of utility functions, forecasting levels of attributes, and calculating expected utilities. Performing these steps obviously depends on having data on stakeholders' preferences and projected/targeted attribute levels. Discussion of such data is well beyond the scope of this chapter and, in light of the nature of the examples, it would be difficult to publish the requisite data.

The needed data can, in many instances, be quite difficult to compile. It can be particularly difficult to relate returns on human effectiveness investments to organizational impacts. Relationships between human and organizational performance are needed. These relationships should answer the following types of questions:

- How do improvements in human performance (e.g., via aiding) translate to increased organizational impacts? Specifically, how does a 2-second improvement in pilot response time due to the VCATS affect mission performance?
- How do improvements of human potential to perform (e.g., via training) translate to actual performance and consequent increased organizational impacts? Specifically, how does increased practice via DMT impact subsequent performance and, in turn, translate to improved mission performance?
- How do improvements in human availability to perform (e.g., via health and safety) translate to actual performance and consequent increased organizational impacts? Specifically, how does prevention of toxic exposure, due to PTOX, affect immediate unit performance and thereby affect mission performance?

These can be difficult questions. However, they are not inherently cost–benefit questions. Instead, they are fundamental system design questions (Sage and Rouse, 1999). If answers are possible, then cost–benefit analyses are more straightforward.

For the VCATS, DMT, and PTOX examples, it may be possible to translate human performance improvements to organizational impacts via mission models. Such models are typically used to determine, for example, the "logistics footprint" needed to support a targeted sortie generation rate or, as another illustration, the combat wins and losses likely with competing defensive measures and countermeasures. Such models can be used, perhaps with extensions, to project the impacts of faster responses due to the VCATS, improved task performance due to DMT, and increased personnel availability due to PTOX.

It is important to note, however, that even if such projections are not available, the multiattribute methodology presented here can still be employed. However, the validity of cost–benefit assessments and predictions will then depend upon subjective perceptions of attribute levels and the relative importance of attributes. Any limitations of this more subjective approach reflect underlying limitations of knowledge rather than inherent limitations of the methodology.

Once $U[U_{\text{user}}, U_{\text{provider}}, U_{\text{customer}}]$ is fully specified, both functionally and in terms of parameters of these functions, one is in the position to project attribute levels (e.g., option values), calculate the expected utility of the alternative investments (e.g., VCATS, DMT, and PTOX), and perform sensitivity analyses. This provides the basis for making investment decisions. There are several ways that these cost–benefit assessments can be used to inform decision making.

The most common way of using expected-utility cost–benefit assessments is to rank order alternative investments in terms of decreasing $U[U_{\text{user}}, U_{\text{provider}}, U_{\text{customer}}]$ and then allocate investment resources from highest ranked to lowest ranked until resources are exhausted. This approach allows the possibility of alternatives with mediocre $U_{\text{customer}}$ making the cut by having substantial $U_{\text{user}}$ and $U_{\text{provider}}$. To avoid this possibility, one can rank order by $U[U_{\text{user}}, U_{\text{provider}}, U_{\text{customer}}]$ all alternatives with $U_{\text{customer}} > U_{\text{co}}$, which implies a minimum acceptable option value.

If resources are relatively unconstrained, one can invest in all alternatives for which $U_{\text{user}} > U_{\text{uo}}$, $U_{\text{provider}} > U_{\text{po}}$ and $U_{\text{customer}} > U_{\text{co}}$. This reflects situations where all stakeholders prefer investment to no investment. Of course, one can also rank order these alternatives by $U[U_{\text{user}}, U_{\text{provider}}, U_{\text{customer}}]$ to determine priorities for investment. However, if resources are truly unconstrained, this rank ordering will not change the resulting investment decisions.

### 17.4.5 Summary

The three examples discussed in this section have portrayed a cross section of human effectiveness investments to enhance HSI, ranging from aiding to training to health-and-safety investments. The discussion has shown how this range of investment alternatives can be fully addressed with an overarching multiattribute utility, multistakeholder cost–benefit formulation. The stakeholder classes of user, provider, and customer are broadly applicable. The classes of attributes discussed also have broad applicability.

These examples have also served to illustrate the merits of a hybrid approach. In particular, option-pricing theory has been used to define the issue of primary interest to customers, ensuring that investments make financial sense, and this issue has then been incorporated into the overall multiattribute formulation. This enabled including in the formulation a substantial degree of objective rigor as well as important subjective attributes and perceptions. As a result, rigor is not sacrificed but instead is balanced with broader, less quantifiable considerations.

It is useful to note that the knowledge capital construct was not employed in the formulation for these three examples, despite the intuitive appeal of the notion that investments in human effectiveness increase knowledge capital (Davenport, 1999). While the formulation reported here could have included increases of knowledge capital as possible benefits, there is no basis for predicting such impacts. Subjective estimates could, of course, be employed. However, this construct is not defined with sufficient crispness to expect reliable estimates from subject matter experts.

The discussion of these examples of human effectiveness investments has served to illustrate the value of an overall cost–benefit formulation. The generality of this formulation allows it to be applied to analyses of a wide variety of HSI investment decisions. The types of information needed to support such analyses are defined by this formulation. While the availability of information remains a potential difficulty, this formulation nevertheless substantially ameliorates the typical problems of comparing ad hoc analyses of competing investments. Also of great importance, this formulation enables cross-stakeholder comparisons and trade-offs that, for the lack of a suitable methodology, are usually ignored or resolved in ad hoc manners.

## 17.5 CONCLUSIONS

It is difficult to make the case for long-term investments that will provide highly uncertain and intangible returns. This chapter has reviewed alternative ways to characterize such investments and presented an overall methodology that incorporates many of the advantages of these alternatives. This methodology has been illustrated in the context of R&D investments in human effectiveness aspects of HSI.

Central to the cost–benefit analysis methodology presented is a multiattribute, multi-stakeholder formulation. This formulation includes nonlinear preference spaces that are not necessarily aligned across stakeholders. The nonlinearities and lack of alignment provide ample opportunities for interesting trade-offs.

It is important to stress the applicability of this methodology to nearer term HSI investments, which may or may not involve R&D. While the time frame will certainly affect choices of attributes (for instance, option values may not be meaningful for near-term investments), the overall cost–benefit methodology remains unchanged. This chapter focused on long-term R&D investments because such analyses are the most difficult to frame and perform.

It is also useful to indicate that cost–benefit analysis, as broadly conceptualized in this chapter, can be a central element in assessment activities related to life-cycle costing (e.g., affordability) and program/contract management (e.g., earned-value management (Earned Value Management Center, 2000)). For the former, attributes reflecting life-cycle costs can easily be incorporated. For the latter, costs and benefits can be tracked and compared to original projections. This does, of course, require that benefits be attributable to ongoing processes and not just outcomes.

This cost–benefit methodology, when coupled with appropriate methods and tools for predicting attribute levels (Sage and Rouse, 1999), can enable cost–benefit predictions and, thereby, support investment decision making. Using attributes such as option values and potentially knowledge capital can make it possible to translate the intuitive appeal of R&D and human effectiveness investments into more tangible measures of value.

Note also that the methodology includes many of the elements necessary to developing a business case for HSI investments. Markets (stakeholders), revenues (benefits), and costs are central issues in business case development and in this methodology. However, this methodology also supports valuation of investments with broader constituencies (e.g., the public) and ranges of issues (e.g., jobs created) than typically considered in business cases.

Finally, we have also found that use of the methodology presented here provides indirect advantages in terms of causing decision-making groups to clarify and challenge underlying assumptions. This helps decision makers avoid being trapped by common delusions that would mislead them relative to likely costs/benefits (Rouse, 1998).

## REFERENCES


Allen, M. S. (2000). *Business Portfolio Management: Valuation, Risk Assessment, and EVA Strategies*. New York: Wiley.

Amram, M., and Kulatilaka, N. (1999). *Real Options: Managing Strategic Investment in an Uncertain World*. Boston: Harvard Business School Press.

Andrews, D. H. (2000). Distributed Mission Training. In W. Karwowski (Ed.), *International Encyclopedia of Ergonomics and Human Factors*. Philadelphia, PA: Taylor & Francis.

Black, F., and Scholes, M. (1973). The Pricing of Options and Corporate Liabilities. *Journal of Political Economy*, 637–659.

Boer, F. P. (1998, September/October). Traps, Pitfalls, and Snares in the Valuation of Technology. *Research Technology Management*, pp. 45–54.

Boer, F. P. (1999). *The Valuation of Technology: Business and Financial Issues in R&D*. New York: Wiley.

Brigham, E. F., and Gapenski, L. C. (1988). *Financial Management: Theory and Practice*. Chicago, IL: Dryden.

Burke, J. (1996). *The Pinball Effect: How Renaissance Water Gardens Made the Carburetor Possible and Other Journeys through Knowledge*. Boston: Little, Brown.

Christensen, C. M. (1997). *The Innovator's Dilemma: When New Technologies Cause Great Firms to Fail*. Boston: Harvard Business School Press.

Cooper, R. G., Edgett, S. J., and Kleinschmidt, E. J. (1998a). *Portfolio Management for New Products*. Reading, MA: Addison-Wesley.

Cooper, R. G., Edgett, S. J., and Kleinschmidt, E. J. (1998b). Best Practices for Managing R&D Portfolios. *Research Technology Management*, *41*(4), 20–33.

Davenport, T. O. (1999). *Human Capital: What it is and Why People Invest It*. San Francisco: Jossey-Bass.

*Economist*. (1999, June 12). A Price on the Priceless: Measuring Intangible Assets. *Economist*, pp. 61–62.

*Earned Value Management Center*. (2000). Available: http://evms.dcmdw.dla.mil.

Fuchs, R., McCarthy, J., Corder, J., Rankine, R., Miller, W., and Gawron, V. (1997). *United States Air Force Expeditionary Forces*. Washington, DC: Air Force Scientific Advisory Board.

Gill, B., Nelson, B., and Spring, S. (1996). Seven Steps to New Product Development. In M. D. Rosenau, Jr. (Ed.), *The PDMA Handbook of New Product Development* (Chapter 2). New York: Wiley.

Gramlich, E. M. (1997). *A Guide to Cost-Benefit Analysis*. Prospect Heights, IL: Waveland.

Hammond, J. S., Keeney, R. L., and Raiffa, H. (1998). *Smart Choices: A Practical Guide to Making Better Decisions*. Boston, MA: Harvard Business School Press.

Keeney, R. L., and Raiffa, H. (1976). *Decisions with Multiple Objectives: Preferences and Value Tradeoffs*. New York: Wiley.

Kirkwood, C. W. (1979). Pareto Optimality and Equity in Social Decision Analysis. *IEEE Transactions on Systems, Man, and Cybernetics*, *9*(2), 89–91.

Layard, R., and Glaister, S. (Eds.). (1994). *Cost-Benefit Analysis*. Cambridge, UK: Cambridge University Press.

Lint, O., and Pennings, E. (1998). R&D as an Option on Market Introduction. *R&D Management*, *28*(4), 279–287.

Luehrman, T. A. (1998, July/August). Investment Opportunities as Real Options. *Harvard Business Review*, pp. 51–67.

Matheson, D., and Matheson, J. (1998). *The Smart Organization: Creating Value through Strategic R&D*. Boston, MA: Harvard Business School Press.

Mintz, S. L. (1998, February). A Better Approach to Estimating Knowledge Capital. *CFO*, pp. 29–37.

Nevins, J. L., and Winner, R. I. (1999, April). *Ford Motor Company's Investment Efficiency Initiative: A Case Study*, Paper P-3311. Alexandria, VA: Institute for Defense Analyses.

Office of Science and Technology Policy (OSTP). (1998, August). *A National Obligation: Planning for Health Preparedness for and Readjustment of the Military, Veterans, and Their Families after Future Deployments*, Presidential Review Directive No. 5. Washington, DC: Office of Science and Technology Policy.

Rastikis, L. (1998). Human-Centered Design Project Revolutionizes Air Combat. *CSERIAC Gateway*, *XIV*(1), 1–6.

Rouse, W. B. (1991). *Design for Success: A Human-Centered Approach to Designing Successful Products and Systems*. New York: Wiley.

Rouse, W. B. (1998). *Don't Jump to Solutions: Thirteen Delusions That Undermine Strategic Thinking*. San Francisco: Jossey-Bass.

Rouse, W. B., and Boff, K. R. (1997). Assessing Cost/Benefits of Human Factors. In G. Salvendy (Ed.), *Handbook of Human Factors and Ergonomics* (Chapter 49). New York: Wiley.

Rouse, W. B., and Boff, K. R. (1998). R&D/Technology Management: A Framework for Putting Technology to Work. *IEEE Transactions on Systems, Man, and Cybernetics*—Part C, *28*(4), 501–515.

Rouse, W. B., Boff, K. R., and Thomas, B. G. S. (1997). Assessing Cost/Benefits of R&D Investments. *IEEE Transactions on Systems, Man, and Cybernetics—Part A*, *27*(4), 389–401.

Rouse, W. B., Kober, N., and Mavor, A. (Eds.). (1997). *The Case for Human Factors in Industry and Government*. Washington, DC: National Academy Press.

Roussel, P. A., Saad, K. N., and Erickson, T. J. (1991). *Third Generation R&D: Managing the Link to Corporate Strategy*. Cambridge, MA: Harvard Business School Press.

Sage, A. P. (1977). *Systems Methodology for Large-Scale Systems*. New York: McGraw-Hill.

Sage, A. P., and Armstrong, J. (2000). *An Introduction to Systems Engineering*. New York: Wiley.

Sage, A. P., and Rouse, W. B. (Eds.). (1999). *Handbook of Systems Engineering and Management*. New York: Wiley.

Strassman, P. A. (1999, September). *Does Knowledge Capital Explain Market/Book Valuations*? Available: www.strassman.com/pubs/km.

## PART IV

# APPLICATIONS

Part I provided the organization, management, and cultural context for HSI applications, Part II provided the HSI framework for applications in the systems engineering and systems acquisition processes, and Part III provided descriptions of the methods and technology available for systems application by human systems integration (HSI) practitioners. The purpose of Part IV is to describe a wide range of HSI applications to systems. Many of the HSI systems applications presented in this part are drawn from military, aviation, and commercial environments that provide representative samples of the types of organizations and cultures HSI professionals are likely to find themselves working in the future. Seven chapters comprise this part, with major applications drawn from the U.S. Army, the U.S. Navy, the Federal Aviation Administration, and small-system developments such as appear with new commercial products. Four of the chapters primarily present the results of past HSI applications and three of the chapters provide general guidance for future HSI applications to system and product design.

Chapter 18 by Booher and Minninger, reviews the results of HSI applied to U.S. Army systems acquisition over the past decade. It explores two different approaches in describing the army's experience with Manpower and Personnel Integration (MANPRINT), the army's HSI program. First, specific system examples are provided for each of 10 HSI principles described in Chapter 1 to illustrate how applying the principles have helped system acquisition programs meet cost, performance, and schedule requirements. The examples cover major systems and modifications, nondevelopmental systems and some small rapidly procured systems to demontrate a wide range of systems design influence from HSI. Second, the chapter describes the results of four case studies of army systems illustrating the large number of performance and cost benefits to the army that have come from applying the HSI factors. Booher and Minninger categorize the benefits into four areas: acquisition process efficiencies, system design improvements, casualty reduction, and cost avoidance. The chapter concludes by projecting the relevance of HSI factors to future weapon systems.

In Chapter 19 Miller, Crowson, and Narkevicius bring the reader a comprehensive description of human characteristics and measures important to system design. In particular, this chapter presents an overview of characteristics, measures, and techniques that exist to describe and quantify a variety of human factors categories, including

*Handbook of Human Systems Integration*, Edited by Harold R. Booher.
ISBN 0-471-02053-2 

anthropometrics, sensation and perception, mental abilities, social abilities, physiological attributes, and operator states under varying environmental conditions. Underlying this discussion of human characteristics and measures for system design is the aim of recognizing, understanding, and accounting for the human performance variance in HSI applications.

Chapter 20 by Osga describes the approach and results of applying a particular type of human-centered design method called a *task-centered design approach* to the Navy Multi-modal Watchstation (MMWS) project. The HSI participants created an integrated set of decision support and user interface designs to support mission execution and management. The project represents a software-based prototype applied within a simulation-based acquisition started early in the conceptual design stage of a shipboard command-and-control system. During the design stage, concepts are tested and refined before they are passed on to advanced engineering development. The designs are validated by iterative performance and usability tests indicating improvements in task response time and accuracy, with lower workload and increased situation awareness. The set of requirements used to generate the MMWS designs can be applied within other mission domains resulting in consistent quality of user support across a variety of shipboard functions and tasks.

In Chapter 21 Pierce and Salas note that the military acquisition process is a highly structured process with milestone decisions at regular intervals between developing a mission needs statement and system fielding. Capabilities are defined and captured in system and operational requirements, and elaborate tests and experiments are conducted to evaluate the equipment and the operational procedures. Even the need to improve the system after it is fielded is formally recognized in a product improvement plan. Yet when information systems are acquired under this process, more often than not, they do not meet the needs of the military command-and-control decision maker. Information systems are particularly weak in meeting human performance requirements that frequently determine success or failure at meeting mission requirements. The contributors believe it is possible, however, to design more reliable and capable information systems through HSI. Their chapter, therefore, provides a description of those issues, concepts, principles, guidelines, and tools available to help integrate the human component into the HSI process for information systems design.

Klesch and Stembler in Chapter 22 raise the question: Why is it that many program managers of new systems recognize the importance of HSI yet tend to trade off HSI considerations in their investment strategy and assume more risk in system performance? They state this approach is particularly true with the training domain, where even promising technologies such as interactive multimedia instruction (IMI) that can reduce training costs are seldom applied to new systems. They acknowledge that the application of training technology to new systems is a difficult issue to resolve not only because of cost but also because of timeliness and quality. This chapter first addresses the reasons that training, and in particular training technology, frequently loses out in new systems trade-off exercises, even though program managers may be aware of the serious performance costs from inadequate training investments. In particular, it examines why promising technologies such as IMI are not fully utilized in new systems training. Then, based on HSI principles, Klesch and Stembler suggest a systems integration approach to training for new systems that can harness both modern technology and traditional means of efficiencies in training. They show how an HSI training production design can lower both cost and risk, thereby helping managers to solve the training dilemma.

Between 1994 and 1998 the National Academy of Sciences' Panel on Human Factors in Air Traffic Control Automation conducted a study that examined the relationship between controllers and automation in a variety of systems under development by the Federal Aviation Administration (FAA) using principles of human-centered automation. Chapter 23 by Mavor and Wickens, reexamines the panel's findings in view of the HSI principles for successful organizational implementation. The primary purpose of the chapter is to illustrate the challenges and benefits of applying a human-centered design philosophy to a large sociotechnical system such as the National Air Traffic Control System. As an example of the mixed results, which frequently come from attempts to implement human factors recommendations into large complex organizations, particular emphasis is placed on one system studied by the Panel—the Center TRACON Automation System (CTAS). In closing, the chapter discusses the types of coordination and integration issues that are frequently associated with harmonizing several systems (some already in existence and some under development) for an organization as complex as the FAA.

In Chapter 24 Rouse considers HSI issues in the context of private-sector new product development (NPD) efforts where market considerations and profit motives drive HSI-related design decisions. The chapter contrasts the characteristics of private- versus public-sector product and system development and discusses product management practices in terms of multistage decision processes and human-centered design. In this chapter, methods and tools are considered for market research, product lines and platforms, product evaluation, NPD project evaluation, and product planning. Results of empirical assessments of best practices are summarized in terms of characteristics of projects, project management, organizations, and individuals.

CHAPTER 18

# Human Systems Integration in Army Systems Acquisition

HAROLD R. BOOHER and JAMES MINNINGER

## 18.1 BACKGROUND

MANPRINT (Manpower and Personnel Integration), the U.S. Army's human systems integration (HSI) program, has been identified as one of the most promising programs ever developed by the military for providing effective human systems performance. (Minninger et al., 1995; Skelton, 1997). This has been supported by other studies [U.S. Army Audit Agency (AAA), 1997; Booher, 1997; 1998; General Officer Steering Committee (GOSC), 1998] that show the vast range and depth of influence that HSI has had upon the army systems whenever its methodologies have been applied. Generally, performance improved, safety increased, and costs were avoided.

In spite of these impressive results, the HSI practitioner often finds it difficult to convince program managers of the full value of the HSI discipline. Part of the difficulty is that the HSI concept is not fully appreciated, even among many practitioners, so the positive benefits that could accrue for a program are never presented in a way that convinces decision makers HSI can make a significant (and affordable) difference in achieving their objectives. Another difficulty is that very few systems throughout the defense and commercial sectors have actually been quantitatively documented for performance and cost benefits resulting from HSI. Finally, it must be realized that the acquisition world has changed such that strategies that worked with past systems may not work with future systems.

This chapter is designed to help the HSI practitioner better formulate arguments that will be convincing to program managers of the need for HSI on future systems. Set within the framework of those HSI factors identified in the literature (Booher, 1996–1999; GOSC, 1998) as crucial organizational and technical principles to the success of HSI programs, the specific army applications provided here should help the reader better understand the importance of the factors and their interactions to a successful systems acquisition

*Handbook of Human Systems Integration*, Edited by Harold R. Booher.
ISBN 0-471-02053-2 

program. A large number of specific examples are provided as supporting evidence for the value of HSI in terms program managers can appreciate such as (1) technology advancements, (2) acquisition process efficiencies, (3) system design enhancements, (4) safety increases, and (5) returns on investment.

## 18.2 HSI SYSTEM SUCCESS FACTORS

A recent army study on HSI success factors identified critical factors important to achieving MANPRINT cost and performance benefits for army systems acquisitions (Booher, 1999). Ten representative army systems were selected and reviewed in this study. Table 18.1 lists the systems reviewed and indicates how well they had met the army's acquisition objectives at the time of the review. Six of the systems were considered successful; two were marginal, because of difficulties meeting soldier requirements within cost, schedule, and performance objectives; one was fielded with reduced performance acceptance (degraded); and one was canceled by the army (failed). Since the study, one of the two marginal systems (the command-and-control vehicle) has also been canceled.

***Factors Identified*** Booher (1999) concluded that 10 HSI factors (listed in Table 18.2) can account for MANPRINT success (or failure) on systems procured by the army. The 10 principal organizational and technical factors hypothesized from the literature as critical to the success of past MANPRINT programs were verified with analyses of the representative systems. Without exception, all of the major development systems adequately adopted all 10 factors. No new top-level factors were identified, and none of the 10 identified were shown to be consistently unimportant on past systems. Consequently, these 10 factors are considered the broad factors that have made MANPRINT successful in the past. The specific examples, which follow in the next two sections, show a large number of examples on army systems that support Booher's conclusions.

**TABLE 18.1 Systems Reviewed for MANPRINT Involvement**

| System | Category | Army Objectives |
|---|---|---|
| 1. Comanche helicopter | ACAT I—full | Successful |
| 2. Longbow Apache helicopter | Major—mod | Successful |
| 3. Javelin Antitank Guided Missile System | ACAT I—full | Successful |
| 4. Multiple Launch Rocket System—Extended Range | Major—mod | Successful |
| 5. Command and Control Vehicle (C2V) | ACAT I—full | Marginal |
| 6. Family of Medium Tactical Vehicles (FMTV) | Major—NDI | Degraded |
| 7. Armored Gun System | Major—NDI | Failed |
| 8. Crusader artillery/resupply | ACAT I—full | Successful |
| 9. Land Warrior | ACAT II | Marginal |
| 10. Nuclear, biological, and chemical (NBC) reconnaissance system (NBCRS—Fox) | ACAT III | Successful |

*Note:* ACAT = army category; ACAT I is highest cost and priority; ACAT II is intermediate cost and priority; ACAT III is relatively low cost and priority; NDI = nondevelopmental item; less than full-scale acquisition process.

**TABLE 18.2 HSI System Success Factors**

1. Top-level support and understanding
2. Human-centered design
3. Source selection
4. Domains integration
5. System documentation integration
6. Quantitative human performance
7. MANPRINT technology
8. Test and evaluation integration
9. Practitioners, skilled, available
10. Education and training: (a) practitioners and (b) nonpractitioners

## 18.3 HSI FACTORS: EXAMPLES FROM ARMY SYSTEMS

The 10 HSI factors can be better appreciated by examining a number of specific examples from 15 army systems (including the 10 systems in Table 18.1). Figure 18.1 is a summary of the HSI factors illustrated by specific system examples in this section. This section is designed to provide a collection of examples arranged by the 10 HSI system factors. The reader who does not need this level of detail may skip to the next section.

### *Factor 1: Top-Level Support*

*Description* This factor is the degree to which top-level management supports HSI concepts and practices for the specific system being developed. Top-level management includes the program manager and the responsible decision makers he or she must report to in achieving program objectives. Because of the rapid and controversial systems engineering trade-offs that often need to be made, it is important that the program manager also understand HSI concepts and data as well as any other systems engineering concepts and data.

| Systems | HSI Factors | | | | | | | | | |
|---|---|---|---|---|---|---|---|---|---|---|
| | 1.TSP | 2. HCD | 3.SS | 4.ODI | 5.SDI | 6.QHP | 7.MT | 8.T&E | 9.PR | 10.ET |
| 1. *Apache Automatic Target Handover System (ATHS)* | | | x | | | | | | | |
| 2. *Apache Longbow* | | x | | | | x | | | | |
| 3. *Armored Gun System (AGS).* | x | x | | | | | | | x | x |
| 4. *Comanche Helicopter* | x | x | x | x | x | x | x | x | x | x |
| 5. *Command and Control Vehicle (C2V).* | | | | | x | x | | | x | x |
| 6. *Crusader Artillery System* | | x | | | | x | x | x | x | x |
| 7. *Family of Medium Tactical Vehicles (FMTV)* | | | | | x | | | | x | x |
| 8. *Fox NBC Reconnaissance Vehicle* | | | | | | x | x | x | x | x |
| 9, *Javelin Antitank Guided Missile System* | | | | x | | | | | | |
| 10. *Land Warrior* | | | | | | | | x | | |
| 11. *Lightweight Howitzer* | | | | | | | x | | x | |
| 12. *Line of Sight - Forward Heavy (LOS-FH) missile system* | | | | | | x | | | | |
| 13. *Stinger Missile System* | | x | | | | x | | | | |
| 14. *T-800 Engine* | | x | | | x | | | | | |
| 15. *Multiple Launch Rocket System-Extended Range* | x | | | x | x | | | x | | |

**Figure 18.1** Systems by HSI factors.

**Example 18.1 Comanche Management** The Army's Comanche is being developed as a lightweight, twin-engine helicopter capable of performing armed reconnaissance and light-attack missions. From the beginning, the Comanche has had a number of ambitious goals including

1. to push the state of the art by incorporating the latest aircraft technologies to enhance its performance in complex missions in a wide range of environments (i.e., night, nap of the earth, and adverse weather conditions),
2. to be one of the most supportable aircraft in the world,
3. to have increased safety measures for aircrew survivability,
4. to achieve the added performance features without unduly increasing operations-and-support (O&S) costs over that to maintain the current reconnaissance and light-attack helicopter fleet.

It was realized by army leadership that the challenges to meet the ambitious performance goals would require major changes in the acquisition and design processes. This was especially true regarding the emphasis to be placed on the human design component. Through the MANPRINT approach, HSI methodology was inserted in the earliest stages of requirements development and carried throughout each subsequent stage of the acquisition process. The Comanche report (Minninger et al., 1995), which documents the results of the program's human-centered approach, is based on a five-year record keeping effort by both the winning contractor, Boeing-Sikorsky, and the Comanche Program Office. These results are without question some of the most impressive ever reported on a major weapon system acquisition.

**Example 18.2 Armored Gun System (AGS) Leadership** Top-level army leadership supported the MANPRINT concerns for soldier performance and survivability, but both government and contractor's program managers did not pay sufficient attention to these concerns until too late in the program. The AGS was designed from a hardware perspective, and crew performance and soldier survivability were at best afterthoughts. However, because MANPRINT reviews were given top-level visibility, the poor application of HSI was highly contributory to the program cancellation in 1996.

**Example 18.3 Multiple Launch Rocket System—Extended Range (MLRS-ER)** The MLRS-ER is an example of a system considered to have relatively simple human interfaces and low manpower, personnel, and training demands, thus suggesting little need for a strong HSI program. However, analyses of this system (AAA, 1997; Booher, 1999) show this system had a good MANPRINT program and was successful with applying several of the HSI factors, in particular: 1.0 for top-level management and organization support, 4.0 for domains integration, 5.0 for system documentation integration, and 8.0 for test and evaluation integration. The MLRS-ER shows that even for a system that appeared to have few human performance issues, HSI top-level support (along with at least some of the other HSI factors) is still necessary for system success.

### *Factor 2: Human-Centered Design*

*Description* Strong emphasis on human-centered design (HCD) begins in the requirements stages. This factor encourages the concept of defining a "system" more broadly than the hardware and software that industrial companies build. Procuring organizations should specify their requirements for a system in such terms as to include operators and maintainers as an inherent part of the "system." These requirements, which include the human element, should be translated quantitatively throughout the design, development, and testing processes in systems engineering measures of effectiveness and performance.

**Example 18.4 Army Stinger Missile System** Numerous system failures have occurred in the past because a system was not defined to include the human. For example, when the U.S. Army Stinger Missile "system" was designed with a "probability of kill" at a certain level, it did so without applying this factor. As a result, the army found actual performance in the hands of the soldier was only one-half of that expected. The designed performance has assumed human performance to be perfect and did not take into account the skill and training level of the operator. If the system probability of kill had been defined as "including the human operator," the procurement process would have been significantly different.

**Example 18.5 Armored Gun System Design** From the beginning, this program did not have an acquisition strategy favorable to making the soldier part of the system design. For example, although expected to fight in the desert and tropical environments with nuclear, biological, and chemical (NBC) gear, none of the escape hatches was wide enough for large soldiers to exit, even without NBC gear, and most of the soldiers could not exit quickly enough if wearing NBC gear. Also, the AGS crew could not be expected to perform well in the cramped and poorly designed workspaces. Driver head clearance when wearing the helmet with the hatch closed was less than 1 inch, so that they would routinely bang their head during motion, and if slumped to avoid the banging, the drivers' field of view was reduced and possibly eliminated altogether. These were only a few of the large number of HSI problems identified by the MANPRINT practitioners.

**Example 18.6 Comanche Cockpit** The crew station design for the Comanche allows the aircrew to set priorities for information criticality at specific points during the conduct of missions. This is unlike previous cockpits where the information was presented in predetermined menus. Overall, the sequence of tasks required to perform mission functions was greatly reduced with this human-centered approach. For example, a sequence for target reporting that previously required 34 procedural steps in the older aircraft (OH-58D) was reduced to only five steps in the Comanche.

### *Factor 3: Source Selection Policy*

*Description* Source selection policy for systems procurement should state that HSI evaluation factors will have the same visibility as technical and cost factors (as a major area) and will be evaluated in all other relevant areas as well. This is a unique evaluation criterion requirement not specified similarly for any other factor. This is because the HSI evaluation must not only show how well the contractor understands the HSI process (visibility) but also show that the contractor will use HSI technology and disciplines in the design of his or her equipment (other relevant areas).

**Example 18.7 Comanche Contractor Selection** The source selection evaluation criteria used in the Comanche program represented a radical departure from past acquisition programs. For example, MANPRINT (including training) was made a separate evaluation area with the same weight as reliability, availability, and maintainability/integrated logistics support (RAM/ILS). MANPRINT and ILS were combined under the same review team so that MANPRINT/ILS had the same weight (35%) as technical. This was made known to industry during the request-for-proposal stage, showing that the government was serious about its commitment to the soldier. With such weighting factors, a contract could be won or lost based on HSI understanding and the proposed approach using HSI methodology.

Early in the competition it was discovered that even more important to effective design (once industry was convinced the government was serious about HSI, which was communicated by showing the major area emphasis) was the additional emphasis on MANPRINT within the technical evaluation criteria. A very high percentage of the technical evaluation areas were also evaluated as having either strong or moderate MANPRINT implications.

The two competing contractor teams were required to commit contractually to the achievement of MANPRINT, supportability, system performance goals, and the overall affordability of the Comanche program. MANPRINT objectives (and HSI methodologies that demonstrated feasibility) tended to both ease the overall manpower requirements for the system and to make more efficient use of available projected manpower than had been done in the past. Because of the unique emphasis in source selection on HCD, MANPRINT/HSI requirements were clearly communicated to the contractor. The contract statement of work (SOW) required the contractor to seek ways to incorporate HSI principles into the operation, support, and maintenance of the aircraft. By adopting HSI objectives as an inherent part of engineering design and development, the contractor was able to integrate soldier capabilities and limitations into the design with an affordable investment. As it turns out, Boeing-Sikorsky won the contract primarily because it received higher MANPRINT/ILS and operability scores than its competitor.

**Example 18.8 Apache Automatic Target Handover System (ATHS)** On the Apache product improvement program (PIP) for ATHS, MANPRINT was one of only two evaluation factors. MANPRINT was 50 percent of the source selection weight and technical was the other 50 percent. The purpose of the ATHS was to automate the function of "handing over" the target once identified and selected by the pilot to the lock-on of the target for delivery of a Hell Fire missile. MANPRINT was evaluated so heavily because of the critical interface with the human operator in the cockpit. As it turns out, this MANPRINT design for the pilot caused a large unanticipated maintenance manpower increase. The wiring in the Apache was so confusing due to the new and old wiring being interwoven that it was estimated that troubleshooting difficulties would require manpower increases costing about $1 million per year. Since MANPRINT is concerned with total system costs, it required the old wiring to be removed. There was great resistance from the program manager, since the Apache program would be paying around $4 million to remove the old wiring as part of the PIP. However, due to the high source selection weighting of MANPRINT, the old wiring was removed. Still another unexpected result came from this issue being resolved in favor of reduced manpower—this time it was additional warfighting capability, which was a windfall for the program manager. The removal of the old wiring reduced the aircraft weight by 16 pounds. This reduced weight translated to either a saving on fuel consumption of $170,000 per year or 14 additional 30-mm rounds that could be carried. The end result of a high MANPRINT source selection factor for the army was improved system target hit capability (the original intent of the PIP) and $16 million cost avoidance (spread over 20 years) in maintenance manpower—all while reducing aircraft weight.

### *Factor 4: Organizational Integration of All HSI Domains*

*Description* A single focus for all HSI domains is necessary if any of the domains is to have substantial influence upon the system being procured. It is important that expertise from each of the HSI domains be provided to the various systems engineering and systems integration working groups. The results of a common focus of all HSI domains to a system acquisition can differ widely from system to system and from feature to feature within a system. Sometimes the domains can provide different perspectives that tend to reinforce each other and, once understood by the program manager, seem to be additive in meeting system objectives. At other times, a domain recommendation may be perceived by the program manager as so low in trade-off decisions that it can only survive with the help of the other domains. In some cases, the domains may create major conflicts for the system because of the differences required among the domains themselves and may not be resolved without compromises among the HSI requirements.

**Example 18.9 Comanche Rotor System Design** The Comanche PENTAFLEX rotor system design provides an excellent lesson learned for industry on the unexpected benefits that can accrue when HSI recommendations are adopted. The Boeing-Sikorsky design team had originally considered a rotor blade design that met government specifications but one for which MANPRINT and ILS contractor personnel had raised maintainability and transportability concerns. Because the team was still in competition with McDonnell Douglas, it was reluctant to expend extra design resources where they were not required by government specifications. Nevertheless, by bringing the full focus of the domains together on this issue, MANPRINT/ILS persevered, and the team decided to develop a new modular design that was easier to maintain, reduced the potential for installation error, and eliminated close-fit tolerance for transportability. The amount of additional effort for the MANPRINT analyses, test and evaluation, and drawing change was 395 man-hours, likely costing the contractor something well below $50,000. However, when a life-cycle cost analysis was conducted later, approximately $150 million was calculated as avoided due to this design improvement. These savings would come primarily from manpower requirements reductions in skill and numbers due to easier and less maintenance on the rotor system and reductions in transportability times.

**Example 18.10 Comanche Tail Rotor Weight Trade-off** During early design the technical advantages of the "fan-in-fin" composite tail rotor (FANTAIL) for flight efficiency were recognized. Moreover, crew and aircraft survivability were also increased with the new FANTAIL design. During the trade-off analysis the FANTAIL design was found to be eight times safer than that of the traditional rotor design. A shroud was added to protect ground crew from the tail rotor. It was known that in the past unprotected tail rotors have contributed to many avoidable accidents on the ground. This was significant for MANPRINT design influence because the shroud added extra weight, which would not have been accepted in the weight trade-off decisions if the safety domain had to argue its case alone. However, because of MANPRINT bringing together the voice of safety, maintenance, and flight operations, weight offsets in other areas allowed the increased weight for ground personnel safety.

**Example 18.11 Javelin Antitank Guided Missile System** The Javelin is an excellent example of a system where manpower versus soldier performance and survivability create conflicting expectations from MANPRINT. The Javelin Weapon System will replace the Dragon as the army and marine corps primary medium Antitank Guided Missile System. Javelin is a man-portable, shoulder-fixed antitank weapon capable of defeating modern and future threat armor. Major improvements over the Dragon are increased range and lethality, increased gunner survivability, reduced launch signature and effects, and reduced maintenance and support requirements. The Javelin program has understood well the role of the soldier in the total system performance. A major difficulty with the Javelin has been conflicting human performance parameters. The weight of the Javelin has always been too heavy for a one-man portable system. Yet one of the domains (soldier survivability) has required increased weight for gunner survivability. The one-man portable requirement has forced technology to reduce weight while providing the survivability advantages. Still another domain, human factors engineering, has added improved human performance features to the system accuracy that increase weight as well. All seven MANPRINT domains participated in the MANPRINT Joint Working Group (MJWG) and relied on the system MANPRINT management plan (SMMP) to bring together issues to effect design and development. A weakness pointed out by the army audit was that the MJWG did not have tasking authority to get issues tested and resolved. (Booher, 1999; AAA, 1997)

### *Factor 5: System Documentation Integration*

*Description* The second integration step of the HSI model applies the information from the first integration directly into the procurement process. The HSI management tool for this principle for the Department of Defense (DoD) is the HSI management plan (HSIMP). The HSIMP is seen as the critical interface document feeding information into all other procurement documents and being fed by them. The quality of information in the HSIMP depends on the quality of personnel assigned to the system joint working groups (SJWGs) and the tools and systems information at their disposal. Some of the critical documents that the HSIMP feeds are the operational requirements document (ORD), request for proposal (RFP), and test-and-evaluation plan (TEMP).

**Example 18.12 T-800 Engine Contractor Request for Proposal** Major advances in maintainability with reductions in manpower, personnel, and training (MPT) were demonstrated in the T-800 engine as a direct result of the government inserting limitations in the RFP. The RFP stated that the design was to have no increase in skills or manpower numbers. This clause was added late in the procurement bidding process but was accepted by industry competitors at "no cost to the government." As a result, impressive design improvements were provided. An early notable example was requiring only six tools for the T-800 organizational maintenance when 136 tools were needed for similar functions on the predecessor engine. *Note that the HSI approach was not to require a reduction in tools, but rather to set limits on the MPT that could result from the contractor's design.*

**Example 18.13 Command-and-Control Vehicle (C2V)** The C2V (now canceled) was to be an improved armored, tracked combat vehicle that would house and transport command-and-control equipment and staff personnel. The improvements desired were (1) speed and mobility to keep up with Abrams tank and Bradley vehicle, (2) conduct operations on the move, and (3) geographic dispersion of command and control. The command-and-control systems in the vehicle would be highly digitized equipment providing a central role for the future battlefield. Several human performance issues were identified as unique to this new equipment. Most importantly, these comprised performance of cognitive tasks and team performance under noise, vibration, and motion. Motion sickness was especially troublesome for many individuals during operations on the move and presented a major human limitation that was not fully considered in the requirements stage. Had this been fully explored, it is likely the requirement for conducting operations on the move would not have been made. This combined with the other human cognitive performance issues while in motion made one of the most important features of the new vehicle—operations on the move—no longer feasible.

**Example 18.14 Family of Medium Tactical Vehicles (FMTV)** The FMTV was singled out by the AAA (1997) as the only program reviewed that did not have MANPRINT properly integrated into the acquisition process. The FMTV is a nondevelopmental item consisting of both the $2\frac{1}{2}$-ton (light medium tactical) and 5-ton (medium tactical) vehicles. Compatible trailers with capacity equal to the prime mover are also included in the FMTV. The system was designed to provide for large reductions in supportability need, to enhance capability and performance, and for multiple and flexible use. Although none of the 10 factors was adequately applied, it is an especially good example of how a program should *not* conduct system documentation integration. The AAA (1997) report found such concerns as: deficiency in documentation to support MPT; MANPRINT issues and concerns were reactive instead of proactive; and the SMMP never addressed issues and concerns. The SMMP prepared at the beginning of the program was not updated to include new issues and concerns found from the prototype hardware (i.e., the SMMP never addressed issues and concerns representing the actual hardware). As a result, the FMTV was fielded with a large number of design flaws. For example, two soldiers are needed to change a tire; female soldiers cannot assemble the electric crane on the $2\frac{1}{2}$-ton truck; and it lacks protection from small arms and fragmentation and has vulnerability to blast overpressure and shock injuries. Other deficiencies include: inability to withstand effects of chemical agents and decontaminates, inadequacy of seat belts and crew seat comfort, poor rollover protection, poor brakes, and high potential for additional training and military occupational specialties (MOSs). Some of the problem may have been because of the newness of MANPRINT. The FMTV program was initiated in 1986 prior to a full understanding of the MANPRINT philosophy by Tank & Automotive Command (TACOM) and the U.S. Army Training and Doctrine Command (TRADOC) systems manager (TSM). In this case both the user and the developer did not understand how to integrate MANPRINT requirements to affect the acquisition process.

### *Factor 6: Quantitative Human Performance*

*Description* The HSI process allows representation of all human factors domains in order to prescribe goals and constraints for the system being procured. Since the human is part of the system and the system is being designed to certain quantifiable specifications, the human aspects should be described quantifiably as well. The U.S. military has compiled performance data for each occupational specialty (based on skill level and training) such that basic tasks can be analyzed quantitatively for proposed weapon system designs. The research community has a very strong role in providing human performance data that comprises cognitive as well as physical performance recorded in human reliability and human error terminology.

**Example 18.15 Stinger Missile System Probability of Kill** In the early 1980s a test was conducted with different skilled soldiers on performance with the U.S. Army Stinger Air Defense System. The Stinger was designed to be held, aimed, and fired by the infantry soldier. The design specification was for a system capable of being fired by the soldier at the enemy aircraft with a probability of kill (aiming, firing, and hitting) 6 out of every 10 enemy aircraft fired upon. Thus, the probability of total system performance was $P_s = 0.6$. Total system performance reliability was $P_s = P_c + P_h$, where $P_c$ is missile component probability and $P_h$ is human operator probability. The accuracy of the missile components themselves ($P_c$) was 0.6, so the gunners' performance would have had to be error free ($P_h = 1.0$). However when actual gunners were tested, it was found that even the best gunners made errors with the system such that the actual system performance reliability with superior gunners was not 1.0 but rather 0.402 ($0.67 \times 0.6$). The average gunner was lower still, having a reliability of 0.51, thus making the total systems performance with them $P_s = 0.306$. In other words, the actual performance the army could expect with its air defense system was only about one-half of what it was designed to do. The requirements document should have stated, "The total systems performance reliability, including the gunner performance reliability, must be $P_s = 0.6$."

**Example 18.16 Line of Sight–Forward Heavy (LOS–FH) Missile System** MANPRINT was introduced in the middle of a programs acquisition process. The LOS–FH was one of those programs. The program manager showed the difference on the program as an example of the increased emphasis on human performance quantification. Before MANPRINT, crew members would be asked questions that provided subjective answers. For example, on one occasion the program manager asked, "How did you feel about information displays used in engaging targets?" Four sample crew member responses were as follows:

1. "Screen was too small and dim."
2. "I feel good about it."

3. “Things were just going too fast.”
4. “That thing kept losing track.”

The resulting database was comprised largely of subjective unquantifiable performance data.

After MANPRINT, the LOS–FH collected human performance data differently. Table 18.3 shows quantified human performance data for two contractor candidates for the system.

**Example 18.17 Command-and-Control Vehicle Human Performance** Quantification of human parameters was an extremely important factor for the C2V. Many of the difficulties with human task performance under noise and motion would not have been identified as early as they have were it not for this factor. Quantitative data for human performance were used to both identify important MANPRINT issues and make design recommendations to improve performance. Some of these efforts included analyzing individual and team performance tasks while the vehicle is in motion, assessing the effects of various shift scenarios, identifying special knowledge requirements' impact on crew members, and recommending design changes to reduce noise.

### *Factor 7: HSI Technology*

*Description* The HSI technology includes three different types of technologies, tools, and techniques: (1) domain unique or common technology shared by one or more domains, (2) technology that allows trade-offs among domains, and (3) technology that aids trade-offs between system capability and affordability. The HSI technology in the hands of highly qualified practitioners will allow better design and development decisions with future systems.

**Example 18.18 Comanche MANPRINT Quantitative Trade Analysis** MANPRINT technology has been productively used in several critical decisions for the Comanche program (Minninger et al., 1995). During the concept exploration phase of the Comanche program (then called LHX, for light-helicopter experimental), a HARDMAN (hardware-versus-

**TABLE 18.3 LOS-FH Human Performance Data**

| | Candidate A | | Candidate B | |
|---|---|---|---|---|
| Event | Recoverable Error | Engagement-Ending Error | Recoverable Error | Engagement-Ending Error |
| Detect | 6 | 57 | 8 | 94 |
| Acquire | 0 | 4 | 1 | 6 |
| Identify | 19 | 9 | 24 | 5 |
| Identify-friend-of-foe (IFF) | 33 | 0 | 135 | 0 |
| Tracking | 13 | 25 | 0 | 10 |
| Ranging | 0 | 0 | 0 | 0 |
| Fire | 0 | 34 | 0 | 31 |
| Slew to cue | 1 | 1 | 4 | 0 |
| | 72 | 129 | 172 | 146 |

manpower) comparability methodology (HCM) study was conducted to provide early estimates of MPT requirements and associated training costs for a family of light helicopters compared to predecessor systems. The HARDMAN results supported the light-helicopter concept as vastly superior for MPT affordability.

The Systems Laboratory, Army Research Institute, employed the crew workload model, a new MANPRINT tool at the time, to determine the degree to which automation would aid one- and two-person crews to conduct the intended missions. The crew workload model demonstrated that without the automation planned for LHX both one- and two-crew cockpit positions were overloaded an excessive number of times for the missions intended. The missions could not be accomplished with either size crew. However, even with automation, the one-person crew was overloaded in 10 critical events. Only a two-crew model with automation predicted no overloads for the LHX missions. The decision to adopt a two-seat design was therefore based on MANPRINT analysis for superior mission performance. This was an important decision, because not only were more flight crew required but also more maintenance personnel. The HARDMAN analysis showed that the two-seat configuration would require 12 percent more maintenance support than the single-seat version due to the additional cockpit equipment.

Altogether, however, a major net reduction of MPT was projected for the army. The manpower capabilities (MANCAP) model (one of six HARDMAN modules) was used to predict about a 25 percent reduction in manpower requirements (primarily maintenance) in the light-infantry division with the introduction of LHX. As manpower requirements became less, so did personnel requirements. For example, MANPRINT analysis showed it would be possible to consolidate maintenance-related MOS from 13 to 4. Still another finding was that the reductions in manpower and numbers of MOS allowed the MPT resource requirements to be reduced on an average of 27 to 39 percent compared to predecessor aircraft.

While showing the overall reductions in MPT requirements was important, still other uses of the MANPRINT technology were demonstrated that utilized HARDMAN's ability to represent the complexity of MPT trade-offs. In maintenance manpower, for example, depot maintenance increased 16 percent for the two-level maintenance concept. (This increase was partially offset by an estimated 6 percent reduction in manpower due to improvements in reliability, availability, and maintainability). Further complexities were revealed for actual operations. While the overall light-helicopter manpower and personnel were less, distribution of personnel was critical since workload requirement could be expected to increase at the unit level. The increases in unit workload were due to increases in operational tempos of the aircraft within the units operating the light helicopter compared to the aircraft it would replace.

**Example 18.19 HSI Modeling and Simulation Program** New human figure modeling tools are continually being advanced as part of the HSI set of tools to answer such questions as workspace layout, egress, and access to equipment in new or modified designs. The advanced human figure models work in combination with advanced simulation methods seeking to reliably predict system mission performance. The Comanche, Crusader, and Fox case studies (Section 18.4) show the importance of HSI to the capability and validity of those simulations directed to questions on system performance, speeded-up acquisition processes, twenty-first-century training techniques, and outcomes in warfighting scenarios.

The HSI modeling and simulation program currently available at the Human Research and Engineering Directorate (HRED) provides a conceptual "build," "test," and "evaluation" that can be performed well before a system is built. Various pieces and their integration on real programs have been demonstrated in the case studies. The human figure model, HARDMAN III, and distributed interactive simulation of small crews were applied to the Fox, whereas HARDMAN III and distributed interactive simulation (DIS) at the Janus level were successfully applied to the Crusader.

**Example 18.20 Lightweight Howitzer Human Figure Modeling and Improved Performance Research Integration Tool (IMPRINT)** The Human Research and Engineering Directorate developed and applied new MANPRINT/HSI technology to the SM777, 155-mm, lightweight, towed howitzer to increase system safety, usability, and efficiency while avoiding costly redesigns and reducing the total cost of ownership. An early HSI evaluation identified numerous operator interface concerns that were corrected with inexpensive fixes. The integration of HSI methods included human figure (HF) modeling, task network models, and a fast azimuth shift tool (FAST). The HF modeling was used to correlate reported operator discomforts with specific crew postures interfacing with the prototype design. Subsystem design alternative related to hand wheels, trails, spades, and fire control were evaluated with the HF modeling effort. Task network models were generated with the IMPRINT for various response functions. The task network modeling results were used by the joint program manager for requirements risk reduction, the training community for crew drill optimization, and the prime contractor to conduct real-time design trade-offs on over two dozen subsystem alternatives. The FAST was used to reduce crew burden and function time for conducting the bold-shift function. The concept and design were rapidly implemented into the final howitzer design.

### *Factor 8: Integrated Test and Evaluation*

*Description* Human systems integration test and evaluation are the final and most reliable factors to assure that the soldier will receive a safe and effective weapon before going into battle. This factor begins by assuring all human performance requirements are fully included in the measures of effectiveness (MOEs) and measures of performance (MOPs) for the test-and-evaluation plan for the system. It is completed when representative users successfully perform the system during the operational test and evaluation (T&E).

**Example 18.21 Crusader Training and Testing Simulator Experiment** At its best, HSI integrated T&E is a continuous activity taking place throughout the system design and development process. The Crusader illustrates how HSI can play a central role in reducing the costs of operational T&E and make training and testing more effective for complex warfighting environments. As an example, Pierce (1996) describes a battlelab experiment where HSI was applied to a combined training and testing simulator for the Crusader operating in a digitized battlefield (see Section 18.4.1). The battlelab experiment showed the value of the simulator as both a trainer for field artillery collective training and as a means of testing alternative Crusader tactics, techniques, and procedures (TTPs).

**Example 18.22 Land Warrior MOEs and MOPs** The Land Warrior program provides a good illustration of the way MOEs and MOPs tie soldier MANPRINT requirements into the T&E program. The T&E requirements will vary over time, but a snapshot of a Land Warrior draft of MOEs/MOPs in 1997 (see Table 18.4) provides the basis for an example of integration with soldier survivability MANPRINT requirements. A few critical issues for the program are broken down into several criteria, then further broken down into MOEs, and finally each MOE comprises a list of MOPs. For Land Warrior there were three critical issues each broken down into criteria ranging in number from 3 to 7. Each criterion has its MOEs described. For critical issue 3 (survivability), the number of MOEs ranged from 4 to 6. Each MOE was further broken down into MOPs. For example, for critical issue 3, Land Warrior had MOPs ranging in number from 3 to 9 for the various MOEs related to that issue.

**TABLE 18.4 Land Warrior T&E Issues, Criteria, MOEs, and MOPs**

*Critical issue 1:* Effectiveness. Is Land Warrior (LW) operationally effective?
*Critical issue 2:* Suitability. Is LW operationally suitable?
*Critical issue 3:* Survivability. Is LW survivable on the modern battlefield?
*Criterion 3.1.* The LW soldier/infantry squad/platoon survivability on the modern battlefield must be equal to or greater than baseline soldier/infantry squad/platoon.
**MOE 3-1-1** Difference in detectability of a LW soldier/unit and a baseline soldier/unit in an operational environment (as stated in ORD para xx)
**MOP 3-1-1-1** Probability of detection due to light by range by mission by soldier/squad/platoon.
*Criterion 3.2.* Determine LW impact on protection afforded a dismounted soldier engaged in close combat.
**MOE 3-2-1** Ability of basic body armor and the modular plates to meet weight goals, dimensions, protection level, and compatibility requirements (as stated in ORD, para xx, and LW spec, para xx)
**MOP 3-2-1-2** Ratings of the fit and comfort associated with wearing body armor with and without front and rear plates
**MOP 3-2-1-8** Coverage of plates of vital organs in the torso region for the 5th to 95th percentile male soldier

**Example 18.23 Fox Vehicle T&E** The Fox vehicle case study (see Section 18.4.2) illustrates the ability of HSI to improve the effectiveness of operational T&E for nonmajor systems. By using an HSI model such as HARDMAN III (newer version called IMPRINT) to obtain operational estimates of measures of performance and effectiveness, the T&E procedures can be conducted much more efficiently.

### Factor 9: Practitioners

*Description* Successful systems need to use highly qualified practitioners on the government side as domain representatives for the system working groups, writers of requirements for SOWs, proposal evaluators, and assessors for the T&E process. Skilled HSI practitioners also need to be employed by the supplier in the research, development, test, and evaluation (RDT&E) of the system. Such individuals need to be conversant with both the technology and operational complexity of the system. Most of the tools and techniques used by the domains and as HSI trade-off methodologies are best applied by experts in their field. Because of the short supply of highly qualified practitioners in HSI, two questions that need to always be asked in assessing program success on this factor are as follows: (1) Were qualified practitioners available? (2) Were the practitioners utilized effectively?

**Example 18.24 Crusader Crew Workload Research** Highly qualified practitioners were both available and utilized effectively on the Crusader. The depth and quality of contributions possible from qualified HSI practitioners are well illustrated by the experience of Pierce (1996). Pierce describes how HSI practitioners using HARDMAN III technology answered system critical research questions on Crusader crew characteristics early in systems design (see Section 18.4.2). The HARDMAN analysis provided design recommendations for optimal crew size for both the Advanced Forward Artillery System (AFAS) and the Forward Area Resupply Vehicle (FARV). It also found the best combination of the Armed Services

Vocational Aptitude Battery (ASVAB), area composites, ASVAB area cutoff scores, and MOSs that would allow enhanced mission performance while not restricting the availability of qualified personnel.

**Example 18.25 Practitioners on Other Successful Programs**

***A. Comanche*** Highly qualified practitioners were available and utilized by both the government and industry. In the government, skilled personnel represented each domain such that a team approach was used. This was true somewhat in industry as well, however; human factors engineering (HFE) and safety were the most heavily utilized in working with ILS personnel in a concurrent engineering environment. Government practitioners tended to specialize in their domains. When the first army acquisition milestone decision meeting for Comanche was held, more than 200 issues were identified from the practitioners of the six domains.

***B. Fox*** Practitioners from HRED using both HFE and MPT domain tools were available but were not utilized appropriately until the program was in danger of being canceled from the adverse operational and test results.

***C. Lightweight Howitzer*** Practitioners were available and fully utilized on this program. The practitioners were fully qualified to conduct an early HFE evaluation, to apply human figure modeling to operator interfaces, to generate task network models using IMPRINT, and to develop, fabricate, and evaluate FAST while applying it to a new howitzer design.

**Example 18.26 Practitioners on Marginal and Failed Programs**

***A. FMTV*** The specification/purchase description for the FMTV addressed only HFE requirements in general (i.e., MIL-STD 1472 requirements). There was no specific requirement for MANPRINT, thus automatically leaving out five practitioner domains. Practitioners for all domains were generally available but not utilized.

***B. AGS*** Highly qualified practitioners were available and utilized by both government and industry, but AGS leadership attempted to keep them from voicing their concerns to top leadership. The MANPRINT leadership was able to override these attempts and get the practitioners' concerns to the top army decision makers.

***C. C2V*** This program started during a period when MANPRINT was receiving reduced emphasis at the top levels. Consequently, inadequate resources were provided for MANPRINT translating to an inadequate number of qualified practitioners being consistently provided for the C2V program. However, qualified practitioners were able to identify and attempt to solve a number of human performance issues for the program. This program did not attempt to hide the MANPRINT issues as did the AGS, but the human performance problems identified were determined too difficult to overcome.

### *Factor 10: Education and Training*

*Description* Human systems integration education and training are essential to assure practitioners are qualified. Moreover, it is important to provide some aspect of HSI for everyone in the system acquisition program, in addition to the practitioners, in order for them to understand the value of HSI in meeting overall system performance, cost, and schedule. Three different levels of HSI education and training are provided. Level 1 is advanced education through formal degree programs in academic settings. Level 2 is specialized practitioner training provided by government or industry short courses. Level 3 is HSI awareness training for nonpractitioners provided as part of government and industry training either in specialized HSI short courses or as part of other courses for non-HSI personnel. (See Chapter 5 for a complete discussion of HSI education and training.)

**Example 18.27 Education and Training in Successful Programs**

***A. Comanche*** All three levels of MANPRINT education and training were fully covered on the Comanche program. Most of the domains had practitioners with advanced degrees who worked issues for all the domains on both the government and industry sides. Specialized MANPRINT training was provided for civilians, military personnel, and industry personnel who participated on the Comanche. Most government individuals associated with the acquisition process but who were not practitioners were trained by the army on the MANPRINT concept.

***B. Crusader*** As with the Comanche, all three levels of MANPRINT education and training have been provided to participants on the Crusader program.

***C. Fox*** MANPRINT and its tools did not receive appropriate visibility among the nonpractitioners. It took being evaluated as "unsuitable" and "ineffective" to gain the necessary visibility.

The fact that highly successful programs in the past have had all three levels of HSI education and training helping to develop knowledge, skills, and abilities for practitioners and knowledge for nonpractitioners suggests a high priority for all three levels to assure success in future systems.

**Example 18.28 Education and Training in Marginal and Failed Programs** The FMTV, AGS, and C2V represent degraded or failed army programs spanning 15 years of MANPRINT. The FMTV was introduced at about the same time MANPRINT was introduced. Then AGS came into the acquisition about the middle of this time period, and the C2V has come most recently. In each of these programs, it was not the lack of available educated and trained HSI practitioners that contributed to their failure. In fact, in two of the cases, it has been the voice of HSI practitioners that has been heard that has helped to eliminate the programs before they were made into problems for the soldiers. The major problem has been nonpractitioners, either within the program or higher in the army acquisition leadership, not fully appreciating the importance of HSI to program decisions. This suggests that the top priority to avoid system failures in the future is increased emphasis on education and training for nonpractitioners involved in the acquisition process.

## 18.4 CASE STUDIES OF SYSTEM BENEFITS

The review of army systems and HSI application has revealed a number of beneficial results that should be especially attractive not only to the HSI practitioner but also to the non-HSI practitioner to better understand the value of HSI in systems engineering and management terminology. Four systems have been studied in detail as case studies by Booher (1997) from this point of view. The systems comprise two aviation systems, Comanche and Apache; one NBC reconnaissance vehicle; and the army's advanced howitzer system, Crusader. Selections from these four case studies are presented here as examples of program benefits in terms of acquisition process efficiencies, system design improvements, casualty reduction, and cost avoidance.

### 18.4.1 Acquisition Process Efficiencies

Two examples of major systems acquisition process efficiencies are provided below: the Comanche acquisition process and the Crusader battlelab experiment.

**Comanche Acquisition Process** The Comanche program provides the best-documented example of HSI influence on the systems acquisition process. The Comanche philosophy was to focus on maximizing the army aviation's battlefield influence by fielding a *totally integrated* weapon system with the appropriate mix of quality soldiers, hardware, and software. To achieve a "total system," as opposed to an "equipment-oriented" perspective, HSI principles were applied to the design and development of the Comanche aircraft. Inherent in such a philosophy of a total system's view was the crucial concept that the soldier is not added to the system but that the soldier (whether aircrew member, maintainer, or support personnel) is an integral part of the system.

This total systems philosophy required a new organization and management process that horizontally integrated the widely disparate MANPRINT, supportability, engineering, and cost disciplines. The horizontal integration of discrete development processes encouraged the breakdown of traditionally organizational barriers and facilitated interaction outside those barriers. In this way, effective design decisions could be made that reflected all participating disciplines. This, of course, is the intention of the modern acquisition improvement concepts with integrated process teams (IPTs). The Comanche program went beyond this, showing that the IPT was most effective because MANPRINT was provided a prominent status. In fact, for integration across disciplines, only the focus on the soldier permitted a true integrating strategy.

Minninger et al. (1995) highlight a number of management initiatives driven by and/or compatible with HSI principles:

- concept exploration and advanced modeling/simulations,
- concurrent engineering (integrated concept teams/integrated process teams),
- source selection and MANPRINT,
- continuous acquisition and life-cycle support (CALS),
- Comanche supportability initiative,
- HFI quantitative trade analyses,
- TRADOC system manager—forward, and
- pilot vehicle interface mechanization and specification.

Several of these initiatives described below illustrate the major influence HSI methodologies had upon the Comanche acquisition process.

*Concept Exploration and Advanced Modeling/Simulations* Long before the current Comanche program during the concept exploration stages for the LHX program, advanced modeling and simulation activities were initiated through the Advanced Rotocraft Technology Integration (ARTI) Program. Pilot workload issues were considered early on as a potential limiting factor to the LHX concept. Advanced simulation was utilized in the study of pilot tasks using a wide-field-of-view helmet-mounted display, electro-optical systems, and very high speed integrated circuit (VHSIC) electronics. Human-driven analyses, computer simulations, and physical mock-ups were used to improve and assess the effectiveness of the aircraft's total system performance. At the time, an important manpower issue was one- versus two-pilot cockpit and a critical training issue was simulation fidelity.

A MANPRINT analysis of pilot tasks was used to reduce the risk of the LHX developmental program and prove the feasibility of a single-pilot scout/attack helicopter as well as general cockpit and architecture design. In order to meet the single-pilot objective, the state of the art had to be pushed to the maximum. As an absolute minimum, not only did human engineering requirements have to be incorporated into the aircraft architecture but also the majority of in-flight functional activities had to be automated. The automated features included detection, recognition, identification, and prioritization of targets; management of noncritical flight control functions; navigation; automatic location reporting; and mission and flight status. The technology thrust was to provide this critical real-time information within the pilot's field of view looking outside the aircraft, so he or she would not have to look down at the control panel. The HSI research showed this was feasible by using sophisticated heads-up/eyes-out displays integrated into the pilot's helmet. The helmet-mounted display also could provide forward-looking infrared (FLI) imagery for target identification and acquisition. The cockpit design also incorporated two integrated multipurpose displays mounted in the control panel.

As part of the modeling and simulation efforts, performance and work loading data were obtained from HSI real-time simulations of flight dynamics, external visual scenes, and responses of mission equipment packages. Flight tests in modified aircraft verified the HSI simulation findings that a pilot could use helmet-mounted and multipurpose displays while performing normal flight tasks.

*Source Selection* See Example 18.7.

*HFI Quantitative Trade Analysis* See Example 18.18.

*TRADOC System Manager—Forward* Prior to the downselect of the contractor team to complete development of the Comanche, the army provided teams of TRADOC soldiers to support the contractors. These teams were composed of aviators and maintenance personnel selected for their experience and ability to communicate "user" information to the contractors during the design phase. Following downselect of the prime contractor, a team of soldiers were provided to the contractors on site as an extension of the Comanche TSM; it became known as the TSM-forward. The TSM-forward was a unique concept in that it was neither a part of the Defense Plant Representative Office (DRPO) or part of the Program Manager's Office (PMO). The objectives of the TSM-forward were to address and prioritize user operational and MANPRINT concerns during the demonstration/validation (DEM/VAL) prototype and subsequent engineering and manufacturing development (EMD) phases. The presence of the TSM-forward in the contractors' facility allowed user's issues and concerns to be identified in a timely manner. As an example, TSM-forward activities with the IPTs reduced the time period to turn around design changes between contractor and government. In one instance, a rotor design change that would routinely have taken 12 months for contractor/government approval was completed in 30 days.

*Integration with Advanced Systems Management* Other new systems management initiatives (e.g., total quality management, concurrent engineering, integrated logistics support) created an environment for Comanche design and development that was compatible with the human-centered approach. As a direct result of these efforts and changes in the acquisition process, more than 500 design improvements were approved to aid in system performance and logistics. These improvements were accomplished while demonstrating projected cost avoidance of $3.29 billion in manpower, personnel, training, and safety. Additionally, 91 fatalities and 116 disabling injuries would be avoided.

***Crusader Battlelab Experiment*** The experiment on Crusader (Pierce, 1996) was conducted in a first of a kind *synthetic environment* comprising real and simulated systems in a complete battlelab environment. The real systems included such tactical digital systems as the Advanced Field Artillery Tactical Data System (AFATDS), Initial Fire Support Automated System (IFSAS), and Fire Direction System (FDS). The simulations were at two levels: the maneuver battle using the DIS-compliant version of Janus and the support processes simulated by the target acquisition and fire support model (TAFSM). World Modeler, an interpreter, created the interface between the two simulations. The Janus simulation was staffed by interactor and player staffs using Crusader scenarios. Crusader characteristics were played in the TAFSM, and soldiers from field artillery units were used to generate and process fire missions, resupply missions, and *tactical coordination and movements*. The HSI personnel at the HRED Ft. Sill Field Element led the experiment to demonstrate the feasibility of the synthetic environment playing war games of complex battle scenarios with full soldier performance data, including battle staff performance.

Forty battalion-level staff from a field artillery unit participated in the battlelab experiment. The scenario selected represented an artillery battalion performing a direct-support role for an attacking brigade and its three task forces. The principal offensive operation was a movement to contact that included a reconnaissance, hasty attack, obstacle breaching, forward passage of lines, and deliberate attack by the maneuver forces. In the experiment, personnel were assigned roles for the maneuver element, the battalion tactical operations center, and each of six platoon operations centers. The event stream was those events that make up a complete command-and-control cycle, including fire mission processing; survivability and tactical displacements; and resupply planning, coordination, and execution. The TAFSM performed fire support officer functions and disseminated instructions to players in tactical message format.

The study examined implications of Crusader systems on command-and-control processes using the event stream. The training/test purpose of the synthetic environment exercise was to stress the unit command-and-control system, to determine what levels of fire support activity stress this system, and where the system is likely to break when these levels of activity occur. The level of activity was varied through fire missions, movements, platoon operations center performance, and the scenario.

Two principal questions about Crusader performance were asked of the first battlelab experiment:

1. Can the Crusader deliver effective fires to defeat the projected threat?
2. Can Crusader ammunition resupply system support the battle (operations tempo OPTEMPO)?

The answer to both questions was in the affirmative, but the experiment provided greater specificity about the relative important of certain TTPs as well as equipment capabilities and limitations. For example, to deliver effective fires, it was discovered that additional command-and-control processors were required at battalion and platoon. The techniques for "shoot and move" were not only confirmed as sound but were also shown necessary to enhance Crusader survivability against counterfires. Additionally, potential fratricide situations were uncovered and tactics were developed to avoid fratricide. Also, for the assurance of effective fires, the experiment found it critical to define specific tactical and technical fire control roles and responsibilities for the platoon centers.

For the resupply system to support the battle OPTEMPO, the experiment confirmed the need for "pooled" resupply vehicles at the platoon level. It was also found that the pooled condition allowed the resupply vehicles operational cycle (rearm, hide, resupply) to keep up with conditions of increased fire mission processing.

The battlelab experiment showed the value of the simulator as a trainer for field artillery collective training and as a means of testing alternative Crusader TTPs. Because of HSI involvement, unit performance can now be observed in the various battle games for systems as complex as the Crusader operating in a digital battlefield. Shortfalls, gaps, and improvements in the warfighting doctrine can be evaluated and used by the army to propose new doctrine for systems such as the Crusader upon fielding.

### 18.4.2 System Design Improvements

Eleven examples of system design improvements from HSI are provided below from the four case study systems: Comanche, Crusader, Apache Longbow, and Fox.

***Comanche System Design Improvements*** The Comanche aircraft has been designed to be the most sophisticated helicopter ever built. It incorporates state-of-the-art technology throughout every component and subsystem of its design. Apart from those disciplines advancing helicopter technology itself, HSI is one of the most important disciplines contributing toward making the Comanche system a highly capable, operable, and supportable weapon system. Figure 18.2 illustrates several of the design features most notably influenced by the MANPRINT design team. The crew station design, the T-800 engine, and the box structure design are selected for further discussion below.

*Crew Station Design* Early simulations and modeling, lessons learned, and user inputs allowed the cockpit truly to be designed from the pilot outward. The objective of the crew station design process was to blend the airframe, computers, sensors, and crew into a low-workload, low-error-rate, high-situation-awareness, and quick-reaction cockpit. The Comanche Human Factors Engineering Group used the army's task analysis/workload (TAWL) methodology to perform analyses of the operator tasks. As a result of the TAWL analyses, designers were able to meet the following crew station design objectives:

- Reduce the number of sequential tasks required to perform mission functions.
- Ensure human performance demands from design do not exceed human performance capabilities.
- Ensure task performance times are acceptable for the mission.
- Ensure that the controls and displays provide adequate interface information to accomplish mission tasks.

More specifically, the TAWL and TAWL Operator Simulation System (TOSS) assisted the design team to simultaneously combine critical target acquisition and attack data with critical flight control data. This information can be displayed to the aircrew through the

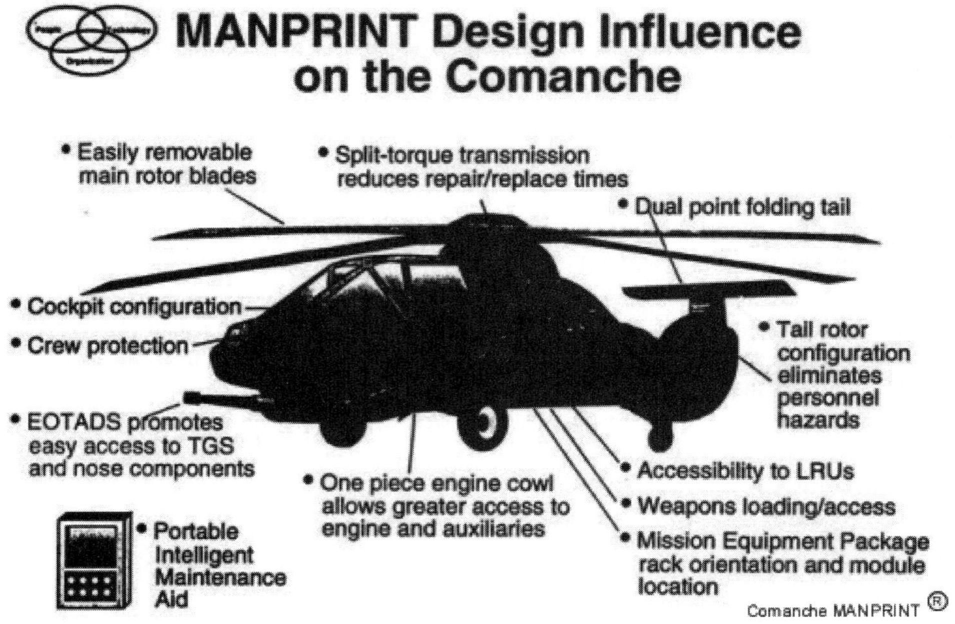


**Figure 18.2** Comanche design improvements.

tactical situation display (TSD) mounted on the display panel or the Helmet Integrated Display Sighting System (HIDSS) attached to the crew member's helmet.

*T-800 Engine* The T-800 engine was the first army development program in which the MANPRINT process played a major role. MANPRINT's visibility allowed ILS and RAM programs to be more effective in influencing the design process and also provided for the integration of soldier capabilities and limitations with system development. During the design and development process, widely varying HSI tools (analyses, models, and mockups) were utilized to improve, validate, and assess the effectiveness of the T-800 system. Benefits were extensive in the areas of MPT as a result of government limitations in the RFP stating the design was to have no increase in skills or manpower numbers. The engine had an extensive number of improvements based on the MPT limitations. The modular design eliminated the need for scheduled overhaul; the elimination of the need for torque wrenches reduces both the number of tools required and the level of maintenance. In designing the engine to be more maintainable, it had become more reliable as well. The increased reliability and maintainability not only decreased the maintenance per operating hour but also reduced overall training burden by as much as 40 percent for comparable engines of the current aircraft fleet. Some of the other many benefits to the T-800 from HSI have been documented by Howington and Goldthwaite (1989), by Booher (1990), and in a 1993 case study held in the army MANPRINT headquarters office (DAPE-MR)

*Box Structure Design* Driven by MANPRINT access requirements to helicopter on-board components, especially in a field environment, an entirely new load-bearing structure was designed for the Comanche. The new box beam structure is a graphite–

epoxy composite material that allows more than 50 percent of the exterior skin to have access doors and panels. Mission equipment packages (MEPs) are accessible for maintenance and/or inspection in a field environment. Several of the access panels open at convenient locations to serve as work platforms, thus eliminating the need for separate ladders or special work platforms. The design and placement of aircraft components, built-in access doors, and convenient work platforms make it possible for fast turnaround of maintenance and loading tasks. By partitioning the Electro-Optical Target Acquisition and Designation System (EOTADS) sensor functions, a 40 percent life-cycle cost avoidance in supply stockage is projected. Loading of the 20-mm gun can be accomplished by one person loading from the side of the aircraft. The feature of adjustable weapon bay doors allows missile ordnance loading in less than 13 minutes with only two personnel.

***Crusader System Design Improvements*** As the Comanche, the Crusader has had a number of system design improvements generated by HSI practitioners. One of the best documented examples is how MANPRINT affected the manpower and personnel design decision for the artillery and resupply systems. At the time, the Crusader was called AFAS-FARV. The general question for HSI was whether the 13B MOSs with regular training, using the AFAS-FARV under sustained operations, could accomplish their mission.

Three specific manpower and personnel questions were asked of the practitioners with their HARDMAN analysis:

1. What is the optimal crew size for the AFAS and the FARV?
2. What combination of ASVAB area composites and area cutoff scores for the AFAS and the FARV results in enhanced mission performance while not restricting the availability of qualified personnel?
3. Is there a basis for selecting an appropriate MOS for the AFAS and the FARV?

To address the crew size question, the HARDMAN analysis team looked at performance of different crew sizes two, three, and four under different environments (Desert Storm, tropical, NE Asia-Korea) under a range of scenarios (standard, rapid fire, direct fire; degraded operations and FARV upload-manual and automatic). The crew's performance was also examined for effects of special stressors such as mission-oriented protective-posture (MOPP) gear, continuous operations, heat, cold, humidity, wind, and noise.

Two of the most significant conclusions on crew size were as follows:

1. With the exception of two-man FARV crews with automatic upload, only three-man crews could perform mission requirements accurately under any of the conditions examined.
2. Automatic upload was essential for FARV. Even a four-man crew could not meet mission performance times in the manual mode. The automatic upload showed either two- or three-man crews consistently met mission performance times.

However,

3. In a desert or tropical environment and after 48 hours of continuous operations, the FARV two- and three-man crews made 40 percent more errors than the four-man crew.

To answer the personnel questions, three-man crews for both AFAS and FARV were assumed. Fewer environments and scenarios were examined and continuous operations were held below 48 hours. Two area composites, field artillery (FA) for the 13B MOS and operations and food (OF) for the 13M MOS were considered. The ASVAB cutoff scores examined were 85, 95, and 105.

The findings supported the following ASVAB area conclusions:

1. For the AFAS—FA for 13B MOS and OF for the 13M MOS perform about the same in normal operations, but the OF area composite crews produced about 34 percent fewer mission aborts than FA-selected crews. The area cutoff scores recommended therefore were FA 95 and OF 85 or OF 95. For the FARV—Increased aptitude was not significant in improving performance.
2. Although the OF area composite teams could perform adequately with lower cutoff scores and better under continuous operations, the difference was not so great as to select a 3M MOS specialty for Crusader. Utilization of personnel from both MOSs could increase the availability of qualified personnel. For the AFAS the standard 13B MOS can perform adequately, so long as the cutoff score is 95.

***Apache Longbow Design Improvements*** Irving et al. (1994) report on a McDonnell Douglas helicopter systems MANPRINT cost savings study conducted on the Longbow Apache. The study covered the four previous years in which the Longbow Apache MANPRINT team participated in the EMD phase where issues were raised throughout the concurrent engineering process. Those issues that were not readily resolved were labeled as problems, issues, and concerns (PICs). An item could become a PIC from recommendation by the army, by failure to comply with documented company or military standards, or by continual refusal by a designer to comply with user-friendly design practices without acceptable rationale. At the time of the study 161 PICs had been documented and 86 had been resolved. Five of the resolved PICs were selected for detailed analysis and are presented as items here as illustrative of typical HSI design issues and methods of resolution.

The five items selected for HSI analysis were (1) seat stroke interference, (2) extended forward avionics bay (EFAB) contour, (3) rotor head access, (4) tail rotor rigging pin, and (5) data rate adapter mounting. The rotor head access and the EFAB contour related to design deficiencies that could have caused loss of life and aircraft had they not been resolved. The remaining three were concerned with maintainer access to components and fasteners and the time and costs involved with difficulties in access.

*Seat Stroke Interference* The Apache is equipped with crash-survivable seats that "stroke" (collapse) during a crash to absorb energy in order to reduce injuries to crew members. The original design for the Longbow Apache included new brackets for the left consoles of both crew stations that reduced the clearance on the left side of the seats and interfered with the stroke. As a result of HSI, the depths of the control panels on the left side were reduced and the Apache Longbow brackets were redesigned to allow the seats to stroke identically to those in the fielded Apache (AH-64A).

*Rotor Head Access* In order to access the rotor head, Apache maintainers have a habit of standing on the engines, the infrared jammer support, and catwalk door hinges. These practices have led to injury and maintenance-induced damage in the AH-64A Apache. A review of lessons learned from the AH-64A brought this issue into the MANPRINT analysis process. The analysis found the Longbow Apache Environmental Control System (ECS) structure would be exposed to damage when used by mechanics as steps and handholds.

As a result of HSI recommendations, the ECS support structures were redesigned to incorporate a work platform. The new platform not only provides maintenance access to the rotor head components but also provides protection to ECS components. The analysis of frequency of repair in the rotor head area showed Apache maintainers might need to access the rotor head area over 92,000 times throughout the fleet life cycle.

*EFAB Contour* The Longbow Apache avionics bays were enlarged over the predecessor system, causing designers to redesign for changes in airflow. On the right side of the aircraft, a fairing was constructed to improve airflow over the top of the wing. Unfortunately, the new design created a safety hazard. If during flight, a foreign object were to be directed down the top of the EFAB, the object would likewise be directed toward the engine inlet and sucked into the engine. The faster the forward aircraft's air speed, the more likely the ingestion of the foreign object. If this were to occur during nap-of-the-earth, an engine failure could result in loss of aircraft and flight crew. As a result of the HSI effort, the fairing was eliminated and replaced by a smaller fairing that diverts air and foreign objects under the wing and outboard rather than into the engine.

*Tail Rotor Rigging Pin* The proposed rigging of the tail rotor flight controls was difficult to access. The maintainer had to insert a pin in the flight control package located below the pilot crew station right console. Two ECS components, a fan, and an evaporator had to be removed to access the rigging pin hole. Also, an additional maintainer MOS was required for removal of ECS components. The human factors redesign was to relocate the fan and evaporator slightly aft to allow access for the rigging pin, eliminating both the access problem and the second maintainer.

*Data Rate Adapter Mounting* Line-replaceable units (LRUs) mounted below the Longbow programmable signal processor are tightly packed. Data rate adapters (DRAs) mounted in this area with fasteners facing inboard could not be removed without first removing adjacent LRUs. By fastening the DRAs to a sheet metal bracket that mounts to a shelf with fasteners facing outward, maintenance was eased.

***Fox Vehicle HSI Modeling*** Illustrated in Figure 18.3, the Fox vehicle (formally the XM93E1, NBC Reconnaissance System) provides one of the clearest examples to date of how the integration of HSI technology from different domains can provide vastly superior results over nonintegrated applications of the same technology.

The Fox is designed to move over terrain possibly having NBC contamination, pick up and analyze the samples, and determine the nearest "clean" area. The Fox was originally designed for operation by a crew of four without consideration for female anthropometrics. The army wished to field the Fox as quickly as possible as an army category (ACAT) III nondevelopmental item (NDI) but with some changes in field operations. The changes included (a) reducing the crew from four soldiers to three soldiers, (b) replacing contractor maintenance with army logistics support (i.e., the soldier), and (c) adding standoff detection capability (an additional soldier task). From a workload perspective, it was

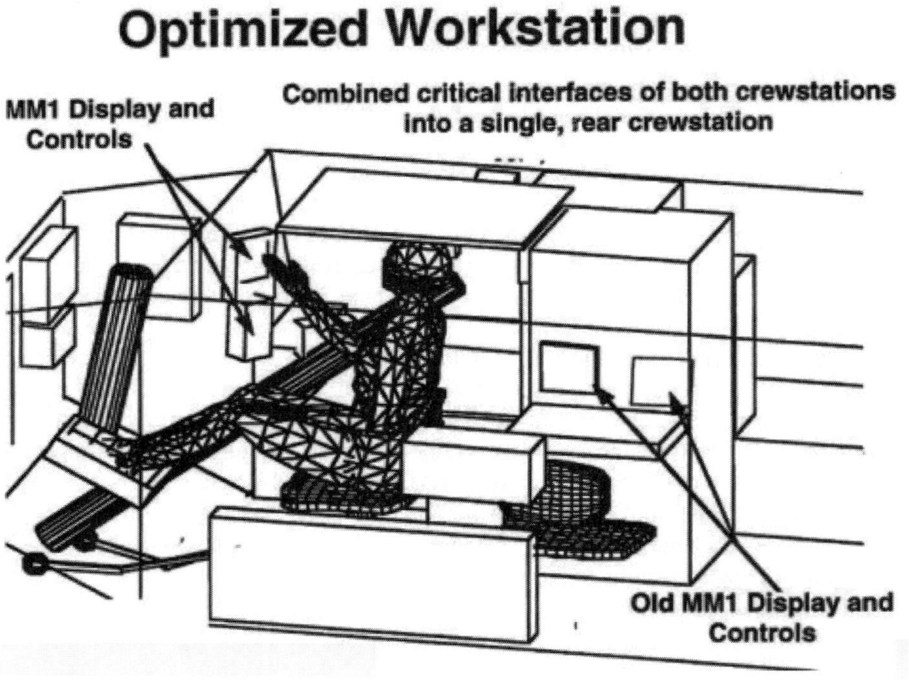


**Figure 18.3** Fox vehicle.

apparent right away that the Fox without design modification would have a serious problem with crew workload. The soldier maintenance and standoff detection would increase the tasks, which would be distributed among fewer soldiers. An excessive workload determination was subsequently confirmed by the Operational Evaluation Command (OEC), which gave the Fox an initial outfit T&E (IOT&E) assessment of "unsuitable and ineffective." The Fox program manager requested HRED of the Army Research Laboratory to assist in making this vehicle effective and affordable. McMahon (1996) describes the strategy utilized by HRED to design a solution based on two different types of HSI modeling capability: a workstation human figure modeling and a HARDMAN III task network modeling.

*Human Figure Modeling* The original four-person crew had two positions at the front of the vehicle, one on the right side and one at the rear. In order to eliminate one of the crew, a workstation design change was required to combine two positions into one. It was decided that the rearward positions could be combined into one by combining the man–machine interfaces. Anthropometric-sized human figure models were created for each of the Fox crew stations. The human figure models of the rear stations showed how the old controls and displays (MM1) for the seat on the right could be combined into a single, rear crew station. The human figure model was also exercised to verify that the design was within the field of view and reach envelope of a 5th percentile female operator.

*HARDMAN III Task Network Modeling* The human figure modeling provided confidence that the two crew stations could be combined into one. It was still a question, however, whether the three crew members would be able to meet mission requirements that included movement to a starting point, taking a spectrum, and then finding the near-side clean area. To accomplish these mission functions, the rear crew member must continually interact with a spectrum monitor, a probe, and sampler wheels. The Operational T&E Command (OPTEC) was not convinced that the functions could be satisfactorily accomplished under conditions of stress and fatigue over long periods of time. The OPTEC test scenario was 24 hours a day for five days a week over two to three months. This test was of such an extent that it was estimated by the program manager to cost between $2 and $4 million, an amount sufficiently large to cancel an ACAT III program. However, OPTEC would allow certain performance model estimates to supplement the operational test and evaluation. By using the HARDMAN III (MAN-SEVAL) model, previously accredited by OPTEC, to obtain system performance estimates, the actual test was reduced to a much more affordable test–4-hour missions, 8 hours a day, for only two to three weeks.

The HARDMAN III model was set up to analyze the Fox operations with inputs on mission definition, crew performance time data, and task workload estimates. Mission definition was stated in terms of functions and subfunctions derived from the Fox mission crew drills. Performance time data came from the Fox IOT&E of fiscal year 1994. The workload assignments for visual, cognitive, psychomotor, and auditory tasks came from subject matter experts using McCracken–Aldrich scale values.

The HARDMAN model verified that the Fox human factors modifications (HFMs) would meet performance requirements in all mission functions. In fact, the overall mission time for HFM showed a 20 percent reduction from the original mission time. It was determined that the modifications not only allowed one soldier to do the combined tasks previously done by two but also improved the soldiers' ability to interact with the monitor, probe, and sampler wheels.

*HSI Tools Interaction* The Fox vehicle demonstrated three significant points about the application of HSI technology. First, HSI technology can make a program successful, even if it is one where only relatively small design modifications are possible. Second, the Fox clearly illustrated that HSI man–machine interfaces and workspace layouts are necessary when attempting to reduce manpower without creating excessive workload. Third, Fox demonstrated the importance of utilizing widely varying HSI tools from the different domain to help the program manager achieve the program mission. The human factors interface technology helped design the optimum solution but would not have been adequate to forego the OPTEC expensive test scenario without the HARDMAN task network modeling. On the other hand, if only network modeling had been done to the original design, little more would have been shown than that OPTEC was correct—that the workload was too excessive to conduct the mission.

### 18.4.3 Safety Improvements

Although safety improvements through HSI design were inherent in all of the case study systems, the Comanche provides the best documented example of how HSI can reduce military casualties.

***Comanche Casualty Reduction*** It is projected that use of the Comanche rather than the OH-58 A/C and AH-1F aircraft will avoid 91 soldiers' deaths over a period of 20 years.

Similarly, use of the Comanche will avoid at least 116 disabling injuries. Nine years of accident/incident data reported to the U.S. Army Safety Center was reviewed for events causing personnel deaths and disabling injuries in the older aircraft. During this period there were 26 AH-1 and 39 OH-58 related fatalities (i.e., fatalities that safety analysis showed could have been prevented by improved design). Also during the 9-year period, there were 23 AH-1 and 63 OH-58 related disabling injuries. Some of the incident types and corresponding design improvements are listed in Table 18.5.

### 18.4.4 Cost Benefits

Three of the systems, the Comanche, Apache Longbow, and Fox, provided clear investment and benefits costs *that could be directly attributed to HSI activities*.

***Comanche Cost Avoidance*** Minninger et al. (1995) fully document their assessment of cost avoidance due to MANPRINT/HSI. Although MANPRINT attributes were closely linked to other disciplines such as ILS and RAM, it was not always possible for the analysis to identify those savings due directly to HSI. However, the cost avoidance documented in that report was entirely from the MANPRINT domains of MPT and safety. It was also recognized that it was the MANPRINT approach with the focus on the soldier and communication to industry through its acquisition process that significantly changed the design process for the contractor. The cost avoidance assumptions and details of the cost avoidance estimate rastionale are provided in Appendix B of Minninger et al. (1995).

The Army Manpower Cost System (AMCOS) model was used to quantify cost avoidance due to the contributing factors of MPT that follow from such items as reduction of number of MOSs, reduction in maintenance levels, and reduced training requirements. The contributing factors for the Comanche were compared to the predecessor systems OH-58 and AH-1 being replaced with the Comanche. In order to standardize comparisons,

**TABLE 18.5 Incident Type and Design Improvements**

| Incident Type | Design Improvements |
|---|---|
| • Aircraft collisions | • Improved outside visibility<br>• Two pilots for all current missions |
| • Aircraft crash | • Improved night vision capabilities<br>• Improved situational awareness<br>• Ground proximity warning system<br>• Improved airframe crash survivability |
| • In-flight break-up | • Strengthened composite airframe<br>• Improved rotor system prevents mast bumping |
| • Engine failure | • Monitoring systems warn of impending failure<br>• Engines can operate 20 minutes after loss of oil<br>• Multiple engines |
| • Loss of tail rotor effectiveness | • Fantail system does not limit flight envelope<br>• Fantail can operate after loss of a blade |
| • Ground accidents | • Work platforms built into the airframe<br>• Fantail shrouded with added safety bars |

identical operational tempos were used for the Comanche and the predecessor systems. It is important to recognize that the systems being replaced would not only require the higher MPT costs but would be unable to perform many of the new capabilities provided by the Comanche. Other analyses such as those described above in determining fielding requirements showing a 25 percent reduction in overall maintenance requirements are not reflected here, because those analyses consider the full MPT needed to make complete use of the Comanche's capabilities.

Safety and soldier survivability estimates were based on safety center mishap data and consideration of those specific Comanche design improvements aimed at eliminating design deficiencies of the Kiowa and Cobra aircraft that safety analyses show could have been prevented by design changes.

The cost avoidance figures due to HSI were broken down into four categories. Manpower showed that 32 percent of the predecessor manpower costs will be avoided in the Comanche equating to $2.67 billion. Personnel and training together will avoid 33 percent of predecessor personnel/training costs or $440 million; safety, health hazards, and soldier survivability costs avoided equate to $180 million. *The total Comanche cost avoidance due to HSI is $3.29 billion.* Since the total costs for MANPRINT on the Comanche (past and projected) is $74.9 million, the return on investment over 20 years is 4390 percent.

***Apache Longbow Cost Benefits*** Irving et al. (1994) report that 80 of the 86 resolved PICs were judged capable of objective analysis for determining quantifiable cost savings or cost avoidance for their customer. The study team felt the five PICs discussed above were a good representation of the range of HSI impacts on the Apache.

*Seat Stroke Interference* Using historical data for class A mishaps, the cost avoidance for the seat stroke interference design correction led to an estimated savings of $2,610,000, not including the loss of crew productivity or the incalculable loss of aviator's lives. This deficiency was resolved by making minor changes to one control panel and a single bracket at a nonrecurring cost of less than $10,000.

*Rotor Head Access* The work platform recommendation for the Apache ECS support structures redesign came as a cost-effective solution to avoid maintenance-induced damage. Assuming the expensive blower or transition duct could be damaged by maintenance personnel to the extent that they would need to be replaced 2 percent of the time, cost avoidance of replacement parts alone (not including aircraft downtime or man-hours to make the repair) would be about $4,577,000. The fleet implementation expense for the maintenance platform will be about $568,000, a return of 8 times the investment.

*EFAB Contour* To avoid the potential hazard of a foreign object being sucked into the engine because of the EFAB contour, HSI recommended a design that diverts air and foreign objects under the wing and outboard rather than into the engine. This hazard was resolved with a nonrecurring cost of approximately $10,000, with a cost avoidance of over $10 million.

*Tail Rotor Rigging Pin* The HSI redesign of the fan and evaporator location to allow access for the rigging pin, which eliminated both the access problem and the second maintainer, was made with an implementation cost of $8000 and reduced manpower costs by about $300,000.

*Data Rate Adapter Mounting* The small change of a bracket that mounts to a shelf with fasteners facing outward cost about $4000 to install but allowed cost savings of over $76,000.

For the five PICs alone the design and implementation costs were $600,000, but the study team found a $16.8 million cost avoidance over the life cycle of the program. They concluded this only represents a small fraction of the total cost savings/avoidance to be realized by the army throughout the Longbow Apache life cycle. The investment in MANPRINT for the entire full-scale development is $2.7 million. Allowing for implementation costs, the five PICs alone will provide a return 5 times (500 percent) the investment into HSI for the program. But if one were to extrapolate to all 80 PICs, the return would amount to over 20 times the same investment, not as high as the Comanche, either in total dollars saved or return on investment, but a number well worth the investment.[1]

***Fox Cost Benefits*** The Fox vehicle demonstrates a number of HSI lessons learned and quantitative cost benefits not realized before. First, as an ACAT III program that is NDI, only relatively small modifications are possible. The Fox clearly demonstrated that HSI human–machine interfaces and workspace layouts are necessary when attempting to reduce manpower without creating excessive workload. Second, Fox demonstrates how widely varying HSI tools can be used to achieve the program mission. The human factors interface technology helped design the optimum solution but would not have been adequate to forego the OPTEC expensive test scenario without the HARDMAN task network modeling. On the other hand, if only network modeling had been done to the original design, little more would have been shown than that OPTEC was correct—that the workload was too excessive to conduct the mission. Finally, not only was the program saved, but it was also done in a very cost-effective manner *that was reflected in the PM budget in the near term*. The estimated cost to the PM for the HFI analyses (which were completed in 60 days) was $60,000. The operational test savings were $2 to $4 million.

## 18.5 HSI FACTORS AND FUTURE WEAPONS SYSTEMS ACQUISITION

There are several difficulties facing the army leadership with a decision to revitalize the old MANPRINT process and apply it to future army systems. Although factors can be identified that were successful in the past, it is fair to question how well these factors will translate to new systems, considering that many of the acquisition processes have changed. For example, it is not known what effect acquisition reform changes such as using integrated concept and process teams or elimination of coordination documents like the system MANPRINT management plan will have on future systems. Additionally, the effect of reducing the numbers of practitioners representing the individual MANPRINT domains may provide a new personnel and training issue that was not a problem in the past.

A second task in the study by Booher (1999) was to evaluate the critical HSI factors for applicability to systems being procured now and in the future under the DoD acquisition reform practices. Starting with the baseline information for the 10 systems in Table 18.1, Booher reanalyzed the 10 HSI success factors in view of 8 relevant acquisition reform factors (ARFs):

R1. rapid acquisition process (RAP);
R2. increased NDI, COTS (INC);
R3. reduced emphasis on SMMPs, MJWGs (RSM);
R4. greater reliance on battle labs, simulation, and modeling (BSM);
R5. ICTs, IPTs, Integrated T&E (INT);
R6. fewer practitioners (FPs);
R7. fewer nonpractitioners (FNPs); and
R8. greater reliance on total system performance (TSP).

Table 18.6 presents the results of a matrix analysis conducted for the 8 ARF factors and the 10 HSI success factors. The HSI success factors for past systems were judged whether they would have been significantly influenced by an ARF, positively (+), negatively (−), or with no change (nc).

The results of the analysis of HSI success factors with the interactions of the ARF are summarized in Table 18.7. The two most important findings were as follows:

1. Two of the acquisition reform factors either have positive or no effect on all the HSI success factors. These are R4 (battlelabs, modeling, and simulation) and R8 (total system performance). MANPRINT is conceptually consistent with front-end decision making that uses human performance data. To the degree that R4 and R8 include human performance parameters on the domains of MANPRINT, better performing and cost-effective systems will be produced.

2. Six of the eight reform factors have a negative effect on many of the HSI success factors. In particular, success factors 4 (domains integration), 5 (system development integration), 6 (quantitative human parameters), 9 (practitioners), and 10 (education and training) are most negatively affected by the reform factors.

Booher (1999) made the following conclusions regarding the HSI success factors' role in future systems acquisition:

1. In general, the 10 success factors for past systems appear to be the same factors that should be part of the process for all army systems of the future.

2. The reform factors provide no basis for the proposition that reduced attention to any of the factors will result in systems successes in the future. Although greater utilization of battlelabs, simulation and modeling, and total system performance are highly compatible with some of the HSI success factors, this does not offset the extremely negative effects of most reform factors on most of the success factors.

3. No tailoring of factors for future systems can be recommended beyond that shown in the HSI success factors. The problem is really how to achieve satisfactory results in the future under seriously degraded conditions. The best solutions appear to be increased emphasis on strong HSI policy for requirements, source selection, and test and evaluation;

**TABLE 18.6 Acquisition Reform Factors Analysis Matrix**

| | Acquisition Reform Factors | | | | | | | |
|---|---|---|---|---|---|---|---|---|
| Success Factors | (1) RAP | (2) NDI | (3) SMP | (4) BSM | (5) INT | (6) FP | (7) FNP | (8) TSP |
| 1. Top-level support | – | – | nc | nc | nc | nc | nc | nc |
| 2. Human-centered design | – | – | nc | + | nc | – | n | + |
| 3. Source selection | nc | nc | nc | nc | – | – | – | + |
| 4. Domains integration | – | nc | – | nc | – | – | – | nc |
| 5. System documents integration | – | nc | – | nc | – | – | – | nc |
| 6. Quantitative human performance | – | – | – | + | – | – | – | + |
| 7. Trade-off methodology | nc | nc | nc | + | + | + | + | + |
| 8. Test and evaluation integration | + | nc | nc | + | + | nc | nc | + |
| 9. Practitioners, skilled, available | – | – | – | nc | – | – | – | nc |
| 10. (a) Education and training (Practitioners) | – | – | – | nc | – | – | nc | nc |
| 10. (b) Education and training (nonpractioners) | – | – | – | nc | – | nc | – | nc |

**TABLE 18.7 HSI Factors and Acquisition Reform Factors Interaction**

| |
|---|
| *Factor 1. Top-level support and understanding:* Most of the acquisition reform factors have little influence on the importance of this factor. The same emphasis from the top will be critical to MANPRINT success on future systems. However, reform factors 1 (rapid acquisition process) and 2 (more NDI/COTS) can have a negative effect because of the pressure to have less coordination and examination of issues at the top levels. |
| *Factor 2. Human-centered design:* This factor will be influenced in several different ways by the ARFs. The rapid acquisition process and more NDI/COTS can be negative for the same reasons as factor 1. Human Centered Design is most influential when considered early in developmental programs, but both these reform factors can act against considering the human in the requirements stages. Fewer practitioners is a negative influence, particularly at the U.S. Army Training and Doctrine Command (TRADOC), again by having fewer informed people to pay adequate attention to user requirements early. Two positive acquisition reform factors, however, are RF-4 (greater attention to battle labs, simulation, and modeling) and RF-8 (total system performance). These provide data and acceptance criteria that are favorable to a human-centered design. |
| *Factor 3. Source selection:* This factor is negatively influenced by three reform factors: RF-5 (integrated teams), RF-6 (fewer practitioners), and RF-7 (fewer non practitioners). This is because each of these reform factors reduce personnel, which means fewer people to fully attend to source selection criteria development and evaluation. As with human-centered design, total system performance (RF-8) helps offset the negative effects of reduced personnel. |
| *Factor 4. Domains integration:* There are no positive aspects of acquisition reform to offset the extensive negative aspects of having inadequate representation from all the MANPRINT domains. *This is a serious problem, considering that this factor was the one factor that was critical or important on all 10 systems evaluated.* It was often this factor that discovered MANPRINT weaknesses in systems such as the AGS and FMTV. |
| *Factor 5. System documentation integration:* As with the domain integration factor, there are no offsetting benefits of acquisition reform that directly and positively make up for the negative impact of weak inputs to the systems documentation for a system. |
| *Factor 6. Quantitative human performance:* Most of the reform factors have a negative effect on this factor. The reform factors of rapid acquisition and emphasis on NDI/COTS provide tendency to gloss over human performance requirements. The reduction in SMMPs and MJWGs, and integrated teams combined with fewer personnel reduce the likelihood of incorporating important quantitative data on human performance. Similar to factor 2, the two positive acquisition reform factors, RF-4 and RF-8, provide data and acceptance criteria that are favorable to quantitative human performance. |
| *Factor 7. MANPRINT technology:* This factor, along with factor 8, are the only HSI success factors that are positively affected from acquisition reform. Five of the eight ARFs show positive influence on this factor. This is because this factor (and the next) most closely coincide with the best intentions of acquisition reform. With emphasis on simulation and modeling and total system performance, MANPRINT technology is indispensable for assuring systems will perform as designed. Although it is always desirable to have more personnel, this factor is the major one to assure informed decisions can be made with fewer people. |
| *Factor 8. Test and evaluation integration:* This is the one factor that must be applied as the last assurance that operational systems will perform as projected. Providing similar benefits and no negatives as HSI factor 7, it is one of the only two factors positively influenced from the rapid acquisition process. This is because T&E soldier performance considerations must now be fully integrated into the earliest stages of system concept and development. Under the old process, T&E of soldier performance could be left to the end. |

(*continued*)

**TABLE 18.7 (*Continued*)**

*Factor 9. Practitioners, skilled, available:* Along with HSI factors 4 and 5, this is one of the most seriously degraded success factors resulting from acquisition reform. This problem is the most central of all the MANPRINT problems, negatively influenced by six of the acquisition reform factors and negatively affecting six other success factors.

*Factor 10(a) Education and training—practitioners:* This factor is negatively affected by most of the acquisition reform factors, because of the increased burden on education and training of those practitioners that are left. The few practitioners left in the human factors engineering domain must be trained to do TRADOC MPT domains; to participate in more MJWGs; and to provide T&E integration. Additionally many of these practitioners are supposed to do human performance research and develop MANPRINT technology, crucial to HSI factors 2, 6, and 7.

*Factor 10(b) Education and training—nonpractitioners:* This factor is negatively affected by most of the acquisition reform factors, because of the increased burden on education and training of individuals in the acquisition system who can be helpful to MANPRINT. Individuals not responsible for MANPRINT can reduce the obstacles impeding the HSI process through better understanding the value of HSI to help meet system performance needs.

increased funding of HSI science and technology; increased funding of HSI practitioner systems support; and increased education and training of both practitioners and nonpractitioners.

## 18.6 SUMMARY AND CONCLUSIONS

The army's experience with HSI/MANPRINT over the past decade is described in two ways. First is a description and explanation of the relevance of each of the 10 HSI factors that the literature and a recent study by Booher (1999) have shown to be crucial to army weapons system success. Thirty-four specific examples from 15 army systems are used in this chapter to describe the HSI success factors. We conclude that 10 HSI factors (listed in Table 18.2) have been major contributors to army systems success (or failure) in the past.

Second is a report of four case studies of army systems (Comanche, Crusader, Apache, and Fox) conducted by Booher (1997) that documents the benefits of HSI to these systems in terms of acquisition process efficiencies, system design improvements, casualty reduction, and cost avoidance.

### 18.6.1 System Benefits from HSI

The four case studies and the other army systems examined for this chapter show the vast range and depth of influence that HSI has had upon the army systems whenever its methodologies have been applied. Generally, performance improved, safety increased, and costs were avoided. The findings of the case studies are summarized for contributions and lessons learned under (1) technology advancements, (2) acquisition process efficiencies, (3) system design enhancements, (4) safety increases, and (5) major returns on investment.

***Technology Advancements*** The Comanche program demonstrates that technologies across the board are advanced rapidly through the influence of HSI. Not only were the human–machine interfaces advanced to take advantage of the state of the art, but literally the entire engine and airframe construction was advanced by the focus on the soldier philosophy. The HSI technology itself is advanced by research focused on an operational environment and the human technology organizational interfaces. New human figure modeling tools such as those employed on the Fox vehicle are continually being advanced as part of the HSI set of tools to answer such questions as workspace layout, egress, and access to equipment in new or modified designs. Critical to the new digitized battle is the HSI advancement in simulation. Human systems integration is the crucial link to the confidence required to make simulations reliable for the environments being simulated. Such simulations cover a vast array of needs for the Objective Force Army. The Comanche, Crusader, and Fox case studies show the importance of HSI to the capability and validity of those simulations directed to questions on systems performance, speeded-up acquisition processes, twenty-first-century training techniques, and outcomes in warfighting scenarios.

***Acquisition Process Efficiencies*** The Comanche illustrated the numerous desirable acquisition processes that were made to work effectively due to HSI influence:

- Advanced modeling and simulation applied to cockpit, engine, and airframe design at early stages of development.

- Unique source selection process—human systems factors evaluated as a separate major area *and* integrated throughout all other areas.
- Human-centered technologies and disciplines drove critical decisions throughout the design process.
- TSM-forward concept—utilized actual Army operators and maintainers to communicate "user" needs and concerns to contractors at contractors' location.
- System performance defined to include operators' and maintainers' performance as well as equipment performance. This definition carried through operational T&E measures of system performance.

The Fox vehicle case study shows that the benefits to the acquisition process are not limited to new systems. The HSI modeling program can be applied anywhere from milestone O up to milestone IV. The Fox vehicle also shows the major benefits to nonmajor systems as well the ability of HSI to improve the effectiveness of operational T&E. The Crusader illustrates how TRADOC can utilize HSI to evaluate operational concepts, improve the criteria for reducing costs of operational T&E, and make training and testing more effective by integrating real and simulated systems in a complete battlelab environment.

***System Design Enhancements*** The case studies indicated clearly that HSI can be applied to enhance system designs appreciably regardless of the stage of development or how large the system is. Longbow Apache HSI made over 160 critical design improvements for the period evaluated. The ACAT-III Fox vehicle could not have performed its mission if HSI had not designed a new workstation. These two systems were, however, modifications of existing systems, so the HSI potential was limited. To appreciate the full impact of HSI potential on system design, the Comanche is without comparison. A few of the improvements are listed in Table 18.8.

**TABLE 18.8 Significant Comanche HSI Design Improvements**

- State-of-the-art crew station design decreasing pilot workload while increasing mission performance.
- Superior modular main rotor blade design with reduced acoustic vibration, automatic rotor tracking, reduced maintenance, greater transportability, and an approximately $150 million manpower life-cycle savings.
- Tail rotor designed to be eight times safer than conventional designs.
- Portable maintenance aid laptop computer to diagnose systems failure, accumulate critical flight and maintenance data, and replace all technical publications.
- Line-replaceable modular design for mission equipment packages for functional partitioning and diagnostics capability.
- Central-box main structure that acts as primary load-bearing carrier for high structural integrity and allows exterior skin with 50% access panels.
- Enhanced drive train with 73% fewer parts than Blackhawk and 62% less than Apache.
- T-800 modular engine design with increased reliability and 40% reduction in maintenance man-hour requirements.
- Tool set with only 50 tools compared to over 150 for other helicopters, with only 22 of the 50 peculiar to Comanche.

***Safety Increases*** Safety was greatly improved by the MANPRINT teams on both the Comanche and the Apache. The Comanche showed 91 lives saved and 116 disabling injuries avoided from HSI designs compared to the predecessor aircraft. The Apache study did not calculate the number of lives and disabling injuries avoided, but two of the five PICs, if they had not been corrected, would have undoubtedly contributed to unnecessary loss of lives and/or disabling injuries.

***Major Returns on Investment*** The three case studies with quantitative analysis of costs and savings make an interesting comparison (see Table 18.9). The Comanche offers both the greatest return on investment and total costs avoided. The Apache Longbow provides a very commendable savings and return on investment. Both Comanche and Apache returns are spread over 20 years. The advantage of the investment in the Apache is that the investment was considerably smaller and the return began earlier as the Longbow started fielding in FY 98. The Fox vehicle is perhaps the most interesting for considering the future army with few new major systems and major modifications. Systems like the Comanche and the Apache represent an acquisition system of the past, not the future. Program managers and training and doctrine system managers should be aware of the tremendous advantages that HSI offers to the smaller but far greater number of systems that can be improved for soldier use as well as saving resources in the near term. The Fox showed that programs can save considerable operational test and evaluation funds if HSI disciplines and technology have played a role in design, modeling, and simulation.

### 18.6.2 HSI and Future Systems

There is every reason to believe that similar benefits from HSI shown with the case studies and the other army system examples can be derived with future weapon systems. We conclude that the 10 HSI success factors for past systems should be made part of the process for military systems acquisition of the future. However, the HSI factors will be more difficult to implement with future weapon systems. Although the utilization of battlelabs, simulation and modeling, and total system performance are highly compatible with two HSI factors (HSI technology and test and evaluation integration), a majority of the reform factors have strong negative effects on most of the HSI success factors.

In view of the projected negative effects of acquisition reform on most of the HSI success factors, it is recommended that the highest priorities for future HSI acquisition organizations should be (a) increased emphasis on strong HSI policy for requirements, source selection, and test and evaluation; (b) increased funding of HSI science and technology; (c) increased funding of HSI practitioner systems support; and (d) increased education and training of both practitioners and nonpractitioners.

**TABLE 18.9 Major Returns on HSI Investment**

| System | Cost Savings ($) | Investment ($) | Return on Investment (%) | Time (years) |
|---|---|---|---|---|
| Comanche | $3.29 \times 10^9$ | $74.9 \times 10^6$ | 4390 | 20 |
| Apache Longbow | $268.8 \times 10^6$ | $12.3 \times 10^6$ | 2180 | 20 |
| Fox | $2\text{–}4 \times 10^6$ | 60,000 | 3300 | 1 |

## NOTE

1. Extrapolating the 5 PICs to 80 increases the cost figures by a multiple of 16. Assuming the 5 PICs are a good representation, $(16.8 \times 10^6) \times 16 = 268.8 \times 10^6$; and $600{,}000 \times 16 = 9.6 \times 10^6$. Combining total design change costs, $(9.6 \times 10^6)$, and MANPRINT costs $(2.7 \times 10^6)$ gives $12.3 \times 10^6$. Dividing savings by costs $(268.8 \times 10^6$ divided by $12.3 \times 10^6)$ equals 21.8, or 2180% return on investment.

## REFERENCES

Booher, H. R. (1996). *U.S. MANPRINT Practices: Potential Applications to UK Human Factors Integration Procedures*. Bristol, England: Human Engineering Limited.

Booher, H. R. (1997, July). *Human Factors Integration: Cost and Performance Benefits on Army Systems*. ARL-CR-341, Aberdeen Proving Ground, MD: Army Research Laboratory.

Booher, H. R. (1998, May). *Human Factors Integration Case Studies*, Final Report. Prepared for Human Engineering Limited, Bristol, England.

Booher, H. R. (1999, May). *Acquisition Process Factors in MANPRINT*, Final Report. Reston, VA: Raytheon Systems Directorate (Instructional Systems Group). Prepared for Army Research Laboratory, Human Engineering and Research Directorate, Aberdeen Proving Ground, MD.

Booher, H. R. (Ed.). (1990). *MANPRINT: An Approach to Systems Integration*, New York: Van Nostrand Reinhold.

General Officer Steering Committee (GOSC). (1998, July). *Report to the General Officer Steering Committee (GOSC) on the Army's Manpower and Personnel Integration (MANPRINT) Program*. Washington, DC: Booz-Allen & Hamilton. Prepared for the Assistant Secretary of the Army (Manpower and Reserve Affairs).

Howington, B., and Goldthwaite, W. (1989, July/August 1989). The T-800 Engine: A MANPRINT Success Story. *MANPRINT Bulletin, 4*(1), 1–3, 7.

Irving, S., Hampton, A., and Cremonese, V. (1994, April 5). *Longbow Apache MANPRINT Cost Savings White Paper*. Mesa, AZ: McDonnell Douglas Helicopter Systems.

McMahon, R. (1996, June). A Quick Response Approach to Improving and Assessing the Operational Performances of the XM93E1 Nuclear, Biological, and Chemical Reconnaissance (NBCRS) through the Use of Modeling and Validation Testing. Paper Presented at 64th Military Operations Research Symposium, Army Research Laboratory, Human Research and Engineering Directerate, Aberdeen Proving Ground, MD.

Minninger, J. E., Skonieczny, J. T., and Yawn, S. R. (1995, January). *MANPRINT/Human Systems Integration Influence on Comanche Design and Development Program*. St. Louis, MO: Analytic Sciences Corporation (TASC).

Pierce, L. G. (1996). Crusader Battlelab Warfighting Experiment. Briefing, Army Research Laboratory, Human Research and Engineering Directorate, Fort Sill Field Element, Fort Sill, OK.

Skelton, I. (1997, October 1) MANPRINT for the U.S. Army. *Congressional Record—House*, pp. H8269–71.

U.S. Army Audit Agency (AAA). (1997, June 10). *Incorporating MANPRINT into Weapons Systems Development*, Audit Report: AA 97-205. Washington, DC: AAA.

CHAPTER 19

# Human Characteristics and Measures in Systems Design

NITA LEWIS MILLER, J. JEFFREY CROWSON, JR. and
JENNIFER MCGOVERN NARKEVICIUS

## 19.1 INTRODUCTION

The ebb and flow of ocean tides, the timing and exact placement of sunrise and sunset, seasonal fluctuations in weather with winter snows, tropical storms, seasonal monsoons, and migratory patterns of whales and many other creatures are natural events with cyclical and rhythmic characteristics. While these events may be difficult to describe with precision, they are quantifiable and predictable. We understand that such complex natural phenomena cannot be described using a single point in time or a single reference. The context and timing of the observation is critical, and a single measurement cannot accurately describe the complexity and dynamic nature of such a system. Similarly, the challenge posed by accurately describing human behavior requires understanding a vast array of conditions impossible to quantify with a few observations. Human behavior can seem mysterious, imprecise, overly complicated, and difficult to replicate. However, like other natural systems, human behavior is quantifiable and often predictable.

The human features prominently in the design of manned systems. However, engineering curricula do not typically address mental and physical characteristics of the human. Without this knowledge, design engineers do not have the tools to quantify the characteristics of the human and therefore often neglect the centrality of the human to systems design. Such human characteristics must be taken into account in the design, testing, and implementation of new technology and are central to human systems integration (HSI) (Booher, 1990).

There is room in the systems engineering design process to include all subsystems including the human component. The human subsystem, like other subsystems, has characteristics and predictable behaviors. Just as the design engineer selects the best materials based on their strengths and weaknesses relative to the design, design engineers

*Handbook of Human Systems Integration*, Edited by Harold R. Booher.
ISBN 0-471-02053-2 

must also design with the properties of the human in mind. For instance, one would not select reinforced concrete as the material for an aircraft skin. It has excellent structural properties but is too heavy for the application. Similarly, one would not purposely design a user interface that does not allow for optimal performance by the human nor design an interface that encourages the operator to make errors and perform slowly. Economic, time, and performance constraints often are the primary drivers in systems design. The HSI concept does not discourage the emphasis on these primary drivers. However, consideration of the human component is critical for most systems to meet realistic cost, schedule, and total system performance requirements.

### 19.1.1 Human System Characteristics

The characteristics of both human and nonhuman components of a system need to be thoroughly evaluated and understood if the benefits of the HSI approach are to be achieved. In an ideal situation, the requirements of the system will flawlessly match the characteristics of the human operator or maintainer, resulting in a one-to-one correspondence between task and person. Designing a system with the characteristics of the target audience in mind increases the likelihood of a superior product through enhanced system performance. Figure 19.1 illustrates such an integrated relationship.

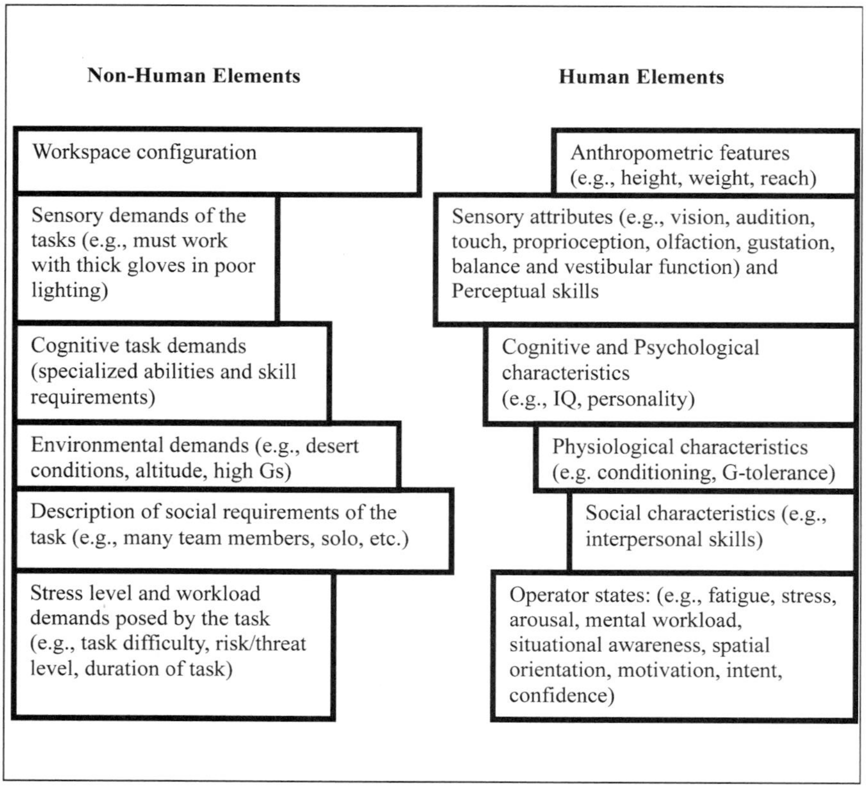


**Figure 19.1** Total system with human and nonhuman elements.

The relationship seen in the two sides of the diagram is analogous to a lock-and-key mechanism, with both human and nonhuman sides having equally important and interdependent qualities. When the two sides map effectively onto each other, this relationship enables a smooth and dynamic boundary between the human and nonhuman components of the system. When the two sides do not mesh effectively, the person must adjust to the nonhuman components through increased personnel skill aptitudes, increased training, or by reduced performance. At best, this mismatch between human and nonhuman components causes additional and unnecessary workload; at worst, it increases the risk for accidents.

This chapter will focus on the right side of Figure 19.1 by describing those characteristics of people that help define them as system components.[1] The chapter will further introduce the reader to sources of information useful in deriving estimates of baseline limits and ranges in the capabilities of people. It will also explore how dynamic shifts in functional capability can occur from highly stressful and complex work environments. Situations that require high cognitive workloads, long work hours, or heightened levels of situation awareness (SA) can cause complete failures in total system performance if the human limitations have not been adequately designed into the system operational expectations.

### 19.1.2 Defining the Human Component of the System

Wherever possible in systems applications, it is important to work with measurable characteristics of the human. This chapter addresses many of the measures used to describe and accommodate the system user, operator, or maintainer. From an HSI standpoint, to adequately define the human component of the system, there must be an engineering understanding of the strengths and limitations of the population of users for which the system is designed. This description needs to define total system performance in such a way that the differences provided by the human component are measurable. Ultimately, knowing more about the human will allow the engineering design team to tailor the system for optimal performance, both from a total system perspective as well as from the perspective of the people who operate or maintain the system.

This chapter seeks to bridge an important gap between engineering and the behavioral sciences for HSI applications. We give an overview of characteristics, measures, and techniques that exist to quantify a variety of human factors categories including anthropometrics, sensation and perception, mental abilities, social abilities, physiological characteristics, and operator states under varying environmental conditions. Underlying this discussion of the primary human factor categories is the aim of recognizing, understanding, and accounting for the variance in human performance.

### 19.1.3 Chapter Overview

There are three primary questions that frame the chapter's discussion of the integration of the human into a system design:

1. How do we describe and measure human characteristics?
2. How do we consider limitations to human capabilities under varying operational states and adverse environments?
3. How do we integrate human components into the system being designed?

Each of these questions is discussed in the sections that follow by including information drawn from the literature and from applications familiar to the authors. A case study is presented in the last section to illustrate how the human considerations discussed can be applied to hypothetical but realistic systems.

## 19.2 HUMAN TRAITS: CHARACTERISTICS OF USERS

Each person has individual characteristics or *traits*, which combine in such a way that each person is unique, distinct from any other individual. People carry these characteristics with them wherever they go. Appreciating the inherent traits of each person is critically important for the systems designer who is creating a new system or modifying an existing system. The indwelling characteristics described in this section refer to traits or features of individuals that tend to remain constant over time. As shown in Table 19.1, these characteristics can be divided into five somewhat distinct categories:

**TABLE 19.1 Human Factors Categories, Characteristics, and Measures**

| Categories | Characteristics | Measure/Technique |
|---|---|---|
| Anthropometrics and physical parameters of the body | Physical dimensions; range of motion (static and dynamic), strength | Anthropometry: Manual and automated methods<br>Published norms and standards for specific populations |
| Sensation and perception | Vision, audition, proprioception, olfaction, gustation, balance, motion | Standardized techniques for sensory threshold testing, just noticeable difference (JND), static and dynamic measures |
| Cognition and psychological attributes | General intelligence, memory, cognitive style, problem-solving skills, and decision making ability | Standardized tests of intelligence, performance, cognitive style, problem solving and decision making |
| Social and personality factors | Personality traits and interactions with other humans; socialization | Measures of personality, social skills, and team performance |
| Physiological factors | Neuronal, electrophysiological, psychophysiological, biochemical, and hormonal | Electroencephalogram (EEG), Electromyogram (EMG), Electrodermal activity (EDA/EDR), electrocardiogram (ECG), heart rate, blood pressure, pupillary response, electrooculogram (EOG), plasma, and salivary cortisol and other hormone levels |

1. Anthropometrics and physical parameters of the body
2. Human sensation and perception
3. Cognition and psychological attributes
4. Social and personality factors
5. Physiological factors

Table 19.1 gives an overview of factors that are frequently used to describe the human component of a system. The first column categorizes the factors, the middle column shows some of the human characteristics that fall into each category, and the last column of the table lists some of the measures and techniques that can be used to quantify those characteristics. In some systems design cases, an accurate representation of the target audience will require that one consider all of these characteristics for all categories, while in other cases, only a subset of one category might be necessary for an adequate description of the human component. For example, designing a military system requires different considerations for the human operators and maintainers (e.g., ease of human interface, reliability, and ease of maintainability) than when designing a similar system for a civilian system where life and death may not hinge on the quality of the human systems interface.

### 19.2.1 Anthropometrics and Physical Parameters of Body

The science of measurement of the human body is known as *anthropometry*. Anthropometry exists as a discipline because people vary considerably in height, weight, body mass, reach, and flexibility. This field encompasses physical and biomechanical traits or characteristics, as well as physical geometry, properties of mass, and human strength capabilities. Anthropometry is used in a wide range of applications, including industrial design, consumer product design, medicine, garment/clothing design, personnel selection, human factors and ergonomics, and office design (Roebuck, 1995). Human dimensions vary independently (e.g., a person with a long torso and short legs may be the same overall height as another person with a short torso and long legs).

Anthropometric measurements are traditionally divided into two areas: static and dynamic. *Static* measurements are passive, physical body dimensions (without motion). Static measurements are typically used to determine size and spacing requirements (e.g., height, weight, distance from elbow to extended finger tip, thigh circumference, or floor area required for a person seated at a desk). *Dynamic* measurements assess motion-related properties including reach, range of motion, endurance, force exertion, and physical strength. Both static and dynamic measurements are used to fit a user to a physical environment and to ensure that control locations are accessible.

An understanding of anthropometry and anthropometric methods is essential for systems design and for the operation of any type of machine, environment, or workplace that involves people. Because ergonomic design principles have been popularized in the mass media, today most designers and engineers know what anthropometry can offer and maintain an awareness and appreciation of how to use anthropometric data and measurements.

Anthropometry is population specific. It is important to identify who will use the system and to use anthropometric data appropriate to that group of users (McDaniel, 1998). These "normalized" databases have tended to focus on U.S. military users and systems (Marras and Kim, 1993). If a commercial system is being designed, these data may not be appropriate; therefore, the design engineers have to become involved in "hands-on" anthropometric measurements. There are a number of population references available, but designers need to be mindful of the limitations of each data set. For example, many databases have been drawn from the U.S. military population, which is mostly white males and may not be applicable to civilian populations or minority groups and women. A few of these widely recognized data sets are:

1. Anthropometry Research Project Staff, *Anthropometric Source Book*, Vols. I, II, and III, 1978
2. *Ergonomic Design for People at Work*, Eastman Kodak Co. 1989
3. U.S. Department of Defense, *Military Handbook, Anthropometry of U.S. Military Personnel*, 1991

A more complete list of anthropometry databases and references is included in the Additional Readings section at the end of this chapter. The measurements reported in these sources are typically obtained using physical measurements of both static and dynamic dimension using a variety of traditional rulers, calipers, goniometers, and anthropometers. Some of these devices are pictured in Figure 19.2.

These tools have not changed markedly over the past 100+ years. Such techniques, still widely used and yielding valid data, are giving way to computer modeling. Computerized anthropometric modeling programs are capable of using traditional measurement data to build complex three-dimensional (3D) models of the human (Vannier and Robinette, 1995). These computer-generated models now allow multidimensional assessments and animations, including virtual reality scenes. There is one important caveat: Such programs are based on data obtained using the traditional mechanical devices, and the availability and expense of digital 3D data will be a limiting factor in their use for the foreseeable future.

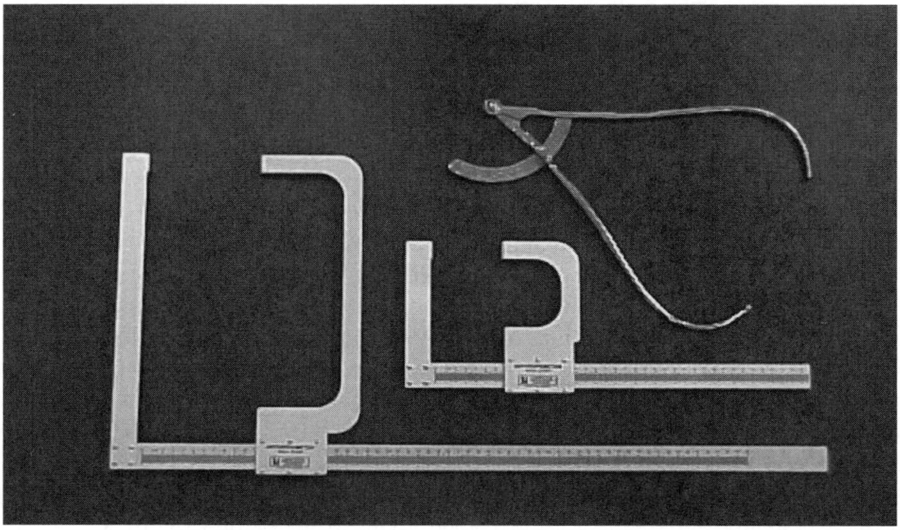

**Figure 19.2** Anthropometric tools. (Courtesy of Lt. Paul Patillo.)

A recent development in anthropometric methods is laser scanning (Bhatia et al., 1994; Vannier and Robinette, 1995). This methodology is very accurate and useful for static measures but is expensive, time-consuming, and resource intensive. However, the use of laser scanning overcomes the issues of measurement accuracy and data entry and results in a more flexible 3D model of the human.

### 19.2.2 Human Sensation and Perception

Humans, like most organisms, have a suite of sensors whose primary responsibility is to glean information about the world. The eyes, ears, nose, tongue, skin, and other sensory organs feed information into the human cognitive and decision-making system. Of particular importance to HSI applications are sensory detection and recognition and the related higher order perceptual functions such as depth perception, auditory localization, and motion perception. These sensory domains include the electromagnetic spectra for vision and audition, and particulate detection in olfaction and gustation. The touch senses, including touch, proprioception, and haptic senses, all are designed to detect pressure. A threshold is the minimal amount of stimulation necessary for a human to detect a particular stimulus (light, sound, taste, odor, pressure, etc.; Ludel, 1978).

The scientific discipline of psychophysics examines the relationship between physical properties of the environment and the detection of those properties by a human (Gescheider, 1997). Signal detection theory is an essential tenet of psychophysics, predicting for each sense and for each situation under investigation when and how people will be able to detect a faint or weak signal against background clutter or noise (Parasuraman et al., 2000; Wickens, 2002). There are no absolute thresholds, and like other sensor systems, response varies with respect to the environment and the individual human being. The ability to detect a weak stimulus (a "signal") is a function of signal strength but is heavily influenced by operator trait and state (e.g., motivation, fatigue, stress, expectations, etc.). Both Boff and Kaufman (1986) and Salvendy (1997) are excellent resources for more detailed information on sensation and other human factors topics.

***Perception and Response Bias*** Each individual has a unique ability to perceive sensory stimuli, and this perception is critical to the resultant responses made by an individual following a sensory stimulus. Responses can often differ radically between people exposed to the same stimuli. Perception varies as a function of many factors and can lead to differences in perception between individuals, providing one type of *response bias*. Issues such as expectancy, fatigue, and stress may contribute to such response bias in an individual. In July of 1988, a tragic example of this type of response bias was seen in the USS *Vincennes* incident in which a passenger airliner was incorrectly identified as a hostile aircraft. The crew mistakenly fired on the airliner, resulting in many civilian casualties (House of Representatives Committee on Armed Services, 1992).

*Vision* The human visual system is a complex system that provides us with information regarding form, color, brightness, and motion. Approximately 80 percent of all information processed by humans is via the visual system. This system, typically conceptualized as an extension of the brain, conveys light energy via chemical, neural, and higher order mental and cognitive processes to the visual centers of the brain for integration, evaluation, and interpretation.

The pupil is the variable opening in the iris, which allows differing amounts of light to enter the eye as light waves, which pass through the flexible lens (which is used to maintain focus). The light waves are focused on the retina, forming an inverted image on the rods and cones. Neural impulses from the retina are then transmitted via the optic nerve to the brain and create a corresponding pattern of nerve impulses in the brain, thereby triggering a series of neural impulses in the brain's visual center. Typical measures of human visual function include measures of visual acuity (both near and distance), contrast sensitivity (both static and dynamic), stereopsis, and tests of color vision. Measuring a person's vision at a single point in time fails to take into account the predictable visual changes that occur with aging. For those seeking further information, excellent treatments of the visual system may be found in Barlow and Mollon (1982), Goldstein (2001), or Regan (2000). Table 19.2 presents a description of the visual system and important design considerations.

*Audition* The human auditory system is the second most important source of information for most individuals. It consists of (1) the ear and associated neuroanatomy, (2) a source of sound, and (3) a transmission medium. The eardrum receives external sounds and transfers them via the middle ear bones to the oval window of the cochlea. The motion is transmitted via fluid-filled canals in the cochlea, which stimulates the cilia within the canals. These cilia, when activated, transmit neuronal impulses via the auditory nerve to the auditory centers of the brain. Sound is typically referred to as both the physical sound that enters the ear and our response to that sound. Hearing is typically used to refer to our subjective response of the auditory system to the sound. This distinction is necessary because our perception of sound does not have an exact linear relationship to the physical sound that enters the ear canal. Sound results from vibrations emanating from a source,

**TABLE 19.2 Human Visual System Parameters and Design Considerations**

| Human Visual System | Parameters | Design Considerations |
|---|---|---|
| *Rods* (black, white, and gray only) > 0.01 lumens per $ft^2$<br>*Cones* (color) < 0.001 lumens per $ft^2$<br>*Minimum visible light intensity*: 1/1,000,000,000 of a lambert ftL<br>*Wavelength of visible light*: 397–723 nm<br>Violet, 397–424 nm<br>Blue, 424–491 nm<br>Green, 491–575 nm<br>Yellow, 575–585 nm<br>Orange, 585–647 nm<br>Red, 647–723 nm | • Visual acuity<br>• Visual field<br>• Depth perception<br>• Motion perception<br>• Feature detectors (lines, curves, circles, etc.)<br>• Color discrimination<br>• Dark adaptation<br>• Absolute threshold<br>• Difference threshold<br>• Flicker-fusion threshold<br>• Stereoscopic vision<br>• Single & double images<br>• Apparent motion<br>• Optical illusions<br>• After-images<br>• Accommodation<br>• Saccadic eye movements | • Light levels (illumination)<br>• Coding<br>• Pattern recognition<br>• Motion detection<br>• 2D/3D convergence<br>• Dim-out (lighting) conditions<br>• Glare/shadows<br>• Diffused light<br>• Direct/indirect light<br>• Aesthetics of color<br>• Transillumination (of control panels)<br>• Color coding |

and emits pressure fluctuations in all directions at a speed that depends on the transmission medium (air, water, etc.). The vibrations are cyclic and consist of frequency, intensity (pressure level), and duration. An excellent source for information on these theories is found in Barlow and Mollon (1982) and Buser and Imbert (1990). Table 19.3 presents a description of the auditory system and important design considerations.

*Haptic Sense: Touch* Nerve endings in the skin and surrounding tissues transmit information regarding our immediate environment. These neurons or nerve cells are specially adapted to transmit information from specialized receptors for pain, pressure, cold, and heat. Specific neural receptors in the skin appear to respond to each of these. The sense of touch helps in our perception of form and is an important source of information for tactile information that is received from knobs and control surface textures. Table 19.4 presents a description of the haptic system and important design considerations.

*Vestibular* Through the otolith and the semicircular canals, the vestibular senses contribute to our sense of stability and give us cues for determining our orientation, self-motion and balance. Normally, the visual system is the dominant sense but it is closely coupled, even hard-wired, to the vestibular system. *Vestibular opportunism* occurs in the absence of visual cues when vestibular inputs must be resolved without their concomitant visual inputs (e.g., a pilot flying in the clouds loses visual references and may become disoriented by trusting a false perception provided by the vestibular system). Motion sickness and the related syndrome of simulator sickness occur when sufficient low-frequency alternating acceleration is transferred to the vertical (z) axis of the body and/or when there is a mismatch between visual, vestibular, and other sensory cues (McCauley and Sharkey, 1992). This cue mismatch or sensory decoupling causes malaise and nausea (Harm, 2002; Flaherty, 1998). In the absence of visual cues, this vestibular opportunism

**TABLE 19.3 Human Auditory System Parameters and Design Considerations**

| Human Auditory System | Parameters | Design Considerations |
|---|---|---|
| Frequencies between: 20 and 20,000 cycles per second.<br>*Minimum intensity*: 5 cycles per second ($\approx 15$ dB) (For most people)<br>*Maximum intensity*: 100,000 cycles per second ($\approx 140$ dB) (pain threshold for most people)<br>*Gender*<br>Men: Better at hearing low-frequency tones<br>Women: Better at hearing high-frequency tones<br>*Loudness* [just noticeable difference (JND)]:<br>$< 20$ dB $=$ 2–6 dB JND<br>$> 20$ dB $= \frac{1}{2}$–1 dB JND | • Frequency<br>• Intensity<br>• Voice recognition<br>• Auditory masking<br>• Auditory fatigue<br>• Vocal intelligibility<br>• Individual differences<br>• Age-related decrements<br>• Gender effects<br>• “White” noise | • Pattern recognition<br>• Tones vs. speech<br>• Signal vs. noise<br>• Intelligibility<br>• Speech distortion<br>• Sound localization<br>• Extraneous noise impact on performance |

**TABLE 19.4 Haptic Sensory System Parameters and Design Considerations**

| Haptic System: The Sense of Touch | Parameters | Design Considerations |
|---|---|---|
| *Skin senses*<br>1. *Pain:*<br>*Mechanical, chemical, thermal, electrical:* Varies with location on body and pain type: (e.g., thermal: 0.21 gram-calories per second per $cm^2$)<br>2. *Pressure:*<br>Inward/outward on skin<br>Vibration: Pressure sensitivity<br>3. *Cold:*<br>4. *Heat:*<br>Skin is a poor conductor of heat and cold. It is possible to achieve partial or complete adaptation to thermal conditions.<br><br>*Pain range:* 0.02 (cornea) to 300 (fingertip) $g/mm^2$<br>*Pressure range:* 2.0 (tip of tongue) to 250 (sole of foot) ($g/mm^2$)<br>*Neuron* (nerve): Electric potential takes between 300 to 1000+ msec to fire (refractory period limits the frequency with which a neuron may fire).<br>**Muscle, Tendon and Ligament Neural Receptors**<br>*Muscle Tissue:* Low efficiency (three fourths of energy released as heat; only one fourth in useful work) | • Reflexes<br>• Reaction time<br>• Muscle tremor<br>• Repetitive movements<br>• Fine vs. gross motor movements<br>• Skin sensitivity | • Tactile feedback |

can result in vestibular illusions or spatial disorientation. Vestibular issues for design consideration arise in moving platforms and in stabililized platforms with peripheral visual cues where decoupling of the two senses may occur (Stoffregen et al., 2002; Hettinger et al., 1990). A description of the vestibular system and design considerations are presented in Table 19.5.

*Gustation and Olfaction* The chemical senses of taste and smell, while important to us from the standpoint of enjoying our daily lives, have experienced limited applicability for human factors engineers and HSI professionals. Exceptions to this rule are the use of an olfactory warning for detection of leaking gas, and the experimental use of wintergreen mint to alert drowsy drivers. Table 19.6 discusses the olfactory and gustatory sensory systems and lists potential design considerations.

### 19.2.3 Cognition and Psychological Attributes

This section addresses mental ability such as intelligence and cognitive processes such as memory function and decision making. All of these human characteristics are interde-

**TABLE 19.5 Vestibular System Parameters and Design Considerations**

| Vestibular System: Sense of Balance and Self Motion | Parameters | Design Considerations |
|---|---|---|
| *Vestibular system*: Gives position and movement of the head<br>Three Semicircular canals: orthogonal to sense angular acceleration<br>Two Otolith Organs: linear accelerometers to indicate head position and orientation<br>*Kinesthesis*: Body awareness in 3D space + joints; muscle contractions;<br>Kinesthetic fibers are large = rapid conduction of information | • Acceleration/ deceleration<br>• Motion sickness<br>• Sopite syndrome<br>• Vestibulo-Ocular Response | • Relationship to other senses (especially vision) |

pendent and interrelated. To maximize overall system effectiveness, these characteristics in the design population should be assessed and considered in systems design.

***Intelligence Quotient*** The cognitive capability of humans, particularly the ability to process information, is known as *intelligence* and is more commonly referred to as an *intelligence quotient*, or *IQ*. In humans, intelligence is not just a process or capacity but is regarded as the intellectual capabilities of one person relative to some standardized population. Psychologists have developed a number of standardized tests to assess individual intelligence and capability for learning (Kaufman, 2000; Snell, 1996). Typically, an IQ score in the range of 70 to 135 is considered to be within the "normal" range. Scores below 70 indicate that the individual may have difficulty functioning in society, while a score over 135 indicates that the individual has exceptional abilities and may function at an intellectually higher level than most other people. While systems designers will probably never administer an IQ test, it is important to consider the intelligence level of the human

**TABLE 19.6 Olfactory and Gustatory Senses and Design Considerations**

| Senses of Taste (Gustation) and Smell (Olfaction) | Design Considerations |
|---|---|
| *Taste:* Gustatory neurons: Proximity receptors (must be stimulated by direct contact)<br>Tongue has areas for detecting *bitter* (most sensitive: Can detect 0.0005 percent solution), *sweet*, *sour*, and *salty*<br>*Smell:* Olfactory neurons: Direct extensions of the brain. Smell involves mechanical and chemical stimulation (e.g., ammonia = smell + pain receptors)<br>Ethyl alcohol: 0.2 of a milligram per liter<br>Vanilla: 1/1,000,000 of a milligram per liter | • Limited design considerations for smell at this time (use of noxious odor in natural gas and wintergreen for drowsy drivers)<br>• No known design considerations for taste at this time |

who will be using and maintaining the system, especially when designing a complex and cognitively demanding system. Systems designers must work closely with personnel selection professionals to ensure that the system being designed is appropriate for the design population.

***Cognition*** Cognition is the exploration of how the human mind processes information, what incoming information it processes, and what it does with that information after it is processed. There are a number of theories about how the brain deals with information, but they are beyond the scope of this chapter. Suffice it to say that there are some widely accepted characteristics of human information processing that are useful, even critical, to the system designer. Again, it is important to work with human factors professionals in an iterative design process to ensure a usable product for the target population. Cognition occurs at the interface between perception and memory. Perception is used to bring information into the system for processing. Memory is used to retain perceived information while determining which to use now, which information to use later, and which to throw out. Bringing information in from external sources is the function of perception; holding on to information while deciding what to do with it is the function of memory.

***Memory, Decision Making, and Cognitive Style*** These three characteristics refer to the way people process, store, and act on information from the environment. *Memory* has a number of characteristics, including limitations on the amount of information that can be held at one time and restrictions in retention and recall of that information. *Decision making* results from the processing of information from the outside world. *Cognitive style*, or the characteristics of the user's response, is vitally important to overall outcome and is essential for the achievement of optimal system performance. For example, one operator may be extremely methodical—slow to respond and careful to avoid errors—while another might respond quickly and impulsively with little regard for errors. Table 19.7 indicates several important design considerations relevant to these three cognitive processing topics.

### 19.2.4 Social and Personality Factors

Individuals have unique social characteristics and needs. Our socialization process begins at birth and continues through the school years into adulthood. In this respect, it can be stated that social skills are not static but are continually undergoing adjustment and development. These individual social skills can have a tremendous impact on an individual's effectiveness and ability to work as a team member and should be considered both in workplace design and in personnel selection. The social environment and physical proximity of individuals in the workplace configuration and the impact of personality within a social system are vitally important considerations for the systems design team. It is understood that people perform better when their personal space and personal needs are taken into consideration. Submariners, for example, need to be selected partially based on their ability to work well in confined spaces and in close proximity to other crew members.

***Personality*** Personality can be thought of as a distinctive pattern of *relatively enduring* behaviors, thoughts, and emotional responses that define who we are and how we interact with other individuals in our environment. Our personalities are generally flexible enough

**TABLE 19.7 Cognitive Characteristics Design Considerations**

| | Design Consideration |
|---|---|
| **Memory Characteristics** | |
| *Attention* | Humans have a limited attention span. They are particularly bad at vigilance tasks requiring sustained attention or monitoring. Humans perform poorly on tasks that are boring. |
| *Chunking* | Humans can learn to group information together in ways that allow them to remember more information. This includes grouping by likeness, location, or association. Most people can hold about $7 \pm 2$ items in memory at one time. |
| *Heuristics/mnemonics* | Heuristics are "rules of thumb" that facilitate recall. Mnemonic devices are mental tricks that allow humans to retain more things in memory than would otherwise be possible. |
| *Forgetting* | Regardless of how frequently things occur, people forget things. Forgetting can lead to errors or unacceptable performance. |
| **Decision-making Characteristics** | |
| *Uncertainty* | Uncertainty requires more cognitive work and places higher demands on memory. The display of information should support a reduction of uncertainty, resulting in faster and more accurate solutions. |
| *Choice* | As the number of choices increases, the cognitive and memory demands also increase exponentially. In a given operational scenario, display of information should reduce the number of choices to the fewest number possible. |
| **Cognitive Style** | |
| *Speed* | The "speed/accuracy trade-off" is a hallmark principle of cognition with the two tasks being inversely related. Speed on a task will vary according to the instructions given to the operator. Speed is exchanged for accuracy with higher speeds related to less accurate performance. |
| *Accuracy* | The human operator will emphasize either speed or accuracy. The decision about which to emphasize should be made in the concept exploration phase of the system design. |
| *Choice* | The more choices the user must make, the longer the response will take. This is an opportunity for decision aids to support the selection of a response by providing supporting information or by limiting the number of choices available. |
| *Errors* | Errors occur. Unfortunately, many errors occur due to suboptimal system design. The effect of errors in each phase of the mission must be evaluated. "Fatal errors" result in redesign of the system, and it is better to identify them in the concept exploration phase than during production. |

to allow us to adapt to a wide range of situations and to the behaviors of those around us. Human factors professionals and design engineers should remain aware of the effects of individual personality on the work environment and on the social environment at work (Holtzman, 2002).

### 19.2.5 Physiological Factors

Human physiology is based on the concept of *homeostasis*, the natural feedback system that continually strives to maintain balance in all cellular processes. Each of us possesses genetic determinants for our individual physiological makeup. These physiological traits include homeostatic functions such as metabolism and immune response. Physiological functions can be measured and quantified using a wide variety of techniques such as the electrical activity of the brain, heart, and skin; pulse and respiratory rates; and biochemical indicators such as hormone levels collected in plasma, salivary, and urine samples. These physiological traits are affected by the various environmental and work conditions to which humans are exposed and as such will fluctuate from baseline levels when exposed to conditions such as fatigue and arousal.

## 19.3 HUMAN STATES: OPERATIONAL AND ENVIRONMENTAL VARIATIONS

This section focuses on the importance of *transitory human states* and the effects these states have on system performance. Covering the depth and breadth of all human operator states is well beyond the scope of a single chapter. However, as examples, the following states and their effects on system performance are discussed: mental workload, fatigue and circadian rhythms, psychological and physiological stress, and SA.

Various techniques including physiological measures, measures of performance (MOPs), and measures of effectiveness (MOEs) are described as they relate to monitoring the state of the human in the system. Table 19.8 lists transitory human operator states and candidate measurement techniques for quantifying these states.

### 19.3.1 Mental Workload

Most systems require some level of mental work by the operator or user. Whether setting the clock on a DVD player or cooking food in a microwave, users perform mental work. To

**TABLE 19.8 Transitory Human Operator States and Candidate Measurement Techniques**

| Operator States | Candidate Measurement Techniques |
|---|---|
| *Mental workload* | Primary and secondary performance measures, subjective workload scales, physiological measures (e.g., EEG, oculography, cardiovascular measures, and respiratory rates) |
| *Circadian rhythms and fatigue* | Actigraphy, temperature, melatonin levels in saliva or plasma samples |
| *Psychological and physiological stress* | Physiological measures such as cortisol levels in saliva, plasma, and urine samples, psychophysiological measures (e.g., EEG, EDA, ECG, and EMG), subjective rating scales (both standardized and task-specific), SME-administered interviews |
| *Situational awareness* | Standardized ratings of SA, subjective questionnaires, SME ratings of SA |

determine the mental workload required by a system, it is necessary first to have a definition of workload. One definition of mental workload is that it is "the amount of cognitive or attentional resources being expended at a given point in time" (Charlton and O'Brien, 2002, p. 98).

Cognitive psychologists and human factors professionals have used a variety of strategies in their attempts to measure mental workload of the human. The intuitive approach to measuring mental workload is to simply query the operator. However, by interrupting the user during the task, the observer has intruded upon and altered the workload being measured. This alteration is referred to as *assessment reactivity*, and the results of such a measurement procedure may be prone to subjective bias. Conversely, assessing workload after the task is completed is also unsatisfactory because one ends up with an overall workload measure for the task but no data regarding workload during the task itself. Certain time segments of a task may be considerably more challenging and could therefore be expected to produce a higher workload. The single measure collected at the end of a task would fail to capture these fluctuations. Further complications are posed by innate individual differences in "workload capacity" [i.e., what is very hard for one person may seem less hard for another (Reid and Nygren, 1988)].

Until the "Vulcan mind-meld" (*Star Trek*, Paramount Pictures, 1970) has been perfected, what is needed is an unobtrusive and noninvasive "window into the brain." In this section, we will examine how mental workload has been measured in the past, how we are currently measuring it, and possible directions for future measurements of mental workload. Traditionally, mental workload has been measured in four ways:

1. Performance on primary task measures
2. Performance on secondary task measures
3. Subjective measures
4. Electrophysiological and psychophysiological measures

The first two methods, both performance measures, rely heavily on information processing models of attention that assume that performance will degrade with increasing workload. For years, theorists have been debating the details of information processing and attentional capacity models, and this debate continues (Charlton and O'Brien, 2002).

***Performance Measures of Mental Workload*** When using primary task measures of performance to evaluate operator workload, operators are given tasks to complete. One task is identified as being relatively more important in the face of other competing tasks. Speed and accuracy of performance on that task is stressed. Operator workload level is derived from performance on that primary task. Using this metric, slower and less accurate performance suggests higher mental workload.

In the secondary task method, performance on the task not identified as important (the task that "suffers" when the operator becomes busy) is considered to be reflective of workload. Poor performance on the secondary task suggests that the workload level is too demanding. Examples of secondary tasks include time estimation, tracking tasks, memory tasks, mental arithmetic, and reaction time. In general, trying to "embed" secondary tasks into a primary task situation is difficult and may result in an artificial and intrusive measure. A comprehensive overview of both primary and secondary task measures, with their strengths and limitations, is found in *Human Performance Measures Handbook*

(Gawron, 2000). This handbook also gives excellent descriptions of a wide variety of human performance measures.

There are problems in using performance as a measure of workload. O'Donnell and Eggemeier (1986) cite four major areas of difficulty: (1) artificially enhanced performance with task underload, (2) a "floor effect" with task overload, (3) the operator's information processing strategy, training, or experience may confound the estimates of mental workload, and (4) issues with generalizing the results to other tasks. Additionally, a single measure of primary task performance may be overly simplistic, especially when the task is complex and multidimensional (Meshkati et al., 1990).

***Subjective Measures of Mental Workload*** Subjective measures of mental workload have high face validity and are possibly the most intuitive and easiest to obtain. Gawron (2000) lists 34 separate subjective rating scales. Some scales are generic measures of workload, while others are designed for specific task domains such as estimating workload in an aviation cockpit.[2]

On the negative side, O'Donnell and Eggemeier (1986) offer these caveats for those using subjective ratings of workload.

- Mental and physical workload can be potentially confounded.
- It is difficult to distinguish the task difficulty from actual workload.
- Subjects cannot accurately rate their level of unconscious or preattentive processing.
- There is a dissociation of subjective ratings and task performance.
- Subjective measures require a well-defined question.
- Subjective ratings are highly dependent on the short-term memory of the rater.

***Physiological Measures of Workload*** In contrast to performance and subjective measures of workload, physiological measures are relatively objective and unbiased. However, these measures, while inherently attractive, may be costly and unwieldy in field settings. In operational settings, psychophysiological measures have been used effectively to monitor the functional state of the operator, to determine the response of the operator to new equipment and/or new procedures, and to determine workload and vigilance levels of the operator (Wilson and Eggemeier, 1991). Wilson (2002, p.128) states, "... an operator's interaction with a system influences their physiology. By monitoring their physiology, we are able to infer the cognitive and emotional demands that the job places on the person."

Central to any discussion of physiological measures is the role of the nervous system in controlling all behavior, including both physical movement and cognitive processes. The nervous system relays information through the "firing" of electrical impulses, which propagates from nerve to nerve. The human nervous system can be divided into the central nervous system, or CNS, comprised of the brain and spinal cord, and the peripheral nervous system, or PNS, which comprises the cranial nerves and all other nerves. Activity from the CNS and the corresponding changes in cellular metabolic function can be monitored using a variety of methods. Table 19.9 describes some of the more important characteristics of these measurements of CNS activity.

**TABLE 19.9 Central Nervous System Physiological Measures**

| Acronym | Name | Transducer Type | Where Conducted? | Expense | Ease of use |
|---|---|---|---|---|---|
| EEG | Electroencephalogram | Scalp electrodes | Field/lab | $ | * |
| ERP | Event-related potentials | Scalp electrodes | Lab | $$ | † |
| MEG | Magnetoencephalogram | Magnetic sensors | Lab | $$$ | ‡ |
| fMRI | Functional magnetic resonance imaging | Magnetic sensors | Lab | $$$ | ‡ |
| PET | Positron emission tomography | Radioactive isotopes and special sensors | Lab/hospital | $$$$ | ‡ |

$ = *Inexpensive*
$$ = *Somewhat expensive*
$$$ = *Expensive*
$$$$ = *Very expensive*
* = *Minimal training required*
† = *Specialized training required*
‡ = *Highly specialized technical training required*

At the level of the CNS, representative measures include:

- Electroencephalogram (EEG)
- Event-related potentials (ERP)
- Magnetoencephalogram (MEG)
- Functional magnetic resonance imaging (fMRI)
- Positron emission tomography (PET)

The last four of these measures require that the subject or operator remain relatively stationary. These measures also have serious operational restrictions posed by the cumbersome and nonportable nature of the apparatus required to record the signals (Center for Position Emission Tomography, 2002). However, recent advances in the EEG recording technology have resulted in portable, human-mounted devices that allow for data collection to be made real-time in field settings.

These five measures of central nervous system activity are not the only indications of nervous system activity, however. There are signals that can be observed peripherally to the brain that are also extremely good indicators of nervous system activity. Quite often, these peripheral measures are less invasive and more useful for field applications. Table 19.10 describes some of the more critical features of these peripheral measures.

At the level of the PNS, representative electrophysiological measures include:

- Electromyogram (EMG)
- Electordermal response (EDR)
- Cardiovascular responses (ECG, heart rate, and heart rate variability; blood pressure, echocardiogram)
- Respiratory rate
- Oculometry (EOG, or electroculogram, pupillary dilation, eye blink rate, and eyelid closure rates)

**TABLE 19.10 Representative Peripheral Measures of Central Nervous System**

| Acronym | Name | Recording Transducer | Where Conducted? | Expense | Ease of Use |
|---|---|---|---|---|---|
| EMG | Electromyogram | Electrode over muscle | Field/lab | $ | * |
| EDA/EDR | Electrodermal activity and response | Skin conductance electrode | Field/lab | $ | * |
| EKG/ECG | Electrocardiogram | ECG electrodes | Field/lab | $ | * |
| HR, HRV | Heart rate and heart rate variability | ECG electrodes | Field/lab | $ | * |
| BP | Blood pressure | Sphygmomano-meter | Lab | $ | |
| RR | Respiratory rate | Spirometer | Field/lab | $ | † |
| EOG | Electrooculogram, blink rate, eyelid closure rate | EOG electrodes placed beside eyes; camera | Field/lab | $ | † |
| SV | Saccadic velocity | Eye reflectance and camera | Lab | $$ | † |
| PD | Pupillary dilation | Eye reflectance and camera | Lab | $ | † |

$ = *Inexpensive*
$$ = *Somewhat expensive*
* = *Minimal training required*
† = *Specialized training required*

Studies have shown that psychophysiological measures have the potential to precede or predict performance decrements (Cacioppo, 2000; Lewis et al., 1988). One remarkable feature of psychophysiological measures is the sensitivity at which they detect alterations and variations in human response. Many studies have demonstrated the utility of these peripheral measures in discriminating between workload levels (Wilson, 2002; Miller and Rokicki, 1996; McCarthy, 1996; Burns et al., 1991). Prodromal indicators of performance decrements would have great usefulness when assessing operator state and workload. One example of an operational military system that is using psychophysiological measures is the use of in-flight EEGs to detect G-induced loss of consciousness in Israeli Air Force pilots. Real-time monitoring of operator state has left the realm of science fiction and has become a reality. Recent advancements in the field of laser Doppler vibrometry hold promise for monitoring many human physiological signals.[3]

### 19.3.2 Circadian Rhythms and Fatigue

Human beings operate on an approximate 24-hour biological clock with a predictable pattern in many parameters of our behavior. For individuals who are adjusted to sleeping nights and working days, many physiological systems slow down in the very early morning hours as can be seen in the predictable drops in body temperature, heart rate, and blood pressure. Although the "normal" human body temperature is 98.6°F, body temperature is just one of many physiological parameters that varies with time of day and with the body cooling off over the course of the night, reaching the coolest point in the early morning hours just before awakening and then beginning to warm up once again. Temperatures reach their peak at about 9:00 p.m. and then repeat this cosinelike cycle. Human performance also changes over the course of a 24-hour period. Performance on many tasks such as reaction time and vigilance mirrors the circadian variations seen in body temperature and other physiological indices (Krueger, 1989; Tilley, 1982). There is a performance trough associated with the circadian nadir occurring around 2:00 p.m. (the "postprandial" dip in performance) and again from 1:00 a.m. to 4:00 a.m. Some cultures acknowledge this performance decrement and accommodate it by providing a designated time for resting, or "siesta", in the early afternoon. Similarly, it is not surprising that many accidents occur in the early morning hours when circadian rhythms are at their nadir, the lowest point of the cycle (e.g., some well-known disasters and the time they occurred include: Chernobyl, 1:23 a.m.; Bhopal, 12:40 a.m.; and Three Mile Island, 4:00 a.m.).

In addition to the substantial differences in performance due strictly to normal circadian variation, fatigue due to sleep deprivation can also be a major source of variance in human performance. Most adult humans require an average of 8 hours of sleep per day. When this requirement is not met, performance can suffer in a most dramatic way. This human performance decrement has been modeled very effectively in computer models such as the SAFTE model (O'Donnell, et al., 1999) [implemented in the Fatique Avoidance Scheduling tool (FAST) computer program, Eddy and Hursh, 2001; Hursh et al., in press] that has been adopted by the Department of Defense (DoD). Sleep inertia, the lethargic feeling that one experiences when awakening from sleep, is also associated with inferior performance (Naitoh et al., 1993; Balkin and Badia, 1988).

A third source of performance variation can be seen in circadian desynchrony when the normal circadian rhythms of an individual are disrupted (see Example 19.1). As anyone who has experienced jet lag will attest, shifts in time zones result in general feelings of malaise and impairment in cognitive functioning. Modern aircraft, with their greatly extended mission durations, have motivated researchers to question how best to manage work and rest cycles during extended missions. Around-the-clock flight operations have become commonplace, both in the civilian and military workplace (DellaRocco; 1999; Caldwell, 1997). In such cases, even limited exposure to normal photic time cues (daylight-darkness) and normal work/social/sleep schedules (day work/night rest and day wake/night sleep cycles) may hamper an individual's circadian inversion and disrupt their sleep patterns. The literature on shift workers is rife with examples of the diminished performance and health risks associated with night shift and swing-shift work schedules (Hossain and Shapiro, 1999).

When assessing operator state, these predictable fluctuations in human performance attributed to circadian rhythms must be considered. The designer should assure that the system can be operated properly, not only during regular business hours (when operated by rested users), but also in the early morning hours when operated by users who have had very little sleep for the preceding week. Table 19.11 lists candidate measures that are frequently used for monitoring circadian rhythms and fatigue.

**Example 19.1 Watch Standing Aboard a U.S. Navy Carrier** In times of combat and military crisis, U.S. Navy (USN) aircrew members are frequently required to fly a tremendous number of night missions. Recently, a number of aircraft carriers have instituted a remarkable adjustment in the work shift work schedule of their entire crew. To accommodate the needs of

**TABLE 19.11 Circadian Rhythm and Fatigue Measures**

| Acronym | Name | Recording Method | Where Conducted? | Expense | Ease of use |
|---|---|---|---|---|---|
| **Actigraph** | Activity levels | Accelerometer worn on wrist | Field/lab | $$ | * |
| **Temp** | Temperature: Oral, axillary, aural or core | Thermometers, digital or analog | Field/lab | $ | * |
| **Melatonin** | Hormone melatonin concentrations | Salivary or plasma level of melatonin | Lab (hard for field) | $$$ | † |
| **Ratings** | Subjective ratings of fatigue | Subjective rating scales (e.g., Stanford Sleepiness Scale, Epworth Sleepiness Scale, POMS) | Field | Minimal | ‡ |

$ = *Inexpensive*
$$ = *Somewhat expensive*
$$$ = *Expensive*
* = *Minimal training required*
† = *Specialized training required*
‡ = *Highly specialized technical training required*

the flight crews, the entire ship's company has shifted its working hours to a night schedule, which can have unexpected ramifications.

Anyone who has crossed several time zones has experienced jet lag and will recognize the difficulty involved when trying to invert the human circadian rhythm. The schedule inversion implemented by these USN aircraft carriers poses a unique question: Can an entire ship's company be successful in inverting individual circadian rhythms in the presence of normal light or photic cues? How do we assess how much rest an individual is getting and how do we determine if they are "fit for duty?"

A proper appreciation of performance decrements seen in individuals whose circadian rhythms are desynchronized serves as a reminder of the importance of adequate rest for all crew members. Watch-standing schedules specifically designed to safeguard against fatigue and promote sleep hygiene are vital. In the near future, field trials of "fitness for duty" batteries, incorporating physiological and performance tests, will determine whether such batteries will be beneficial to commanders and supervisors.

### 19.3.3 Psychological and Physiological Stress

Stress is defined as a process by which we receive information and then respond to an event, either real or imagined. Stressors can be both positive and negative. Stressors can have strong motivating or empowering properties, although stressors can also have detrimental effects on our psychological and physical health. Long-term severe stress can lead to a host of medical problems, and severe stress can even contribute to (if not cause) death (Lazarus and Folkman, 1984).

All living organisms respond to stress. The term *stress* (or *stressor*) is used in a variety of ways and can impact an organism individually or in combination. Stress can be induced by *physical or environmental* conditions (e.g., heat, cold, noise, illumination, motion, etc.). It may also be caused by *psychological* pressures (e.g., anxiety, anger/hostility, the "fight or flight" syndrome, threat perception, etc.), and it is also associated with *physiological* factors (e.g., sleep loss, fatigue, mental or physical workload, etc., Wickens, 1996).

The *stress response* system has been eloquently operationalized by Seyle (1937, 1975) using his general adaptation syndrome (GAS), which is divided into three distinct phases: (1) the alarm reaction (mobilize resources), (2) the resistance stage (cope with the stressor), and (3) exhaustion (reserves of energy depleted). In response to chronic stress, these reactions can lead to physical ailments such as hypertension, heart disease, stroke, ulcers, and even death. Seyle's pioneering work on the effects of stress on organisms, and the publication of his "stressful life events scale," represented an attempt to quantify a wide range of typical stressors, both positive and negative.

The "fight or flight" is the characteristic reaction of organisms when exposed to an emergency or to a life-threatening situation (Cannon, 1994). This response offers obvious survival advantages for an organism. When confronted with a life-threatening event, we will instinctively respond by (a) fighting to protect ourselves, or if fighting is not an option, we will (b) remove ourselves from the vicinity by fleeing. But over the millennia, humans have had to face fewer and fewer truly life-threatening or emergency events in daily living. Unfortunately, our anatomy and physiology have not adapted to this less threatening life-style. We still respond to life-threatening events, but we also respond to stressors with this same response. This hard-wired fight or flight reaction is at the center of most stress-induced physiological responses and is a causal factor in many physical changes. The following (partial) list of responses is well known:

- Increase in sympathetic nervous activity
- Increase in blood pressure
- Rapid heart rate
- Release of red blood cells/blood coagulant
- Increase in epinephrine
- Surge of adrenaline
- Change in acid/alkaline blood balance
- Redistribution of blood supply (from viscera and skin to brain and muscles)

There is also a wide range of psychological emotions associated with the fight or flight response, and most of these center on thoughts related to escape and the availability of means to protect oneself. During periods of very high stress, such as in combat, individuals faced with life-threatening situations often report a "narrowing" of perception or attention, a clarity of thought processes, and an ability to "hyper focus." Alternately, some soldiers under fire report thought patterns that are so jumbled and disordered that clear thought and action is almost impossible. Different people can (and often will) respond differently to the same stressor (Hancock and Desmond, 2001).

While excessive stress has the potential to cause physical and psychological harm, the opposite extreme—too little or nonexistent levels of stress—is also not good. A moderate amount of stress can be a good motivator, increasing the arousal level of the individual. The level of arousal plays an important role in motivation and task-oriented behaviors. This inverted U-shaped curve describes the effects of stress on the organism and is typically referred to as the *Yerkes–Dodson law*. (For a review of Yerkes–Dodson law, see Teigen, 1994.) At the left end of the inverted U there is so little stress that there is no motivation, while at the right-hand side of the inverted U, too much stress exists. The optimal amount of stress lies somewhere in between, and varies greatly by individual, as well as by the importance of a task, the time allowed for the task, and a host of other variables.

The ability to assess the impact of stressors as a function of their physiological and cognitive impacts on an individual is an important factor for a systems design engineer to consider (Hockey, 1983). Something as simple as requiring an individual to work in an intensely stressful environment for only a limited and prescribed period of time may prove to have extremely beneficial effects on long-term employee health and cognitive functioning. This "exposure-based" approach of allowing workers to work for a limited time in excessively stressful environments is used in many high-stress occupations, the best-known example being air traffic controllers (Stokes and Kite, 1994).

***Environmental Stressors*** Stress takes many forms, including a class of stressors termed *environmental stressors* (Banderet et al., 1987; Kanki, 1996). Environmental stressors are those factors in our physical environment that can have a negative effect on our ability to function physically or cognitively. These factors exist around us, in one form or another, all the time. Such factors include:

- Excessive heat (and/or) excessive cold
- Climate (humidity)
- Illumination (both quality and intensity)
- Motion (on a vehicle; vibration from machinery; g forces)

- Noise (excessive noise requires hearing protection)
- Quality of air (ventilation odors; particulate matter requires breathing protection; high altitude requires supplemental oxygen; underwater diving requires pressurized air)
- Air pressure (very little at high altitudes)
- Water pressure (very high pressure while diving underwater)
- Social (working alone vs. part of a team or group)

When one or more of these variables leaves the normal range, we begin to notice and become aware of the way they affect our ability to function (Griffin, 1997).

***Background Stressors*** Finally, there is the concept of daily hassles. Daily hassles are those everyday, relatively minor "background stressors" that have a cumulative and annoying effect on us (Hahn and Smith, 1999). This additive effect can lead to physiological and psychological effects that parallel the effects of regular stressors described above. For example, imagine if a system has small warning lights and audible alarms. The warning lights are not to signify emergency conditions, but rather to alert an operator that attention is needed on some time-critical process or function. As such, the warning lights are designed to help allow the operator to attend to systems other than the one on which he/she may be concentrating. However, if these small alarms, not particularly annoying by themselves, start lighting up and sounding every few minutes, the operator must reset the equipment, silencing the warnings, and then continue normal activities. If these false alarms continue, the operator is likely to find a way to permanently silence them. The cumulative effects of such seemingly small hassles build until the operator experiences significant stress. Systems engineers and designers should be reminded that if such systems are designed to continuously alert an operator, the cumulative effects of those warnings can contribute to the stress they were designed to ameliorate.

### 19.3.4 Situation Awareness

Situation awareness (SA) can be described as an awareness or knowledge of what is going on around you. The definition of SA also takes into account the accuracy of the information, as well as how much the individual believes the information to be accurate. There is a tremendous amount of information available in the world but not all of that information is available to or perceived by the human operator. Figure 19.3 illustrates how information about the state of the world is sequentially filtered to yield the most basic level of human operator SA (Pew, 2000).

The top level of the figure is "ground truth" which consists of each piece of information and every data point that is available about the world. Sensed Truth or Potential SA is the next level in the diagram and includes all information that is sensed, whether directly or indirectly. Of course, as our sensor capabilities improve, we know more and more about the state of the world around us but it can be argued that we will never know everything about the world. The difference between the first two levels represents the inability or limitations of our sensor capability.

Operator SA is at the bottom level of this cone of awareness and is a subset of sensed truth or potential SA. All information that has been sensed or detected is available at this level for the human operator. Of course, due to limitations on human information

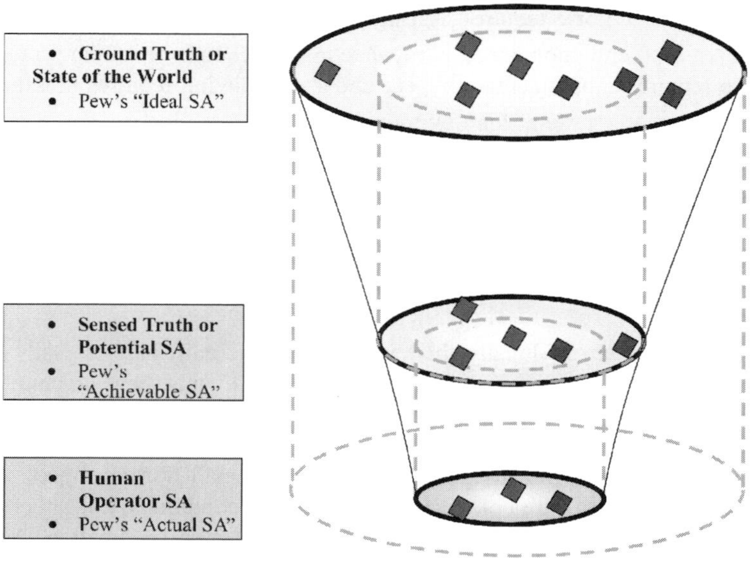


**Figure 19.3** Depiction of situation awareness.

processing, the human operator will not perceive everything detected by the sensors. Endsley and Garland (2000) describe SA as a three-tiered process of perception, comprehension and projection. These three stages are described as Levels 1, 2 and 3 SA. In Figure 19.3 the lower level represents perception or Level 1 SA.

To acquire and maintain SA requires an awareness of what is going on around you, followed by the ability to make judgments concerning things to which you should attend. This process is a type of "cognitive filter," which allows an individual to make decisions on the relative importance of environmental features. Depending on the situation and context, we may use all of our senses in assessing our environment, while at other times we may use only a subset of the available sensory modalities. We may use one sensory modality to such an extent that we ignore other senses. For example, a pilot has to learn to maintain SA, not only using the eyes and ears, but by attending both physically and cognitively to important information, while concurrently ignoring or "filtering" irrelevant information. SA is more than just keeping a mental picture: It is a dynamic process whereby an individual maintains environmental awareness and is also aware of what to ignore in the environment (Hartman and Secrist, 1991; Endsley, 1995; Endsley and Garland, 2000).

***Aids to Enhance Situation Awareness*** Situation awareness is an internal state that is acquired and maintained by the individual and therefore differs from many other factors that a systems design engineer considers. Individuals vary considerably in their ability to acquire and maintain SA. Various types of equipment have been used in attempting to "design in" aids to SA for the system operator. Such aids help an individual acquire and maintain SA, while at the same time enhancing their ability to attend to other functions or problems.

One well-known and very successful example is use of the HUD ("Head Up Display") in a cockpit. This system helps a pilot maintain SA by allowing an awareness of the world outside while simultaneously maintaining an awareness of vital flight parameters. From this perspective, a HUD can be conceived as more than simply a means to view flight parameters, but rather as a SA-enhancement device (Will, 2000; Leger et al., 1999). However, HUDs have also been found to significantly *decrease* SA when an individual's attention is overly focused on the information presented on the HUD, thus diverting attention away from other critical (often dangerous) environmental cues (Weiner, 1989, 1990, 1993).

***Quantifying Situation Awareness*** While attempts to quantify SA have been met with mixed results, a critical consideration is how to accurately measure an individual's SA (Vidulich, Stratton, and Wilson, 1994). Pritchett and Hansman (2000) divide SA measures into 3 categories: knowledge-based, verbalization measures and performance-based. The strengths and limitations of the various measures are succinctly described in tabular form in their chapter. For more in-depth information on these measures, the reader is referred to Endsley and Garland (2000) and Gawron (2000).

***Enhancing Situation Awareness*** The workplace configuration and the cognitive task demands on the human should be taken into consideration when designing a SA enhancement system. There does not appear to be "too much" SA. However, the danger lies in assuming that "more is better." Endsley offers a list of 50 principles for enhancing SA in systems design. A design should always take into account the *desirability* and *need* by a person to enhance or modify the level of SA. It is important to determine the optimal level (and type) of information delivery that will *enhance* SA, while not inadvertently obscuring or clouding existing SA. Poorly designed equipment or interfaces, the delivery of too much unimportant information, and poorly presented information can hinder the development and maintenance of SA. An experienced operator will have learned to attend to various cues regarding the state of the environment and equipment, some consciously and some at the level of the sub-conscious and unconscious level. That "funny feeling" that individuals often use to describe their sense of an impending equipment failure can be explained by the various cues that are constantly being processed by the user.

***The Temporal Nature of Situation Awareness*** One variable that significantly impacts SA is time. SA is time-sensitive. The salience or "newness" of SA information is inexorably tied to the length of time that has elapsed since it was last refreshed and the rate of change of relevant information in the task at hand. Periodic updates are essential to maintaining SA. The rate of updates to SA information varies considerably, and is typically a function of task requirements and the ability of an individual to receive periodic information updates. These updates may need to occur frequently (e.g., demanding air traffic control tasks), or more methodically (e.g., monitoring of aircraft instruments on autopilot at cruising altitude). One important factor that greatly influences the time it takes an operator to acquire and maintain SA is the level of familiarity the operator has with a particular task or situation. We are able to respond most rapidly to situations with which we have the most familiarity (i.e., over-learned behaviors). Our ability to respond to new situations is heavily impacted by our knowledge of previous, similar situations. Additionally, there is a learning curve that occurs while attempting to master a new system. Before an individual will acquire high levels of SA, they must have achieved mastery of the new system.

***New Frontiers in Situation Awareness*** Recently, the concept of SA has been extended beyond previous definitions. One new technological area that involves SA, but relies entirely on synthetic sources of information, is the emerging field of "remote" SA. This involves the ability of an operator to acquire and maintain SA when operating an unmanned aerial vehicle (UAV) via remote control. This technology, also called "tele-robotics," relies on remote sensors to provide the information necessary for the operator of the UAV to acquire and maintain SA. As with any new technology, many questions remain unanswered. Can a machine be equipped with an appropriate set of sensory devices that transmit vital "real time" SA information back to the operator? And how well does the operator receive such information, selectively attend to it, and form a reasonable mental model of the situation? Such "cutting edge" questions are germane to the discussion of the role of human systems integration, cognitive task demands, and workplace configuration in the situation awareness of the "human-in-the-loop."

## 19.4 HUMAN SYSTEMS INTERFACES

This section addresses the critical juncture between the human and the machine and gives guidance for systems designers who are striving to optimize total system performance while making the most effective use of the human. These guidelines are divided into four sections:

- Workspace design and anthropometric considerations
- HSI considerations for the design of displays
- Task allocation: man versus machine
- Social issues and team performance

### 19.4.1 Workspace Design and Anthropometric Considerations

From a human systems design perspective, it is important to focus the design on the population of individuals who will be the end users. The importance of designing with the user in mind is no more readily apparent than in work space design issues. A worker in a poorly designed work environment will be less productive, more error prone, more injury prone, and eventually may no longer be able to work due to repetitive strain injuries (RSI). Optimal work space design increases worker productivity, enhances safety, contributes to a reduction in worker errors, reduces work-related injuries, and lowers personnel turnover.

Several criteria typify developing new work space environments. These include *the use of anthropometric data*. These data require more than a simple "look-up table" approach. Anthropometric data may be easily misinterpreted or misapplied. If used inappropriately or carelessly, or if interpreted incorrectly, these data may do more harm than good. The same thoughtful, methodical, and comprehensive approach one would use in the design and selection of material for a project should be applied when using anthropometric data.

***Thoughtful Application of Anthropometric Data*** Determining which data table to use and applying those data judiciously is not enough. A systems designer should keep in mind several key points before using such data. These points reflect the system under design and the potential user population. Similarly, if data for a particular design dimension do not exist, the same care should be taken in acquiring the necessary data.

*Know the Population of Users under Consideration* Who will be using the system? What characteristics do they posses? What are the human user requirements (physical, cognitive, skill level, etc.)? What kind of physical demands will be placed on the user in the work space? What will the operating environment be like?

*Determine Essential Body Dimensions and Which Dimensions Are Most Important for Design* Similarly, which dimensions are practical, given economic, weight, size, durability, and maintainability considerations? When referring to anthropometric tables/charts, always use recent data (individuals and general populations *do* change over time). Do not extrapolate or attempt to correlate dimensions from existing data unless you are aware of how doing so may skew the data. There is always the potential for complex and unforeseen interactions between measurements.

*Determine Percentage of Population Accommodated* This consideration may vary depending on the use of the system. Consumer products require much wider population variances than military systems, for instance. While it is obviously impractical to fit a design or process to 100 percent of the population, it is important to know in the conceptual design phase who comprises the population of users and what percentage of them you intend to accommodate.

*Design to the Portion of the Population That Has Been Selected for Accommodation* Use the average of that population as a midpoint and develop a design that is adjustable or modifiable for the variance of that population. Again, for commercial systems, the populations have larger variances than military populations. Make necessary adjustments using anthropometric tables for reach and clearance envelopes, seat adjustment envelopes (transportation equipment), etc. Like all design trade-offs, the more flexible the design becomes in one area, the more likely certain constraints will appear in other areas of the design.

Use prototyping, mockup, simulation, and computer-aided design tools in the concept exploration phases to determine if human user requirements are met. Early design verification for user fit will save time and money in later engineering phases. (Gawron et al., 2002). Other appropriate considerations in the design of a system and its user environment include:

*Determine the Controls and Tools Necessary to Perform System Tasks* Primary and secondary controls and tools must be inside the reach envelope for the population selected. Tools, controls, or parts that the worker must reach and use most often should be placed within the primary reach envelope, while tools, controls, and parts used less frequently should be placed progressively further away (in secondary or tertiary reach envelopes).

*Determine the Physical Characteristics and Clearances (e.g., leg, head, arm, etc.) for Users of System* Comfortable seating, adequate legroom and headroom, sufficiently wide passageways to/from the work location should all be planned. Many populations include special-needs members such as users with limited mobility, sensory limitations, and/or pregnancy.

Physical characteristics of the work surface must take into consideration such factors as height, depth, clearances, and inclination of the work surface. Overhead or side tool storage must be placed to allow access with minimal disruption to productivity.

Maintenance of the system requires accessibility to maintainers. Designing for maintenance requires that the system be usable and accessible to both its operator and maintainer (see Example 19.2).

**Example 19.2 Engine Room Habitability** The engine room crew of a nuclear aircraft carrier has to contend with issues involving the "habitability" domain—that is, those issues associated with living, sleeping, and eating within the confines of the systems the crew is operating and maintaining.

The design engineer in this context is commonly referred to as a marine architect, namely, one who designs ships. A major challenge for the marine architect charged with designing a turbine engine compartment deep within a modern aircraft carrier would be habitability. Because space is at a premium, it is no small task to design a space large enough to hold the engines and ancillary equipment, but to also ensure that the crew members who will work within this environment are able to safely and effectively perform their assigned tasks.

Another issue is the high stress levels likely to be a factor for the engine room crew. The work environment is inherently stressful (high heat, extremely loud, close quarters; physically uncomfortable work positions). The unpredictable nature of equipment malfunction and the requisite necessity to work at odd hours or "on call" to operate and/or repair equipment can add to this problem. Inability to "get away" from work—most engine room crews' sleep and eat in relatively close proximity to their duty station—combined with working very long hours can induce high levels of stress and fatigue. Due to the nature of the work location (far below decks), individuals may not see daylight for a week or more. Crewmembers also must maintain a high level of SA due to the dangerous environment.

### 19.4.2 HSI Considerations for Design of Displays

Many, if not all, modern systems involve the display of information. Complex presentation of information has been designed into modern weapons systems, power generation plants, and desktop workstations. The goal of any display is to optimize the performance of the person using the system while allowing for a reduction in errors (Woodson and Conover, 1966).

***Visual Display of Information*** Visual displays should be designed so that users are able to easily and quickly ascertain the state of the system at a glance. While good displays facilitate accurate transfer of information from system to the human, poor displays may contribute to accidents and errors. Poorly designed displays may make it difficult for the user to quickly and accurately detect a problem or determine and implement a solution. Designing a usable visual display does not require extensive knowledge of vision theory and the supporting brain and cognitive functions. There are principles that can be applied by the design engineer in planning and developing a system that people can use successfully. Table 19.12 lists many of the factors that should be considered when designing a system that requires the use of visual information.

*Ambient lighting* must be adequate to allow the person to see the task at hand. For some jobs, supplemental *task lighting* must be added so people can see what they are doing. For example, kitchen designers provide an overhead (ambient) fixture, but also include task lighting over the stove, in the oven, and over work surfaces. Detailed work may require

**TABLE 19.12 Visual Characteristics Useful for Human Systems Design Consideration**

*Visual primacy:* As the dominant sensory system for humans, vision is used for orientation, intake of information, and verification of other senses. In the absence of visual input, other sensory systems become more important.

*Light levels*: Ambient light levels have a profound effect on visual functioning. Light levels that are too low inhibit detection and color vision while light levels that are too bright, e.g., direct glare, can be equally disruptive to vision.

*Edge detection*: The visual system of humans has built-in "edge detectors" that allow for immediate recognition and detection of edges.

*Motion detection:* Built-in motion detectors allow for immediate and automatic recognition of movement and are obviously important in many technologies.

*Pattern recognition:* The human visual system has an excellent ability for pattern recognition, which is particularly useful in monitoring tasks or for off-center vision. It is automatic and allows for the rapid integration of dissimilar visual elements into a cohesive whole. For example, on an aircraft display, symbology is used to facilitate rapid recognition of visual targets.

*Coding:* Visual design of information can be coded to give more information in less space. The addition of color, shape, or grouping to indicate another dimension is an excellent practice that tells the user more in less space.

*Gestalt:* Visual information should be grouped to ensure that similar items are processed as a unit. Control panels that have related controls "boxed" together using linear demarcation facilitate human performance and allow the operator to quickly detect when one gauge is out of range.

more light. There should also be adequate contrast between the area of attention and the background. Kroemer and Grandjean (1997) provide excellent guidance for the placement of light for visual work.

Because people tend to identify things that are placed close together as a group, gauges, dials, or other items that are closely related should be located in close proximity to one another (Chapanis et al., 1963). Displays should be grouped according to use. For instance, in an aircraft, all the displays having to do with engine health should be located together. The displays should also be arranged to allow "quick looks." Many aircraft displays are installed so that the indicators for normal operations are in the same position (e.g., 12 o'clock) for all the displays in a group. *Display consistency* allows the operator to quickly glance at a display to determine whether the system is functioning properly. Displays should be directly linked with their controls by putting control actuators (knobs, dials, etc.) on or close to the display. There should be little or no delay or lag in the display/control interface. Displays need to be properly lighted to ensure readability in all lighting conditions that may be encountered. In an aircraft, the system should accommodate for light levels ranging from night light, low light, low sun angle light, to bright sunlight. There should also be good contrast and readability in all lighting levels. Human Engineering Design Criteria for Military Systems, Equipment and Facilities (MIL-STD-1472) is a useful resource for basic design. The MIL-STD 1472 (currently in version F) contains some basic considerations that apply to all systems in which visual information is presented.

There are a number of other important visual principles related to presentation of information (Tufte, 1983, 1990). Some important design considerations include redundancy, alarm and caution signals, and display type. *Redundancy* is the presence of information in more than one place and/or in more than one sensory modality. *Color coding* is an example of different modes, while the presence of a digital and analog clock on a display is an example of information in more than one place and in more than one mode. Another consideration is that while many people have good color vision, an inherited deficiency in color vision (color blindness) exists in about 10 percent of the population, primarily males.

Redundancy helps people gather information by presenting data in more than one way. *Alarms and cautions* have a range of urgency and can be presented in multiple ways. For example, alarm or caution information may be presented in red (color coding) or flashing (to catch the user's attention) or coupled with another sensory modality to ensure that the user identifies a problem and initiates a solution. *Display type* includes both the method of information presentation (e.g., digital vs. analog) or means of presentation (e.g., CRT, AMLCD, plasma, etc.). The method of presentation should be carefully considered to ensure that the user receives the information in the most usable way. For instance, while digital speedometers in cars were tried for a short while, it was difficult to control speed because digital speedometers show state information. Speed control requires trend information (e.g., is the vehicle accelerating or decelerating?) As has been shown, vision is a complex but critical information channel whose use should be optimized to ensure peak user performance.

***Auditory Display of Information*** As with vision, since so many modern systems use auditory cues to convey information, this section focuses on general design principles critical for the system designer to know about the auditory characteristics of the human. Table 19.13 lists auditory characteristics that are of primary consideration by systems engineering design teams.

Typically, vision and audition function together. For example, a radio call alerts a pilot to air traffic and the pilot begins a visual search. Auditory information is processed serially while visual information can be processed in a parallel fashion. Auditory signals can range from simple (e.g., warning horns) to complex (e.g., speech). The auditory channel is well suited to the presentation of imperative information, such as warnings or cautions. But for most tasks, audition should supplement visually presented information, rather than being

**TABLE 19.13 Auditory Characteristics Useful for Human Systems Design Consideration**

| |
|---|
| *Auditory localization:* Humans are very good at determining the direction of a sound source, with the exception of sounds generated on the exact centerline, i.e., directly in front or behind the operator. |
| *3D auditory capability:* The occurrence of a sound in three-dimensional space allows a user to receive information other than the location of a sound. This feature of the human auditory system can be used to aid a user who is overloaded with visual input. |
| *Pattern recognition:* Auditory pattern recognition is very similar to visual pattern recognition. This process occurs automatically and allows for the rapid integration of dissimilar auditory elements into a cohesive whole (e.g., music recognition). |
| *Tones vs. speech:* To be effective, alarms need to be audible and distinctive in the operating environment. Alarms do not have to be transmitted verbally as long as their intent is conveyed. |

the primary source of information. Auditory displays of information should be limited to short messages or information that requires an immediate response. The auditory display alerts the user to make use of the visual display for more complete and amplifying information. While verbal displays may provide more information, they may also take longer to present the information than if it was presented visually. MIL-STD-1472 provides guidance for the use and design of verbal displays.

People with normal hearing exhibit temporal proximity, that is, tones close together in time are perceived to be together. People also exhibit auditory similarity based on the pitch of the sounds (i.e., sounds that have similar pitch are perceived as a group). Humans have the ability to localize sounds in three dimensions [i.e., they can determine the direction of the source of a sound (Proctor and Van Zandt, 1994)]. Three-dimensional displays provide location information. The localization of the signal alerts the user to the position of a threat or other sound and is very different from the information conveyed visually.

***Haptic Sensory Display of Information*** Virtual environments have increasingly relied on the insertion of haptic cues to enhance the user's sense of immersion in the virtual environment. In particular, one journal, *Presence*, has a wealth of information on haptics and their use in virtual environments. Staying abreast with developments in this rapidly changing field can be challenging but rewarding for those wanting to include senses other than just vision and audition in virtual environments. In aviation, user presentation of information using the touch sensory modality has focused on the presentation of attitude, proximity, and spatial mapping of information [e.g., U.S. Navy research on a vibro-tactile suit to display aircraft attitude information to the pilot (Rupert et al., 1994; Rupert, 2000)]. The touch senses do not transmit highly specific information as occurs more commonly in the visual and auditory senses. This characteristic does not in any way discount the utility of the touch senses for a more general display of information, improving SA, or for redirecting operator attention. Table 19.14 illustrates the touch characteristics that are important for consideration in HSI applications.

**TABLE 19.14 Touch Characteristics Useful for Human Systems Design Consideration**

*Proprioception:* An awareness of the position of our body joints relative to each other and to our body. This includes awareness of the position of the body in space and with respect to objects in the environment.

*Sensitivity variability:* Sensors are more closely spaced in some areas of the skin than in other areas (e.g., the skin on the tips of the fingers has many more receptors than does the skin on the back.). Thus, if the operator is required to detect small patterns, it would be better to use the tips of the fingers than the skin on the back. One example of this is the use of the fingers to read Braille letters in visually impaired individuals.

*Haptic:* When touching an object, we respond to the shape and feel of a manipulated object. Shape coding of controls uses the haptic sense to impart more information for operation of the control without visual input (blind mapping). Vibro-tactal display devices capitalize on haptic sense. Another haptic design consideration is control actuation feedback.

*Kinesthetic:* Our ability to sense the relative motion and speed of movement of our limbs. There are no practical design considerations for kinesthesia at this time.

### 19.4.3 Task Allocation—Humans versus Machines

Any comprehensive discussion of the human component of a total system should include a section comparing human versus machine or nonhuman abilities, highlighting the relative merits of the two subsystems. Task analysis is a method used to determine which part of a system should perform different types of work.[4] It ensures that each task is assigned to a specific part of the system being designed and helps to illustrate what characteristics are most appropriate for the successful completion of a task. For example, if a requirement-driven task is "monitoring a fuel level," the task of monitoring the fuel level may be levied on a computer subsystem, while the task of "monitoring the monitoring of the fuel level" could be assigned to the mission commander, a human subsystem. Task analysis can also be used to model the distribution of tasks across a system and/or team and determine the effect of that distribution. Task allocation can be reiterated until optimal performance is achieved.

There are a number of task types that are better left to people or the human subsystem, while other tasks are better left to machine subsystems (Parasuraman, 2000; Parasuraman et al., 1996; Parasuraman and Riley, 1997). Table 19.15 lists tasks that are performed better by humans and tasks that are performed better by machines.

### 19.4.4 Social Issues and Team Performance

While working with other people is generally beneficial, desirable, and often necessary, it is also not without its problems. The vast majority of us are intimately tied to our social environment. We are heavily influenced by social factors, and this social environment exists completely within our physical environment. We think and work differently as a function of our social environment. Yet we have become so accustomed to living and working within this social climate, that we are rarely conscious of it. A pervasive theme running throughout this section is to make the reader aware of the influence of factors that are easily overlooked—in this case, the social factor.

One very important factor that is often ignored in the planning, design, and operation of many complex systems is the social nature of the human user. Social interactions are a very important part of our lives, and because these interactions so strongly influence our behavior, we need to be aware of the processes and social dynamics that influence the operator/user. Both the physical workplace configuration and the cognitive task demands placed on the human operator/user should be considered within this social milieu.

***Social Interactions*** It is important to consider the social processes that occur when an individual in a system interacts with other people. Social interactions refer to the subtle yet pervasive verbal and nonverbal interactive styles all humans exhibit. These interactions are important whenever two or more people are required to function as a team, and become even more important when that team is responsible for controlling highly sophisticated, technologically intense, and potentially dangerous equipment (e.g., nuclear power plant, chemical factory or oil refinery, air traffic control operations, etc.).

***Social Comparison*** People assess each other constantly. From a human factors perspective, this social comparison can influence our actions in ways that defy the attempts of engineers to accommodate them in the system design. Pilots or others in high-risk, high-

**TABLE 19.15 Task Allocation to Appropriate Subsystem**

| Example of Tasks Performed Better by Humans | Example of Tasks Performed Better by Machines |
|---|---|
| Detection of certain forms of very low energy levels | Monitoring of humans and machines |
| Sensitivity to an extremely wide variety of stimuli | Performing routine, repetitive, or very precise operations |
| Perceiving patterns and making generalizations about them | Responding very quickly to control signals |
| Detecting signals in high noise levels | Exerting great force, smoothly and with precision |
| Ability to profit from experience and alter course of action | Insensitivity to extraneous factors |
| Ability to react to unexpected low-probability events | Ability to process many different things simultaneously |
| Applying originality in solving problems | Deductive processes |
| Ability for fine manipulation, especially where misalignment appears unexpectedly | Ability to repeat operations very rapidly, continuously, and precisely the same way over a long period of time |
| Ability to perform even when overloaded | Operating in environments that are hostile or dangerous to humans or are beyond human tolerance |
| Ambiguity resolution | Continuous collection of data to support decision making |
| Ability to reason inductively | Deductive reasoning |
| Tasks requiring high motivation or involving strong emotions | Consistent reasoning across all cases |
| Remembering exceptional cases | Remembering all cases (and the probability of each) |
| Following hunches/flexibility | Consistent application of rules to situations or cases |

tech occupations may perform their job poorly or in an unsafe manner in attempting to gain the approval (or admiration) of those around them, or perhaps to flaunt their "mastery" of a complex skill.

***Diffusion of Responsibility*** The attribution of responsibility can be ascribed to an individual's deference to the "person in charge." Unfortunately, this communication is often made in non-verbal ways, which can leave the senior individual under the assumption that "...if I do something wrong or unsafe, he/she (e.g., the junior individual) will tell me about it," when in fact the junior individual may remain quiet out of respect (or fear) of the senior individual. The subordinate may also feel it is "not their place" to alert their superior to a potentially dangerous situation. Such unstated assumptions can have disastrous consequences (Evans, 2000).

Computers provide assistance to the operator in most modern complex systems, but as long as humans are an integral part of such complex systems, the potential for human error will always be present (Parasuraman et al., 1996; Wiegmann and Shappell, 2001). Failure to take into account the personality, social interaction style, and the ability of individuals to effectively communicate and work together *as a team* during critical aspects of any complex operation is to invite disaster (see Examples 19.3 and 19.4).

**Example 19.3 Crew (or Cockpit) Resource Management** To appreciate why flight crew training and coordination is so important, one can look at safety statistics that demonstrate that approximately 70 per cent of all major airplane accidents were caused by aircrew mistakes (O'Hare and Roscoe, 1990). Crew resource management (CRM) was developed to mitigate some of the more dangerous social aspects of human performance in a complex system, specifically to improve performance in multiplace aircraft (Weiner et al., 1993). Performance of the multiplace aircraft system is dependent on the crew working *together* to detect and solve problems as they occur in flight operations. Crew coordination is combined with other decision and performance aids such as checklists and instrument configuration changes. The CRM approach is often associated with checklist design but goes well beyond the checklist. Crew coordination focuses on determining a process for performance, and following that process, to ensure safe performance and successful mission completion (Brown et al., 1991).

**Example 19.4 Automation and Fight Emergencies** One of the issues in modern aircraft flight decks is the misallocation of crew and equipment resources. The current generation of computer-controlled "fly-by-wire" aircraft has sophisticated instrumentation and onboard computers capable of flying the entire route from take-off to landing. While this reduces the workload on the crew, the design of these automated systems are not based on the performance strengths and limitations of the system subcomponents (the computer and the human operator). Therefore, although workload is reduced in regular operations, the operators are forced to perform monitoring tasks, which humans perform poorly. In emergencies, the operators, who have been tasked to monitor the system, may not have the information needed to resolve the emergency and are required to "catch up" to the rest of the system when time is short.

## 19.5 CASE STUDY

For all systems that include the human as an operator and user, there are certain considerations that are universal. Perhaps most obvious is the physical consideration that involves fitting the human *into the system*, whether it is in a crew station, control room, or cockpit. Sensory, perceptual, and cognitive considerations are also required for all systems: The operator must have the ability to sense and process information from the system to make decisions about how to control it. The following case study illustrates the importance of human consideration in HSI. Although based on a real military system and tasks, the case study is hypothetical and not meant to be used as guidelines for actual systems.

***Unmanned Aerial Vehicle [UAV(x)] System*** This case study is for a hypothetical UAV system design that we will term UAV(x). The DoD has actually invested heavily in both the technology and capabilities of UAVs. As shown in the war in Afghanistan, UAVs can function in a wide variety of roles *without endangering the life of the human operating the system* (i.e., a pilot). Other benefits of the UAV besides pilot safety are weight savings and a larger payload for sensors. The UAV itself is only one component of a very complex system. It requires many of the same things that a piloted aircraft requires, including a runway, a hanger for maintenance, the ability of maintenance personnel to readily access major components for repair/replacement, etc. Unlike a piloted aircraft, however, it also requires a "cockpit" apart from the UAV itself, typically at a remote site to allow the "pilot" (or operator) to "fly" the aircraft from a distant location.

Table 19.16 provides an engineering description of the UAV(x), which includes an example of the operational and hardware requirements that the engineering community might provide for such a system. Usually the engineering community will also provide the human interface requirements in very general language. As shown in Table 19.16, the "pilot" is expected to "fly" the UAV for the duration of its flight, and should have flight experience and good SA. The HSI program manager should identify special issues to be included early in the engineering design process (see part IV of Table 19.16).

The UAV operator selection is a critical HSI issue. How do we select and train an operator for the UAV system? What human characteristics and capabilities are required to operate a UAV? How long a shift can an operator work without impairment in performance? Are there standards that could be developed and applied to operators of UAVs? Who will make the best UAV operators?—Experienced pilots/aviators *or* specially trained operators with extensive training in UAV operations?

**TABLE 19.16 UAV(x) Operational and HSI Description**

*I.* *Operational Requirements:* Ability to taxi, take off, fly, and land like a manned, fixed wing aircraft; ability to fly to specific coordinates and loiter (either manually or by ground-based operator direction) or via autopilot; good fuel efficiency to remain on station for extended periods; ability to respond to operator commands and send operational and avionics data back to ground operator; speed of vehicle not a primary design consideration; weight of payload and endurance.

*II.* *Hardware Requirements:* Fuselage with wings; tail assembly; propulsion system (jet or propeller); avionics and communications bay; fuel tanks; landing gear (fixed or retractable); hydraulic system (if needed); wiring and piping; payload bay(s): video camera(s), still (digital) camera, IR sensor, electronic warfare (EW) offensive capacity, offensive missile capacity, defensive systems (EW; chaff, IR flares; etc.); IFF and/or transponder for identification in combat/controlled airspace; GPS receiver to aid in localization; payload bay designed to optimize quick equipment change with minimal down time; low noise level of vehicle designed to avoid detection; use of lightweight material imperative; shape of fuselage incorporates "Stealth" technology to reduce radar return.

*III.* *Human Interface Requirements:* Control station (flight deck) must allow for full flight control of the UAV from takeoff, to mission control, to landing; control station must have the ability to transmit data to the UAV and receive data from the UAV in real time; video image(s) must be received and displayed as clearly as possible; flight controls must be suitable for wide range of operations and should be similar (where possible) to flight controls on regular aircraft. The individual(s) who operate or "fly" the UAV remotely will likely have some flight experience and should have good hand–eye coordination, good situation awareness, and good mechanical/electronics abilities.

*IV.* *Special HSI Issues*

1. Skill requirements for UAV operators
2. Ability to "fly" vehicle from remote location
3. Remote station multitask displays design
4. Operator fatigue for long period flights
5. Vehicle recovery and repair time

The military has considerable experience in selecting, classifying, and training pilots and aviators over the last century. Pilots and aviators enter flight training schools and are quickly advanced into flying specialized military aircraft based on their skills and performance in flight schools—or else they are moved into nonflying activities. But regular flight school revolves around one major fact—the pilot is actually in the aircraft he or she is flying. But with a UAV, the opposite is true, the pilot/operator flies the aircraft remotely via sensors and avionics information relayed via data links from the "flight deck" to/from the UAV. Although there are obvious similarities, the operation of a UAV is considerably different from the operation of a regular manned aircraft. Examples of operational issues with the UAV, which are not issues with manned aircraft, include the ability to maintain SA, ability to respond to vestibular cues (aircraft motion), ability to view the world in 3D and ability to respond to events quickly without time delay in the data link.

Because UAVs are capable of staying airborne for extremely long periods of time (Global Hawk can remain airborne for over 4 days), issues of operator fatigue, crew rest, appropriate handoff between crews, and the decay in performance seen over time with vigilance tasks are critical considerations for this system. Monitoring the fluctuations of operator efficiency, as evidenced by operator states, will ensure optimal system performance.

A special document might be prepared for any new system listing the HSI characteristics and quantification methods of special characteristics. As shown in Table 19.17, critical human characteristics for the UAV(x) include the target audience description (operators and maintainers), human factors design of the remote control station, and special maintenance requirements for the system. For measures of many of these characteristics, the Armed Services Vocational Aptitude Battery (ASVAB) provides the skill categories (CAT I, II, III, IV) (see Chapter 11), and MIL-STD-1472 provides the general standards for human factors engineering of crew stations and maintenance (see Chapter 7).

## 19.6 SUMMARY AND CONCLUSIONS

As is the case with all natural systems, the complexity of the physical forces and the resulting interaction of the organisms within the systems are vital elements in understanding how systems operate and can be designed for optimal performance. In this chapter, we have pointed out the need to consider the characteristics of the human components within our "man-made" systems engineering ecosystem.

Understanding the strengths and limitations of the human operator and maintainer is imperative to the systems designer. Humans have measurable psychological and physical characteristics; some of these characteristics are traits, innate and relatively unchanging; others are transitory states that may vary according to a range of conditions. The ability to quantify these human parameters provides a tremendous advantage to the systems designer. Design trade-offs made with an understanding of these human characteristics are more likely to result in a superior product with improved systems performance than one where these characteristics are not a priority. Ultimately, understanding the salient human considerations can allow the design engineer to tailor a system to effective and efficient performance, both from a total systems perspective as well as that of the human user.

**TABLE 19.17 UAV(x) HSI Characteristics and Quantification Methods**

| Human Characteristics | Quantification Methods |
|---|---|
| 1. Target audience description | |
| • Operator skill requirements | |
| • High aptitude—CAT II | • Military ASVAB scores |
| • Visual acuity—20/20 correctable | • Snellen eye chart |
| • High hand–eye coordination | |
| • Good 3D spatial perception | • FAA—air traffic control tests |
| • Flight training | |
| • Operator anthropometric limits | • Body height, leg length, arm length |
| • Male/female 5th–95th percentile | |
| • Maintainer skill requirements | • Military ASVAB scores |
| • Average aptitude—CAT III | |
| • Avionics qualified | |
| • Number of personnel | |
| • Operators—1 per vehicle | |
| • Maintainers—1 per four vehicles | |
| 2. Remote Control Station | |
| • Workstation design for one operator; layout lighting, controls, displays, alarms, seating in accordance with MIL-STD-1492. | • MIL-STD-1472 |
| • Multitask displays capable of representing 3D spatial images of aircraft environment in form easily detected and processed by operator. | • NASA/FAA standards |
| • Ability to quickly acquire and maintain SA using data provided from UAV | • SA measures for distant SA via data link need to be developed. |
| 3. Maintenance | |
| • Parts removal and repairs to be 80% organizational level. | |
| • Vehicle maintenance hatches, equipment bays easily accessible; equipment within each bay easily removed/replaced with minimal UAV down time; reconfigurable software must allow for modifications as new hardware is added or removed. | • MIL-STD-1472 |
| • Remote station flight deck and related electronics designed for easy access to all components | |

This chapter presents an overview of human characteristics that should be considered within a total systems perspective. It is pointed out that individuals can be defined and characterized using a variety of criteria, including broad categories of "trait" and "state." *Human traits*, or those characteristics of the user that tend to be static and unchanging, are described along with corresponding measurement techniques. *Human states*, or those characteristics that vary based on individual responses to operational and/or environmental conditions are also described. Such states may be complex responses to environmental conditions and/or demands, or they may entail individual reactivity to internal processes. In a section on human–system interfaces we provide some guidelines for bridging crucial junctures between human and machine. Guidelines such as these should be useful to the systems engineer seeking to optimize system performance through effectively integrating the human into the system design. Finally, a case study on the design and operation of a hypothetical UAV system is used to illustrate some of the HSI lessons learned throughout the chapter.

If the human component of the systems are to perform to their optimum, it is recommended that human factors professionals be on the design team and basic iterative human factors design principles be used in all phases of the systems engineering process, especially during the concept development phase. To obtain HSI support from certified human factors professionals, the Human Factors and Ergonomics Society (HFES), the Board of Certification of Professional Ergonomists (BCPE), and the International Ergonomics Association (IEA) are useful resources. The web sites for these organizations have been listed at the end of the reference section.

## NOTES

1. Although the domains for manpower, personnel, and training (MPT) are important to a complete description of the human component, MPT descriptions and issues are not covered in this chapter. Chapters 11 and 12 cover the MPT characteristics important for system engineering and management issues.
2. Table 19, page 103, of Gawron (2000) lists these measures, along with estimates of reliability, task time, and ease of scoring.
3. Personal communication, John Rohrbaugh, June 2002.
4. See Chapters 10, 11, 13, and 20 for details on task analysis.

## REFERENCES

Balkin, T. J., and Badia, P. (1988). Relationship between Sleep Inertia and Sleepiness: Cumulative Effects of Four Nights of Sleep Disruption/Restriction on Performance Following Abrupt Nocturnal Awakenings. *Biological Psychology, 27*(3), 245–258.

Banderet, L. E., Shukitt, B. L., Crohn, E. A., Burse, R. L., and Roberts, D. E. (1987). *Effects of Various Environmental Stressors on Cognitive Performance*. U.S. Army Research Institute of Environmental Medicine. Report No. AD-A177587. (NTIS No. HC A02/MF A01). Natick, MA: U.S. Army.

Barlow, H. B., and Mollon, J. D. (1982). *The Senses*. Cambridge, MA: Cambridge University Press.

Bhatia, G. H., Smith, K. E., Commean, P. K., Whitestone, J. J., and Vannier, M. W. (1994). Design of a multi-sensor optical scanner (of the human body). In *Proceedings of the Society of Photo-Optical Instrumentation Engineers: Sensor Fusion VII Meeting*, Boston, MA, SPIE Proceedings, Vol. 2355 (pp. 262–273). Bellingham WA: Society of Photo-Optical Instrumentation Engineers.

Boff, K. R., and Kaufman, J. P. (Eds.). (1986). *Handbook of Perception and Human Performance*, Vol. 2: Cognitive Processes and Performance. New York: Wiley.

Booher, H. R. (1990). *MANPRINT: An Approach to Systems Integration*. New York: Van Nostrand Reinhold.

Brown, C. E., Boff, K. R., and Swierenga, S. J. (1991). Cockpit resource management—A social psychological perspective. In *Proceedings of the International Symposium on Aviation Psychology, 6th* Columbus, OH, Vol. 1 (pp. 398–403). Columbus, OH: Ohio State University.

Burns, J. W., Werchan, P. M., Fanton, J. W., and Dollins, A. B. (1991). Performance Recovery Following +Gz-Induced Loss of Consciousness. *Aviation, Space, and Environmental Medicine*, *62*, 615–617.

Buser, P., and Imbert, M. (1992). *Audition* (Trans. R. H. Kay). Cambridge, MA: MIT Press.

Cacioppo, J. T., Tassinary, L. G., and Berntson, G. G. (Eds.). (2000). *Handbook of Psychophysiology*, 2nd ed. New York: Cambridge University Press.

Caldwell, Jr., J. A. (1997). Fatigue in the Aviation Environment—An Overview of the Causes and Effects as Well as Recommended Countermeasures. *Aviation, Space, and Environmental Medicine*, *68*(10), 932–938.

Cannon, B. (1994). Walter Bradford Cannon: Reflections on the Man and His Contributions. *International Journal of Stress Management*, *1*(2), 145–158.

Center For Positron Emission Tomography (CPET). (2002). *Writings on Positron Emission Tomography*. Available: http://www.nucmed.buffalo.edu/petdef.htm (retrieved May 2002).

Chapanis, A., Cook, III, J. S., Lund, M. W., and Morgan, C. T. (Eds.). (1963). *Human Engineering Guide to Equipment Design*. New York: McGraw-Hill. Sponsored by the Joint Army-Navy-Air Force Steering Committee.

Charlton, S. G., and O'Brien, T. G. (Eds.). (2002). *Handbook of Human Factors Testing and Evaluation*, 2nd ed. Mahwah, NJ: Lawrence Erlbaum.

DellaRocco, P. S. (1999). *The Role of Shift Work and Fatigue in Air Traffic Control Operational Errors and Incidents*. NASA No. 19990025333; DOT/FAA/AM-99/2. Civil Aeromedical Institute (CAMI), Oklahoma City, OK: DOT/FAA.

Eddy, D. R., and Hursh, S. R. (2001). Fatigue Avoidance Scheduling Tool (FAST). SBIR Phase I Final Report, *Human Effectiveness Directorate, Biodynamics and Protection Division, Flight Motion Effects Branch*. Air Force Research Laboratory (AFRL-HE-BR-TR-2001-0140). Brooks AFB, TX: USAF.

Endsley, M. R. (1995). Toward a Theory of Situation Awareness in Dynamic Systems. *Human Factors. Special Issue: Situation Awareness*, *37*(1), 32–64.

Endsley, M. R., and Garland, D. J. (Eds). (2000). *Situation Awareness Analysis and Measurement*. Mahwah, NJ: Lawrence Erlbaum.

Evans, D. (2000). Blindly Following the Computer. *Avionics*, *24*(3), 42–43.

Flaherty, D. E. *Sopite Syndrome in Operational Flight Training*. Master's thesis, Naval Postgraduate School. NASA No. 19990018574; AD-A354942. (NTIS No. AD-A354942). Monterey, CA: U.S. Navy. Retrieved December 10, 2002 from Defense Technical Information Center (DTIC): http:/handle.dtic.mil/100.2/ADA354942.

Gawron, V. J. (2000). *Human Performance Measures Handbook*. Mahway, NJ: Lawrence Erlbaum.

Gawron, V. J., Dennison, T. W., and Biferno, M. A. (2002). Mock-Ups, Models, Simulations, and Embedded Testing. In S. G. Charlton and T. G. O'Brien (Eds.), *Handbook of Human Factors Testing and Evaluation*, 2nd ed. (pp. 181–223). Mahwah, NJ: Lawrence Erlbaum.

Gescheider, G. A. (1997). *Psychophysics: The Fundamentals*. Mahwah, NJ: Lawrence Erlbaum.

Goldstein, E. B. (Ed.). (2001). *Blackwell Handbook of Perception*. Malden, MA: Blackwell.

Griffin, M. (1997). Vibration and Motion. In G. Salvendy (Ed.), *Handbook of Human Factors and Ergonomics*. New York: Wiley.

Hahn, S. E., and Smith, C. S. (1999). Daily Hassles and Chronic Stressors: Conceptual and Measurement Issues. *Stress Medicine*, *15*(2), 89–101.

Hancock, P. A., and Desmond, P. A. (Eds.). (2001). *Stress, Workload, and Fatigue*. Mahwah, NJ: Lawrence Erlbaum.

Harm, D. L. (2002). Motion Sickness Neurophysiology, Physiological Correlates, and Treatment. In K. M. Stanley (Ed.), *Handbook of Virtual Environments: Design, Implementation, and Applications. Human Factors and Ergonomics* (pp. 637–661). Mahwah, NJ: Lawrence Erlbaum.

Hartman, B. O., and Secrist, G. E. (1991). Situational Awareness Is More Than Exceptional Vision. *Aviation, Space, and Environmental Medicine*, *62*(11), 1084–1089.

Hettinger, L. J., Kennedy, R. S., and McCauley, M. E. (1990). Motion and Human Performance. *Motion and Space Sickness* (pp. 411–441) (A93-55929 24-52). Boca Raton, FL: CRC Press.

Hockey, R. (Ed.). (1983). *Stress and Fatigue in Human Performance*. New York: Wiley.

Holtzman, W. H. (2002). Personality Theory and Assessment: Current and Timeless Issues. In H. I. Braun and D. N. Jackson (Eds.), *The Role of Constructs in Psychological and Educational Measurement*. Mahwah, NJ: Lawrence Erlbaum.

Hossain, J. M., and Shapiro, C. M. (1999). Considerations and Possible Consequences of Shift Work. *Journal of Psychosomatic Research. Special Issue: Sleep and Fatigue*, *47*(4), 293–296.

House of Representatives Committee on Armed Services. (1992, July 21). The July 3, 1988 Attack by the Vincennes on an Iranian Aircraft. 102d Congress, Second Session.

Hursh, S. R., Redmond, D. P., Johnson, M. L., Thorne, D. R., Belenky, G., Balkin, T. J., Storm, W. F., Miller, J. C., and Eddy, D. R. (in press). Fatigue Models for Applied Research in War Fighting. *Aviation, Space, and Environmental Medicine*.

Kanki, B. G. (1996). Stress and Aircrew Performance. In J. E. Driskell and E. Salas (Eds.), *Stress and Human Performance* (pp. 127–162). Mahwah, NJ: Lawrence Erlbaum.

Kaufman, A. S. (2000). Tests of Intelligence. In R. J. Sternberg (Ed.), *Handbook of Intelligence* (pp. 445–476). New York: Cambridge University.

Kroemer, K. H. E., and Grandjean, E. (1997). *Fitting the Task to the Human: A Textbook of Occupational Ergonomics*, 5th ed. Philadelphia, PA: Taylor and Francis.

Krueger, G. P. (1989). Sustained Work, Fatigue, Sleep Loss and Performance: A Review of the Issues. *Work and Stress*, *3*(2), 129–141.

Lazarus, R. S., and Folkman, S. (1984). *Stress, Appraisal and Coping*. New York: Springer.

Leger, A., Aymeric, B., Audrezet, H., and Alba, P. (October 1999). Human Factors Associated with HUD-based Hybrid Landing Systems—SEXTANT's experience. In *AIAA and SAE 1999 World Aviation Conference*, San Francisco, CA. AIAA Paper 99-5512; SAE Paper 1999-01-5512. Reston, VA: American Institute of Aeronautics and Astronautics (AIAA); Warrendale, PA: Society of Automotive Engineers (SAE).

Lewis, N. L., McGovern, J. B., Miller, J. C., Eddy, D. R., Forster, E. M. (1988). EEG Indices of G-induced Loss of Consciousness (G-LOC). In *Advisory Group for Aerospace Research and Development (AGARD), Electric and Magnetic Activity of the Central Nervous System: Research and Applications in Aerospace Medicine*. (SEE N88-27683 21-51). (NTIS No. HC A18/MF A01). Brooks AFB, TX: AGARD.

Ludel, J. (1978). *Introduction to Sensory Processes*. San Francisco: W.H. Freeman.

Marras, W. S., and Kim, J. Y. (1993). Anthropometry of Industrial Populations. *Ergonomics*, *36*(4), 371–378.

McCarthy, G. W. (1996). G-Induced Loss of Consciousness (GLOC)—An Aviation Psychology Challenge. *Human Performance in Extreme Environments*, *1*(2), 42–43.

McCauley, M. E. and Sharkey, T. J. (1992). Cybersickness—Perceptions of Self-motion in Virtual Environments. *Presence: Teleoperators and Virtual Environments*, *1*(3), 311–318.

McDaniel, J. W. (1998). Design Requirements for Human Strength: Asking the Right Question. In S. Kuman, (Ed.), *Advances in Occupational Ergonomics and Safety* 2 (pp. 345–348). Amsterdam, The Netherlands: IOS Press.

Meshkati, N., Hancock, P. A., and Rahimi, M. (1990). Techniques in Mental Workload. In J. R. Wilson and E. N. Corlett (Eds.), *Evaluation of Human Work: A Practical Ergonomics Methodology* (pp. 605–627). Philadelphia, PA: Taylor and Francis.

Miller, J. C., and Rokicki, S. M. (1996). Psychophysiological Test Methods and Procedures. In T. G. O'Brien and S. G. Charlton (Eds.), *Handbook of Human Factors Testing and Evaluation*. Hillsdale, NJ: Lawrence Erlbaum.

Naitoh, P. Kelley, T., and Babkoff, H. (1993). Sleep Inertia: Best Time Not to Wake Up? *Chronobiology International*, *10*(2), 109–118.

O'Donnell, R. D., and Eggemeier, T. D. (1986). Workload Assessment Methodology. In K. R. Boff, L. Kaufman and J. Thomas (Eds.), *Handbook of Perception and Human Performance,* Vol. 2: *Cognitive Processes and Performance* (pp. 42.1–42.49). New York: Wiley.

O'Donnell, R. D., Moise, S., Smith, R., Cardenas, R., and Eddy, D. (1999). Development of the Situation Awareness Flight Training and Simulation Evaluation (SAFTE) System: Final Development, Initial Test, and Documentation of the System. *NTI, Inc., Report No. AL/XX-TR-1999-XXXX: ADA372633*. Dayton, OH: NTI.

O'Hare, D., and Roscoe, S. (1990). *Flight Deck Performance: The Human Factor*. Ames, IA: Iowa State University Press.

Parasuraman, R. (2000). Designing Automation for Human Use: Empirical Studies and Quantitative Models. *Ergonomics. Special Issue: Ergonomics for the New Millennium*, *43*(7), 931–951.

Parasuraman, R., Masalonis, A. J., and Hancock, P. A. (2000). Fuzzy Signal Detection Theory: Basic Postulates and Formulas for Analyzing Human and Machine Performance. *Human Factors: Special Issue*, *42*, 636–659.

Parasuraman, R., Molloy, R., Moulousa, M., and Hilburn, B. (1996). Monitoring of Automated Systems. In *Automation and Human Performance: Theory and Applications*, (Collected works) (pp. 91–115). Mahwah, NJ: Lawrence Erlbaum.

Parasuraman, R., and Riley, V. (1997). Humans and Automation: Use, Misuse, Disuse, Abuse. *Human Factors*, *39*(2), 230–253.

Pew, R. W. (2000). The State of Situation Awareness Measurement: Heading Toward the Next Century. In M. R. Endsley and D. J. Garland (Eds.), *Situation Awareness Analysis and Measurement*. Mahwah, NJ: Lawrence Erlbaum.

Pritchett, A. R., and Hansman, R. J. (2000). Use of Testable Responses for Performance-Based Measurement of Situation Awareness. In M. R. Endsley and D. J. Garland (Eds.), *Situation Awareness Analysis and Measurement*. Mahwah, NJ: Lawrence Erlbaum.

Proctor, R. W., and Van Zandt, T. (1994). *Human Factors in Simple and Complex Systems*. Boston: Allyn and Bacon.

Regan, D. (2000). *Human Perception of Objects*. Sunderland, MA: Sinauer Associates.

Reid, G. B., and Nygren, T. E. (1988). The Subjective Workload Assessment Technique: A Scaling Procedure for Measuring Mental Workload. In P. A. Hancock and N. Meshkati (Eds.), *Human Mental Workload*. Amsterdam: North Holland.

Roebuck, J. A. (1995). *Anthropometrics Methods: Designing to Fit the Human Body*. Santa Monica, CA: Human Factors and Ergonomics Society.

Rupert, A. H. (2000). Tactile Situation Awareness System—Proprioceptive Prostheses for Sensory Deficiencies. *Aviation Space, and Environmental Medicine, Section 2, Supplement*, *71*(9), A92–A99.

Rupert, A. H., Guedry, F. E., and Reschke, M. F. (1994). The Use of a Tactile Interface to Convey Position and Motion Perceptions. In *Advisory Group for Aerospace Research and Development (AGARD), Virtual Interfaces: Research and Applications*. (SEE N94-37261 12-53). Johnson Space Center, Houston, TX: AGARD.

Salvendy, G. (Ed.). (1997). *Handbook of Human Factors and Ergonomics*. New York: Wiley.

Seyle, H. (1937). Further Evidence in Support of the Alarm Reaction Theory of Adrenal Insufficiency. *American Journal Physiology*, *119*, 400–401.

Seyle, H. (1975). Stress without Distress. *Vie medicale au Canada francais*, *4*(8), 964–968.

Snell, J. C. (1996). The Bell Curve, IQ, and Assessment Testing. *Journal of Instructional Psychology, 23*(4), 305–306.

Stoffregen, T. A., Draper, M. H., Kennedy, R. S., and Compton, D. (2002). Vestibular Adaptation and Aftereffects. In K. M. Stanney (Ed.), *Handbook of Virtual Environments: Design Implementation, and Applications. Human Factors and Ergonomics* (pp. 773–790). Mahwah, NJ: Lawrence Erlbaum.

Stokes, A. F., and Kite, K. (1994). *Flight Stress: Stress, Fatigue and Performance in Aviation.* Brookfield, VT: Ashgate Aviation.

Tilley, A. J. (1982). The Sleep and Performance of Shift Workers. *Human Factors, 24*(6), 629–641.

Teigen, K. H. (1994). Yerkes-Dodson: A Law for All Seasons. *Theory and Psychology, 4*(4), 525–547.

Tufte, E. R. (1983). *The Visual Display of Quantitative Information*. Cheshire, CT: Graphics.

U.S. Department of Defense (DOD). (1989). *Military Standard: Human Engineering Design Criteria for Military Systems. Equipment and Facilities*. Report: MIL-STD-1472D. (NTIS No. MILSTD1472D: ADA2814010). Washington, DC: DOD.

Tufte, E. R. (1990). *Envisioning Information*. Cheshire, CT: Graphics.

Vannier, M. W. and Robinette, K. M. (1995). Three Dimensional Anthropometry. In *Proceedings of the Biomedical Visualization Conference*, Atlanta, GA, (pp. 2–8). Piscataway, NJ: Institute of Electrical and Electronics Engineers (IEEE).

Vidulich, M. A., Stratton, M., and Wilson, G. (1994). Performance-Based and Physiological Measures of Situational Awareness. *Aviation, Space, and Environmental Medicine, 65*(5 Suppl.), A7–12.

Weiner, E. L. (1989). Reflections on Human Error—Matters of Life and Death. In *Proceedings of the Human Factors Society, 33rd Annual Meeting*, Denver, CO, Vol. 1 (pp. 1–7). Santa Monica, CA: Human Factors Society.

Weiner, E. L. (1990). Potential Benefits and Hazards of Increased Reliance on Cockpit Automation. In *Safety at Sea and in the Air—Taking Stock Together: Proceedings of the Royal Aeronautical Society*, London, UK (pp. 15-1–15.7). London: Royal Aeronautical Society.

Weiner, E. L. (1993). Life in the Second Decade of the Glass Cockpit. In *Proceedings of the International Symposium on Aviation Psychology, 7th*, Columbus, OH, Vol. 1 (pp. 1–7). Columbus, OH: Ohio State University.

Weiner, E. L., Kanki, B. G., and Helmreich, R. L. (Eds.). (1993). *Cockpit Resource Management.* San Diego, CA: Academic.

Wickens, C. D. (1996). Designing for Stress. In J. E. Driskell and E. Salas (Eds.), *Stress and Human Performance*. Mahwah, NJ: Lawrence Erlbaum.

Wickens, T. D. (2002). *Elementary Signal Detection Theory.* London: Oxford University Press.

Wiegmann, D. A., and Shappell, S. A. (2001). Human Error Perspectives in Aviation. *International Journal of Aviation Psychology, 11*(4), 341–357.

Will, B. (2000). HUD Operational Effectiveness. Situational Awareness on the Flight Deck. The Current and Future Contribution by Systems and Equipment. In *Proceedings of the Royal Aeronautical Society*, London, UK (pp. 1-34–1.48). London: Royal Aeronautical Society.

Wilson, G. F. (1992, November). Cardiorespiratory Measures and Their Role in Studies of Performance. *Biological Psychology, 34*, 2–3.

Wilson, G. F. (2002). Psychophysiological Test Methods and Procedures. In S. G. Charlton and T. G. O'Brien (Eds.), *Handbook of Human Factors Testing and Evaluation*, (2nd ed.) (pp. 127–156). Mahwah, NJ: Lawrence Erlbaum.

Wilson, G. F., and Eggemeier, F. T. (1991). Physiological Measures of Workload in Multi-Task Environments. In D. Damos (Ed.), *Multiple-Task Performance* (pp. 329–360). London: Taylor and Francis.

Woodson, W. E., and Conover, D. W. (1966). *Human Engineering Guide for Equipment Designers*, 2nd ed. Berkeley: University of California Press.

## ADDITIONAL READING

### General

Casey, S. (1998). *Set Phasers on Stun: And Other True Tales of Design, Technology, and Human Error*, 2nd ed. Santa Barbara, CA: Aegean.

Damon, A., Stoudt, H. W, and McFarland, R. A. (1966). *The Human Body in Equipment Design*. Cambridge, MA: Harvard University Press.

Drefuss, H. (1971). *The Measure of Man: Human Factors in Design*, 2nd ed. New York: Whitney Library of Design.

Driskell, J. E., and Salas, E. (Eds.). (1996). *Stress and Human Performance*. Mahwah, NJ: Lawrence Erlbaum.

EIA Engineering Bulletin (EIAEB). (2002). *Human Engineering—Principles and Practice (HEB1)*. Arlington, VA: EIAEB.

Gould, S. J. (1996). *The Mismeasure of Man*. New York: Norton.

Gregory, R. L. (1997). *Eye and Brain*. Princeton, NJ: Princeton University Press.

Lachman, R., Lachman, J. L., and Butterfield, E. C. (1979). *Cognitive Psychology and Information Processing: An Introduction*. Hillsdale, N.J.: Lawrence Erlbaum.

Norman, D. A. (1988). *The Psychology of Everyday Things*. New York: Basic Books.

Pew, R. W., and Mavor, A. S. (Eds.). (1996). *Modeling Human and Organizational Behavior: Application to Military Situations*. Washington, DC: National Academy Press.

Robinson, J. P., Shaver, P. R., and Wrightsman, L. S. (1991). *Measures of Personality and Psychological Attitudes*. San Diego: Academic.

Sanders, M. S., and McCormick, E. J. (1993). *Human Factors in Engineering Design*. New York: McGraw-Hill.

Schiffman, H. R. (2001). *Sensation and Perception: An Integrated Approach*, 5th ed. New York: Wiley.

Schultz, D., and Schultz, S. E. (2002). *Psychology and Work Today: An introduction to Industrial and Organizational Psychology*, 8th ed. Englewood Cliffs, NJ: Prentice-Hall.

Schultz, W., and Oskamp, S. (2000). *Social Psychology: An Applied Perspective*. Englewood Cliffs, NJ: Prentice-Hall.

Sternberg, R. J. (1990). *Handbook of Human Intelligence*. New York: Cambridge University Press.

Weimer, J. (1995). *Research Techniques in Human Engineering*. Upper Saddle River, NJ: Prentice-Hall PTR.

Wickens, C. D., Gordon, S. E., and Liu, Y. (1997). *An Introduction to Human Factors Engineering*. New York: Addison-Wesley.

### Anthropometry Database Sources

Anthropometry Research Project Staff. (Eds.). (1978a). *Anthropometric Source Book*, Vol. I: *Anthropometry for Designers*, NASA Reference Publication 1024. Houston: NASA Scientific and Technical Information Office.

Anthropometry Research Project Staff. (Eds.). (1978b). *Anthropometric Source Book*, Vol. II: *A Handbook of Anthropometric Data*, NASA Reference Publication 1024. Houston: NASA Scientific and Technical Information Office.

Anthropometry Research Project Staff. (Eds.). (1978c). *Anthropometric Source Book*, Vol. III: *Annotated Bibliography of Anthropometry*, NASA Reference Publication 1024. Houston: NASA Scientific and Technical Information Office.

Chapanis, A. (Ed.). (1975). *Ethnic Variables in Human Factors Engineering*. Baltimore: Johns Hopkins University Press.

Eastman Kodak Company. (1989). *Ergonomic Design for People at Work*: Vol. 1. New York: Wiley.

Eastman Kodak Company. (1989). *Ergonomic Design for People at Work*: Vol. 2. New York: Wiley.

Garrett, J. W., and Kennedy, K. W. (1971). *A Collation of Anthropometry*, AAMRL TR-68-1. Wright-Patterson Air Force Base, OH: Aerospace Medical Research Laboratory.

Gordon, C. C., Churchill, T., Clauser, C. E., Bradtmiller, B., McConville, J. T., Tebbetts, I., and Walker, R. A. (1989a). *1988 Anthropometric Survey of U.S. Army Personnel: Summary Statistics Interim Report*, Technical Report Natick/TR-89/O27. Natick, MA: U.S. Army Natick Research, Development, and Engineering Center.

Gordon, C. C., Churchill, T., Clauser, C. E., Bradtmiller, B., McConville, J. T., Tebbetts, I., and Walker, R. A. (1989b). *1988 Anthropometric Survey of U.S. Army Personnel: Methods and Summary Statistics*, Technical Report Natick/TR-89/O27. Natick, MA: U.S. Army Natick Research, Development, and Engineering Center.

National Aeronautics and Space Administration (NASA). (1986). *Man-Systems Integration Standards*, Vol. I NASA-STD-3000. Houston: NASA.

U.S. Department of Defense. (1991). *Military Handbook, Anthropometry of U.S. Military Personnel* (Metric), DOD-HDBK-743A. Natick, MA: U.S. Army Natick Research and Development Labs.

## Websites

Board of Certification of Professional Ergonomics: http://www.bcpe.org.

Human Factors and Ergonomics Society: http://hfes.org.

International Ergonomics Association: http://www.iea.cc.

CHAPTER 20

# Human-Centered Shipboard Systems and Operations

GLENN A. OSGA

## 20.1 BACKGROUND

One of the primary principles of successful human systems integration (HSI) in systems engineering and management is utilizing a human-centered design (HCD) approach throughout the systems acquisition process (Chapters 1, 10, and 18). Several other chapters (Chapters 4, 6, 7, and 9, in particular) have pointed out the need to establish HSI requirements early in the process, if the HCD principle is to be fully effective. Unfortunately, system design requirements based upon human capabilities and limitations may not be considered early in the design process, leading to costly changes during implementation. Often, new systems simply evolve from past systems approaches using established procedural and design methods.

The designer may rely on the user during the requirements stage to consider the human component, but user input must be carefully considered in that it can maintain previous designer flaws relative to human performance. User input and design qualities must be abstracted into basic task requirements. Unless the methods and procedures used in establishing requirements are specifically analyzed for impact on human performance and efficiency, neither the user or the designer is likely to fully recognize the effect the design will have on the human component when the system is fielded.

A major requirement for improved user interface and decision support aboard ships has arisen from the need for crew size optimization. Optimization must be achieved without sacrifice of performance, mission risk, and without crew overload. Crew optimization in future ships has been recognized as a significant cost factor and therefore has become a performance capability objective for newer classes of ships [Naval Sea Systems Command (NAVSEA), 1996, 1997]. When the U.S. Navy required a drastic reduction of crew size from 350 to 95 personnel on DD 21 ships, it recognized the need to use HSI principles for equipment design requirements and design solutions to successfully achieve mission objectives (Bush et. al., 1999).

*Handbook of Human Systems Integration*, Edited by Harold R. Booher.
ISBN 0-471-02053-2 

Consequently the Multimodal Watchstation (MMWS) project was conceived as a risk-reduction research effort to create concept designs that aid in HSI with optimized crews.[1] The concept designs also demonstrated a *task-centered approach* to requirements determination during the system definition stage, without major restrictions imposed by current design practice.

### 20.1.1 Multimodal Watchstation Project

As an example of the early stages of the design process and its products, MMWS represents the conceptual design stages of engineering, before full-scale development is attempted. The purpose of concept definition is not to create a product for final delivery or fielding but to investigate innovative features that are hypothesized to improve human performance and training. This process further refines requirements and guidelines that are then transferred into advanced engineering model development. The reader must recognize, however, that the primary MMWS project focus is on software-based decision aids and not on watchstation hardware or display technology. The hardware design is totally driven by available commercial display and control technologies, with some innovation in how the technologies are integrated and used by the software, together with ergonomic features for the physical configuration. As display technologies improved over the project life, the watchstation was also modified to take advantage of these changes. The primary focus of the MMWS design project was simulation-based design, in which a user interface simulation was constructed to test and refine requirements.

The conceptual design process included the identification of critical tasks within one of the two broad mission domains and the specification of task requirements based on task characteristics and job design. This evolutionary approach allowed for technology insertion and improvements over the 4-year MMWS concept design cycle, with operator involvement in all stages of the design process.

Over the 4-years to complete the project, requirements were generated using a task-centered design approach from which alternative design concepts were developed. The design concepts were subjected to a series of usability tests and team performance evaluations to verify that both human performance and training objectives could be met. Performance and workload measures were collected with reduced crews relative to today's systems estimating the potential impact on crew size optimization.

The iterative design process resulted in a mission execution and management system prototype capable of simulating work activity typical of navy command and control information centers and designed for meeting mission goals for both land attack and air defense operations. The warfighting functions supported by the MMWS are the same as current command and control centers but offer reduced workload and workload distribution capabilities among team members that may enable crew size optimization.

The work discussed in this chapter applies directly to the ship command center information systems design and does not imply that crew size is reduced for other ship operational functions as a result. Decision support systems, cooperative automation, and effective displays are enablers of optimized crews but do not directly reduce crew size, unless other operational methods are changed.

The path from requirements to effective display and interface design is multidimensional. If requirements omit major work factors that contribute workload or performance risk, the resulting design solution is at risk. There is a degree of art and innovation in design not easily quantified, and it is likely that multiple design solutions can work to achieve acceptable performance as well. Despite this first impression that one design example and set of requirements only serve as a loose connection with other designs, we are seeing a broader use of decision support "components" that are modified across diverse mission areas, without the need to "reinvent the wheel" for new task requirements. The specific design properties that address the task-centered requirements identified in the MMWS project also apply to other mission areas and work settings such as ship propulsion and engineering tasks (Osga, 2001).

Thus, an important lesson learned to retain throughout the chapter discussion is that the *type of requirements* and *type of tasks* covered are stable within the task-centered approach across diverse systems. This stability allows for modification of various design components to "fine-tune" results for various missions, such as defensive, strike warfare, and ship engineering control. In all cases, the human has a need to project mission events ahead in time in order to visualize the upcoming processes and anticipate potential results. It is then possible to enact mission solutions based on lower stress planned responses rather than on surprised and late reactions to failures.

The advantage of HSI research and development in the early conceptual process is that design ideas of varying risk can be combined and tested, with the ability to accept or reject design solutions based on iterative modeling or human performance testing. This design process will vary between every project based on the cost and expertise of the design team. Important qualities of the MMWS decision aids were defined through years of focused research related to the air defense task domain. The project allowed the integration of various design concepts and techniques in a common design approach. Important innovations were newly derived, however, based on task support areas not previously addressed.

### 20.1.2 Chapter Overview

This chapter presents a *conceptual design process* based on the experience with the MMWS project. A significant part of this process lies in the definition of tasks and establishment of key requirements. An HCD focus characterizes tasks in an information system work space according to task qualities and dynamic properties. This *task-centered approach* drives design thinking toward solving users' needs across a broader spectrum of task types and dynamics than is typically considered by systems designers.

The chapter is divided into the following sections that describe how HSI requirements were defined and design solutions for these requirements were addressed in the MMWS project:

- Task-centered approach
- Task coverage requirements
- Human support task requirements
- Dynamic task requirements
- Design by task requirement
- Other design qualities
- Benefits of task-centered design

## 20.2 TASK-CENTERED APPROACH

The task-centered approach fits into a conceptual design process, as shown in Figure 20.1. The process is iterative and cyclic, meaning that not all system design components and features are fully developed at the same time and fully output to another design processing stage.

First, mission and task requirements are derived from design reference missions (DRMs), which capture the future use of the system into a time-based story depicting system use. The quality of the DRMs is critical for system requirements and definition of scope.

Mission tasks are then derived as part of an iterative function allocation process, with levels of desired automation considered for each task. The function allocation may not be entirely fixed in a dynamic system in which the user can vary function allocation. This phase of processing produces the DRM, task definitions, task flows, and decision points to describe the task domain.

The human–computer interface (HCI) design is developed and validated, as shown in the lower right circle in Figure 20.1, with input from related discovery research and other decision support tools that may be modified to fit the current mission focus. Important design requirements (beyond just the mission task requirements) that feed this part of the design process are discussed later in the chapter.

The software validation process and prototyping are conducted to verify that computational methods can be found that are reliable, accurate, and serve the information needs of the tasks. This prototyping process can be separate from the HCI prototyping and requirements definition thus enabling HCI designers and software engineers to coordinate work in a parallel process. The HCI prototypes, whether in paper, slide show, or simulation may be subject to repeated usability tests before they are submitted to the software prototyping process. As usability data is collected and HCI requirements mature and are better specified over time, they serve as input to the software validation process.

An important facet of this approach is that in large complex systems, requirements and design are not fully described as in a hierarchical noniterative approach but that testing and refinement can occur over time for "pieces" of the system. The software architecture is also designed to accommodate successive improvements over time. The chapter content primarily covers the "task analysis" and "HCI design and validate" parts of this design process.

Before proceeding further into the design process example with MMWS, several important terms used throughout the chapter should be defined. Several basic definitions related to "tasks" are needed to better understand the *task-centered approach* as a design process. These are *task, mission task, task description, job design*, and *task definition.*

A *task* is a goal-oriented work activity component of a job. The task may be accomplished manually, automatically, or some combination of the two. The composite of all tasks for a given job description accounts for all workload during a prescribed work period.

*Mission tasks* are workload producing components typically addressed in military system specifications involving human control elements.

*A task description* represents a taxonomic description of labeled work activities. This creates a written description of a definable process by which human and machine cooperate at achieving a work-related goal.

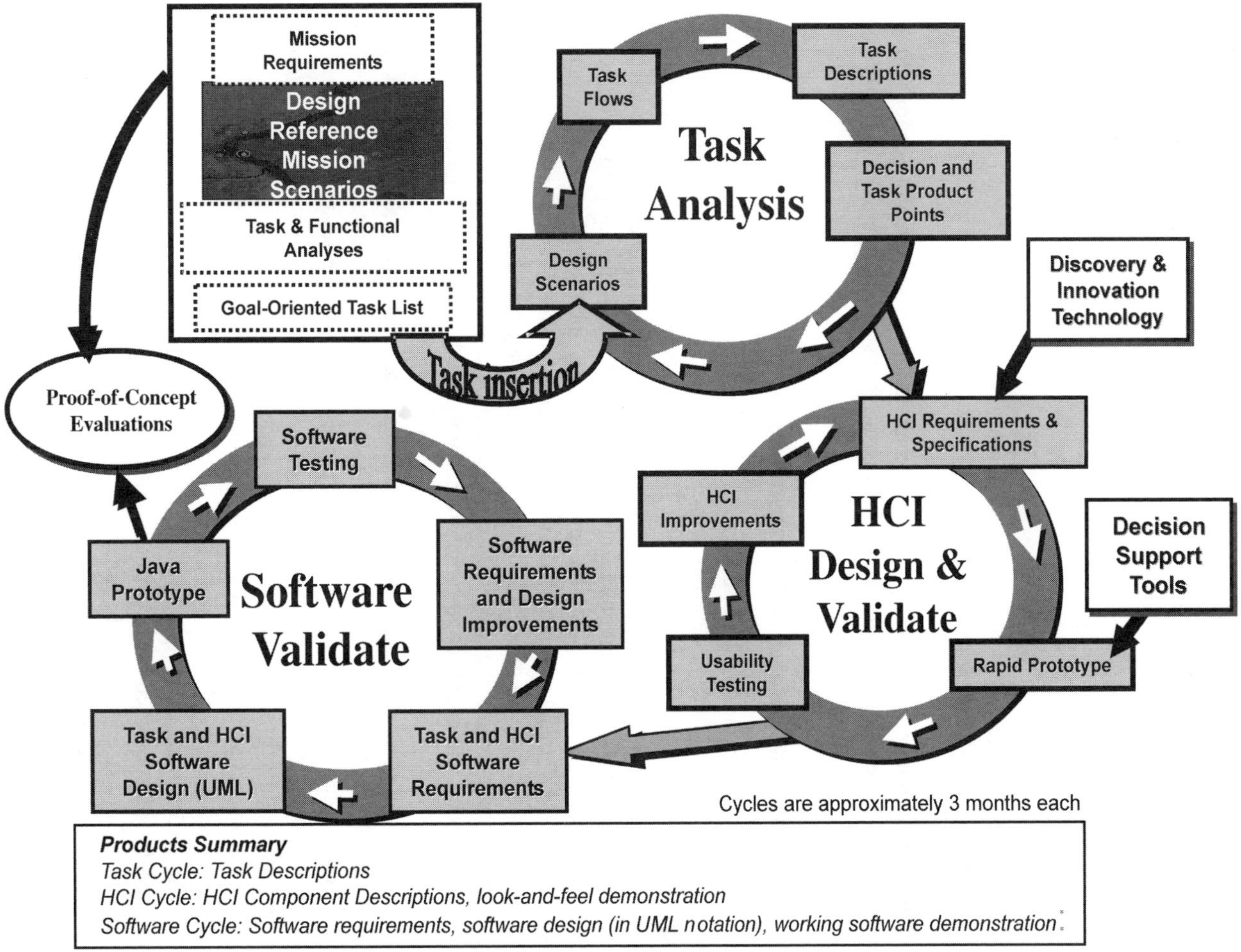


**Figure 20.1** Conceptual design process incorporating a task-centered approach.

A *job design* is a collection of tasks defined as a set of related work goals to which a human operator is assigned to complete.

*Task definition* is the process of putting defined labels on a set of work activities. There is no unified, agreed approach within the human-engineering discipline on task labels. Typically it is up to the system designer and architect to define a hierarchy and level of detail judged appropriate for the design problem.

The *task-centered approach* is an analytical HSI design process broadly comprising two components:

- Defining HSI requirements within defined task domains
- Creating task-centered designs supporting task goal achievement

Defining HSI requirements is the focus of the remainder of this section and Sections 20.3, 20.4, and 20.5. Creating task-centered designs is the focus of Sections 20.6 and 20.7.

### 20.2.1 Establishing Key HSI Requirements

The premise is that a task-centered design focus during the systems engineering process provides a mechanism to fully describe the work environment in a manner that establishes a comprehensive set of design requirements. These requirements are structured to cover the various types of tasks that compose the majority of workload sources at the tactical watchstation. Another important premise is that design attention must be paid to the major sources of workload whether they be mission, computer-interface, human information processing, or work management tasks. These tasks also operate in a dynamic work cycle with defined phases. Much of today's design focus in legacy ship systems is on a narrow subset of processes within the task work cycle, leaving the system operators unsupported to use their visual and cognitive resources to carry the workload through the unsupported task phases.

A process that is key to successful HSI is adequate description of the system task and work environment. A design concept is simply a set of design hypotheses matching design solutions to the task requirements, with the hypotheses being made relative to the human performance outcome in both training and operational results. A task-centered approach can be used to directly tie requirements to task characteristics.

The most important concepts utilized in establishing key HSI requirements are:

1. *Task Coverage* The quantity of tasks and the qualities of task requirements addressed by the designer relative to the entire workload environment within the job design. A major concern here is the breadth of the tasks in job design. If a task is not even considered or recognized by the designer, then there can be no hypothesized design solution. Most of the major types of tasks (e.g., mission, computer, work management, and human support tasks, along with the various types of support requirements for situation awareness, attention management, and decision making) can be developed through the method of task definition. These types of requirements can be thought of as "static" task requirements, in the sense that task qualities can be described independent of time or sequence.

2. *Task Dynamics* The life cycle of a task within a dynamic task decision process. The dynamic properties of tasks are described with reference to how these dynamic

properties create additional design requirements. For example, human memory is subject to decay over time. Task interruptions are affected by time. Task deadlines and parallel tasks affect workload. The dynamic pacing and workload requirements of the job environment are not readily visible when each task is analyzed independently and out of the timing context.

3. *Task Goals* The detailed level focus of system design needed to create the task product in an efficient and cost-effective manner.

Task coverage and task dynamics will be covered in more detail in the following sections, but since the process of establishing key HSI requirements starts with task goals, they are described in the following section.

### 20.2.2 Definition of Task Goals

The goal of many mission tasks is to create a product (e.g., an order, control action, message, awareness update), but a task-centered design goal is often a process. For example, to reduce human cognitive workload, the task goal may be to move tasks from requiring knowledge-based aptitudes toward the skill or rule-based performance (Rasmussen, 1986). For such tasks the goal would be to relieve the user of tedious rule- or skill-based steps, by capturing such processes within algorithms or computational resources that result in the presentation of "draft" task products for human verification and delivery. Thus, as designers we shift the human role in system interaction from repetitious lower level task rules toward a role of monitoring and directing automated processes that produce "draft" products for human inspection.

The purpose of a task in the context of human work is for the useful completion of work related to mission goals. The definition of task goals is critical in early design stages in that they describe any gaps between the conceptualized system products and the task goals that the human must support. As we see in the *task definition process* described in the next section, the scope of these goals may be broad or narrow depending on the vision of the designer and the acknowledgment of the types of tasks and their associated goals.

Goals can usually be stated in hierarchical terms, and the designer must make important design decisions as to what type of functional system support to provide toward the attainment of task goals at various hierarchical levels. See Example 20.1.

**Example 20.1 Message Preparation Requirements** The task of message preparation and delivery to meet a mission requirement may be supported across various subtask steps of information collection, message creation, editing, formatting, and delivery. A word processor may support the functional goal of writing text in the message format with various features to support text editing. Consistency across messages may be supported by display views that show proper message format. The system could collect the proper information for the message and prepare a draft message if the form and content is known. The same function can be supported with only a basic word processor, forcing the user to collect information and type in the message based on training and task expertise. With minimum support, the user must collect the information from displays, form the message content in their head, and verbally speak the message to the receiver while perhaps writing notes with no system support for creation, editing, and delivery. In each example, the task goal remains the same for the user—deliver a correct, timely, and succinct message as soon as it is required—but the requirements for system support would vary tremendously depending on how much automation and task product drafting the system is required to support the user. Task-centered design attempts to identify requirements related to task goals and provide support through all task phases leading to the goal accomplishment.

Many task performance goals could be derived from mission performance requirements. If the requirement is set at five messages per minute or no more than 5 percent error in message content, the contractor and designer must determine a human–system solution that meets performance goals. Without the specification of performance goals, design becomes a somewhat arbitrary process of debate between the human factors engineer and project management on what level of system support is proper or needed. Unfortunately, the definition of performance requirements and design solutions often does not involve the human component. When focused correctly technology support could often provide improved performance with improved HSI, resulting in more accurate adjustment of task performance goals and mission requirements based on those improvements.

The research process should play an important role in the setting of task goals. In an impartial lab setting, task goals are not artificially held back or restricted by business practices, risk aversion, or loyalty to a specific product or approach. Research must identify tasks and goals representative of operational requirements. Results should provide "honest broker" evaluation of a system's qualities and its potential to support task goals. Setting of task goals represents common metrics of quality across disparate design approaches, and as such, the goals create the opportunity to specify performance objectives as metrics of system success.

Task goals provide a focus for design of task products, and task coverage defines the breadth of design support toward the attainment of the variety of task goals. Sections 20.3, 20.4, and 20.5 describe how HSI task coverage and task dynamic requirements were developed for the MMWS in support of various task goals.

## 20.3 TASK COVERAGE REQUIREMENTS

Task coverage represents the amount of work activity that the designer supports with system features. In the MMWS project, task coverage represented a comprehensive view of the work requirements for an air defense mission application, and a "task to be covered" was defined as a segment of a job activity with the following attributes:

- Varying in time from seconds to hours, or the entire watch period (6 hours or more).
- Supportable by computer-based aids, (e.g., not work activities such as physically connecting cables or cleaning and maintenance).
- Supportable by various levels of automation, which may include user selectable or fixed automation levels. Thus, levels of task supervision and user–system task sharing are dynamic.
- May vary from structured, rigid protocols to open-ended user-defined sequences. Following Rasmussen's (1986) hierarchy, tasks may include skill-, rule-, or knowledge-based behaviors. Many tasks in the air defense warfare area had defined procedures and structured protocols and could be defined as rule-based tasks.

Task coverage is strongly influenced by designer *vision, cost, and time*. Often the *potential* system support is a result of design *vision* to see how *potential* technology, algorithms, and computational methods can be utilized in support of tasks. During early design concept formulation, user task activities must be discussed and judgments rendered as to whether the activity is "too difficult" or "too costly" to support. Revised design vision might appear weeks or months later during the task definition and analysis process. The level of task support may be increased with upgraded versions of the software and system. These successive improvements create a requirement for a flexible software architecture to allow for expanded user support while task requirements evolve. Thus, during the early stages in the conceptual design process, it is imperative to identify as complete a concept of task coverage as possible, while avoiding premature narrowing of design focus because an immediate solution is not readily available.

A significant step in the *task coverage analysis process* is to identify all the critical task goals in the job domain. *Task definitions* are strongly related to task goals. These goals are a starting point from which to create task definitions. For example, the goal to create and send a new track report message in MMWS is covered under the task with the same name "create new track report." Other goals such as "review rules of engagement" have the goal of updating short-term memory during the mission progression with information that affects other task decisions. Thus, task goals may be to produce defined products such as a mission order, mission report, message, plan, etc., or the goal may be to update and reinforce information storage supporting situation awareness in the human cognitive processor.

Subject experts are able to define concrete task products relatively easily but appear to have much more difficulty with goals that are related to cognitive processing. This process can be aided by task walkthroughs or observations of task progress with experts used to explain these processes. The task definition process must be as thorough as possible to ensure that complete coverage is considered, and the level of functional support within the coverage area is based on deliberate design decisions on how much workload to allocate to human or machine to cooperatively achieve the goals.

For MMWS design purposes five main classes of tasks were defined as requiring support by the watchstation design. These classes were:

1. Mission tasks
2. Human support tasks
3. Work space computer management and control tasks
4. Work management tasks
5. Communication tasks

These classes represent a total task coverage approach to the MMWS workload. Within each of the classes of tasks, the process of task definition was applied to create task labels that made sense to both designer and subject expert, and were explainable and defendable to software engineers and project management.

### 20.3.1 Task Definition Process

The process of task definition may differ significantly if the system design problem is an upgrade to an existing task process or if the system is a new or innovative concept. With system upgrades there may be increased pressure from current users and management to minimize impact on training or documentation costs. With a new system there is less legacy and design tradition to affect the task definitions. Designers must analyze the current task process carefully and define task elements at different levels. The designer must look for inefficient or awkward methods in current procedures. These factors affect the manner in which the task taxonomy is described and then formulated for the new system.

Subject matter experts (SMEs) will most likely describe their tasks within the scope of the current work environment. Some air defense warfare tasks were initially omitted in MMWS because the current design provided little or no support for those tasks; therefore they were not part of the software description or documentation. Without some introspection and observation of current task processes in action, these undocumented tasks remain invisible to the designer. Also, the goal of a task can be obscured by lack of full documentation of the current systems' task process. For example, the SME might guide the task definition and process for a communications task toward designing refinements to support that process, hence focusing the design discussion on improving today's process. With tasks related to voice communications, the SME drove the design discussion toward better communication methods or technologies. Further analysis on the *task product* changed the task definition by revealing a potential improvement in performance by focusing on the preparation and delivery of the tasks' communication products.

**Example 20.2 Human Factors and Computer Science Collaboration During Task Definition** The MMWS design hypothesized that perhaps a voice message can be prepared automatically and sent digitally without requiring the human to both conceive and verbalize the report through a communication channel. The human factors engineer estimated a reduction in workload if this technology can be implemented while the computer scientist determined if there was sufficient task structure to automate the message preparation. When the task requirements are stated first, it allows greater flexibility in the design discussion than if the design solution is discussed first.

A key lesson learned during the MMWS project was to define task "products" first or early in the task definition stage. If a product cannot be clearly defined, the task concept may be a candidate for being "reduced" to a task "step" or "subtask" within another defined task area.

With system design changes, tasks may become obsolete or technology can change the entire nature of the task or process. Tasks may evolve from totally manual to partially automated to produce "intermediate" products in a multistage complex process. Other tasks have easily defined beginning and end points and concise products. There are no clearly defined quantitative solutions to defining the best "size" and workload of the work activity within a given task label. The function of "job design" is to subgroup tasks and work in a manner that allows user work pacing and rest in manageable cycles. Typically, the more concise and smaller the task with respect to time, complexity, and steps, the less likely the task will be interrupted, left incomplete, or subject to forgetting. Fewer tasks relate to a more manageable design problem, software effort, and user training effort. Smaller tasks with simple procedural steps produce easier training challenges that may be presented in a building-block fashion. The ability to combine smaller task products into larger outcomes also facilitates training and instruction. Guidelines for the purpose of defining task size and complexity in MMWS include the definition of a task unit that is:

A "trainable unit" (e.g., a cohesive and related set of goals)

A reasonable software design/build module (e.g., a related set of computations and results)

A reasonable grouping of information and goals (e.g., related information and logical process flow)

The task definition process may generate confusion about how to define the differences between the qualities of "tasks" and "functions." A function is *what* is done and a task is *how* it gets done. For example, a function could be described as "mission planning" whereas the task would be to "prepare the strike mission plan." The task also has associated subtasks such as "receive and review ops orders," "review auto assignment of weapons to targets," or "review and edit schedule." The auto assignment task is supported by the system functions that calculate the best weapon–target pairing. These types of functions were termed "task services" in MMWS. Tasks were defined as processes involving various levels of human intervention, while functions could be either fully automated computations or have human involvement.

During the process of defining MMWS tasks, a taxonomy of work tasks was created and then evolved over time as tasks were defined, created, deleted, combined, or separated. This happened over a period of several months as the work environment and task products became better defined.

This evolution was necessary as the design team completed the initial task description process. This definition and refinement process was supported by teams, comprising SMEs, human factors engineers, and computer scientists. The role of each discipline was valuable as the computer scientist was concerned about how to computationally model the task; the SME carried the perspective of real-world constraints, procedures and operations; and the human factors engineer presented the perspective of design impact on human performance.

The task definition process must also remain flexible through early design stages. Some work activities initially defined as tasks requiring human processing later became more fully automated after further study found analysis methods to create reliable task products without human intervention. These tasks then became task services in support of other tasks. Thus, function allocation was not a one-time event, but instead involved the creation of a set of hypotheses that were refined as task methods were created and matured. The process of defining each of the major task classes is described in the following sections.

### 20.3.2 Task Properties

The next step in the task coverage analysis process was to conduct a task definition exercise for all of the tasks within the five classes covered. This results in a requirements definition that fully considers unique task properties. A special section follows later for human support tasks. Communication tasks are combined with mission tasks in this discussion. The other three classes, mission, work space computer management and control, and work management tasks are covered in this section.

Mission Tasks are the workload producing components that are typically addressed in naval system specifications involving human control elements. The mission tasks provide a structure for analysis and development of design approaches toward each task goal.

The MMWS analysis for mission tasks was difficult because current naval systems have minimal task design documentation (Osga, 1989). Another difficulty confronting the analyst was the sparse information regarding future shipboard tasks. The analyst, therefore, had to take the following steps:

1. Abstract information from current task methods
2. Project design and technology properties forward in time for new tasks
3. Reanalyze the newly designed task structure

Air defense mission tasks initial listings of mission tasks were reviewed (Osga, 1989) as were function analyses for land attack and air missions [Naval Surface Warfare Center (NSWC), 1997, 1998]. An initial set of tasks was derived when a design reference scenario was developed. The reference scenario served to focus the larger set of possible functions and tasks down to a manageable set under the scope of the project. The types of mission tasks analyzed were:

- Visually identify (VID) all unknown air contacts within a defined area of responsibility (AOR).
- Escort air contacts from threat country with aircraft-carrier-based defensive counter air (DCA).
- Issue warnings to threat country aircraft.
- Conduct positive identification of air contacts unable to VID by correlating indications and warning, electronic emissions, profile, point of origin or initial detection, air tasking order and interrogate friend or foe signal (IFF) received.
- Convey internal communications and external communications with air warfare commander, DCA, and carrier.
- Conduct weapon engagement in self-defense.

Within these types of tasks, mission task labels were defined as a "verb–noun" phrase for consistency. Task verbs such as "prepare . . . , check . . . , deliver . . . , review . . . , order . . . , issue . . . " are descriptive of the type of work activity being performed. The task noun can indicate the product of the task, e.g., a "level 1 query, level II warning, new track report, engagement order, etc."

Work space computer management and control tasks involve the workload that is inherently part of operating within the computing environment. In a graphic user interface (GUI) environment, workload may be induced by tasks such as searching for files, organizing windows, de-cluttering displays, moving and navigating between windows, etc. If the designer is satisfied with accepting an off-the-shelf GUI without consideration to impact on workload or performance, then the system design is accepting a given level of performance. Typical computer control tasks include the selection of displayed objects, resizing of windows, movement of objects, copy and paste text between windows, search for windows, files, objects, etc. The MMWS design included changes to the standard "desktop" approach based on previous research conducted with similar task conditions (Osga, 1995.) The HCI design features were used to reduce workload for computer work space management.

Work management tasks include the management of work activities with regard to work sequence, task prioritizing, multiple tasks time sharing, and scheduling. This work includes the transition between tasks and decisions regarding activity prioritization, e.g., recognizing what is important to do now versus what can be delayed. Individual work management includes decisions regarding multiple task time sharing—when to shift resources from one task to another. Experts, based on training and experience, create a background of work environment knowledge that contains individualized patterns and rules about task start, break, and stop points.

**Example 20.3 Expertise in Work Management Tasks** An expert knows through repeated task experience that "step 3" in the process is a good time to pause for rest or to shift attention to another task because of on-the-job knowledge gained. The expert anticipates that the next step requires a process that takes several minutes of other system resources (e.g., automation or another user). Thus, the expert gains experience on the systems' *timing* and *dynamics* and can better schedule attention resources during multitasking work events.

Another important aspect of work management is the knowledge of when to begin tasks or sometimes when to terminate them prematurely. Tasks appropriate in one context may be deleted during another. The system design versus cost trade-off for work management tasks is the cost of training and developing user expertise developed over time with reliable repetition versus the cost of developing system support features that aid in reliable work management.

## 20.4 HUMAN SUPPORT TASK REQUIREMENTS

Human support task requirements are one of the major classes of tasks included in the total task coverage of the MMWS. Special features needing attending in the development of human support task requirements are:

1. Maintaining situation awareness (SA)
2. Attention management
3. Decision making
4. Working memory
5. Task interruption
6. Supervisory control
7. Ergonomics

### 20.4.1 Maintaining Situation Awareness

Situation Awareness is a human process of information collection, filtering and storage, interpretation, and reaction. System aiding can be provided for various types of SA sampling, storage, and retrieval activities. Jones and Endsley (2000) refer to three levels of SA as:

***Level 1: Perception of Elements in Environment*** Elements are perceived within a volume of time and space. Tasks utilize visual search, filtering of important task information from peripheral visual "noise," and auditory sampling from multiple circuits. All are part of the sampling process to continually update human awareness.

***Level 2: Comprehension of Their Meaning*** Implies that information presented is compared to the current and near-term goal states of mission tasks and activities to determine the significance of events relative to goals. The result includes decisions to start, delay, or cancel task activities.

***Level 3: Projection of Their Status in Near Future*** This level implies that there is a temporal nature to decision making and that activities may be launched or altered based on projections into the near-term future. In air defense missions, this implies issuing warnings or reports or deciding that such reports are not warranted. There is evidence that users build a story (or mental model) based on the operating environment, expected events, observed events, and compare this to past experiences in their decision-making training or operational experiences (Klein, 1993).

Problems in mission performance may appear when errors occur in SA, producing a mismatch between the user's mental model of the situation and the actual situation. Jones and Endsley (2000) refer to "representational errors" when information has been correctly perceived but the significance of various pieces of information is not properly understood, meaning problems with level 2 SA. In air defense, environment cues that an aircraft is potentially hostile may be overlooked in favor of evidence that it is a commercial airliner, if the event was unexpected that a hostile should be in that location or if there is speed, altitude, or position data suggesting "commercial air." The system requirement, therefore, is to provide information in a manner that *prompts* the user to be attentive and aware of conflicting task data—providing track identification evidence both for and against the current track identification state. The designer must also create methods to support user understanding and projection of events in the near future.

### 20.4.2 Attention Management Requirements

Attention Management is a critical support activity invoking human cognitive and visual skills. With increasing knowledge and skill of the work environment, the human processor is better able to determine the importance of work events and to disregard background stimulus noise. Human attention resources are limited and are quickly forced to time-share multiple events (visual, auditory) in most tactical situations. The human processor, while conducting task A, must preattend and be ready for task B. Experts develop "work habits" as methods for moving attention resources between task activities. The degree to which the system effectively supports these processes reduces human–system performance variance across the spectrum of users who possess varying degrees of visual search and attention management skills. A significant requirement is to guide attention at an appropriate level of intrusion into the user's work focus using both visual and auditory stimuli.

### 20.4.3 Decision-Making Requirements

Decision making in naval warfare varies according to the nature of the warfare environment (e.g., rules of engagement, operational orders, battlegroup firing status, and type of warfare: anti–air defense, air offense and strike, land attack strike). In recent years, much attention has been paid to the single-ship, single-threat scenario following the USS

*Vincennes* incident in 1988. During the 1990s, the Tactical Decision Making Under Stress (TADMUS) program studied air defense tasks with respect to the identification process and displayed information to support decision making with ambiguous ID information (Morrison et al., 1997). The *Vincennes* incident review showed that data within the combat system (such as decreasing altitude of an aircraft) does not translate directly to information for system users if the displays and information presentation do not clearly present *historical trend data*.

When operating in an international battle group under defined doctrine and rules of engagement, the decision-maker's actions must conform to operational rules. As the situation changes from peacetime to hostilities, the decision rules change but the response methods with each task should remain stable. A watchstation design human interface must support all types of tactical situations, ranging from a "single possible threat in peacetime" situation to the "multiple threat hot war situation."

Important requirements must be addressed with respect to information gathering and management for decision support. With an increased system functional role in information gathering and synthesis, the nature of task activity for the user changes from the current "information gathering mode" to the "monitor of intelligent automation" mode. Table 20.1 presents a summary of the changing nature of decision support requirements following a trend away from "manual" information gathering systems toward increased automation for information management. MMWS represents significant progress toward addressing the requirements listed in Table 20.1 but does not completely satisfy them.

Increased dependence on automation support for task information processing will in turn change the information requirements to support the task process. Warfighter needs for "explanation" facilities and supporting information will evolve as systems become more capable of reliable support for multiple tasks throughout the detect-to-engage process. Freeman et al. (1997) identify a "dual-process" model in which decision makers employ either critical thinking or rapid recognition process depending upon the time, stakes, and familiarity of the task and decision context. This requirement for support of both types of decision processes indicates that MMWS information displays must support each decision strategy. See Example 20.4.

**Example 20.4 Change in Automation Trust and Decision Processes over Time** A reliable identification method that reports a track identification based upon comparing the track information to battle group identification parameters may initially be monitored or even questioned by users (using critical thinking) who will visually check each ID parameter to see if it conforms with the battlegroup rules. If the method is highly reliable (e.g., the system repeatedly creates IDs that match the battle group ID rules), the occurrence of such user checks will diminish (e.g., they will shift toward using rapid recognition processing).

Thus, the decision-making requirements for MMWS were stated as: (1) provide task summary products for rapid recognition processing and (2) provide task amplifying information to aid in critical thinking processes when time is plentiful, stakes are high, and familiarity with the task is low.

The requirements listed in Table 20.1 change the user's role from the current paradigm of checking details of incoming tactical data streams using vigilance and cognitive storage/recall skills (critical thinking detailed), to the role of strategic decision making for more global mission goals (critical thinking globally). The user checks and confirms or denies mission actions recommended by automation support (rapid processing). With

**TABLE 20.1 Comparison of Decision Support Methods, Today and in Future**

| Mission Function | Decision-Making Today | Today's Decision Support Methods | Future Decision Support Requirements |
|---|---|---|---|
| Response planning | • Battle orders and op orders form basis for plans.<br>• User burden on integration of planned responses with system responses.<br>• Selection of ownship plan based on battle group protocols, ownship role, anticipated events, and recent experience.<br>• User burden for plan recall and match with tactical situation. | • Requires heavy working memory loading–storage and retrieval.<br>• No integration of plans with events.<br>• Little or no auto support for planning, testing of plans.<br>• Support for simulated testing of response plans during training. | • Plan templates available for air defense, land attack response tasks.<br>• Ability to compare planned responses to tactical events to trigger tasks.<br>• Quality check of plans—safety, weaponeering, communications, coordination.<br>• Present visual summary of plans. |
| Detection | • User vigilance of tactical events plays key role. "Hooking and looking" to spot key events as quickly as possible.<br>• Workload often delegated to small part of team. | • Systems record new track positions and kinematics or electromagnetic returns and visually display symbols.<br>• Auditory or visual alert (blinking) provided to enhance detected info. | • Matching of detected events to planned task responses to events.<br>• Summarize info. of detected event and task consequences. |
| Identification | • User recall of ID factors and manual comparison to system proposed ID.<br>• Single ID check without recheck during track life, requires user monitoring and updating. | • Initial track data is compared to basis ID factors.<br>• ID doctrine difficult to impart to system or explain. | • Clearly summarize all ID factors and comparison to current status whether for/against or unknown in support of ID.<br>• Check ID basis regularly and trigger appropriate tasks if changes warrant. |

| | | | |
|---|---|---|---|
| Monitoring | • User vigilance of tactical events plays key role. "Hooking and looking" to spot key events as quickly as possible.<br>• Time lag or delay of multiple decision support items forces note taking or memory loading.<br>• Workload distributed across team in unpredictable fashion.<br>• Workload pacing forced by tactical actions (external pacing).<br>• Priority of visual search/monitoring based on user expectations. | • The track plot or electromagnetic data log presents current tactical events.<br>• Limited history or projection of tactical events/actions. | • System monitoring of tactical events and triggering of attention direction cues.<br>• Maintain history of events associated with a chain of related tasks (by track, target, or mission function).<br>• Guide attention to most critical events and thus most critical decision actions.<br>• Guide workload distribution to avoid "tunnel vision" and narrow focus based on erroneous expectations. |
| Response actions<br>Queries<br>Warnings<br>Illumination<br>Jamming<br>DCA intercept<br>Weapon engage<br>Chaff and counter | • Compare tactical events with plans stored in various formats (paper, notes, viewgraphs) or stored in user memory.<br>• Recall similar events on specific track/event in recent history.<br>• Recall and execute steps to complete response procedures. | • Provides manual response methods.<br>• Users must construct many task products manually, e.g., verbal reports constructed by combining pieces of data.<br>• System does not log task occurrences.<br>• Procedures not consistent across similar tasks—training burden increased. | • Provide automatic task responses proposed by system with supervision and approval of task responses.<br>• Provide full explanation of task responses proposed by system.<br>• Provide consistent, redundant, and easily performed methods and procedures. |

reliable system performance, the user digs into the background information only if workload allows, or the user questions the result, or the decision has serious mission/safety consequences. Freeman and Cohen (1998) discuss the decision tasks and critical decision points for anti–air warfare in the context of several currently fielded systems. MacMillan et al. (1997) define critical decision points for air defense warfare in a review of current air defense methods for initial MMWS design. The watchstation mission task information was designed to address many of these critical decision points.

### 20.4.4 Working Memory Requirements

Tasks that evolve over time involve human cognitive processes in short-term or working memory. Information that is processed by visual and attention systems is temporarily stored while the task goal is active. Problems in storage and retrieval may appear when stored information content is similar (Fowler, 1980).

**Example 20.5 Information Lost in Working Memory** A previously mentioned course change with track 7150 is confused and reported as a change in track 7157 or a few moments later a course heading of 310 is recalled as 030. A common theme in "number reversal" during voice communications may be lack of common visual cues at each end of the conversation to augment the visual information. In today's command centers, information such as electronic warfare emissions may only be available to a single user at a specific workstation designed to receive that information, and results may be only transferred by voice to decision makers. Without immediate note taking upon delivery, such information is easily degraded in or lost in working memory, perhaps within 10 to 20 seconds after arrival (Wickens, 1987).

System design features must be used to unburden working memory, including storage and retrieval of visual information to augment transient auditory information. Also, a combination of numeric and spatial information presentation for track objects supports both verbal and spatial working memory storage.

**Example 20.6 Compensation for Short-Term Working Memory Overload by Note Taking** An example of short-term task overload in working memory may be manifested in either the repetition or forgetting of task events. Without computer assistance, system users try to compensate for memory limitations through note taking. Observations of operators in action indicate that some write down notes and others do not (Hildebrand, 2000), but there is no known data on the effects of note taking on task outcomes.

System design features that can alleviate memory loading include features that keep track of information changes over time and clearly present past, present, and planned task events. The MMWS Response Planning/Manager is an example of a design feature that provides a visual time review of tasks planned, proposed, in-progress, and accomplished. In today's systems this can only be accomplished by note taking or recall from short-term memory.

### 20.4.5 Task Interruption Requirements

Mission demands for real-time multitasking require increased designer focus on supportive workstation tools that take into account human limitations resulting from multitasking and the disruptive effects of interrupting ongoing tasks. McFarlane (1997) refers to task interruption as a process of coordination between human and machine. To support this coordination workstation, designs must include software that accounts for task interruption and subsequent user refocusing. The MMWS alleviates short-term interruption effects by:

1. Providing low-workload task reaccess using multiple screens
2. Keeping relevant information together for the task without user burden
3. Providing visual cues of changes and highlighting of information changed since the user last checked the information

Another important design requirement is the method of design for interruption alerts that change user task and attention focus. Software methods must be developed that consider the context of the interruption across multiple events. An event may appear to be important to require an interruption of the user when considered in isolation but not be worthy of interruption when other life-threatening tasks are also present. Thus, the requirement for context-sensitive alerting is present. The design approach of whether to interrupt abruptly or provide more subtle interruption cues depends on mission context and time criticality of the task events, taken in context within the mission focus. This requires interruption in the form of visual, auditory, or haptic cues that inform the user beyond the simple auditory alert buzzer used in many systems today. A multilevel alerting system has been proposed for the MMWS project, allowing a wide range of interruption possibilities (Obermayer, 1998). The proposed system has yet to be fully implemented in MMWS and tested with workstation operators.

### 20.4.6 Supervisory Control Requirements

The human role working in cooperation with automation is variable, ranging from: (1) task monitoring with hands off to (2) task confirmation at the last procedural step to (3) full involvement through multiple task steps. Task management automation removes much of the user burden of task initiation, and other aids reduce the workload in completing each task procedural step. The degree of user involvement for task supervision may be dictated by task external pacing. With more work time available to focus per task, there is more opportunity for hands-on work, but with less time available there is an increased need for intermittent task focus by way of supervision of the automation. The workstation must easily support both types of work allowing the users to adapt to changes in workload. MMWS provides several important features to aid in visual search and information acquisition during supervisory control tasks. The design requirement is to provide relevant cues for high-level information visual scanning, supporting decisions regarding the need to drill down into detailed information content. The requirements are made more complex by studies that suggest the type of work tasks assigned to the crew member should vary over short periods of time.

**Example 20.7 Continuous Skill-Based Tasks and Alertness** Neerincx (1999) reports that just 10 minutes of continuous skill-based performance tasks can decrease performance. As workstations and automation can be made more capable, a newer design challenge emerges to vary tasks in ways that maintain crew alertness throughout the typical 6-hour watch session.

### 20.4.7 Ergonomic Requirements

The comfort and safety of the system operator is a critical design requirement for the watchstation design. The console display pedestal must provide a comfortable work environment for body statures ranging from the 2.5 percent; female through 97.5 percent male size. Visual and reach envelopes for displays and touch-screen interaction must be considered such that the user can rest the elbow on the desktop surface comfortably and reach the majority of the display surfaces. Input devices must consider the motion effects of sea conditions and the possibility that the watchstation could be moving with a rough sea state. Another important consideration affecting the pedestal design is team arrangements and the ability for team members to see each other. The taller, bulky consoles of today prohibit face-to-face interactions and force arrangements of crews into rows and aisles of equipment. Ergonomic touch, reach, visibility, and team interaction requirements led to an MMWS design approach that accommodates the physical statures listed, provides controls suitable for various sea conditions, and allows close-proximity team interactions.

## 20.5 DYNAMIC TASK REQUIREMENTS

Many of the static or qualitative requirements for user task support discussed in the previous sections do not account for the *dynamic* stages and processes involved in task completion, and the effects of time and job pacing on task progression and the human information processor. Dynamic task properties include task and job attributes that change over time due to the rate of information change, the loading of tasks, and the context of multiple competing tasks occurring in the same time continuum. User qualities of fatigue and alertness change over time also. Dynamically varying data sets and information support tasks have a time context within the changing tactical environment. The degree to which the system designers can capture the context and relevancy of the task information with respect to current operations could make systems more responsive to the users' current needs. This section discusses these important qualities of tasks in a task-centered design approach from the perspective of task timing and the dynamic life cycle of tasks.

### 20.5.1 Dynamic Task Timing and Pacing

Frequency and repetition of task events within the job context play a role shaping the system design concepts. Task pacing, vigilance loading, and multiple task timing are important factors requiring design support.

***Externally Paced versus Internal Pacing*** Tasks that are externally paced cause work-induced stress. The tactical work environment creates this stressful pacing since system users have no control over the pace at which the enemy or other tactical entities decide to operate. By nature, the goal of an adversary is to overwhelm the opponent. But designers can consider design options to mitigate the effects of external task pacing. For example, tasks can be designed such that their workload could be distributed across team members if it increases to unacceptable levels. Workload distribution is prohibited in today's systems due to strict assignment of tasks with specialized operating modes in each workstation. The specialized training of today and lack of ability to flex workload decreases survivability, given that removal of key consoles or positions offers no replacement. System tools and architectures for information delivery that allow for distributed workload and for assignment of any task at any station maximize survivability across the entire ship. These design properties associated with workload distribution should be factored into ship-level survivability and failure mode analyses. In MMWS the task management system is designed to reduce externally paced workload on any given individual by allowing tasks to be distributed among the team members. Voice and auditory information tasks are often externally paced, and the design can support increased user control over the pacing of tasks through digital storage of audio, thereby allowing the task to become internally paced.

***Vigilance and Situation Awareness*** Vigilance tasks over periods of time create boredom and induce fatigue. The designer should consider system support in eliminating vigilance tasks wherever possible. A warfare tactic sometimes used involves human vigilance and perception. An adversary may repeat training exercises for many days or months, until the exercise becomes the perceived "norm." When an attack is planned during a training exercise, the initial reaction is that "it's another training exercise." Then the initial stages of attack are less perceptible from the routine exercises, which evolve to not be of high interest. The MMWS design focuses on system support to reduce vigilance workload by automatic detection and triggering of tasks. The design attempts reduce or eliminate errors and risk emanating from situations where human detection is the sole method for beginning a task process. The design will need to support human manipulation and editing of the thresholds and external conditions desired for task triggering. Given the complexity of these external conditions, the interface design will present a challenging problem.

***Long- and Short-Term Work*** Task time frames may vary from seconds to minutes to hours. Task times will vary across mission domains such as air and land attack. Some tasks may be time-on-target, requiring scheduling of task goals at a specific time in the future. The design should consider the varying time properties of tasks and require the system to support multiple active tasks with different time frames—with the ability to quickly move between these tasks to update situation awareness or to easily refocus attention on a task product.

### 20.5.2 Dynamic Task Life Cycle

Another important set of task properties relate to the task life cycle. In the previous section we spoke briefly of the concept of task initiation, with respect to task management work activities. Definition of additional states in a task life cycle are necessary to fully support the work process through the life cycle. The life cycle phases in a task may be defined as follows:

*Initiation*: The task supports part of an ongoing mission activity. The task processing is invisible to the user and only seen as part of a list of possible tasks that might occur or that are planned for in the future.

*Activation*: The task goal becomes active and requires servicing by the user or system.

*Assign*: The task is assigned to human(s) or system (e.g., if automated).

*Execution*: A task product is prepared or response activities are conducted.

*Completion*: The product is finalized, delivered, and delivery is confirmed. Task events or activities are completed and results are confirmed. The instance of the task is complete. The general task then goes into the same pending state as during the initiation phase.

*Retire*: There is no longer any anticipated near-term need for the task, such that any processing or system activity associated with it can be ended.

Events monitored while progressing through the life cycle can trigger decision support tools or "behind-the-scenes" automation and services. It is important to recognize that today's tactical system designs focus almost entirely on execution. In automated doctrine systems such as found in the AEGIS ship system, we find some predefined activation facilities based on tactical events. But many of these system services are seldom used as they only apply to tactical events that seldom occur.

These phases of task activity generate requirements to support the human processor's activity including:

*Initiation*: Informing the human that a task goal has become active and that work is scheduled to be done.

*Orientation*: Guiding the human visual and auditory processors to and through the information required to process the task, such as reviewing a proposed task product and the amplifying information to support the product.

*Decision*: Supporting the decision process for one or several task steps ending in approval, delay, or canceling of the task product and task goal. Providing the summary information, basis for results, recommendations, etc. for the task decision.

*Execution and Product Delivery*: Final preparations or confirmation of task products and approval of their delivery or execution.

*Confirmation*: Clearly defined indication that the task process is initiated or products are now sent and delivered.

*Transition*: User decision process to move to another task activity and selection by the user of the next task activity to focus current attention.

These phases of task processing can be compared to the command and control ($C^2$) process models such as the observe–orient–decide–act (OODA) decision processing loop as defined by John Boyd or Lawson's $C^2$ process model (see Allard, 1996). Lawson's model—sense–process–compare–decide–act—more closely represents MMWS functions but still is incomplete with respect to defining confirmation and transition. The designer is faced with the challenge of determining what system support affords the human information and decision processing in *each* of the process steps (e.g., what can the human do reliably and what can automation do?; how do both cooperate to achieve task goals?). While some designers might interpret "observe" or "compare" as an inherently human function, in MMWS the human first observes information that has been "sensed," "processed," and "compared" (as in Lawson's model) to the desired state of tasks, supported by automation in the form of task rules and heuristics.

***User Decision Paths*** There are several decision paths the user may take after initial information processing. The task completion process may vary according to workload and task risk or mission criticality. If the user understands the task, product, and goal, and judges the quality of the product to be sufficient, the user may move to final execution and satisfaction of the immediate task goal. However, the initial decision "action" may be a decision to spend further dwell time on the information to decide whether it requires further processing. Several factors may affect this decision. The "newness" or novelty of the task in the current work context will likely affect task processing strategies. A new task usually warrants more investigation by the user and longer orientation times. Workload and task priority will also drive the decision strategy for orientation and review of task products. A familiar and repeated task will require less orientation. The user strategy for familiar and repeated tasks will lean toward a "naturalistic" process of reviewing the task information and quickly confirming the draft task product and deciding if the task is both timely and required in the current mission context. The expert user will recognize that the pattern of information and results drafted by the system for a task meet the current requirements either for approval, delay, or cancellation. Task risk is another important factor in the user's decision process on how much attention and cognitive processing to allocate to the task. In MMWS the initial orientation phase involved a visual review of a task draft product, the context of the task, including user judgment on whether the task is to be completed, delayed, deleted, or shed (passed to another team member).

***Confirmation*** The process step of "confirmation" is omitted from the $C^2$ process models but in MMWS the requirement was addressed to provide feedback to the user about task processing beyond the immediate task execution action. The warfighter's visual and aural senses must receive immediate confirmation (visual or auditory) that the system is executing the task commands. Confirmation information of task completion must also be persistent (able to be revisited) to guard against possible degradation of confirmation information within working memory. This loss or interference of confirmation information retrieval could lead to task duplication. This requirement is addressed in the design of the response planner/manager display in MMWS.

***Transition*** Task transition is critical but not accounted for in legacy system requirements nor addressed in the $C^2$ models by Lawson or Boyd. Delays or inefficient decisions during this stage of processing can decrease performance reaction time on critical mission tasks. Without system assistance the user is forced into an intertask workload demand to recall mission activities in progress, decide whether to scan for new tasks or recall a previously incomplete task, and then gather information to check the status and relative importance of events to prioritize the next task action. There can be search paths and strategies that lead to diminished results and further waste workload. St. John and Osga (1999) showed that transition strategies for selection of tasks could be improved by providing task priority selection cues to users for selection between mission and time-critical tasks.

### 20.5.3 Task Management Requirements

A goal of the system design is to match the dynamic task life cycle to human information processing and decision requirements. These must be matched for each of the major life-cycle phases in task processing. Table 20.2 shows the major stages in the task life cycle paired with human information requirements. The requirements in this table are repeated in Table 20.4, with design options listed to address these requirements. The process of "task management" addresses a set of requirements that afford the focus of user attention

**TABLE 20.2 System Requirements Related to Task Life Cycle**

| Software State | Mission State | Human State | System Requirements |
|---|---|---|---|
| *Automation initiated* to detect task trigger events. Access provided to manually trigger tasks. | Task pending as part of a planned mission process (e.g., air defense). Ship on-station, with orders, assigned mission role and responsibility. | • Assigned role in team with pending task responsibilities.<br>• Aware of pending tasks and goals.<br>• Developing SA relative to mission goals and responsibilities. | • Present task plans.<br>• Assign tasks to match user roles and responsibilities.<br>• Provide practice and rehearsal functions.<br>• Monitor events for task triggers. |
| *Activate task* processing and produce draft products and information sets to explain products. | A. Task activated due to mission events and/or scheduled time.<br>B. Task activated by human decision. | A. Same as above—awaiting information cues.<br>B. Decide to manually initiate task or react to verbal orders to activate. | • Calculate task information and draft products.<br>• Provide controls to easily launch task for manual activation. |
| *Assign task*: Provide information on task to assigned watchstation. | Task goal is active and pending assignment to individual or team based on naval operational rules and procedures. | • Become aware of task goal being activated.<br>• Discontinue previous activity to launch task.<br>• Launch task information presentation. | • Determine which team member gets task assignment (by preassignment, current workload, or team leader).<br>• Provide appropriate visual and aural attention cues to guide user to task launching. |
| *Launch and execute task* | Task goal awaiting execution and completion. | *Initiation (launch) task*<br>• Become *aware* of task goal being activated.<br>• Suspend or end other task activity to start task.<br>• Launch task information presentation.<br>*Orientation*: Gain SA for task.<br>*Decision*: to execute, delay, cancel task.<br>*Execution*: Perform actions appropriate to satisfy task goals. | • Provide controls to launch task.<br>• Provide flexible methods to account for various user processing strategies.<br>• Supervisory displays—display for quick assessment.<br>• Summarize information to orient user and speed.<br>• Provide decision support and produce "draft" task products for review execution. |

| | | | |
|---|---|---|---|
| | | | • Task product summaries—packaging for execution, delivery appropriate for automation approval level. |
| A. *Suspend (delay) task*<br>B. *Complete task*<br>C. *Cancel task* | A. Mission timing or system constraints allow only partial completion.<br>B. Task goals are met and products delivered.<br>C. A task is triggered but in current mission context is not needed. | A. *Suspend* task after partial completion due to mission, timing, resource, or system constraints. Store task awareness in short-term memory (STM).<br>B. *Delivery*: Complete final step of task product delivery. Store task awareness in STM.<br>C. *Cancel* task because the product and result of task is not currently required to meet mission goals. Store task awareness in STM. | A. Record state of task when suspended. Continue task processing if appropriate. Monitor task state and inform user if appropriate when to reengage task.<br>B. Conduct final task processing and provide feedback that task executed properly—message sent, product delivered.<br>C. Provide function to cancel a task and remove it from the display, and record in any historical task documentation that task was canceled by user. |
| A. *Awaiting task execution*<br>B. *Awaiting task trigger events* | A. New tasks are pending execution.<br>B. No current mission response activities. Information changing. | A. *Transition*: Pause and rest or decision to select another task.<br>B. User proceeds to update SA and mission monitor tasks, await new tasks, or look for manual initiation opportunities. | A. Provide direction and cues to the next most important task to be executed.<br>B. Provide general SA information, update on important events since last SA check. |

**TABLE 20.3 Key Task Characteristics Related to Task Management Requirements**

| Task Characteristics: Tasks . . . | Design Requirement: System Should . . . |
|---|---|
| May have definable start/stop schedules. | Monitor concurrent loading and make schedules visible to user. |
| Have definable goals. | Monitor progress toward goals; offer assistance if needed; report progress toward goals; allow user to modify or create new goals. |
| Are grouped as parts of overall job role. | Provide visual indication of task assignments and task "health." |
| May be user and/or system invoked. | Indicate who has task responsibility. Invoke and "offer" tasks when possible. |
| Have information and control requirements. | Minimize workload to access info. or controls. |
| Are mission or computer control focused. | Provide full top-down task flow and status for mission tasks with consistent, short multimodal procedures. |
| May involve varying levels of automation from full manual to partial to fully automated. | Provide visual indication of automation state with supervisory indicators. |
| May require one or many databases. | Do not require the user to know which database for any task. Direct queries automatically. |
| May require one or many software applications. | Require user to know the tasks, not multiple applications; integrate information across the job versus application. |
| Will require attention shift between multiple tasks in foreground and background (parallel). | Provide attention management and minimize workload to shift between task focus. |
| Have definable cognitive, visual, and motor workload components. | Use task estimates for workload distribution and monitoring among crew members. |
| Will likely be interrupted. | Provide assistance to reorient progress and resources to minimize working memory load. |
| Should be consistent from training to field. | Provide consistent terms, content, goals throughout. |
| Will evolve as missions, systems evolve over the life cycle of the ship. | Support reconfiguration of task groupings and addition of new tasks as systems are upgraded. |
| May be individual or collaborative. | Support close proximity and distant collaboration via visual and auditory tools. |

throughout the task life cycle. Endsley and Garland (2000) indicate that, in "general aviation" pilots, task management, including ability to accurately assess the importance and severity of events and tasks is an important component of level 2 SA (see Section 20.4.1). In MMWS a design focus on task management requirements led to definition of task characteristics (see Meister, 1985) and projected (estimated) characteristics for a future naval system as shown in Table 20.3 (Osga, 1997). The need for visual feedback and guidance for task management listed in the right column of Table 20.3 led to the development of a task management support function in MMWS.

**TABLE 20.4 System Design Related to Task Life Cycle**

| Software State | Design Approach Examples | Human State | System Requirements |
|---|---|---|---|
| *Automation initiated* to detect task trigger events. Access available to user to manually trigger tasks. | • Summary displays of task plans by track (mission focus).<br>• Summary of tracks awaiting task processing.<br>• User task assignments.<br>• Tools to support task planning and practice. | • Assigned role in team with pending task responsibilities.<br>• Aware of pending tasks and goals.<br>• Developing SA relative to mission goals and responsibilities. | • Present task plans for user inspection/editing.<br>• Provide practice and rehearsal functions<br>• Monitor events for task triggers. |
| *Activate task* processing and produce draft products and information sets to explain products.<br>*Assign Task*: Provide information on task to assigned watchstation. | Access methods (VABs, voice, menus) for manual task launch. | A. Same as above.<br>B. Decide to manually initiate task or react to verbal orders to activate.<br>• Become aware of task goal being activated.<br>• Discontinue previous activity to launch task.<br>• Launch task information presentation. | • Calculate task information and draft products.<br>• Provide controls to easily launch task for manual activation.<br>• Determine which team member gets task assignment (by pre-assignment, current workload, or team leader).<br>• Provide appropriate visual and aural attention cues to guide user to task launching. |
| *Launch and execute task* | • TACSIT icons, TM icons, track popup menu, RPM task bar.<br>• Information set design.<br>• Color coding standards across ID and information assessments.<br>• Attention cues such as track fill or no-fill.<br>• Task "draft" products.<br>• Recommendations for best task solutions. | *Initiation (launch) task*<br>• Become *aware* of task goal being activated.<br>• Suspend or end other task activity to start task.<br>• Launch task information presentation.<br>*Orientation*: Gain SA for task.<br>*Decision*: to execute, delay, cancel task.<br>*Execution*: Perform actions appropriate to satisfy task goals. | • Provide controls to launch task.<br>• Provide flexible methods to account for various user processing strategies.<br>• Supervisory displays—display for quick assessment.<br>• Summarize information to quickly orient user.<br>• Provide decision support and produce "draft" task products for review execution. |

(*continued*)

**TABLE 20.4 (*Continued*)**

| Software State | Design Approach Examples | Human State | System Requirements |
|---|---|---|---|
| | • Concise product displays with "one-touch, one-command" delivery methods.<br>• Redundant point or hands-on-keypad/trackball methods.<br>• Communications setup for task in case voice conference required. | | • Task product summaries—packaging for execution, delivery appropriate for automation approval level. |
| A. *Suspend task*<br>B. *Complete task*<br>C. *Cancel task* | A. Brief summary indicator of task progress. Indicator of when to return to task if predicted. Attention cues to return to task.<br>B. Clear and concise feedback of task successful completion. Confirmation of final product delivery. Storage of task instance for future reference.<br>C. A function to delete task is provided with the task product or task icon. | A. *Suspend* task after partial completion due to mission, timing, resource, or system constraints. Store task awareness in short-term memory (STM.)<br>B. *Delivery*: Complete final step of task product delivery. Store task awareness in STM<br>C. *Cancel* task because the product and result of task is not currently required to meet mission goals. Do not store in STM. | A. Record state of task when suspended. Continue task processing if appropriate. Monitor task state and inform user if appropriate when to reengage task.<br>B. Conduct final task processing and provide feedback that task executed properly—message sent, product delivered.<br>C. Provide function to cancel a task and remove it from the display, and record in any historical task documentation that task was canceled by user. |
| *Transition*<br>A. *Awaiting task execution*<br>B. *Awaiting task trigger events—SA maintenance* | A. Task instance icons on task summary display. Concise listing of most important tasks on the queue. Track instances shown (a small window per track) within a task type, peripheral task indicators (near TACSIT window) shown across task types.<br>B. Provide access and cues to SA updates and important information changes. | A. Pause and rest or decision to select another task.<br>B. Proceed to update SA and await next task assignment. | A. Provide direction and cues to the next most important task to be executed.<br>B. Provide general SA information, update on important events since last SA check. |

## 20.6 DESIGN BY TASK REQUIREMENTS

The previous sections described how HSI provided assistance to the MMWS project using a task-centered approach. In particular, the HSI process focused the designer on providing user support through the task life cycle, with the critical contribution of establishing both static and dynamic requirements for the four major task categories (mission, human support, work management, and workspace computer management and control). These sections covered the first major component of the task-centered design (TCD) process—establishing HSI requirements. This section and the next cover the MMWS experience in the second major component of the TCD design process—creating TCDs.

The creation of design concepts to address the requirements for MMWS included several key inputs:

1. *Experience and Lessons Learned for Similar Systems with Similar Tasks* Previous research projects with similar tasks provided design input by supplying HCI tool "components" that supported computer interaction tasks (Osga, 1995). Decision support study results provided a basis for decision support methods (Morrison et al., 1997).

2. *Innovation and Creative Design Solutions* The general philosophy of designing the watchstation to support task goals (e.g., "task-centered" design) was a central theme for innovation within each critical task area. The dynamic task life cycle, as described in previous sections, is supported by system functions that account for human capabilities in visual search, cognition, memory, and training issues.

3. *Traceability of Requirements to Design Results* Requirement lists were generated and used to focus concept design toward methods to address these requirements. Traceability is particularly critical in new design, when management seeks an explanation of what requirement the design addresses.

4. *Iterative Testing of Design Concepts with Users* All requirements identified were not addressed in the initial concept design. Iterative testing was a critical part of the design methodology and focused the results on products that worked with the navy user population.

**Example 20.8 Rapid Prototype Refinement of Design Requirements** The design concepts were captured in task description documents and design descriptions. They were then turned into working models using the Macromedia™ Director authoring software. This software provided a rapid prototyping method to support usability testing. A parallel development team created a JAVA-based software version, as requirements and design were further stabilized. In this manner the Rapid Prototype version consistently fed design requirements to the JAVA programming team as usability tests were completed.

A summary of the MMWS display design is shown in Figure 20.2. The four-screen watchstation is shown with an "information set" assigned to each of the top three screens and the bottom center screen containing the Task Manager display with other windows. Each of these components is described in further detail together with the requirements that were addressed with the design features. This description includes how the design addressed the requirements of the task life cycle and decision support, attention management, task management, user navigation, and ergonomics.

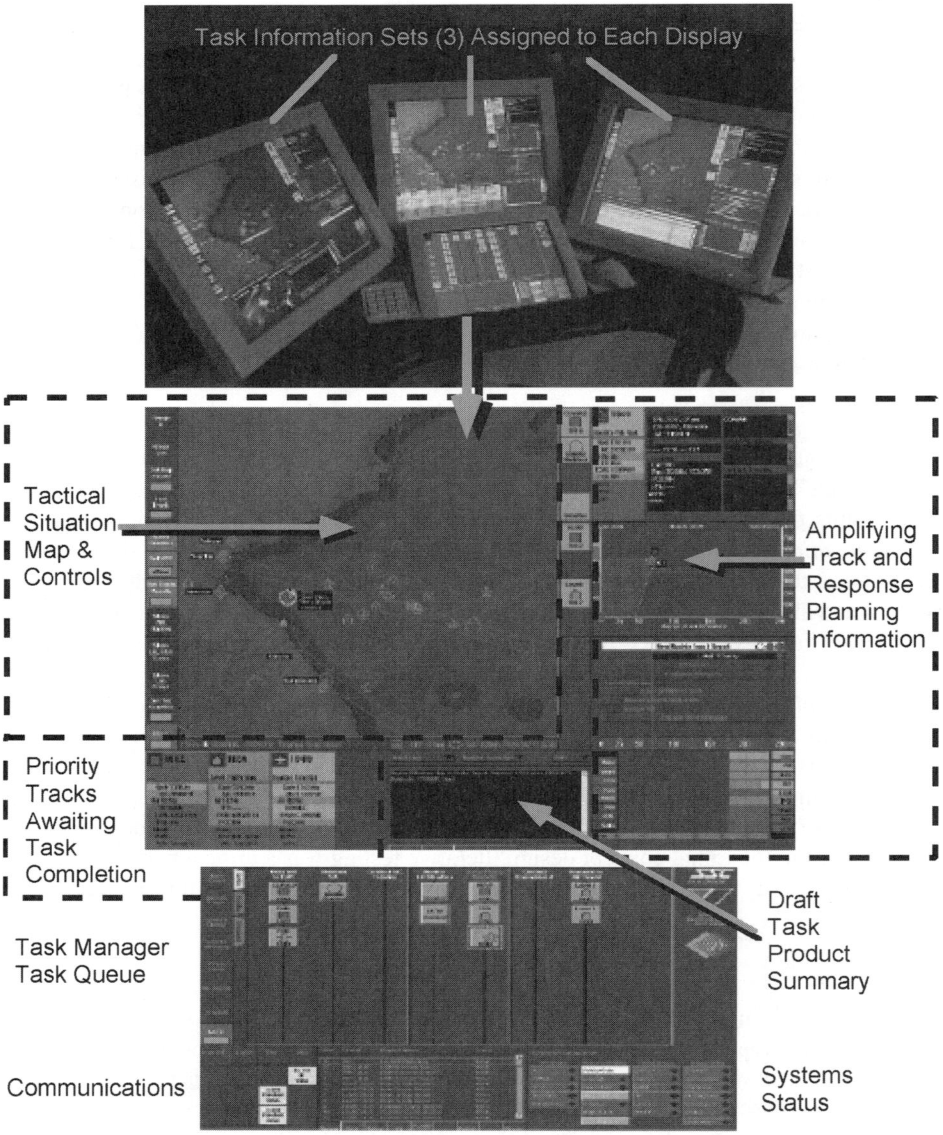


**Figure 20.2** MMWS display layout and task information sets.

Table 20.4 summarizes many of the design properties of MMWS in relation to user support through the stages of the task life cycle. Each of these task phases and design attributes are discussed in further detail in the following sections.

## 20.6.1 Task Initiation Design

Task *initiation* is defined as the initial processing of task triggering information and ends with the start of the next phase of calculations for draft task products. This processing of task information is invisible to the end user. The user is brought into the loop at the end of the initiation process, when the system identifies the presence of a task to the user.

The following task initiation requirements are addressed by various design attributes of the watchstation.

1. *Present Task Plans for User Inspection/Editing* The MMWS presents task plans using several views: (a) Top-level iconic view of all tasks, (b) graphic view of assigned tasks (coded by assignment to the user or automation), (c) graphic view of plans within a task (detailed by track if appropriate), and (d) iconic view of tracks within a task focus area (sorted by simple ID priority). The Task Manager display column headings (see Fig. 20.3) shows the current tasks assigned to the warfare team. The response Planner/Manager display was designed to allow user inspection of task plans (see Fig. 20.4).

2. *Provide Practice and Rehearsal Functions* The requirement to support task response planning and practice was not addressed in the current MMWS design, and the plan was fixed for the test scenario operational area. This design did not allow any flexibility in editing task plans during the mission simulation. This requirement allows the user to cognitively rehearse mission responses and adapt the responses to different operational areas and conditions.

3. *Monitor Events for Task Triggers* The MMWS simulation was designed to monitor simulated shipboard databases for events and information changes, using rule-based event triggers. Tasks are initiated in response to predetermined events, using simple mechanisms and rules. The task description documents generated for each task contained details of prescribed task triggers (Osga et al., 2002b).

***Task Initiation Design Summary*** Task initiation requirements play an important part in the life-cycle task process. If the user or system does not initiate a task, the goal is

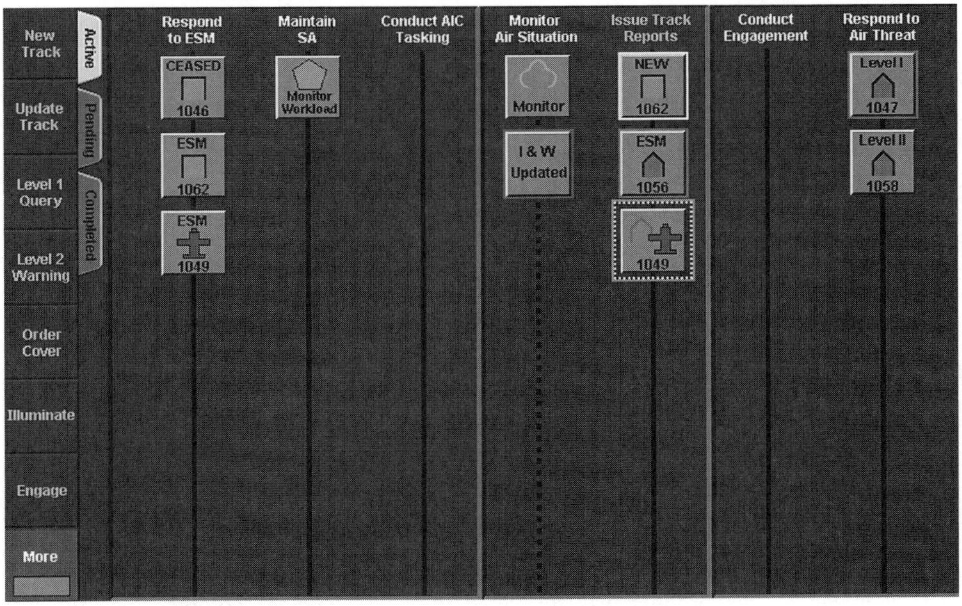


**Figure 20.3** Task management icon list display for air defense tasks.

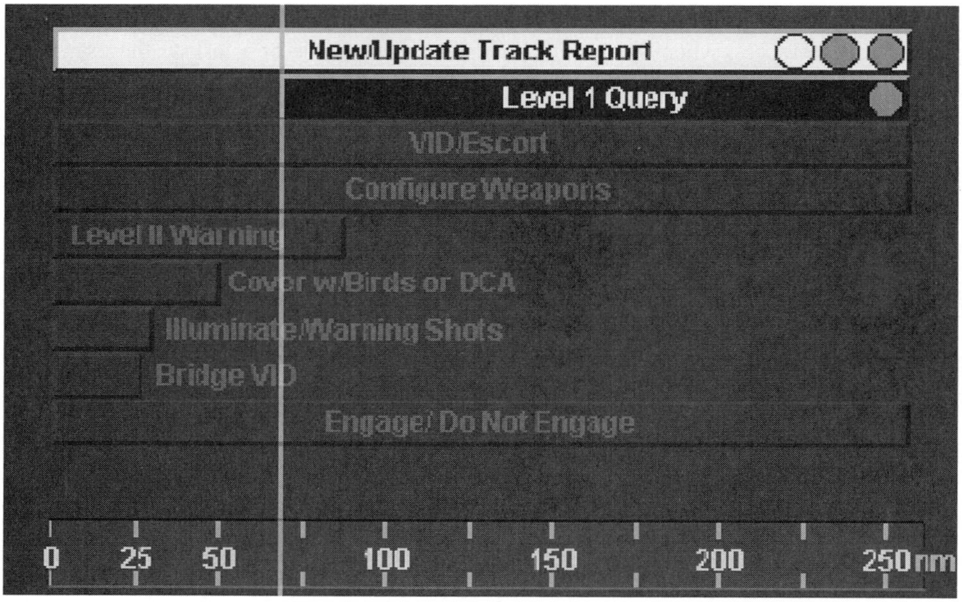


**Figure 20.4** Response planner decision support tool.

not obtained. These requirements were addressed in the MMWS prototype by using embedded task triggers for all air defense warfare (ADW) tasks within the scope of the current test mission problem. The triggers were fixed and not editable by end users, but they followed a battle response plan agreed to by SMEs as reasonable and following accepted practice with fleet methods for the scenario. Task inspection information and response plans were provided using several iconic and graphic display formats.

### 20.6.2 Task Activation and Assignment Design

Task activation may follow initiation and starts the process of finalizing the task product and meeting the immediate task goal. Activation can be either manually performed by a human action or automated in a fielded system. In MMWS software and design, activation was manually performed in one software version and had automatic assistance in a second version. Requirements during activation and assignment are as follows:

1. *Calculate Task Information and Draft Products* When a task was triggered, software mechanisms were set in motion to create task products. These products included draft messages such as new/updated reports, queries, and warnings. The design philosophy was that the system would attempt to create a "draft" product in best format possible, allowing for user inspection and approval of the draft. The current software design did not address user editing of draft products. Some tasks did not involve products for delivery, but for inspection, such as an update to operational orders or rules of engagement that required user cognizance. The product was formatted with text changes colored since the last inspection performed by the user.

2. *Determine Which Team Member Gets Task Assignment* The initial ADW design did not address assignment by workload. The tasks were preassigned as designated by SMEs' judgment of appropriate assignments. The limiting factor on task assignment was related to the monitoring of ship audio circuits. The various circuits needed an assigned operator to monitor replies from external sources—other ships, aircraft, etc. The assignment of a single person to a single circuit work strategy significantly limited workload distribution and task assignment for tasks associated with communications events. This also prohibited the distribution and leveling of workload across the team as originally planned for MMWS. While there was considerable controversy among the MMWS design team as to how communications might be handled in the future, the limitation of workload distribution represented a worse-case design condition basis that communications external to the ship would be handled using today's voice technology. Members of the design team could envision digital messaging and transfer information to and from the ship in ways that would lessen the workload restrictions for some types of messages such as "new" or "update track" reports. Other messages such as directions to aircraft or warnings to aircraft were determined to require an operator dedicated to getting the replies from the external aircraft. The task demands for external communications must be given serious consideration in determining workload distribution aboard future ships.

3. *Provide Appropriate Visual and Aural Attention Cues to Guide User to Task Launching* When a task was initiated, three display events occurred: (1) An icon was presented on the task manager (see Fig. 20.3). (2) An icon could appear on the peripheral task indicators if the task was at the top of the queue for that task category. (3) An instance of the task could appear as a small amplifying information summary window in the list of windows for a task category (see "priority tracks awaiting completion" in Fig. 20.2). Aural cues were used in usability studies to represent different task attributes, and it was determined that they did not add benefit to task launching performance while creating unnecessary distraction. Auditory cues were delegated to a supportive role if the task response exceeded a certain time limit and urgency requirement.

### 20.6.3 Task Execution Design

During execution the users' attention processes are focused on the task requirement when a decision has been made to begin task execution. Task execution involves the process of supporting control actions and decisions relevant to satisfying the task goal(s). Execution includes the user launching the task to populate displays and windows with the task information set, and then the user monitoring or executing the task as appropriate. The final step to execution would be delivery or cancellation of the task product. Execution could also be delayed and then restarted at a future time.

The MMWS design included multiple displays to allow the user to easily time-share display allocation between concurrent tasks without requiring changes to a single display to transition between tasks. The need for task time-sharing varies according to mission demands, and at times of low workload, a single display may suffice. The three displays were considered supportive in a high workload environment. They were also selected and positioned on the basis of ergonomic requirements (see Section 20.4.7).

**Example 20.9 Flexible Control Methods to Launch Tasks** To aid in quick performance reaction and reduce visual search, redundant methods were provided to launch tasks. These methods were based on user cognitive and visual strategies envisioned for task processing.

Several methods, including task icons, task bars, and pop-up windows, were provided to launch a task. Several of these task launch methods provided a similar support strategy to launch a sequence of task events allowing the operator to maintain visual focus on a single display area to accomplish a sequence of tasks. These methods allowed the user to work within a task type, "task family," or to move between task families and types.

Quick assessment and flow through task processing is done by making visual search and visual work flow through the task efficient. Visual work must flow within a display and flow across displays. In the design of display layouts there are no perfect answers, but instead there are many layouts that could foster effective task flow. The MMWS design supports a user strategy of continued work within a task (single display), quick sampling of the larger work activity (Task Manager and three displays), and switching rapidly between tasks (visual shift to primary displays). The workload induced by a display visual shift, combined with common formats and common placement of similar information (such as task products), would be less than that required to access, remember, and locate commands/menus to navigate between tasks. This simple visual shift between tasks should be less disruptive to cognitive processing of higher level mission activities.

**Example 20.10 Supervisory Displays** There are several supervisory "layers" provided in MMWS design to aid in fast assessment. The highest layer is across an entire display, where differences in color provide visual cues for conflicting or homogeneous information on ID. Supervision requires visual and cognitive processes to first sample information and second to decide when to dig deeper into a task processing. A key information issue is the urgency and mission-critical nature of the task or information. Information that is neither urgent nor mission critical is left for future processing while urgent or critical information is given attention. The first layer of user processing is by position and color. For example, position coding has task family positions constant in the Task Manager (TM) display and task icons placed and coded on the TM list according to urgency; whereas color coding is used to aid in quick scanning such as for conflicting ID information by using multiple hues. Other design methods include:

- *Summarize information to quickly orient user.* Kellmeyer and Osga (2000) report that the Basis of Assessment window (see Fig. 20.5), with its color coding and consistent summary of ID information, is one of the most useful information summary displays on the MMWS.
- Provide decision support and produce "draft" task products for review before execution.
- Provide task product summaries—ready for execution, delivery appropriate for automation approval level.
- If task is suspended, record state of task when suspended. Continue task processing if appropriate. Monitor task state and inform user if appropriate when to reengage task.
- Conduct final task processing and provide feedback that task executed properly—message sent, product delivered.
- Provide function to cancel a task and remove it from the display and record in any historical task documentation that task was canceled by user.

### 20.6.4 Task Transition Design

Task *transition* design includes support for decisions about work strategy and direction of attention toward available task opportunities. Transition involves a change of immediate

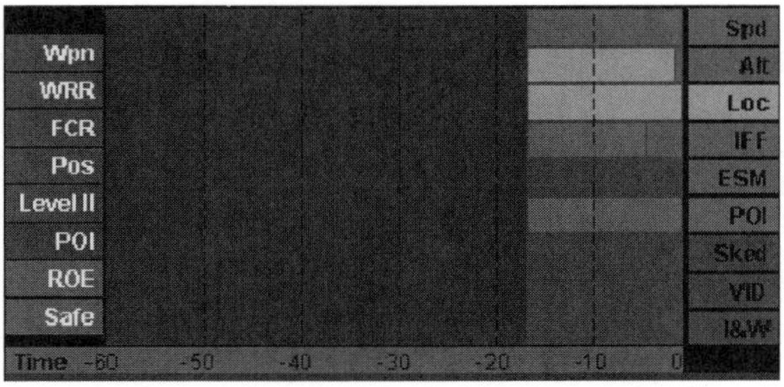


**Figure 20.5** ID basis of assessment display. Right side of window shows ID history parameters and colored bars indicate change over time. Left side shows current threat positive for selected track.

user focus from a specific task goal toward identifying the broader scope of task goals to be accomplished, followed by a decision whether to continue sampling for task opportunities or to begin to work to accomplish a specific task goal.

1. *Provide Direction and Cues to the Next Most Important Task to Be Executed* Several visual cues were used to provide information on the remaining tasks to be executed. The coding methods are shown in Table 20.5. On the tactical display window, symbols were filled if an incomplete task was remaining and unfilled if no tasks were pending. Thus, if New Track Report task was currently selected, all filled symbols shown were those pending a new track report. If Monitor Air task was selected, all pending tasks were shown for suspect and unknown tracks. In the periphery of the tactical display the task icons were listed showing the top task in each task family, and the Amplifying-Information windows showed a sorted list of tracks within the selected task. Table 20.6 lists the triggers and

**TABLE 20.5 Visual Cues to Aid Task Transition**

| Display Location | Type of Visual Cue | Comments |
|---|---|---|
| Tactical situation map | Filled symbols | Indicate task in queue awaiting processing. If monitor air situation then only suspect or unknown symbols with pending tasks filled. |
| Tactical situation display peripheral area | Task icon | Show top task icon from each task family. |
| Tactical situation periphery | Amplifying information windows | Show sorted windows for top 7 tracks in the selected task family. |
| Response planner/manager (RPM) | Show next suggested task with highlighted text on task bar. | A circle appears on the bar if someone on the team activates a task and it is in progress. |
| Task manager (TM) | Task icon with time or urgency color border on the task icon. | Task icon border colors were used. (See Table 20.6 for coding rules by task type.) |

**TABLE 20.6 Visual Cues for Task Urgency/Latency**

| Task Type | Visual Cue Trigger | Type of Cue (lower to higher urgency shown) |
|---|---|---|
| New track report | 2 minutes—no response | Yellow border on task icon |
| Update track report | 3 minutes—no response | Orange border on task icon |
| I&W updated | 5 minutes—no response | Red border on task icon |
| ATO updated | | |
| ROE updated | | |
| Level I query | Longer range from ownship | Yellow border on task icon |
| Level II warning | Medium range from ownship | Orange border on task icon |
| | Close range from ownship | Red border on task icon |
| ESM tasks | No cues used | No colored borders used |
| Maintain workload | | |
| Monitor air situation | | |

visual cues associated with tasks that had a late response or an increase in urgency due to the position and heading of the track in relation to friendly ships.

2. *Provide General Situation Awareness Information, Update on Important Events Since Last User Information Check* Within the limited air defense task domain studied, several tasks were included to provide an update to situation awareness and changing information. The system provided updates to the indications & warnings (I&W) status, air tasking order, air warfare situation representation (SITREP) report, ship equipment status, and rules of engagement (ROE) as information changed for these documents. Information that changed since the last user update was shown using an alternate color in the window.

## 20.7 SPECIAL DESIGN QUALITIES

There are a number of design qualities stimulated by the HSI process that were integrated into the product such that the overall design produced shows a strong focus on HCD qualities including:

- Design for decision support
- Design for attention management
- Task manager design concepts
- Design for user navigation and selection
- Design for user ergonomics

### 20.7. 1 Design for Decision Support

The decision support design principles used in MMWS were:

1. Bring the information to the decision and summarize it.
2. Clearly show any ambiguity or conflicting information with regard to the decision.
3. Provide assistance in the timing, planning, and scheduling of decisions.

**TABLE 20.7 Coding Methods Used in RPM Display for Decision Support**

| |
|---|
| Coding for task name on task "bar" |
| Gray text—task not yet recommended for this type of track and its kinematics. |
| White text—task may be recommended at future point if track maintains same ID and same kinematics. |
| Black text—task is completed already if task bar is white. |
| Coding for task completion status |
| Black bar—task completed. |
| White bar—task has been created (system or operator). |
| Gray bar—task not initiated. |
| Coding to keep record of occurences of task for track |
| Open circle—task currently in process or pending. |
| Green circle—task has been completed. |
| No circle with white bar—indicates that the task was probably deleted by an operator. |

An example of these design principles were shown with the information sets that provide the task information for each task goal, with color-coding used to show ambiguous or conflicting information related to the track ID involved in the task decision. Also, the TM and response planner manager (RPM) displays provided work strategy decision support mechanisms. Further, visual coding rules were used in the RPM display to provide decision support information on work strategy to the user as summarized in Table 20.7.

### 20.7.2 Design for Attention Management

Attention management is the process of system support to guide human resources such that those resources are allocated in an efficient manner to the most critical or urgent task activities. In situations where no time-urgent or mission-urgent tasks are in the queue to be done, attention should be guided toward information relevant to pending and future task goals. Attention management should be handled carefully, due to issues discussed earlier concerning task interruption. In MMWS, a layered approach to management included (1) visual cues and (2) alerts (visual and auditory) that supplement the visual cues. The primary visual cues guide work flow and resource allocation between and within tasks. Specific cues guide attention within a task. Many of these visual cues have been presented in earlier sections on design for task initiation and execution. In addition to capturing visual and auditory channels when needed, the system must foster smooth and efficient flow toward completing the work activity and then through task transition. The following sections discuss two attention mechanisms in MMWS: task prioritization within the task management functions and alerting mechanisms. Two examples are presented.

**Example 20.11 Task Prioritization** Task prioritization schemes were proposed but not fully implemented in the MMWS software during the project time frame. A priority scheme was proposed with four levels ranked from highest to lowest priority: (1) mission critical and time critical; (2) mission critical but not time critical; (3) time critical but not mission critical; and (4) neither time critical nor mission critical. This task prioritization scheme was not effective by itself, and another variable came into play that did not allow preassignment of a "rank" to a task type. The object or track involved in a task could make that task change between levels of mission or time criticality. Thus, a new track report for a track identified as a commercial air at some distance was level 4, while the same report for a suspect closing to the battle group might be level 1. Then, a more elaborate prioritization was proposed based on various track ID parameters (Hildebrand, 1999). The detailed prioritization methods were not implemented in the current MMWS software, and a simple scheme of first-in, first-out was selected with the most recent task instance shown at the top of the display for each task group. As expected, in comments from users following tests, users did not approve this simple prioritization method. Further research is warranted on best methods to prioritize and rank tasks, including methods on how to update the task priority rankings as these priorities change in real time.

**Example 20.12 Attention Management Cueing Methods** Attention cueing supports the process of bringing the user's attention to critical issues or problem tasks. Cues were described earlier to guide task progress and transition. Other cueing support was provided to indicate late or delayed tasks and information changes within tasks. The cues were numbered from low to high, ranging from a low amount of visual and auditory stimulus to progressively higher amounts of stimuli. Figure 20.6 shows the visual appearance of several graphic cues. The first and primary-type visual cues notify the user of task initiation and presence, with icons and visual indicators. Higher levels of cue stimulation add additional visual cues in a change of color for the TM icon border. These cues were time-based and appeared within a certain period after no response for a presented task. Higher intensity cues also involved the use of audio cues and blinking of the standard alert icon (a small triangle). The icons appeared in static form as shown on a button or window as shown in Figure 20.6 and then could become blinking after no response for a given period. Lower priority alerts could be delayed if higher priority alerts were present. The relative priority of multiple alerts across tasks becomes an important issue when workload increases.

### 20.7.3 Task Manager Design Concepts

Design concepts related to task management requirements are listed in Table 20.8. Many, but not all, requirements were addressed in the current design.

***Task Manager Summary Window Format Design*** In order to address requirements related to depiction of task state information, formats were designed to depict tasks currently active in the work queue. Early concepts addressing air defense task progress

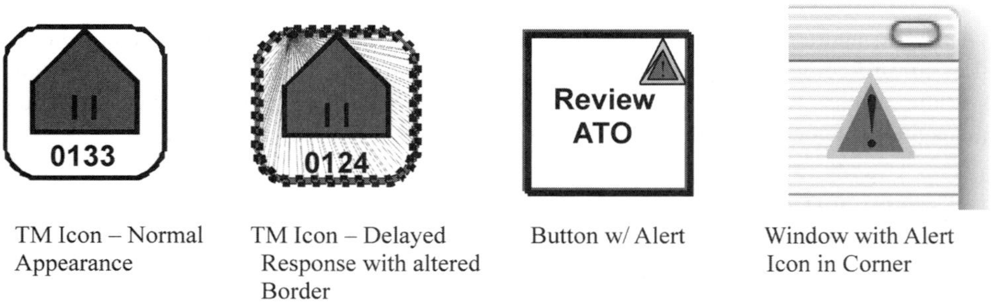


**Figure 20.6** Examples of visual alert cues (low priority) used in task manager icons, buttons, and windows.

**TABLE 20.8 Key MMWS Design Concepts Related to Task Management Requirements**

| MMWS Design Concept Basis | Design Requirement |
|---|---|
| RPM—individual threat response summary. TM display—composite workload and task icons. | Monitor concurrent loading and make schedules visible to user. |
| RPM—range based, single threat summary. TM display—task summary display. No user modification in current design. | Monitor progress toward goals; offer assistance if needed; report progress toward goals; allow user to modify or create new goals. |
| TM display and workload indicators. | Provide visual indication of task assignments and task "health." |
| TM display—task assignment summary. MMWS context and event monitoring to support task initiation. | Indicate who has task responsibility. Invoke and "offer" tasks when possible. |
| Multiple display surfaces—maximize visual work space (within 5–95% reach envelope for touch). | Minimize workload to access info. or controls. |
| TM expand/contract task list and task filters. | Provide full top-down task flow and status for mission tasks with consistent, short multimodal procedures. |
| Earlier TM designs indicated automation type. Removed for ADW when automation was fixed for testing. Added for land attack. | Provide visual indication of automation state with supervisory indicators. |
| Information sets provide information automatically for task. | Do not require the user to know which database for any task. Direct queries automatically. |
| Apply consistent procedures across different tasks. | Require user to know the tasks, not multiple applications; integrate information across the job versus application. |
| Multiple displays allow simple visual shift between tasks. Task priority visual cues. Tasks assigned to columns in similar groupings. Task columns match display assignment. | Provide attention management and minimize workload to shift between task focus. |
| Workload distribution summary display shows relative loading among crew members. | Use task estimates for workload distribution and monitoring among crew members. |
| Highlight changed information when task is "dormant." Reminders and notes tied to tasks. | Provide assistance to reorient progress and resources to minimize working memory load. |
| Consistent task design across multiple tasks. | Provide consistent terms, content, goals throughout. |
| Task groupings fixed in current design. Future support should provide flexibility. | Support reconfiguration of task groupings and addition of new tasks as systems are upgraded. |

were created in 1989 and reported in Osga (1995). Design concepts for the RPM display from the TADMUS project were also reviewed (Kelly et al., 1996; Morrison et al., 1997) and from research efforts following TADMUS (Manes et al., 1999; St. John et al., 1999).

The RPM display was used to depict planned response actions in air defense warfare showing task duration and deadlines related to individual air threats. Additional information was required beyond the single-threat RPM focus to address task situations with multiple threats and multiple mission activities. The TM display was created to provide a view of all tasks planned or in progress. The TM air defense display format differed for long-term tasks such as mission plans or execution of events that occurred over many minutes and involved multiple steps in their sequence.

Usability testing results (Kellmeyer and Osga, 2000) indicated that visual depiction of time, automation, and deadline with display scrolling on the task manager window were not beneficial during high workload periods. Information concerning task deadlines and schedules was not needed in fast-paced air defense tasks. The users simply wanted to see current work in the queue and process the task as quickly as possible. Figure 20.3 shows the simpler TM display with task icons. Simple icons were found to be sufficient for air defense mission task depiction.

### 20.7.4 Design for User Navigation and Selection

Two important design features include methods to navigate through task procedures and for selection of objects or functions. Five multimodal selection methods are:

1. *Redundant Touch and Trackball Cursor Movement* The watchstation provided several redundant methods with which to navigate the four-screen work space. Methods employed were touch, trackball for full cursor navigation, and partial navigation with keypad. Gross movements were aided by touch. With this method it was impossible to visually lose the cursor since it would always appear where the screen was touched. Moving the cursor large distances between all screens was easily done with touch. Fine selection movements to select tracks, icons, and other GUI objects were done with either touch or trackball. Selection of tracks was aided by the advanced hooking algorithm (Osga, 1991).

2. *Navigation on Task Manager with Keypad* Navigation on the task manager was also supported by the keypad. The arrow keys could be used to move between task icons on the TM and the ENTER key used to select an icon (or the select button on trackball). The user could proceed through most tasks with one hand on keypad and one on the trackball without ever touching the screens or reaching to hook a track. The default task product window and DONE or SEND function would gain cursor focus at the end of a task allowing the select function on the trackball to be used to complete the task.

3. *Track Search with Keypad* Tracks could be hooked and located using the search function with the numeric keypad. With the NumLock set in the "on" position, the user typed a four-digit number in the keypad. A virtual keypad appeared in the top right corner of the tactical plot that showed what was being typed and on which plot the track would be hooked. When the ENTER function was selected, the track with that number would be hooked. Usability test feedback provided positive results for all the methods used for navigation.

4. *Prehook Selection Methods and Information* Prehook information refers to track information obtained by moving the cursor near the track object, before a selection action is made. A dashed circle indicated the track that will be hooked when a select action is made and shows a small set of summary information about the track. When the select (hook) action is made, the circle changes to solid and other auxiliary windows present the amplifying information for that track. A select action used either the left trackball button or

**TABLE 20.9 Popup Menus and Methods**

| Type of Menu | Method Accessed | Notes |
|---|---|---|
| Track context popup | Right trackball button | Cursor near track on map |
| Map context popup | Right trackball button | Cursor over map—not near track |
| Track list popup | Left trackball button, depress and hold | Cursor near track |
| Auxiliary window list popup | Left trackball button, depress and hold | Cursor over an unused part of window—not over button |

a tap on the screen with the finger. Dragging the finger or moving the trackball showed the prehook indicator as the cursor was moved.

5. *Function Selection Methods* Methods used to select functions included variable action buttons (VABs) and popup windows including the track contextual menu, tactical situation (TACSIT) map menu, track declutter menu, and auxiliary window context menu. Table 20.9 indicates how each pop-up menu was activated.

### 20.7.5 Task Procedure Design

The MMWS job design contains a set of repeatable procedures designed such that tasks could be launched by several methods. This approach allows the user to adopt multiple task flow strategies during task transition. The user scans for task opportunities, starts the task using several alternate methods, scans the task products and information sets, and makes a task decision and transition to the next task. Table 20.10 compares procedures for

**TABLE 20.10 MMWS Task Procedure Design Summary**

| Procedure Step | Basic MMWS Method[a] | Enhanced MMWS Method |
|---|---|---|
| Scan for task opportunities | Tactical symbol color coding for ID | Color coding and Task manager icons |
| Start task | Variable action button | Task manager icon<br>Track pull-down menu<br>Tactical display peripheral icon<br>Mini-Amp info. selection (if user stays within same task for repeated tracks)<br>Manual Variable action button |
| Collect information for task | Visual scanning | Decision support information sets |
| Task decision | Send order, message, report, or read/comprehend information | Send prepared order, message, report, or read/comprehend information |
| Task transition | Visually scan and wait | Review next Task manager icon |

[a]"Basic MMWS" refers to the version with limited decision aids while the "Enhanced" version contained the full set of decision support aids.

the Basic and Enhanced versions of MMWS. This simple procedural method was able to service many different types of tasks, and training was streamlined due to the consistency across task types.

### 20.7.6 Design for User Ergonomics

In the spring of 1998, the NEC Corporation began producing flat-panel color liquid-crystal displays with a much wider viewing angle. These displays were selected for an upgrade to the MMWS console configuration. An Elographics guided-acoustic wave touch screen was also selected. Initial foam-core mockups of the MMWS pedestal were constructed to evaluate reach envelopes. When the larger NEC displays became available at a 20-inch size, the design was altered to accommodate them. Three displays were placed in the optimum reach/viewing envelope with adequate resolution and display area to accommodate multiple tasks. A desktop version of the MMWS was used for usability testing prior to construction of the display pedestal. The final configuration is shown in Figure 20.7.

## 20.8 BENEFITS OF TASK-CENTERED DESIGN

The benefits of the design approach are seen with results from individual and group performance testing (Osga et al., 2002a). Individual and group performance tests were conducted with naval fleet operators. Performance gains were found for both speed and

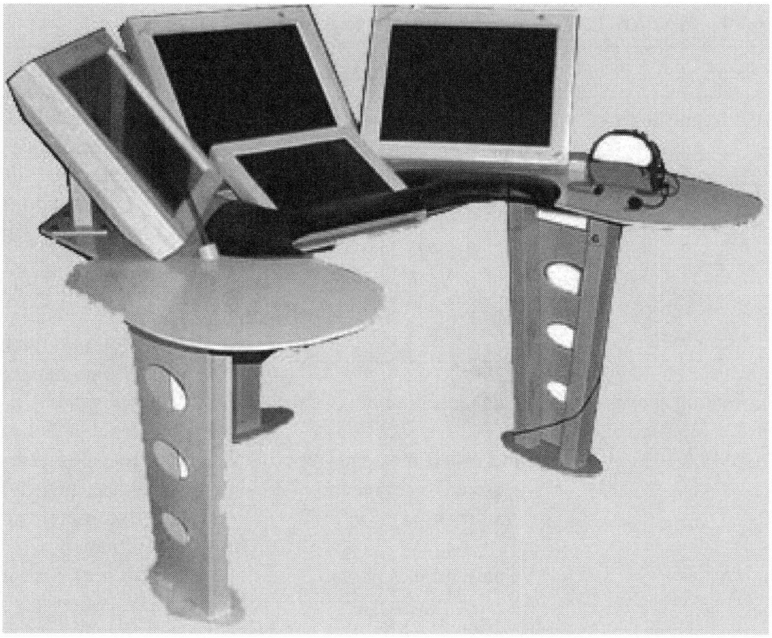

**Figure 20.7** MMWS pedestal design.

accuracy with improvement in SA and workload. Training was also simplified relative to the training requirements for similar systems currently in operation.

### 20.8.1 Performance Testing

A team performance test was conducted comparing shipboard nine-member crews using today's equipment and methods (legacy team) to five-member crews using the MMWS configuration (HCD team). Eight ship crew teams were tested using the scenario aboard AEGIS-class ships at pier-side or in land-based training sites. Six MMWS crews were tested with the basic-capability (BC1) MMWS and two teams with the enhanced-capability (EC2) MMWS. The BC1 version lacked some of the dynamic decision aids, whereas the EC2 version contained the full spectrum of aids. A realistic air defense scenario was prepared containing both low- and high-density track periods to stimulate various levels of tasks required. The scenario test used role players who acted the part of aircraft and other ships in the battlegroup. The role players were positioned in another room separate from the test teams, using voice communications simulating battlegroup operations. The AEGIS teams had eight air defense members plus an air intercept controller, responsible for vectoring aircraft. The MMWS teams had four members with a combination of duties assigned to the smaller crew size (see Fig. 20.8). Teams were instructed to conduct air defense warfare tasks in accordance with the rules of engagement and operational plans briefed during training and as practiced during the training exercised preceding the test. Primary operational tasks were:

- Visually identify (VID) all unknown air contacts within a defined area of responsibility (AOR).
- Escort air contacts from threat country with aircraft-carrier-based friendly aircraft.
- Issue warnings to threat country aircraft.
- Make positive identification of air contacts unable to VID by correlating indications and warning, electronic emissions, profile, point of origin or initial detection, air tasking order, and electronic data received.
- Conduct internal communications and external communications with battlegroup commanders and aircraft.
- Engage in self-defense.

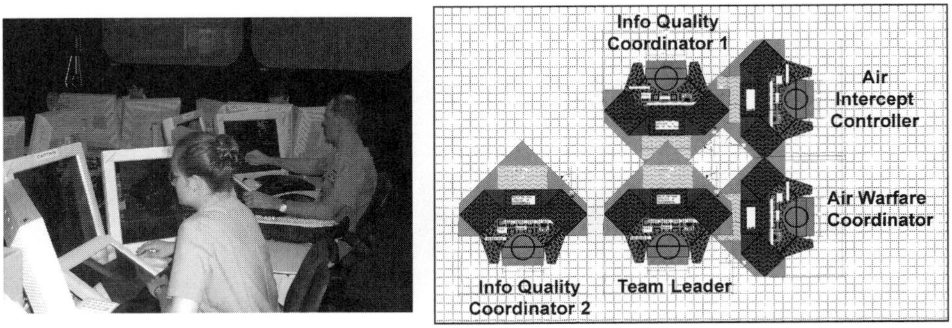


**Figure 20.8** Integrated Command Environment Lab. MMWS team performance testing (left) and team positions (right).

- Verify positive communications and communication equipment check for departing strike force aircraft.

Results for time and accuracy of reporting new tracks to the battlegroup are shown in Figures 20.9 and 20.10. There was a large decrease in performance variability from the AEGIS crews to MMWS versions BC1 and EC2. The results are shown for the first and second half of the scenario test period, with the first half being the lower workload period. Note that performance variance decreases for the Basic and Enhanced MMWS design in both the low and high workload periods. The low, medium, and high ranked tracks within 25 critical scenario events are shown, with indication that MMWS teams were better able to balance their workload among the types of scenario events.

The "overall" score shows a summary of all scenario periods with a similar decrease in variance. The high variance of results with legacy system teams requires a large number of subjects (greater than 20 teams calculated) to allow for inferential statistics. The low variance of performance with MMWS indicates that an increased homogeneity of response may be possibly a result of the design features guiding user information processing through the task cycle.

Figure 20.10 indicates that fewer MMWS users missed performing the report tasks, with only one report missed by the two MMWS EC teams tested. There were fewer missed tasks in the first and second scenario periods, with reduced performance variance. The legacy system relies on poorly coded graphic displays with a burden on human visual search tasking to locate and define task opportunities.

The MMWS provides enhanced visual cues for task initiation yielding fewer missed task opportunities. Table 20.11 shows SA results for a few of the critical scenario events. Track number 132 was a critical event where evidence was built over several minutes that the track might have hostile intentions toward the friendly forces. *The track eventually*

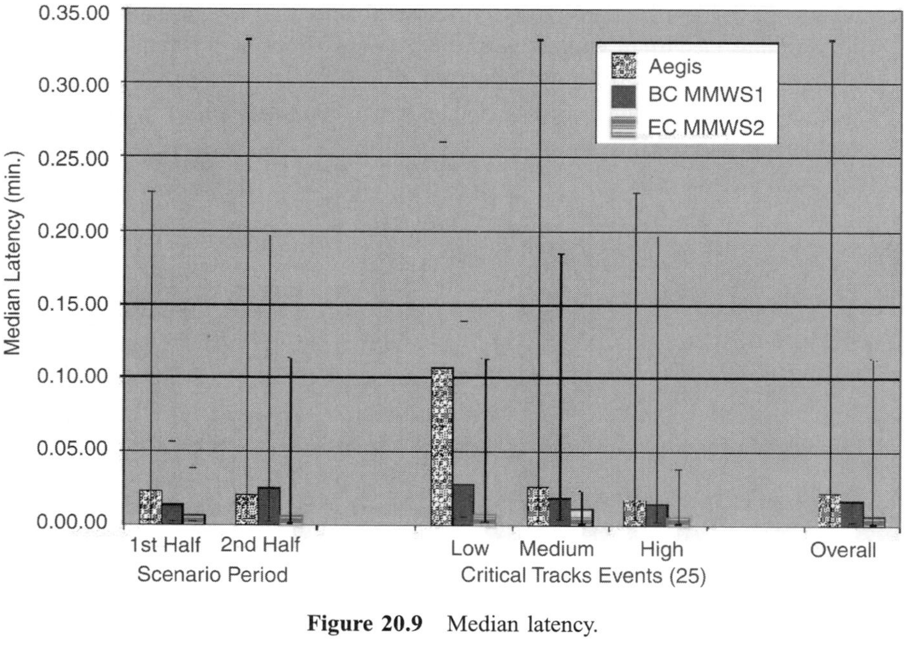


**Figure 20.9** Median latency.

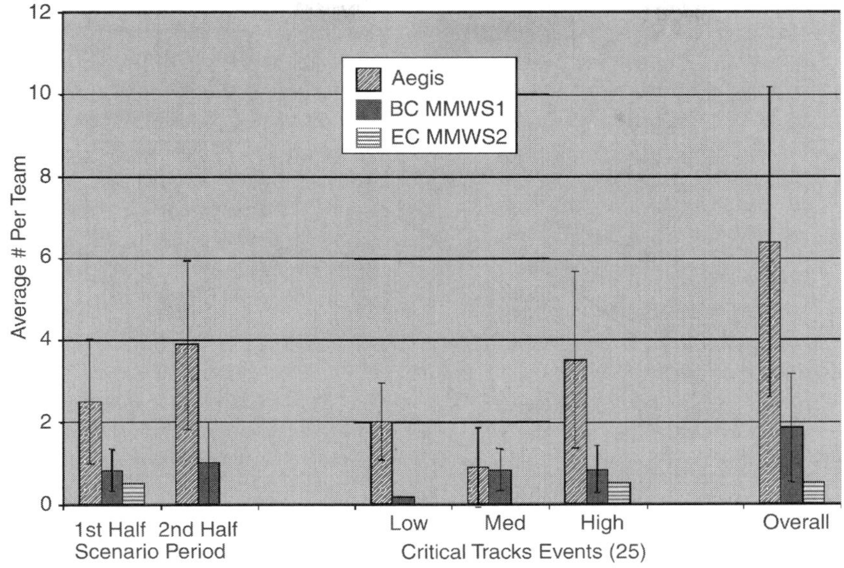


**Figure 20.10** Averaged number of missed new track reports.

*attacks friendly forces*. Note that all the MMWS teams followed the information changes about the track represented by kinematic cues (course, speed, altitude, position) and exhibited markedly improved SA as evidenced by their preparations in issuing queries or warnings leading up to the time of attack. In comparison, most of the AEGIS teams using the legacy equipment missed key kinematic events, and few teams issued queries or warnings and responded with last second engagement responses after the attack. Thus, the engagement outcome may be successful with legacy systems, but the risk is higher due to shortened reaction times with lower SA. Figures 20.9 and 20.10 represent a small subset of data collected, and further testing is required to replicate results with larger sample sizes. The team testing results correlated very well with the speed and accuracy results obtained with the same tasks and scenario with individual operators during usability testing.

Workload was measured by ratings of subject experts who observed the crew members and by crew members themselves during scenario breaks. Figure 20.11 presents the results of the expert raters. Although the raters were not condition blind, considerable time passed between the legacy system data collection and MMWS collections (one year). *Results*

**TABLE 20.11 Summary of AEGIS and MMWS Responses to Critical Track Number TN 132**

| Teams | Kinematics Detected | Query/Warnings Issued | Engage Antiship Missile (ASM) after attacked |
|---|---|---|---|
| AEGIS teams | 1 of 8 | 2 of 8 | 7 of 8 |
| MMWS BC1 | 6 of 6 | 6 of 6 | 6 of 6 |
| MMWS EC2 | 2 of 2 | 2 of 2 | 2 of 2 |

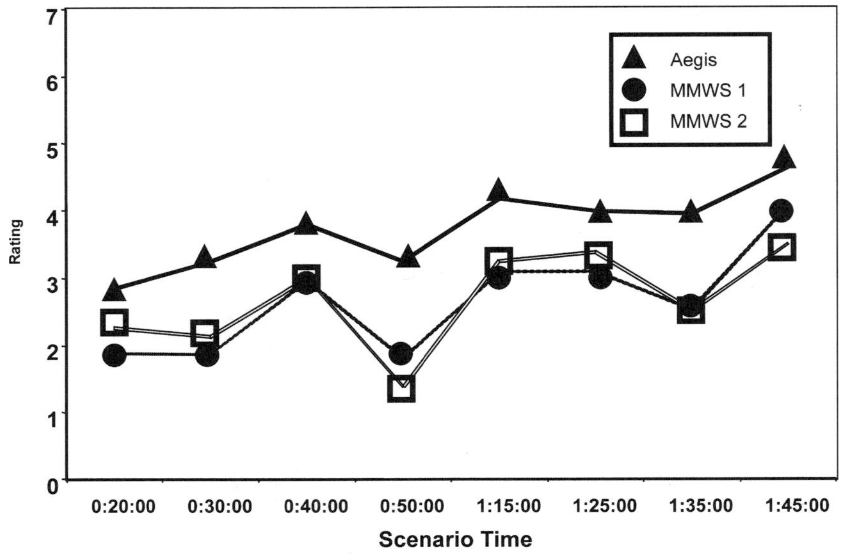


**Figure 20.11** Subject expert ratings of workload (1 = low, 7 = high) over entire scenario period for MMWS and ship board systems.

*indicate that despite the smaller teams used with MMWS, the crews were not overloaded in comparison to the larger crews using the legacy system.*

### 20.8.2 Training Results

Training requirements for ship crews included knowledge and skills applied across several task domains: (1) warfighting and mission, (2) individual responsibility and team role, (3) system Command and Control ($C^2$), (4) verbal communications, and (5) work strategy, planning, and prioritization. Subjects used in team testing were experts in the mission domain and required no training in mission tasks. They were skilled in communications methods and vocabulary used today. Training was required in system $C^2$. The watchstation training required a minimum of 1 to 2 hours for simple usability studies and tasks. Approximately 6 to 8 hours of training were required for full team testing. Teams intact from ships had previous experience of working together as a unit. Teams composed of training personnel or instructors were familiar with individual tasks but not as working together as a unit.

*Both teams performed well with no detected difference in results*. Results indicated that despite being challenged by new symbols, graphics, operating procedures, and display formats, that the crews using MMWS performed as well or better than the larger intact shipboard teams. The TCD plays an important role in facilitating training by providing a design focus on simple procedures across many tasks. Most MMWS tasks could be performed using identical procedural steps, allowing for simple procedural knowledge training that could be extrapolated across many work activities. Personnel commented following training that the watchstation and associated displays and tasks were easy to learn and could condense the longer training courses of today's workstations into a shorter time period.

## 20.9 SUMMARY AND CONCLUSIONS

Evidence from performance studies supports the hypothesis that the MMWS design may improve mission performance and reduce mission risk. Training complexity and burden are also significantly reduced. While there appears to be a performance gain from the Basic- to Enhanced-capability MMWS, there is still too little data to make firm conclusions. The TM, decision aids, and dynamic RPM in the Enhanced MMWS version appear to reduce performance variance and possibly improve decision reaction time and reduce missed tasks. The task-centered approach focused the design effort on critical tasks needed to complete a complex mission scenario. This approach directed the design cost toward the necessary display and control elements to get the "core" work done.

The cost benefit of these results, as well as the potential for crew size optimization due to lower workload and improved task execution, project a significant role for the application of task-centered human engineering in future work environments. These results apply across various task domains in other mission areas and in ship propulsion and control systems.

A central design theme in MMWS was the evolution of the human role in many $C^2$ tasks from being a manual preparation of task products to the supervisor and reviewer of draft task products. The human is better able to allocate resources to planning and strategy tasks that are difficult for the machine, and the machine off-loads the rule-based tasks from the human, with a reliable and repeatable result. The challenge then exists to make these machine assistants increasingly flexible and pliable under a variety of task conditions and demands, while keeping the human informed to monitor, supervise, and approve task activities.

### 20.9.1 HSI Principles

Clearly, the focus on HCD in the MMWS design illustrates an example of principle 2, described in Chapter 1. But what of the other principles? The context of where MMWS fits in the design process also illustrates the relevance of several other HSI principles. Certain principles apply more to the early concept design phase during research and development (R&D) whereas others are more appropriate during later stages of design.

*Leadership* (principle 1) is critical to the viability of any project and program from concept through fielding. For innovative R&D concepts, leadership is necessary to see a state of design beyond what exists today. The ability to sell this "vision" to leadership in the procurement and funding allocation roles is critical. In the case of MMWS, navy leadership puts forth a vision of reduced crews on ships, driven by cost and budgeting realities, as well as recruiting and personnel projections. This in turn led to the requirement for improvement in human engineering and crew workload. The conceptual design phase of MMWS required leadership with a sense of vision that HSI methods and processes could be improved relative to the state-of-art for today. Project leaders had to be convinced that this goal was worthwhile as part of a global crew reduction HSI strategy.

The MMWS project had an interesting and unplanned benefit for the *source selection process* (principle 3) and *documentation integration* (principle 5). Military requirements policymakers are increasing both the content and strength of verbiage applied to HSI in procurement documents for future systems. The recently released Land-Attack Training Guidance document (Chief of Naval Operations, 2001) is a good example. Notably, it requires program executive offices and program managers to plan and budget for HSI support activities during design and procurement. The document also states that "System operation and watchstanding requirements may be reduced through . . . Enhanced system ergonomic and Human Centered Designs that improve the performance and efficiency of watchstanders, especially in the areas of information management and operator interfaces . . . [and] Use of multi-modal watch stations that permit task sharing and optimize workload within the watch team." The document also specifies working–level integrated product teams (WIPTs) that specifically include HSI and HCD as prime considerations. From the government point of view, the systems acquisition documentation should include greater emphasis on HCD. From the contractors point of view, contract awards should be given to those with best HCD technical approach.

*HSI technologies* (principle 7) are recognized to be fast moving targets with commercial hardware advancements occurring in rapid succession. An important goal of HSI, therefore, must be to provide guidance with regard to HCI architecture. Systems that place HCI functions in a software layer as either independent or plug-in components allow for further upgrades and adaptation as technology quickly evolves. The concept of TCD fits the plug-and-play architecture very well as task components are upgraded and added through an evolutionary approach. The software design process also benefits from the testing and debugging afforded by a modular approach to architecture.

*Testing and performance evaluation* (principle 8) is a critical part of the design process. The process of evolutionary design and usability testing differs from the more conventional hierarchical linear design method that includes user testing at the end of the design process. With iterative design, risk is mitigated by usability testing starting with early conceptual walkthroughs on paper or by creating low-fidelity simulations in general-purpose presentation tools such as Microsoft PowerPoint before any code is written and while requirements are in formation. While the team performance tests were useful in the MMWS design process, the numerous usability tests through 2 years of multiple software versions held the most value for risk reduction. Design ideas were very much changed or discarded that had looked good on paper but failed due to a combination of dynamic task demands and lower than expected utility with operators. Programs that delay user testing and hands-on interaction until later stages of design incur unnecessary risk with respect to user performance and acceptance.

The use of *highly qualified human factors practitioners* (principle 9) contributed strongly to the MMWS design process. The design requirements were stated and held as design goals by a qualified Ph.D. human factors professional. There were occasions where the project team considered a design path directed toward a solution that was expedient for software risk or acceptance of a commercial product solution that did not appear to support human performance in a desirable manner. A qualified professional can screen the design options and select options based on HCD goals and performance improvement. Many HSI aspects of the design process are invisible to the nonqualified engineer who might be placed in charge of HSI by program management. One of the most difficult issues in using checklists or guidance information is the comparison of the task conditions represented in the guidelines to the task conditions in the current problem. The designer must recognize whether task differences are meaningful and what aspects of human performance are affected by these differences. Another important facet of professional support is the evolving literature and technologies surrounding the HCI. This fast-paced evolution requires dedicated professionals to keep abreast of changes relevant to any engineering project.

### 20.9.2 Navy HSI Capability Maturity

In general, it can be said that within the navy, the underlying government procurement organization structure is trying to enhance recognition of HSI considerations during design. In most program execution offices, however, the HSI responsibilities still are buried at a level far down in the organization hierarchy and typically as a collateral duty. The procurement officer may be in the hardware display or information systems component of the project. The prevalent conception among the engineering community is that the HSI issues revolve around display formats or use of color formatting at a superficial design level. Human factors professionals are not consulted during the system requirements definition phases or other early design processes.

Moreover, even though there is increased recognition that usability is a system design requirement having great importance, there is little R&D or development funding to follow through in improving HSI nor are there penalties for HSI ignorance. Design problems are often passed along as issues that the training community must address when the system is fielded. The Department of Defense (DoD) engineering community still attacks the myriad of problems in complex information and $C^2$ systems from a network and hardware architecture perspective, with HSI narrowly seen as a problem of maintaining consistency in the graphics user interface (GUI). Performance goals or requirements are not quantified, leaving no specific human performance requirements with which to test design success or failure. Thus, currently, the many components of HSI do not drive broader design solutions, and the main tenants of task coverage and dynamic task life-cycle support discussed in this chapter are not widely known or considered.

However, design success stories such as MMWS should increase the education and awareness level of management, while increasing the awareness of the user community that improved HSI is feasible. If user-centered design processes and successful results increase the number of visible system successes, particularly with respect to system life-cycle costs, the prospects for improved HSI during the design process will increase.

## NOTE

1. MMWS was conceived by the Space & Naval Warfare Systems Center, San Diego and supported under the DD21/ONR Manning Affordability Program executed through the Office of Navy Research, Arlington, Va.

## REFERENCES


Allard, K. (1996). *Command, Control and the Common Defense*, rev. ed. Chapter 6 "Tactical Command and Control of American Armed Forces Problems of Modernization," Section "Some Conceptual Models of Command and Control" Washington, DC: National Defense University, pp 153–160.

Bush, C. T., Bost, J. R., Hamburger, P. S., and Malone, T. B. (1999, April 14). Optimizing Manning on DD21. In *Association of Scientist and Engineers Proceedings*. Arlington, VA.

Chief of Naval Operations. (2001, January 26). *Surface Combatant Land Attack Warfare Training Requirements Document*. Washington, DC: CNO Land Attack Capstone Organization.

Endsley, M. R., and Garland, D. I. (2000 July 31–Aug 4). Pilot Situation Awareness Training in General Aviation. In *Proceedings of the International Ergonomics Society/Human Factors & Ergonomics Society 2000 Congress* pp. 2–357 to 2–359, San Diego, CA: HFES.

Fowler, F. D. (1980) Air Traffic Control Problems: A Pilot's View. *Human Factors*, *22*(6), 645–653.

Freeman, J., and Cohen, M. A. (1998). *Critical Decision Analysis of Aspects of Naval Anti-Air Warfare*, Tech Report 98-2. Arlington, VA: Cognitive Technologies.

Freeman, J. T., Cohen, M. S., Serfaty, D., Thompson, B., and Bresnick, T. (1997). *Training in Information Management for Army Brigade and Battalion Staff: Methods and Preliminary Findings*, Technical Report 1073. Ft. Knox, KY: U.S. Army Research Institute for the Behavioral and Social Sciences Armored Forces Research Unit.

Hildebrand, G. (1999, November 24). Project Memorandum. *Track Prioritization by Task*. pp 1–2.

Hildebrand, G. (2000, September 26). Personal correspondence. pp 1–2.

Jones, D., and Endsley, M. (2000). Overcoming Representational Errors in Complex Environments. *Human Factors*, *42*(3), 367–378.

Kellmeyer, D., and Osga, G. (2000, July 31–Aug 4). Usability Testing and Analysis of Advanced Multimodal Watchstation Functions. In *Proceedings of the International Ergonomics Society/ Human Factors & Ergonomics Society 2000 Congress*, San Diego, CA: HFES.

Kelly, R. T., Morrison, J. G., & Hutchins, S. G. (1996 September 22–26). Impact of Naturalistic Decision Support on Tactical Situation Awareness. *Proceedings of the 40th Human Factors and Ergonomics Society Annual Meeting*, Philadelphia, PA: HFES.

Klein, G. (1993). A Recognition-Primed Decision (RPD) Model of Rapid Decision Making. In G. A. Klein, J. Orasanu, R. Calderwood, and C. E. Zsambok (Eds.), *Decision Making in Action: Models and Methods*, Norwood, NJ: Ablex.

MacMillan, J., Serfaty, D., Cohen, M., Freeman F., Klein, G., and Thordsen, M. (1997). Advanced Multimodal Watchstation Quick Look Critical Decisions in the AMMWS Air Dominance Scenario. Unpublished Technical Report. Prepared by Decision Spectrum Group for NSWC-DD CSACT Laboratory, Naval Surface Warfare Center, Dahlgren, VA.

Manes, D. I., St. John, M., and Smith, C. A. P. (1999). Response Planner & Manager: Human Computer Interface Issues and Display Design. Unpublished Technical Report. Pacific Science & Engineering Group, San Diego, CA.

McFarlane, D. C. (1997). *Interruption of People in Human-Computer Interaction: A General Unifying Definition of Human Interruption and Taxonomy*, Report NRL/FR/5510-97-9870, Washington, DC: Naval Research Laboratory.

Meister, D. (1985). *Behavioral Foundations of System Development*, 2nd ed. Malabar, FL: Robert E. Drieger.

Morrison, J. G., Kelly, R. T., Moore, R. A., and Hutchins, S. G. (1997, April 14–17). Tactical Decision Making Under Stress (TADMUS)—Decision Support System. Paper presented at the IRIS National Symposium on Sensor and Data Fusion, MIT Lincoln Laboratory, Lexington, MA.

Naval Sea Systems Command (NAVSEA). (1996). *SC-21 Concept of Operations (CONOPs) DD 21 Ship Requirements*, Draft Rev. (3) 12/17/96. Washington, DC: NAVSEA.

Naval Sea Systems Command (NAVSEA). (1997). *Operational Requirements Document (ORD) for Land Attack Destroyer DD 21*, Document 479-86-97 (Unclass version). Washington, DC: NAVSEA.

Naval Surface Warfare Center (NSWC). (1997). SC-21 *Combat Information Center Top—Down Function Analysis*, Technical Report (unnumbered). Dahlgren, VA: NSWC, Basic Commerce & Industries, Planning Consultants, Carlow International.

Naval Surface Warfare Center (NSWC). (1998). *S&T Mission Function Analysis.* Technical Report (unnumbered). Dahlgren, VA: NSWC, Basic Commerce & Industries, Planning Consultants, Carlow International.

Neerincx, M. A. (1999) Optimising Cognitive Task Load in Naval Ship Control Centres *Proceedings of the Twelvth Ship Control Systems Symposium*, 19–21 October, The Hague, Netherlands.

Obermayer, R. W. (1998). Human Computer Interaction Design Guidelines for an Alert Warning System and Attention Allocation System of the Multi-Modal Watchstation. Unpublished Technical Report. Pacific Science and Engineering Group, San Diego, CA.

Osga, G. A. (1989). *Measurement, Modeling and Analysis of Human Performance with Combat Information Center Consoles*, Technical Document 1465. San Diego; CA: Naval Ocean Systems Center.

Osga, G. (1991). Using Enlarged Target Area and Constant Visual Feedback to Aid Cursor Pointing Tasks. In *Proceedings of the 35th Human Factors and Ergonomics Society Annual Meeting* (pp. 369–373). San Francisco, CA: HFES

Osga, G. (1995, February). *Combat Information Center Human-Computer Interface Design Studies*, Technical Document 2822. San Diego, CA: Naval Command Control and Ocean Surveillance Center RDT&E Division.

Osga, G. (1997, February). Task-Centered Design. Briefing presented at the Second Multimodal Watchstation Architecture Working Group, San Diego CA: Naval Ocean Systems Center.

Osga, G. (2001). *Human-System Integration Review Distributed Network Control System for YP 679*, Technical Note 1815. San Diego, CA: Space & Naval Warfare Systems Center.

Osga, G., Van Orden, K., Campbell, N., Kellmeyer, D., and Lulue, D. (2002a). *Design and Evaluation of Warfighter Task Support Methods in a Multi-Modal Watchstation*, Technical Report 1874. San Diego, CA: Space & Naval Warfare Command Systems Center.

Osga, G., Van Orden, K., Campbell, N., Kellmeyer, D., and Lulue, D. (2002b). *Task Description Documents for Air Defense Warfare Design Support in the Multimodal Watchstation*, Technical Note 3130. San Diego, CA: Space & Naval Warfare Systems Center.

Rasmussen, J. (1986). *Information Processing and Human-Machine Interaction: An Approach to Cognitive Engineering*. Amsterdam: Elsevier.

St. John, M., Manes, D. I., Moore, R. A., and Smith, C. A. P. (1999). Development of a Naval Air Warfare Decision Support Interface Using Rapid Prototyping Techniques. In *Proceedings of the 1999 Command and Control Research and Technology Symposium*. Washington, DC: National Defense University.

St. John, M., and Osga, G. (1999, October). Supervision of Concurrent Tasks Using a Dynamic Task Status Display. In *Proceedings of the 43rd Human Factors & Ergonomics Society Annual Meeting* (pp.168–172), Houston, TX: HFES.

Wickens, C. D. (1987). Information Processing, Decision-Making, and Cognition. In G. Salvendy (Ed.), *Handbook of Human Factors*, 1st ed. (p. 81). New York: Wiley.

CHAPTER 21

# Linking Human Performance Principles to Design of Information Systems

LINDA G. PIERCE and EDUARDO SALAS

## 21.1 BACKGROUND

In 1988, the U.S. Navy accidentally shot down Iran Air flight 655. Due to a trivial design decision, the naval computer did not provide a summary of crucial information about position, heading, and altitude on a single screen. Related information was displayed on different consoles, which made it difficult for the operator to interpret. The operator was under time pressure to diagnose the aircraft as friend or foe, which was mistakenly identified as a military flight.

In December 2001, a "friendly-fire" accident occurred in the Afghanistan war where 3 U.S. Special Forces soldiers and 20 others were injured. A 2000-pound satellite-guided bomb landed on a battalion command post instead of the Taliban post. A precision lightweight global positioning system receiver was used to calculate the Taliban's coordinates for a B-52 attack. However, the controller did not realize that after he changed the device's battery, the device was programmed to automatically display the coordinates for its own location rather than the enemy's target location.

These are just a couple of examples illustrating how information systems have performed inadequately in the recent past, but the prospect of acceptable information systems performance is even more discouraging for the future. In a presentation of a concept for future military operations, Becker (2002) provides a bleak assessment of our readiness to respond (p. 54):

> We have an unmatched ability to gather information about the environment, the adversary, and ourselves, but this information is not always the critical information needed. Furthermore, we lack the collaborative planning and command and control systems to use this information to enable decision superiority. We have precision weapons that can hit an aim point with great accuracy, but our planning is limited in its ability to consistently produce the desired

*Handbook of Human Systems Integration*, Edited by Harold R. Booher.
ISBN 0-471-02053-2 

> operational effect. Our services bring great capabilities to each domain, but continuing interoperability problems, insufficient joint training, and the lack of a fully coherent joint command and control system limit their ability to perform routinely and effectively in integrated joint action.

Why are we in this position? In 2000, a panel chaired by the Deputy Under Secretary of the Army (Operations Research) reviewed how the U.S. Army was acquiring information systems as part of the digitization program to modernize army operations. The assessment was initiated because the army's efforts to acquire information systems were behind schedule and had not resolved training deficiencies noted during advanced warfighting experiments. The charge to the panel was to review the current process and make recommendations for improvement. The panel looked at the acquisition process from requirements definition through testing and fielding.

Deficiencies noted by the army digitization review panel included a lack of horizontal integration across information systems and between information systems and weapon platforms [U.S. Department of Defense (DoD), 2000b]. These deficiencies were noted for all echelons. Information systems were being acquired to support requirements within battlefield functional areas such as artillery, maneuver, and intelligence, with little attention given to integration across functional areas, within the army, or across the services. This finding extends to a lack of interoperability with other government agencies, multinational partners, and the international community, which may be especially problematic in current operations in Bosnia-Herzegovina (B-H), Kosovo, and Afghanistan. The panel suggested that additional work should be invested in integration efforts, developing a capstone requirements document to outline the *system-of-systems* requirements, defining the information processing and transport requirements in an operational architecture, and developing joint training methods. In short, a comprehensive system-of-systems approach was recommended.

The panel noted that requirements were often refined during early user testing, but while recognizing the importance of user input, especially in developing tactics, techniques, and procedures (TTPs), there was a need to synchronize user refinements with system-of-systems requirements identified by system and program managers. The panel concluded that a system-of-systems view must be maintained, that while the effectiveness of each system was important for battlefield success, it was clearly not sufficient. Further, a spiral development process was required such that users were involved with developers in a develop–refine–develop process. The spiral development process should facilitate the systematic development and testing of TTPs, system-of-systems operational architectures, training methods and tools, and system refinements.

How people, teams, and human organizations use and adapt to new technology, procedures, organizational structures, and environmental complexities remains at the heart of the system-of-systems performance issue. Greater understanding of the relationship among the human, the technology, and the mission and a systematic assessment framework are required. Without these things it is difficult, if not impossible, to judge the relevance and utility of emerging information systems and to guide combat developers in design and integration of information systems to meet the complexity that will be inherent in future operational requirements.

In light of the above, how can information systems be acquired that meet the needs of the military command-and-control decision makers? How should the military acquisition system account for human performance? Can the findings from human performance research help guide the design and delivery of information systems? Our motivation in writing this chapter was to answer these questions.

The purpose of this chapter is to identify the major issues, concepts, principles, guidelines, and tools of human performance that are relevant to information systems design and operation. This is accomplished by reviewing the U.S. Army's *cognitive engineering of the digital battlefield* project and the U.S. Navy's *tactical decision-making under stress* project. These projects are briefly described below. Examples from the projects are used throughout the chapter to illustrate the concepts derived from them. The human performance data from these projects are considered along with the state-of-the-art literature on human performance to provide guidance for information systems designers and decision makers.

### 21.1.1 Cognitive Engineering of the Digital Battlefield

In the latter part of the twentieth century, the U.S. Army became heavily involved in the acquisition of information technology to "digitize the forces." The concept was that information technology would improve *situation awareness* of own and enemy forces for *more timely and accurate decision making* and improve battlefield performance. Initially, situation awareness tended to be considered as synonymous with information acquisition, with relatively less effort devoted to understanding how commanders and teams would actually use the information to make decisions and to assessing the implications for doctrine, organizations, training, leadership, and soldiers. A focus on the technology rather than the human was due in part to the significant technical challenges that needed to be solved and in part to limited understanding of the importance of the interaction between the human and the computer. A lack of scientifically valid, relevant, and practical methods to assess human performance in battle command further interfered with efforts to examine the interdependencies among the human, the technology, and the mission. Very quickly, however, it became apparent that human performance issues had to be considered if information technology was to have the intended effect.

The impact on performance of an army research and development focus on the data-driven, information management aspects of battle command has been documented in numerous reviews of the army's digitization process (U.S. Army Audit Agency, 2001; DoD, 2000a,b; Grynovicki et al., 2001). Creating automated devices has generally been much easier than using them. New automation has often failed to produce expected gains because system designers treated the operator as just another switch or sensor. Regarding operators as mechanical components simplified system design but overlooked the active and highly complex nature of human information processing (Beck et al., 2002).

The Army Research Laboratory Human Research and Engineering Directorate (ARL HRED) in collaboration with the Army Research Institute (ARI) planned and executed the cognitive engineering of the digital battlefield science and technology objective (CE STO) to improve army battle command through integration of the human dimension in information system acquisition. This five-year research program was designed to close the gap that was growing between the army's ability to generate and distribute megabytes of data across the battlefield and the soldiers' ability to cognitively assimilate and translate these data into situation awareness for effective decision making or common ground for adaptable, distributed teamwork. This work was aimed at developing models of human performance to guide investments in information technology and methods and tools for assessing and improving command decision making and teamwork.

While the gap still exists, the concepts in decision making and teamwork defined during the course of the CE STO have promoted innovation in training and information system design, helping the army realize the potential of digitization. Specific areas of investigation included leader and team learning for adaptable performance; collaborative decision making by multinational and dispersed teams; visualization techniques for timely decision making in uncertain, rapidly evolving situations; and methods for assessing human performance in battle command and comparison of alternative systems, organizations, or procedures.

Defining features of the CE STO were reliance on a practice-centered approach to research using both theory and application to define and test solutions (Woods and Christoffersen, 2001); collaboration among researchers from the government, academia, and industry; and use of military experts to implement and validate proposed solutions. Methods of inquiry ranged from highly controlled laboratory experiments to field studies, with each area informing the next in a spiral process of iterative advances in science and application.

### 21.1.2 Tactical Decision Making Under Stress (TADMUS)

Several incidents led to the launching of the TADMUS program. First was the USS *Stark* incident. The USS *Stark* commander failed to identify an enemy ship allowing his vessel to be hit by two missiles from an Iraqi *Mirage* on May 17, 1987. Thirty-seven crew members lost their lives. Several months later, the USS *Samuel B. Roberts*, shortly after being commissioned, struck an Iranian mine. The ship was repaired and survived. The investigations of these two incidents pointed to problems in system displays and crew training.

One of the most salient examples of the need for TADMUS was the *Vincennes* incident mentioned above. Rear Admiral Fogarty (1988), who conducted the investigation and published the report on the incident, identified several contributing factors to the accident in his official report about the incident. The incident was ultimately attributed to human error, in general, by the official Fogarty report (Collyer and Malecki, 1998). The accident was blamed, in large part, on operator stress, task fixation, and unconscious distortion due to expectancy bias and scenario fulfillment. Klein (1989) and others expanded on the findings and said blaming human error was too simplistic. There were certain factors reported by Fogarty that could not be filed under the heading of simple human error. For example, inadequate displays led crew members to believe the aircraft was descending rather than ascending. Systems then available, such as the *AEGIS* (the Navy's most sophisticated battle management system at the time), were often ambiguous in identifying the position and intentions of aircraft.

The TADMUS objectives were to define what problems navy tactical teams face—from designing tactical decision-making performance measures to determining the effect of stress on tactical decision making and, lastly, to developing and then testing principles for training, decision support, displays, and simulation. TADMUS scientists accomplished this by applying notions from several different research areas, including human–computer interaction, human factors, and naturalistic decision making to the design of training, performance measurement, and decision support systems (see Cannon-Bowers and Salas, 1998).

### 21.1.3 Chapter Overview

The CE STO and TADMUS programs produced a large amount of data on human performance and information systems. Selected findings from these two research programs are summarized and used to illustrate

- human performance issues in information systems operations,
- human performance concepts and principles applicable to information systems, and
- guidelines and tools for information systems design.

## 21.2 HUMAN PERFORMANCE ISSUES

Table 21.1 lists eight of the most troublesome issues facing human systems integration (HSI) in information systems operations. Each of these issues will be described in this section.

### 21.2.1 Learning *How* to Think, Not *What* to Think

The U.S. Army is moving from a battlefield approach that emphasizes planning to a more flexible, execution-based focus. Operational transformation is being spurred by advances in information technology with the promise of nearly perfect situation awareness and by the increased breadth and complexity of missions military leaders and teams must be prepared to address (e.g., warfighting, combating terrorism, peacekeeping, humanitarian assistance). They are expected to rapidly assess, continually monitor, and appropriately adapt in evolving, ambiguous situations. A level of expertise not generally seen across the force structure will be required, and an innovative training approach is needed to help these men and women develop their thinking skills earlier and more thoroughly.

Learning how to think, not what to think, means practicing to be adaptive. Immersion in multiple, realistic, challenging, and cognitively complex situations and iteration, performance assessment, and scaffolding are key elements of an adaptive learning model developed by Ross et al. (1999). The adaptive learning model was applied and refined in a series of Army experiments (Lussier et al., 2000; Ross, 2000), and while the tenants of adaptive learning are theoretically sound and intuitively appealing, the process is not well

**TABLE 21.1 Human Performance Issues in Information Systems**

- Learning *how* to think, not *what* to think
- Leader mindsets constraining flexibility
- Difficulty managing uncertainty
- Degraded situation awareness
- Problems with team coordination
- Inadequate information filters
- Abuse, misuse, and disuse of automation
- Inadequate human performance assessments

supported by currently available training technologies or information systems. Efforts to use the adaptive learning model to design and evaluate training methods and tools and to define requirements for information systems have been initiated, but additional work is required to link the resulting methods, tools, and systems to adaptable operational performance.

### 21.2.2 Leader Mindsets Constraining Flexibility

In applying the adaptive learning model, the army CE STO study team worked with an active-duty army unit as they prepared for and then completed a one-year deployment as the sustainment force (SFOR), multinational division north [MND(N)] in B-H. The CE STO study team found a *warfighting mindset* to be a primary barrier preventing adequate preparation of the unit for peacekeeping prior to deployment (Klein and Pierce, 2001; Pierce and Klein, 2002; Pierce and Pomranky, 2001). The investigators noted that a warfighting mindset tended to limit diversity in team decision making, was uncomfortable with ambiguity, and did not promote training in the peacekeeping art of negotiations, persuasion, and influence (see Example 21.1).

Leader mindsets are also a problem with the selection of technology. Because of this, technological barriers were seen in the use of systems designed to support decision making in major theater-of-war conflicts, tracking enemy and friendly forces, and developing and executing warfighting solutions such as movement to contact or defense in sector. It is evident in our sociotechnical culture that technology and engineering thrusts are preferred over human-centered ones by the acquisition community. This "technology mindset" is not unique with information systems. Information systems, however, because of the need for rapid changes and complex human–machine and human–human communications, quickly reveal failures in total system performance due to poor designs for human performance.

**Example 21.1 SFOR Peacekeeping Activities** A primary mission of the army is to fight and win the nation's wars. Given this requirement, some within the army leadership have perceived peacekeeping as a detractor from that mission. Because of this mindset, time devoted to deployment preparation was strictly controlled in the deploying SFOR unit. In addition, personnel continued to rotate into and out of the unit less than 30 days before deployment with training events rarely including intact unit teams. Key team members from civil affairs, multinational forces, and the international community were grossly under-represented throughout the predeployment training cycle. The result was a lack of understanding of the roles and functions of team members in peacekeeping that was difficult to overcome once deployed.

Although SFOR unit actions were predominately nonlethal, designed to maintain the steady state to allow freedom of movement and support the nonviolent return of displaced persons to their prewar homes, training and operational planning continued to emphasize high threat events such as how to respond to criminal activities and violent demonstrations. One of the final training events, the mission rehearsal exercise was a complex series of interrelated events designed and orchestrated by the exercise developers to introduce unit personnel to a range of potential threats. Few opportunities to learn or plan for high probability functions and practice interpersonal skills such as negotiations, persuasion, and influence were provided. This focus reflected a force protection philosophy and an assumption that a warfighting mindset would assure the safety of the unit for the first 30 days of deployment and that the functions and tasks specific to peacekeeping could be learned on the job. Our observations in B-H of deployed forces and interviews with selected leaders and teams generally revealed that this philosophy was not fully supported, although most believed that preparation and planning to handle crisis events were required, but not to the exclusion of training for the more likely events. Further defining or measuring success, which was key to learning and commitment, was difficult for the unit once it had deployed. *The unit practiced what they were good at—using force to respond to threats.* The unit did not fully appreciate the novelty and ambiguity—the uncertainty—inherent in adapting warfighting skills to a peacekeeping mission until after deployment.

### 21.2.3 Difficulty Managing Uncertainty

Within the military, the rise in uncertainty is due in part to the loss of a defining peer threat in the breakup of the Soviet Union and the emerging more ill-defined asymmetric threats as well as the advances in technology, especially information technology and the data deluge that has resulted. The latter cause is due in part to a focus on information acquisition rather than technology to support decision makers in their interpretation and use of information. Often, in the case of uncertain environments, tactical leaders have tended to overplan. Overplanning involves trying to anticipate all the possible situations the unit is likely to face, defining appropriate responses in as much detail as possible, and maintaining resources to implement plans and contingency plans. [See Lipshitz and Strauss (1997) for a review of coping with uncertainty in the field.]

In the SFOR environment, many military leaders advocated a strategy of "tactical patience" to manage uncertainty. The notion was to delay action, or in the words of one highly experienced senior mentor, to "not rush to failure." In peacekeeping, situations were thought to develop more slowly than in warfighting. In observing the unit and interviewing key decision makers, the CE STO study team suggested that applying a strategy of tactical patience might have encouraged overplanning and interfered with unit adaptability.

Managing uncertainty by withholding decisions or actions is problematic to army operations on at least two levels. The first is the tendency for inaction to move the team from a proactive to a reactive stance, perhaps missing opportunities to influence or control situations. The second concern in applying a strategy of tactical patience is the false assumption that uncertainty can be reduced if the team waits long enough. This is not likely in highly ambiguous situations, where uncertainty exists not only in what is happening but also in how best to respond. If the task is cognitively complex or novel, there is no guarantee that the decision makers will know the best response (Beck, 1997). Tactical patience may have been due to poor preparation in peacekeeping skills and a lack of decision support systems.

As stated previously, information systems have been designed to primarily address the acquisition of information rather than the interpretation or use of that information in decision making or teamwork. Further, this has primarily been the acquisition of information for warfighting, not peacekeeping, and certainly not to enable adaptability. Procuring systems for information acquisition in warfighting provides only a small portion of the situation awareness requirements.

### 21.2.4 Degraded Situation Awareness

Situation awareness is defined as a global representation of the current and future situation. Decisions about what actions to be taken are byproducts of the situation awareness that precedes the selection of that action (Hutchins, 1996). The *USS Stark*, the *USS Vincennes*, and the Afghanistan friendly-fire incidents all took place when operators were under stress and in degraded situation awareness conditions. The *Stark* commander failed to identify an enemy; the *Vincennes* operator mistakenly detected an enemy action from a nonthreatening aircraft; and the controller gave a B-52 the coordinates for the friendly location rather than the enemy's target location under conditions where none of them was adequately aware of the true situation before making his decision. The fault for such degraded situation awareness may be any number of things—from faulty, too much, or too little information; lack of training; or poor design of displays—but when the decision maker is not fully aware of what is going on, the consequences are often tragic.

### 21.2.5 Problems with Team Coordination

Uncertainty and situation awareness were considered from the perspective of the individual; however, many performance failures are not the result of a lack of individual skills or machine failures, but, rather, are caused by the inability of team members (human or automated) to coordinate their actions (see Example 21.2). The empirical literature is filled with examples that illustrate the importance of team functions. For instance, Terborg et al. (1976) discovered that only 3 percent of the variance in the performance of land survey teams could be attributed to differences in individual skill levels. Similarly, Jones (1974) reported that measures of individual skills only accounted for 35 percent of the success of basketball teams. This leaves most of the variance to be explained by teamwork and other factors such as coaching.

**Example 21.2 Aircraft Disasters from Teamwork Breakdown** Some of the most graphic examples of the need to understand group process variables come from examinations of aircraft disasters. In 1978, a flight crew became preoccupied with a minor mechanical problem, allowing the airplane to run out of gas. The U.S. National Transportation Safety Board (NTSB, 1979) attributed the crash to a breakdown in teamwork. Poor teamwork may also have led to the crash of an Air Florida jet into the Potomac River bridge in 1982. As the plane awaited takeoff, the copilot repeatedly, but deferentially, reminded the captain of dangerous ice accumulation. The NTSB (1982) report implied that the disaster might have been averted if the copilot had more forcefully conveyed his misgivings to the captain. Analyses of multicrew aircraft accidents (Cooper et al., 1979; Foushee, 1984) have clearly established that communication and coordination failures contributed to most crashes.

### 21.2.6 Inadequate Information Filters

In the design and acquisition of information systems for military applications, problems of situation awareness especially within and between teams are highlighted in the lack of system interoperability. Interoperability includes the integration of information systems into networks with filters and communication strategies that either promote or hinder team performance. The team's approach to information filtering determines what data are transmitted and what data are not transmitted. Filtering systems are designed to provide decision makers with access to relevant information and prevent the passage of unimportant information. The process of setting filters demands an understanding of the relationship between data and performance (Beck, 1997). Any minimally effective filtering system must pass vital data and exclude the most extraneous data. Most filtering controversies concern the treatment of information that is neither critical nor extraneous. This large middle ground of information could be of value to decision makers but is of secondary importance. Permeable filters send both primary and secondary information to decision makers. Restrictive filters transmit only critical or primary information. The most appropriate filter, permeable or restrictive, will be determined by the situation. Filtering strategies that work well in one situation may have catastrophic effects on performance if the amount of data or the environment changes.

Social conditions within most organizations promote permeable filters. Most people want as much information as possible before making important decisions (Cialdini, 2000). There is also evidence that most people have an inflated view of their skill at processing information (see, e.g., Carver et al., 1980; Zuckerman, 1979). Thus, it is reasonable to hypothesize that many military commanders tend to overestimate their information processing capacities, often asking for more data than they can efficiently analyze (Beck, 1997). In observing a unit conducting their mission rehearsal exercise prior to deployment to B-H, the commander was observed to order his subordinates to tell him everything and he would decide what was and was not important. This order was in response to the commander's perception that a key piece of information was not passed and his recognition that the novelty and ambiguity inherent in a peacekeeping mission made defining his critical information requirements more difficult. He quickly became overloaded with data and had to reexamine his decision requirements!

### 21.2.7 Abuse, Misuse, and Disuse of Automation

Advances in information systems have increased the role of automation in teamwork. This evolution in user system interaction was recognized by Halpin (1984) and described in a three-stage process. In the first stage, users monitored systems, without an ability to interact with or control the data presented. Interactive computer systems led to the second phase, which was characterized by more user control of what data were displayed. The third stage heralded the introduction of the computer as an interactive partner or team member. Guidelines for human–system collaboration, however, have lagged behind the capability. For example, automation is not always the best option for improving system performance. Automation can improve efficiency, performance, and system productivity. Automation can also reduce workload and operational costs. However, people do some things better than machines. In some circumstances, human expert ability greatly surpasses that of automated systems. One such area is in adaptive performance. People, especially experts who have experience with a number of different situations, can display enormous flexibility (or adaptability) in performance.

Parasuraman et al. (2000) have developed a four-stage model to guide automation decisions in design. The model includes full automation to manual operations in information acquisition, information analysis, decision and action selection, and action implementation. Parasuraman and Riley (1997) refer to the automation of functions that should not be automated as abuse. Human operators can become *complacent* and rely too heavily on automated systems if they are *highly reliable but not perfect.* Complacency results when operators overtrust a system (Parasuraman and Riley, 1997). By not requiring human operators to be actively involved with a system, the operator may be convinced that the system will not fail, especially if it had not failed in the past. The Three-Mile Island nuclear power plant accident is an example of what can happen when operators misuse automation. Mistakes resulting from misuse or overreliance on automation are one of the causes of automation disuse or underutilization.

Dzindolet et al. (2001b) considered the Battlefield Combat Identification System (BCIS) acquisition (see Example 21.3) within the context of the literature on *automation reliance*. In a series of experiments, they examined the impact of system reliability on reliance and found that more reliable automation did not necessarily produce greater reliance or better performance by human–machine teams. The relationship between decision maker and decision aid was much more complex with cognitive biases toward and against automation and self-serving biases affecting reliance (Beck et al., 2002).

In subsequent work, the impact of trust and motivation on automation use decisions was added to define a comprehensive model of cognitive, social, and motivational influences on automation use decisions (Dzindolet et al., 2001a, 2002). Field reports have been collected indicating that trust in automation is affecting reliance, especially as battle intensity increases. Despite apparent acceptance of automation in training, in actual combat or during highly realistic battle rehearsals, there has been a tendency for soldiers to turn off their automation. They were not fighting as they were trained—their objectives had changed. In training, soldiers work to improve processes, often including integration of automation. In combat, priorities change. Automation that is hard to use or does not clearly provide an advantage to the individual over manual operations will be discarded (Beck, 1997).

**Example 21.3 Battlefield Combat Identification System** The BCIS was proposed to assist in vehicle-to-vehicle identification, and the Combat Identification for the Dismounted Soldier (CIDS) performed a similar function between individual soldiers. The concept was that a system able to identify other friendly systems would reduce fratricide. The BCIS provided the ability to "interrogate" a potential target by sending a microwave or laser signal that, if returned, identified the target as a "friend." Unanswered signals produced an "unknown" response. Early limited user tests of the BCIS indicated that it did indeed increase the ability of soldiers to identify friendly vehicles and reduce fratricide. However, the results were not supported by more realistic assessments in advanced warfighting experiments (Grynovicki et al., 2001) in which fratricide, even with the BCIS, continued to be a significant problem. Further, the literature on collaboration between humans and automation indicates that human operators do not appropriately rely on automated decision-making aids, depending, instead, on the situation, they may underutilize (disuse) or overly rely (misuse) on these aids (cf. Parasuraman and Riley, 1997; Dzindolet et al., 1999). In the case of the BCIS, receipt of an unknown response from an unanswered signal presents two potential dangers. The first is that the commander's dilemma will remain unresolved, slow his reaction to threat, and increase the chance of his own destruction. The other hazard is that soldiers, especially during battle, will be too quick to treat an unknown signal as an enemy. If that happens, the BCIS could increase fratricide.

## 21.2.8 Inadequate Human Performance Assessments

Assessing human performance in HSI requires new measurement approaches. In laboratory-based research, cognitive scientists typically measure such simple features of behavior as response time, accuracy of the response, or type of response. However, as cognitive engineers examine complex behavior in fields of practice, these simple measures may not accurately reflect the behavior. In other words, to study cognition in the wild (e.g., Hutchins, 1996), HSI investigators must first create and validate new methodological approaches. These new methods must represent the complex behavior in a way that accurately reflects the behavior and that also makes the data comprehensible and useful for making inferences about the cognitive processes underlying the behavior.

## 21.3 HUMAN PERFORMANCE CONCEPTS AND PRINCIPLES

Recognizing some of the major issues facing HSI in the operation of information systems, we now turn to those concepts and principles of human performance that show the most promise toward helping develop information system design guidelines. There are four major concepts derived from army and navy research-and-development (R&D) studies that can greatly aid in establishing how HSI can enhance information system design and use:

- adaptive performance,
- situation awareness,
- information presentation, and
- performance assessment.

Table 21.2 lists a number of theoretical and empirically based human performance principles. Categorized under the four concepts, 10 human performance principles are described in this section.[1]

### 21.3.1 Adaptive Performance

***What Is Adaptive Performance and Why Is It Important?*** Military leaders and teams must have the training and technology required to respond to the full spectrum of military missions. These missions range from warfighting to peacekeeping and humanitarian assistance, with the continuum characterized by complexity and ambiguity. Although there is a core set of military skills that are required across the spectrum, each point along the continuum requires specific knowledge, skills, and abilities (KSAs). An overarching skill is adaptability or the ability to rapidly assess the situation and make the right decisions based on that assessment, monitor the situation for changing requirements, and respond appropriately. In addition, advances in information technology continue to challenge adaptive decision makers and teams with more and more, often unprocessed, information received more quickly than ever before.

Decision making has and will likely remain primarily a human responsibility. With advances in sensor and information technologies, decision uncertainty will shift from knowing what is happening to understanding the situation and knowing what to do. The situations are too varied, and the variables are too many to prescribe behavior or to rely on automation. Information systems certainly may be designed to better support their use, but command decision making demands expert leaders and teams who expect and prepare to respond to change.

***P.1 Learning to Be Adaptive Requires Practice*** Learning to be adaptive is different from overlearning (Driskell and Johnston, 1998). Overlearning refers to skills

**TABLE 21.2 Human Performance Principles**

**Adaptive performance**
- P.1. Learning to be adaptive requires practice.
- P.2. Adaptability is a way of managing uncertainty.
- P.3. User and computer adaptability is a social activity.

**Situation awareness**
- P.4. Situation awareness is affected by time and operator involvement with the system.
- P.5. Shared situation awareness promotes team adaptability.

**Information presentation**
- P.6. Decision support systems should be designed to work within the constraints of cognitive processing capabilities.
- P.7. Individuals rely on heuristics to make decisions and the decisions contain biases.
- P.8. Method of communication used affects workload and performance.

**Performance assessment**
- P.9. Improvement in system performance requires knowing what to measure and how to measure human performance.
- P.10. Human performance must be assessed as part of system performance.

that were deliberately overtrained beyond initial proficiency often to reduce errors, especially under stress. Overlearned skills can actually lead to rigid responding, the antithesis of adaptability. Ross (2000) described the process of learning to be adaptive as one in which the learner is allowed to grapple with tough problems and to learn to appreciate ambiguity and disequilibrium as part of the learning process, receiving guidance only as necessary to move to a higher level of understanding. This type of guidance or feedback is given in the form of scaffolding, allowing the learner to move from one level of understanding to the next. Feedback enhances—but does not replace—the intellectual struggle. Researchers have proposed that individuals and teams can learn to be adaptive through this dynamic process of immersion, iteration, performance assessment, and scaffolding. Researchers across the services have begun development of training tools designed to promote adaptability.

***P.2 Adaptability Is a Way of Managing Uncertainty*** It is commonplace for organizations to assert that they want to encourage their teams to be adaptive as a way of managing uncertainty (Klein and Pierce, 2001). Uncertainty and the need for adaptability have always been a part of military operations. As described previously, however, information technology has changed what the military does and how it is done, increasing the need for adaptability. The choice has become not whether or not the team should be adaptive but rather how well or quickly the team will adapt.

***P.3 User and Computer Adaptability Is a Social Activity*** Cognition must be examined within the larger sociotechnical system in which it is embedded. Within this system, the smallest team is the user and the computer. Social psychology therefore may be applied to the acquisition of information systems. As examples, social psychology concepts that should be considered in information system acquisition include diffusion of responsibility, social loafing, social facilitation, collaborative debate, trust, and coordination (see, e.g., Beck and Pierce, 1996). Adaptation occurs in the context of transactions; adaptation and adaptive responses are dynamic, evolving, and dependent on the situation. In human–computer collaboration, adaptability is a requirement for both the human and the computer.

### 21.3.2 Situation Awareness

Decision making in naturalistic environments consists of "what is going on?" and "what do you do about it?" *"What is going on?" refers to situation assessment. "What do you do about it?" refers to actually choosing what action to take.* These decisions reflect situation awareness (Hutchins, 1996). Therefore, human performance researchers and designers must concern themselves with the process of situation assessment and awareness.

***P.4 Situation Awareness Is Affected by Time and Operator Involvement with the System*** *Time is important in establishing awareness of the current and future situation.* Past and present information is critical to establishing situational awareness in both teams and individuals (Hutchins et al., 1996). Users need the past to understand the present, and past and present events are used to predict the future. The user must also be aware of changes over time, which may tax working memory.

*Situation awareness* can lead to increased performance if the operator is involved with the system (Niessen et al., 1999). In designing systems, work schedules, and work tasks, vigilance and its demands on human operators must be taken into consideration. Human performance research demonstrates that human operators must be involved with the operation of the system, even with automation, in order to perform successfully and consistently. Situation assessment is based not only on anticipation of future events but also on the evaluation of further information processing requirements.

***P.5 Shared Situation Awareness Promotes Team Adaptability*** In interacting with the environment, others, and the artifacts of technology, people form internal, mental models of themselves and the things with which they are interacting. These models provide predictive and explanatory power for understanding the interaction. Mental models are important cognitive tools for human factors (Langan-Fox et al., 2000). An individual's mental model in large part determines how they behave and how they will react to novel, ambiguous situations. Teams have shared mental models that help them predict and coordinate their behavior. Shared mental models that include an understanding of the situation, the team's resources, and the other team members' roles and needs improve team effectiveness by permitting implicit coordination to occur (Entin and Serfaty, 1999). This is especially important during periods of high workload. One way to form adaptive teams is to ensure all systems and displays have common capabilities to increase the likelihood of a shared mental model. Although it is sometimes difficult to determine what is "common," if done successfully, teams or dependent individual operators are better able to visualize the mission as a whole, which leads to better and shared situational awareness. Goals, resources, plans, actions, and progress reports can be shared to improve shared knowledge and mental models.

### 21.3.3 Information Presentation

The information processing capabilities of humans should be a primary factor in designing displays. The way the data are presented (e.g., size, color, and organization) has an impact on human performance. If critical data are displayed outside of the operator's normal scan area, then they may not be perceived. If a display presents data across time or relies on the operator to track several pieces of data, then the operator's short-term memory capacity may be exceeded and key data may be forgotten. If systems are designed without taking these limitations into account, then operators may not be able to use the system. However, when human capabilities are folded into the design, positive outcomes such as increased accuracy, fewer errors, and less time needed in interaction with the system can result. Designing to support operator pattern recognition may reduce information processing and memory overload and may provide users with more time to make decisions about emerging problem situations.

***P.6 Decision Support Systems Should Be Designed to Work within Constraints of Cognitive Processing Capabilities*** Decision support systems may reduce cognitive workload of the user by reducing the amount of information processing that must be done, by reducing working memory requirements, and by assisting users by properly allocating the limited cognitive resources that they have available. Unfortunately, however, decision support systems are often created to provide operators too much data and not enough information. Operators are often given data that are raw and must be processed and evaluated in order to be helpful (Morrison and Moore, 1999). The extra processing required can increase workload and distract the user from the task at hand. Information must be presented in a way that is appropriate to the given situation and not add to the workload of the user.

Human performance can be improved by providing decision support that does not require dependence on previous contact data and a vast amount of information obtained in past training and experience (Hutchins et al., 1996). By providing users with more information in order to support their decision making, users with limited attentional capacity, often found in complex situations, have the tools to perform better. This enables users to rely on recognition rather than recall, which has traditionally been found to be better and less demanding.

***P.7 Individuals Rely on Heuristics to Make Decisions and Decisions Contain Biases*** Diagnoses begin as an initial provisional hypothesis is formed and more evidence is sought to either confirm or disprove it. We use heuristics to help us find the "true" state. Heuristics help save cognitive resources and enable human operators to make rapid decisions. However, heuristics may contain biases and direct us away from the true state.

In the work related to the BCIS described previously, these biases were further described as appraisal and action biases. Appraisal biases occurred when operators misjudged their own competence relative to an automated alternative. People misused automation aids when the perceived utility of the aid was overestimated and disused aids when the perceived utility was underestimated. Self-serving biases, an illusion of control of chance events, and the availability heuristic have all been used to explain disuse. The bias toward action was observed in operators who accurately assessed the reliability of both team members but chose to ignore the automated aid even though probability of success would be less. The result was nonrational automation use decisions (Beck et al., 2002). There is some evidence that expertise may encourage disuse (Sheridan et al., 1983). Experts believe they have less need for decision aids and so ignore potentially useful information. Analyses and knowledge of heuristics and biases can help direct training, procedural guidelines, and design guidelines.

### P.8 Method of Communication Used Affects Workload and Performance

*There is an optimal amount of perceptual information that can be attended to. Perception of information is optimum when enough is presented to stimulate attention but not so much as to cause cognitive overload.* Users should not be required to take in too much perceptual information at once. If the human capacity for perceptual information is exceeded, the result can be a decrement in performance. Perceptual load also affects haptic tasks. Excessive amounts of visual, auditory, or tactile information can lead to failure to process most, if not all, of the information provided.

When an individual is required to do too many tasks at once or when task demand is too high, there is high workload. Performance declines during periods of very low and high workload (Beck, 1997). Performance is optimal with some, but not too much workload. This principle is based upon the idea that task performance is based on the amount of resources available to dedicate to the task.

When a person is aroused, the manner in which attention is distributed depends on what attentional resources are available. This principle is based on the Yerkes–Dodson law (Yerkes and Dodson, 1908) and the inverted-U shape of the arousal function. Systems that must be operated during times of increased arousal must compensate for decreased performance by the user. It is accurate to say that a little arousal is a good thing, but too much can cause a decline in performance.

The function describing the actual relationship between the amount of information received and the value of that information is also curvilinear (Beck, 1997). At first, *increments in the quantity of information received leads to better performance, until an inflection point is reached. Thereafter, further increases in the amount of information causes a decrease in team performance.* Deterioration in the performance of either an individual or team resulting from excessive information input is referred to as cognitive overload. The old premise that more information is better has proven false. The quantity of data that can be processed without exceeding the inflection point will depend on a variety of factors, including the size and skill of the team and the communication links between teammates.

Different users or operators require different levels of detail, fidelity, and depth of information. Some operators need overview data that are concise and provides them with the essential but general information. Some only need recommendations for action. Others need more detailed discussions of issues in order to operate effectively. Some operators require orders and directions. A user's duties and standing within the team and the organization often determine what type of information is needed. Information presented so that the viewer can rely on pattern recognition rather than recall has traditionally been found to be better and less demanding.

The method of communication used affects performance. Some multimedia presentations of information result in better performance of human operators on memory tasks (Najjar, 1998). Verbal information leads to better memory of small amounts of information. Text appears to be better for remembering information for longer periods of time. However, if an operator's visual channel is already occupied, verbal information is better.

*Graphic presentations reduce the mental computation required of users to complete a task.* The proverbial expression "a picture is worth a thousand words" is certainly true for human processing of display information. Users have been found to easily understand "cartoon" graphics (Moore and Averett, 1999), and graphic presentation has been found to be preferable to text-based presentations. One reason is that graphic presentations allow users to skip mental steps when performing a task. Instead of having to visually compare mental images, users can use the graphic on the screen to make their decision. The image is already generated for them. Users can use the perceptual processes that are less demanding by looking at the screen. This saves the user cognitive resources to devote to more logical operations.

*Color-coding conventions reduce errors in communications.* Most systems offer users definable color settings. However, if each system within a network has different coding conventions users may not be able to read, see, or understand each other's displays (Moore and Averett, 1999). This failure between users can lead to many errors because of missed or misunderstood communication of information. Research recommends following human factors guidelines regarding colors and using common color settings across systems, thus enabling free and open communication of ideas. Based on that same idea, researchers have found that operators need to share a common picture across systems and units (Moore and Averett, 1999).

### 21.3.4 Performance Assessment

Adequate assessments of human performance need to be based on reliable, valid measures of performance. In the CE STO research program to improve command decision making and teamwork, the study team worked under the premise that "you could not improve what you could not understand and you could not understand what you could not measure." Two principles of performance assessment were found most beneficial to information systems assessments.

***P.9 Improvement in System Performance Requires Knowing What to Measure and How to Measure Human Performance*** The CE STO study team first observed warfighting exercises to assess who was making what decisions and how and then went on to conduct critical decision interviews and to describe battle command processes based on their observations and interviews. Their descriptions were operationalized in human performance models of battle command (Knapp et al., 1997a–c; Middlebrooks et al., 1999; Plott, 1999; Wojcik and Plott, 2001). These models included a depiction of the functions and tasks performed, the flow of information, and the decisions made by personnel within a command organization using various types of information systems and under different environmental conditions. Model outcomes included workload, information bottlenecks, and decision quality. This process is being refined to explore human-system performance in future concepts.

***P.10 Human Performance Must Be Assessed as Part of System Performance*** Advances in simulation systems have enabled the use of synthetic task environments to assess human performance as a part of system performance. Human performance metrics are being developed to focus on the human system within the larger mission and ensure that neither the human interface nor the technical capabilities of the system drive acquisition decisions; rather it is the interaction among the soldier, the system, and the mission—the system of systems—that is considered.

## 21.4 GUIDELINES AND TOOLS FOR SYSTEM DESIGNERS

With an understanding of the major issues, concepts, and principles of human–system performance in information systems, we can now focus on recommended guidelines and tools for information systems acquisition and design. Organized under the four categories below, these guidelines and tools are summarized in Table 21.3

- acquisition process decision recommendations;
- tactical decision making—aiding and support;
- adaptability design for training and operations; and
- performance assessment.

### 21.4.1 Acquisition Process Decision Recommendations

For cognitive engineering to have the greatest impact on information system acquisition, it must be integrated throughout the acquisition cycle. As described in other chapters, this includes such things as describing the user during requirements determination and determining automation and human–system interface needs. We add the need to design for adaptability as a top-level recommendation. The acquisition of information systems requires a sociotechnical approach in which the information system is considered a team member within a system of systems. This approach supports the use of HSI to assure the success of information systems in meeting the expectations of the designer and the requirements of the user.

***G.1 Describe the Users*** Understanding the intended users is the first step in system design. Factors such as level of expertise, culture, cognitive ability, and personality should be considered when designing systems. User profiles provide information about user capabilities and limitations. Cognitive task analyses (CTAs), help determine what the user needs from the system to ensure that the system supports the major task sequences. Other important information needed to describe the user is education and reading level and the experience they have in performing the job.

*T.1 Cognitive Task Analyses* The CTAs are conducted to understand and define the cognitive requirements of individuals and teams on jobs and tasks (see Schraagen et al., 2000; Vincente, 1999). This is done mainly through interviews and observations. One specific way to conduct CTAs is through a *knowledge audit*, which uses structured interviews, often in connection with other CTA methods (Klein and Militello, 2001). Related to CTA are manpower, personnel, and training (MPT) trade-off analyses, analyses that may also be used to describe the user. (See Chapters 8 and 11 for MPT trade-off analyses.)

***G.2 Determine Automation Needs*** It is critical that automation should not be utilized without carefully considering its drawbacks. *Automation should not be used*

**TABLE 21.3 HSI Information System Guidelines and Tools**

| Guidelines | Tools |
|---|---|
| *I. Acquisition Process Decision Recommendations* | |
| G.1 Describe the users | T.1 Cognitive task analyses (Chapters 10, 20); MPT trade-off analyses (Chapters 8, 11) |
| G.2 Determine automation needs | T.2 Function and allocation analyses (Chapters 10, 13, 20) |
| G.3 Determine human–system interface needs | T.3 Task and workload analyses (Chapters 10, 13, 20) |
| | T.4 Breakdown analyses |
| G.4 Design for adaptability | |
| *II. Tactical Decision Making—Aiding and Support* | |
| G.5 Design for how users view tasks | |
| G.6 Design automation to improve team performance | T.5 Automation use decision model |
| G.7 Design a common shared picture for teams | |
| G.8 Display states of uncertainty | T.6 Advanced tactical architecture for combat knowledge systems (ATACKS) |
| G.9 Rely on experts in special circumstances | |
| G.10 Design displays to present tailored information and support operator pattern recognition | |
| G.11 Design displays to provide feedback and active practice | |
| *III. Adaptability Design for Training and Operations* | |
| G.12 Apply adaptive learning design requirements. | T.7 Advanced cognitive understanding simulation (ACUSIM) |
| | T.8 Think like a commander (TLAC) |
| | T.9 Simulations for adaptability (SFOR Adapt) |
| G.13 Utilize naturalistic decision-making concepts | T.10 Embedded decision and team performance measures |
| G.14 Apply team training Strategies | T.11 Team dimensional training (TDT) |
| | T.12 Team adaptation and coordination training (TACT) |
| | T.13 Stress exposure training (SET) |
| | T.14 Stress inoculation training (SIT) |
| *IV. Performance Assessment* | |
| G.15 Apply human performance models | T.15 Command, control, and communications tactically reliable assessment of combat environments (C3 TRACE) (see also Chapter 11, IMPRINT) |

**TABLE 21.3** ***(Continued)***

| Guidelines | Tools |
|---|---|
| G.16 Utilize simulation-based training and assessment methods | |
| G.17 Measure individual and team processes | T.16 The behavioral observation booklet (BOB) |
| | T.17 Anti–air warfare team performance index (ATPI) |
| | T.18 Anti–air teamwork observation measure (ATOM) |

*simply because it can* (Wickens and Hollands, 2000). Human factors research has shown that completely automated systems often leave the user with nothing to do except to monitor the system, which may lead to decrements in attention and reduced operator vigilance in monitoring problem situations. Operators may get lost within modes (Sarter and Woods, 1995) or overly trust the automation and assume that problem situations are being handled by the automation (Parasuraman and Riley, 1997). Users may not be able to problem solve or troubleshoot if they do not understand where they are in the system, or if a function has been automated, users' skills may deteriorate and they may not be able to take manual control back from the system. Generally, performance is better if operators play an active role in using the system, with automation supporting their needs, rather than operating a highly automated system.

*T.2 Function and Allocation Analyses* Chapters 10, 13, and 20 describe function and allocation analyses that are useful in determining system automation needs.

***G.3 Determine Human–System Interface Needs*** Advances in information system technology have increased the likelihood that individuals and teams will rely on technology as if it were another team member (Halpin, 1984; Noah and Halpin, 1986). Without an understanding of how teams operate, information systems may be designed to be bad team members allowing breakdowns in performance to occur. A breakdown occurs when the user becomes aware of the system, rather than being able to focus on the task at hand (Scrivener et al., 1993). Questions such as "What is that thing doing now or why is it doing that?" exemplify the breakdown.

*T.3 Task and Workload Analyses* Chapters 10, 13, and 20 describe task and workload analyses that are useful in determining human–system interface needs.

*T.4 Breakdown Analyses* Breakdown analyses are geared to identify, diagnose, and remedy breakdowns between user and task, user and tool, user and environment, and user and user. In addition to the user and tool breakdowns, the communication breakdowns Scrivener et al. (1993) proposed as most likely between user and user can be applied to better understand user and information system interactions.

***G.4 Design for Adaptability*** A top-level recommendation for information systems is to design for adaptability. The system design should be adjustable for different users' needs, including expert, novice, and any other users with different capabilities and needs who are intended to use the system. *Adaptable systems are usable by both novices and experts.* Novices and experts have different mental models, experience, and access to information on which to base their decisions (Kalyuga et al., 1998; Schlechter et al., 1998). Further, individuals can be novices in one area and experts in another. However, displays and system features can help novices perform more comparably with experts. By providing novice users with additional (text) information, they will have access to information similar to that held by the experts. In addition, training and tasks can be designed to reflect the level of expertise of the operator. A proper adaptive design can ensure all users' needs are met. Further, when the system is adaptive to all user needs, each user will have a more accurate mental model and more accurate expectations of system performance (Duncan et al., 1996).

### 21.4.2 Tactical Decision Making—Aiding and Support

Decision making involves the process of detecting a pattern of cues in a situation, assessing what is going on, choosing a response for dealing with it, and determining the degree that the response successfully dealt with the situation. Further, the consequences of decisions in tactical situations are great, and the decision-making process is often performed under time stress. Ignoring the information requirements of operators or how they need that information to be presented can be catastrophic. Users read displays to gather information in order to make decisions. By understanding the user's task, the designer ensures that systems are built to support operators in their decision making. Specifically, HSI drives what and how information is displayed, based on the criticality and frequency that the operator needs information. Decision makers must be able to access the information they need, when they need it, where they need it, and in the form they need it to make effective decisions (Morrison et al., 2000).

***G.5 Design for How Users View Tasks*** System design should be based on how users view tasks and how they learn, explore, navigate, and predict future system states (Morrison et al., 2000). When the user's mental model is incorporated into system design, the result is more accurate expectations and greater understanding of what the system is doing at various points in time.

***G.6 Design Automation to Improve Team Performance*** Automated decision aids must be designed as collaborative systems, and the impact of automation on other team members must be considered. For example, the high incidence of fratricide during Operation Desert Storm led Secretary of Defense William Perry to charge the services to enhance their capabilities to distinguish friendly, enemy, neutral, and noncombatant (Doton, 1996). The army responded to the Secretary's challenge by developing combat identification devices to identify friendly vehicles and individual soldiers. However, as described previously, in operation these devices did not deliver the expected gain in performance. The system was not designed to consider the biases of the human decision maker.

*T.5 Automation Use Decision Model* A *comprehensive model of cognitive, social, and motivational influences on automation use decisions* (Dzindolet et al., 2001a, 2002) grew out of the research on automation reliance. This model has been applied in the laboratory and has provided evidence that *training decision makers to better understand when aids were likely to make a mistake encourages more appropriate automation use* (Dzindolet et al., 2000). St. John et al. (2000) propose a "trust-but-verify" strategy with a two-stage model of conditional trust and qualitative verification. Conditional trust involved knowing when the automation should be trusted, and qualitative verification uses the automation as a guide or even an input to manual decision making. Findings on automation reliance must be validated in the field as a next step. While laboratory findings are consistent with anecdotes from the field, the model needs to be tested under stressful, battlefield conditions to determine how stress will interact with automation use.

***G.7 Design a Common Shared Picture for Teams*** In tasks that require teamwork, it is important that the system support team performance. To accomplish this, the system should support the team members' ability to monitor each other and to create a common picture across the team—a shared mental model (Salas et al., 1997). This picture can then help team members see the same problem situations, so that all team members have a common understanding. Further, it may facilitate the ability to predict what other team members should be doing at different points in time. In large, distributed, tactical teams, it is even more difficult to monitor what other units and team members are doing. Common displays that compile relevant information can increase the chances that all team members are seeing a common problem situation or share a team mental model. A good team mental model may enable team members to anticipate the information needs of other teammates and engage in implicit coordination by "pushing" information rather than waiting for it to be "pulled" via time-consuming requests (Entin and Seraty, 1999). Implicit coordination occurs when team members have an awareness of the roles and functions of other team members to include their information needs and are able to predict and respond in a timely manner to those needs. Common displays enable a shared team mental model and implicit coordination.

***G.8 Display States of Uncertainty*** Various presentation methods have been considered with an objective to reduce or at least understand the impact of *cognitive biases on decision making* (Barnes et al., 2000b). The ultimate purpose of decision aids or visualization techniques is to increase the commander's ability to understand the battle dynamics, consider options, make decisions, and predict outcomes. As evolving systems become more sophisticated, the display of states of uncertainty and the concomitant cognitive biases will require innovative cognitive engineering solutions. Findings from this preliminary work are being used to design aids to enhance situation awareness in nonconventional situations such as B-H, Kosovo, and Afghanistan with a goal of increasing understanding of historical trends, political changes, ethnic conceptions, and changing perceptions of various combatant and noncombatant groups (Zacharias and Hudlicka, 2001).

*T.6 Advanced Tactical Architecture for Combat Knowledge System (ATACKS)* The ATACKS provides a simulation environment to evaluate new visualization concepts for commander and staff decision-making. The ATACKS operates on a standard personal computer and is composed of visualization tools and decision support drivers (Suantak et al., 2001). The visualization tools portray three-dimensional standard and nonconventional military symbols, important terrain and urban features, and realistic animated behaviors for the objects depicted. Decision support has been demonstrated using genetic algorithms as well as more conventional rule-based algorithms. The ATACKS is applicable to both a major theater of war and support and stability operation (SASO) environments and is being used to evaluate visualization and decision-aiding tools for the Army Future Combat System (FCS).

***G.9 Rely on Experts in Special Circumstances*** Overall, though, it is important to remember that since all possible circumstances cannot be anticipated, even with the advantage of decision support, the *expert's abilities* and intuition are indispensable (Hutchins et al., 1996). In some circumstances, human expert ability surpasses that of technology or automated systems. One such example is flexibility, especially for experts who have experience with a number of different situations. Automated systems simply do not have the capabilities to exhibit adaptability, creativity, and commonsense knowledge that are an important part of human performance. Furthermore, human operators are much more able to incorporate experience online, use analogical reasoning, and maintain a broader focus than machines. Experts are even more adept at incorporating these capabilities into their performance.

***G.10 Design Displays to Present Tailored Information and Support Operator Pattern Recognition*** After designers understand what information is required, the next step is to ensure that it is presented in a way that can be perceived and remembered. Displays should illustrate data in a way that operators can use to make decisions. The data that are displayed should be useful, in that raw data should be analyzed and condensed so that patterns in the data are clear and calculations and analysis are minimized. Performance will be negatively impacted if the display does not match operator needs, information processing capabilities, and expectations.

A good display design provides relevant information tailored to the situation rather than data that require interpretation (Morrison and Moore, 1999). Operators are often given raw data that must be processed and evaluated in order to be helpful in decision making. The extra processing required can increase workload and distract the user from the task at hand. Information must be presented in a way that is appropriate to the given situation and not add to the workload of the user.

In addition, system design should support user tasks and capabilities. In many cases, users may not be able to glean pertinent information from a large number of system inputs and outputs. Designing to support operator pattern recognition may reduce information processing and memory overload and may provide users with more time to make decisions about emerging problem situations (Klein, 1997).

***G.11 Design Displays to Provide Feedback and Active Practice*** Embedded system features should include *active practice and feedback.* Practice is essential for effective learning, but other tools, such as feedback, are also necessary (Johnston et al., 1997; Ross et al., 1999). Feedback should be provided to ensure human operators and teams understand their performance to avoid repeating mistakes and errors. Feedback should be given relatively quickly after an action to ensure it is understood and applied. Feedback, when used effectively, is also good for such affective components as morale and efficacy. Systems and training should include a feedback component for human operators and teams. Feedback and active practice encourage adaptable performance.

### 21.4.3 Adaptability Design for Training and Operations

Information on adaptive learning, naturalistic decision making, and team training are useful in developing design requirements for adaptive systems. The following guidelines and tools are recommended for adaptability design for training and operations of future information systems.

***G.12 Apply Adaptive Learning Design Requirements*** Practicing to be adaptive in multiple, realistic, challenging, and cognitively complex situations enhances learning to think adaptively. To meet the challenge, adaptive learning models have been developed (see Ross, 2000; Ross et al., 1999; Lussier et al., 2000).

The multiple training tools developed under the CE STO were based on the same underlying model of learning and are complementary, with the primary difference being the mission focus, which varied across the spectrum from warfighting to peacekeeping and humanitarian assistance. Using a student-centered process, learners improve through sustained exploration and practice geared to their unique requirements. The following simulation tools for adaptive learning were developed under the CE STO.

*T.7 Advanced Cognitive Understanding Simulation (ACUSIM)* A training tool for battle staffs, ACUSIM provides a forum for staff officers and staff teams to plan and execute battle operations with embedded performance tips, available online coaching, and faster than real-time implementation (Ross, 2000).

*T.8 Think Like Commander (TLAC)* The Army Research Institute developed TLAC to promote deliberate practice of thinking skills required for command in warfighting and responding to threatening situations that could be encountered in small-scale contingencies and peace enforcement (Lussier et al., 1997).

*T.9 Simulations for Adaptability (SFOR Adapt)* This is a program of instruction that includes two automated tools, facilitation guides, and performance measures. The automated tools are PC-based and Web-hosted. It was developed to prepare soldiers for SASO, a general and inclusive term for operations other than war, including peace enforcement, peacekeeping, and humanitarian assistance. It is characterized by cooperation between military forces and civilian agencies working together to establish or promote regional stability (Pierce and Pomranky, 2001).

***G.13 Utilize Naturalistic Decision-Making Concepts*** Naturalistic decision-making concepts, such as recognition-primed decision making and explanation-based reasoning should be considered in training and information system design to improve decision-making performance, especially in novel situations.

*Recognition-primed decision making* assumes that people form a new but tentative representation when confronted with a novel situation. The representation or hypothesis is based on past experiences that seem to be similar to the present situation. This representation contains observed situation data and is the basis for future expectations about what will happen. Incoming data can confirm the representation and enforce people's observations. If incoming data conflict with the current representation, additional data are gathered to refine or dispel it.

*Explanation-based reasoning* is another, less common form of decision making that involves the selection and evaluation of plausible hypotheses. This decision-making process is often employed when the situation is novel and ambiguous. As in recognition-primed decision making, explanation-based reasoning is quick, concise, and done online. Both strategies are improved by team training in realistic situations.

*T.10 Embedded Decision and Team Performance Measures* Measurement of team processes and performance is important for providing team members with feedback and represents a critical component of the adaptive learning model and event-based practice. *Event-based practice* is an approach to training in which practice is embedded within the task environment, and training objectives, exercise design, performance measurement, and feedback are linked (Cannon-Bowers and Salas, 1997; Dwyer et al., 1997, 1999; Johnston et al., 1997). A scenario is designed with cues and events that cause individual and team tasks to be performed. Measures of individual and team performance are derived from these events and embedded within the scenario.

The TADMUS and CE STO researchers have used embedded decision and team performance measures to examine situation awareness and team adaptability. Performance is improved by providing teams an opportunity to practice in various situations and environments and learn skills that increase the team's ability to confront novel situations. Decision and team performance measures may be embedded in training tools or information systems. An event-based approach to training with embedded decision and team performance measures facilitates the development of naturalistic decision-making skills and expertise.

***G.14 Apply Team Training Strategies*** A number of training strategies are currently available:

- team dimensional training (TDT),
- team adaptation and coordination training (TACT),
- stress exposure training (SET), and
- stress inoculation training (SIT).

*T.11 Team Dimensional Training* This involves the teaching of teamwork skills and knowledge through guided self-correction (Smith-Jenstch et al., 1998). In guided team self-correction, a facilitator works to keep discussions focused, keep the climate positive, facilitate active participation through encouragement, facilitate self-correction through modeling effective feedback between team members, and provide instructions on how to give constructive, useful feedback. This facilitator is provided with a debriefing based on the focus of that particular self-correction exercise that outlines specific questions that can be asked to encourage useful team discussions. Self-correction gives team members a guide to how they should interact and what topics should be discussed while enforcing shared knowledge structures and mental models.

*T.12 Team Adaptation and Coordination Training* This strategy was developed to enhance a team's ability to adapt their coordination strategies to changes in workload and stress. It uses the intermediate feedback loop from the theoretical framework for team adaptation (Serfaty et al., 1998), which contains adaptive coordination skills. It entails the training of adaptive skills, coordination skills, and team exercise of shared situation assessment procedures that apply to the appropriate situations. In research, TACT was found to alter communication patterns according to the training strategies used. Teams who were trained with TACT have exhibited implicit coordination (i.e., team members' ability to maintain performance even under high stress and workload when team members cannot really communicate) (Entin and Serfaty, 1999) and more adaptability (Cannon-Bowers and Salas, 1998).

*T.13 Stress Exposure Training* This is an integrated model of stress training comprising three stages of training: (1) information is provided regarding stress and its effects, (2) behavioral and cognitive skills are acquired, and (3) skills are demonstrated in an environment that approximates the real-world setting (Driskell and Johnston, 1998). It was influenced by a cognitive-behavioral approach to stress training.

*T.14 Stress Inoculation Training* This is designed to provide coping skills after stressful situations. It emerged from research regarding cognitive and affective research on cognitive behavior modeling (Meichenbaum, 1996). It attempts to increase familiarity with the environment and boost confidence in a team's ability to learn.

### 21.4.4 Performance Assessment

Models and simulations provide a forum for the assessment of human performance in concept evaluation and training. Using models and simulations, system designers have a unique opportunity to base HSI decisions on human performance data that approximate performance in the "real world" but allow for more control of implementation and environmental conditions. Based on task analysis and performance data, designers can uncover and correct errors early in system acquisition. Performance assessment can provide designers information to improve HSI and operators with tools to improve performance for future interactions with the system. The importance of models and simulations to HSI and training is captured in the notion that you cannot train what you cannot understand and you cannot understand what you cannot model.

***G.15 Apply Human Performance Models*** Task network models provide a method to efficiently conduct "what if" analyses to further concept exploration and derive the most valuable alternatives for assessment in the more costly, less well-controlled synthetic environment. Applications include the development of task network models to examine the impact of digitization on brigade command and control (Knapp et al., 1997a–c) and command and control of fires and effects (Plott, 1999; Wojcik and Plott, 1999). Building on this work, task network models are being developed to define requirements for both the future combat system and the unmanned combat, armed rotorcraft. Human performance modeling technologies can be employed to understand how to design *decision support systems* to optimize information flow and human performance (see Zachary et al., 2001).

*T.15 Command, Control, and Communications Tactically Reliable Assessment of Combat Environments (C3 TRACE)* This builds on other human performance models developed by the Army Research Laboratory [e.g., Improved Performance Research

Integration (IMPRINT)]. It allows a user to evaluate and propose high-payoff operational and system architectures for use in human-in-the-loop experiments. Human performance data captured during the experiments are then used to refine the model for use in an interactive cycle of model–test–model. The C3 TRACE has been refined through several applications to include an interface that supports what-if-type analyses by the user community. It may also be used to highlight potential workload or performance challenges to drive changes in system or organization design or training.

***G.16 Utilize Simulation-Based Training and Assessment Methods*** When using simulations for training, operators are often provided unstructured practice supported by somewhat general after action reviews that are separated in time from task performance. This method of practice does not provide detailed information about common errors or tasks that are taking up unreasonable amounts of time. Timely, relevant feedback is a critical aspect of human performance that is often overlooked by system designers. Both training theory and empirical testing have supported the value of providing performance feedback to operators (Cannon-Bowers and Salas, 1997).

The HSI tools listed in Section 21.4.3 for adaptability design for training and operations are unlike most simulation techniques. For example, these tools provide a cost-efficient platform that individuals and teams can use to practice decision making and teamwork to master the advanced learning stage for adaptable performance. The products allow learners to practice in cognitively complex and immersive, synthetic task environments with process- and outcome-based performance feedback available throughout the process and tailored to the needs of the learner. The products are usable by colocated or distributed teams.

***G.17 Measure Individual and Team Processes*** Several measures are available to evaluate individual and team processes. Three in particular are

- the behavioral observation booklet (BOB),
- anti–air warfare performance index (ATPI), and
- anti–air teamwork observation measure (ATOM).

*T.16 Behavioral Observation Booklet* This is used to *measure individual processes within teams* in order to evaluate how team members perform on task-specific jobs (Hall et al., 1993). It is formatted like a critical incident report, with events and expected actions listed. An observer marks whether or not the expected action was taken by the individual team members.

*T.17 Anti–Air Warfare Team Performance Index* This provides a measure of outcomes of teams using an anchored rating scale. Raters evaluate outcomes based on a scale of 0 to 3 (Zachary et al., 1991).

*T.18 Anti–Air Teamwork Observation Measure* In ATOM, 11 teamwork behaviors are categorized under 4 major dimensions. Raters record detailed information regarding team performance during a scenario using an ATOM work sheet that contains a time line with prescripted events. For a final score, raters give each event a rating based on their detailed notes (Johnston et al., 1997).

## 21.5 CONCLUSION

The primary conclusion that may be drawn from this chapter is that processes for acquiring information systems must be revised to capitalize on rapid advances in automation and to meet emerging requirements, especially in the area of adaptable performance. Examples were presented that demonstrated both the weakness of the current approach in information system acquisition and the importance of an approach that considers the interaction among the human, the technology, and the mission, especially when designing systems for use in highly uncertain, dynamic, and information-rich environments. Vincente (2002) stated that adaptation to change and novelty will be the primary role of the human in system performance as more "routine activities that are well understood, and which can be reduced to a set of rules or an algorithm are increasingly automated with computer technology" (p. 62). The implications of this evolution in automation to the design of information systems have been highlighted throughout this chapter.

We reviewed human performance issues in information systems operations, defined human performance concepts and principles applicable to the design of information systems, and derived or identified guidelines and tools to improve information systems design. The result of our review was a framework to move from automating the routine to designing for adaptable system performance. Based on this framework, we began the work of defining methods and identifying technology to aid in the process. The framework evolved during two R&D programs in which theory was applied to training and system design to improve battle command decision making and teamwork using research methods that ranged from the laboratory to field experiments but emphasized the importance of a multimethod, iterative approach.

Based on this work, we know with some confidence how to model and measure battle command performance (see Grynovicki et al., 2001; Knapp et al., 1997a–c; Middlebrooks et al., 1999; Plott, 1999; Wojcik and Plott, 2001), to design and deliver training (see Ross et al., 1999; Salas and Cannon-Bowers, 2001), to optimize human–computer interaction (see Dzindolet et al., 2002; Parasuraman and Riley, 1997; Parasuraman et al., 2000), and to use visualization aids to increase situation awareness and improve decision making (see Barnes et al., 2000a,b). Our challenge now is to use and expand the framework and our knowledge of human performance, to test and refine proposed tools, or to develop tools where none are indicated and to use the framework in collaboration with users and technologists to design and field the next generation of information systems promoting adaptable system performance.

## NOTE

1. The 10 human performance principles discussed in this chapter should not be confused with the 10 HSI principles of Chapter 1.

## REFERENCES


Barnes, M. J., Pierce, L. G., Wickens, C. D., Dzindolet, M., and Rozenblit, J. (2000a). Human Performance Issues in Battlefield Visualization. In *Proceedings of the 22nd Army Science Conference* (pp. 795–802). Baltimore, MD: Assistant Secretary of the Army for Acquisitions, Logistics, and Technology.

Barnes, M. J., Wickens, C. D., and Smith, M. (2000b). Visualizing Uncertainty in an Automated National Missile Defense Simulation Environment. In *Proceedings of the 4th Annual FedLab Symposium: Advanced Displays and Interactive Displays*, (pp. 117–122). Adelphi, MD: U.S. Army Research Laboratory.

Beck, H. P. (1997). Application of Group Processes to the C2 Information Flow and Workload Model. Unpublished manuscript.

Beck, H. P., Dzindolet, M. T., and Pierce, L. G. (2002) Applying a Decision-Making Model to Understand Misuse, Disuse, and Appropriate Automation Use. In E. Salas (Ed.), *Advances in Human Performance and Cognitive Engineering Research*, Vol. 2. Greenwich, CT: JAI Press.

Beck, H. P., and Pierce, L. G. (1996). *The Impact of Selected Group Processes on the Coordination and Motivation of Army Teams*, ARL-CR- 292. Aberdeen Proving Ground, MD: U.S. Army Research Laboratory.

Becker, J. J. (February, 2002). Operational Concept Found: Rapid Decisive Operations as a Joint Operational Concept. *Army*, pp. 49–57.

Cannon-Bowers, J. A., and Salas, E. (1997). Teamwork Competencies: The Intersection of Team Member Knowledge, Skills, and Attitudes. In H. F. O'Neil (Eds.), *Assessment and Measurement of Team Performance: Theory, Research, and Applications* (pp. 45–62). Hillsdale, NJ: Lawrence Erlbaum Associates.

Cannon-Bowers, J. A., and Salas, E. (1998). Individual and Team Decision Making under Stress: Theoretical Underpinnings. In J. A. Cannon-Bowers and E. Salas (Eds.), *Decision Making Under Stress: Implications for Individual and Team Training* (pp. 17–38). Washington, DC: American Psychological Association.

Carver, C. S., DeGregorio, E., and Gillis, R. (1980). Ego-Defensive Bias in Attribution among Two Categories of Observers. *Personality and Social Psychology Bulletin, 6*, 44–50.

Cialdini, R. B. (2000). *Influence: Science and Practice*. Boston, MA: Allyn & Bacon.

Collyer, S. C., and Malecki, G. S. (1998). Tactical Decision Making Under Stress: History and Overview. In J. A. Cannon-Bowers and E. Salas (Eds.), *Decision Making Under Stress: Implications for Individual and Team Training* (pp. 3–16). Washington, DC: American Psychological Association.

Cooper, G. E., White, M. E., and Lauber, J. K., (Eds.). (1979, June). *Resource Management on the Flight Deck*, NASA No. CP-2120, NTIS No. N80-22283. Moffett Field, CA: NASA-Ames Research Center.

Doton, L. (1996, Winter). Integrating Technology to Reduce Fratricide. *Acquisition Review Quarterly*, pp. 1–18.

Driskell, J. E., and Johnston, J. H. (1998). Stress Exposure Training. In J. A. Cannon-Bowers and E. Salas (Eds.), *Making Decisions Under Stress: Implications for Individual and Team Training* (pp. 191–217). Washington, DC: American Psychological Association.

Duncan, P. C., Rouse, W. B., Johnston, J. H., Cannon-Bowers, J.A., Salas, E., and Burns, J. (1996). Training Teams Working in Complex systems: A Mental Model-Based Approach. *Human/Technology Interaction in Complex Systems, 9*, 173–231.

Dwyer, D. J., Fowlkes, J. E., Oser, R. L., and Salas, E. (1997). Team Performance Measurement in Distributed Environments: The TARGETs Methodology. In M. T. Brannick, E. Salas, and

C. Prince (Eds.), *Team Performance Assessment and Measurement: Theory, Methods, and Applications* (pp. 137–153). Mahwah, NJ: Lawrence Erlbaum Associates.

Dwyer, D. J., Oser, R. L., Salas, E., and Fowlkes, J. E. (1999). Performance Measurement in Distributed Environments: Initial Results and Implications for Training. *Military Psychology, 11*, 189–213.

Dzindolet, M. T., Beck, H. P., Pierce, L. G., and Dawe, L. A. (1999). Bias Towards Automation Leads to Underutilization of Automated Aids (Abstract). In *Proceedings of the 11th Annual Meeting of the American Psychological Society.*

Dzindolet, M. T., Beck, H. P., Pierce, L. G., and Dawe, L. A. (2001a). *A Framework of Automation Use*, ARL-TR-2412. Aberdeen Proving Ground, MD: U.S. Army Research Laboratory.

Dzindolet, M. T., Pierce, L. G., Beck, H. P., and Dawe, L. A. (2002). The Perceived Utility of Human and Automated Aids in a Visual Detection Task. *Human Factors, 44*, 79–94.

Dzindolet, M. T., Pierce, L. G., Beck, H. P., Dawe, L. A., and Anderson, B. W. (2001b). Predicting Misuse and Disuse of Combat Identification Systems. *Military Psychology, 13*(3), 147–164.

Dzindolet, M. T., Pierce, L. G., Dawe, L. A., Peterson, S., and Beck, H. P. (2000). Building Trust in Automation (Abstract). In *Proceedings of the 2000 Human Performance, Situation Awareness & Automation Conference.*

Entin, E. E., and Serfaty, D. (1999). Adaptive Team Coordination. *Human Factors, 41*, 312–325.

Fogarty, W. M. (1988). *Formal Investigation into the Circumstances Surrounding the Downing of a Commercial Airliner by the U.S.S. Vincennes* (CG 49) *on 3 July 1988.* Unclassified Letter, Series 1320 of July 28, 1988, to Commander in Chief, U.S. Central Command.

Foushee, H. C. (1984). Dyads and Triads at 35,000 Feet: Factors Affecting Group Process and Aircrew Performance. *American Psychologist, 39*, 885–893.

Grynovicki, J. O., Kysor, K. P., and Murphy, J. (2001). *ARL Insight About Human Factors Issues Evaluated During the Joint Contingency Force War-Fighting Experiment*, ARL-TR-2424. Aberdeen Proving Ground, MD: U.S. Army Research Laboratory.

Hall, J. K., Dwyer, D. J., Cannon-Bowers, J. A., Salas, E., and Volpe, C. E. (1993). Toward Assessing Team Tactical Decision Making Under Stress: The Development of a Methodology for Structuring Team Training Scenarios. In *Proceedings of the 15th Annual Interservice/Industry Training Systems Conference*, (pp. 87–98).

Halpin, S. M. (1984). A Proposal for an Intelligent Interface in Man-Machine Systems. In *Proceedings of the 23th IEEE Conference on Decision and Control* (pp. 592–595).

Hoffman, R. R., Crandall, B., and Shadbolt, N. (1998). Use of the Critical Decision Method to Elicit Expert Knowledge: A Case Study in the Methodology of Cognitive Task Analysis. *Human Factors, 40*(2), 254–276.

Hutchins, S. (1996). *Principles for Intelligent Decision Aiding*, TADMUS Technical Report No.1718. San Diego, CA: Space and Naval Warfare Systems Center.

Hutchins, S. G., Morrison, J. G., and Kelly, R. T. (1996, June). Principles for Aiding Complex Military Decision Making. In *Proceedings of the Second International Symposium On Command and Control Research and Technology.*

Johnston, J. H., Smith-Jenstch, K. A., and Cannon-Bowers, J. A. (1997). Performance Measurement Tools for Enhancing Team Decision Making. In M. T. Brannick, E. Salas, and C. Prince (Eds.), *Team Performance Assessment and Measurement: Theory, Methods, and Applications* (pp. 311–327). Mahwah, NJ: Lawrence Erlbaum Associates.

Jones, M. B. (1974). Regressing Group on Individual Effectiveness. *Organizational Behavior and Human Performance, 11*, 426–451.

Kalyuga, S., Chandler, P., and Sweller, J. (1998). Levels of Expertise and Instructional Design. *Human Factors, 40*(1), 1–17.

Klein, G. A. (1989). Do Decision Biases Explain Too Much? *Human Factors Society Bulletin, 32*, 1–3.

Klein, G.A. (1992). *Decision Making in Complex Military Environments*. Klein Associates. Technical Report for the Naval Command, Control and Ocean Surveillance Center, San Diego, CA.

Klein, G. A. (1997). Developing Expertise in Decision Making. *Thinking & Reasoning, 3*(4), 337–352.

Klein, G., and Militello, L. (2001). Some Guidelines for Conducting a Cognitive Task Analysis. In E. Salas (Ed.), *Advances in Human Performance and Cognitive Engineering Research*, Vol. 1 (pp. 163–199). Ukraine: Elsevier Science/JAI Press.

Klein, G., and Pierce, L. (2001). Adaptive Teams. In *Proceedings of the 6th International Command and Control Research & Technology Symposium*.

Knapp, B. G., Johnson, J., Barnette, D. B., Wojciechowski, J., Kilduff, P., and Swoboda, J. (1997a). Modeling Maneuver Battalion C2 Operations of a Current Army Command Post for a Force on Force Scenario—Baseline Model, Baseline model delivery paper to U.S. Army Armor Center and School, U.S. Army Research Laboratory, Aberdeen Proving Ground, MD.

Knapp, B. G., Johnson, J., Barnette, D. B., Wojciechowski, J., Kilduff, P., and Swoboda, J. (1997b). Modeling Maneuver Battalion C2 Operations of a Force XXI Equipped Army Command Post for a Force on Force Scenario—Traditional Model, Traditional C2V battalion TOC model delivery paper to U.S. Army Armor Center and School, U.S. Army Research Laboratory, Aberdeen Proving Ground, MD.

Knapp, B. G., Johnson, J., Barnette, D. B., Wojciechowski, J., Kilduff, P., and Swoboda, J. (1997c). Modeling Maneuver Battalion C2 Operations of a Force XXI Equipped Army Command Post for a Force on Force Scenario—Integrated Model, Integrated C2V battalion TOC model delivery paper to U.S. Army Armor Center and School, U.S. Army Research Laboratory, Aberdeen Proving Ground, MD.

Langan-Fox, J., Code, S., and Langfield-Smith, K. (2000). Team Mental Models: Techniques, Methods, and Analytic Approaches. *Human Factors, 42*, 242–271.

Lipshitz, R., and Strauss, O. (1997). Coping with Uncertainty: A Naturalistic Decision-Making Analysis. *Organizational Behavior and Human Decision Processes, 69*(2), 149–163.

Lussier, J., Michel, R., and Frame, A. (1997). *NTC-CD System: Recreating the NTC Experience*, ARL Study Report 97-02. Alexandria, VA: U.S. Army Research Institute for the Behavioral and Social Sciences.

Lussier, J. W., Ross, K. G., and Mayes, B. (2000). Coaching Techniques for Adaptive Thinking. In *Proceedings of the 2000 Interservice/Industry Training, Simulation, and Education Conference* (Abstract).

Meichenbaum, D. (1996). Stress Inoculation Training for Coping with Stressors. *Clinical Psychologist, 49*, 4–7.

Middlebrooks, S. E., Knapp, B. G., Barnette, B. D., Bird, C. A., Johnson, J. M., Kilduff, P. W., Schipani, S. P., Swoboda, J. C., Wojciechowski, J. Q., Tillman, B. W., Ensing, A. R., Archer, S. G., Archer, R. D., and Plott, B. M. (1999). *CoHOST (Computer Modeling of Human Operator System Tasks) Computer Simulation Models to Investigate Human Performance Task and Workload Conditions in a U.S. Army Heavy Maneuver Battalion Tactical Operations Center*, ARL-TR-1994. Aberdeen Proving Ground, MD: U.S. Army Research Laboratory.

Moore, R. A., and Averett, M. G. (1999, June). Identifying and Addressing User Needs: A Preliminary Report on the Command and Control Requirements for CJTF Staff. In *Proceedings of the 5th International Symposium on Command & Control Research & Technology*.

Morrison, J. G., Kelly, R. T., Moore, R. A., and Hutchins, S. G. (2000). Implications of Decision-Making Research for Decision Support and Displays. In J. A. Cannon-Bowers and E. Salas (Eds.),

*Making Decisions Under Stress: Implications for Individual and Team Training*. Washington, DC: American Psychological Association.

Morrison, J. G., and Moore, R. A. (1999, June). Design Evaluation and Technology Transition: Moving Ideas from the Drawing Board to the Fleet. In *Proceedings of the 5th International Symposium on Command and Control Research and Technology*.

Najjar, L. J. (1998). Principles of Educational Multimedia User Interface Design. *Human Factors, 40*(2), 311–323.

National Transportation Safety Board. (1979, June). *Aircraft Accident Report,* NTSB Report No. AAR-79-07, NTIS No. UB/E/104-007. Washington, DC: National Transportation Safety Board Bureau of Accident Investigation.

National Transportation Safety Board. (1982, August). *Aircraft Accident Report*, NTSB Report No. AAR-82-08, NTIS No. PB82-910408. Washington, DC: National Transportation Safety Board Bureau of Accident Investigation.

Niessen, C., Eyferth, K., and Bierwagen, T. (1999). Modeling Cognitive Processes of Experienced Air Traffic Controllers. *Ergonomics, 42*(11), 1507–1520.

Noah, W. W., and Halpin, S. M. (1986). Adaptive User Interfaces for Planning and Decision Aids in C3I. *IEEE Transactions on Systems, Man and Cybernetics, 16*, 909–918.

Parasuraman, R., and Riley, V. (1997). Humans and Automation: Use, Misuse, Disuse, Abuse. *Human Factors, 39*, 230–253.

Parasuraman, R., Sheridan, T. B., and Wickens, C. D. (2000). A Model for Types and Levels of Human Interaction with Automation. *IEEE Transactions on Systems, Man, and Cybernetics—Part A: Systems and Humans, 30,* 286–297.

Pierce, L. G., and Klein, G. (2002). Preparing and Supporting Adaptable Leaders and Teams for Support and Stability Operations. In *Proceedings of the Defense Analysis Seminar XI*.

Pierce, L. G., and Pomranky, R. A. (2001). The Chameleon Project for Adaptable Commanders and Teams. In *Proceedings on the 45th Human Factors and Ergonomics Society Annual Meeting*. (Abstract).

Plott, B. (1999). *Final Technical Report for Field Artillery Battalion Tactical Operations Center (FA BN TOC) Model Support*. Boulder, CO: Micro Analysis and Design.

Ross, K. G. (2000, September/October). Training Adaptive Leaders—Are We Ready? *Field Artillery*, pp. 15–19.

Ross, K. G., Pierce, L. G., and Baehr, M. (1999). *Revitalizing Battle Staff Training*, ARL-TR-2079. Aberdeen Proving Ground, MD: U.S. Army Research Laboratory.

Salas, E., and Cannon-Bowers, J. A. (2001). The Science of Training: A Decade of Progress. *Annual Review of Psychology, 52,* 471–499.

Salas, E., Cannon-Bowers, J. A., and Johnston, J. H. (1997). How Can You Turn a Team of Experts into an Expert Team?: Emerging Training Strategies. In C. E. Zsambok and G. Klein (Eds.), *Naturalistic Decision Making* (pp. 359–370). Mahwah, NJ: Lawrence Erlbaum Associates.

Salas, E., and Klein, G. (Eds.). (2001). *Linking Expertise and Naturalistic Decision Making*. Mahwah, NJ: Lawrence Erlbaum Associates.

Sarter, N.S., and Woods, D.D. (1995). How in the World Did We Ever Get into That Mode? Mode Error and Awareness in Supervisory Control. *Human Factors, 37*(1), 5–19.

Schraagen, J. M., Chipman, S. F., and Shalin, V. L. (2000). *Cognitive Task Analysis*. Mahwah, NJ: Lawrence Erlbaum Associates.

Scrivener, S. A., Urquijo, S. P., and Palmen, H. K. (1993). The Use of Breakdown Analysis in Synchronous CSCW System Design. *International Journal of Man-Machine Studies, 31*, 517–534.

Serfaty, D., Entin, E. E., and Johnston, J. H. (1998). Team Coordination Training. In J. A. Cannon Bowers and E. Salas (Eds.), *Making Decisions Under Stress: Implications for Individual and Team Training* (pp. 221–245). Washington, DC: American Psychological Association.

Sheridan, T. B., Vamos, T., and Aida, S. (1983). Adapting Automation to Man, Culture, and Society. *Automation, 19*, 605–612.

Shlechter, T. M., Zaccaro, S. J., and Burke, C. S. (1998). Toward an understanding of Shared Mental Models Associated with Proficient Team Performance. Paper presented at the American Psychological Society Annual Meeting, Washington, DC.

Smith-Jentsch, K. A., Zeisig, R. L., Acton, B., and McPherson, J. A. (1998). Team Dimensional Training: A Strategy for Guided Team Self-Correction. In J. Cannon Bowers and E. Salas (Eds.), *Making Decisions Under Stress: Implications for Individual and Team Training* (pp. 271–297). Washington, DC: American Psychological Association.

St. John, M., Oonk, H. M., and Osga, G. A. (2000). Designing Displays for Command and Control Supervision: Contextualizing Alerts and "Trust But Verify" Automation. In *Proceedings of the Human Factors and Ergonomics Society 44th Annual Meeting* (pp. 646–649). Santa Monica, CA: Human Factors and Ergonomics Society.

Suantak, L., Momen, F., Rozenblit, J., Barnes, M., and Fichtl, T. (2001). Intelligent Decision Support for Support and Stability Operations (SASO) through Symbolic Visualization. In *Proceedings of the 2001 IEEE International Conference on Systems, Man and Cybernetics*, (pp. 2927–2931).

Terborg, J. R., Castore, C. H., and DeNinno, J. A. (1976, May). A Longitudinal Field Investigation of the Impact of Group Composition on Group Performance and Cohesion. Paper presented at the meeting of the Midwestern Psychological Association, Chicago, IL.

U.S. Army Audit Agency. (2001). *Joint Contingency Force Advanced Warfighting Experiment*, Audit Report: AA 01-133. Washington, DC: Department of the Army.

U.S. Department of Defense (DoD). (March, 1998). *Department of Defense Design Criteria Standard: Human Engineering*, MIL-STD-1472F. Washington, DC: DoD.

U.S. Department of Defense (DoD). (2000a). *DOT&E FY00 annual report*, Director, Operational Test & Evaluation. Washington, DC: DoD.

U.S. Department of Defense (DoD). (2000b). *Initial Report of the Army Digitization Review Panel, 8 August 2000*, Director, Operational Test & Evaluation. Washington, DC: DoD.

Vincente, K. (1999). *Cognitive Work Analysis: Toward Safe, Productive, and Healthy Computer-Based Work.* Hillsdale, NJ: Lawrence Erlbaum Associates.

Vincente, K. (2002). Ecological Interface Design. *Human Factors, 44*, 62–78.

Wickens, C. D., and Hollands, J. G. (2000). *Engineering Psychology and Human Performance* (pp. 538–550). Upper Saddle River, NJ: Prentice-Hall.

Wojcik, T., and Plott, B. (1999). *Final Technical Report for Task Network Modeling of Fires Effects Coordination Cell for an Interim Division*. Boulder, CO: Micro Analysis and Design.

Wojcik, T., and Plott, B. (2001). *Final Technical Report for Task Network Modeling of Fires Effects Coordination Cell for an Interim Division*. Unpublished manuscript.

Woods, D. D., and Christoffersen, K. (2001). Balancing Practice-Centered Research and Design. In M. McNeese and M. A. Vidulich (Eds.), *Cognitive Systems Engineering in Military Aviation Domains*. Wright-Patterson AFB, OH: Human Systems Information Analysis Center.

Yerkes, R. M., and Dodson, J. D. (1908). The Relation of Strength of Stimulus to Rapidity of Habit Formation. *Journal of Comparative Neurology and Psychology 18*, 459–482.

Zacharias, G., and Hudlicka, E. (2001). *Visualization Aids for Stability and Support Operations*, CRA Contractor Report 599. Cambridge, MA: Charles Rivers Analytics.

Zachary, W., Campbell, G. E., Laughery, K. R., Glen, F., and Cannon-Bowers, J. A. (2001). The Application of Human Modeling Technology to the Design, Evaluation and Operation of

Complex Systems. In E. Salas (Ed.), *Advances in Human Performance and Cognitive Engineering Research*, Vol. 1 (pp. 201–250). Ukraine: Elsevier Science/JAI Press.

Zsambok, C. E., and Klein, G. (Eds.) (1997). *Naturalistic Decision Making*. Mahwah, NJ: Lawrence Erlbaum Associates.

Zuckerman, M. (1979). Attribution of Success and Failure Revisited Or: The Motivational Bias Is Alive and Well in Attribution Theory. *Journal of Personality, 47*, 245–287.

## CHAPTER 22

# Human Systems Integration and Training for New Systems

JOHN KLESCH and WILLIAM STEMBLER

## 22.1 INTRODUCTION

Program managers face a great dilemma in deciding how to incorporate training into new systems. They are aware of training's importance and do not wish to neglect it, but it is viewed as a high-cost element that can be deferred most easily without sacrificing overall program objectives, so there is a strong incentive to neglect it. There are three overriding reasons for this dilemma. First is a management misperception in the relative value of training compared to other system requirements. On the one hand, training is one of the critical elements in any new hardware or software system. Training is needed not only by operators and maintainers but also by leaders, supervisors, and managers involved in any new system. As a critical element, it is expected that training would play an equal role with all other critical elements of the system to maximize the gain on capital investment for the system as a whole. On the other hand, training is seen as a convenient means to reduce rising costs in the overall development program or to avoid schedule slippages that may threaten the program. This reason stems from poor management practices and can only be resolved by smart business decisions that will not allow training or any other critical element to be the bill payer for program funding and scheduling problems.

The second reason for the program manager's dilemma is the perceived high initial costs of human systems integration (HSI) requirements. Although many program managers of new systems recognize the importance of HSI, they tend to trade off HSI considerations in their investment strategy and assume more risk because of upfront costs. This invariably leads to less than optimum performance of the system initially and, in the long run, may lead to a serious loss of return or even failure of the entire investment. This reason is more legitimate than the first but should be resolved in favor of HSI so long as the costs are justified in terms of total system performance and affordability trade-offs.

*Handbook of Human Systems Integration*, Edited by Harold R. Booher.
ISBN 0-471-02053-2 

The third reason for the dilemma is meeting the total requirement for training personnel to operate, maintain, and manage new systems. This reason is the soundest reason for the program manager's dilemma and will be dealt with extensively in this chapter. An introduction to the personnel training requirements that any new systems must meet to assure both initial training and sustainment training is provided below. This will be followed by an introduction to the problems facing the program manager when utilizing training technology to help meet these requirements.

### 22.1.1 Personnel Training Requirements

There are three major requirements for personnel training on new systems:

1. *Personnel Must Be Trained Effectively* The training for operators and maintainers of new systems must first and foremost be relevant. Students must be able to acquire the skills needed to perform all of the tasks necessary to meet the mission or goals and objectives of the system. They must reach a degree of proficiency as quickly as possible so they can initiate operations with the new systems. Since there will be very few others to turn to for assistance on this brand-new system, they must get as close to mastery as possible. They must also have some means to sustain that proficiency once it is achieved. Errors on new systems are frequently critical because that which becomes obvious over time became obvious because the errors led to serious failures. For example, placing a simple but critical filter incorrectly into an army tank has led to repeated engine failures that continued until training was refined.
2. *Personnel Must Be Trained Immediately* New equipment training must be done for all organizations as soon as they receive the equipment. Preferably, personnel are trained before the systems arrive so they can be immediately productive. Any delay means downtime of the system or risk of damage if operated or maintained by unqualified personnel.
3. *Personnel Must Be Trained Efficiently* Most program managers are faced with rising costs, potential cost overruns, and a tight budget that focuses on hardware. In order to train efficiently, they must develop the training as cheaply and quickly as possible and devise means to also deliver training inexpensively to keep the cost per student in line with the budget allocated.

Nowhere is the difficulty of meeting personnel training requirements better illustrated than in the Department of Defense (DoD) where dozens of systems are fielded each year that generate a like number of new system training requirements. Program managers, who are to acquire these systems, are carefully screened and educated in university-level programs. These university programs and curricula are painstakingly designed to ensure program managers are equipped to do a good job. The managers are taught to carefully consider the requirement; study the various alternatives to meet this requirement; perhaps engage in some prototype testing using mockups, simulators, and breadboard models; and, finally, obtain independent field testing and evaluation of the hardware. They understand the need for operators and maintainers to perform with some degree of proficiency if the system is to approach its design capability. Nevertheless, training is seldom developed or tested with the same rigor and priority that the hardware receives. Consequently, the goal of proficient performance is seldom achieved, and as a result, the training burden continues to grow as the systems mature.

In the civilian sector, the computer industry and electronic equipment industry provide prime examples of the mismatch between training requirements and new system operations. Extremely powerful software systems are seldom used to anywhere near their potential. Meanwhile, help desks are required to maintain a 24-hour, 7-day-a-week (24/7) service for users who frequently do not know how to operate the basic features because they have not been trained adequately, if at all. Nearly everyone, at some time or another, is expected to sit down and start using a computer system regardless of his or her experience with the system. It is left up to the employee to learn, usually haphazardly, on the job how the actual application program operates. While there is no accurate way to measure the loss of productivity each time this occurs, the fact that it is nearly a universal experience, even among experts, tells us the loss is significant, if only on the basis of a very expensive item of equipment and software not being used to capacity. Retail companies invest huge amounts in automation of sales, linking it inextricably with inventory and accounting. But, they fail to train their clerks to use the equipment and programs properly and efficiently (Compton, 1999; Stapp, 2001).[1] The net result (especially during holidays and sales days) is long lines and lost sales as sales personnel wait for a supervisor and customers finally leave in frustration. Amazingly, the retailer may later invest in expensive programs to determine why customer loyalty is waning and still more expensive programs to attempt to win them back. Did they save money by shortcutting training?

### 22.1.2 Training Compromises

To paraphrase a common saying among experienced trainers and managers of new systems, "Personnel must be trained effectively, immediately, and efficiently: Pick any two!" Faced with the problem of considering quality, time, and cost simultaneously, program managers almost always are forced to trade off at least one of the three. Because of the tight requirements to meet performance standards, the trade-off is usually between cost and time. Typically the program managers resort to traditional training methods that appear to allow them to meet time requirements for the first units, minimally meet performance standards, and stay within the initial budget. The subsequent system owner, who is responsible for life-cycle costs, however, may sooner or later be confronted with a huge sustainment bill to pay, so this initial balance of *effectiveness, immediacy, and efficiency* can be very short term in duration.

Instructional developers sometimes must minimize development time by covering only basic information formally, expecting the instructor to fill in the gaps as they are revealed during training. Time on actual equipment also may be minimized and available photographs or drawings used in their place. If necessary, managers may raise the instructor-to-student ratio or simply cut down on the amount of time spent in training. The latter is done by reducing the amount of time spent on a task or by eliminating entire tasks from the curriculum.

Follow-on units subsequently face long periods of expensive train-up time as shortcomings are revealed and replacement personnel attempt to learn the system. Revised, "beefed-up" training programs are later produced, duplicating much of the original effort. While new personnel will receive the improved training, those already in the field will still require updated training. However, to be efficient, this training is tailored to their needs to avoid training in what they already know. In essence, this requires yet another expensive training program to be generated. In addition to these high costs of training, there is the immediate cost of downtime that can be experienced with any failures of the new system when first fielded, which in manufacturing can bring about a crisis and in the case of a military unit disaster.

The above paragraphs describe typical methods that allow the manager to keep a program moving but almost always introduce numerous delays and higher costs to correct deficiencies later. This raises the question of whether the three requirements could be better met with improved training technology.

### 22.1.3 Training Technology Limitations

Technology and, in particular, information technology improvements are systematically being applied to new systems, yet the training for new systems has not received the same treatment. While many training techniques and approaches have been developed, tested, and refined, they all fall short in one way or another when applied to new systems. For example, one of the more promising training technologies, interactive multimedia instruction (IMI), which is now being routinely and successfully applied to other training needs, is not being applied to new systems.

The primary difficulties in applying IMI to new systems in the past have been the following:

1. It is very expensive to develop IMI and becomes too expensive when many changes are required, which is always the case in the development of new systems.
2. It is very time consuming in the development stage.
3. It has not shown sufficient flexibility to meet the fielding schedules of new systems.
4. Metrics do not exist for assessing training impact over the system life cycle so managers and decision makers have not been able to justify additional expense over traditional, conventional training techniques.

In the sections that follow, we will examine the strengths and weaknesses of IMI for new systems and explore ways that a new system can utilize the strengths of this technology while retaining some of the advantages of more conventional training methods.

### 22.1.4 Chapter Overview

The chapter has three primary objectives:

- to examine how technology has been applied in the past to solve training problems;
- to evaluate the advantages and disadvantages of IMI for improving training effectiveness, immediacy, and efficiency; and
- to describe an HSI process that combines IMI and conventional training techniques into a cost-effective process to meet new system training requirements.

## 22.2. HSI TRAINING TECHNOLOGY APPLICATIONS

Over the last decade great strides have been made in addressing training needs through the use of computers and communication systems. Computers have provided a number of advantages over traditional systems for training. They can store vast quantities of information; provide graphics of various types, ranging from line-drawn diagrams to photographs, video motion, and animation; and include interaction capability. Communications have likewise had a dramatic effect on training. Video teletraining (VTT) has made it possible to bring live training to multiple remote sites. Telephone lines have made long-distance audio presentations and interactions available anywhere in the world. The Internet with its websites has made possible the sending and retrieval of information and, when harnessed in support of computer training, has made interaction of students with vast amounts and types of information an everyday reality. So how has technology specifically improved training in effectiveness, immediacy, and efficiency for new systems?

### 22.2.1 Technology and Training Effectiveness

The overriding question with training effectiveness has to do with how well students can perform tasks when they have completed training. Students' skills to perform critical tasks never before encountered can be developed and sustained by technology in a number of ways. First, the stimuli presented in the actual task can be reproduced and enhanced for careful study and examination. Being able to see objects clearly and as often as necessary will make the task easier to understand and thereby to perform. For example, removal of complex or delicate parts of equipment, such as an aircraft radar antenna or a computer display unit in a tank, can now be illustrated on a monitor that shows task details in a linear sequence. Graphics, rather than sketches or very cluttered photographs, can show the removal steps in three dimensions, showing components as they are removed and focus on critical elements by blinking, highlighting, or other graphic techniques. Figure 22.1, for example, shows how graphic cues can be used to highlight the location of a control box on an item of equipment and at the same time magnify the control for observation.

These capabilities provide numerous opportunities for the developer to make training more effective. The opportunity to inject various whole/part learning paradigms is quickly evident. With computer technology each student can be shown a whole procedure step by step, then brought back to focus on part tasks of that procedure that can be learned independently. Within the part tasks the students can be required to interact by answering questions and demonstrating that they can pick out relevant cues. Finally, each student can

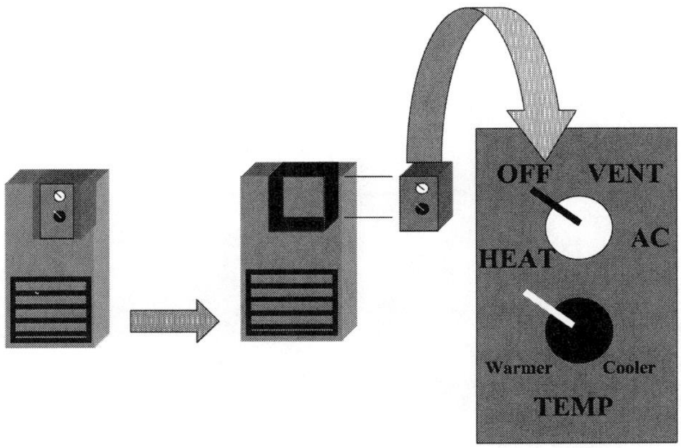


**Figure 22.1** Graphic cues used to highlight an item.

be asked to step through the entire sequence with the student in control to demonstrate that he or she understands the entire procedure. In the case of tasks where the real-world task is also done on the computer, the instructor can have immediate evidence of training effectiveness and transfer. When external or mechanical work is required, there is the opportunity to learn the part tasks first in simulation. There will be far less uncertainty in performance when the students work on real equipment. Students will also be likely to perform with fewer errors and require far less time to reach mastery. Supervisors can use the same training program to learn the new system and later monitor the less experienced trainees' progress (Kulik, 1994; Metzko et al., 1996; Howard, 1997; Parchman et al., 2000).

As a specific example, consider that the critical task of learning to identify friendly or enemy combat systems has always been done using two-dimensional "flashcards." With technology we can now teach students to identify and distinguish between combat systems by presenting images at different distances, in different attitudes, and different environments. Now the soldiers can learn at what range they can identify friend or foe with certainty by having the computer-driven graphics present images that simulate known distances. Students also can view comparisons that highlight silhouettes and distinctive characteristics. Next, they can see the differences when the cues are gradually faded or completely removed. See Figure 22.2 for an example of fading cues as darkness approaches. The stimuli can also be degraded to simulate dust, twilight, camouflage, or other real-world conditions.

Moreover, these same combat systems can be immediately shown in a total desert environment with appropriate camouflage or mountainous, snowy terrain as background with another type of camouflage. Still further, students can now learn what the systems look like while they are moving, as in the case of a mobile system. Students can see how the appearance of the system will change as it draws closer or moves farther away either directly or at an oblique angle. Computer graphics can also now be used to aid the soldier in learning to identify thermal heat signatures of friendly and enemy systems again using accurate simulations. These skills can be absolutely critical in conducting night operations

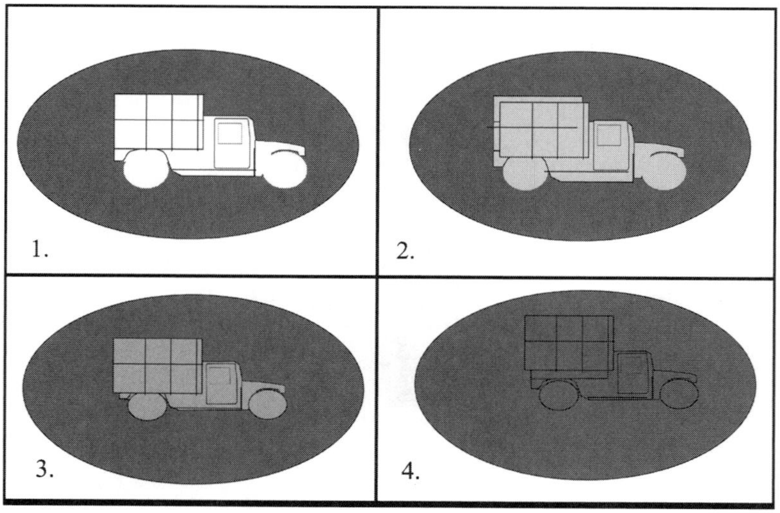


**Figure 22.2** Graphics used to diminish cues.

(and avoiding false targets such as animals, trees, and outhouses!). These are a few examples of many ways improvements can be made in training effectiveness with technology to support new systems. Electronic performance support systems (EPSSs) are under development that will be embedded in many new systems that will use similar techniques to aid actual operation. A training support system that replicates that EPSS can be used anywhere outside the system to enable both less expensive initial skill acquisition as well as sustainment when students are in a school or otherwise distant environment from the actual new system hardware.

### 22.2.2 Technology and Training Immediacy

The most important question for training immediacy has to do with how quickly we can get the training to all the students who require training. Technology is now used to improve training immediacy in several ways. Instructors videotape the live training sessions and can offer the tape to students who were absent when the new system training was presented. This can be repeated for as many additional sessions as required. Where properly equipped facilities exist it is possible to do live VTT via long lines or satellite to long-distance multiple locations, although some scheduling problems can arise. Placing the training session on a CD-ROM or on a website makes training accessible and available most anywhere in the world. Moreover, changes or safety alerts can be supplied via the Internet in a matter of minutes, if required, as opposed to hours, days, weeks, or even months.

### 22.2.3 Technology and Training Efficiency

The primary question for training efficiency has to do with how we can lower the costs of production, reproduction, and delivery. Technology has been applied in numerous ways to reduce both time and cost of lesson production. For example, computer word programs permit the developer to make corrections quite easily to plans of instruction, lesson materials, and other handouts. Application of templates to maintain a standardized "look and feel" is done by simple electronic command. Vugraphs can be quickly built and modified by accessing easy-to-use programs. Graphics and photographs can now be electronically imported for easy insertion into the vugraphs or handouts. Vugraphs can be stored and projected electronically and hence are easily updated. A demonstration on actual equipment can be more quickly captured on videotape than was done previously with film. Immediate editing and updating with video technology also serve to lower costs of development. Modern copy machines make reproduction of even color diagrams and photos within reasonable costs.

For delivery, accommodating more students via VTT can reduce the cost per student trained. Using VTT to bring the training to widely dispersed students will also greatly reduce travel expenses and time spent in traveling. Using websites on the Internet or using CD-ROMs to provide timely updates everywhere at once can lower distribution costs. Mailing a CD is much cheaper than mailing heavy boxes of printed materials.

## 22.3 TRAINING REQUIREMENTS AND IMI

In the previous section, we presented several technologies that can aid the developer and the customer. In this section, we now look at IMI, which has incorporated many of the technologies listed above, to determine how training can be made more effective, immediate, and efficient. Being a combination of useful technologies, one might expect it to be the best of the best. If it can be the best, why is it not used more often? Before one can judge the merits of employing a new technology, it is important to examine more closely that which it is to replace. What are the specific areas where improvement can be made? What are the expected goals for the new technology? How does one determine if the goals are achieved with the new technology? Is it really better, faster, and cheaper? And if not, why not, and can anything be done to improve it? By answering these questions, one can systematically design an improvement process using the new technology.

### 22.3.1 Training Effectiveness

The advantages and disadvantages of traditional standup instruction compared to IMI for training effectiveness are considered below.

***Traditional Standup Instruction and Training Effectiveness*** Traditional standup instruction remains the methodology of choice for new systems simply because the systems are so prone to last-minute design changes that expending any effort to create IMI has not been viable. Even use of video technology is not considered favorably because it must be carefully scripted, then filmed under close-to-ideal conditions that are rarely available in the laboratory or production lines. Moreover, traditional standup instruction is a proven technique. Since the time of Socrates, the teacher-and-pupil, one-on-one pedagogical approach has been known for its effectiveness. Given a subject matter expert with a modicum of teaching skills, a set of vugraphs, possibly the actual equipment, and a plan of instruction, one can expect to obtain some success. The instructor can respond to the student on the spot. He or she can detect which student is struggling and offer more help. More importantly, growth in skill and knowledge can occur when a student explores ideas and concepts with a receptive, experienced trainer. In the military, for example, the instructors are trained to respond in just this manner. Conveying leadership skills as well as technical "tricks of the trade" are two examples where live instruction excels.

The reality for new systems, however, is that the session is more than likely to become just a lecture with possible questions and answers. The forgetting curve is steep under these circumstances, and unless the developer or instructor has the time or inclination to make detailed handouts, the students must rely on a good memory or the class leader who took good personal notes. The resulting problems impacting effectiveness on the job are well known and tough to completely overcome. Serious errors can be committed or new personnel may be restricted to an observer/helper role for a substantial period of time. Parts may be replaced that do not need to be replaced and secondary malfunctions induced. The reasons are sometimes very evident. With new equipment the actual hardware is not likely to be initially available for any kind of hands-on training. The instructor may demonstrate procedures, but the usual problems of being able to see the procedure or see it more than once leave the student far removed from the desired recognition of cues, stimuli, and opportunity to practice responses. Consistency in the instructor's responses or among different instructors sometimes poses a problem for effectiveness. Video taping of critical procedures can be used, but quality shooting, scripting, and narration are rarely available in the time allocated for new system training. When time is available, the cost for professional production would equal the entire training development budget. While even poor-quality taping done in-house to save dollars will have instructional value, the point is simply that the instruction is far from optimized for effectiveness. The most important drawback is that traditional standup instructional classes are not optimized to elicit responses from students, especially all the students. Even if a good instructor asks frequent questions, each student does not have to answer each time. Thus, learning is not being reinforced and the effects, if not later sustained, will diminish significantly.

***Advantages of IMI to Training Effectiveness*** The power of IMI begins with its ability to (a) provide a useful demonstration of the task and (b) allow students to view the demonstration as many times as necessary until they are comfortable with the sequence. To do this, IMI may incorporate a variety of graphics and graphic techniques that aid the student in visualization of the task. For new equipment this becomes crucial if the student must begin training on equipment that may not have reached the production line or the location where he or she is to work. There are many graphic alternatives that IMI can use, usually in combination. Still photographs of actual equipment can be used to provide initial cues and context. Figure 22.3, for example, shows the engine and then provides focus on the distributor. Graphic information may also be provided using line drawings, video clips, or animations depending on the complexity of the task.

In addition to graphics, text can also be presented in a variety of ways to keep student interest in the content and allow focus on critical material or objects that the instructor wishes to call to the attention of the student. In Figure 22.4 the example on the left uses numbered paragraphs to help the student logically group the electrical and mechanical components. The example on the right uses the same idea but with a bullet structure to reduce the number of words and to provide additional emphasis on the key items by singling them out.

The verbal content and the graphics can be supplemented and enhanced by the addition of audio that can provide additional or reinforcing verbal information via narrative and

**Figure 22.3** Example of still photos.

1. Mechanical and electrical controls are used to regulate the heating and air conditioning system.

2. Mechanical controls such as grilles and louvers can direct air flow and regulate the amount of air coming into the unit or being discharged from the unit.

3. Electrical controls allow the user to select the type of air conditioning desired such as cooling , ventilation or heating and the temperature of the area .

Mechanical systems regulate the HVAC system air flow:

•Louvers direct air flow

•Grilles regulate the amount of air entering or leaving the unit

Electrical systems allow selection of mode and temperature to be maintained:

• Mode selector switch- allows heating , cooling, ventilation, or Off modes

•Temperature control thermostat allows selection of desired temperature

**Figure 22.4** Examples of text presentation.

aural cues (e.g., sounds of the equipment under normal and failing conditions) to aid depiction of the task and the opportunity to practice discriminating among aural cues when the task requires it.

What follows next is even more powerful because IMI can then use the same procedures to allow (or require) the student to perform the task (now *you* do it). Often this hands-on practice is not possible with traditional instruction because of time, expense, availability of equipment, danger, and/or the availability of an instructor to monitor performance as it occurs. In contrast to traditional methods, each student can be required to respond to each element of the task. The lesson can be designed so that the student cannot advance until all elements are covered. If the student does not respond correctly, he or she can be guided or sent into a remediation sequence. As illustrated in Figure 22.5, the students may be told, based on their responses, that they have identified part of the answer but that they still must acquire additional information. They are directed by the screen and the highlighted navigation arrow to return to the previous information presented. Further, IMI can allow students to practice as much as they like until they are comfortable with their performance.

Interactive multimedia instruction can be designed to show and use a variety of feedback mechanisms and schedules to achieve optimum student performance. Novice personnel can be given early and frequent feedback by eliciting small chunks of information that they have just been exposed to in the lesson. As the lesson and student progress, fewer single-item feedback questions can be posed and summative questions requiring them to link several sequences of information can then be introduced. Figure 22.6 illustrates a test of the student's knowledge of both the sequence in which a task is to be performed and the location of the items addressed in the task. When desirable, scoring can be used as progress or diagnostic indicators and also for recordkeeping purposes. (Have you done it correctly? Do you need to review again? Have you met all the criteria?)

***Disadvantages of IMI to Training Effectiveness*** Authoring systems are frequently complex to use and maintain and are sometimes proprietary, making it difficult

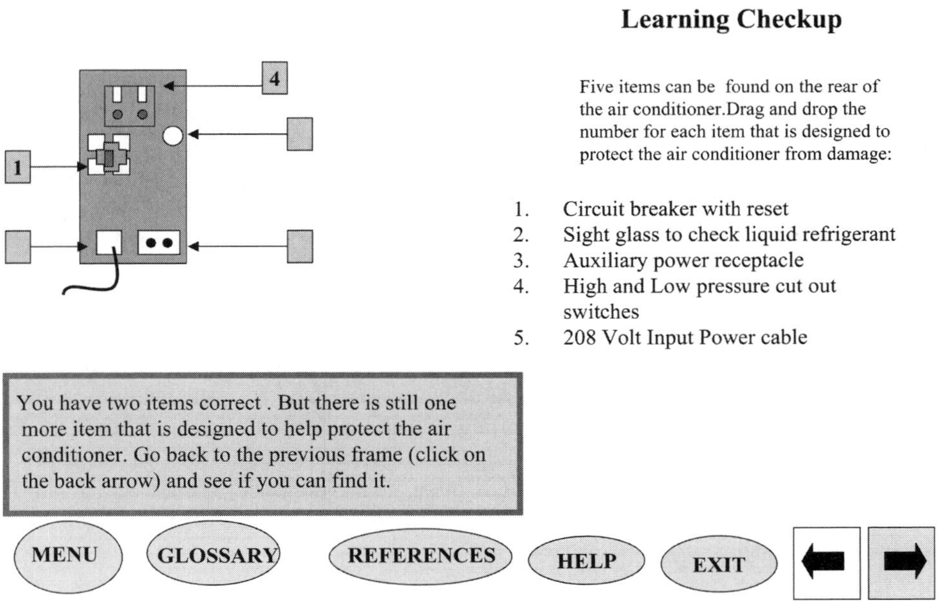


**Figure 22.5** Example of remediation direction.

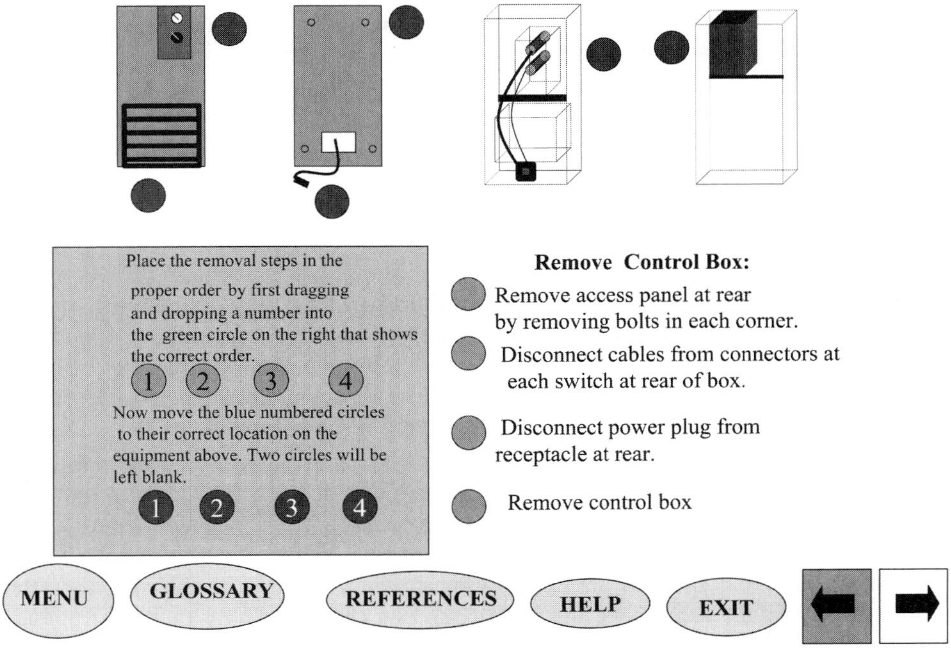


**Figure 22.6** Linking sequences.

to disperse without purchasing site licenses for all recipients. Animations, photographs, video clips, and other high-density graphics quickly generate problems for Web-based IMI over the Internet because of the size of the files created. In order to produce really effective IMI, the course developer must prepare storyboards, create appropriate graphics, and create appropriate and useful audio. Then he or she must decide on the right mix. If the right mix among the three media is not found, the effort may only confound the student. The "look and feel" between developers' lessons can vary widely, but if an IMI course is to be prepared quickly to meet a deadline, the production manager must use multiple developers. The developer could therefore wind up with lessons that are different enough that the students are distracted or, in worse cases, completely frustrated because the requirements and presentations differ so much within the same course or perhaps within the same lesson. Listings of some common problems that arise in the design of IMI are provided in Table 22.1. Some of the problems can be fixed by establishing standards (indicated by "S" in the table) and ensuring that the designers all adhere to them. For example, establishing conventions for navigation through a program or consistent placement of text can solve some problems. For other items, judgment (coded as "J" in the table) must be used that comes from experience based on comments or observations from trials with the lessons.

### 22.3.2 Training Immediacy

When customers buy a new item, they want the training for the new item immediately upon receipt or preferably even before receipt if that were possible. That way new systems are placed in immediate use with personnel who are immediately productive. There are two major difficulties facing the training manager in meeting these customer preferences. First, organizations that are spread over vast geographic areas pose challenging logistics problems because to be cost effective, the objectives are to train everyone at once and minimize the period during which two separate systems (old and new) must be maintained. Second, since engineering or other system deficiencies are likely to induce changes in the system until the last possible moment before fielding, the data for training materials are always out of date. This requires the training manager to begin training development early with the best available data but at the same time have a flexible system that can accommodate last-minute changes.

***Traditional Standup Instruction and Training Immediacy*** Traditional standup instruction provides training immediacy very well. This is accomplished by training an instructor or instructor team, devising a plan of instruction or syllabus that relies on reproducible handouts for the students, and providing some vugraphs and possible training aids for the instructor. Vugraphs and handouts usually can be changed rapidly, so the only waste is in the outdated materials that are replaced by updates. The problems begin to arise when either the students or the instructors must travel in order to receive training. In addition to adjustment of personnel schedules, the airlines, hotels, adequate training sites, local transportation, reservations, and eating arrangements all must be quickly scheduled at each place. The more students, the more systems, or the more geographically separated students are from the instructors, the more likely the chances are for major disruptions. The situation only worsens when weather changes, labor strikes occur, unexpected illness or other emergencies occur, or simple mistakes occur. The result is personnel miss training and do not acquire the immediate skill capability required.

**TABLE 22.1 IMI Media Problems**

| IMI Problems | Media | Category |
|---|---|---|
| Media competes with rather than complements other media. | Narration may be distracting if not tied tightly to text or graphic. Multimedia animation may provide only "entertainment" and lessens the educational value. | J |
| Too busy, therefore stimuli are distracting. | Text can be in too many formats or a poor format for providing focus on key information. Graphics can be too complicated to quickly grasp relevant cues. | J, S |
| Too much information is presented in a single frame. | Typically a problem with text but can occur with graphics and even narration if too lengthy. | J |
| Information is too large for the screen or too small to be seen for required detail. | Usual problems with graphics. Can be a serious problem if video only is used for fine detail work without adequate lighting, angles, and lenses. | **S**, J |
| Sounds used are distracting. | Audio sounds may be irrelevant and become annoying after a while instead of adding to the learning environment. | J |
| File is too large for Web delivery. | Delay in downloading may inhibit learning process or be unusable. | **S**, J |
| Frames are very complicated and expensive to produce. | Animations, details, and colors may not be required for conveying the skills required. May impact budget and cost effectiveness of media. | J |
| Student finds it difficult to follow from one frame to another. | Frames may not be consistent in layout and navigational tools may not be present or adequate. | S |
| Student gets confused from one lesson to another. | No common guidelines between designers and writers. | S |

*Note:* Fix categorized as J = judgment and S = standards (bold indicates it is more important).

***Advantages of IMI to Training Immediacy*** Interactive multimedia instruction can get in-depth asynchronous training to everybody, everywhere quickly by using CD-ROMs to store and deliver the training. With CD-ROMs there is a great opportunity to include substantial amounts of video and other graphics. The speed of distributing the courseware can even be made greater if the courseware is designed for use over the Internet. Being able to reach all of the students at the same time almost regardless of location is an extremely powerful advantage over other methods that are constrained by the availability of students, facilities, and instructors.

***Disadvantages of IMI to Training Immediacy*** At the present time, IMI cannot be used practically for Web-based training if the files are too large (because of the graphics containing long video sequences, bitmap photos, or heavily graphic animation). If last-minute changes are required, it can cause a significant delay in the release of the courseware while updates are being made. If CD-ROMs are used, new CDs may have to be cut. If not done well, IMI will still need to provide a substantial period for students to contact instructors, retake the course, or seek other assistance such as peer help. Longitudinal studies are needed to determine if the absence of a human will have a significant impact on long-term retention of material compared to live instruction. Investigations show that certain students will not be comfortable in an unstructured environment where an instructor is not present (Abell, 2000; George et al., 2001; Lee et al., 1996).

### 22.3.3 Training Efficiency

To adequately address efficiency of the training methods, it is important to examine the strengths and weaknesses of both development and delivery. Costs must be addressed considering both monies expended (what does it cost to develop an hour lesson of instruction) and performance efficiency. To accomplish the latter, a measure of efficiency (MOE) is needed that compares student and instructor time spent against expected student performance. This allows the training manager to assess the potential value added by each student for the amount of investment.

***Costs Associated with Development*** In this category we consider the costs associated with the development of the product to where it is ready to be used by either an instructor or a student. Actual costs will vary depending on location of the work and the workforce, the nature of the topic under development, the requirement for subject matter technical expertise, and the technical approach used to prepare the courseware. The first two factors are self-evident while the latter factor refers to the method in which training development personnel are assigned tasks and how they function.

***Costs of Traditional Standup Instruction*** Standup Instruction can be prepared quickly because it is typically prepared by experienced staff that work from a previous plan of instruction (POI) built for similar systems. They may likely get access to early draft materials from the system developer such as draft technical manuals (TMs). Following an outline they can easily prepare sufficient vugraphs to conduct a class. Under conscientious management, the developers will be given access to engineers and other subject matter experts (SMEs) to develop the initial materials. They may be able to photograph the equipment or mockups in the shop. The costs associated with this development can be as low as 40 hours development time for 1 hour of instruction.

Vugraphs prepared to support this type of training typically contain entirely too much text to be effective as a visual media. So the reliance and success of the technique is on the dynamics of the instructor. On the audio side, the narrative is presented simultaneously with what appears on the screen. If the narrative and screen cues are totally unrelated, you now have inefficient media, because they are competing for the student's attention. If the instructor is dynamic, he or she will first call attention to a major teaching point on the screen and then proceed to support, enlarge, and reinforce the teaching point with examples, anecdotes, and questions to stimulate the students' responses. However, the keyword is *if*. To make instructors consistent, you must take time to develop a script or detailed outline in your plan of instruction. This takes time and additional funds and is not always effective. We have all witnessed the instructor who, not wishing to miss any major points, stands behind a podium and reads a script verbatim. The students' time is now not being used efficiently; some may be snoozing or daydreaming.

If vugraphs are used with good graphics, they take time to develop, and the cost goes up. If videotaping is used to avert the need for drawings, it takes time to prepare and shoot a proper script. If nonprofessionals do the videotaping, it will usually be less than perfect in terms of camera angles, jumpiness, shadows, and general composition. If professionals are used, it is expensive, especially if you need the work done quickly as you would for a new system. It also can no longer be readily updated so the speed advantage is taken away.

***Costs Associated with Development of IMI*** The major drawback for using IMI is the initial cost of development. Good, highly interactive IMI has been known to require 300 hours of development time per hour of instruction, which can mean costs ranging from $20,000 to $40,000. The more animation and interaction that are inserted, the higher the cost. Interactive multimedia instruction is usually referred to in terms of levels of interactivity supported by the IMI from simple page-turner (level 1) to exceptionally complex interactivity based on an intelligent learning model that adjusts to each student (level 4).

Another major reason IMI is not used more often to support new systems is the total length of time that can be required to prepare it for delivery and use. A single course of 130 instructional hours can take up to two years from the time of task analysis to final delivery of a useful product. A painstaking sequential process is generally pursued following task analysis to first develop outline storyboards, decide on the right mix of media, develop the media, go through the editing and review process, add narration and other audio, and assemble each lesson. Then the lessons are staffed allowing a certain number of weeks for review. After all comments are received, the entire development process is repeated to accommodate the changes. If hardware changes occur at the last moment, it is very difficult, if not impossible, to recover. When the lessons are totally prepared they are loaded on the server if they are to be part of a Web-based course. Validation is then accomplished with target audience students using the materials directly from the Web.

***Costs and Savings Associated with Time Spent in Training*** Given that two courses using different media have been validated as teaching the required skills and knowledge, time spent in training can be used as an MOE of each media to deliver training to an individual (e.g., the time it takes the average student to get through the course). The efficiency of each medium can also be determined in terms of meeting the needs of all students collectively (e.g., the time it takes to get all students up to standard). Finally, the time required for each media to sustain the acquired skills can be compared.

***Costs and Savings Associated with Traditional Standup Instruction*** Instructor-led live instruction can save costs because it can be easily changed in place to accommodate a specific target audience. This can reduce the amount of time spent in training if the majority of students indicate by either pretest or in-class discussion that they already know certain materials. The instructor can reduce the amount of instruction required or even bypass certain lessons or parts of lessons. Because this method is live, it is also flexible, and the instructor can improve on the course as he or she goes from group to group and discovers deficiencies or ways to improve the instruction.

Traditional live instruction can incur significant costs, however, since it may require that a great deal of time be spent traveling to each site for instructors or to a training site for the students. Travel may consume an entire day and, if delays are encountered, can result in fatigue before training has even started. Frequently, equipment must be carried along, which at a minimum is expensive and inconvenient to transport. The equipment may also get damaged and may not work upon arriving at the site. These indirect costs can be significant and must be covered in any study of cost–benefit analysis. Experience has shown that another significant cost to traditional standup instruction is that if instructors visit a work site they will have to repeat that visit periodically because of turnover of personnel. The alternative is to develop a train-the-trainer package, which adds to the cost of the program and represents another set of documentation that must be kept current. Traditional training is also "lockstep" in mode. This means that, on the average, unless the student population is unusually homogeneous, the brightest students will be spending hours in the classroom unproductively while attention or time is given to the slower students. The temptation will usually be to leave the slower students behind to catch up later as best possible. This latter group will, therefore, not spend as many productive hours in the classroom as needed to meet the standards of training. These students must be retrained, and the costs must be added to the program.

If traditional instructional methods are used, especially in connection with VTT, to squeeze more students into each class to drive down the cost per student, this must be approached very cautiously. There are limits to the number of students that can be attended to simultaneously to give each the opportunity and requirement to respond. This has been found to be about 8 to16 students per site, especially for technical subjects, and with VTT no more than three or four sites, even if only 8 students are located at each site. With any higher number of students or sites, efficiency quickly turns to inefficiency because the unattended students fail to learn to mastery. They will need to be retrained or learn on the job and risk inadequate performance and remain unproductive while learning. These costs too must be entered into any cost–benefit analysis.

***Costs and Savings Associated with IMI*** If properly executed, IMI can be very cost effective. Cost savings have been validated in many different areas. For example, many studies have shown that the time required for training a task to a measured standard can often be reduced by 20 to 30 percent over a traditional means of instruction. As a hypothetical example, consider that 100 employees earning \$100 a day are being trained together in a course requiring five days or 40 hours. If the same performance can be achieved in 20 percent less time, that would equal 8 hours or one day. In pay that amounts to \$100 per student and a total of \$10,000 for the group.

Another type of savings validated is travel time. In our hypothetical example, if the students were delivered IMI at their company location or their home, they could save on average \$1000 per person in total travel costs (e.g., transportation, food, and lodging). That would amount to a total of \$100,000 in savings for that cost. If you consider that they are sometimes paid for traveling on travel days, that could amount to another \$100 to \$200 per student and an additional \$10,000 to \$20,000 for the total group. The bottom line is that if you trained four or five such groups annually, you would recoup your investment within the first year and get all of your personnel trained effectively. Moreover, they could use the same courseware to sustain their skills over time.

A third type of savings documented, though not as universally studied as the above factors, is the time and cost of training equipment and materials saved when students are learning to operate or repair an item of equipment. This is for the hands-on phase after completing the knowledge phase of training. In studies of marksmanship, trainees require fewer rounds of live ammunition to qualify. In studies of both operation and maintenance, IMI-trained students make fewer mistakes, thus reducing wear and tear on the end item (Ross and Yoder, 1999; Throne and Lickteig, 1997).

Note that these are cost savings associated only with training. Because more students can be brought to greater competency quickly, the cost savings due to improved performance on the job can be well above that expected with more conventional training. Moreover, students will be able to access the training for refresher training at any time and therefore sustain their proficiency, whereas conventionally trained personnel can be expected to have a steeper decline in retention, particularly for tasks infrequently performed. These are perceived advantages that have not been adequately tested. Metrics do not exist for assessing training impact over the life cycle of a system. While strong rational arguments can be made for the efficacy of IMI, the fact remains that we do not know the long-term impact of the technique. It can legitimately be questioned as to whether IMI alone can ever provide a satisfactory solution. Long-term studies are clearly required. The few that have been done comparing any types of media and techniques typically show no significant difference over time. Many other factors such as motivation account for long-term performance. It has been suggested that the question can be partially answered by channeled metrics that can be used to attribute performance to a particular method. Examples of such metrics would be the need for refresher training looked at in terms of elapsed time (how soon is it required based on performance tests and work records), frequency (how often is it needed), and length (how long of a training period is needed).

In summary, there is mounting evidence that traditional instruction may not be as effective, immediate, or efficient as needed by new systems when compared with the possibilities offered by new IMI technology. Although there are clear, short-term cost-effectiveness advantages, some real issues remain to be clarified in terms of long-term training and cost effectiveness.

## 22.4 HSI APPLIED TO TRAINING DEVELOPMENT PROCESS

As powerful as IMI appears to be, equally powerful are some of the drawbacks that prevent it from being applied to new systems. The challenge is to reengineer the training development process such that a large IMI project can be completed on time, within a reasonable budget, and still be able to reap the outstanding benefits of IMI that have been listed above. In this concluding section we present a new paradigm for the production of IMI. The procedures described below are based on a commercial program called The Courseware Factory.[2] It is in essence a marriage of the old concepts of mass production and efficiency studies with new concepts of information management made possible by powerful computers, servers, database technology, software such as Hyper Text Markup Language (HTML), and the Internet. UTOPIA[3] is a unique tool for authoring and data management that has made execution of this paradigm particularly reliable and cost effective. The specific examples shown here are based on UTOPIA; however, the principles illustrated can be applied using other similar tools to achieve the same general benefits.

The principles used by UTOPIA are illustrated with the following topics:

- mass production,
- analysis of the bottlenecks,
- reengineering the design and development process,
- new tools for the designers and developers of IMI, and
- additional tools and methods.

### 22.4.1 Mass Production

One of the tenets of Henry Ford's mass production process is use of an assembly line that is laid out in logical, usually linear order that supports continuous production. Specialized activities are carried out at each of the workstations, and the number and type of workstations are determined by the classic and proven instructional systems design (ISD) model. Typically, a training developer is assigned to perform each of the ISD functions shown in Figure 22.7 for the assigned lesson. Functions would be performed linearly with the cycle beginning again following student or customer feedback.

The training production process for an IMI production facility would begin with a similar streamlined approach adding technology specialists to the process. Figure 22.8 shows the role of each of the specialists in the production sequence. Interactive multimedia instruction requires specially trained graphic artists who prepare new types of graphics such as animation made feasible by computer software. If the IMI is to be Web based, personnel are required who have training in that specialized area. Audio specialists and narrators are other special skills required for comprehensive IMI.

### 22.4.2 Analysis of the Bottlenecks

F. B. Gilbreth's principle of breaking a task down to its most basic activities to study how it can be done more efficiently is applied to the above classic model. There are at least two

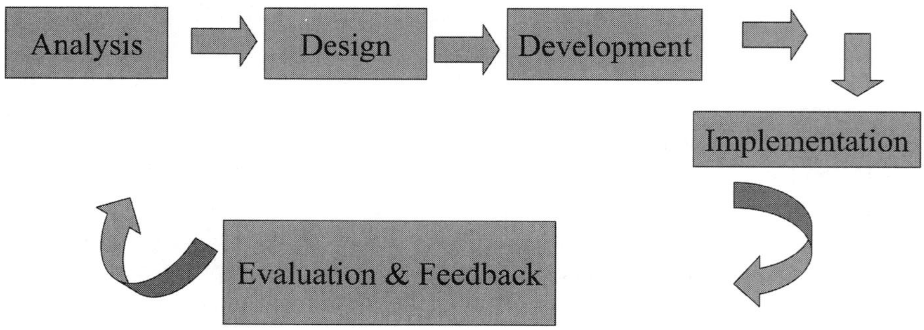


**Figure 22.7** Instructional systems design model.

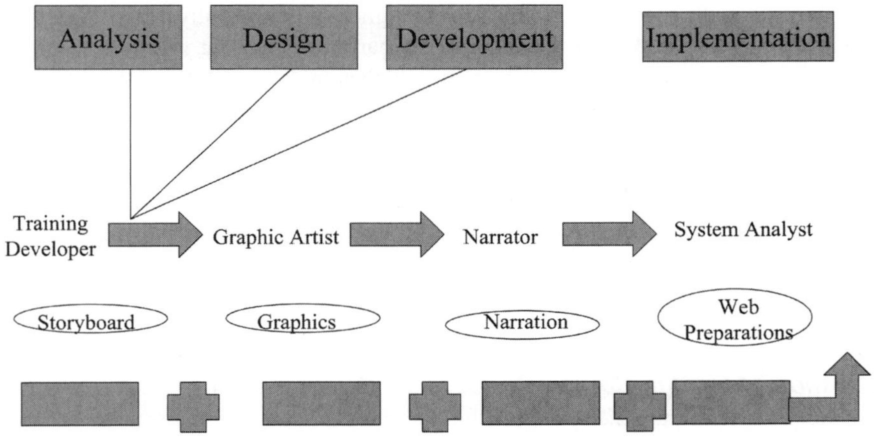


**Figure 22.8** ISD model applied to IMI production.

major bottlenecks that can occur in the production of IMI of any scale when the linear model is followed: (1) the process itself and (2) evaluation of the product.

***Linearity of the Development Process*** The process itself can be a major bottleneck because, as noted earlier, it has been developed linearly, which generates queues. Instructional designers must generate storyboards, usually starting with the output of a task analysis that lists the task, the condition, the standard that must be achieved, and then the task elements or details. The graphics specialists, narrators, and system analysts usually wait until this process is done to take their turn to act on the product. Figure 22.9 shows the linear flow of data as modified by each specialist.

When the courseware designer has completed a few draft storyboards, he or she will discuss the required visuals item by item with the graphics designer. When the graphics are completed, the screens would be produced and would have to be checked by the designer. Next, the narration would be prepared and added. When the lesson designer has written the narration, the narrator can then proceed. Note that if the narration is to be synchronized

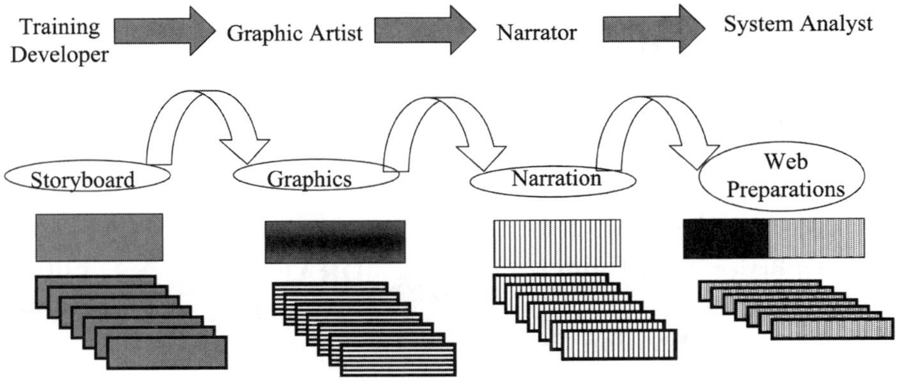


**Figure 22.9** Linear development model—production stage.

with the action in the text and graphics, the designer must carefully script the cues and work with the narrator and audio engineer. If the narrator is in a far removed studio when recording the narration and discovers apparent glitches in the script, there must be either a temporary halt or a scheduled update to the session; this increases the bottleneck. When the narration is done and added to the screens, the designer must again check the product. Finally, the product is turned over to a systems analyst who applies appropriate coding and adds other details so the product is ready for testing for transmission from the website. If the analyst uncovers problems such as size of files being too large to move over the Internet, graphics not fitting properly on the screen, text being too small or runs off the screen and cannot be seen, the product must be returned to the designers, graphic artists, or others for an appropriate fix.

***Application of Quality Control Functions*** The second major bottleneck is evaluation of the product. Figure 22.10 shows that there are at least two major layers of quality control review that enter the production queue. These reviews, if performed, can create serious bottlenecks for the manager.

The supervisor may at any time wish to (and should) either spot check or perform a complete review of each lesson. Serious additional bottlenecks arise if you expect the designer and developer to continue work on new tasks and attend at the same time to the supervisor's review. If designers or supervisors have all the materials in their possession to do the evaluation, it means that others do not have access, and the bottlenecks continue. Multiple copies or a shared drive can be used but must be carefully orchestrated and controlled by the designer to ensure all inputs are correlated and entered into a master copy. This process consumes a great deal of time if done correctly, and meanwhile the designer is not able to move forward on new material development. Still

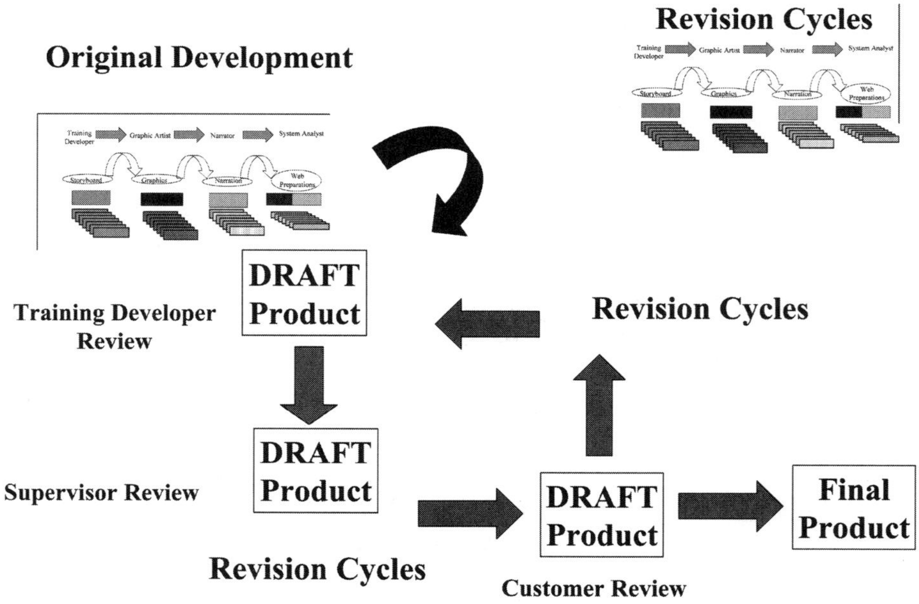


**Figure 22.10** Linear development model quality control stage.

other serious bottlenecks arise if the supervisor evaluates the lessons of several designers and developers who are all working on lessons that belong to the same course and discovers that the styles are too different to be used in the same course. If some of the designers must reformat their lessons to resemble the other lessons, the bottlenecks continue.

At various points along the way the users or customers of the product also expect to review the product. When they do review the product, they again initiate their own internal queue going from department to department and evaluators and supervisor. When they return the product, they expect to see the changes that they made and have an accounting of the disposition of their comments and whether or not they were resolved. The product or answers must then be prepared by the development staff and returned to the customer. It can readily be seen why IMI production is a very time-consuming process and why to attempt to do it on a massive scale has run into serious difficulties.

To overcome these problems, HSI engineering principles can be applied to all stages of the process. First, the design and development process can be reengineered. Second, tools and methods can be developed to assist in the production and control of products that can lead to standardization and high quality. These methods must still be coupled with training and job performance support materials for all employees and supervisors to ensure proper execution.

### 22.4.3 Reengineering the Design and Development Process

Methods to overcome the bottlenecks are based on certain principles that, while traditional and time honored, lend themselves well to solving the new problems of development of technology-based training. These are small-team concept versus individual effort, synchronization of multiple teams working in parallel, and continuous quality control.

***Small Team versus Individuals*** The workstations are broken down to provide individual teams consisting of a training designer to develop or author the lesson and a graphic artist who is also assigned other application responsibilities, such as loading the storyboards, graphics, and narration into the database. Together they have the responsibility to generate the first stage of the lessons.

***Synchronization of Multiple Teams*** Placed with these personnel are other teams who also produce lessons for the same course. Assigned to the entire group of teams are Web designers who load the material and test it for suitability for transmission over the Internet with major browsers to ensure usability as Web-based instruction. In order to standardize the output across lessons, one lesson author and one graphics specialist are designated as leads in their respective specialties. The entire group is assigned a chief who can work directly with the customer to determine both product and schedule. This organization could be nested and replicated as necessary to accommodate large production efforts. Figure 22.11 depicts such an organization sharing a networked database. This organization allows for efficient use of specialized skills as well as a free flow of information in the production of IMI.

***Principles of Quality Control*** Quality control must be done as a major, multilevel, continuous process and not as an end-state process when it is too late, a "lip service" check that is worse than no checks, or an after-the-fact check after poor products are

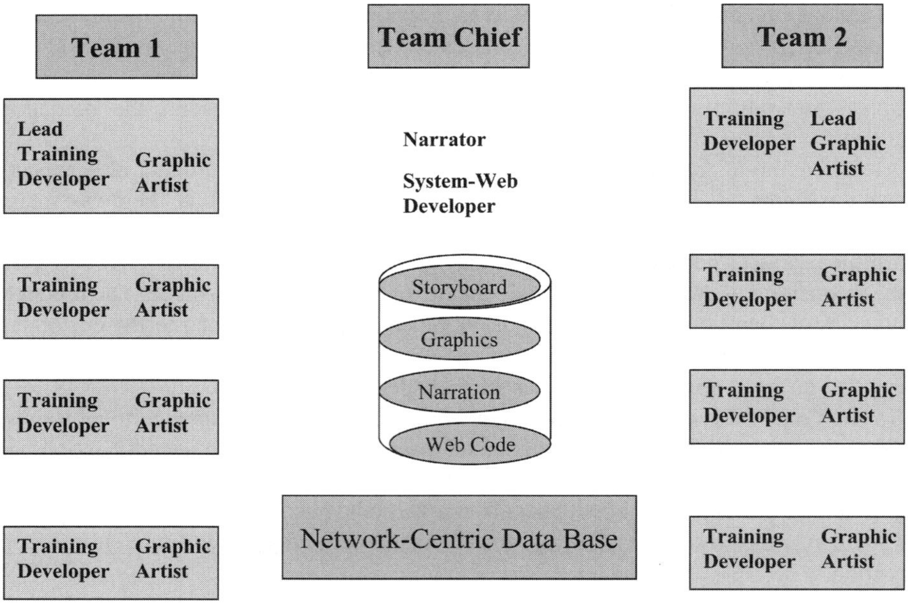


**Figure 22.11** A model production assembly.

delivered. Beginning at the very top, there must be a production manager who watches both quality of products and schedules across the entire assembly. For each lesson and course there are two additional levels of checks that can be done to assure standardization and quality: the lead designers (content and graphics) and the group chief who oversees all lessons going into a course. In addition, the system developer could serve as a further quality control check if items do not load on the server properly or cannot be retrieved properly through the website.

To continue the analogy to a mass production factory, the requirement is to create a process to produce a series of lessons that all meet certain standards (effective in conveying the skills and knowledge to the criteria required) that have the same "look and feel" across all lessons to relieve the student from having to learn new navigational skills or encountering confusing formats because the lessons were designed by different authors. These same lessons must be usable on the Internet (efficient). Each course must be custom designed to meet a particular customer's needs and delivered on time as requested (immediate).

### 22.4.4 New Tools and Processes

New tools are being continuously developed that have been made possible as a result of the information revolution, including use of authoring systems that use a network-centric common database, automate the generation of products, provide on-line access, and leverage the use of templates.

***Common-Use Database*** Figure 22.12 illustrates how a database should be designed to allow multiple authorized users simultaneous access to the database. With simultaneous access to a central database, the lesson designer can be working on one storyboard while

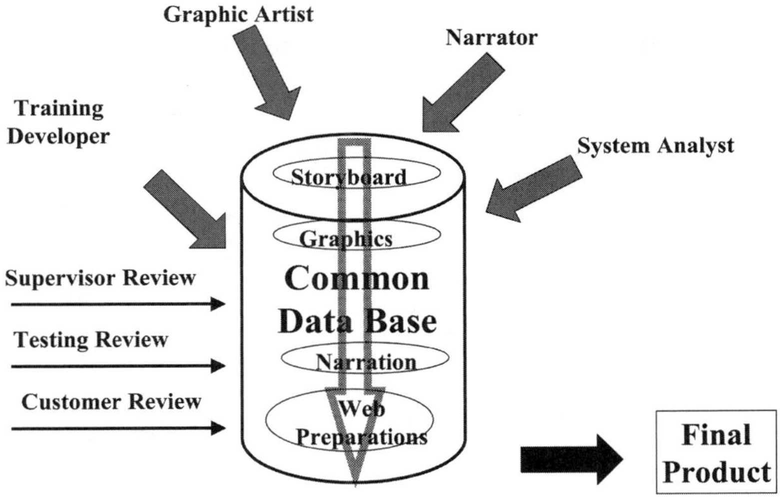


**Figure 22.12** Network-centric nonlinear development model.

the graphics designer is accessing another storyboard to see what graphic has been requested. The lead designers can call up still other storyboards to check on progress and quality. At the same time the designer is preparing storyboards he or she can also be preparing instructions for the graphic artist describing what is needed by way of graphics. This is done on the same document accessed on the database. Thus, the supervisor and other reviewers can simultaneously use the same data.

***Use of Templates*** To generate a consistent product efficiently, there are tools to assist team members to perform their roles. The tools consist of an authoring tool featuring automated protocols and templates making data entry and retrieval into and out of a database very efficient. The database should contain templates for a variety of types of frames and graphics that will serve as a job aid, serve to generate a consistent product, and allow for speed in application, as will be illustrated later. With a single click the lesson designer or graphic artist or supervisor can cause the storyboard to be formatted for a preview or with another single click the entire lesson (in whatever status it may be in at the time—text only, text and graphic, text, graphic, and audio) can be generated and viewed as if a student were taking the lesson. It should be able to be refreshed at anytime. This will make the process ideally suited to handle last-minute changes that arise with new systems. Additional features and functions of these recommended tools are described below in the context of the roles they support.

Using object-oriented programming, lead designers and chiefs can select certain templates based on customer input and target audience descriptions, which are then required for use by everyone working on individual lessons. This enables the design team to quickly generate a consistent product. One way to be consistent is to require graphics to be generally placed into a certain quadrant of the screen for all lessons. For example, all or most graphics may be on top or bottom of the screen or left half or right half, so the student always knows where to expect the graphic and likewise the text to occur. Figure 22.13 illustrates two frames that have the graphic in the same location (in this case, the upper two-thirds of each frame is reserved for the graphic).

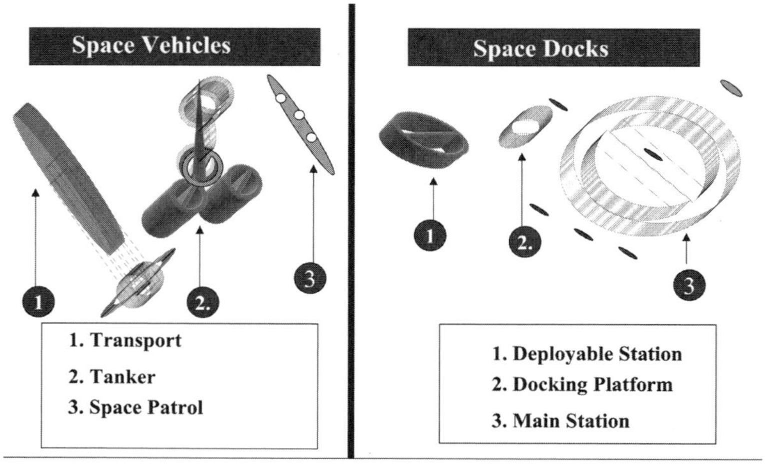


**Figure 22.13** Examples of "upper$\frac{2}{3}$" graphic layouts.

A useful template here would allow the author/designer to display text he or she is generating for a screen to see if the text will fit the remaining space, given that space is reserved for the graphic. Note that the actual graphic does not need to be created in order to do this; only a template need be used. An example is provided below where this particular authoring tool automatically generates a screen so the designer can see how the layout will appear to the student. The designer can immediately make changes to ensure that the text fits properly and is displayed properly. In Figure 22.14 the illustration depicts a

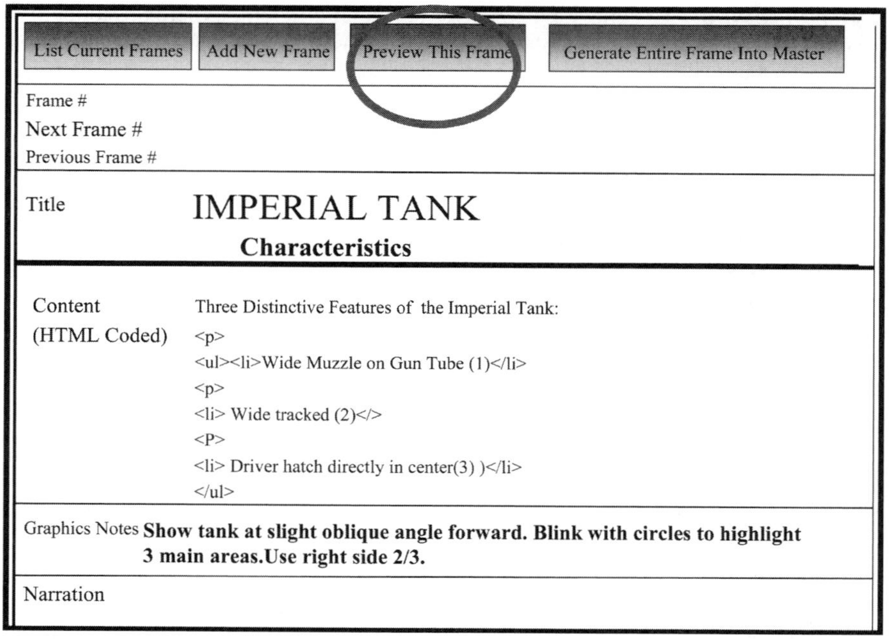


**Figure 22.14** Net-centric authoring screen layout.

**IMPERIAL TANK**

**Characteristics**

Three Distinctive Features of the Imperial Tank:

- **Wide Muzzle on Gun Tube (1)**
- **Wide tracked (2)**
- **Driver hatch directly in center (3)**

**Figure 22.15** Composed screen with graphic added.

comprehensive storyboard that is Web based. The designer can type in the content with code. Then with a click a preview of the screen can be called up to determine if both text and graphic will fit properly (see Fig. 22.15).

This process alone saves countless hours when graphics are being prepared because it saves both designer time and artist time and shortstops false starts that may have to be changed later when either the graphics or text prove to be too large or not usable for the screen. Likewise, at the same time the designer also enters the draft narrative on the same document that will supplement and enhance the graphic and text. Figure 22.16 depicts the same storyboard as above but with narrative added.

This is ideal because the developer is focused on the learning strategy and can immediately capture the narrative while still thinking through the process. This again saves time so that the developer does not have to refocus his or her thoughts later and try to create the narrative at a later time. The same tool allows the designer to assign numerical codes to each storyboard so that sequencing can be done, interaction can be supported, and navigation can be supported. Once the numerical tags are assigned, the tool takes over. The tool generates the text screens, the new screen with addition of the graphics, and a narration report that becomes the actual script to be read by the narrator.

***Tools for Graphic Designer*** The graphic artists could draw a variety of catalogued graphics from a library and also prepare new graphics using contract-furnished illustrations. When required, graphic artists and developers can digitally photograph the items required at customer sites. These photographs and line drawings can then be changed and enhanced with software to other forms such as vector graphics to enhance the graphic and use it for animation. In this way animation sequences can be devised that use less file space than photographs and make highly interactive Web-based IMI practical. Figure 22.17 shows stages of a simple line drawing being enhanced to more closely depict real

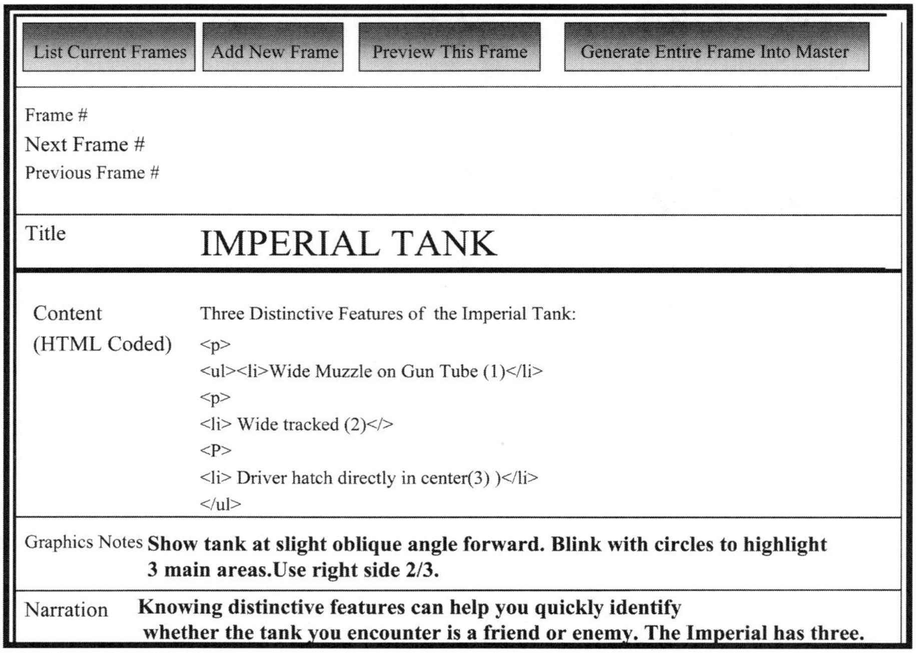


**Figure 22.16** Authoring screen layout with narrative.

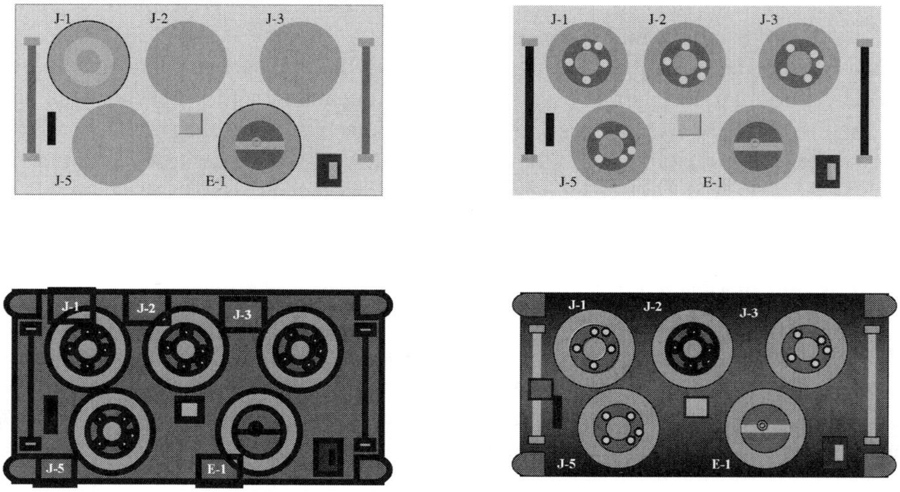


**Figure 22.17** Example of enhanced graphics.

equipment without having a large file size required for the illustration. This is currently critical in Web-based training that must reach users with a wide range of receiving capabilities.

***Tools for Narrators and Audio Engineers*** Being in-house, the narrator, audio engineer, and lesson designer can quickly meet to go over any problem areas whenever they arise and are able to fix them on the spot. The audio engineer may also wish to maintain a library of copyright–free music as well as sound effects that could be used to enhance a lesson where required. He or she can advise the designer of both the availability and suitability of audio cues and stimuli beyond narration. In-house personnel can be used as the narrators for the draft lesson as well as the audio recording and editing tasks. This method allows them to aid the audio engineer as to the intent of the narrative and allows them to serve as an immediate quality control for the designer.

The draft lesson with narration is made accessible to the customer to ensure their preferences are met. When the audio is sufficiently ironed out, a professional narrator is brought in to complete the narration. The lesson designer is usually present in the event of any questions. Because the system is stylized and standardized, there are rarely major problem areas that are required to be fixed.

***Software Application/System Developer/Web Developer*** The system developer loads the draft narration so the lesson designer can actually audit the lesson storyboard by storyboard to determine if any changes were required before the same script is turned over to the customer for review. Later, the developer reloads the narrative when the professional narrator has completed the lesson.

The system developer will next load the entire lesson outside the internal firewall and determine if any problems are encountered when two major browsers are used to access the lesson. When the developer is satisfied that the lesson can be delivered, it is ready for release to the customer who can review the product.

***Role of User*** When feasible, the user should be involved throughout the development period. Representatives of the target audience can provide required insights as to skills already attained, aspects of the work environment that may influence how a skill is conveyed or practiced, and how the developing formats meet their needs and expectations. Still other user representatives may provide the subject matter expertise that ensure the lessons are technically correct before they get to the field for testing. When it is not feasible to have the users available, the developer should seek surrogates from within his or her own staff and ensure that the development process gets back to the users frequently.

***In-House Evaluators/Independent Testers*** In addition to the group leader doing spot checks to ensure the look and feel are following the customers' preferences and are consistent, the production manager, subject matter experts, and formally assigned editors review the lesson. Clarifications to text, graphics, audio, or inherent training strategy may be requested. This may be done with only a sample or, if spot checks show the product as high quality, essentially an alpha version of the entire lesson. In this latter instance the product may simultaneously be released to the customer in order to speed the review process.

***Streamlining the Review and Follow-up Process*** The same database that allowed the lesson to be built can now be used to support the review process. The special tool supports free play review by multiple authorized (password-protected) reviewers simultaneously worldwide. They can review the entire lesson frame by frame and with a single click be able to submit a comment, correction, or suggested change that is automatically cataloged by frame number and identifies not only the comment but also the comment owner and the date of the comment. The tool generates a special report on-demand available again to authorized users who can view all the comments. The comments can be reported and sorted by date, user, or subcategories of the course or lesson. The lesson designer, or if deferred to the supervisor for sensitive comments, can then generate the deficiency report (DR) and respond to each item right on the same report. He or she can correct the database and report what action was taken in the report. Typical items to include in a DR file for both management and archival purposes are shown in Figure 22.18.

A quality control person (evaluator/tester) will then verify that the correction was made and that the response was appropriate to the comment. Thus, a customer or supervisor or any reviewer can determine the status of their comments as to whether or not the problem was understood, who made the correction, and whether or not the correction was validated and when. There is, of course, opportunity to simply comment in reply with no further action when a satisfactory answer or rationale can be given as to why something was presented in a certain way. Sometimes the comment is simply laudatory in nature, which helps designers to know they are meeting their goal. At any rate, the lesson will not be released until all such comments have been satisfied and validated. The speed, precision, and simplicity of this process again make it ideal to support new systems that continue to be refined.

| New Comments | Fix recommended | MGMT Review | Customer Accepted Fix |
|---|---|---|---|

| | Submitted By: ____________ Date ____________ |
|---|---|
| Fixed By | ____________ Date ____________ |
| Reviewed By | ____________ Date ____________ |
| Approved By | ____________ Date ____________ |
| Title | |
| Content | |
| Graphic | |
| Narration | |

**Figure 22.18** Deficiency reports of resolution.

***Test Questions and Exercises*** To ensure training effectiveness, a large pool of test items are prepared for the customers' approval based on the task analysis and target audience assessment. In-house personnel and later target audience personnel are administered selected test items as a pretest that cover all critical tasks. Given that the students are expected to not know the majority of questions, a pool of items in the same category including some identical questions are administered as a posttest to validate that the lesson has conveyed certain knowledge required to perform certain tasks. Treatment of training effectiveness is beyond the scope of this chapter. It is important to note, however, that such testing of relevant skills and knowledge is an absolute requirement to ensure training effectiveness.

Throughout the lessons about every fourth frame seeks to elicit a response from the student to ensure they have acquired the knowledge of those immediate frames. The tests and exercises given at the end of the lesson topic are also used as a means of reinforcing that knowledge. Figure 22.19 shows an example of an end of a section within a lesson on the setup of an item of equipment. The task requires that the correct cables and connectors be mated and that the ground wire be attached first to prevent a hazardous condition. The student would be required to know each of the cables and connectors in order to be able to pass the test. The student's ability to complete the procedure without error would indicate mastery of the knowledge required to perform the task completely and safely.

## 22.5 SUMMARY AND CONCLUSIONS

An HSI approach to new systems training can provide a process whereby personnel can be trained effectively, immediately, and cost effectively. The new technology provided by IMI is essential to this new process, provided the strengths are utilized and the weaknesses

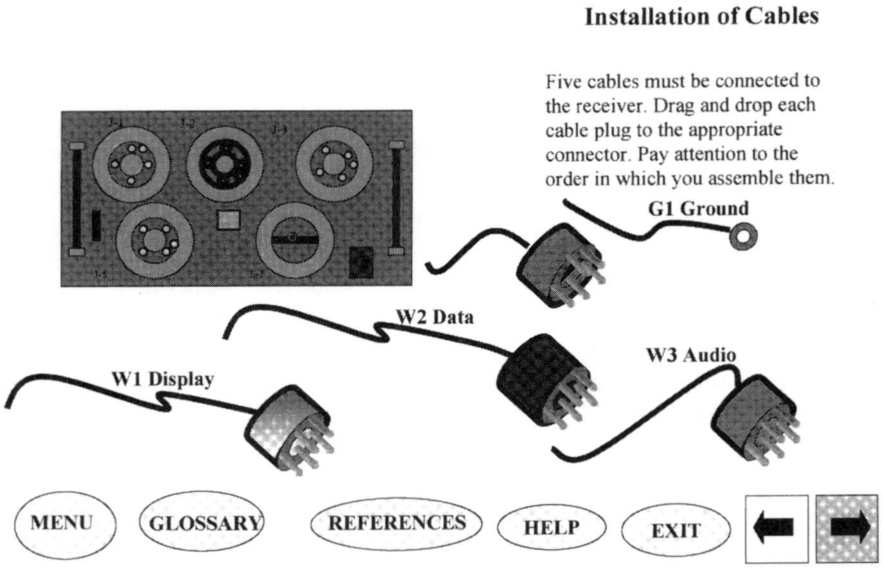


**Figure 22.19** Example of end of section test.

controlled. The chapter discusses the major process changes and concepts that can make IMI a viable candidate to support new systems training. These changes and concepts are summarized under the three critical requirements for personnel training on new systems:

1. Personnel must be trained effectively. This requires an IMI production process having the following:

- continuous and timely quality control with production supervisor oversight and incremental testing and review;
- IMI designed to use the proper mix of audio, text, and a full range of graphics;
- IMI designed from the student's perspective, including, as a minimum, a) a useful demonstration of the task understandable by the student, (b) capability for the student to perform the task, (c) capability for student practice, and (d) feedback of student performance; and
- experienced personnel leading the project, staff adherence to a common look and feel, and building competency among all production staff.

2. Personnel must be trained immediately. This means training development must be completed on time with the fielding of the new equipment despite last-minute changes. This is feasible for the following resons:

- It is now possible to simultaneously work and review draft training materials using the same database.
- Team and multiple team concepts can be applied.
- The DR process provides for rapid simultaneous worldwide reviews and fixes.
- The acceptance process can be streamlined, allowing the customer to see the changes requested and the supervisors to have complete oversight of the progress.
- Internal testing of the first draft greatly reduces need for subsequent changes.
- The product can be placed quickly outside the internal company network and be made available, with password protection, over the Internet for customer review. This will allow customers to review the product at their leisure, home site, or even their home. Access to the Internet and one of the two major browsers is all that is required.

3. Personnel must be trained efficiently (i.e., at least cost). The IMI production and delivery process can be cost effective by introducing actions like those that follow:

- Documenting the process with audit trails to show the customer the basis for any designs as well as provide a strict accounting for time and effort spent in development.
- Using procedures that provide ability to work concurrently, rather than linearly. This includes use of templates, reuse of data including graphics, and having free access to a common database.
- Training and sustaining acquired skills for many personnel simultaneously worldwide via CD ROM or Web-based training.

In conclusion, we have found that by determining and analyzing the bottlenecks of the training development and delivery process for new systems, it is possible to define a new set of efficient training requirements. These new requirements can be fully met through reengineering of the development process and by providing new tools and methods that are now cost effective because of improving information technology. Using these methods and tools as part of an HSI approach to new systems training, it is now possible to develop and deliver effective training materials on time and at reasonable, competitive costs.

## NOTES

1. Stapp (2001) was a major source for the background literature in this chapter. Compton (1999), cited in Stapp (2001) annotated bibliography, states that historically over half of all sales and field force automation projects have failed.
2. *The Courseware Factory*™ process was devised by the coauthor, William A. Stembler, in 1996.
3. UTOPIA is a proprietary tool developed for Computer Sciences Corporation by Dave MacLuskie.

## REFERENCES


Abell, M. (2000, December 7). Soldiers as Distance Learners: What Army Trainers Need to Know. Paper presented at the Interservice/Industry Training, Simulation and Education Conference, Orlando, FL.

Compton, J. (1999). CRM Training by the Book, Sales and Field Force Automation—The Executive's Guide to Customer Relationship Management. Cited in Stapp, K. M. (2001). *Benefits and Costs of Distance Learning: A Perspective from the Distance Learning Literature Since 1995*, Report Number AB-01-025. Department of the Army, TRADOC Analysis Center—White Sands Missile Range (TRAC-WSMR, NM)

George, E. L., Bretl, D., and Jackson, G. (2001). *MOS 92A Distance Learning Training Effectiveness Analysis. DC*: TRADOC Analysis Center—White Sands Missile Range (TRAC-WSMR).

Howard, F. S. (1997). *Distance Learning Annotated Bibliography.* Washington, DC: White Sands Missile Range, Department of the Army, TRAC-WSMR, NM.

Kulik, J. A. (1994). Meta-Analytic Studies of Findings on Computer-Based Instruction. In E. L. Baker and F. O'Neil (Eds.), *Technology Assessment in Education and Training* (pp. 9–33). Hillsdale, NJ: Lawrence Erlbaum Associates.

Lee, A. Y., Gillan, D. J., and Harrison, C. L. (1996). Assessing the Effectiveness of a Multimedia-Based Lab for Upper Division Psychology Students. *Behavior Research Methods, Instruments, and Computers, 28* (2), 295–299.

Metzko, J., Redding, G. A., and Fletcher, J. D. (1996). *Distance Learning and the Reserve Components*, Washington, DC: Institute for Defense Analysis (IDA).

Parchman, S. W., Ellis, J. A., Christinaz, D., and Vogel, M. (2000). An Evaluation of Three Computer-Based Instructional Strategies in Basic Electricity and Electronics Training. *Military Psychology, 12*(1), 73–87.

Ross, K., and Yoder, M. K. R. (1999). Producing Computer Literacy for the Digitized Battlespace of the Future. Paper presented at the Interservice/Industry Training, Simulation and Education Conference, Orlando FL.

Stapp, K. M. (2001). *Benefits and Costs of Distance Learning: A Perspective from the Distance Learning Literature Since 1995*, Report Number AB-01-025. Department of the Army , TRADOC Analysis Center—White Sands Missile Range (TRAC-WSMR) NM.

Throne, M. H., and Lickteig, C. W. (1997). *Training Computer Skills for the Future Battlefield: A Review and Annotated Bibliography*, Research Product 97-15. Alexandria, VA: U.S. Army Research Institute for the Behavioral and Social Sciences.

CHAPTER 23

# Air Traffic Control and Human Factors Integration

ANNE MAVOR and CHRISTOPHER WICKENS

## 23.1 INTRODUCTION

This chapter draws heavily on the reports of the National Academy of Sciences' Panel on Human Factors in Air Traffic Control Automation.[1,2] Over a four-year period (1994 to 1998), the panel reviewed the air traffic control (ATC) system from a human factors perspective and assessed future automation alternatives as they related to the role of the human operator in ensuring safety and efficiency. Two reports were published: *Flight to the Future* (Wickens et al., 1997) and *The Future of Air Traffic Control* (Wickens et al., 1998).

The ATC system and its development and management by the Federal Aviation Administration (FAA) provide an excellent opportunity for examining the role of human factors and human-centered design in a safety-critical, complex system. The American airspace system is impressive in its capacity and safety. In addition to maintaining safety, the ATC system is charged with the efficient flow of traffic from origin to destination. The joint goals of safety and efficiency are accomplished by controllers through an intricate series of procedures, judgments, plans, decisions, communications, and coordinated activities. The success of the current system is demonstrated by the safe and rapid response of the system immediately following the terrorist attacks on September 11. Within 3 hours, all aircraft flying over the United States were safely landed (Bond, 2001).

The primary purpose of this chapter is to illustrate the challenges and benefits of applying human factors integration (HFI) to a large complex system. This will be accomplished by using the national ATC system as a case example. To help the reader better understand the issues underlying HFI applications to ATC, some general background is provided below on the ATC system.

*Handbook of Human Systems Integration*, Edited by Harold R. Booher.
ISBN 0-471-02053-2 

### 23.1.1 Air Traffic Control System

The most familiar aspects of the ATC system to the public are the communication and coordination between the pilot and the controller. However, many operations take place out-of-sight as standard procedures. For example, the task of ATC includes several phases: ground operations from the gate to the taxiway to the runway, takeoff and climb operations to reach a cruising altitude, cross-country flight to the destination, approach and landing operations at the destination, and finally, taxi back to the gate (or other point of unloading). The traffic to be controlled includes not only commercial flights but also corporate, military, and general aviation flights. Three general classes of controllers accomplish control, each resident in different sorts of control facilities. First, ground and local controllers (both referred to as tower controllers) handle aircraft on the taxiways and runways. Second, radar controllers handle aircraft from their takeoff to their cruising path at the origin (departure control) and return them through their approach at the destination (approach control) through the busy airspace surrounding airport facilities. This region is referred to as a terminal radar control approach (TRACON). Third, en-route controllers working at the air route traffic control center manage the flow of traffic along the airways between the TRACON areas.

The functions of ATC have evolved from a few crude navigation aids for pilots to a technologically sophisticated system using satellites, wireless digital communications, various forms of radar, and high-speed, high-capacity computers both on the ground and in the aircraft. For many flights, only relatively passive monitoring of the flight path is needed during the cruise portion of the flight. However, during taxiing and departure and arrival—that is, in the near vicinity of an air terminal—safe passage is likely to require several instructions for change of path or altitude from the ground-based controller to the pilot. In any case, it is the responsibility of the controller to oversee all movements of the aircraft to ensure avoidance of collisions with other aircraft or obstacles. The fulfillment of this responsibility has become increasingly complex over time. The main source of complexity is in the growth in the volume of flights and the diversity of aircraft.

Complexity raises the demand on controller–machine workload, which suggests a number of HFI issues in any attempt to handle the increased workload with the right combination of human operators and new technology. HFI issues could include, for example, the following questions:

- Should the number of controllers be increased or decreased?
- What special needs for increased team training are required?
- How much automation can be introduced to simplify controller workload?

This last question is of special concern to the committee because increased automation is frequently the solution offered by the engineering community for handling increased complexity in human operations. However, the goals of safety and efficiency required by the FAA cannot be met by simply automating those features that are capable of being automated.

### 23.1.2 Automation and the Goals of Safety and Efficiency

Even given the very low accident rate in commercial and private aviation, the need remains to strive for greater safety levels: This is a clearly articulated implication of the "zero-accident" philosophy of the FAA and of current research programs of the National Aeronautics and Space Administration (NASA). These research activities typically incorporate human factors concerns and often are directed explicitly at human factors questions. Solutions for improved air traffic safety have been explored in a number of areas, including automation, changing procedures, improving training and selection of staff, and introducing technological modernization programs that do not involve automation per se.

As noted above, the topic of particular interest to the panel was whether human factors applied to decisions about automation. The approach was driven by the philosophy of human-centered automation defined as follows (Wickens et al., 1998, p. 2):

> The choice of what to automate should be guided by the need to compensate for human vulnerabilities and to exploit human strengths. The development of the automated tools should proceed with the active involvement of both users and trained human factors practitioners. The evaluation of such tools should be carried out with human-in-the-loop simulation and careful experimental design. The introduction of these tools into the workplace should proceed gradually, with adequate attention given to user training, to facility differences, and to user requirements. The operational experience from initial introduction should be very carefully monitored, with mechanisms in place to respond rapidly to the lessons learned from the experiences.

Automation has the capability both to compensate for human information processing vulnerabilities and to better support and exploit human strengths. Controllers, such as human operators in other complex domains, are vulnerable in the following areas:

- monitoring for and detection of unexpected low-frequency events,
- expectancy-driven perceptual processing,
- extrapolation of complex four-dimensional trajectories, and
- use of working memory to either carry out complex cognitive problem solving and planning or temporarily retain information.

In contrast to these vulnerabilities, when controllers are provided with accurate and enduring (i.e., visual rather than auditory) information, they can be very effective in solving problems, and if such problem solving demands creativity or access to knowledge from more distantly related domains, their problem-solving ability can clearly exceed that of automation. Furthermore, to the extent that accurate and enduring information is shared among multiple operators (i.e., other controllers, dispatchers, and pilots), their collaborative skills in problem solving and negotiation represent important human strengths to be preserved. In many respects, the automated capabilities of data storage, presentation, and communications can facilitate these strengths.

A considerable amount of automation has already been applied to ATC tasks for the en-route, TRACON, and tower environments, and future automation is likely to be significant for all environments. This automation has been applied to support controller tasks across all levels of cognitive complexity. However, the application of highly automated features, which often virtually replace controller actions, has to date been largely reserved for tasks of lower cognitive complexity. When automation has been applied to tasks of higher cognitive complexity, the automation was used to provide assistance to controllers.

In its second report, the panel provided an analysis of human factors issues associated with several ATC automation efforts. These analyses are instructive in that they lay out the system functions, the context for development, and the human factors issues associated with design, testing, and implementation.

The analysis of the Center TRACON Automation System (CTAS)—designed to provide support for controllers in the "TRACON" region, surrounding major airports—was particularly illustrative of how HFI principles were incorporated by the FAA at various acquisition stages.

This chapter first reports the panel analysis for CTAS as an example of how HFI was incorporated into the design and development of an automated ATC system. Next, we consider how well the CTAS has been implemented to date and speculate on what this example might illustrate in terms of the HSI principles for organizational maturity. Finally, we discuss the types of coordination and integration issues that are frequently associated with harmonizing several systems (some already in existence and some under development) for an organization as complex as the FAA.

We focus on the CTAS for three reasons. First, the CTAS supports controllers in the approach phase of flight, which appears to be the greatest source of bottleneck in the national airspace system, as well as the most dangerous phase of flight, as defined by risk of accident. Second, based on the panel report, the CTAS appears to have a particularly positive record of human factors input in its development. Third, there have been a number of agency system integration difficulties in acquiring a fully deployed system throughout TRACON facilities, in spite of the strong operational need and positive human factors features inherent in the CTAS design.

Many other ATC automation systems were also considered in the panel's report, such as those that support planning for en-route controllers, those that address issues of safety on conflict avoidance on the runway surface, or those that address communications between pilots and air traffic controllers. The interested reader should consult this report (Wickens et al., 1998) for more detail on these systems.

## 23.2 HFI IN THE DEVELOPMENT OF AN AUTOMATED ATC SYSTEM

The main impetus toward the development of the CTAS was the desire to maximize the use of capacity in airport arrivals and landings. Limitations in prediction of trajectories and weather led to spaces on the final approaches that were not occupied by an aircraft, thus creating delays and not meeting the actual capabilities of an airport's true capacity. In the 1980s, NASA and the FAA Technical Center began an in-house research-and-development project to develop the software tools for achieving this optimization (Erzberger and Tobias, 1986), working closely with controllers and human factors professionals to create a fielded system. During the mid-1990s, this system has received several field tests at Dallas–Fort Worth International Airport and the Denver Airport and Center. It was also being installed at Schiphol Airport in Amsterdam, the Netherlands. Two components of the CTAS (the traffic management advisor and the final approach spacing tool), described below, are currently being installed by the FAA at a larger number of airports as part of its Freeflight Phase 1 program (Nordwall, 2001). The future status of the decent advisor is unclear.

### 23.2.1 Center TRACON Automation System

The primary objective of the CTAS is to assist the air traffic controller in optimizing the traffic flow in the terminal area (Erzberger et al., 1993). Delays are reduced and flight paths are flown in a more economical fashion so that potential fuel savings are estimated to range from 45 to 135 kg per landing (Scott, 1994). These benefits are accomplished by providing assistance in prediction, planning, and control in both routine and unexpected circumstances (e.g., changes in runway configuration). The CTAS is also capable of providing advice to controllers regarding particular airline preferences. The CTAS is comprised of three separate components, each supporting different classes of ATC personnel, located in different facilities, and coordinating different phases of the approach:

1. The traffic management advisor (TMA) supports the TRACON and en route traffic management controllers, primarily in developing an optimal plan, to assign each aircraft a scheduled time of arrival at a downstream point, such as a final approach fix or runway threshold, and a sequence of arrival relative to other aircraft approaching the terminal area. The TMA begins to compute these for inbound aircraft at a point about 200 miles or 45 minutes from the final approach. The plan is designed to optimize the overall flow of the set of aircraft as well as the fuel consumption of each individual aircraft. At the same time, it accounts for various constraints on runway availability and aircraft maneuverability. The plan is also accompanied by an assessment of flight path changes to be implemented in order to accomplish the plan. A set of three displays assists the traffic management coordinator in evaluating the plan. These include a time line of scheduled and estimated times of arrivals for the aircraft, a listing of alternative runway configurations, and a load graph that indicates the anticipated traffic load across designated points in the airspace in 15-minute increments. The displays can be presented in large-screen formats for group viewing. The actual implementation of the plan generated by the controller with the assistance of the TMA is carried out by the other two elements of the CTAS, the descent advisor and the final approach spacing tool.
2. The descent advisor (DA) provides controllers at the final sector of the en route center with advice on proper speed, altitude, and (occasionally) heading control necessary to accomplish the plan generated by the TMA. The critical algorithm underlying the DA is a four-dimensional predictor that is individually tailored for each aircraft, based on that aircraft's type and preferred maneuver, along with local atmospheric data. This predictor generates a set of possible trajectories for the aircraft to implement the TMA plan. The DA then provides the controller with a set of advisories regarding speed, top of descent point, and descent speed. In cases in which these parameters are not sufficient to accomplish the plan, path-stretching advisories are offered that advise lateral maneuvers. The DA also contains a conflict probe that will monitor for possible conflicts up to 20 minutes ahead. If such conflicts are detected, it will offer resolution advisories based initially on speed and altitude changes. If none of these is feasible, lateral maneuvers will be offered as a solution.
3. The final approach spacing tool (FAST) is the corresponding advisory tool designed to support the TRACON controller in implementing the TMA plan by issuing speed and heading advisories and runway assignments necessary to maintain optimal spacing between aircraft of different classes (Davis et al., 1994; Lee and Davis, 1995). An important secondary function of the final-approach spacing tool is its ability to rapidly adjust to—and reschedule on the basis of—unexpected events such as a missed approach or a sudden unexpected runway closure. Like the DA, the controller receives advice in the fourth line of the data tag and also has access to time lines. The final approach spacing tool exists in two versions: The passive FAST provides only aircraft sequence and runway assignments, and the active FAST includes speed and heading advisories.

### 23.2.2 Human Factors Implementation

Human factors have played a relatively important role in the maturation of the CTAS, from concept, to laboratory prototype, to simulation, to field test (Erzberger and Tobias, 1986; Tobias et al., 1989; Harwood et al., 1998). From 1992 to 1997, approximately 30,000 person-hours of human factors expertise have been devoted to CTAS development and fielding. In part, the successful implementation of the human factors input was a result of the fact that the development took place at NASA laboratories, with ready access to human factors professionals and active participation of controllers in developing the specifications. The development was not under constraints related to contract delivery time or required specifications. Human factors implementation was also facilitated in part by the frequent input of controllers to the design concepts of functions at all phases and frequent human-in-the-loop evaluations at varying levels of simulation fidelity. The controller's input was filtered by human factors professionals (Lee and Davis, 1995; Harwood et al., 1998).

Another important factor is that these evaluations (and system changes based thereon) continued as the system was field tested at the Dallas and Denver facilities (Harwood et al., 1998). In particular, developers realized the need for extensive input from a team of controllers at the facility in order to tailor the system to facility-specific characteristics. The introduction process was quite time consuming, taking place over several years. This proved necessary (and advantageous) both in order to secure inputs from controllers at all levels and also in order for human factors professionals and engineers on the design team to thoroughly familiarize themselves with the culture and operating procedures at the Denver and Dallas–Fort Worth facilities; this, in turn, was necessary in order for the trust of the operational controllers to be gained and for the CTAS advisories to be employed successfully.

It is also important to note that the system was designed to have a minimal effect on the existing automated systems and procedures. Finally, the CTAS was presented to controllers with the philosophy that it is an advisory aid, designed to improve their capabilities, rather than as an automation replacement. That is, nothing in the CTAS qualitatively alters the way in which controllers implement their control over the aircraft.

### 23.2.3 HFI Issues

The integration of human factors into the system design process requires that a team of knowledgeable specialists undertake a set of analytical steps. The results of these analyses can be used to help evaluate alternative design features as they are proposed for inclusion in the evolving system. The analytic steps carried out in the CTAS design effort is summarized below.

***Cognitive Task Analysis*** A cognitive task analysis reveals that the CTAS supports the controller's task in three critical respects. First, its four-dimensional predictive capabilities compensate for difficulties that the unaided controller will have in predicting and visualizing the long-term (i.e., 5-minute) implications of multiple, complex, speed-varying trajectories subjected to various constraints, such as fuel consumption, winds, and runway configuration. Second, its interactive planning and scheduling capabilities allow multiple solutions to be evaluated off-line, with the graphics feedback available in the time lines, to facilitate the choice of plans. Here also the system supports the workload-intensive aspects of planning (Johannsen and Rouse, 1983; Tulga and Sheridan, 1980), particularly prevalent when multiple plans need be compared. Finally, the CTAS, particularly the final-approach spacing tool, supports the controller's ability to deal with the high workload imposed by unexpected and complex events, characterized, for example, by a missed approach or unanticipated runway closure. The first and second of these tasks primarily affect the efficiency of system performance, whereas the latter has direct and beneficial safety implications.

***Workload*** A stated objective of the CTAS is that it will not increase controller workload; indeed, field tests of the system reveal that this criterion has been met (Harwood et al., 1998). As noted above, the CTAS has the potential to reduce workload during the "spikes" imposed by unexpected scheduling and spacing requirements due to a missed approach or closed runway. However, it is also the case that workload may be shifted somewhat with the introduction of the CTAS. Relying on an added channel of display information, rather than the controller's own mental judgment, may impose an increase in visual workload. In fact, any new set of procedures (such as those associated with the CTAS) would be likely to impose some transient workload increase.

Finally, although not yet reported, a tool such as the CTAS does have the potential of advising maneuvers that create an airspace considerably more complex than that viewed under unaided conditions (Wyndemere, 1996). In such a case, controller monitoring and perceptual workload may be increased by the controller's effort to maintain a full level of situation awareness of the more complex airspace.

***Training*** The general approach to training for the CTAS is to first provide simulation and then provide a shadowing of the real traffic off-line in the system. In the shadowing mode, CTAS elements provide the advice, and the controller can compare clearances that he or she might provide on the basis of that advice with clearances more typical of an unadvised controller and evaluate the differences (Lee and Davis, 1995). The controller can then determine the rationale behind the automated advisory. This builds confidence that the computer can provide advice to maintain separation. One might anticipate the need for some training of pilots regarding the CTAS, not because procedures are altered, but because the nature of the clearances and instructions may be changed, relative to the more standardized, space-based approaches (i.e., using the standard terminal arrival system) in a non-CTAS facility.

***Communication and Coordination*** Because of the philosophy by which the TMA plans are implemented via the DA and the final-approach spacing tool advisories, the CTAS imposes a relatively heavy communication load between operators and facilities. This is supported via digital data transfer rather than voice communications. Furthermore, the philosophy of repeated displays across different environments supports greater communications and coordination between operators, in that these can better support a shared situation awareness of the implications of different schedules. The extent to which ground–air communications are altered by the CTAS remains unclear. At least one field study of the final-approach spacing tool (Harwood et al., 1998) carried out at the Dallas Airport over a six-month period indicated that the system imposed no increase in overall communications, although the nature of the communications was altered somewhat, involving more messages pertaining to runway assignments and sequencing.

### 23.2.4 Automation Issues

In its report and review of the literature, the panel identified a number of important cognitive issues and lessons learned pertaining to automation of systems in other domains, particularly automation on the flight deck (see reviews by Billings, 1996a, b; Parasuraman and Riley, 1997; Parasuraman et al., 2000). It then applied these to several proposed ATC automation tools and to the CTAS in particular.

The CTAS remains sufficiently recent in its introduction that there has not been time to identify specific human factors automation issues on the basis of operational experience (e.g., operational errors or aviation safety reporting system incidents). However, analysis of system capabilities does suggest at least some that might surface.

***Mode Errors*** The CTAS contains some multimode operations. For example, with the DA, controllers can choose a route intercept or a waypoint capture mode for individual aircraft as well as one of three possible speed control modes for all aircraft (Erzberger and Nedell, 1989). However, the system appears to be designed so that different modes are prominently displayed, and active decisions must be carried out to change modes, so that mode errors would appear to be very unlikely.

***Mistrust*** There is a possibility that the advice offered by the CTAS could be initially mistrusted by controllers if it differs substantially from the way in which control is typically accomplished. Accordingly, trust must be carefully built through careful training with both simulated and live traffic. Indeed, Harwood et al. (1998) noted an increase in controller confidence after they had used the system (and relied on the final-approach spacing tool advice) with live traffic. This provided the opportunity to see the real improvement achieved in traffic flow (13 percent).

***Overtrust and Complacency*** Currently the philosophy of system implementation safeguards against undue complacency. This is because controllers must still give the actual clearances orally, as they would in a nonaided situation. Hence, they remain more likely to actively think about those clearances, for example, than they would in a system in which CTAS-advised clearances could be relayed via data link with a simple keystroke. Complacency is not generally recognized as a concern until an incident of automation failure occurs, in which the human's failure to intervene or resume control appropriately is attributed to such complacency. No such incidents have been observed with the CTAS. The advice-giving algorithms were thoroughly tested and in operational trials have yet to fail; alternatively, if inappropriate advice was ever provided, controllers were sufficiently noncomplacent that they chose to ignore it.

Past experience with other systems indicates that systems can fail in ways that cannot be foreseen in advance (e.g., the software does not anticipate a particular unusual circumstance). Furthermore, despite the design philosophy that appears to keep the controller a relatively active participant in the control loop, it is also the case that the primary objective of the CTAS is to increase the efficiency (and therefore saturation) of the terminal airspace. Such circumstance would make recovery more difficult should problems emerge for which the CTAS would be unable to offer reliable advice.

***Skill Degradation*** As with complacency, so with skill degradation: The CTAS has not been used long enough to determine whether this is an issue. Yet, it is easy to imagine circumstances in which controllers increasingly begin to rely on CTAS advice, relaying this as instructions to pilots, losing the skills at selecting maneuvers on their own. This may be more problematic still, to the extent that the maneuvers recommended by the CTAS are qualitatively different from those that would previously have been issued by unaided controllers. At this time, a clear tabulation of maneuver differences with and without the CTAS has not been carried out.

***Organization*** The organizational implications of the CTAS remain uncertain. A strength of the system is that it is designed to be advisory only; by not directly affecting required procedures, the negative impact on organizational functioning should be minimized. However, it is possible that subtle shifts in authority from the R-side controller to the D-side (who is more likely to have direct access to CTAS advisories) could have unpredictable consequences. We explore these consequences further in the discussion of conflict probes in the following section.

### 23.2.5 Conclusions Regarding CTAS

The CTAS appears to be a well-conceived automation concept addressing a valid concern of the less automated system and designed with an appropriate philosophy that is based on automated advice giving rather than automation-based control. As such, it is characterized by a relatively low level of automation that accordingly diminishes (but does not eliminate) the extent of concern for complacency. Finally, the CTAS has been developed and introduced gradually in a manner sensitive to human factors issues and to the importance of filtered controller input into the functioning of the system. Careful human factors monitoring of the system's field use should be continued.

The CTAS has the potential to radically alter the procedures of pilot–ATC coordination, pilot choices, and flight plans. Yet there are other systems in the airspace that also have similar impacts, such as the pilot's flight management system or the digital "datalink" communications between ground and air. The panel saw the vital need to "harmonize" any new automation system, such as the CTAS, with other systems currently in existence, under development, or proposed—an issue we address in the following section.

Unfortunately, in spite of the promising results of early HFI reviews and evaluations by NASA, during field tests at Dallas–Fort Worth and Denver (Harwood et al., 1998), the CTAS has yet to realize long-term success in system integration. Subsequent to its transfer from a developmental version to a fully deployed ATC system, implementation in TRACON facilities has been limited, difficult, and expensive. Of the three major CTAS components, one (TMA) has been partially implemented and another (FAST) has to this date seen, adoption at only two sites.

When we look at the 10 HSI principles of Chapter 1, we may find a clue for the mixed results of the CTAS as part of the national ATC system. The early positive results the panel's report relied upon to indicate an HFI success with the CTAS were based upon many of the 10 principles being utilized appropriately. For example, there was top-level support (1), a focus on the operator as a central design philosophy (2), an integrated approach to system documentation (4), proper use of HFI technology (7), and application of the appropriate human factors skills throughout the design and development process and initial field evaluations (9).

However, it appears that at least two principles—quantitative human performance (6) and test and evaluation (8)—were weak throughout the process of system acquisition by the FAA. While much was done to demonstrate, evaluate, and revise designs, there were few quantitatively established endpoints for human–system performance. "Goodness" was defined by user acceptance, which was generally nonquantifiable in total system performance terms, and the user acceptance criteria tended to vary throughout the program's development.

## 23.3 HARMONIZATION OF MULTIPLE SYSTEMS

The idea of harmonization or integration is central to the process of systems engineering. Basically, integration refers to the compatibility of the components of a particular system. Electronic and electromechanical systems, in particular, may generate many instances in which some elements or components are unable to communicate with each other. The development of such systems to the point of effective utility is a challenge for engineer designers. In 1998, the developmental pipeline contained a series of substantial ATC subsystems that were proposed for inclusion in the national airspace system. The panel's report concluded that more research was needed to determine if the new subsystems could perform as well as expected and whether they fit together to make an effective total system. At that point in time, subsystems had been developed in relative isolation from one another and from the overall modernization program. For example, the specifications for the Standard Terminal Automation Replacement System (STARS) and Display System Replacement (DSR) required that developers provide an architecture that would allow future plug-in of preplanned product enhancements. However, no human factors analysis of how these enhancements would be integrated with one another or with the STARS and DSR baselines was evident.

The lack of evidence of a unifying human factors analysis for advanced automation products in order to guide their integration into complementary workstation designs or procedures is exemplified by the CTAS. Although NASA's in-house scientists and their supporting contractors were also working on projects such as cockpit automation and data link in ATC, the role of data link with respect to the CTAS and the potential constraints of data link on the CTAS did not receive significant attention of CTAS researchers.

In general, tests that determine inter-subsystem compatibility should immediately follow the tests that demonstrate subsystem performance. When the compatibility between pairs of subsystems has been established and possible sources of confusion resulting from conflicting sensors, databases, and algorithms have been identified, the assembly should be enlarged to include other innovations until the subsystems that must be used together have been included in an overall test. At each stage, the evaluation should include a comparison against the base case represented by the current operational system.

Typically, despite careful analysis and validation efforts, not all human errors can be predicted. The human–computer interface is not the only source of error; new systems can introduce new sources of error (e.g., mode, logic, and procedural errors). This may be especially true when a given system will be integrated within a set of existing systems or when systems in parallel development will be implemented together because such integration can produce unexpected and unintended consequences. Reason and Zapf (1994) note that testing components in isolation and then putting them together open the opportunity for previously unidentified "resident pathogens" to strike. Systems designed without consideration of the implementation context risk incorporating such error-inducing features as computer interface logic that conflicts with that of other systems, information that unnecessarily duplicates (and possibly conflicts with) that provided by other systems, information whose interpretation or use requires data from other remotely located systems, information that confuses what is offered by other systems, alarms that distract the user from those of other systems, and disruption of team work flow (Miller et al., 1996). It should be noted that even field testing can miss unanticipated errors caused by combining new systems if systems planned for simultaneous implementation are not field tested together. For example, the parallel runway approach system, which is ground based, uses distance as a separation algorithm while a proposed cockpit-based system uses time-to-contact for separation. The ground-based system relies on radar; the airborne system uses the Global Positioning System (GPS). These differences in technology are important should a redundant system involving both air and ground alerts be considered. If technologies are different and they provide conflicting advice, which should be followed?

Such cumulative or interactive effects must be taken into account throughout a system development process that anticipates the integration of system elements. In addition, since controller training, sector staffing, operational procedures, control room conditions, and equipment maintenance affect system effectiveness, system development and testing should include attention to how these context factors affect controller tasks, workload, and performance during use of the system under development and test (Grossberg, 1994). Modeling and analytical techniques as well as prototyping and simulation are all important methodologies for examining possible interactions between new technologies and the equipment and procedural contexts into which they are introduced.

While these techniques do not obviate the need for operational validation with controller-in-the-loop simulation in the actual ATC context, the panel's report recognized the critical importance of valid human performance models in the particular area of the human response to unexpected failures in otherwise highly reliable automation systems along with the impact of that response on system safety. Statistically reliable data regarding such responses from empirical studies are extremely difficult to obtain because, by definition, such events must be rare to be unexpected; if they are rare, there will be a very small sample size of observations per subject (Wickens, 2001). Yet, the complexity of the real-world systems involved inhibits the design of experiments using a large number of subjects. Valid models thus become vital in predicting the impact of system failures.

## 23.4 NATIONAL AIRSPACE SYSTEM: AN ORGANIZATIONAL HFI EXAMPLE

The panel's reported conclusions on the requirements for effective management of human factors for the national airspace system illustrate the types of challenges and benefits inherent in attempts to fully integrate human factors into an organization's systems acquisition process.

One conclusion from the panel's report was that effectiveness of human factors activities requires coordination and oversight by a central human factors management within the organization. In reaching that conclusion for the FAA, the report considered requirements for an effective human factors program. Although this conclusion applied to all human factors activities within the agency, the report focused on the development and acquisition of automation systems for ATC. As an example, the specific activities of a centralized human factors program management reported as necessary included

- coordinating communication of human factors performance data across integrated product teams and between researchers, developers, users, and testers in the United States as well as in other countries;
- developing and monitoring human factors program plans;
- monitoring and guiding the activities of contractors' human factors representatives;
- developing policies and procedures for the application of human factors to the development, test, and implementation of automated systems;
- evaluating the qualifications and performance of human factors specialists; and
- guiding the trade-offs pertaining to cost and schedule of human factors activities.

Human factors management plays a key role in identifying the appropriate mix of research and test methods that support system development. Human factors management should interface with engineering and program management personnel within an agency and with its support contractors to ensure that human performance requirements drive specifications and that hardware and software developers are responsive to the human performance requirements. Poor alternatives are the unfortunate situations in which human factors specialists become documenters of previously written computer code for the human–computer interface or in which training is expected to compensate for poor design.

It is also the role of human factors management to remind program managers, as necessary, that good human factors is a "pay now or pay more later" proposition. By the time a system reaches late stages of development or testing, major design commitments have been made, resources have been spent, and there is reduced motivation to discover design flaws that threaten deployment schedules. An example pointed out by the panel was the many downstream adjustments required in the STARS system.

It is not unusual for system designers or program managers to request that human factors specialists devise improved training programs to compensate for discovered design problems after system designs are frozen. Training, however, is not considered a substitute for effective design (reliance on training will not prevent errors if the design itself is inadequate), and flawed systems often require redesign despite improved training methods. Systems in which human factors are not properly addressed may require costly redesign after inadequacies are discovered (Grossberg, 1994; Stein and Wagner, 1994).

An effective human factors program presumes the activity of knowledgeable human factors specialists. In addition, it is important that researchers, system developers, and developers of policies, procedures, and regulations share appreciation of the importance of human factors activities and understanding of fundamental human factors principles. There are several avenues by which a systems acquisition organization can pursue development of human factors understanding throughout the agency as well as the enhancement of human factors expertise:

- The human factors management function, as stated above, includes coordination of information sharing between researchers and system developers. One appropriate vehicle is a human factors newsletter broadly disseminated within and beyond the agency to summarize studies, lessons learned, and issues raised by fielded systems, as the FAA has done after fielding the pilot based automation aid for collision avoidance, known as the TCAS system (Wickens et al., 1998).
- Widespread appreciation of fundamental human factors principles requires education of those within the agencies who perform research, support system development and testing, and establish regulations and procedures.

- Government acquisition programs have generally relied on development contractors and subcontractors to perform human factors activities. Qualifications of good human factors specialists, however, are often not made clear during the hiring of personnel by contractors. One function of the human factors management is to review the qualifications of human factors specialists hired by contractors.
- It is important to work toward an agency infrastructure in which some human factors training is provided to personnel and program managers at all levels of the organization (and to contract teams).
- The federal government increasingly supports integrated product teams with well-trained human factors specialists assigned to the team. It is important that these specialists be responsible to human factors management within the agency as well as to project managers.

## 23.5 CONCLUSION

In conclusion, the nation's ATC system is the prototype of a complex, high-risk system whose effectiveness and safety have the potential to benefit from well-conceived human-centered automation availed by advanced technology. Yet, the final goal of integrating such technology effectively is a lengthy process requiring many steps: careful task and workload analysis of operator needs and demands, good human factors in design and evaluation, effective training and cautious introduction of technology into the workplace, harmonization with other existing systems and procedures, consideration of the sorts of errors that can be committed, and the ways in which low-frequency events could seriously jeopardize the safety of the new technology. Finally, successful integration requires the full commitment and support to human factors of top-level managers in the organization who are responsible for design, acquisition, and deployment. Fulfilling all of these steps is a difficult challenge, but it is one that we believe will underlie the safe adoption of new technology and the satisfaction of the controllers who must supervise that technology.

## NOTES

1. The views expressed in this chapter do not reflect the views of the National Academy of Sciences or the National Research Council.
2. Information for this chapter was adapted with permission from Wickens et al. (1998), *The Future of Air Traffic Control*, Copyright 1998 by the National Academy of Sciences. Courtesy of the National Academy Press, Washington, DC.

## REFERENCES

Billings, C. E. (1996a). *Human-Centered Aviation Automation: Principles and Guidelines*, NASA TM 110381. Moffett Field, CA: NASA Ames Research Center.

Billings, C. E. (1996b). *Aviation Automation: The Search for a Human-Centered Approach*. Mahweh, NJ: Erlbaum.

Bond, D. (2001, December 17). Crisis at Herndon: 11 Airplanes Astray. *Aviation Week and Space Technology*, pp. 96–99.

Davis, T. J., Krzeczowski, K. J., and Bergh, C. (1994). The Final Approach Spacing Tool. In *Proceedings of the 13th IFAC Symposium on Automatic Control in Aerospace-Aerospace Control'94* (pp. 70–76). Palo Alto, CA.

Erzberger, H., Davis, T. J., and Green, S. M. (1993). Design of Center-TRACON Automation System. In *Proceedings of the AGARD Guidance and Control Panel 56th Symposium on Machine Intelligence in Air Traffic Management* (pp. 1–12). pp. 11.1 to 11.12.

Erzberger, H., and Nedell, W. (1989). *Design of Automated System for Management of Arrival Traffic*. NASA Technical Memorandum 102201. Moffett Field, CA: Ames Research Center.

Erzberger, H., and Tobias, L. (1986). A Time-Based Concept for Terminal-Area Traffic Management. In *AGARD Conference Proceedings 410: Efficient Conduct of Individual Flights and Air Traffic* (pp. 1–14). Neuilly-sur-Seine, France: Advisory Group for Aerospace Research and Development. (pp. 52-1 to 52-14).

Grossberg, M. (1994, October/December). Issues in Operational Test and Evaluation of Air Traffic Control Systems. *Journal of Air Traffic Control*, pp. 4–5.

Harwood, K., Sanford, B., and Lee, K. (1998). Developing ATC Automation in the Field: It Pays to Get Your Hands Dirty. *Air Traffic Control Quarterly*, *6*(1), 45–70.

Johannsen, G., and Rouse, W. B. (1983). Studies of Planning Behavior of Aircraft Pilots in Normal, Abnormal, and Emergency Situations. *IEEE Transactions on Systems, Man, and Cybernetics*, *SMC-13*(3), 267–278.

Lee, K. K., and Davis,T. J. (1995). *The Development of the Final Approach Spacing Tool (FAST): A Cooperative Controller-Engineer Design Approach*, NASA Technical Memorandum 110359. Moffett Field, CA: NASA Ames Research Center.

Miller, D. L., Wolfman, G. J., Volanth, A. J., and Mullins, R. T. (1996, April). Systems Integration, User Interface Design, and Tower Air Traffic Control. *IEEE AES Systems Magazine*, pp. 22–25.

Nordwall, B. D. (2001, November 13) Free Flight Benefits Anticipated as FAA Deploys Controller Aids. *Aviation Week and Space Technology*, pp. 50–51.

Parasuraman, R., and Riley, A. (1997). Humans and Automation: Use, Misuse, Disuse, Abuse. *Human Factors*, *39*, 230–253.

Parasuraman, R., Sheridan, T., and Wickens, C. D. (2000). A Model for Types and Levels of Human Interaction with Automations. *IEEE Transactions on Systems, Man, and Cybernetics, Part A: Systems and Humans*, *3*(3), 286–297.

Reason, J. T., and Zapf, D. (Eds.). (1994). Errors, Error Detection and Error Recovery. *Applied Psychology: An International Review*, *43*(4), 427–584 (special issue).

Scott, W. B. (1994, October 17). CTAS Tests Confirm Traffic Flow Benefits. *Aviation Week and Space Technology*. pp. 1–11.

Stein, E. S., and Wagner, D. (1994). A Psychologist's View of Validating Aviation Systems. In J. A. Wise, V. D. Hopkin, and D. J. Garland (Eds.), *Human Factors Certification of Advanced Aviation Technologies* (pp. 45–52). Daytona Beach, FL: Embry-Riddle Aeronautical University Press.

Tobias, L., Völckers, U., and Erzberger, H. (1989). Controller Evaluations of the Descent Advisor Automation Aid. In *AIAA Guidance, Navigation and Control Conference Proceedings* (pp. 1609–1618). Washington, DC: American Institute of Aeronautics and Astronautics.

Tulga, M. K., and Sheridan, T. (1980). Dynamic Decisions and Workload in Multi-Task Supervisory Control. *IEEE Transactions on Systems, Man, and Cybernetics*, *SMC-10*(5), 217–231.

Wickens, C. D. (2001). Attention to Safety and the Psychology of Surprise. Keynote Address. In *Proceedings of the 11th International Symposium on Aviation Psychology*. Columbus OH: Ohio State University.

Wickens, C. D., Mavor, A. S., and McGee, J. P. (Eds.). (1997). *Flight to the Future: Human Factors in Air Traffic Control*. Panel on Human Factors in Air Traffic Control Automation, National

Research Council, Commission on Behavioral and Social Sciences and Education. Washington, DC: National Academy Press.

Wickens, C. D., Mavor, A. S., Parasuraman, R., and McGee, J. P. (Eds.). (1998). *The Future of Air Traffic Control.* Panel on Human Factors in Air Traffic Control Automation, National Research Council, Commission on Behavioral and Social Sciences and Education. Washington, DC: National Academy Press.

Wyndemere (1996). *An Evaluation of Air Traffic Control Complexity: Final Report*, Contract Number NAS 2-14284. Boulder, CO: Wyndemere.

**CHAPTER 24**

# Human Systems Integration and New Product Development

WILLIAM B. ROUSE

## 24.1 INTRODUCTION

This chapter focuses on the nature of human systems integration (HSI) within new product development (NPD), primarily in the private sector. The discussions in this chapter draw upon experiences in product domains ranging from aviation to appliances, computers to communications, and drugs to data warehouses. Hundreds of new product planning engagements form this experience base.

It is essential at the outset to note that few, if any, of these experiences involved explicit use of the phrase "human systems integration." While human-related issues are often primary in NPD, this set of issues and how they are addressed are rarely labeled "HSI." However, as this chapter delineates in some detail, the HSI philosophy permeates much of NPD.

### 24.1.1 Overarching Design Objectives

The primary emphasis in this chapter is on human-related issues in NPD, how these issues interact with other issues, and how inherent trade-offs can be addressed. Elsewhere (Rouse, 1991, 2000a), I have argued that these issues and trade-offs are driven by three overarching design objectives:

- Enhancing human abilities
- Overcoming human limitations
- Fostering human acceptance

Enhancing human abilities dictates that humans' capabilities in the roles of interest be identified, understood, and cultivated. For example, people tend to have excellent pattern

*Handbook of Human Systems Integration*, Edited by Harold R. Booher.
ISBN 0-471-02053-2 

recognition abilities. NPD should take advantage of these abilities—for instance, by using displays of information that enable users to respond on a pattern recognition basis rather than requiring more analytical evaluation of the information.

Overcoming human limitations requires that limits be identified and appropriate compensatory mechanisms devised. A good illustration of an apparent human limitation is the occasional exhibition of behaviors deemed to be "human errors." Humans are fairly flexible information processors, which is usually important to successful system operation. However, this flexibility can lead to innovative behaviors that are erroneous in the sense that undesirable consequences are likely to occur. Such "errors" often reflect a mismatch between the requirements of an unforeseen situation and humans' natural inclinations.

One way of dealing with this problem is to eliminate innovations, perhaps via interlocks and rigid procedures. However, this is akin to the proverbial "throwing out the baby with the bath water." Instead, mechanisms are needed to compensate for undesirable consequences without precluding innovations. Creatively addressing this need may provide considerable market advantage if done well.

Fostering human acceptance dictates that stakeholders' preferences and concerns be explicitly considered in the design process. This requires, of course, that stakeholders be identified and their concerns, values, and perceptions be assessed. Ideally, this will provide a basis for delighting primary stakeholders and gaining the support of secondary stakeholders.

These three overarching objectives lead to design attributes that potentially affect achieving these objectives for each class of stakeholder. The resulting multiattribute, multistakeholder nature of design problems can be quite difficult to address. This is further complicated by typical NPD contexts where there are no contractual requirements and constraints. There are only markets for which you may or may not correctly assess project needs and devise appropriate solutions. As discussed later in this chapter, the nature of risks in such environments is quite different from typical public system design and development efforts.

### 24.1.2 Human Systems Integration of New Product Development

Before pursuing HSI for NPD in more depth, it is useful to consider briefly HSI *of* NPD. Many of the methods and tools discussed in this chapter resulted from taking this perspective (Cody et al., 1995; Rouse and Boff, 1998a). Specifically, many of the concepts presented in this chapter emerged from focusing on enhancing designers' abilities, overcoming designers' limitations, and fostering designers' acceptance.

This series of studies included observations, interviews, and other data collection involving hundreds of designers over roughly 10 years. Overall, conclusions of these studies are presented in terms of typical design environments, common design challenges, and implications for design methods and tools. Later discussions of methods and tools illustrate direct benefits of the knowledge gained from these studies.

Two aspects of typical NPD environments are of particular note. First, are the pervasive and substantial impacts of market/business drivers on the technology/design envelope. Consequently, design issues, including HSI, are often highly influenced by the competition-determined, cost-sensitive business environment. The primary goals are to win the competition and make a profit rather than design a perfect product.

The second and related environmental factor is the inherent multiattribute, multi-stakeholder, time-pressured, information-rich problem solving and decision making.

Consequently, most design decisions are framed and made much too quickly to allow for model building and experimentation. There is little time to peruse archival information that might result in a better product.

Two design challenges are common. First is the need to understand high-impact uncertainties in terms of probability distributions of impacts and criteria for decision making. Expected values are not adequate when uncertainties are large. Design solutions need to function, and the investment needed to create such functionality needs to make sense—for plus and minus 2 to 3 sigma of the distributions of uncertain variables.

The second challenge concerns the difficulty of cross-disciplinary representation, manipulation, and (quasi-) optimization of problems, requirements, and solutions. Often, several of the multiple stakeholders in NPD involve the various disciplines contributing to creating design solutions. The abilities to seamlessly integrate the perspective of these different disciplines are often central to design success.

Considering design methods and tools, these studies indicated the importance of spiral discovery, prototyping, and evaluation processes versus traditional waterfall approaches. The practice of freezing requirements and then procuring solutions results in many immediately obsolete design solutions. Much more flexibility and adaptability are needed to consistently create competitive new products in the private sector.

A second methods and tools consideration relates to the need for targeted, specialized tools rather than monolithic approaches with, ideally, compatible representations and information flows across component tools. This conclusion recognizes the reality that there is not and will not be a "megamodel" that crosses all issues and disciplines, accessing all the world's data. Instead, frameworks that include a range of targeted tools are much more likely to be valued.

This brief tangent into HSI *of* NPD provides some philosophical underpinnings of the remainder of this chapter. Not only does HSI need to address abilities, limitations, and preferences of systems operators and maintainers; but, whatever prescriptions for HSI that we advocate have to be feasible for the humans that do HSI work, that is, designers. Thus, we should be "willing to take our own medicine."

### 24.1.3 Chapter Overview

The next section provides a fairly detailed comparison of private- and public-sector product/system development, primarily to explicate the significant differences between these sectors and the implications for how HSI issues are framed and addressed in NPD. This sets the stage for consideration of NPD management processes, both in terms of general multistage decision processes and more HSI-specific human-centered design. Methods and tools for supporting these processes are next discussed. Finally, best practices relative to product realization processes are summarized.

## 24.2 PRIVATE VERSUS PUBLIC DEVELOPMENT

The primary differences between product development in the private and public sectors are elaborated in Table 24.1. In this section, these differences are first elaborated and then an overarching explanation is offered for why these differences are manifested.

Both the private and public sectors are driven by customers' needs and desires, but the level of specificity relative to these needs and desires differs substantially. In the public

**TABLE 24.1 Comparison of Private- and Public-Sector Product/System Development**

| Comparison | Private Sector | Public Sector |
|---|---|---|
| Driving force | Market needs and opportunities | Procurement plans and budgets |
| Competition | Continued new offerings and players | Little once production contract won |
| Customers | Many potential sales opportunities | Few potential sales opportunities |
| Customers' requirements | Typically researched and inferred | Publicly specified |
| Customers' budgets | Seldom openly available | Publicly available information |
| Production runs | Often very large | Usually relatively small |
| Product life cycles | Usually relatively short | Often very long |
| Risks | Usually borne by developer | Usually borne by customer |
| Rewards | Determined by marketplace | Controlled by customer |
| Sales of new offerings | Brand and relationship loyalty key | Usually competitively bid |

sector, procurement plans and budgets are publicly available information[1]—customers' intentions are quite explicit, often many years in advance, although political factors can derail these plans. In the private sector, needs and opportunities must be inferred from buying patterns and possibly technology trends—customers' intentions may be only vaguely articulated and very open to change. Hence, the market uncertainties in the private sector are substantially greater than in the public sector.

It is, however, very important to recognize the possibility that well-documented and communicated requirements for public systems can nevertheless be ill-conceived. Such a starting point provides ample opportunities for cost overruns, schedule slippages, and much finger pointing. Ensuring that requirements are well-founded is important to successful HSI in all domains, whether private or public (Sage and Rouse, 1999).

Winning a public-sector contract to provide products and systems often assures a long stream of revenues, perhaps decades in the case of defense systems. In the private sector, new offerings and players emerge more frequently and customers may switch to these providers if the costs/benefits are better. Thus in the private sector, winning provides only temporary advantage. On the other hand, losing results is only a temporary disadvantage. Thus, overall private-sector market relationships are much more responsive to change than in the public sector.

There are usually only a few possible customers for public systems. If, for instance, the Federal Aviation Administration does not buy your air traffic control system, to whom else can you sell this system? As another example, if the U.S. military does not buy your defense system, there are unlikely to be many foreign military sales. In contrast, the private sector typically has many potential customers, ranging from 20 to 30 airlines that buy commercial aircraft to millions of people who buy automobiles and computers.

Public-sector customers' requirements are usually quite specific and publicly available. All potential providers compete to offer the most cost-effective way to meet these requirements. Private-sector vendors have to research and infer requirements, often because customers do not really know what they want—hence, market research such as described by Blattberg et al. (1994) is pursued. As a result, alternative solutions tend not to have the same functions and features. Eventual winning solutions may have significant competitive advantages due to proprietary functionality, performance characteristics, etc.

Budgetary information is publicly available for public-sector customers. Thus, providers know what spending is expected and when this spending is projected. Information on private-sector customers' budgets is seldom openly available. In fact, there may be no budget items in many areas. Sales of products in the private sector may, therefore, depend on arguing the costs/benefits of possible solutions and, in effect, creating needs that were not previously perceived.

Public-sector production runs tend to be relatively small, typically ranging from hundreds (e.g., aircraft) to thousands (e.g., small defense systems). In contrast, production runs of hundreds of thousands to millions are not uncommon in the private sector. This enables amortization of research and development (R&D) costs over many more units. It also results in significantly greater cost savings as providers move down production learning curves. Consequently, prices for private-sector products tend to decrease, often as quality also improves.

The product life cycles for public-sector products tend to be quite long, with many defense systems, for example, remaining in use for several decades. The private sector often experiences product life cycles as short as a few months to as long as a few years. The provider who gets to market first, and makes it down the production learning curve the fastest, tends to capture market share and realize the best margins. Of course, there is also the risk of getting to market too quickly, or getting there with the wrong sets of functions and features.

The risks associated with creation of public-sector products and systems are usually assumed by the customer—who often is the only customer. In the private sector, such risks are assumed by the developers of the product. If they are too early, too expensive, or off target in terms of functions and features, they must accept the consequences.

Those who accept risks often earn the greatest rewards. Thus, public-sector profit margins are often quite modest and explicitly controlled by customers. Private-sector profits are determined by the marketplace and are typically not visible to customers, at least not in terms of profit per unit. Of course, if this were not the case, then private-sector product developers would be unlikely to accept the inherent risks.

Sale of new offerings in the public sector usually involves an open competitive bid, despite superior past performance, service, etc. Brand and relationship loyalty provides much more advantage in the private sector, often resulting in sales of new offerings without competition. In fact, loyalty may be such that customers no longer even consider the possibility of alternative ways to meet their needs.

The impact of the differences summarized in Table 24.1 can be considered in the context of a typical product/system life-cycle model such as shown in Figure 24.1 (Patterson, 1999; Sage, 1995). New product development in the private sector involves much more early uncertainty and risk, particularly in terms of market and competitive factors rather than technology. Thus, more effort and time are invested in concept definition, requirements analysis, and conceptual design, partly because concepts and requirements must be evaluated to ensure likely market acceptance. If, for example, a product has major usability deficiencies, private-sector markets cannot be, in effect, forced to buy the product until these deficiencies are remedied.

Across the whole product life cycle, private-sector NPD usually involves many more hypotheses and tests, regularly trying to catch bad ideas quickly—"bad" meaning things that will not sell or may sell but lead to warranty or product liability problems.

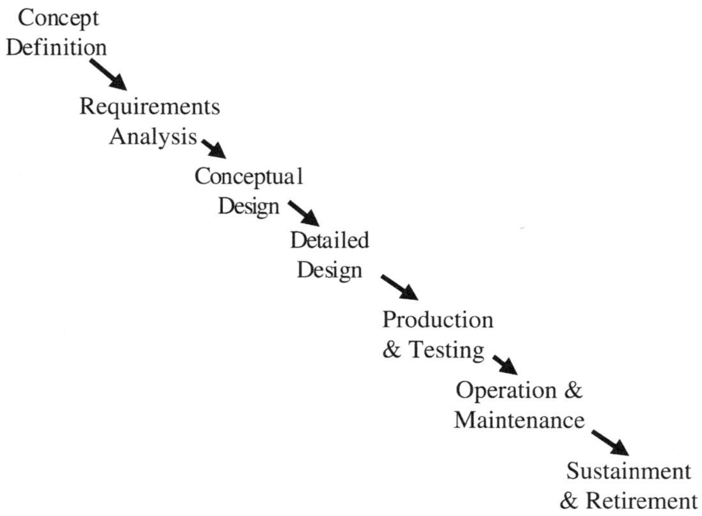


**Figure 24.1** Elements of simplified product/system life cycle.

Consequently, there is much more reliance on "spiral" models of development (Boehm, 1988), rather than the "waterfall" model depicted in Figure 24.1. The requirements analysis, conceptual design, detailed design, and production and testing phases of the life cycle are repeated relatively quickly, each time learning and refining the product concept to improve the design and spiral toward a good design.

Best practices for public-sector system development do include evolutionary waterfall models and spiral models [Sage, 1995, 1999; U.S. Department of Defense (DoD), 1996], often discussed in terms of integrated product and process development (IPPD). Thus, public system developers are often well aware of many, and perhaps most, of NPD best practices. Unfortunately, complications of procurement processes, as well as political processes, can extend and distort life cycles in ways that undermine the benefits of these practices.

Faster development processes, as well as less customer control, result in private-sector products evolving much more quickly. This enables faster technology upgrades and insertion of new technologies. Lessons learned in operations and maintenance are quickly fed back to the evolution of new releases of the product or, more appropriately, the evolving product family. Consequently, private-sector products and systems are much more likely to include the latest, leading-edge technologies.

This blessing for private-sector customers comes with the curse—for developers—of greatly reduced sustainability of competitive advantage. Substantial revenues for spare parts and maintenance of decades-old products and systems are rare in the private sector. Competitive displacement of older technologies tends to be merciless in all but near-monopoly private-sector markets (e.g., commercial aviation). Thus, products usually cannot be viewed as "loss leaders" for downstream recurring revenues.

One very major exception to this assertion involves products where there are substantial ongoing service components. Automobile maintenance is a good example where innovations in products are, to a great extent, intended to attract and retain buyers who will avail themselves of the ongoing services for the products. Thus, in this case, the automobile dealer does not have to make very large margins on the initial product sale.

Summarizing the distinctions drawn in this section, private-sector NPD differs from public-sector product/system development in terms of uncertainties, risks, and rewards. Private sector NPD involves much greater uncertainty and many more risks. However, the *potential* rewards are much greater, due to both large unit volumes and much greater profits per unit.

These differences strongly affect how HSI issues are pursued. Human-related concerns—such as the possibility of creating products with wrong function/feature sets or products with major usability problems—receive substantial attention because the consequences to the developer are so substantial. On the other hand, concerns such as long-term health and safety considerations receive much less attention, mostly because they do not affect near-term sales but also due to discounting of long-term impacts in general.

To a great extent, differences in addressing HSI issues are determined by who suffers the consequences of being wrong. Public-sector product/system development projects are typically sold before development begins; thus, the sale does not depend on how HSI issues are addressed, although it may depend on having a plan for addressing these issues. In contrast, private-sector products/systems are sold after they are developed; customers who are not satisfied with how HSI concerns have been addressed can simply choose not to make purchases. In general, things get done when sales are dependent on them!

The differences summarized thus far are related to both the nature of products and systems created in the private and public sectors and to inherent dissimilarities between these sectors. Figure 24.2 suggests an overarching explanation for these differences. A large percentage of NPD in the private sector involves creating standard products (solutions) for a relatively homogeneous market (stakeholders) that only has scrutiny over the end product. In contrast, a large portion of systems developed in the public sector, for which HSI has received the most attention, involve tailored solutions for which there are a wide range of stakeholders—users, customers, employees, politicians, etc.—who

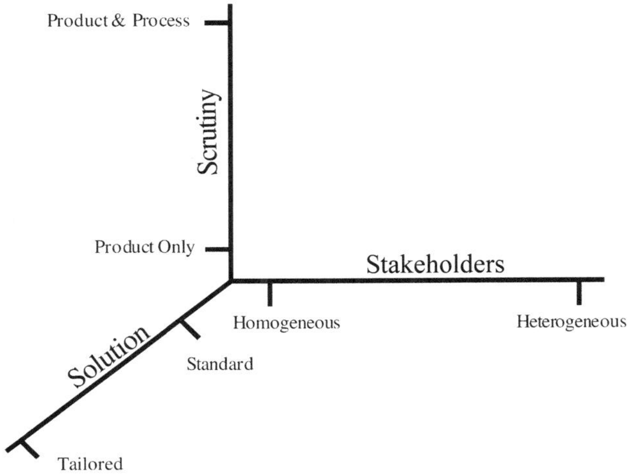


**Figure 24.2** Differences underlying HSI products and systems.

have considerable scrutiny over both system characteristics and the process whereby the system is designed and developed.

As a consequence, many NPD best practices are difficult to implement in the public sector, despite the fact that practitioners are well aware of these practices. For example, it is often the case that the military end user (the warfighter) will dictate design decisions, including design changes that adversely affect budget and schedule. Congress may, for instance, preempt design changes that would adversely affect developers in favored congressional districts. And, of course, the whole federal procurement process can complicate and extend the acquisition process in ways in direct opposition to NPD best practices.

It is also useful to note that private-sector system development efforts involving tailored solutions for heterogeneous stakeholders who have considerable scrutiny of both product and process can encounter some, if not all, of the same difficulties encountered in the public sector. For example, Fryer (1999) reports that at least 90 percent of enterprise resource planning (ERP) projects end up late or over budget, often taking 6 to 7 years or more to realize positive returns. This tends to result in rather large "expectations gaps," often created by vendors of ERP systems. Thus, dissimilarities between the private and public sectors provide only a partial explanation of the differences summarized in Table 24.1.

## 24.3 PRODUCT MANAGEMENT PROCESSES

The uncertainties and risks associated with NPD in the private sector result in many more concepts being pursued than eventually make it to the marketplace. The ratio of initial ideas to market innovations through the product development "funnel" ranges from 3,000 : 1 (Stevens and Burley, 1997) to 10,000 : 1 (Nichols, 1994). Statistics for particular companies commonly reflect 200 to 300 funded projects for each product eventually introduced to the market.

### 24.3.1 Multi-Stage Decision Processes

Companies who manage this winnowing process with firm "go/no-go" decision points perform much better than those who are more ad hoc in their approach. Cooper (1998) has pioneered the formalization of multistage decision processes for making these decisions. An example of a multistage decision process is depicted in Figure 24.3.

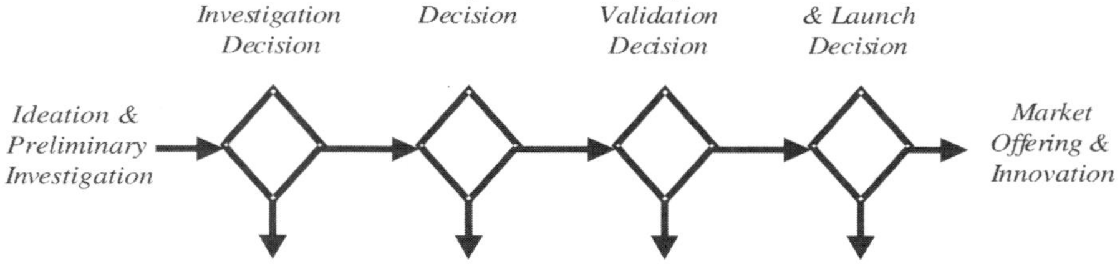


**Figure 24.3** Typical multistage decision process for new product development.

At each stage, NPD projects must pass specified criteria to move on to the next stage. Not passing these criteria results in projects being killed, shelved, or possibly retained at the earlier stage. Of great importance, project managers know exactly what the criteria are for each stage long before having to satisfy these criteria.

Investment levels increase substantially with each stage, adding significant potential value to the concepts being pursued by projects. The stages are designed to eliminate projects where the value added downstream will not justify the costs. The criteria change with each stage, shifting emphasis from strategic relevance and technical feasibility to economic return and risk management. For attributes that are relevant across stages, "hurdle rates," or decision thresholds, increase with each stage.

Investments in all the projects in the funnel are managed as portfolios (Cooper et al., 1998–2000). Plots of returns versus risks are typically used to represent portfolios, where return might be represented as net present value (NPV) or net option value (NOV), and risk might be characterized as the probability that NPV or NOV falls below some criterion, for example, zero (Rouse, 2000b). A common goal is to create an "efficient" portfolio such that each project in the portfolio has the minimum risk for a given level of return or the maximum return for a given level of risk. However, other nonfinancial criteria, for example, "strategic criticality," may result in elements of portfolios that are inefficient from a purely financial perspective.

The nature of multistage decision making just depicted results in many more "failures" than experienced for public systems. The vast majority of all ideas do not make it to applications. However, these nonsuccesses are viewed as simply part of the process of bringing new products to market.

Put differently, NPD in the private sector is much less risk averse than in the public sector. Emphasis is on assessing and managing risks rather than eliminating them. Of course, the consequences of risks are spread across many companies. In the public sector, the government absorbs most risks and nonsuccesses can result in endless scrutiny. Further, companies who develop products and systems for the public sector face the risk that their one and only customer will no longer be able to justify an acquisition, resulting in substantial loss of revenue although seldom significant loss of capital.

### 24.3.2 Human-Centered Design

Human-centered design (HCD) is a multistaged process focused on HSI aspects of NPD (Rouse, 1991, 2000a). The HCD concept emerged from the recognition that product success usually depends on the support of a range of stakeholders beyond the users of the product, hence the phrase human-centered rather than user-centered design. Relative to commercial aircraft, this idea is captured by the expression, "Pilots may fly 'em, but they don't build 'em or buy 'em." Thus, designing an aircraft cockpit to ensure pilot (user) acceptance does not at all guarantee that the airlines will want to buy this aircraft, that the airframe manufacturers will want to produce it, or that the regulatory authorities will certify its use. Human-centered design is a process of ensuring that the concerns, values, and perceptions of all stakeholders in a design effort are considered and balanced.

Stakeholders and their roles in the success of a product or system often include:

- Users—use solutions
- Customers—buy solutions

- Technical—evaluate and possibly regulate solutions
- Providers—invest in creating solutions
- Suppliers—provide elements of solutions
- Maintainers—troubleshoot and repair solutions
- Distributors—sell and deliver solutions

Some of these stakeholders represent several other stakeholders. For example, providers include functional stakeholders such as marketing, sales, finance, engineering, manufacturing, etc.

An overarching principle of HCD concerns the need to delight primary stakeholders and gain the support of secondary stakeholders to have a successful product or system. This reflects the need to convince the user stakeholder to want the solution, the customer stakeholder to buy the solution, and so on. Within the private sector, customers cannot typically be forced to accept a solution by a procurement function. Thus, all stakeholders have to want the solution or at least not argue against its purchase.

The HCD approach provides a framework for pursuing this principle. The core of this framework involves a set of measurement issues that are progressively addressed within a four-phase process. In the remainder of this section, relevant portions of this overall framework are presented.

What do successful products and systems have in common? The fact that people buy and use them is certainly a common attribute. However, sales are not a very useful measure for designers[2]. In particular, using *lack* of sales as a way to uncover poor design choices is akin to using airplane crashes as a method of identifying design flaws; this method works, but the feedback provided is a bit late.

The question, therefore, is one of determining what can be measured early that is indicative of subsequent poor sales. In other words, what can be measured early to find out if the product or system is unlikely to fly? If this can be done early, it should be possible to change the characteristics of the product or system to avoid the predicted failure.

The overall HCD methodology involves seven classes of measures. However, only three of these classes are useful for differentiating private-sector and public-sector approaches to HSI: viability, acceptability, and validity. The other four classes of measures (i.e., testing, verification, demonstration, and evaluation) are similarly addressed in both domains, at least for those organizations that employ leading-edge design methods and tools.

Measures of viability, acceptability, and validity emphasize quite different issues than measures related to testing, verification, demonstration, and evaluation. In fact, viability, acceptability, and validity are the driving issues in HCD. Thus the *last* concern is, "Does it run?" The *first* concern is, "What matters?" or "What constitutes benefits and costs?" Viability, acceptability, and validity are defined as follows:

- *Viability* Are the benefits of system use sufficiently greater than the costs? While this question cannot be answered empirically prior to having a design, one can determine how the question is likely to be answered. How do stakeholders characterize benefits? Are they looking for speed, throughput, an easier job, or appealing surroundings? What influences their perceptions of these characteristics? How do stakeholders characterize costs? Is it simply purchase price or do costs include maintenance and perhaps training? Are all the costs monetary?

- *Acceptability* Do organizations/individuals use the system? This is also a question that cannot be answered definitively prior to having the results of design. However, one can determine in advance the factors that are likely to influence the answer. Most of these factors relate to the extent to which a product or system fits into an organization's philosophy, technology, work practices, etc. Also important can be the extent to which successful adoption of the product involves substantial organizational change.
- *Validity* Does the product or system solve the problem? This, of course, leads to the question, What is the problem? How would you know if the problem was solved or not solved? The answer depends on how stakeholders perceive their problems. Quite possibly, they perceive an effectiveness or efficiency shortfall rather than a need for a new product or system.

These three broad classes of measurement issues are addressed in HCD using an overall approach to measurement that balances the allocation of resources among the issues of concern at each stage of design. This approach integrates intermediate measurement results in a way that provides maximal benefit to the evolution of the design product. This is accomplished by embedding measurement in an NPD process involving the following four phases:

- *Naturalist Phase* This phase involves researching market needs, that is, understanding the domains and tasks of stakeholders from the perspective of individuals, the organization, and the environment. This understanding not only includes people's activities but also prevalent values and attitudes relative to productivity, technology, and change in general. Evaluative assessments of interest include identification of difficult and easy aspects of tasks, barriers to and potential avenues of improvement, and the relative leverage of the various stakeholders in the organization.
- *Marketing Phase* Once one understands the domain and tasks of current and potential stakeholders, as well as their needs and desires, one is in a position to conceptualize alternative products or systems to support these people. Product concepts can be used for evaluative market research in the sense of determining how users react to the concepts. Stakeholder's reactions are needed relative to validity, acceptability, and viability. In other words, one wants to determine whether or not people perceive a product concept as solving an important problem in an acceptable way and at a reasonable cost.
- *Engineering Phase* One is now in a position to begin trade-offs between desired conceptual functionality and technological reality. Most traditional research, development, test, and evaluation occur in this phase. Technology development will usually have been pursued prior to and in parallel with the naturalist and marketing phases. This will have at least partially ensured that the product concepts shown to stakeholders will be technologically and economically feasible. However, one now must be very specific about how desired functionality is to be provided, what performance is possible, and the time and dollars necessary to provide it.
- *Sales and Service Phase* As this phase begins, the product should have successfully been tested, verified, demonstrated, and evaluated. From a measurement point of view, the focus is now on validity, acceptability, and viability. It is also at this point that one ensures that implementation conditions are consistent with the assumptions under-

lying the design basis of the product or system. Also important is identification of new market opportunities. This information is compiled by sales people in the process of selling solutions and by service people in the process of providing ongoing support.

Table 24.2 illustrates typical measures of viability, acceptability, and validity and shows how these three classes of measurement issues should be organized or sequenced in the four phases. *Framing* an issue denotes the process of determining what an issue means within a particular context and defining the variables to be measured. *Planning* is concerned with devising a sequence of steps and schedule for making measurements. *Refining* involves using initial results to modify the plan or even rethink issues and variables. Finally, *completing* is the process of making outcome measurements and interpreting results. Elsewhere, I discuss a wide range of methods and tools for making the measurements associated with the four phases of HCD (Rouse, 1991, 2000a).

It is useful to note the philosophical compatibility of HCD as outlined here with the aforementioned integrated product and process development (Sage 1995, 1999; DoD, 1996), although IPPD tends to be most focused on the engineering phase of HCD. In general, HCD emphasizes very early, upstream involvement, where HSI problems are best avoided and continual, downstream involvement where problems can be remedied and new opportunities identified.

Table 24.2 provides a useful context in which to discuss typical measurement problems. There are two classes of problems of interest. The first class is *planning too late* where, for example, failure to plan for assessing acceptability issues can preclude measurement prior to putting a product into use. The second class of problems is *executing too early* where, for instance, product demonstrations are performed prior to resolving validity issues and potentially lead to negative initial impressions of a product or system—such impressions can later be difficult to change.

It is important to note the role of technology in the HCD process. As depicted in Figure 24.4, technology is pursued in parallel with the four phases of the design process. In fact, technology feasibility, development, and refinement often consume the greatest share of the resources in a product or system design effort. However, technology should not drive the design process. The HCD objectives should drive the process, and technology should support these objectives.

The HCD process was formulated to avoid HSI failures. In private sector NPD, such failures can lead to lack of product sales, substantial recall/warranty costs, and possible lawsuits. Meeting requirements may be a sufficient condition for technical success but is by no means sufficient for market success. The success of NPD in the marketplace depends on competitively enhancing human abilities, overcoming human limitations, and fostering human acceptance. The HCD approach provides a systematic means for achieving these ends.

## 24.4 METHODS AND TOOLS

Design philosophies and frameworks are often adopted and utilized to the extent that methods and tools are available to support their use. This section outlines a variety of methods and tools commonly employed for NPD in the private sector. Also described are methods and tools specifically created to support HCD.

**TABLE 24.2 Measures vs. Phases in Human-Centered Design**

| Attributes | | Phase | | | |
|---|---|---|---|---|---|
| | | Naturalist | Marketing | Engineering | Sales and Service |
| Viability | Benefits | Frame | Plan | Refine | Complete |
| | Performance improvements | Frame | Plan | Refine | Complete |
| | Quality improvements | Frame | Plan | Refine | Complete |
| | Etc. | Frame | Plan | Refine | Complete |
| | Costs | Frame | Plan | Refine | Complete |
| | Acquisition costs | Frame | Plan | Refine | Complete |
| | Operating costs | Frame | Plan | Refine | Complete |
| | Etc. | Frame | Plan | Refine | Complete |
| Acceptability | Matches operating philosophy | Frame | Plan | Refine | Complete |
| | Compatible with existing infrastructure | Frame | Plan | Refine | Complete |
| | Consistent with training practices | Frame | Plan | Refine | Complete |
| | Etc. | Frame | Plan | Refine | Complete |
| Validity | Meets technical specifications | | Frame/plan | Refine/complete | |
| | Complies with relevant standards | | Frame/plan | Refine/complete | |
| | Passes usability evaluation | | Frame/plan | Refine/complete | |
| | Etc. | | Frame/plan | Refine/complete | |

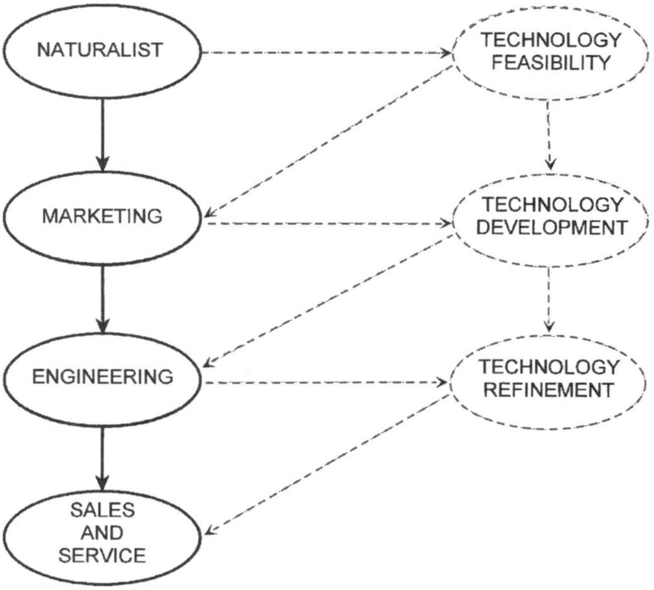


**Figure 24.4** Role of technology in human-centered design.

### 24.4.1 Market Research

It should be apparent from the foregoing discussion that the private sector has to make much greater investments than the public sector in researching market needs, customer preferences, and possible product requirements. Private-sector markets do not publish their requirements. Indeed, these markets often do not know what they want until they see it.

There is a wide variety of methods and tools for market research, ranging from secondary sources such as newspapers, magazines, and the Internet, to information gathering via questionnaires, interviews, and prototyping (Rouse, 1991, 2000a). The various alternatives have advantages and disadvantages depending on the phase or stage of the product realization life cycle.

Many companies have in recent years created or purchased databases of customer characteristics and purchasing behaviors. They mine and model this data to form and test hypotheses regarding preferences and responses to, for example, alternative marketing programs (Blattberg et al., 1994). Gensch and colleagues (1990) discuss development of such models and present data that show the financial returns from making NPD decisions with the support of such models.

### 24.4.2 Product Lines and Platforms

Market research is used to drive choices of product functions and features that are likely to provide compelling value propositions to the stakeholders (Rouse, 2000b). Increasingly, such decisions concern product families rather than just "point" solutions. Thus, NPD involves systematic thinking about product lines (Brownsword and Clements, 1996), product platforms (Robertson and Ulrich, 1998), evolutionary architectures (Rouse, 1991), and the modularity needed to enable these families of products (Baldwin and Clark, 1997). This has been a common practice in the automobile industry for many decades. However, it is being extended to products such as software (Brownsword and Clements, 1996) and computers (Baldwin and Clark, 1997).

### 24.4.3 Product Evaluation

Despite extensive market research, it is still quite possible to create products that people find difficult to use. Many companies employ formal usability evaluation processes to minimize chances of significant difficulties (Henneman, 1999). Such evaluations are motivated by risks of difficult-to-use products not selling in the private sector and possible safety problems in both the private and public sectors.

Product evaluation usually continues after introduction to the marketplace and during ongoing sales and service. Such evaluative efforts typically focus on rapid remediation of problems and responses to competitive threats and opportunities. Of course, follow-up of products and systems in use is common in both private and public sectors. The difference in the private sector, however, is the speed with which opportunities can be converted to business advantages and profitable outcomes.

### 24.4.4 NPD Project Evaluation

An important contributor to NPD success is the nature of the design and development process. Best practices in this area are reviewed later in this chapter. Relative to methods and tools, it is nevertheless useful to mention at this point NewProd, an NPD project benchmarking tool [Cooper, 1985; Product Development Institute (PDI), 2000]. This software tool is used to evaluate proposals for new product initiatives. Several 0–1 scales are used to characterize NPD projects. These ratings are combined to create an overall score that is compared to the scores of roughly 200 actual, historical NPD projects to predict probability of success.

The obvious limitations of this tool include the idiosyncrasies of historical linkages and the focus on a single attribute—probability of success. A finer-grained characterization can be needed to diagnose and remediate project deficiencies. Later discussion considers other types of project attributes.

### 24.4.5 Decision Support Tools

There is a variety of decision support and project management tools available to support NPD (Rouse, 2000b). Most of these tools are quite general in nature, aimed at providing support in a wide range of decision-making situations or managing projects in general. As such, the advice and online help provided are not premised on particular types of decisions being addressed.

The Product Planning Advisor (PPA) was specifically developed to support HCD as outlined earlier [Enterprise Support Systems (ESS), 2000a].[3] This tool helps in developing and manipulating models of the form depicted in Figure 24.5. The purpose of such models is identification of products and systems that delight primary stakeholders and ensure the support of secondary stakeholders, while also assuring competitive advantages for those who invest in creating these products and systems.

The portion of Figure 24.5 labeled "What the Market Wants" relates to characterizations of stakeholders and their issues and concerns in terms of attributes related to viability,

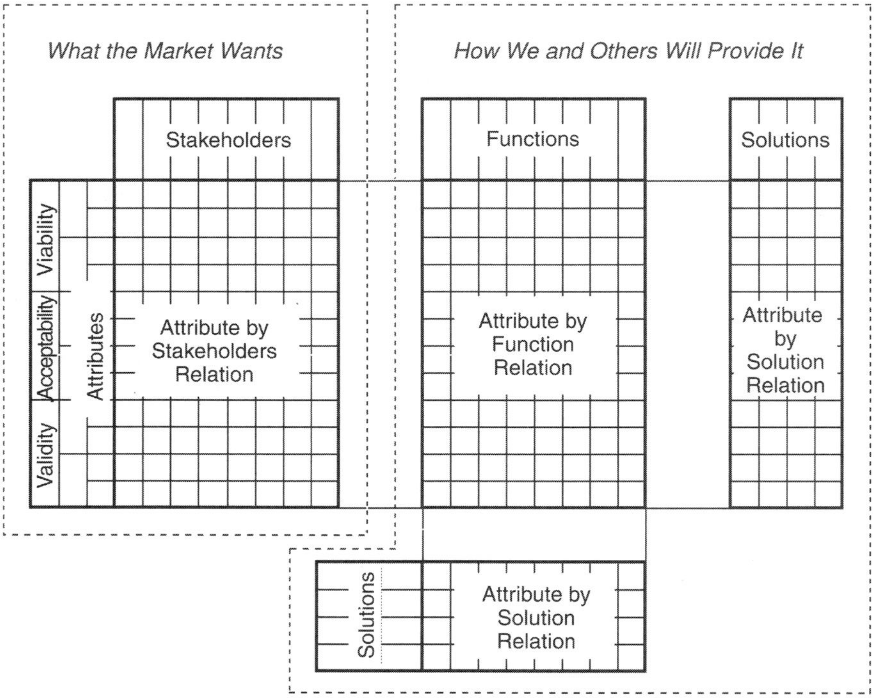


**Figure 24.5** Multistakeholder, multiattribute product planning framework.

acceptability, and validity. The cells of this matrix include stakeholders' utility functions that map attributes to relative preferences. Each cell relates to one stakeholder and one attribute.

The portion labeled "How We and Others Will Provide It" includes characterizations of product and system functionality as they relate to attributes of interest to stakeholders. The elements of the functions versus attributes matrix include ratings of the strength of relationships—either positive or negative—between functions and attributes. This representation is similar to quality function deployment (Hauser and Clausing, 1988), albeit much simplified. This enables providing users of PPA with product improvement advice in terms of functional improvements most likely to support achieving relative competitive advantage.

Solutions are "composed" of collections of functions, as indicated in the lower matrix in Figure 24.5. This often includes both solutions provided—or being entertained—by users of PPA and solutions provided or potentially provided by competitors. This enables competitive analyses and positioning of potential market offerings relative to competitors, possibly with different competitors in different market segments.

The right-most matrix in Figure 24.5 represents solutions versus attributes. The cells of this matrix include the attribute values for each solution. These values may be based on empirical measurements, market requirements, or stakeholder perceptions. In the latter case, multiple attributes may be needed to characterize differences among different stakeholders' perceptions of particular variables.

Summarizing, the overall characterization in Figure 24.5 involves an object-oriented model—in terms of the underlying software—of the multistakeholder, multiattribute nature of how multifunction solutions compete for stakeholders' perceptions of value and, hopefully, influence their subsequent purchase decisions. The PPA helps to both create such models and manipulate these models to determine the most competitive market offerings.

The representation summarized in Figure 24.5 underlies the PPA shown in Figure 24.6. Use of this tool begins with defining goals of the NPD effort, which usually includes both market and organizational goals. The second through fourth steps, as well as associated substeps, involve defining the rows and columns of the matrices depicted in Figure 24.5.

The fifth step, "Assess Solutions," enables using the underlying models to calculate the expected utility of the alternative solutions, for each or all solutions, attributes, and stakeholders. A "How to Improve?" feature within this step performs sensitivity analyses relative to each attribute, rank orders attributes by impact on overall utility, and provides guidance on the functions needing attention to achieve improvements. A "What If?" feature enables assessing the impacts of particular combinations of attribute values, stakeholder preferences, and relative stakeholder importance.

Typical use of PPA involves initial representation of one or two stakeholders, attributes, etc., and subsequent "growing" of representations via insights gained from sensitivity and "What If?" analyses. This approach is much more useful than attempting to "complete" a model before performing analyses, which tends to result in overly complex models from which intuitions can be difficult to glean.

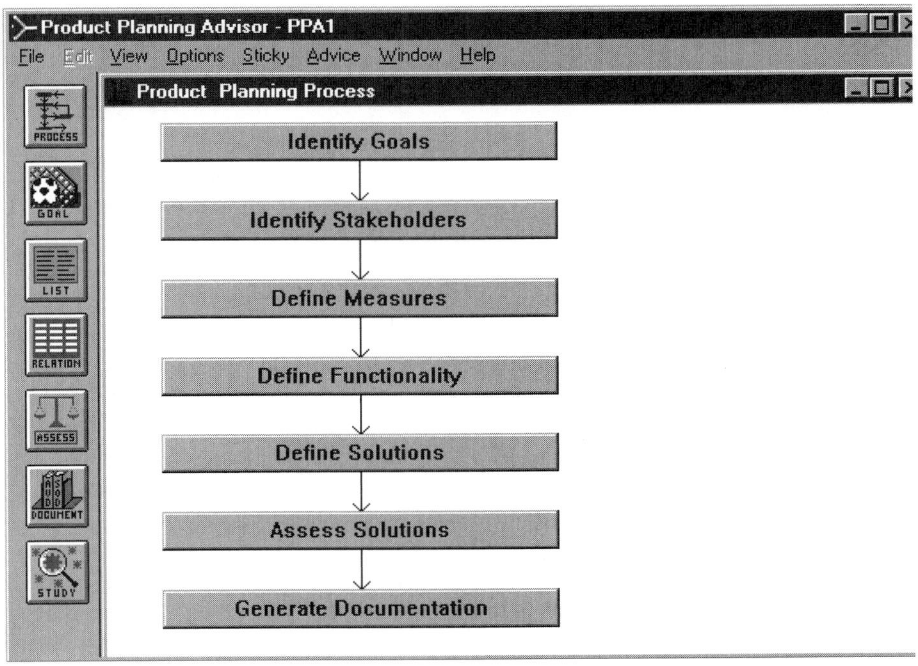


**Figure 24.6** Product planning advisor (ESS, 2000a).

The following examples of how PPA has been used illustrate the ways in which this tool is applied and the types of insights that are gained. In particular, these examples depict trade-offs across stakeholders and how the impacts of assumptions can be explored. It is important to note that while these three examples show how NPD teams reached counterintuitive conclusions, use of PPA does not always result in such conclusions.

***Digital Signal Processor*** Members of the NPD team began this effort convinced that they already knew the best function/feature set with which to delight the market. The marketing manager, however, insisted that they test their intuitions using PPA. After developing the models in Figure 24.5 and using them for competitive analyses, the team concluded that assumptions regarding stakeholders' preferences for three particular attributes, as well as the values of these attributes, were critical to the original intuitions being correct. Attempts to support these assumptions by talking with stakeholders, especially end users and customers, resulted in the conclusions that all three assumptions were unsupportable. The team subsequently pursued a different product concept.

***Automobile Engine*** A team working on new emission control systems decided to evaluate an earlier technology investment using PPA. The team members compared the chosen approach to four other candidates that had been rejected with the earlier decision. Development and use of the models shown in Figure 24.5 resulted in the conclusion that the chosen approach was the worst among the five original candidates. This surprising conclusion led to an in-depth exploration of the assumptions built into their PPA models. This exploration resulted in further support for these assumptions. Reviewing these results, the team leader realized that the earlier decision had not fully considered the impact of the alternatives on the manufacturing stakeholder. The earlier choice had been of high utility to customers and other stakeholders but very complex to manufacture. As a result of this insight, a new approach was adopted.

***Medical Imaging System*** An NPD team had developed an advanced concept for medical imaging that members argued would enable their company to enter a very crowded market, where a couple of brand name companies currently dominated. They used PPA to assess the market advantages of their concept relative to the offerings of the market leaders. Initial results showed a considerably greater market utility for their potential offering. Attention then shifted to the likely reactions of the market leaders to the introduction of this advanced product. The team's expectation was that the leaders would have to invest in 2 years of R&D to catch up with the new technology embodied in their offering. However, using the "How to Improve?" feature for PPA models of the competitors' offerings resulted in the conclusion that the best strategy for the market leaders was to reduce prices significantly. The team had not anticipated this possibility—someone said, "That's not fair!" This caused the team to reconsider the firmness of its revenue projections, in terms of both number of units sold and price per unit.

These three examples serve to illustrate types of human-related issues in NPD. The first example depicted the impact of unsupportable assumptions regarding the preferences of primary stakeholders. The next example showed how the concerns of a secondary stakeholder could affect the attractiveness of a solution. The last example demonstrated how the likely reactions of competitors impact the possible market advantages of a product or system. Taken together, these three examples clearly illustrate how a human-centered orientation helps to avoid creating solutions that some stakeholders may want but other stakeholders will not buy.

### 24.4.6 Summary

This section has discussed selected methods and tools for NPD in the private sector. These methods and tools are also applicable in the public sector. However, due to the differences between these domains, as summarized in Table 24.1, these methods and tools are much more important elements of risk reduction for private-sector NPD. In particular, when the risks of product failure fall completely on developers, there is strong motivation to employ methods and tools that will help in gaining understanding of these risks and evaluate alternative approaches to managing risks.

## 24.5 BEST PRACTICES

Thus far, several NPD methods and tools have been discussed that are judged by many as best practices. Multistage decision processes are pervasive among new product developers, and spiral development processes are also quite common. The HCD concept exploits these concepts for HSI-intensive design contexts.

Market research, often involving compilation of extensive databases, is also key to NPD. The notion of product lines and platforms enabling evolutionary product families is also common, as are formal product evaluation processes. Evaluation of NPD projects is often framed by the particular nature of the multistage decision process employed.

The PPA represents an instance of a structured decision framing and decision-making process. Many research studies have shown that such structured processes tend to result in well-informed decisions with better outcomes (Rouse, 2000b). It is important to emphasize, however, that such tools are best used to guide decision making, *not* dictate decisions.

Beyond these practices, there has been a wealth of studies of the utility of alternative new product realization processes, as well as various characteristics of these processes (Balachandra and Friar, 1999; Cooper, 1999; Rosenau, 1996; Rouse, 2000b, Rouse and Boff, 1998b). The foci of these studies generally fall in categories of project characteristics, project management practices, organizational characteristics, and individual characteristics. The remainder of this section provides a brief review of the highlights of the findings of these many studies.

### 24.5.1 Project Characteristics

Not surprisingly, the nature of a NPD project influences its likely success (Rouse and Boff, 1998b). For example, Tatikonda and Rosenthal (2000) investigated the impacts on project success of technology novelty and project complexity. They found that technology novelty is strongly associated with poor unit cost and time-to-market results, and project complexity is strongly associated with poor unit cost outcomes. They also found that novel process technology presents more problems than novel product technology. With regard to project complexity, they found that the newness of project objectives is more problematic than other aspects of project complexity.

The implications of these findings need to include uncertainties of costs and time to market in financial valuations of novel and/or complex projects (ESS, 2000b; Rouse et al., 2000). These uncertainties might be included in Monte Carlo analyses focused on the probability distributions for projected returns. Risks could then be quantified in terms of, for instance, probability that net values are less than zero. Projects could then be compared in return versus risk portfolio plots.

A wide variety of other project characteristics have been the focus of numerous studies. Recent analyses of this wealth of studies have summarized the impacts of company characteristics, management practices, market characteristics, motivation/incentives, product/project characteristics, team skills/abilities, organizational structure, technologies involved, and time issues on various measures of NPD technical and/or market success (Rouse and Boff, 1998b; Rouse et al., 1998). This immense literature is quite rich, although it tends to rely heavily on subjective perceptions assessed via questionnaires and interviews. Nevertheless, the availability of such data for private-sector NPD efforts far exceeds that available for public-sector endeavors.

### 24.5.2 Project Management Practices

Beyond the inherent nature of NPD undertakings, project management practices can also impact likely success. Lester (1998) summarizes several practices found to be valuable for project management:

- Cross-functional teams are ideally suited for NPD assignments.
- NPD organizations should be focused on adding value to the efforts of NPD teams by helping, supporting, and guiding.
- The NPD process should provide strategy and fundamental operational guidelines.
- Teams and organizations are more effective if they share a common understanding of the NPD process.

Thus, to the extent that NPD projects are dominated by one or two disciplines, do not have clear strategic guidance and objectives, and do not employ commonly understood processes, the likely success of these projects may be compromised.

Regarding teams, McDonough (2000) reports the results of a survey of NPD professionals asking them about success factors in NPD. Almost all respondents employ cross-functional teams, with a primary motivation of improving time to market. Establishing clear, unchanging goals, team leadership, and cooperation were most frequently mentioned as the factors influencing team success. Thus, the mere presence of teams, without other factors being addressed appropriately, does not ensure success.

Jassawalla and Sashittal (2000) provide another view of NPD teams, particularly team leaders. Summarizing the literature on leadership of NPD teams, they conclude that effective team leaders should:

- Clearly communicate the organization's expectations.
- Foster high levels of communication.
- Create a climate that raises morale and energizes the team.
- Take responsibility for the team's goals.

- Guide and share burdens.
- Interface with key external constituents.
- Enjoy high levels of autonomy and support from superiors.
- Involve all functional groups from the initiating stages.
- Balance both technical and human interaction issues.
- Reduce destructive conflict.

Based on interviews of 40 managers in 10 high-tech companies, Jassawalla and Sashittal (2000) suggest that carefully selected team leaders endowed with high levels of autonomy be likely to:

- Ensure commitment.
- Build transparency.
- Function as facilitators.
- Strengthen human relations.
- Foster learning.

Lynn and colleagues (1998) discuss NPD teams and learning. They provide several suggestions for what teams can do to learn more. First, do not expect to get it right the first time, and do not fire the scouts who are exploring new territory. It is important to embrace new information and build on it. Finally, teams should be structured to learn. This involves both refining current ways of doing things and determining new procedures.

Tabrizi and Walleigh (1997) discuss three sets of practices that distinguish companies that are able to successfully launch next-generation products—gleaned from examination of 28 next-generation product projects at 14 leading high-tech companies. They conclude that practices surrounding product strategy are very important, including creating a clear map of the company's product stream for the next 2 years and using it to manage all aspects of the company's development activities, generating a seamless product strategy; that is, one that leaves no holes for competitors to exploit, and collecting, interpreting, and assimilating good information about the market.

With regard to project organization, Tabrizi and Walleigh (1997) emphasize being willing to turn the development of new-platform products over to business units created solely for that purpose, knowing how to choose the optimum number of NPD team members—with the right mix of skills—for product definition over the course of the product development process, and matching other product development resources—such as shifting workloads of engineers and marketers—to the cyclical demands of the process. When executing during the definition stage, they suggest tracking progress and sustaining urgency, developing early prototypes, and using development partnerships.

The representative set of studies discussed here make it quite clear that project management practices are central to success. The nature of NPD teams, how they are guided, and how progress is measured are central issues.

### 24.5.3 Organizational Characteristics

Organizational characteristics beyond the nature of the NPD effort and project management practices can also impact likely success. For instance, Englund and Graham (1999) provide a compilation of good practices for top management. They assert that a successful top manager exhibits the following:

- Knows that projects without strategic emphasis often end in failure
- Develops an upper management team to oversee project selection
- Focuses on the goals of what an organization should do before limiting choices by considering only what the organization is capable of doing
- Works to develop a system of projects and links them to organizational strategy
- Guides the development of consistent criteria that are used to prioritize projects
- Selects projects based on comparative priority ranking contributions to strategy
- Reduces the total number of projects to minimize possible disruption
- Knows that a system of projects utilizes a common resource pool and that the pool may be abused without cooperation across the organization
- Develops a system to manage the resource pool and reward interdepartmental cooperation
- Allows unallocated capacity in a resource pool for emergencies and creativity
- Believes in the power generated by a learning organization
- Creates a model for linking projects to strategy and supports it with authenticity and integrity (p.64)

There can be significant interactions across the categories of variables discussed here. Swink's (2000) recently reported study provides a good illustration. This report presents the results of a survey of NPD projects across most manufacturing industries. Questions were focused on impacts of design integration (i.e., coordination of product and process design activities) and top management support of NPD projects. Design integration was found to be positively associated with higher design quality but not better financial performance. Design integration was also found to be a factor in achieving NPD time goals but only for projects with high technological innovation and/or high levels of uncertainty. Top management support was generally associated with better time-based performance, design quality, and financial performance. However, high levels of top management support were found to be ineffective in securing good financial performance with high levels of technological innovativeness. It is suggested that this might be attributed to managers' likely greater understanding of incremental versus disruptive innovation.

### 24.5.4 Individual Characteristics

The characteristics of NPD team members also impact projects. Perhaps no other aspect of such characteristics has been explored more than the role of champions. Markham and Griffin's (1998) study of almost 400 firms found the impact of champion support for projects tends to be indirect through program success, innovation strategy, process implementation, and program success. While champions significantly improve project performance, it is difficult to demonstrate any measurable overall effects at the company level, although positive effects are manifested through the just-mentioned mediating variables.

Another oft-discussed individual characteristic is entrepreneurship. Roberts (1991) studied high-tech entrepreneurs in several hundred companies over a 25-year period. Of particular interest are his conclusions regarding why entrepreneurs often encounter difficulty attracting investors to their NPD visions. Plans that inadequately reflect market issues and support for financial assumptions are typical shortfalls of NPD plans. Best practice approaches to NPD avoid such flaws.

Bhide (1996) notes that entrepreneurs' great ideas often do not lead to great performance. In discussing entrepreneurs' abilities to deal with people issues, he concluded that:

> When entrepreneurs neglect to articulate organizational norms and instead hire employees mainly for their technical skills and credentials, their organizations develop a culture by chance rather than by design. The personalities and values of the first wave of employees shape a culture that may not serve the founders' goals and strategies. Once a culture is established, it is difficult to change. (p. 129)

The roles of individual knowledge and skills in NPD have also received considerable attention. Mascitelli (2000) argues for taking advantage of the creative power of tacit knowledge to foster breakthrough innovations, defined in terms of creative and original market offerings. He suggests that breakthrough innovators, who are often highly capable generalists, tend to "see" solutions without the conscious ability to explain their visions. Their knowledge is, therefore, very difficult to codify and share—in other words, this knowledge is tacit. His suggested mechanisms for harnessing this tacit knowledge include:

- Engendering deep personal commitments—tacit knowledge flows from the pull of emotional commitment and deep personal involvement
- Using prototypes as catalysts for breakthrough thinking—enables a physically active approach to learning and experimentation
- Sharing tacit knowledge face-to-face—sharing relies more on showing than telling

The obvious implication is that managers need to support the development and maintenance of these mechanisms.

Thus championship, entrepreneurship, and knowledge and skills are individual characteristics that play important roles in NPD efforts. These characteristics can be both great strengths and substantial weaknesses, especially if the problematic aspects are not anticipated.

These findings for private-sector NPD are quite consistent with those of a much more modest study of determinants of success of public-sector R&D projects (Rouse and Boff, 1994). Senior managers from several military R&D organizations were asked to explain factors that differentiated past project successes from failures. The one factor upon which all interviewees agreed was the association of project success with the efforts of champions who displayed strong entrepreneurial orientations. It was often noted that the commitments of champions of successful projects approached irrationality in light of the organizational hurdles that had to be addressed and lack of monetary incentives and rewards for success.

### 24.5.5 Summary

It is beyond the scope of this chapter to provide a complete review of the wealth of studies of the types of factors and variables just discussed. Sources such as those noted earlier—

Balachandra and Friar (1999), Cooper (1999), Rosenau (1996), Rouse and Boff (1998b), and Rouse (2000b)—provide, collectively, a thorough review. In fact, the breadth and depth of this work are sufficient to warrant creation of "knowledge maps" of relationships among primary variables and the secondary variables that affect these relationships (Rouse et al., 1998).

It is useful to consider how the best practices outlined in this section might differ for private and public sectors. This immediately raises the question of the types of outcomes the best practices are intended to influence. In general, technical success and market success are the outcomes sought when attempting to identify potential best practices for private-sector NPD. This search for best practices involves finding those practices associated with the highest degrees of technical and market success.

Technical success in the private and public sectors is likely to be much more similar than is market success. Put in terms of HCD issues, validity is much more likely to be similar in two domains than are acceptability and viability. Thus, those private-sector best practices associated with maximizing technical success are likely to also be applicable in the public sector. In contrast, practices aimed at maximizing market success may not transition well, or even meaningfully, from private to public sectors.

In terms of HSI issues, the key distinction is between design practices that are indirectly regulated by market forces versus practices that are directly regulated by government dictates. In the former, HSI shortfalls are absorbed by the developer in terms of lost sales, warranty costs, etc. In the latter, HSI shortfalls are borne by the customers, often because they failed to specify or regulate some requirement or process.

## 24.6 CONCLUSIONS

This chapter has considered HSI issues in the context of private-sector NPD efforts where market considerations and profit motives drive design decisions. The characteristics of private versus public-sector product and system development were contrasted.

Product management practices were discussed in terms of multistage decision processes and HCD. Methods and tools considered included those for market research, product lines and platforms, product evaluation, NPD project evaluation, and product planning. Results of empirical assessments of best practices were summarized in terms of characteristics of projects, project management, organizations, and individuals.

Overall, competitive pressures, as well as providers having to absorb most risks, result in private-sector NPD being very sensitive to HSI issues. The possibilities of products not selling, having to be recalled, and leading to legal suits provide strong motivations for paying attention to and resolving HSI issues. On the other hand, for new, innovative products and services, the marketplace is often quite forgiving with regard to HSI limitations.

To the extent that public-sector products and systems are similar to private-sector offerings, one may be able to rely on similar competitive forces to ensure responsiveness to HSI issues. On the other hand, for public systems procurements that are sufficiently unique to require significant deviations from off-the-shelf, private-sector solutions, HSI compliance may have to be a regulatory requirement.

Overall, the issues in private- and public-sector NPD are quite similar. However, the motivations for pursuing and resolving these issues tend to be rather different. The HCD concept applies equally well in both domains, but the nature of the stakeholders—beyond users—is very different. Consequently, common practices tend to differ significantly, although one could quite reasonably argue for similar best practices.

## NOTES

1. An obvious exception to this generalization concerns systems associated with national security where a "need to know" may be necessary to gain access to information. Those involved with proposing solutions inherently have this need.
2. One can clearly learn from baselines of previously successful products, especially if they are similar to the product currently being developed. However, the concern here is with assessing likely success as product development proceeds.
3. The author openly acknowledges a principal role in development of this software tool and an ongoing role in its application and sales.

## REFERENCES

Balachandra, R., and Friar, J. H. (1999). Managing New Product Development Processes the Right Way. *Information Knowledge Systems Management, 1*(1), 33–43.

Baldwin, C. Y., and Clark, K. B. (1997, September/October). Managing in an Age of Modularity. *Harvard Business Review*, pp. 84–93.

Bhide, A. (1996). The Questions Every Entrepreneur Should Ask. *Harvard Business Review, 74*(6), 120–130.

Blattberg, R. C., Glazer, R., and Little, J. D. C. (Eds.). (1994). *The Marketing Information Revolution*. Boston, MA: Harvard Business School Press.

Boehm, B. W. (1988, May). A Spiral Model of Software Development and Enhancement. *IEEE Computer, 21*(5), 61–72.

Brownsword, L., and Clements, P. (1996). *A Case Study in Successful Product Line Development*, ESC-TR-96-016. Pittsburgh, PA: Carnegie Mellon University, Software Engineering Institute.

Cody, W. J., Rouse, W. B., and Boff, K. R. (1995). Designers' Associates: Intelligent Support for Information Access and Utilization in Design, Vol. 7 (pp. 173–260). In W. B. Rouse (Ed.), *Human/Technology Interaction in Complex Systems*. Greenwich, CT: JAI Press.

Cooper, R. G. (1985). Selecting Winning New Product Projects: Using the NewProd System. *Journal of Product Innovation Management, 2*, 34–44.

Cooper, R. G. (1998). *Product Leadership: Creating and Launching Superior New Products*. Reading, MA: Perseus Books.

Cooper, R. G. (1999). The Invisible Success Factors in Product Innovation. *Journal of Product Innovation Management, 16*(2), 115–133.

Cooper, R. G., Edgett, S. J., and Kleinschmidt, E. J. (1998). *Portfolio Management for New Products*. Reading, MA: Addison-Wesley.

Cooper, R. G., Edgett, S. J., and Kleinschmidt, E. J. (1999). New Product Portfolio Management: Practices and Performance. *Journal of Product Innovation Management, 16*(4), 333–351.

Cooper, R. G., Edgett, S. J., and Kleinschmidt, E. J. (2000, March/April). New Problems, New Solutions: Making Portfolio Management More Effective. *Research Technology Management, 43*(2), 18–33.

Englund, R. L., and Graham, R. J. (1999). Linking Projects to Strategy. *Journal of Product Innovation Management, 16*(1), 52–64.

Enterprise Support Systems (ESS). (2000a). *Product Planning Advisor.* Available: http://ess-advisors.com/software.htm.

Enterprise Support Systems (ESS). (2000b). *Technology Investment Advisor.* Available: http://www.ess-advisors.com/software.htm.

Fryer, B. (1999, September). The ROI Challenge: Can You Produce a Positive Return on Investment from ERP? *CFO*, pp. 85–90.

Gensch, D. H., Aversa, N., and Moore, S. P. (1990). A Choice-Modeling Market Information System That Enabled ABB Electric to Expand Its Market Share. *Interfaces, 20*, 6–25.

Hauser, J. R., and Clausing, D. (1988, May/June). The House of Quality. *Harvard Business Review*, pp. 63–73.

Henneman, R. L. (1999). Design for Usability: Process, Skills, and Tools. *Information Knowledge Systems Management, 1*(2), 133–144.

Jassawalla, A. R., and Sashittal, H. C. (2000, Winter). Strategies of Effective New Product Team Leaders. *California Management Review, 42*(2), 34–51.

Lester, D. H. (1998). Critical Success Factors for New Product Development. *Research Technology Management, 41*(1), 36–43.

Lynn, G. S., Mazzuca, M., Morone, J. G., and Paulson, A. S. (1998). Learning Is the Critical Success Factor in Developing Truly New Products. *Research Technology Management, 41*(3), 45–51.

Markham, S. K., and Griffin, A. (1998). The Breakfast of Champions: Associations between Champions and Product Development Environments, Practices, and Performance. *Journal of Product Innovation Management, 15*(5), 436–454.

Mascitelli, R. (2000). Harnessing Tacit Knowledge to Achieve Breakthrough Innovation. *Journal of Product Development Management, 17*(3), 179–193.

McDonough, E. F., III. (2000). An Investigation of Factors Contributing to the Success of Cross-Functional Teams. *Journal of Product Innovation Management, 17*(3), 221–235.

Nichols, N. A. (1994, January/February). Scientific Management at Merck: An Interview with CFO Judy Lewent. *Harvard Business Review*, pp. 88–99.

Patterson, F. G., Jr. (1999). Systems Engineering Life Cycles. In A. P. Sage and W. B. Rouse (Eds.), *Handbook of Systems Engineering and Management* (Chapter 1). New York: Wiley.

Product Development Institute (PDI). (2000). *NewProd.* Available: http://www.prod-dev.com.

Roberts, E. B. (1991). *Entrepreneurs in High Technology: Lessons from MIT and Beyond.* New York: Oxford University Press.

Robertson, D., and Ulrich, K. (1998, Summer). Planning for Product Platforms. *Sloan Management Review*, pp. 19–31.

Rosenau, M. D., Jr. (Ed.). (1996). *The PDMA Handbook of New Product Development.* New York: Wiley.

Rouse, W. B. (1991). *Design for Success: A Human-Centered Approach to Designing Successful Products and Systems.* New York: Wiley.

Rouse, W. B. (2000a). Human-Centered Product Planning and Design. In G. Salvendy (Ed.), *Handbook of Industrial Engineering*, 2nd ed., New York: Wiley.

Rouse, W. B. (2000b). *Essential Challenges of Strategic Management.* New York: Wiley.

Rouse, W. B., and Boff, K. R. (1994, December). *Technology Transfer from R&D to Applications.* Wright-Patterson AFB: Armstrong Research Laboratory.

Rouse, W. B., and Boff, K. R. (1998a). Packaging Human Factors for Designers. *Ergonomics in Design, 6*(1), 11–17.

Rouse, W. B., and Boff, K. R. (1998b). R&D/Technology Management: A Framework for Putting Technology to Work. *IEEE Transactions on Systems, Man, and Cybernetics—Part C, 28*(4), 501–515.

Rouse, W. B., Howard, C. H., Carns, W., and Prendergast, J. (2000, Spring). An Options-Based Approach to Technology Strategy. *Information Knowledge Systems Management, 2*(1), 63–81.

Rouse, W. B., Thomas, B. G. S., and Boff, K. R. (1998). Knowledge Maps for Knowledge Mining: Application to R&D/Technology Management. *IEEE Transactions on Systems, Man, and Cybernetics—Part C, 28*(3), 309–317.

Sage, A. P. (1995). *Systems Management.* New York: Wiley.

Sage, A. P. (1999). Systems Reengineering. In A. P. Sage and W. B. Rouse (Eds.), *Handbook of Systems Engineering and Management* (Chapter 23). New York: Wiley.

Sage, A. P., and Rouse, W. B. (Eds.). (1999). *Handbook of Systems Engineering and Management.* New York: Wiley.

Stevens, G. A., and Burley, J. (1997, May/June). 3000 Raw Ideas = 1 Commercial Success! *Research Technology Management, 40*(3), 16–27.

Swink, M. (2000). Technological Innovativeness as a Moderator of New Product Design Innovation and Top Management Support. *Journal of Product Innovation Management, 17*(3), 208–220.

Tabrizi, B., and Walleigh, R. (1997, November/December). Defining Next-Generation Products: An Inside Look. *Harvard Business Review, 75*(6), 116–124.

Tatikonda, M., and Rosenthal, S. R. (2000). Technology Novelty, Project Complexity, and Product Development Project Execution Success: A Deeper Look at Task Uncertainty in Product Innovation. *IEEE Transactions on Engineering Management, 47*(1), 74–87.

U.S. Department of Defense. (1996). *Guide to Integrated Product and Process Development.* Washington, DC: U.S. Department of Defense, Office of the Undersecretary of Defense for Acquisition and Technology.

# AFTERWORD

## I INTRODUCTION

In 1997 Congressman Ike Skelton made an important statement to Congress on MANPRINT (reprinted in the Appendix). He was concerned that the turbulence of government "downsizing" and Department of Defense acquisition reform would eliminate an Army program which had all the features of exactly the kinds of programs that the nation's leaders should encourage—high performance, safety, and cost savings with new weapon systems developments. His statement not only had the effect of preserving the Army program, but also of encouraging the other branches of the military to seek out Human Systems Integration (HSI) benefits for their systems acquisition programs as well. Soon thereafter, as shown by the contributions to the *Handbook of Systems Integration* from both military and non-military sources, we found a growing interest to the degree that might be called a sociotechnical cultural revolution (at least among engineering and business communities) on how we should acquire and implement complex technological systems. This technological cultural revolution, which would focus all decisions concerning the acquisition of new systems on the people who must use them, is spreading not only throughout the military and aerospace industries, but is receiving consideration from other fields and endeavors which rely on complex systems to fulfill their missions.

The *Handbook* chapters present the state of the art for this new systems approach which begins and ends with fully integrating people, technology, and organizations for a common purpose which is almost always to promote the well being of the nation's citizenry whether at work, on the battlefield, in classrooms, or as patients. While one of the principal objectives of using the HSI approach is to increase the financial health of an organization, it seeks ways to accomplish financial goals without sacrificing the human resource upon which the success of the endeavor ultimately depends. According to Paul O'Neill's business philosophy while CEO of ALCOA, when business focuses on the safety, health, and well being of its people, business success will follow.[1] In the long run, this success could far exceed any short term gains an organization might be tempted to acquire at the expense of its people.

In ending his statement, Mr. Skelton posed a challenge to his colleagues that we would like to address here. He concluded:

> The possible applications for ... [HSI] go far beyond the military in our constantly evolving technological-based society... It would seem our medical and educational systems could benefit from a technological development and management process which focuses on the end user. One may wonder what a difference it would make if these systems were made to operate primarily for the doctor and the patient or the teacher and the learner rather than fitting these individuals to the system as an afterthought.

**TABLE 1 Comparison of Critical System Factors on Four Activities**

| Factors | Military | Health Care | Education | National Security |
|---|---|---|---|---|
| Users | Soldiers, Operators, Maintainers | Doctors, Nurses, Patients | Teachers, Students, Parents | FBI, CDC, etc. Fire fighters, Police, Civilians |
| Mission Issues | 1. Weapons Systems acquisition<br>2. Manpower, personnel, training costs<br>3. Combat readiness | 1. Health care quality and resource capacity<br>2. Health care costs<br>3. Supply of Health care professionals (manpower)<br>4. Emergency preparedness | 1. Students' educational quality<br>2. Teachers' quality<br>3. Education Delivery systems and facilities<br>4. Education costs | 1. Terrorism defense<br>2. Emergency readiness<br>3. Security Costs |
| Technology Complexity | Extremely high | Moderate | Low | High |
| Organizational Complexity | Extremely high | High | Moderate | Extremely high |
| Operational Complexity | Extremely high during war; moderate otherwise | Extremely high —Continuously | High | Extremely high |

***Note:*** Complexity ratings: Low, Moderate, High, Extremely high

Accordingly, we have picked three activities of high national interest that we believe have the potential to receive major benefits from applying the HSI approach to their systems decisions; particularly in making far better use of their resources than has happened in the past. They are health care, education, and national security. Table 1 provides a broad comparison of these three activities with military systems for five factors: users, mission issues, technological complexity, organizational complexity, and operational complexity. These five factors help provide a framework for evaluating the degree to which HSI benefits found in military systems might transfer to the other activities.

It is quickly seen that all four activities have costs issues associated with accomplishing their missions. One of the promises of HSI is that system performance objectives can be met with projected affordable costs. The problem with the health care and educational activities is that costs can increase exorbitantly while falling behind on meeting performance objectives. National security costs could be potentially unlimited without meeting desired objectives. None of the other activities has the level of technological complexity that the military requires, but all look to technology to assist in meeting their missions in a cost effective manner. For the organizational factor, only national security has the same level of complexity as the military. Military operational complexity is extremely high during war, but most of the time is fairly moderate. Similarly, depending on the level of threat, national security can range from moderately "alert" to extremely heightened states of emergency. The greatest daily operational complexity probably resides in health care because of the requirements to deliver a wide variety of highly specialized therapies for a large spectrum of illnesses under conditions which involve continuous, but unscheduled, patient emergencies. Further, like the military, the health care system must be prepared to deal with casualties of disasters. The operational complexity in education is high, primarily because of conditions placed on the education system by many disparate competing entities without common goals.

## II HEALTH CARE

Healthcare is expensive, and national health expenditures in the United States continued to climb (at a rate of 6.9 percent, increased from 5.7 percent in 1999) in the year 2000, amounting to nearly $1.3 trillion and 13.2 percent of the gross domestic product (GDP). They are projected to total $2.8 trillion, grow at an annual rate of 7.3 percent, and reach 17.0 percent of the GDP by 2011.[2] Delivering state of the art medical care today generally requires the simultaneous and ongoing provision of many different, and often very specialized, services (various medical professional evaluations with treatment planning, diagnostic testing, monitoring, delivery of pharmaceuticals and other therapeutic interventions etc.). This necessitates a greater degree of teamwork than might have been necessary in the past (when there were fewer services and therapies to be rendered). Well reasoned, thoughtful application of medical knowledge to clinical conditions is a prerequisite of good care. However, clinical outcomes now, more than ever, depend upon a host of factors beyond good clinical judgment, some of which include the availability of timely, coordinated, safe and reliable services. With few exceptions, the delivery of medical care has become dependent upon a health care 'system' of services to provide our complex and technological standard of medical care.

In the recent, highly publicized Institute of Medicine report "To Err is Human: Building a Safer Health System,"[3] it was estimated that adverse events as a result of active medical errors (errors of commission) occurred among 3–4 percent of hospitalized patients. One in ten of these adverse events resulted in death, and at least half of these errors were thought to be preventable according to current standards of care. It has been estimated that 44 to 98 thousand deaths per year in the United States occur as a result of medical injury, with associated direct health care costs totaling 9–15 billion dollars a year. These estimates do not include care delivered outside of the hospital setting or instances where effective medical care may have been withheld (errors of omission in medical care are likely to represent a much larger issue for the effectiveness of health care delivery). Some have challenged the exact figures presented in the Institute of Medicine report, contending that the adverse event rate is overestimated.[4] Whatever the case, there is an increasing acknowledgement that medical errors do occur; they are often preventable; they are costly; and they significantly threaten the safety and quality of American health care. With the complexity of the healthcare enterprise, a systems approach to evaluating, managing and designing care processes seems logical.

The culmination of technological, organizational, and operational complexity for the health care system is best illustrated by a snap shot of any moment in a hospital environment. The following four exhibits describe a typical hour in the emergency room, the medicine floor, the surgery floor, and the administration office of an urban hospital.[5]

**Exhibit 1 – Emergency room (ER)** It's 5 PM on a Friday during flu season. The ER is busy. One of the nurses scheduled to work this shift had injured her back lifting a patient the day before, and a receptionist called out sick. The ER manager was unable to find anyone who could cover this shift for them. The ER triages the people with flu symptoms to an urgent care setting. Today, this unit is overwhelmed. The ER will help to accommodate them when they can. However, the ER is unable to help those with more minor illnesses on this day, because it is full to capacity with a number of critically ill patients to care for.

One patient is a middle-aged executive who is having chest pain, and whose blood pressure suddenly drops. It becomes immediately clear that he is not having a heart attack but is suffering from a ruptured thoracic aortic aneurysm. The operating rooms are busy with other urgent and elective cases; the surgeons and anesthesiologists are all presently occupied.

Another patient is a young woman who suddenly fainted and is in a coma. She had been two blocks from a university hospital, but that hospital was presently so overwhelmed with patients that the ambulance had to be diverted to a community hospital. The patient had to be placed on a respirator, and testing revealed an intracranial bleed. The patient needed a level of neurosurgical care that would best be performed at a university hospital. The ER doctor called two different university hospitals, but neither had beds available in their neurosurgical intensive care units (ICUs), so could not accept the patient in transfer.

In another room is an obese woman who could not communicate or move her right side as a result of a previous stroke. A hired caregiver who brought her in reports the patient complained of abdominal pain. The caregiver could give some limited information about the patient's past medical history, but was unable to list the patient's medications or name the patient's primary physician. In any case, medical offices were closing for the day, and without looking at the patient's office record, the patient's own physician, let alone a covering physician, would probably not be able to list the medications either. The caregiver reported that the patient had multiple previous hospitalizations at another local hospital, and received primary care in that hospital's medical teaching clinic. The caregiver recalled that the patient had been previously hospitalized at this institution as well, however, a medical record could not be located. A medical technician is scrambling to find cables for a cardiac monitor, because another patient presents an irregular pulse and dizziness. Bloodwork is drawn and barcoded labels are placed on the specimen tubes. The labels are jamming in the bar code labeling machine, so another technician is trying to solve this problem. He is able to get them to print now, but the labels are slightly out of alignment, so the complete code is not being printed correctly. He's still working on this.

There is a patient with a broken bone in another room. No one has been able to keep up with re-stocking the rooms, and the cast material specified by the physician has run out. Someone needs to go to the hospital's central supply to get more. The police have just brought a disheveled, agitated and unruly patient to the ER for evaluation. He has a history of mental illness, and smells of alcohol. He has not been going to work and is now charged with assaulting his wife. The thoracic surgeons finished a cardiac bypass operation they had been involved with, and rush to the ER to take the man with the aneurysm to the operating room. The husband of the young woman with the intracranial bleed has also rushed to the ER. He is shocked and bewildered to see his wife on a ventilator. A nurse hurries to provide him with information and comfort. The ER doctor has done a physical exam, which required the help of two nurses to move and position the obese, stroke disabled woman with the abdominal pain. While the doctor documents the exam in the chart and is ordering studies, a nurse, with the help of a medical technician to hold the patient in position, is placing a urinary catheter with sterile preparation and draping to assess for bladder obstruction. Another nurse is preparing to draw blood from the patient's arm. A respiratory therapist is administering nebulized treatments, begun in triage, to a patient with asthma in another room. The registrar is waiting outside this patient's room to get the necessary information to register the patient, and create a chart. The doctor has already seen this patient, but will wait to document his findings once the chart is created.

A patient in another room presented with an upper gastro-intestinal (GI) bleed that seems to have stabilized, and is ready to be transferred to the medical intensive care unit (ICU). The ER unit clerk has been paging the ICU doctor who is listed as being on-call. There has been no answer. The patient will remain in the ER until an ICU bed can be made ready. The patient looks comfortable, is surrounded by family members, and is receiving a blood transfusion. His daughter is ringing the bell for his nurse, because he wants something to eat. In the medical laboratory, a laboratory technician is trying to scan the barcode label on a tube of blood. He complains that the way it was placed won't allow scanning because of the rounded surface of the tube. He enters the patient's information into his computer to create a new label and proceeds to peel off the previous label. If he has more than one label on the tube it won't fit into the centrifuge, which is a required step in the processing of specimens. He looses his grip on the tube. It falls to the floor and breaks open, with blood splattering on his clothes and elsewhere. It is 6 PM. The ER waiting room is still full. The unit clerk announces to everyone in the ER that the computer system used for patient registration, test requisition, test results reporting, and billing has just gone down. The ICU physician who is on-call pays a spontaneous visit to the ER to see if anyone needs help. This physician is not who was listed as on-call for this night. The unit clerk says hello, and takes a pen to correct the on-call list that had been faxed by the department of medicine.

**Exhibit 2 – Medicine floor.** Up on the medicine floor, a patient seems to be choking. A doctor is summoned, who inspects the airway and asks the nurse for a suction kit that can be hooked up to the wall suction unit. The nurse runs to the supply room, but says she doesn't regularly work at this hospital since she's an agency nurse, and can't find it. The unit clerk pages other nurses for help. Another nurse says that the suction units are no longer stocked on the floors and have to be ordered from the hospital's central supply. The doctor is getting angry and asks the nurse why such a decision had been made. She expresses similar frustration and says that the nurses had fought this decision when made two months ago by a hospital administrator. They decide that they have to break into a code blue cart, which is a locked cart of supplies for emergencies. It apparently costs hundreds of dollars to break into a code cart, but what else is there to do right now? Just then, another nurse comes running with an unused suction kit that she has taken from another patient's room. They suction oral secretions from the choking patient, who recovers. Everyone says they ought to write out an incident report, and they vow to, but right now they have too many other things to do.

A physician receives a call from a nurse to report that a diabetic patient had a dangerously low blood sugar measurement. She reports that the patient was given orange juice mixed with sugar to drink. The blood sugar was measured using a battery operated glucose meter. The physician asks what physical manifestations of the low blood sugar the patient showed. The nurse replies that the patient had no physical symptoms whatsoever, and was alert and oriented throughout. The physician suspects that the metered measurement was incorrect because the clinical status of the patient did not support a diagnosis of low blood sugar. A repeat blood sugar measurement is requested in one hour. The physician is called in one hour with a report that the blood sugar is now too high.

**Exhibit 3. Surgery floor.** On the surgery floor, a patient who came in for an elective hip surgery was being readied for discharge to home. The patient was an elderly woman with a number of chronic conditions for which she took a number of medications. One of her usual medications, for high blood pressure, was not specifically available in the hospital pharmacy. Another medication with similar pharmacologic properties had been substituted during her hospitalization so that she could continue to receive uninterrupted treatment for her hypertension. The surgery resident wrote discharge instructions as well as prescriptions for all the medications she had received from the hospital formulary. The patient assumed that the substitution blood pressure medication was something new she had to take, since she was not familiar with its name. She had that prescription filled at the hospital outpatient pharmacy on her way out (instead of her usual pharmacy), and began to take the 'new' medication in addition to her usual regimen of medicines at home. Two days later she presented, as an outpatient, to her primary physician complaining of lightheadedness and fatigue. Her pulse and blood pressure were abnormally low. Fortunately, she had brought all her medication bottles with her for the appointment, making it a simple matter to determine that she was suffering from an overdose of antihypertensive medication.

**Exhibit 4. Administration office.** The hospital administrators are meeting to discuss strategic planning for the upcoming fiscal year. A few years ago, with excess hospital bed capacity, they had focused on attracting primary care doctors to the hospital community in an effort to boost hospital patient volumes. There weren't many gains in primary care doctors, yet, today, they find themselves dealing with hospital overcrowding. To make matters worse, most of the increases in hospital admissions have been medical admissions, which aren't reimbursed by payers (both private and public) as highly as surgical procedures are. The revenue stream has increased, but is becoming dominated by less profitable business. To make matters worse, this trend of increasing medical admissions is hampering the hospital's ability to accommodate the more profitable surgical business. With the hospital beds being filled with sick medical patients (for diagnoses like: pneumonia, congestive heart failure, emphysema etc.) there were fewer beds available to schedule patients for elective procedures (for example: joint replacement surgery, elective hysterectomy or gallbladder removal, cosmetic surgery etc.), which were increasingly being delayed, and sometimes cancelled. A few of the staff specialty surgeons have abandoned the hospital for free standing surgical centers. The administrators are not talking about expanding the number of beds at this hospital, because the beds are already there. A whole floor was shut down during the last decade because of previous downsizing. Even if they wanted to open this floor back up, they can't because, despite massive recruitment efforts, there aren't enough nurses available to staff them.

## HSI Applied to Health Care

***How can HSI help improve the health care system?*** There are a number of areas with common ground between the military HSI advances and applications to health care. For example consider costs containment, human and systems performance analysis and testing for delivery of complex services, patient safety, technological procurement and deployment strategies, dealing with manpower shortages, and achieving common organizational objectives.

When we look at Table 1, we see a different set of users from what HSI has focused in the military, but the idea of focusing on the user is still the driving concept. Doctors and nurses are skilled professionals whereas patients represent the diverse population of our society. The techniques for describing characteristics of the human, for conducting task analyses of work environments, and human-technology interfaces are covered throughout the *Handbook* in a number of chapters and need not be restricted to military personnel. Methods for analyzing costs of HSI applications and seeing their benefits in terms of safety and system performance are provided in Chapters 17 and 18. Federal Government personnel and manufacturers concerned with improving medical devices might read Chapters 7, 13, and 24 to better understand the processes for procuring products with a user focus.

**TABLE 2 Health Care Priorities**

**HOSPITALS**

Management and Organization:

1. Studies to determine most cost effective way for hospital network to a) meet increasing demands for patient services, b) reduce operations costs, c) improve medical outcomes, and d) increase delivery service effectiveness
2. Risk analyses to predict sources and effects of human and system error on patient care outcomes.
3. Development of data bases, data collection standards, and data evaluation processes for reliable health care delivery information used in decision making.

Operational Processes:

1. Mapping of current hospital processes and identification of resources to provide patient health care.
2. Identification of major impediments in current processes
3. Identify key areas to improve or redesign current processes
4. Provide user oriented measures of effectiveness for processes.

Product Design:

1. Technology requirements for hospital procedures and processes
2. Standards for medical devices development and usage.

**FEDERAL GOVERNMENT**

Management and Organization:

1. Laws, policies, and procedures that focus on safe, affordable, and effective health care practices and technology
2. National risk pool data to reflect projected costs of care; including analysis of demographics and access issues

Operational Processes:

1. Comprehensive human factors study of medical workplace and health care delivery system

Product Design:

1. Human factors criteria for safe and effective medical devices design and operation
2. Government sponsored medical research that focuses on design and development of new human factors technology; e.g., noninvasive surgery
3. Research and development of job performance aids for health care providers

**MEDICAL DEVICE MANUFACTURERS**

Product Design:

1. Develop and implement user centered design procedures
2. Integrate contributions from multiple skilled disciplines in human sciences and technology.
3. Apply human systems integration tools in the design and development of products.
4. Test and evaluate human performance, safety, and usability of products.

One of the greatest problems facing the health care system is the large number of disparate organizations that can be considered stakeholders in health care delivery. Hospitals, the federal government, health maintenance organizations, insurance companies, state governments, medical device manufactures, and universities/medical schools, the pharmaceutical industry, employers and patients are all part of the health care system. All of these organizations interact in ways to create the daily operational complexity in health care delivery. For leaders of health care organizations, the philosophy, cultural change requirements, and benefits of HSI are described in Chapters 1, 2, 3, 18, and 23. To better appreciate the HSI process for handling complex systems design, Chapters 4, and 6-10 are recommended reading. Two concepts are critical for success; these being 1) defining requirements for user involvement in all early decisions and 2) testing any ideas to be assured of their true value in the work environment.

For studies to determine tradeoffs among personnel, training and human factors design and systems performance see the methods described in Chapters 11, 12, and 13. For quantitative definition of users and their task environments see Chapters 19 and 20. For safety and health factors and methods, see Chapters 14 and 15.

It is realized that solutions to the health care crisis will need to create systems involving all the stakeholders, but to begin to help the doctors, nurses, and patients in a hospital environment we have reviewed three areas, the federal government, the hospital, and medical device manufacturers for priorities. Table 2 lists some of the highest priorities we would suggest for health care system improvement based on the HSI approach.

## III EDUCATION

In FY 1996 our States and territories spent collectively over 255 billion dollars on elementary and secondary education. In FY 2000 that figure had grown to over 300 billion dollars (an increase of nearly 20 percent).[6] The Federal government, through the Department of Education, contributed another 40 billion in FY 2000 (which was a 33 percent increase over its FY 90 expenditures in constant dollars).[7] Yet in FY 2000 about 37 percent of fourth graders continued to read below a basic achievement level on the National Assessment of Educational Progress (NAEP) standardized test; a trend that has remained stable throughout the past decade.[8] Since the ability to read is such a critical skill that underlies most all of academic endeavors and our everyday life skills as well (from reading tax forms and insurance policies to instructions for operation of household items and even safe cooking) why do we seem unable to do better? Funding is always a consideration but the above statistics show that even with substantial increases in funding, success continues to elude the educational system.

The following two exhibits help illustrate the intricacies of the problem and why simply spending more money has not resulted in long lasting solutions for education.

**Exhibit 5. First grade classroom.** Mrs. Foster has thirty five children in her first grade class. This is her first day and their first day. They will begin to learn to read using simple familiar words and repetitive sounds that will enable them to "see" and "hear" the effects of different letters of the alphabet. Hence a sequence of letters C A T may be written or shown in a graphic display followed by the pronunciation. This sequence of letters may be followed by the sequence H A T and B A T. It will be an explicit assumption and expectation that the children have already learned to recognize the individual letters of the alphabet. Simple enough and straightforward it would seem. Still Mrs. Foster feels apprehensive.

Soon she discovers that many do not already know their ABCs. Also there is one child who seems to have a hearing impairment and yet another who has trouble "seeing" the letter sequence correctly. Even several of those who have the ability to see and say aloud "A B C " when presented with the spelling and pronunciation of "CAT" appear puzzled on this first day of school. "C" simply does not sound like "CAT". The problem becomes worse when she attempts to show and pronounce the word "S E E" since five of the children recently immigrated to this country and are more familiar with the Spanish "S I" looking and meaning something different. Some of the children are having trouble staying awake and two of the boys already appear likely to be disciplinary problems.

The school has just acquired new computers for the teacher and children to use in helping them learn to read, but there are no instructions provided with the computers. Mrs. Foster knows very little about computers and is especially unfamiliar with anything other than the reading programs, yet she is expected to use them to help with math and geography as well. The computers will sit unused for the next several months.

Several special sessions (during lunch and after the children go home) are called by the school principal to discuss administrative matters. At these Mrs. Foster learns she is to keep records on the students, noting their progress and shortcomings. She must also provide her own logistics support including identifying and acquiring supplies, equipment, and other resources. There will also be additional duties assigned as the year progresses.

In the afternoon, Mrs. Foster assesses her class for potential in developing writing skills. Quickly she becomes aware that Paulo and Mary are having a difficult time printing and she can sense their embarrassment. When she goes home that evening, although tired and a bit discouraged, she resolves to work harder tomorrow to keep up with the goals she has set for herself.

**Exhibit 6. Efforts to improve the education system.** Many efforts have been made to improve and reform our education "system" and new methods continue to be studied. One such major longitudinal study is reported by Berends et al.[9] in which the RAND Corporation gathered data on the project of the New American Schools (NAS), launched in 1991, which was designed to address whole-school reform. Approximately 185 schools partnered with special design teams in 14 separate districts throughout the nation. By 1995 over 500 schools had entered the project. The project initially consisted of theory-based new learning designs with no external intervention.

The RAND study investigators quickly discovered that external support for implementation of the designs would be required if the designs were to be implemented at all. Schools could not simply be "given" the programs without some guidance, help, and impetus to move forward. One of RAND'S major observations is that the learning designs were in a continuous process of adaptation often to the detriment of a major premise of the project, that of unification. Secondly, it was observed that barriers to achieving high degrees of implementation arose related to poverty, achievement, and school and district climate. School "capacity" to absorb the new learning designs proved to be an important factor. "Principal leadership (communicating expectations to the staff, securing critical resources for the school, talking with teachers about instruction) proved to be an important contributor across all studies." There was an indication that teachers bore a significant cost of the design reforms, particularly when other reforms were being attempted simultaneously. While student gains in mathematics and reading were observed in approximately 50 percent of the participating schools the findings suggested "that weak implementation will lead to weak impacts on student performance."

## HSI Applied to Education

What do we know about HSI that can be applied to the problems faced by Mrs. Foster and her students? What kinds of help from an HSI perspective might a new look at education provide in the future and what are some of the major considerations that need to be taken into account with such a new look?

Broadly the principles of HSI can be employed to improve the educational system by:

- Developing educational programs, organizations, facilities, materials, and technology that are people focused as opposed to school or district focused.

- Placing emphasis on measurement, access, and utility of performance data.
- Developing educational strategies, models, and tools that are adaptable to local and individual needs.

HSI principles begin with a focus on the person rather than the institution or the technology. Table 1 shows these people to be primarily teachers and students. As with the health care system users, the *Handbook* techniques for describing the user characteristics, conducting task analyses, and defining human-technology interfaces can be useful for the education system users. Federal, regional, and local educators may read Chapters 7, 13, 22, and 24 to better understand the processes for procuring educational products with the user in mind. Methods for analyzing costs of HSI applications and seeing their benefits in quantitative versus subjective terms can be found in Chapters 17, 18, and 22.

Also as with the health care system, there are a large number of disparate organizations, including the Federal government, state and local school systems and boards, and parent teacher organizations that influence how any particular classroom will be utilized. From an HSI point of view all such organizations need a consistent and rational basis for deciding on how to design, staff, and operate a classroom. This means that suggestions and programs that affect the school organization and physical plant as well as educational methods and materials should be made focused first and foremost on the student, teacher, and parent. Decision makers for educational institutions may be stimulated from reading Chapters 1,2,3, 18 and 22 to consider the value of seeking a cultural change in education based on a user centered systems approach.

Next HSI urges the development and adoption of overall models and strategies that include the educational needs of the target audience; in this case it would be all the K-12 student population. To address this problem, a systematic approach will be required that can assess the traditional model of education and perhaps suggest alternative models of K-12 education to determine possible solutions at the national level. Successful programs such as Headstart indicate that a truly comprehensive model would apply even before students enter kindergarten.

No one "new" method or program (especially at the national level) will be able to address the complexities of providing adequate education for all. It is safe to say that schools, teachers, parents and students will need tools tailored to address unique regional and local requirements. These will need to be tested to assure they actually achieve the performance expected. Chapters 4, 6–8, and 22 describe the central two concepts (user requirements based on user involvement in early decision making and testing educational methods in the classroom) that are critical for success.

Perhaps the greatest contribution in the near term from an HSI approach would be in aiding educators to make better decisions about technology as an aid to learning. Many schools have embraced technology as a means of enhancing student education. Such programs can be very costly however, and few are adequate in helping all students learn. At best most have experienced varying degrees of success. In some cases, the programs may actually inhibit learning.

Developing and applying the right technology is important, however. (The chapters to read for more information on the process of developing technology from an HSI perspective are 12, 13, and 22.) For example, an "early up front needs analysis" can provide valuable insights and possible solutions, many of which may require application of technology. Design requirements for continuous, flexible, adaptation to student's immediate and changing needs may be imposed on any technology based solutions. Tools for assessment, collaboration and support can likewise be built into future solutions. Integration of schoolhouse, external environments and home, through distributed techniques made possible by technology can present continuous and expert support to all students.

Crawl/walk/run learning paradigms can be developed for individual students guided by continuous diagnostics. Individualized presentation of instruction based on sensory preferences and strengths or weaknesses can adapt to meet and support each student's learning style. Tools that permit collaboration of any two or all three of these groups with each other and with other peers can be very valuable in providing tailored, adapted learning and learning support activities for the student. A few recommended classroom educational tools are listed in Table 3.

**TABLE 3 Classroom Educational Tools**

**Tools for teachers and students**

- Tools of assessment evaluation and feedback for both teacher and students. (Diagnostics)
- Automated Management Systems (Record Keeping, Planning)

**Tools for students**

- Programs that can adjust using artificial intelligence (AI) to students' sensory preferences and strengths
- Computer software that employs voice recognition, speech synthesis software converting text to speech.
- Combination visual and audio programs with diminished cues capability that allow both machine and student to "show and tell."

## IV NATIONAL SECURITY

After September 11, 2001 it is no longer possible in the U.S. to go about our daily activities with the same confidence we may have had before that date. Our national security from terrorist attacks is threatened in nearly every aspect of our society. Our transportation systems, communications, infrastructure, energy systems, air, water, centers of business and entertainment, even our homes are vulnerable – none is immune from the threat of terrorism.

The government of the United States has responded to the terrorist events of 2001 by heightening security across a broad array of industries (e.g., nuclear, aviation) and national assets, and has formed a new federal Department of Homeland Security. A 2002 report sponsored by the Brookings Institute[10] provides a four-point plan for enhancing national security against terrorist events that focuses on (1) perimeter defense at the border to prevent infiltration by terrorist and/or potential terrorist weaponry, (2) detection of internal threats and the securing of potential terrorist weaponry, (3) identification and defense of key sites within the country, and (4) providing those involved in responding to an attack that may nevertheless occur with the tools to effectively respond to and contain it. The proposed federal cost for the added security is around $45 billion annually compared to less than $20 billion in 2001.[11] State, local and private sectors would need to bear higher costs for homeland security as well. One of the most fundamental challenges will be structuring the federal government in ways that will make the responsible agencies capable of addressing national security threats efficiently and effectively.

The following two exhibits provide examples of the cross-organizational difficulties facing those responsible for preventing, preparing for, and reacting to terrorist acts.

**Exhibit 7. Bioterrorism events.** O'Toole provides a stark illustration of the extreme demands that would be placed on public health, safety, and defense resources in the event of a bioterrorist attack specifically, one in which the smallpox virus is used as the primary "weapon."[12] According to O'Toole's scenario, the smallpox virus is released by a terrorist group at a public ceremony attended by the vice-president in a major northeastern city. Although FBI informants later report there were rumors that "something happened" during this event, there is no awareness within the government of the smallpox release. Within two weeks, a small number of patients begin to present themselves at emergency rooms with signs and symptoms such as fever, backache, headache, chills, vomiting and influenza-like symptoms. For the most part, these patients are instructed to return home, rest in bed, take ibuprofen, and drink plenty of fluids.

Within several more days, these signs and symptoms begin to worsen. Of particular note is the appearance of vesicular rashes that are initially interpreted as indicating the presence of chickenpox. Further testing, based on recommendations from an infectious disease specialist, reveals the presence of the smallpox virus, and approximately two weeks after the initial release of the virus a contagious disease emergency is declared. At this point, a complex chain of events is set in motion involving the FBI, local police, the Centers for Disease Control, and local hospital administrators and healthcare personnel. Within hours of the emergency declaration, the issue has risen to the level of the National Security Council and White House, and has begun to attract the attention of local and national media.

O'Toole's detailed scenario goes on to illustrate the serious organizational challenges that arise as the nation attempts to respond to the crisis. The logistical problems involved in distributing limited supplies of smallpox vaccine to those most in need, identifying, locating, and isolating infected individuals, managing the flow of information to maintain public order in the face of rising public fear, concern, and civil unrest, and coordinating overall command and control activities at the national and local level are all highlighted in stark terms. Written to "stimulate review of institutional capacities for rapid communication and coordinated action in the wake of attack."[13]

Perceived shortcomings in the nature of the command and control structure needed to respond to a significant bioterrorist attack have been the topic of several recent articles.[14] Rosen et al. argue that current federal emergency response plans are well-suited to respond to "limited" disasters, but ill-suited to respond effectively to "unlimited" disasters.[15] The latter are defined as a disaster that spreads relatively spontaneously and indefinitely, last for periods of weeks to months. Rosen et al. cite a communicable bioweapon such as smallpox, plague, or influenza as an example of a man-made, unlimited disaster.

To illustrate their point, Rosen et al. cite results of two major exercises that simulated the release of bioweapons in the United States.[16] The first exercise, referred to as TOPOFF,[17] simulated the release of plague in the Denver area, and resulted in the collapse of the Colorado public health system after six days. The second exercise referenced by Rosen et al was the "Dark Winter" simulation, which included a scenario in which smallpox was released in Oklahoma.[18] The Dark Winter exercise resulted in the following major findings: (1) a bioterrorist attack on the United States would clearly threaten vital nation security interests, (2) current organizational structures and capabilities are not well suited for managing the results of such an attack, (3) there is no surge capability in the US healthcare and public health systems that could manage the results of such an attack, (4) managing the media response to a biowarfare attack would be a major challenge at all levels of government, and (5) containing the spread of disease will present significant ethical, political, cultural, operational, and legal challenges.[19]

**Exhibit 8. Fire and police communications on 9/11/01.** The events of September 11, 2001, revealed some clear deficiencies in the functional integration of the nation's sociotechnical defense assets. For instance, a long history of cultural distrust and animosity between the New York City Police and Fire Departments contributed directly to the occurrence of needless deaths and impeded effective performance of rescue operations. After the south tower of the World Trade Center had collapsed, police helicopters hovered in the air above the area of the remaining north tower. To the pilots and observers on board these aircraft, the imminent collapse of the north tower was apparent, and an immediate evacuation of all personnel was ordered.

"Those clear warnings, captured on police radio tapes, were transmitted 21 minutes before the building fell, and officials say they were relayed to police officers, most of whom managed to escape. Yet most firefighters never heard those warnings, or earlier orders to get out. Their radio system failed frequently that morning. Even if the radio network had been reliable, it was not linked to the police system. And the police and fire commanders guiding the rescue efforts did not talk to one another during the crisis. Cut off from critical information, at least 121 firefighters, most in striking distance of safety, died when the north tower fell."[20]

## HSI Applied to National Security

A fundamental assumption of an HSI approach to national security preparedness is that an effective response to a large-scale terrorist event in the future will of necessity involve smooth, functional coordination between a number of different government and private sector entities. The success or failure of this response will largely be a function of the degree to which organizational and technical assets are designed (or redesigned) to effectively transmit the right information to the right people in support of the right response at the right time.

The observation of heroic human effort thwarted by the presence of technical and organizational shortcomings is an old story. Seldom, however, has it been as dramatically illustrated as it was on September 11, 2001. A very important point illustrated by this view of the events of 9–11 is that individual organizations can have all the well engineered, highly usable systems they want (though typically they do not) but if the demands of the situation call for effective interaction between organizations in terms of their personnel and their technology, then the technical assets of one specific group, no matter how well-designed, will almost certainly not be sufficient to overcome the more serious short-comings of poor inter-organizational operability. Unfortunately (1) individual organizations all too often have technical systems that fail to effectively support human performance under emergency, high stress situations, even within their own limited domains (2) these systems are not interoperable with those of other organizations with whom they must cooperate in order to generate an effective response to an emergency and/or high stress event, and (3) the organizations themselves are often loathe to cooperate with one another due to cultural animosities.

If we might focus on a particular type of terrorist event, such as bioterrorism, it will be easier to illustrate the applicability of the HSI approach to national security. For example, an HSI approach to bioterrorism preparedness would focus first and foremost on the identification of current shortcomings (of the type listed above) in the nation's ability to respond to a bioterrorist event. These shortcomings are, in many respects, very similar to those that prevent the optimal operation of complex, multidisciplinary sociotechnical systems such as those in the military. (Because of this the *Handbook* in general appears directly relevant to the design of National Security systems. Also a special chapter on Personnel Survivability is included; it has little application to the other two activities reviewed here, but the threats and methods discussed in Chapter 16 bear directly on designing systems which most operate in response to such hostile events as might occur from bioterrorism.) Table 4 lists some of the most pressing priorities we believe need to be considered for bioterrorism preparedness and for which the HSI approach provides a realistic method to help assure the nation is prepared for such events.

**TABLE 4 HSI Priorities for Bioterrorism Events**

- Clear identification of the human-system performance requirements needed to successfully address a broad array of bioterrorist events. An analysis of this sort would be intended to illuminate key deficiencies, such as those described by O'Toole (1999) and Rosen et al (2002), in the nation's bioterrorism response capability, and to identify those that can be addressed by means of HSI principles and techniques.
- Identification of deficiencies in manpower capabilities, and formulation of a plan to provide "surge" capacity to the public health system in the event of a bioterrorist attack.
- Identification of human-machine/computer system deficiencies, including issues related to the interoperability of such systems across the diverse agencies that would be called upon to cooperate in response to a bioterrorist attack.
- Identification of deficiencies, bottlenecks, and other problems related to cultural issues occurring at the interface of difference agencies whose coordinated efforts will be required to achieve a successful response.
- Identification of training issues and development of a plan to train individuals and agencies in the necessary, coordinated activities.

Clearly there are other issues, those related to other HSI concerns and those of a more purely medical nature, which must be addressed to devise a successful bioterrorism response capability. HSI is not a panacea, and cannot successfully address all the problems that need to be resolved. However, an HSI approach to bioterrorism preparedness would provide an appropriate, coherent framework within which to pursue such a program in light of its overarching emphasis on human-systems performance objectives, and its emphasis on breaking down cultural and functional barriers that exist that between those organizations whose effective cooperation will be required.

## IV SUMMARY AND CONCLUSIONS

If we examine HSI for its most distinctive features we can summarize them as follows:

1. HSI introduces organizational cultural change from one that makes people fit systems to one that makes systems fit people.
2. HSI introduces a safety culture for anticipating, preventing, and minimizing effects of system error.
3. HSI conducts risk and cost/benefit analyses to compare systems integration approaches with varying degrees of human performance considerations.

4. HSI provides ways to measure effectiveness of operational performance that includes human performance.
5. HSI provides the skills and tools needed to design systems that are user-oriented.

As we further examine the advances made with the HSI approach described in this *Handbook* and consider the three areas above of pressing national and international interest and urgency, we would like to leave the reader five thoughts on:

- Target Audiences
- Technology Needs
- Decision-maker Needs
- HSI Processes
- HSI resources

***Target Audiences.*** The HSI target audiences for military and commercial systems are primarily operators and maintainers. The common aspect of the military audience with that of the health, education, and security systems is that all are professionals who work with our technical systems. For these individuals, there is almost a direct transfer of HSI knowledge to doctors, nurses, teachers, police, and fire fighters who also are professionals and must work with the systems of their trade. What is primarily different from the HSI audiences studied in the past and those of the three future arenas are the non-professional target audiences–patients, students, and every-day civilians going about their business where our technological systems are imposed upon them for better or worse. It is with these latter audiences, the ultimate reason for health, education, and security systems that the greatest needs exist for furthering HSI knowledge.

***Technology Needs.*** Advances in technology will continue to be one of the primary forces driving our sociotechnical culture into the future. The HSI approach simply provides processes and methods to help assure technology is selected, designed and implemented in such a way that it makes the most of what the decision-makers, designers and implementers intend. When people are the central focus of technology design and usage, system performance can be enhanced, systems can be used more safely, and overall system resources can be conserved.

***Decision-maker Needs.*** In all the activities considered by the *Handbook* contributors, the keys to system improvements are in the hands of decision-makers leaders who can decide to apply the processes and methods of HSI. To the degree leaders of organizations decide to implement some of the concepts outlined in the *Handbook*, many of the benefits promised by the HSI approach can become reality for their organizations. The two most important decisions for all the activities discussed are 1) defining requirements for new systems in terms of the human user and 2) testing the usefulness of the systems in performance terms. Decision makers need to be aware, however, of the HSI approach and its benefits and methods presented in terms that can be understood economically. Providing the economic case for HSI tailored to decision-maker criteria is the greatest challenge for the systems engineering and human systems integration communities.

***HSI Processes and Methods.*** When decision-makers understand HSI concepts and see the benefits applicable to their activities, they will need something specific to apply. This *Handbook* presents the state of the art for HSI processes and methods and as such is recommended as the primary guidance document for this purpose.

***HSI Resources.*** The quality of HSI implementation is dependant upon the availability and utilization of HSI professionals. As with the *Handbook* for processes and methods, the contributors are good sources for additional information on locating qualified HSI professionals. Additionally the Human Factors and Ergonomics Society with over 5000 members is a good source for identifying individuals skilled in various aspects of HSI. The U.S. has over 60 academic institutions for human factors and ergonomics. As the number of systems engineering and operations research institutions who teach human systems integration increase, the availability of HSI professionals will also increase. There are currently sufficient HSI resources to meet current demands. The challenge will be to meet increases in demand as more decision-makers join the HSI sociotechnical cultural revolution.

Harold R. Booher
Catherine A. Booher
Lawrence J. Hettinger
John Klesch

## NOTES

1. Paul O'Neill was key note speaker at the forum, "Developing a National Policy Agenda for Improving Patient Safety" convened by American Hospital Association, Joint Commission on Accreditation of Healthcare Organizations, and the National Patient Safety Foundation, held at National Press Club, Washington, DC July 15, 1999.
2. Source: Centers for Medicare and Medicaid Services, formerly the Health Care Finance Administration http://cms.hhs.gov/statistics/nhe/projections-2001/highlights.asp.)
3. Kohn, L., Corrigan, J., and Donaldson, M. (Eds.) (2000). *To Err is Human: Building a Safer Health System*, Washington DC: National Academy Press.
4. See for example, Sox Jr. H. and Woloshin S., (Nov-Dec 2000) "How many deaths are due to medical error? Getting the number right," Eff Clin Pract 3(6):277-83; Hayward, R. and Hofer, T. (Jul 25, 2001). "Estimating hospital deaths due to medical errors: preventability is in the eye of the reviewer," JAMA 286(4):415-20.
5. Hospital examples drawn from professional experiences and observations made by C. Booher, M.D. during a study, "Patient Centered Systems Analysis in a Hospital Setting" for the Food and Drug Administration; Center for Devices and Radiological Health Order No. FDA 269689-00-99-GH02. Final Report July 7, 2000.
6. Source: Website; US Department of Education, National Center for Educational Statistics (NCES); original source: US Department of Education, National Center for Education Statistics, Common Core of Data, "Early Estimates of Public Elementary/Secondary Education Survey," 1999-2000 and "National Public Education Financial Survey", 1995-96 through 1997-98.
7. Source: Website; US Department of Education, National Center for Educational Statistics (NCES); original source: U.S. Department of Education: Office of the Under Secretary, unpublished data, and National Center for Education Statistics, compiled from data appearing

in U.S. Office of Management and Budget, Budget of the United States Government, fiscal years (FY) 1982–2001 (selected years); National Science Foundation, Federal Funds for Research and Development, FY 1980–2000 (selected years); and unpublished data obtained from various federal agencies.

8. Source: Website; US Department of Education, National Center for Educational Statistics (NCES); original source U.S. Department of Education, NCES. (2001). The Nation's Report Card: Fourth-Grade Reading 2000 (NCES 2001499).
9. Source: Mark Berends, Susan J. Bodilly, Sheila Nataraj Kirby, Facing the Challenges of Whole-School Reform: New American Schools After a Decade (Santa Monica, CA.: RAND, MR-1498-EDU, 2002).
10. O'Hanlon, M.; Orszag, I.; Daalder, I.; Destler, I.; Gunter, L.; Litan, R.; and Steinberg, J. (2002). *Protecting the American Homeland,* Washington, DC: Brookings Press.
11. Ibid.
12. O'Toole, T., (2001), Smallpox: An attack scenario, *Emerging Infectious Diseases*, 5, 540-546.
13. Ibid, p. 540.
14. For examples, see Kun, L.G., & Bray, D.A., (2002), Information infrastructure tools for bioterrorism preparedness, *IEEE Engineering in Medicine and Biology*, 21(5), 69-85; Popovich, M.L., Henderson, J.M., & Stinn, J., (2002), Information technology in the age of emergency public health response, *IEEE Engineering in Medicine and Biology*, 21(5), 48-55; Rosen, J., Grigg, E., Lanier, J., McGrath, S., Lillibridge, S., Sargent, D., & Koop, C.E. (2002), The future of command and control for disaster response, *IEEE Engineering in Medicine and Biology*, 21(5), 56-68.
15. Op. cit., Rosen et al., 2002.
16. Ibid.
17. See also Inglesby, T.V. (2000), Lessons from TOPOFF, *The Second National Symposium on Medical and Public Health Response to Bioterrorism Speaker Transcript*, Washington, DC. Available at: http://www.hopkins-biodefense.org/sympcast/transcripts/trans_ingl.html.
18. See also ANSER, (2001), "Dark winter: Summary." Available at http://www.homelandsecurity.org/darkwinter.index.cfm.
19. Op cit., Rosen et al, (2002); ANSER, (2001).
20. New York Times, July 7, 2002.

# APPENDIX

Statement by Congressman Ike Skelton, October 1, 1997, Congressional Record-House, (H8269-H8271)

MANPRINT for the U.S. Army

(Mr. SKELTON asked and was given permission to address the House for 1 minute.)

Mr. SKELTON. Mr. Speaker, today, it is my pleasure to share with my colleagues a good news story, one about our Nation's military and, in particular, our Army. It involves a materiel acquisition program first developed in the 1980's for Army soldiers. It is called MANPRINT, which stands for manpower and personnel integration.

The MANPRINT program objective is to improve the performance of Army weapons and equipment through a man-machine total systems approach. That is, MANPRINT focuses on the interrelationship of the soldier and his or her weapon or equipment and the human requirements for maximizing system performance. In a nutshell, it does not make any difference if there is a tank that is capable of firing 10 rounds per minute if its crew can only operate it at three rounds per minute. Regardless of its technical capabilities, the tank is a three-round-per-minute tank due to the human factors that limit its output. This is the kind of problem MANPRINT addresses.

MANPRINT is an umbrella term that refers to seven disciplines that are critical to optimizing the man-machine, total-system approach. They are manpower, personnel, training, human factors engineering, system safety, health hazards, and soldier survivability. The central idea is to integrate considerations of these domains continuously into the acquisition process.

Thanks to MANPRINT the Army now has a vastly increased confidence that its new systems will perform as expected in the hands of its soldiers-and, at the same time, save lives and dollars. As I will explain later, MANPRINT has, in fact, already saved hundreds of soldiers' lives and billions of dollars. It has returned thousands of percent on a trickle of investment dollars. It is, or should be, a governmental downsizer's dream come true. Moreover, in this day of increased reliance on technology, we are only beginning to explore the ramifications the Army's concept could have for our entire society.

There is an element of urgency associated with this Army program, however, and the very real danger that we could repeat mistakes of the past—the type where U.S. inventors or progressive thinkers create great ideas which we fail to appreciate and implement. Instead, other countries capitalize on them. You will recall the Dr. W. Edward Deming's ideas on quality were ignored in this country in the 1950's and then successfully adopted by the Japanese. We may be on the verge of committing such a mistake with the Army's MANPRINT program. The Army resources devoted to MANPRINT have been continually slashed during the drawdown. At the same time, the United Kingdom has picked up on the U.S. Army's idea and is already in the process of implementing it throughout all services in the royal force. Moreover, as the Japanese recognized, Deming's quality ideas applied to all technology, not just defense. Not surprisingly, the British are starting MANPRINT programs in the Departments of Trade and Industry as well.

In order to reduce the likelihood of our making the same error with MANPRINT as we did with Deming's quality management, I want to make sure my colleagues are familiar with this highly successful soldier-oriented concept for the design, development, manufacturing, and fielding of the Army's newest weapon's systems.

## ARMY ACQUISITION PROGRAMS LED TO ADOPTION OF MANPRINT

I am sure that many of you recall the manpower and readiness problems that plagued the Army force modernization program in the early 1980's. It seemed that whenever a new system was put into the hands of the soldier, actual field performance often failed to match the standards predicted during its development. The Stinger anti-aircraft missile, for example, was designed to hit incoming aircraft better than 6 percent of the time. But if it had been placed in service as originally designed, it would actually have achieved hits only 30 percent of the time when operated by soldiers in combat units. The Stinger's problems were eventually corrected. But the problems of soldier utilization were so great in the Division Air Defense Gun, known as the DIVAD or Sergeant York, that the program had to be canceled. In the case of the Dragon anti-tank missile, that soldier's nightmare is still in the Army's inventory.

In addition to unacceptable performance from new systems, the Army experienced problems in crew performance. When the Army replaced an existing system with a newer, more technologically complex system, the newer system often generated requirements for soldiers of a higher level of skill and for more soldiers per system. The Army personnel system simply could not provide enough soldiers of the caliber required to operate and maintain such sophisticated systems.

The Army's first study on what to do about the disappointing performance and unaffordable manpower costs of new weapons systems and equipment was conducted by retired Generals Walter T. Kerwin and George S. Blanchard in 1980. In examining the Army's concerns about the mobilization, readiness and sustainability of new systems, the report concluded that it was primarily a lack of consideration of the human in the system that was causing the problem. Human performance assessments either were not done or were too late to influence weapons design. Supporting the Kerwin and Blanchard findings, the General Accounting Office [GAO] published reports in 1981 and 1985 attributing 50 percent of equipment failures to human error. GAO, too, stressed the need for integrating into the acquisition process human disciplines, such as, in particular, manpower, personnel and training needs.

The recommendations for a new soldier-oriented approach to systems acquisition were taken very seriously in the mid-1980's. With the full support of the entire Army leadership, military and civilian, Gen. Maxwell Thurman, as the Vice Chief of Staff, directed that an entirely new approach to systems acquisition be adopted by the Army, one which required that systems fit the soldiers rather than that the soldier—through selection or training—fit the systems.

This new concept also affected industry because, as we all know, defense contractors actually design and develop Army systems. In the mid-eighties, the concept required a radical change in the way contractors did business. To successfully compete in the new Army acquisition process, industry had to focus on the human element and design systems that fit soldier's needs and capabilities. In the MANPRINT process, human parameters are specified in the same manner as any other component of the system. System performance is measured with the humans quantitative performance included as an inherent part of the total system performance. No longer could performance in the laboratory be extrapolated as satisfying the requirements of performance in the field.

The MANPRINT philosophy and examples of the array of concepts inherent in MANPRINT are documented in a book, 'MANPRINT: An Approach to Systems Integration' (Van Nostrand Reinhold, 1990), edited by Dr. Harold R. Booher, who was the first senior Army civilian official appointed to direct the Army's MANPRINT program.

## COMANCHE AND MANPRINT

Nowhere has the new soldier-oriented partnership between Government and industry been more visible than on the Army's Light Helicopter Experimental [LHX] program. Better known to us today as the Comanche, the LHX in 1986 was the Army's true experimental program, testing where it was possible to introduce cutting-edge technology into its inventory without running headlong into the problems of unsatisfactory performance and runaway personnel costs. Even opponents of Comanche cannot ignore the great advances achieved in this program beyond the standard of normal acquisition practices.

Perhaps the first indication that MANPRINT was not only viable but could revolutionize the military's procurement process was the successful development of the Comanche's T-800 engine. The MANPRINT approach fostered hundreds of design improvements affecting both maintenance and reliability. In one striking example, the tool kit for the organization mechanic was reduced from 134 tools to only 6. The trunk-sized caster tool kit used on other helicopters was reduced to a canvass pouch half the size of a rolled-up newspaper. Furthermore, this reduction cost Government and industry nothing and will save taxpayer dollars.

For the Comanche itself, MANPRINT resulted in more than 500 design improvements in system performance and logistics. The cockpit was designed outward, from the pilot seat, using simulations and modeling, lessons learned from previous aircraft programs, and user inputs. In addition, when fielded, the Comanche would allow the aircrew to select what information is needed during missions. The result is an anticipated system with a much improved pilot-crew workload. A typical performance benefit is illustrated in the reduced number of steps it takes for the pilot to acquire a target. The OH-58D Kiowa Warrior required 34; the Comanche, 5.

Incorporation of MANPRINT considerations during Comanche development also introduced entirely new concepts to the acquisition process. The source selection competition included MANPRINT in all evaluation areas. It became impossible for a company to win the contract without a plan to integrate MANPRINT in the design, development, and manufacture of Comanche. In addition, seasoned maintenance personnel and other soldiers with field experience in operational units were assigned to the contractor's plant as representatives of the users in the operating commands. These soldiers were invaluable in fitting the machine to the operator. For example, they completed a rotor design change in 30 days that would otherwise have taken 12 months to achieve contractor-Government approval.

MANPRINT was also responsible for technological advances. To provide for easy maintenance to aircraft components, Comanche was built around a box-like, load-bearing keel. In most helicopters, the load is carried by the external skin. In Comanche, the load-bearing keel made it possible to locate easy-access panels almost anywhere on the aircraft. Consequently, maintenance personnel can easily reach all of the internal components. In this case, a maintenance requirement drove the technological design, which in turn resulted in an aerodynamic improvement.

In another instance MANPRINT and transport considerations suggested the need for an improved rotor blade removal capability. The contractor design team already had a rotor blade design which met Government specifications and was concerned about the added expense. Nevertheless, because of soldier concerns, MANPRINT prevailed. A new blade was designed at a cost of approximately $60,000. Life cycle cost calculations have indicated that the new blade will remain easier to manufacture and should save approximately $150 million in personnel, maintenance, and transport costs from the original design.

From the outset soldier safety has been a major design objective. Safety experts studied more than two decades of helicopters accident reports to determine how the designers could make Comanche a safer aircraft. As a result of their efforts, the Comanche's safety-related design features are projected—when compared to other helicopters such as the OH-58 Kiowa and AH-1F Cobra—to save 91 soldiers lives and avoid at least 116 disabling injuries.

A 1995 report by the Analytic Sciences Corp: Minninger, et al: documents the performance improvements and savings on Comanche attributable to MANPRINT. The report found Comanche cost avoidance in manpower, personnel, training, and safety to be a whopping $3.29 billion. This return resulted from a design investment of approximately 4 percent of the Comanche R&D budget. Calculated as a return on design investment, MANPRINT in the Comanche program yielded over an 8,000-percent return. Moreover, if the costs of the remaining MANPRINT disciplines—health hazards and soldiers survivability—are included in the calculation, the return on investment for the entire program remains well over 4000 percent.

## MANPRINT APPLIED TO OTHER ARMY SYSTEMS

MANPRINT is not only limited to new or major acquisition systems. It works with systems already in the inventory as well. In 1994, McDonnell Douglas conducted a study covering

4 years of MANPRINT design improvements on Longbow Apache. More than 80 MANPRINT problems, issues, and concerns were identified and resolved. Each of them yielded an improvement either for the operator or the maintainer of the aircraft. Once again, improved human performance proved cost effective. From a $2.7 million investment, a return in manpower and safety costs reached $268 million, approximately a 2,000-percent return on investment.

The Fox vehicle modification is an illustrative example of MANPRINT's contribution to smaller, less visible acquisition programs. The Army uses the Fox—a mobile sensing module built into an eight-wheeled armored vehicle—as a nuclear, biological, and chemical reconnaissance system for identifying contaminated areas. In a recent system improvement project, the Army wanted to reduce the crew from four soldiers to three. But operational evaluators labeled the vehicle, when operated by three soldiers, 'unsuitable and ineffective.' The program appeared doomed because it was out of money and time. But MANPRINT experts, using two different types of integration models, redesigned the Fox and it was subsequently shown to be fully effective in its projected missions. The MANPRINT effort cost $60,000 and was completed in a short time; additional operational testing was avoided and the Army saved $2 to $4 million from projected program costs while removing on crew member requirement from each vehicle.

## MANPRINT VIABILITY TODAY

A recent Army Audit Agency [AAA] report evaluated how the Army, after its radical downsizing, is 'incorporating MANPRINT into weapon systems development.' The good news is that nine Army weapons systems were evaluated and all but one were considered to have incorporated MANPRINT adequately. Based on the AAA's audit assessment, the Army can expect positive MANPRINT results in such current programs as Land Warrior, Javelin, and Extended Range Multiple Launch Rocket System. The Command and Control Vehicle program and several nondevelopmental programs examined by AAA, including the Embedded Global Positioning System/Inertial Navigation System, also include good MANPRINT initiatives. Because of MANPRINT, the Army can have increased confidence in many of the systems it will be fielding in the not-too-distant future.

The Army cannot rest on its laurels, however. Several developments cloud the future of MANPRINT.

First, the AAA report noted that not all systems under development have incorporated MANPRINT. The now-canceled Armored Gun System is an example in the recent past of a program in which MANPRINT considerations were purposely rejected. It is not a coincidence that the Army canceled the program.

Second, the new DOD acquisition system may make it easier to omit MANPRINT from programs. The new system rightly attempts to give program managers more latitude by removing regulations that previously proved too restrictive. But this new-found freedom in itself may make it more difficult in the future to ensure an appropriate incorporation of MANPRINT. It would be very unfortunate if an unintended consequence of streamlining the acquisition process proved to be a reduced emphasis on MANPRINT.

That need not be the case, as the AAA report points out. The new acquisition system, if approached correctly, affords the opportunity for greater integration of people-oriented concerns into the acquisition process. If the 'unbound' program managers appreciate the value of optimizing the man-machine interface, they are free under the new system to tailor their programs to incorporate people-oriented considerations. Consequently, a major effort is needed to adapt MANPRINT to the new acquisition process.

A third concern is the erosion of the MANPRINT program in recent years as the Army has experienced the drawdown. The Army made a commitment to understand and incorporate the features that optimize man-machine performance in the mid-1980's but until recently has been in danger of returning to old ways. MANPRINT personnel have been reduced 55 percent while the active Army has come down approximately 37 percent. The AAA audit report concluded that the Army's training process, which started out so well in 1986, is now inadequate. Career paths no longer identify MANPRINT as important. Nor does MANPRINT always play as prominent a role in source selection as in some programs, such as Comanche. Finally, the technology resources devoted to the research and development needed to advance the state of the art for quantitative tradeoffs of manpower, personnel skills, and training have shrunk significantly.

Fortunately, thanks to the AAA audit report, Army leadership has been reminded that MANPRINT is a golden nugget and seems determined that it must be revitalized. A panel of senior officers has been working for several months to ensure that the wounds inflicted on the program by the drawdown are not fatal and that MANPRINT recovers its health.

In closing I want to congratulate the Army for developing MANPRINT and for continuing to support the program in a time of very scarce resources.

I also want to suggest that the Army's approach to systems integration is relevant to the other military departments, to the entire Department of Defense, and probably to the remainder of the Government. Acquisition reform seeks to advance technology while holding down procurement costs. Downsizing seeks to ensure essential Government functions are accomplished with a minimum of staff. MANPRINT can be an essential ingredient in both initiatives. With respect to the military, it ensures that the weapons and equipment supporting a reduced force structure will perform as expected on the battlefield.

But the possible applications for MANPRINT go far beyond the military in our constantly evolving technological-based society. Our regulatory agencies like the Federal Aviation Agency, the Nuclear Regulatory Commission, the Food and Drug Administration should push this concept to the forefront with the systems and equipment they regulate. Also it would seem our medical and educational systems could benefit from a technological development and management process which focuses on the end user. One may wonder what a difference it would make it these systems were made to operate primarily for the doctor and the patient or the teacher and the learner rather than fitting these individuals to the system as an afterthought. We have not been in such an enviable position to take advantage of a technological cultural change since Deming's total quality management. Let's not miss our opportunity this time around.

# AUTHOR INDEX

# SUBJECT INDEX